Electronic Principles

Seventh Edition

Albert Malvino

David J. Bates

McGraw Hill **Higher Education**

Boston   Burr Ridge, IL   Dubuque, IA   Madison, WI   New York   San Francisco   St. Louis
Bangkok   Bogotá   Caracas   Kuala Lumpur   Lisbon   London   Madrid   Mexico City
Milan   Montreal   New Delhi   Santiago   Seoul   Singapore   Sydney   Taipei   Toronto

Higher Education

ELECTRONIC PRINCIPLES, SEVENTH EDITION

Some ancillaries, including electronic and print components, may not be available to customers outside the United States.

This book is printed on acid-free paper.

6 7 8 9 0 DOW/DOW 10 9 8 7 6 5 4 3 2

ISBN-13   978–0–07–297527–7
ISBN-10   0–07–297527–X

Publisher, Career Education: *David T. Culverwell*
Publisher, Trades and Engineering Technology: *Thomas E. Casson*
Managing Developmental Editor: *Jonathan Plant*
Director of Development: *Kristine Tibbetts*
Editorial Coordinator: *Lindsay M. Roth*
Executive Marketing Manager: *James F. Connely*
Senior Project Manager: *Kay J. Brimeyer*
Senior Production Supervisor: *Sherry L. Kane*
Lead Media Project Manager: *Audrey A. Reiter*
Associate Media Producer: *Christina Nelson*
Designer: *Laurie B. Janssen*
Cover/Interior Designer: *Christopher Reese*
(USE) Cover Image: *Stocksearch/Alamy*
Senior Photo Research Coordinator: *John C. Leland*
Photo Research: *David Tietz*
Supplement Producer: *Tracy L. Konrardy*
Compositor: *Interactive Composition Corporation*
Typeface: *10/12 Times New Roman*
Printer: *R R Donnelley, Willard, Ohio*

Library of Congress Cataloging-in-Publication Data

Malvino, Albert Paul.
  Electronic principles / Albert Paul Malvino, David J. Bates. — 7th ed.
    p. cm.
  Rev. ed. of: Malvino electronic principles. 6th ed. © 1999.
  Includes index.
  ISBN 978–0–07–297527–7 — ISBN 0–07–297527–X (alk. paper)
  1. Electronics. I. Bates, David J. II. Malvino, Albert Paul. Malvino electronic principles. III. Title.

TK7816.M25   2007
621.381—dc22   2006041092
                                                                        CIP

www.mhhe.com

About the Authors

Albert P. Malvino *was an electronics technician while serving in the U.S. Navy from 1950 to 1954. He graduated from the University of Santa Clara Summa Cum Laude in 1959 with a B.S. degree in Electrical Engineering. For the next five years, he worked as an electronics engineer at Microwave Laboratories and at Hewlett-Packard while earning his MSEE from San Jose State University in 1964. He taught at Foothill College for the next four years and was awarded a National Science Foundation Fellowship in 1968. After receiving a Ph.D. in Electrical Engineering from Stanford University in 1970, Dr. Malvino embarked on a full-time writing career. He has written 10 textbooks that have been translated into 20 foreign languages with over 108 editions. Dr. Malvino is currently a consultant and designs microcontroller circuits for SPD-Smart™ windows. In addition, he is writing educational software for electronics technicians and engineers. He also serves on the Board of Directors of Research Frontiers Incorporated. His website address is www.malvino.com.*

David J. Bates *is an instructor in the Electronic Technologies Department of Western Technical College located in La Crosse, Wisconsin. Along with working as an electronic servicing technician and as an electrical engineering technician, he has over 25 years of teaching experience.*

Credentials include an A.S. degree in Industrial Electronics Technology, B.S. degree in Industrial Education, and an M.S. degree in Vocational/Technical Education. Certifications include an FCC GROL license, A+ certification as a computer hardware technician, and a Journeyman Level certification as a Certified Electronics Technician (CET) by the International Society of Certified Electronics Technicians (ISCET). David J. Bates is presently a certification administrator (CA) for ISCET and has served as a member of the ISCET Board of Directors, along with serving as a Subject Matter Expert (SME) of basic electronics for the National Coalition for Electronics Education (NCEE).

David J. Bates is also a co-author of "Basic Electricity" a text-lab manual by Zbar, Rockmaker, and Bates.

Dedication

Electronic Principles, 7th ed. is dedicated to my family and friends who have been so supportive during this process and especially to Jackie, whose patience and skills helped put this all together.

Contents

Preface

Electronic Principles, seventh edition, continues its tradition as a clearly explained, in-depth introduction to electronic semiconductor devices and circuits. This textbook is intended for students who are taking their first course in linear electronics. The prerequisites are a DC/AC Circuits course, algebra, and some trigonometry.

Electronic Principles provides essential understanding of semiconductor device characteristics, testing, and the practical circuits in which they are found. The text provides clearly explained concepts—written in an easy-to-read conversational style—establishing the foundation needed for understanding the operation and troubleshooting of electronic systems. Practical circuit examples and troubleshooting exercises are found throughout the chapters.

New to This Edition

Based on the feedback from extensive reviewing and course research, the seventh edition of *Electronic Principles* contains enhanced material on a variety of electronic devices and circuits, including:

- additional PNP transistor coverage,
- basic Bipolar Junction Transistor (BJT) voltage-divider circuit design
- increased ac load line analysis of BJT power amplifiers
- power E-MOSFET and D-MOSFET biasing
- Insulated Gate Bipolar Transistors (IGBTs)
- R/2R ladder D/A converters
- Function generator integrated circuits
- Class-D amplifiers

Starting in Chapter 1, "Introduction," the **T-shooter** troubleshooting exercise feature has been simplified for easier student use and understanding. This practical feature is integrated throughout the remaining chapters. In Chapter 2, "Semiconductors," one of many **"Summary Tables"** has been added to provide concept reinforcement and a convenient information resource. Starting in Chapter 3, "Diode Theory," **data sheets** for specific semiconductor devices will be found within the chapter as they are discussed. Chapters 4 through 6 have additional content regarding the testing of diodes and transistors using DMMs, VOMs, and semiconductor curve tracers. While Chapter 10, "Voltage Amplifiers," retains its coverage of the common-emitter (CE) amplifier, Chapter 11, "CC and CB Amplifiers," focuses on common-collector (CC) or emitter follower, common-base (CB), and Darlington amplifier configurations. Chapter 12, "Power Amplifiers," has been modified to include class A, B, AB and C power amplifiers, along with additional ac load line analysis. In Chapter 13, "JFETs," modifications have been made to the order of JFET biasing and biasing techniques. In Chapter 14, "MOSFETs," depletion-mode and power enhancement-mode MOSFET amplifiers have been added along with MOSFET testing. Chapter 15, "Thyristors," includes more detail in RC phase-shift control, SCR testing, and introduces IGBTs. Frequency analysis of FET stages has been added to Chapter 16, "Frequency Effects." In Chapter 20, "Linear Op-Amp Circuits," the circuit operation of the R/2R ladder D/A converter has been included. Chapter 22, "Nonlinear Op-Amp Circuits," now includes the basic operation of a class-D amplifier. Also, Chapter 23, "Oscillators," has an additional section on function generator ICs, including the XR-2206.

Guided Tour

Learning Features

Many new learning features have been incorporated into the seventh edition of *Electronic Principles*. These learning features, found throughout the chapters, include:

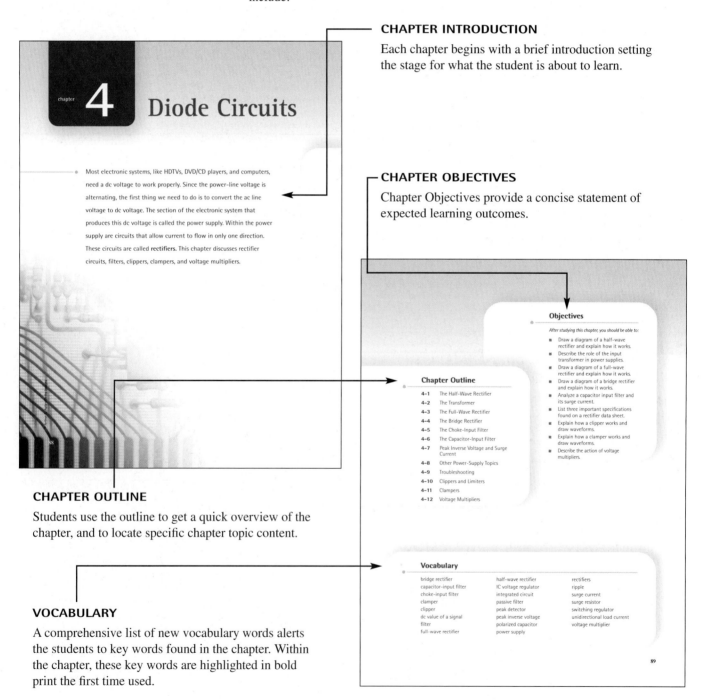

chapter 4 Diode Circuits

Most electronic systems, like HDTVs, DVD/CD players, and computers, need a dc voltage to work properly. Since the power-line voltage is alternating, the first thing we need to do is to convert the ac line voltage to dc voltage. The section of the electronic system that produces this dc voltage is called the power supply. Within the power supply are circuits that allow current to flow in only one direction. These circuits are called **rectifiers.** This chapter discusses rectifier circuits, filters, clippers, clampers, and voltage multipliers.

CHAPTER INTRODUCTION

Each chapter begins with a brief introduction setting the stage for what the student is about to learn.

CHAPTER OBJECTIVES

Chapter Objectives provide a concise statement of expected learning outcomes.

Objectives

After studying this chapter, you should be able to:

- Draw a diagram of a half-wave rectifier and explain how it works.
- Describe the role of the input transformer in power supplies.
- Draw a diagram of a full-wave rectifier and explain how it works.
- Draw a diagram of a bridge rectifier and explain how it works.
- Analyze a capacitor input filter and its surge current.
- List three important specifications found on a rectifier data sheet.
- Explain how a clipper works and draw waveforms.
- Explain how a clamper works and draw waveforms.
- Describe the action of voltage multipliers.

Chapter Outline

4-1 The Half-Wave Rectifier
4-2 The Transformer
4-3 The Full-Wave Rectifier
4-4 The Bridge Rectifier
4-5 The Choke-Input Filter
4-6 The Capacitor-Input Filter
4-7 Peak Inverse Voltage and Surge Current
4-8 Other Power-Supply Topics
4-9 Troubleshooting
4-10 Clippers and Limiters
4-11 Clampers
4-12 Voltage Multipliers

CHAPTER OUTLINE

Students use the outline to get a quick overview of the chapter, and to locate specific chapter topic content.

Vocabulary

bridge rectifier	half-wave rectifier	rectifiers
capacitor-input filter	IC voltage regulator	ripple
choke-input filter	integrated circuit	surge current
clamper	passive filter	surge resistor
clipper	peak detector	switching regulator
dc value of a signal	peak inverse voltage	unidirectional load current
filter	polarized capacitor	voltage multiplier
full-wave rectifier	power supply	

89

VOCABULARY

A comprehensive list of new vocabulary words alerts the students to key words found in the chapter. Within the chapter, these key words are highlighted in bold print the first time used.

EXAMPLES

Each chapter contains worked-out Examples that demonstrate important concepts or circuit operation, including circuit analysis, applications, troubleshooting, and basic design.

PRACTICE PROBLEMS

Students can obtain critical feedback by performing the Practice Problems that immediately follow most Examples. Answers to these problems are found at the end of each chapter.

GOOD TO KNOW

Good To Know statements, found in the margins, provide interesting added insights to topics being presented.

MULTISIM

Students can "bring to life" many of the circuits found in each chapter. A CD containing MultiSim files is included with the textbook; with these files students can change the value of circuit components and instantly see the effects, using realistic Tektronix and Agilent simulation instruments. Troubleshooting skills can be developed by inserting circuit faults and making circuit measurements. Students new to computer simulation software will find a MultiSim Primer in the appendix.

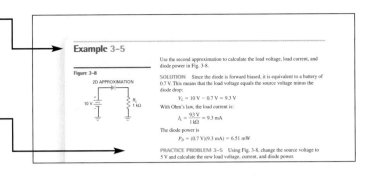

Example 3-5

Figure 3-8

2D APPROXIMATION

Use the second approximation to calculate the load voltage, load current, and diode power in Fig. 3-8.

SOLUTION Since the diode is forward biased, it is equivalent to a battery of 0.7 V. This means that the load voltage equals the source voltage minus the diode drop:

$$V_L = 10\ V - 0.7\ V = 9.3\ V$$

With Ohm's law, the load current is:

$$I_L = \frac{9.3\ V}{1\ k\Omega} = 9.3\ mA$$

The diode power is

$$P_D = (0.7\ V)(9.3\ mA) = 6.51\ mW$$

PRACTICE PROBLEM 3-5 Using Fig. 3-8, change the source voltage to 5 V and calculate the new load voltage, current, and diode power.

Figure 6-10 Set of collector curves.

GOOD TO KNOW

When displayed on a curve tracer, the collector curves in Fig. 6-10 actually have a slight upward slope as V_{CE} increases. This rise is the result of the base region becoming slightly smaller as V_{CE} increases. (As V_{CE} increases, the CB depletion layer widens, thus narrowing the base.) With a smaller base region, there are fewer holes available for recombination. Since each curve represents a constant base current, the effect looks like an increase in collector current.

DATA SHEETS

Full and partial component data sheets are provided for many semiconductor devices; key specifications are examined and explained. Complete data sheets of these devices can be found on the Internet.

Example 4-1

Figure 4-3 shows a half-wave rectifier that you can build on the lab bench or on a computer screen with MultiSim. An oscilloscope is across the 1 kΩ. This will show us the half-wave load voltage. Also, a multimeter is across the 1 kΩ to read the dc load voltage. Calculate the theoretical values of peak load voltage and the dc load voltage. Then, compare these values to the readings on the oscilloscope and the multimeter.

SOLUTION Figure 4-3 shows an ac source of 10 V and 60 Hz. Schematic diagrams usually show ac source voltages as effective or rms values. Recall that the *effective value* is the value of a dc voltage that produces the same heating effect as the ac voltage.

Figure 4-3 Lab example of half-wave rectifier.

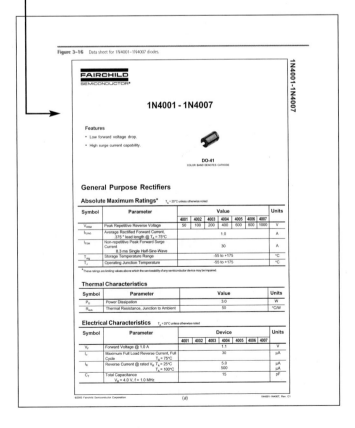

Figure 3-16 Data sheet for 1N4001–1N4007 diodes.

FAIRCHILD SEMICONDUCTOR®

1N4001-1N4007

1N4001 - 1N4007

Features
- Low forward voltage drop.
- High surge current capability.

DO-41
COLOR BAND DENOTES CATHODE

General Purpose Rectifiers

Absolute Maximum Ratings* $T_A = 25°C$ unless otherwise noted

Symbol	Parameter	Value							Units
		4001	4002	4003	4004	4005	4006	4007	
V_{RRM}	Peak Repetitive Reverse Voltage	50	100	200	400	600	800	1000	V
$I_{O(AV)}$	Average Rectified Forward Current, .375 " lead length @ $T_A = 75°C$				1.0				A
I_{FSM}	Non-repetitive Peak Forward Surge Current 8.3 ms Single Half-Sine-Wave				30				A
T_{stg}	Storage Temperature Range				-55 to +175				°C
T_J	Operating Junction Temperature				-55 to +175				°C

*These ratings are limiting values above which the serviceability of any semiconductor device may be impaired.

Thermal Characteristics

Symbol	Parameter	Value	Units
P_D	Power Dissipation	3.0	W
$R_{θJA}$	Thermal Resistance, Junction to Ambient	50	°C/W

Electrical Characteristics $T_A = 25°C$ unless otherwise noted

Symbol	Parameter	Device							Units
		4001	4002	4003	4004	4005	4006	4007	
V_F	Forward Voltage @ 1.0 A				1.1				V
I_{rr}	Maximum Full Load Reverse Current, Full Cycle $T_A = 75°C$				30				µA
I_R	Reverse Current @ rated V_R $T_A = 25°C$				5.0				µA
	$T_A = 100°C$				500				µA
C_T	Total Capacitance $V_R = 4.0\ V,\ f = 1.0\ MHz$				15				pF

©2003 Fairchild Semiconductor Corporation

(a)

1N4001-1N4007, Rev. C1

COMPONENT PHOTOS

Photos of actual electronic devices bring students closer to the device being studied.

SUMMARY TABLES

Summary Tables have been included at important points within many chapters. Students use these tables as an excellent review of important topics, and as a convenient information resource.

COMPONENT TESTING

Students will find clear descriptions of how to test individual electronic components using common equipment such as digital multimeters (DMMs).

Figure 7-22 Phototransistor. (a) Open base gives maximum sensitivity; (b) variable base resistor changes sensitivity; (c) typical phototransistor.

(a)

(b)

(c)
© Brian Moskau/Brian Moskau Photography

concentrate on the thermally produced carriers in the collector diode. Visualize the reverse current produced by these carriers as an ideal current source in parallel with the collector-base junction of an ideal transistor (Fig. 7-21b).

Because the base lead is open, all the reverse current is forced into the base of the transistor. The resulting collector current is:

$$I_{CEO} = \beta_{dc}I_R$$

where I_R is the reverse minority-carrier current. This says that the collector current is higher than the original reverse current by a factor of β_{dc}.

The collector diode is sensitive to light as well as heat. In a phototransistor, light passes through a window and strikes the collector-base junction. As the light increases, I_R increases, and so does I_{CEO}.

Phototransistor versus Photodiode

The main difference between a phototransistor and a photodiode is the current gain β_{dc}. The same amount of light striking both devices produces β_{dc} times more current in a phototransistor than in a photodiode. The increased sensitivity of a phototransistor is a big advantage over that of a photodiode.

Figure 7-22a shows the schematic symbol of a phototransistor. Notice the open base. This is the usual way to operate a phototransistor. You can control the sensitivity with a variable base return resistor (Fig. 7-22b), but the base is usually left open to get maximum sensitivity to light.

The price paid for increased sensitivity is reduced speed. A phototransistor is more sensitive than a photodiode, but it cannot turn on and off as fast. A photodiode has typical output currents in microamperes and can switch on and off in nanoseconds. The phototransistor has typical output currents in milliamperes but switches on and off in microseconds. A typical phototransistor is shown in Fig. 7-22c.

Optocoupler

Figure 7-23a shows an LED driving a phototransistor. This is a much more sensitive optocoupler than the LED-photodiode discussed earlier. The idea is straightforward. Any changes in V_S produce changes in the LED current, which changes the current through the phototransistor. In turn, this produces a changing voltage across the collector-emitter terminals. Therefore, a signal voltage is coupled from the input circuit to the output circuit.

Again, the big advantage of an optocoupler is the electrical isolation between the input and output circuits. Stated another way, the common for the input circuit is different from the common for the output circuit. Because of this,

Figure 7-23 (a) Optocoupler with LED and phototransistor; (b) optocoupler IC

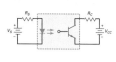

(a)

(b)
© Brian Moskau/Brian Moskau Photography

Transistor Fundamentals

249

Summary Table 8-1 (continued)

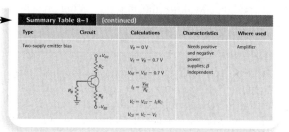

Type	Circuit	Calculations	Characteristics	Where used
Two-supply emitter bias		$V_B = 0$ V	Needs positive and negative power supplies; β independent	Amplifier
		$V_E = V_B - 0.7$ V		
		$V_{RE} = V_{EE} - 0.7$ V		
		$I_E = \dfrac{V_{RE}}{R_E}$		
		$V_C = V_{CC} - I_C R_C$		
		$V_{CE} = V_C - V_E$		

Figure 8-12 (a) Base bias; (b) emitter-feedback bias.

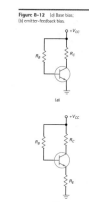

(a)

(b)

8-5 Other Types of Bias

In this section, we will discuss some other types of bias. A detailed analysis of these types of bias is not necessary because they are rarely used in new designs. But you should at least be aware of their existence in case you see them on a schematic diagram.

Emitter-Feedback Bias

Recall our discussion of base bias (Fig. 8-12a). This circuit is the worst when it comes to setting up a fixed Q point. Why? Since the base current is fixed, the collector current varies when the current gain varies. In a circuit like this, the Q point moves all over the load line with transistor replacement and temperature change.

Historically, the first attempt at stabilizing the Q point was **emitter-feedback bias**, shown in Fig. 8-12b. Notice that an emitter resistor has been added to the circuit. The basic idea is this: If I_C increases, V_E increases, causing V_B to increase. More V_B means less voltage across R_B. This results in less I_B, which opposes the original increase in I_C. It's called *feedback* because the change in emitter voltage is being fed back to the base circuit. Also, the feedback is called *negative* because it opposes the original change in collector current.

Emitter-feedback bias never became popular. The movement of the Q point is still too large for most applications that have to be mass-produced. Here are the equations for analyzing the emitter-feedback bias:

$$I_E = \frac{V_{CC} - V_{BE}}{R_E + R_B/\beta_{dc}} \tag{8-17}$$

$$V_E = I_E R_E \tag{8-18}$$

$$V_B = V_E + 0.7 \text{ V} \tag{8-19}$$

$$V_C = V_{CC} - I_C R_C \tag{8-20}$$

The intent of emitter-feedback bias is to **swamp out** the variations in β_{dc}; that is, R_E should be much greater than R_B/β_{dc}. If this condition is satisfied,

---

Figure 7-15 npn transistor.

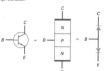

Out-of-Circuit Tests

A transistor is commonly tested using a DMM set to the diode test range. Figure 7-15 shows how an *npn* transistor resembles two back-to-back diodes. Each *pn* junction can be tested for normal forward and reverse biased readings. The collector to emitter can also be tested and should result in an overrange indication with either DMM polarity connection. Since a transistor has three leads, there are six DMM polarity connections possible. These are shown in Fig. 7-16a. Notice that only two polarity connections result in approximately a 0.7 V reading. Also important to note here is that the base lead is the only connection common to both 0.7 V readings and it requires a (+) polarity connection. This is also shown in Fig. 7-16b.

A *pnp* transistor can be tested using the same technique. As shown in Fig. 7-17, the *pnp* transistor also resembles two back-to-back diodes. Again, using the DMM in the diode test range, Fig. 7-18a and 7-18b show the results for a normal transistor.

Many DMMs have a special β_{dc} or h_{FE} test function. By placing the transistor's leads into the proper slots, the forward current gain is displayed. This current gain is for a specified base current or collector current and V_{CE}. You can check the DMM's manual for the specific test condition.

Another way to test transistors is with an ohmmeter. You can begin by measuring the resistance between the collector and the emitter. This should be very high in both directions because the collector and emitter diodes are back to

Figure 7-16 NPN DMM Readings (a) Polarity connections; (b) pn junction readings.

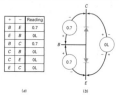

+	−	Reading
B	E	0.7
E	B	OL
B	C	0.7
C	B	OL
C	E	OL
E	C	OL

(a)

(b)

Transistor Fundamentals

245

Summary

SEC. 8-1 VOLTAGE-DIVIDER BIAS

The most famous circuit based on the emitter-bias prototype is called voltage-divider bias. You can recognize it by the voltage divider in the base circuit.

SEC. 8-2 ACCURATE VDB ANALYSIS

The key idea is for the base current to be much smaller than the current through the voltage divider. When this condition is satisfied, the voltage divider holds the base voltage almost constant and equal to the unloaded voltage out of the voltage divider. This produces a solid Q point under all operating conditions.

SEC. 8-3 VDB LOAD LINE AND Q POINT

The load line is drawn through saturation and cutoff. The Q point lies on the load line with the exact location determined by the biasing. Large variations in current gain have almost no effect on the Q point because this type of bias sets up a constant value of emitter current.

SEC. 8-4 TWO-SUPPLY EMITTER BIAS

This design uses two power supplies: one positive and the other negative. The idea is to set up a constant value of emitter current. The circuit is a variation of the emitter-bias prototype discussed earlier.

SEC. 8-5 OTHER TYPES OF BIAS

This section introduced negative feedback, a phenomenon that exists when an increase in an output quantity produces a decrease in an input quantity. It is a brilliant idea that led to voltage-divider bias. The other types of bias cannot use enough negative feedback, so they fail to attain the performance level of voltage-divider bias.

SEC. 8-6 TROUBLESHOOTING

Troubleshooting is an art. Because of this, it cannot be reduced to a set of rules. You learn troubleshooting mostly from experience.

SEC. 8-7 PNP TRANSISTORS

These *pnp* devices have all currents and voltages reversed from their *npn* counterparts. They may be used with negative power supplies; more commonly, they are used with positive power supplies in an upside-down configuration.

Troubleshooting

▌▌▌ **MultiSim** Use Fig. 8-30 for the remaining problems.

8-35 Find Trouble 1.

8-36 Find Trouble 2.

8-37 Find Troubles 3 and 4.

8-38 Find Troubles 5 and 6.

8-39 Find Troubles 7 and 8.

8-40 Find Troubles 9 and 10.

8-41 Find Troubles 11 and 12.

Figure 8-30

Trouble	MEASUREMENTS			
	V_B (V)	V_E (V)	V_C (V)	R_2 (Ω)
OK	1.8	1.1	6	OK
T1	10	9.3	9.4	OK
T2	0.7	0	0.1	OK
T3	1.8	1.1	10	OK
T4	2.1	2.1	2.1	OK
T5	0	0	10	OK
T6	3.4	2.7	2.8	∞
T7	1.83	1.212	10	OK
T8	0	0	10	OK
T9	1.1	0.4	0.5	OK
T10	1.1	0.4	10	OK
T11	0	0	0	OK
T12	1.83	0	10	OK

Job Interview Questions

1. Draw a VDB circuit. Then, tell me all the steps in calculating the collector-emitter voltage. Why does this circuit have a very stable Q point?

2. Draw a TSEB circuit and tell me how it works. What happens to the collector current when the transistor is replaced or the temperature changes?

3. Describe a few other kinds of bias. What can you tell me about their Q points?

4. What are the two types of feedback biasing, and why were they developed?

5. What is the primary type of biasing used with discrete bipolar transistor circuits?

6. Should transistors being used as switching circuits be biased in the active region? If not, what two points associated with the load line are important with switching circuits?

7. In a VDB circuit, the base current is not small compared to the current through the voltage divider. What is the shortcoming of this circuit? What should be changed to correct it?

8. What is the most commonly used transistor biasing configuration? Why?

9. Draw a VDB circuit using an *npn* transistor. Label directions of divider, base, emitter, and collector currents.

10. What is wrong with a VDB circuit in which R_1 and R_2 are 100 times greater than R_E?

Self-Test Answers

1.	d	11.	b	21.	c
2.	a	12.	a	22.	a
3.	a	13.	c	23.	d
4.	d	14.	c	24.	b
5.	b	15.	c	25.	b
6.	b	16.	a	26.	c
7.	b	17.	b	27.	b
8.	a	18.	a	28.	c
9.	c	19.	d	29.	a
10.	a	20.	a	30.	d

Practice Problem Answers

8-1 $V_B = 2.7$ V;
$V_E = 2$ mA;
$V_C = 7.78$ V;
$V_{CE} = 5.78$ V

8-2 $V_{CE} = 5.85$ V;
Very close to the predicted value

8-4 $R_E = 1$ kΩ;
$R_C = 4$ kΩ;
$R_2 = 700$ Ω (680);
$R_1 = 3.4$ kΩ (3.3k)

8-5 $V_{CE} = 6.96$ V

8-6 $V_{CE} = 7.05$ V

8-7 For 8-19a:
$V_B = 2.16$ V;
$V_E = -1.46$ V;
$V_C = -6.73$ V;
$V_{CE} = -5.27$ V

For 8-19b:
$V_B = 9.84$ V;
$V_E = 10.54$ V;
$V_C = 5.27$ V;
$V_{CE} = -5.27$ V

Problems

SEC. 6-3 TRANSISTOR CURRENTS

6-1 A transistor has an emitter current of 10 mA and a collector current of 9.95 mA. What is the base current?

6-2 The collector current is 10 mA, and the base current is 0.1 mA. What is the current gain?

6-3 A transistor has a current gain of 150 and a base current of 30 μA. What is the collector current?

6-4 If the collector current is 100 mA and the current gain is 65, what is the emitter current?

SEC. 6-5 THE BASE CURVE

6-5 ▌▌▌ **MultiSim** What is the base current in Fig. 6-20?

(a)

(b)

Figure 6-21

6-6 ▌▌▌ **MultiSim** If the current gain decreases from 200 to 100 in Fig. 6-20, what is the base current?

6-7 If the 470 kΩ of Fig. 6-20 has a tolerance of ±5 percent, what is the maximum base current?

SEC. 6-6 COLLECTOR CURVES

6-8 ▌▌▌ **MultiSim** A transistor circuit similar to Fig. 6-20 has a collector supply voltage of 20 V, a collector resistance of 1.5 kΩ, and a collector current of 6 mA. What is the collector-emitter voltage?

6-9 If a transistor has a collector current of 100 mA and a collector-emitter voltage of 3.5 V, what is its power dissipation?

SEC. 6-7 TRANSISTOR APPROXIMATIONS

6-10 What are the collector-emitter voltage and the transistor power dissipation in Fig. 6-20? (Give answers for the ideal and the second approximation.)

6-11 Figure 6-21a shows a simpler way to draw a transistor circuit. It works the same as the circuits already discussed. What is collector-emitter voltage? The transistor power dissipation? (Give answers for the ideal and the second approximation.)

6-12 When the base and collector supplies are equal, the transistor can be drawn as shown in Fig. 6-21b. What is the collector voltage in this circuit? The transistor power? (Give answers for the ideal and the second approximation.)

Figure 6-20

SEC. 6-8 READING DATA SHEETS

6-13 What is the storage temperature range of a 2N3904?

6-14 What is the minimum h_{FE} for a 2N3904 for a collector current of 1 mA and a collector-emitter voltage of 1 V?

6-15 A transistor has a power rating of 1 W. If the collector-emitter voltage is 10 V and the collector current is 120 mA, what happens to the transistor?

6-16 A 2N3904 has a power dissipation of 625 mW without a heat sink. If the ambient temperature is 65°C, what happens to the power rating?

SEC. 6-10 TROUBLESHOOTING

6-17 ▌▌▌ **MultiSim** In Fig. 6-20, does the collector-emitter voltage increase, decrease, or remain the same for each of these troubles?

a. 470 kΩ is shorted

b. 470 kΩ is open

c. 820 Ω is shorted

d. 820 Ω is open

e. No base supply voltage

f. No collector supply

Student Resources

In addition to the fully updated text, a number of student learning resources have been developed to aid readers in their understanding of electronic principles and applications.

- **Student CD-ROM,** provided with the textbook, provides MultiSim simulation files for circuits covered in text examples and problems. They are arranged by chapter for easy reference.
- **Experiments Manual** to accompany Electronic Principles, correlated to the textbook, provides a full array of hands-on labs; MultiSim "pre-lab" routines are included for those wanting to integrate computer simulation, and files for these are included on the bound-in **Experiments Manual CD ROM.**
- **Online Learning Center (OLC)** website (Student Site) contains a wealth of student features, including extra review questions, links to industry sites, circuit & component lists, and activities based around key terms.

Instructor Resources

- **Instructor's Manual** provides printed solutions and teaching suggestions for the text and Experiments Manual.
- **Instructor Productivity Center CD ROM,** bound with the Instructor's Manual, provides prepared **instructional PowerPoint** for all chapters in the text; **Electronic Testbanks** with additional review questions for each chapter that can be arranged, edited, and modified to fit class needs; and **e-Instruction's Classroom Performance System (CPS),** an active in-class learning system using hand-held consoles.
- **Online Learning Center** website (Instructor Site, password-protected) includes the Instructor's Manual and PowerPoint slides online, links to industry and educational websites, Parts Kit list for experiments, and electronic testbanks online.
- **Experiments Manual,** to accompany Electronic Principles, correlated to the textbook, with lab follow-up information included in the textbook Instructor's Manual, Instructor Productiviy Center and Instructor side of the Online Learning Center website.
- **Visual Calculator For Electronics** software allows you to analyze over 140 basic electronics circuits with the ability to display any of the 1500 equations used in the calculations. With Visual Calculator you can substitute standard resistor values to see the effects on circuit operation, view load lines and other graphs, along with viewing data sheets for many of the components. This software can be used by instructors to help teach and demonstrate electronic circuit operation. Students can use the software to review for midterms and final exams, check answers to homework problems, and get answers faster than by any other method. Visual Calculator can be found at http://www.malvino.com.

Acknowledgements

The production of *Electronic Principles,* seventh edition, is truly a team effort. It requires the hard work and professional dedication of a large number of people. Thank you to everyone at McGraw-Hill Higher Education who contributed to this edition, especially Tom Casson, Jonathan Plant, Lindsay Roth, Kay Brimeyer, and Carol Kromminga. Special thanks should also go out to Pat Hoppe whose insights, careful review, and tremendous work on the MultiSim files has been a significant contribution to this textbook.

Thanks to everyone whose comments and suggestions were extremely valuable in the development of this edition. This includes those who took the time to respond to surveys prior to manuscript development and those who carefully reviewed the revised material. Every survey and review was carefully examined and has contributed greatly to this edition. Here is a list of the reviewers who helped make this edition comprehensive and relevant.

Current Edition Reviewers

Ron Barrier
  Rowan Cabarrus Community College, NC

Adrien Berthiaume
  Northern Essex Community College, MA

M. C. Greenfield
  Indiana State University, IN

Craig Hill
  Erie Institute of Technology, PA

Patrick Hoppe
  Gateway Technical College, WI

Paul Kiser
  National Institute of Technology, WV

Dan Lookadoo
  New River Community College, VA

William Murray
  Broome Community College, NY

Rina Mazzucco
  Mesa Community College, AZ

Rajappa Papannareddy
  Purdue University, IN

Ken White
  Lakeland Community College, OH

Survey Respondents

Ben Bartlett
  College of Southern Idaho, ID

Michele J. Chance
  Rowan-Cabarrus Community College, NC

Walter O. Craig, III
  Southern University, LA

Sheila Donchoo
  Southern Polytechnic State University, GA

James A. Duru
  Essex County College, NJ

William Eaton
  Hinds Community College, MS

Udezei F. Edgal
  North Carolina A&T State University, NC

Glen Elliott
  Cambria County Area Community College, PA

Fred Etcheverry
  Hartnell College, CA

Jim Fiore
  Mohawk Valley Community College, NY

Rex Fisher
  Brigham Young University, ID

John E. Fitzen
  Idaho State University, ID

George Fredericks
  Northeast State Technical Community College, TN

G. J. Gerard
  Gateway Community Technical College, CT

Albert Gerth
  Corning Community College, NY

Melvin G. Gomez
  Green River Community College, WA

James Henderson
  Arkansas State University, AR

George Hendricks
  Gaston College, NC

Larry Hoffman
Purdue University, IN

David A. Kruse
Lane Community College, OR

Daniel Landiss
St. Louis Community College, MO

M. David Luneau, Jr.
University of Arkansas, AR

Richard McKinney
Nashville State Technical Community College, TN

Paul Nelson
College of the Sequoias, CA

Robert Peeler
Lamar State College, TX

Nasser H. Rashidi
Virginia State University, VA

Steven D. Rice
University of Montana, MT

Robert J. Scoff
University of Memphis, TN

Ron Tinckham
Santa Fe Community College, FL

Anthony Webb
Missouri Tech, MO

Harold Wiebe
Northern Kentucky University, KY

Michael Wilson
Kansas State University, KS

Manuscript Reviewers

Abraham Falsafi
National Institute of Technology, WV

Mohamed Haj-Mohamadi
North Carolina A&T University, NC

Patrick Hoppe
Gateway Technical College, WI

John Lindsey
Kansas Community and Technical College System, KS

Jim Ramming
Vatterott College, MO

Vince Vasco
Pittsburgh Technical Institute, PA

Electronic Principles

chapter 1

Introduction

The topics in this chapter include formulas, voltage sources, current sources, two circuit theorems, and troubleshooting. Although some of the discussion will be review, you will find new ideas that can make it easier for you to understand semiconductor devices and to serve as a framework for the rest of the textbook.

Objectives

After studying this chapter, you should be able to:

- Name the three types of formulas and explain why each is true.

- Explain why approximations are often used instead of exact formulas.

- Define an ideal voltage source and an ideal current source.

- Describe how to recognize a stiff voltage source and a stiff current source.

- State Thevenin's theorem and apply it to a circuit.

- State Norton's theorem and apply it to a circuit.

- List two facts about an open device and two facts about a shorted device.

Chapter Outline

Vocabulary

cold-solder joint

definition

derivation

duality principle

formula

ideal (first) approximation

law

Norton current

Norton resistance

open device

second approximation

shorted device

solder bridge

theorem

Thevenin resistance

Thevenin voltage

third approximation

troubleshooting

1-1 The Three Kinds of Formulas

A **formula** is a rule that relates quantities. The rule may be an equation, an inequality, or other mathematical description. You will see many formulas in this book. Unless you know why each one is true, you may become confused as they accumulate. Fortunately, there are only three ways formulas can come into existence. Knowing what they are will make your study of electronics more logical and satisfying.

The Definition

When you study electricity and electronics, you have to memorize new words like *current, voltage,* and *resistance.* However, a verbal explanation of these words is not enough. Why? Because your idea of current must be mathematically identical to everyone else's. The only way to get this identity is with a **definition,** a formula invented for a new concept.

Here is an example of a definition. In your earlier course work, you learned that capacitance equals the charge on one plate divided by the voltage between plates. The formula looks like this:

$$C = \frac{Q}{V}$$

This formula is a definition. It tells you what capacitance C is and how to calculate it. Historically, some researcher made up this definition and it became widely accepted.

Here is an example of how to create a new definition out of thin air. Suppose we are doing research on reading skills and need some way to measure reading speed. Out of the blue, we might decide to define *reading speed* as the number of words read in a minute. If the number of words is W and the number of minutes is M, we could make up a formula like this:

$$S = \frac{W}{M}$$

In this equation, S is the speed measured in words per minute.

To be fancy, we could use Greek letters: ω for words, μ for minutes, and σ for speed. Our definition would then look like this:

$$\sigma = \frac{\omega}{\mu}$$

This equation still translates to speed equals words divided by minutes. When you see an equation like this and know that it is a definition, it is no longer as impressive and mysterious as it initially appears to be.

In summary, *definitions are formulas that a researcher creates.* They are based on scientific observation and form the basis for the study of electronics. They are simply accepted as facts. It's done all the time in science. A definition is true in the same sense that a word is true. Each represents something we want to talk about. When you know which formulas are definitions, electronics is easier to understand. Because definitions are starting points, all you need to do is to understand and memorize them.

The Law

A **law** is different. It summarizes a relationship that already exists in nature. Here is an example of a law:

$$f = K \frac{Q_1 Q_2}{d^2}$$

where f = force
K = a constant of proportionality, $9(10^9)$
Q_1 = first charge
Q_2 = second charge
d = distance between charges

This is Coulomb's law. It says that the force of attraction or repulsion between two charges is directly proportional to the charges and inversely proportional to the square of the distance between them.

This is an important equation, for it is the foundation of electricity. But where does it come from? And why is it true? To begin with, all the variables in this law existed before its discovery. By experiment, Coulomb was able to prove that the force was directly proportional to each charge and inversely proportional to the square of the distance between the charges. Coulomb's law is an example of a relationship that exists in nature. Although earlier researchers could measure f, Q_1, Q_2, and d, Coulomb discovered the law relating the quantities and wrote a formula for it.

Before discovering a law, someone may have a hunch that such a relationship exists. After a number of experiments, the researcher writes a formula that summarizes the discovery. When enough people confirm the discovery through experiments, the formula becomes a law. *A law is true because you can verify it with an experiment.*

The Derivation

Given an equation like this:

$$y = 3x$$

we can add 5 to both sides to get:

$$y + 5 = 3x + 5$$

The new equation is true because both sides are still equal. There are many other operations like subtraction, multiplication, division, factoring, and substitution that preserve the equality of both sides of the equation. For this reason, we can derive many new formulas using mathematics.

A **derivation** *is a formula that we can get from other formulas.* This means that we start with one or more formulas and, using mathematics, arrive at a new formula not in our original set of formulas. A derivation is true because mathematics preserves the equality of both sides of every equation between the starting formula and the derived formula.

For instance, Ohm was experimenting with conductors. He discovered that the ratio of voltage to current was a constant. He named this constant *resistance* and wrote the following formula for it:

$$R = \frac{V}{I}$$

This is the original form of Ohm's law. By rearranging it, we can get:

$$I = \frac{V}{R}$$

This is a derivation. It is the original form of Ohm's law converted to another equation.

Here is another example. The definition for capacitance is:

$$C = \frac{Q}{V}$$

We can multiply both sides by V to get the following new equation:

$$Q = CV$$

This is a derivation. It says that the charge on a capacitor equals its capacitance times the voltage across it.

What to Remember

Why is a formula true? There are three possible answers. To build your understanding of electronics on solid ground, classify each new formula in one of these three categories:

> Definition: A formula invented for a new concept
> Law: A formula for a relationship in nature
> Derivation: A formula produced with mathematics

1-2 Approximations

We use approximations all the time in everyday life. If someone asks you how old you are, you might answer 21 (ideal). Or you might say 21 going on 22 (second approximation). Or, maybe, 21 years and 9 months (third approximation). Or, if you want to be more accurate, 21 years, 9 months, 2 days, 6 hours, 23 minutes, and 42 seconds (exact).

The foregoing illustrates different levels of approximation: an ideal approximation, a second approximation, a third approximation, and an exact answer. The approximation to use will depend on the situation. The same is true in electronics work. In circuit analysis, we need to choose an approximation that fits the situation.

The Ideal Approximation

Did you know that 1 ft of AWG 22 wire that is 1 in from a chassis has a resistance of 0.016 Ω, an inductance of 0.24 μH, and a capacitance of 3.3 pF? If we had to include the effects of resistance, inductance, and capacitance in every calculation for current, we would spend too much time on calculations. This is why everybody ignores the resistance, inductance, and capacitance of connecting wires in most situations.

The **ideal approximation,** sometimes called the **first approximation,** is the simplest equivalent circuit for a device. For instance, the ideal approximation of a piece of wire is a conductor of zero resistance. This ideal approximation is adequate for everyday electronics work.

The exception occurs at higher frequencies, where you have to consider the inductance and capacitance of the wire. Suppose 1 in of wire has an inductance of 0.24 μH and a capacitance of 3.3 pF. At 10 MHz, the inductive reactance is 15.1 Ω, and the capacitive reactance is 4.82 kΩ. As you see, a circuit designer can no longer idealize a piece of wire. Depending on the rest of the circuit, the inductance and capacitive reactances of a connecting wire may be important.

As a guideline, we can idealize a piece of wire at frequencies under 1 MHz. This is usually a safe rule of thumb. But it does not mean that you can be careless about wiring. In general, keep connecting wires as short as possible, because at some point on the frequency scale, those wires will begin to degrade circuit performance.

When you are troubleshooting, the ideal approximation is usually adequate because you are looking for large deviations from normal voltages and currents. In this book, we will idealize semiconductor devices by reducing them to simple equivalent circuits. With ideal approximations, it is easier to analyze and understand how semiconductor circuits work.

The Second Approximation

The ideal approximation of a flashlight battery is a voltage source of 1.5 V. The **second approximation** adds one or more components to the ideal approximation. For instance, the second approximation of a flashlight battery is a voltage source of 1.5 V and a series resistance of 1 Ω. This series resistance is called the *source* or *internal* resistance of the battery. If the load resistance is less than 10 Ω, the load voltage will be noticeably less than 1.5 V because of the voltage drop across the source resistance. In this case, accurate calculations must include the source resistance.

The Third Approximation and Beyond

The **third approximation** includes another component in the equivalent circuit of the device. Chapter 3 will give you an example of the third approximation when we discuss semiconductor diodes.

Even higher approximations are possible with many components in the equivalent circuit of a device. Hand calculations using these higher approximations can become very difficult and time-consuming. Because of this, computers using circuit simulation software are often used. For instance, MultiSim by Electronics Workbench (EWB) and PSpice are commercially available computer programs that use higher approximations to analyze semiconductor circuits. Many of the circuits and examples in this book can be analyzed and demonstrated using this type of software.

Conclusion

Which approximation to use depends on what you are trying to do. If you are troubleshooting, the ideal approximation is usually adequate. For many situations, the second approximation is the best choice because it is easy to use and does not require a computer. For higher approximations, you should use a computer and a program like MultiSim.

1–3 Voltage Sources

An *ideal dc voltage source* produces a load voltage that is constant. The simplest example of an ideal dc voltage source is a perfect battery, one whose internal resistance is zero. Figure 1-1a shows an ideal voltage source connected to a variable load resistance of 1 Ω to 10 MΩ. The voltmeter reads 10 V, exactly the same as the source voltage.

Figure 1-1b shows a graph of load voltage versus load resistance. As you can see, the load voltage remains fixed at 10 V when the load resistance changes from 1 Ω to 1 MΩ. In other words, an ideal dc voltage source produces a constant load voltage, regardless of how small or large the load resistance is. With an ideal voltage source, only the load current changes when the load resistance changes.

Figure 1–1 (*a*) Ideal voltage source and variable load resistance; (*b*) load voltage is constant for all load resistances.

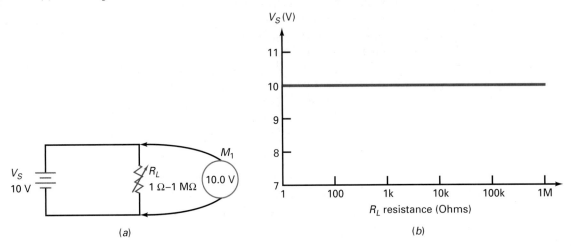

(*a*)   (*b*)

Second Approximation

An ideal voltage source is a theoretical device; it cannot exist in nature. Why? When the load resistance approaches zero, the load current approaches infinity. No real voltage source can produce infinite current because a real voltage source always has some internal resistance. The second approximation of a dc voltage source includes this internal resistance.

Figure 1-2*a* illustrates the idea. A source resistance R_S of 1 Ω is now in series with the ideal battery. The voltmeter reads 5 V when R_L is 1 Ω. Why? Because the load current is 10 V divided by 2 Ω, or 5 A. When 5 A flows through the source resistance of 1 Ω, it produces an internal voltage drop of 5 V. This is why the load voltage is only half of the ideal value, with the other half being dropped across the internal resistance.

Figure 1-2*b* shows the graph of load voltage versus load resistance. In this case, the load voltage does not come close to the ideal value until the load resistance is much greater than the source resistance. But what does *much greater* mean? In other words, when can we ignore the source resistance?

Figure 1–2 (*a*) Second approximation includes source resistance; (*b*) load voltage is constant for large load resistances.

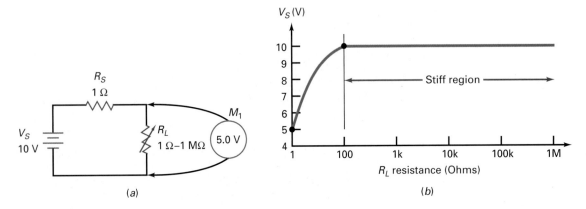

(*a*)   (*b*)

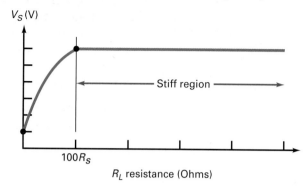

Figure 1–3 Stiff region occurs when load resistance is large enough.

Stiff Voltage Source

Now is the time when a new definition can be useful. So, let us invent one. We can ignore the source resistance when it is at least 100 times smaller than the load resistance. Any source that satisfies this condition is a *stiff voltage source*. As a definition,

$$\text{Stiff voltage source: } R_S < 0.01R_L \tag{1-1}$$

This formula defines what we mean by a *stiff voltage source*. The boundary of the inequality (where $<$ is changed to $=$) gives us the following equation:

$$R_S = 0.01R_L$$

Solving for load resistance gives the minimum load resistance we can use and still have a stiff source:

$$R_{L(\text{min})} = 100R_S \tag{1-2}$$

In words, the minimum load resistance equals 100 times the source resistance.

Equation (1-2) is a derivation. We started with the definition of a stiff voltage source and rearranged it to get the minimum load resistance permitted with a stiff voltage source. As long as the load resistance is greater than $100R_S$, the voltage source is stiff. When the load resistance equals this worst-case value, the calculation error from ignoring the source resistance is 1 percent, small enough to ignore in a second approximation.

Figure 1-3 visually summarizes a stiff voltage source. The load resistance has to be greater than $100R_S$ for the voltage source to be stiff.

Example 1–1

The definition of a stiff voltage source applies to ac sources as well as to dc sources. Suppose an ac voltage source has a source resistance of 50 Ω. For what load resistance is the source stiff?

SOLUTION   Multiply by 100 to get the minimum load resistance:

$$R_L = 100R_S = 100(50 \; \Omega) = 5 \; \text{k}\Omega$$

As long as the load resistance is greater than 5 kΩ, the ac voltage source is stiff and we can ignore the internal resistance of the source.

A final point. Using the second approximation for an ac voltage source is valid only at low frequencies. *At high frequencies, additional factors such as lead inductance and stray capacitance come into play.* We will deal with these high-frequency effects in a later chapter.

PRACTICE PROBLEM 1–1   If the ac source resistance in Example 1-1 is 600 Ω, for what load resistance is the source stiff?

GOOD TO KNOW

A well-regulated power supply is a good example of a stiff voltage source.

GOOD TO KNOW

At the output terminals of a constant current source, the load voltage V_L increases in direct proportion to the load resistance.

1-4 Current Sources

A dc voltage source produces a constant load voltage for different load resistances. A *dc current source* is different. It produces a constant load current for different load resistances. An example of a dc current source is a battery with a large source resistance (Fig. 1-4a). In this circuit, the source resistance is 1 MΩ and the load current is:

$$I_L = \frac{V_S}{R_S + R_L}$$

When R_L is 1 Ω in Fig. 1-4a, the load current is:

$$I_L = \frac{10\ \text{V}}{1\ \text{MΩ} + 1\ \text{Ω}} = 10\ \mu\text{A}$$

In this calculation, the small load resistance has an insignificant effect on the load current.

Figure 1-4b shows the effect of varying the load resistance from 1 Ω to 1 MΩ. In this case, the load current remains constant at 10 μA over a large range. It is only when the load resistance is greater than 10 kΩ that a noticeable drop-off occurs in load current.

Figure 1–4   (*a*) Simulated current source with a dc voltage source and a large resistance; (*b*) load current is constant for small load resistances.

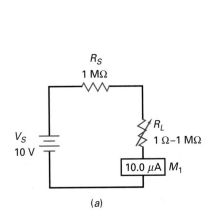

(a)

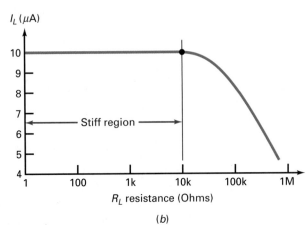

(b)

Stiff Current Source

Here is another definition that will be useful, especially with semiconductor circuits. We will ignore the source resistance of a current source when it is at least 100 times larger than the load resistance. Any source that satisfies this condition is a *stiff current source*. As a definition:

$$\text{Stiff current source: } R_S > 100R_L \tag{1-3}$$

The upper boundary is the worst case. At this point:

$$R_S = 100R_L$$

Solving for load resistance gives the maximum load resistance we can use and still have a stiff current source:

$$R_{L(\text{max})} = 0.01R_S \tag{1-4}$$

In words: The maximum load resistance equals $\frac{1}{100}$ of the source resistance.

Equation (1-4) is a derivation, because we started with the definition of a stiff current source and rearranged it to get the maximum load resistance. When the load resistance equals this worst-case value, the calculation error is 1 percent, small enough to ignore in a second approximation.

Figure 1-5 shows the stiff region. As long as the load resistance is less than $0.01R_S$, the current source is stiff.

Schematic Symbol

Figure 1-6a is the schematic symbol of an ideal current source, one whose source resistance is infinite. This ideal approximation cannot exist in nature, but it can exist mathematically. Therefore, we can use the ideal current source for fast circuit analysis, as in troubleshooting.

Figure 1-6a is a visual definition: It is the symbol for a current source. When you see this symbol, it means that the device produces a constant current I_S. It may help to think of a current source as a pump that pushes out a fixed number of coulombs per second. This is why you will hear expressions like "The current source pumps 5 mA through a load resistance of 1 kΩ."

Figure 1-6b shows the second approximation. The internal resistance is in parallel with the ideal current source, not in series as it was with an ideal voltage source. Later in this chapter we will discuss Norton's theorem. You will then see why the internal resistance must be in parallel with the current source. Table 1-1 will help you understand the differences between a voltage source and a current source.

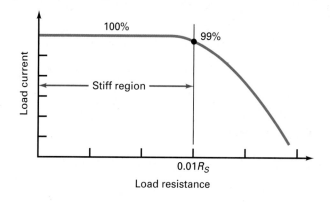

Figure 1–5   Stiff region occurs when load resistance is small enough.

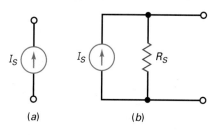

Figure 1–6   (*a*) Schematic symbol of a current source; (*b*) second approximation of a current source.

Table 1–1	Properties of Voltage and Current Sources	
Quantity	**Voltage Source**	**Current Source**
R_S	Typically low	Typically high
R_L	Greater than $100R_S$	Less than $0.01R_S$
V_L	Constant	Depends on R_L
I_L	Depends on R_L	Constant

Example 1-2

A current source of 2 mA has an internal resistance of 10 MΩ. Over what range of load resistance is the current source stiff?

SOLUTION Since this is a current source, the load resistance has to be small compared to the source resistance. With the 100:1 rule, the maximum load resistance is:

$$R_{L(max)} = 0.01(10 \text{ M}\Omega) = 100 \text{ k}\Omega$$

The stiff range for the current source is a load resistance from 0 to 100 kΩ.

Figure 1-7 summarizes the solution. In Fig. 1-7a, a current source of 2 mA is in parallel with 10 MΩ and a variable resistor set to 1 Ω. The ammeter measures a load current of 2 mA. When the load resistance changes from 1 Ω to 1 MΩ, as shown in Fig. 1-7b, the source remains stiff up to 100 kΩ. At this point, the load current is down about 1 percent from the ideal value. Stated another way, 99 percent of the source current passes through the load resistance. The other 1 percent passes through the source resistance. As the load resistance continues to increase, load current continues to decrease.

Figure 1-7 Solution.

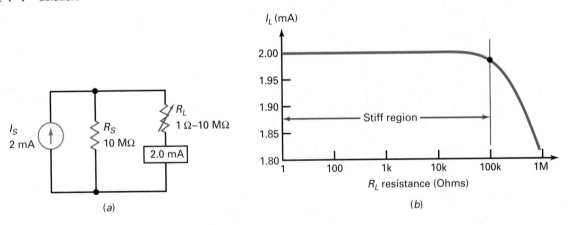

(a)

(b)

PRACTICE PROBLEM 1-2 What is the load voltage in Fig. 1-7a when the load resistance equals 10 kΩ?

Example 1-3

When you analyze transistor circuits, you will visualize a transistor as a current source. In a well-designed circuit, the transistor will act like a stiff current source, so that you can ignore its internal resistance. Then you can calculate the load voltage. For instance, if a transistor is pumping 2 mA through a load resistance of 10 kΩ, the load voltage is 20 V.

1-5 Thevenin's Theorem

Every once in a while, somebody makes a big breakthrough in engineering and carries all of us to a new high. A French engineer, M. L. Thevenin, made one of these quantum leaps when he derived the circuit theorem named after him: Thevenin's theorem.

Definition of Thevenin Voltage and Resistance

A **theorem** is a statement that we can prove mathematically. Because of this, it is not a definition or a law. So, we classify it as a derivation. Recall the following ideas about Thevenin's theorem from earlier courses. In Fig. 1-8a, the **Thevenin voltage** V_{TH} is defined as the voltage across the load terminals when the load resistor is open. Because of this, the Thevenin voltage is sometimes called the *open-circuit voltage*. As a definition:

$$\text{Thevenin voltage: } V_{TH} = V_{OC} \tag{1-5}$$

The **Thevenin resistance** is defined as the resistance that an ohmmeter measures across the load terminals of Fig. 1-8a when all sources are reduced to zero and the load resistor is open. As a definition:

$$\text{Thevenin resistance: } R_{TH} = R_{OC} \tag{1-6}$$

With these two definitions, Thevenin was able to derive the famous theorem named after him.

There is a subtle point in finding the Thevenin resistance. Reducing a source to zero has different meanings for voltage and current sources. When you reduce a voltage source to zero, you are effectively replacing it by a short because that's the only way to guarantee zero voltage when a current flows through the voltage source. When you reduce a current source to zero, you are effectively replacing it by an open because that's the only way you can guarantee zero current when there is a voltage across the current source. To summarize:

To zero a voltage source, replace it by a short.
To zero a current source, replace it by an open.

The Derivation

What is Thevenin's theorem? Look at Fig. 1-8a. This black box can contain any circuit with dc sources and linear resistances. (A *linear resistance* does not change with increasing voltage.) Thevenin was able to prove that no matter how

Figure 1-8 (*a*) Black box has a linear circuit inside of it; (*b*) Thevenin circuit.

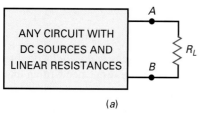

(a)

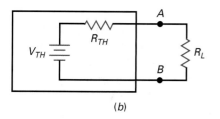

(b)

complicated the circuit inside the black box of Fig. 1-8a was, it would produce exactly the same load current as the simple circuit of Fig. 1-8b. As a derivation:

$$I_L = \frac{V_{TH}}{R_{TH} + R_L} \tag{1-7}$$

Let the idea sink in. Thevenin's theorem is a powerhouse tool. Engineers and technicians use the theorem constantly. Electronics could not possibly be where it is today without the Thevenin theorem. It not only simplifies calculations, it enables us to explain circuit operation that would be impossible to explain with only Kirchhoff equations.

Example 1-4

‖‖ MultiSim

What are the Thevenin voltage and resistance in Fig. 1-9a?

SOLUTION   First, calculate the Thevenin voltage. To do this, you have to open the load resistor. Opening the load resistance is equivalent to removing it from the circuit, as shown in Fig. 1-9b. Since 8 mA flows through 6 kΩ in series with 3 kΩ, 24 V will appear across the 3 kΩ. With no current through the 4 kΩ, 24 V will appear across the *AB* terminals. Therefore:

$$V_{TH} = 24 \text{ V}$$

Second, get the Thevenin resistance. Reducing a dc source to zero is equivalent to replacing it by a short, as shown in Fig. 1-9c. If we connect an ohmmeter across the *AB* terminals of Fig. 1-9c, what will it read?

It will read 6 kΩ. Why? Because looking back into the *AB* terminals with the battery shorted, the ohmmeter sees 4 kΩ in series with a parallel connection of 3 kΩ and 6 kΩ. We can write:

$$R_{TH} = 4 \text{ k}\Omega + \frac{3\,\text{k}\Omega \times 6\,\text{k}\Omega}{3\,\text{k}\Omega + 6\,\text{k}\Omega} = 6\,\text{k}\Omega$$

The product over sum of 3 kΩ and 6 kΩ is 2 kΩ, which, added to 4 kΩ, gives 6 kΩ.

Again, we need a new definition. Parallel connections occur so often in electronics that most people use a shorthand notation for them. From now on, we will use the following notation:

$$\| = \text{in parallel with}$$

Whenever you see two vertical bars in an equation, it means *in parallel with*. In industry, you will see the foregoing equation for Thevenin resistance written like this:

$$R_{TH} = 4 \text{ k}\Omega + (3 \text{ k}\Omega \,\|\, 6 \text{ k}\Omega) = 6 \text{ k}\Omega$$

Most engineers and technicians know that the vertical bars mean *in parallel with*. So, they automatically use product over sum or reciprocal method to calculate the equivalent resistance of 3 kΩ and 6 kΩ.

Figure 1-10 shows the Thevenin circuit with a load resistor. Compare this simple circuit with the original circuit of Fig. 1-9a. Can you see how much

Figure 1-9 (a) Original circuit; (b) open-load resistor to get Thevenin voltage; (c) reduce source to zero to get Thevenin resistance.

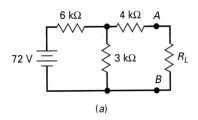

(a)

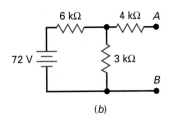

(b)

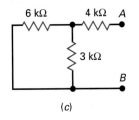

(c)

Figure 1–10 Thevenin circuit for Fig. 1-9a.

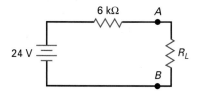

easier it will be to calculate the load current for different load resistances? If not, the next example will drive the point home.

PRACTICE PROBLEM 1–4 Using Thevenin's theorem, what is the load current in Fig. 1-9a for the following values of R_L: 2 kΩ, 6 kΩ, and 18 kΩ?

 If you really want to appreciate the power of Thevenin's theorem, try calculating the foregoing currents using the original circuit of Fig. 1-9a and any other method.

Example 1–5

A *breadboard* is a circuit often built with solderless connections without regard to the final location of parts to prove the feasibility of a design. Suppose you have the circuit of Fig. 1-11a breadboarded on a lab bench. How would you measure the Thevenin voltage and resistance?

SOLUTION Start by replacing the load resistor with a multimeter, as shown in Fig. 1-11b. After you set the multimeter to read volts, it will indicate 9 V. This is the Thevenin voltage. Next, replace the dc source by a short (Fig. 1-11c). Set the multimeter to read ohms, and it will indicate 1.5 kΩ. This is the Thevenin resistance.

 Are there any sources of error in the foregoing measurements? Yes, the one thing to watch out for is the input impedance of the multimeter when voltage is measured. Because this input impedance is across the measured terminals, a small current flows through the multimeter. For instance, if you use a moving-coil multimeter, the typical sensitivity is 20 kΩ per volt. On the 10-V range, the voltmeter has an input resistance of 200 kΩ. This will load the circuit down slightly and decrease the load voltage from 9 to 8.93 V.

 As a guideline, the input impedance of the voltmeter should be at least 100 times greater than the Thevenin resistance. Then, the loading error is less than 1 percent. *To avoid loading error, use a field-effect transistor (FET) input or digital multimeter (DMM) instead of a moving-coil multimeter.* The input impedance of these instruments is at least 10 MΩ, which usually eliminates loading error.

Figure 1–11 (a) Circuit on lab bench; (b) measuring Thevenin voltage; (c) measuring Thevenin resistance.

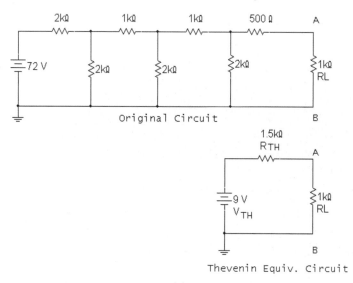

Figure 1-11 (continued)

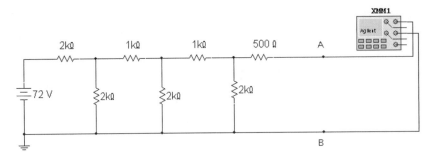

Measuring the Thevenin Voltage

(*b*)

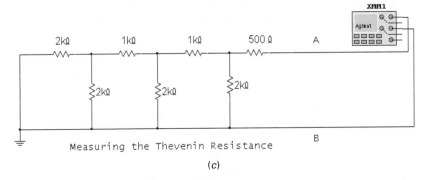

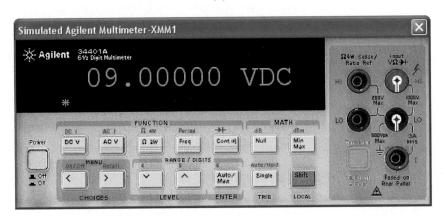

Measuring the Thevenin Resistance

(*c*)

1-6 Norton's Theorem

Recall the following ideas about Norton's theorem from earlier courses. In Fig. 1-12a, the Norton current I_N is defined as the load current when the load resistor is shorted. Because of this, the **Norton current** is sometimes called the *short-circuit current*. As a definition:

$$\text{Norton current: } I_N = I_{SC} \tag{1-8}$$

The **Norton resistance** is the resistance that an ohmmeter measures across the load terminals when all sources are reduced to zero and the load resistor is open. As a definition:

$$\text{Norton resistance: } R_N = R_{OC} \tag{1-9}$$

Since Thevenin resistance also equals R_{OC}, we can write:

$$R_N = R_{TH} \tag{1-10}$$

This derivation says that Norton resistance equals Thevenin resistance. If you calculate a Thevenin resistance of 10 kΩ, you immediately know that the Norton resistance equals 10 kΩ.

Basic Idea

What is Norton's theorem? Look at Fig. 1-12a. This black box can contain any circuit with dc sources and linear resistances. Norton proved that the circuit inside the black box of Fig. 1-12a would produce exactly the same load voltage as the simple circuit of Fig. 1-12b. As a derivation, Norton's theorem looks like this:

$$V_L = I_N(R_N \| R_L) \tag{1-11}$$

In words: The load voltage equals the Norton current times the Norton resistance in parallel with the load resistance.

Earlier we saw that Norton resistance equals Thevenin resistance. But notice the difference in the location of the resistors: Thevenin resistance is always in series with a voltage source; Norton resistance is always in parallel with a current source.

Note: If you are using electron flow, keep the following in mind. In industry, the arrow inside the current source is almost always drawn in the direction of conventional current. The exception is a current source drawn with a dashed arrow instead of a solid arrow. In this case, the source pumps electrons in the direction of the dashed arrow.

Figure 1-12 (*a*) Black box has a linear circuit inside of it; (*b*) Norton circuit.

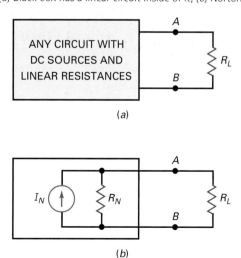

Figure 1-13 Duality principle: Thevenin theorem implies Norton theorem and vice versa. (*a*) Converting Thevenin to Norton; (*b*) converting Norton to Thevenin.

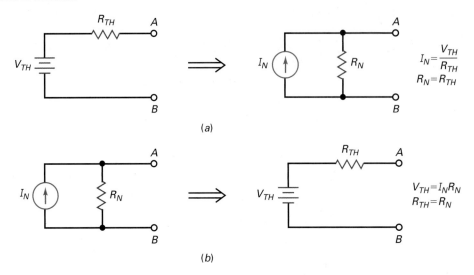

$$I_N = \frac{V_{TH}}{R_{TH}}$$
$$R_N = R_{TH}$$

(*a*)

$$V_{TH} = I_N R_N$$
$$R_{TH} = R_N$$

(*b*)

The Derivation

Norton's theorem can be derived from the **duality principle.** It states that for any theorem in electrical circuit analysis there is a dual (opposite) theorem in which one replaces the original quantities with dual quantities. Here is a brief list of dual quantities:

Voltage ⟷ Current
Voltage source ⟷ Current source
Series ⟷ Parallel
Series resistance ⟷ Parallel resistance

Figure 1-13 summarizes the duality principle as it applies to Thevenin and Norton circuits. It means that we can use either circuit in our calculations. As you will see later, both equivalent circuits are useful. Sometimes, it is easier to use Thevenin. At other times, we use Norton. It depends on the specific problem. Summary Table 1-1 shows the steps for getting the Thevenin and Norton quantities.

Summary Table 1-1	Thevenin and Norton Values	
Process	**Thevenin**	**Norton**
Step 1	Open the load resistor.	Short the load resistor.
Step 2	Calculate or measure the open-circuit voltage. This is the Thevenin voltage.	Calculate or measure the short-circuit current. This is the Norton current.
Step 3	Short voltage sources and open current sources.	Short voltage sources, open current sources, and open load resistor.
Step 4	Calculate or measure the open-circuit resistance. This is the Thevenin resistance.	Calculate or measure the open-circuit resistance. This is the Norton resistance.

Relationships between Thevenin and Norton Circuits

We already know that the Thevenin and Norton resistances are equal in value, but different in location: Thevenin resistance is in series with a voltage source, and Norton resistance is in parallel with a current source.

We can derive two more relationships, as follows. We can convert any Thevenin circuit to a Norton circuit, as shown in Fig. 1-13a. The proof is straightforward. Short the AB terminals of the Thevenin circuit, and you get the Norton current:

$$I_N = \frac{V_{TH}}{R_{TH}} \qquad \qquad \text{(1-12)}$$

This derivation says that the Norton current equals the Thevenin voltage divided by the Thevenin resistance.

Similarly, we can convert any Norton circuit to a Thevenin circuit, as shown in Fig. 1-13b. The open-circuit voltage is:

$$V_{TH} = I_N R_N \qquad \qquad \text{(1-13)}$$

This derivation says that the Thevenin voltage equals the Norton current times the Norton resistance.

Figure 1-13 summarizes the equations for converting either circuit into the other.

Example 1-6

Suppose that we have reduced a complicated circuit to the Thevenin circuit shown in Fig. 1-14a. How can we convert this to a Norton circuit?

Figure 1-14   Calculating Norton current.

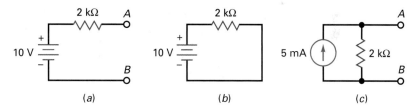

(a)   (b)   (c)

SOLUTION   Use Eq. (1-12) to get:

$$I_N = \frac{10\text{V}}{2\text{k}\Omega} = 5 \text{ mA}$$

Figure 1-14c shows the Norton circuit.

Most engineers and technicians forget Eq. (1-12) soon after they leave school. But they always remember how to solve the same problem using Ohm's law. Here is what they do. Look at Fig. 1-14a. Visualize a short across the AB terminals, as shown in Fig. 1-14b. The short-circuit current equals the Norton current:

$$I_N = \frac{10\text{V}}{2\text{k}\Omega} = 5 \text{ mA}$$

This is the same result, but calculated with Ohm's law applied to the Thevenin circuit. Figure 1-15 summarizes the idea. This memory aid will help you calculate the Norton current, given the Thevenin circuit.

Figure 1-15 A memory aid for Norton current.

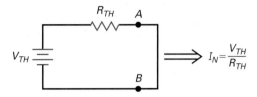

PRACTICE PROBLEM 1-6 If the Thevenin resistance of Fig. 1-14*a* is 5 kΩ, determine the Norton current value.

1-7 Troubleshooting

Troubleshooting means finding out why a circuit is not doing what it is supposed to do. The most common troubles are opens and shorts. Devices like transistors can become open or shorted in a number of ways. One way to destroy any transistor is by exceeding its maximum-power rating.

Resistors become open when their power dissipation is excessive. But you can get a shorted resistor indirectly as follows. During the stuffing and soldering of printed-circuit boards, an undesirable splash of solder may connect two nearby conducting lines. Known as a **solder bridge,** this effectively shorts any device between the two conducting lines. On the other hand, a poor solder connection usually means no connection at all. This is known as a **cold-solder joint** and means that the device is open.

Besides opens and shorts, anything is possible. For instance, temporarily applying too much heat to a resistor may permanently change the resistance by several percent. If the value of resistance is critical, the circuit may not work properly after the heat shock.

And then there is the troubleshooter's nightmare: the intermittent trouble. This kind of trouble is very difficult to isolate because it appears and disappears. It may be a cold-solder joint that alternately makes and breaks a contact, or a loose cable connector, or any similar trouble that causes on-again, off-again operation.

An Open Device

Always remember these two facts about an **open device:**

> *The current through an open device is zero.*
> *The voltage across it is unknown.*

The first statement is true because an open device has infinite resistance. No current can exist in an infinite resistance. The second statement is true because of Ohm's law:

$$V = IR = (0)(\infty)$$

In this equation, zero times infinity is mathematically indeterminate. You have to figure out what the voltage is by looking at the rest of the circuit.

A Shorted Device

A shorted device is exactly the opposite. Always remember these two statements about a **shorted device:**

> *The voltage across a shorted device is zero.*
> *The current through it is unknown.*

The first statement is true because a shorted device has zero resistance. No voltage can exist across zero resistance. The second statement is true because of Ohm's law:

$$I = \frac{V}{R} = \frac{0}{0}$$

Zero divided by zero is mathematically meaningless. You have to figure out what the current is by looking at the rest of the circuit.

Procedure

Normally, you measure voltages with respect to ground. From these measurements and your knowledge of basic electricity, you can usually deduce the trouble. After you have isolated a component as the top suspect, you can unsolder or disconnect the component and use an ohmmeter or other instrument for confirmation.

Figure 1–16 Voltage divider and load used in troubleshooting discussion.

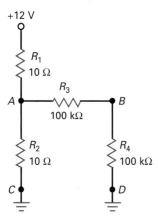

Normal Values

In Fig. 1-16, a stiff voltage divider consisting of R_1 and R_2 drives resistors R_3 and R_4 in series. Before you can troubleshoot this circuit, you have to know what the normal voltages are. The first thing to do, therefore, is to work out the values of V_A and V_B. The first is the voltage between A and ground. The second is the voltage between B and ground. Because R_1 and R_2 are much smaller than R_3 and R_4 (10 Ω versus 100 kΩ), the stiff voltage at A is approximately +6 V. Furthermore, since R_3 and R_4 are equal, the voltage at B is approximately +3 V. When this circuit is trouble-free, you will measure 6 V between A and ground, and 3 V between B and ground. These two voltages are the first entry of Table 1-2.

R_1 Open

When R_1 is open, what do you think happens to the voltages? Since no current can flow through the open R_1, no current can flow through R_2. Ohm's law tells us the voltage across R_2 is zero. Therefore, $V_A = 0$ and $V_B = 0$, as shown in Table 1-2 for R_1 open.

R_2 Open

When R_2 is open, what happens to the voltages? Since no current can flow through the open R_2, the voltage at A is pulled up toward the supply voltage. Since R_1 is much smaller than R_3 and R_4, the voltage at A is approximately 12 V. Since R_3 and R_4 are equal, the voltage at B becomes 6 V. This is why $V_A = 12$ V and $V_B = 6$ V, as shown in Table 1-2 for an R_2 open.

Table 1-2	Troubles and Clues	
Trouble	V_A	V_B
Circuit OK	6 V	3 V
R_1 open	0	0
R_2 open	12 V	6 V
R_3 open	6 V	0
R_4 open	6 V	6 V
C open	12 V	6 V
D open	6 V	6 V
R_1 shorted	12 V	6 V
R_2 shorted	0	0
R_3 shorted	6 V	6 V
R_4 shorted	6 V	0

Remaining Troubles

If ground C is open, no current can pass through R_2. This is equivalent to an open R_2. This is why the trouble C open has $V_A = 12$ V and $V_B = 6$ V in Table 1-2.

You should work out all of remaining entries in Table 1-2, making sure that you understand why each voltage exists for the given trouble.

Example 1-7

In Fig. 1-16, you measure $V_A = 0$ and $V_B = 0$. What is the trouble?

SOLUTION   Look at Table 1-2. As you can see, two troubles are possible: R_1 open or R_2 shorted. Both of these produce zero voltage at points A and B. To isolate the trouble, you can disconnect R_1 and measure it. If it measures open, you have found the trouble. If it measures OK, then R_2 is the trouble.

PRACTICE PROBLEM 1-7   What could the possible troubles be if you measure $V_A = 12$ V and $V_B = 6$ V in Fig. 1-16?

Summary

SEC. 1-1 THE THREE KINDS OF FORMULAS

A *definition* is a formula invented for a new concept. A *law* is a formula for a relation in nature. A *derivation* is a formula produced with mathematics.

SEC. 1-2 APPROXIMATIONS

Approximations are widely used in industry. The ideal approximation is useful for troubleshooting. The second approximation is useful for preliminary circuit calculations. Higher approximations are used with computers.

SEC. 1-3 VOLTAGE SOURCES

An ideal voltage source has no internal resistance. The second approximation of a voltage source has an internal resistance in series with the source. A stiff voltage source is defined as one whose internal resistance is less than $^1/_{100}$ of the load resistance.

SEC. 1-4 CURRENT SOURCES

An ideal current source has an infinite internal resistance. The second approximation of a current source has a large internal resistance in parallel with the source. A *stiff current source* is defined as one whose internal resistance is more than 100 times the load resistance.

SEC. 1-5 THEVENIN'S THEOREM

The *Thevenin voltage* is defined as the voltage across an open load. The *Thevenin resistance* is defined as the resistance an ohmmeter would measure with an open load and all sources reduced to zero. Thevenin proved that a Thevenin equivalent circuit will produce the same load current as any other circuit with sources and linear resistances.

SEC. 1-6 NORTON'S THEOREM

The Norton resistance equals the Thevenin resistance. The Norton current equals the load current when the load is shorted. Norton proved that a Norton equivalent circuit produces the same load voltage as any other circuit with sources and linear resistances. Norton current equals Thevenin voltage divided by Thevenin resistance.

SEC. 1-7 TROUBLESHOOTING

The most common troubles are shorts, opens, and intermittent troubles. A short always has zero voltage across it; the current through a short must be calculated by examining the rest of the circuit. An open always has zero current through it; the voltage across an open must be calculated by examining the rest of the circuit. An intermittent trouble is an on-again, off-again trouble that requires patient and logical troubleshooting to isolate it.

Definitions

(1-1)   Stiff voltage source:

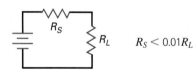

$$R_S < 0.01R_L$$

(1-3)   Stiff current source:

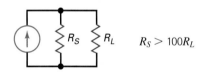

$$R_S > 100R_L$$

(1-5)   Thevenin voltage:

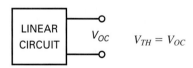

$$V_{TH} = V_{OC}$$

(1-6)   Thevenin resistance:

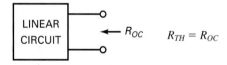

$$R_{TH} = R_{OC}$$

(1-8)   Norton current:

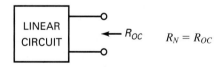

$$I_N = I_{SC}$$

(1-9)   Norton resistance:

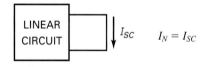

$$R_N = R_{OC}$$

Derivations

(1-2)   Stiff voltage source:

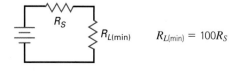

$$R_{L(min)} = 100R_S$$

(1-4)   Stiff current source:

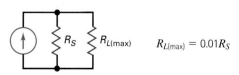

$$R_{L(max)} = 0.01R_S$$

(1-7)   Thevenin's theorem:

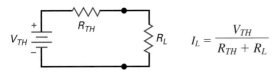

$$I_L = \frac{V_{TH}}{R_{TH} + R_L}$$

(1-10)   Norton resistance:

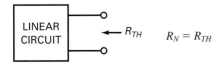

$$R_N = R_{TH}$$

(1-11)   Norton's theorem:

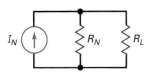

$$V_L = I_N(R_N \| R_L)$$

(1-12)   Norton current:

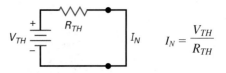

$$I_N = \frac{V_{TH}}{R_{TH}}$$

(1-13)   Thevenin voltage:

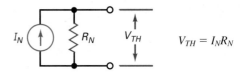

$$V_{TH} = I_N R_N$$

Student Assignments

1. **An ideal voltage source has**
   a. Zero internal resistance
   b. Infinite internal resistance
   c. A load-dependent voltage
   d. A load-dependent current

2. **A real voltage source has**
   a. Zero internal resistance
   b. Infinite internal resistance
   c. A small internal resistance
   d. A large internal resistance

3. **If a load resistance is 100 Ω, a stiff voltage source has a resistance of**
   a. Less than 1 Ω
   b. At least 10 Ω
   c. More than 10 kΩ
   d. Less than 10 kΩ

4. **An ideal current source has**
   a. Zero internal resistance
   b. Infinite internal resistance
   c. A load-dependent voltage
   d. A load-dependent current

5. **A real current source has**
   a. Zero internal resistance
   b. Infinite internal resistance
   c. A small internal resistance
   d. A large internal resistance

6. **If a load resistance is 100 Ω, a stiff current source has a resistance of**
   a. Less than 1 Ω
   b. Less than 1 Ω

   c. Less than 10 kΩ
   d. More than 10 kΩ

7. **The Thevenin voltage is the same as the**
   a. Shorted-load voltage
   b. Open-load voltage
   c. Ideal source voltage
   d. Norton voltage

8. **The Thevenin resistance is equal in value to the**
   a. Load resistance
   b. Half the load resistance
   c. Internal resistance of a Norton circuit
   d. Open-load resistance

9. **To get the Thevenin voltage, you have to**
   a. Short the load resistor
   b. Open the load resistor
   c. Short the voltage source
   d. Open the voltage source

10. **To get the Norton current, you have to**
    a. Short the load resistor
    b. Open the load resistor
    c. Short the voltage source
    d. Open the current source

11. **The Norton current is sometimes called the**
    a. Shorted-load current
    b. Open-load current
    c. Thevenin current
    d. Thevenin voltage

12. **A solder bridge**
    a. May produce a short
    b. May cause an open
    c. Is useful in some circuits
    d. Always has high resistance

13. **A cold-solder joint**
    a. Always has low resistance
    b. Shows good soldering technique
    c. Usually produces an open
    d. Will cause a short circuit

14. **An open resistor has**
    a. Infinite current through it
    b. Zero voltage across it
    c. Infinite voltage across it
    d. Zero current through it

15. **A shorted resistor has**
    a. Infinite current through it
    b. Zero voltage across it
    c. Infinite voltage across it
    d. Zero current through it

16. **An ideal voltage source and an internal resistance are examples of the**
    a. Ideal approximation
    b. Second approximation
    c. Higher approximation
    d. Exact model

17. **Treating a connecting wire as a conductor with zero resistance is an example of the**
    a. Ideal approximation
    b. Second approximation
    c. Higher approximation
    d. Exact model

18. **The voltage out of an ideal voltage source**

   a. Is zero

   b. Is constant

   c. Depends on the value of load resistance

   d. Depends on the internal resistance

19. **The current out of an ideal current source**

   a. Is zero

   b. Is constant

   c. Depends on the value of load resistance

   d. Depends on the internal resistance

20. **Thevenin's theorem replaces a complicated circuit facing a load by an**

   a. Ideal voltage source and parallel resistor

   b. Ideal current source and parallel resistor

   c. Ideal voltage source and series resistor

   d. Ideal current source and series resistor

21. **Norton's theorem replaces a complicated circuit facing a load by an**

   a. Ideal voltage source and parallel resistor

   b. Ideal current source and parallel resistor

   c. Ideal voltage source and series resistor

   d. Ideal current source and series resistor

22. **One way to short a device is**

   a. With a cold-solder joint

   b. With a solder bridge

   c. By disconnecting it

   d. By opening it

23. **Derivations are**

   a. Discoveries

   b. Inventions

   c. Produced by mathematics

   d. Always called theorems

Problems

SEC. 1–3 VOLTAGE SOURCES

1-1 A given voltage source has an ideal voltage of 12 V and an internal resistance of 0.1 Ω. For what values of load resistance will the voltage source appear stiff?

1-2 A load resistance may vary from 270 Ω to 100 kΩ. For a stiff voltage source to exist, what is the largest internal resistance the source can have?

1-3 The internal output resistance of a function generator is 50 Ω. For what values of load resistance does the generator appear stiff?

1-4 A car battery has an internal resistance of 0.04 Ω. For what values of load resistance does the car battery appear stiff?

1-5 The internal resistance of a voltage source equals 0.05 Ω. How much voltage is dropped across this internal resistance when the current through it equals 2 A?

1-6 In Fig. 1-17, the ideal voltage is 9 V and the internal resistance is 0.4 Ω. If the load resistance is zero, what is the load current?

Figure 1-17

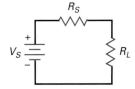

SEC. 1–4 CURRENT SOURCES

1-7 Suppose a current source has an ideal current of 10 mA and an internal resistance of 10 MΩ. For what values of load resistance will the current source appear stiff?

1-8 A load resistance may vary from 270 Ω to 100 kΩ. If a stiff current source drives this load resistance, what is the internal resistance of the source?

1-9 A current source has an internal resistance of 100 kΩ. What is the largest load resistance if the current source must appear stiff?

1-10 In Fig. 1-18, the ideal current is 20 mA and the internal resistance if 200 kΩ. If the load resistance equals zero, what does the load current equal?

Figure 1-18

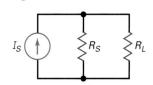

1-11 In Fig. 1-18, the ideal current is 5 mA and the internal resistance is 250 kΩ. If the load resistance is 10 kΩ, what is the load current? Is this a stiff current source?

SEC. 1–5 THEVENIN'S THEOREM

1-12 What is the Thevenin voltage in Fig. 1-19? The Thevenin resistance?

Figure 1-19

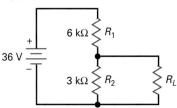

1-13 Use Thevenin's theorem to calculate the load current in Fig. 1-19 for each of these load resistances: 0, 1 kΩ, 2 kΩ, 3 kΩ, 4 kΩ, 5 kΩ, and 6 kΩ.

1-14 The voltage source of Fig. 1-19 is decreased to 18 V. What happens to the Thevenin voltage? To the Thevenin resistance?

1-15 All resistances are doubled in Fig. 1-19. What happens to the Thevenin voltage? To the Thevenin resistance?

SEC. 1–6 NORTON'S THEOREM

1-16 A circuit has a Thevenin voltage of 12 V and a Thevenin resistance of 3 kΩ. What is the Norton circuit?

1-17 A circuit has a Norton current of 10 mA and a Norton resistance of 10 kΩ. What is the Thevenin circuit?

1-18 What is the Norton circuit for Fig. 1-19?

SEC. 1–7   TROUBLESHOOTING

1-19 Suppose the load voltage of Fig. 1-19 is 36 V. What is wrong with R_1?

1-20 The load voltage of Fig. 1-19 is zero. The battery and the load resistance are OK. Suggest two possible troubles.

1-21 If the load voltage is zero in Fig. 1-19 and all resistors are normal, where does the trouble lie?

1-22 In Fig. 1-19, R_L is replaced with a voltmeter to measure the voltage across R_2. What input resistance must the voltmeter have to prevent meter loading?

Critical Thinking

1-23 Suppose we temporarily short the load terminals of a voltage source. If the ideal voltage is 12 V and the shorted load current is 150 A, what is the internal resistance of the source?

1-24 In Fig. 1-17, the ideal voltage is 10 V and the load resistance is 75 Ω. If the load voltage equals 9 V, what does the internal resistance equal? Is the voltage source stiff?

1-25 Somebody hands you a black box with a 2-kΩ resistor connected across the exposed load terminals. How can you measure the Thevenin voltage?

1-26 The black box in Prob. 1-25 has a knob on it that allows you to reduce all internal voltage and current sources to zero. How can you measure the Thevenin resistance?

1-27 Solve Prob. 1-13. Then solve the same problem without using Thevenin's theorem. After you are finished, comment on what you have learned about the Thevenin theorem.

1-28 You are in the laboratory looking at a circuit like the one shown in Fig. 1-20. Somebody challenges you to find the Thevenin circuit driving the load resistor. Describe an experimental procedure for measuring the Thevenin voltage and the Thevenin resistance.

1-29 Design a hypothetical current source using a battery and a resistor. The current source must meet the following specifications: It must supply a stiff 1 mA of current to any load resistance between 0 and 1 kΩ.

1-30 Design a voltage divider (similar to the one in Fig. 1-19) that meets these specifications: Ideal source voltage is 30 V, open-load voltage is 15 V, and Thevenin resistance is equal to or less than 2 kΩ.

1-31 Design a voltage divider like the one in Fig. 1-19 so that it produces a stiff 10 V to all load resistances greater than 1 MΩ. Use an ideal voltage of 30 V.

Figure 1–20

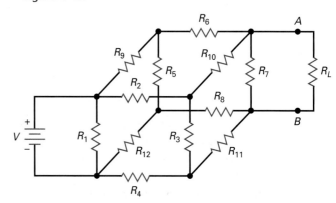

1-32 Somebody hands you a D-cell flashlight battery and a digital multimeter (DMM). You have nothing else to work with. Describe an experimental method for finding the Thevenin equivalent circuit of the flashlight battery.

1-33 You have a D-cell flashlight battery, a DMM, and a box of different resistors. Describe a method that uses one of the resistors to find the Thevenin resistance of the battery.

1-34 Calculate the load current in Fig. 1-21 for each of these load resistances: 0, 1 kΩ, 2 kΩ, 3 kΩ, 4 kΩ, 5 kΩ, and 6 kΩ.

Figure 1–21

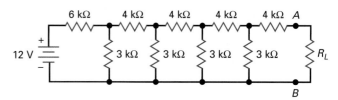

Troubleshooting

1-35 |III| **MultiSim** Using Fig. 1-22 and its troubleshooting table, find the circuit troubles for conditions 1 to 8. The troubles are one of the resistors open, one of the resistors shorted, an open ground, or no supply voltage.

Figure 1–22   Troubleshooting.

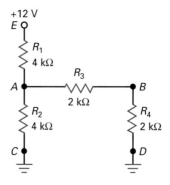

Condition	V_A	V_B	V_E	Condition	V_A	V_B	V_E
Normal	4 V	2 V	12 V	Trouble 5	6 V	3 V	12 V
Trouble 1	12 V	6 V	12 V	Trouble 6	6 V	6 V	12 V
Trouble 2	0 V	0 V	12 V	Trouble 7	0 V	0 V	0 V
Trouble 3	6 V	0 V	12 V	Trouble 8	3 V	0 V	12 V
Trouble 4	3 V	3 V	12 V				

Job Interview Questions

A job interviewer can quickly tell whether your learning is skin-deep or whether you really understand electronics. Interviewers do not always ask neat and tidy questions. Sometimes, they leave out data to see how you handle the question. When you interview for a job, the interviewer might ask you questions like the following.

1. What is the difference between a voltage source and a current source?
2. When do you have to include the source resistance in your calculations for load current?
3. If a device is modeled as a current source, what can you say about the load resistance?
4. What does a stiff source mean to you?
5. I have a circuit breadboarded on my lab bench. Tell me what measurements I can make to get the Thevenin voltage and Thevenin resistance.

6. What would be an advantage of a 50 Ω voltage source compared to a 600 Ω voltage source?
7. How are the Thevenin resistance and "cold cranking amperes" of a car battery related?
8. Someone tells you that a voltage source is heavily loaded. What do you think this means?
9. Which approximation does the technician normally use when performing initial troubleshooting procedures? Why?
10. When troubleshooting an electronics system, you measure a dc voltage of 9.5 V at a test point where the schematic diagram says it should be 10 V. What should you infer from this reading? Why?
11. What are some of the reasons for using a Thevenin or Norton circuit?
12. What is the value of the Thevenin and Norton theorems in bench testing?

Self–Test Answers

1.	a	5.	d	9.	b	13.	c	17.	a	21.	b
2.	c	6.	d	10.	a	14.	d	18.	b	22.	b
3.	a	7.	b	11.	a	15.	b	19.	b	23.	c
4.	b	8.	c	12.	a	16.	b	20.	c		

Practice Problem Answers

1-1 60 kΩ

1-2 $V_L = 20$ V

1-4 3 mA when $R_L = 2$ kΩ; 2 mA $R_L = 6$ kΩ; 1 mA $R_L = 18$ kΩ

1-6 $I_N = 2$ mA

1-7 Either R_2 open, C open, or R_1 shorted

2 Semiconductors

To understand how diodes, transistors, and integrated circuits work, you first have to study semiconductors: materials that are neither conductors nor insulators. Semiconductors contain some free electrons, but what makes them unusual is the presence of holes. In this chapter, you will learn about semiconductors, holes, and other related topics.

Objectives

After studying this chapter, you should be able to:

- Recognize, at the atomic level, the characteristics of good conductors and semiconductors.

- Describe the structure of a silicon crystal.

- List the two types of carriers and name the type of impurity that causes each to be a majority carrier.

- Explain the conditions that exist at the *pn* junction of an unbiased diode, a forward-biased diode, and a reverse-biased diode.

- Describe the types of breakdown current caused by excessive reverse voltage across a diode.

Chapter Outline

Vocabulary

ambient temperature
avalanche effect
barrier potential
breakdown voltage
conduction band
covalent bond
depletion layer
diode
doping
extrinsic semiconductor

forward bias
free electron
hole
intrinsic semiconductor
junction diode
junction temperature
majority carriers
minority carriers
n-type semiconductor
p-type semiconductor

pn junction
recombination
reverse bias
saturation current
semiconductor
silicon
surface-leakage current
thermal energy

2-1 Conductors

Copper is a good conductor. The reason is clear when we look at its atomic structure (Fig. 2-1). The nucleus of the atom contains 29 protons (positive charges). When a copper atom has a neutral charge, 29 electrons (negative charges) circle the nucleus like planets around the sun. The electrons travel in distinct *orbits* (also called *shells*). There are 2 electrons in the first orbit, 8 electrons in the second, 18 in the third, and 1 in the outer orbit.

Stable Orbits

The positive nucleus of Fig. 2-1 attracts the planetary electrons. The reason why these electrons are not pulled into the nucleus is the centrifugal (outward) force created by their circular motion. This centrifugal force is exactly equal to the inward pull of the nucleus, so that the orbit is stable. The idea is similar to a satellite that orbits the earth. At the right speed and height, a satellite can remain in a stable orbit above the earth.

The larger the orbit of an electron, the smaller the attraction of the nucleus. In a larger orbit, an electron travels more slowly, producing less centrifugal force. The outermost electron in Fig. 2-1 travels very slowly and feels almost no attraction to the nucleus.

The Core

In electronics, all that matters is the outer orbit. It is called the *valence orbit*. This orbit controls the electrical properties of the atom. To emphasize the importance of the valence orbit, we define the *core* of an atom as the nucleus and all the inner orbits. For a copper atom, the core is the nucleus (+29), and the first three orbits (−28).

The core of a copper atom has a net charge of +1 because it contains 29 protons and 28 inner electrons. Figure 2-2 can help in visualizing the core and the valence orbit. The valence electron is in a large orbit around a core that has a net charge of only +1. Because of this, the inward pull felt by the valence electron is very small.

Figure 2-1 Copper atom.

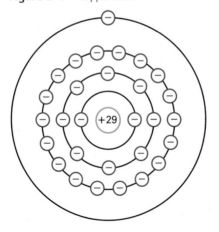

Figure 2-2 Core diagram of copper atom.

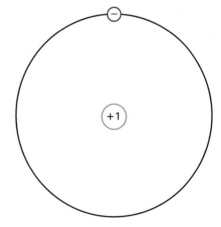

Free Electron

Since the attraction between the core and the valence electron is very weak, an outside force can easily dislodge this electron from the copper atom. This is why we often call the valence electron a **free electron.** This is also why copper is a good conductor. The slightest voltage causes the free electrons to flow from one atom to the next. The best conductors are silver, copper, and gold. All have a core diagram like Fig. 2-2.

Example 2-1

Suppose an outside force removes the valence electron of Fig. 2-2 from a copper atom. What is the net charge of the copper atom? What is the net charge if an outside electron moves into the valence orbit of Fig. 2-2?

SOLUTION   When the valence electron leaves, the net charge of the atom becomes +1. Whenever an atom loses one of its electrons, it becomes positively charged. We call a positively charged atom a *positive ion.*

When an outside electron moves into the valence orbit of Fig. 2-2, the net charge of the atom becomes −1. Whenever an atom has an extra electron in its valence orbit, we call the negatively charged atom a *negative ion.*

2-2 Semiconductors

The best conductors (silver, copper, and gold) have one valence electron, whereas the best insulators have eight valence electrons. A **semiconductor** is an element with electrical properties between those of a conductor and those of an insulator. As you might expect, the best semiconductors have four valence electrons.

Germanium

Germanium is an example of a semiconductor. It has four electrons in the valence orbit. Many years ago, germanium was the only material suitable for making semiconductor devices. But these germanium devices had a fatal flaw (their excessive reverse current, discussed in a later section) that engineers could not overcome. Eventually, another semiconductor named **silicon** became practical and made germanium obsolete in most electronic applications.

Silicon

Next to oxygen, silicon is the most abundant element on the earth. But there were certain refining problems that prevented the use of silicon in the early days of semiconductors. Once these problems were solved, the advantages of silicon (discussed later) immediately made it the semiconductor of choice. Without it, modern electronics, communications, and computers would be impossible.

Figure 2-3 (*a*) Silicon atom; (*b*) core diagram.

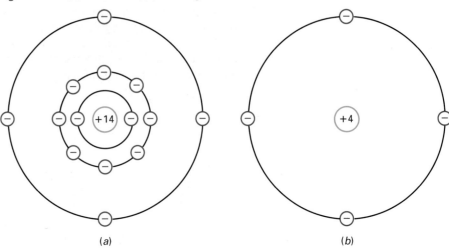

(a)   (b)

GOOD TO KNOW

Another common semiconductor element is carbon (C), which is used mainly in the production of resistors.

An isolated silicon atom has 14 protons and 14 electrons. As shown in Fig. 2-3*a*, the first orbit contains 2 electrons and the second orbit contains 8 electrons. The 4 remaining electrons are in the valence orbit. In Fig. 2-3*a*, the core has a net charge of +4 because it contains 14 protons in the nucleus and 10 electrons in the first two orbits.

Figure 2-3*b* shows the core diagram of a silicon atom. The 4 valence electrons tell us that silicon is a semiconductor.

Example 2-2

What is the net charge of the silicon atom in Fig. 2-3*b* if it loses one of its valence electrons? If it gains an extra electron in the valence orbit?

SOLUTION   If it loses an electron, it becomes a positive ion with a charge of +1. If it gains an extra electron, it becomes a negative ion with a charge of −1.

2-3 Silicon Crystals

When silicon atoms combine to form a solid, they arrange themselves into an orderly pattern called a *crystal.* Each silicon atom shares its electrons with four neighboring atoms in such a way as to have eight electrons in its valence orbit. For instance, Fig. 2-4*a* shows a central atom with four neighbors. The shaded circles represent the silicon cores. Although the central atom originally had four electrons in its valence orbit, it now has eight.

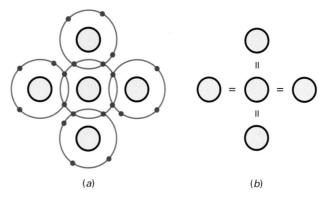

Figure 2-4 (*a*) Atom in crystal has four neighbors; (*b*) covalent bonds.

(*a*)　　　　　　　　　　　(*b*)

Covalent Bonds

Each neighboring atom shares an electron with the central atom. In this way, the central atom has four additional electrons, giving it a total of eight electrons in the valence orbit. The electrons no longer belong to any single atom. Each central atom and its neighbors share the electrons. The same idea is true for all the other silicon atoms. In other words, every atom inside a silicon crystal has four neighbors.

In Fig. 2-4*a*, each core has a charge of $+4$. Look at the central core and the one to its right. These two cores attract the pair of electrons between them with equal and opposite force. This pulling in opposite directions is what holds the silicon atoms together. The idea is similar to tug-of-war teams pulling on a rope. As long as both teams pull with equal and opposite force, they remain bonded together.

Since each shared electron in Fig. 2-4*a* is being pulled in opposite directions, the electron becomes a bond between the opposite cores. We call this type of chemical bond a **covalent bond.** Figure 2-4*b* is a simpler way to show the concept of the covalent bonds. In a silicon crystal, there are billions of silicon atoms, each with eight valence electrons. These valence electrons are the covalent bonds that hold the crystal together—that give it solidity.

Valence Saturation

Each atom in a silicon crystal has eight electrons in its valence orbit. These eight electrons produce a chemical stability that results in a solid piece of silicon material. No one is quite sure why the outer orbit of all elements has a predisposition toward having eight electrons. When eight electrons do not exist naturally in an element, there seems to be a tendency for the element to combine and share electrons with other atoms so as to have eight electrons in the outer orbit.

There are advanced equations in physics that partially explain why eight electrons produce chemical stability in different materials, but no one knows the reason why the number eight is so special. It is one of those laws like the law of gravity, Coulomb's law, and other laws that we observe but cannot fully explain.

When the valence orbit has eight electrons, it is *saturated* because no more electrons can fit into this orbit. Stated as a law:

Valence saturation: $n = 8$　　　　　　　　　　　　　　　　　(2-1)

In words, *the valence orbit can hold no more than eight electrons.* Furthermore, the eight valence electrons are called *bound electrons* because they are tightly held by the atoms. Because of these bound electrons, a silicon crystal is almost a perfect insulator at room temperature, approximately 25°C.

The Hole

The **ambient temperature** is the temperature of the surrounding air. When the ambient temperature is above absolute zero (−273°C), the heat energy in this air causes the atoms in a silicon crystal to vibrate. The higher the ambient temperature, the stronger the mechanical vibrations become. When you pick up a warm object, the warmth you feel is the effect of the vibrating atoms.

In a silicon crystal, the vibrations of the atoms can occasionally dislodge an electron from the valence orbit. When this happens, the released electron gains enough energy to go into a larger orbit, as shown in Fig. 2-5a. In this larger orbit, the electron is a free electron.

But that's not all. The departure of the electron creates a vacancy in the valence orbit called a **hole** (see Fig. 2-5a). This hole behaves like a positive charge because the loss of the electron produces a positive ion. The hole will attract and capture any electron in the immediate vicinity. The existence of holes is the critical difference between conductors and semiconductors. Holes enable semiconductors to do all kinds of things that are impossible with conductors.

At room temperature, thermal energy produces only a few holes and free electrons. To increase the number of holes and free electrons, it is necessary to *dope* the crystal. More is said about this in a later section.

Recombination and Lifetime

In a pure silicon crystal, **thermal** (heat) **energy** creates an equal number of free electrons and holes. The free electrons move randomly throughout the crystal. Occasionally, a free electron will approach a hole, feel its attraction, and fall into it. **Recombination** is the merging of a free electron and a hole (see Fig. 2-5b).

The amount of time between the creation and disappearance of a free electron is called the *lifetime*. It varies from a few nanoseconds to several microseconds, depending on how perfect the crystal is and other factors.

Main Ideas

At any instant, the following is taking place inside a silicon crystal:

1. Some free electrons and holes are being created by thermal energy.
2. Other free electrons and holes are recombining.
3. Some free electrons and holes exist temporarily, awaiting recombination.

Figure 2-5 (a) Thermal energy produces electron and hole; (b) recombination of free electron and hole.

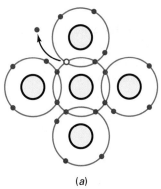

(a)

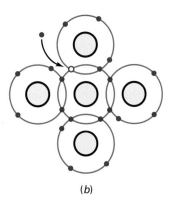

(b)

Example 2-3

If a pure silicon crystal has 1 million free electrons inside it, how many holes does it have? What happens to the number of free electrons and holes if the ambient temperature increases?

SOLUTION   Look at Fig. 2-5a. When heat energy creates a free electron, it automatically creates a hole at the same time. Therefore, a pure silicon crystal always has the same number of holes and free electrons. If there are 1 million free electrons, there are 1 million holes.

A higher temperature increases the vibrations at the atomic level, which means that more free electrons and holes are created. But no matter what the temperature is, a pure silicon crystal has the same number of free electrons and holes.

2-4 Intrinsic Semiconductors

An **intrinsic semiconductor** is a pure semiconductor. A silicon crystal is an intrinsic semiconductor if every atom in the crystal is a silicon atom. At room temperature, a silicon crystal acts like an insulator because it has only a few free electrons and holes produced by thermal energy.

Flow of Free Electrons

Figure 2-6 shows part of a silicon crystal between charged metallic plates. Assume that thermal energy has produced a free electron and a hole. The free electron is in a large orbit at the right end of the crystal. Because of the negatively charged plate, the free electron is repelled to the left. This free electron can move from one large orbit to the next until it reaches the positive plate.

Flow of Holes

Notice the hole at the left of Fig. 2-6. This hole attracts the valence electron at point A. This causes the valence electron to move into the hole.

When the valence electron at point A moves to the left, it creates a new hole at point A. The effect is the same as moving the original hole to the right. The new hole at point A can then attract and capture another valence electron. In this way, valence electrons can travel along the path shown by the arrows. This means the hole can move the opposite way, along path A-B-C-D-E-F, acting the same as a positive charge.

Figure 2-6 Hole flow through a semiconductor.

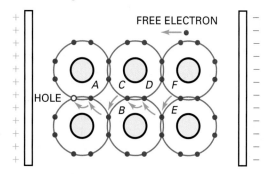

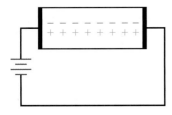

2-5 Two Types of Flow

Figure 2-7 shows an intrinsic semiconductor. It has the same number of free electrons and holes. This is because *thermal energy produces free electrons and holes in pairs.* The applied voltage will force the free electrons to flow left and the holes to flow right. When the free electrons arrive at the left end of the crystal, they enter the external wire and flow to the positive battery terminal.

On the other hand, the free electrons at the negative battery terminal will flow to the right end of the crystal. At this point, they enter the crystal and recombine with holes that arrive at the right end of the crystal. In this way, a steady flow of free electrons and holes occurs inside the semiconductor. Note that there is no hole flow outside the semiconductor.

In Fig. 2-7, *the free electrons and holes move in opposite directions.* From now on, we will visualize the current in a semiconductor as the combined effect of the two types of flow: the flow of free electrons in one direction and the flow of holes in the other direction. Free electrons and holes are often called *carriers* because they carry a charge from one place to another.

2-6 Doping a Semiconductor

One way to increase conductivity of a semiconductor is by **doping.** This means adding impurity atoms to an intrinsic crystal to alter its electrical conductivity. A doped semiconductor is called an **extrinsic semiconductor.**

Increasing the Free Electrons

How does a manufacturer dope a silicon crystal? The first step is to melt a pure silicon crystal. This breaks the covalent bonds and changes the silicon from a solid to a liquid. To increase the number of free electrons, *pentavalent atoms* are added to the molten silicon. Pentavalent atoms have five electrons in the valence orbit. Examples of pentavalent atoms include arsenic, antimony, and phosphorus. Because these materials *will donate an extra electron* to the silicon crystal, they are often referred to as *donor impurities.*

Figure 2-8a shows how the doped silicon crystal appears after it cools down and re-forms its solid crystal structure. A pentavalent atom is in the center, surrounded by four silicon atoms. As before, the neighboring atoms share an electron with the central atom. But this time, there is an extra electron left over. Remember that each pentavalent atom has five valence electrons. Since only eight electrons can fit into the valence orbit, the extra electron remains in a larger orbit. In other words, it is a free electron.

Each pentavalent or donor atom in a silicon crystal produces one free electron. This is how a manufacturer controls the conductivity of a doped semiconductor. The more impurity that is added, the greater the conductivity. In this way, a semiconductor may be lightly or heavily doped. A lightly doped semiconductor has a high resistance, whereas a heavily doped semiconductor has a low resistance.

Increasing the Number of Holes

How can we dope a pure silicon crystal to get an excess of holes? By using a *trivalent impurity,* one whose atoms have only three valence electrons. Examples include aluminum, boron, and gallium.

Figure 2-8b shows a trivalent atom in the center. It is surrounded by four silicon atoms, each sharing one of its valence electrons. Since the trivalent atom originally had only three valence electrons and each neighbor shares one

Figure 2-8 (*a*) Doping to get more free electrons; (*b*) doping to get more holes.

• FREE ELECTRON

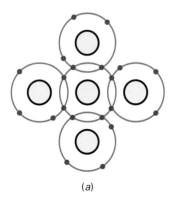

(*a*)

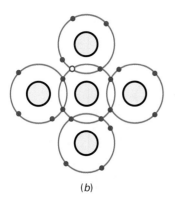

(*b*)

electron, only seven electrons are in the valence orbit. This means that a hole exists in the valence orbit of each trivalent atom. A trivalent atom is also called an *acceptor atom* because each hole it contributes can accept a free electron during recombination.

Points to Remember

Before manufacturers can dope a semiconductor, they must produce it as a pure crystal. Then, by controlling the amount of impurity, they can precisely control the properties of the semiconductor. Historically, pure germanium crystals were easier to produce than pure silicon crystals. This is why the earliest semiconductor devices were made of germanium. Eventually, manufacturing techniques improved and pure silicon crystals became available. Because of its advantages, silicon has become the most popular and useful semiconductor material.

Example 2-4

A doped semiconductor has 10 billion silicon atoms and 15 million pentavalent atoms. If the ambient temperature is 25°C, how many free electrons and holes are there inside the semiconductor?

SOLUTION   Each pentavalent atom contributes one free electron. Therefore, the semiconductor has 15 million free electrons produced by doping. There will be almost no holes by comparison because the only holes in the semiconductor are those produced by heat energy.

PRACTICE PROBLEM 2–4   As in Example 2-4, if 5 million trivalent atoms are added instead of pentavalent atoms, how many holes are there inside the semiconductor?

2-7 Two Types of Extrinsic Semiconductors

A semiconductor can be doped to have an excess of free electrons or an excess of holes. Because of this, there are two types of doped semiconductors.

n-Type Semiconductor

Silicon that has been doped with a pentavalent impurity is called an ***n*-type semiconductor,** where the *n* stands for negative. Figure 2-9 shows an *n*-type semiconductor. Since the free electrons outnumber the holes in an *n*-type semiconductor, the free electrons are called the **majority carriers** and the holes are called the **minority carriers.**

Because of the applied voltage, the *free electrons move to the left* and the *holes move to the right*. When a hole arrives at the right end of the crystal, one of the free electrons from the external circuit enters the semiconductor and recombines with the hole.

Figure 2-9   *n*-type semiconductor has many free electrons.

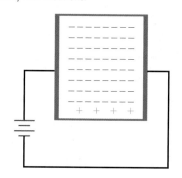

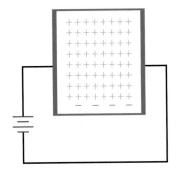

Figure 2-10 *p-type semiconductor has many holes.*

The free electrons shown in Fig. 2-9 flow to the left end of the crystal, where they enter the wire and flow on to the positive terminal of the battery.

p–Type Semiconductor

Silicon that has been doped with a trivalent impurity is called a ***p*-type semiconductor,** where the *p* stands for positive. Figure 2-10 shows a *p*-type semiconductor. Since holes outnumber free electrons, the holes are referred to as the majority carriers and the free electrons are known as the minority carriers.

Because of the applied voltage, the *free electrons move to the left* and the *holes move to the right.* In Fig. 2-10, the holes arriving at the right end of the crystal will recombine with free electrons from the external circuit.

There is also a flow of minority carriers in Fig. 2-10. The free electrons inside the semiconductor flow from right to left. Because there are so few minority carriers, they have almost no effect in this circuit.

2–8 The Unbiased Diode

By itself, a piece of *n*-type semiconductor is about as useful as a carbon resistor; the same can be said for a *p*-type semiconductor. But when a manufacturer dopes a crystal so that one-half of it is *p*-type and the other half is *n*-type, something new comes into existence.

The border between *p*-type and *n*-type is called the ***pn* junction.** The *pn* junction has led to all kinds of inventions including diodes, transistors, and integrated circuits. Understanding the *pn* junction enables you to understand all kinds of semiconductor devices.

The Unbiased Diode

As discussed in the preceding section, each trivalent atom in a doped silicon crystal produces one hole. For this reason, we can visualize a piece of *p*-type semiconductor as shown on the left side of Fig. 2-11. Each circled minus sign is the trivalent atom, and each plus sign is the hole in its valence orbit.

Similarly, we can visualize the pentavalent atoms and free electrons of an *n*-type semiconductor as shown on the right side of Fig. 2-11. Each circled plus sign represents a pentavalent atom, and each minus sign is the free electron it contributes to the semiconductor. Notice that each piece of semiconductor material is *electrically neutral because the number of pluses and minuses is equal.*

A manufacturer can produce a single crystal with *p*-type material on one side and *n*-type on the other side, as shown in Fig. 2-12. The junction is the border where the *p*-type and the *n*-type regions meet, and **junction diode** is another name for a *pn* crystal. The word **diode** is a contraction of two electrodes, where *di* stands for "two."

Figure 2-11 Two types of semiconductor.

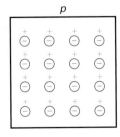
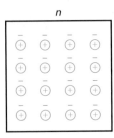

Figure 2-12 The *pn* junction.

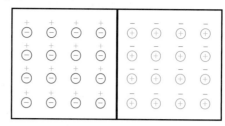

The Depletion Layer

Because of their repulsion for each other, the free electrons on the *n* side of Fig. 2-12 tend to diffuse (spread) in all directions. Some of the free electrons diffuse across the junction. When a free electron enters the *p* region, it becomes a minority carrier. With so many holes around it, this minority carrier has a short lifetime. Soon after entering the *p* region, the free electron recombines with a hole. When this happens, the *hole disappears* and the *free electron becomes a valence electron*.

Each time an electron diffuses across a junction, it creates a pair of ions. When an electron leaves the *n* side, it leaves behind a pentavalent atom that is short one negative charge; this pentavalent atom becomes a positive ion. After the migrating electron falls into a hole on the *p* side, it makes a negative ion out of the trivalent atom that captures it.

Figure 2-13*a* shows these ions on each side of the junction. The circled plus signs are the positive ions, and the circled minus signs are the negative ions. The ions are fixed in the crystal structure because of covalent bonding, and they cannot move around like free electrons and holes.

Each pair of positive and negative ions at the junction is called a *dipole*. The creation of a dipole means that one free electron and one hole have been taken out of circulation. As the number of dipoles builds up, the region near the junction is emptied of carriers. We call this charge-empty region the **depletion layer** (see Fig. 2-13*b*).

Barrier Potential

Each dipole has an electric field between the positive and negative ions. Therefore, if additional free electrons enter the depletion layer, the electric field tries to push these electrons back into the *n* region. The strength of the electric field increases with each crossing electron until equilibrium is reached. To a first approximation, this means that the electric field eventually stops the diffusion of electrons across the junction.

Figure 2-13 (*a*) Creation of ions at junction; (*b*) depletion layer.

IONS

DEPLETION LAYER

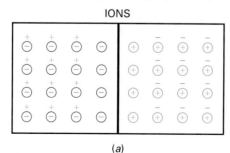
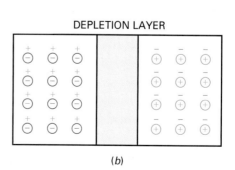

(a)   (b)

In Fig. 2-13a, the electric field between the ions is equivalent to a difference of potential called the **barrier potential.** At 25°C, the barrier potential equals approximately 0.3 V for germanium diodes and 0.7 V for silicon diodes.

2-9 Forward Bias

Figure 2-14 shows a dc source across a diode. The negative source terminal is connected to the *n*-type material, and the positive terminal is connected to the *p*-type material. This connection produces what is called **forward bias.**

Flow of Free Electrons

In Fig. 2-14, the battery pushes holes and free electrons toward the junction. If the battery voltage is less than the barrier potential, the free electrons do not have enough energy to get through the depletion layer. When they enter the depletion layer, the ions will push them back into the *n* region. Because of this, there is no current through the diode.

When the dc voltage source is greater than the barrier potential, the battery again pushes holes and free electrons toward the junction. This time, the free electrons have enough energy to pass through the depletion layer and recombine with the holes. If you visualize all the holes in the *p* region moving to the right and all the free electrons moving to the left, you will have the basic idea. Somewhere in the vicinity of the junction, these opposite charges recombine. Since free electrons continuously enter the right end of the diode and holes are being continuously created at the left end, there is a continuous current through the diode.

The Flow of One Electron

Let us follow a single electron through the entire circuit. After the free electron leaves the negative terminal of the battery, it enters the right end of the diode. It travels through the *n* region until it reaches the junction. When the battery voltage is greater than 0.7 V, the free electron has enough energy to get across the depletion layer. Soon after the free electron has entered the *p* region, it recombines with a hole.

In other words, the free electron becomes a valence electron. As a valence electron, it continues to travel to the left, passing from one hole to the next until it reaches the left end of the diode. When it leaves the left end of the diode, a new hole appears and the process begins again. Since there are billions of electrons taking the same journey, we get a continuous current through the diode. A series resistor is used to limit the amount of forward current.

||| MultiSim **Figure 2-14** Forward bias.

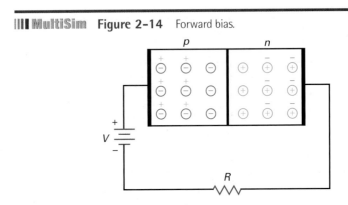

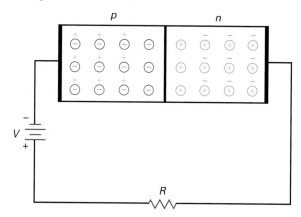

What to Remember

Current flows easily in a forward-biased diode. As long as the applied voltage is greater than the barrier potential, there will be a large continuous current in the circuit. In other words, if the source voltage is greater than 0.7 V, a silicon diode allows a continuous current in the forward direction.

2-10 Reverse Bias

Turn the dc source around and you get Fig. 2-15. This time, the negative battery terminal is connected to the *p* side, and the positive battery terminal to the *n* side. This connection produces what is called **reverse bias.**

Depletion Layer Widens

The negative battery terminal attracts the holes, and the positive battery terminal attracts the free electrons. Because of this, holes and free electrons flow away from the junction. Therefore, the depletion layer gets wider.

How wide does the depletion layer get in Fig. 2-16*a*? When the holes and electrons move away from the junction, the newly created ions increase the difference of potential across the depletion layer. The wider the depletion layer, the greater the difference of potential. The depletion layer stops growing when its difference of potential equals the applied reverse voltage. When this happens, electrons and holes stop moving away from the junction.

Figure 2-16 (*a*) Depletion layer; (*b*) increasing reverse bias widens depletion layer.

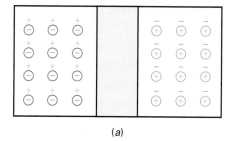

(*a*)

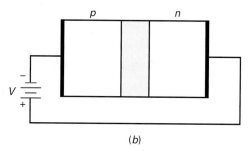

(*b*)

Figure 2-17 Thermal production of free electron and hole in depletion layer produces reverse minority saturation current.

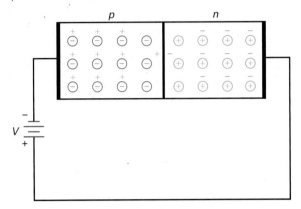

Sometimes the depletion layer is shown as a shaded region like that of Fig. 2-16b. The width of this shaded region is proportional to the reverse voltage. *As the reverse voltage increases, the depletion layer gets wider.*

Minority-Carrier Current

Is there any current after the depletion layer stabilizes? Yes. A small current exists with reverse bias. Recall that thermal energy continuously creates pairs of free electrons and holes. This means that a few minority carriers exist on both sides of the junction. Most of these recombine with the majority carriers. But those inside the depletion layer may exist long enough to get across the junction. When this happens, a small current flows in the external circuit.

Figure 2-17 illustrates the idea. Assume that thermal energy has created a free electron and hole near the junction. The depletion layer pushes the free electron to the right, forcing one electron to leave the right end of the crystal. The hole in the depletion layer is pushed to the left. This extra hole on the *p* side lets one electron enter the left end of the crystal and fall into a hole. Since thermal energy is continuously producing electron-hole pairs inside the depletion layer, a small continuous current flows in the external circuit.

The reverse current caused by the thermally produced minority carriers is called the **saturation current.** In equations, the saturation current is symbolized by I_S. The name *saturation* means that we cannot get more minority-carrier current than is produced by the thermal energy. In other words, *increasing the reverse voltage will not increase the number of thermally created minority carriers.*

Surface-Leakage Current

Besides the thermally produced minority-carrier current, does any other current exist in a reverse-biased diode? Yes. A small current flows on the surface of the crystal. Known as the **surface-leakage current,** it is caused by surface impurities and imperfections in the crystal structure.

What to Remember

The reverse current in a diode consists of a minority-carrier current and a surface-leakage current. In most applications, the reverse current in a silicon diode is so small that you don't even notice it. The main idea to remember is this: *Current is approximately zero in a reverse-biased silicon diode.*

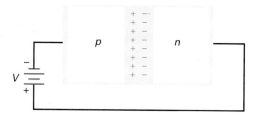

2-11 Breakdown

Diodes have maximum voltage ratings. There is a limit to how much reverse voltage a diode can withstand before it is destroyed. If you continue increasing the reverse voltage, you will eventually reach the **breakdown voltage** of the diode. For many diodes, breakdown voltage is at least 50 V. The breakdown voltage is shown on the *data sheet* for the diode. We will discuss data sheets in Chap. 3.

Once the breakdown voltage is reached, a large number of the minority carriers suddenly appears in the depletion layer and the diode conducts heavily.

Where do the carriers come from? They are produced by the **avalanche effect** (see Fig. 2-18), which occurs at higher reverse voltages. Here is what happens. As usual, there is a small reverse minority-carrier current. When the reverse voltage increases, it forces the minority carriers to move more quickly. These minority carriers collide with the atoms of the crystal. When these minority carriers have enough energy, they can knock valence electrons loose, producing free electrons. These new minority carriers then join the existing minority carriers to collide with other atoms. The process is geometric, because one free electron liberates one valence electron to get two free electrons. These two free electrons then free two more electrons to get four free electrons. The process continues until the reverse current becomes huge.

Figure 2-19 shows a magnified view of the depletion layer. The reverse bias forces the free electron to move to the right. As it moves, the electron gains speed. The larger the reverse bias, the faster the electron moves. If the high-speed electron has enough energy, it can bump the valence electron of the first atom into a larger orbit. This results in two free electrons. Both of these then accelerate and go on to dislodge two more electrons. In this way, the number of minority carriers may become quite large and the diode can conduct heavily.

The breakdown voltage of a diode depends on how heavily doped the diode is. With rectifier diodes (the most common type), the breakdown voltage is usually greater than 50 V. Summary Table 2-1 illustrates the difference between a forward- and reverse-biased diode.

Figure 2-19   The process of avalanche is a geometric progression: 1, 2, 4, 8, . . .

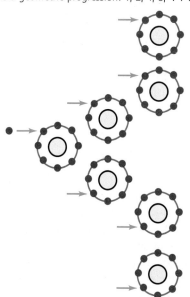

2-12 Energy Levels

To a good approximation, we can identify the total energy of an electron with the size of its orbit. That is, we can think of each radius of Fig. 2-20*a* as equivalent to an energy level in Fig. 2-20*b*. Electrons in the smallest orbit are on the first energy level; electrons in the second orbit are on the second energy level; and so on.

Higher Energy in Larger Orbit

Since an electron is attracted by the nucleus, extra energy is needed to lift an electron into a larger orbit. When an electron is moved from the first to the second orbit, it gains potential energy with respect to the nucleus. Some of the external forces that can lift an electron to higher energy levels are heat, light, and voltage.

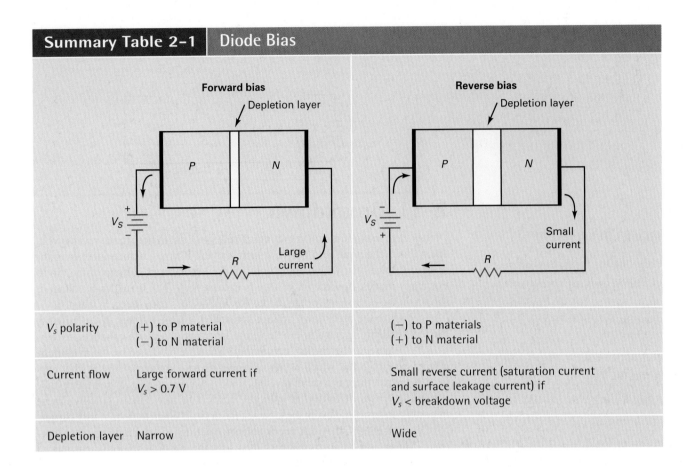

	Forward bias	Reverse bias
V_s polarity	(+) to P material (−) to N material	(−) to P materials (+) to N material
Current flow	Large forward current if $V_s > 0.7$ V	Small reverse current (saturation current and surface leakage current) if $V_s <$ breakdown voltage
Depletion layer	Narrow	Wide

For instance, assume that an outside force lifts the electron from the first to the second orbit in Fig. 2-20a. This electron has more potential energy because it is farther from the nucleus (Fig. 2-20b). It is like an object above the earth: The higher the object, the greater its potential energy with respect to the earth. If released, the object falls farther and does more work when it hits the earth.

Falling Electrons Radiate Light

After an electron has moved into a larger orbit, it may fall back to a lower energy level. If it does, it will give up its extra energy in the form of heat, light, and other radiation.

In a *light-emitting diode (LED)*, the applied voltage lifts the electrons to higher energy levels. When these electrons fall back to lower energy levels, they give off light. Depending on the material used, the light is red, green, orange, or blue. Some LEDs produce infrared radiation (invisible), which is useful in burglar alarm systems.

Energy Bands

When a silicon atom is isolated, the orbit of an electron is influenced only by the charges of the isolated atom. This results in energy levels like the lines of Fig. 2-20b. But when silicon atoms are in a crystal, the orbit of each electron is also influenced by the charges of many other silicon atoms. Since each electron has a unique position inside the crystal, no two electrons see exactly the same pattern of surrounding charges. Because of this, the orbit of each electron is different; or, to put it another way, the energy level of each electron is different.

Figure 2-21 shows what happens to the energy levels. All electrons in first orbit have slightly different energy levels because no two electrons see

Figure 2-20 Energy level is proportional to orbit size. (*a*) Orbits; (*b*) energy levels.

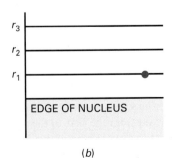

(a)

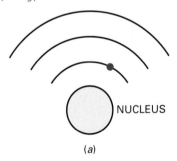

(b)

Figure 2-21 Intrinsic semiconductor and its energy bands.

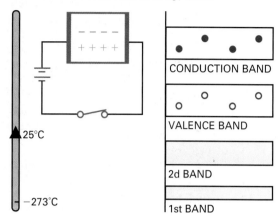

exactly the same charge environment. Since there are billions of first-orbit electrons, the slightly different energy levels form a cluster or *band* of energy. Similarly, the billions of second-orbit electrons, all with slightly different energy levels, form the second energy band—and so on for remaining bands.

Another point. As you know, thermal energy produces a few free electrons and holes. The holes remain in the valence band, but the free electrons go to the next-higher energy band, which is called the **conduction band.** This is why Fig. 2-21 shows a conduction band with some free electrons and a valence band with some holes. When the switch is closed, a small current exists in the pure semiconductor. The free electrons move through the conduction band, and holes move through the valence band.

n-Type Energy Bands

Figure 2-22 shows the energy bands for an *n*-type semiconductor. As you would expect, the majority carriers are the free electrons in the conduction band, and the minority carriers are the holes in the valence band. Since the switch is closed in Fig. 2-22, the majority carriers flow to the left, and the minority carriers flow to the right.

p-Type Energy Bands

Figure 2-23 shows the energy bands for a *p*-type semiconductor. Here you see a reversal of the carrier roles. Now, the majority carriers are the holes in the valence

Figure 2-22 *n*-type semiconductor and its energy bands.

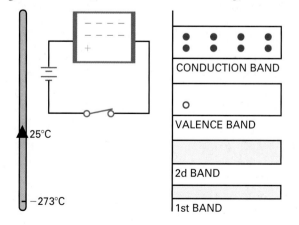

Figure 2-23 *p*-type semiconductor and its energy bands.

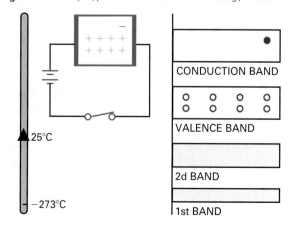

band, and the minority carriers are the free electrons in the conduction band. Since the switch is closed in Fig. 2-23, the majority carriers flow to the right, and the minority carriers flow to the left.

2-13 The Energy Hill

To understand more advanced types of semiconductor devices, you will need to know how energy levels control the action of a *pn* junction.

Before Diffusion

Assuming an abrupt junction (one that suddenly changes from *p* to *n* material), what does the energy diagram look like? Figure 2-24*a* shows the energy bands before electrons have diffused across the junction. The *p* side has many holes in the valence band, and the *n* side has many electrons in the conduction band. But why are the *p* bands slightly higher than the *n* bands?

The *p* side has trivalent atoms with a core charge of +3, shown in Fig. 2-24*b*. On the other hand, the *n* side has pentavalent atoms with a core charge of +5 (Fig. 2-24*c*). A +3 core attracts an electron less than a +5 core does. Therefore, the orbits of a trivalent atom (*p* side) are slightly larger than those of a pentavalent atom (*n* side). This is why the *p* bands of Fig. 2-24*a* are slightly higher than the *n* bands.

An abrupt junction like that of Fig. 2-24*a* is an idealization because the *p* side cannot suddenly end where the *n* side begins. A manufactured diode has a gradual change from one material to the other. For this reason, Fig. 2-25*a* is a more realistic energy diagram of a junction diode.

At Equilibrium

When the diode is first formed, there is no depletion layer (Fig. 2-25*a*). In this case, free electrons will diffuse across the junction. In terms of energy levels, this means that the electrons near the top of the *n* conduction band move across the junction, as previously described. Soon after crossing the junction, a free electron will recombine with a hole. In other words, the electron will fall from the conduction band to the valence band. As it does, it emits heat, light, and other radiation. This recombination not only creates the depletion layer, it also changes the

Figure 2-24   (*a*) Energy bands of abrupt junction before diffusion; (*b*) *p*-type atom has larger orbits, equivalent to higher energy level; (*c*) *n*-type atom has smaller orbits, equivalent to lower energy level.

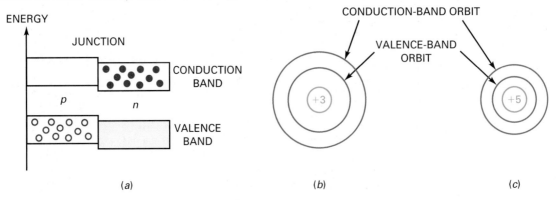

Figure 2-25 (a) Energy bands before diffusion; (b) energy bands after depletion layer is formed; (c) p-type atom before diffusion has smaller orbit; (d) p-type atom after diffusion has larger orbit, equivalent to higher energy level.

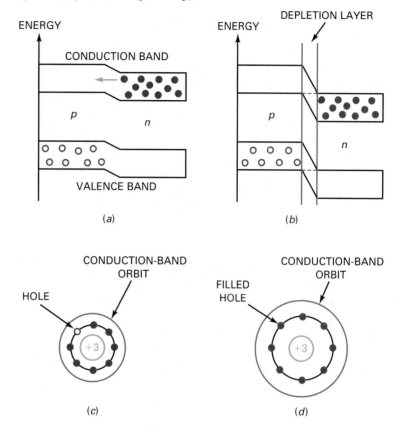

energy levels at the junction by increasing the energy level difference between the *p* and *n* bands.

Figure 2-25*b* shows the energy diagram after the depletion layer is created. The *p* bands have moved up with respect to the *n* bands. As you can see, the bottom of each *p* band is level with the top of the corresponding *n* band. This means that electrons on the *n* side no longer have enough energy to get across the junction. What follows is a simplified explanation of why the *p* band moves up.

Figure 2-25*c* shows a conduction-band orbit around one of the trivalent atoms before diffusion has occurred. When an electron diffuses across the junction, it falls into the hole of a trivalent atom (Fig. 2-25*d*). This extra electron in the valence orbit will push the conduction-band orbit farther away from the trivalent atom, as shown in Fig. 2-25*d*. Therefore, any new electrons coming into this area will need more energy than before to travel in a conduction-band orbit. Stated another way, the larger conduction-band orbit means that the energy level has increased. This is equivalent to saying that the *p* bands move up with respect to the *n* bands after the depletion layer has built up.

At equilibrium, conduction-band electrons on the *n* side travel in orbits not quite large enough to match the *p* side orbits (Fig. 2-25*b*). In other words, electrons on the *n* side do not have enough energy to get across the junction. To an electron trying to diffuse across the junction, the path it must travel looks like a hill, an energy hill (see Fig. 2-25*b*). The electron cannot climb this hill unless it receives energy from an outside source. This energy source may be a voltage source, but it can also be heat, light, or other radiation. Do not think of the energy

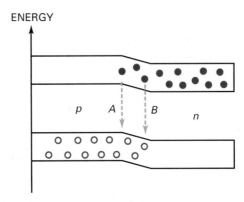

hill as a "physical" hill. Instead, think of it as the necessary higher energy level for the valence electrons to "rise" to before they can cross the depletion layer.

Forward Bias

Forward bias lowers the energy hill (see Fig. 2-26). In other words, the battery increases the energy level of the free electrons; this is equivalent to forcing the *n* band upward. Because of this, free electrons have enough energy to enter the *p* region. Soon after entering the *p* region, they fall into holes (path *A*). As valence electrons, they continue moving toward the left end of the crystal; this is equivalent to holes moving toward the junction.

Some holes penetrate the *n* region as shown in Fig. 2-26. In this case, conduction-band electrons can follow recombination path *B*. Regardless of where the recombination takes place, the result is the same. A steady stream of free electrons moves toward the junction and falls into holes near the junction. The captured electrons (now valence electrons) move left in a steady stream through the holes in the *p* region. In this way, we get a continuous flow of electrons through the diode.

Incidentally, when free electrons fall from the conduction band to the valence band, they radiate their excess energy in the form of heat and light. With an ordinary diode, the radiation is heat energy, which serves no useful purpose. But with an LED, the radiation can be light such as red, green, blue, or orange. LEDs are widely used as visual indicators on electronic instruments, computer keyboards, and consumer equipment.

2-14 Barrier Potential and Temperature

The **junction temperature** is the temperature inside a diode, right at the *pn* junction. The *ambient temperature* is different. It is the temperature of the air outside the diode, the air that surrounds the diode. When the diode is conducting, the junction temperature is higher than the ambient temperature because of the heat created by recombination.

The barrier potential depends on the junction temperature. An increase in junction temperature creates more free electrons and holes in the doped regions. As these charges diffuse into the depletion layer, it becomes narrower. This means that there is *less barrier potential at higher junction temperatures.*

Before continuing, we need to define a symbol:

$$\Delta = \text{the change in} \qquad\qquad (2\text{-}2)$$

The Greek letter Δ (delta) stands for "the change in." For instance, ΔV means the change in voltage, and ΔT means the change in temperature. The ratio $\Delta V/\Delta T$ stands for the change in voltage divided by the change in temperature.

Now we can state a rule for estimating the change in barrier potential: *The barrier potential of a silicon diode decreases by 2 mV for each degree Celsius rise.*

As a derivation:

$$\frac{\Delta V}{\Delta T} = -2\ \text{mV}/°\text{C} \qquad\qquad (2\text{-}3)$$

By rearranging:

$$\Delta V = (-2\ \text{mV}/°\text{C})\ \Delta T \qquad\qquad (2\text{-}4)$$

With this, we can calculate the barrier potential at any junction temperature.

Example 2-5

Assuming a barrier potential of 0.7 V at an ambient temperature of 25°C, what is the barrier potential of a silicon diode when the junction temperature is 100°C? At 0°C?

SOLUTION   When the junction temperature is 100°C, the change in barrier potential is:

$$\Delta V = (-2\ \text{mV}/°\text{C})\ \Delta T = (-2\ \text{mV}/°\text{C})(100°\text{C} - 25°\text{C}) = -150\ \text{mV}$$

This tells us that the barrier potential decreases 150 mV from its room temperature value. So, it equals:

$$V_B = 0.7\ \text{V} - 0.15\ \text{V} = 0.55\ \text{V}$$

When the junction temperature is 0°C, the change in barrier potential is:

$$\Delta V = (-2\ \text{mV}/°\text{C})\ \Delta T = (-2\ \text{mV}/°\text{C})(0°\text{C} - 25°\text{C}) = 50\ \text{mV}$$

This tells us that the barrier potential increases 50 mV from its room temperature value. So, it equals:

$$V_B = 0.7\ \text{V} + 0.05\ \text{V} = 0.75\ \text{V}$$

PRACTICE PROBLEM 2–5   What would be the barrier potential in Example 2-5 when the junction temperature is 50°C?

2-15 Reverse-Biased Diode

Let's discuss a few advanced ideas about a reverse-biased diode. To begin with, the depletion layer changes in width when the reverse voltage changes. Let us see what this implies.

Transient Current

When the reverse voltage increases, holes and electrons move away from the junction. As the free electrons and holes move away from the junction, they leave positive and negative ions behind. Therefore, the depletion layer gets wider. The greater the reverse bias, the wider the depletion layer becomes. While the depletion layer is adjusting to its new width, a current flows in the external circuit. This transient current drops to zero after the depletion layer stops growing.

The amount of time the transient current flows depends on the RC time constant of the external circuit. It typically happens in a matter of nanoseconds. Because of this, you can ignore the effects of the transient current below approximately 10 MHz.

Reverse Saturation Current

As discussed earlier, forward-biasing a diode raises the n band and allows free electrons to cross the junction. Reverse bias has the opposite effect: It widens the depletion layer and lowers the n band, as shown in Fig. 2-27.

Here is the energy viewpoint on reverse saturation. Suppose that thermal energy creates a hole and free electron inside the depletion layer, as shown in Fig. 2-27. The free electron at A and the hole at B can now contribute to reverse current. Because of the reverse bias, the free electron will move to the right, effectively pushing an electron out of the right end of the diode. Similarly, the hole will move to the left. This extra hole on the p side lets an electron enter the left end of the crystal.

The higher the junction temperature, the greater the saturation current. A useful approximation to remember is this: I_S doubles for each 10°C rise. As a derivation,

$$\text{Percent } \Delta I_S = 100\% \text{ for a 10°C increase} \tag{2-5}$$

In words, the change in saturation current is 100 percent for each 10°C rise in temperature. If the changes in temperature are less than 10°C, you can use this equivalent rule:

$$\text{Percent } \Delta I_S = 7\% \text{ per °C} \tag{2-6}$$

In words, the change in saturation current is 7 percent for each Celsius degree rise. This 7 percent solution is a close approximation of the 10° rule.

Figure 2-27 Thermal energy produces free electron and hole inside depletion layer.

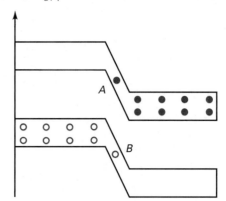

Figure 2-28 (*a*) Atoms on the surface of a crystal have no neighbors; (*b*) surface of crystal has holes.

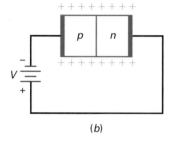

(*a*)

(*b*)

Silicon versus Germanium

In a silicon atom, the distance between the valence band and the conduction band is called the *energy gap*. When thermal energy produces free electrons and holes, it has to give the valence electrons enough energy to jump into the conduction band. The larger the energy gap, the more difficult it is for thermal energy to produce electron-hole pairs. Fortunately, silicon has a large energy gap; this means that thermal energy does not produce many electron-hole pairs at normal temperatures.

In a germanium atom the valence band is much closer to the conduction band. In other words, germanium has a much smaller energy gap than silicon has. For this reason, thermal energy produces many more electron-hole pairs in germanium devices. This is the fatal flaw mentioned earlier. The excessive reverse current of germanium devices precludes their widespread use in modern computers, consumer electronics, and communications circuits.

Surface-Leakage Current

We discussed surface-leakage current briefly in Sec. 2-10. Recall that it is a reverse current on the surface of the crystal. Here is an explanation of why surface-leakage current exists. Suppose that the atoms at the top and bottom of Fig. 2-28*a* are on the surface of the crystal. Since these atoms have no neighbors, they have only six electrons in the valence orbit, implying two holes in each surface atom. Visualize these holes along the surface of the crystal shown in Fig. 2-28*b*. Then you can see that the skin of a crystal is like a *p*-type semiconductor. Because of this, electrons can enter the left end of the crystal, travel through the surface holes, and leave the right end of the crystal. In this way, we get a small reverse current along the surface.

The surface-leakage current is directly proportional to the reverse voltage. For instance, if you double the reverse voltage, the surface-leakage current I_{SL} doubles. We can define the surface-leakage resistance as follows:

$$R_{SL} = \frac{V_R}{I_{SL}} \tag{2-7}$$

Example 2-6

A silicon diode has a saturation current of 5 nA at 25°C. What is the saturation current at 100°C?

SOLUTION The change in temperature is:

$$\Delta T = 100°C - 25°C = 75°C$$

With Eq. (2-5), there are seven doublings between 25°C and 95°C:

$$I_S = (2^7)(5\text{ nA}) = 640\text{ nA}$$

With Eq. (2-6), there are an additional 5° between 95°C and 100°C:

$$I_S = (1.07^5)(640\text{ nA}) = 898\text{ nA}$$

PRACTICE PROBLEM 2-6 Using the same diode as in Example 2-6, what would be the saturation current at 80°C?

Example 2-7

If the surface-leakage current is 2 nA for a reverse voltage of 25 V, what is the surface-leakage current for a reverse voltage of 35 V?

SOLUTION   There are two ways to solve this problem. First, calculate the surface-leakage resistance:

$$R_{SL} = \frac{25 \text{ V}}{2 \text{ nA}} = 12.5(10^9) \; \Omega$$

Then, calculate the surface-leakage current at 35 V as follows:

$$I_{SL} = \frac{35 \text{ V}}{12.5(10^9) \; \Omega} = 2.8 \text{ nA}$$

Here is a second method. Since surface-leakage current is directly proportional to reverse voltage:

$$I_{SL} = \frac{35 \text{ V}}{25 \text{ V}} \, 2 \text{ nA} = 2.8 \text{ nA}$$

PRACTICE PROBLEM 2-7   In Example 2-7, what is the surface-leakage current for a reverse voltage of 100 V?

Summary

SEC. 2-1 CONDUCTORS

A neutral copper atom has only one electron in its outer orbit. Since this single electron can be easily dislodged from its atom, it is called a *free electron*. Copper is a good conductor because the slightest voltage causes free electrons to flow from one atom to the next.

SEC. 2-2 SEMICONDUCTORS

Silicon is the most widely used semiconductor material. An isolated silicon atom has four electrons in its outer or valence orbit. The number of electrons in the valence orbit is the key to conductivity. Conductors have one valence electron, semiconductors have four valence electrons, and insulators have eight valence electrons.

SEC. 2-3 SILICON CRYSTALS

Each silicon atom in a crystal has its four valence electrons plus four more electrons that are shared by the neighboring atoms. At room temperature, a pure silicon crystal has only a few thermally produced free electrons and holes. The amount of time between the creation and recombination of a free electron and a hole is called the *lifetime*.

SEC. 2-4 INTRINSIC SEMICONDUCTORS

An intrinsic semiconductor is a pure semiconductor. When an external voltage is applied to the intrinsic semiconductor, the free electrons flow toward the positive battery terminal and the holes flow toward the negative battery terminal.

SEC. 2-5 TWO TYPES OF FLOW

Two types of carrier flow exist in an intrinsic semiconductor. First, there is the flow of free electrons through larger orbits (conduction band). Second, there is the flow of holes through smaller orbits (valence band).

SEC. 2-6 DOPING A SEMICONDUCTOR

Doping increases the conductivity of a semiconductor. A doped semiconductor is called an *extrinsic semiconductor*. When an intrinsic semiconductor is doped with pentavalent (donor) atoms, it has more free electrons than holes. When an intrinsic semiconductor is doped with trivalent (acceptor) atoms, it has more holes than free electrons.

SEC. 2-7 TWO TYPES OF EXTRINSIC SEMICONDUCTORS

In an *n*-type semiconductor the free electrons are the majority carriers, and the holes are the minority carriers. In a *p*-type semiconductor the holes are the majority carriers, and the free electrons are the minority carriers.

SEC. 2-8 THE UNBIASED DIODE

An unbiased diode has a depletion layer at the *pn* junction. The ions in this depletion layer produce a barrier potential. At room temperature, this barrier potential is approximately 0.7 V for a silicon diode and 0.3 V for a germanium diode.

SEC. 2-9 FORWARD BIAS

When an external voltage opposes the barrier potential, the diode is forward-biased. If the applied voltage is greater than the barrier potential, the current is large. In other words, current flows easily in a forward-biased diode.

SEC. 2-10 REVERSE BIAS

When an external voltage aids the barrier potential, the diode is reverse-biased. The width of the depletion layer increases when the reverse voltage increases. The current is approximately zero.

SEC. 2-11 BREAKDOWN

Too much reverse voltage will produce either avalanche or zener effect. Then, the large breakdown current destroys the diode. In general, diodes are never operated in the breakdown region. The only exception is the zener diode, a special-purpose diode discussed in a later chapter.

SEC. 2-12 ENERGY LEVELS

The larger the orbit, the higher the energy level of an electron. If an outside force raises an electron to a higher energy level, the electron will emit energy when it falls back to its original orbit.

SEC. 2-13 THE ENERGY HILL

The barrier potential of a diode looks like an energy hill. Electrons attempting to cross the junction need to have enough energy to climb this hill. An external voltage source that forward-biases the diode gives electrons the energy required to pass through the depletion layer.

SEC. 2-14 BARRIER POTENTIAL AND TEMPERATURE

When the junction temperature increases, the depletion layer becomes narrower and the barrier potential decreases. It will decrease approximately 2 mV for each degree Celsius increase.

SEC. 2-15 REVERSE-BIASED DIODE

There are three components of reverse current in a diode. First, there is the transient current that occurs when the reverse voltage changes. Second, there is the minority-carrier current, also called the *saturation current* because it is independent of the reverse voltage. Third, there is the surface-leakage current. It increases when the reverse voltage increases.

Definitions

(2-2) Δ = the change in

(2-7) $R_{SL} = \dfrac{V_R}{I_{SL}}$

Laws

(2-1)   Valence saturation: $n = 8$

Derivations

(2-3) $\dfrac{\Delta V}{\Delta T} = -2 \text{ mV/°C}$

(2-4) $\Delta V = (-2 \text{ mV/°C}) \, \Delta T$

(2-5)   Percent ΔI_S = 100% for a 10°C increase

(2-6)   Percent ΔI_S = 7% per°C

Student Assignments

1. The nucleus of a copper atom contains how many protons?
   a. 1
   b. 4
   c. 18
   d. 29

2. The net charge of a neutral copper atom is
   a. 0
   b. +1
   c. −1
   d. +4

3. Assume the valence electron is removed from a copper atom. The net charge of the atom becomes
   a. 0
   b. +1
   c. −1
   d. +4

4. The valence electron of a copper atom experiences what kind of attraction toward the nucleus?
   a. None
   b. Weak
   c. Strong
   d. Impossible to say

5. How many valence electrons does a silicon atom have?
   a. 0
   b. 1
   c. 2
   d. 4

6. Which is the most widely used semiconductor?
   a. Copper
   b. Germanium
   c. Silicon
   d. None of the above

7. How many protons does the nucleus of a silicon atom contain?
   a. 4
   b. 14
   c. 29
   d. 32

8. Silicon atoms combine into an orderly pattern called a
   a. Covalent bond
   b. Crystal
   c. Semiconductor
   d. Valence orbit

9. An intrinsic semiconductor has some holes in it at room temperature. What causes these holes?
   a. Doping
   b. Free electrons
   c. Thermal energy
   d. Valence electrons

10. When an electron is moved to a higher orbit level, its energy level with respect to the nucleus
    a. Increases
    b. Decreases
    c. Remains the same
    d. Depends on the type of atom

11. The merging of a free electron and a hole is called
    a. Covalent bonding
    b. Lifetime
    c. Recombination
    d. Thermal energy

12. At room temperature an intrinsic silicon crystal acts approximately like
    a. A battery
    b. A conductor
    c. An insulator
    d. A piece of copper wire

13. The amount of time between the creation of a hole and its disappearance is called
    a. Doping
    b. Lifetime
    c. Recombination
    d. Valence

14. The valence electron of a conductor can also be called a
    a. Bound electron
    b. Free electron
    c. Nucleus
    d. Proton

15. A conductor has how many types of flow?
    a. 1
    b. 2
    c. 3
    d. 4

16. A semiconductor has how many types of flow?
    a. 1
    b. 2
    c. 3
    d. 4

17. When a voltage is applied to a semiconductor, holes will flow
    a. Away from the negative potential
    b. Toward the positive potential
    c. In the external circuit
    d. None of the above

18. For semiconductor material, its valence orbit is saturated when it contains
    a. 1 electron
    b. Equal (+) and (−) ions
    c. 4 electrons
    d. 8 electrons

19. In an intrinsic semiconductor, the number of holes
    a. Equals the number of free electrons
    b. Is greater than the number of free electrons
    c. Is less than the number of free electrons
    d. None of the above

20. Absolute zero temperature equals
    a. −273°C
    b. 0°C
    c. 25°C
    d. 50°C

21. At absolute zero temperature an intrinsic semiconductor has
    a. A few free electrons
    b. Many holes
    c. Many free electrons
    d. No holes or free electrons

22. At room temperature an intrinsic semiconductor has
    a. A few free electrons and holes
    b. Many holes
    c. Many free electrons
    d. No holes

23. The number of free electrons and holes in an intrinsic semiconductor decreases when the temperature
    a. Decreases
    b. Increases
    c. Stays the same
    d. None of the above

24. The flow of valence electrons to the right means that holes are flowing to the
    a. Left
    b. Right
    c. Either way
    d. None of the above

25. Holes act like
    a. Atoms
    b. Crystals
    c. Negative charges
    d. Positive charges

26. Trivalent atoms have how many valence electrons?
    a. 1
    b. 3
    c. 4
    d. 5

27. An acceptor atom has how many valence electrons?
    a. 1
    b. 3
    c. 4
    d. 5

28. If you wanted to produce a **p**-type semiconductor, which of these would you use?
   a. Acceptor atoms
   b. Donor atoms
   c. Pentavalent impurity
   d. Silicon

29. Electrons are the minority carriers in which type of semiconductor?
   a. Extrinsic
   b. Intrinsic
   c. *n*-type
   d. *p*-type

30. How many free electrons does a **p**-type semiconductor contain?
   a. Many
   b. None
   c. Only those produced by thermal energy
   d. Same number as holes

31. Silver is the best conductor. How many valence electrons do you think it has?
   a. 1
   b. 4
   c. 18
   d. 29

32. Suppose an intrinsic semiconductor has 1 billion free electrons at room temperature. If the temperature drops to 0°C, how many holes are there?
   a. Fewer than 1 billion
   b. 1 billion
   c. More than 1 billion
   d. Impossible to say

33. An external voltage source is applied to a **p**-type semiconductor. If the left end of the crystal is positive, which way do the majority carriers flow?
   a. Left
   b. Right
   c. Neither
   d. Impossible to say

34. Which of the following doesn't fit in the group?
   a. Conductor
   b. Semiconductor
   c. Four valence electrons
   d. Crystal structure

35. Which of the following is approximately equal to room temperature?
   a. 0°C
   b. 25°C
   c. 50°C
   d. 75°C

36. How many electrons are there in the valence orbit of a silicon atom within a crystal?
   a. 1
   b. 4
   c. 8
   d. 14

37. Negative ions are atoms that have
   a. Gained a proton
   b. Lost a proton
   c. Gained an electron
   d. Lost an electron

38. Which of the following describes an **n**-type semiconductor?
   a. Neutral
   b. Positively charged
   c. Negatively charged
   d. Has many holes

39. A **p**-type semiconductor contains holes and
   a. Positive ions
   b. Negative ions
   c. Pentavalent atoms
   d. Donor atoms

40. Which of the following describes a **p**-type semiconductor?
   a. Neutral
   b. Positively charged
   c. Negatively charged
   d. Has many free electrons

41. As compared to a germanium diode, a silicon diode's reverse saturation current is
   a. Equal at high temperatures
   b. Lower
   c. Equal at lower temperatures
   d. Higher

42. What causes the depletion layer?
   a. Doping
   b. Recombination
   c. Barrier potential
   d. Ions

43. What is the barrier potential of a silicon diode at room temperature?
   a. 0.3 V
   b. 0.7 V
   c. 1 V
   d. 2 mV per degree Celsius

44. When comparing the energy gap of germanium and silicon atoms, a silicon atom's energy gap is
   a. About the same
   b. Lower
   c. Higher
   d. Unpredictable

45. In a silicon diode the reverse current is usually
   a. Very small
   b. Very large
   c. Zero
   d. In the breakdown region

46. While maintaining a constant temperature, a silicon diode has its reverse-bias voltage increased. The diode's saturation current will
   a. Increase
   b. Decrease
   c. Remain the same
   d. Equal its surface-leakage current

47. The voltage where avalanche occurs is called the
   a. Barrier potential
   b. Depletion layer
   c. Knee voltage
   d. Breakdown voltage

48. The energy hill of diode's **pn** junction will decrease when the diode is
   a. Forward biased
   b. First formed
   c. Reverse biased
   d. Not conducting

49. When the reverse voltage decreases from 10 to 5 V, the depletion layer
   a. Becomes smaller
   b. Becomes larger
   c. Is unaffected
   d. Breaks down

50. When a diode is forward-biased, the recombination of free electrons and holes may produce

a. Heat

b. Light

c. Radiation

d. All of the above

51. A reverse voltage of 10 V is across a diode. What is the voltage across the depletion layer?

a. 0 V

b. 0.7 V

c. 10 V

d. None of the above

52. The energy gap in a silicon atom is the distance between the valence band and the

a. Nucleus

b. Conduction band

c. Atom's core

d. Positive ions

53. The reverse saturation current doubles when the junction temperature increases

a. 1°C

b. 2°C

c. 4°C

d. 10°C

54. The surface-leakage current doubles when the reverse voltage increases

a. 7%

b. 100%

c. 200%

d. 2 mV

Problems

2-1 What is the net charge of a copper atom if it gains two electrons?

2-2 What is the net charge of a silicon atom if it gains three valence electrons?

2-3 Classify each of the following as conductor or semiconductor:

a. Germanium

b. Silver

c. Silicon

d. Gold

2-4 If a pure silicon crystal has 500,000 holes inside it, how many free electrons does it have?

2-5 A diode is forward biased. If the current is 5 mA through the *n* side, what is the current through each of the following?

a. *p* side

b. External connecting wires

c. Junction

2-6 Classify each of the following as *n*-type or *p*-type semiconductors:

a. Doped by acceptor atoms

b. Crystal with pentavalent impurities

c. Majority carriers are holes

d. Donor atoms were added to crystal

e. Minority carriers are free electrons

2-7 A designer will be using a silicon diode over a temperature range of 0° to 75°C. What are the minimum and maximum values of barrier potential?

2-8 If a silicon diode has a saturation current of 10 nA at 25° to 75°C, what are the minimum and maximum values of saturation current?

2-9 A diode has a surface-leakage current of 10 nA when the reverse voltage is 10 V. What is the surface-leakage current if the reverse voltage is increased to 100 V?

Critical Thinking

2-10 A silicon diode has a reverse current of 5 μA at 25°C and 100 μA at 100°C. What are values of the saturation current and the surface-leakage current at 25°C?

2-11 Devices with *pn* junctions are used to build computers. The speed of computers depends on how fast a diode can be turned off and on. Based on what you have learned about reverse bias, what can we do to speed up a computer?

Job Interview Questions

A team of experts in electronics created these questions. In most cases, the text provides enough information to answer all questions. Occasionally, you may come across a term that is not familiar. If this happens, look up the term in a technical dictionary. Also, a question may appear that is not covered in this text. In this case, you may wish to do some library research.

1. Tell me why copper is a good conductor of electricity.
2. How does a semiconductor differ from a conductor? Include sketches in your explanation.

3. Tell me all you know about holes and how they differ from free electrons. Include some drawings.
4. Give me the basic idea of doping semiconductors. I want to see some sketches that support your explanation.
5. Show me, by drawing and explaining the action, why current exists in a forward-biased diode.
6. Tell me why a very small current exists in a reverse-biased diode.
7. A reverse-biased semiconductor diode will break down under certain conditions. I want you to describe avalanche in enough detail so that I can understand it.
8. I want to know why a light-emitting diode produces light. Tell me about it.
9. Do holes flow in a conductor? Why or why not? What happens to holes when they reach the end of a semiconductor?
10. What is surface leakage current?
11. Why is recombination important in a diode?
12. How does extrinsic silicon differ from intrinsic silicon, and why is the difference important?
13. In your own words, describe the action that takes place when the *pn* junction is initially created. Your discussion should include the formation of the depletion layer.
14. In a *pn* junction diode, which of the charge carriers move? Holes or free electrons?

Self-Test Answers

1.	d	19.	a	37.	c
2.	a	20.	a	38.	a
3.	b	21.	d	39.	b
4.	b	22.	a	40.	a
5.	d	23.	a	41.	b
6.	c	24.	a	42.	b
7.	b	25.	d	43.	b
8.	b	26.	b	44.	c
9.	c	27.	b	45.	a
10.	a	28.	a	46.	c
11.	c	29.	d	47.	d
12.	c	30.	c	48.	a
13.	b	31.	a	49.	a
14.	b	32.	a	50.	d
15.	a	33.	b	51.	c
16.	b	34.	a	52.	b
17.	d	35.	b	53.	d
18.	d	36.	c	54.	b

Practice Problem Answers

2-4 Approximately 5 million holes

2-5 $V_B = 0.65$ V

2-6 $I_S = 224$ nA

2-7 $I_{SL} = 8$ nA

3 Diode Theory

This chapter continues our study of diodes. After discussing the diode curve, we look at approximations of a diode. We need approximations because exact analysis is very tedious and time-consuming in most situations. For instance, an ideal approximation is usually adequate for troubleshooting, and a second approximation gives quick and easy solutions in many cases. Beyond this, we can use a third approximation for better accuracy or a computer solution for almost exact answers.

Objectives

After studying this chapter, you should be able to:

- Draw a diode symbol and label the anode and cathode.
- Draw a diode curve and label all significant points and areas.
- Describe the ideal diode.
- Describe the second approximation.
- Describe the third approximation.
- List four basic characteristics of diodes shown on a data sheet.
- Describe how to test a diode using a DMM and VOM.

Chapter Outline

Vocabulary

anode	knee voltage	nonlinear device
bulk resistance	linear device	ohmic resistance
cathode	load line	power rating
ideal diode	maximum forward current	up-down analysis

3-1 Basic Ideas

An ordinary resistor is a **linear device** because the graph of its current versus voltage is a straight line. A diode is different. It is a **nonlinear device** because the graph of its current versus voltage is not a straight line. The reason is the barrier potential. When the diode voltage is less than the barrier potential, the diode current is small. When the diode voltage exceeds the barrier potential, the diode current increases rapidly.

The Schematic Symbol and Case Styles

Figure 3-1a shows the schematic symbol of a diode. The p side is called the **anode,** and the n side the **cathode.** The diode symbol looks like an arrow that points from the p side to the n side, from the anode to the cathode. Figure 3-1b shows some of the many typical diode case styles. Many, but not all, diodes have the cathode lead (K) identified by a colored band.

Basic Diode Circuit

Figure 3-1c shows a diode circuit. In this circuit, the diode is forward biased. How do we know? Because the positive battery terminal drives the p side through a resistor, and the negative battery terminal is connected to the n side. With this connection, the circuit is trying to push holes and free electrons toward the junction.

In more complicated circuits, it may be difficult to decide whether the diode is forward biased. Here is a guideline. Ask yourself this question: Is the external circuit pushing current in the *easy direction* of flow? If the answer is yes, the diode is forward biased.

What is the easy direction of flow? If you use conventional current, the easy direction is the same direction as the diode arrow. If you prefer electron flow, the easy direction is the other way.

Figure 3-1   Diode. (*a*) Schematic symbol; (*b*) diode case styles; (*c*) forward bias.

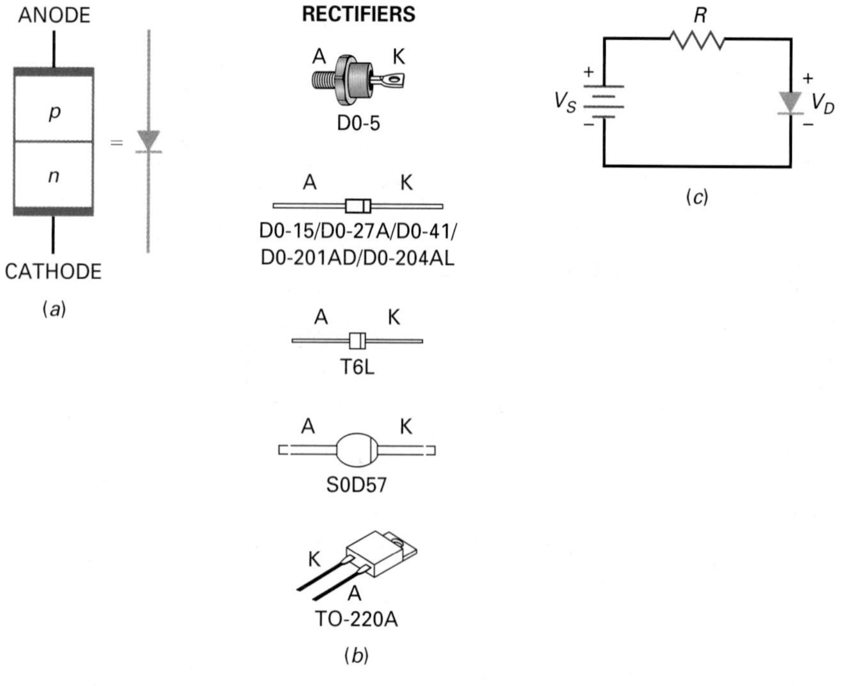

Figure 3-2 Diode curve.

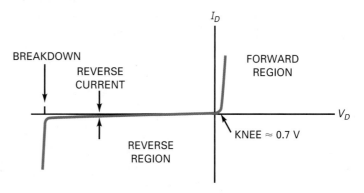

When the diode is part of a complicated circuit, we also can use Thevenin's theorem to determine whether it is forward biased. For instance, assume that we have reduced a complicated circuit with Thevenin's theorem to get Fig. 3-1c. We would know that the diode is forward biased.

The Forward Region

Figure 3-1c is a circuit that you can set up in the laboratory. After you connect this circuit, you can measure the diode current and voltage. You can also reverse the polarity of the dc source and measure diode current and voltage for reverse bias. If you plot the diode current versus the diode voltage, you will get a graph that looks like Fig. 3-2.

This is a visual summary of the ideas discussed in the preceding chapter. For instance, when the diode is forward biased, there is no significant current until the diode voltage is greater than the barrier potential. On the other hand, when the diode is reverse biased, there is almost no reverse current until the diode voltage reaches the breakdown voltage. Then, avalanche produces a large reverse current, destroying the diode.

Knee Voltage

In the forward region, the voltage at which the current starts to increase rapidly is called the **knee voltage** of the diode. The knee voltage equals the barrier potential. Analysis of diode circuits usually comes down to determining whether the diode voltage is more or less than the knee voltage. If it's more, the diode conducts easily. If it's less, the diode conducts poorly. We define the knee voltage of a silicon diode as:

$$V_K \approx 0.7 \text{ V} \tag{3-1}$$

(*Note:* The symbol ≈ means "approximately equal to.")

Even though germanium diodes are rarely used in new designs, you may still encounter germanium diodes in special circuits or in older equipment. For this reason, remember that the knee voltage of a germanium diode is approximately 0.3 V. This lower knee voltage is an advantage and accounts for the use of a germanium diode in certain applications.

Bulk Resistance

Above the knee voltage, the diode current increases rapidly. This means that small increases in the diode voltage cause large increases in diode current. After the barrier potential is overcome, all that impedes the current is the **ohmic resistance** of the *p* and *n* regions. In other words, if the *p* and *n* regions were two separate pieces of semiconductor, each would have a resistance that you could measure with an ohmmeter, the same as an ordinary resistor.

The sum of the ohmic resistances is called the **bulk resistance** of the diode. It is defined as:

$$R_B = R_P + R_N \tag{3-2}$$

The bulk resistance depends on the size of the *p* and *n* regions, and how heavily doped they are. Often, the bulk resistance is less than 1 Ω.

Maximum DC Forward Current

If the current in a diode is too large, the excessive heat can destroy the diode. For this reason, a manufacturer's data sheet specifies the maximum current a diode can safely handle without shortening its life or degrading its characteristics.

The **maximum forward current** is one of the maximum ratings given on a data sheet. This current may be listed as I_{max}, $I_{F(max)}$, I_O, etc., depending on the manufacturer. For instance, a 1N456 has a maximum forward current rating of 135 mA. This means that it can safely handle a continuous forward current of 135 mA.

Power Dissipation

You can calculate the power dissipation of a diode the same way as you do for a resistor. It equals the product of diode voltage and current. As a formula:

$$P_D = V_D I_D \tag{3-3}$$

The **power rating** is the maximum power the diode can safely dissipate without shortening its life or degrading its properties. In symbols, the definition is:

$$P_{max} = V_{max} I_{max} \tag{3-4}$$

where V_{max} is the voltage corresponding to I_{max}. For instance, if a diode has a maximum voltage and current of 1 V and 2 A, its power rating is 2 W.

Example 3-1

IIII MultiSim

Is the diode of Fig. 3-3*a* forward biased or reverse biased?

SOLUTION   The voltage across R_2 is positive; therefore, the circuit is trying to push current in the easy direction of flow. If this is not clear, visualize the Thevenin circuit facing the diode as shown in Fig. 3-3*b*. In this series circuit, you can see that the dc source is trying to push current in the easy direction of flow. Therefore, the diode is forward biased.

Whenever in doubt, reduce the circuit to a series circuit. Then, it will be clear whether the dc source is trying to push current in the easy direction or not.

Figure 3-3

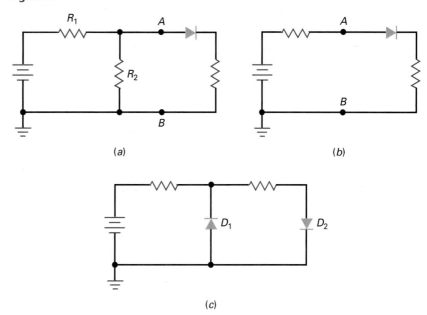

(a)

(b)

(c)

PRACTICE PROBLEM 3–1 Are the diodes of Fig. 3-3c forward biased or reverse biased?

Example 3-2

A diode has a power rating of 5 W. If the diode voltage is 1.2 V and the diode current is 1.75 A, what is the power dissipation? Will the diode be destroyed?

SOLUTION

$$P_D = (1.2 \text{ V})(1.75 \text{ A}) = 2.1 \text{ W}$$

This is less than the power rating, so the diode will not be destroyed.

PRACTICE PROBLEM 3–2 Referring to Example 3-2, what is the diode's power dissipation if the diode voltage is 1.1 V and the diode current is 2 A?

3-2 The Ideal Diode

Figure 3-4 shows a detailed graph of the forward region of a diode. Here you see the diode current I_D versus diode voltage V_D. Notice how the current is approximately zero until the diode voltage approaches the barrier potential. Somewhere in the vicinity of 0.6 to 0.7 V, the diode current increases. When the diode voltage is greater than 0.8 V, the diode current is significant and the graph is almost linear.

Depending on how a diode is doped and its physical size, it may differ from other diodes in its maximum forward current, power rating, and other characteristics. If we need an exact solution, we would have to use the graph of

Figure 3-4 Graph of forward current.

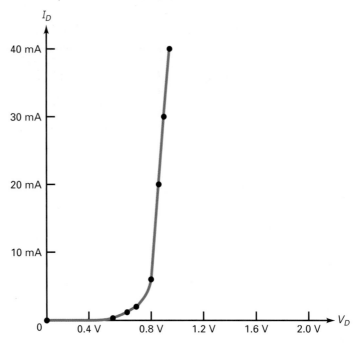

the particular diode. Although the exact current and voltage points will differ from one diode to the next, the graph of any diode is similar to Fig. 3-4. All silicon diodes have a knee voltage of approximately 0.7 V.

Most of the time, we do not need an exact solution. This is why we can and should use approximations for a diode. We will begin with the simplest approximation, called an **ideal diode.** In the most basic terms, what does a diode do? It conducts well in the forward direction and poorly in the reverse direction. Ideally, a diode acts like a perfect conductor (zero resistance) when forward biased and like a perfect insulator (infinite resistance) when reverse biased.

Figure 3-5*a* shows the current-voltage graph of an ideal diode. It echoes what we just said: zero resistance when forward biased and infinite resistance when reverse biased. It is impossible to build such a device, but this is what manufacturers would produce if they could.

Is there any device that acts like an ideal diode? Yes. An ordinary switch has zero resistance when closed and infinite resistance when open. Therefore, an ideal diode acts like a switch that closes when forward biased and opens when reverse biased. Figure 3-5*b* summarizes the switch idea.

Figure 3-5 (*a*) Ideal diode curve; (*b*) ideal diode acts like a switch.

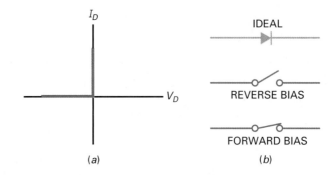

Example 3-3

Use the ideal diode to calculate the load voltage and load current in Fig. 3-6a.

SOLUTION   Since the diode is forward biased, it is equivalent to a closed switch. Visualize the diode as a closed switch. Then, you can see that all of the source voltage appears across the load resistor:

$$V_L = 10 \text{ V}$$

With Ohm's law, the load current is:

$$I_L = \frac{10 \text{ V}}{1 \text{ k}\Omega} = 10 \text{ mA}$$

PRACTICE PROBLEM 3-3   In Fig. 3-6a, find the ideal load current if the source voltage is 5 V.

Example 3-4

Calculate the load voltage and load current in Fig. 3-6b using an ideal diode.

Figure 3-6

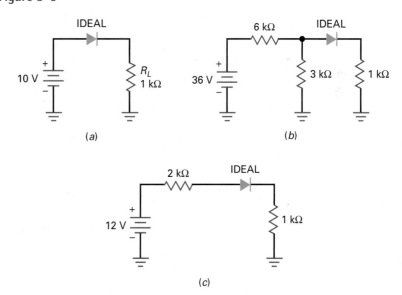

(a)   (b)

(c)

SOLUTION   One way to solve this problem is to Thevenize the circuit to the left of the diode. Looking from the diode back toward the source, we see a voltage divider with 6 kΩ and 3 kΩ. The Thevenin voltage is 12 V, and the Thevenin resistance is 2 kΩ. Figure 3-6c shows the Thevenin circuit driving the diode. (If you have any problem understanding this, review Example 1-3.)

Now that we have a series circuit, we can see that the diode is forward biased. Visualize the diode as a closed switch. Then, the remaining calculations are:

$$I_L = \frac{12 \text{ V}}{3 \text{ k}\Omega} = 4 \text{ mA}$$

and

$$V_L = (4 \text{ mA})(1 \text{ k}\Omega) = 4 \text{ V}$$

You don't have to use Thevenin's theorem. You can analyze Fig. 3-6*b* by visualizing the diode as a closed switch. Then, you have 3 kΩ in parallel with 1 kΩ, equivalent to 750 Ω. Using Ohm's law, you can calculate a voltage drop of 32 V across the 6 kΩ. The rest of the analysis produces the same load voltage and load current.

PRACTICE PROBLEM 3-4 Using Fig. 3-6*b*, change the 36 V source to 18 V and solve for the load voltage and load current using an ideal diode.

GOOD TO KNOW

When you troubleshoot a circuit that contains a silicon diode that is supposed to be forward biased, a diode voltage measurement much greater than 0.7 V means that the diode has failed and is in fact open.

3-3 The Second Approximation

The ideal approximation is all right in most troubleshooting situations. But we are not always troubleshooting. Sometimes, we want a more accurate value for load current and load voltage. This is where the *second approximation* comes in.

Figure 3-7*a* shows the graph of current versus voltage for the second approximation. The graph says that no current exists until 0.7 V appears across the diode. At this point, the diode turns on. Thereafter, only 0.7 V can appear across the diode, no matter what the current.

Figure 3-7*b* shows the equivalent circuit for the second approximation of a silicon diode. We think of the diode as a switch in series with a barrier potential of 0.7 V. If the Thevenin voltage facing the diode is greater than 0.7 V, the switch will close. When conducting, then the diode voltage is 0.7 V for any forward current.

On the other hand, if the Thevenin voltage is less than 0.7 V, the switch will open. In this case, there is no current through the diode.

Figure 3-7 (*a*) Diode curve for second approximation; (*b*) equivalent circuit for second approximation.

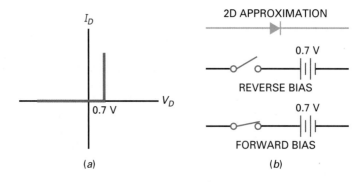

(*a*)   (*b*)

Example 3-5

Figure 3-8

2D APPROXIMATION

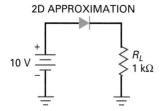

Use the second approximation to calculate the load voltage, load current, and diode power in Fig. 3-8.

SOLUTION   Since the diode is forward biased, it is equivalent to a battery of 0.7 V. This means that the load voltage equals the source voltage minus the diode drop:

$$V_L = 10\text{ V} - 0.7\text{ V} = 9.3\text{ V}$$

With Ohm's law, the load current is:

$$I_L = \frac{9.3\text{ V}}{1\text{ k}\Omega} = 9.3\text{ mA}$$

The diode power is

$$P_D = (0.7\text{ V})(9.3\text{ mA}) = 6.51\text{ mW}$$

PRACTICE PROBLEM 3–5   Using Fig. 3-8, change the source voltage to 5 V and calculate the new load voltage, current, and diode power.

Example 3-6

Calculate the load voltage, load current, and diode power in Fig. 3-9a using the second approximation.

Figure 3-9   (a) Original circuit; (b) simplified with Thevenin's theorem.

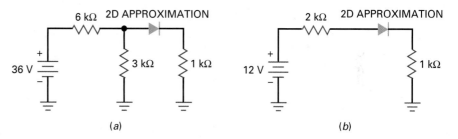

(a)   (b)

SOLUTION   Again, we will Thevenize the circuit to the left of the diode. As before, the Thevenin voltage is 12 V and the Thevenin resistance is 2 kΩ. Figure 3-9b shows the simplified circuit.

Since the diode voltage is 0.7 V, the load current is:

$$I_L = \frac{12\text{ V} - 0.7\text{ V}}{3\text{ k}\Omega} = 3.77\text{ mA}$$

The load voltage is:

$$V_L = (3.77\text{ mA})(1\text{ k}\Omega) = 3.77\text{ V}$$

and the diode power is:

$$P_D = (0.7\text{ V})(3.77\text{ mA}) = 2.64\text{ mW}$$

PRACTICE PROBLEM 3–6   Repeat Example 3-6 using 18 V as the voltage source value.

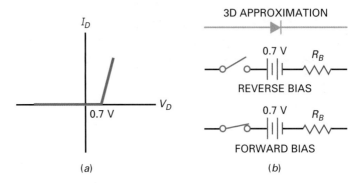

Figure 3-10 (*a*) Diode curve for third approximation; (*b*) equivalent circuit for third approximation.

3-4 The Third Approximation

In the *third approximation* of a diode, we include the bulk resistance R_B. Figure 3-10a shows the effect that R_B has on the diode curve. After the silicon diode turns on, the voltage increases linearly with an increase in current. The greater the current, the larger the diode voltage because of the voltage drop across the bulk resistance.

The equivalent circuit for the third approximation is a switch in series with a barrier potential of 0.7 V and a resistance of R_B (see Fig. 3-10b). When the diode voltage is larger than 0.7 V, the diode conducts. During conduction, the total voltage across the diode is:

$$V_D = 0.7 \text{ V} + I_D R_B \tag{3-5}$$

Often, the bulk resistance is less than 1 Ω, and we can safely ignore it in our calculations. A useful guideline for ignoring bulk resistance is this definition:

$$\textbf{Ignore bulk: } R_B < 0.01 R_{TH} \tag{3-6}$$

This says to ignore the bulk resistance when it is less than 1/100 of the Thevenin resistance facing the diode. When this condition is satisfied, the error is less than 1 percent. The third approximation is rarely used by technicians because circuit designers usually satisfy Eq. (3-6).

Example 3-7

The 1N4001 of Fig. 3-11a has a bulk resistance of 0.23 Ω. What is the load voltage, load current, and diode power?

SOLUTION   Replacing the diode by its third approximation, we get Fig. 3-11b. The bulk resistance is small enough to ignore because it is less than 1/100 of the load resistance. In this case, we can use the second approximation to solve the problem. We already did this in Example 3-6, where we found a load voltage, load current, and diode power of 9.3 V, 9.3 mA, and 6.51 mW.

Figure 3–11

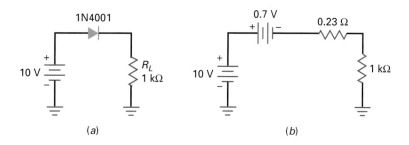

(a)   (b)

Example 3-8

||| MultiSim

Repeat the preceding example for a load resistance of 10 Ω.

SOLUTION   Figure 3-12a shows the equivalent circuit. The total resistance is:

$$R_T = 0.23\ \Omega + 10\ \Omega = 10.23\ \Omega$$

The total voltage across R_T is:

$$V_T = 10\ \text{V} - 0.7\ \text{V} = 9.3\ \text{V}$$

Therefore, the load current is:

$$I_L = \frac{9.3\ \text{V}}{10.23\ \Omega} = 0.909\ \text{A}$$

The load voltage is:

$$V_L = (0.909\ \text{A})(10\ \Omega) = 9.09\ \text{V}$$

Figure 3–12

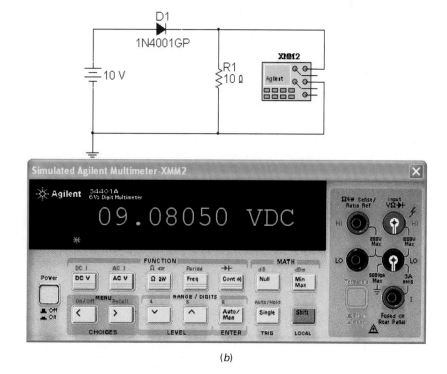

(a)

(b)

To calculate the diode power, we need to know the diode voltage. We can get this in either of two ways. We can subtract the load voltage from the source voltage:

$$V_D = 10 \text{ V} - 9.09 \text{ V} = 0.91 \text{ V}$$

or we can use Eq. (3-5):

$$V_D = 0.7 \text{ V} + (0.909 \text{ A})(0.23 \text{ }\Omega) = 0.909 \text{ V}$$

The slight difference in the last two answers is caused by rounding. The diode power is:

$$P_D = (0.909 \text{ V})(0.909 \text{ A}) = 0.826 \text{ W}$$

Two more points. First, the 1N4001 has a maximum forward current of 1 A and a power rating of 1 W, so the diode is being pushed to its limits with a load resistance of 10 Ω. Second, the load voltage calculated with the third approximation is 9.09 V, which is in very close agreement with the MultiSim load voltage of 9.08 V (see Fig. 3-12b).

Summary Table 3-1 illustrates the differences between the three diode approximations.

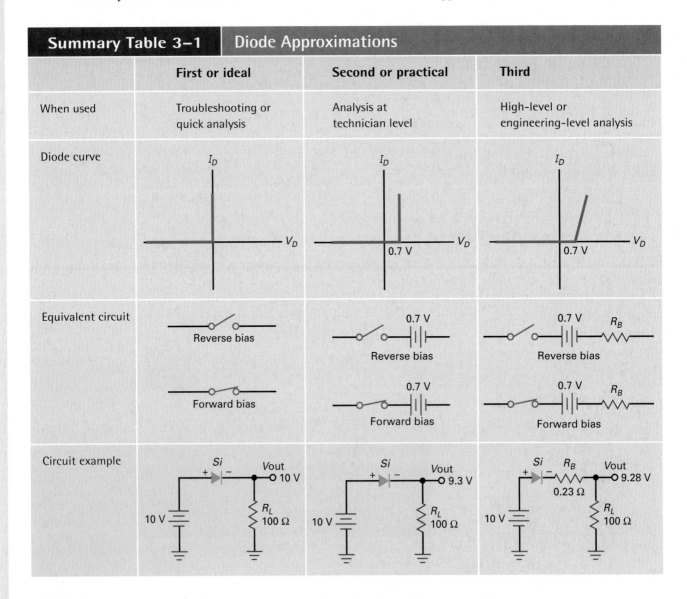

Summary Table 3–1	Diode Approximations		
	First or ideal	**Second or practical**	**Third**
When used	Troubleshooting or quick analysis	Analysis at technician level	High-level or engineering-level analysis
Diode curve			
Equivalent circuit			
Circuit example			

PRACTICE PROBLEM 3-8   Repeat Example 3-8 using 5 V as the voltage source value.

3-5 Troubleshooting

You can quickly check the condition of a diode with an ohmmeter on a medium-to-high resistance range. Measure the dc resistance of the diode in either direction, and then reverse the leads and measure the dc resistance again. The forward current will depend on which ohmmeter range is used, which means that you get different readings on different ranges.

The main thing to look for, however, is a high ratio of reverse to forward resistance. For typical silicon diodes used in electronics work, the ratio should be higher than 1000:1. Remember to use a high enough resistance range to avoid the possibility of diode damage. Normally, the R × 100 or R × 1K ranges will provide proper safe measurements.

Using an ohmmeter to check diodes is an example of go/no-go testing. You're really not interested in the exact dc resistance of the diode; all you want to know is whether the diode has a low resistance in the forward direction and a high resistance in the reverse direction. Diode troubles are indicated for any of the following: extremely low resistance in both directions (diode shorted); high resistance in both directions (diode open); somewhat low resistance in the reverse direction (called a *leaky diode*).

When set to the ohms or resistance function, most digital multimeters (DMMs) do not have the required voltage and current output capability to properly test *pn*-junction diodes. Most DMMs do, however, have a special diode test range. When the meter is set to this range, it supplies a constant current of approximately 1 mA to whatever device is connected to its leads. When forward biased, the DMM will display the *pn*-junction's forward voltage V_F shown in Fig. 3-13*a*. This forward voltage will generally be between 0.5 V and 0.7 V for normal silicon *pn*-junction diodes. When the diode is reverse biased by the test leads, the meter will give an overrange indication such as "OL" or "1" on the display as shown in Fig. 3-13*b*. A shorted diode would display a voltage of less than 0.5 V in both directions. An open diode would be indicated by an overrange display in both directions. A leaky diode would display a voltage less than 2.0 V in both directions.

Figure 3-13 (*a*) DMM diode forward test.

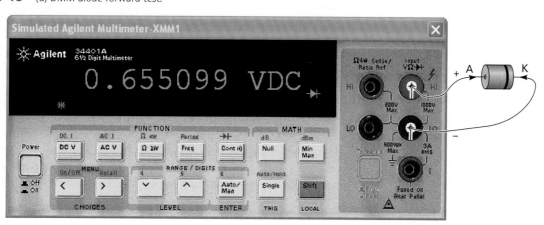

(*a*)

Figure 3-13 (*b*) DMM diode reverse test.

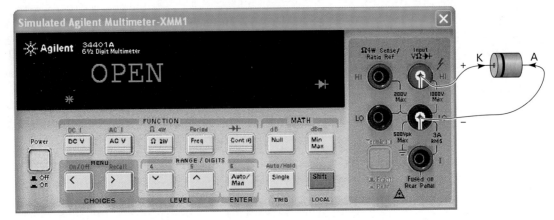

(*b*)

Example 3-9

Figure 3-14   Troubleshooting a circuit.

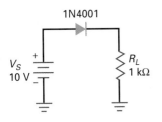

Figure 3-14 shows the diode circuit analyzed earlier. Suppose something causes the diode to burn out. What kind of symptoms will you get?

SOLUTION   When a diode burns out, it becomes an open circuit. In this case, the current drops to zero. Therefore, if you measure the load voltage, the voltmeter will indicate zero.

Example 3-10

Suppose the circuit of Fig. 3-14 is not working. If the load is not shorted, what is the trouble?

SOLUTION   Many troubles are possible. First, the diode could be open. Second, the supply voltage could be zero. Third, one of the connecting wires could be open.

How do you find the trouble? Measure the voltages to isolate the defective component. Then disconnect any suspected component and test its resistance. For instance, you could measure the source voltage first and the load voltage second. If there is source voltage but no load voltage, the diode may be open. An ohmmeter or DMM test will tell. If the diode passes the ohmmeter or DMM test, check the connections because there's nothing else to account for having source voltage but no load voltage.

If there is no source voltage, the power supply is defective or a connection between the supply and the diode is open. Power-supply troubles are common. Often, when electronics equipment is not working, the trouble is in the power supply. This is why most troubleshooters start by measuring the voltages out of the power supply.

Figure 3-15 Up-down analysis of a circuit.

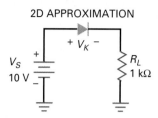

2D APPROXIMATION

3-6 Up-Down Circuit Analysis

There is nothing quite like **up-down analysis** to help you understand circuits. The idea is this: Any circuit has independent variables (like source voltages and branch resistances) and dependent variables (like voltages across resistors, currents, and powers). When an independent variable increases, each of the dependent variables will usually respond by increasing or decreasing. If you understand how the circuit works, you will be able to predict whether a dependent variable will increase or decrease.

Here is how it works for Fig. 3-15. A voltage V_S of 10 V is applied to a diode in series with a load resistance R_L of 1 kΩ. In the second approximation of a diode, there are three independent variables for this circuit: V_S, R_L, and V_K. We are including the knee voltage as an independent variable because it may be slightly different from the ideal value of 0.7 V. There are five dependent variables as follows: V_L, I_L, P_D, P_L, and P_T. These are the load voltage, load current, diode power, load power, and total power.

Suppose the source voltage V_S increases slightly, say 10 percent. How will each of the dependent variables respond? Will each go up (U), go down (D), or show no change (N)? Here are some of the thoughts that might pass through your mind as you solve this problem:

> *In the second approximation, the diode has a voltage drop of 0.7 V. If the source voltage increases slightly, the diode drop is still 0.7 V, which means that the load voltage has to increase. If the load voltage increases, the load current increases. An increase in load current means that the diode power and load power increase. The total power is the sum of diode power and load power, so total power must increase.*

The first row of Table 3-1 summarizes the effect of a small increase in source voltage. As you can see, each dependent variable increases.

What do you think happens when the load resistance of Fig. 3-15 increases slightly? Since the diode voltage is constant in the second approximation, the load voltage shows no change, but the load current will go down. This implies less diode power, load power, and total power. The second row of Table 3-1 summarizes this case.

Finally, consider the effect of knee voltage. If the knee voltage increases slightly in Fig. 3-15, the dependent variables decrease, except for the diode power, as shown in the third row of Table 3-1.

Look at Fig. 3-25 (at the end of the chapter). How do you use this to find dependent changes?

The way you can practice up-down analysis for the circuit is by selecting one independent variable (V_S, R_1, R_2, R_3, or V_K). Next, select any dependent variable (V_A, V_B, V_C, I_1, etc.). Then try to figure out whether the dependent variable goes up, goes down, or shows no change.

Table 3-1	Up-Down Analysis				
	V_L	I_L	P_D	P_L	P_T
V_S increase	U	U	U	U	U
R_L increase	N	D	D	D	D
V_K increase	D	D	U	D	D

For instance, how does an increase in knee voltage affect the current in R_3? In Fig. 3-25, a stiff voltage divider drives the diode in series with the 100 kΩ. Therefore, a slight increase in knee voltage will decrease the voltage across the 100 kΩ. Then, Ohm's law tells us that I_3 should decrease.

A final point: Do not use a calculator for up-down circuit analysis. This would defeat the whole purpose of this type of thinking. Up-down circuit analysis is similar to troubleshooting because the emphasis is on logic rather than equations. The purpose of up-down analysis is to train your mind to get in touch with the circuit action. It does this by forcing you to think about how the different parts of the circuit interact.

3-7 Reading a Data Sheet

A data sheet, or specification sheet, lists important parameters and operating characteristics for semiconductor devices. Also, essential information such as case styles, pinouts, testing procedures, and typical applications can be obtained from a component's data sheet. Semiconductor manufacturers generally provide this information in data books or from the manufacturer's website. This information can also be found on the Internet by companies that specialize in cross-referencing or component substitution.

Much of the information on a manufacturer's data sheet is obscure and of use only to circuit designers. For this reason, we will discuss only those entries on the data sheet that describe quantities in this book.

Reverse Breakdown Voltage

Let us start with the data sheet for a 1N4001, a rectifier diode used in power supplies (circuits that convert ac voltage to dc voltage). Figure 3-16 shows a data sheet for the 1N4001 to 1N4007 series of diodes: seven diodes that have the same forward characteristics but differ in their reverse characteristics. We are interested in the 1N4001 member of this family. The first entry under "Absolute Maximum Ratings" is this:

	Symbol	1N4001
Peak Repetitive Reverse Voltage	V_{RRM}	50 V

The breakdown voltage for this diode is 50 V. This breakdown occurs because the diode goes into avalanche when a huge number of carriers suddenly appears in the depletion layer. With a rectifier diode like the 1N4001, breakdown is usually destructive.

With the 1N4001, a reverse voltage of 50 V represents a destructive level that a designer avoids under all operating conditions. This is why a designer includes a *safety factor*. There is no absolute rule on how large to make the safety factor because it depends on too many design factors. A conservative design would use a safety factor of 2, which means never allowing a reverse voltage of more than 25 V across the 1N4001. A less-conservative design might allow as much as 40 V across the 1N4001.

On other data sheets, reverse breakdown voltage may be designated *PIV*, *PRV*, or *BV*.

Figure 3-16   Data sheet for 1N4001–1N4007 diodes.

1N4001 - 1N4007

Features

- Low forward voltage drop.

- High surge current capability.

DO-41
COLOR BAND DENOTES CATHODE

General Purpose Rectifiers

Absolute Maximum Ratings* T_A = 25°C unless otherwise noted

Symbol	Parameter	Value							Units
		4001	4002	4003	4004	4005	4006	4007	
V_{RRM}	Peak Repetitive Reverse Voltage	50	100	200	400	600	800	1000	V
$I_{F(AV)}$	Average Rectified Forward Current, .375 " lead length @ T_A = 75°C				1.0				A
I_{FSM}	Non-repetitive Peak Forward Surge Current 8.3 ms Single Half-Sine-Wave				30				A
T_{stg}	Storage Temperature Range				-55 to +175				°C
T_J	Operating Junction Temperature				-55 to +175				°C

*These ratings are limiting values above which the serviceability of any semiconductor device may be impaired.

Thermal Characteristics

Symbol	Parameter	Value	Units
P_D	Power Dissipation	3.0	W
$R_{\theta JA}$	Thermal Resistance, Junction to Ambient	50	°C/W

Electrical Characteristics T_A = 25°C unless otherwise noted

Symbol	Parameter	Device							Units
		4001	4002	4003	4004	4005	4006	4007	
V_F	Forward Voltage @ 1.0 A				1.1				V
I_{rr}	Maximum Full Load Reverse Current, Full Cycle T_A = 75°C				30				μA
I_R	Reverse Current @ rated V_R T_A = 25°C T_A = 100°C				5.0 500				μA μA
C_T	Total Capacitance V_R = 4.0 V, f = 1.0 MHz				15				pF

(a)   1N4001-1N4007, Rev. C1

Figure 3-16 (continued)

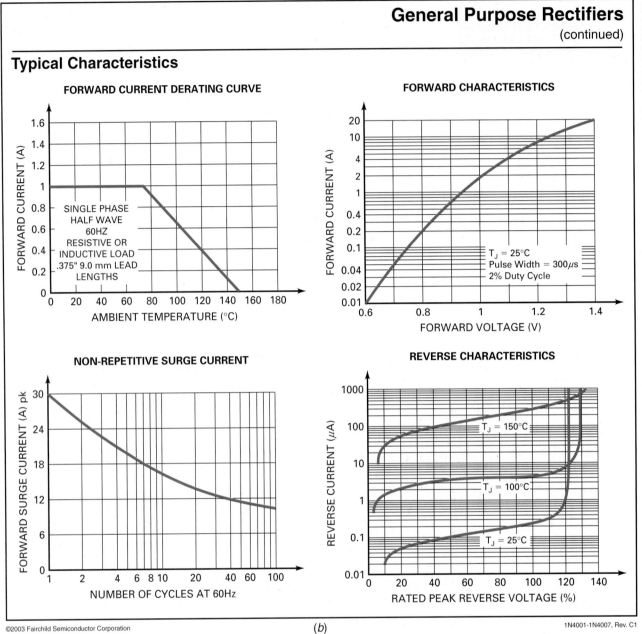

General Purpose Rectifiers

(continued)

Typical Characteristics

FORWARD CURRENT DERATING CURVE

SINGLE PHASE
HALF WAVE
60HZ
RESISTIVE OR
INDUCTIVE LOAD
.375" 9.0 mm LEAD
LENGTHS

FORWARD CHARACTERISTICS

$T_J = 25°C$
Pulse Width $= 300\mu s$
2% Duty Cycle

NON-REPETITIVE SURGE CURRENT

REVERSE CHARACTERISTICS

$T_J = 150°C$

$T_J = 100°C$

$T_J = 25°C$

©2003 Fairchild Semiconductor Corporation

(b)

1N4001-1N4007, Rev. C1

Maximum Forward Current

Another entry of interest is average rectified forward current, which looks like this on the data sheet:

	Symbol	Value
Average Rectified Forward Current @ $T_A = 75°C$	$I_{F(AV)}$	1 A

This entry tells us that the 1N4001 can handle up to 1 A in the forward direction when used as a rectifier. You will learn more about average rectified forward current in the next chapter. For now, all you need to know is that 1 A is the level of forward current when the diode burns out because of excessive power dissipation. On other data sheets, the average current may be designated as I_o.

Again, a designer looks upon 1 A as the absolute maximum rating of the 1N4001, a level of forward current that should not even be approached. This is why a safety factor would be included—possibly a factor of 2. In other words, a reliable design would ensure that the forward current is less than 0.5 A under all operating conditions. Failure studies of devices show that the lifetime of a device decreases the closer you get to the maximum rating. This is why some designers use a safety factor of as much as 10:1. A really conservative design would keep the maximum forward current of the 1N4001 at 0.1 A or less.

Forward Voltage Drop

Under "Electrical Characteristics" in Fig. 3-16, the first entry shown gives you these data:

Characteristic and Conditions	Symbol	Maximum Value
Forward Voltage Drop $(i_F) = 1.0$ A, $T_A = 25°C$	v_F	1.1 V

As shown in Fig. 3-16 on the chart titled "Forward Characteristics," the typical 1N4001 has a forward voltage drop of 0.93 V when the current is 1 A and the junction temperature is 25°C. If you test thousands of 1N4001s, you will find that a few will have as much as 1.1 V across them when the current is 1 A.

Maximum Reverse Current

Another entry on the data sheet that is worth discussing is this one:

Characteristic and Conditions	Symbol	Typical Value	Maximum Value
Reverse Current	I_R		
$T_A = 25°C$		0.05 μA	10 μA
$T_A = 100°C$		1.0 μA	50 μA

This is the reverse current at the maximum reverse dc rated voltage (50 V for a 1N4001). At 25°C, the typical 1N4001 has a maximum reverse current of 5.0 μA. But notice how it increases to 500 μA at 100°C. Remember that this reverse current includes thermally produced saturation current and surface-leakage current. You can see from these numbers that temperature is important. A design that requires a reverse current of less than 5.0 μA will work fine at 25°C with a typical 1N4001, but will fail in mass production if the junction temperature reaches 100°C.

3-8 How to Calculate Bulk Resistance

When you are trying to analyze a diode circuit accurately, you will need to know the bulk resistance of the diode. Manufacturers' data sheets do not usually list the bulk resistance separately, but they do give enough information to allow you to calculate it. Here is the derivation for bulk resistance:

$$R_B = \frac{V_2 - V_1}{I_2 - I_1} \tag{3-7}$$

where V_1 and I_1 are the voltage and current at some point at or above the knee voltage; V_2 and I_2 are the voltage and current at some higher point on the diode curve.

For instance, the data sheet of a 1N4001 gives a forward voltage of 0.93 V for a current of 1 A. Since this is a silicon diode, it has a knee voltage of approximately 0.7 V and a current of approximately zero. Therefore, the values to use are $V_2 = 0.93$ V, $I_2 = 1$ A, $V_1 = 0.7$ V, and $I_1 = 0$. Substituting these values into equation, we get a bulk resistance of:

$$R_B = \frac{V_2 - V_1}{I_2 - I_1} = \frac{0.93\text{V} - 0.7\text{ V}}{1\text{ A} - 0\text{ A}} = \frac{0.23\text{ V}}{1\text{ A}} = 0.23\ \Omega$$

Incidentally, the diode curve is a graph of current versus voltage. The bulk resistance equals the inverse of the slope above the knee. The greater the slope of the diode curve, the smaller the bulk resistance. In other words, the more vertical the diode curve is above the knee, the lower the bulk resistance.

3-9 DC Resistance of a Diode

If you take the ratio of total diode voltage to total diode current, you get the *dc resistance* of the diode. In the forward direction, this dc resistance is symbolized by R_F; in the reverse direction, it is designated R_R.

Forward Resistance

Because the diode is a nonlinear device, its dc resistance varies with the current through it. For example, here are some pairs of forward current and voltage for a 1N914: 10 mA at 0.65 V, 30 mA at 0.75 V, and 50 mA at 0.85 V. At the first point, the dc resistance is:

$$R_F = \frac{0.65\text{ V}}{10\text{ mA}} = 65\ \Omega$$

At the second point:

$$R_F = \frac{0.75\text{ V}}{30\text{ mA}} = 25\ \Omega$$

And at the third point:

$$R_F = \frac{0.85\text{ mV}}{50\text{ mA}} = 17\ \Omega$$

Notice how the dc resistance decreases as the current increases. In any case, the forward resistance is low compared to the reverse resistance.

Reverse Resistance

Similarly, here are two sets of reverse current and voltage for a 1N914: 25 nA at 20 V; 5 μA at 75 V. At the first point, the dc resistance is:

$$R_R = \frac{20\text{ V}}{25\text{ nA}} = 800\text{ M}\Omega$$

At the second point:

$$R_R = \frac{75\text{ V}}{5\ \mu\text{A}} = 15\text{ M}\Omega$$

Notice how the dc resistance decreases as we approach the breakdown voltage (75 V).

DC Resistance versus Bulk Resistance

The dc resistance of a diode is different from the bulk resistance. The dc resistance of a diode equals the bulk resistance *plus* the effect of the barrier potential. In other words, the dc resistance of a diode is its total resistance, whereas the bulk resistance is the resistance of only the *p* and *n* regions. For this reason, the dc resistance of a diode is always greater than bulk resistance.

3-10 Load Lines

This section is about the **load line,** a tool used to find the exact value of diode current and voltage. Load lines are useful with transistors, so a detailed explanation will be given later in the transistor discussions.

Equation for the Load Line

How can we find the exact diode current and voltage in Fig. 3-17*a*? The current through the resistor is:

$$I_D = \frac{V_S - V_D}{R_s} \tag{3-8}$$

Because of the series circuit, this current is the same through the diode.

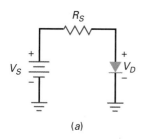

Figure 3-17 Load-line analysis.

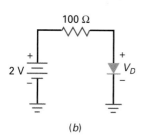

An Example

If the source voltage is 2 V and the resistance is 100 Ω as shown in Fig. 3-17*b*, then Eq. (3-8) becomes:

$$I_D = \frac{2 - V_D}{100} \tag{3-9}$$

Equation (3-9) is a linear relationship between current and voltage. If we plot this equation, we will get a straight line. For instance, let V_D equal zero. Then:

$$I_D = \frac{2\,V - 0\,V}{100\,\Omega} = 20\;mA$$

Plotting this point ($I_D = 20$ mA, $V_D = 0$) gives the point on the vertical axis of Fig. 3-18. This point is called *saturation* because it represents maximum current with 2 V across 100 Ω.

Here's how to get another point. Let V_D equal 2 V. Then Eq. (3-9) gives:

$$I_D = \frac{2\,V - 2\,V}{100\,\Omega} = 0$$

When we plot this point ($I_D = 0$, $V_D = 2$ V), we get the point shown on the horizontal axis (Fig. 3-18). This point is called *cutoff* because it represents minimum current.

By selecting other voltages, we can calculate and plot additional points. Because Eq. (3-9) is linear, all points will lie on the straight line shown in Fig. 3-18. The straight line is called the *load line.*

The Q Point

Figure 3-18 shows the load line and a diode curve. The point of intersection, known as the *Q* point, represents a simultaneous solution between the diode curve and the load line. In other words, the *Q* point is the only point on the graph that works for both the diode and the circuit. By reading the coordinates of the *Q* point, we get a current of 12.5 mA and a diode voltage of 0.75 V.

Figure 3–18 *Q* point is the intersection of the diode curve and the load line.

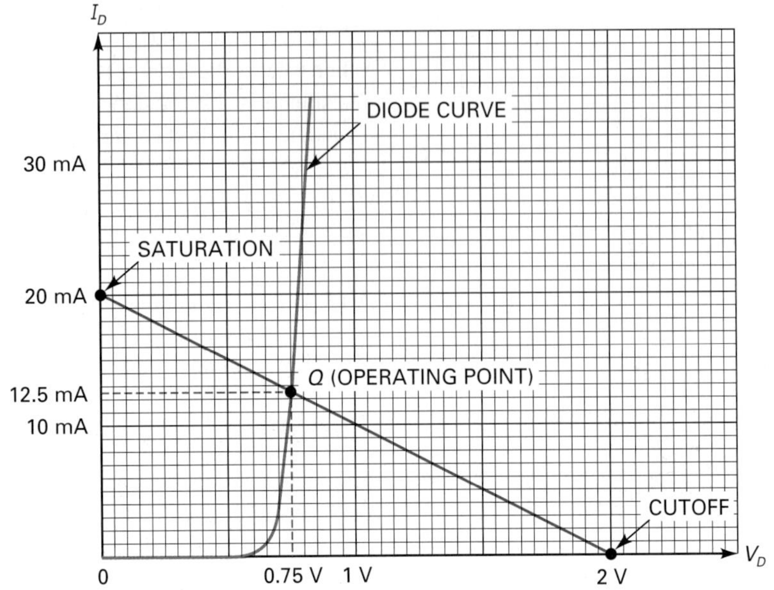

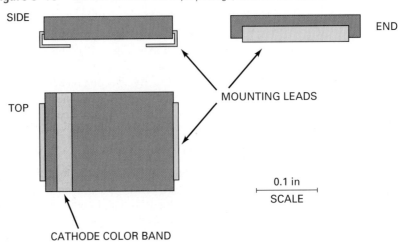

Incidentally, the *Q* point has no relationship to the figure of merit of a coil. In the present discussion, *Q* is an abbreviation for *quiescent,* which means "at rest." The quiescent or *Q* point of semiconductor circuits is discussed in later chapters.

3–11 Surface-Mount Diodes

Surface-mount (SM) diodes can be found anywhere there is a need for diode applications. SM diodes are small, efficient, and relatively easy to test, remove, and replace on the circuit board. Although there are a number of SM package styles, two basic styles dominate the industry: SM (surface mount) and SOT (small outline transistor).

The SM package has two L-bend leads and a colored band on one end of the body to indicate the cathode lead. Figure 3-19 shows a typical set of dimensions. The length and width of the SM package are related to the current rating of

Figure 3–19 The two-terminal SM-style package, used for SM diodes.

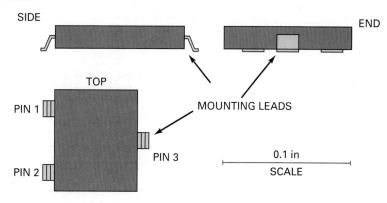

Figure 3-20 The SOT-23 is a three-terminal transistor package commonly used for SM diodes.

SIDE

END

TOP

PIN 1

MOUNTING LEADS

PIN 3

PIN 2

0.1 in

SCALE

the device. The larger the surface area, the higher the current rating. So an SM diode rated at 1 A might have a surface area given by 0.181 by 0.115 in. The 3 A version, on the other hand, might measure 0.260 by 0.236 in. The thickness tends to remain at about 0.103 in for all current ratings.

Increasing the surface area of an SM-style diode increases its ability to dissipate heat. Also, the corresponding increase in the width of the mounting terminals increases the thermal conductance to a virtual heat sink made up of the solder joints, mounting lands, and the circuit board itself.

SOT-23 packages have three gull-wing terminals (see Fig. 3-20). The terminals are numbered counterclockwise from the top, pin 3 being alone on one side. However, there are no standard markings indicating which two terminals are used for the cathode and the anode. To determine the internal connections of the diode, you can look for clues printed on the circuit board, check the schematic diagram, or consult the diode manufacturer's data book. Some SOT-style packages include two diodes, which have a common-anode or common-cathode connection at one of the terminals.

Diodes in SOT-23 packages are small, no dimension being greater than 0.1 in. Their small size makes it difficult to dissipate larger amounts of heat, so the diodes are generally rated at less than 1 A. The small size also makes it impractical to label them with identification codes. As with many of the tiny SM devices, you have to determine the PIN from other clues on the circuit board and schematic diagram.

Summary

SEC. 3–1   BASIC IDEAS

A diode is a nonlinear device. The knee voltage, approximately 0.7 V for a silicon diode, is where the forward curve turns upward. The bulk resistance is the ohmic resistance of the p and n regions. Diodes have a maximum forward current and a power rating.

SEC. 3–2   THE IDEAL DIODE

This is the first approximation of a diode. The equivalent circuit is a switch that closes when forward biased and opens when reverse biased.

SEC. 3–3   THE SECOND APPROXIMATION

In this approximation, we visualize a silicon diode as a switch in series with a knee voltage of 0.7 V. If the Thevenin voltage facing the diode is greater than 0.7 V, the switch closes.

SEC. 3-4 THE THIRD APPROXIMATION

We seldom use this approximation because bulk resistance is usually small enough to ignore. In this approximation, we visualize the diode as a switch in series with a knee voltage and a bulk resistance.

SEC. 3-5 TROUBLESHOOTING

When you suspect that a diode is the trouble, remove it from the circuit and use an ohmmeter to measure its resistance in each direction. You should get a high resistance one way and a low resistance the other way, at least 1000:1 ratio. Remember to use a high enough resistance range when testing a diode, to avoid possible diode damage. A DMM will display 0.5–0.7 V when a diode is forward biased and an overrange indication when it is reverse biased.

SEC. 3-6 UP–DOWN CIRCUIT ANALYSIS

No calculation is required in this type of circuit analysis. All you are after is *up*, *down*, or *no change*. When you know beforehand how a dependent variable should respond to an increase in an independent variable, you will be more successful at troubleshooting, analysis, and design.

SEC. 3-7 READING A DATA SHEET

Data sheets are useful to a circuit designer and may be useful to a repair technician for selecting a substitute device, which is sometimes required. Diode data sheets from different manufacturers contain similar information, but different symbols are used to indicate different operating conditions. Diode data sheets may list the following: breakdown voltage (V_R, V_{RRM}, V_{RWM}, PIV, PRV, BV), maximum forward current ($I_{F(max)}$, $I_{F(av)}$, I_0), forward voltage drop ($V_{F(max)}$, V_F), and maximum reverse current $I_{R(max)}$, I_{RRM}).

SEC. 3-8 HOW TO CALCULATE BULK RESISTANCE

You need two points in the forward region of the third approximation. One point can be 0.7 V with zero current. The second point comes from the data sheet at a large forward current where both a voltage and a current are given.

SEC. 3-9 DC RESISTANCE OF A DIODE

The dc resistance equals the diode voltage divided by the diode current at some operating point. This resistance is what an ohmmeter will measure. DC resistance has limited application, aside from telling you that it is small in the forward direction and large in the reverse direction.

SEC. 3-10 LOAD LINES

The current and voltage in a diode circuit have to satisfy both the diode curve and Ohm's law for the load resistor. These are two separate requirements that graphically translate to the intersection of the diode curve and the load line.

SEC. 3-11 SURFACE-MOUNT DIODES

Surface-mount diodes are often found on modern electronics circuits boards. These diodes are small, efficient, and typically found either as an SM (surface mount) or an SOT (small outline transistor) case style.

Definitions

(3-1) Silicon knee voltage:

$$V_K \approx 0.7 \text{ V}$$

(3-2) Bulk resistance:

$$R_B = R_P + R_N$$

(3-4) Maximum power dissipation

$$P_{max} = V_{max} I_{max}$$

(3-6) Ignore bulk:

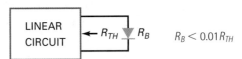

$$R_B < 0.01 R_{TH}$$

Derivations

(3-3) Diode power dissipation:

$$P_D = V_D I_D$$

(3-5) Third approximation:

$$V_D = 0.7 \text{ V} + I_D R_B$$

(3-7) Bulk resistance:

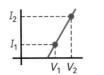

$$R_B = \frac{V_2 - V_1}{I_2 - I_1}$$

Student Assignments

1. When the graph of current versus voltage is a straight line, the device is referred to as
   a. Active
   b. Linear
   c. Nonlinear
   d. Passive

2. What kind of device is a resistor?
   a. Unilateral
   b. Linear
   c. Nonlinear
   d. Bipolar

3. What kind of a device is a diode?
   a. Bilateral
   b. Linear
   c. Nonlinear
   d. Unipolar

4. How is a nonconducting diode biased?
   a. Forward
   b. Inverse
   c. Poorly
   d. Reverse

5. When the diode current is large, the bias is
   a. Forward
   b. Inverse
   c. Poor
   d. Reverse

6. The knee voltage of a diode is approximately equal to the
   a. Applied voltage
   b. Barrier potential
   c. Breakdown voltage
   d. Forward voltage

7. The reverse current consists of minority-carrier current and
   a. Avalanche current
   b. Forward current
   c. Surface-leakage current
   d. Zener current

8. How much voltage is there across the second approximation of a silicon diode when it is forward biased?
   a. 0
   b. 0.3 V
   c. 0.7 V
   d. 1 V

9. How much current is there through the second approximation of a silicon diode when it is reverse biased?
   a. 0
   b. 1 mA
   c. 300 mA
   d. None of the above

10. How much forward diode voltage is there with the ideal-diode approximation?
    a. 0
    b. 0.7 V
    c. More than 0.7 V
    d. 1 V

11. The bulk resistance of a 1N4001 is
    a. 0
    b. 0.23 Ω
    c. 10 Ω
    d. 1 kΩ

12. If the bulk resistance is zero, the graph above the knee becomes
    a. Horizontal
    b. Vertical
    c. Tilted at 45°
    d. None of the above

13. The ideal diode is usually adequate when
    a. Troubleshooting
    b. Doing precise calculations
    c. The source voltage is low
    d. The load resistance is low

14. The second approximation works well when
    a. Troubleshooting
    b. Load resistance is high
    c. Source voltage is high
    d. All of the above

15. The only time you have to use the third approximation is when
    a. Load resistance is low
    b. Source voltage is high
    c. Troubleshooting
    d. None of the above

16. **MultiSim** How much load current is there in Fig. 3-21 with the ideal diode?
    a. 0   c. 12 mA
    b. 11.3 mA   d. 25 mA

Figure 3-21

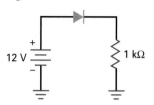

17. **MultiSim** How much load current is there in Fig. 3-21 with the second approximation?
    a. 0
    b. 11.3 mA
    c. 12 mA
    d. 25 mA

18. **MultiSim** How much load current is there in Fig. 3-21 with the third approximation?
    a. 0
    b. 11.3 mA
    c. 12 mA
    d. 25 mA

19. **MultiSim** If the diode is open in Fig. 3-21, the load voltage is
    a. 0
    b. 11.3 V
    c. 20 V
    d. −15 V

20. **MultiSim** If the resistor is ungrounded in Fig. 3-21, the voltage measured with a DMM between the top of the resistor and ground is closest to
    a. 0   c. 20 V
    b. 12 V   d. −15 V

21. **MultiSim** The load voltage measures 12 V in Fig. 3-21. The trouble may be
    a. A shorted diode
    b. An open diode
    c. An open load resistor
    d. Too much supply voltage

22. Using the third approximation in Fig. 3-21, how low must R_L be before the diode's bulk resistance must be considered?
    a. 1 Ω c. 23 Ω
    b. 10 Ω d. 100 Ω

Problems

SEC. 3–1 BASIC IDEAS

3-1 A diode is in series with 220 Ω. If the voltage across the resistor is 6 V, what is the current through the diode?

3-2 A diode has a voltage of 0.7 V and a current of 100 mA. What is the diode power?

3-3 Two diodes are in series. The first diode has a voltage of 0.75 V and the second has a voltage of 0.8 V. If the current through the first diode is 400 mA, what is the current through the second diode?

SEC. 3–2 THE IDEAL DIODE

3-4 In Fig. 3-22a, calculate the load current, load voltage, load power, diode power, and total power.

3-5 If the resistor is doubled in Fig. 3-22a, what is the load current?

3-6 In Fig. 3-22b, calculate the load current, load voltage, load power, diode power, and total power.

3-7 If the resistor is doubled in Fig. 3-22b, what is the load current?

3-8 If the diode polarity is reversed in Fig. 3-22b, what is the diode current? The diode voltage?

Figure 3-22

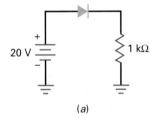

(a)

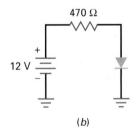

(b)

SEC. 3–3 THE SECOND APPROXIMATION

3-9 In Fig. 3-22a, calculate the load current, load voltage, load power, diode power, and total power.

3-10 If the resistor is doubled in Fig. 3-22a, what is the load current?

3-11 In Fig. 3-22b, calculate the load current, load voltage, load power, diode power, and total power.

3-12 If the resistor is doubled in Fig. 3-22a, what is the load current?

3-13 If the diode polarity is reversed in Fig. 3-22b, what is the diode current? The diode voltage?

SEC. 3–4 THE THIRD APPROXIMATION

3-14 In Fig. 3-22a, calculate the load current, load voltage, load power, diode power, and total power. ($R_B = 0.23\ \Omega$)

3-15 If the resistor is doubled in Fig. 3-22a, what is the load current? ($R_B = 0.23\ \Omega$)

3-16 In Fig. 3-22b, calculate the load current, load voltage, load power, diode power, and total power. ($R_B = 0.23\ \Omega$)

3-17 If the resistor is doubled in Fig. 3-22b, what is the load current? ($R_B = 0.23\ \Omega$)

3-18 If the diode polarity is reversed in Fig. 3-22b, what is the diode current? The diode voltage?

SEC. 3–5 TROUBLESHOOTING

3-19 Suppose the voltage across the diode of Fig. 3-23a is 5 V. Is the diode open or shorted?

Figure 3-23

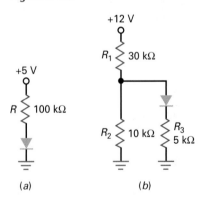

(a)   (b)

3-20 Something causes R to short in Fig. 3-23a. What will the diode voltage be? What will happen to the diode?

3-21 You measure 0 V across the diode of Fig. 3-23a. Next you check the source voltage and it reads +5 V with respect to ground. What is wrong with the circuit?

3-22 In Fig. 3-23b, you measure a potential of +3 V at the junction of R_1 and R_2. (Remember, potentials are always with respect to ground.) Next you measure 0 V at the junction of the diode and the 5-kΩ resistor. Name some possible troubles.

3-23 The forward and reverse DMM diode test reading is 0.7 V and 1.8 V. Is this diode good?

3-24 Which diode would you select in the 1N4000 series if it has to withstand a peak repetitive reverse voltage of 300 V?

3-25 The data sheet shows a band on one end of the diode. What is the name of this band? Does the diode arrow of the schematic symbol point toward or away from this band?

3-26 Boiling water has a temperature of 100°C. If you drop a 1N4001 into a pot of boiling water, will it be destroyed or not? Explain your answer.

Critical Thinking

3-27 Here are some diodes and their worst-case specifications:

Diode	I_F	I_R
1N914	10 mA at 1 V	25 nA at 20 V
1N4001	1 A at 1.1 V	10 μA at 50 V
1N1185	10 A at 0.95 V	4.6 mA at 100 V

Calculate the forward and the reverse resistance for each of these diodes.

3-28 In Fig. 3-23a, what value should R be to get a diode current of approximately 20 mA?

3-29 What value should R_2 be in Fig. 3-23b to set up a diode current of 0.25 mA?

3-30 A silicon diode has a forward current of 500 mA at 1 V. Use the third approximation to calculate its bulk resistance.

3-31 Given a silicon diode with a reverse current of 5 μa at 25°C and 100 μA at 100°C, calculate the surface leakage current.

3-32 The power is turned off and the upper end of R_1 is grounded in Fig. 3-23b. Now you use an ohmmeter to read the forward and reverse resistance of the diode. Both readings are identical. What does the ohmmeter read?

3-33 Some systems, like burglar alarms and computers, use battery backup just in case the main source of power should fail. Describe how the circuit of Fig. 3-24 works.

Figure 3-24

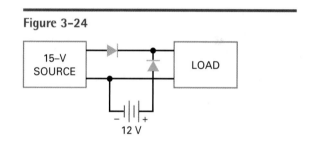

Up-Down Circuit Analysis

Use Fig. 3-25 for the remaining problems. Assume increases of approximately 10 percent in the independent variable and use the second approximation of a diode.

For each independent variable increase, determine what the respective dependent variable will do. Will it go up (U),

down (D), or show no change (N)? Refer to Sec. 3-6, Up-Down Circuit Analysis, to review this procedure.

3-34 Predict the response of each dependent variable in the row labeled V_S. Check your answers. Then, answer the

Figure 3-25 Up-down circuit analysis.

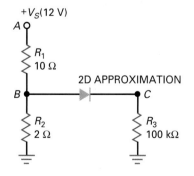

	DEPENDENT VARIABLES								
	V_A	V_B	V_C	I_1	I_2	I_3	P_1	P_2	P_3
V_S									
R_1									
R_2									
R_3									
V_K									

Independent Variables

following questions as simply and directly as possible. What effect does an increase in source voltage have on the dependent variables of the circuit?

3-35 Predict the response of each dependent variable in the row labeled R_1. Check your answers. Then summarize your findings in one or two sentences.

3-36 Predict the response of each dependent variable in the row labeled R_2. Check your answers. List the dependent variables that decrease. Explain why these

variables decrease, using Ohm's law or similar basic ideas.

3-37 Predict the response of each dependent variable in the row labeled R_3. List the dependent variables that show no change. Explain why these variables show no change.

3-38 Predict the response of each dependent variable in the row labeled V_K. List the dependent variables that decrease. Explain why these variables decrease.

Job Interview Questions

For the following questions, whenever possible draw circuits, graphs, or any figures that will help illustrate your answers. If you can combine words and pictures in your explanations, you are more likely to understand what you are talking about. Also, if you have privacy, pretend that you are at an interview and speak out loud. This practice will make it easier later, when the interview actually takes place.

1. Have you ever heard of an ideal diode? If so, tell me what it is and when you would use it.
2. One of the approximations for a diode is the second approximation. Tell me what the equivalent circuit is and when a silicon diode conducts.
3. Draw the diode curve and explain the different parts of it.
4. A circuit on my lab bench keeps destroying a diode every time I connect a new one. If I have a data sheet for the diode, what are some of the quantities I need to check?

5. In the most basic terms, describe what a diode acts like when it is forward biased and when it is reverse biased.
6. What is the difference between the typical knee voltage of a germanium diode and a silicon diode?
7. What would be a good technique for a technician to use to determine the current through a diode without breaking the circuit?
8. If you suspect that there is a defective diode on a circuit board, what steps would you take to determine whether it is actually defective?
9. For a diode to be useful, how much larger should the reverse resistance be than the forward resistance?
10. How might you connect a diode to prevent a second battery from discharging in a recreational vehicle, and yet still allow it to charge from the alternator?
11. What instruments can you use to test a diode in or out of a circuit?
12. Describe the operation of a diode in detail. Include majority and minority carriers in your discussion.

Self-Test Answers

1.	b	9.	a	17.	b
2.	b	10.	a	18.	b
3.	c	11.	b	19.	a
4.	d	12.	b	20.	b
5.	a	13.	a	21.	a
6.	b	14.	d	22.	c
7.	c	15.	a		
8.	c	16.	c		

Practice Problem Answers

3-1 D_1 is reverse biased; D_2 is forward biased

3-2 $P_D = 2.2$ W

3-3 $I_L = 5$ mA

3-4 $V_L = 2$ V;
$I_L = 2$ mA

3-5 $V_L = 4.3$ V;
$I_L = 4.3$ mA;
$P_D = 3.01$ mW

3-6 $I_L = 1.77$ mA;
$V_L = 1.77$ V;
$P_D = 1.24$ mW

3-8 $R_T = 10.23$ Ω;
$I_L = 420$ mA;
$V_L = 4.2$ V;
$P_D = 335$ mW

4

Diode Circuits

Most electronic systems, like HDTVs, DVD/CD players, and computers, need a dc voltage to work properly. Since the power-line voltage is alternating, the first thing we need to do is to convert the ac line voltage to dc voltage. The section of the electronic system that produces this dc voltage is called the power supply. Within the power supply are circuits that allow current to flow in only one direction. These circuits are called **rectifiers.** This chapter discusses rectifier circuits, filters, clippers, clampers, and voltage multipliers.

Objectives

After studying this chapter, you should be able to:

- Draw a diagram of a half-wave rectifier and explain how it works.
- Describe the role of the input transformer in power supplies.
- Draw a diagram of a full-wave rectifier and explain how it works.
- Draw a diagram of a bridge rectifier and explain how it works.
- Analyze a capacitor input filter and its surge current.
- List three important specifications found on a rectifier data sheet.
- Explain how a clipper works and draw waveforms.
- Explain how a clamper works and draw waveforms.
- Describe the action of voltage multipliers.

Chapter Outline

Vocabulary

bridge rectifier

capacitor-input filter

choke-input filter

clamper

clipper

dc value of a signal

filter

full-wave rectifier

half-wave rectifier

IC voltage regulator

integrated circuit

passive filter

peak detector

peak inverse voltage

polarized capacitor

power supply

rectifiers

ripple

surge current

surge resistor

switching regulator

unidirectional load current

voltage multiplier

Figure 4-1 (a) Ideal half-wave
rectifier; (b) on positive half cycle;
(c) on negative half cycle.

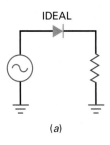

IDEAL

(a)

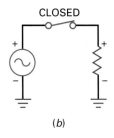

CLOSED

(b)

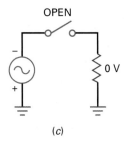

OPEN

(c)

4-1 The Half-Wave Rectifier

Figure 4-1a shows a **half-wave rectifier** circuit. The ac source produces a sinusoidal voltage. Assuming an ideal diode, the positive half cycle of source voltage will forward bias the diode. Since the switch is closed, as shown in Fig. 4-1b, the positive half cycle of source voltage will appear across the load resistor. On the negative half cycle, the diode is reverse biased. In this case, the ideal diode will appear as an open switch, as shown in Fig. 4-1c, and no voltage appears across the load resistor.

Ideal Waveforms

Figure 4-2a shows a graphical representation of the input voltage waveform. It is a sine wave with an instantaneous value of v_{in} and a peak value of $V_{p(in)}$. A pure sinusoid like this has an average value of zero over one cycle because each instantaneous voltage has an equal and opposite voltage half a cycle later. If you measure this voltage with a dc voltmeter, you will get a reading of zero because a dc voltmeter indicates the average value.

In the half-wave rectifier of Fig. 4-2b, the diode is conducting during the positive half cycles but is nonconducting during the negative half cycles. Because of this, the circuit clips off the negative half cycles, as shown in Fig. 4-2c. We call a waveform like this a *half-wave signal.* This half-wave voltage produces a **unidirectional load current.** This means that it flows in only one direction. If the diode were reversed, the output pulses would be negative.

A half-wave signal like the one in Fig. 4-2c is a pulsating dc voltage that increases to a maximum, decreases to zero, and then remains at zero during the negative half cycle. This is not the kind of dc voltage we need for electronics equipment. What we need is a constant voltage, the same as you get from a

Figure 4-2 (a) Input to half-wave rectifier; (b) circuit; (c) output of half-wave rectifier.

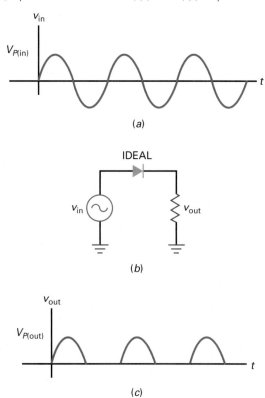

(a)

IDEAL

(b)

(c)

battery. To get this kind of voltage, we need to **filter** the half-wave signal (discussed later in this chapter).

When you are troubleshooting, you can use the ideal diode to analyze a half-wave rectifier. It's useful to remember that the peak output voltage equals the peak input voltage:

$$\textbf{Ideal half wave: } V_{p(out)} = V_{p(in)} \tag{4-1}$$

DC Value of Half–Wave Signal

The **dc value of a signal** is the same as the average value. If you measure a signal with a dc voltmeter, the reading will equal the average value. In basic courses the dc value of a half-wave signal is derived. The formula is:

$$\textbf{Half wave: } V_{dc} = \frac{V_p}{\pi} \tag{4-2}$$

The proof of this derivation requires calculus because we have to work out the average value over one cycle.

Since $1/\pi \approx 0.318$, you may see Eq. (4-2) written as:

$$V_{dc} \approx 0.318 V_p$$

When the equation is written in this form, you can see that the dc or average value equals 31.8 percent of the peak value. For instance, if the peak voltage of the half-wave signal is 100 V, the dc voltage or average value is 31.8 V.

Output Frequency

The output frequency is the same as the input frequency. This makes sense when you compare Fig. 4-2c with Fig. 4-2a. Each cycle of input voltage produces one cycle of output voltage. Therefore, we can write:

$$\textbf{Half wave: } f_{out} = f_{in} \tag{4-3}$$

We will use this derivation later with filters.

Second Approximation

We don't get a perfect half-wave voltage across the load resistor. Because of the barrier potential, the diode does not turn on until the ac source voltage reaches approximately 0.7 V. When the peak source voltage is much greater than 0.7 V, the load voltage will resemble a half-wave signal. For instance, if the peak source voltage is 100 V, the load voltage will be very close to a perfect half-wave voltage. If the peak source voltage is only 5 V, the load voltage will have a peak of only 4.3 V. When you need to get a better answer, use this derivation:

$$\textbf{2d half wave: } V_{p(out)} = V_{p(in)} - 0.7 \textbf{ V} \tag{4-4}$$

Higher Approximations

Most designers will make sure that the bulk resistance is much smaller than the Thevenin resistance facing the diode. Because of this, we can ignore bulk resistance in almost every case. If you must have better accuracy than you can get with the second approximation, you should use a computer and a circuit simulator like MultiSim.

Example 4–1

Figure 4-3 shows a half-wave rectifier that you can build on the lab bench or on a computer screen with MultiSim. An oscilloscope is across the 1 kΩ. This will show us the half-wave load voltage. Also, a multimeter is across the 1 kΩ to read the dc load voltage. Calculate the theoretical values of peak load voltage and the dc load voltage. Then, compare these values to the readings on the oscilloscope and the multimeter.

SOLUTION Figure 4-3 shows an ac source of 10 V and 60 Hz. Schematic diagrams usually show ac source voltages as effective or rms values. Recall that the *effective value* is the value of a dc voltage that produces the same heating effect as the ac voltage.

Figure 4-3 Lab example of half-wave rectifier.

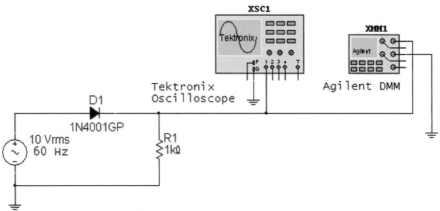

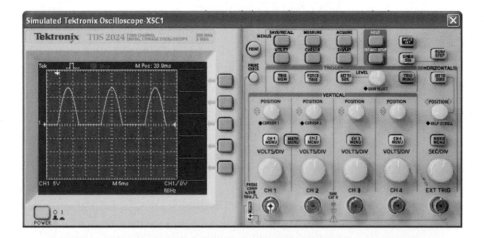

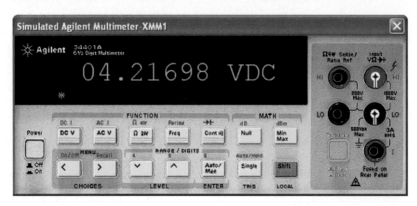

Since the source voltage is 10 V rms, the first thing to do is calculate the peak value of the ac source. You know from earlier courses that the rms value of a sine wave equals:

$$V_{rms} = 0.707V_p$$

Therefore, the peak source voltage in Fig. 4-3 is:

$$V_p = \frac{V_{rms}}{0.707} = \frac{10 \text{ V}}{0.707} = 14.1 \text{ V}$$

With an ideal diode, the peak load voltage is:

$$V_{p(out)} = V_{p(in)} = 14.1 \text{ V}$$

The dc load voltage is:

$$V_{dc} = \frac{V_p}{\pi} = \frac{14.1 \text{ V}}{\pi} = 4.49 \text{ V}$$

With the second approximation, we get a peak load voltage of:

$$V_{p(out)} = V_{p(in)} - 0.7 \text{ V} = 14.1 \text{ V} - 0.7 \text{ V} = 13.4 \text{ V}$$

and a dc load voltage of:

$$V_{dc} = \frac{V_p}{\pi} = \frac{13.4 \text{ V}}{\pi} = 4.27 \text{ V}$$

Figure 4-3 shows you the values that an oscilloscope and a multimeter will read. Channel 1 of the oscilloscope is set at 5 V per major division (5 V/Div). The half-wave signal has a peak value between 13 and 14 V, which agrees with the result from our second approximation. The multimeter also gives good agreement with theoretical values, because it reads approximately 4.22 V.

PRACTICE PROBLEM 4–1 Using Fig. 4-3, change the ac source voltage to 15 V. Calculate the second approximation dc load voltage V_{dc}.

4-2 The Transformer

Power companies in the United States supply a nominal line voltage of 120 V rms and a frequency of 60 Hz. The actual voltage coming out of a power outlet may vary from 105 to 125 V rms, depending on the time of day, the locality, and other factors. Line voltage is too high for most of the circuits used in electronics equipment. This is why a transformer is commonly used in the power-supply section of almost all electronics equipment. The transformer steps the line voltage down to safer and lower levels that are more suitable for use with diodes, transistors, and other semiconductor devices.

Basic Idea

Earlier courses discussed the transformer in detail. All we need in this chapter is a brief review. Figure 4-4 shows a transformer. Here you see line voltage applied to the primary winding of a transformer. Usually, the power plug has a third prong to ground the equipment. Because of the turns ratio N_1/N_2, the secondary voltage is stepped down when N_1 is greater than N_2.

Phasing Dots

Recall the meaning of the phasing dots shown at the upper ends of the windings. Dotted ends have the same instantaneous phase. In other words, when a positive half cycle appears across the primary, a positive half cycle appears across the

Figure 4-4 Half-wave rectifier with transformer.

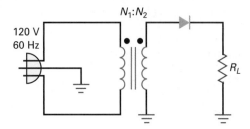

secondary. If the secondary dot were on the ground end, the secondary voltage would be 180° out of phase with the primary voltage.

On the positive half cycle of primary voltage, the secondary winding has a positive half sine wave across it and the diode is forward biased. On the negative half cycle of primary voltage, the secondary winding has a negative half cycle and the diode is reverse-biased. Assuming an ideal diode, we will get a half-wave load voltage.

Turns Ratio

Recall from your earlier course work the following derivation:

$$V_2 = \frac{V_1}{N_1/N_2} \qquad (4\text{-}5)$$

This says that the secondary voltage equals the primary voltage divided by the turns ratio. Sometimes you will see this equivalent form:

$$V_2 = \frac{N_2}{N_1} V_1$$

This says that the secondary voltage equals the inverse turns ratio times the primary voltage.

You can use either formula for rms, peak values, and instantaneous voltages. Most of the time, we will use Eq. (4-5) with rms values because ac source voltages are almost always specified as rms values.

The terms *step up* and *step down* are also encountered when dealing with transformers. These terms always relate the secondary voltage to the primary voltage. This means that a step-up transformer will produce a secondary voltage that is larger than the primary, and a step-down transformer will produce a secondary voltage that is smaller than the primary.

Example 4-2

What are the peak load voltage and dc load voltage in Fig. 4-5?

Figure 4-5

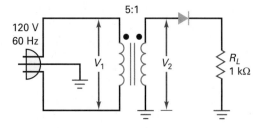

SOLUTION The transformer has a turns ratio of $5:1$. This means that the rms secondary voltage is one-fifth of the primary voltage:

$$V_2 = \frac{120 \text{ V}}{5} = 24 \text{ V}$$

and the peak secondary voltage is:

$$V_p = \frac{24 \text{ V}}{0.707} = 34 \text{ V}$$

With an ideal diode, the peak load voltage is:

$$V_{p(\text{out})} = 34 \text{ V}$$

The dc load voltage is:

$$V_{\text{dc}} = \frac{V_p}{\pi} = \frac{34 \text{ V}}{\pi} = 10.8 \text{ V}$$

With the second approximation, the peak load voltage is:

$$V_{p(\text{out})} = 34 \text{ V} - 0.7 \text{ V} = 33.3 \text{ V}$$

and the dc load voltage is:

$$V_{\text{dc}} = \frac{V_p}{\pi} = \frac{33.3 \text{ V}}{\pi} = 10.6 \text{ V}$$

PRACTICE PROBLEM 4–2 Using Fig. 4-5, change the transformer's turns ratio to $2:1$ and solve for the ideal dc load voltage.

4–3 The Full–Wave Rectifier

Figure 4-6a shows a **full-wave rectifier** circuit. Notice the grounded center tap on the secondary winding. The full-wave rectifier is equivalent to two half-wave rectifiers. Because of the center tap, each of these rectifiers has an input voltage equal to half the secondary voltage. Diode D_1 conducts on the positive half cycle, and diode D_2 conducts on the negative half cycle. As a result, the rectified load current flows during both half cycles. The full-wave rectifier acts the same as two back-to-back half-wave rectifiers.

Figure 4-6b shows the equivalent circuit for the positive half cycle. As you see, D_1 is forward biased. This produces a positive load voltage as indicated by the plus-minus polarity across the load resistor. Figure 4-6c shows the equivalent circuit for the negative half cycle. This time, D_2 is forward biased. As you can see, this also produces a positive load voltage.

During both half cycles, the load voltage has the same polarity and the load current is in the same direction. The circuit is called a *full-wave rectifier* because it has changed the ac input voltage to the pulsating dc output voltage shown in Fig. 4-6d. This waveform has some interesting properties that we will now discuss.

Figure 4-6 (*a*) Full-wave rectifier; (*b*) equivalent circuit for positive half cycle; (*c*) equivalent circuit for negative half cycle; (*d*) full-wave output.

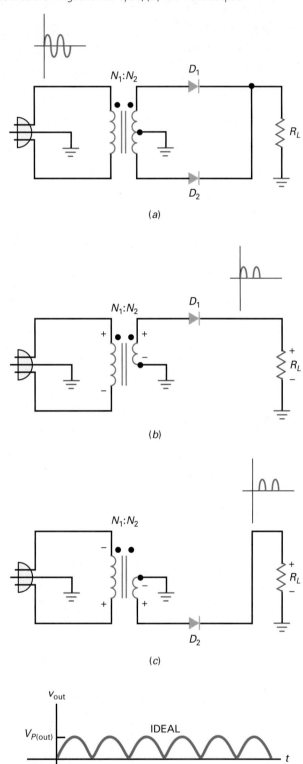

DC or Average Value

Since the full-wave signal has twice as many positive cycles as the half-wave signal, the dc or average value is twice as much, given by:

$$\text{Full wave: } V_{dc} = \frac{2V_p}{\pi} \tag{4-6}$$

Since $2/\pi = 0.636$, you may see Eq. (4-6) written as:

$$V_{dc} \approx 0.636V_p$$

In this form, you can see that the dc or average value equals 63.6 percent of the peak value. For instance, if the peak voltage of the full-wave signal is 100 V, the dc voltage or average value is 63.6 V.

Output Frequency

With a half-wave rectifier, the output frequency equals the input frequency. But with a full-wave rectifier, something unusual happens to the output frequency. The ac line voltage has a frequency of 60 Hz. Therefore, the input period equals:

$$T_{in} = \frac{1}{f} = \frac{1}{60 \text{ Hz}} = 16.7 \text{ ms}$$

Because of the full-wave rectification, the period of the full-wave signal is half the input period:

$$T_{out} = 0.5(16.7 \text{ ms}) = 8.33 \text{ ms}$$

(If there is any doubt in your mind, compare Fig. 4-6d to Fig. 4-2c.) When we calculate the output frequency, we get:

$$f_{out} = \frac{1}{T_{out}} = \frac{1}{8.33 \text{ ms}} = 120 \text{ Hz}$$

The frequency of the full-wave signal is double the input frequency. This makes sense. A full-wave output has twice as many cycles as the sine-wave input has. The full-wave rectifier inverts each negative half cycle, so that we get double the number of positive half cycles. The effect is to double the frequency. As a derivation:

$$\text{Full wave: } f_{out} = 2f_{in} \tag{4-7}$$

Second Approximation

Since the full-wave rectifier is like two back-to-back half-wave rectifiers, we can use the second approximation given earlier. The idea is to subtract 0.7 V from the ideal peak output voltage. The following example will illustrate the idea.

Example 4-3

||| MultiSim

Figure 4-7 shows a full-wave rectifier that you can build on lab bench or on a computer screen with MultiSim. Channel 1 of the oscilloscope displays the primary voltage (the sine wave), and channel 2 displays the load voltage (the full-wave signal). Calculate the peak input and output voltages. Then compare the theoretical values to the measured values.

SOLUTION

The peak primary voltage is:

$$V_{p(1)} = \frac{V_{rms}}{0.707} = \frac{120 \text{ V}}{0.707} = 170 \text{ V}$$

Figure 4–7 Lab example of full-wave rectifier.

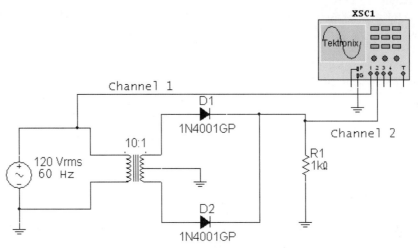

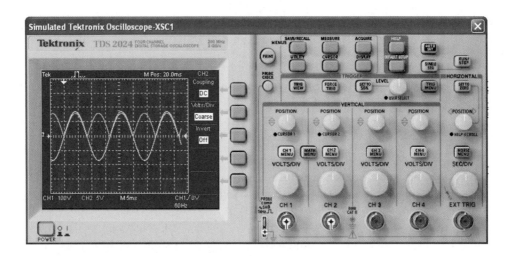

Because of the 10:1 step-down transformer, the peak secondary voltage is:

$$V_{p(2)} = \frac{V_{p(1)}}{N_1/N_2} = \frac{170\text{ V}}{10} = 17\text{ V}$$

The full-wave rectifier acts like two back-to-back half-wave rectifiers. Because of the center tap, the input voltage to each half-wave rectifier is only half the secondary voltage:

$$V_{p(\text{in})} = 0.5(17\text{ V}) = 8.5\text{ V}$$

Ideally, the output voltage is:

$$V_{p(\text{out})} = 8.5\text{ V}$$

Using the second approximation:

$$V_{p(\text{out})} = 8.5\text{ V} - 0.7\text{ V} = 7.8\text{ V}$$

Now, let's compare the theoretical values with the measured values. The sensitivity of channel 1 is 100 V/Div. Since the sine-wave input reads approximately 1.7 divisions, its peak value is approximately 170 V. Channel 2 has a sensitivity of 5 V/Div. Since the full-wave output reads approximately 1.4 Div, its peak value is approximately 7 V. Both input and output readings are in reasonable agreement with theoretical values.

Once again, notice that the second approximation improves the answer only slightly. If you were troubleshooting, the improvement would not be of much value. If something was wrong with the circuit, the chances are that the full-wave output would be drastically different from the ideal value of 8.5 V.

PRACTICE PROBLEM 4–3 Using Fig. 4-7, change the transformer's turns ratio to 5:1 and calculate the V_p (in) and V_p (out) second approximation values.

Example 4-4

IIII MultiSim

If one of the diodes in Fig. 4-7 were open, what would happen to the different voltages?

SOLUTION   If one of the diodes is open, the circuit reverts to a half-wave rectifier. In this case, half the secondary voltage is still 8.5 V, but the load voltage will be a half-wave signal rather than a full-wave signal. This half-wave voltage will still have a peak of 8.5 V (ideally) or 7.8 V (second approximation).

4-4 The Bridge Rectifier

Figure 4-8*a* shows a **bridge rectifier** circuit. The bridge rectifier is similar to a full-wave rectifier because it produces a full-wave output voltage. Diodes D_1 and D_2 conduct on the positive half cycle, and D_3 and D_4 conduct on the negative half cycle. As a result, the rectified load current flows during both half cycles.

Figure 4-8*b* shows the equivalent circuit for the positive half cycle. As you can see, D_1 and D_2 are forward biased. This produces a positive load voltage as indicated by the plus-minus polarity across the load resistor. As a memory aid, visualize D_2 shorted. Then, the circuit that remains is a half-wave rectifier, which we are already familiar with.

Figure 4-8*c* shows the equivalent circuit for the negative half cycle. This time, D_3 and D_4 are forward biased. This also produces a positive load voltage. If you visualize D_3 shorted, the circuit looks like a half-wave rectifier. So the bridge rectifier acts like two back-to-back half-wave rectifiers.

During both half cycles, the load voltage has the same polarity and the load current is in the same direction. The circuit has changed the ac input voltage to the pulsating dc output voltage shown in Fig. 4-8*d*. Note the advantage of this type of full-wave rectification over the center-tapped version in the previous section: *The entire secondary voltage can be used.*

Fig. 4-8*e* shows bridge rectifier packages that contain all four diodes.

Average Value and Output Frequency

Because a bridge rectifier produces a full-wave output, the equations for average value and output frequency are the same as given for a full-wave rectifier:

$$V_{dc} = \frac{2V_p}{\pi}$$

Figure 4-8 (*a*) Bridge rectifier; (*b*) equivalent circuit for positive half cycle; (*c*) equivalent circuit for negative half cycle; (*d*) full-wave output; (*e*) bridge rectifier packages.

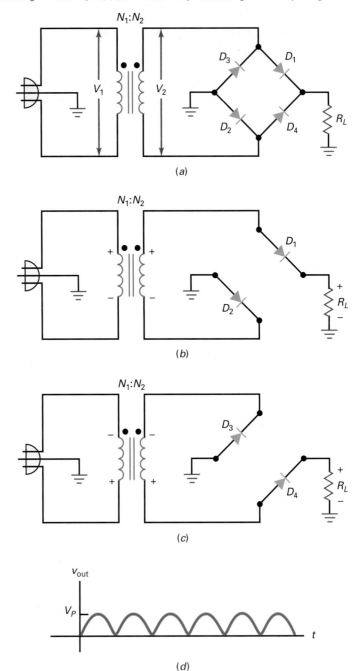

and

$$f_{\text{out}} = 2f_{\text{in}}$$

The average value is 63.6 percent of the peak value, and the output frequency is 120 Hz, given a line frequency of 60 Hz.

One advantage of a bridge rectifier is that all the secondary voltage is used as the input to the rectifier. Given the same transformer, we get twice as much peak voltage and twice as much dc voltage with a bridge rectifier as with a full-wave rectifier. Doubling the dc output voltage compensates for having to use two extra diodes. As a rule, you will see *the bridge rectifier used a lot more than the full-wave rectifier.*

Figure 4-8 (continued)

(e)

Incidentally, the full-wave rectifier was in use for many years before the bridge rectifier was used. For this reason, it has retained the name *full-wave rectifier* even though a bridge rectifier also has a full-wave output. To distinguish the full-wave rectifier from the bridge rectifier, some literature may refer to a full-wave rectifier as a *conventional full-wave rectifier,* a *two-diode full-wave rectifier,* or a *center-tapped full-wave rectifier.*

Second Approximation and Other Losses

Since the bridge rectifier has two diodes in the conducting path, the peak output voltage is given by:

$$\text{2d bridge: } V_{p(\text{out})} = V_{p(\text{in})} - 1.4 \text{ V} \tag{4-8}$$

As you can see, we have to subtract two diode drops from the peak to get a more accurate value of peak load voltage. Summary Table 4-1 compares the three rectifiers and their properties.

Summary Table 4–1	Unfiltered Rectifiers*		
	Half–wave	**Full–wave**	**Bridge**
Number of diodes	1	2	4
Rectifier input	$V_{p(2)}$	$0.5V_{p(2)}$	$V_{p(2)}$
Peak output (ideal)	$V_{p(2)}$	$0.5V_{p(2)}$	$V_{p(2)}$
Peak output (2d)	$V_{p(2)} - 0.7$ V	$0.5V_{p(2)} - 0.7$ V	$V_{p(2)} - 1.4$ V
DC output	$V_{p(\text{out})}/\pi$	$2V_{p(\text{out})}/\pi$	$2V_{p(\text{out})}/\pi$
Ripple frequency	f_{in}	$2f_{\text{in}}$	$2f_{\text{in}}$

*$V_{p(2)}$ = peak secondary voltage; $V_{p(\text{out})}$ = peak output voltage.

Example 4-5

Calculate the peak input and output voltages in Fig. 4-9. Then, compare the theoretical values to the measured values.

Notice the circuit uses a bridge rectifier package.

SOLUTION The peak primary and secondary voltages are the same as in Example 4-3:

$V_{p(1)} = 170$ V

$V_{p(2)} = 17$ V

With a bridge rectifier, all of the secondary voltage is used as the input to the rectifier. Ideally, the peak output voltage is:

$V_{p(\text{out})} = 17$ V

Figure 4-9 Lab example of bridge rectifier.

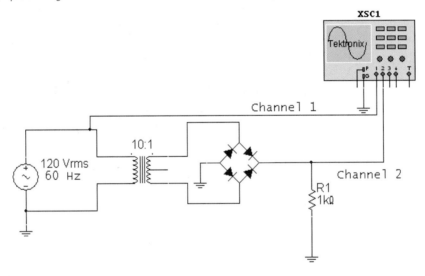

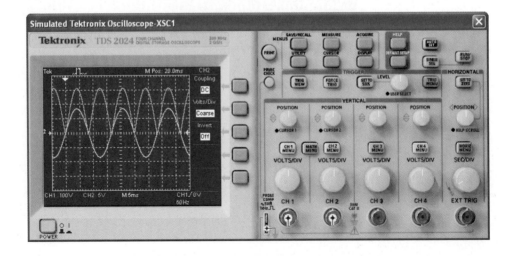

To a second approximation:

$$V_{p(\text{out})} = 17 \text{ V} - 1.4 \text{ V} = 15.6 \text{ V}$$

Now, let's compare the theoretical values with the measured values. The sensitivity of channel 1 is 100 V/Div. Since the sine-wave input reads approximately 1.7 Div, its peak value is approximately 170 V. Channel 2 has a sensitivity of 5 V/Div. Since the half-wave output reads approximately 3.2 Div, its peak value is approximately 16 V. Both input and output readings are approximately the same as the theoretical values.

PRACTICE PROBLEM 4–5 As in Example 4-5, calculate the ideal and second approximation $V_p(\text{out})$ values using a 5:1 transformer turns ratio.

4–5 The Choke-Input Filter

At one time, the choke-input filter was widely used to filter the output of a rectifier. Although not used much anymore because of its cost, bulk, and weight, this type of filter has instructional value and helps make it easier to understand other filters.

Basic Idea

Look at Fig. 4-10a. This type of filter is called a **choke-input filter.** The ac source produces a current in the inductor, capacitor, and resistor. The ac current in each component depends on the inductive reactance, capacitive reactance, and the resistance. The inductor has a reactance given by:

$$X_L = 2\pi f L$$

The capacitor has a reactance given by:

$$X_C = \frac{1}{2\pi f C}$$

As you learned in previous courses, the choke (or inductor) has the primary characteristic of opposing a change in current. Because of this, a choke-input filter ideally reduces the ac current in the load resistor to zero. To a second approximation, it reduces the ac load current to a very small value. Let us find out why.

The first requirement of a well-designed choke-input filter is to have X_C at the input frequency be much smaller than R_L. When this condition is satisfied, we can ignore the load resistance and use the equivalent circuit of Fig. 4-10b. The second requirement of a well-designed choke-input filter is to have X_L be much greater than X_C at the input frequency. When this condition is satisfied, the ac

Figure 4–10 (a) Choke-input filter; (b) ac equivalent circuit.

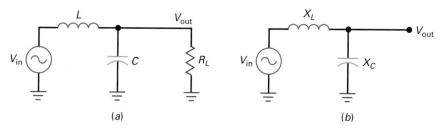

output voltage approaches zero. On the other hand, since the choke approximates a short circuit at 0 Hz and the capacitor approximates an open at 0 Hz, the dc current can be passed to the load resistance with minimum loss.

In Fig. 4-10b, the circuit acts like a reactive voltage divider. When X_L is much greater than X_C, almost all the ac voltage is dropped across the choke. In this case, the ac output voltage equals:

$$V_{\text{out}} \approx \frac{X_C}{X_L} V_{\text{in}} \qquad \qquad \textbf{(4-9)}$$

For instance, if $X_L = 10 \text{ k}\Omega$, $X_C = 100 \text{ }\Omega$, and $V_{\text{in}} = 15 \text{ V}$, the ac output voltage is:

$$V_{\text{out}} \approx \frac{100 \text{ }\Omega}{10 \text{ k}\Omega} 15 \text{ V} = 0.15 \text{ V}$$

In this example, the choke-input filter reduces the ac voltage by a factor of 100.

Filtering the Output of a Rectifier

Figure 4-11a shows a choke-input filter between a rectifier and a load. The rectifier can be a half-wave, full-wave, or bridge type. What effect does the choke-

---

Figure 4–11   (a) Rectifier with choke-input filter; (b) rectifier output has dc and ac components; (c) dc equivalent circuit; (d) filter output is direct current with small ripple.

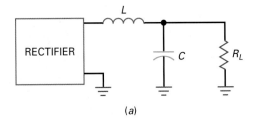

(a)

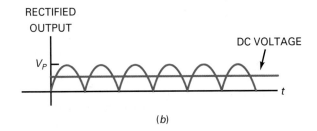

(b)

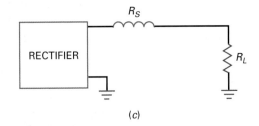

(c)

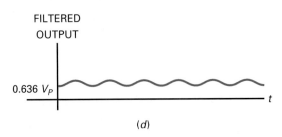

(d)

input filter have on the load voltage? The easiest way to solve this problem is to use the superposition theorem. Recall what this theorem says: If you have two or more sources, you can analyze the circuit for each source separately and then add the individual voltages to get the total voltage.

The rectifier output has two different components: a dc voltage (the average value) and an ac voltage (the fluctuating part), as shown in Fig. 4-11b. Each of these voltages acts like a separate source. As far as the ac voltage is concerned, X_L is much greater than X_C, and this results in very little ac voltage across the load resistor. Even though the ac component is not a pure sine wave, Eq. (4-9) is still a close approximation for the ac load voltage.

The circuit acts like Fig. 4-11c as far as dc voltage is concerned. At 0 Hz, the inductive reactance is zero and the capacitive reactance is infinite. Only the series resistance of the inductor windings remains. Making R_S much smaller than R_L causes most of the dc component to appear across the load resistor.

That's how a choke-input filter works: Almost all of the dc component is passed on to the load resistor, and almost all of the ac component is blocked. In this way, we get an almost perfect dc voltage, one that is almost constant, like the voltage out of a battery. Figure 4-11d shows the filtered output for a full-wave signal. The only deviation from a perfect dc voltage is the small ac load voltage shown in Fig. 4-11d. This small ac load voltage is called **ripple.** With an oscilloscope, we can measure its peak-to-peak value.

Main Disadvantage

A **power supply** is the circuit inside electronics equipment that converts the ac input voltage to an almost perfect dc output voltage. It includes a rectifier and a filter. The trend nowadays is toward low-voltage, high-current power supplies. Because line frequency is only 60 Hz, large inductances have to be used to get enough reactance for adequate filtering. But large inductors have large winding resistances, which create a serious design problem with large load currents. In other words, too much dc voltage is dropped across the choke resistance. Furthermore, bulky inductors are not suitable for modern semiconductor circuits, where the emphasis is on lightweight designs.

Switching Regulators

One important application does exist for the choke-input filter. A **switching regulator** is a special kind of power supply used in computers, monitors, and an increasing variety of equipment. The frequency used in a switching regulator is much higher than 60 Hz. Typically, the frequency being filtered is above 20 kHz. At this much higher frequency, we can use much smaller inductors to design efficient choke-input filters. We will discuss the details in a later chapter.

4-6 The Capacitor-Input Filter

The choke-input filter produces a dc output voltage equal to the average value of the rectified voltage. The **capacitor-input filter** produces a dc output voltage equal to the peak value of the rectified voltage. This type of filter is the most widely used in power supplies.

Basic Idea

Figure 4-12a shows an ac source, a diode, and a capacitor. The key to understanding a capacitor-input filter is understanding what this simple circuit does during the first quarter cycle.

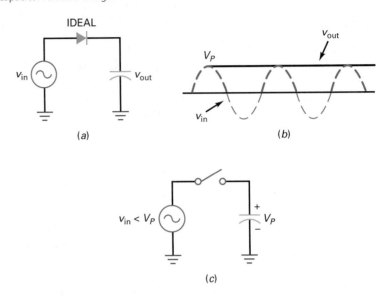

Initially, the capacitor is uncharged. During the first quarter cycle of Fig. 4-12b, the diode is forward biased. Since it ideally acts like a closed switch, the capacitor charges, and its voltage equals the source voltage at each instant of the first quarter cycle. The charging continues until the input reaches its maximum value. At this point, the capacitor voltage equals V_p.

After the input voltage reaches the peak, it starts to decrease. As soon as the input voltage is less than V_p, the diode turns off. In this case, it acts like the open switch of Fig. 4-12c. During the remaining cycles, the capacitor stays fully charged and the diode remains open. This is why the output voltage of Fig. 4-12b is constant and equal to V_p.

Ideally, all that the capacitor-input filter does is charge the capacitor to the peak voltage during the first quarter cycle. This peak voltage is constant, the perfect dc voltage we need for electronics equipment. There's only one problem: There is no load resistor.

Effect of Load Resistor

For the capacitor-input filter to be useful, we need to connect a load resistor across the capacitor, as shown in Fig. 4-13a. As long as the R_LC time constant is much greater than the period, the capacitor remains almost fully charged and the load voltage is approximately V_p. The only deviation from a perfect dc voltage is the small ripple seen in Fig. 4-13b. The smaller the peak-to-peak value of this ripple, the more closely the output approaches a perfect dc voltage.

Between peaks, the diode is off and the capacitor discharges through the load resistor. In other words, the capacitor supplies the load current. Since the capacitor discharges only slightly between peaks, the peak-to-peak ripple is small. When the next peak arrives, the diode conducts briefly and recharges the capacitor to the peak value. A key question is: What size should the capacitor be for proper operation? Before discussing capacitor size, consider what happens with the other rectifier circuits.

Full-Wave Filtering

If we connect a full-wave or bridge rectifier to a capacitor-input filter, the peak-to-peak ripple is cut in half. Figure 4-13c shows why. When a full-wave voltage is

Figure 4-13 (a) Loaded capacitor-input filter; (b) output is direct current with small ripple; (c) full wave output has less ripple.

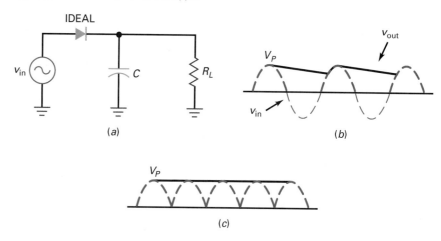

(a)

(b)

(c)

applied to the *RC* circuit, the capacitor discharges for only half as long. Therefore, the peak-to-peak ripple is half the size it would be with a half-wave rectifier.

The Ripple Formula

Here is a derivation we will use to estimate the peak-to-peak ripple out of any capacitor-input filter:

$$V_R = \frac{I}{fC} \qquad\qquad (4\text{-}10)$$

where V_R = peak-to-peak ripple voltage
I = dc load current
f = ripple frequency
C = capacitance

This is an approximation, not an exact derivation. We can use this formula to estimate the peak-to-peak ripple. When a more accurate answer is needed, one solution is to use a computer with a circuit simulator like MultiSim.

For instance, if the dc load current is 10 mA and the capacitance is 200 μF, the ripple with a bridge rectifier and a capacitor-input filter is:

$$V_R = \frac{10 \text{ mA}}{(120 \text{ Hz})(200 \text{ }\mu\text{F})} = 0.417 \text{ V pp}$$

In using this derivation, remember two things. First, the ripple is in peak-to-peak (pp) voltage. This is useful because you normally measure ripple voltage with an oscilloscope. Second, the formula works with half-wave or full-wave voltages. Use 60 Hz for half-wave, and 120 Hz for full-wave.

You should use an oscilloscope for ripple measurements if one is available. If not, you can use an ac voltmeter, although there will be a significant error in the measurement. Most ac voltmeters are calibrated to read the rms value of a sine wave. Since the ripple is not a sine wave, you may get a measurement error of as much as 25 percent, depending on the design of the ac voltmeter. But this should be no problem when you are troubleshooting, since you will be looking for much larger changes in ripple.

If you do use an ac voltmeter to measure the ripple, you can convert the peak-to-peak value given by Eq. (4-10) to an rms value using the following formula for a sine wave:

$$V_{rms} = \frac{V_{pp}}{2\sqrt{2}}$$

Dividing by 2 converts the peak-to-peak value to a peak value, and dividing by $\sqrt{2}$ gives the rms value of a sine wave with the same peak-to-peak value as the ripple voltage.

Exact DC Load Voltage

It is difficult to calculate the exact dc load voltage in a bridge rectifier with a capacitor-input filter. To begin with, we have the two diode drops that are subtracted from the peak voltage. Besides the diode drops, an additional voltage drop occurs, as follows: The diodes conduct heavily when recharging the capacitor because they are on for only a short time during each cycle. This brief but large current has to flow through the transformer windings and the bulk resistance of the diodes. In our examples, we will calculate either the ideal output or the output with the second approximation of a diode, remembering that the actual dc voltage is slightly lower.

Example 4–6

What is the dc load voltage and ripple in Fig. 4-14?

SOLUTION   The rms secondary voltage is:

$$V_2 = \frac{120 \text{ V}}{5} = 24 \text{ V}$$

The peak secondary voltage is:

$$V_p = \frac{24 \text{ V}}{0.707} = 34 \text{ V}$$

Assuming an ideal diode and small ripple, the dc load voltage is:

$$V_L = 34 \text{ V}$$

To calculate the ripple, we first need to get the dc load current:

$$I_L = \frac{V_L}{R_L} = \frac{34 \text{ V}}{5 \text{ k}\Omega} = 6.8 \text{ mA}$$

Figure 4-14   Half-wave rectifier and capacitor-input filter.

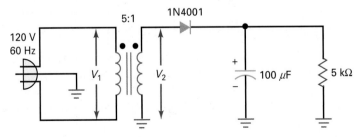

Figure 4-15 Full-wave rectifier and capacitor-input filter.

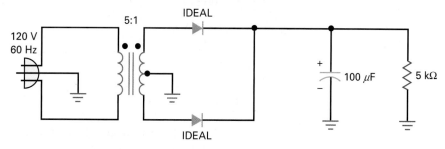

Now we can use Eq. (4-10) to get:

$$V_R = \frac{6.8 \text{ mA}}{(60 \text{ Hz})(100 \text{ } \mu\text{F})} = 1.13 \text{ V pp} \approx 1.1 \text{ V pp}$$

We rounded the ripple to two significant digits because it is an approximation and cannot be accurately measured with an oscilloscope with greater precision.

Here is how to improve the answer slightly: There is about 0.7 V across a silicon diode when it is conducting. Therefore, the peak voltage across the load will be closer to 33.3 V than to 34 V. The ripple also lowers the dc voltage slightly. So the actual dc load voltage will be closer to 33 V than to 34 V. But these are minor deviations. Ideal answers are usually adequate for troubleshooting and preliminary analysis.

A final point about the circuit. The plus sign on the filter capacitor indicates a **polarized capacitor,** one whose plus side must be connected to the positive rectifier output. In Fig. 4-15, the plus sign on the capacitor case is correctly connected to the positive output voltage. You must look carefully at the capacitor case when you are building or troubleshooting a circuit to find out whether it is polarized or not.

Power supplies often use polarized electrolytic capacitors because this type can provide high values of capacitance in small packages. As discussed in earlier courses, *electrolytic capacitors must be connected with the correct polarity* to produce the oxide film. If an electrolytic capacitor is connected in opposite polarity, *it becomes hot and may explode.*

Example 4–7

IIII MultiSim

What is the dc load voltage and ripple in Fig. 4-15?

SOLUTION   Since the transformer is 5:1 step-down like the preceding example, the peak secondary voltage is still 34 V. Half this voltage is the input to each half-wave section. Assuming an ideal diode and small ripple, the dc load voltage is:

$$V_L = 17 \text{ V}$$

The dc load current is:

$$I_L = \frac{17 \text{ V}}{5 \text{ k}\Omega} = 3.4 \text{ mA}$$

Now, Eq. (4-10) gives:

$$V_R = \frac{3.4 \text{ mA}}{(120 \text{ Hz})(100 \text{ } \mu\text{F})} = 0.283 \text{ V pp} \approx 0.28 \text{ V pp}$$

Because of the 0.7 V across the conducting diode, the actual dc load voltage will be closer to 16 V than to 17 V.

PRACTICE PROBLEM 4–7   Using Fig. 4-15, change R_L to 2 kΩ and calculate the new ideal dc load voltage and ripple.

Example 4-8

▌▌▌ MultiSim

What is the dc load voltage and ripple in Fig. 4-16? Compare the answers with those in the two preceding examples.

SOLUTION   Since the transformer is 5:1 step-down as in the preceding example, the peak secondary voltage is still 34 V. Assuming an ideal diode and small ripple, the dc load voltage is:

$$V_L = 34 \text{ V}$$

The dc load current is:

$$I_L = \frac{34 \text{ V}}{5 \text{ k}\Omega} = 6.8 \text{ mA}$$

Now, Eq. (4-10) gives:

$$V_R = \frac{6.8 \text{ mA}}{(120 \text{ Hz})(100 \text{ } \mu\text{F})} = 0.566 \text{ V pp} \approx 0.57 \text{ V pp}$$

Because of the 1.4 V across two conducting diodes and the ripple, the actual dc load voltage will be closer to 32 V than to 34 V.

We have calculated the dc load voltage and ripple for the three different rectifiers. Here are the results:

Half wave: 34 V and 1.13 V

Full wave: 17 V and 0.288 V

Bridge: 34 V and 0.566 V

For a given transformer, the bridge rectifier is better than the half-wave rectifier because it has less ripple, and it's better than the full-wave rectifier because it produces twice as much output voltage. Of the three, *the bridge rectifier has emerged as the most popular.*

Figure 4-16   Bridge rectifier and capacitor-input filter.

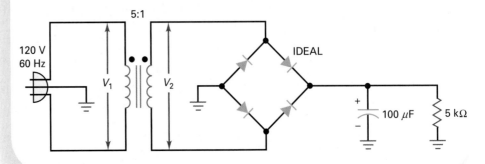

Example 4-9

Figure 4-17 shows the values measured with MultiSim. Calculate the theoretical load voltage and ripple and compare them to the measured values.

Figure 4-17   Lab example of bridge rectifier and capacitor-input filter.

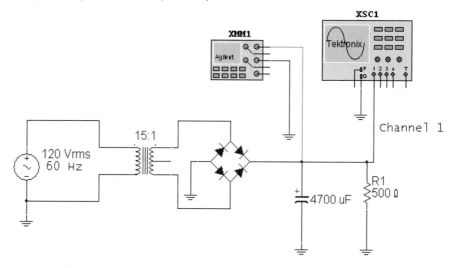

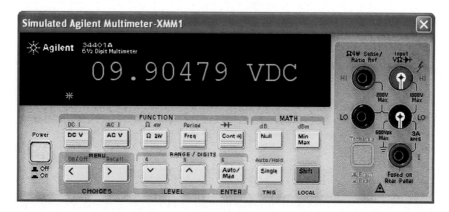

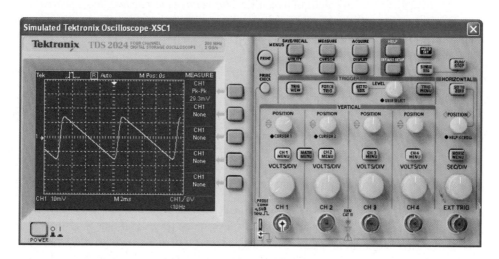

The transformer is a 15 : 1 step-down, so the rms secondary voltage is:

$$V_2 = \frac{120\text{ V}}{15} = 8\text{ V}$$

and the peak secondary voltage is:

$$V_p = \frac{8\text{ V}}{0.707} = 11.3\text{ V}$$

Let's use the second approximation of the diodes to get the dc load voltage:

$$V_L = 11.3\text{ V} - 1.4\text{ V} = 9.9\text{ V}$$

To calculate the ripple, we first need to get the dc load current:

$$I_L = \frac{9.9\text{ V}}{500\ \Omega} = 19.8\text{ mA}$$

Now, we can use Eq. (4-10) to get:

$$V_R = \frac{19.8\text{ mA}}{(120\text{ Hz})(4700\ \mu\text{F})} = 35\text{ mV pp}$$

In Fig. 4-17, a multimeter reads a dc load voltage of 9.9 V.

Channel 1 of the oscilloscope is set to 10 mV/Div. The peak-to-peak ripple is approximately 2.9 Div and the measured ripple is 29.3 mV. This is less than the theoretical value of 35 mV, which emphasizes the point made earlier. Equation (4-10) is to be used for *estimating* ripple. If you need more accuracy, use computer simulation software.

PRACTICE PROBLEM 4–9   Change the capacitor value in Fig. 4-17 to 1,000 μF. Calculate the new V_R value.

4–7 Peak Inverse Voltage and Surge Current

The **peak inverse voltage (PIV)** is the maximum voltage across the nonconducting diode of a rectifier. *This voltage must be less than the breakdown voltage of the diode; otherwise, the diode will be destroyed.* The peak inverse voltage depends on the type of rectifier and filter. The worst case occurs with the capacitor-input filter.

As discussed earlier, data sheets from various manufacturers use many different symbols to indicate the maximum reverse voltage rating of a diode. Sometimes, these symbols indicate different conditions of measurement. Some of the data sheet symbols for the maximum reverse voltage rating are PIV, PRV, V_B, V_{BR}, V_R, V_{RRM}, V_{RWM}, and $V_{R(max)}$.

Half–Wave Rectifier with Capacitor–Input Filter

Figure 4-18a shows the critical part of a half-wave rectifier. This is the part of the circuit that determines how much reverse voltage is across the diode. The rest of the circuit has no effect and is omitted for the sake of clarity. In the worse case, the peak secondary voltage is on the negative peak and the capacitor is fully charged

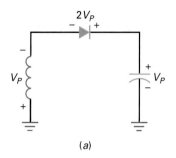

(a)

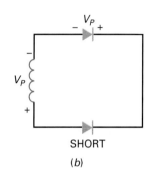

SHORT

(b)

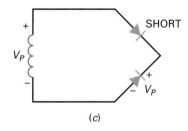

(c)

with a voltage of V_p. Apply Kirchhoff's voltage law, and you can see right away that the peak inverse voltage across the nonconducting diode is:

$$\textbf{PIV} = \textbf{2}V_p \qquad (4\text{-}11)$$

For instance, if the peak secondary voltage is 15 V, the peak inverse voltage is 30 V. As long as the breakdown voltage of the diode is greater than this, the diode will not be damaged.

Full-Wave Rectifier with Capacitor–Input Filter

Figure 4-18b shows the essential part of a full-wave rectifier needed to calculate the peak inverse voltage. Again, the secondary voltage is at the negative peak. In this case, the lower diode acts like a short (closed switch) and the upper diode is open. Kirchhoff's law implies:

$$\textbf{PIV} = V_p \qquad (4\text{-}12)$$

Bridge Rectifier with Capacitor–Input Filter

Figure 4-18c shows part of a bridge rectifier. This is all you need to calculate the peak inverse voltage. Since the upper diode is shorted and the lower one is open, the peak inverse voltage across the lower diode is:

$$\textbf{PIV} = V_p \qquad (4\text{-}13)$$

Another advantage of the bridge rectifier is that it has the lowest peak inverse voltage for a given load voltage. To produce the same load voltage, the full-wave rectifier would need twice as much secondary voltage.

Surge Resistor

Before the power is turned on, the filter capacitor is uncharged. At the first instant the power is applied, this capacitor looks like a short. Therefore, the initial charging current may be very large. All that exists in the charging path to impede the current is the resistance of the transformer windings and the bulk resistance of the diodes. The initial rush of current when the power is turned on is called the **surge current.**

Ordinarily, the designer of the power supply will select a diode with enough current rating to withstand the surge current. The key to the surge current is the size of the filter capacitor. Occasionally, a designer may decide to use a **surge resistor** rather than select another diode.

Figure 4-19 illustrates the idea. A small resistor is inserted between the bridge rectifier and the capacitor-input filter. Without the resistor, the surge current might destroy the diodes. By including the surge resistor, the designer reduces the surge current to a safe level. Surge resistors are not used very often and are mentioned just in case you see one used in a power supply.

Figure 4-19   Surge resistor limits surge current.

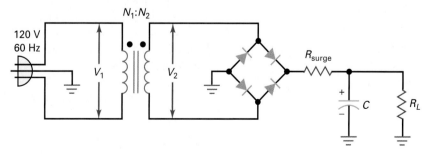

Example 4–10

What is the peak inverse voltage in Fig. 4-19 if the turns ratio is 8:1? A 1N4001 has a breakdown voltage of 50 V. Is it safe to use a 1N4001 in this circuit?

SOLUTION   The rms secondary voltage is:

$$V_2 = \frac{120 \text{ V}}{8} = 15 \text{ V}$$

The peak secondary voltage is:

$$V_p = \frac{15 \text{ V}}{0.707} = 21.2 \text{ V}$$

The peak inverse voltage is:

$$\text{PIV} = 21.2 \text{ V}$$

The 1N4001 is more than adequate, since the peak inverse voltage is much less than the breakdown voltage of 50 V.

PRACTICE PROBLEM 4–10   Using Fig. 4-19, change the transformer's turns ratio to 2:1. Which 1N4000 series of diodes should you use?

4-8 Other Power–Supply Topics

You have a basic idea of how power-supply circuits work. In the preceding sections, you have seen how an ac input voltage is rectified and filtered to get a dc voltage. There are a few additional ideas you need to know about.

Commercial Transformers

The use of turns ratios with transformers applies only to ideal transformers. Iron-core transformers are different. In other words, the transformers you buy from a parts supplier are not ideal because the windings have resistance, which produces power losses. Furthermore, the laminated core has eddy currents, which produce additional power losses. Because of these unwanted power losses, the turns ratio is only an approximation. In fact, the data sheets for transformers rarely list the turns ratio. Usually, all you get is the secondary voltage at a rated current.

For instance, Fig. 4-20*a* shows an F-25X, an industrial transformer whose data sheet gives only the following specifications: for a primary voltage of 115 V ac, the secondary voltage is 12.6 V ac when the secondary current is 1.5 A. If the secondary current is less than 1.5 A in Fig. 4-20*a*, the secondary voltage will be more than 12.6 V ac because of lower power losses in the windings and laminated core.

If it is necessary to know the primary current, you can estimate the turns ratio of a real transformer by using this definition:

$$\frac{N_1}{N_2} = \frac{V_1}{V_2} \tag{4-14}$$

Figure 4-20 (a) Rating on real transformer; (b) calculating fuse current.

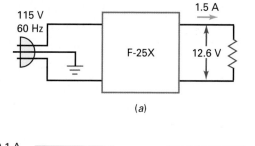

(a)

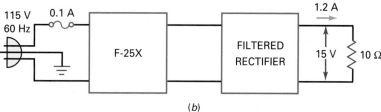

(b)

For instance, the F25X has $V_1 = 115$ V and $V_2 = 12.6$ V. The turns ratio at the rated load current of 1.5 A is:

$$\frac{N_1}{N_2} = \frac{115}{12.6} = 9.13$$

This is an approximation because the calculated turns ratio decreases when the load current decreases.

Calculating Fuse Current

When troubleshooting, you may need to calculate the primary current to determine whether a fuse is adequate or not. The easiest way to do this with a real transformer is to assume that the input power equals the output power: $P_{in} = P_{out}$. For instance, Fig. 4-20b shows a fused transformer driving a filtered rectifier. Is the 0.1-A fuse adequate?

Here is how to estimate the primary current when troubleshooting. The output power equals the dc load power:

$$P_{out} = VI = (15 \text{ V})(1.2 \text{ A}) = 18 \text{ W}$$

Ignore the power losses in the rectifier and the transformer. Since the input power must equal the output power:

$$P_{in} = 18 \text{ W}$$

Since $P_{in} = V_1 I_1$, we can solve for the primary current:

$$I_1 = \frac{18 \text{ W}}{115 \text{ V}} = 0.156 \text{ A}$$

This is only an estimate because we ignored the power losses in the transformer and rectifier. The actual primary current will be higher by about 5 to 20 percent because of these additional losses. In any case, the fuse is inadequate. It should be at least 0.25 A.

Slow-Blow Fuses

Assume that a capacitor-input filter is used in Fig. 4-20b. If an ordinary 0.25-A fuse is used in Fig. 4-20b, it will blow out when you turn the power on. The reason is the surge current, described earlier. Most power supplies use a slow-blow fuse, one that can temporarily withstand overloads in current. For instance, a 0.25-A slow-blow fuse can withstand

> 2 A for 0.1 s
> 1.5 A for 1 s
> 1 A for 2 s

and so on. With a slow-blow fuse, the circuit has time to charge the capacitor. Then, the primary current drops down to its normal level with the fuse still intact.

Calculating Diode Current

Whether a half-wave rectifier is filtered or not, the average current through the diode has to equal the dc load current because there is only one path for current. As a derivation:

$$\text{Half wave: } I_{\text{diode}} = I_{\text{dc}} \tag{4-15}$$

On the other hand, the average current through a diode in the full-wave rectifier equals only half the dc load current because there are two diodes in the circuit, each sharing the load. Similarly, each diode in a bridge rectifier has to withstand an average current of half the dc load current. As a derivation:

$$\text{Full wave: } I_{\text{diode}} = 0.5I_{\text{dc}} \tag{4-16}$$

Summary Table 4-2 compares the properties of the three capacitor-input filtered rectifiers.

Reading a Data Sheet

Refer to the data sheet of the 1N4001 in Chap. 3, Fig. 3-16. The maximum peak repetitive reverse voltage, V_{RRM} on the data sheet, is the same as the peak inverse voltage discussed earlier. The data sheet says that the 1N4001 can withstand a voltage of 50 V in the reverse direction.

The average rectified forward current—$I_{F(\text{av})}$, $I_{(\text{max})}$, or I_0—is the dc or average current through the diode. For a half-wave rectifier, the diode current equals the dc load current. For a full-wave or bridge rectifier, it equals half the dc load current. The data sheet says that a 1N4001 can have a dc current of 1 A, which means that the dc load current can be as much as 2 A in a bridge rectifier.

Summary Table 4–2	Capacitor-Input Filtered Rectifiers*		
	Half-wave	**Full-wave**	**Bridge**
Number of diodes	1	2	4
Rectifier input	$V_{p(2)}$	$0.5V_{p(2)}$	$V_{p(2)}$
DC output (ideal)	$V_{p(2)}$	$0.5V_{p(2)}$	$V_{p(2)}$
DC output (2d)	$V_{p(2)} - 0.7$ V	$0.5V_{p(2)} - 0.7$ V	$V_{p(2)} - 1.4$ V
Ripple frequency	f_{in}	$2f_{\text{in}}$	$2f_{\text{in}}$
PIV	$2V_{p(2)}$	$V_{p(2)}$	$V_{p(2)}$
Diode current	I_{dc}	$0.5I_{\text{dc}}$	$0.5I_{\text{dc}}$

*$V_{p(2)}$ = peak secondary voltage; $V_{p(\text{out})}$ = peak output voltage; I_{dc} = dc load current.

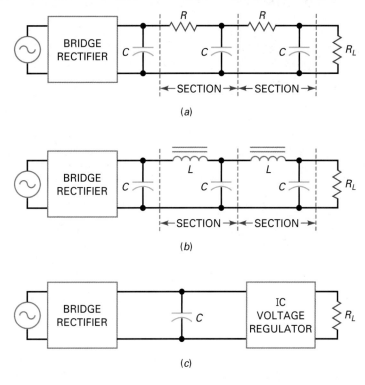

Figure 4–21 (*a*) *RC* filtering; (*b*) *LC* filtering; (*c*) voltage-regulator filtering.

Notice also the surge-current rating I_{FSM}. The data sheet says that a 1N4001 can withstand 30 A during the first cycle when the power is turned on.

RC Filters

Before the 1970s, **passive filters** (*R*, *L*, and *C* components) were often connected between the rectifier and the load resistance. Nowadays, you rarely see passive filters used in semiconductor power supplies, but there might be special applications, such as audio power amplifiers, in which you might encounter them.

Figure 4-21*a* shows a bridge rectifier and a capacitor-input filter. Usually, a designer will settle for a peak-to-peak ripple of as much as 10 percent across the filter capacitor. The reason for not trying to get even lower ripple is because the filter capacitor would become too large. Additional filtering is then done by *RC* sections between the filter capacitor and the load resistor.

The *RC* sections are examples of a passive filter, one that uses only *R*, *L*, or *C* components. By deliberate design, *R* is much greater than X_C at the ripple frequency. Therefore, the ripple is reduced before it reaches the load resistor. Typically, *R* is at least 10 times greater than X_C. This means that each section attenuates (reduces) the ripple by a factor of at least 10. The disadvantage of an *RC* filter is the loss of dc voltage across each *R*. Because of this, the *RC* filter is suitable only for very light loads (small load current or large load resistance).

LC Filter

When the load current is large, the *LC* filters of Fig. 4-21*b* are an improvement over *RC* filters. Again, the idea is to drop the ripple across the series components, in this case, the inductors. By making X_L much greater than X_C, we can reduce the ripple to a very low level. The dc voltage drop across the inductors is much smaller than it is across the resistors of *RC* sections because the winding resistance is smaller.

The *LC* filter was very popular at one time. Now, it's becoming obsolete in typical power supplies because of the size and cost of inductors. For low-voltage

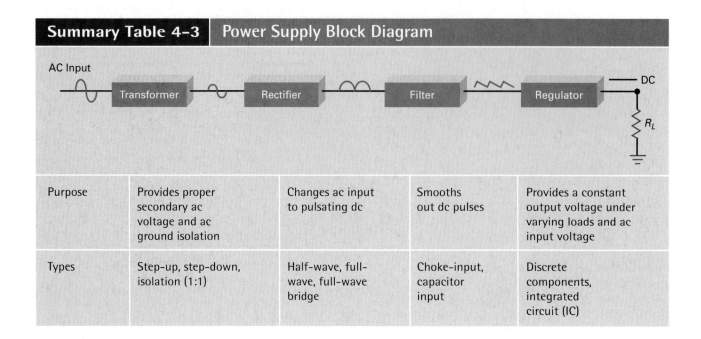

Purpose	Provides proper secondary ac voltage and ac ground isolation	Changes ac input to pulsating dc	Smooths out dc pulses	Provides a constant output voltage under varying loads and ac input voltage
Types	Step-up, step-down, isolation (1:1)	Half-wave, full-wave, full-wave bridge	Choke-input, capacitor input	Discrete components, integrated circuit (IC)

GOOD TO KNOW

A filter made of an inductor placed in between two capacitors is often called a pi (π) filter.

power supplies, the *LC* filter has been replaced by an **integrated circuit (IC).** This is a device that contains diodes, transistors, resistors, and other components in a miniaturized package to perform a specific function.

Figure 4-21*c* illustrates the idea. An **IC voltage regulator,** one type of integrated circuit, is between the filter capacitor and the load resistor. This device not only reduces the ripple, it also holds the output voltage constant. We will discuss IC voltage regulators in a later chapter. Because of their low cost, IC voltage regulators are now the standard method used for ripple reduction.

Summary Table 4-3 breaks the power supply down into functional blocks.

4-9 Troubleshooting

Almost every piece of electronics equipment has a power supply, typically a rectifier driving a capacitor-input filter followed by a voltage regulator (discussed later). This power supply produces the dc voltages needed by transistors and other devices. If a piece of electronics equipment is not working properly, start your troubleshooting with the power supply. More often than not, *equipment failure is caused by troubles in the power supply.*

Procedure

Assume that you are troubleshooting the circuit of Fig. 4-22. You can start by measuring the dc load voltage. It should be approximately the same as the peak secondary voltage. If not, there are two possible courses of action.

First, if there is no load voltage, you can use a floating VOM or DMM to measure the secondary voltage (ac range). The reading is the rms voltage across the secondary winding. Convert this to peak value. You can estimate the peak value by adding 40 percent to the rms value. If this is normal, the diodes may be defective. If there is no secondary voltage, either the fuse is blown or the transformer is defective.

Second, if there is dc load voltage, but it is lower than it should be, look at the dc load voltage with an oscilloscope and measure the ripple. A peak-to-peak ripple around 10 percent of the ideal load voltage is reasonable. The ripple may be

Figure 4-22 Troubleshooting.

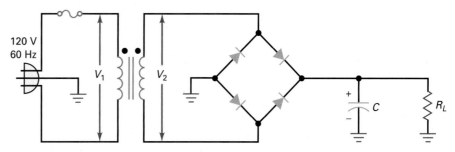

somewhat more or less than this, depending on the design. Furthermore, the ripple frequency should be 120 Hz for a full-wave or bridge rectifier. If the ripple is 60 Hz, one of the diodes may be open.

Common Troubles

Here are the most common troubles that arise in bridge rectifiers with capacitor-input filters:

1. If the fuse is open, there will be no voltages anywhere in the circuit.
2. If the filter capacitor is open, the dc load voltage will be low because the output will be an unfiltered full-wave signal.
3. If one of the diodes is open, the dc load voltage will be low because there will be only half-wave rectification. Also, the ripple frequency will be 60 Hz instead of 120 Hz. If all diodes are open, there will be no output.
4. If the load is shorted, the fuse will be blown. Possibly, one or more diodes may be ruined or the transformer may be damaged.
5. Sometimes the filter capacitor becomes leaky with age, and this reduces the dc load voltage.
6. Occasionally, shorted windings in the transformer reduce the dc output voltage. In this case, the transformer often feels very warm to the touch.
7. Besides these troubles, you can have solder bridges, cold-solder joints, bad connections, and so on.

Summary Table 4-4 lists these troubles and their symptoms.

Summary Table 4-4	Typical Troubles for Capacitor–Input Filtered Bridge Rectifier					
	V_1	V_2	$V_{L(dc)}$	V_R	f_{ripple}	Scope on Output
Fuse blown	Zero	Zero	Zero	Zero	Zero	No output
Capacitor open	OK	OK	Low	High	120 Hz	Full-wave signal
One diode open	OK	OK	Low	High	60 Hz	Half-wave ripple
All diodes open	OK	OK	Zero	Zero	Zero	No output
Load shorted	Zero	Zero	Zero	Zero	Zero	No output
Leaky capacitor	OK	OK	Low	High	120 Hz	Low output
Shorted windings	OK	Low	Low	OK	120 Hz	Low output

Example 4–11

When the circuit of Fig. 4-23 is working normally, it has an rms secondary voltage of 12.7 V, a load voltage of 18 V, and a peak-to-peak ripple of 318 mV. If the filter capacitor is open, what happens to the dc load voltage?

Figure 4-23

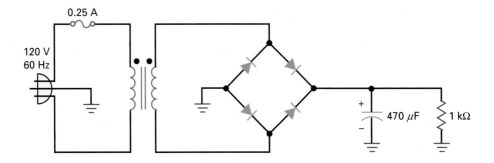

SOLUTION   With an open filter capacitor, the circuit reverts to a bridge rectifier with no filter capacitor. Because there is no filtering, an oscilloscope across the load will display a full-wave signal with a peak value of 18 V. The average value is 63.6 percent of 18 V, which is 11.4 V.

Example 4–12

Suppose the load resistor of Fig. 4-23 is shorted. Describe the symptoms.

SOLUTION   A short across the load resistor will increase the current to a very high value. This will blow out the fuse. Furthermore, it is possible that one or more diodes will be destroyed before the fuse blows. Often, when one diode shorts, it will cause the other rectifier diodes to also short. Because of the blown fuse, all voltages will measure zero. When you check the fuse visually or with an ohmmeter, you will see that it is open.

   With the power off, you should check the diodes with an ohmmeter to see whether any of them have been destroyed. You should also measure the load resistance with an ohmmeter. If it measures zero or very low, you have more troubles to locate.

   The trouble could be a solder bridge across the load resistor, incorrect wiring, or any number of possibilities. Fuses do occasionally blow out without a permanent short across the load. But the point is this: *When you get a blown fuse, check the diodes for possible damage and the load resistance for a possible short.*

   A troubleshooting exercise at the end of the chapter has eight different troubles, including open diodes, filter capacitors, shorted loads, blown fuses, and open grounds.

4-10 Clippers and Limiters

The diodes used in low-frequency power supplies are *rectifier diodes*. These diodes are optimized for use at 60 Hz and have power ratings greater than 0.5 W. The typical rectifier diode has a forward current rating in amperes. Except for power supplies, rectifier diodes have little use because most circuits inside electronics equipment are running at much higher frequencies.

Figure 4-24 (*a*) Positive clipper; (*b*) output waveform.

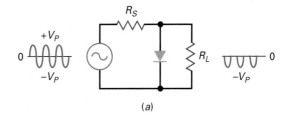

(*a*)

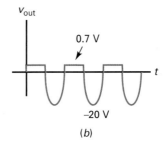

(*b*)

Small-Signal Diodes

In this section, we will be using *small-signal diodes*. These diodes are optimized for use at high frequencies and have power ratings less than 0.5 W. The typical small-signal diode has a current rating in milliamperes. It is this smaller and lighter construction that allows the diode to work at higher frequencies.

The Positive Clipper

A **clipper** is a circuit that removes either positive or negative parts of a waveform. This kind of processing is useful for signal shaping, circuit protection, and communications. Figure 4-24*a* shows a *positive clipper*. The circuit removes all the positive parts of the input signal. This is why the output signal has only negative half cycles.

Here is how the circuit works: During the positive half cycle, the diode turns on and looks like a short across the output terminals. Ideally, the output voltage is zero. On the negative half cycle, the diode is open. In this case, a negative half cycle appears across the output. By deliberate design, the series resistor is much smaller than the load resistor. This is why the negative output peak is shown as $-V_p$ in Fig. 4-24*a*.

To a second approximation, the diode voltage is 0.7 V when conducting. Therefore, the clipping level is not zero, but 0.7 V. For instance, if the input signal has a peak value of 20 V, the output of the clipper will look like Fig. 4-24*b*.

Defining Conditions

Small-signal diodes have a smaller junction area than rectifier diodes because they are optimized to work at higher frequencies. As a result, they have more bulk resistance. The data sheet of a small-signal diode like the 1N914 lists a forward current of 10 mA at 1 V. Therefore, the bulk resistance is:

$$R_B = \frac{1 \text{ V} - 0.7 \text{ V}}{10 \text{ mA}} = 30 \text{ }\Omega$$

Why is bulk resistance important? Because the clipper will not work properly unless the series resistance R_S is much greater than the bulk resistance. Furthermore, the clipper won't work properly unless the series resistance R_S is

Figure 4-25 (*a*) Negative clipper; (*b*) output waveform.

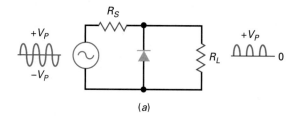

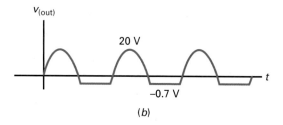

much smaller than the load resistance. For a clipper to work properly, we will use this definition:

$$\text{Stiff clipper: } 100R_B < R_S < 0.01R_L \qquad (4\text{-}17)$$

This says that the series resistance must be 100 times greater than the bulk resistance and 100 times smaller than the load resistance. When a clipper satisfies these conditions, we call it a *stiff clipper.* For instance, if the diode has a bulk resistance of 30 Ω, the series resistance should be at least 3 kΩ and the load resistance should be at least 300 kΩ.

The Negative Clipper

If we reverse the polarity of the diode as shown in Fig. 4-25*a*, we get a *negative clipper.* As you would expect, this removes the negative parts of the signal. Ideally, the output waveform has nothing but positive half cycles.

The clipping is not perfect. Because of the diode *offset voltage* (another way of saying *barrier potential*), the clipping level is at −0.7 V. If the input signal has a peak of 20 V, the output signal will look like Fig. 4-25*b*.

The Limiter or Diode Clamp

The clipper is useful for waveshaping, but the same circuit can be used in a totally different way. Take a look at Fig. 4-26*a*. The normal input to this circuit is a signal with a peak of only 15 mV. Therefore, the normal output is the same signal because neither diode is turned during the cycle.

What good is the circuit if the diodes don't turn on? Whenever you have a sensitive circuit, one that cannot have too much input, you can use a positive-negative *limiter* to protect its input, as shown in Fig. 4-26*b*. If the input signal tries to rise above 0.7 V, the output is limited to 0.7 V. On the other hand, if the input signal tries to drop below −0.7 V, the output is limited to −0.7 V. In a circuit like this, normal operation means that the input signal is always smaller than 0.7 V in either polarity.

An example of a sensitive circuit is the *op amp,* an IC that will be discussed in later chapters. The typical input voltage to an op amp is less than 15 mV. Voltages greater than 15 mV are unusual, and voltages greater than 0.7 V are abnormal. A limiter on the input side of an op amp will prevent excessive input voltage from being accidentally applied.

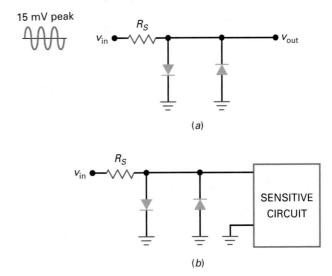

Figure 4-26 (a) Diode clamp; (b) protecting a sensitive circuit.

A more familiar example of a sensitive circuit is a moving-coil meter. By including a limiter, we can protect the meter movement against excessive input voltage or current.

The limiter of Fig. 4-26a is also called a *diode clamp*. The term suggests clamping or limiting the voltage to a specified range. With a diode clamp, the diodes remain off during normal operation. The diodes conduct only when something is abnormal, when the signal is too large.

Biased Clippers

The reference level (same as the clipping level) of a positive clipper is ideally zero, or 0.7 V to a second approximation. What can we do to change this reference level?

In electronics, *bias* means applying an external voltage to change the reference level of a circuit. Figure 4-27a is an example of using bias to change the reference level of a positive clipper. By adding a dc voltage source in series with the diode, we can change the clipping level. The new V must be less than V_p for normal operation. With an ideal diode, conduction starts as soon as the input

Figure 4-27 (a) Biased positive clipper; (b) biased negative clipper.

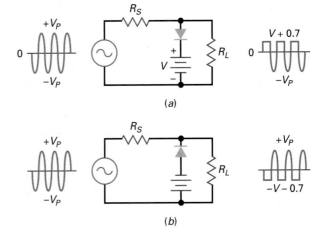

Figure 4-28   Biased positive-negative clipper.

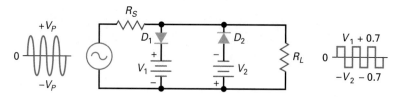

voltage is greater than V. To a second approximation, it starts when the input voltage is greater than $V + 0.7$ V.

Figure 4-27b shows how to bias a negative clipper. Notice that the diode and battery have been reversed. Because of this, the reference level changes to $-V - 0.7$ V. The output waveform is negatively clipped at the bias level.

Combination Clipper

We can combine the two biased clippers as shown in Fig. 4-28. Diode D_1 clips off positive parts above the positive bias level, and diode D_2 clips off parts below the negative bias level. When the input voltage is very large compared to the bias levels, the output signal is a *square wave,* as shown in Fig. 4-28. This is another example of the signal shaping that is possible with clippers.

Variations

Using batteries to set the clipping level is impractical. One approach is to add more silicon diodes because each produces a bias of 0.7 V. For instance, Fig. 4-29a shows three diodes in a positive clipper. Since each diode has an offset of around 0.7 V, the three diodes produce a clipping level of approximately $+2.1$ V. The application does not have to be a clipper (waveshaping). We can use the same circuit as a diode clamp (limiting) to protect a sensitive circuit that cannot tolerate more than a 2.1-V input.

Figure 4-29b shows another way to bias a clipper without batteries. This time, we are using a voltage divider (R_1 and R_2) to set the bias level. The bias level is given by:

$$V_{bias} = \frac{R_2}{R_1 + R_2} V_{dc} \tag{4-18}$$

In this case, the output voltage is clipped or limited when the input is greater than $V_{bias} + 0.7$ V.

Figure 4-29c, shows a biased diode clamp. It can be used to protect sensitive circuits from excessive input voltages. The bias level is shown as $+5$ V. It can be any bias level you want it to be. With a circuit like this, a destructively large voltage of $+100$ V never reaches the load because the diode limits the output voltage to a maximum value of $+5.7$ V.

Sometimes a variation like Fig. 4-29d is used to remove the offset of the limiting diode D_1. Here is the idea: Diode D_2 is biased slightly into forward conduction, so that it has approximately 0.7 V across it. This 0.7 V is applied to 1 kΩ in series with D_1 and 100 kΩ. This means that diode D_1 is on the verge of conduction. Therefore, when a signal comes in, diode D_1 conducts near 0 V.

Figure 4-29   (a) Clipper with three offset voltages; (b) voltage divider biases clipper; (c) diode clamp protects above 5.7 V; (d) diode D_2 biases D_1 to remove offset voltage.

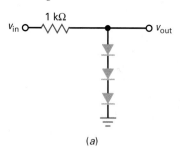

(a)

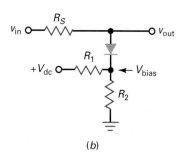

(b)

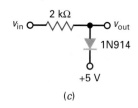

(c)

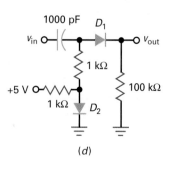

(d)

4-11 Clampers

The diode clamp, which was discussed in the preceding section, protects sensitive circuits. The **clamper** is different, so don't confuse the similar-sounding names. A clamper adds a dc voltage to the signal.

Positive Clamper

Figure 4-30a shows the basic idea for a positive clamper. When a positive clamper has a sine-wave input, it adds a positive dc voltage to the sine wave. Stated another way, the positive clamper shifts the ac reference level (normally zero) up to a dc level. The effect is to have an ac voltage centered on a dc level. This means that each point on the sine wave is shifted upward, as shown on the output wave.

Figure 4-30b shows an equivalent way of visualizing the effect of a positive clamper. An ac source drives the input side of the clamper. The Thevenin voltage of the clamper output is the superposition of a dc source and an ac source. The ac signal has a dc voltage of V_p added to it. This is why the entire sine wave of Fig. 4-30a has shifted upward so that it has a positive peak of $2V_p$ and a negative peak of zero.

Figure 4-31a is a positive clamper. Ideally, here is how it works. The capacitor is initially uncharged. On the first negative half cycle of input voltage, the diode turns on (Fig. 4-31b). At the negative peak of the ac source, the capacitor has fully charged and its voltage is V_p with the polarity shown.

Slightly beyond the negative peak, the diode shuts off (Fig. 4-31c). The $R_L C$ time constant is deliberately made much larger than the period T of the signal. We will define *much larger* as at least 100 times greater:

$$\textbf{Stiff clamper: } R_L C > 100T \qquad \textbf{(4-19)}$$

For this reason, the capacitor remains almost fully charged during the off time of the diode. To a first approximation, the capacitor acts like a battery of V_p volts. This is why the output voltage in Fig. 4-31a is a positively clamped signal. Any clamper that satisfies Eq. (4-19) is called a *stiff clamper.*

Figure 4-30 (a) Positive clamper shifts waveform upward; (b) positive clamper adds a dc component to signal.

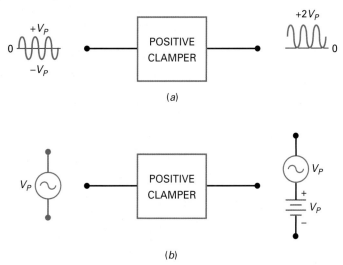

(a)

(b)

Figure 4-31 (*a*) Ideal positive clamper; (*b*) at the positive peak; (*c*) beyond the positive peak; (*d*) clamper is not quite perfect.

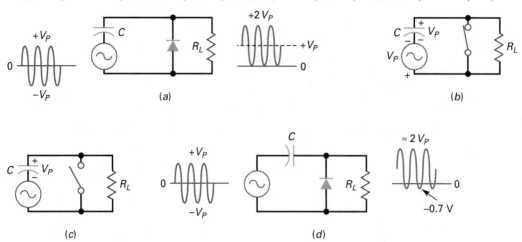

The idea is similar to the way a half-wave rectifier with a capacitor-input filter works. The first quarter cycle charges the capacitor fully. Then, the capacitor retains almost all of its charge during subsequent cycles. The small charge that is lost between cycles is replaced by diode conduction.

In Fig. 4-31*c*, the charged capacitor looks like a battery with a voltage of V_p. This is the dc voltage that is being added to the signal. After the first quarter cycle, the output voltage is a positively clamped sine wave with a reference level of zero; that is, it sits on a level of 0 V.

Figure 4-31*d* shows the circuit as it is usually drawn. Since the diode drops 0.7 V when conducting, the capacitor voltage does not quite reach V_p. For this reason, the clamping is not perfect, and the negative peaks have a reference level of −0.7 V.

Negative Clamper

What happens if we turn the diode in Fig. 4-31*d* around? We get the negative clamper of Fig. 4-32. As you can see, the capacitor voltage reverses, and the circuit becomes a negative clamper. Again, the clamping is less than perfect because the positive peaks have a reference level of 0.7 V instead of 0 V.

As a memory aid, notice that the diode points in the direction of shift. In Fig. 4-32, the diode points downward, the same direction as the shift of the sine wave. This tells you that it's a negative clamper. In Fig. 4-31*a*, the diode points up, the waveform shifts up, and you have positive clamper.

Figure 4-32 Negative clamper.

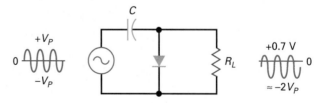

Figure 4-33 Peak-to-peak detector.

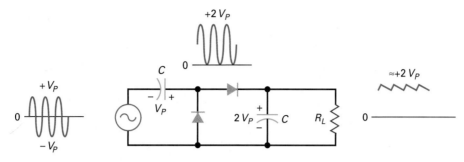

Both positive and negative clampers are widely used. For instance, television receivers use a clamper to change the reference level of video signals. Clampers are also used in radar and communication circuits.

A final point. The less than perfect clipping and clamping discussed so far are no problem. After we discuss op amps, we will look again at clippers and clampers. At that time, you will see how easy it is to eliminate the barrier-potential problem. In other words, we will look at circuits that are almost perfect.

Peak-to-Peak Detector

A half-wave rectifier with a capacitor-input filter produces a dc output voltage approximately equal to the peak of the input signal. When the same circuit uses a small-signal diode, it is called a **peak detector.** Typically, peak detectors operate at frequencies that are much higher than 60 Hz. The output of a peak detector is useful in measurements, signal processing, and communications.

If you cascade a clamper and a peak detector, you get a *peak-to-peak detector* (see Fig. 4-33). As you can see, the output of a clamper is used as the input to a peak detector. Since the sine wave is positively clamped, the input to the peak detector has a peak value of $2V_p$. This is why the output of the peak detector is a dc voltage equal to $2V_p$.

As usual, the RC time constant must be much greater than the period of the signal. By satisfying this condition, you get good clamping action and good peak detection. The output ripple will therefore be small.

One application is in measuring nonsinusoidal signals. An ordinary ac voltmeter is calibrated to read the rms value of an ac signal. If you try to measure a nonsinusoidal signal, you will get an incorrect reading with an ordinary ac voltmeter. However, if the output of a peak-to-peak detector is used as the input to a dc voltmeter, it will indicate the peak-to-peak voltage. If the nonsinusoidal signal swings from -20 to $+50$ V, the reading is 70 V.

4–12 Voltage Multipliers

A peak-to-peak detector uses small-signal diodes and operates at high frequencies. By using rectifier diodes and operating at 60 Hz, we can produce a new kind of power supply called a *voltage doubler.*

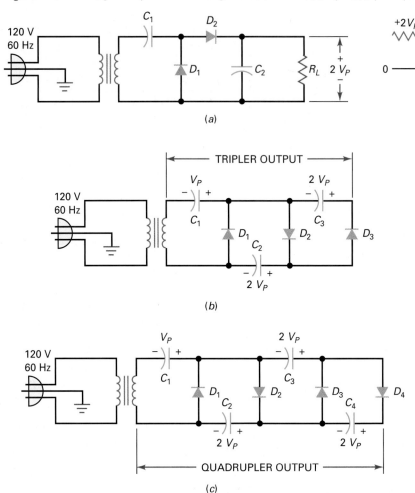

Figure 4-34 Voltage multipliers with floating loads. (*a*) Doubler; (*b*) tripler; (*c*) quadrupler.

Voltage Doubler

Figure 4-34*a* is a *voltage doubler.* The configuration is the same as a peak-to-peak detector, except that we use rectifier diodes and operate at 60 Hz. The clamper section adds a dc component to the secondary voltage. The peak detector then produces a dc output voltage that is 2 times the secondary voltage.

Why bother using a voltage doubler when you can change the turns ratio to get more output voltage? The answer is that you don't need to use a voltage doubler at lower voltages. The only time you run into a problem is when you are trying to produce very high dc output voltages.

For instance, line voltage is 120 V rms, or 170 V peak. If you are trying to produce 3400 V dc, you will need to use a 1:20 step-up transformer. Here is where the problem comes in. Very high secondary voltages can be obtained only with bulky transformers. At some point, a designer may decide that it would be simpler to use a voltage doubler and a smaller transformer.

Voltage Tripler

By connecting another section, we get the *voltage tripler* of Fig. 4-34b. The first two sections act like a doubler. At the peak of the negative half cycle, D_3 is forward-biased. This charges C_3 to $2V_p$ with the polarity shown in Fig. 4-34b. The tripler output appears across C_1 and C_3. The load resistance can be connected across the tripler output. As long as the time constant is long, the output equals approximately $3V_p$.

Voltage Quadrupler

Figure 4-34c is a *voltage quadrupler* with four sections in *cascade* (one after another). The first three sections are a tripler, and the fourth makes the overall circuit a quadrupler. The first capacitor charges to V_p. All others charge to $2V_p$. The quadrupler output is across the series connection of C_2 and C_4. We can connect a load resistance across the quadrupler output to get an output of $4V_p$.

Theoretically, we can add sections indefinitely, but the ripple gets much worse with each new section. Increased ripple is another reason why **voltage multipliers** (doublers, triplers, and quadruplers) are not used in low-voltage power supplies. As stated earlier, voltage multipliers are almost always used to produce high voltages, well into the hundreds or thousands of volts. Voltage multipliers are the natural choice for high-voltage and low-current devices like the cathode-ray tube (CRT) used in television receivers, oscilloscopes, and computer monitors.

Variations

All of the voltage multipliers shown in Fig. 4-34 use load resistances that are *floating*. This means that neither end of the load is grounded. Figure 4-35a, b, and c shows variations of the voltage multipliers. Figure 4-35a merely adds grounds to Fig. 4-34a. On the other hand, Fig. 4-35b and c are redesigns of the tripler (Fig. 4-34b) and quadrupler (Fig. 4-34c). In some applications, you may see floating-load designs used (such as in the CRT); in others, you may see the grounded-load designs used.

Full–Wave Voltage Doubler

Figure 4-35d shows a full-wave voltage doubler. On the positive half cycle of the source, the upper capacitor charges to the peak voltage with the polarity shown. On the next half cycle, the lower capacitor charges to the peak voltage with the indicated polarity. For a light load, the final output voltage is approximately $2V_p$.

The voltage multipliers discussed earlier are half-wave designs; that is, the output ripple frequency is 60 Hz. On the other hand, the circuit of Fig. 4-35d is called a *full-wave voltage doubler* because one of the output capacitors is being charged during each half cycle. Because of this, the output ripple is 120 Hz. This ripple frequency is an advantage because it is easier to filter. Another advantage of the full-wave doubler is that the PIV rating of the diodes need only be greater than V_p.

Figure 4-35 Voltage multipliers with grounded loads, except full-wave doubler. (a) Doubler; (b) tripler; (c) quadrupler; (d) Full-wave doubler.

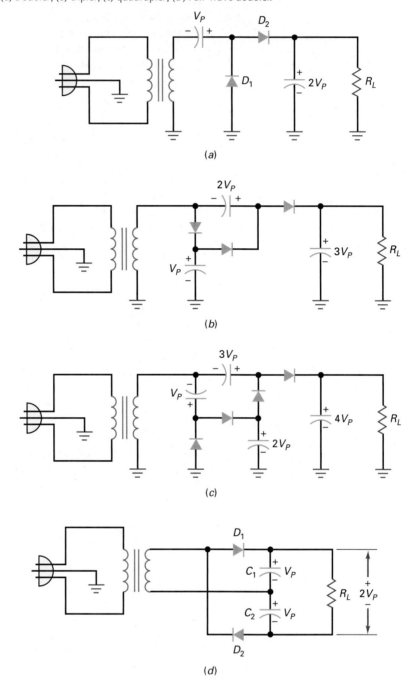

(a)

(b)

(c)

(d)

Summary

SEC. 4–1 THE HALF–WAVE RECTIFIER

The half-wave rectifier has a diode in series with a load resistor. The load voltage is a half-wave output. The average or dc voltage out of a half-wave rectifier equals 31.8 percent of the peak voltage.

SEC. 4–2 THE TRANSFORMER

The input transformer is usually a step-down transformer in which the voltage steps down and the current steps up. The secondary voltage equals the primary voltage divided by the turns ratio.

SEC. 4-3 THE FULL-WAVE RECTIFIER

The full-wave rectifier has a center-tapped transformer with two diodes and a load resistor. The load voltage is a full-wave signal whose peak value is half the secondary voltage. The average or dc voltage out of a full-wave rectifier equals 63.6 percent of the peak voltage, and the ripple frequency equals 120 Hz instead of 60 Hz.

SEC. 4-4 THE BRIDGE RECTIFIER

The bridge rectifier has four diodes. The load voltage is a full-wave signal with a peak value equal to the secondary voltage. The average or dc voltage out of a half-wave rectifier equals 63.6 percent of the peak voltage, and the ripple frequency equals 120 Hz.

SEC. 4-5 THE CHOKE-INPUT FILTER

The choke-input filter is an LC voltage divider in which the inductive reactance is much greater than the capacitive reactance. The type of filter allows the average value of the rectified signal to pass through to the load resistor.

SEC. 4-6 THE CAPACITOR-INPUT FILTER

This type of filter allows the peak value of the rectified signal to pass through to the load resistor. With a large capacitor, the ripple is small, typically less than 10 percent of the dc voltage. The capacitor-input filter is the most widely used filter in power supplies.

SEC. 4-7 PEAK INVERSE VOLTAGE AND SURGE CURRENT

The peak inverse voltage is the maximum voltage that appears across the nonconducting diode of a rectifier circuit. This voltage must be less than the breakdown voltage of the diode. The surge current is the brief and large current that exists when the power is first turned on. It is brief and large because the filter capacitor must charge to the peak voltage during the first cycle or, at most, during the first few cycles.

SEC. 4-8 OTHER POWER-SUPPLY TOPICS

Real transformers usually specify the secondary voltage at a rated load current. To calculate the primary current, you can assume that the input power equals the output power. Slow-blow fuses are typically used to protect against the surge current. The average diode current in a half-wave rectifier equals the dc load current. In a full-wave or bridge rectifier, the average current in any diode is half the dc load current. LC filters and LC filters may occasionally be used to filter the rectified output.

SEC. 4-9 TROUBLESHOOTING

Some of the measurements that can be made with a capacitor-input filter are the dc output voltage, the primary voltage, the secondary voltage, and the ripple. From these, you can usually deduce the trouble. Open diodes reduce the output voltage to zero. An open filter capacitor reduces the output to the average value of the rectified signal.

SEC. 4-10 CLIPPERS AND LIMITERS

A clipper shapes the signal. It clips off positive or negative parts of the signal. The limiter or diode clamp protects sensitive circuits from too much input.

SEC. 4-11 CLAMPERS

The clamper shifts a signal positively or negatively by adding a dc voltage to the signal. The peak-to-peak detector produces a load voltage equal to the peak-to-peak value.

SEC. 4-12 VOLTAGE MULTIPLIERS

The voltage doubler is a redesign of the peak-to-peak detector. It uses rectifier diodes instead of small-signal diodes. It produces an output equal to 2 times the peak value of the rectified signal. Voltage triplers and quadruplers multiply the input peak by factors of 3 and 4. Very high voltage power supplies are the main use of voltage multipliers.

Definitions

(4-14) Turns ratio:

$$\frac{N_1}{N_2} = \frac{V_1}{V_2}$$

(4-17) Stiff clipper:

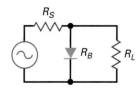

$$100R_B < R_S < 0.01R_L$$

(4-19) Stiff clamper:

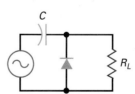

$$R_LC > 100T$$

Derivations

(4-1)   Ideal half-wave:

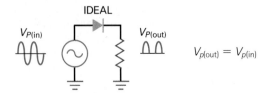

$$V_{p(out)} = V_{p(in)}$$

(4-2)   Half-wave:

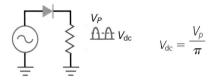

$$V_{dc} = \frac{V_p}{\pi}$$

(4-3)   Half-wave:

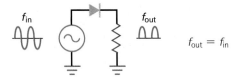

$$f_{out} = f_{in}$$

(4-4)   2d half-wave:

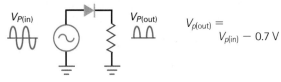

$$V_{p(out)} = V_{p(in)} - 0.7\ V$$

(4-5)   Ideal transformer:

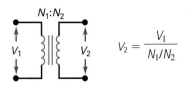

$$V_2 = \frac{V_1}{N_1/N_2}$$

(4-6)   Full-wave:

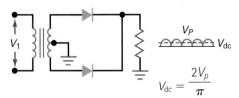

$$V_{dc} = \frac{2V_p}{\pi}$$

(4-7)   Full-wave:

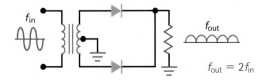

$$f_{out} = 2f_{in}$$

(4-8)   2d bridge:

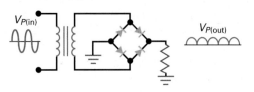

$$V_{p(out)} = V_{p(in)} - 1.4\ V$$

(4-9)   Choke-input filter:

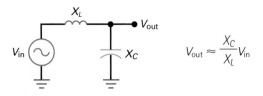

$$V_{out} \approx \frac{X_C}{X_L} V_{in}$$

(4-10)   Peak-to-peak ripple:

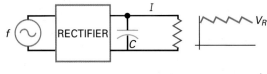

$$V_R = \frac{I}{fC}$$

(4-11)   Half-wave:

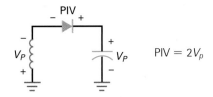

$$PIV = 2V_p$$

(4-12)   Full-wave:

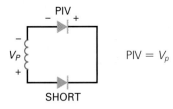

$$PIV = V_p$$

(4-13)   Bridge:

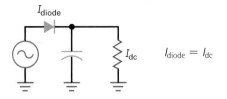

$$PIV = V_p$$

(4-15)   Half-wave:

$$I_{diode} = I_{dc}$$

(4-16)   Full-wave and bridge:

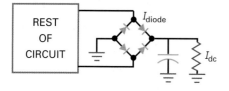

$$I_{diode} = 0.5I_{dc}$$

(4-18)   Biased clipper:

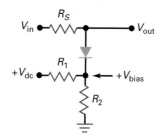

$$V_{bias} = \frac{R_2}{R_1 + R_2}V_{dc}$$

Student Assignments

1. If $N_1/N_2 = 4$, and the primary voltage is 120 V, what is the secondary voltage?

   a. 0 V

   b. 30 V

   c. 60 V

   d. 480 V

2. In a step-down transformer, which is larger?

   a. Primary voltage

   b. Secondary voltage

   c. Neither

   d. No answer possible

3. A transformer has a turns ratio of 2:1. What is the peak secondary voltage if 115 V rms is applied to the primary winding?

   a. 57.5 V

   b. 81.3 V

   c. 230 V

   d. 325 V

4. With a half-wave rectified voltage across the load resistor, load current flows for what part of a cycle?

   a. 0°

   b. 90°

   c. 180°

   d. 360°

5. Suppose line voltage may be as low as 105 V rms or as high as 125 rms in a half-wave rectifier. With a 5:1 step-down transformer, the minimum peak load voltage is closest to

   a. 21 V

   b. 25 V

   c. 29.7 V

   d. 35.4 V

6. The voltage out of a bridge rectifier is a

   a. Half-wave signal

   b. Full-wave signal

   c. Bridge-rectified signal

   d. Sine wave

7. If the line voltage is 115 V rms, a turns ratio of 5:1 means the rms secondary voltage is closest to

   a. 15 V

   b. 23 V

   c. 30 V

   d. 35 V

8. What is the peak load voltage in a full-wave rectifier if the secondary voltage is 20 V rms?

   a. 0 V

   b. 0.7 V

   c. 14.1 V

   d. 28.3 V

9. We want a peak load voltage of 40 V out of a bridge rectifier. What is the approximate rms value of secondary voltage?

   a. 0 V

   b. 14.4 V

   c. 28.3 V

   d. 56.6 V

10. With a full-wave rectified voltage across the load resistor, load current flows for what part of a cycle?

    a. 0°

    b. 90°

    c. 180°

    d. 360°

11. What is the peak load voltage out of a bridge rectifier for a secondary voltage of 12.6 V rms? (Use second approximation.)

    a. 7.5 V

    b. 16.4 V

    c. 17.8 V

    d. 19.2 V

12. If line frequency is 60 Hz, the output frequency of a half-wave rectifier is

    a. 30 Hz

    b. 60 Hz

    c. 120 Hz

    d. 240 Hz

13. If line frequency is 60 Hz, the output frequency of a bridge rectifier is

    a. 30 Hz

    b. 60 Hz

    c. 120 Hz

    d. 240 Hz

14. With the same secondary voltage and filter, which has the most ripple?

    a. Half-wave rectifier

    b. Full-wave rectifier

    c. Bridge rectifier

    d. Impossible to say

15. With the same secondary voltage and filter, which produces the least load voltage?

    a. Half-wave rectifier

    b. Full-wave rectifier

    c. Bridge rectifier

    d. Impossible to say

16. If the filtered load current is 10 mA, which of the following has a diode current of 10 mA?

   a. Half-wave rectifier
   b. Full-wave rectifier
   c. Bridge rectifier
   d. Impossible to say

17. If the load current is 5 mA and the filter capacitance is 1000 μF, what is the peak-to-peak ripple out of a bridge rectifier?

   a. 21.3 pV
   b. 56.3 nV
   c. 21.3 mV
   d. 41.7 mV

18. The diodes in a bridge rectifier each have a maximum dc current rating of 2 A. This means the dc load current can have a maximum value of

   a. 1 A
   b. 2 A
   c. 4 A
   d. 8 A

19. What is the PIV across each diode of a bridge rectifier with a secondary voltage of 20 V rms?

   a. 14.1 V
   b. 20 V
   c. 28.3 V
   d. 34 V

20. If the secondary voltage increases in a bridge rectifier with a capacitor-input filter, the load voltage will

   a. Decrease
   b. Stay the same
   c. Increase
   d. None of these

21. If the filter capacitance is increased, the ripple will

   a. Decrease
   b. Stay the same
   c. Increase
   d. None of these

22. A circuit that removes positive or negative parts of a waveform is called a

   a. Clamper
   b. Clipper
   c. Diode clamp
   d. Limiter

23. A circuit that adds a positive or negative dc voltage to an input sine-wave is called a

   a. Clamper
   b. Clipper
   c. Diode clamp
   d. Limiter

24. For a clamper circuit to operate properly, its R_LC time constant should be

   a. Equal to the period T of the signal
   b. > 10 times the period T of the signal
   c. > 100 times the period T of the signal
   d. < 10 times the period T of the signal

25. Voltage multipliers are circuits best used to produce

   a. Low voltage and low current
   b. Low voltage and high current
   c. High voltage and low current
   d. High voltage and high current

Problems

SEC. 4–1 THE HALF-WAVE RECTIFIER

4-1 |||| **MultiSim** What is the peak output voltage in Fig. 4-36*a* if the diode is ideal? The average value? The dc value? Sketch the output waveform.

Figure 4–36

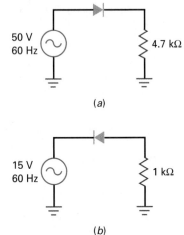

(a)

(b)

4-2 |||| **MultiSim** Repeat the preceding problem for Fig. 4-36*b*.

4-3 |||| **MultiSim** What is the peak output voltage in Fig. 4-36*a* using the second approximation of a diode? The average value? The dc value? Sketch the output waveform.

4-4 |||| **MultiSim** Repeat the preceding problem for Fig. 4-36*b*.

SEC. 4–2 THE TRANSFORMER

4-5 If a transformer has a turns ratio of 6 : 1, what is the rms secondary voltage? The peak secondary voltage? Assume a primary voltage of 120 V rms.

4-6 If a transformer has a turns ratio of 1 : 12, what is the rms secondary voltage? The peak secondary voltage? Assume a primary voltage 120 V rms.

4-7 Calculate the peak output voltage and the dc output voltage in Fig. 4-37 using an ideal diode.

4-8 Calculate the peak output voltage and the dc output voltage in Fig. 4-37 using the second approximation.

Figure 4-37

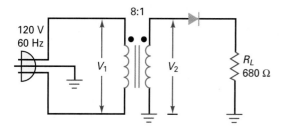

SEC. 4-3 THE FULL-WAVE RECTIFIER

4-9 A center-tapped transformer with 120 V input has a turns ratio of 4:1. What is the rms voltage across the upper half of the secondary winding? The peak voltage? What is the rms voltage across the lower half of the secondary winding?

4-10 |||| MultiSim What is the peak output voltage in Fig. 4-38 if the diodes are ideal? The average value? The dc value? Sketch the output waveform.

4-11 |||| MultiSim Repeat the preceding problem using the second approximation.

Figure 4-38

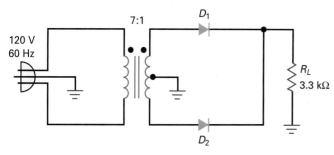

SEC. 4-4 THE BRIDGE RECTIFIER

4-12 |||| MultiSim In Fig. 4-39, what is the peak output voltage if the diodes are ideal? The average value? The dc value? Sketch the output waveform.

4-13 |||| MultiSim Repeat the preceding problem using the second approximation.

4-14 If the line voltage in Fig. 4-39 varies from 105 to 125 V rms, what is the minimum dc output voltage? The maximum?

SEC. 4-5 THE CHOKE–INPUT FILTER

4-15 A half-wave signal with a peak of 20 V is the input to a choke-input filter. If $X_L = 1$ kΩ and $X_C = 25$ Ω, what is the approximate peak-to-peak ripple across the capacitor?

4-16 A full-wave signal with a peak of 14 V is the input to a choke-input filter. If $X_L = 2$ kΩ and $X_C = 50$ Ω, what is the approximate peak-to-peak ripple across the capacitor?

SEC. 4-6 THE CAPACITOR–INPUT FILTER

4-17 What is the dc output voltage and ripple in Fig. 4-40a? Sketch the output waveform.

4-18 In Fig. 4-40b, calculate the dc output voltage and ripple.

4-19 What happens to the ripple in Fig. 4-40a if the capacitance value is reduced to half?

4-20 In Fig. 4-40a, what happens to the ripple if the resistance is reduced to 500 Ω?

4-21 What is the dc output voltage in Fig. 4-41? The ripple? Sketch the output waveform.

4-22 If the line voltage decreases to 105 V in Fig. 4-41, what is the dc output voltage?

SEC. 4-7 PEAK INVERSE VOLTAGE
AND SURGE CURRENT

4-23 What is the peak inverse voltage in Fig. 4-41?

4-24 If the turns ratio changes to 3:1 in Fig. 4-41, what is the peak inverse voltage?

SEC. 4-8 OTHER POWER–SUPPLY TOPICS

4-25 An F-25X replaces the transformer of Fig. 4-41. What is the approximate peak voltage across the secondary winding? The approximate dc output voltage? Is the transformer being operated at its rated output current? Will the dc output voltage be higher or lower than normal?

4-26 What is the primary current in Fig. 4-41?

4-27 What is the average current through each diode in Fig. 4-40a and 4-40b?

4-28 What is the average current through each diode of Fig. 4-41?

Figure 4-39

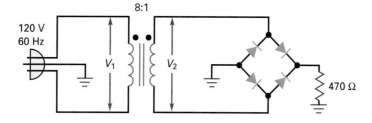

Figure 4-40

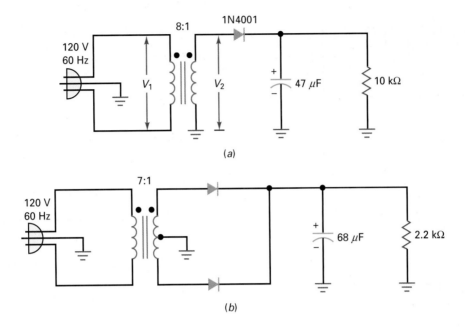

(a)

(b)

Figure 4-41

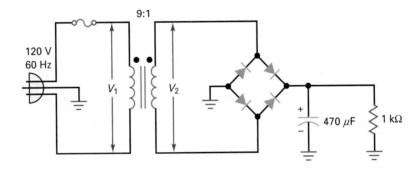

SEC. 4-9 TROUBLESHOOTING

4-29 If the filter capacitor in Fig. 4-41 is open, what is the dc output voltage?

4-30 If only one diode in Fig. 4-41 is open, what is the dc output voltage?

4-31 If somebody builds the circuit of Fig. 4-41 with the electrolytic capacitor reversed, what kind of trouble is likely to happen?

4-32 If the load resistance of Fig. 4-41 opens, what changes will occur in the output voltage?

SEC. 4-10 CLIPPERS AND LIMITERS

4-33 In Fig. 4-42a, sketch the output waveform. What is the maximum positive voltage? The maximum negative?

4-34 Repeat the preceding problem for Fig. 4-42b.

4-35 The diode clamp of Fig. 4-42c protects the sensitive circuit. What are the limiting levels?

4-36 In Fig. 4-42d, what is maximum positive output voltage? Maximum negative output voltage? Sketch the output waveform.

4-37 If the sine wave of Fig. 4-42d is only 20 mV, the circuit will act as a diode clamp instead of a biased clipper. In this case, what is the protected range of output voltage?

SEC. 4-11 CLAMPERS

4-38 In Fig. 4-43a, sketch the output waveform. What is the maximum positive voltage? The maximum negative?

4-39 Repeat the preceding problem for Fig. 4-43b.

4-40 Sketch the output waveform of the clamper and final output in Fig. 4-43c. What is the dc output voltage with ideal diodes? To a second approximation?

Figure 4-42

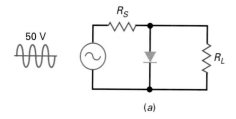

(a)

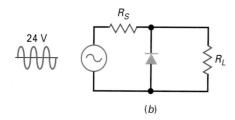

(b)

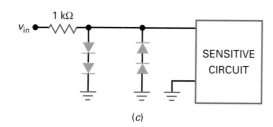

(c)

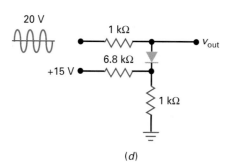

(d)

Figure 4-43

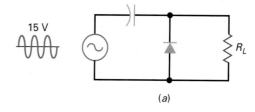

(a)

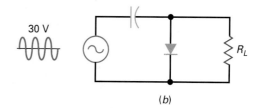

(b)

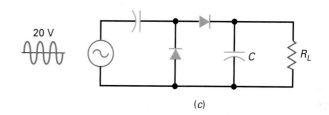

(c)

SEC. 4–12 VOLTAGE MULTIPLIERS

4-41 Calculate the dc output voltage in Fig. 4-44a.

4-42 What is the tripler output in Fig. 4-44b?

4-43 What is the quadrupler output in Fig. 4-44c?

Critical Thinking

4-44 If one of the diodes in Fig. 4-41 shorts, what will the probable result be?

4-45 The power supply of Fig. 4-45 has two output voltages. What are their approximate values?

4-46 A surge resistor of 4.7 Ω is added to Fig. 4-45. What is the maximum possible value of surge current?

4-47 A full-wave voltage has a peak value of 15 V. Somebody hands you a book of trigonometry tables, so that you can look up the value of a sine wave at intervals of 1°. Describe how you could prove that the average value of a full-wave signal is 63.6 percent of the peak value.

4-48 For the switch position shown in Fig. 4-46, what is the output voltage? If the switch is thrown to the other position, what is the output voltage?

4-49 If V_{in} is 40 V rms in Fig. 4-47 and the time constant RC is very large compared to the period of the source voltage, what does V_{out} equal? Why?

Figure 4-44

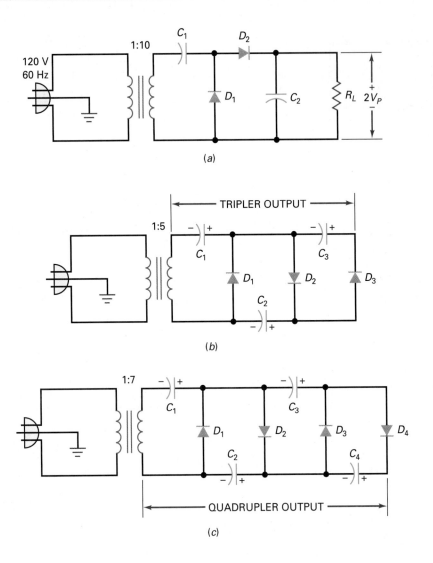

(a)

(b)

(c)

Figure 4-45

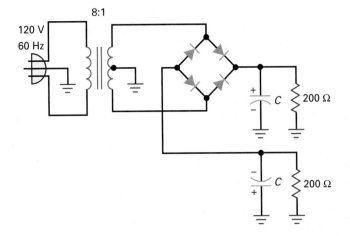

Figure 4-46

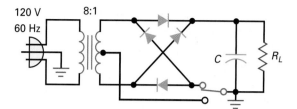

Figure 4-47

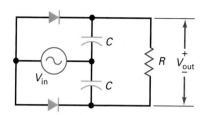

Troubleshooting

4-50 Figure 4-48 shows a bridge rectifier circuit with normal circuit values and eight troubles—T1–T8. Find all eight troubles.

Figure 4-48   Troubleshooting

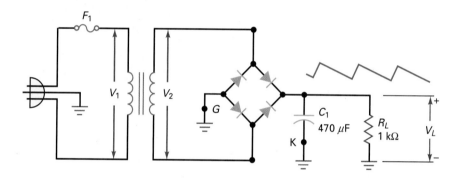

TROUBLESHOOTING							
	V_1	V_2	V_L	V_R	f	R_L	C_1	F_1
ok	115	12.7	18	0.3	120	1k	ok	ok
T1	115	12.7	11.4	18	120	1k	∞	ok
T2	115	12.7	17.7	0.6	60	1k	ok	ok
T3	0	0	0	0	0	0	ok	∞
T4	115	12.7	0	0	0	1k	ok	ok
T5	0	0	0	0	0	1k	ok	∞
T6	115	12.7	18	0	0	∞	ok	ok
T7	115	0	0	0	0	1k	ok	ok
T8	0	0	0	0	0	1k	0	∞

Job Interview Questions

1. Here's a pencil and paper. Tell me how a bridge rectifier with a capacitor-input filter works. In your explanation, I expect to see a schematic diagram and waveforms at different points in the circuit.

2. Suppose there's a bridge rectifier with a capacitor-input filter on my lab bench. It's not working. Tell me how you would troubleshoot it. Indicate the kind of instruments you would use and how you would isolate common troubles.

3. Excessive current or voltage can destroy the diodes in a power supply. Draw a bridge rectifier with a capacitor-input filter and tell me how current or voltage can destroy a diode. Do the same for excessive reverse voltage.

4. Tell me everything you know about clippers, clampers, and diode clamps. Show me typical waveforms, clipping levels, clamping levels, and protection levels.

5. I want you to tell me how a peak-to-peak detector works. Then, I want you to tell me in what ways a voltage doubler is similar to a peak-to-peak detector and in what ways it differs from a peak-to-peak detector.

6. What is the advantage of using a bridge rectifier in a power supply as opposed to using a half-wave or a full-wave rectifier? Why is the bridge rectifier more efficient than the others?

7. In what power-supply application might I prefer to use an *LC*-type filter instead of the *RC*-type? Why?

8. What is the relationship between a half-wave rectifier and a full-wave rectifier?

9. Under what circumstances is it appropriate to use a voltage multiplier as part of a power supply?

10. A dc power supply is supposed to have an output of 5 V. You measure exactly 5 V out of this supply using a dc voltmeter. Is it possible for the power supply to still have a problem? If so, how would you troubleshoot it?

11. Why would I use a voltage multiplier instead of a transformer with a higher turns ratio and an ordinary rectifier?

12. List the advantages and disadvantages of the *RC* filter and *LC* filter.

13. While troubleshooting a power supply, you find a resistor burned black. A measurement shows that the resistor is open. Should you replace the resistor and turn the supply back on? If not, what should you do next?

14. For a bridge rectifier, list three possible faults and what the symptoms of each would be.

Self-Test Answers

1.	b	10.	d	19.	c
2.	a	11.	b	20.	c
3.	b	12.	b	21.	a
4.	c	13.	c	22.	b
5.	c	14.	a	23.	a
6.	b	15.	b	24.	c
7.	b	16.	a	25.	c
8.	c	17.	d		
9.	c	18.	c		

Practice Problem Answers

4-1 $V_{dc} = 6.53$ V

4-2 $V_{dc} = 27$ V

4-3 $V_{p(\text{in})} = 12$ V;
$V_{p(\text{out})} = 11.3$ V

4-5 $V_{p(\text{out})}$ ideal = 34 V;
$2^{\text{nd}} = 32.6$ V

4-7 $V_L = 17$ V;
$V_R = 0.71$ Vpp

4-9 $V_R = 0.165$ Vpp

4-10 1N4002 or 1N4003 for safety
factor of 2

chapter 5

Special-Purpose Diodes

Rectifier diodes are the most common type of diode. They are used in power supplies to convert ac voltage to dc voltage. But rectification is not all that a diode can do. Now we will discuss diodes used in other applications. The chapter begins with the zener diode, which is optimized for its breakdown properties. Zener diodes are very important because they are the key to voltage regulation. The chapter also covers optoelectronic diodes, Schottky diodes, varactors, and other diodes.

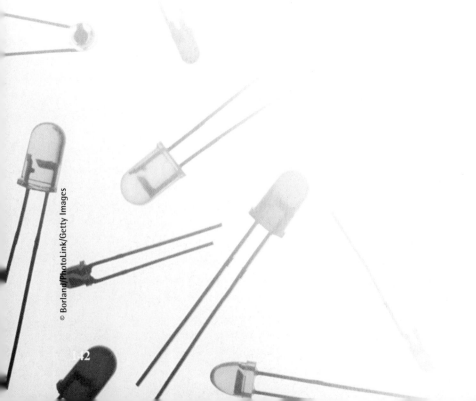

Objectives

After studying this chapter, you should be able to:

■ Show how the zener diode is used and calculate various values related to its operation.

■ List several optoelectronic devices and describe how each works.

■ Recall two advantages Schottky diodes have over common diodes.

■ Explain how a varactor works.

■ State a primary use of the varistor.

■ List four items of interest to the technician found on a zener diode data sheet.

■ List and describe the basic function of other semiconductor diodes.

Chapter Outline

Vocabulary

back diode
common-anode
common-cathode
current-regulator diode
derating factor
laser diode
leakage region
light-emitting diode (LED)
negative resistance

optocoupler
optoelectronics
photodiode
PIN diode
preregulator
Schottky diode
seven-segment display
step-recovery diode
temperature coefficient

tunnel diode
varactor
varistor
zener diode
zener effect
zener regulator
zener resistance

5-1 The Zener Diode

Small-signal and rectifier diodes are never intentionally operated in the breakdown region because this may damage them. A **zener diode** is different; it is a silicon diode that the manufacturer has optimized for operation in the breakdown region. The zener diode is the backbone of voltage regulators, circuits that hold the load voltage almost constant despite large changes in line voltage and load resistance.

I-V Graph

Figure 5-1a shows the schematic symbol of a zener diode; Fig. 5-1b is an alternative symbol. In either symbol, the lines resemble a z, which stands for "zener." By varying the doping level of silicon diodes, a manufacturer can produce zener diodes with breakdown voltages from about 2 to over 1000 V. These diodes can operate in any of three regions: forward, leakage, and breakdown.

Figure 5-1c shows the I-V graph of a zener diode. In the forward region, it starts conducting around 0.7 V, just like an ordinary silicon diode. In the **leakage region** (between zero and breakdown), it has only a small reverse current. In a zener diode, the breakdown has a very sharp knee, followed by an almost vertical increase in current. Note that the voltage is almost constant, approximately equal to V_Z over most of the breakdown region. Data sheets usually specify the value of V_Z at a particular test current I_{ZT}.

Figure 5-1c also shows the maximum reverse current I_{ZM}. As long as the reverse current is less than I_{ZM}, the diode is operating within its safe range. If the current is greater than I_{ZM}, the diode will be destroyed. To prevent excessive reverse current, a *current-limiting resistor* must be used (discussed later).

Zener Resistance

In the third approximation of a silicon diode, the forward voltage across a diode equals the knee voltage plus the additional voltage across the bulk resistance.

Figure 5-1 Zener diode. (*a*) Schematic symbol; (*b*) alternative symbol; (*c*) graph of current versus voltage; (*d*) typical zener diodes.

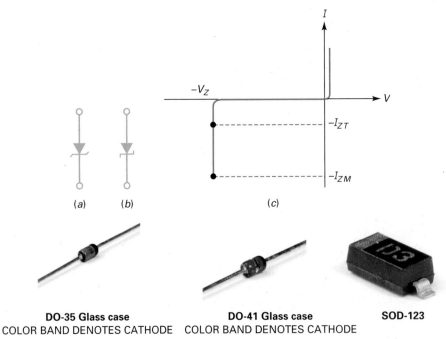

(a) **(b)** **(c)**

DO-35 Glass case
COLOR BAND DENOTES CATHODE

DO-41 Glass case
COLOR BAND DENOTES CATHODE

SOD-123

(d)

© Brian Moeskau/Brian Moeskau Photography   © Brian Moeskau/Brian Moeskau Photography   © Brian Moeskau/Brian Moeskau Photography

Similarly, in the breakdown region, the reverse voltage across a diode equals the breakdown voltage plus the additional voltage across the bulk resistance. In the reverse region, the bulk resistance is referred to as the **zener resistance**. This resistance equals the inverse of the slope in the breakdown region. In other words, the more vertical the breakdown region, the smaller the zener resistance.

In Fig. 5-1c, the zener resistance means that an increase in reverse current produces a slight increase in reverse voltage. The increase in voltage is very small, typically only a few tenths of a volt. This slight increase may be important in design work, but not in troubleshooting and preliminary analysis. Unless otherwise indicated, our discussions will ignore the zener resistance. Fig. 5-1(d) shows typical zener diodes.

Zener Regulator

A zener diode is sometimes called a *voltage-regulator diode* because it maintains a constant output voltage even though the current through it changes. For normal operation, you have to reverse bias the zener diode, as shown in Fig. 5-2a. Furthermore, to get breakdown operation, the source voltage V_S must be greater than the zener breakdown voltage V_Z. A series resistor R_S is always used to limit the zener current to less than its maximum current rating. Otherwise, the zener diode will burn out like any device with too much power dissipation.

Figure 5-2b shows an alternative way to draw the circuit with grounds. Whenever a circuit has grounds, you can measure voltages with respect to ground.

For instance, suppose you want to know the voltage across the series resistor of Fig. 5-2b. Here is the one way to find it when you have a built-up circuit. First, measure the voltage from the left end of R_S to ground. Second, measure the voltage from the right end of R_S to ground. Third, subtract the two voltages to get the voltage across R_S. If you have a floating VOM or DMM, you can connect directly across the series resistor.

Figure 5-2c shows the output of a power supply connected to a series resistor and a zener diode. This circuit is used when you want a dc output voltage that is less than the output of the power supply. A circuit like this is called a *zener voltage regulator,* or simply a **zener regulator.**

Ohm's Law Again

In Fig. 5-2, the voltage across the series or current-limiting resistor equals the difference between the source voltage and the zener voltage. Therefore, the current through the resistor is:

$$I_S = \frac{V_S - V_Z}{R_S} \tag{5-1}$$

Figure 5-2 Zener regulator. (*a*) Basic circuit; (*b*) same circuit with grounds; (*c*) power supply drives regulator.

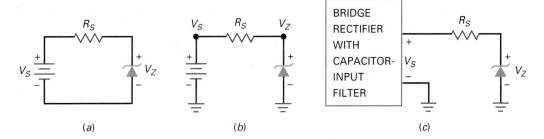

(a)   (b)   (c)

Figure 5-3 Ideal approximation of a zener diode.

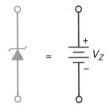

Once you have the value of series current, you also have the value of zener current. This is because Fig. 5-2 is a series circuit. Note that I_S must be less than I_{ZM}.

Ideal Zener Diode

For troubleshooting and preliminary analysis, we can approximate the breakdown region as vertical. Therefore, the voltage is constant even though the current changes, which is equivalent to ignoring the zener resistance. Figure 5-3 shows the ideal approximation of a zener diode. This means that a zener diode operating in the breakdown region ideally acts like a battery. In a circuit, it means that you can mentally replace a zener diode by a voltage source of V_Z, provided the zener diode is operating in the breakdown region.

Example 5-1

Suppose the zener diode of Fig. 5-4a has a breakdown voltage of 10 V. What are the minimum and maximum zener currents?

Figure 5-4 Example.

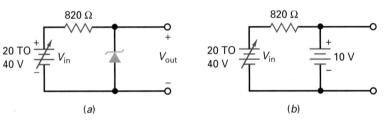

(a) (b)

SOLUTION The applied voltage may vary from 20 to 40 V. Ideally, a zener diode acts like the battery shown in Fig. 5-4b. Therefore, the output voltage is 10 V for any source voltage between 20 and 40 V.

The minimum current occurs when the source voltage is minimum. Visualize 20 V on the left end of the resistor and 10 V on the right end. Then you can see that the voltage across the resistor is 20 V − 10 V, or 10 V. The rest is Ohm's law:

$$I_S = \frac{10\ V}{820\ \Omega} = 12.2\ \text{mA}$$

The maximum current occurs when the source voltage is 40 V. In this case, the voltage across the resistor is 30 V, which gives a current of

$$I_S = \frac{30\ V}{820\ \Omega} = 36.6\ \text{mA}$$

In a voltage regulator like Fig. 5-4a, the output voltage is held constant at 10 V, despite the change in source voltage from 20 to 40 V. The larger source voltage produces more zener current, but the output voltage holds rock-solid at 10 V. (If the zener resistance is included, the output voltage increases slightly when the source voltage increases.)

PRACTICE PROBLEM 5-1 Using Fig. 5-4, what is the zener current I_S if $V_{in} = 30$ V?

5-2 The Loaded Zener Regulator

Figure 5-5a shows a *loaded* zener regulator, and Fig. 5-5b shows the same circuit with grounds. The zener diode operates in the breakdown region and holds the load voltage constant. Even if the source voltage changes or the load resistance varies, the load voltage will remain fixed and equal to the zener voltage.

Breakdown Operation

How can you tell whether the zener diode of Fig. 5-5 is operating in the breakdown region? Because of the voltage divider, the Thevenin voltage facing the diode is:

$$V_{TH} = \frac{R_L}{R_S + R_L} V_S \tag{5-2}$$

This is the voltage that exists when the zener diode is disconnected from the circuit. This Thevenin voltage has to be greater than the zener voltage; otherwise, breakdown cannot occur.

Series Current

Unless otherwise indicated, in all subsequent discussions we assume that the zener diode is operating in the breakdown region. In Fig. 5-5, the current through the series resistor is given by:

$$I_S = \frac{V_S - V_Z}{R_S} \tag{5-3}$$

This is Ohm's law applied to the current-limiting resistor. It is the same whether or not there is a load resistor. In other words, if you disconnect the load resistor, the current through the series resistor still equals the voltage across the resistor divided by the resistance.

Load Current

Ideally, the load voltage equals the zener voltage because the load resistor is in parallel with the zener diode. As an equation:

$$V_L = V_Z \tag{5-4}$$

This allows us to use Ohm's law to calculate the load current:

$$I_L = \frac{V_L}{R_L} \tag{5-5}$$

Figure 5-5   Loaded zener regulator. (*a*) Basic circuit; (*b*) practical circuit.

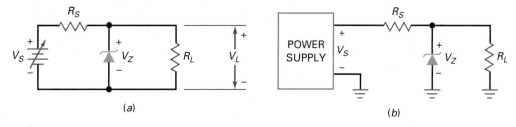

(*a*)   (*b*)

Zener Current

With Kirchhoff's current law:

$$I_S = I_Z + I_L$$

The zener diode and the load resistor are in parallel. The sum of their currents has to equal the total current, which is the same as the current through the series resistor.

We can rearrange the foregoing equation to get this important formula:

$$I_Z = I_S - I_L \tag{5-6}$$

This tells you that the zener current no longer equals the series current, as it does in an unloaded zener regulator. Because of the load resistor, the zener current now equals the series current minus the load current.

Table 5-1 summarizes the steps in the analysis of a loaded zener regulator. You start with the series current, followed by the load voltage and load current, and finally the zener current.

Zener Effect

When the breakdown voltage is greater than 6 V, the cause of the breakdown is the avalanche effect, discussed in Chap. 2. The basic idea is that minority carriers are accelerated to high enough speeds to dislodge other minority carriers, producing a chain or avalanche effect that results in a large reverse current.

The zener effect is different. When a diode is heavily doped, the depletion layer becomes very narrow. Because of this, the electric field across the depletion layer (voltage divided by distance) is very intense. When the field strength reaches approximately 300,000 V/cm, the field is intense enough to pull electrons out of their valence orbits. The creation of free electrons in this way is called the **zener effect** (also known as *high-field emission*). This is distinctly different from the avalanche effect, which depends on high-speed minority carriers dislodging valence electrons.

When the breakdown voltage is less than 4 V, only the zener effect occurs. When the breakdown voltage is greater than 6 V, only the avalanche effect occurs. When the breakdown voltage is between 4 and 6 V, both effects are present.

The zener effect was discovered before the avalanche effect, so all diodes used in the breakdown region came to be known as zener diodes. Although you may occasionally hear the term *avalanche diode,* the name *zener diode* is in general use for all breakdown diodes.

Table 5–1	Analyzing a Loaded Zener Regulator	
	Process	**Comment**
Step 1	Calculate the series current, Eq. (5-3)	Apply Ohm's law to R_S
Step 2	Calculate the load voltage, Eq. (5-4)	Load voltage equals diode voltage
Step 3	Calculate the load current, Eq. (5-5)	Apply Ohm's law to R_L
Step 4	Calculate the zener current, Eq. (5-6)	Apply the current law to the diode

Temperature Coefficients

When the ambient temperature changes, the zener voltage will change slightly. On data sheets the effect of temperature is listed under the **temperature coefficient,** which is defined as the change in breakdown voltage per degree of increase. The temperature coefficient is negative for breakdown voltages less than 4 V (zener effect). For instance, a zener diode with a breakdown voltage of 3.9 V may have a temperature coefficient of -1.4 mV/°C. If temperature increases by 1°, the breakdown voltage decreases by 1.4 mV.

On the other hand, the temperature coefficient is positive for breakdown voltages greater than 6 V (avalanche effect). For instance, a zener diode with a breakdown voltage of 6.2 V may have a temperature coefficient of 2 mV/°C. If the temperature increases by 1°, the breakdown voltage increases by 2 mV.

Between 4 and 6 V, the temperature coefficient changes from negative to positive. In other words, there are zener diodes with breakdown voltages between 4 and 6 V in which the *temperature coefficient equals zero.* This is important in some applications when a solid zener voltage is needed over a large temperature range.

Example 5-2

||| MultiSim

Is the zener diode of Fig. 5-6*a* operating in the breakdown region?

Figure 5-6   Example.

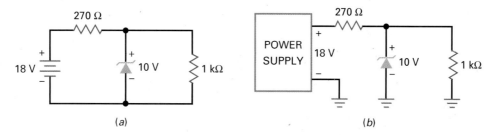

(a)   (b)

SOLUTION   With Eq. (5-2):

$$V_{TH} = \frac{1 \text{ k}\Omega}{270 \ \Omega + 1 \text{ k}\Omega}(18 \text{ V}) = 14.2 \text{ V}$$

Since this Thevenin voltage is greater than the zener voltage, the zener diode is operating in the breakdown region.

Example 5-3

||| MultiSim

What does the zener current equal in Fig. 5-6*b*?

SOLUTION   You are given the voltage on both ends of the series resistor. Subtract the voltages, and you can see that 8 V is across the series resistor. Then Ohm's law gives:

$$I_S = \frac{8 \text{ V}}{270 \ \Omega} = 29.6 \text{ mA}$$

Since the load voltage is 10 V, the load current is:

$$I_L = \frac{10\text{ V}}{1\text{ k}\Omega} = 10\text{ mA}$$

The zener current is the difference between the two currents:

$$I_Z = 29.6\text{ mA} - 10\text{ mA} = 19.6\text{ mA}$$

PRACTICE PROBLEM 5-3   Using Fig. 5-6b, change the power supply to 15 V and calculate I_S, I_L, and I_Z.

Example 5-4

IIII MultiSim

What does the circuit of Fig. 5-7 do?

Figure 5-7   Example.

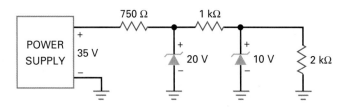

SOLUTION   This is an example of a **preregulator** (the first zener diode) driving a zener regulator (the second zener diode). First, notice that the preregulator has an output voltage of 20 V. This is the input to the second zener regulator, whose output is 10 V. The basic idea is to provide the second regulator with a well-regulated input, so that the final output is extremely well regulated.

Example 5-5

IIII MultiSim

What does the circuit of Fig. 5-8 do?

Figure 5-8   Zener diodes used for waveshaping.

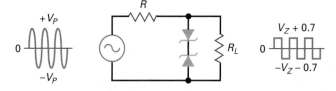

SOLUTION   In most applications, zener diodes are used in voltage regulators where they remain in the breakdown region. But there are exceptions. Sometimes zener diodes are used in waveshaping circuits like Fig. 5-8.

Notice the back-to-back connection of two zener diodes. On the positive half cycle, the upper diode conducts and the lower diode breaks down. Therefore, the output is clipped as shown. The clipping level equals the zener voltage (broken-down diode) plus 0.7 V (forward-biased diode).

On the negative half cycle, the action is reversed. The lower diode conducts, and the upper diode breaks down. In this way, the output is almost a square wave. The larger the input sine wave, the better-looking the output square wave.

PRACTICE PROBLEM 5-5   In Fig. 5-8, the V_Z for each diode is 3.3 V. What would the voltage across R_L be?

Example 5-6

Briefly describe the circuit action for each of the circuits in Fig. 5-9.

Figure 5-9 Zener applications. (*a*) Producing nonstandard output voltages; (*b*) using a 6-V relay in a 12-V system; (*c*) using a 6-V capacitor in a 12-V system.

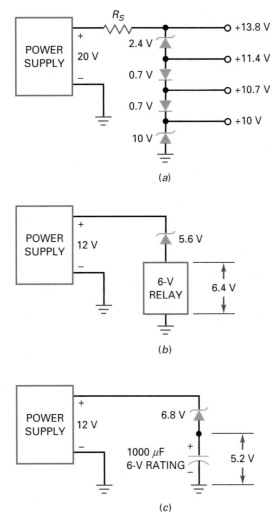

(a)

(b)

(c)

SOLUTION Figure 5-9*a* shows how zener diodes and ordinary silicon diodes can produce several dc output voltages, given a 20-V power supply. The bottom diode produces an output of 10 V. Each silicon diode is forward biased, producing outputs of 10.7 V and 11.4 V, as shown. The top diode has a breakdown voltage of 2.4 V, giving an output of 13.8 V. With other combinations of zener and silicon diodes, a circuit like this can produce different dc output voltages.

If you try to connect a 6-V relay to a 12-V system, you will probably damage the relay. It is necessary to drop some of the voltage. Figure 5-9*b* shows one way to do this. By connecting a 5.6-V zener diode in series with the relay, only 6.4 V appears across the relay, which is usually within the tolerance of the relay's voltage rating.

Large electrolytic capacitors often have small voltage ratings. For instance, an electrolytic capacitor of 1000 μF may have a voltage rating of only 6 V. This means that the maximum voltage across the capacitor should be less than 6 V. Figure 5-9*c* shows a workaround solution in which a 6-V electrolytic capacitor is used with a 12-V power supply. Again, the idea is to use a zener diode to drop some of the voltage. In this case, the zener diode drops 6.8 V, leaving only 5.2 V across the capacitor. This way, the electrolytic capacitor can filter the power supply and still remain with its voltage rating.

5-3 Second Approximation of a Zener Diode

Figure 5-10a shows the second approximation of a zener diode. A zener resistance is in series with an ideal battery. The total voltage across the zener diode equals the breakdown voltage plus the voltage drop across the zener resistance. Since R_Z is relatively small in a zener diode, it has only a minor effect on the total voltage across the zener diode.

Effect on Load Voltage

How can we calculate the effect of the zener resistance on the load voltage? Figure 5-10b shows a power supply driving a loaded zener regulator. Ideally, the load voltage equals the breakdown voltage V_Z. But in the second approximation, we include the zener resistance as shown in Fig. 5-10c. The additional voltage drop across R_Z will slightly increase the load voltage.

Since the zener current flows through the zener resistance in Fig. 5-10c, the load voltage is given by:

$$V_L = V_Z + I_Z R_Z$$

As you can see, the change in the load voltage from the ideal case is:

$$\Delta V_L = I_Z R_Z \tag{5-7}$$

Usually, R_Z is small, so the voltage change is small, typically in tenths of a volt. For instance, if $I_Z = 10$ mA and $R_Z = 10\ \Omega$, then $\Delta V_L = 0.1$ V.

Figure 5-10 Second approximation of a zener diode. (a) Equivalent circuit; (b) power supply drives zener regulator; (c) zener resistance included in analysis.

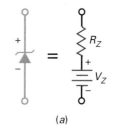

(a)

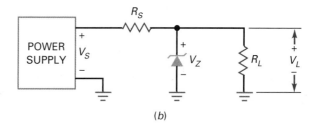

(b)

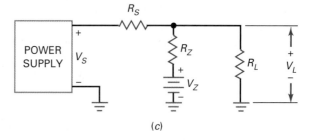

(c)

Figure 5-11 Zener regulator reduces ripple. (a) Complete ac equivalent circuit; (b) simplified ac equivalent circuit.

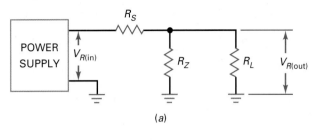

(a)

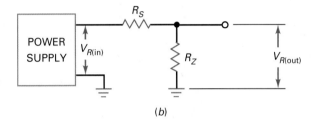

(b)

Effect on Ripple

As far as ripple is concerned, we can use the equivalent circuit shown in Fig. 5-11a. In other words, the only components that affect the ripple are the three resistances shown. We can simplify this even further. In a typical design, R_Z is much smaller than R_L. Therefore, the only two components that have a significant effect on ripple are the series resistance and zener resistance shown in Fig. 5-11b.

Since Fig. 5-11b is a voltage divider, we can write the following equation for the output ripple:

$$V_{R(\text{out})} = \frac{R_Z}{R_S + R_Z} V_{R(\text{in})}$$

Ripple calculations are not critical; that is, they don't have to be exact. Since R_S is always much greater than R_Z in a typical design, we can use this approximation for all troubleshooting and preliminary analysis:

$$V_{R(\text{out})} \approx \frac{R_Z}{R_S} V_{R(\text{in})} \tag{5-8}$$

Example 5-7

The zener diode of Fig. 5-12 has a breakdown voltage of 10 V and a zener resistance of 8.5 Ω. Use the second approximation to calculate the load voltage when the zener current is 20 mA.

Figure 5-12   Loaded zener regulator.

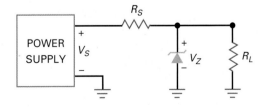

SOLUTION The change in load voltage equals the zener current times the zener resistance:

$$\Delta V_L = I_Z R_Z = (20 \text{ mA})(8.5 \; \Omega) = 0.17 \text{ V}$$

To a second approximation, the load voltage is:

$$V_L = 10 \text{ V} + 0.17 \text{ V} = 10.17 \text{ V}$$

PRACTICE PROBLEM 5-7 Use the second approximation to calculate the load voltage of Fig. 5-12 when $I_Z = 12$ mA.

Example 5-8

In Fig. 5-12, $R_S = 270 \; \Omega$, $R_Z = 8.5 \; \Omega$, and $V_{R(\text{in})} = 2$ V. What is the approximate ripple voltage across the load?

SOLUTION The load ripple approximately equals the ratio of R_Z to R_S, times the input ripple:

$$V_{R(\text{out})} \approx \frac{8.5 \, \Omega}{270 \, \Omega} 2 \text{ V} = 63 \text{ mV}$$

PRACTICE PROBLEM 5-8 Using Fig. 5-12, what is the approximate load ripple voltage if $V_{R \, (\text{in})} = 3$ V?

Example 5-9

The zener regulator of Fig. 5-13 has $V_Z = 10$ V, $R_S = 270 \; \Omega$, and $R_Z = 8.5 \; \Omega$, the same values used in Examples 5-7 and 5-8. Describe the measurements being made in this MultiSim circuit analysis.

||| **MultiSim** **Figure 5-13** MultiSim analysis of ripple in zener regulator.

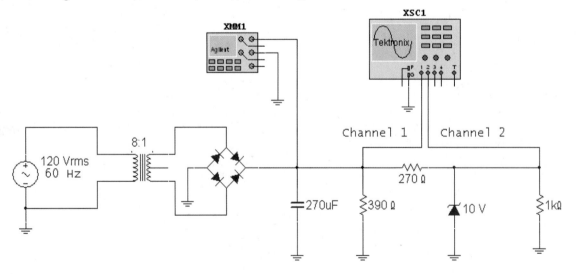

Figure 5-13 (continued)

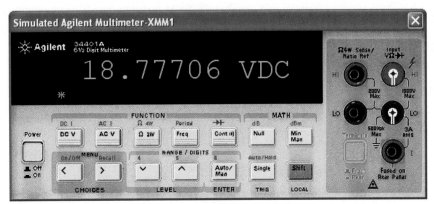

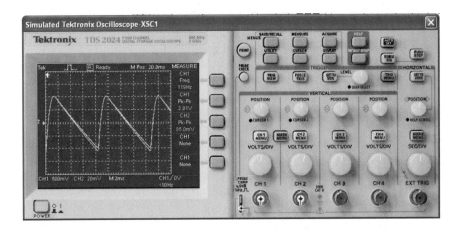

SOLUTION If we calculate the voltages in Fig. 5-13 using the methods discussed earlier, we will get the following results. With an 8:1 transformer, the peak secondary voltage is 21.2 V. Subtract two diode drops, and you get a peak of 19.8 V across the filter capacitor. The current through the 390-Ω resistor is 51 mA, and the current through R_S is 36 mA. The capacitor has to supply the sum of these two currents, which is 87 mA. With Eq. (4-10), this current results in a ripple across the capacitor of approximately 2.7 V pp. With this, we can calculate the ripple out of the zener regulator, which is approximately 85 mV pp.

Since the ripple is large, the voltage across the capacitor swings from a high of 19.8 V to a low of 17.1 V. If you average these two values, you get 18.5 V as the approximate dc voltage across the filter capacitor. This lower dc voltage means that the input and output ripple calculated earlier will also be lower. As discussed in the preceding chapter, calculations like these are only estimates because the exact analysis has to include higher-order effects.

Now, let us look at the MultiSim measurements, which are almost exact answers. The multimeter reads 18.78 V, very close to the estimated value of 18.5 V. Channel 1 of the oscilloscope shows the ripple across the capacitor. It is approximately 2 V pp, somewhat less than the estimated 2.7 V pp, but still in reasonable agreement. And finally, the output ripple of the zener regulator is approximately 85 mV pp (channel 2).

5-4 Zener Drop-Out Point

For a zener regulator to hold the output voltage constant, the zener diode must remain in the breakdown region under all operating conditions. This is equivalent to saying that there must be zener current for all source voltages and load currents.

Worst-Case Conditions

Figure 5-14a shows a zener regulator. It has the following currents:

$$I_S = \frac{V_S - V_Z}{R_S} = \frac{20\ V - 10\ V}{200\ \Omega} = 50\ mA$$

$$I_L = \frac{V_L}{R_L} = \frac{10\ V}{1\ k\Omega} = 10\ mA$$

and

$$I_Z = I_S - I_L = 50\ mA - 10\ mA = 40\ mA$$

Now, consider what happens when the source voltage decreases from 20 to 12 V. In the foregoing calculations, you can see that I_S will decrease, I_L will remain the same, and I_Z will decrease. When V_S equals 12 V, I_S will equal 10 mA, and $I_Z = 0$. At this low source voltage, the zener diode is about to come out of the breakdown region. If the source decreases any further, regulation will be lost. In other words, the load voltage will become less than 10 V. Therefore, a low source voltage can cause the zener circuit to fail to regulate.

Another way to get a loss of regulation is by having too much load current. In Fig. 5-14a, consider what happens when the load resistance decreases from 1 kΩ to 200 Ω. When the load resistance is 200 Ω, the load current increases to 50 mA and the zener current decreases to zero. Again, the zener diode is about to come out of breakdown. Therefore, a zener circuit will fail to regulate if the load resistance is too low.

Finally, consider what happens when R_S increases from 200 Ω to 1 kΩ. In this case, the series current decreases from 50 to 10 mA. Therefore, a high series resistance can bring the circuit out of regulation.

Figure 5-14 Zener regulator. (a) Normal operation; (b) worst-case conditions at drop-out point.

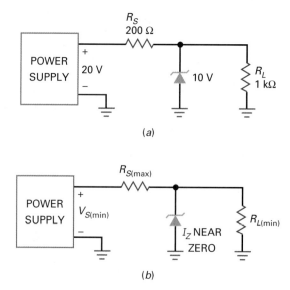

Figure 5-14*b* summarizes the foregoing ideas by showing the worst-case conditions. When the zener current is near zero, the zener regulation is approaching a drop-out or failure condition. By analyzing the circuit for these worst-case conditions, it is possible to derive the following equation:

$$R_{S(max)} = \left(\frac{V_{S(min)}}{V_Z} - 1\right) R_{L(min)} \tag{5-9}$$

An alternative form of this equation is also useful:

$$R_{S(max)} = \frac{V_{S(min)} - V_Z}{I_{L(max)}} \tag{5-10}$$

These two equations are useful because you can check a zener regulator to see whether it will fail under any operating conditions.

Example 5-10

A zener regulator has an input voltage that may vary from 22 to 30 V. If the regulated output voltage is 12 V and the load resistance varies from 140 Ω to 10 kΩ, what is the maximum allowable series resistance?

SOLUTION Use Eq. (5-9) to calculate the maximum series resistance as follows:

$$R_{S(max)} = \left(\frac{22\,V}{12\,V} - 1\right) 140\ \Omega = 117\ \Omega$$

As long as the series resistance is less than 117 Ω, the zener regulator will work properly under all operating conditions.

PRACTICE PROBLEM 5-10 Using Example 5-10, what is the maximum allowable series resistance if the regulated output voltage is 15 V?

Example 5-11

A zener regulator has an input voltage ranging from 15 to 20 V and a load current ranging from 5 to 20 mA. If the zener voltage is 6.8 V, what is the maximum allowable series resistance?

SOLUTION Use Eq. (5-10) to calculate the maximum series resistance as follows:

$$R_{S(max)} = \frac{15\,V - 6.8\,V}{20\ mA} = 410\ \Omega$$

If the series resistance is less than 410 Ω, the zener regulator will work properly under all conditions.

PRACTICE PROBLEM 5-11 Repeat Example 5-11 using a zener voltage of 5.1 V.

5-5 Reading a Data Sheet

Figure 5-15 shows the data sheets for the 1N957B and 1N4728A series of zener diodes. Refer to these data sheets during the following discussion. Again, most of the information on a data sheet is for designers, but there are a few items that even troubleshooters and testers will want to know about.

Maximum Power

The power dissipation of a zener diode equals the product of its voltage and current:

$$P_Z = V_Z I_Z \qquad (5\text{-}11)$$

For instance, if $V_Z = 12$ V and $I_Z = 10$ mA, then

$$P_Z = (12 \text{ V})(10 \text{ mA}) = 120 \text{ mW}$$

As long as P_Z is less than the power rating, the zener diode can operate in the breakdown region without being destroyed. Commercially available zener diodes have power ratings from ¼ to more than 50 W.

For example, the data sheet for the 1N957B series lists a maximum power rating of 500 mW. A safe design includes a safety factor to keep the power dissipation well below this 500-mW maximum. As mentioned elsewhere, safety factors of 2 or more are used for conservative designs.

Maximum Current

Data sheets often include the *maximum current* a zener diode can handle without exceeding its power rating. I_{ZM} for a 1N961B is shown to be 32 mA. If this value is not listed, the maximum current can be found as follows:

$$I_{ZM} = \frac{P_{ZM}}{V_Z} \qquad (5\text{-}12)$$

where I_{ZM} = maximum rated zener current
P_{ZM} = power rating
V_Z = zener voltage

For example, the 1N4742A has a zener voltage of 12 V and a 1 W power rating. Therefore, it has a maximum current rating of

$$I_{ZM} = \frac{1 \text{ W}}{12 \text{ V}} = 83.3 \text{ mA}$$

If you satisfy the current rating, you automatically satisfy the power rating. For instance, if you keep the maximum zener current less than 83.3 mA, you are also keeping the maximum power dissipation less than 1 W. If you throw in the safety factor of 2, you don't have to worry about a marginal design blowing the diode.

Tolerance

Most zener diodes will have a suffix A, B, C, or D to identify the zener voltage tolerance. Because these suffix markings are not always consistent, be sure to identify any special notes included on the zener's data sheet that indicate that specific tolerance. For instance, the data sheet for the 1N4728A series shows its tolerance to equal ±5 percent while the 1N957B series also has a tolerance of ±5 percent. A suffix of C generally indicates ±2 percent, D ±1 percent, and no suffix ±20 percent.

Figure 5-15(*a*)   Zener data sheets. (Copyright of Fairchild Semiconductor. Used by Permission).

SEMICONDUCTOR®

Zeners
1N957B - 1N991B

Absolute Maximum Ratings * T_A = 25˚C unless otherwise noted

Tolerance = 5%

Symbol	Parameter	Value	Units
P_D	Power Dissipation @ TL ≤ 75°C, Lead Length = 3/8"	500	mW
	Derate above 75°C	4.0	mW/°C
T_J, T_{STG}	Operating and Storage Temperature Range	-65 to +200	°C

* These ratings are limiting values above which the serviceability of the diode may be impaired.

DO-35 Glass case
COLOR BAND DENOTES CATHODE

Electrical Characteristics T_A = 25˚C unless otherwise noted

Device	V_Z (Volts) (Note 1)				Z_Z (Ω) (Note 2)			I_R @ V_R		I_{ZM} (mA) (Note 3)
	Min.	Typ.	Max.	@ I_Z (mA)	Z_Z @ I_Z	Z_{ZK} @ I_{ZK} Ω	mA	μA	Volts	
1N957B	6.46	6.8	7.14	18.5	4.5	700	1.0	150	5.2	47
1N958B	7.125	7.5	7.875	16.5	5.5	700	0.5	75	5.7	42
1N959B	7.79	8.2	8.61	15	6.5	700	0.5	50	6.2	38
1N960B	8.645	9.1	9.555	14	7.5	700	0.5	25	6.9	35
1N961B	9.5	10	10.5	12.5	8.5	700	0.25	10	7.6	32
1N962B	10.45	11	11.55	11.5	9.5	700	0.25	5	8.4	28
1N963B	11.4	12	12.6	10.5	11.5	700	0.25	5	9.1	26
1N964B	12.35	13	13.65	9.5	13	700	0.25	5	9.9	24
1N965B	14.25	15	15.75	8.5	16	700	0.25	5	11.4	21
1N966B	15.2	16	16.8	7.8	17	700	0.25	5	12.2	19
1N967B	17.1	18	18.9	7.0	21	750	0.25	5	13.7	17
1N968B	19	20	21	6.2	25	750	0.25	5	15.2	15
1N969B	20.9	22	23.1	5.6	29	750	0.25	5	16.7	14
1N970B	22.8	24	25.2	5.2	33	750	0.25	5	18.2	13
1N971B	25.652	27	28.35	4.6	41	750	0.25	5	20.6	11
1N972B	8.5	30	31.5	4.2	49	1000	0.25	5	22.8	10
1N973B	31.35	33	34.65	3.8	58	1000	0.25	5	25.1	9.2
1N974B	34.2	36	37.8	3.4	70	1000	0.25	5	27.4	8.5
1N975B	37.05	39	40.95	3.2	80	1000	0.25	5	29.7	7.8
1N976B	40.85	43	45.15	3.0	93	1500	0.25	5	32.7	7.0
1N977B	44.65	47	49.35	2.7	105	1500	0.25	5	35.8	6.4
1N978B	48.45	51	53.55	2.5	125	1500	0.25	5	38.8	5.9
1N979B	53.2	56	58.8	2.2	150	2000	0.25	5	42.6	5.4
1N980B	58.9	62	65.1	2.0	185	2000	0.25	5	47.1	4.9
1N981B	64.6	68	71.4	1.8	230	2000	0.25	5	51.7	4.5

Notes:
1. Zener Voltage (V_Z) Measurement
   Nominal zener voltage is measured with the device junction in the thermal equilibrium at the lead temperature (T_L) at 30°C ± 1°C and 3/8" lead length.
2. Zener Impedance (Z_Z) Derivation
   Z_{ZT} and Z_{ZK} are measured by dividing the ac voltage drop across the device by the ac current applied. The specified limits are for $I_{Z(ac)}$ = 0.1 $I_{Z(dc)}$ with the ac frequency = 60Hz.
3. Maximum Zener Current Ratings (I_{ZM})
   The maximum current handling capability on a worst case basis is limited by the actual zener voltage at the operation point and the power derating curve.

1N957B - 1N991B, Rev. E2

Figure 5-15(b) (continued)

Zeners
1N4728A - 1N4764A

Absolute Maximum Ratings * T_A = 25°C unless otherwise noted

Tolerance = 5%

Symbol	Parameter	Value	Units
P_D	Power Dissipation @ TL ≤ 50°C, Lead Length = 3/8"	1.0	W
	Derate above 50°C	6.67	mW/°C
T_J, T_STG	Operating and Storage Temperature Range	-65 to +200	°C

* These ratings are limiting values above which the serviceability of the diode may be impaired.

DO-41 Glass case
COLOR BAND DENOTES CATHODE

Electrical Characteristics T_A = 25°C unless otherwise noted

Device	V_Z (V) @ I_Z (Note 1) Min.	Typ.	Max.	Test Current I_Z (mA)	Z_Z @ I_Z (Ω)	Z_ZK @ I_ZK (Ω)	I_ZK (mA)	I_R (µA)	V_R (V)
1N4728A	3.315	3.3	3.465	76	10	400	1	100	1
1N4729A	3.42	3.6	3.78	69	10	400	1	100	1
1N4730A	3.705	3.9	4.095	64	9	400	1	50	1
1N4731A	4.085	4.3	4.515	58	9	400	1	10	1
1N4732A	4.465	4.7	4.935	53	8	500	1	10	1
1N4733A	4.845	5.1	5.355	49	7	550	1	10	1
1N4734A	5.32	5.6	5.88	45	5	600	1	10	2
1N4735A	5.89	6.2	6.51	41	2	700	1	10	3
1N4736A	6.46	6.8	7.14	37	3.5	700	1	10	4
1N4737A	7.125	7.5	7.875	34	4	700	0.5	10	5
1N4738A	7.79	8.2	8.61	31	4.5	700	0.5	10	6
1N4739A	8.645	9.1	9.555	28	5	700	0.5	10	7
1N4740A	9.5	10	10.5	25	7	700	0.25	10	7.6
1N4741A	10.45	11	11.55	23	8	700	0.25	5	8.4
1N4742A	11.4	12	12.6	21	9	700	0.25	5	9.1
1N4743A	12.35	13	13.65	19	10	700	0.25	5	9.9
1N4744A	14.25	15	15.75	17	14	700	0.25	5	11.4
1N4745A	15.2	16	16.8	15.5	16	700	0.25	5	12.2
1N4746A	17.1	18	18.9	14	20	700	0.25	5	13.7
1N4747A	19	20	21	12.5	22	700	0.25	5	15.2
1N4748A	20.9	22	23.1	11.5	23	750	0.25	5	16.7
1N4749A	22.8	24	25.2	10.5	25	750	0.25	5	18.2
1N4750A	25.65	27	28.35	9.5	35	750	0.25	5	20.6
1N4751A	28.5	30	31.5	8.5	40	1000	0.25	5	22.8
1N4752A	31.35	33	34.65	7.5	45	1000	0.25	5	25.1
1N4753A	34.2	36	37.8	7	50	1000	0.25	5	27.4
1N4754A	37.05	39	40.95	6.5	60	1000	0.25	5	29.7
1N4755A	40.85	43	45.15	6	70	1500	0.25	5	32.7
1N4756A	44.65	47	49.35	5.5	80	1500	0.25	5	35.8
1N4757A	48.45	51	53.55	5	95	1500	0.25	5	38.8

1N4728A - 1N4764A, Rev. F2

Zener Resistance

The zener resistance (also called *zener impedance*) may be designated R_{ZT} or Z_{ZT}. For instance, the 1N961B has a zener resistance of 8.5 Ω measured at a test current of 12.5 mA. As long as the zener current is beyond the knee of the curve, you can use 8.5 Ω as the approximate value of the zener resistance. But note how the zener resistance increases at the knee of the curve (700 Ω). The point is this: Operation should be at or near the test current, if at all possible. Then you know that the zener resistance is relatively small.

The data sheet contains a lot of additional information, but it is primarily aimed at designers. If you do get involved in design work, then you have to read the data sheet carefully, including the notes that specify how quantities were measured.

Derating

The **derating factor** shown on a data sheet tells you how much you have to reduce the power rating of a device. For instance, the 1N4728A series has a power rating of 1 W for a lead temperature of 50°C. The derating factor is given as 6.67 mW/°C. This means that you have to subtract 6.67 mW for each degree above 50°C. Even though you may not be involved in design, you have to be aware of the effect of temperature. If it is known that the lead temperature will be above 50°C, the designer has to derate or reduce the power rating of the zener diode.

5-6 Troubleshooting

Figure 5-16 shows a zener regulator. When the circuit is working properly, the voltage between A and ground is +18 V, the voltage between B and ground is +10 V, and the voltage between C and ground is +10 V.

Unique Symptoms

Now, let's discuss what can go wrong with the circuit. When a circuit is not working as it should, a troubleshooter usually starts by measuring voltages. These voltage measurements give clues that help isolate the trouble. For instance, suppose these voltages are measured:

$$V_A = +18 \text{ V} \qquad V_B = +10 \text{ V} \qquad V_C = 0$$

Here is what may go through a troubleshooter's mind after measuring the foregoing voltages:

> *What if the load resistor were open? No, the load voltage would still be 10 V. What if the load resistor were shorted? No, that would pull B and C down to ground, producing 0 V. All right, what if the connecting wire between B and C were open? Yes, that would do it.*

This trouble produces unique symptoms. The only way you can get this set of voltages is with an open connection between B and C.

Ambiguous Symptoms

Not all troubles produce unique symptoms. Sometimes, two or more troubles produce the same set of voltages. Here is an example. Suppose the troubleshooter measures these voltages:

$$V_A = +18 \text{ V} \qquad V_B = 0 \qquad V_C = 0$$

What do you think the trouble is? Think about this for a few minutes. When you have an answer, read what follows.

Figure 5-16 Troubleshooting a zener regulator.

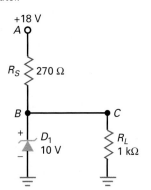

Here is a way that a troubleshooter might find the trouble. The thinking goes like this:

> *I've got voltage at A, but not at B and C. What if the series resistor were open? Then no voltage could reach B or C, but I would still measure 18 V between A and ground. Yes, the series resistor is probably open.*

At this point, the troubleshooter would disconnect the series resistor and measure its resistance with an ohmmeter. Chances are that it would be open. But suppose it measures OK. Then the troubleshooter's thinking continues like this:

> *That's strange. Well, is there any other way I can get 18 V at A and 0 V at B and C? What if the zener diode were shorted? What if the load resistor were shorted? What if a solder splash were between B or C and ground? Any of these will produce the symptoms I'm getting.*

Now the troubleshooter has more possible troubles to check out. Eventually, he or she will find the trouble.

When components burn out, they may become open, but not always. Some semiconductor devices can develop internal shorts, in which case they are like zero resistances. Other ways to get shorts include a solder splash between traces on a printed-circuit board, a solder ball touching two traces, and so on. Because of this, you must include what-if questions in terms of shorted components, as well as open components.

Table of Troubles

Table 5-2 shows the possible troubles for the zener regulator of Fig. 5-16. In working out the voltages, remember this: A shorted component is equivalent to a resistance of zero, and an open component is equivalent to a resistance of infinity. If you have trouble calculating with 0 and ∞, use 0.001 Ω and 1000 MΩ. In other words, use a very small resistance for a short and a very large resistance for an open.

In Fig. 5-16, the series resistor R_S may be shorted or open. Let us designate these troubles as R_{SS} and R_{SO}. Similarly, the zener diode may be shorted or open, symbolized by D_{1S} and D_{1O}. Also the load resistor may be shorted or open, R_{LS} and R_{LO}. Finally, the connecting wire between B and C may be open, designated BC_O.

Table 5–2	Zener Regulator Troubles and Symptoms			
Trouble	V_A, V	V_B, V	V_C, V	**Comments**
None	18	10	10	No trouble.
R_{SS}	18	18	18	D_1 and R_L may be open.
R_{SO}	18	0	0	
D_{1S}	18	0	0	R_S may be open.
D_{1O}	18	14.2	14.2	
R_{LS}	18	0	0	R_S may be open.
R_{LO}	18	10	10	
BC_O	18	10	0	
No supply	0	0	0	Check power supply.

In Table 5-2, the second row shows the voltages when the trouble is R_{SS}, a shorted series resistor. When the series resistor is shorted in Fig. 5-16, 18 V appears at B and C. This destroys the zener diode and possibly the load resistor. For this trouble, a voltmeter measures 18 V at points A, B, and C. This trouble and its voltages are shown in Table 5-2.

If the series resistor were open in Fig. 5-16, the voltage could not reach point B. In this case, B and C would have zero voltage, as shown in Table 5-2. Continuing like this, we can get the remaining entries of Table 5-2.

In Table 5-2, the comments indicate troubles that might occur as a direct result of the original short circuits. For instance, a shorted R_S will destroy the zener diode and may also open the load resistor. It depends on the power rating of the load resistor. A shorted R_S means that there's 18 V across 1 kΩ. This produces a power of 0.324 W. If the load resistor is rated at only 0.25 W, it may open.

Some of the troubles in Table 5-2 produce unique voltages, and others produce ambiguous voltages. For instance, the voltages for R_{SS}, D_{1O}, BC_O, and No supply are unique. If you measure these unique voltages, you can identify the trouble without breaking into the circuit to make an ohmmeter measurement.

On the other hand, all the other troubles in Table 5-2 produce ambiguous voltages. This means that two or more troubles can produce the same set of voltages. If you measure a set of ambiguous voltages, you will need to break into the circuit and measure the resistance of the suspected components. For instance, suppose you measure 18 V at A, 0 V at B, and 0 V at C. The troubles that can produce these voltages are R_{SO}, D_{1S}, and R_{LS}.

Zener diodes can be tested in a variety of ways. A DMM, set to the diode range, will enable the diode to be tested for being open or shorted. A normal reading will be approximately 0.7 V in the forward-biased direction and an open (overrange) indication in the reverse-biased direction. This test, though, will not indicate if the zener diode has the proper breakdown voltage V_Z.

A semiconductor curve tracer, shown in Fig. 5-17, will accurately display the zener's forward/reverse-biased characteristics. If a curve tracer is not available, a simple test is to measure the voltage drop across the zener diode while connected in a circuit. The voltage drop should be close to its rated value.

Figure 5-17 Curve Tracer.

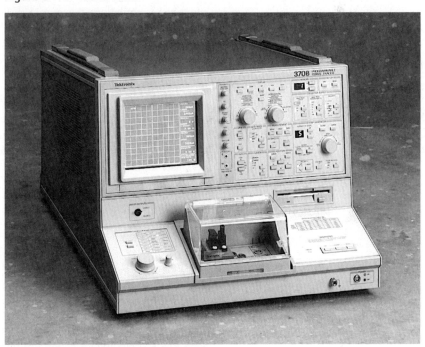

Special-Purpose Diodes

Figure 5-18 (*a*) Zener regulator circuit; (*b*) load lines.

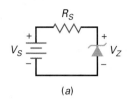

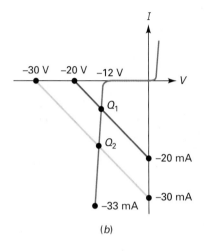

(a)

(b)

5-7 Load Lines

The current through the zener diode of Fig. 5-18*a* is given by

$$I_Z = \frac{V_S - V_Z}{R_S}$$

Suppose $V_S = 20$ V and $R_S = 1$ kΩ. Then the foregoing equation reduces to:

$$I_Z = \frac{20 - V_Z}{1000}$$

We get the saturation point (vertical intercept) by setting V_Z equal to zero and solving for I_Z to get 20 mA. Similarly, to get the cutoff point (horizontal intercept), we set I_Z equal to zero and solve for V_Z to get 20 V.

Alternatively, you can get the ends of the load line as follows. Visualize Fig. 5-18*a* with $V_S = 20$ V and $R_S = 1$ kΩ. With the zener diode shorted, the maximum diode current is 20 mA. With the diode open, the maximum diode voltage is 20 V.

Suppose the zener diode has a breakdown voltage of 12 V. Then its graph appears as shown in Fig. 5-18*b*. When we plot the load line for $V_S = 20$ V and $R_S = 1$ kΩ, we get the upper load line with an intersection point of Q_1. The voltage across the zener diode will be slightly more than the knee voltage at breakdown because the curve slopes slightly.

To understand how voltage regulation works, assume that the source voltage changes to 30 V. Then the zener current changes to:

$$I_Z = \frac{30 - V_Z}{1000}$$

This implies that the ends of the load line are 30 mA and 30 V, as shown in Fig. 5-18*b*. The new intersection is at Q_2. Compare Q_2 with Q_1, and you can see that there is more current through the zener diode, but approximately the same zener voltage. Therefore, even though the source voltage has changed from 20 to 30 V, the zener voltage is still approximately equal to 12 V. This is the basic idea of voltage regulation; the output voltage has remained almost constant even though the input voltage has changed by a large amount.

5-8 Optoelectronic Devices

Optoelectronics is the technology that combines optics and electronics. This field includes many devices based on the action of a *pn* junction. Examples of optoelectronic devices are **light-emitting diodes (LEDs)**, photodiodes, optocouplers, and laser diodes. Our discussion begins with the LED.

Light-Emitting Diode

Figure 5-19*a* shows a source connected to a resistor and an LED. The outward arrows symbolize the radiated light. In a forward-biased LED, free electrons cross the junction and fall into holes. As these electrons fall from a higher to a lower energy level, they radiate energy. In ordinary diodes, this energy is radiated in the form of heat. But in an LED, the energy is radiated as light. LEDs made from different elements have the ability to radiate energy across a wide wavelength spectrum. LEDs have replaced incandescent lamps in many applications because of their low voltage, long life, and fast on-off switching.

Figure 5-19 LED indicator. (*a*) Basic circuit; (*b*) practical circuit; (*c*) typical LEDs.

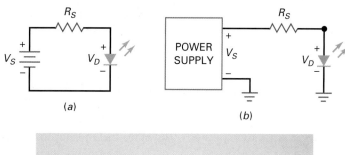

(a) (b)

Red Green

(c)

By using elements like gallium, arsenic, and phosphorus, a manufacturer can produce LEDs that radiate red, green, yellow, blue, orange, or infrared (invisible). LEDs that produce visible radiation are useful with instruments, calculators, and so on. The infrared LED finds applications in burglar alarm systems, remote controls, CD players, and other devices requiring invisible radiation.

LED Voltage and Current

The resistor of Fig. 5-19*b* is the usual current-limiting resistor that prevents the current from exceeding the maximum current rating of the diode. Since the resistor has a node voltage of V_S on the left and a node voltage of V_D on the right, the voltage across the resistor is the difference between the two voltages. With Ohm's law, the series current is:

$$I_S = \frac{V_S - V_D}{R_S} \qquad (5\text{-}13)$$

For most commercially available LEDs, the typical voltage drop is from 1.5 to 2.5 V for currents between 10 and 50 mA. The exact voltage drop depends on the LED current, color, tolerance, and so on. Unless otherwise specified, we will use a nominal drop of 2 V when troubleshooting or analyzing the LED circuits in this book. Fig. 5-19*c* shows typical LEDs.

LED Brightness

The brightness of an LED depends on the current. When V_S is much greater than V_D in Eq. (5-13), the brightness of the LED is approximately constant. For instance, a TIL222 is a green LED with a forward voltage of between 1.8 (minimum) and 3 V (maximum), for a current of 25 mA. If a circuit like Fig. 5-19*b* is mass-produced using a TIL222, the brightness of the LED will be almost constant if V_S is much greater than V_D. If V_S is only slightly more than V_D, the LED brightness will vary noticeably from one circuit to the next.

The best way to control the brightness is by driving the LED with a current source. This way, the brightness is constant because the current is constant. When we discuss transistors (they act like current sources), we will show how to use a transistor to drive an LED.

Breakdown Voltage

LEDs have very low breakdown voltages, typically between 3 and 5 V. Because of this, they are easily destroyed if reverse biased with too much voltage. When troubleshooting an LED circuit in which the LED will not light, check the polarity of the LED connection to make sure that it is forward biased.

An LED is often used to indicate the presence of power-line voltage into equipment. In this case, a rectifier diode may be used in parallel with the LED to prevent reverse-bias destruction of the LED. An example of using a rectifier diode to protect an LED is given later.

Seven-Segment Display

Figure 5-20a shows a **seven-segment display.** It contains seven rectangular LEDs (A through G). Each LED is called a *segment* because it forms part of the character being displayed. Figure 5-20b is a schematic diagram of the seven-segment display. External series resistors are included to limit the currents to safe levels. By grounding one or more resistors, we can form any digit from 0 through 9. For instance, by grounding A, B, and C, we get a 7. Grounding A, B, C, D, and G produces a 3.

A seven-segment display can also display capital letters A, C, E, and F, plus lowercase letters b and d. Microprocessor trainers often use seven-segment displays that show all digits from 0 through 9, plus A, b, C, d, E, and F.

The seven-segment indicator of Fig. 5-20b is referred to as the **common-anode** type because all anodes are connected together. Also available is the **common-cathode** type, in which all cathodes are connected together.

Photodiode

As previously discussed, one component of reverse current in a diode is the flow of minority carriers. These carriers exist because thermal energy keeps dislodging valence electrons from their orbits, producing free electrons and holes in the process. The lifetime of the minority carriers is short, but while they exist, they can contribute to the reverse current.

When light energy bombards a *pn* junction, it can dislodge valence electrons. The more light striking the junction, the larger the reverse current in a diode. A **photodiode** has been optimized for its sensitivity to light. In this diode, a window lets light pass through the package to the junction. The incoming light

Figure 5-20 Seven-segment indicator. (*a*) Physical layout of segments; (*b*) schematic diagram.

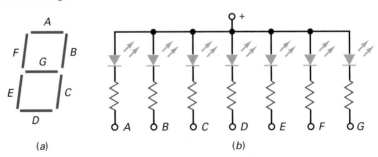

(a)   (b)

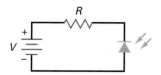

produces free electrons and holes. The stronger the light, the greater the number of minority carriers and the larger the reverse current.

Figure 5-21 shows the schematic symbol of a photodiode. The arrows represent the incoming light. Especially important, the source and the series resistor reverse bias the photodiode. As the light becomes brighter, the reverse current increases. With typical photodiodes, the reverse current is in the tens of microamperes.

Optocoupler

An optocoupler (also called an *optoisolator*) combines an LED and a photodiode in a single package. Figure 5-22 shows an optocoupler. It has an LED on the input side and a photodiode on the output side. The left source voltage and the series resistor set up a current through the LED. Then the light from the LED hits the photodiode, and this sets up a reverse current in the output circuit. This reverse current produces a voltage across the output resistor. The output voltage then equals the output supply voltage minus the voltage across the resistor.

When the input voltage is varying, the amount of light is fluctuating. This means that the output voltage is varying in step with the input voltage. This is why the combination of an LED and a photodiode is called an **optocoupler.** The device can couple an input signal to the output circuit. Other types of optocouplers use phototransistors, photothyristors, and other photo devices in their output circuit side. These devices will be discussed in later chapters.

The key advantage of an optocoupler is the electrical isolation between the input and output circuits. With an optocoupler, the only contact between the input and the output is a beam of light. Because of this, it is possible to have an insulation resistance between the two circuits in the thousands of megohms. Isolation like this is useful in high-voltage applications in which the potentials of the two circuits may differ by several thousand volts.

Laser Diode

In an LED, free electrons radiate light when falling from higher energy levels to lower ones. The free electrons fall randomly and continuously, resulting in light waves that have every phase between 0 and 360°. Light that has many different phases is called *noncoherent light.* An LED produces noncoherent light.

A **laser diode** is different. It produces a *coherent light.* This means that all the light waves are *in phase with each other.* The basic idea of a laser diode is to use a mirrored resonant chamber that reinforces the emission of light waves at a single frequency of the same phase. Because of the resonance, a laser diode produces a narrow beam of light that is very intense, focused, and pure.

Laser diodes are also known as *semiconductor lasers.* These diodes can produce visible light (red, green, or blue) and invisible light (infrared). Laser diodes are used in a large variety of applications. They are used in telecommunications, data communications, broadband access, industrial, aerospace, test and measurement, medical and defense industries. They are also used in laser printers and consumer products requiring large-capacity optical disk systems, such as

Figure 5-22 Optocoupler combines an LED and a photodiode.

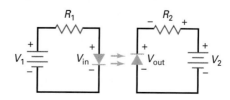

compact disk (CD) and digital video disk (DVD) players. In broadband communication, they are used with fiber-optic cables to increase the speed of the Internet.

A *fiber-optic cable* is analogous to a stranded wire cable, except that the strands are thin flexible fibers of glass or plastic that transmit light beams instead of free electrons. The advantage is that much more information can be sent through a fiber-optic cable than through a copper cable.

New applications are being found as the lasing wavelength is pushed lower into the visible spectrum with visible laser diodes (VLDs). Also, near-infrared diodes are being used in machine vision systems, sensors, and security systems.

Example 5-12

Figure 5-23*a* shows a voltage-polarity tester. It can be used to test a dc voltage of unknown polarity. When the dc voltage is positive, the green LED lights up. When the dc voltage is negative, the red LED lights up. What is the approximate LED current if the dc input voltage is 50 V and the series resistance is 2.2 kΩ?

Figure 5-23   (*a*) Polarity indicator; (*b*) continuity tester.

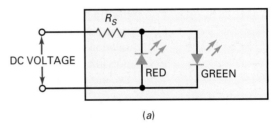

(*a*)

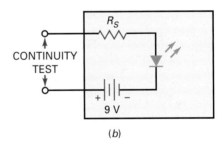

(*b*)

SOLUTION   We will use a forward voltage of approximately 2 V for either LED. With Eq. (5-13):

$$I_S = \frac{50\text{ V} - 2\text{ V}}{2.2\text{ k}\Omega} = 21.8\text{ mA}$$

Example 5-13   ‖‖‖ MultiSim

Figure 5-23*b* is a continuity tester. After you turn off all the power in a circuit under test, you can use this circuit to check for the continuity of cables, connectors, and switches. How much LED current is there if the series resistance is 470 Ω?

SOLUTION When the input terminals are shorted (continuity), the internal 9-V battery produces an LED current of:

$$I_S = \frac{9\text{ V} - 2\text{ V}}{470\ \Omega} = 14.9\text{ mA}$$

PRACTICE PROBLEM 5-13 Using Fig. 5-23, what value series resistor should be used to produce 21 mA of LED current?

Example 5-14

LEDs are often used to indicate the existence of ac voltages. Figure 5-24 shows an ac voltage source driving an LED indicator. When there is ac voltage, there is LED current on the positive half cycles. On the negative half cycles, the rectifier diode turns on and protects the LED from too much reverse voltage. If the ac source voltage is 20 V rms and the series resistance is 680 Ω, what is the average LED current? Also, calculate the approximate power dissipation in the series resistor.

Figure 5-24 Low ac voltage indicator.

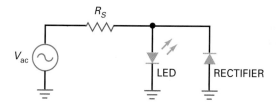

SOLUTION The LED current is a rectified half-wave signal. The peak source voltage is 1.414 × 20 V, which is approximately 28 V. Ignoring the LED voltage drop, the approximate peak current is:

$$I_S = \frac{28\text{ V}}{680\ \Omega} = 41.2\text{ mA}$$

The average of the half-wave current through the LED is:

$$I_S = \frac{41.2\text{ mA}}{\pi} = 13.1\text{ mA}$$

Ignore the diode drops in Fig. 5-24; this is equivalent to saying that there is a short to ground on the right end of the series resistor. Then the power dissipation in the series resistor equals the square of the source voltage divided by the resistance:

$$P = \frac{(20\text{ V})^2}{680\ \Omega} = 0.588\text{ W}$$

As the source voltage in Fig. 5-24 increases, the power dissipation in the series resistor may increase to several watts. This is a disadvantage because a high-wattage resistor is too bulky and wasteful for most applications.

PRACTICE PROBLEM 5-14 If the ac input voltage of Fig. 5-24 is 120 V and the series resistance is 2 kΩ, find the average LED current and approximate series resistor power dissipation.

Example 5-15

The circuit of Fig. 5-25 shows an LED indicator for the ac power line. The idea is basically the same as in Fig. 5-24, except that we use a capacitor instead of a resistor. If the capacitance is 0.68 μF, what is the average LED current?

Figure 5-25   High ac voltage indicator.

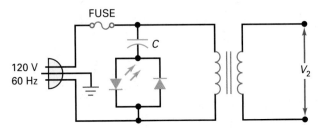

SOLUTION   Calculate the capacitive reactance:

$$X_C = \frac{1}{2\pi fC} = \frac{1}{2\pi(60\ \text{Hz})(0.68\ \mu\text{F})} = 3.9\ \text{k}\Omega$$

Ignoring the LED voltage drop, the approximate peak LED current is:

$$I_S = \frac{170\ \text{V}}{3.9\ \text{k}\Omega} = 43.6\ \text{mA}$$

The average LED current is:

$$I_S = \frac{43.6\ \text{mA}}{\pi} = 13.9\ \text{mA}$$

What advantage does a series capacitor have over a series resistor? Since the voltage and current in a capacitor are 90° out of phase, there is no power dissipation in the capacitor. If a 3.9-kΩ resistor were used instead of a capacitor, it would have a power dissipation of approximately 3.69 W. Most designers would prefer to use a capacitor, since it's smaller and ideally produces no heat.

Example 5-16

What does the circuit of Fig. 5-26 do?

Figure 5-26   Blown-fuse indicator.

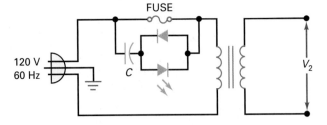

SOLUTION   This is a *blown-fuse indicator.* If the fuse is OK, the LED is off because there is approximately zero voltage across the LED indicator. On the other hand, if the fuse is open, some of the line voltage appears across the LED indicator and the LED lights up.

5-9 The Schottky Diode

As frequency increases, the action of small-signal rectifier diodes begins to deteriorate. They are no longer able to switch off fast enough to produce a well-defined half-wave signal. The solution to this problem is the *Schottky diode*. Before describing this special-purpose diode, let us look at the problem that arises with ordinary small-signal diodes.

Charge Storage

Figure 5-27*a* shows a small-signal diode, and Fig. 5-27*b* illustrates its energy bands. As you can see, conduction-band electrons have diffused across the junction and traveled into the *p* region before recombining (path *A*). Similarly, holes have crossed the junction and traveled into the *n* region before recombination occurs (path *B*). The greater the lifetime, the farther the charges can travel before recombination occurs.

For instance, if the lifetime equals 1 μs, free electrons and holes exist for an average of 1 μs before recombination takes place. This allows the free electrons to penetrate deeply into the *p* region, where they remain temporarily stored at the higher energy band. Similarly, the holes penetrate deeply into the *n* region, where they are temporarily stored in the lower energy band.

The greater the forward current, the larger the number of charges that have crossed the junction. The greater the lifetime, the deeper the penetration of these charges and the longer the charges remain in the high and low energy bands. The temporary storage of free electrons in the upper energy band and holes in the lower energy band is referred to as *charge storage*.

Charge Storage Produces Reverse Current

When you try to switch a diode from on to off, charge storage creates a problem. Why? Because if you suddenly reverse-bias a diode, the stored charges will flow in the reverse direction for a while. The greater the lifetime, the longer these charges can contribute to reverse current.

For example, suppose a forward-biased diode is suddenly reverse-biased, as shown in Fig. 5-28*a*. Then a large reverse current can exist for a while because of the flow of stored charges in Fig. 5-28*b*. Until the stored charges either cross the junction or recombine, the reverse current will continue.

Figure 5-27 Charge storage. (*a*) Forward bias creates stored charges; (*b*) stored charges in high- and low-energy bands.

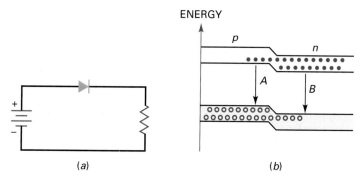

(*a*)   (*b*)

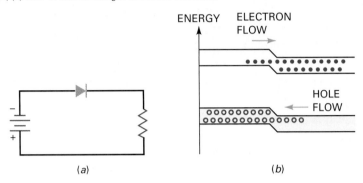

(a)   (b)

Reverse Recovery Time

The time it takes to turn off a forward-biased diode is called the *reverse recovery time t_{rr}*. The conditions for measuring t_{rr} vary from one manufacturer to the next. As a guide, t_{rr} is the time it takes for the reverse current to drop to 10 percent of the forward current.

For instance, the 1N4148 has a t_{rr} of 4 ns. If this diode has a forward current of 10 mA and it is suddenly reverse-biased, it will take approximately 4 ns for the reverse current to decrease to 1 mA. Reverse recovery time is so short in small-signal diodes that you don't even notice its effect at frequencies below 10 MHz or so. It's only when you get well above 10 MHz that you have to take t_{rr} into account.

Poor Rectification at High Frequencies

What effect does reverse recovery time have on rectification? Take a look at the half-wave rectifier shown in Fig. 5-29*a*. At low frequencies the output is a half-wave rectified signal. As the frequency increases well into megahertz, however, the output signal begins to deviate from the half-wave shape, as shown in Fig. 5-29*b*. Some reverse conduction (called *tails*) is noticeable near the beginning of the reverse half cycle.

The problem is that the reverse recovery time has become a significant part of the period, allowing conduction during the early part of the negative half cycle. For instance, if $t_{rr} = 4$ ns and the period is 50 ns, the early part of the reverse half cycle will have tails similar to those shown in Fig. 5-29*b*. As the frequency continues to increase, the rectifier becomes useless.

Figure 5-29 Stored charges degrade rectifier behavior at high frequencies. (*a*) Rectifier circuit with ordinary small-signal diode; (*b*) tails appear on negative half cycles at higher frequencies.

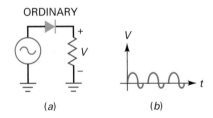

(a)   (b)

Eliminating Charge Storage

The solution to the problem of tails is a special-purpose device called a **Schottky diode.** This kind of diode uses a metal such as gold, silver, or platinum on one side of the junction and doped silicon (typically n-type) on the other side. Because of the metal on one side of the junction, the Schottky diode has no depletion layer. The lack of a depletion layer means that there are *no stored charges at the junction.*

When a Schottky diode is unbiased, free electrons on the n side are in smaller orbits than are the free electrons on the metal side. This difference in orbit size is called the *Schottky barrier,* approximately 0.25 V. When the diode is forward biased, free electrons on the n side can gain enough energy to travel in larger orbits. Because of this, free electrons can cross the junction and enter the metal, producing a large forward current. Since the metal has no holes, there is no charge storage and no reverse recovery time.

Hot-Carrier Diode

The Schottky diode is sometimes called a *hot-carrier diode.* This name came about as follows. Forward bias increases the energy of the electrons on the n side to a higher level than that of the electrons on the metal side of the junction. This increase in energy inspired the name *hot carrier* for the n-side electrons. As soon as these high-energy electrons cross the junction, they fall into the metal, which has a lower-energy conduction band.

High-Speed Turnoff

The lack of charge storage means that the Schottky diode can switch off faster than an ordinary diode can. In fact, a Schottky diode can easily rectify frequencies above 300 MHz. When it is used in a circuit like Fig. 5-30a, the Schottky diode produces a perfect half-wave signal like Fig. 5-30b even at frequencies above 300 MHz.

Figure 5-30a shows the schematic symbol of a Schottky diode. Notice the cathode side. The lines look like a rectangular S, which stands for *Schottky.* This is how you can remember the schematic symbol.

Applications

The most important application of Schottky diodes is in digital computers. The speed of computers depends on how fast their diodes and transistors can turn on and off. This is where the Schottky diode comes in. Because it has no charge storage, the Schottky diode has become the backbone of low-power Schottky TTLs, a group of widely used digital devices.

A final point. Since a Schottky diode has a barrier potential of only 0.25 V, you may occasionally see it used in low-voltage bridge rectifiers because you subtract only 0.25 V instead of the usual 0.7 V for each diode when using the second approximation. In a low-voltage supply, this lower diode voltage drop is an advantage.

Figure 5-30 Schottky diodes eliminate tails at high frequencies. (*a*) Circuit with Schottky diode; (*b*) half-wave signal at 300 MHz.

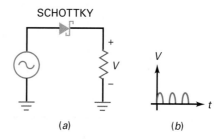

Figure 5-31 Varactor. (*a*) Doped regions are like capacitor plates separated by a dielectric; (*b*) ac equivalent circuit; (*c*) schematic symbol; (*d*) graph of capacitance versus reverse voltage.

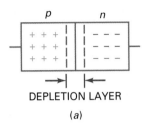

p   *n*

DEPLETION LAYER

(*a*)

C_T

(*b*)

(*c*)

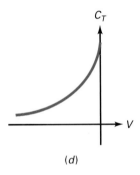

C_T

V

(*d*)

5-10 The Varactor

The **varactor** (also called the *voltage-variable capacitance, varicap, epicap,* and *tuning diode*) is widely used in television receivers, FM receivers, and other communications equipment because it can be used for electronic tuning.

Basic Idea

In Fig. 5-31*a*, the depletion layer is between the *p* region and the *n* region. The *p* and *n* regions are like the plates of a capacitor, and the depletion layer is like the dielectric. When a diode is reverse-biased, the width of the depletion layer increases with the reverse voltage. Since the depletion layer gets wider with more reverse voltage, the capacitance becomes smaller. It's as though you moved apart the plates of a capacitor. The key idea is that capacitance is controlled by reverse voltage.

Equivalent Circuit and Symbol

Figure 5-31*b* shows the ac equivalent circuit for a reverse-biased diode. In other words, as far as an ac signal is concerned, the varactor acts the same as a variable capacitance. Figure 5-31*c* shows the schematic symbol for a varactor. The inclusion of a capacitor in series with the diode is a reminder that a varactor is a device that has been optimized for its variable-capacitance properties.

Capacitance Decreases at Higher Reverse Voltages

Figure 5-31*d* shows how the capacitance varies with reverse voltage. This graph shows that the capacitance gets smaller when the reverse voltage gets larger. The really important idea here is that reverse dc voltage controls capacitance.

How is a varactor used? It is connected in parallel with an inductor to form a parallel resonant circuit. This circuit has only one frequency at which maximum impedance occurs. This frequency is called the *resonant frequency*. If the dc reverse voltage to the varactor is changed, the resonant frequency is also changed. This is the principle behind electronic tuning of a radio station, a TV channel, and so on.

Varactor Characteristics

Because the capacitance is voltage-controlled, varactors have replaced mechanically tuned capacitors in many applications such as television receivers and automobile radios. Data sheets for varactors list a reference value of capacitance measured at a specific reverse voltage, typically -3 V to -4 V. Figure 5-32 shows a partial data sheet for a MV209 varactor diode. It lists a reference capacitance C_t of 29 pF at -3 V.

In addition to providing the reference value of capacitance, data sheets normally list a capacitance ratio, C_R, or tuning range associated with a voltage range. For example, along with the reference value of 29 pF, the data sheet of a MV209 shows a minimum capacitance ratio of 5:1 for a voltage range of -3 V to -25 V. This means that the capacitance, or tuning range, decreases from 29 to 6 pF when the voltage varies from -3 V to -25 V.

The tuning range of a varactor depends on the doping level. For instance, Fig. 5-33*a* shows the doping profile for an abrupt-junction diode (the ordinary type of diode). The profile shows that the doping is uniform on both sides of the junction. The tuning range of an abrupt-junction diode is between 3:1 and 4:1.

Device	C_t, Diode Capacitance V_R = 3.0 Vdc, f = 1.0 MHz pF			Q, Figure of Merit V_R = 3.0 Vdc f = 50 MHz	C_R, Capacitance Ratio C_3/C_{25} f = 1.0 MHz (Note 1)	
	Min	Nom	Max	Min	Min	Max
MMBV109LT1, MV209	26	29	32	200	5.0	6.5

1. C_R is the ratio of C_t measured at 3 Vdc divided by C_t measured at 25 Vdc.

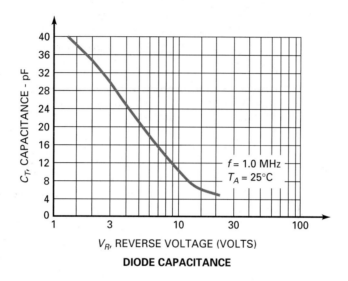

DIODE CAPACITANCE

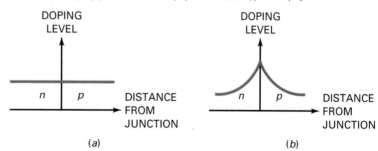

Figure 5–33 Doping profiles. (*a*) Abrupt junction; (*b*) hyperabrupt junction.

To get larger tuning ranges, some varactors have a *hyperabrupt junction,* one whose doping profile looks like Fig. 5-33*b*. This profile tells us that the doping level increases as we approach the junction. The heavier doping produces a narrower depletion layer and a larger capacitance. Furthermore, changes in reverse voltage have more pronounced effects on capacitance. A hyperabrupt varactor has a tuning range of about 10:1, enough to tune an AM radio through its frequency range of 535 to 1605 kHz. (Note: You need a 10:1 range because the resonant frequency is inversely proportional to the square root of capacitance.)

Example 5-17

What does the circuit of Fig. 5-34a do?

SOLUTION   As mentioned in Chap. 1, a transistor is a semiconductor device that acts like a current source. In Fig. 5-34a, the transistor pumps a fixed number of milliamperes into the resonant LC tank circuit. A negative dc voltage reverse-biases the varactor. By varying this dc control voltage, we can vary the resonant frequency of the LC circuit.

As far as the ac signal is concerned, we can use the equivalent circuit shown in Fig. 5-34b. The coupling capacitor acts like a short circuit. An ac current source drives a resonant LC tank circuit. The varactor acts like variable capacitance, which means that we can change the resonant frequency by changing the dc control voltage. This is the basic idea behind the tuning of radio and television receivers.

Figure 5-34   Varactors can tune resonant circuits. (*a*) Transistor (current source) drives tuned LC tank; (*b*) ac equivalent circuit.

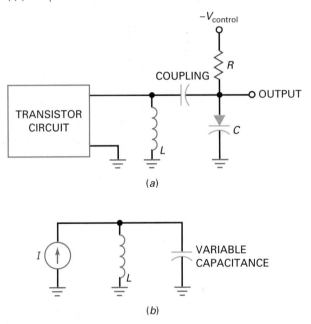

5-11 Other Diodes

Besides the special-purpose diodes discussed so far, there are others you should know about. Because they are so specialized, only a brief description follows.

Varistors

Lightning, power-line faults, and transients can pollute the ac line voltage by superimposing dips and spikes on the normal 120 V rms. *Dips* are severe voltage drops lasting microseconds or less. *Spikes* are very brief overvoltages up to 2000 V

Figure 5-35 (*a*) Varistor protects primary from ac line transients; (*b*) current-regulator diode.

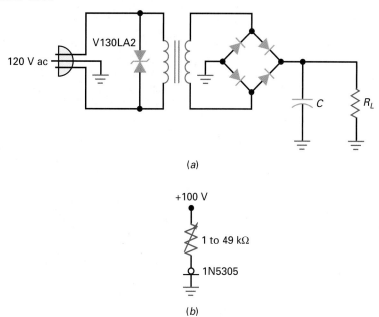

(a)

(b)

or more. In some equipment, filters are used between the power line and the primary of the transformer to eliminate the problems caused by ac line transients.

One of the devices used for line filtering is the **varistor** (also called a *transient suppressor*). This semiconductor device is like two back-to-back zener diodes with a high breakdown voltage in both directions. Varistors are commercially available with breakdown voltages from 10 to 1000 V. They can handle peak transient currents in the hundreds or thousands of amperes.

For instance, a V130LA2 is a varistor with a breakdown voltage of 184 V (equivalent to 130 V rms) and a peak current rating of 400 A. Connect one of these across the primary winding as shown in Fig. 5-35*a*, and you don't have to worry about spikes. The varistor will clip all spikes at the 184-V level and protect your power supply.

Current-Regulator Diodes

These are diodes that work in a way exactly opposite to the way zener diodes work. Instead of holding the voltage constant, these diodes hold the current constant. Known as **current-regulator diodes** (or *constant-current diodes*), these devices keep the current through them fixed when the voltage changes. For example, the 1N5305 is a constant-current diode with a typical current of 2 mA over a voltage range of 2 to 100 V. Figure 5-35*b* shows the schematic symbol of a current-regulator diode. In Fig. 5-35*b*, the diode will hold the load current constant at 2 mA even though the load resistance is varied from 1 to 49 kΩ.

Step-Recovery Diodes

The **step-recovery diode** has the unusual doping profile shown in Fig. 5-36*a*. This graph indicates that the density of carriers decreases near the junction. This unusual distribution of carriers causes a phenomenon called *reverse snap-off*.

Figure 5-36*b* shows the schematic symbol for a step-recovery diode. During the positive half cycle, the diode conducts like any silicon diode. But during the negative half cycle, reverse current exists for a while because of the stored charges, and then suddenly drops to zero.

Figure 5-36 Step-recovery diode. (*a*) Doping profile shows less doping near junction; (*b*) circuit rectifying an ac input signal; (*c*) snap-off produces a positive voltage step rich in harmonics.

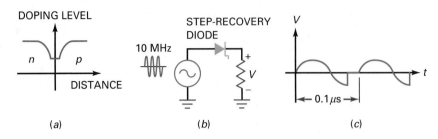

(*a*) (*b*) (*c*)

Figure 5-36*c* shows the output voltage. It's as though the diode conducts reverse current for a while, and then suddenly snaps open. This is why the step-recovery diode is also known as a *snap diode*. The sudden step in current is rich in harmonics and can be filtered to produce a sine wave of a higher frequency. (*Harmonics* are multiples of the input frequency like $2f_{in}$, $3f_{in}$, and $4f_{in}$.) Because of this, step-recovery diodes are useful in frequency multipliers, circuits whose output frequency is a multiple of the input frequency.

Back Diodes

Zener diodes normally have breakdown voltages greater than 2 V. By increasing the doping level, we can get the zener effect to occur near zero. Forward conduction still occurs around 0.7 V, but now reverse conduction (breakdown) starts at approximately −0.1 V.

A diode with a graph like Fig. 5-37*a* is called a **back diode** because it conducts better in the reverse than in the forward direction. Figure 5-34*b* shows a sine wave with a peak of 0.5 V driving a back diode and a load resistor. (Notice that the zener symbol is used for the back diode.) The 0.5 V is not enough to turn on the diode in the forward direction, but it is enough to break down the diode in the reverse direction. For this reason, the output is a half-wave signal with a peak of 0.4 V, as shown in Fig. 5-37*b*.

Back diodes are occasionally used to rectify weak signals with peak amplitudes between 0.1 and 0.7 V.

Tunnel Diodes

By increasing the doping level of a back diode, we can get breakdown to occur at 0 V. Furthermore, the heavier doping distorts the forward curve, as shown in Fig. 5-38*a*. A diode with this graph is called a **tunnel diode.**

Figure 5-38*b* shows the schematic symbol for a tunnel diode. This type of diode exhibits a phenomenon known as **negative resistance.** This means that

Figure 5-37 Back diode. (*a*) Breakdown occurs at −0.1 V; (*b*) circuit rectifying weak ac signal.

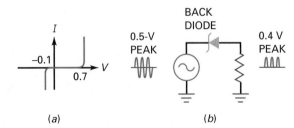

(*a*) (*b*)

Chapter 5

Figure 5-38 Tunnel diode. (*a*) Breakdown occurs at 0 V; (*b*) schematic symbol.

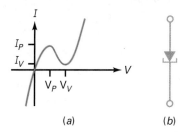

an increase in forward voltage produces a decrease in forward current, at least over the part of the graph between V_P and V_V. The negative resistance of tunnel diodes is useful in high-frequency circuits called *oscillators*. These circuits are able to generate a sinusoidal signal, similar to that produced by an ac generator. But unlike the ac generator that converts mechanical energy to a sinusoidal signal, an oscillator converts dc energy to a sinusoidal signal. Later chapters will show you how to build oscillators.

PIN Diodes

A **PIN diode** is a semiconductor device that operates as a variable resistor at RF and microwave frequencies. Figure 5-39*a* shows its construction. It consists of an intrinsic (pure) semiconductor material sandwiched between p-type and n-type materials. Figure 5-39*b* shows the schematic symbol for the PIN diode.

When the diode is forward biased, it acts like a current-controlled resistance. Figure 5-39*c* shows how the PIN diode's series resistance R_S decreases as its forward current increases. When reverse biased, the PIN diode acts like a fixed capacitor. The PIN diode is widely used in RF and microwave modulator circuits.

Table of Devices

Summary Table 5-1 lists all the special-purpose devices in this chapter. The zener diode is useful in voltage regulators, the LED as a dc or an ac indicator, the seven-segment indicator in measuring instruments, and so on. You should study the table and remember the ideas it contains.

Figure 5-39 PIN diode. (*a*) Construction; (*b*) schematic symbol; (*c*) series resistance.

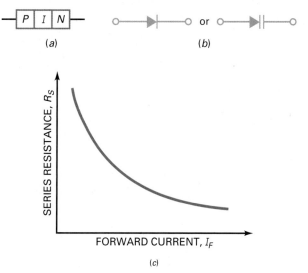

Summary Table 5-1 | Special-Purpose Devices

Device	Key Idea	Application
Zener diode	Operates in breakdown region	Voltage regulators
LED	Emits noncoherent light	DC or ac indicators
Seven-segment indicator	Can display numbers	Measuring instruments
Photodiode	Light produces minority carriers	Light detectors
Optocoupler	Combines LED and photodiode	Input/output isolators
Laser diode	Emits coherent light	CD/DVD players, broadband communications
Schottky diode	Has no charge storage	High-frequency rectifiers (300 MHz)
Varactor	Acts like variable capacitance	TV and receiver tuners
Varistor	Breaks down both ways	Line-spike protectors
Current-regulator diode	Holds current constant	Current regulators
Step-recovery diode	Snaps off during reverse conduction	Frequency multipliers
Back diode	Conducts better in reverse	Weak-signal rectifiers
Tunnel diode	Has a negative-resistance region	High-frequency oscillators
PIN diode	Controlled resistance	Microwave communications

Summary

SEC. 5-1 THE ZENER DIODE

This is a special diode optimized for operation in the breakdown region. Its main use is in voltage regulators—circuits that hold the load voltage constant. Ideally, a reverse-biased zener diode is like a perfect battery. To a second approximation, it has a bulk resistance that produces a small additional voltage.

SEC. 5-2 THE LOADED ZENER REGULATOR

When a zener diode is in parallel with a load resistor, the current through the current-limiting resistor equals the sum of the zener current and the load current. The process for analyzing a zener regulator consists of finding the series current, load current, and zener current (in that order).

SEC. 5-3 SECOND APPROXIMATION OF A ZENER DIODE

In the second approximation, we visualize a zener diode as a battery of V_Z and a series resistance of R_Z. The current through R_Z produces an additional voltage across the diode, but this voltage is usually small. You need zener resistance in order to calculate ripple reduction.

SEC. 5-4 ZENER DROP-OUT POINT

A zener regulator will fail to regulate if the zener diode comes out of breakdown. The worst-case conditions occur for minimum source voltage, maximum series resistance, and minimum load resistance. For the zener regulator to work properly under all operating conditions, there must be zener current under the worst-case conditions.

SEC. 5–5 READING A DATA SHEET

The most important quantities on the data sheet of zener diodes are the zener voltage, the maximum power rating, the maximum current rating, and the tolerance. Designers also need the zener resistance, the derating factor, and a few other items.

SEC. 5–6 TROUBLESHOOTING

Troubleshooting is an art and a science. Because of this, you can learn only so much from a book. The rest has to be learned from direct experience with circuits in trouble. Because troubleshooting is an art, you have to ask "What if?" often and feel your way to a solution.

SEC. 5–7 LOAD LINES

The intersection of the load line and the zener diode graph is the Q point. When the source voltage changes, a different load line appears with a different Q point. Although the two Q points may have

different currents, the voltages are almost identical. This is a visual demonstration of voltage regulation.

SEC. 5–8 OPTOELECTRONIC DEVICES

The LED is widely used as an indicator on instruments, calculators, and other electronic equipment. By combining seven LEDs in a package, we get a seven-segment indicator. Another important optoelectronic device is the optocoupler, which allows us to couple a signal between two isolated circuits.

SEC. 5–9 THE SCHOTTKY DIODE

The reverse recovery time is the time it takes a diode to shut off after it is suddenly switched from forward to reverse bias. This time may be only a few nanoseconds, but it places a limit on how high the frequency can be in a rectifier circuit. The Schottky diode is a special diode with almost zero reverse recovery time. Because of this, the Schottky diode

is useful at high frequencies where short switching times are needed.

SEC. 5–10 THE VARACTOR

The width of the depletion layer increases with the reverse voltage. This is why the capacitance of a varactor can be controlled by the reverse voltage. A common application is remote tuning of radio and television sets.

SEC. 5–11 OTHER DIODES

Varistors are useful as transient suppressors. Constant-current diodes hold the current, rather than the voltage, constant. Step-recovery diodes snap off and produce a step voltage that is rich in harmonics. Back diodes conduct better in the reverse direction than in the forward direction. Tunnel diodes exhibit negative resistance, which can be used in high-frequency oscillators. PIN diodes use a forward-biased control current to change its resistance in RF and microwave communication circuits.

Derivations

(5-3)   Series current:

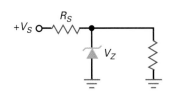

$$I_S = \frac{V_S - V_Z}{R_S}$$

(5-4)   Load voltage:

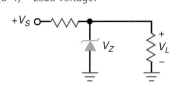

$$V_L = V_Z$$

(5-5)   Load current:

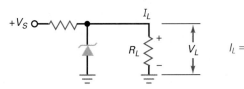

$$I_L = \frac{V_L}{R_L}$$

(5-6)   Zener current:

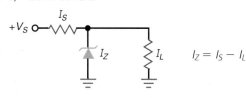

$$I_Z = I_S - I_L$$

(5-7)   Change in load voltage:

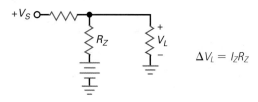

$$\Delta V_L = I_Z R_Z$$

(5-8)   Output ripple:

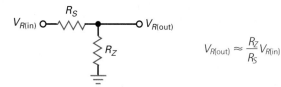

$$V_{R(out)} \approx \frac{R_Z}{R_S} V_{R(in)}$$

(5-9)   Maximum series resistance:

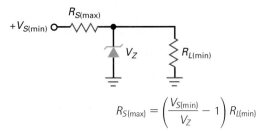

$$R_{S(max)} = \left(\frac{V_{S(min)}}{V_Z} - 1 \right) R_{L(min)}$$

(5-10)   Maximum series resistance:

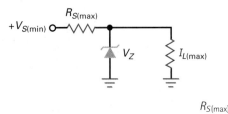

$$R_{S(max)} = \frac{V_{S(min)} - V_Z}{I_{L(max)}}$$

(5-13)   LED current:

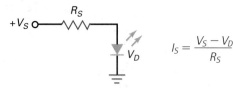

$$I_S = \frac{V_S - V_D}{R_S}$$

Student Assignments

1. **What is true about the breakdown voltage in a zener diode?**
   a. It decreases when current increases.
   b. It destroys the diode.
   c. It equals the current times the resistance.
   d. It is approximately constant.

2. **Which of these is the best description of a zener diode?**
   a. It is a rectifier diode.
   b. It is a constant-voltage device.
   c. It is a constant-current device.
   d. It works in the forward region.

3. **A zener diode**
   a. Is a battery
   b. Has a constant voltage in the breakdown region
   c. Has a barrier potential of 1 V
   d. Is forward biased

4. **The voltage across the zener resistance is usually**
   a. Small
   b. Large
   c. Measured in volts
   d. Subtracted from the breakdown voltage

5. **If the series resistance increases in an unloaded zener regulator, the zener current**
   a. Decreases
   b. Stays the same
   c. Increases
   d. Equals the voltage divided by the resistance

6. **In the second approximation, the total voltage across the zener diode is the sum of the breakdown voltage and the voltage across the**
   a. Source
   b. Series resistor
   c. Zener resistance
   d. Zener diode

7. **The load voltage is approximately constant when a zener diode is**
   a. Forward biased
   b. Reverse biased
   c. Operating in the breakdown region
   d. Unbiased

8. **In a loaded zener regulator, which is the largest current?**
   a. Series current
   b. Zener current

   c. Load current
   d. None of these

9. **If the load resistance increases in a zener regulator, the zener current**
   a. Decreases
   b. Stays the same
   c. Increases
   d. Equals the source voltage divided by the series resistance

10. **If the load resistance decreases in a zener regulator, the series current**
    a. Decreases
    b. Stays the same
    c. Increases
    d. Equals the source voltage divided by the series resistance

11. **When the source voltage increases in a zener regulator, which of these currents remains approximately constant?**
    a. Series current
    b. Zener current
    c. Load current
    d. Total current

12. If the zener diode in a zener regulator is connected with the wrong polarity, the load voltage will be closest to
    a. 0.7 V
    b. 10 V
    c. 14 V
    d. 18 V

13. When a zener diode is operating above its power-rated temperature
    a. It will immediately be destroyed
    b. You must decrease its power rating
    c. You must increase its power rating
    d. It will not be affected

14. Which of the following will not indicate a zener diode's breakdown voltage?
    a. In-circuit voltage drop
    b. Curve tracer
    c. Reverse-bias test circuit
    d. DMM

15. At high frequencies, ordinary diodes don't work properly because of
    a. Forward bias
    b. Reverse bias
    c. Breakdown
    d. Charge storage

16. The capacitance of a varactor diode increases when the reverse voltage across it
    a. Decreases
    b. Increases
    c. Breaks down
    d. Stores charges

17. Breakdown does not destroy a zener diode, provided the zener current is less than the
    a. Breakdown voltage
    b. Zener test current
    c. Maximum zener current rating
    d. Barrier potential

18. As compared to a silicon rectifier diode, an LED has a
    a. Lower forward voltage and lower breakdown voltage
    b. Lower forward voltage and higher breakdown voltage
    c. Higher forward voltage and lower breakdown voltage
    d. Higher forward voltage and higher breakdown voltage

19. To display the digit 0 in a seven-segment indicator,
    a. C must be off
    b. G must be off
    c. F must be on
    d. All segments must be lighted

20. A photodiode is normally
    a. Forward biased
    b. Reverse biased
    c. Neither forward nor reverse biased
    d. Emitting light

21. When the light decreases, the reverse minority-carrier current in a photodiode
    a. Decreases
    b. Increases
    c. Is unaffected
    d. Reverses direction

22. The device associated with voltage-controlled capacitance is a
    a. Light-emitting diode
    b. Photodiode
    c. Varactor diode
    d. Zener diode

23. If the depletion layer width decreases, the capacitance
    a. Decreases
    b. Stays the same
    c. Increases
    d. Is variable

24. When the reverse voltage decreases, the capacitance
    a. Decreases
    b. Stays the same
    c. Increases
    d. Has more bandwidth

25. The varactor is usually
    a. Forward biased
    b. Reverse biased
    c. Unbiased
    d. Operated in the breakdown region

26. The device to use for rectifying a weak ac signal is a
    a. Zener diode
    b. Light-emitting diode
    c. Varistor
    d. Back diode

27. Which of the following has a negative-resistance region?
    a. Tunnel diode
    b. Step-recovery diode
    c. Schottky diode
    d. Optocoupler

28. A blown-fuse indicator uses a
    a. Zener diode
    b. Constant-current diode
    c. Light-emitting diode
    d. PIN diode

29. To isolate an output circuit from an input circuit, which is the device to use?
    a. Back diode
    b. Optocoupler
    c. Seven-segment indicator
    d. Tunnel diode

30. The diode with a forward voltage drop of approximately 0.25 V is the
    a. Step-recovery diode
    b. Schottky diode
    c. Back diode
    d. Constant-current diode

31. For typical operation, you need to use reverse bias with a
    a. Zener diode
    b. Photodiode
    c. Varactor
    d. All of the above

32. As the forward current through a PIN diode decreases, its resistance
    a. Increases
    b. Decreases
    c. Remains constant
    d. Cannot be determined

Problems

SEC. 5-1 THE ZENER DIODE

5-1 |||| **MultiSim** An unloaded zener regulator has a source voltage of 24 V, a series resistance of 470 Ω, and a zener voltage of 15 V. What is the zener current?

5-2 If the source voltage in Prob. 5-1 varies from 24 to 40 V, what is the maximum zener current?

5-3 If the series resistor of Prob. 5-1 has a tolerance of ±5 percent, what is the maximum zener current?

SEC. 5-2 THE LOADED ZENER REGULATOR

5-4 |||| **MultiSim** If the zener diode is disconnected in Fig. 5-40, what is the load voltage?

Figure 5-40

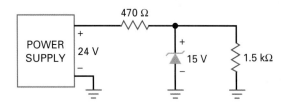

5-5 |||| **MultiSim** Calculate all three currents in Fig. 5-40.

5-6 Assuming a tolerance of ±5 percent in both resistors of Fig. 5-40, what is the maximum zener current?

5-7 Suppose the supply voltage of Fig. 5-40 can vary from 24 to 40 V. What is the maximum zener current?

5-8 The zener diode of Fig. 5-40 is replaced with a 1N963B. What are the load voltage and the zener current?

5-9 Draw the schematic diagram of a zener regulator with a supply voltage of 20 V, a series resistance of 330 Ω, a zener voltage of 12 V, and a load resistance of 1 kΩ. What are the load voltage and the zener current?

SEC. 5-3 SECOND APPROXIMATION OF A ZENER DIODE

5-10 The zener diode of Fig. 5-40 has a zener resistance of 14 Ω. If the power supply has a ripple of 1 V pp, what is the ripple across the load resistor?

5-11 During the day, the ac line voltage changes. This causes the unregulated 24-V output of the power supply to vary from 21.5 to 25 V. If the zener resistance is 14 Ω, what is the voltage change over the foregoing range?

SEC. 5-4 ZENER DROP-OUT POINT

5-12 Assume the supply voltage of Fig. 5-40 decreases from 24 to 0 V. At some point along the way, the zener diode will stop regulating. Find the supply voltage where regulation is lost.

5-13 In Fig. 5-40, the unregulated voltage out of the power supply may vary from 20 to 26 V and the load resistance may vary from 500 Ω to 1.5 kΩ. Will the zener regulator fail under these conditions? If so, what value should the series resistance be?

5-14 The unregulated voltage in Fig. 5-40 may vary from 18 to 25 V, and the load current may vary from 1 to 25 mA. Will the zener regulator stop regulating under these conditions? If so, what is the maximum value for R_S?

5-15 What is the minimum load resistance that may be used in Fig. 5-40 without losing zener regulation?

SEC. 5-5 READING A DATA SHEET

5-16 A zener diode has a voltage of 10 V and a current of 20 mA. What is the power dissipation?

5-17 A 1N968 has 5 mA through it. What is the power?

5-18 What is the power dissipation in the resistors and zener diode of Fig. 5-40?

5-19 The zener diode of Fig. 5-40 is a 1N4744A. What is the minimum zener voltage? The maximum?

5-20 If the lead temperature of a 1N4736A zener diode rises to 100°C, what is the diode's new power rating?

SEC. 5-6 TROUBLESHOOTING

5-21 In Fig. 5-40, what is the load voltage for each of these conditions?

   a. Zener diode shorted

   b. Zener diode open

   c. Series resistor open

   d. Load resistor shorted

5-22 If you measure approximately 18.3 V for the load voltage of Fig. 5-40, what do you think the trouble is?

5-23 You measure 24 V across the load of Fig. 5-40. An ohmmeter indicates the zener diode is open. Before replacing the zener diode, what should you check for?

5-24 In Fig. 5-41, the LED does not light. Which of the following are possible troubles?

   a. V130LA2 is open.

   b. Ground between two left bridge diodes is open.

   c. Filter capacitor is open.

   d. Filter capacitor is shorted.

   e. 1N5314 is open.

   f. 1N5314 is shorted.

Figure 5-41

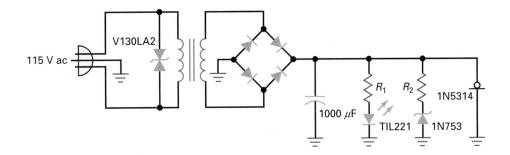

SEC. 5-8 OPTOELECTRONIC DEVICES

5-25 ||| **MultiSim** What is the current through the LED of Fig. 5-42?

5-26 If the supply voltage of Fig. 5-42 increases to 40 V, what is the LED current?

5-27 If the resistor is decreased to 1 kΩ, what is the LED current in Fig. 5-42?

5-28 The resistor of Fig. 5-42 is decreased until the LED current equals 13 mA. What is the value of the resistance?

Figure 5-42

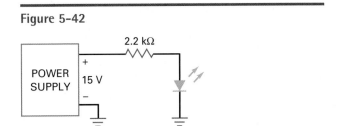

Critical Thinking

5-29 The zener diode of Fig. 5-40 has a zener resistance of 14 Ω. What is the load voltage if you include R_Z in your calculations?

5-30 The zener diode of Fig. 5-40 is a 1N4744A. If the load resistance varies from 1 to 10 kΩ, what is the minimum load voltage? The maximum load voltage? (Use the second approximation.)

5-31 Design a zener regulator to meet these specifications: Load voltage is 6.8 V, source voltage is 20 V, and load current is 30 mA.

5-32 A TIL312 is a seven-segment indicator. Each segment has a voltage drop between 1.5 and 2 V at 20 mA. The supply voltage is +5 V. Design a seven-segment display circuit controlled by on-off switches that has a maximum current drain of 140 mA.

5-33 The secondary voltage of Fig. 5-41 is 12.6 V rms when the line voltage is 115 V rms. During the day the power line varies by ±10 percent. The resistors have tolerances of ±5 percent. The 1N4733A has a tolerance of ±5 percent and a zener resistance of 7 Ω. If R_2 equals 560 Ω, what is the maximum possible value of the zener current at any instant during day?

5-34 In Fig. 5-41, the secondary voltage is 12.6 V rms, and diode drops are 0.7 V each. The 1N5314 is a constant-current diode with a current of 4.7 mA. The LED current is 15.6 mA, and the zener current is 21.7 mA. The filter capacitor has a tolerance of ±20 percent. What is the maximum peak-to-peak ripple?

5-35 Figure 5-43 shows part of a bicycle lighting system. The diodes are Schottky diodes. Use the second approximation to calculate the voltage across the filter capacitor.

Figure 5-43

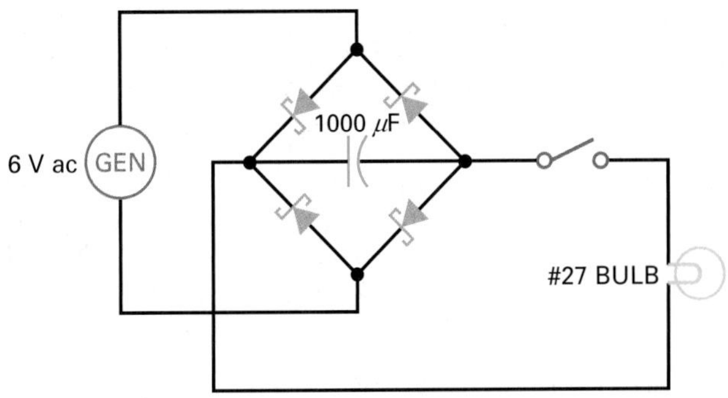

6 V ac (GEN) 1000 μF #27 BULB

Troubleshooting

|||| MultiSim The troubleshooting table shown in Fig. 5-44 lists the voltage values at each respective circuit point and the condition of the diode D_1, for circuit troubles T_1 through T_8. The first row displays what values would be found under normal operating conditions.

5-36 Find Troubles 1 to 4 in Fig. 5-44.

5-37 Find Troubles 5 to 8 in Fig. 5-44.

Figure 5-44 Troubleshooting.

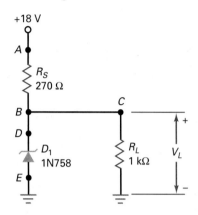

	V_A	V_B	V_C	V_D	D_1
OK	18	10.3	10.3	10.3	OK
T1	18	0	0	0	OK
T2	18	14.2	14.2	0	OK
T3	18	14.2	14.2	14.2	∞
T4	18	18	18	18	∞
T5	0	0	0	0	OK
T6	18	10.5	10.5	10.5	OK
T7	18	14.2	14.2	14.2	OK
T8	18	0	0	0	0

Job Interview Questions

1. Draw a zener regulator. Then explain to me how it works and what its purpose is.

2. I have a power supply that produces an output of 25 V dc. I want three regulated outputs of approximately 15 V, 15.7 V, and 16.4 V. Show me a circuit that will produce these outputs.

3. I have a zener regulator that stops regulating during the day. The ac line voltage in my area varies from 105 to 125 V rms. Also, the load resistance on the zener regulator varies from 100 Ω to 1 kΩ. Tell me some of the possible reasons why the zener regulator fails during the day.

4. This morning, I was breadboarding an LED indicator. After I connected the LED and turned on the power, the LED did not light up. I checked the LED and discovered that it was open. I tried another LED and got the same

results. Tell me some of the possible reasons why this happened.

5. I have heard that a varactor can be used to tune a television receiver. Tell me the basic idea of how it tunes a resonant circuit.

6. Why would an optocoupler be used in an electronic circuit?

7. Given a standard plastic-dome LED package, name two ways to identify the cathode.

8. Explain the differences, if any, between a rectifier diode and a Schottky diode.

9. Draw a circuit like Fig. 5-4a, except replace the dc source by an ac source with a peak value of 40 V. Draw the graph of output voltage for a zener voltage of 10 V.

Self-Test Answers

1.	d	12.	a	23.	c
2.	b	13.	b	24.	c
3.	b	14.	d	25.	b
4.	a	15.	d	26.	d
5.	a	16.	a	27.	a
6.	c	17.	c	28.	c
7.	c	18.	c	29.	b
8.	a	19.	b	30.	b
9.	c	20.	b	31.	d
10.	b	21.	a	32.	a
11.	c	22.	c		

Practice Problem Answers

5-1 $I_S = 24.4$ mA

5-3 $I_S = 18.5$ mA;
$I_L = 10$ mA;
$I_Z = 8.5$ mA

5-5 $V_{RL} = 8$ Vpp square-wave

5-7 $V_L = 10.1$ V

5-8 $V_{R \, (out)} = 94$ mVpp

5-10 $R_{S \, (max)} = 65 \, \Omega$

5-11 $R_{S \, (max)} = 495 \, \Omega$

5-13 $R_S = 330 \, \Omega$

5-14 $I_S = 27$ mA;
$P = 7.2$ W

6

Bipolar Junction Transistors

In 1951, William Schockley invented the first **junction transistor,** a semiconductor device that can amplify (enlarge) electronic signals such as radio and television signals. The transistor has led to many other semiconductor inventions including the **integrated circuit (IC),** a small device that contains thousands of miniaturized transistors. Because of the IC, modern computers and other electronic miracles are possible.

This chapter introduces the **bipolar junction transistor (BJT),** the kind that uses both free electrons and holes. The word *bipolar* is an abbreviation for "two polarities." Following chapters will explore how this BJT can be used as an amplifier and as a switch.

Objectives

After studying this chapter, you should be able to:

- Describe the relationships among the base, emitter, and collector currents of a bipolar junction transistor.
- Draw a diagram of the CE circuit and label each terminal, voltage, and resistance.
- Draw a hypothetical base curve and a set of collector curves, labeling both axes.
- Label the three regions of operation on a bipolar junction transistor collector curve.
- Calculate the respective CE transistor current and voltage values using the ideal transistor and the second transistor approximation.
- List several bipolar junction transistor ratings that might be used by a technician.

Chapter Outline

Vocabulary

active region

base

bipolar junction transistor (BJT)

breakdown region

collector

collector diode

common emitter (CE)

current gain

cutoff region

dc alpha

dc beta

emitter

emitter diode

h parameters

heat sink

integrated circuit (IC)

junction transistor

power transistors

saturation region

small-signal transistors

surface-mount transistors

switching circuit

thermal resistance

6-1 The Unbiased Transistor

A transistor has three doped regions, as shown in Fig. 6-1. The bottom region is called the **emitter,** the middle region is the **base,** and the top region is the **collector.** In an actual transistor, the base region is much thinner as compared to the collector and emitter regions. The transistor of Fig. 6-1 is an *npn device* because there is a *p* region between two *n* regions. Recall that the majority carriers are free electrons in *n*-type material and holes in *p*-type material.

Transistors are also manufactured as *pnp* devices. A *pnp* transistor has an *n* region between two *p* regions. To avoid confusion between the *npn* and the *pnp* transistors, our early discussions will focus on the *npn* transistor.

Doping Levels

In Fig. 6-1, the emitter is heavily doped. On the other hand, the base is lightly doped. The doping level of the collector is intermediate, between the heavy doping of the emitter and the light doping of the base. The collector is physically the largest of the three regions.

Emitter and Collector Diodes

The transistor of Fig. 6-1 has two junctions: one between the emitter and the base, and another between the collector and the base. Because of this, a transistor is like two back-to-back diodes. The lower diode is called the *emitter-base diode,* or simply the **emitter diode.** The upper diode is called the *collector-base diode,* or the **collector diode.**

Before and After Diffusion

Figure 6-1 shows the transistor regions before diffusion has occurred. As discussed in Chap. 2, free electrons in the *n* region will diffuse across the junction and recombine with the holes in the *p* region. Visualize the free electrons in each *n* region crossing the junction and recombining with holes.

The result is two depletion layers, as shown in Fig. 6-2. For each of these depletion layers, the barrier potential is approximately 0.7 V at 25°C for a silicon transistor (0.3 V at 25°C for a germanium transistor). As before, we emphasize silicon devices because they are now more widely used than germanium devices.

Figure 6-1 Structure of a transistor.

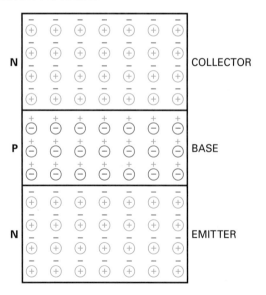

Figure 6-2   Depletion layers.

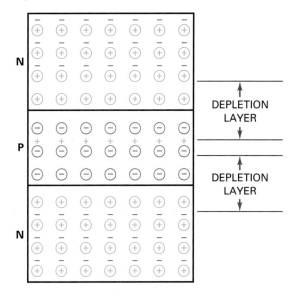

6-2 The Biased Transistor

An unbiased transistor is like two back-to-back diodes. Each diode has a barrier potential of approximately 0.7 V. When you connect external voltage sources to the transistor, you will get currents through the different parts of the transistor.

Emitter Electrons

Figure 6-3 shows a biased transistor. The minus signs represent free electrons. The heavily doped emitter has the following job: to emit or inject its free electrons into the base. The lightly doped base also has a well-defined purpose: to pass emitter-injected electrons on to the collector. The collector is so named because it collects or gathers most of the electrons from the base.

Figure 6-3 is the usual way to bias a transistor. The left source V_{BB} of Fig. 6-3 forward-biases the emitter diode, and the right source V_{CC} reverse-biases the collector diode. Although other biasing methods are possible, forward-biasing the emitter diode and reverse-biasing the collector diode produce the most useful results.

Figure 6-3   Biased transistor.

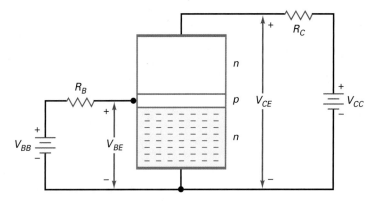

Base Electrons

At the instant that forward bias is applied to the emitter diode of Fig. 6-3, the electrons in the emitter have not yet entered the base region. If V_{BB} is greater than the emitter-base barrier potential in Fig. 6-3, emitter electrons will enter the base region, as shown in Fig. 6-4. Theoretically, these free electrons can flow in either of two directions. First, they can flow to the left and out of the base, passing through R_B on the way to the positive source terminal. Second, the free electrons can flow into the collector.

Which way will the free electrons go? Most will continue on to the collector. Why? Two reasons: the base is *lightly doped* and *very thin*. The light doping means that the free electrons have a long lifetime in the base region. The very thin base means that the free electrons have only a short distance to go to reach the collector. For these two reasons, almost all the emitter-injected electrons pass through the base to the collector.

Only a few free electrons will recombine with holes in the lightly doped base of Fig. 6-4. Then, as valence electrons, they will flow through the base resistor to the positive side of the V_{BB} supply.

Collector Electrons

Almost all the free electrons go into the collector, as shown in Fig. 6-5. Once they are in the collector, they feel the attraction of the V_{CC} source voltage. Because of this, the free electrons flow through the collector and through R_C until they reach the positive terminal of the collector supply voltage.

Figure 6-4   Emitter injects free electrons into base.

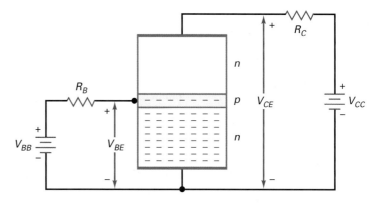

Figure 6-5   Free electrons from base flow into collector.

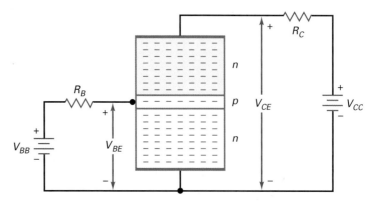

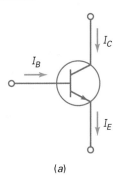

Figure 6-6 Three transistor currents. (a) Conventional flow; (b) electron flow; (c) *pnp* currents.

(a)

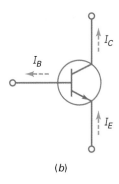

(b)

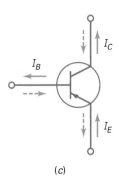

(c)

Here's a summary of what's going on: In Fig. 6-5, V_{BB} forward biases the emitter diode, forcing the free electrons in the emitter to enter the base. The thin and lightly doped base gives almost all these electrons enough time to diffuse into the collector. These electrons flow through the collector, through R_C, and into the positive terminal of the V_{CC} voltage source.

6-3 Transistor Currents

Figures 6-6a and 6-6b show the schematic symbol for an *npn* transistor. If you prefer conventional flow, use Fig. 6-6a. If you prefer electron flow, use Fig. 6-6b. In Fig. 6-6, there are three different currents in a transistor: emitter current I_E, base current I_B, and collector current I_C.

How the Currents Compare

Because the emitter is the source of the electrons, it has the largest current. Since most of the emitter electrons flow to the collector, the collector current is almost as large as the emitter current. The base current is very small by comparison, *often less than 1 percent of the collector current.*

Relation of Currents

Recall Kirchhoff's current law. It says that the sum of all currents into a point or junction equals the sum of all currents out of the point or junction. When applied to a transistor, Kirchhoff's current law gives us this important relationship:

$$I_E = I_C + I_B \tag{6-1}$$

This says that the emitter current is the sum of the collector current and the base current. Since the base current is so small, the collector current approximately equals the emitter current:

$$I_C \approx I_E$$

and the base current is much smaller than the collector current:

$$I_B \ll I_C$$

(*Note:* $\ll$ means *much smaller than.*)

Figure 6-6 (c) shows the schematic symbol for a *pnp* transistor and its currents. Notice that the current directions are opposite that of the *npn*. Again notice that Eq. (6-1) holds true for the *pnp* transistor currents.

Alpha

The **dc alpha** (symbolized α_{dc}) is defined as the dc collector current divided by the dc emitter current:

$$\alpha_{dc} = \frac{I_C}{I_E} \tag{6-2}$$

Since the collector current almost equals the emitter current, the dc alpha is slightly less than 1. For instance, in a low-power transistor the dc alpha is typically greater than 0.99. Even in a high-power transistor, the dc alpha is typically greater than 0.95.

Beta

The **dc beta** (symbolized β_{dc}) of a transistor is defined as the ratio of the dc collector current to the dc base current:

$$\beta_{dc} = \frac{I_C}{I_B} \tag{6-3}$$

The dc beta is also known as the **current gain** because a small base current controls a much larger collector current.

The current gain is a major advantage of a transistor and has led to all kinds of applications. For low-power transistors (under 1 W), the current gain is typically 100 to 300. High-power transistors (over 1 W) usually have current gains of 20 to 100.

Two Derivations

Equation (6-3) may be rearranged into two equivalent forms. First, when you know the value of β_{dc} and I_B, you can calculate the collector current with this derivation:

$$I_C = \beta_{dc}I_B \tag{6-4}$$

Second, when you have the value of β_{dc} and I_C, you can calculate the base current with this derivation:

$$I_B = \frac{I_C}{\beta_{dc}} \tag{6-5}$$

Example 6-1

A transistor has a collector current of 10 mA and a base current of 40 μA. What is the current gain of the transistor?

SOLUTION   Divide the collector current by the base current to get:

$$\beta_{dc} = \frac{10\text{ mA}}{40\ \mu\text{A}} = 250$$

PRACTICE PROBLEM 6-1   What is the current gain of the transistor in Example 6-1 if its base current is 50 μA?

Example 6-2

A transistor has a current gain of 175. If the base current is 0.1 mA, what is the collector current?

SOLUTION   Multiply the current gain by the base current to get:

$$I_C = 175(0.1\text{ mA}) = 17.5\text{ mA}$$

PRACTICE PROBLEM 6-2   Find I_C in Example 6-2 if $\beta_{dc} = 100$.

Example 6-3

A transistor has a collector current of 2 mA. If the current gain is 135, what is the base current?

SOLUTION Divide the collector current by the current gain to get:

$$I_B = \frac{2 \text{ mA}}{135} = 14.8 \ \mu\text{A}$$

PRACTICE PROBLEM 6-3 If $I_C = 10$ mA in Example 6-3, find the transistor's base current.

6-4 The CE Connection

There are three useful ways to connect a transistor: with a CE (common emitter), a CC (common collector), or a CB (common base). The CC and CB connections are discussed in later chapters. In this chapter, we will focus on the CE connection because it is the most widely used.

Common Emitter

In Fig. 6-7a, the common or ground side of each voltage source is connected to the emitter. Because of this, the circuit is called a **common emitter (CE)** connection. The circuit has two loops. The left loop is the base loop, and the right loop is the collector loop.

Figure 6-7 CE connection. (*a*) Basic circuit; (*b*) circuit with grounds.

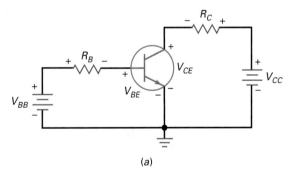

(a)

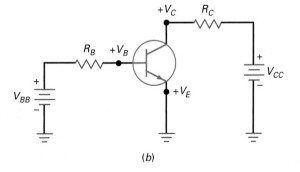

(b)

In the base loop, the V_{BB} source forward biases the emitter diode with R_B as a current-limiting resistance. By changing V_{BB} or R_B, we can change the base current. Changing the base current will change the collector current. In other words, *the base current controls the collector current.* This is important. It means that a small current (base) controls a large current (collector).

In the collector loop, a source voltage V_{CC} reverse biases the collector diode through R_C. The supply voltage V_{CC} must reverse bias the collector diode as shown, or else the transistor won't work properly. Stated another way, the collector must be positive in Fig. 6-7a to collect most of the free electrons injected into the base.

In Fig. 6-7a, the flow of base current in the left loop produces a voltage across the base resistor R_B with the polarity shown. Similarly, the flow of collector current in the right loop produces a voltage across the collector resistor R_C with the polarity shown.

Double Subscripts

Double-subscript notation is used with transistor circuits. When the subscripts are the same, the voltage represents a source (V_{BB} and V_{CC}). When the subscripts are different, the voltage is between the two points (V_{BE} and V_{CE}).

For instance, the subscripts of V_{BB} are the same, which means that V_{BB} is the base voltage source. Similarly, V_{CC} is the collector voltage source. On the other hand, V_{BE} is the voltage between points B and E, between the base and the emitter. Likewise, V_{CE} is the voltage between points C and E, between the collector and the emitter.

Single Subscripts

Single subscripts are used for node voltages, that is, voltages between the subscripted point and ground. For instance, if we redraw Fig. 6-7a with grounds, we get Fig. 6-7b. Voltage V_B is the voltage between the base and ground, voltage V_C is the voltage between the collector and ground, and voltage V_E is the voltage between the emitter and ground. (In this circuit, V_E is zero.)

You can calculate a double-subscript voltage of different subscripts by subtracting its single-subscript voltages. Here are three examples:

$$V_{CE} = V_C - V_E$$

$$V_{CB} = V_C - V_B$$

$$V_{BE} = V_B - V_E$$

This is how you could calculate the double-subscript voltages for any transistor circuit: Since V_E is zero in this CE connection (Fig. 6-7b), the voltages simplify to:

$$V_{CE} = V_C$$

$$V_{CB} = V_C - V_B$$

$$V_{BE} = V_B$$

6-5 The Base Curve

What do you think the graph of I_B versus V_{BE} looks like? It looks like the graph of an ordinary diode as shown in Fig. 6-8a. And why not? This is a forward-biased emitter diode, so we would expect to see the usual diode graph of current versus voltage. What this means is that we can use any of the diode approximations discussed earlier.

Figure 6-8 (*a*) Diode curve; (*b*) example.

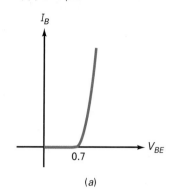

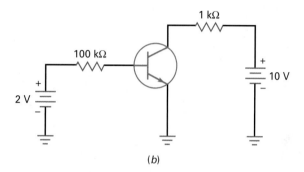

Applying Ohm's law to the base resistor of Fig. 6-7*b* gives this derivation:

$$I_B = \frac{V_{BB} - V_{BE}}{R_B} \tag{6-6}$$

If you use an ideal diode, $V_{BE} = 0$. With the second approximation, $V_{BE} = 0.7$ V.

Most of the time, you will find the second approximation to be the best compromise between the speed of using the ideal diode and accuracy of higher approximations. All you need to remember for the second approximation is that V_{BE} is 0.7 V, as shown in Fig. 6-8*a*.

Example 6–4   IIII MultiSim

Use the second approximation to calculate the base current in Fig. 6-8*b*. What is the voltage across the base resistor? The collector current if $\beta_{dc} = 200$?

SOLUTION The base source voltage of 2 V forward biases the emitter diode through a current-limiting resistance of 100 kΩ. Since the emitter diode has 0.7 V across it, the voltage across the base resistor is:

$$V_{BB} - V_{BE} = 2\text{ V} - 0.7\text{ V} = 1.3\text{ V}$$

The current through the base resistor is:

$$I_B = \frac{V_{BB} - V_{BE}}{R_B} = \frac{1.3\text{ V}}{100\text{ k}\Omega} = 13\ \mu\text{A}$$

With a current gain of 200, the collector current is:

$$I_C = \beta_{dc}I_B = (200)(13~\mu A) = 2.6~mA$$

PRACTICE PROBLEM 6–4 Repeat Example 6-4 using a base source voltage $V_{BB} = 4$ V.

6-6 Collector Curves

In Fig. 6-9a, we already know how to calculate the base current. Since V_{BB} forward biases the emitter diode, all we need to do is calculate the current through the base resistor R_B. Now, let us turn our attention to the collector loop.

We can vary V_{BB} and V_{CC} in Fig. 6-9a to produce different transistor voltages and currents. By measuring I_C and V_{CE}, we can get data for a graph of I_C versus V_{CE}.

For instance, suppose we change V_{BB} as needed to get $I_B = 10~\mu A$. With this fixed value of base current, we can now vary V_{CC} and measure I_C and V_{CE}. Plotting the data gives the graph shown in Fig. 6-9b. (*Note:* this graph is for a 2N3904, a widely used low-power transistor. With other transistors, the numbers may vary but the shape of the curve will be similar.)

When V_{CE} is zero, the collector diode is not reverse biased. This is why the graph shows a collector current of zero when V_{CE} is zero. When V_{CE} increases from zero, the collector current rises sharply in Fig. 6-9b. When V_{CE} is a few tenths of a volt, the collector current becomes *almost constant* and equal to 1 mA.

Figure 6-9 (*a*) Basic transistor circuit; (*b*) collector curve.

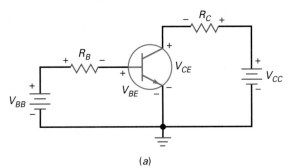

(*a*)

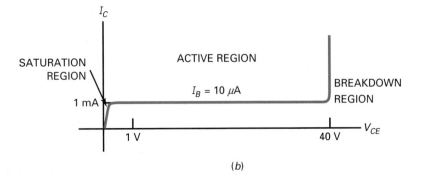

(*b*)

The constant-current region in Fig. 6-9b is related to our earlier discussions of transistor action. After the collector diode becomes reverse biased, it is gathering all the electrons that reach its depletion layer. Further increases in V_{CE} cannot increase the collector current. Why? Because the collector can collect only those free electrons that the emitter injects into the base. The number of these injected electrons depends only on the base circuit, not on the collector circuit. This is why Fig. 6-9b shows a constant collector current between a V_{CE} of less than 1 V to a V_{CE} of more than 40 V.

If V_{CE} is greater than 40 V, the collector diode breaks down and normal transistor action is lost. The transistor is not intended to operate in the breakdown region. For this reason, one of the maximum ratings to look for on a transistor data sheet is the collector-emitter breakdown voltage $V_{CE(\text{max})}$. If the transistor breaks down, it will be destroyed.

Collector Voltage and Power

Kirchhoff's voltage law says that the sum of voltages around a loop or closed path is equal to zero. When applied to the collector circuit of Fig. 6-9a, Kirchhoff's voltage law gives us this derivation:

$$V_{CE} = V_{CC} - I_C R_C \tag{6-7}$$

This says that the collector-emitter voltage equals the collector supply voltage minus the voltage across the collector resistor.

In Fig. 6-9a, the transistor has a power dissipation of approximately:

$$P_D = V_{CE} I_C \tag{6-8}$$

This says that the transistor power equals the collector-emitter voltage times the collector current. This power dissipation causes the junction temperature of the collector diode to increase. The higher the power, the higher the junction temperature.

Transistors will burn out when the junction temperature is between 150 and 200°C. One of the most important pieces of information on a data sheet is the maximum power rating $P_{D(\text{max})}$. The power dissipation given by Eq. (6-8) must be less than $P_{D(\text{max})}$. Otherwise, the transistor will be destroyed.

Regions of Operation

The curve of Fig. 6-9b has different regions where the action of a transistor changes. First, there is the region in the middle where V_{CE} is between 1 and 40 V. This represents the normal operation of a transistor. In this region, the emitter diode is forward biased, and the collector diode is reverse biased. Furthermore, the collector is gathering almost all the electrons that the emitter has sent into the base. This is why changes in collector voltage have no effect on the collector current. This region is called the **active region.** Graphically, the active region is the horizontal part of the curve. In other words, the collector current is *constant* in this region.

Another region of operation is the **breakdown region.** The transistor should never operate in this region because it will be destroyed. Unlike the zener diode, which is optimized for breakdown operation, a transistor is not intended for operation in the breakdown region.

Third, there is the early rising part of the curve, where V_{CE} is between 0 V and a few tenths of a volt. This sloping part of the curve is called the **saturation region.** In this region, the collector diode has insufficient positive voltage to collect all the free electrons injected into the base. In this region, the base current I_B is larger than normal and the current gain β_{dc} is smaller than normal.

GOOD TO KNOW

When displayed on a curve tracer, the collector curves in Fig. 6-10 actually have a slight upward slope as V_{CE} increases. This rise is the result of the base region becoming slightly smaller as V_{CE} increases. (As V_{CE} increases, the CB depletion layer widens, thus narrowing the base.) With a smaller base region, there are fewer holes available for recombination. Since each curve represents a constant base current, the effect looks like an increase in collector current.

Figure 6-10 Set of collector curves.

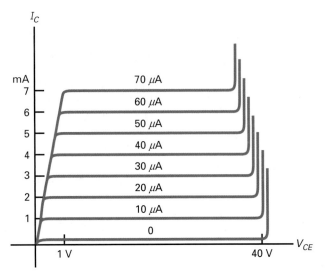

More Curves

If we measure I_C and V_{CE} for $I_B = 20$ μA, we can plot the second curve of Fig. 6-10. The curve is similar to the first curve, except that the collector current is 2 mA in the active region. Again, the collector current is constant in the active region.

When we plot several curves for different base currents, we get a set of collector curves like those in Fig. 6-10. Another way to get this set of curves is with a *curve tracer* (a test instrument that can display I_C versus V_{CE} for a transistor). In the active region of Fig. 6-10, each collector current is 100 times greater than the corresponding base current. For instance, the top curve has a collector current of 7 mA and a base current of 70 μA. This gives a current gain of:

$$\beta_{dc} = \frac{I_C}{I_B} = \frac{7\,\text{mA}}{70\,\mu\text{A}} = 100$$

If you check any other curve, you get the same result: a current gain of 100.

With other transistors, the current gain may be different from 100, but the shape of the curves will be similar. All transistors have an active region, a saturation region, and a breakdown region. The active region is the most important because amplification (enlargement) of signals is possible in the active region.

Cutoff Region

Figure 6-10 has an unexpected curve, the one on the bottom. This represents a fourth possible region of operation. Notice that the base current is zero, but there still is a small collector current. On a curve tracer, this current is usually so small that you cannot see it. We have exaggerated the bottom curve by drawing it larger than usual. This bottom curve is called the **cutoff region** of the transistor, and the small collector current is called the *collector cutoff current*.

Why does the collector cutoff current exist? Because the collector diode has reverse minority-carrier current and surface-leakage current. In a well-designed circuit, the collector cutoff current is small enough to ignore. For instance, a 2N3904 has a collector cutoff current of 50 nA. If the actual collector current is 1 mA, ignoring a collector cutoff current of 50 nA produces a calculation error of less than 5 percent.

Recap

A transistor has four distinct operating regions: *active, cutoff, saturation,* and *breakdown.* Transistors operate in the active region when they are used to amplify weak signals. Sometimes, the active region is called the *linear region* because changes in the input signal produce proportional changes in the output signal. The saturation and cutoff regions are useful in digital and computer circuits, referred to as **switching circuits.**

Example 6-5

The transistor of Fig. 6-11a has $\beta_{dc} = 300$. Calculate I_B, I_C, V_{CE}, and P_D.

Figure 6-11 Transistor circuit. (*a*) Basic schematic diagram; (*b*) circuit with grounds; (*c*) simplified schematic diagram.

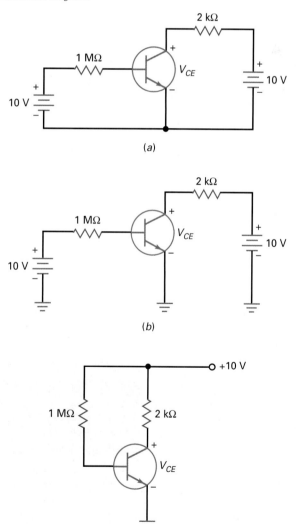

(a)

(b)

(c)

SOLUTION Figure 6-11*b* shows the same circuit with grounds. The base current equals:

$$I_B = \frac{V_{BB} - V_{BE}}{R_B} = \frac{10 \text{ V} - 0.7 \text{ V}}{1 \text{ M}\Omega} = 9.3 \ \mu\text{A}$$

The collector current is:

$$I_C = \beta_{dc}I_B = (300)(9.3 \ \mu\text{A}) = 2.79 \text{ mA}$$

and the collector-emitter voltage is:

$$V_{CE} = V_{CC} - I_C R_C = 10 \text{ V} - (2.79 \text{ mA})(2 \text{ k}\Omega) = 4.42 \text{ V}$$

The collector power dissipation is:

$$P_D = V_{CE}I_C = (4.42 \text{ V})(2.79 \text{ mA}) = 12.3 \text{ mW}$$

Incidentally, when both the base and the collector supply voltages are equal, as in Fig. 6-11*b*, you usually see the circuit drawn in the simpler form of Fig. 6-11*c*.

PRACTICE PROBLEM 6-5 Change R_B to 680 kΩ and repeat Example 6-5.

Example 6-6

||| MultiSim

Figure 6-12 shows a transistor circuit built on a computer screen with MultiSim. Calculate the current gain of the 2N4424.

Figure 6-12 MultiSim circuit for calculating current gain of 2N4424.

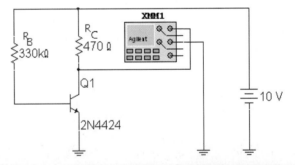

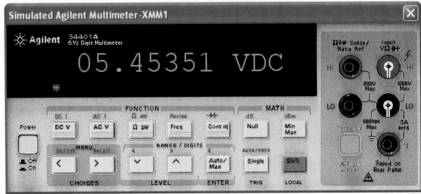

SOLUTION First, get the base current as follows:

$$I_B = \frac{10 \text{ V} - 0.7 \text{ V}}{330 \text{ k}\Omega} = 28.2 \ \mu\text{A}$$

Next, we need the collector current. Since the multimeter indicates a collector-emitter voltage of 5.45 V (rounded to three places), the voltage across the collector resistor is:

$$V = 10 \text{ V} - 5.45 \text{ V} = 4.55 \text{ V}$$

Since the collector current flows through the collector resistor, we can use Ohm's law to get the collector current:

$$I_C = \frac{4.55 \text{ V}}{470 \ \Omega} = 9.68 \text{ mA}$$

Now, we can calculate the current gain:

$$\beta_{\text{dc}} = \frac{9.68 \text{ mA}}{28.2 \ \mu\text{A}} = 343$$

The 2N4424 is an example of transistor with a high current gain. The typical range of β_{dc} for small-signal transistors is 100 to 300.

PRACTICE PROBLEM 6-6 Using MultiSim, change the base resistor of Fig. 6-12 to 560 kΩ and calculate the current gain of the 2N4424.

6-7 Transistor Approximations

Figure 6-13a shows a transistor. A voltage V_{BE} appears across the emitter diode, and a voltage V_{CE} appears across the collector-emitter terminals. What is the equivalent circuit for this transistor?

Ideal Approximation

Figure 6-13b shows the ideal approximation of a transistor. We visualize the emitter diode as an ideal diode. In this case, $V_{BE} = 0$. This allows us to calculate base current quickly and easily. This equivalent circuit is often useful for troubleshooting when all we need is a rough approximation of base current.

As shown in Fig. 6-13b, the collector side of the transistor acts like a current source that pumps a collector current of $\beta_{\text{dc}}I_B$ through the collector resistor. Therefore, after you calculate the base current, you can multiply by the current gain to get the collector current.

The Second Approximation

Figure 6-13c shows the second approximation of a transistor. This is more commonly used because it may improve the analysis significantly when the base-supply voltage is small.

This time we use the second approximation of a diode when calculating base current. For silicon transistors, this means that $V_{BE} = 0.7$ V. (For germanium transistors, $V_{BE} = 0.3$ V.) With the second approximation, the base and collector currents will be slightly less than their ideal values.

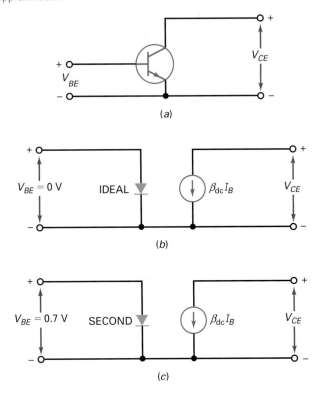

(a)

(b)

(c)

Higher Approximations

The bulk resistance of the emitter diode becomes important only in high-power applications in which the currents are large. The effect of bulk resistance in the emitter diode is to increase V_{BE} to more than 0.7 V. For instance, in some high-power circuits, the V_{BE} across the base-emitter diode may be more than 1 V.

Likewise, the bulk resistance of the collector diode may have a noticeable effect in some designs. Besides emitter and collector bulk resistances, a transistor has many other higher-order effects that make hand calculations tedious and time-consuming. For this reason, calculations beyond the second approximation should use a computer solution.

Example 6-7

What is the collector-emitter voltage in Fig. 6-14? Use the ideal transistor.

Figure 6-14 Example.

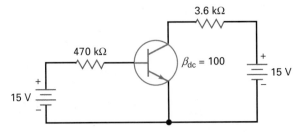

SOLUTION An ideal emitter diode means that:

$$V_{BE} = 0$$

Therefore, the total voltage across R_B is 15 V. Ohm's law tells us that:

$$I_B = \frac{15 \text{ V}}{470 \text{ k}\Omega} = 31.9 \text{ } \mu\text{A}$$

The collector current equals the current gain times the base current:

$$I_C = 100(31.9 \text{ } \mu\text{A}) = 3.19 \text{ mA}$$

Next, we calculate the collector-emitter voltage. It equals the collector supply voltage minus the voltage drop across the collector resistor:

$$V_{CE} = 15 \text{ V} - (3.19 \text{ mA})(3.6 \text{ k}\Omega) = 3.52 \text{ V}$$

In a circuit like Fig. 6-14, knowing the value of the emitter current is not important, so most people would not calculate this quantity. But since this is an example, we will calculate the emitter current. It equals the sum of the collector current and the base current:

$$I_E = 3.19 \text{ mA} + 31.9 \text{ } \mu\text{A} = 3.22 \text{ mA}$$

This value is extremely close to the value of the collector current, which is another reason for not bothering to calculate it. Most people would say that the emitter current is approximately 3.19 mA, the value of the collector current.

Example 6-8

IIII MultiSim

What is the collector-emitter voltage in Fig. 6-14 if you use the second approximation?

SOLUTION In Fig. 6-14, here is how you would calculate the currents and voltages, using the second approximation. The voltage across the emitter diode is:

$$V_{BE} = 0.7 \text{ V}$$

Therefore, the total voltage across R_B is 14.3 V, the difference between 15 and 0.7 V. The base current is:

$$I_B = \frac{14.3 \text{ V}}{470 \text{ k}\Omega} = 30.4 \text{ } \mu\text{A}$$

The collector current equals the current gain times the base current:

$$I_C = 100(30.4 \text{ } \mu\text{A}) = 3.04 \text{ mA}$$

The collector-emitter voltage equals:

$$V_{CE} = 15 \text{ V} - (3.04 \text{ mA})(3.6 \text{ k}\Omega) = 4.06 \text{ V}$$

The improvement in this answer over the ideal answer is about half a volt: 4.06 versus 3.52 V. Is this half a volt important? It depends on whether you are troubleshooting, designing, and so on.

Example 6-9

Suppose you measure a V_{BE} of 1 V. What is the collector-emitter voltage in Fig. 6-14?

SOLUTION   The total voltage across R_B is 14 V, the difference between 15 and 1 V. Ohm's law tells us that the base current is:

$$I_B = \frac{14 \text{ V}}{470 \text{ k}\Omega} = 29.8 \text{ } \mu\text{A}$$

The collector current equals the current gain times the base current:

$$I_C = 100(29.8 \text{ } \mu\text{A}) = 2.98 \text{ mA}$$

The collector-emitter voltage equals:

$$V_{CE} = 15 \text{ V} - (2.98 \text{ mA})(3.6 \text{ k}\Omega) = 4.27 \text{ V}$$

Example 6-10

What is the collector-emitter voltage in the three preceding examples if the base supply voltage is 5 V?

SOLUTION   With the ideal diode:

$$I_B = \frac{5 \text{ V}}{470 \text{ k}\Omega} = 10.6 \text{ } \mu\text{A}$$

$$I_C = 100 \, (10.6 \text{ } \mu\text{A}) = 1.06 \text{ mA}$$

$$V_{CE} = 15 \text{ V} - (1.06 \text{ mA})(3.6 \text{ k}\Omega) = 11.2 \text{ V}$$

With the second approximation:

$$I_B = \frac{4.3 \text{ V}}{470 \text{ k}\Omega} = 9.15 \text{ } \mu\text{A}$$

$$I_C = 100(9.15 \text{ } \mu\text{A}) = 0.915 \text{ mA}$$

$$V_{CE} = 15 \text{ V} - (0.915 \text{ mA})(3.6 \text{ k}\Omega) = 11.7 \text{ V}$$

With the measured V_{BE}:

$$I_B = \frac{4 \text{ V}}{470 \text{ k}\Omega} = 8.51 \text{ } \mu\text{A}$$

$$I_C = 100(8.51 \text{ } \mu\text{A}) = 0.851 \text{ mA}$$

$$V_{CE} = 15 \text{ V} - (0.851 \text{ mA})(3.6 \text{ k}\Omega) = 11.9 \text{ V}$$

This example allows you to compare the three approximations for the case of low base supply voltage. As you can see, all answers are within a volt of each other. This is the first clue as to which approximation to use. If you are troubleshooting this circuit, the ideal analysis will probably be adequate. But if you are designing the circuit, you might want to use a computer solution because of its accuracy. Summary Table 6-1 illustrates the difference between the ideal and second transistor approximations.

PRACTICE PROBLEM 6-10   Repeat Example 6-10 using a base supply voltage of 7 V.

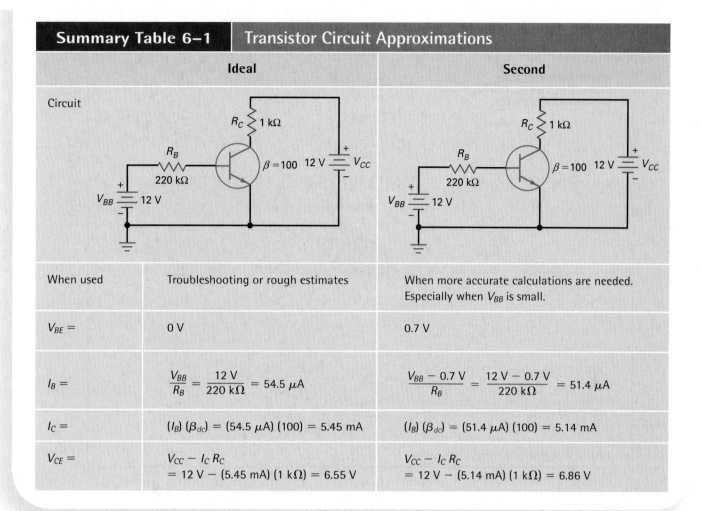

	Ideal	Second
Circuit		
When used	Troubleshooting or rough estimates	When more accurate calculations are needed. Especially when V_{BB} is small.
$V_{BE} =$	0 V	0.7 V
$I_B =$	$\dfrac{V_{BB}}{R_B} = \dfrac{12 \text{ V}}{220 \text{ k}\Omega} = 54.5 \ \mu\text{A}$	$\dfrac{V_{BB} - 0.7 \text{ V}}{R_B} = \dfrac{12 \text{ V} - 0.7 \text{ V}}{220 \text{ k}\Omega} = 51.4 \ \mu\text{A}$
$I_C =$	$(I_B)(\beta_{dc}) = (54.5 \ \mu\text{A})(100) = 5.45 \text{ mA}$	$(I_B)(\beta_{dc}) = (51.4 \ \mu\text{A})(100) = 5.14 \text{ mA}$
$V_{CE} =$	$V_{CC} - I_C R_C$ $= 12 \text{ V} - (5.45 \text{ mA})(1 \text{ k}\Omega) = 6.55 \text{ V}$	$V_{CC} - I_C R_C$ $= 12 \text{ V} - (5.14 \text{ mA})(1 \text{ k}\Omega) = 6.86 \text{ V}$

6-8 Reading Data Sheets

Small-signal transistors can dissipate less than a watt; **power transistors** can dissipate more than a watt. When you look at a data sheet for either type of transistor, you should start with the maximum ratings because these are the limits on the transistor currents, voltages, and other quantities.

Breakdown Ratings

In the data sheet shown in Fig. 6-15, the following maximum ratings of a 2N3904 are given:

$$V_{CEO} \quad 40 \text{ V}$$
$$V_{CBO} \quad 60 \text{ V}$$
$$V_{EBO} \quad 6 \text{ V}$$

These voltage ratings are reverse breakdown voltages, and V_{CEO} is the voltage between the collector and the emitter with the base open. The second rating is V_{CBO}, which stands for the voltage from collector to base with the emitter open. Likewise, V_{EBO} is the maximum reverse voltage from emitter to base with the collector open. As usual, a conservative design never allows voltages to get even close to the foregoing maximum ratings. If you recall, even getting close to maximum ratings can shorten the lifetime of some devices.

Figure 6-15(a)   2N3904 data sheet.

SEMICONDUCTOR ™

2N3904   MMBT3904   PZT3904

TO-92

C
B E

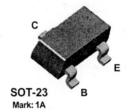

C

E

SOT-23
Mark: 1A

B

C

E

C

B

SOT-223

NPN General Purpose Amplifier

This device is designed as a general purpose amplifier and switch.
The useful dynamic range extends to 100 mA as a switch and to
100 MHz as an amplifier.

Absolute Maximum Ratings* $T_A = 25°C$ unless otherwise noted

Symbol	Parameter	Value	Units
V_{CEO}	Collector-Emitter Voltage	40	V
V_{CBO}	Collector-Base Voltage	60	V
V_{EBO}	Emitter-Base Voltage	6.0	V
I_C	Collector Current - Continuous	200	mA
T_J, T_{stg}	Operating and Storage Junction Temperature Range	-55 to +150	°C

*These ratings are limiting values above which the serviceability of any semiconductor device may be impaired.

NOTES:
1) These ratings are based on a maximum junction temperature of 150 degrees C.
2) These are steady state limits. The factory should be consulted on applications involving pulsed or low duty cycle operations.

Thermal Characteristics $T_A = 25°C$ unless otherwise noted

Symbol	Characteristic	Max			Units
		2N3904	*MMBT3904	**PZT3904	
P_D	Total Device Dissipation	625	350	1,000	mW
	Derate above 25°C	5.0	2.8	8.0	mW/°C
$R_{\theta JC}$	Thermal Resistance, Junction to Case	83.3			°C/W
$R_{\theta JA}$	Thermal Resistance, Junction to Ambient	200	357	125	°C/W

*Device mounted on FR-4 PCB 1.6" X 1.6" X 0.06."

**Device mounted on FR-4 PCB 36 mm X 18 mm X 1.5 mm; mounting pad for the collector lead min. 6 cm².

2N3904/MMBT3904/PZT3904, Rev A

Figure 6–15(*b*)   (continued)

NPN General Purpose Amplifier
(continued)

Electrical Characteristics T_A = 25°C unless otherwise noted

Symbol	Parameter	Test Conditions	Min	Max	Units
OFF CHARACTERISTICS					
$V_{(BR)CEO}$	Collector-Emitter Breakdown Voltage	I_C = 1.0 mA, I_B = 0	40		V
$V_{(BR)CBO}$	Collector-Base Breakdown Voltage	I_C = 10 μA, I_E = 0	60		V
$V_{(BR)EBO}$	Emitter-Base Breakdown Voltage	I_E = 10 μA, I_C = 0	6.0		V
I_{BL}	Base Cutoff Current	V_{CE} = 30 V, V_{EB} = 3V		50	nA
I_{CEX}	Collector Cutoff Current	V_{CE} = 30 V, V_{EB} = 3V		50	nA
ON CHARACTERISTICS*					
h_{FE}	DC Current Gain	I_C = 0.1 mA, V_{CE} = 1.0 V	40		
		I_C = 1.0 mA, V_{CE} = 1.0 V	70		
		I_C = 10 mA, V_{CE} = 1.0 V	100	300	
		I_C = 50 mA, V_{CE} = 1.0 V	60		
		I_C = 100 mA, V_{CE} = 1.0 V	30		
$V_{CE(sat)}$	Collector-Emitter Saturation Voltage	I_C = 10 mA, I_B = 1.0 mA		0.2	V
		I_C = 50 mA, I_B = 5.0 mA		0.3	V
$V_{BE(sat)}$	Base-Emitter Saturation Voltage	I_C = 10 mA, I_B = 1.0 mA	0.65	0.85	V
		I_C = 50 mA, I_B = 5.0 mA		0.95	V
SMALL SIGNAL CHARACTERISTICS					
f_T	Current Gain - Bandwidth Product	I_C = 10 mA, V_{CE} = 20 V, f = 100 MHz	300		MHz
C_{obo}	Output Capacitance	V_{CB} = 5.0 V, I_E = 0, f = 1.0 MHz		4.0	pF
C_{ibo}	Input Capacitance	V_{EB} = 0.5 V, I_C = 0, f = 1.0 MHz		8.0	pF
NF	Noise Figure	I_C = 100 μA, V_{CE} = 5.0 V, R_S =1.0kΩ,f=10 Hz to 15.7kHz		5.0	dB
SWITCHING CHARACTERISTICS					
t_d	Delay Time	V_{CC} = 3.0 V, V_{BE} = 0.5 V,		35	ns
t_r	Rise Time	I_C = 10 mA, I_{B1} = 1.0 mA		35	ns
t_s	Storage Time	V_{CC} = 3.0 V, I_C = 10mA		200	ns
t_f	Fall Time	I_{B1} = I_{B2} = 1.0 mA		50	ns

*Pulse Test: Pulse Width ≤ 300 μs, Duty Cycle ≤ 2.0%

Spice Model

NPN (Is=6.734f Xti=3 Eg=1.11 Vaf=74.03 Bf=416.4 Ne=1.259 Ise=6.734 Ikf=66.78m Xtb=1.5 Br=.7371 Nc=2 Isc=0 Ikr=0 Rc=1 Cjc=3.638p Mjc=.3085 Vjc=.75 Fc=.5 Cje=4.493p Mje=.2593 Vje=.75 Tr=239.5n Tf=301.2p Itf=.4 Vtf=4 Xtf=2 Rb=10)

Maximum Current and Power

Also shown in the data sheet are these values:

$$I_C \qquad 200 \text{ mA}$$
$$P_D \qquad 625 \text{ mW}$$

Here I_C is the maximum dc collector current rating. This means that a 2N3904 can handle up to 200 mA of direct current, provided the power rating is not exceeded. The next rating, P_D, is the maximum power rating of the device. This power rating depends on whether any attempt is being made to keep the transistor cool. If the transistor is not being fan-cooled and does not have a heat sink (discussed below), its case temperature T_C will be much higher than the ambient temperature T_A.

In most applications, a small-signal transistor like the 2N3904 is not fan-cooled and it does not have a heat sink. In this case, the 2N3904 has a power rating of 625 mW when the ambient temperature T_A is 25°C.

The case temperature T_C is the temperature of the transistor package or housing. In most applications, the case temperature will be higher than 25°C because the internal heat of the transistor increases the case temperature.

The only way to keep the case temperature at 25°C when the ambient temperature is 25°C is by fan-cooling or by using a large heat sink. If fan cooling or a large heat sink is used, it is possible to reduce the temperature of the transistor case to 25°C. For this condition, the power rating can be increased to 1.5 W.

Derating Factors

As discussed in Chap. 5, the derating factor tells you how much you have to reduce the power rating of a device. The derating factor of the 2N3904 is given as 5 mW/°C. This means that you have to reduce the power rating of 625 mW by 5 mW for each degree above 25°C.

Heat Sinks

One way to increase the power rating of a transistor is to get rid of the internal heat faster. This is the purpose of a **heat sink** (a mass of metal). If we increase the surface area of the transistor case, we allow the heat to escape more easily into the surrounding air. For instance, Fig. 6-16a shows one type of heat sink. When this is pushed onto the transistor case, heat radiates more quickly because of the increased surface area of the fins.

Figure 6-16b shows another approach. This is the outline of a power-tab transistor. A metal tab provides a path out of the transistor for heat. This metal tab can be fastened to the chassis of electronic equipment. Because the chassis is a massive heat sink, heat can easily escape from the transistor to the chassis.

Large power transistors like Fig. 6-16c have the collector connected to the case to let heat escape as easily as possible. The transistor case is then fastened to the chassis. To prevent the collector from shorting to chassis ground, a thin insulating washer and heat-conducting compound is used between the transistor case and the chassis. The important idea here is that heat can leave the transistor more rapidly, which means that the transistor has a higher power rating at the same ambient temperature. Sometimes, the transistor is fastened to a large heat sink with fins; this is even more efficient in removing heat from the transistor.

No matter what kind of heat sink is used, the purpose is to lower the case temperature because this will lower the internal or junction temperature of the transistor. The data sheet includes other quantities called **thermal resistances.** These allow a designer to work out the case temperature for different heat sinks.

Figure 6-16 (a) Push-on heat sink; (b) power-tab transistor; (c) power transistor with collector connected to case.

(a)

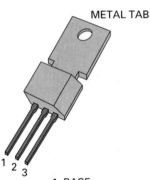

METAL TAB

1 2 3
TO-220

1. BASE
2. COLLECTOR
3. EMITTER

(b)

TO-204AA (TO-3)
CASE 1-07

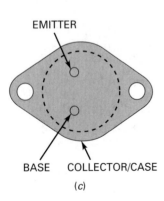

EMITTER

BASE   COLLECTOR/CASE

(c)

Current Gain

In another system of analysis called **h parameters,** h_{FE} rather than β_{dc} is defined as the symbol for current gain. The two quantities are equal:

$$\beta_{dc} = h_{FE} \tag{6-9}$$

Remember this relation because data sheets use the symbol h_{FE} for the current gain.

In the section labeled "On Characteristics," the data sheet of a 2N3904 lists the values of h_{FE} as follows:

I_C, mA	Min. h_{FE}	Max. h_{FE}
0.1	40	–
1	70	–
10	100	300
50	60	–
100	30	–

The 2N3904 works best when the collector current is in the vicinity of 10 mA. At this level of current, the minimum current gain is 100 and the maximum current gain is 300. What does this mean? It means that if you mass-produce a circuit using 2N3904s and a collector current of 10 mA, some of the transistors will have a current gain as low as 100, and some will have a current gain as high as 300. Most of the transistors will have a current gain in the middle of this range.

Notice how the minimum current gain decreases for collector currents that are less than or greater than 10 mA. At 0.1 mA, the minimum current gain is 40. At 100 mA, the minimum current gain is 30. The data sheet shows only the minimum current gain for currents different from 10 mA because the minimum values represent the worst case. Designers usually do worst-case design; that is, they figure out how the circuit will work when the transistor characteristics such as current gain are at their worst case.

Example 6-11

A 2N3904 has $V_{CE} = 10$ V and $I_C = 20$ mA. What is the power dissipation? How safe is this level of power dissipation if the ambient temperature is 25°C?

SOLUTION Multiply V_{CE} by I_C to get:

$$P_D = (10\text{ V})(20\text{ mA}) = 200\text{ mW}$$

Is this safe? If the ambient temperature is 25°C, the transistor has a power rating of 625 mW. This means that the transistor is well within its power rating.

As you know, a good design includes a safety factor to ensure a longer operating life for the transistor. Safety factors of 2 or more are common. A safety factor of 2 means that the designer would allow up to half of 625 mW, or 312 mW. Therefore, a power of only 200 mW is very conservative, provided the ambient temperature stays at 25°C.

Example 6-12

How safe is the level of power dissipation if the ambient temperature is 100°C in Example 6-11?

SOLUTION First, work out the number of degrees that the new ambient temperature is above the reference temperature of 25°C. Do this as follows:

$$100°C - 25°C = 75°C$$

Sometimes, you will see this written as:

$$\Delta T = 75°C$$

where Δ stands for "difference in." Read the equation as the difference in temperature equals 75°C.

Now, multiply the derating factor by the difference in temperature to get:

$$(5 \text{ mW/°C})(75°C) = 375 \text{ mW}$$

You often see this written as:

$$\Delta P = 375 \text{ mW}$$

where ΔP stands for the difference in power. Finally, you subtract the difference in power from the power rating at 25°C:

$$P_{D(\text{max})} = 625 \text{ mW} - 375 \text{ mW} = 250 \text{ mW}$$

This is the power rating of the transistor when the ambient temperature is 100°C.

How safe is this design? The transistor is still all right because its power is 200 mW compared with the maximum rating of 250 mW. But we no longer have a safety factor of 2. If the ambient temperature were to increase further, or if the power dissipation were to increase, the transistor could get dangerously close to the burnout point. Because of this, a designer might redesign the circuit to restore the safety factor of 2. This means changing circuit values to get a power dissipation of half of 250 mW, or 125 mW.

PRACTICE PROBLEM 6-12 Using a safety factor of 2, could you safely use the 2N3904 transistor of Example 6-12 if the ambient temperature were 75°C?

6-9 Surface-Mount Transistors

Surface-mount transistors are usually found in a simple three-terminal, gull-wing package. The SOT-23 package is the smaller of the two and is used for transistors rated in the milliwatt range. The SOT-223 is the larger package and is used when the power rating is about 1 W.

Figure 6-17 shows a typical SOT-23 package. Viewed from the top, the terminals are numbered in a counterclockwise direction, with terminal 3 the lone terminal on one side. The terminal assignments are fairly well standardized for bipolar transistors: 1 is the base, 2 is the emitter, and 3 is the collector. (The usual terminal assignments for FETs: 1 is the drain, 2 is the source, and 3 is the gate.)

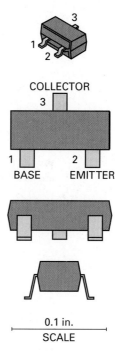

Figure 6-18 The SOT-223 package is designed to dissipate the heat generated by transistors operating in the 1-W range.

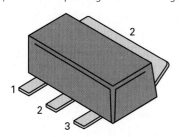

TOP

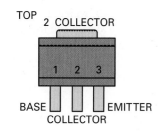

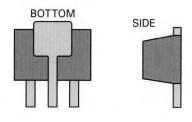

The SOT-223 package is designed to dissipate the heat generated by transistors operating in the 1-W range. This package has a larger surface area than the SOT-23; this increases its ability to dissipate heat. Some of the heat is dissipated from the top surface, and much is carried away by the contact between the device and the circuit board below. The special feature of the SOT-223 case, however, is the extra collector tab that extends from the side opposite the main terminals. The bottom view in Fig. 6-18 shows that the two collector terminals are electrically identical.

The standard terminal assignments are different for the SOT-23 and SOT-223 packages. The three terminals located on one edge are numbered in sequence, from left to right as viewed from the top. Terminal 1 is the base, 2 is the collector (electrically identical to the large tab at the opposite edge), and 3 is the emitter. This is also shown in the data sheet of Fig. 6-15.

The SOT-23 packages are too small to have any standard part identification codes printed on them. Usually the only way to determine the standard identifier is by noting the part number printed on the circuit board and then consulting the parts list for the circuit. SOT-223 packages are large enough to have identification codes printed on them, but these codes are rarely standard transistor identification codes. The typical procedure for learning more about a transistor in an SOT-223 package is the same as for the smaller SOT-23 configurations.

Occasionally a circuit uses SOIC packages that house multiple transistors. The SOIC package resembles the tiny dual-inline package commonly used for ICs and the older feed-through circuit board technology. The terminals on the SOIC, however, have the gull-wing shape required for SM technology.

6-10 Troubleshooting

Figure 6-19 shows a common-emitter circuit with grounds. A base supply of 15 V forward-biases the emitter diode through a resistance of 470 kΩ. A collector supply of 15 V reverse biases the collector diode through a resistance of 1 kΩ. Let us

Figure 6-19 Troubleshooting a circuit.

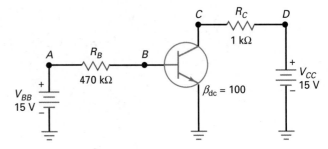

use the ideal approximation to find the collector-emitter voltage. The calculations are as follows:

$$I_B = \frac{15\,\text{V}}{470\,\text{k}\Omega} = 31.9\ \mu\text{A}$$

$$I_C = 100(31.9\ \mu\text{A}) = 3.19\ \text{mA}$$

$$V_{CE} = 15\,\text{V} - (3.19\,\text{mA})(1\,\text{k}\Omega) = 11.8\,\text{V}$$

Common Troubles

If you are troubleshooting a circuit like Fig. 6-19, one of the first things to measure is the collector-emitter voltage. It should have a value in the vicinity of 11.8 V. Why don't we use the second or third approximation to get a more accurate answer? Because resistors usually have a tolerance of at least ±5 percent, which causes the collector-emitter voltage to differ from your calculations, no matter what approximation you use.

In fact, when troubles come, they are usually big troubles like shorts or opens. Shorts may occur because of damaged devices or solder splashes across resistors. Opens may occur when components burn out. Troubles like these produce big changes in currents and voltages. For instance, one of the most common troubles occurs when no supply voltage reaches the collector. This could happen in a number of ways, such as a trouble in the power supply itself, an open lead between the power supply and the collector resistor, an open collector resistor, and so on. In any of these cases, the collector voltage of Fig. 6-19 will be approximately zero because there is no collector supply voltage.

Another possible trouble is an open base resistor, which drops the base current to zero. This forces the collector current to drop to zero and the collector-emitter voltage to rise to 15 V, the value of the collector supply voltage. An open transistor has the same effect.

How Troubleshooters Think

The point is this: Typical troubles cause big deviations in transistor currents and voltages. Troubleshooters are seldom looking for differences in tenths of a volt. They are looking for voltages that are radically different from the ideal values. This is why the ideal transistor is useful as a starting point in troubleshooting. Furthermore, it explains why many troubleshooters don't even use calculators to find the collector-emitter voltage.

If they don't use calculators, what do they do? They mentally estimate the value of the collector-emitter voltage. Here is the thinking of an experienced troubleshooter while estimating the collector-emitter voltage in Fig. 6-19.

> *The voltage across the base resistor is about 15 V. A base resistance of 1 MΩ would produce a base current of about 15 μA. Since 470 kΩ is half of 1 MΩ, the base current is twice as much, approximately 30 μA.*

Table 6-1	Troubles and Symptoms		
Trouble	V_B, V	V_C, V	Comments
None	0.7	12	No trouble
R_{BS}	15	15	Transistor blown
R_{BO}	0	15	No base or collector current
R_{CS}	0.7	15	
R_{CO}	0.7	0	
No V_{BB}	0	15	Check supply and lead
No V_{CC}	0.7	0	Check supply and lead

A current gain of 100 gives a collector current of about 3 mA. When this flows through 1 kΩ, it produces a voltage drop of 3 V. Subtracting 3 V from 15 V leaves 12 V across the collector-emitter terminals. So, V_{CE} should measure in the vicinity of 12 V, or else there is something wrong in this circuit.

A Table of Troubles

As discussed in Chap. 5, a shorted component is equivalent to a resistance of zero, whereas an open component is equivalent to a resistance of infinity. For instance, the base resistor R_B may be shorted or open. Let us designate these troubles by R_{BS} and R_{BO}. Similarly, the collector resistor may be shorted or open, symbolized by R_{CS} and R_{CO}.

Table 6-1 shows a few of the troubles that could occur in a circuit like Fig. 6-19. The voltages were calculated using the second approximation. When the circuit is operating normally, you should measure a collector voltage of approximately 12 V. If the base resistor were shorted, +15 V would appear at the base. This large voltage would destroy the emitter diode. The collector diode would probably open as a result, forcing the collector voltage to go to 15 V. The trouble R_{BS} and its voltages are shown in Table 6-1.

If the base resistor were open, there would be no base voltage or current. Furthermore, the collector current would be zero, and the collector voltage would increase to 15 V. The trouble R_{BO} and its voltages are shown in Table 6-1. Continuing like this, we can get the remaining entries of the table.

Figure 6-19 is repeated here for reference.

Figure 6-19 Troubleshooting a circuit.

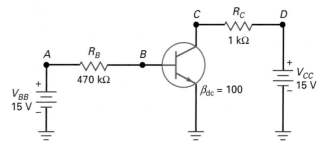

Summary

SEC. 6-1 THE UNBIASED TRANSISTOR

A transistor has three doped regions: an emitter, a base, and a collector. A *pn* junction exists between the base and the emitter; this part of the transistor is called the emitter diode. Another *pn* junction exists between the base and the collector; this part of the transistor is called the collector diode.

SEC. 6-2 THE BIASED TRANSISTOR

For normal operation, you forward bias the emitter diode and reverse bias the collector diode. Under these conditions, the emitter sends free electrons into the base. Most of these free electrons pass through the base to the collector. Because of this, the collector current approximately equals the emitter current. The base current is much smaller, typically less than 5 percent of the emitter current.

SEC. 6-3 TRANSISTOR CURRENTS

The ratio of the collector current to the base current is called the current gain, symbolized as β_{dc} or h_{FE}. For low-power transistors, this is typically 100 to 300. The emitter current is the largest of the three currents, the collector current is almost as large, and the base current is much smaller.

SEC. 6-4 THE CE CONNECTION

The emitter is grounded or common in a CE circuit. The base-emitter part of a transistor acts approximately like an ordinary diode. The base-collector part acts like a current source that is equal to β_{dc} times the base current. The transistor has an active region, a saturation region, a cutoff region, and a breakdown region. The active region is used in linear amplifiers. Saturation and cutoff are used in digital circuits.

SEC. 6-5 THE BASE CURVE

The graph of base current versus base-emitter voltage looks like the graph of an ordinary diode. Because of this, we can use any of the three diode approximations to calculate the base current. Most of the time, the ideal and the second approximation are all that is necessary.

SEC. 6-6 COLLECTOR CURVES

The four distinct operating regions of a transistor are the active region, the saturation region, the cutoff region, and the breakdown region. When it is used as an amplifier, the transistor operates in the active region. When it is used in digital circuits, the transistor usually operates in the saturation and cutoff regions. The breakdown region is avoided because the risk of transistor destruction is too high.

SEC. 6-7 TRANSISTOR APPROXIMATIONS

Exact answers are a waste of time in most electronics work. Almost everybody uses approximations because the answers are adequate for most applications. The ideal transistor is useful for basic troubleshooting. The third approximation is needed for precise design. The second approximation is a good compromise for both troubleshooting and design.

SEC. 6-8 READING DATA SHEETS

Transistors have maximum ratings on their voltages, currents, and powers. Small-signal transistors can dissipate 1 W or less. Power transistors can dissipate more than 1 W. Temperature can change the value of the transistor characteristics. Maximum power decreases with a temperature increase. Also, current gain varies greatly with temperature.

SEC. 6-9 SURFACE-MOUNT TRANSISTORS

Surface-mount transistors (SMTs) are found in a variety of packages. A simple three-terminal gull-wing package is common. Some SMTs are packaged in styles that can dissipate more than 1 W of power. Other surface-mount devices may contain (house) multiple transistors.

SEC. 6-10 TROUBLESHOOTING

When troubles arise, they usually produce large changes in transistor voltages. This is why ideal analysis is usually adequate for troubleshooters. Furthermore, many troubleshooters spurn the use of calculators because it slows down their thinking. The best troubleshooters learn to mentally estimate the voltages they want to measure.

Definitions

(6-2) DC alpha:

$$\alpha_{dc} = \frac{I_C}{I_E}$$

(6-3) DC beta (current gain)

$$\beta_{dc} = \frac{I_C}{I_B}$$

Derivations

(6-1) Emitter current:

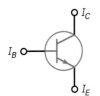

$$I_E = I_C + I_B$$

(6-4) Collector current:

$$I_C = \beta_{dc} I_B$$

(6-5) Base current:

$$I_B = \frac{I_C}{\beta_{dc}}$$

(6-6) Base current:

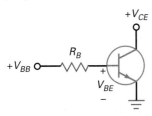

$$I_B = \frac{V_{BB} - V_{BE}}{R_B}$$

(6-7) Collector-emitter voltage:

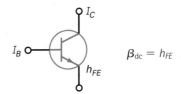

$$V_{CE} = V_{CC} - I_C R_C$$

(6-8) CE power dissipation:

$$P_D = V_{CE} I_C$$

(6-9) Current gain:

$$\beta_{dc} = h_{FE}$$

Student Assignments

1. **A transistor has how many *pn* junctions?**
   a. 1
   b. 2
   c. 3
   d. 4

2. **What is one important thing transistors do?**
   a. Amplify weak signals
   b. Rectify line voltage
   c. Step-down voltage
   d. Emit light

3. **Who invented the first junction transistor?**
   a. Bell
   b. Faraday
   c. Marconi
   d. Schockley

4. **In an *npn* transistor, the majority carriers in the emitter are**
   a. Free electrons
   b. Holes
   c. Neither
   d. Both

5. **The barrier potential across each silicon depletion layer is**
   a. 0
   b. 0.3 V
   c. 0.7 V
   d. 1 V

6. **The emitter diode is usually**
   a. Forward biased
   b. Reverse biased
   c. Nonconducting
   d. Operating in the breakdown region

7. For normal operation of the transistor, the collector diode has to be
   a. Forward biased
   b. Reverse biased
   c. Nonconducting
   d. Operating in the breakdown region

8. The base of an *npn* transistor is thin and
   a. Heavily doped
   b. Lightly doped
   c. Metallic
   d. Doped by a pentavalent material

9. Most of the electrons in the base of an *npn* transistor flow
   a. Out of the base lead
   b. Into the collector
   c. Into the emitter
   d. Into the base supply

10. Most of the electrons in the base of an *npn* transistor do not recombine because they
    a. Have a long lifetime
    b. Have a negative charge
    c. Must flow through the base
    d. Flow out of the base

11. Most of the electrons that flow through the base will
    a. Flow into the collector
    b. Flow out of the base lead
    c. Recombine with base holes
    d. Recombine with collector holes

12. The beta of a transistor is the ratio of the
    a. Collector current to emitter current
    b. Collector current to base current
    c. Base current to collector current
    d. Emitter current to collector current

13. Increasing the collector supply voltage will increase
    a. Base current
    b. Collector current
    c. Emitter current
    d. None of the above

14. The fact that there are many free electrons in a transistor emitter region means the emitter is
    a. Lightly doped
    b. Heavily doped
    c. Undoped
    d. None of the above

15. In a normally biased *npn* transistor, the electrons in the emitter have enough energy to overcome the barrier potential of the
    a. Base-emitter junction
    b. Base-collector junction
    c. Collector-base junction
    d. Recombination path

16. In a *pnp* transistor, the major carriers in the emitter are
    a. Free electrons
    b. Holes
    c. Neither
    d. Both

17. What is the most important fact about the collector current?
    a. It is measured in milliamperes.
    b. It equals the base current divided by the current gain.
    c. It is small.
    d. It approximately equals the emitter current.

18. If the current gain is 100 and the collector current is 10 mA, the base current is
    a. 10 μA   c. 1 A
    b. 100 μA   d. 10 A

19. The base-emitter voltage is usually
    a. Less than the base supply voltage
    b. Equal to the base supply voltage
    c. More than the base supply voltage
    d. Cannot answer

20. The collector-emitter voltage is usually
    a. Less than the collector supply voltage
    b. Equal to the collector supply voltage
    c. More than the collector supply voltage
    d. Cannot answer

21. The power dissipated by a transistor approximately equals the collector current times
    a. Base-emitter voltage
    b. Collector-emitter voltage
    c. Base supply voltage
    d. 0.7 V

22. A small collector current with zero base current is caused by the leakage current of the
    a. Emitter diode
    b. Collector diode
    c. Base diode
    d. Transistor

23. A transistor acts like a diode and a
    a. Voltage source
    b. Current source
    c. Resistance
    d. Power supply

24. If the base current is 100 mA and the current gain is 30, the emitter current is
    a. 3.33 mA
    b. 3 A
    c. 3.1 A
    d. 10 A

25. The base-emitter voltage of an ideal transistor is
    a. 0
    b. 0.3 V
    c. 0.7 V
    d. 1 V

26. If you recalculate the collector-emitter voltage with the second approximation, the answer will usually be
    a. Smaller than the ideal value
    b. The same as the ideal value
    c. Larger than the ideal value
    d. Inaccurate

27. In the active region, the collector current is not changed significantly by
    a. Base supply voltage
    b. Base current
    c. Current gain
    d. Collector resistance

28. The base-emitter voltage of the second approximation is
    a. 0
    b. 0.3 V
    c. 0.7 V
    d. 1 V

29. If the base resistor is open, what is the collector current?
    a. 0
    b. 1 mA
    c. 2 mA
    d. 10 mA

30. When comparing the power dissipation of a 2N3904 transistor to the PZT3904 surface-mount version, the 2N3904
    a. Can handle less power
    b. Can handle more power
    c. Can handle the same power
    d. Is not rated

Problems

SEC. 6-3 TRANSISTOR CURRENTS

6-1 A transistor has an emitter current of 10 mA and a collector current of 9.95 mA. What is the base current?

6-2 The collector current is 10 mA, and the base current is 0.1 mA. What is the current gain?

6-3 A transistor has a current gain of 150 and a base current of 30 μA. What is the collector current?

6-4 If the collector current is 100 mA and the current gain is 65, what is the emitter current?

SEC. 6-5 THE BASE CURVE

6-5 IIII MultiSim What is the base current in Fig. 6-20?

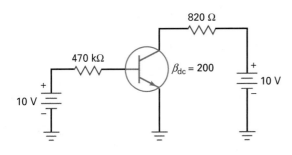

Figure 6-20

6-6 IIII MultiSim If the current gain decreases from 200 to 100 in Fig. 6-20, what is the base current?

6-7 If the 470 kΩ of Fig. 6-20 has a tolerance of $\pm$5 percent, what is the maximum base current?

SEC. 6-6 COLLECTOR CURVES

6-8 IIII MultiSim A transistor circuit similar to Fig. 6-20 has a collector supply voltage of 20 V, a collector resistance of 1.5 kΩ, and a collector current of 6 mA. What is the collector-emitter voltage?

6-9 If a transistor has a collector current of 100 mA and a collector-emitter voltage of 3.5 V, what is its power dissipation?

SEC. 6-7 TRANSISTOR APPROXIMATIONS

6-10 What are the collector-emitter voltage and the transistor power dissipation in Fig. 6-20? (Give answers for the ideal and the second approximation.)

6-11 Figure 6-21a shows a simpler way to draw a transistor circuit. It works the same as the circuits already discussed. What is collector-emitter voltage? The transistor power dissipation? (Give answers for the ideal and the second approximation.)

6-12 When the base and collector supplies are equal, the transistor can be drawn as shown in Fig. 6-21b. What is the collector-emitter voltage in this circuit? The transistor power? (Give answers for the ideal and the second approximation.)

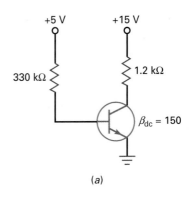

(a)

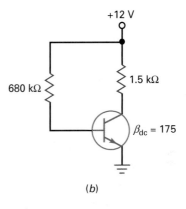

(b)

Figure 6-21

SEC. 6-8 READING DATA SHEETS

6-13 What is the storage temperature range of a 2N3904?

6-14 What is the minimum h_{FE} for a 2N3904 for a collector current of 1 mA and a collector-emitter voltage of 1 V?

6-15 A transistor has a power rating of 1 W. If the collector-emitter voltage is 10 V and the collector current is 120 mA, what happens to the transistor?

6-16 A 2N3904 has a power dissipation of 625 mW without a heat sink. If the ambient temperature is 65°C, what happens to the power rating?

SEC. 6-10 TROUBLESHOOTING

6-17 IIII MultiSim In Fig. 6-20, does the collector-emitter voltage increase, decrease, or remain the same for each of these troubles?

   a. 470 kΩ is shorted

   b. 470 kΩ is open

   c. 820 Ω is shorted

   d. 820 Ω is open

   e. No base supply voltage

   f. No collector supply

Critical Thinking

6-18 What is the dc alpha of a transistor that has a current gain of 200?

6-19 What is the current gain of a transistor with a dc alpha of 0.994?

6-20 Design a CE circuit to meet these specifications: $V_{BB} = 5$ V, $V_{CC} = 15$ V, $h_{FE} = 120$, $I_C = 10$ mA, and $V_{CE} = 7.5$ V.

6-21 In Fig. 6-20, what value base resistor would be needed so $V_{CE} = 6.7$ V?

6-22 A 2N3904 has a power rating of 350 mW at room temperature (25°C). If the collector-emitter voltage is 10 V, what is the maximum current that the transistor can handle for an ambient temperature of 50°C?

6-23 Suppose we connect an LED in series with the 820 Ω of Fig. 6-20. What does the LED current equal?

6-24 What is the collector-emitter saturation voltage of a 2N3904 when the collector current is 50 mA? Use the data sheet.

Up–Down Analysis

Use Fig. 6-22 for the remaining problems. Assume increases of approximately 10 percent in the independent variable, and use the second approximation of the transistor. A response should be an N (no change) if the change in a dependent variable is so small that you would have difficulty measuring it.

6-25 Predict the response of each dependent variable in the row labeled V_{BB}. Then answer the following question as simply and directly as possible. What effect does an increase in the base supply voltage have on the dependent variables of the circuit?

6-26 Predict the response of each dependent variable in the row labeled V_{CC}. Then summarize your findings in one or two sentences.

6-27 Predict the response of each dependent variable in the row labeled R_B. List the dependent variables that decrease. Explain why these variables decrease, using Ohm's law or similar basic ideas.

6-28 Predict the response of each dependent variable in the row labeled R_C. List the dependent variables that show no change. Explain why these variables show no change.

6-29 Predict the response of each dependent variable in the row labeled β_{dc}. List the dependent variables that decrease. Explain why these variables decrease.

Figure 6-22 Up-down analysis.

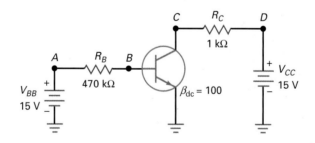

UP-DOWN ANALYSIS

		DEPENDENT VARIABLES								
		V_A	V_B	V_C	V_D	I_B	I_C	P_B	P_C	P_D
INDEPENDENT VARIABLES	V_{BB}									
	V_{CC}									
	R_B									
	R_C									
	β_{dc}									

Job Interview Questions

1. I want you to draw an *npn* transistor showing the *n* and *p* regions. Then I want you to bias the transistor properly and tell me how it works.
2. Draw a set of collector curves. Then, using these curves, show me where the four operating regions of a transistor are.
3. Draw the two equivalent circuits (ideal and second approximation) to represent a transistor that is operating in the active region. Then, tell me when and how you would use these circuits to calculate the transistor currents and voltages.
4. Draw a transistor circuit with a CE connection. Now, what kind of troubles can you get with a circuit like this and what measurements would you make to isolate each trouble?
5. When looking at a schematic diagram that shows *npn* and *pnp* transistors, how can you identify each type? How can you tell the direction of electron (or conventional) flow?
6. Name a test instrument that can display a set of collector curves, I_C versus V_{CE}, for a transistor.
7. What is the formula for transistor power dissipation? Knowing this relationship, where on the load line would you expect the power dissipation to be maximum?
8. What are the three currents in a transistor, and how are they related?
9. Draw an *npn* and a *pnp* transistor. Label all currents and show directions of flow.
10. Transistors may be connected in any of the following configurations: common emitter, common collector, and common base. Which is the most common configuration?

Self-Test Answers

1.	b	11.	a	21.	b
2.	a	12.	b	22.	b
3.	d	13.	d	23.	b
4.	a	14.	b	24.	c
5.	c	15.	a	25.	a
6.	a	16.	b	26.	c
7.	b	17.	d	27.	d
8.	b	18.	b	28.	c
9.	b	19.	a	29.	a
10.	a	20.	a	30.	a

Practice Problem Answers

6-1 $\beta_{dc} = 200$

6-2 $I_C = 10$ mA

6-3 $I_B = 74.1$ μA

6-4 $V_B = 0.7$ V;
$I_B = 33$ μA;
$I_C = 6.6$ mA

6-5 $I_B = 13.7$ μA;
$I_C = 4.11$ mA;
$V_{CE} = 1.78$ V;
$P_D = 7.32$ mW

6-6 $I_B = 16.6$ μA;
$I_C = 5.89$ mA;
$\beta_{dc} = 355$

6-10 Ideal: $I_B = 14.9$ μA;
$I_C = 1.49$ mA;
$V_{CE} = 9.6$ V
Second: $I_B = 13.4$ μA;
$I_C = 1.34$ mA;
$V_{CE} = 10.2$ V

6-12 $P_{D\,(max)} = 375$ mW. Not within the safety factor of 2.

chapter 7

Transistor Fundamentals

There are two basic ways to set up the operating point of a transistor: base bias and emitter bias. Base bias produces a fixed value of base current, whereas emitter bias produces a fixed value of emitter current. Base bias is most useful in switching circuits, whereas emitter bias is predominant in amplifying circuits. This chapter discusses base bias, emitter bias, switching circuits, and optoelectronic circuits.

Objectives

After studying this chapter, you should be able to:

- State why base bias does not work well in amplifying circuits.
- Identify the saturation point and the cutoff point for a given base-biased circuit.
- Calculate the Q point for a given base-biased circuit.
- Draw an emitter bias circuit and explain why it works well in amplifying circuits.
- Demonstrate how to do out-of-circuit and in-circuit transistor tests.

Chapter Outline

Vocabulary

amplifying circuit
base bias
correction factor
cutoff point
emitter bias

hard saturation
load line
phototransistor
quiescent point
saturation point

soft saturation
switching circuit
two-state circuit

7-1 Variations in Current Gain

The current gain β_{dc} of a transistor depends on three factors: the transistor, the collector current, and the temperature. For instance, when you replace a transistor with another of the same type, the current gain usually changes. Likewise, if the collector current or temperature changes, the current gain will change.

Worst and Best Case

As a concrete example, the data sheet of a 2N3904 lists a minimum h_{FE} of 100 and a maximum of 300 when the temperature is 25°C and the collector current is 10 mA. If we build thousands of circuits with 2N3904s, some of the transistors will have a current gain as low as 100 (worst case), and others will have a current gain as high as 300 (best case).

Figure 7-1 shows the graphs of a 2N3904 for the worst case (minimum h_{FE}). Look at the middle curve, the current gain for an ambient temperature of 25°C. When the collector current is 10 mA, the current gain is 100, the worst case for a 2N3904. (In the best case, a few 2N3904s have a current gain of 300 at 10 mA and 25°C.)

Effect of Current and Temperature

When the temperature is 25°C (the middle curve), the current gain is 50 at 0.1 mA. As the current increases from 0.1 mA to 10 mA, h_{FE} increases to a maximum of 100. Then, it decreases to less than 20 at 200 mA.

Also notice the effect of temperature. When the temperature decreases, the current gain is less (the bottom curve). On the other hand, when the temperature increases, h_{FE} increases over most of the current range (the top curve).

Main Idea

As you can see, transistor replacement, collector-current changes, or temperature changes can produce large changes in h_{FE} or β_{dc}. At a given temperature, a 3:1 change is possible when a transistor is replaced. When the temperature varies, an additional 3:1 variation is possible. And when the current varies, more than a 3:1 variation is possible. In summary, the 2N3904 may have a current gain of less than 10 to more than 300. Because of this, any design that depends on a precise value of current gain will fail in mass production.

Figure 7-1 Variation of current gain.

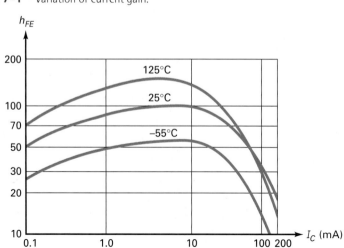

7-2 The Load Line

Figure 7-2a shows the CE connection discussed in Chap. 6. Given the values of R_B and β_{dc}, we can calculate collector current I_C and collector voltage V_{CE} using the methods of the preceding chapter.

Base Bias

The circuit of Fig. 7-2a is an example of **base bias,** which means setting up a *fixed value of base current.* For instance, if $R_B = 1$ MΩ, the base current is 14.3 μA (second approximation). Even with transistor replacements and temperature changes, the base current remains fixed at approximately 14.3 μA under all operating conditions.

If $\beta_{dc} = 100$ in Fig. 7-2a, the collector current is approximately 1.43 mA and the collector-emitter voltage is:

$$V_{CE} = V_{CC} - I_C R_C = 15 \text{ V} - (1.43 \text{ mA})(3 \text{ k}\Omega) = 10.7 \text{ V}$$

Therefore, the quiescent or Q point in Fig. 7-2a is:

$$I_C = 1.43 \text{ mA} \qquad \text{and} \qquad V_{CE} = 10.7 \text{ V}$$

Graphical Solution

We can also find the Q point using a graphical solution based on the transistor **load line,** a graph of I_C versus V_{CE}. In Fig. 7-2a, the collector-emitter voltage is given by:

$$V_{CE} = V_{CC} - I_C R_C$$

Solving for I_C gives:

$$I_C = \frac{V_{CC} - V_{CE}}{R_C} \tag{7-1}$$

If we graph this equation (I_C versus V_{CE}), we will get a straight line. This line is called the *load line* because it represents the effect of the load on I_C and V_{CE}.

For instance, substituting the values of Fig. 7-2a into Eq. (7-1) gives:

$$I_C = \frac{15 \text{ V} - V_{CE}}{3 \text{ k}\Omega}$$

Figure 7-2 Base bias. (*a*) Circuit; (*b*) load line.

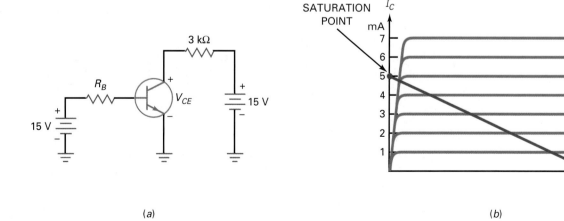

(*a*)

(*b*)

This equation is a linear equation; that is, its graph is a straight line. (*Note:* A *linear equation* is any equation that can be reduced to the standard form of $y = mx + b$.) If we graph the foregoing equation on top of the collector curves, we get Fig. 7-2b.

The ends of the load line are the easiest to find. When $V_{CE} = 0$ in the load-line equation (the foregoing equation):

$$I_C = \frac{15 \text{ V}}{3 \text{ k}\Omega} = 5 \text{ mA}$$

The values, $I_C = 5$ mA and $V_{CE} = 0$, plot as the upper end of the load line in Fig. 7-2b.

When $I_C = 0$, the load-line equation gives:

$$0 = \frac{15 \text{ V} - V_{CE}}{3 \text{ k}\Omega}$$

or

$$V_{CE} = 15 \text{ V}$$

The coordinates, $I_C = 0$ and $V_{CE} = 15$ V plot as the lower end of the load line in Fig. 7-2b.

Visual Summary of All Operating Points

Why is the load line useful? Because it contains every possible operating point for the circuit. Stated another way, when the base resistance varies from zero to infinity, it causes I_B to vary, which makes I_C and V_{CE} to vary over their entire ranges. If you plot the I_C and V_{CE} values for every possible I_B value, you will get the load line. Therefore, the load line is a visual summary of *all the possible transistor operating points*.

The Saturation Point

When the base resistance is too small, there is too much collector current, and the collector-emitter voltage drops to approximately zero. In this case, the transistor goes into *saturation*. This means that the collector current has increased to its maximum possible value.

The **saturation point** is the point in Fig. 7-2b where the load line intersects the saturation region of the collector curves. Because the collector-emitter voltage V_{CE} at saturation is very small, the saturation point is almost touching the upper end of the load line. From now on, we will approximate the saturation point as the upper end of the load line, bearing in mind that there is a slight error.

The saturation point tells you the maximum possible collector current for the circuit. For instance, the transistor of Fig. 7-3a goes into saturation when the collector current is approximately 5 mA. At this current, V_{CE} has decreased to approximately zero.

There is an easy way to find the current at the saturation point. Visualize a short between the collector and emitter to get Fig. 7-3b. Then V_{CE} drops to zero. All the 15 V from the collector supply will be across the 3 kΩ. Therefore, the current is:

$$I_C = \frac{15 \text{ V}}{3 \text{ k}\Omega} = 5 \text{ mA}$$

You can apply this "mental short" method to any base-biased circuit.

Here is the formula for the saturation current in base-biased circuits:

$$I_{C(\text{sat})} = \frac{V_{CC}}{R_C} \tag{7-2}$$

GOOD TO KNOW

When a transistor is saturated, further increases in base current produce no further increases in collector current.

Figure 7-3 Finding the ends of the load line. (*a*) Circuit; (*b*) calculating collector saturation current; (*c*) calculating collector-emitter cutoff voltage.

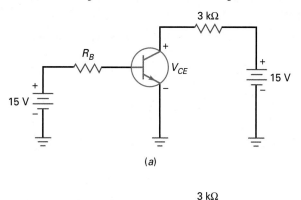

(a)

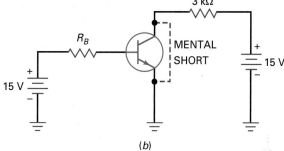

(b)

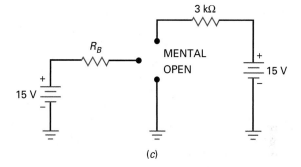

(c)

This says that the maximum value of the collector current equals the collector supply voltage divided by the collector resistance. It is nothing more than Ohm's law applied to the collector resistor. Figure 7-3*b* is a visual reminder of this equation.

The Cutoff Point

The **cutoff point** is the point at which the load line intersects the cutoff region of the collector curves in Fig. 7-2*b*. Because the collector current at cutoff is very small, the cutoff point almost touches the lower end of the load line. From now on, we will approximate the cutoff point as the lower end of the load line.

The cutoff point tells you the maximum possible collector-emitter voltage for the circuit. In Fig. 7-3*a*, the maximum possible V_{CE} is approximately 15 V, the collector supply voltage.

There is a simple process for finding the cutoff voltage. Visualize the transistor of Fig. 7-3*a* as an open between the collector and the emitter (see Fig. 7-3*c*). Since there is no current through the collector resistor for this open condition, all the 15 V from the collector supply will appear between the collector-emitter terminals. Therefore, the voltage between the collector and the emitter will equal 15 V:

$$V_{CE\text{(cutoff)}} = V_{CC} \tag{7-3}$$

Example 7-1

What are the saturation current and the cutoff voltage in Fig. 7-4a?

SOLUTION   Visualize a short between the collector and emitter. Then:

$$I_{C(sat)} = \frac{30\ V}{3\ k\Omega} = 10\ mA$$

Next, visualize the collector-emitter terminals open. In this case:

$$V_{CE(cutoff)} = 30\ V$$

Figure 7-4   Load lines when collector resistance is the same. (*a*) With a collector supply of 30 V; (*b*) with a collector supply of 9 V; (*c*) load lines have same slope.

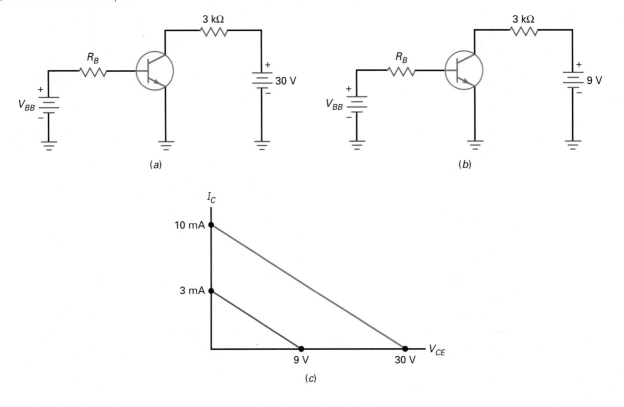

Example 7-2

Calculate the saturation and cutoff values for Fig. 7-4b. Draw the load lines for this and the preceding example.

SOLUTION   With a mental short between the collector and emitter:

$$I_{C(sat)} = \frac{9\ V}{3\ k\Omega} = 3\ mA$$

A mental open between the collector and emitter gives:

$$V_{CE(cutoff)} = 9\ V$$

Figure 7-4c shows the two load lines. Changing the collector supply voltage while keeping the same collector resistance produces two load lines of the same slope but with different saturation and cutoff values.

PRACTICE PROBLEM 7-2   Find the saturation current and cutoff voltage of Fig. 7-2a if the collector resistor is 2 kΩ and V_{CC} is 12 V.

Example 7-3

What are the saturation current and the cutoff voltage in Fig. 7-5a?

SOLUTION   The saturation current is:

$$I_{C(\text{sat})} = \frac{15\text{ V}}{1\,\text{k}\Omega} = 15\text{ mA}$$

The cutoff voltage is:

$$V_{CE(\text{cutoff})} = 15\text{ V}$$

Figure 7-5   Load lines when collector voltage is the same. (a) With a collector resistance of 1 kΩ; (b) with a collector resistance of 3 kΩ; (c) smaller R_C produces steeper slope.

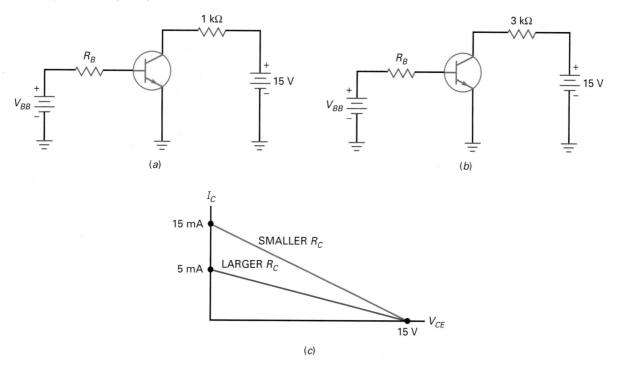

Example 7-4

Calculate the saturation and cutoff values for Fig. 7-5b. Then, compare the load lines for this and the preceding example.

SOLUTION   The calculations are as follows:

$$I_{C(\text{sat})} = \frac{15\text{ V}}{3\,\text{k}\Omega} = 5\text{ mA}$$

and

$$V_{CE(cutoff)} = 15 \text{ V}$$

Figure 7-5c shows the two load lines. Changing the collector resistor with the same collector supply voltage produces load lines of different slopes but the same cutoff values. Also, notice that a smaller collector resistance produces a larger slope (steeper or closer to vertical). This happens because the slope of the load line is equal to the reciprocal of the collector resistance:

$$\text{Slope} = \frac{1}{R_C}$$

PRACTICE PROBLEM 7-4 Using Fig. 7-5b, what happens to the circuit's load line if the collector resistor is changed to 5 kΩ?

7-3 The Operating Point

Every transistor circuit has a load line. Given any circuit, work out the saturation current and the cutoff voltage. These values are plotted on the vertical and horizontal axes. Then draw a line through these two points to get the load line.

Plotting the Q Point

Figure 7-6a shows a base-biased circuit with a base resistance of 500 kΩ. We get the saturation current and cutoff voltage by the process given earlier. First, visualize a short across the collector-emitter terminals. Then all the collector supply voltage appears across the collector resistor, which means that the saturation current is 5 mA. Second, visualize the collector-emitter terminals open. Then there is no current, and all the supply voltage appears across the collector-emitter terminals, which means that the cutoff voltage is 15 V. If we plot the saturation current and cutoff voltage, we can draw the load line shown in Fig. 7-6b.

Figure 7-6 Calculating the Q point. (a) Circuit; (b) change in current gain changes Q point.

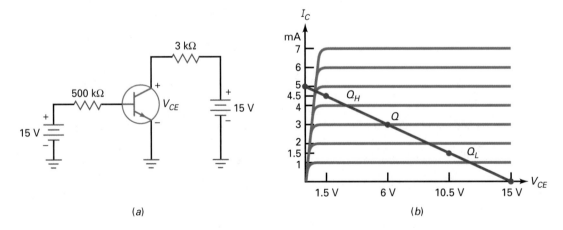

(a)

(b)

Let us keep the discussion simple for now by assuming an ideal transistor. This means that all the base supply voltage will appear across the base resistor. Therefore, the base current is:

$$I_B = \frac{15 \text{ V}}{500 \text{ k}\Omega} = 30 \text{ } \mu\text{A}$$

We cannot proceed until we have a value for the current gain. Suppose the current gain of the transistor is 100. Then the collector current is:

$$I_C = 100(30 \text{ } \mu\text{A}) = 3 \text{ mA}$$

This current flowing through 3 kΩ produces a voltage of 9 V across the collector resistor. When we subtract this from the collector supply voltage, we get the voltage across the transistor. Here are the calculations:

$$V_{CE} = 15 \text{ V} - (3 \text{ mA})(3 \text{ k}\Omega) = 6 \text{ V}$$

By plotting 3 mA and 6 V (the collector current and voltage), we get the operating point shown on the load line of Fig. 7-6b. The operating point is labeled Q because this point is often called the **quiescent point.** (*Quiescent* means quiet, still, or resting.)

Why the Q Point Varies

We assumed a current gain of 100. What happens if the current gain is 50? If it is 150? To begin, the base current remains the same because the current gain has no effect on the base current. Ideally, the base current is fixed at 30 μA. When the current gain is 50:

$$I_C = 50(30 \text{ } \mu\text{A}) = 1.5 \text{ mA}$$

and the collector-emitter voltage is:

$$V_{CE} = 15 \text{ V} - (1.5 \text{ mA})(3 \text{ k}\Omega) = 10.5 \text{ V}$$

Plotting the values gives the low point Q_L shown in Fig. 7-6b.
If the current gain is 150, then:

$$I_C = 150(30 \text{ } \mu\text{A}) = 4.5 \text{ mA}$$

and the collector-emitter voltage is:

$$V_{CE} = 15 \text{ V} - (4.5 \text{ mA})(3 \text{ k}\Omega) = 1.5 \text{ V}$$

Plotting these values gives the high point Q_H point shown in Fig. 7-6b.
The three Q points of Fig. 7-6b illustrate how sensitive the operating point of a base-biased transistor is to changes in β_{dc}. When the current gain varies from 50 to 150, the collector current changes from 1.5 to 4.5 mA. If the changes in current gain were much greater, the operating point could be driven easily into saturation or cutoff. In this case, an amplifying circuit would become useless because of the loss of current gain outside the active region.

The Formulas

The formulas for calculating the Q point are as follows:

$$I_B = \frac{V_{BB} - V_{BE}}{R_B} \tag{7-4}$$

$$I_C = \beta_{dc}I_B \tag{7-5}$$

$$V_{CE} = V_{CC} - I_C R_C \tag{7-6}$$

Example 7-5

Suppose the base resistance of Fig. 7-6a is increased to 1 MΩ. What happens to the collector-emitter voltage if β_{dc} is 100?

SOLUTION Ideally, the base current would decrease to 15 μA, the collector current would decrease to 1.5 mA, and the collector-emitter voltage would increase to:

$$V_{CE} = 15 - (1.5 \text{ mA})(3 \text{ k}\Omega) = 10.5 \text{ V}$$

To a second approximation, the base current would decrease to 14.3 μA, and the collector current would decrease to 1.43 mA. The collector-emitter voltage would increase to:

$$V_{CE} = 15 - (1.43 \text{ mA})(3 \text{ k}\Omega) = 10.7 \text{ V}$$

PRACTICE PROBLEM 7-5 If the β_{dc} value of Example 7-5 changed to 150 due to a temperature change, find the new value of V_{CE}.

7-4 Recognizing Saturation

There are two basic kinds of transistor circuits: **amplifying** and **switching.** With amplifying circuits, the Q point must remain in the active region under all operating conditions. If it does not, the output signal will be distorted on the peak where saturation or cutoff occurs. With switching circuits, the Q point usually switches between saturation and cutoff. How switching circuits work, what they do, and why they are used will be discussed later.

Impossible Answers

Assume that the transistor of Fig. 7-7a has a breakdown voltage greater than 20 V. Then, we know that it is not operating in the breakdown region. Furthermore, we can tell at a glance that the transistor is not operating in the cutoff region because of the biasing voltages. What is not immediately clear, however, is whether the transistor is operating in the active region or the saturation region. It must be operating in one of these regions. But which?

Figure 7-7 (a) Base-biased circuit; (b) load line.

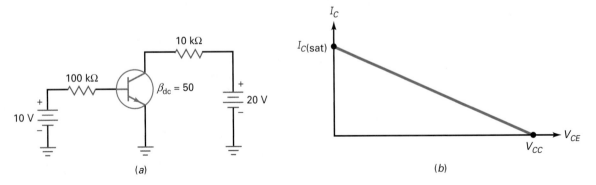

(a)

(b)

Troubleshooters and designers often use the following method to determine whether a transistor is operating in the active region or the saturation region. Here are the steps in using the method:

1. Assume that the transistor is operating in the active region.
2. Carry out the calculations for currents and voltages.
3. If an impossible result occurs in any calculation, the assumption is false.

An impossible answer means that the transistor is saturated. Otherwise, the transistor is operating in the active region.

Saturation–Current Method

For instance, Fig. 7-7a shows a base-biased circuit. Start by calculating the saturation current:

$$I_{C(\text{sat})} = \frac{20\,\text{V}}{10\,\text{k}\Omega} = 2\,\text{mA}$$

The base current is ideally 0.1 mA. Assuming a current gain of 50 as shown, the collector current is:

$$I_C = 50(0.1\,\text{mA}) = 5\,\text{mA}$$

The answer is impossible because the collector current cannot be greater than the saturation current. Therefore, the transistor cannot be operating in the active region; it must be operating in the saturation region.

Collector–Voltage Method

Suppose you want to calculate V_{CE} in Fig. 7-7a. Then you can proceed like this: The base current is ideally 0.1 mA. Assuming a current gain of 50 as shown, the collector current is:

$$I_C = 50(0.1\,\text{mA}) = 5\,\text{mA}$$

and the collector-emitter voltage is:

$$V_{CE} = 20\,\text{V} - (5\,\text{mA})(10\,\text{k}\Omega) = -30\,\text{V}$$

This result is impossible because the collector-emitter voltage cannot be negative. So the transistor cannot be operating in the active region; it must be operating in the saturation region.

Current Gain Is Less in Saturation Region

When you are given the current gain, it is usually for the active region. For instance, the current gain of Fig. 7-7a is shown as 50. This means that the collector current will be 50 times the base current, provided the transistor is operating in the active region.

When a transistor is saturated, the current gain is less than the current gain in the active region. You can calculate the saturated current gain as follows:

$$\beta_{\text{dc(sat)}} = \frac{I_{C(\text{sat})}}{I_B}$$

In Fig. 7-7a, the saturated current gain is

$$\beta_{\text{dc(sat)}} = \frac{2\,\text{mA}}{0.1\,\text{mA}} = 20$$

Hard Saturation

A designer who wants a transistor to operate in the saturation region under all conditions often selects a base resistance that produces a current gain of 10. This is called **hard saturation,** because there is more than enough base current to saturate the transistor. For example, a base resistance of 50 kΩ in Fig. 7-7a will produce a current gain of:

$$\beta_{dc} = \frac{2 \text{ mA}}{0.2 \text{ mA}} = 10$$

For the transistor of Fig. 7-7a, it takes only

$$I_B = \frac{2 \text{ mA}}{50} = 0.04 \text{ mA}$$

to saturate the transistor. Therefore, a base current of 0.2 mA will drive the transistor deep into saturation.

Why does a designer use hard saturation? Recall that the current gain changes with collector current, temperature variation, and transistor replacement. To make sure that the transistor does not slip out of saturation at low collector currents, low temperatures, and so on, the designer uses hard saturation to ensure transistor saturation under all operating conditions.

From now on, *hard saturation* will refer to any design that makes the saturated current gain approximately 10. **Soft saturation** will refer to any design in which the transistor is barely saturated, that is, in which the saturated current gain is only a little less than the active current gain.

Recognizing Hard Saturation at a Glance

Here is how you can quickly tell whether a transistor is in hard saturation. Often, the base supply voltage and the collector supply voltage are equal: $V_{BB} = V_{CC}$. When this is the case, a designer will use the 10 : 1 rule, which says to make the base resistance approximately 10 times as large as the collector resistance.

Figure 7-8a was designed by using the 10 : 1 rule. Therefore, whenever you see a circuit with a 10 : 1 ratio (R_B to R_C), you can expect it to be saturated.

Figure 7-8   (a) Hard saturation; (b) load line.

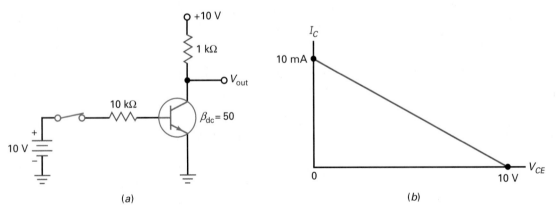

(a)

(b)

Example 7-6

Suppose the base resistance of Fig. 7-7a is increased to 1 MΩ. Is the transistor still saturated?

SOLUTION Assume the transistor is operating in the active region, and see whether a contradiction arises. Ideally, the base current is 10 V divided by 1 MΩ, or 10 μA. The collector current is 50 times 10 μA, or 0.5 mA. This current produces 5 V across the collector resistor. Subtract 5 from 20 V to get:

$$V_{CE} = 15 \text{ V}$$

There is no contradiction here. If the transistor were saturated, we would have calculated a negative number or, at most, 0 V. Because we got 15 V, we know that the transistor is operating in the active region.

Example 7-7

Suppose the collector resistance of Fig. 7-7a is decreased to 5 kΩ. Does the transistor remain in the saturation region?

SOLUTION Assume the transistor is operating in the active region, and see whether a contradiction arises. We can use the same approach as in Example 7-6, but for variety, let us try the second method.

Start by calculating the saturation value of the collector current. Visualize a short between the collector and the emitter. Then you can see that 20 V will be across 5 kΩ. This gives a saturated collector current of:

$$I_{C(\text{sat})} = 4 \text{ mA}$$

The base current is ideally 10 V divided by 100 kΩ, or 0.1 mA. The collector current is 50 times 0.1 mA, or 5 mA.

There is a contradiction. The collector current cannot be greater than 4 mA because the transistor saturates when I_C = 4 mA. The only thing that can change at this point is the current gain. The base current is still 0.1 mA, but the current gain decreases to:

$$\beta_{\text{dc(sat)}} = \frac{4\,\text{mA}}{0.1\,\text{mA}} = 40$$

This reinforces the idea discussed earlier. A transistor has two current gains, one in the active region and another in the saturation region. The second is equal to or smaller than the first.

PRACTICE PROBLEM 7–7 If the collector resistance of Fig. 7-7a is 4.7 kΩ, what value of base resistor would produce hard saturation using the 10:1 design rule?

7-5 The Transistor Switch

Base bias is useful in *digital circuits* because these circuits are usually designed to operate at saturation and cutoff. Because of this, they have either low output voltage or high output voltage. In other words, none of the Q points between saturation and cutoff are used. For this reason, variations in the Q point don't matter, because the transistor remains in saturation or cutoff when the current gain changes.

Here is an example of using a base-biased circuit to switch between saturation and cutoff. Figure 7-8a shows an example of a transistor in hard saturation. Therefore, the output voltage is approximately 0 V. This means the Q point is at the upper end of the load line (Fig. 7-8b).

When the switch opens, the base current drops to zero. Because of this, the collector current drops to zero. With no current through the 1 kΩ, all the collector supply voltage will appear across the collector-emitter terminals. Therefore, the output voltage rises to +10 V. Now, the Q point is at the lower end of load line (see Fig. 7-8b).

The circuit can have only two output voltages: 0 or +10 V. This is how you can recognize a digital circuit. It has only two output levels: low or high. The exact values of the two output voltages are not important. All that matters is that you can distinguish the voltages as low or high.

Digital circuits are often called *switching circuits* because their Q point switches between two points on the load line. In most designs, the two points are saturation and cutoff. Another name often used is **two-state circuits,** referring to the low and high outputs.

Example 7-8

The collector supply voltage of Fig. 7-8a is decreased to 5 V. What are the two values of the output voltage? If the saturation voltage $V_{CE(sat)}$ is 0.15 V and the collector leakage current I_{CEO} is 50 nA, what are the two values of the output voltage?

SOLUTION   The transistor switches between saturation and cutoff. Ideally, the two values of output voltage are 0 and 5 V. The first voltage is the voltage across the saturated transistor, and the second voltage is the voltage across the cutoff transistor.

If you include the effects of saturation voltage and collector leakage current, the output voltages are 0.15 and 5 V. The first voltage is the voltage across the saturated transistor, which is given as 0.15 V. The second voltage is the collector-emitter voltage with 50 nA flowing through 1 kΩ:

$$V_{CE} = 5 \text{ V} - (50 \text{ nA})(1 \text{ k}\Omega) = 4.99995 \text{ V}$$

which rounds to 5 V.

Unless you are a designer, it is a waste of time to include the saturation voltage and the leakage current in your calculations of switching circuits. With switching circuits, all you need is two distinct voltages: one low and the other high. It doesn't matter whether the low voltage is 0, 0.1, 0.15 V, and so on. Likewise, it doesn't matter whether the high voltage is 5, 4.9, or 4.5 V. All that usually matters in the analysis of switching circuits is that you can distinguish the low voltage from the high voltage.

PRACTICE PROBLEM 7–8   If the circuit of Fig. 7-8a used 12 V for its collector and base supply voltage, what are the two values of switched output voltage? ($V_{CE(sat)} = 0.15$ V and $I_{CEO} = 50$ nA)

7-6 Emitter Bias

Digital circuits are the type of circuits used in computers. In this area, base bias and circuits derived from base bias are useful. But when it comes to amplifiers, we need circuits whose Q points are immune to changes in current gain.

Figure 7-9 shows **emitter bias.** As you can see, the resistor has been moved from the base circuit to the emitter circuit. That one change makes all the difference in the world. The Q point of this new circuit is now rock-solid. When the current gain changes from 50 to 150, the Q point shows almost no movement along the load line.

Basic Idea

The base supply voltage is now applied directly to the base. Therefore, a troubleshooter will read V_{BB} between the base and ground. The emitter is no longer grounded. Now the emitter is above the ground and has a voltage given by:

$$V_E = V_{BB} - V_{BE} \qquad (7\text{-}7)$$

If V_{BB} is more than 20 times V_{BE}, the ideal approximation will be accurate. If V_{BB} is less than 20 times V_{BE}, you may want to use the second approximation. Otherwise your error will be more than 5 percent.

Finding the Q Point

Let us analyze the emitter-biased circuit of Fig. 7-10. The base supply voltage is only 5 V, so we use the second approximation. The voltage between the base and ground is 5 V. From now on, we refer to this base-to-ground voltage as the *base voltage,* or V_B. The voltage across the base-emitter terminals is 0.7 V. We refer to this voltage as the *base-emitter voltage,* or V_{BE}.

Figure 7-9 Emitter bias.

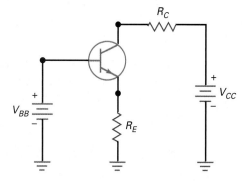

Figure 7-10 Finding the Q point.

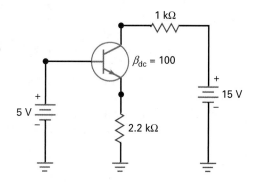

The voltage between the emitter and ground is called the *emitter voltage.* It equals:

$$V_E = 5\text{ V} - 0.7\text{ V} = 4.3\text{ V}$$

This voltage is across the emitter resistance, so we can use Ohm's law to find the emitter current:

$$I_E = \frac{4.3\text{ V}}{2.2\text{ k}\Omega} = 1.95\text{ mA}$$

This means that the collector current is 1.95 mA to a close approximation. When this collector current flows through the collector resistor, it produces a voltage drop of 1.95 V. Subtracting this from the collector supply voltage gives the voltage between the collector and ground:

$$V_C = 15\text{ V} - (1.95\text{ mA})(1\text{ k}\Omega) = 13.1\text{ V}$$

From now on, we will refer to this collector-to-ground voltage as the *collector voltage.*

This is the voltage a troubleshooter would measure when testing a transistor circuit. One lead of the voltmeter would be connected to the collector, and the other lead would be connected to ground. If you want the collector-emitter voltage, you have to subtract the emitter voltage from the collector voltage as follows:

$$V_{CE} = 13.1\text{ V} - 4.3\text{ V} = 8.8\text{ V}$$

So, the emitter-biased circuit of Fig. 7-10 has a Q point with these coordinates: $I_C = 1.95$ mA and $V_{CE} = 8.8$ V.

The collector-emitter voltage is the voltage used for drawing load lines and for reading transistor data sheets. As a formula:

$$V_{CE} = V_C - V_E \tag{7-8}$$

Circuit Is Immune to Changes in Current Gain

Here is why emitter bias excels. The Q point of an emitter-biased circuit is immune to changes in current gain. The proof lies in the process used to analyze the circuit. Here are the steps we used earlier:

1. Get the emitter voltage.
2. Calculate the emitter current.
3. Find the collector voltage.
4. Subtract the emitter from the collector voltage to get V_{CE}.

At no time do we need to use the current gain in the foregoing process. Since we don't use it to find the emitter current, collector current, and so on, the exact value of current gain no longer matters.

By moving the resistor from the base to the emitter circuit, we force the base-to-ground voltage to equal the base supply voltage. Before, almost all this supply voltage was across the base resistor, setting up a *fixed base current.* Now, all this supply voltage minus 0.7 V is across the emitter resistor, setting up a *fixed emitter current.*

Minor Effect of Current Gain

The current gain has a minor effect on the collector current. Under all operating conditions, the three currents are related by:

$$I_E = I_C + I_B$$

which can be rearranged as:

$$I_E = I_C + \frac{I_C}{\beta_{\text{dc}}}$$

Solve this for the collector current, and you get:

$$I_C = \frac{\beta_{dc}}{\beta_{dc} + 1} I_E \qquad (7\text{-}9)$$

The quantity that multiplies I_E is called a **correction factor.** It tells you how I_C differs from I_E. When the current gain is 100, the correction factor is:

$$\frac{\beta_{dc}}{\beta_{dc} + 1} = \frac{100}{100 + 1} = 0.99$$

This means that the collector current is equal to 99 percent of the emitter current. Therefore, we get only a 1 percent error when we ignore the correction factor and say that the collector current equals the emitter current.

Example 7-9   ‖‖ MultiSim

What is the voltage between the collector and ground in the MultiSim Fig. 7-11? Between the collector and the emitter?

SOLUTION   The base voltage is 5 V. The emitter voltage is 0.7 V less than this, or:

$$V_E = 5\text{ V} - 0.7\text{ V} = 4.3\text{ V}$$

‖‖ MultiSim   **Figure 7-11**   Meter Values.

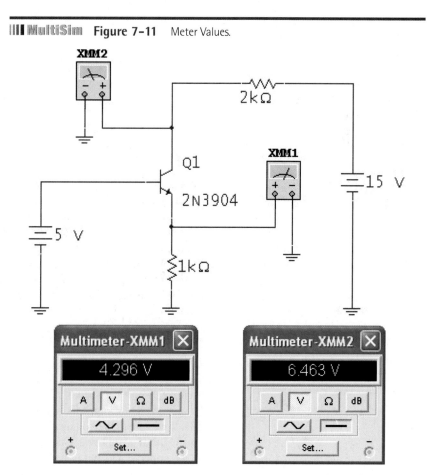

This voltage is across the emitter resistance, which is now 1 kΩ. Therefore, the emitter current is 4.3 V divided by 1 kΩ, or:

$$I_E = \frac{4.3 \text{ V}}{1 \text{ k}\Omega} = 4.3 \text{ mA}$$

The collector current is approximately equal to 4.3 mA. When this current flows through the collector resistance (now 2 kΩ), it produces a voltage of:

$$I_C R_C = (4.3 \text{ mA})(2 \text{ k}\Omega) = 8.6 \text{ V}$$

When you subtract this voltage from the collector supply voltage, you get:

$$V_C = 15 \text{ V} - 8.6 \text{ V} = 6.4 \text{ V}$$

This voltage value is very close to the value measured by the MultiSim meter. Remember, this is the voltage between the collector and ground. This is what you would measure when troubleshooting.

Unless you have a voltmeter with a high input resistance and a floating ground lead, you should not attempt to connect a voltmeter directly between the collector and the emitter because this may short the emitter to ground. If you want to know the value of V_{CE}, you should measure the collector-to-ground voltage, then measure the emitter-to-ground voltage, and subtract the two. In this case:

$$V_{CE} = 6.4 \text{ V} - 4.3 \text{ V} = 2.1 \text{ V}$$

PRACTICE PROBLEM 7-9 IIII MultiSim Decrease the base supply voltage of Fig. 7-11 to 3 V. Predict and measure the new value of V_{CE}.

7-7 LED Drivers

You have learned that base-biased circuits set up a fixed value of base current, and emitter-biased circuits set up a fixed value of emitter current. Because of the problem with current gain, base-biased circuits are normally designed to switch between saturation and cutoff, whereas emitter-biased circuits are usually designed to operate in the active region.

In this section, we discuss two circuits that can be used as LED drivers. The first circuit uses base bias, and the second circuit uses emitter bias. This will give you a chance to see how each circuit performs in the same application.

Base-Biased LED Driver

The base current is zero in Fig. 7-12a, which means that the transistor is at cutoff. When the switch of Fig. 7-12a closes, the transistor goes into hard saturation. Visualize a short between the collector-emitter terminals. Then the collector supply voltage (15 V) appears across the series connection of the 1.5 kΩ and the LED. If we ignore the voltage drop across the LED, the collector current is ideally 10 mA. But if we allow 2 V across the LED, then there is 13 V across the 1.5 kΩ, and the collector current is 13 V divided by 1.5 kΩ, or 8.67 mA.

Figure 7-12 (*a*) Base-biased; (*b*) emitter-biased.

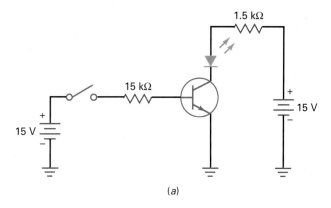

(*a*)

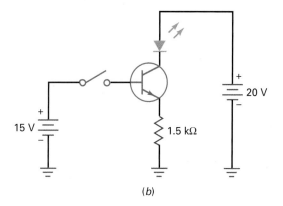

(*b*)

There is nothing wrong with this circuit. It makes a fine LED driver because it is designed for hard saturation, where the current gain doesn't matter. If you want to change the LED current in this circuit, you can change either the collector resistance or the collector supply voltage. The base resistance is made 10 times larger than the collector resistance because we want hard saturation when the switch is closed.

Emitter–Biased LED Driver

The emitter current is zero in Fig. 7-12*b*, which means that the transistor is at cut-off. When the switch of Fig. 7-12*b* closes, the transistor goes into the active region. Ideally, the emitter voltage is 15 V. This means that we get an emitter current of 10 mA. This time, the LED voltage drop has no effect. It doesn't matter whether the exact LED voltage is 1.8, 2, or 2.5 V. This is an advantage of the emitter-biased design over the base-biased design. The LED current is independent of the LED voltage. Another advantage is that the circuit doesn't require a collector resistor.

The emitter-biased circuit of Fig. 7-12*b* operates in the active region when the switch is closed. To change the LED current, you can change the base supply voltage or the emitter resistance. For instance, if you vary the base supply voltage, the LED current varies in direct proportion.

Example 7-10

We want 25 mA of LED current when the switch is closed in Fig. 7-12b. How can we do it?

SOLUTION One solution is to increase the base supply. We want 25 mA to flow through the emitter resistance of 1.5 kΩ. Ohm's law tells us that the emitter voltage has to be:

$$V_E = (25 \text{ mA})(1.5 \text{ k}\Omega) = 37.5 \text{ V}$$

Ideally, $V_{BB} = 37.5$ V. To a second approximation, $V_{BB} = 38.2$ V. This is a bit high for typical power supplies. But the solution is workable if the particular application allows this high a supply voltage.

A supply voltage of 15 V is common in electronics. Therefore, a better solution in most applications is to decrease the emitter resistance. Ideally, the emitter voltage will be 15 V, and we want 25 mA through the emitter resistor. Ohm's law gives:

$$R_E = \frac{15 \text{ V}}{25 \text{ mA}} = 600 \text{ }\Omega$$

The nearest standard value with a tolerance of 5 percent is 620 Ω. If we use the second approximation, the resistance is:

$$R_E = \frac{14.3 \text{ V}}{25 \text{ mA}} = 572 \text{ }\Omega$$

The nearest standard value is 560 Ω.

PRACTICE PROBLEM 7-10 In Fig. 7-12b, what value of R_E is needed to produce an LED current of 21 mA?

Example 7-11

What does the circuit of Fig. 7-13 do?

Figure 7-13 Base-biased LED driver.

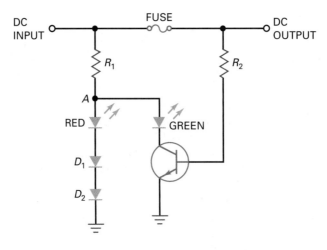

SOLUTION This is a blown-fuse indicator for a dc power supply. When the fuse is intact, the transistor is base-biased into saturation. This turns on the green LED to indicate that all is OK. The voltage between point *A* and ground is approximately 2 V. This voltage is not enough to run on the red LED. The two series diodes (D_1 and D_2) prevent the red LED from turning on because they require a drop of 1.4 V to conduct.

When the fuse blows, the transistor goes into cutoff, turning off the green LED. Then, the voltage of point *A* is pulled up toward the supply voltage. Now there is enough voltage to turn on the two series diodes and the red LED to indicate a blown fuse. Summary Table 7-1 illustrates the differences between base bias and emitter bias.

Summary Table 7–1	Base Bias Vs. Emitter Bias	

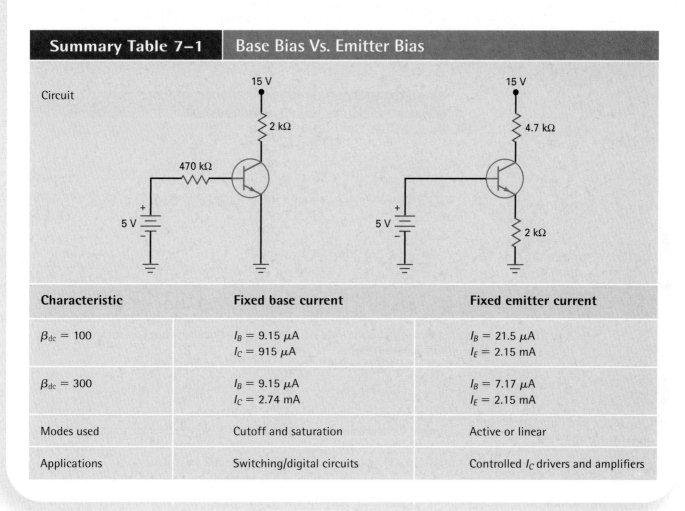

Circuit		

Characteristic	Fixed base current	Fixed emitter current
$\beta_{dc} = 100$	$I_B = 9.15 \ \mu A$ $I_C = 915 \ \mu A$	$I_B = 21.5 \ \mu A$ $I_E = 2.15 \ mA$
$\beta_{dc} = 300$	$I_B = 9.15 \ \mu A$ $I_C = 2.74 \ mA$	$I_B = 7.17 \ \mu A$ $I_E = 2.15 \ mA$
Modes used	Cutoff and saturation	Active or linear
Applications	Switching/digital circuits	Controlled I_C drivers and amplifiers

7-8 The Effect of Small Changes

Earlier chapters introduced up-down analysis, which is helpful to anyone trying to understand circuits. For the up-down analysis of Fig. 7-14, a small change means a change of approximately 10 percent (the tolerance of some resistors).

For instance, Fig. 7-14 shows an emitter-biased circuit with these circuit values:

$$V_{BB} = 2 \text{ V} \qquad V_{CC} = 15 \text{ V} \qquad R_E = 130 \ \Omega \qquad R_C = 470 \ \Omega$$

Figure 7-14 Up-down analysis.

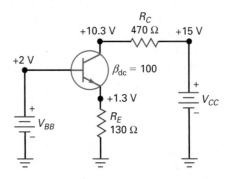

Table 7-1	Up-Down Analysis					
	V_E	I_E	I_B	I_C	V_C	V_{CE}
V_{BB} increase	U	U	U	U	D	D
V_{CC} increase	N	N	N	N	U	U
R_E increase	N	D	D	D	U	U
R_C increase	N	N	N	N	D	D

These are the independent variables of the circuit (often called the *circuit values*) because their values are independent of one another; changing one of them has no effect on the others.

The remaining voltages and currents are as follows:

$$V_E = 1.3 \text{ V} \quad V_C = 10.3 \text{ V} \quad I_B = 99 \text{ } \mu\text{A} \quad I_C = 9.9 \text{ mA} \quad I_E = 10 \text{ mA}$$

Each of these is called a *dependent variable* because its value may change when one of the independent variables changes. If you really understand how a circuit works, you can tell whether a dependent variable will increase, decrease, or remain the same when an independent variable increases.

For instance, suppose V_{BB} increases by about 10 percent in Fig. 7-14. Will V_C increase, decrease, or remain the same? It will decrease. Why? Because an increase in the base supply voltage will increase the emitter current, increase the collector current, increase the voltage across the collector resistor, and decrease the collector voltage.

Table 7-1 shows the effects of small increases in the independent variables of Fig. 7-14. We use U for up, D for down, and N for no change (change of less than 1 percent). These results assume the second approximation. By studying this table and asking why the changes occur, you can improve your understanding of how this circuit works.

7-9 Troubleshooting

Many things can go wrong with a transistor. Since it contains two diodes, exceeding any of the breakdown voltages, maximum currents, or power ratings can damage either or both diodes. The troubles may include shorts, opens, high leakage currents, and reduced β_{dc}.

Figure 7-15 *npn* transistor.

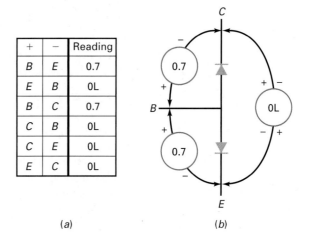

Out-of-Circuit Tests

A transistor is commonly tested using a DMM set to the diode test range. Figure 7-15 shows how an *npn* transistor resembles two back-to-back diodes. Each *pn* junction can be tested for normal forward and reverse biased readings. The collector to emitter can also be tested and should result in an overrange indication with either DMM polarity connection. Since a transistor has three leads, there are six DMM polarity connections possible. These are shown in Fig. 7-16*a*. Notice that only two polarity connections result in approximately a 0.7 V reading. Also important to note here is that the base lead is the only connection common to both 0.7 V readings and it requires a (+) polarity connection. This is also shown in Fig. 7-16*b*.

     A *pnp* transistor can be tested using the same technique. As shown in Fig. 7-17, the *pnp* transistor also resembles two back-to-back diodes. Again, using the DMM in the diode test range, Fig. 7-18*a* and 7-18*b* show the results for a normal transistor.

     Many DMMs have a special β_{dc} or h_{FE} test function. By placing the transistor's leads into the proper slots, the forward current gain is displayed. This current gain is for a specified base current or collector current and V_{CE}. You can check the DMM's manual for the specific test condition.

     Another way to test transistors is with an ohmmeter. You can begin by measuring the resistance between the collector and the emitter. This should be very high in both directions because the collector and emitter diodes are back to

Figure 7-16 *NPN* DMM Readings (*a*) Polarity connections; (*b*) *pn* junction readings.

+	−	Reading
B	E	0.7
E	B	OL
B	C	0.7
C	B	OL
C	E	OL
E	C	OL

(*a*)

(*b*)

Figure 7-17 *PNP* transistor.

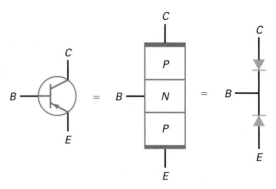

Figure 7-18 *PNP* DMM readings (*a*) polarity connections; (*b*) *pn* junction readings.

+	−	Readings
B	E	0L
E	B	0.7
B	C	0L
C	B	0.7
C	E	0L
E	C	0L

(*a*)

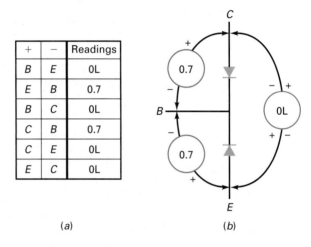

(*b*)

back in series. One of the most common troubles is a collector-emitter short, produced by exceeding the power rating. If you read zero to a few thousand ohms in either direction, the transistor is shorted and should be replaced.

Assuming that the collector-emitter resistance is very high in both directions (in megohms), you can read the reverse and forward resistances of the collector diode (collector-base terminals) and the emitter diode (base-emitter terminals). You should get a high reverse/forward ratio for both diodes, typically more than 1000:1 (silicon). If you do not, the transistor is defective.

Even if the transistor passes the ohmmeter tests, it still may have some faults. After all, the ohmmeter tests each transistor junction only under dc conditions. You can use a curve tracer to look for more subtle faults, such as too much leakage current, low β_{dc}, or insufficient breakdown voltage. A transistor being tested with a curve tracer is shown in Fig. 7-19. Commercial transistor testers are also available; these check the leakage current, current gain β_{dc}, and other quantities.

In-Circuit Tests

The simplest in-circuit tests are to measure transistor voltages with respect to ground. For instance, measuring the collector voltage V_C and the emitter voltage V_E is a good start. The difference $V_C - V_E$ should be more than 1 V but less than V_{CC}. If the reading is less than 1 V in an amplifier circuit, the transistor may be shorted. If the reading equals V_{CC}, the transistor may be open.

The foregoing test usually pins down a dc trouble if one exists. Many people include a test of V_{BE}, done as follows: Measure the base voltage V_B and the

Figure 7-19 Transistor curve tracer test courtesy of Tektronix.

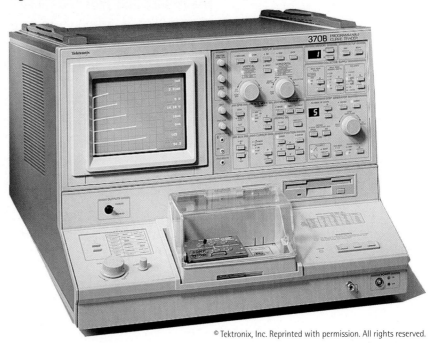

emitter voltage V_E. The difference of these readings is V_{BE}, which should be 0.6 to 0.7 V for small-signal transistors operating in the active region. For power transistors, V_{BE} may be 1 V or more because of the bulk resistance of the emitter diode. If the V_{BE} reading is less than approximately 0.6 V, the emitter diode is not being forward-biased. The trouble could be in the transistor or in the biasing components.

Some people include a cutoff test, performed as follows: Short the base-emitter terminals with a jumper wire. This removes the forward bias on the emitter diode and should force the transistor into cutoff. The collector-to-ground voltage should equal the collector supply voltage. If it does not, something is wrong with the transistor or the circuitry.

Care should be taken when doing this test. If another device or circuit is directly connected to the collector terminal, be sure that the increase in collector-to-ground voltage will not cause any damage.

A Table of Troubles

As discussed in Chap. 6, a shorted component is equivalent to a resistance of zero, and an open component is equivalent to a resistance of infinity. For instance, the emitter resistor may be shorted or open. Let us designate these troubles by R_{ES} and R_{EO}, respectively. Similarly, the collector resistor may be shorted or open, symbolized by R_{CS} and R_{CO}, respectively.

When a transistor is defective, anything can happen. For instance, one or both diodes may be internally shorted or open. We are going to limit the number of possibilities to the most likely defects as follows: a collector-emitter short (*CES*) will represent all three terminals shorted together (base, collector, and emitter), and a collector-emitter open (*CEO*) stands for all three terminals open. A base-emitter open (*BEO*) means that the base-emitter diode is open, and a collector-base open (*CBO*) means that the collector-base diode is open.

Table 7-2 shows a few of the troubles that could occur in a circuit like Fig. 7-20. The voltages were calculated by using the second approximation. When the circuit is operating normally, you should measure a base voltage of 2 V, an emitter voltage of 1.3 V, and a collector voltage of approximately 10.3 V. If the

Figure 7-20 (Repeated for reference.) Up-down analysis.

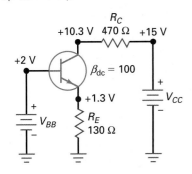

Table 7-2	Troubles and Symptoms			
Trouble	V_B, V	V_E, V	V_C, V	**Comments**
None	2	1.3	10.3	No trouble
R_{ES}	2	0	15	Transistor destroyed (CEO)
R_{EO}	2	1.3	15	No base or collector current
R_{CS}	2	1.3	15	
R_{CO}	2	1.3	1.3	
No V_{BB}	0	0	15	Check supply and lead
No V_{CC}	2	1.3	1.3	Check supply and lead
CES	2	2	2	All transistor terminals shorted
CEO	2	0	15	All transistor terminals open
BEO	2	0	15	Base-emitter diode open
CBO	2	1.3	15	Collector-base diode open

emitter resistor were shorted, $+2$ V would appear across the emitter diode. This large voltage would destroy the transistor, possibly producing a collector-emitter open. This trouble R_{ES} and its voltages are shown in Table 7-2.

If the emitter resistor were open, there would be no emitter current. Furthermore, the collector current would be zero, and the collector voltage would increase to 15 V. This trouble R_{EO} and its voltages are shown in Table 7-2. Continuing like this, we can get the remaining entries of the table.

Notice the entry for no V_{CC}. This is worth commenting on. Your initial instinct might be that the collector voltage is zero, because there is no collector supply voltage. But that is not what you will measure with a voltmeter. When you connect a voltmeter between the collector and ground, the base supply will set up a small forward current through the collector diode in series with the voltmeter. Since the base voltage is fixed at 2 V, the collector voltage is 0.7 V less than this. Therefore, the voltmeter will read 1.3 V between the collector and ground. In other words, the voltmeter completes the circuit to ground because the voltmeter looks like a very large resistance in series with the collector diode.

Figure 7-21 (*a*) Transistor with open base; (*b*) equivalent circuit.

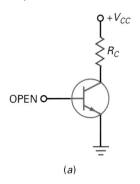

(*a*)

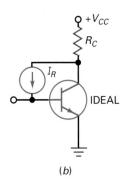

(*b*)

7-10 More Optoelectronic Devices

As mentioned earlier, a transistor with an open base has a small collector current consisting of thermally produced minority carriers and surface leakage. By exposing the collector junction to light, a manufacturer can produce a **phototransistor,** a device that has more sensitivity to light than a photodiode.

Basic Idea of Phototransistors

Figure 7-21*a* shows a transistor with an open base. As mentioned earlier, a small collector current exists in this circuit. Ignore the surface-leakage component, and

Figure 7-22 Phototransistor.
(*a*) Open base gives maximum sensitivity;
(*b*) variable base resistor changes
sensitivity; (*c*) typical phototransistor.

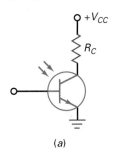

(*a*)

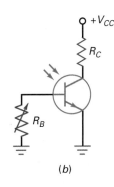

(*b*)

(*c*)

concentrate on the thermally produced carriers in the collector diode. Visualize the reverse current produced by these carriers as an ideal current source in parallel with the collector-base junction of an ideal transistor (Fig. 7-21*b*).

Because the base lead is open, all the reverse current is forced into the base of the transistor. The resulting collector current is:

$$I_{CEO} = \beta_{dc} I_R$$

where I_R is the reverse minority-carrier current. This says that the collector current is higher than the original reverse current by a factor of β_{dc}.

The collector diode is sensitive to light as well as heat. In a phototransistor, light passes through a window and strikes the collector-base junction. As the light increases, I_R increases, and so does I_{CEO}.

Phototransistor versus Photodiode

The main difference between a phototransistor and a photodiode is the current gain β_{dc}. The same amount of light striking both devices produces β_{dc} times more current in a phototransistor than in a photodiode. The increased sensitivity of a phototransistor is a big advantage over that of a photodiode.

Figure 7-22*a* shows the schematic symbol of a phototransistor. Notice the open base. This is the usual way to operate a phototransistor. You can control the sensitivity with a variable base return resistor (Fig. 7-22*b*), but the base is usually left open to get maximum sensitivity to light.

The price paid for increased sensitivity is reduced speed. A phototransistor is more sensitive than a photodiode, but it cannot turn on and off as fast. A photodiode has typical output currents in microamperes and can switch on and off in nanoseconds. The phototransistor has typical output currents in milliamperes but switches on and off in microseconds. A typical phototransistor is shown in Fig. 7-22*c*.

Optocoupler

Figure 7-23*a* shows an LED driving a phototransistor. This is a much more sensitive optocoupler than the LED-photodiode discussed earlier. The idea is straightforward. Any changes in V_S produce changes in the LED current, which changes the current through the phototransistor. In turn, this produces a changing voltage across the collector-emitter terminals. Therefore, a signal voltage is coupled from the input circuit to the output circuit.

Again, the big advantage of an optocoupler is the electrical isolation between the input and output circuits. Stated another way, the common for the input circuit is different from the common for the output circuit. Because of this,

Figure 7-23   (*a*) Optocoupler with LED and phototransistor; (*b*) optocoupler *IC.*

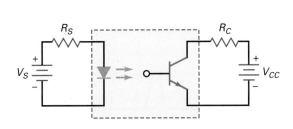

(*a*)

(*b*)

no conductive path exists between the two circuits. This means that you can ground one of the circuits and float the other. For instance, the input circuit can be grounded to the chassis of the equipment, while the common of the output side is ungrounded. Figure 7-23*b* shows a typical optocoupler *IC*.

An Example

The 4N24 optocoupler of Fig. 7-24*a* provides isolation from the power line and detects zero crossings of line voltage. The graph of Fig. 7-24*b* shows how the collector current is related to the LED current. Here is how you can calculate the peak output voltage from the optocoupler:

The bridge rectifier produces a full-wave current through the LED. Ignoring diode drops, the peak current through the LED is:

$$I_{LED} = \frac{1.414(115 \text{ V})}{16 \text{ k}\Omega} = 10.2 \text{ mA}$$

The saturated value of the phototransistor current is:

$$I_{C(sat)} = \frac{20 \text{ V}}{10 \text{ k}\Omega} = 2 \text{ mA}$$

Figure 7-24*b* shows the static curves of phototransistor current versus LED current for three different optocouplers. With a 4N24 (top curve), an LED current of 10.2 mA produces a collector current of approximately 15 mA when the load resistance is zero. In Fig. 7-24*a*, the phototransistor current never reaches 15 mA because the phototransistor saturates at 2 mA. In other words, there is more

Figure 7-24 (*a*) Zero-crossing detector; (*b*) optocoupler curves; (*c*) output of detector.

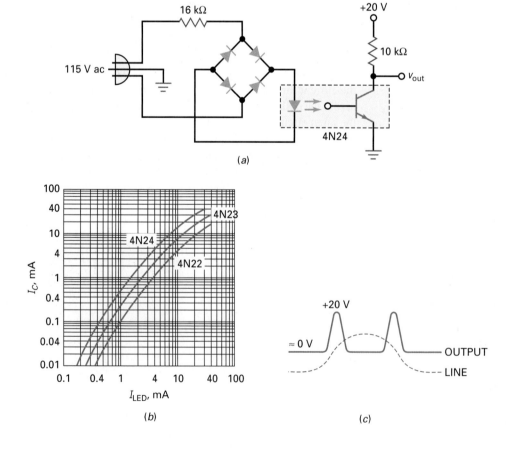

than enough LED current to produce saturation. Since the peak LED current is 10.2 mA, the transistor is saturated during most of the cycle. At this time, the output voltage is approximately zero, as shown in Fig. 7-24c.

The zero crossings occur when the line voltage is changing polarity, from positive to negative, or vice versa. At a zero crossing, the LED current drops to zero. At this instant, the phototransistor becomes an open circuit, and the output voltage increases to approximately 20 V, as indicated in Fig. 7-24c. As you can see, the output voltage is near zero most of the cycle. At the zero crossings, it increases rapidly to 20 V and then decreases to the baseline.

A circuit like Fig. 7-24a is useful because it does not require a transformer to provide isolation from the line. The photocoupler takes care of this. Furthermore, the circuit detects zero crossings, desirable in applications where you want to synchronize some other circuit to the frequency of the line voltage.

Summary

SEC. 7–1 VARIATIONS IN CURRENT GAIN

The current gain of a transistor is an unpredictable quantity. Because of manufacturing tolerances, the current gain of a transistor may vary over as much as a 3 : 1 range when you change from one transistor to another of the same type. Changes in the temperature and the collector current produce additional variations in the dc gain.

SEC. 7–2 THE LOAD LINE

The dc load line contains all the possible dc operating points of a transistor circuit. The upper end of the load line is called saturation, and the lower end is called cutoff. The key step in finding the saturation curent is to visualize a short between the collector and the emitter. The key step to finding the cutoff voltage is to visualize an open between the collector and emitter.

SEC. 7–3 THE OPERATING POINT

The operating point of the transistor is on the dc load line. The exact location of this point is determined by the collector current and the collector-emitter voltage. With base bias, the Q point moves whenever any of the circuit values change.

SEC. 7–4 RECOGNIZING SATURATION

The idea is to assume that the *npn* transistor is operating in the active region.

If this leads to a contradiction (such as negative collector-emitter voltage or collector current greater than saturation current), then you know that the transistor is operating in the saturation region. Another way to recognize saturation is by comparing the base resistance to the collector resistance. If the ratio is in the vicinity of 10 : 1, the transistor is probably saturated.

SEC. 7–5 THE TRANSISTOR SWITCH

Base bias tends to use the transistor as a switch. The switching action is between cutoff and saturation. This type of operation is useful in digital circits. Another name for switching circuits is two-state circuits.

SEC. 7–6 EMITTER BIAS

Emitter bias is virtually immune to changes in current gain. The process for analyzing emitter bias is to find the emitter voltage, emitter current, collector voltage, and collector-emitter voltage. All you need for this process is Ohm's law.

SEC. 7–7 LED DRIVERS

A base-biased LED driver uses a saturated or cutoff transistor to control the current through an LED. An emitter-biased LED driver uses the active region and cutoff

to control the current through the LED.

SEC. 7–8 THE EFFECT OF SMALL CHANGES

Useful to both troubleshooters and designers is the ability to predict the direction of change for a dependent voltage or current when one of the circuit values changes. When you can do this, you can better understand what happens for different troubles and can more easily analyze circuits.

SEC. 7–9 TROUBLESHOOTING

You can use a DMM or ohmmeter to test a transistor. This is best done with the transistor disconnected from the circuit. When the transistor is in the circuit with the power on, you can measure its voltages, which are clues to possible troubles.

SEC. 7–10 MORE OPTOELECTRONIC DEVICES

Because of its β_{dc}, the phototransistor is more sensitive to light than a photodiode. Combined with an LED, the phototransistor gives us a more sensitive optocoupler. The disadvantage with the phototransistor is that it responds more slowly to changes in light intensity than a photodiode.

Derivations

(7-1) Load-line analysis:

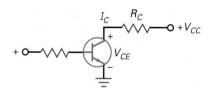

$$I_C = \frac{V_{CC} - V_{CE}}{R_C}$$

(7-2) Saturation current (base bias):

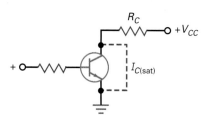

$$I_{C(sat)} = \frac{V_{CC}}{R_C}$$

(7-3) Cutoff voltage (base bias)

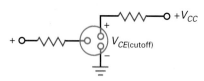

$$V_{CE(cutoff)} = V_{CC}$$

(7-4) Base current:

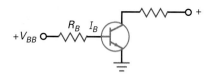

$$I_B = \frac{V_{BB} - V_{BE}}{R_B}$$

(7-5) Current gain:

$$I_C = \beta_{dc} I_B$$

(7-6) Collector-emitter voltage:

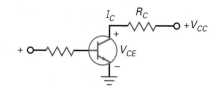

$$V_{CE} = V_{CC} - I_C R_C$$

(7-7) Emitter voltage:

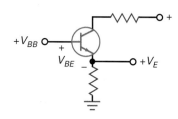

$$V_E = V_{BB} - V_{BE}$$

(7-8) Collector-emitter voltage:

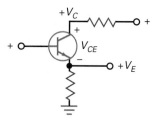

$$V_{CE} = V_C - V_E$$

(7-9) Insensitivity of I_C to β_{dc}

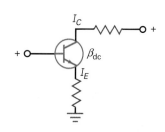

$$I_C = \frac{\beta_{dc}}{\beta_{dc} + 1} I_E$$

Student Assignments

1. The current gain of a transistor is defined as the ratio of the collector current to the
   a. Base current
   b. Emitter current
   c. Supply current
   d. Collector current

2. The graph of current gain versus collector current indicates that the current gain
   a. Is constant
   b. Varies slightly
   c. Varies significantly
   d. Equals the collector current divided by the base current

3. When the collector current increases, what does the current gain do?
   a. Decreases
   b. Stays the same
   c. Increases
   d. Any of the above

4. As the temperature increases, the current gain
   a. Decreases
   b. Remains the same
   c. Increases
   d. Can be any of the above

5. When the base resistor increases, the collector voltage will probably
   a. Decrease
   b. Stay the same
   c. Increase
   d. Do all of the above

6. If the base resistor is very small, the transistor will operate in the
   a. Cutoff region
   b. Active region
   c. Saturation region
   d. Breakdown region

7. Ignoring the bulk resistance of the collector diode, the collector-emitter saturation voltage is
   a. 0
   b. A few tenths of a volt
   c. 1 V
   d. Supply voltage

8. Three different Q points are shown on a load line. The upper Q point represents the
   a. Minimum current gain
   b. Intermediate current gain
   c. Maximum current gain
   d. Cutoff point

9. If a transistor operates at the middle of the load line, a decrease in the base resistance will move the Q point
   a. Down   c. Nowhere
   b. Up   d. Off the load line

10. If a transistor operates at the middle of the load line, a decrease in the current gain will move the Q point
    a. Down
    b. Up
    c. Nowhere
    d. Off the load line

11. If the base supply voltage increases, the Q point moves
    a. Down
    b. Up
    c. Nowhere
    d. Off the load line

12. Suppose the base resistor is open. The Q point will be
    a. In the middle of the load line
    b. At the upper end of the load line
    c. At the lower end of the load line
    d. Off the load line

13. If the base supply voltage is disconnected, the collector-emitter voltage will equal
    a. 0 V
    b. 6 V
    c. 10.5 V
    d. Collector supply voltage

14. If the base resistor has zero resistance, the transistor will probably be
    a. Saturated
    b. In cutoff
    c. Destroyed
    d. None of the above

15. If the collector resistor opens in a base-biased circuit, the load line will become
    a. Horizontal   c. Useless
    b. Vertical   d. Flat

16. The collector current is 1.5 mA. If the current gain is 50, the base current is
    a. 3 μA
    b. 30 μA
    c. 150 μA
    d. 3 mA

17. The base current is 50 μA. If the current gain is 100, the collector current is closest in value to
    a. 50 μA
    b. 500 μA
    c. 2 mA
    d. 5 mA

18. When the Q point moves along the load line, V_{CE} decreases when the collector current
    a. Decreases
    b. Stays the same
    c. Increases
    d. Does none of the above

19. When there is no base current in a transistor switch, the output voltage from the transistor is
    a. Low
    b. High
    c. Unchanged
    d. Unknown

20. A circuit with a fixed emitter current is called
    a. Base bias
    b. Emitter bias
    c. Transistor bias
    d. Two-supply bias

21. The first step in analyzing emitter-based circuits is to find the
    a. Base current
    b. Emitter voltage
    c. Emitter current
    d. Collector current

22. If the current gain is unknown in an emitter-biased circuit, you cannot calculate the
    a. Emitter voltage
    b. Emitter current
    c. Collector current
    d. Base current

23. If the emitter resistor is open, the collector voltage is
    a. Low
    b. High
    c. Unchanged
    d. Unknown

24. If the collector resistor is open, the collector voltage is
    a. Low
    b. High
    c. Unchanged
    d. Unknown

25. When the current gain increases from 50 to 300 in an emitter-biased circuit, the collector current
    a. Remains almost the same
    b. Decreases by a factor of 6
    c. Increases by a factor of 6
    d. Is zero

26. If the emitter resistance increases, the collector voltage
    a. Decreases
    b. Stays the same
    c. Increases
    d. Breaks down the transistor

27. If the emitter resistance decreases, the
    a. Q point moves up
    b. Collector current decreases
    c. Q point stays where it is
    d. Current gain increases

28. When using a DMM to test a transistor, an approximate reading of 0.7 V will be found with how many polarity connections?

   a. One

   b. Two

   c. Three

   d. None

29. What DMM polarity connection is needed on an *npn* transistor's base to get a 0.7 V reading?

   a. Positive

   b. Negative

   c. Either positive or negative

   d. Unknown

30. When testing an *npn* transistor using an ohmmeter, the collector-emitter resistance will be low when

   a. The collector is positive in respect to the emitter

   b. The emitter is positive in respect to the collector

   c. The transistor is normal

   d. The transistor is defective

31. The major advantage of a phototransistor as compared to a photodiode is its

   a. Response to higher frequencies

   b. AC operation

   c. Increased sensitivity

   d. Durability

Problems

SEC. 7–1 VARIATIONS IN CURRENT GAIN

7-1 Refer to Fig. 7-1. What is the current gain of a 2N3904 when the collector current is 100 mA and the junction temperature is 125°C?

7-2 Refer to Fig. 7-1. The junction temperature is 25°C, and the collector current is 1.0 mA. What is the current gain?

SEC. 7–2 THE LOAD LINE

7-3 Draw the load line for Fig. 7-25a. What is the collector current at the saturation point? The collector-emitter voltage at the cutoff point?

Figure 7-25

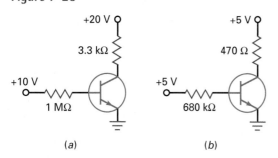

    (a)   (b)

7-4 If the collector supply voltage is increased to 25 V in Fig. 7-25a, what happens to the load line?

7-5 If the collector resistance is increased to 4.7 kΩ in Fig. 7-25a, what happens to the load line?

7-6 If the base resistance of Fig. 7-25a is reduced to 500 kΩ, what happens to the load line?

7-7 Draw the load line for Fig. 7-25b. What is the collector current at the saturation point? The collector-emitter voltage at the cutoff point?

7-8 If the collector supply voltage is doubled in Fig. 7-25b, what happens to the load line?

7-9 If the collector resistance is increased to 1 kΩ in Fig. 7-25b, what happens to the load line?

SEC. 7–3 THE OPERATING POINT

7-10 In Fig. 7-25a, what is the voltage between the collector and ground if the current gain is 200?

7-11 The current gain varies from 25 to 300 in Fig. 7-25a. What is the minimum voltage from the collector to ground? The maximum?

7-12 The resistors of Fig. 7-25a have a tolerance of ±5 percent. The supply voltages have a tolerance of ±10 percent. If the current gain can vary from 50 to 150, what is the minimum possible voltage from the collector to ground? The maximum?

7-13 In Fig. 7-25b, what is the voltage between the collector and ground if the current gain is 150?

7-14 The current gain varies from 100 to 300 in Fig. 7-25b. What is the minimum voltage from the collector to ground? The maximum?

7-15 The resistors of Fig. 7-25b have a tolerance of ±5 percent. The supply voltages have a tolerance of ±10 percent. If the current gain can vary from 50 to 150, what is the minimum possible voltage from the collector to ground? The maximum?

SEC. 7–4 RECOGNIZING SATURATION

7-16 In Fig. 7-25a, use the circuit values shown unless otherwise indicated. Determine whether the transistor is saturated for each of these changes:

   a. $R_B = 33\ k\Omega$ and $h_{FE} = 100$

   b. $V_{BB} = 5\ V$ and $h_{FE} = 200$

   c. $R_C = 10\ k\Omega$ and $h_{FE} = 50$

   d. $V_{CC} = 10\ V$ and $h_{FE} = 100$

7-17 In Fig. 7-25b, use the circuit values shown unless otherwise indicated. Determine whether the transistor is saturated for each of these changes:

    a. $R_B = 51$ kΩ and $h_{FE} = 100$

    b. $V_{BB} = 10$ V and $h_{FE} = 500$

    c. $R_C = 10$ kΩ and $h_{FE} = 100$

    d. $V_{CC} = 10$ V and $h_{FE} = 100$

SEC. 7-5 THE TRANSISTOR SWITCH

7-18 The 680 kΩ in Fig. 7-25b is replaced by 4.7 kΩ and a series switch. Assuming an ideal transistor, what is the collector voltage if the switch is open? What is the collector voltage if the switch is closed?

7-19 Repeat Prob. 7-18, except use $V_{CE(sat)} = 0.2$ V and $I_{CEO} = 100$ nA.

SEC. 7-6 EMITTER BIAS

7-20 IIII MultiSim What is the collector voltage in Fig. 7-26a? The emitter voltage?

Figure 7-26

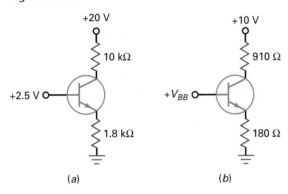

(a)   (b)

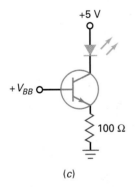

(c)

7-21 IIII MultiSim If the emitter resistor is doubled in Fig. 7-26a, what is the collector-emitter voltage?

7-22 IIII MultiSim If the collector supply voltage is decreased to 15 V in Fig. 7-26a, what is the collector voltage?

7-23 IIII MultiSim What is the collector voltage in Fig. 7-26b if $V_{BB} = 2$ V?

7-24 IIII MultiSim If the emitter resistor is doubled in Fig. 7-26b, what is the collector-emitter voltage for a base supply voltage of 2.3 V?

7-25 IIII MultiSim If the collector supply voltage is increased to 15 V in Fig. 7-26b, what is the collector-emitter voltage for $V_{BB} = 1.8$ V?

SEC. 7-7 LED DRIVERS

7-26 IIII MultiSim If the base supply voltage is 2 V in Fig. 7-26c, what is the current through the LED?

7-27 IIII MultiSim If $V_{BB} = 1.8$ V in Fig. 7-26c, what is the LED current? The approximate V_C?

SEC. 7-8 THE EFFECT OF SMALL CHANGES

Use the letters U (up), D (down), and N (no change) for your answers in the following problems.

7-28 The base supply voltage of Fig. 7-27a decreases by 10 percent. What happens to the base current, collector current, and collector voltage?

Figure 7-27

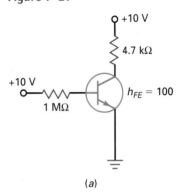

(a)

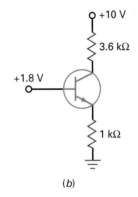

(b)

7-29 The base resistance of Fig. 7-27a decreases by 10 percent. What happens to the base current, collector current and collector voltage?

7-30 The collector resistance of Fig. 7-27a increases by 10 percent. What happens to the base current, collector current, and collector voltage?

7-31 The collector supply voltage of Fig. 7-27a increases by 10 percent. What happens to the base current, collector current, and collector voltage?

7-32 The base supply voltage of Fig. 7-27b decreases by 10 percent. What happens to the base current, collector current, and collector voltage?

7-33 The emitter resistance of Fig. 7-27b increases by 10 percent. What happens to the emitter current, collector current, and collector voltage?

7-34 The collector resistance of Fig. 7-27b increases by 10 percent. What happens to the emitter current, collector current, and collector voltage?

7-35 The collector supply voltage of Fig. 7-27b increases by 10 percent. What happens to the emitter current, collector current, and collector voltage?

SEC. 7-9 TROUBLESHOOTING

7-36 A voltmeter reads 10 V at the collector of Fig. 7-27a. What are some of the troubles that can cause this high reading?

7-37 What if the ground on the emitter is open in Fig. 7-27a? What will a voltmeter read for the base voltage? For the collector voltage?

7-38 A dc voltmeter measures a very low voltage at the collector of Fig. 7-27a. What are some of the possible troubles?

7-39 A voltmeter reads 10 V at the collector of Fig. 7-27b. What are some of the troubles that can cause this high reading?

7-40 What if the emitter resistor is open in Fig. 7-27b? What will a voltmeter read for the base voltage? For the collector voltage?

7-41 A dc voltmeter measures 1.1 V at the collector of Fig. 7-27b. What are some of the possible troubles?

Critical Thinking

7-42 You have built the circuit of Fig. 7-27a, and it is working normally. Now your job is to destroy the transistor. In other words, you are trying to find the ways in which to ruin the transistor. What will you try?

7-43 A first-year electronics student invents a new circuit. It works quite well when the current gain is between 90 and 110. Outside this range, it fails. The student plans to mass-produce the circuit by hand selecting 2N3904s that have the right current gain, and asks for your advice. What are some of the things you would say?

7-44 A student swears that a base-biased circuit with a load line that is not straight can be built and is willing to bet you $50 that it can be done. Should you take the bet? Explain your answer.

7-45 A student wants to measure the collector-emitter voltage in Fig. 7-27b and so connects a voltmeter between the collector and emitter. What does the voltmeter read? (*Note:* There are many right answers.)

7-46 What is the collector current of Q_2 in Fig. 7-28a?

7-47 In Fig. 7-28a, the first transistor has a current gain of 100, and the second transistor has a current gain of 50. What is the base current in the first transistor?

7-48 What is the current through the LED of Fig. 7-28b if $V_{BB} = 0$? If $V_{BB} = 10$ V?

7-49 The zener diode of Fig. 7-28b is replaced by a 1N4736. What is the LED current when $V_{BB} = 0$?

7-50 What is the maximum possible value of current through the 2 kΩ of Fig. 7-29a?

7-51 Figure 7-29b applies to the 4N33 of Fig. 7-29a. If the voltage across the 2 kΩ is 2 V, what is the value of V_{BB}?

7-52 The LED is open in Fig. 7-29a, and $V_{BB} = 3$ V. A voltmeter is connected between the collector of the 2N3904 and ground. What does the voltmeter read?

7-53 A DMM has an input resistance of 10 MΩ. The DMM is connected between the collector of Fig. 7-25a and

ground. If the 3.3 kΩ collector resistor is open, what does the DMM read?

7-54 Redesign a transistor switch similar to Fig. 7-27a for hard saturation and to meet these specifications:

$$V_{CC} = 15 \text{ V} \quad V_{BB} = 0 \text{ V and 15 V} \quad I_{C(sat)} = 5 \text{ mA}$$

7-55 Using Fig. 7-27b, change the value of the collector resistor (3.6 kΩ) so $V_{CE} = 6.6$ V.

Figure 7-28

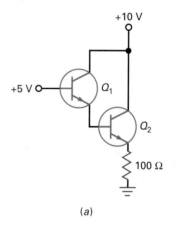

(a)

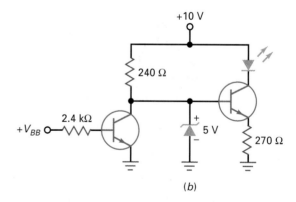

(b)

Figure 7-29

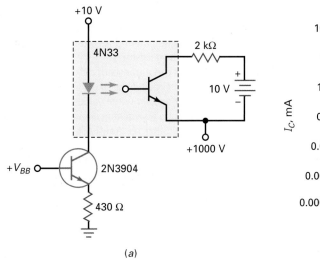

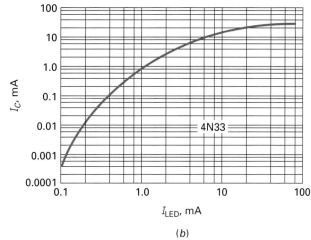

(a)

(b)

Up-Down Analysis

Use Fig. 7-30 for the remaining problems. Assume increases of approximately 10 percent in the independent variable, and use the second approximation of the transistor. A response should be an N (no change) if the change in a dependent variable is so small that you would have difficulty measuring it. For instance, you probably would find it difficult to measure a change of less than 1 percent. To a troubleshooter, a change like this is usually considered to be no change at all.

7-56 Try to predict the response of each dependent variable in the row labeled V_{BB}. Then answer the following question as simply and directly as possible. What effect does an increase in the base-supply voltage have on the dependent variables of the circuit?

7-57 Predict the response of each dependent variable in the row labeled V_{CC}. Then summarize your findings in one or two sentences.

7-58 Predict the response of each dependent variable in the row labeled R_E. List the dependent variables that decrease. Explain why these variables decrease, using Ohm's law or similar basic ideas.

7-59 Predict the response of each dependent variable in the row labeled R_C. List the dependent variables that show no change. Explain why these variables show no change.

Figure 7-30 Up-down analysis.

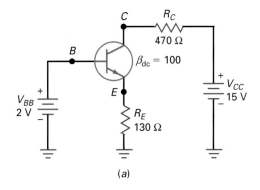

(a)

INDEPENDENT VARIABLES	10% increase	DEPENDENT VARIABLES					
		V_B	V_E	V_C	I_E	I_C	I_B
	V_{BB}						
	V_{CC}						
	R_E						
	R_C						

(b)

Job Interview Questions

1. Draw a base-biased circuit. Then, tell me how to calculate the collector-emitter voltage. Why is this circuit likely to fail in mass production if a precise value of current gain is needed?
2. Draw another base-biased circuit. Draw a load line for the circuit and tell me how to calculate the saturation and cutoff points. Discuss the effects of a changing current gain on the location of the Q point.
3. What is the difference between base bias and emitter bias? In what kind of circuits is each useful?
4. Draw an emitter-biased circuit and tell me how it works. What happens to the collector current when the transistor is replaced or the temperature changes?
5. Tell me how you would test a transistor out of a circuit. What kind of tests can you do while the transistor is in a circuit with the power on?
6. What is an optocoupler and what are its advantages? I want to see a drawing of the device and an explanation of how it works.
7. What effect does temperature have on current gain?
8. What is the primary application of a base-biased circuit?
9. What piece of test equipment does a technician use for a preliminary test of a transistor?
10. What kind of faults can a transistor curve tracer detect?
11. Draw a base-biased circuit. Now, explain three ways to saturate this circuit and show the values.
12. When troubleshooting a switching circuit with a voltmeter, how can you tell whether the transistor is in saturation or cutoff?
13. Which transistor would tend to saturate more with less base current: a transistor with a large R_C or one with a small R_C?
14. When a base-biased transistor is used as a switch, what is the transistor doing?

Self-Test Answers

1.	a	12.	c	23.	b
2.	b	13.	d	24.	a
3.	d	14.	c	25.	a
4.	d	15.	a	26.	c
5.	c	16.	b	27.	a
6.	c	17.	d	28.	b
7.	a	18.	c	29.	a
8.	c	19.	b	30.	d
9.	b	20.	b	31.	c
10.	a	21.	b		
11.	b	22.	d		

Practice Problem Answers

7-2 $I_{C(sat)} = 6$ mA;
$V_{CE(cutoff)} = 12$ V

7-4 $I_{C(sat)} = 3$ mA;
The slope would decrease.

7-5 $V_{CE} = 8.25$ V

7-7 $R_B = 47$ kΩ

7-8 $V_{CE} = 11.999$ V and 0.15 V

7-9 $V_{CE} = 8.1$ V

7-10 $R_E = 680$ Ω

8

Transistor Biasing

A **prototype** is a basic circuit design that can be modified to get more advanced circuits. Base bias is a prototype used in the design of switching circuits. Emitter bias is a prototype used in the design of amplifying circuits. In this chapter, we emphasize emitter bias and the practical circuits that can be derived from it.

Objectives

After studying this chapter, you should be able to:

- Draw a diagram of a voltage-divider bias circuit.

- Calculate the divider current, base voltage, emitter voltage, emitter current, collector voltage, and collector-emitter voltage for an *npn* VDB circuit.

- Determine how to draw the load line and calculate the Q point for a given VDB circuit.

- Design a VDB circuit using design guidelines.

- Draw a two-supply emitter bias circuit and calculate V_{RE}, I_E, V_C, and V_{CE}.

- Compare several different types of bias and describe how well each works.

- Calculate the Q point of *pnp* VDB circuits.

- Troubleshoot transistor-biasing circuits.

Chapter Outline

Vocabulary

collector-feedback bias

emitter-feedback bias

firm voltage divider

prototype

self-bias

stage

stiff voltage divider

swamp out

two-supply emitter bias (TSEB)

voltage-divider bias (VDB)

8-1 Voltage-Divider Bias

Figure 8-1a shows the most widely used biasing circuit. Notice that the base circuit contains a voltage divider (R_1 and R_2). Because of this, the circuit is called **voltage-divider bias (VDB).**

Figure 8-1 Voltage-divider bias. (a) Circuit; (b) voltage divider; (c) simplified circuit.

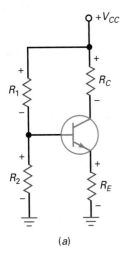

(a)

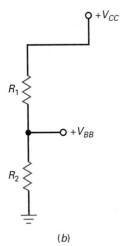

(b)

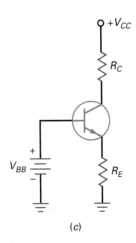

(c)

Simplified Analysis

For troubleshooting and preliminary analysis, use the following method. In any well-designed VDB circuit, the base current is much smaller than the current through the voltage divider. Since the base current has a negligible effect on the voltage divider, we can mentally open the connection between the voltage divider and the base to get the equivalent circuit of Fig. 8-1b. In this circuit, the output of the voltage divider is

$$V_{BB} = \frac{R_2}{R_1 + R_2} V_{CC}$$

Ideally, this is the base-supply voltage as shown in Fig. 8-1c.

As you can see, voltage-divider bias is really emitter bias in disguise. In other words, Fig. 8-1c is an equivalent circuit for Fig. 8-1a. This is why VDB sets up a fixed value of emitter current, resulting in a solid Q point that is independent of the current gain.

There is an error in this simplified approach, and we will discuss it in the next section. The crucial point is this: In any well-designed circuit, the error in using Fig. 8-1c is very small. In other words, a designer deliberately chooses circuit values so that Fig. 8-1a acts like Fig. 8-1c.

Conclusion

After you calculate V_{BB}, the rest of the analysis is the same as discussed earlier for emitter bias in Chap. 7. Here is a summary of the equations you can use to analyze VDB:

$$V_{BB} = \frac{R_2}{R_1 + R_2} V_{CC} \tag{8-1}$$

$$V_E = V_{BB} - V_{BE} \tag{8-2}$$

$$I_E = \frac{V_E}{R_E} \tag{8-3}$$

$$I_C \approx I_E \tag{8-4}$$

$$V_C = V_{CC} - I_C R_C \tag{8-5}$$

$$V_{CE} = V_C - V_E \tag{8-6}$$

These equations are based on Ohm's and Kirchhoff's laws. Here are the steps in the analysis:

1. Calculate the base voltage V_{BB} out of the voltage divider.
2. Subtract 0.7 V to get the emitter voltage (use 0.3 V for germanium).
3. Divide by the emitter resistance to get the emitter current.
4. Assume that the collector current is approximately equal to the emitter current.

5. Calculate the collector-to-ground voltage by subtracting the voltage across the collector resistor from the collector supply voltage.
6. Calculate the collector-emitter voltage by subtracting the emitter voltage from the collector voltage.

Since these six steps are logical, they should be easy to remember. After you analyze a few VDB circuits, the process becomes automatic.

Example 8-1

Figure 8-2 Example.

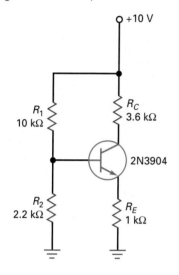

What is the collector-emitter voltage in Fig. 8-2?

SOLUTION The voltage divider produces an unloaded output voltage of:

$$V_{BB} = \frac{2.2 \text{ k}\Omega}{10 \text{ k}\Omega + 2.2 \text{ k}\Omega} 10 \text{ V} = 1.8 \text{ V}$$

Subtract 0.7 V from this to get:

$$V_E = 1.8 \text{ V} - 0.7 \text{ V} = 1.1 \text{ V}$$

The emitter current is:

$$I_E = \frac{1.1 \text{ V}}{1 \text{ k}\Omega} = 1.1 \text{ mA}$$

Since the collector current almost equals the emitter current, we can calculate the collector-to-ground voltage like this:

$$V_C = 10 \text{ V} - (1.1 \text{ mA})(3.6 \text{ k}\Omega) = 6.04 \text{ V}$$

The collector-emitter voltage is:

$$V_{CE} = 6.04 - 1.1 \text{ V} = 4.94 \text{ V}$$

Here is an important point: The calculations in this preliminary analysis do not depend on changes in the transistor, the collector current, or the temperature. This is why the Q point of this circuit is stable, almost rock-solid.

PRACTICE PROBLEM 8-1 Change the power supply voltage of Fig. 8-2 from 10 V to 15 V and solve for V_{CE}.

Example 8-2

Discuss the significance of Fig. 8-3, which shows a MultiSim analysis of the same circuit analyzed in the preceding example.

SOLUTION This really drives the point home. Here we have an almost identical answer using a computer to analyze the circuit. As you can see, the voltmeter reads 6.03 V (rounded to 2 places). Compare this to 6.04 V in the preceding example, and you can see the point. A simplified analysis has produced essentially the same result as a computer analysis.

You can expect this kind of close agreement whenever a VDB circuit has been well designed. After all, the whole point of VDB is to act like emitter bias to virtually eliminate the effects of changing the transistor, collector current, or temperature.

Figure 8-3 MultiSim example.

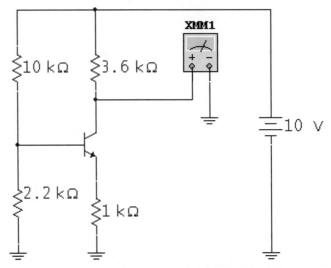

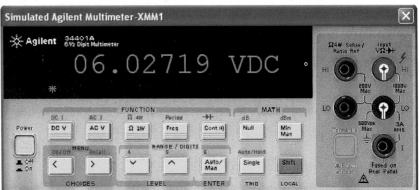

PRACTICE PROBLEM 8-2 Using MultiSim, change the supply voltage of Fig. 8-3 to 15 V and measure V_{CE}. Compare your measured value to the answer of Practice Problem 8-1.

8-2 Accurate VDB Analysis

What is a well-designed VDB circuit? It is one in which *the voltage divider appears stiff to the input resistance of the base.* The meaning of the last sentence needs to be discussed.

Source Resistance

Chapter 1 introduced the idea of a stiff voltage source:

Stiff voltage source: $R_S < 0.01R_L$

When this condition is satisfied, the load voltage is within 1 percent of the ideal voltage. Now, let us extend this idea to the voltage divider.

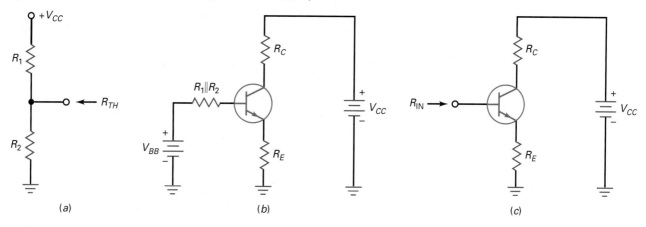

What is the Thevenin resistance of the voltage divider in Fig. 8-4*a*? Looking back into the voltage divider with V_{CC} grounded, we see R_1 in parallel with R_2. As an equation:

$$R_{TH} = R_1 \| R_2$$

Because of this resistance, the output voltage of the voltage divider is not ideal. A more accurate analysis includes the Thevenin resistance, as shown in Fig. 8-4*b*. The current through this Thevenin resistance reduces the base voltage from the ideal value of V_{BB}.

Load Resistance

How much less than ideal is the base voltage? The voltage divider has to supply the base current in Fig. 8-4*b*. Put another way, the voltage divider sees a load resistance of R_{IN}, as shown in Fig. 8-4*c*. For the voltage divider to appear stiff to the base, the 100 : 1 rule:

$$R_S < 0.01 R_L$$

translates to:

$$R_1 \| R_2 < 0.01 R_{IN} \tag{8-7}$$

A well-designed VDB circuit will satisfy this condition.

Stiff Voltage Divider

If the transistor of Fig. 8-4*c* has a current gain of 100, its collector current is 100 times greater than the base current. This implies that the emitter current is also 100 times greater than the base current. When seen from the base side of the transistor, the emitter resistance R_E appears to be 100 times larger. As a derivation:

$$R_{IN} = \beta_{dc} R_E \tag{8-8}$$

Therefore, Eq. (8-7) may be written as:

$$\textbf{Stiff voltage divider: } R_1 \| R_2 < 0.01 \beta_{dc} R_E \tag{8-9}$$

Whenever possible, a designer selects circuit values to satisfy this 100 : 1 rule because it will produce an ultrastable *Q* point.

Firm Voltage Divider

Sometimes a stiff design results in such small values of R_1 and R_2 that other problems arise (discussed later). In this case, many designers compromise by using this rule:

$$\text{Firm voltage divider: } R_1 \| R_2 < 0.1\beta_{dc}R_E \qquad (8\text{-}10)$$

We call any voltage divider that satisfies this 10 : 1 rule a **firm voltage divider.** In the worst case, using a firm voltage divider means that the collector current will be approximately 10 percent lower than the stiff value. This is acceptable in many applications because the VDB circuit still has a reasonably stable Q point.

A Closer Approximation

If you want a more accurate value for the emitter current, you can use the following derivation:

$$I_E = \frac{V_{BB} - V_{BE}}{R_E + (R_1 \| R_2)/\beta_{dc}} \qquad (8\text{-}11)$$

This differs from the stiff value because $(R_1 \| R_2)/\beta_{dc}$ is in the denominator. As this term approaches zero, the equation simplifies to the stiff value.

Equation (8-11) will improve the analysis, but it is a fairly complicated formula. If you have a computer and need a more accurate analysis obtained with the stiff analysis, you should use MultiSim or an equivalent circuit simulator.

Example 8-3

Figure 8-5 Example.

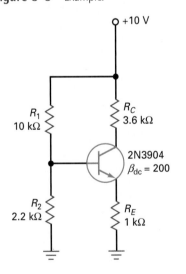

Is the voltage divider of Fig. 8-5 stiff? Calculate the more accurate value of emitter current using Eq. (8-11).

SOLUTION Check to see whether the 100:1 rule has been used:

$$\text{Stiff voltage divider: } R_1 \| R_2 < 0.01\beta_{dc}R_E$$

The Thevenin resistance of the voltage divider is:

$$R_1 \| R_2 = 10 \text{ k}\Omega \| 2.2 \text{ k}\Omega = \frac{(10 \text{ k}\Omega)(2.2 \text{ k}\Omega)}{10 \text{ k}\Omega + 2.2 \text{ k}\Omega} = 1.8 \text{ k}\Omega$$

The input resistance of the base is:

$$\beta_{dc}R_E = (200)(1 \text{ k}\Omega) = 200 \text{ k}\Omega$$

and one-hundredth of this is:

$$0.01\beta_{dc}R_E = 2 \text{ k}\Omega$$

Since 1.8 kΩ is less than 2 kΩ, the voltage divider is stiff.
With Eq. (8-11), the emitter current is

$$I_E = \frac{1.8 \text{ V} - 0.7 \text{ V}}{1 \text{ k}\Omega + (1.8 \text{ k}\Omega)/200} = \frac{1.1 \text{ V}}{1 \text{ k}\Omega + 9 \text{ }\Omega} = 1.09 \text{ mA}$$

This is extremely close to 1.1 mA, the value we get with the simplified analysis.

The point is this: You don't have to use Eq. (8-11) to calculate emitter current when the voltage divider is stiff. Even when the voltage divider is firm, the use of Eq. (8-11) will improve the calculation for emitter current only by at most 10 percent. Unless otherwise indicated, from now on all analysis of VDB circuits will use the simplified method.

8-3 VDB Load Line and Q Point

Because of the stiff voltage divider in Fig. 8-6, the emitter voltage is held constant at 1.1 V in the following discussion.

The Q Point

The Q point was calculated in Sec. 8-1. It has a collector current of 1.1 mA and a collector-emitter voltage of 4.94 V. These values are plotted to get the Q point shown in Fig. 8-6. Since voltage-divider bias is derived from emitter bias, the Q point is virtually immune to changes in current gain. One way to move the Q point in Fig. 8-6 is by varying the emitter resistor.

For instance, if the emitter resistance is changed to 2.2 kΩ, the collector current decreases to:

$$I_E = \frac{1.1\ \text{V}}{2.2\ \text{k}\Omega} = 0.5\ \text{mA}$$

The voltages change as follows:

$$V_C = 10\ \text{V} - (0.5\ \text{mA})(3.6\ \text{k}\Omega) = 8.2\ \text{V}$$

and

$$V_{CE} = 8.2\ \text{V} - 1.1\ \text{V} = 7.1\ \text{V}$$

Therefore, the new Q point will be Q_L and will have coordinates of 0.5 mA and 7.1 V.

On the other hand, if we decrease the emitter resistance to 510 Ω, the emitter current increases to:

$$I_E = \frac{1.1\ \text{V}}{510\ \Omega} = 2.15\ \text{mA}$$

Figure 8-6 Calculating the Q point.

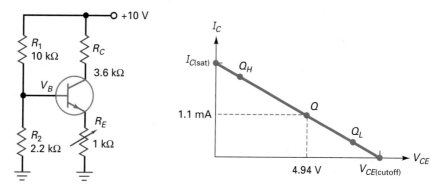

and the voltages change to:

$$V_C = 10\text{ V} - (2.15\text{ mA})(3.6\text{ k}\Omega) = 2.26\text{ V}$$

and

$$V_{CE} = 2.26\text{ V} - 1.1\text{ V} = 1.16\text{ V}$$

In this case, the Q point shifts to a new position at Q_H with coordinates of 2.15 mA and 1.16 V.

Q Point in Middle of Load Line

V_{CC}, R_1, R_2, and R_C control the saturation current and the cutoff voltage. A change in any of these quantities will change $I_{C(\text{sat})}$ and/or $V_{CE(\text{cutoff})}$. Once the designer has established the values of the foregoing variables, the *emitter resistance* is varied to set the Q point at any position along the load line. If R_E is too large, the Q point moves into the cutoff point. If R_E is too small, the Q point moves into saturation. Some designers set the Q point at the middle of the load line.

VDB Design Guideline

Fig. 8-7 shows a VDB circuit. This circuit will be used to demonstrate a simplified design guideline to establish a stable Q point. This design technique is suitable for most circuits, but it is only a guideline. Other design techniques can be used.

Before starting the design, it is important to determine the circuit requirements or specifications. The circuit is normally biased for V_{CE} to be at a midpoint value with a specified collector current. You also need to know the value of V_{CC} and the range of β_{dc} for the transistor being used. Also, be sure the circuit will not cause the transistor to exceed its power dissipation limits.

Start by making the emitter voltage approximately one-tenth of the supply voltage:

$$V_E = 0.1\,V_{CC}$$

Next, calculate the value of R_E to set up the specified collector current:

$$R_E = \frac{V_E}{I_E}$$

Since the Q point needs to be at approximately the middle of the dc load line, about 0.5 V_{CC} appears across the collector-emitter terminals. The remaining 0.4 V_{CC} appears across the collector resistor; therefore:

$$R_C = 4\,R_E$$

Next, design for a stiff voltage divider using the 100 : 1 rule:

$$R_{TH} \leq 0.01\,\beta_{dc}\,R_E$$

Usually, R_2 is smaller than R_1. Therefore, the stiff voltage divider equation can be simplified to:

$$R_2 \leq 0.01\,\beta_{dc}\,R_E$$

You may also choose to design for a firm voltage divider by using the 10 : 1 rule:

$$R_2 \leq 0.1\,\beta_{dc}\,R_E$$

In either case, use the minimum-rated β_{dc} value at the specified collector current.

Finally, calculate R_1 by using proportion:

$$R_1 = \frac{V_1}{V_2}\,R_2$$

Figure 8-7 VDB design.

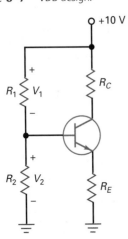

Example 8-4

For the circuit shown in Fig. 8-7, design the resistor values to meet these specifications:

$$V_{CC} = 10 \text{ V} \qquad V_{CE} \text{ @ midpoint}$$

$$I_C = 10 \text{ mA} \qquad \text{2N3904's } \beta_{dc} = 100\text{–}300$$

SOLUTION First, establish the emitter voltage by:

$$V_E = 0.1 \, V_{CC}$$

$$V_E = (0.1)(10 \text{ V}) = 1 \text{ V}$$

The emitter resistor is found by:

$$R_E = \frac{V_E}{I_E}$$

$$R_E = \frac{1 \text{ V}}{10 \text{ mA}} = 100 \, \Omega$$

The collector resistor is:

$$R_C = 4 \, R_E$$

$$R_C = (4)(100 \, \Omega) = 400 \, \Omega \text{ (use 390 } \Omega)$$

Next, choose either a stiff or firm voltage divider. A stiff value of R_2 is found by:

$$R_2 \leq 0.01 \, \beta_{dc} \, R_E$$

$$R_2 \leq (0.01)(100)(100 \, \Omega) = 100 \, \Omega$$

Now, the value of R_1 is:

$$R_1 = \frac{V_1}{V_2} R_2$$

$$V_2 = V_E + 0.7 \text{ V} = 1 \text{ V} + 0.7 \text{ V} = 1.7 \text{ V}$$

$$V_1 = V_{CC} - V_2 = 10 \text{ V} - 1.7 \text{ V} = 8.3 \text{ V}$$

$$R_1 = \left(\frac{8.3 \text{ V}}{1.7 \text{ V}} \right)(100 \, \Omega) = 488 \, \Omega \text{ (use 490 } \Omega)$$

PRACTICE PROBLEM 8-4 Using the given VDB design guidelines, design the VDB circuit of Fig. 8-7 to meet these specifications:

$$V_{CC} = 10 \text{ V} \qquad V_{CE} \text{ @ midpoint} \qquad \text{stiff voltage divider}$$

$$I_C = 1 \text{ mA} \qquad \beta_{dc} = 70\text{-}200$$

8-4 Two-Supply Emitter Bias

Some electronic equipment has a power supply that produces both positive and negative supply voltages. For instance, Fig. 8-8 shows a transistor circuit with two power supplies: $+10$ and -2 V. The negative supply forward biases the emitter

Figure 8-8 Two-supply emitter bias.

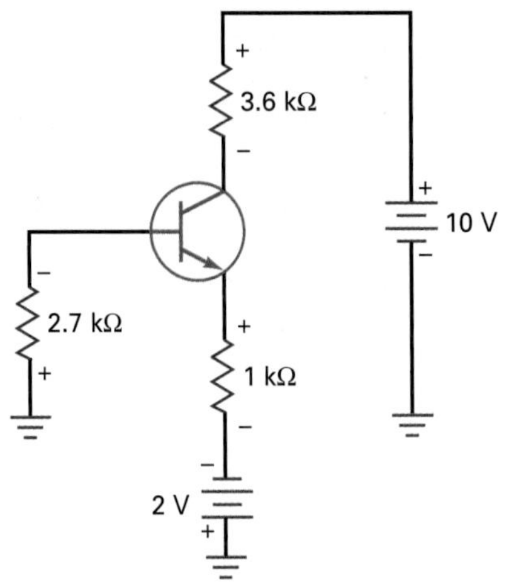

Figure 8-9 Redrawn TSEB circuit.

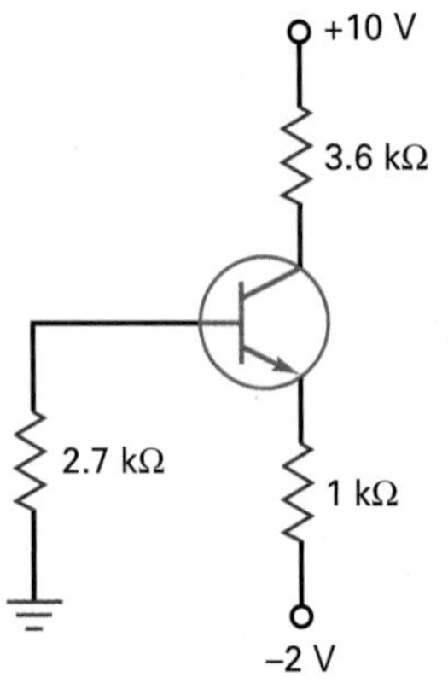

diode. The positive supply reverse biases the collector diode. This circuit is derived from emitter bias. For this reason, we refer to it as **two-supply emitter bias (TSEB).**

Analysis

The first thing to do is to redraw the circuit as it usually appears on schematic diagrams. This means deleting the battery symbols, as shown in Fig. 8-9. This is necessary on schematic diagrams because there usually is no room for battery symbols on complicated diagrams. All the information is still on the diagram, except that it is in condensed form. That is, a negative supply voltage of −2 V is applied to the bottom of the 1 kΩ, and a positive supply voltage of +10 V is applied to the top of the 3.6-kΩ resistor.

When this type of circuit is correctly designed, the base current will be small enough to ignore. This is equivalent to saying that the base voltage is approximately 0 V, as shown in Fig. 8-10.

The voltage across the emitter diode is 0.7 V, which is why −0.7 V is shown on the emitter node. If this is not clear, stop and think about it. There is a plus-to-minus drop of 0.7 V in going from the base to the emitter. If the base voltage is 0 V, the emitter voltage must be −0.7 V.

In Fig. 8-10, the emitter resistor again plays the key role in setting up the emitter current. To find this current, apply Ohm's law to the emitter resistor as follows: The top of the emitter resistor has a voltage of −0.7 V, and the bottom has a voltage of −2 V. Therefore, the voltage across the emitter resistor equals the difference between the two voltages. To get the right answer, subtract the more negative value from the more positive value. In this case, the more negative value is −2 V, so:

$$V_{RE} = -0.7\text{ V} - (-2\text{ V}) = 1.3\text{ V}$$

Once you have found the voltage across the emitter resistor, calculate the emitter current with Ohm's law:

$$I_E = \frac{1.3\text{ V}}{1\text{ k}\Omega} = 1.3\text{ mA}$$

This current flows through the 3.6 kΩ and produces a voltage drop that we subtract from +10 V as follows:

$$V_C = 10\text{ V} - (1.3\text{ mA})(3.6\text{ k}\Omega) = 5.32\text{ V}$$

The collector-emitter voltage is the difference between the collector voltage and the emitter voltage:

$$V_{CE} = 5.32\text{ V} - (-0.7\text{ V}) = 6.02\text{ V}$$

GOOD TO KNOW

When transistors are biased using well-designed voltage-divider or emitter-bias configurations, they are classified as beta-independent circuits because the values of I_C and V_{CE} are unaffected by changes in the transistor's beta.

Figure 8-10 Base voltage is ideally zero.

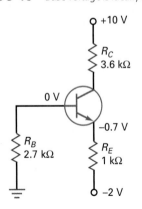

When two-supply emitter bias is well designed, it is similar to voltage-divider bias and satisfies this 100 : 1 rule:

$$R_B < 0.01\beta_{dc}R_E \qquad (8\text{-}12)$$

In this case, the simplified equations for analysis are:

$$V_B \approx 0 \qquad (8\text{-}13)$$

$$I_E = \frac{V_{EE} - 0.7\ \text{V}}{R_E} \qquad (8\text{-}14)$$

$$V_C = V_{CC} - I_C R_C \qquad (8\text{-}15)$$

$$V_{CE} = V_C + 0.7\ \text{V} \qquad (8\text{-}16)$$

Base Voltage

One source of error in the simplified method is the small voltage across the base resistor of Fig. 8-10. Since a small base current flows through this resistance, a negative voltage exists between the base and ground. In a well-designed circuit, this base voltage is less than -0.1 V. If a designer has to compromise by using a larger base resistance, the voltage may be more negative than -0.1 V. If you are troubleshooting a circuit like this, the voltage between the base and ground should produce a low reading; otherwise, something is wrong with the circuit.

Example 8-5

IIII MultiSim

What is the collector voltage in Fig. 8-10 if the emitter resistor is increased to 1.8 kΩ?

SOLUTION   The voltage across the emitter resistor is still 1.3 V. The emitter current is:

$$I_E = \frac{1.3\ \text{V}}{1.8\ \text{k}\Omega} = 0.722\ \text{mA}$$

The collector voltage is:

$$V_C = 10\ \text{V} - (0.722\ \text{mA})(3.6\ \text{k}\Omega) = 7.4\ \text{V}$$

PRACTICE PROBLEM 8-5   Change the emitter resistor of Fig. 8-10 to 2 kΩ and solve for V_{CE}.

Example 8-6

A **stage** is a transistor and the passive components connected to it. Figure 8-11 shows a three-stage circuit using two-supply emitter bias. What are the collector-to-ground voltages for each stage in Fig. 8-11?

SOLUTION   To begin with, ignore the capacitors because they appear as open circuits to dc voltage and currents. Then, we are left with three isolated transistors, each using two-supply emitter bias.

The first stage has an emitter current of:

$$I_E = \frac{15\ \text{V} - 0.7\ \text{V}}{20\ \text{k}\Omega} = \frac{14.3\ \text{V}}{20\ \text{k}\Omega} = 0.715\ \text{mA}$$

and a collector voltage of:

$$V_C = 15\ \text{V} - (0.715\ \text{mA})(10\ \text{k}\Omega) = 7.85\ \text{V}$$

Since the other stages have the same circuit values, each has a collector-to-ground voltage of approximately 7.85 V. Summary Table 8-1 illustrates the four main types of bias circuits.

PRACTICE PROBLEM 8-6   Change the supply voltages of Fig. 8-11 to $+12$ V and -12 V. Then, calculate V_{CE} for each transistor.

Figure 8-11   Three-stage circuit.

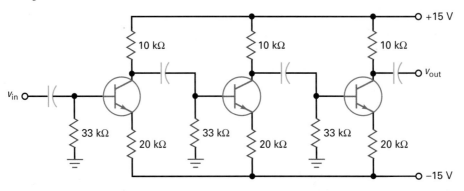

Summary Table 8–1		Main Bias Circuits		
Type	**Circuit**	**Calculations**	**Characteristics**	**Where used**
Base bias		$I_B = \dfrac{V_{BB} - 0.7\text{ V}}{R_B}$ $I_C = \beta I_B$ $V_{CE} = V_{CC} - I_C R_C$	Few parts; β dependent; fixed base current	Switch; digital
Emitter bias		$V_E = V_{BB} - 0.7\text{ V}$ $I_E = \dfrac{V_E}{R_E}$ $V_C = V_C - I_C R_C$ $V_{CE} = V_C - V_E$	Fixed emitter current; β independent	I_C driver; amplifier
Voltage divider bias		$V_B = \dfrac{R_2}{R_1 + R_2} V_{CC}$ $V_E = V_B - 0.7\text{ V}$ $I_E = \dfrac{V_E}{R_E}$ $V_C = V_{CC} - I_C R_C$ $V_{CE} = V_C - V_E$	Needs more resistors; β independent; needs only one power supply	Amplifier

Type	Circuit	Calculations	Characteristics	Where used
Two-supply emitter bias		$V_B \approx 0\ \text{V}$ $V_E = V_B - 0.7\ \text{V}$ $V_{RE} = V_{EE} - 0.7\ \text{V}$ $I_E = \dfrac{V_{RE}}{R_E}$ $V_C = V_{CC} - I_C R_C$ $V_{CE} = V_C - V_E$	Needs positive and negative power supplies; β independent	Amplifier

Figure 8-12   (a) Base bias; (b) emitter-feedback bias.

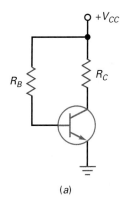

(a)

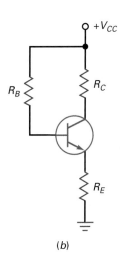

(b)

8-5 Other Types of Bias

In this section, we will discuss some other types of bias. A detailed analysis of these types of bias is not necessary because they are rarely used in new designs. But you should at least be aware of their existence in case you see them on a schematic diagram.

Emitter-Feedback Bias

Recall our discussion of base bias (Fig. 8-12a). This circuit is the worst when it comes to setting up a fixed Q point. Why? Since the base current is fixed, the collector current varies when the current gain varies. In a circuit like this, the Q point moves all over the load line with transistor replacement and temperature change.

Historically, the first attempt at stabilizing the Q point was **emitter-feedback bias,** shown in Fig. 8-12b. Notice that an emitter resistor has been added to the circuit. The basic idea is this: If I_C increases, V_E increases, causing V_B to increase. More V_B means less voltage across R_B. This results in less I_B, which opposes the original increase in I_C. It's called *feedback* because the change in emitter voltage is being fed back to the base circuit. Also, the feedback is called *negative* because it opposes the original change in collector current.

Emitter-feedback bias never became popular. The movement of the Q point is still too large for most applications that have to be mass-produced. Here are the equations for analyzing the emitter-feedback bias:

$$I_E = \frac{V_{CC} - V_{BE}}{R_E + R_B/\beta_{dc}} \tag{8-17}$$

$$V_E = I_E R_E \tag{8-18}$$

$$V_B = V_E + 0.7\ \text{V} \tag{8-19}$$

$$V_C = V_{CC} - I_C R_C \tag{8-20}$$

The intent of emitter-feedback bias is to **swamp out** the variations in β_{dc}; that is, R_E should be much greater than R_B/β_{dc}. If this condition is satisfied,

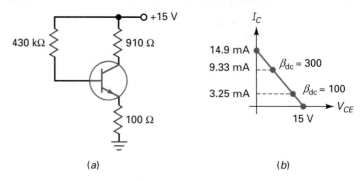

Eq. (8-17) will be insensitive to changes in β_{dc}. In practical circuits, however, a designer cannot select R_E large enough to swamp out the effects of β_{dc} without cutting off the transistor.

Figure 8-13*a* shows an example of an emitter-feedback bias circuit. Figure 8-13*b* shows the load line and the *Q* points for two different current gains. As you can see, a 3:1 variation in current gain produces a large variation in collector current. The circuit is not much better than base bias.

Collector–Feedback Bias

Figure 8-14*a* shows **collector-feedback bias** (also called **self-bias**). Historically, this was another attempt at stabilizing the *Q* point. Again, the basic idea is to feed back a voltage to the base in an attempt to neutralize any change in collector

Figure 8–14 (*a*) Collector-feedback bias; (*b*) example; (*c*) *Q* point is less sensitive to changes in current gain.

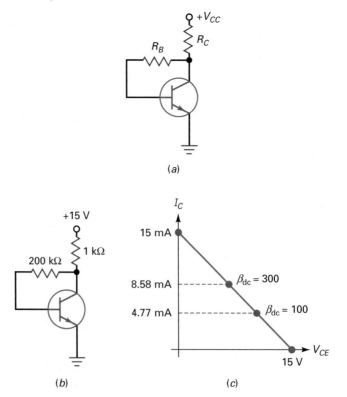

current. For instance, suppose the collector current increases. This decreases the collector voltage, which decreases the voltage across the base resistor. In turn, this decreases the base current, which opposes the original increase in collector current.

Like emitter-feedback bias, collector-feedback bias uses negative feedback in an attempt to reduce the original change in collector current. Here are the equations for analyzing collector-feedback bias:

$$I_E = \frac{V_{CC} - V_{BE}}{R_C + R_B/\beta_{dc}} \tag{8-21}$$

$$V_B = 0.7 \text{ V} \tag{8-22}$$

$$V_C = V_{CC} - I_C R_C \tag{8-23}$$

The Q point is usually set near the middle of the load line by using a base resistance of:

$$R_B = \beta_{dc} R_C \tag{8-24}$$

Figure 8-14b shows an example of collector-feedback bias. Figure 8-14c shows the load line and the Q points for two different current gains. As you can see, a 3:1 variation in current gain produces less variation in collector current than emitter-feedback (see Fig. 8-13b).

Collector-feedback bias is more effective than emitter-feedback bias in stabilizing the Q point. Although the circuit is still sensitive to changes in current gain, it is used in practice because of its simplicity.

Collector- and Emitter-Feedback Bias

Emitter-feedback bias and collector-feedback bias were the first steps toward a more stable bias for transistor circuits. Even though the idea of negative feedback is sound, these circuits fall short because there is not enough negative feedback to do the job. This is why the next step in biasing was the circuit shown in Fig. 8-15. The basic idea is to use both emitter and collector feedback to try to improve the operation.

As it turns out, more is not always better. Combining both types of feedback in one circuit helps but still falls short of the performance needed for mass production. If you come across this circuit, here are the equations for analyzing it:

$$I_E = \frac{V_{CC} - V_{BE}}{R_C + R_E + R_B/\beta_{dc}} \tag{8-25}$$

$$V_E = I_E R_E \tag{8-26}$$

$$V_B = V_E + 0.7 \text{ V} \tag{8-27}$$

$$V_C = V_{CC} - I_C R_C \tag{8-28}$$

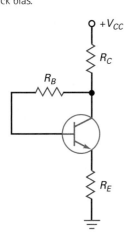

Figure 8-15 Collector-emitter feedback bias.

8-6 Troubleshooting

Let us discuss troubleshooting voltage-divider bias because this biasing method is the most widely used. Figure 8-16 shows the VDB circuit analyzed earlier. Table 8-1 lists the voltages for the circuit when it is simulated with MultiSim. The voltmeter used to make the measurements has an input impedance of 10 MΩ.

Figure 8-16 Troubleshooting.

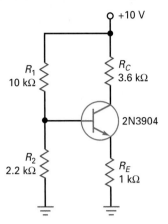

Table 8-1				Troubles and Symptoms
Trouble	V_B	V_E	V_C	**Comment**
None	1.79	1.12	6	No trouble
R_{1S}	10	9.17	9.2	Transistor saturated
R_{1O}	0	0	10	Transistor cutoff
R_{2S}	0	0	10	Transistor cutoff
R_{2O}	3.38	2.68	2.73	Reduces to emitter-feedback bias
R_{ES}	0.71	0	0.06	Transistor saturated
R_{EO}	1.8	1.37	10	10-MΩ voltmeter reduces V_E
R_{CS}	1.79	1.12	10	Collector resistor shorted
R_{CO}	1.07	0.4	0.43	Large base current
CES	2.06	2.06	2.06	All transistor terminals shorted
CEO	1.8	0	10	All transistor terminals open
No V_{CC}	0	0	0	Check supply and leads

Unique Troubles

Often, an open or shorted component produces unique voltages. For instance, the only way to get 10 V at the base of the transistor in Fig. 8-16 is with a shorted R_1. No other shorted or open component can produce the same result. Most of the entries in Table 8-1 produce a unique set of voltages, so you can identify them without breaking into the circuit to make further tests.

Ambiguous Troubles

Two troubles in Table 8-1 do not produce unique voltages: R_{1O} and R_{2S}. They both have measured voltages of 0, 0, and 10 V. With ambiguous troubles like this, the troubleshooter has to disconnect one of the suspected components and use an ohmmeter or other instrument to test it. For instance, we could disconnect R_1 and measure its resistance with an ohmmeter. If it's open, we have found the trouble. If it's OK, then R_2 is shorted.

Voltmeter Loading

Whenever you use a voltmeter, you are connecting a new resistance to a circuit. This resistance will draw current from the circuit. If the circuit has a large resistance, the voltage being measured will be lower than normal.

For instance, suppose the emitter resistor is open in Fig. 8-16. The base voltage is 1.8 V. Since there can be no emitter current with an open emitter resistor, the unmeasured voltage between the emitter and ground must also be 1.8 V. When you measure V_E with a 10-MΩ voltmeter, you are connecting 10 MΩ between the emitter and ground. This allows a small emitter current to flow, which produces a voltage across the emitter diode. This is why $V_E = 1.37$ V instead of 1.8 V for R_{EO} in Table 8-1.

Figure 8-17 *PNP* transistor.

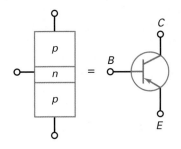

Figure 8-18 *PNP* currents.

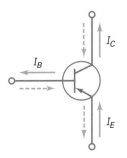

Figure 8-19 *PNP* circuit. (*a*) Negative supply; (*b*) positive supply.

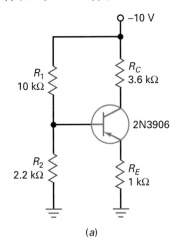

(*a*)

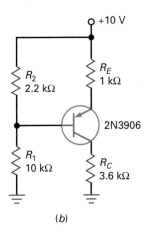

(*b*)

8-7 *PNP* Transistors

To this point, we have concentrated on bias circuits using *npn* transistors. Many circuits also use *pnp* transistors. This type of transistor is often used when the electronics equipment has a negative power supply. Also, *pnp* transistors are used as complements to *npn* transistors when dual (positive and negative) power supplies are available.

Figure 8-17 shows the structure of a *pnp* transistor along with its schematic symbol. Because the doped regions are of the opposite type, we have to turn our thinking around. Specifically, holes are the majority carriers in the emitter instead of free electrons.

Basic Ideas

Briefly, here is what happens at the atomic level: The emitter injects holes into the base. The majority of these holes flow on to the collector. For this reason, the collector current is almost equal to the emitter current.

Figure 8-18 shows the three transistor currents. Solid arrows represent conventional current, and dashed arrows represent electron flow.

Negative Supply

Figure 8-19*a* shows voltage-divider bias with a *pnp* transistor and a negative supply voltage of -10 V. The 2N3906 is the complement of the 2N3904; that is, its characteristics have the same absolute values as those of the 2N3904, but all currents and voltage polarities are reversed. Compare this *pnp* circuit with the *npn* circuit of Fig. 8-16. The only differences are the supply voltages and the transistors.

The point is this: Whenever you have a circuit with *npn* transistors, you can often use the same circuit with a negative power supply and *pnp* transistors.

Because of using a negative supply voltage, producing negative circuit values, care needs to be taken when doing circuit calculations. The steps in determining the Q point of Fig. 8-19*a* would be as follows:

$$V_B = \frac{R_2}{R_1 + R_2} V_{CC} = \frac{2.2\text{ k}\Omega}{10\text{ k}\Omega + 2.2\text{ k}\Omega}(-10\text{ V}) = -1.8\text{ V}$$

With a *pnp* transistor, the base-emitter junction will be forward biased when V_E is 0.7 V above V_B. Therefore,

$$V_E = V_B + 0.7\text{ V}$$

$$V_E = -1.8\text{ V} + 0.7\text{ V}$$

$$V_E = -1.1\text{ V}$$

Next, determine the emitter and collector currents:

$$I_E = \frac{V_E}{R_E} = \frac{-1.1\text{ V}}{1\text{ k}\Omega} = 1.1\text{ mA}$$

$$I_C \approx I_E = 1.1\text{ mA}$$

Now, solve for the collector and collector-emitter voltage values:

$$V_C = -V_{CC} + I_C R_C$$

$$V_C = -10\text{ V} + (1.1\text{ mA})(3.6\text{ k}\Omega)$$

$$V_C = -6.04\text{ V}$$

$$V_{CE} = V_C - V_E$$

$$V_{CE} = -6.04\text{ V} - (-1.1\text{ V}) = -4.94\text{ V}$$

Positive Power Supply

Positive power supplies are used more often in transistor circuits than negative power supplies. Because of this, you often see *pnp* transistors drawn upside down, as shown in Fig. 8-19*b*. Here is how the circuit works: The voltage across R_2 is applied to the emitter diode in series with the emitter resistor. This sets up the emitter current. The collector current flows through R_C, producing a collector-to-ground voltage. For troubleshooting, you can calculate V_C, V_B, and V_E as follows:

1. Get the voltage across R_2.
2. Subtract 0.7 V to get the voltage across the emitter resistor.
3. Get the emitter current.
4. Calculate the collector-to-ground voltage.
5. Calculate the base-to-ground voltage.
6. Calculate the emitter-to-ground voltage.

Example 8–7   ‖‖‖ MultiSim

Calculate the three transistor voltages for the *pnp* circuit of Fig. 8-19*b*.

SOLUTION   Start with the voltage across R_2. We can calculate this voltage using the voltage-divider equation:

$$V_2 = \frac{R_2}{R_1 + R_2} V_{EE}$$

Alternatively, we can calculate the voltage in a different way: Get the current through the voltage divider and then multiply by R_2. The calculation looks like this:

$$I = \frac{10\text{ V}}{12.2\text{ k}\Omega} = 0.82\text{ mA}$$

and

$$V_2 = (0.82\text{ mA})(2.2\text{ k}\Omega) = 1.8\text{ V}$$

Next, subtract 0.7 V from the foregoing voltage to get the voltage across the emitter resistor:

$$1.8\text{ V} - 0.7\text{ V} = 1.1\text{ V}$$

Then, calculate the emitter current:

$$I_E = \frac{1.1\text{ V}}{1\text{ k}\Omega} = 1.1\text{ mA}$$

When the collector current flows through the collector resistor, it produces a collector-to-ground voltage of:

$$V_C = (1.1\text{ mA})(3.6\text{ k}\Omega) = 3.96\text{ V}$$

The voltage between the base and ground is:

$$V_B = 10\text{ V} - 1.8\text{ V} = 8.2\text{ V}$$

The voltage between the emitter and ground is:

$$V_E = 10\text{ V} - 1.1\text{ V} = 8.9\text{ V}$$

PRACTICE PROBLEM 8–7   For both circuits, Fig. 8-19*a* and 8-19*b*, change the power supply voltage from 10 V to 12 V and calculate V_B, V_E, V_C, and V_{CE}.

Summary

SEC. 8-1 VOLTAGE-DIVIDER BIAS

The most famous circuit based on the emitter-bias prototype is called voltage-divider bias. You can recognize it by the voltage divider in the base circuit.

SEC. 8-2 ACCURATE VDB ANALYSIS

The key idea is for the base current to be much smaller than the current through the voltage divider. When this condition is satisfied, the voltage divider holds the base voltage almost constant and equal to the unloaded voltage out of the voltage divider. This produces a solid Q point under all operating conditions.

SEC. 8-3 VDB LOAD LINE AND Q POINT

The load line is drawn through saturation and cutoff. The Q point lies on the load line with the exact location determined by the biasing. Large variations in current gain have almost no effect on the Q point because this type of bias sets up a constant value of emitter current.

SEC. 8-4 TWO-SUPPLY EMITTER BIAS

This design uses two power supplies: one positive and the other negative. The idea is to set up a constant value of emitter current. The circuit is a variation of the emitter-bias prototype discussed earlier.

SEC. 8-5 OTHER TYPES OF BIAS

This section introduced negative feedback, a phenomenon that exists when an increase in an output quantity produces a decrease in an input quantity. It is a brilliant idea that led to voltage-divider bias. The other types of bias cannot use enough negative feedback, so they fail to attain the performance level of voltage-divider bias.

SEC. 8-6 TROUBLESHOOTING

Troubleshooting is an art. Because of this, it cannot be reduced to a set of rules. You learn troubleshooting mostly from experience.

SEC. 8-7 *PNP* TRANSISTORS

These *pnp* devices have all currents and voltages reversed from their *npn* counterparts. They may be used with negative power supplies; more commonly, they are used with positive power supplies in an upside-down configuration.

VDB Derivations

(8-1)   Base voltage:

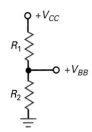

$$V_{BB} = \frac{R_2}{R_1 + R_2} V_{CC}$$

(8-2)   Emitter voltage:

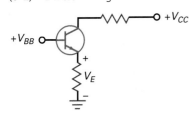

$$V_E = V_{BB} - V_{BE}$$

(8-3)   Emitter current:

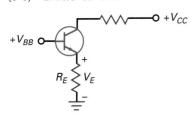

$$I_E = \frac{V_E}{R_E}$$

(8-4)   Collector current:

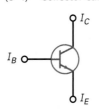

$$I_C \approx I_E$$

(8-5)   Collector voltage:

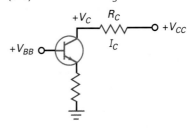

$$V_C = V_{CC} - I_C R_C$$

(8-6)   Collector-emitter voltage:

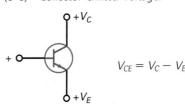

$$V_{CE} = V_C - V_E$$

Transistor Biasing

279

TSEB Derivations

(8-13) Base voltage:

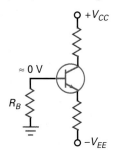

$V_B \approx 0$

(8-15) Collector voltage (TSEB)

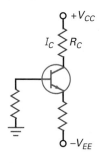

$V_C = V_{CC} - I_C R_C$

(8-14) Emitter current:

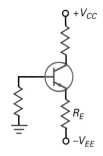

$$I_E = \frac{V_{EE} - 0.7\ \text{V}}{R_E}$$

(8-16) Collector-emitter voltage (TSEB)

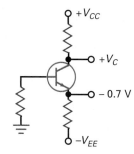

$V_{CE} = V_C + 0.7\ \text{V}$

Student Assignments

1. For the emitter bias, the voltage across the emitter resistor is the same as the voltage between the emitter and the
   a. Base
   b. Collector
   c. Emitter
   d. Ground

2. For emitter bias, the voltage at the emitter is 0.7 V less than the
   a. Base voltage
   b. Emitter voltage
   c. Collector voltage
   d. Ground voltage

3. With voltage-divider bias, the base voltage is
   a. Less than the base supply voltage
   b. Equal to the base supply voltage
   c. Greater than the base supply voltage
   d. Greater than the collector supply voltage

4. VDB is noted for its
   a. Unstable collector voltage
   b. Varying emitter current
   c. Large base current
   d. Stable Q point

5. With VDB, an increase in collector resistance will
   a. Decrease the emitter voltage
   b. Decrease the collector voltage
   c. Increase the emitter voltage
   d. Decrease the emitter current

6. VDB has a stable Q point like
   a. Base bias
   b. Emitter bias
   c. Collector-feedback bias
   d. Emitter-feedback bias

7. VDB needs
   a. Only three resistors
   b. Only one supply
   c. Precision resistors
   d. More resistors to work better

8. VDB normally operates in the
   a. Active region
   b. Cutoff region
   c. Saturation region
   d. Breakdown region

9. The collector voltage of a VDB circuit is not sensitive to changes in the
   a. Supply voltage
   b. Emitter resistance
   c. Current gain
   d. Collector resistance

10. If the emitter resistance decreases in a VDB circuit, the collector voltage
    a. Decreases   c. Increases
    b. Stays the same   d. Doubles

11. Base bias is associated with
    a. Amplifiers
    b. Switching circuits
    c. Stable Q point
    d. Fixed emitter current

12. If the emitter resistance is reduced by one-half in a VDB circuit, the collector current will
    a. Double
    b. Drop in half
    c. Remain the same
    d. Increase

13. If the collector resistance decreases in a VDB circuit, the collector voltage will
    a. Decrease
    b. Stay the same
    c. Increase
    d. Double

14. The Q point of a VDB circuit is
    a. Hypersensitive to changes in current gain
    b. Somewhat sensitive to changes in current gain
    c. Almost totally insensitive to changes in current gain
    d. Greatly affected by temperature changes

15. The base voltage of two-supply emitter bias (TSEB) is
    a. 0.7 V
    b. Very large
    c. Near 0 V
    d. 1.3 V

16. If the emitter resistance doubles with TSEB, the collector current will
    a. Drop in half
    b. Stay the same
    c. Double
    d. Increase

17. If a splash of solder shorts the collector resistor of TSEB, the collector voltage will
    a. Drop to zero
    b. Equal the collector supply voltage
    c. Stay the same
    d. Double

18. If the emitter resistance decreases with TSEB, the collector voltage will
    a. Decrease
    b. Stay the same
    c. Increase
    d. Equal the collector supply voltage

19. If the base resistor opens with TSEB, the collector voltage will
    a. Decrease
    b. Stay the same
    c. Increase slightly
    d. Equal the collector supply voltage

20. In TSEB, the base current must be very
    a. Small
    b. Large
    c. Unstable
    d. Stable

21. The Q point of TSEB does not depend on the
    a. Emitter resistance
    b. Collector resistance
    c. Current gain
    d. Emitter voltage

22. The majority carriers in the emitter of a *pnp* transistor are
    a. Holes
    b. Free electrons
    c. Trivalent atoms
    d. Pentavalent atoms

23. The current gain of a *pnp* transistor is
    a. The negative of the *npn* current gain
    b. The collector current divided by the emitter current
    c. Near zero
    d. The ratio of collector current to base current

24. Which is the largest current in a *pnp* transistor?
    a. Base current
    b. Emitter current
    c. Collector current
    d. None of these

25. The currents of a *pnp* transistor are
    a. Usually smaller than *npn* currents
    b. Opposite *npn* currents
    c. Usually larger than *npn* currents
    d. Negative

26. With *pnp* voltage-divider bias, you must use
    a. Negative power supplies
    b. Positive power supplies
    c. Resistors
    d. Grounds

27. With a TSEB *pnp* circuit using a negative V_{CC} supply, the emitter voltage is
    a. Equal to the base voltage
    b. 0.7 V higher than the base voltage
    c. 0.7 V lower than the base voltage
    d. Equal to the collector voltage

28. In a well-designed VDB circuit, the base current is
    a. Much larger than the voltage divider current
    b. Equal to the emitter current
    c. Much smaller than the voltage divider current
    d. Equal to the collector current

29. In a VDB circuit, the base input resistance R_{IN} is
    a. Equal to $\beta_{dc} R_E$
    b. Normally smaller than R_{TH}
    c. Equal to $\beta_{dc} R_C$
    d. Independent of β_{dc}

30. In a TSEB circuit, the base voltage is approximately zero when
    a. The base resistor is very large
    b. The transistor is saturated
    c. β_{dc} is very small
    d. $R_B < 0.01 \beta_{dc} R_E$

Problems

SEC. 8-1 VOLTAGE-DIVIDER BIAS

8-1 |||| MultiSim What is the emitter voltage in Fig. 8-20? The collector voltage?

8-2 |||| MultiSim What is the emitter voltage in Fig. 8-21? The collector voltage?

8-3 |||| MultiSim What is the emitter voltage in Fig. 8-22? The collector voltage?

8-4 |||| MultiSim What is the emitter voltage in Fig. 8-23? The collector voltage?

8-5 All resistors in Fig. 8-22 have a tolerance of ±5 percent. What is the lowest possible value of the collector voltage? The highest?

8-6 The power supply of Fig. 8-23 has a tolerance of ±10 percent. What is the lowest possible value of the collector voltage? The highest?

Figure 8-20

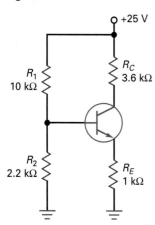

Figure 8-21

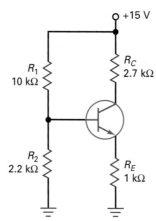

Figure 8-22

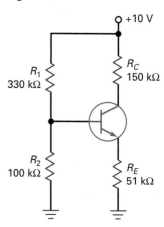

SEC. 8-3 VDB LOAD LINE AND Q POINT

8-7 What is the Q point for Fig. 8-20?

8-8 What is the Q point for Fig. 8-21?

8-9 What is the Q point for Fig. 8-22?

8-10 What is the Q point for Fig. 8-23?

8-11 All resistors in Fig. 8-22 have a tolerance of ±5 percent. What is the lowest value of the collector current? The highest?

8-12 The power supply of Fig. 8-23 has a tolerance of ±10 percent. What is the lowest possible value of the collector current? The highest?

SEC. 8-4 TWO-SUPPLY EMITTER BIAS

8-13 What is the emitter current in Fig. 8-24? The collector voltage?

8-14 If all resistances are doubled in Fig. 8-24, what is the emitter current? The collector voltage?

8-15 All resistors in Fig. 8-24 have a tolerance of ±5 percent. What is the lowest possible value of the collector voltage? The highest?

SEC. 8-5 OTHER TYPES OF BIAS

8-16 Does the collector voltage increase, decrease, or remain the same in Fig. 8-23 for small changes in each of the following?

  a. R_1 increases   d. R_C decreases

  b. R_2 decreases   e. V_{CC} increases

  c. R_E increases   f. β_{dc} decreases

8-17 Does the collector voltage increase, decrease, or remain the same in Fig. 8-25 for small increases in each of the following circuit values?

  a. R_1 d. R_C

  b. R_2 e. V_{CC}

  c. R_E f. β_{dc}

Figure 8-23

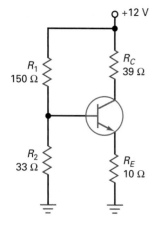

Figure 8-24

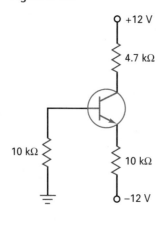

Figure 8-25

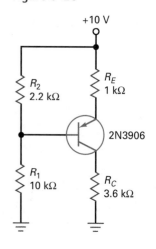

SEC. 8-6 TROUBLESHOOTING

8-18 What is the approximate value of the collector voltage in Fig. 8-23 for each of these troubles?

   a. R_1 open

   b. R_2 open

   c. R_E open

   d. R_C open

   e. Collector-emitter open

8-19 What is the approximate value of the collector voltage in Fig. 8-25 for each of these troubles?

   a. R_1 open

   b. R_2 open

   c. R_E open

   d. R_C open

   e. Collector-emitter open

SEC. 8-7 *PNP* TRANSISTORS

8-20 What is the collector voltage in Fig. 8-25?

8-21 What is the collector-emitter voltage in Fig. 8-25?

8-22 What is the collector saturation current in Fig. 8-25? The collector-emitter cutoff voltage?

8-23 What is the emitter voltage in Fig. 8-26? The collector voltage?

Figure 8-26

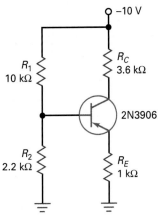

Critical Thinking

8-24 Somebody has built the circuit of Fig. 8-23, except for changing the voltage divider as follows: $R_1 = 150 \text{ k}\Omega$ and $R_2 = 33 \text{ k}\Omega$. The builder cannot understand why the base voltage is only 0.8 V instead of 2.16 V (the ideal output of the voltage divider). Can you explain what is happening?

8-25 Somebody builds the circuit of Fig. 8-23 with a 2N3904. What do you have to say about that?

8-26 A student wants to measure the collector-emitter voltage in Fig. 8-23, and so connects a voltmeter between the collector and the emitter. What does it read?

8-27 You can vary any circuit value in Fig. 8-23. Name all the ways you can think of to destroy the transistor.

8-28 The power supply of Fig. 8-23 has to supply current to the transistor circuit. Name all the ways you can think of to find this current.

8-29 Calculate the collector voltage for each transistor of Fig. 8-27. (*Hint:* Capacitors are open to direct current.)

8-30 The circuit of Fig. 8-28*a* uses silicon diodes. What is the emitter current? The collector voltage?

8-31 What is the output voltage in Fig. 8-28*b*?

8-32 How much current is there through the LED of Fig. 8-29*a*?

8-33 What is the LED current in Fig. 8-29*b*?

8-34 We want the voltage divider of Fig. 8-22 to be stiff. Change R_1 and R_2 as needed without changing the Q point.

Figure 8-27

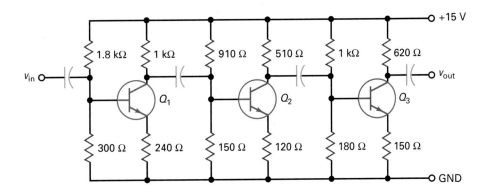

Figure 8-28

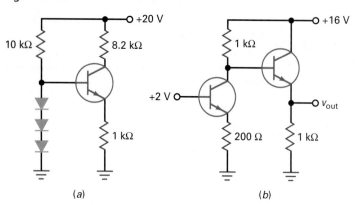

(a)   (b)

Figure 8-29

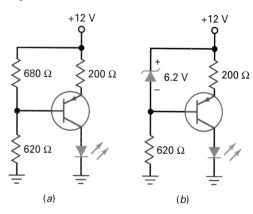

(a)   (b)

Troubleshooting

MultiSim Use Fig. 8-30 for the remaining problems.

8-35 Find Trouble 1.

8-36 Find Trouble 2.

8-37 Find Troubles 3 and 4.

8-38 Find Troubles 5 and 6.

8-39 Find Troubles 7 and 8.

8-40 Find Troubles 9 and 10.

8-41 Find Troubles 11 and 12.

Figure 8-30

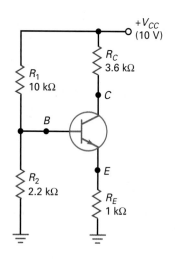

MEASUREMENTS

Trouble	V_B (V)	V_E (V)	V_C (V)	R_2 (Ω)
OK	1.8	1.1	6	OK
T1	10	9.3	9.4	OK
T2	0.7	0	0.1	OK
T3	1.8	1.1	10	OK
T4	2.1	2.1	2.1	OK
T5	0	0	10	OK
T6	3.4	2.7	2.8	∞
T7	1.83	1.212	10	OK
T8	0	0	10	0
T9	1.1	0.4	0.5	OK
T10	1.1	0.4	10	OK
T11	0	0	0	OK
T12	1.83	0	10	OK

Job Interview Questions

1. Draw a VDB circuit. Then, tell me all the steps in calculating the collector-emitter voltage. Why does this circuit have a very stable Q point?
2. Draw a TSEB circuit and tell me how it works. What happens to the collector current when the transistor is replaced or the temperature changes?
3. Describe a few other kinds of bias. What can you tell me about their Q points?
4. What are the two types of feedback biasing, and why were they developed?
5. What is the primary type of biasing used with discrete bipolar transistor circuits?
6. Should transistors being used as switching circuits be biased in the active region? If not, what two points associated with the load line are important with switching circuits?
7. In a VDB circuit, the base current is not small compared to the current through the voltage divider. What is the shortcoming of this circuit? What should be changed to correct it?
8. What is the most commonly used transistor biasing configuration? Why?
9. Draw a VDB circuit using an *npn* transistor. Label directions of divider, base, emitter, and collector currents.
10. What is wrong with a VDB circuit in which R_1 and R_2 are 100 times greater than R_E?

Self-Test Answers

1.	d	11.	b	21.	c
2.	a	12.	a	22.	a
3.	a	13.	c	23.	d
4.	d	14.	c	24.	b
5.	b	15.	c	25.	b
6.	b	16.	a	26.	c
7.	b	17.	b	27.	b
8.	a	18.	a	28.	c
9.	c	19.	d	29.	a
10.	a	20.	a	30.	d

Practice Problem Answers

8-1 $V_B = 2.7$ V;
$V_E = 2$ mA;
$V_C = 7.78$ V;
$V_{CE} = 5.78$ V

8-2 $V_{CE} = 5.85$ V;
Very close to the predicted value

8-4 $R_E = 1$ kΩ;
$R_C = 4$ kΩ;
$R_2 = 700$ Ω (680);
$R_1 = 3.4$ kΩ (3.3k)

8-5 $V_{CE} = 6.96$ V

8-6 $V_{CE} = 7.05$ V

8-7 For 8-19*a*:
$V_B = 2.16$ V;
$V_E = -1.46$ V;
$V_C = -6.73$ V;
$V_{CE} = -5.27$ V

For 8-19*b*:
$V_B = 9.84$ V;
$V_E = 10.54$ V;
$V_C = 5.27$ V;
$V_{CE} = -5.27$ V

AC Models

After a transistor has been biased with the Q point near the middle of the load line, we can couple a small ac voltage into the base. This will produce an ac collector voltage. The ac collector voltage looks like the ac base voltage, except that it's a lot bigger. In other words, the ac collector voltage is an *amplified* version of the ac base voltage.

The invention of amplifying devices, first vacuum tubes and later transistors, was crucial to the evolution of electronics. Without amplification, there would be no radio, no television, and no computers.

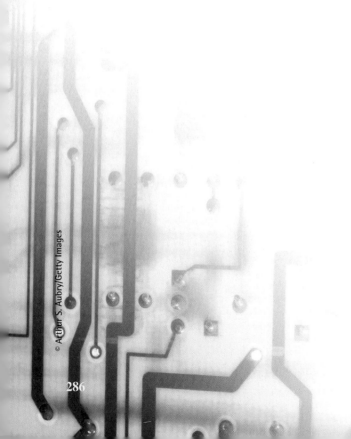

Objectives

After studying this chapter, you should be able to:

- Draw a transistor amplifier and explain how it works.
- Describe what coupling and bypass capacitors are supposed to do.
- Give examples of ac shorts and ac grounds.
- Use the superposition theorem to draw the dc and ac equivalent circuits.
- Define small-signal operation and why it may be desirable.
- Draw an amplifier that uses VDB. Then, draw its ac equivalent circuit.

Chapter Outline

Vocabulary

ac current gain
ac emitter resistance
ac equivalent circuit
ac ground
ac short
bypass capacitor

CB amplifier
CC amplifier
CE amplifier
coupling capacitor
dc equivalent circuit
distortion

Ebers-Moll model
π model
small-signal amplifiers
superposition theorem
T model
voltage gain

9-1 Base-Biased Amplifier

In this section, we will discuss a base-biased amplifier. Although a base-biased amplifier is not useful in mass production, it has instructional value because its basic ideas can be used to build more complicated amplifiers.

Coupling Capacitor

Figure 9-1a shows an ac voltage source connected to a capacitor and a resistor. Since the impedance of the capacitor is inversely proportional to frequency, the capacitor effectively blocks dc voltage and transmits ac voltage. When the frequency is high enough, the capacitive reactance is much smaller than the resistance. In this case, almost all the ac source voltage appears across the resistor. When used in this way, the capacitor is called a **coupling capacitor** because it couples or transmits the ac signal to the resistor. Coupling capacitors are important because they allow us to couple an ac signal into an amplifier without disturbing its Q point.

For a coupling capacitor to work properly, its reactance must be much smaller than the resistance at the *lowest frequency of the ac source*. For instance, if the frequency of the ac source varies from 20 Hz to 20 kHz, the worst case occurs at 20 Hz. A circuit designer will select a capacitor whose reactance at 20 Hz is much smaller than the resistance.

How small is small? As a definition:

Good coupling: $X_C < 0.1R$ (9-1)

In words: The reactance should be at least 10 times smaller than the resistance at the lowest frequency of operation.

When the 10:1 rule is satisfied, Fig. 9-1a can be replaced by the equivalent circuit of Fig. 9-1b. Why? The magnitude of impedance in Fig. 9-1a is given by:

$$Z = \sqrt{R^2 + X_C^2}$$

When you substitute the worst-case into this, you get:

$$Z = \sqrt{R^2 + (0.1R)^2} = \sqrt{R^2 + 0.01R^2} = \sqrt{1.01R^2} = 1.005R$$

Since the impedance is within half of a percent of R at the lowest frequency, the current in Fig. 9-1a is only half a percent less than the current in Fig. 9-1b. Since any well-designed circuit satisfies the 10:1 rule, we can approximate all coupling capacitors as an **ac short** (Fig. 9-1b).

A final point about coupling capacitors: Since dc voltage has a frequency of zero, the reactance of a coupling capacitor is infinite at zero frequency. Therefore, we will use these two approximations for a capacitor:

1. For dc analysis, the capacitor is open.
2. For ac analysis, the capacitor is shorted.

Figure 9-1 (a) Coupling capacitor; (b) capacitor is an ac short; (c) dc open and ac short.

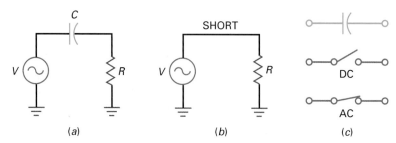

Figure 9-1c summarizes these two important ideas. Unless otherwise stated, all the circuits we analyze from now on will satisfy the 10:1 rule, so that we can visualize a coupling capacitor as shown in Fig. 9-1c.

Example 9-1

Using Fig. 9-1a, if $R = 2$ kΩ and the frequency range is from 20 Hz to 20 kHz, find the value of C needed to act as a good coupling capacitor.

SOLUTION Following the 10:1 rule, X_C should be ten times smaller than R at the lowest frequency.

Therefore,

$$X_C < 0.1 \, R \text{ at 20 Hz}$$

$$X_C < 200 \, \Omega \text{ at 20 Hz}$$

Since $X_C = \dfrac{1}{2\pi f C}$

by rearrangement, $C = \dfrac{1}{2\pi f X_C} = \dfrac{1}{(2\pi)(20 \text{ Hz})(200 \, \Omega)}$

$$C = 39.8 \, \mu\text{F}$$

PRACTICE PROBLEM 9-1 Using Example 9-1, find the value of C when the lowest frequency is 1 kHz and R is 1.6 kΩ.

DC Circuit

Figure 9-2a shows a base-biased circuit. The dc base voltage is 0.7 V. Because 30 V is much greater than 0.7 V, the base current is approximately 30 V divided by 1 MΩ, or:

$$I_B = 30 \, \mu\text{A}$$

With a current gain of 100, the collector current is:

$$I_C = 3 \text{ mA}$$

and the collector voltage is:

$$V_C = 30 \text{ V} - (3 \text{ mA})(5 \text{ k}\Omega) = 15 \text{ V}$$

So, the Q point is located at 3 mA and 15 V.

Amplifying Circuit

Figure 9-2b shows how to add components to build an amplifier. First, a coupling capacitor is used between an ac source and the base. Since the coupling capacitor is open to direct current, the same dc base current exists with or without the capacitor and ac source. Similarly, a coupling capacitor is used between the collector and the load resistor of 100 kΩ. Since this capacitor is open to direct current,

Figure 9-2 (*a*) Base bias; (*b*) base-biased amplifier.

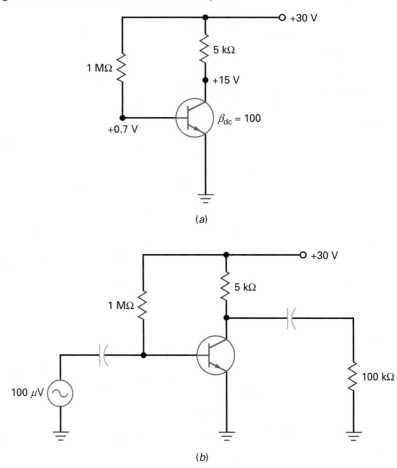

(*a*)

(*b*)

the dc collector voltage is the same with or without the capacitor and load resistor. The key idea is that the coupling capacitors prevent the ac source and load resistance from changing the Q point.

In Fig. 9-2*b*, the ac source voltage is 100 μV. Since the coupling capacitor is an ac short, all the ac source voltage appears between the base and the ground. This ac voltage produces an ac base current that is added to the existing dc base current. In other words, the total base current will have a dc component and an ac component.

Figure 9-3*a* illustrates the idea. An ac component is superimposed on the dc component. On the positive half cycle, the ac base current adds to the 30 μA of dc base current, and on the negative half cycle it subtracts from it.

The ac base current produces an amplified variation in collector current because of the current gain. In Fig. 9-3*b*, the collector current has a dc component of 3 mA. Superimposed on this is an ac collector current. Since this amplified collector current flows through the collector resistor, it produces a varying voltage across the collector resistor. When this voltage is subtracted from the supply voltage, we get the collector voltage shown in Fig. 9-3*c*.

Again, an ac component is superimposed on a dc component. The collector voltage is swinging sinusoidally above and below the dc level of +15 V. Also, the ac collector voltage is *inverted*, 180° out of phase with the input voltage. Why? On the positive half cycle of ac base current, the collector current increases, producing more voltage across the collector resistor. This means that there is less voltage between the collector and ground. Similarly, on the negative half cycle the

Figure 9–3 DC and ac components. (*a*) Base current; (*b*) collector current; (*c*) collector voltage.

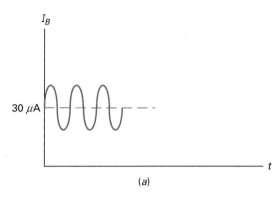

(*a*)

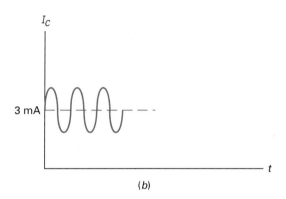

(*b*)

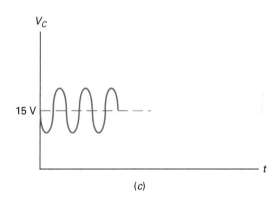

(*c*)

collector current decreases. Since there is less voltage across the collector resistor, the collector voltage increases.

Voltage Waveforms

Figure 9-4 shows the waveforms for a base-biased amplifier. The ac source voltage is a small sinusoidal voltage. This is coupled into the base, where it is superimposed on the dc component of +0.7 V. The variation in base voltage produces sinusoidal variations in base current, collector current, and collector voltage. The total collector voltage is an inverted sine wave superimposed on the dc collector voltage of +15 V.

Notice the action of the output coupling capacitor. Since it is open to direct current, it blocks the dc component of collector voltage. Since it is shorted to alternating current, it couples the ac collector voltage to the load resistor. This is why the load voltage is a pure ac signal with an average value of zero.

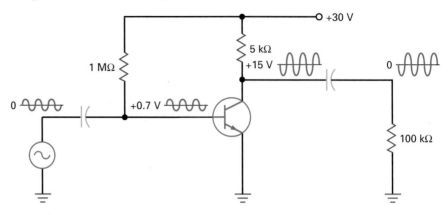

Figure 9-4 Base-biased amplifier with waveforms.

Voltage Gain

The **voltage gain** of an amplifier is defined as the ac output voltage divided by the ac input voltage. As a definition:

$$A_V = \frac{v_{out}}{v_{in}} \tag{9-2}$$

For instance, if we measure an ac load voltage of 50 mV with an ac input voltage of 100 μV, the voltage gain is:

$$A_V = \frac{50 \text{ mV}}{100 \text{ } \mu\text{V}} = 500$$

This says that the ac output voltage is 500 times larger than the ac input voltage.

Calculating Output Voltage

We can multiply both sides of Eq. (9-2) by v_{in} to get this derivation:

$$v_{out} = A_V v_{in} \tag{9-3}$$

This is useful when you want to calculate the value of v_{out}, given the values of A_V and v_{in}.

For instance, the triangular symbol shown in Fig. 9-5a is used to indicate an amplifier of any design. Since we are given an input voltage of 2 mV and a voltage gain of 200, we can calculate an output voltage of:

$$v_{out} = (200)(2 \text{ mV}) = 400 \text{ mV}$$

Calculating Input Voltage

We can divide both sides of Eq. (9-3) by A_V to get this derivation:

$$v_{in} = \frac{v_{out}}{A_V} \tag{9-4}$$

This is useful when you want to calculate the value of v_{in}, given the values v_{out} and A_V. For instance, the output voltage is 2.5 V in Fig. 9-5b. With a voltage gain of 350, the input voltage is:

$$v_{in} = \frac{2.5 \text{ V}}{350} = 7.14 \text{ mV}$$

Figure 9-5 (a) Calculating output voltage; (b) calculating input voltage.

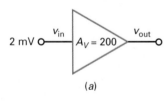

(a)

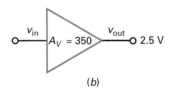

(b)

Figure 9-6 (*a*) Bypass capacitor;
(*b*) point *E* is an ac ground.

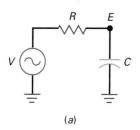

(*a*)

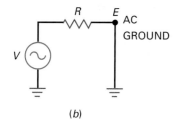

(*b*)

9-2 Emitter-Biased Amplifier

The base-biased amplifier has an unstable Q point. For this reason, it is not used much as an amplifier. Instead, an emitter-biased amplifier (either VDB or TSEB) with its stable Q point is preferred.

Bypass Capacitor

A **bypass capacitor** is similar to a coupling capacitor because it appears open to direct current and shorted to alternating current. But it is not used to couple a signal between two points. Instead, it is used to create an **ac ground.**

Figure 9-6*a* shows an ac voltage source connected to a resistor and a capacitor. The resistance R represents the Thevenin resistance as seen by the capacitor. When the frequency is high enough, the capacitive reactance is much smaller than the resistance. In this case, almost all the ac source voltage appears across the resistor. Stated another way, point E is effectively shorted to ground.

When used in this way, the capacitor is called a *bypass capacitor* because it bypasses or shorts point E to ground. A bypass capacitor is important because it allows us to create an ac ground in an amplifier without disturbing its Q point.

For a bypass capacitor to work properly, its reactance must be much smaller than the resistance at the *lowest frequency of the ac source.* The definition for good bypassing is identical to that for good coupling:

$$\text{Good bypassing: } X_C < 0.1R \tag{9-5}$$

When this rule is satisfied, Fig. 9-6*a* can be replaced by the equivalent circuit of Fig. 9-6*b*.

Example 9-2

Figure 9-7

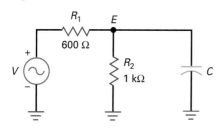

In Fig. 9-7, the input frequency of V is 1 kHz. What value of C is needed to effectively short point E to ground?

SOLUTION First, find the Thevenin resistance as seen by the capacitor C.

$$R_{TH} = R_1 \| R_2$$

$$R_{TH} = 600 \ \Omega \| 1 \ \text{k}\Omega = 375 \ \Omega$$

Next, X_C should be ten times smaller than R_{TH}. Therefore, $X_C < 37.5 \ \Omega$ at 1 kHz. Now solve for C by

$$C = \frac{1}{2\pi f X_C} = \frac{1}{(2\pi)(1 \ \text{kHz})(37.5 \ \Omega)}$$

$$C = 4.2 \ \mu\text{F}$$

PRACTICE PROBLEM 9-2 In Fig. 9-7, find the value of C needed if R is 50 Ω.

Figure 9-8 VDB amplifier with waveforms.

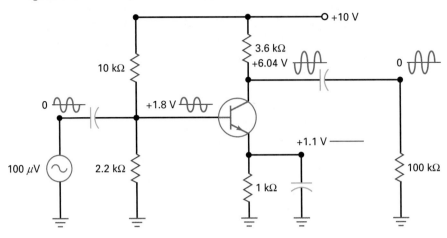

VDB Amplifier

Figure 9-8 shows a voltage-divider-biased (VDB) amplifier. To calculate the dc voltages and currents, mentally open all capacitors. Then, the transistor circuit simplifies to the VDB circuit analyzed in Chap. 8. The quiescent or dc values for this circuit are:

$$V_B = 1.8 \text{ V}$$

$$V_E = 1.1 \text{ V}$$

$$V_C = 6.04 \text{ V}$$

$$I_C = 1.1 \text{ mA}$$

As before, we use a coupling capacitor between the source and base, and another coupling capacitor between the collector and the load resistance. We also need to use a bypass capacitor between the emitter and ground. Without this capacitor, the ac base current would be much smaller. But with the bypass capacitor, we get a much larger voltage gain. The mathematical details of why this happens are discussed in the next chapter.

In Fig. 9-8, the ac source voltage is 100 μV. This is coupled into the base. Because of the bypass capacitor, all of this ac voltage appears across the base-emitter diode. The ac base current then produces an amplified ac collector voltage, as previously described.

VDB Waveforms

Notice the voltage waveforms in Fig. 9-8. The ac source voltage is a small sinusoidal voltage with an average value of zero. The base voltage is an ac voltage superimposed on a dc voltage of +1.8 V. The collector voltage is an amplified and inverted ac voltage superimposed on the dc collector voltage of +6.04 V. The load voltage is the same as the collector voltage, except that it has an average value of zero.

Notice also the voltage on the emitter. It is a pure dc voltage of +1.1 V. There is no ac emitter voltage because emitter is at ac ground, a direct result of using a bypass capacitor. This is important to remember because it is useful in troubleshooting. If the bypass capacitor were to open, an ac voltage would appear between the emitter and ground. This symptom would immediately point to the open bypass capacitor as the unique trouble.

Discrete versus Integrated Circuits

The VDB amplifier of Fig. 9-8 is the standard way to build a discrete transistor amplifier. *Discrete* means that all components like resistors, capacitors, and transistors are separately inserted and connected to get the final circuit. A *discrete circuit* differs from an *integrated circuit (IC),* in which all the components are simultaneously created and connected on a *chip,* a piece of semiconductor material. Later chapters will discuss the *op amp,* an IC amplifier that produces voltage gains of more than 100,000.

TSEB Circuit

Figure 9-9 shows a two-supply emitter bias (TSEB) amplifier. We analyzed the dc part of the circuit in Chap. 8 and calculate these quiescent voltages:

$$V_B \approx 0 \text{ V}$$

$$V_E = -0.7 \text{ V}$$

$$V_C = 5.32 \text{ V}$$

$$I_C = 1.3 \text{ mA}$$

Figure 9-9 shows two coupling capacitors and an emitter bypass capacitor. The ac operation of the circuit is similar to that of a VDB amplifier. We couple a signal into the base. The signal is amplified to get the collector voltage. The amplified signal is then coupled to the load.

Notice the waveforms. The ac source voltage is a small sinusoidal voltage. The base voltage has a small ac component riding on a dc component of approximately 0 V. The total collector voltage is an inverted sine wave riding on the dc collector voltage of +5.32 V. The load voltage is the same amplified signal with no dc component.

Again, notice the pure dc voltage on the emitter, a direct result of using the bypass capacitor. If the bypass capacitor were to open, an ac voltage would appear at the emitter. This would greatly reduce the voltage gain. Therefore, when troubleshooting an amplifier with bypass capacitors, remember that all ac grounds should have zero ac voltage.

Figure 9-9 TSEB amplifier with waveforms.

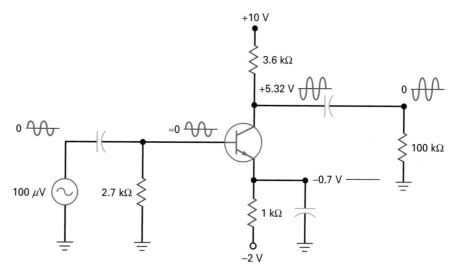

9-3 Small-Signal Operation

Figure 9-10 shows the graph of current versus voltage for the base-emitter diode. When an ac voltage is coupled into the base of a transistor, an ac voltage appears across the base-emitter diode. This produces the sinusoidal variation in V_{BE} shown in Fig. 9-10.

Instantaneous Operating Point

When the voltage increases to its positive peak, the instantaneous operating point moves from Q to the upper point shown in Fig. 9-10. On the other hand, when the sine wave decreases to its negative peak, the instantaneous operating point moves from Q to the lower point.

The total base-emitter voltage of Fig. 9-10 is an ac voltage centered on a dc voltage. The size of the ac voltage determines how far the instantaneous point moves away from the Q point. Large ac base voltages produce large variations, whereas small ac base voltages produce small variations.

Distortion

The ac voltage on the base produces the ac emitter current shown in Fig. 9-10. This ac emitter current has the same frequency as the ac base voltage. For instance, if the ac generator driving the base has a frequency of 1 kHz, the ac emitter current has a frequency of 1 kHz. The ac emitter current also has approximately the same shape as the ac base voltage. If the ac base voltage is sinusoidal, the ac emitter current is approximately sinusoidal.

The ac emitter current is not a perfect replica of the ac base voltage because of the curvature of the graph. Since the graph is curved upward, the positive half cycle of ac emitter current is elongated (stretched) and the negative half cycle is compressed. This stretching and compressing of alternate half cycles is called **distortion.** It is undesirable in high-fidelity amplifiers because it changes the sound of voice and music.

Reducing Distortion

One way to reduce distortion in Fig. 9-10 is by keeping the ac base voltage small. When you reduce the peak value of the base voltage, you reduce the movement of the instantaneous operating point. The smaller this swing or variation, the less the curvature in the graph. If the signal is small enough, the graph appears to be linear.

Figure 9-10   Distortion when signal is too large.

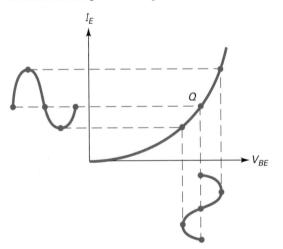

Figure 9-11   Definition of small-signal operation.

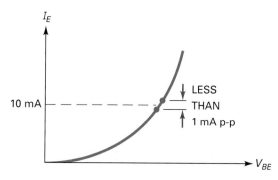

Why is this important? Because there is negligible distortion for a small signal. When the signal is small, the changes in ac emitter current are almost directly proportional to the changes in ac base voltage because the graph is almost linear. In other words, if the ac base voltage is a small enough sine wave, the ac emitter current will also be a small sine wave with no noticeable stretching or compression of half cycles.

The 10 Percent Rule

The total emitter current shown in Fig. 9-10 consists of a dc component and an ac component, which can be written as:

$$I_E = I_{EQ} + i_e$$

where I_E = the total emitter current
I_{EQ} = the dc emitter current
i_e = the ac emitter current

To minimize distortion, the peak-to-peak value of i_e must be small compared to I_{EQ}. Our definition of small-signal operation is:

Small signal: $i_{e(pp)} < 0.1I_{EQ}$ (9-6)

This says that the ac signal is small when the peak-to-peak ac emitter current is less than 10 percent of the dc emitter current. For instance, if the dc emitter current is 10 mA, as shown in Fig. 9-11, the peak-to-peak emitter current should be less than 1 mA in order to have small-signal operation.

From now on, we will refer to amplifiers that satisfy the 10 percent rule as **small-signal amplifiers.** This type of amplifier is used at the front end of radio and television receivers because the signal coming in from the antenna is very weak. When coupled into a transistor amplifier, a weak signal produces very small variations in emitter current, much less than the 10 percent rule requires.

Example 9-3

Using Fig. 9-9, find the maximum small signal emitter current.

SOLUTION: First, find the Q point emitter current, I_{EQ}.

$$I_{EQ} = \frac{V_{EE} - V_{BE}}{R_E} \qquad I_{EQ} = \frac{2\text{ V} - 0.7\text{ V}}{1\text{ k}\Omega} \qquad I_{EQ} = 1.3\text{ mA}$$

Then solve for the small signal emitter current $i_{e(pp)}$

$$i_{e(pp)} < 0.1 \, I_{EQ}$$

$$i_{e(pp)} = (0.1)(1.3 \text{ mA})$$

$$i_{e(pp)} = 130 \, \mu A_{pp}$$

PRACTICE PROBLEM 9-3 Using Fig. 9-9, change R_E to 1.5 kΩ and calculate the maximum small signal emitter current.

9-4 AC Beta

The current gain in all discussions up to this point has been *dc current gain.* This was defined as:

$$\beta_{dc} = \frac{I_C}{I_B} \tag{9-7}$$

The currents in this formula are the currents at the Q point in Fig. 9-12. Because of the curvature in the graph of I_C versus I_B, the dc current gain depends on the location of the Q point.

Definition

The **ac current gain** is different. It is defined as:

$$\beta = \frac{i_c}{i_b} \tag{9-8}$$

In words, the ac current gain equals the ac collector current divided by the ac base current. In Fig. 9-12, the ac signal uses only a small part of the graph on both sides

Figure 9-12 AC current gain equals ratio of changes.

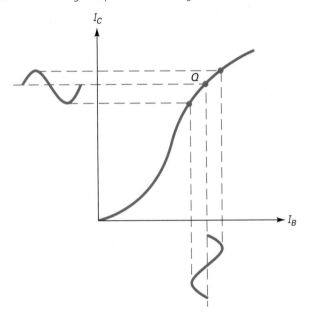

of the Q point. Because of this, the value of the ac current gain is different from the dc current gain, which uses almost all of the graph.

Graphically, β equals the slope of the curve at the Q point in Fig. 9-12. If we were to bias the transistor to a different Q point, the slope of the curve would change, which means that β would change. In other words, the value of β depends on the amount of dc collector current.

On data sheets, β_{dc} is listed as h_{FE} and β is shown as h_{fe}. Notice that capital subscripts are used with dc current gain, and lowercase subscripts with ac current gain. The two current gains are comparable in value, not differing by a large amount. For this reason, if you have the value of one, you can use the same value for the other in preliminary analysis.

Notation

To keep dc quantities distinct from ac quantities, it is standard practice to use capital letters and subscripts for dc quantities. For instance, we have been using:

I_E, I_C, and I_B for the dc currents
V_E, V_C, and V_B for the dc voltages
V_{BE}, V_{CE}, and V_{CB} for the dc voltages between terminals

For ac quantities, we will use lowercase letters and subscripts as follows:

i_e, i_c, and i_b for the ac currents
v_e, v_c, and v_b for the ac voltages
v_{be}, v_{ce}, and v_{cb} for the ac voltages between terminals

Also worth mentioning is the use of capital R for dc resistances and lowercase r for ac resistances. The next section will discuss ac resistance.

9–5 AC Resistance of the Emitter Diode

Figure 9-13 shows a graph of current versus voltage for the emitter diode. When a small ac voltage is across the emitter diode, it produces the ac emitter current shown. The size of this ac emitter current depends on the location of the Q point. Because of the curvature, we get more peak-to-peak ac emitter current when the Q point is higher up the graph.

Figure 9-13 AC resistance of emitter diode.

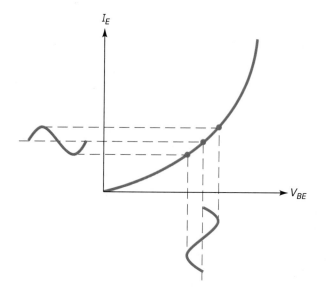

Definition

As discussed in Sec. 9-3, the total emitter current has a dc component and an ac component. In symbols:

$$I_E = I_{EQ} + i_e$$

where I_{EQ} is the dc emitter current and i_e is the ac emitter current.

In a similar way, the total base-emitter voltage of Fig. 9-13 has a dc component and an ac component. Its equation can be written as:

$$V_{BE} = V_{BEQ} + v_{be}$$

where V_{BEQ} is the dc base-emitter voltage and v_{be} is the ac base-emitter voltage.

In Fig. 9-13, the sinusoidal variation in V_{BE} produces a sinusoidal variation in I_E. The peak-to-peak value of i_e depends on the location of the Q point. Because of the curvature in the graph, a fixed v_{be} produces more i_e as the Q point is biased higher up the curve. Stated another way, the ac resistance of the emitter diode decreases when the dc emitter current increases.

The **ac emitter resistance** of the emitter diode is defined as:

$$r_e' = \frac{v_{be}}{i_e} \tag{9-9}$$

This says that the ac resistance of the emitter diode equals the ac base-emitter voltage divided by the ac emitter current. The prime (') in r_e' is a standard way to indicate that the resistance is inside the transistor.

For instance, Fig. 9-14 shows an ac base-emitter voltage of 5 mV pp. At the given Q point, this sets up an ac emitter current of 100 μA pp. The ac resistance of the emitter diode is:

$$r_e' = \frac{5 \text{ mV}}{100 \, \mu\text{A}} = 50 \, \Omega$$

As another example, assume that a higher Q point in Fig. 9-14 has $v_{be} = 5$ mV and $i_e = 200 \, \mu$A. Then, the ac resistance decreases to:

$$r_e' = \frac{5 \text{ mV}}{200 \, \mu\text{A}} = 25 \, \Omega$$

The point is this: The ac emitter resistance always decreases when the dc emitter current increases, because v_{be} is essentially a constant value.

Figure 9-14   Calculating r_e'.

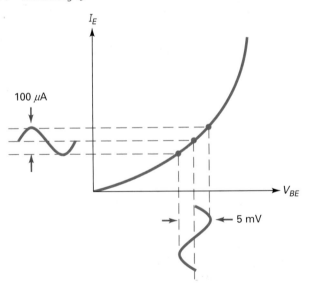

Formula for AC Emitter Resistance

With solid-state physics and calculus, it is possible to derive the following remarkable formula for the ac emitter resistance:

$$r'_e = \frac{25 \text{ mV}}{I_E} \tag{9-10}$$

This says that the ac resistance of the emitter diode equals 25 mV divided by the dc emitter current.

This formula is remarkable because of its simplicity and the fact that it applies to all transistor types. It is widely used in industry to calculate a preliminary value for the ac resistance of the emitter diode. The derivation assumes small-signal operation, room temperature, and an abrupt rectangular base-emitter junction. Since commercial transistors have gradual and nonrectangular junctions, there will be some deviations from Eq. (9-10). In practice, almost all commercial transistors have an ac emitter resistance between 25 mV/I_E and 50 mV/I_E.

The reason r'_e is important is because it determines the voltage gain. The smaller it is, the higher the voltage gain. Chapter 10 will show you how to use r'_e to calculate the voltage gain of a transistor amplifier.

Example 9-4   IIII MultiSim

What does r'_e equal in the base-biased amplifier of Fig. 9-15a?

Figure 9-15   (a) Base-biased amplifier; (b) VDB amplifier; (c) TSEB amplifier.

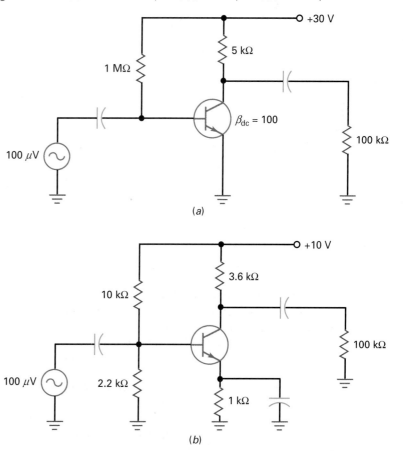

Figure 9-15 (continued)

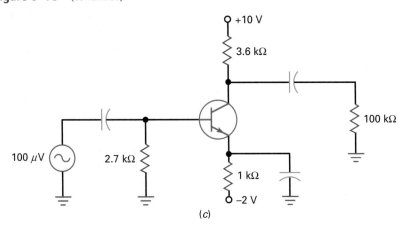

(c)

SOLUTION Earlier, we calculated a dc emitter current of approximately 3 mA for this circuit. With Eq. (9-10), the ac resistance of the emitter diode is:

$$r'_e = \frac{25 \text{ mV}}{3 \text{ mA}} = 8.33 \ \Omega$$

Example 9-5

IIII MultiSim

In Fig. 9-15b, what does r'_e equal?

SOLUTION We analyzed this VDB amplifier earlier and calculated a dc emitter current of 1.1 mA. The ac resistance of the emitter diode is:

$$r'_e = \frac{25 \text{ mV}}{1.1 \text{ mA}} = 22.7 \ \Omega$$

Example 9-6

IIII MultiSim

What is the ac resistance of the emitter diode for the two-supply emitter-bias amplifier of Fig. 9-15c?

SOLUTION From an earlier calculation, we got a dc emitter current of 1.3 mA. Now, we can calculate the ac resistance of the emitter diode:

$$r'_e = \frac{25 \text{ mV}}{1.3 \text{ mA}} = 19.2 \ \Omega$$

PRACTICE PROBLEM 9–6 Using Fig. 9-15c, change the V_{EE} supply to -3 V and calculate r'_e.

9-6 Two Transistor Models

To analyze the ac operation of a transistor amplifier, we need an ac equivalent circuit for a transistor. In other words, we need a model for the transistor that simulates how it behaves when an ac signal is present.

The T Model

One of the earliest ac models was the **Ebers-Moll model** shown in Fig. 9-16. As far as a small ac signal is concerned, the emitter diode of a transistor acts like an ac resistance r_e' and the collector diode acts like a current source i_c. Since the Ebers-Moll model looks like a T on its side, the equivalent circuit is also called the **T model.**

When analyzing a transistor amplifier, we can replace each transistor by a T model. Then, we can calculate value of r_e' and other ac quantities like voltage gain. The details are discussed in the next chapter.

When an ac input signal drives a transistor amplifier, an ac base-emitter voltage v_{be} is across the emitter diode, as shown in Fig. 9-17a. This produces an ac base current i_b. The ac voltage source has to supply this ac base current, so that the transistor amplifier will work properly. Stated another way, the ac voltage source is loaded by the input impedance of the base.

Figure 9-17b illustrates the idea. Looking into the base of the transistor, the ac voltage source sees an input impedance $z_{in(base)}$. At low frequencies, this impedance is purely resistive and defined as:

$$z_{in(base)} = \frac{v_{be}}{i_b} \tag{9-11}$$

Applying Ohm's law to the emitter diode of Fig. 9-17a, we can write:

$$v_{be} = i_e r_e'$$

Figure 9-16 T model of a transistor.

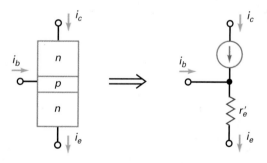

Figure 9-17 Defining the input impedance of the base.

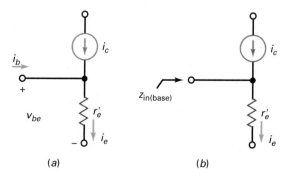

(a)   (b)

Figure 9-18 π model of a transistor.

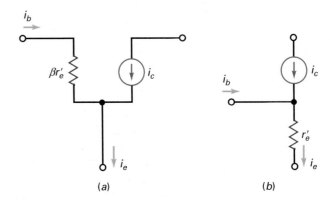

(a)   (b)

Substitute this equation into the preceding one to get:

$$z_{\text{in(base)}} = \frac{v_{be}}{i_b} = \frac{i_e r_e'}{i_b}$$

Since $i_e \approx i_c$, the foregoing equation simplifies to:

$$z_{\text{in(base)}} = \beta r_e' \qquad (9\text{-}12)$$

This equation tells us that the input impedance of the base is equal to the ac current gain times the ac resistance of the emitter diode.

The π Model

Figure 9-18a shows the π **model** of a transistor. It's a visual representation of Eq. (9-12). The π model is easier to use than the T model (Fig. 9-18b) because the input impedance is not obvious when you look at the T model. On the other hand, the π model clearly shows that an input impedance of $\beta r_e'$ will load the ac voltage source driving the base.

Since the π and T models are ac equivalent circuits for a transistor, we can use either one when analyzing an amplifier. Most of the time, we will use the π model. With some circuits like the differential amps in Chap. 17, the T model gives a better insight into the circuit action. Both models are widely used in industry.

9-7 Analyzing an Amplifier

Amplifier analysis is complicated because both dc and ac sources are in the same circuit. To analyze amplifiers, we can calculate the effect of the dc sources, and then the effect of the ac sources. When using the superposition theorem in this analysis, the effect of each source acting alone is added to get the total effect of all sources acting simultaneously.

The DC Equivalent Circuit

The simplest way to analyze an amplifier is to split the analysis into two parts: a dc analysis and an ac analysis. In the dc analysis, we calculate the dc voltages and currents. To do this, we mentally open all capacitors. The circuit that remains is the **dc equivalent circuit.**

With the dc equivalent circuit, you can calculate the transistor currents and voltages as needed. If you are troubleshooting, approximate answers are adequate. The most important current in the dc analysis is the dc emitter current. This is needed to calculate r_e' for the ac analysis.

Figure 9-19 DC voltage source is an ac short.

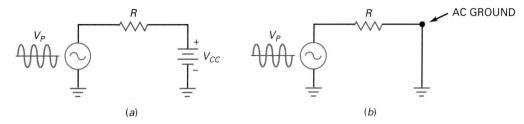

(a) (b)

AC Effect of a DC Voltage Source

Figure 9-19a shows a circuit with ac and dc sources. What is the ac current in a circuit like this? As far as the ac current is concerned, the dc voltage source acts like an ac short, as shown in Fig. 9-19b. Why? Because a dc voltage source has a constant voltage across it. Therefore, any ac current flowing through it cannot produce an ac voltage across it. If no ac voltage can exist, the dc voltage source is equivalent to an ac short.

Another way to understand the idea is to recall the **superposition theorem** discussed in basic electronics courses. In applying superposition to Fig. 9-19a, we can calculate the effect of each source acting separately while the other is reduced to zero. Reducing the dc voltage source to zero is equivalent to shorting it. Therefore, to calculate the effect of the ac source in Fig. 9-19a, we can short the dc voltage source.

From now on, we will short all dc voltage sources when analyzing the ac operation of an amplifier. As shown in Fig. 9-19b, this means that each dc voltage supply point acts like an ac ground.

AC Equivalent Circuit

After analyzing the dc equivalent circuit, the next step is to analyze the **ac equivalent circuit.** This is the circuit that remains after you have mentally shorted all capacitors and dc voltage sources. The transistor can be replaced by either the π model or the T model. In the next chapter, we will show the mathematical details of ac analysis. For the remainder of this chapter, let us focus on how to get the ac equivalent circuit for the three amplifiers discussed up to now: base-biased, VDB, and TSEB.

Base–Biased Amplifier

Figure 9-20a is a base-biased amplifier. After mentally opening all capacitors and analyzing the dc equivalent circuit, we are ready for the ac analysis. To get the ac equivalent circuit, we short all capacitors and dc voltage sources. Then, the point labeled $+V_{CC}$ is an ac ground.

Figure 9-20b shows the ac equivalent circuit. As you can see, the transistor has been replaced by its π model. In the base circuit, the ac input voltage appears across R_B in parallel with $\beta r_e'$. In the collector circuit, the current source pumps an ac current of i_c through R_C in parallel with R_L.

VDB Amplifier

Figure 9-21a is a VDB amplifier, and Fig. 9-21b is its ac equivalent circuit. As you can see, all capacitors have been shorted, the dc supply point has become an ac ground, and the transistor has been replaced by its π model. In the base circuit, the ac input voltage appears across R_1 in parallel with R_2 in parallel with $\beta r_e'$. In the collector circuit, the current source pumps an ac current of i_c through R_C in parallel with R_L.

Figure 9-20 (a) Base-biased amplifier; (b) ac equivalent circuit.

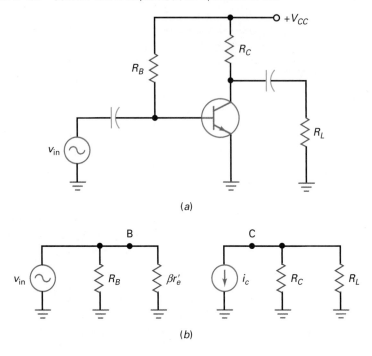

(a)

(b)

Figure 9-21 (a) VDB amplifier; (b) ac equivalent circuit.

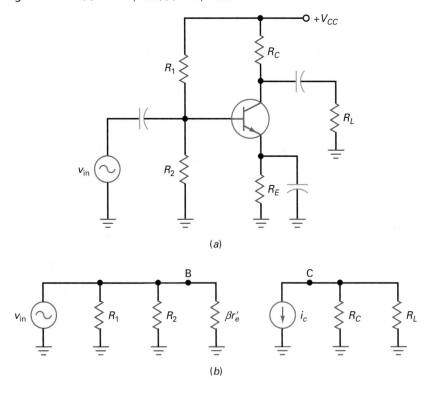

(a)

(b)

TSEB Amplifier

Our last example is the two-supply emitter-bias circuit of Fig. 9-22a. After analyzing the dc equivalent circuit, we can draw the ac equivalent circuit of Fig. 9-22b. Again, all capacitors are shorted, the dc source voltage becomes an ac ground, and the transistor is replaced by its π model. In the base circuit, the ac

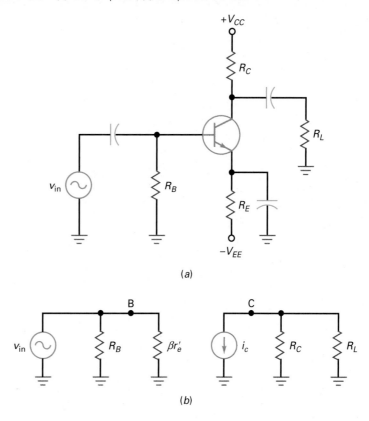

Figure 9-22 (*a*) TSEB amplifier; (*b*) ac equivalent circuit.

(a)

(b)

input voltage appears across R_B in parallel with $\beta r_e'$. In the collector circuit, the current source pumps an ac current of i_c through R_C in parallel with R_L.

CE Amplifiers

The three different amplifiers of Figs. 9-20, 9-21, and 9-22 are examples of a **common-emitter (CE) amplifier.** You can recognize a CE amplifier immediately because its emitter is at ac ground. With a CE amplifier, the ac signal is coupled into the base, and the amplified signal appears at the collector.

Two other basic transistor amplifiers are possible. **The common-base (CB) amplifier** and the **common-collector (CC) amplifier.** The CB amplifier has its base at ac ground, and the CC amplifier has its collector at ac ground. They are useful in some applications, but are not as popular as the CE amplifier. Later chapters discuss the CB and CC amplifiers.

Main Ideas

The foregoing method of analysis works for all amplifiers. You start with the dc equivalent circuit. After calculating the dc voltages and currents, you analyze the ac equivalent circuit. The crucial ideas in getting the ac equivalent circuit are:

1. Short all coupling and bypass capacitors.
2. Visualize all dc supply voltages as ac grounds.
3. Replace the transistor by its π or T model.
4. Draw the ac equivalent circuit.

Subsequent chapters will use this method to calculate the voltage gain, input impedance, and other characteristics of amplifiers.

The process of using superposition to analyze a VDB circuit is shown in Summary Table 9-1.

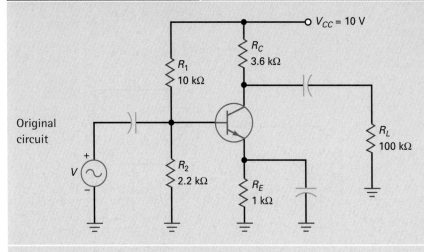

Original circuit

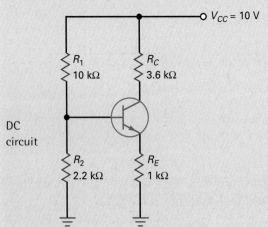

DC circuit

- Open all coupling and bypass capacitors.
- Redraw the circuit.
- Solve the dc circuit's Q point:
  $V_B = 1.8$ V
  $V_E = 1.1$ V
  $I_E = 1.1$ mA
  $V_{CE} = 4.94$ V

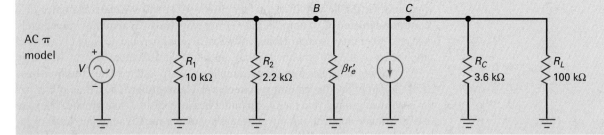

AC π model

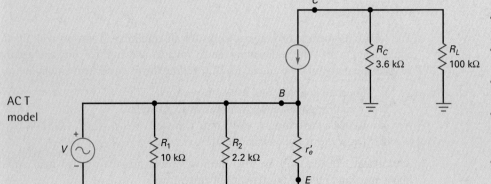

AC T model

- Short all coupling and bypass capacitors.
- Visualize all dc supply voltages as ac grounds.
- Replace the transistor by its π or T model.
- Draw the ac equivalent circuit.

- $r'_e = \dfrac{25\ \text{mV}}{I_{EQ}} = 22.7\ \Omega$

9-8 AC Quantities on the Data Sheet

Refer to the partial data sheet of a 2N3904 in Fig. 9-23 during the following discussion. The ac quantities appear in the section labeled "Small-Signal Characteristics." In this section, you will find four new quantities labeled h_{fe}, h_{ie}, h_{re}, and h_{oe}. These are called h parameters. What are they?

H Parameters

When the transistor was first invented, an approach known as the h parameters was used to analyze and design transistor circuits. This mathematical approach models the transistor on what is happening at its terminals without regard for the physical processes taking place inside the transistor.

A more practical approach is the one we are using. It is called the r' parameter method, and it uses quantities like β and r'_e. With this approach, you can use Ohm's law and other basic ideas in the analysis and design of transistor circuits. This is why the r' parameters are better suited to most people.

This does not mean that the h parameters are useless. They have survived on data sheets because they are easier to measure than r' parameters. When you read data sheets, therefore, don't look for β, r'_e, and other r' parameters. You won't find them. Instead, you will find h_{fe}, h_{ie}, h_{re}, and h_{oe}. These four h parameters give useful information when translated into r' parameters.

Relations Between R and H Parameters

For instance, h_{fe} given in the "Small-Signal Characteristics" section of the data sheet is identical to the ac current gain. In symbols this is represented as:

$$\beta = h_{fe}$$

The data sheet lists a minimum h_{fe} of 100 and a maximum of 400. Therefore, β may be as low as 100 or as high as 400. These values are for a collector current of 1 mA and a collector-emitter voltage of 10 V.

Another h parameter is the quantity h_{ie}, equivalent to the input impedance. The data sheets give a minimum h_{ie} of 1 kΩ and a maximum of 10 kΩ. The quantity h_{ie} is related to r' parameters like this:

$$r'_e = \frac{h_{ie}}{h_{fe}} \tag{9-13}$$

For instance, the maximum values of h_{ie} and h_{fe} are 10 kΩ and 400. Therefore:

$$r'_e = \frac{10 \text{ k}\Omega}{400} = 25 \ \Omega$$

The last two h parameters, h_{re} and h_{oe}, are not needed for troubleshooting and basic design.

Other Quantities

Other quantities listed under "Small-Signal Characteristics" include f_T, C_{ibo}, C_{obo}, and NF. The first, f_T, gives information about the high-frequency limitations on a 2N3904. The second and third quantities, C_{ibo} and C_{obo}, are the input and output capacitances of the device. The final quantity, NF, is the noise figure; it indicates how much noise the 2N3904 produces.

The data sheet of a 2N3904 includes a lot of graphs, which are worth looking at. For instance, the graph on the data sheet labelled *current gain* shows that h_{fe} increases from approximately 70 to 160 when the collector current increases from 0.1 mA to 10 mA. Notice that h_{fe} is approximately 125 when the

2N3903, 2N3904

ELECTRICAL CHARACTERISTICS (T$_A$ = 25°C unless otherwise noted)

Characteristic		Symbol	Min	Max	Unit
SMALL–SIGNAL CHARACTERISTICS					
Current–Gain–Bandwidth Product (I_C = 10 mAdc, V_{CE} = 20 Vdc, f = 100 MHz) 2N3903 2N3904		f_T	250 300	– –	MHz
Output Capacitance (V_{CB} = 0.5 Vdc, I_E = 0, f = 1.0 MHz)		C_{obo}	–	4.0	pF
Input Capacitance (V_{EB} = 0.5 Vdc, I_C = 0, f = 1.0 MHz)		C_{ibo}	–	8.0	pF
Input Impedance (I_C = 1.0 mAdc, V_{CE} = 10 Vdc, f = 1.0 kHz) 2N3903 2N3904		h_{ie}	1.0 1.0	8.0 10	kΩ
Voltage Feedback Ratio (I_C = 1.0 mAdc, V_{CE} = 10 Vdc, f = 1.0 kHz) 2N3903 2N3904		h_{re}	0.1 0.5	5.0 8.0	×10^{-4}
Small–Signal Current Gain (I_C = 1.0 mAdc, V_{CE} = 10 Vdc, f = 1.0 kHz) 2N3903 2N3904		h_{fe}	50 100	200 400	–
Output Admittance (I_C = 1.0 mAdc, V_{CE} = 10 Vdc, f = 1.0 kHz)		h_{oe}	1.0	40	μmhos
Noise Figure (I_C = 100 μAdc, V_{CE} = 5.0 Vdc, R_S = 1.0 kΩ, f = 1.0 kHz) 2N3903 2N3904		NF	– –	6.0 5.0	dB

H PARAMETERS
V_{CE} = 10 Vdc, f = 1.0 kHz, T_A = 25°C

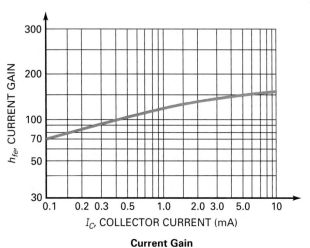

Current Gain

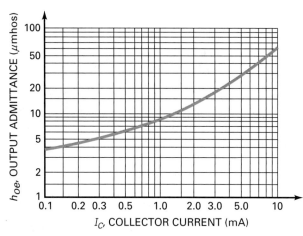

Output Admittance

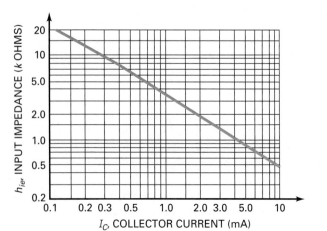

Input Impedance

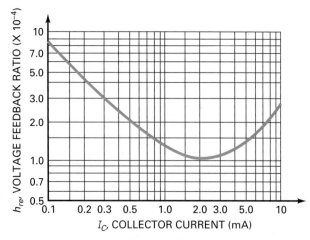

Voltage Feedback Ratio

collector current is 1 mA. This graph is for a typical 2N3904 at room temperature. If you recall that the minimum and maximum h_{fe} values were given as 100 and 400, then you can see that h_{fe} will have a large variation in mass production. Also worth remembering is that h_{fe} changes with temperature.

Take a look at the graph labelled Input Impedance on the data sheet of the 2N3904. Notice how h_{ie} decreases from approximately 20 kΩ to 500 Ω when the collector current increases from 0.1 mA to 10 mA. Equation (9-13) tells us how to calculate r'_e. It says to divide h_{ie} by h_{fe} to get r'_e. Let's try it. If you read the value of h_{fe} and h_{ie} at a collector current of 1 mA from the graphs on the data sheet, you get these approximate values: $h_{fe} = 125$ and $h_{ie} = 3.6$ kΩ. With Eq. (9-13):

$$r'_e = \frac{3.6 \text{ k}\Omega}{125} = 28.8 \ \Omega$$

The ideal value of r'_e is:

$$r'_e = \frac{25 \text{ mV}}{1 \text{ mA}} = 25 \ \Omega$$

Summary

SEC. 9–1 BASE–BIASED AMPLIFIER

Good coupling occurs when the reactance of the coupling capacitor is much smaller than the resistance at the lowest frequency of the ac source. In a base-biased amplifier, the input signal is coupled into the base. This produces an ac collector voltage. The amplified and inverted ac collector voltage is then coupled to the load resistance.

SEC. 9–2 EMITTER–BIASED AMPLIFIER

Good bypassing occurs when the reactance of the coupling capacitor is much smaller than the resistance at the lowest frequency of the ac source. The bypassed point is an ac ground. With either a VDB or a TSEB amplifier, the ac signal is coupled into the base. The amplified ac signal is then coupled to the load resistance.

SEC. 9–3 SMALL–SIGNAL OPERATION

The ac base voltage has a dc component and an ac component. These set up dc and ac components of emitter current. One way to avoid excessive distortion is to use small-signal operation. This means keeping the peak-to-peak ac emitter current less than one-tenth of the dc emitter current.

SEC. 9–4 AC BETA

The ac beta of a transistor is defined as the ac collector current divided by the ac base current. The values of the ac beta usually differ only slightly from the values of the dc beta. When troubleshooting, you can use the same value for either beta. On data sheets, h_{FE} is equivalent to β_{dc}, and h_{fe} is equivalent to β.

SEC. 9–5 AC RESISTANCE OF THE EMITTER DIODE

The base-emitter voltage of a transistor has a dc component V_{BEQ} and an ac component v_{be}. The ac base-emitter voltage sets up an ac emitter current of i_e. The ac resistance of the emitter diode is defined as v_{be} divided by i_e. With mathematics, we can prove that the ac resistance of the emitter diode equals 25 mV divided by dc emitter current.

SEC. 9–6 TWO TRANSISTOR MODELS

As far as ac signals are concerned, a transistor can be replaced by either of two equivalent circuits: the π model or the T model. The π model indicates that the input impedance of the base is $\beta r'_e$.

SEC. 9–7 ANALYZING AN AMPLIFIER

The simplest way to analyze an amplifier is to split the analysis into two parts: a dc analysis and an ac analysis. In the dc analysis, the capacitors are open. In the ac analysis, the capacitors are shorted and the dc supply points are ac grounds.

SEC. 9–8 AC QUANTITIES ON THE DATA SHEET

The h parameters are used on data sheets because they are easier to measure than r' parameters. The r' parameters are easier to use in analysis because we can use Ohm's law and other basic ideas. The most important quantities are the data sheet are h_{fe} and h_{ie}. They can be easily converted into β and r'_e.

Definitions

(9-1) Good coupling:

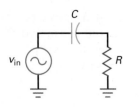

$X_C < 0.1R$

(9-2) Voltage gain:

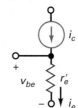

$A_V = \dfrac{v_{out}}{v_{in}}$

(9-5) Good bypassing:

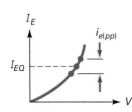

$X_C < 0.1R$

(9-6) Small signal:

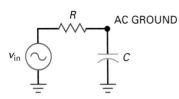

$i_{e(pp)} < 0.1I_{EQ}$

(9-7) DC current gain:

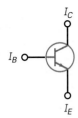

$\beta_{dc} = \dfrac{I_C}{I_B}$

(9-8) AC current gain:

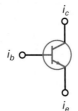

$\beta = \dfrac{i_c}{i_b}$

(9-9) AC resistance:

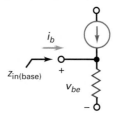

$r_e' = \dfrac{v_{be}}{i_e}$

(9-11) Input impedance:

$z_{in(base)} = \dfrac{v_{be}}{i_b}$

Derivations

(9-3) AC output voltage:

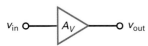

$v_{out} = A_v v_{in}$

(9-4) AC input voltage:

$v_{in} = \dfrac{v_{out}}{A_v}$

(9-10)   AC resistance:

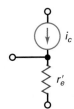

$$r'_e = \frac{25 \text{ mV}}{I_E}$$

(9-12)   Input impedance:

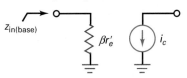

$$z_{\text{in(base)}} = \beta r'_e$$

Student Assignments

1. **For dc, the current in a coupling circuit is**
   a. Zero
   b. Maximum
   c. Minimum
   d. Average

2. **The current in a coupling circuit for high frequencies is**
   a. Zero   c. Minimum
   b. Maximum   d. Average

3. **A coupling capacitor is**
   a. A dc short
   b. An ac open
   c. A dc open and an ac short
   d. A dc short and an ac open

4. **In a bypass circuit, the top of a capacitor is**
   a. An open
   b. A short
   c. An ac ground
   d. A mechanical ground

5. **The capacitor that produces an ac ground is called a**
   a. Bypass capacitor
   b. Coupling capacitor
   c. DC open
   d. AC open

6. **The capacitors of a CE amplifier appear to be**
   a. Open to ac
   b. Shorted to dc
   c. Open to supply voltage
   d. Shorted to ac

7. **Reducing all dc sources to zero is one of the steps in getting the**
   a. DC equivalent circuit
   b. AC equivalent circuit
   c. Complete amplifier circuit
   d. Voltage-divider biased circuit

8. **The ac equivalent circuit is derived from the original circuit by shorting all**
   a. Resistors
   b. Capacitors
   c. Inductors
   d. Transistors

9. **When the ac base voltage is too large, the ac emitter current is**
   a. Sinusoidal
   b. Constant
   c. Distorted
   d. Alternating

10. **In a CE amplifier with a large input signal, the positive half cycle of the ac emitter current is**
    a. Equal to the negative half cycle
    b. Smaller than the negative half cycle
    c. Larger than the negative half cycle
    d. Equal to the negative half cycle

11. **AC emitter resistance equals 25 mV divided by the**
    a. Quiescent base current
    b. DC emitter current
    c. AC emitter current
    d. Change in collector current

12. **To reduce the distortion in a CE amplifier, reduce the**
    a. DC emitter current
    b. Base-emitter voltage
    c. Collector current
    d. AC base voltage

13. **If the ac voltage across the emitter diode is 1 mV and the ac emitter current is 100 μA, the ac resistance of the emitter diode is**
    a. 1 Ω
    b. 10 Ω
    c. 100 Ω
    d. 1 kΩ

14. **A graph of ac emitter current versus ac base–emitter voltage applies to the**
    a. Resistor
    b. Emitter diode
    c. Collector diode
    d. Power supply

15. **The output voltage of a CE amplifier is**
    a. Amplified
    b. Inverted
    c. 180° out of phase with the input
    d. All of the above

16. **The emitter of a CE amplifier has no ac voltage because of the**
    a. DC voltage on it
    b. Bypass capacitor
    c. Coupling capacitor
    d. Load resistor

17. **The voltage across the load resistor of a capacitor-coupled CE amplifier is**
    a. DC and ac
    b. DC only
    c. AC only
    d. Neither dc nor ac

18. The ac collector current is approximately equal to the
   a. AC base current
   b. AC emitter current
   c. AC source current
   d. AC bypass current

19. The ac emitter current times the ac emitter resistance equals the
   a. DC emitter voltage
   b. AC base voltage
   c. AC collector voltage
   d. Supply voltage

20. The ac collector current equals the ac base current times the
   a. AC collector resistance
   b. DC current gain
   c. AC current gain
   d. Generator voltage

21. When the emitter resistance R_E doubles, the ac emitter resistance
   a. Increases
   b. Decreases
   c. Remains the same
   d. Cannot be determined

Problems

SEC. 9–1 BASE–BIASED AMPLIFIER

9-1 ‖‖ MultiSim In Fig. 9-24, what is the lowest frequency at which good coupling exists?

Figure 9-24

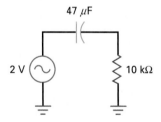

47 μF

2 V   10 kΩ

9-2 ‖‖ MultiSim If the load resistance is changed to 1 kΩ in Fig. 9-24, what is the lowest frequency for good coupling?

9-3 ‖‖ MultiSim If the capacitor is changed to 100 μF in Fig. 9-24, what is the lowest frequency for good coupling?

9-4 If the lowest input frequency of Fig. 9-24 is 100 Hz, what C value is required for good coupling?

SEC. 9–2 EMITTER–BIASED AMPLIFIER

9-5 In Fig. 9-25, what is the lowest frequency at which good bypassing exists?

9-6 If the series resistance is changed to 10 kΩ in Fig. 9-25, what is the lowest frequency for good bypassing?

Figure 9-25

R_1 A

2.2 kΩ

3 V R_2 220 μF
      10 kΩ

9-7 If the capacitor is changed to 47 μF in Fig. 9-25, what is the lowest frequency for good bypassing?

9-8 If the lowest input frequency of Fig. 9-25 is 1 kHz, what C value is required for effective bypassing?

SEC. 9–3 SMALL–SIGNAL OPERATION

9-9 If we want small-signal operation in Fig. 9-26, what is the maximum allowable ac emitter current?

9-10 The emitter resistor of Fig. 9-26 is doubled. If we want small-signal operation in Fig. 9-26, what is the maximum allowable ac emitter current?

SEC. 9–4 AC BETA

9-11 If an ac base current of 100 μA produces an ac collector current of 15 mA, what is the ac beta?

9-12 If the ac beta is 200 and the ac base current is 12.5 μA, what is the ac collector current?

9-13 If the ac collector current is 4 mA and the ac beta is 100, what is the ac base current?

Figure 9-26

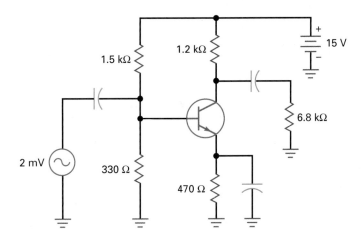

SEC. 9-5 AC RESISTANCE OF THE EMITTER DIODE

9-14 IIII MultiSim What is the ac resistance of the emitter diode in Fig. 9-26?

9-15 IIII MultiSim If the emitter resistance of Fig. 9-26 is doubled, what is the ac resistance of the emitter diode?

SEC. 9-6 TWO TRANSISTOR MODELS

9-16 What is the input impedance of the base in Fig. 9-26 if $\beta = 200$?

9-17 If the emitter resistance is doubled in Fig. 9-26, what is the input impedance of the base with $\beta = 200$?

9-18 If the 1.2 kΩ resistance is changed to 680 Ω in Fig. 9-26, what is the input impedance of the base if $\beta = 200$?

SEC. 9-7 ANALYZING AN AMPLIFIER

9-19 IIII MultiSim Draw the ac equivalent circuit for Fig. 9-26 with $\beta = 150$.

9-20 Double all the resistances in Fig. 9-26. Then, draw the ac equivalent circuit for an ac current gain of 300.

SEC. 9-8 AC QUANTITIES ON THE DATA SHEET

9-21 What are the minimum and maximum values listed under "Small-Signal Characteristics" in Fig. 9-23 for the h_{fe} of a 2N3903? For what collector current are these values given? For what temperature are these values given?

9-22 Refer to the data sheet of a 2N3904 for the following. What is the typical value of r'_e that you can calculate from the h parameter if the transistor operates at a collector current of 5 mA? Is this smaller or larger than the ideal value of r'_e calculated with 25 mV/I_E?

Critical Thinking

9-23 Somebody has built the circuit of Fig. 9-24. The builder cannot understand why a very small dc voltage is measured across the 10 kΩ when the source is 2 V at zero frequency. Can you explain what is happening?

9-24 Assume that you are in the laboratory testing the circuit of Fig. 9-25. As you increase the frequency of the generator, the voltage at node A decreases until it becomes too small to measure. If you continue to increase the frequency well above 10 MHz, the voltage at node A begins to increase. Can you explain why this happens?

9-25 In the rule for good coupling, R represents all the resistance that is in series with the coupling capacitor. With this hint in mind, what is the lowest frequency for good coupling in Fig. 9-27a?

9-26 What is the lowest frequency for good bypassing in Fig. 9-27b? (*Hint:* Thevenin resistance.)

9-27 In the two-stage amplifier of Fig. 9-28, what is the input impedance of the first base if the ac current gain is 250? If the second transistor has $\beta = 100$, what is the input impedance of the second base?

9-28 Draw the ac equivalent circuit for Fig. 9-28 using $\beta = 200$ for both transistors.

9-29 In Fig. 9-26, the Thevenin resistance seen by the bypass capacitor is 30 Ω. If the emitter is supposed to be ac ground over a frequency range of 20 Hz to 20 kHz, what size should the bypass capacitor be?

Figure 9-27

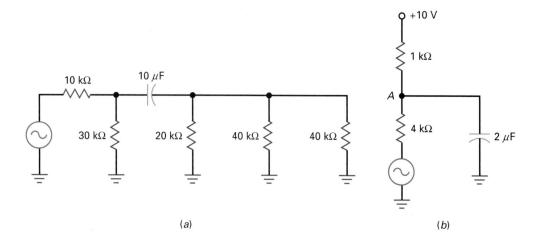

(a)   (b)

Figure 9-28

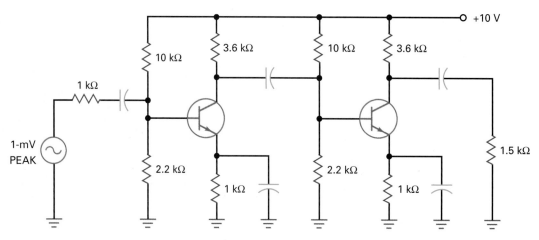

Job Interview Questions

1. Why are coupling and bypass capacitors used?
2. I want you to draw a base-biased amplifier with waveforms. Then, I want you to explain how the circuit amplifies and why the different waveforms have the dc and ac voltage levels you have shown in your drawing.
3. I want you to draw a VDB amplifier with waveforms. Then, explain the different waveforms.
4. Tell me everything you know about the ac resistance of the emitter diode.
5. Explain what *small-signal operation* means. Including drawings in your discussion.
6. Draw the two ac models of a transistor that were discussed in this chapter. Explain how to use them.
7. Why is it important to bias a transistor near the middle of the ac load line?
8. Why are ac models used for transistors, and which are the two most commonly used?
9. Compare and contrast coupling and bypass capacitors.
10. What is the difference between β and β_{dc}?
11. If you have a VDB circuit and the collector resistor opens, what would happen to the ac output voltage?

Self–Test Answers

1.	a	8.	b	15.	d
2.	b	9.	c	16.	b
3.	c	10.	c	17.	c
4.	c	11.	b	18.	b
5.	a	12.	d	19.	b
6.	d	13.	b	20.	c
7.	b	14.	b	21.	a

Practice Problem Answers

9-1 $C = 1\ \mu F$

9-2 $C = 33\ \mu F$

9-3 $i_{e(pp)} = 86.7\ \mu A_{pp}$

9-6 $r'_e = 28.8\ \Omega$

chapter 10

Voltage Amplifiers

This chapter continues the discussion of CE amplifiers and shows you how to calculate the voltage gain and the ac voltages from the circuit values. This is important when troubleshooting because you can measure the ac voltages to see whether they are in reasonable agreement with theoretical values. This chapter also discusses input impedance, multistage amplifiers, and negative feedback. The CC and CB amplifier configurations will be discussed in Chap. 11.

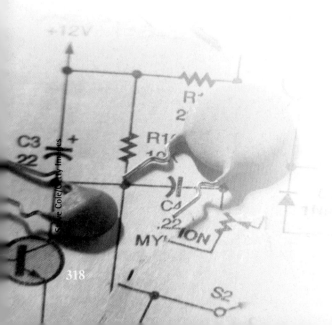

Objectives

After studying this chapter, you should be able to:

- Discuss the important characteristics of the CE amplifier.
- Show how to calculate and predict the voltage gain of a CE amplifier.
- Explain how the swamped amplifier works and list three of its advantages.
- Draw a diagram of a two-stage CE amplifier.
- Describe two capacitor-related problems that can occur in the CE amplifier.
- Troubleshoot CE amplifier circuits.

Chapter Outline

Vocabulary

ac collector resistance
ac emitter feedback
cascading
feedback resistor

multistage amplifier
swamped amplifier
swamping
total voltage gain

two-stage feedback
voltage gain

10-1 Voltage Gain

Figure 10-1a shows a voltage-divider-biased (VDB) amplifier. **Voltage gain** was defined as the ac output voltage divided by the ac input voltage. With this definition, we can derive another equation for voltage gain that is useful in troubleshooting.

Derived from the π Model

Figure 10-1b shows the ac equivalent circuit using the π model of the transistor. The ac base current i_b flows through the input impedance of the base ($\beta r'_e$). With Ohm's law, we can write:

$$v_{in} = i_b \beta r'_e$$

In the collector circuit, the current source pumps an ac current i_c through the parallel connection of R_C and R_L. Therefore, the ac output voltage equals:

$$v_{out} = i_c(R_C \parallel R_L) = \beta i_b(R_C \parallel R_L)$$

Figure 10-1 (a) CE amplifier; (b) ac equivalent circuit with π model; (c) ac equivalent circuit with T model.

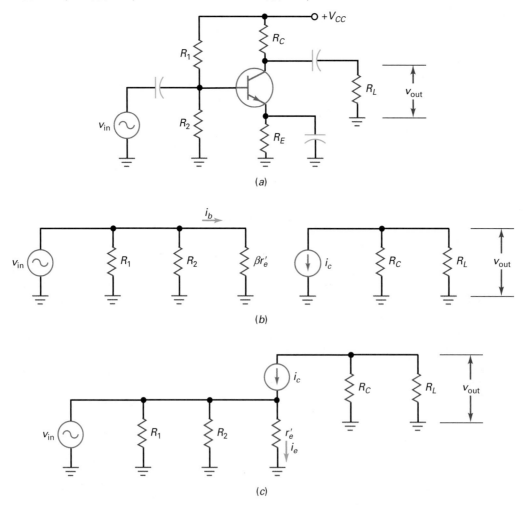

Now, we can divide v_{out} by v_{in} to get:

$$A_V = \frac{v_{out}}{v_{in}} = \frac{\beta i_b (R_C \parallel R_L)}{i_b \beta r_e'}$$

which simplifies to:

$$A_V = \frac{(R_C \parallel R_L)}{r_e'} \qquad (10\text{-}1)$$

AC Collector Resistance

In Fig. 10-1b, the total ac load resistance seen by the collector is the parallel combination of R_C and R_L. This total resistance is called the **ac collector resistance,** symbolized r_c. As a definition:

$$r_c = R_C \parallel R_L \qquad (10\text{-}2)$$

Now, we can rewrite Eq. (10-1) as:

$$A_V = \frac{r_c}{r_e'} \qquad (10\text{-}3)$$

In words: The voltage gain equals the ac collector resistance divided by the ac resistance of the emitter diode.

Derived from the T Model

Either transistor model gives the same results. Later, we will use the T model when analyzing differential amplifiers. For practice, let us derive the equation for voltage gain using the T model.

Figure 10-1c shows the ac equivalent circuit using the T model of the transistor. The input voltage v_{in} appears across r_e'. With Ohm's law, we can write:

$$v_{in} = i_e r_e'$$

In the collector circuit, the current source pumps an ac current i_c through the ac collector resistance. Therefore, the ac output voltage equals:

$$v_{out} = i_c r_c$$

Now, we can divide v_{out} by v_{in} to get:

$$A_V = \frac{v_{out}}{v_{in}} = \frac{i_c r_c}{i_e r_e'}$$

Since $i_c \approx i_e$, we can simplify the equation to get:

$$A_V = \frac{r_c}{r_e'}$$

This is the same equation derived with the π model. It applies to all CE amplifiers because all have an ac collector resistance of r_c and an emitter diode with an ac resistance of r_e'.

Example 10-1

What is the voltage gain in Fig. 10-2a? The output voltage across the load resistor?

Figure 10-2   (a) Example of VDB amplifier; (b) example of TSEB amplifier.

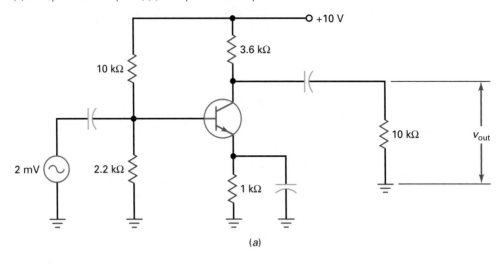

(a)

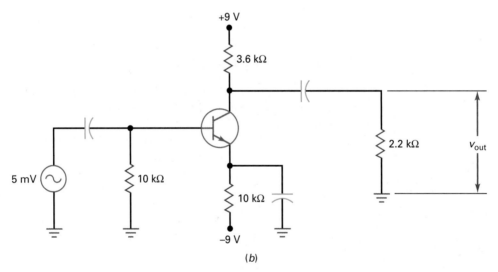

(b)

SOLUTION   The ac collector resistance is:

$$r_c = R_C \| R_L = (3.6 \text{ k}\Omega \| 10 \text{ k}\Omega) = 2.65 \text{ k}\Omega$$

In Example 9-2, we calculated an r'_e of 22.7 Ω. So, the voltage gain is:

$$A_V = \frac{r_c}{r'_e} = \frac{2.65 \text{ k}\Omega}{22.7 \text{ }\Omega} = 117$$

The output voltage is:

$$v_{out} = A_V v_{in} = (117)(2 \text{ mV}) = 234 \text{ mV}$$

PRACTICE PROBLEM 10-1   Using Fig. 10-2a, change R_L to 6.8 kΩ and find A_V.

Example 10-2

What is the voltage gain in Fig. 10-2b? The output voltage across the load resistor?

SOLUTION   The ac collector resistance is:

$$r_c = R_C \,\|\, R_L = (3.6\ \text{k}\Omega \,\|\, 2.2\ \text{k}\Omega) = 1.37\ \text{k}\Omega$$

The dc emitter current is approximately:

$$I_E = \frac{9\ \text{V} - 0.7\ \text{V}}{10\ \text{k}\Omega} = 0.83\ \text{mA}$$

The ac resistance of the emitter diode is:

$$r_e' = \frac{25\ \text{mV}}{0.83\ \text{mA}} = 30\ \Omega$$

The voltage gain is:

$$A_V = \frac{r_c}{r_e'} = \frac{1.37\ \text{k}\Omega}{30\ \Omega} = 45.7$$

The output voltage is:

$$v_\text{out} = A_V v_\text{in} = (45.7)(5\ \text{mV}) = 228\ \text{mV}$$

PRACTICE PROBLEM 10-2   In Fig. 10-2b, change the emitter resistor R_E from 10 kΩ to 8.2 kΩ and calculate the new output voltage, v_out.

10-2 The Loading Effect of Input Impedance

Up to now, we have assumed an ideal ac voltage source, one with zero source resistance. In this section, we will discuss how the input impedance of an amplifier can load down the ac source, that is, reduce the ac voltage appearing across the emitter diode.

Input Impedance

In Fig. 10-3a, an ac voltage source v_g has an internal resistance of R_G. (The subscript g stands for "generator," a synonym for *source*.) When the ac generator is not stiff, some of the ac source voltage is dropped across its internal resistance. As a result, the ac voltage between the base and ground is less than ideal.

The ac generator has to drive the input impedance of the stage $z_\text{in(stage)}$. This input impedance includes the effects of the biasing resistors R_1 and R_2, in parallel with the input impedance of the base $z_\text{in(base)}$. Figure 10-3b illustrates the idea. The input impedance of the stage equals:

$$z_\text{in(stage)} = R_1 \,\|\, R_2 \,\|\, \beta r_e'$$

Figure 10-3 CE amplifier. (*a*) Circuit; (*b*) ac equivalent circuit; (*c*) effect of input impedance.

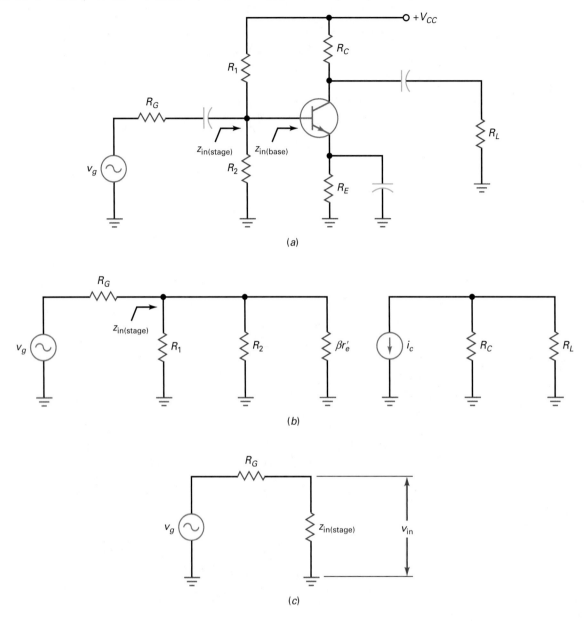

(*a*)

(*b*)

(*c*)

Equation for Input Voltage

When the generator is not stiff, the ac input voltage v_{in} of Fig. 10-3*c* is less than v_g. With the voltage-divider theorem, we can write:

$$v_{in} = \frac{z_{in(stage)}}{R_G + z_{in(stage)}} v_g \qquad (10\text{-}4)$$

This equation is valid for any amplifier. After you calculate or estimate the input impedance of the stage, you can determine what the input voltage is. Note: The generator is stiff when R_G is less than $0.01 z_{in(stage)}$.

Example 10-3

In Fig. 10-4, the ac generator has an internal resistance of 600 Ω. What is the output voltage in Fig. 10-4 if $\beta = 300$?

Figure 10-4 Example.

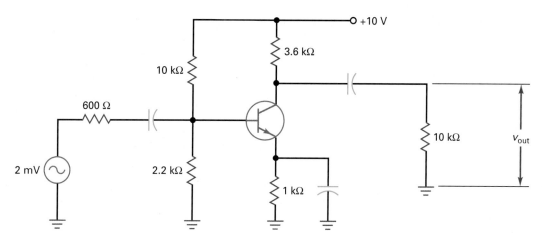

SOLUTION Here are two quantities calculated in earlier examples: $r_e' = 22.7\ \Omega$ and $A_V = 117$. We will use these values in solving the problem.

When $\beta = 300$, the input impedance of the base is:

$$z_{\text{in(base)}} = (300)(22.7\ \Omega) = 6.8\ \text{k}\Omega$$

The input impedance of the stage is:

$$z_{\text{in(stage)}} = 10\ \text{k}\Omega \parallel 2.2\ \text{k}\Omega \parallel 6.8\ \text{k}\Omega = 1.42\ \text{k}\Omega$$

With Eq. (10-4), we can calculate the input voltage:

$$v_{\text{in}} = \frac{1.42\ \text{k}\Omega}{600\ \Omega + 1.42\ \text{k}\Omega}\, 2\ \text{mV} = 1.41\ \text{mV}$$

This is the ac voltage that appears at the base of the transistor, equivalent to the ac voltage across the emitter diode. The amplified output voltage equals:

$$v_{\text{out}} = A_V v_{\text{in}} = (117)(1.41\ \text{mV}) = 165\ \text{mV}$$

PRACTICE PROBLEM 10-3 Change the R_G value of Fig. 10-4 to 50 Ω and solve for the new amplified output voltage.

Example 10-4

Repeat the preceding example for $\beta = 50$.

SOLUTION When $\beta = 50$, the input impedance of the base decreases to:

$$z_{\text{in(base)}} = (50)(22.7\ \Omega) = 1.14\ \text{k}\Omega$$

The input impedance of the stage decreases to:

$$z_{\text{in(stage)}} = 10\ \text{k}\Omega \parallel 2.2\ \text{k}\Omega \parallel 1.14\ \text{k}\Omega = 698\ \Omega$$

With Eq. (10-4), we can calculate the input voltage:

$$v_{\text{in}} = \frac{698\ \Omega}{600\ \Omega + 698\ \Omega}\, 2\ \text{mV} = 1.08\ \text{mV}$$

The output voltage equals:

$$v_{out} = A_V v_{in} = (117)(1.08 \text{ mV}) = 126 \text{ mV}$$

This example illustrates how the ac current gain of the transistor can change the output voltage. When β decreases, the input impedance of the base decreases, the input impedance of the stage decreases, the input voltage decreases, and the output voltage decreases.

PRACTICE PROBLEM 10-4 Using Fig. 10-4, change the β value to 400 and calculate the output voltage.

10-3 Multistage Amplifiers

To get more voltage gain, we can create a **multistage amplifier** by **cascading** two or more amplifier stages. This means using the output of the first stage as the input to a second stage. In turn, the output of the second stage can be used as the input to the third stage, and so on.

Voltage Gain of First Stage

Figure 10-5a shows a two-stage amplifier. The amplified and inverted signal out of the first stage is coupled to the base of the second stage. The amplified and inverted output of the second stage is then coupled to the load resistance. The

Figure 10-5 (a) Two-stage amplifier; (b) ac equivalent circuit.

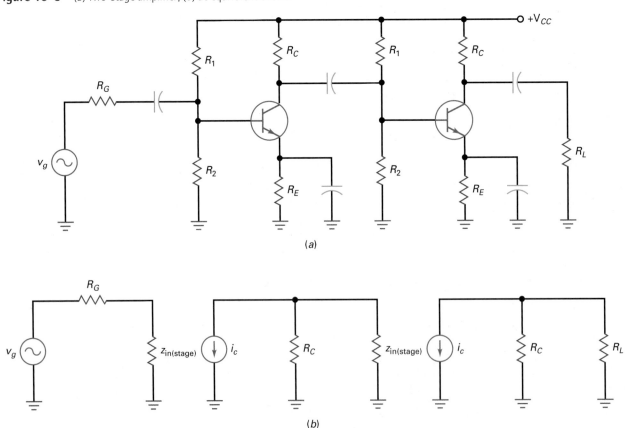

(a)

(b)

signal across the load resistance is in phase with the generator signal. The reason is that each stage inverts the signal by 180°. Therefore, two stages invert the signal by 360°, equivalent to 0° (in phase).

Voltage Gain of First Stage

Fig. 10-5b shows the ac equivalent circuit. Note that the input impedance of the second stage loads down the first stage. In other words, the z_{in} of the second stage is in parallel with the R_C of the first stage. The ac collector resistance of the first stage is

$$\text{First stage:} \quad r_c = R_C \| z_{in(stage)}$$

The voltage gain of the first stage is:

$$A_{V_1} = \frac{R_C \| z_{in(stage)}}{r_e'}$$

Voltage Gain of Second Stage

The ac collector resistance of the second stage is:

$$\text{Second stage:} \quad r_c = R_C \| R_L$$

and the voltage gain is:

$$A_{V_2} = \frac{R_C \| R_L}{r_e'}$$

Total Voltage Gain

The **total voltage gain** of the amplifier is given by the product of the individual gains:

$$A_V = (A_{V_1})(A_{V_2}) \tag{10-5}$$

For instance, if each stage has a voltage gain of 50, the overall voltage gain is 2500.

Example 10–5

What is the ac collector voltage in the first stage of Fig. 10-6? The ac output voltage across the load resistor?

Figure 10-6 Example.

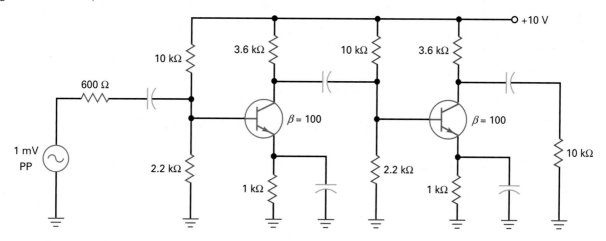

SOLUTION The input impedance of the first base is:

$$z_{\text{in(base)}} = (100)(22.7\ \Omega) = 2.27\ \text{k}\Omega$$

The input impedance of the first stage is:

$$z_{\text{in(stage)}} = 10\ \text{k}\Omega \parallel 2.2\ \text{k}\Omega \parallel 2.27\ \text{k}\Omega = 1\ \text{k}\Omega$$

The input signal to the first base is:

$$v_{\text{in}} = \frac{1\ \text{k}\Omega}{600\ \Omega + 1\ \text{k}\Omega}\ 1\ \text{mV} = 0.625\ \text{mV}$$

The input impedance of the second base is the same as the first stage:

$$z_{\text{in(stage)}} = 10\ \text{k}\Omega \parallel 2.2\ \text{k}\Omega \parallel 2.27\ \text{k}\Omega = 1\ \text{k}\Omega$$

This input impedance is the load resistance on the first stage. In other words, the ac collector resistance of the first stage is:

$$r_c = 3.6\ \text{k}\Omega \parallel 1\ \text{k}\Omega = 783\ \Omega$$

The voltage gain of the first stage is:

$$A_{V_1} = \frac{783\ \Omega}{22.7\ \Omega} = 34.5$$

Therefore, the ac collector voltage in the first stage is:

$$v_c = A_{V_1} v_{\text{in}} = (34.5)(0.625\ \text{mV}) = 21.6\ \text{mV}$$

The ac collector resistance of the second stage is:

$$r_c = 3.6\ \text{k}\Omega \parallel 10\ \text{k}\Omega = 2.65\ \text{k}\Omega$$

and the voltage gain is:

$$A_{V_2} = \frac{2.65\ \text{k}\Omega}{22.7\ \Omega} = 117$$

Therefore, the ac output voltage across the load resistor is:

$$v_{\text{out}} = A_{V_2} v_{b_2} = (117)(21.6\ \text{mV}) = 2.52\ \text{V}$$

Another way to calculate the final output voltage is by using the overall voltage gain:

$$A_V = (34.5)(117) = 4037$$

The ac output voltage across the load resistor is:

$$v_{\text{out}} = A_V v_{\text{in}} = (4037)(0.625\ \text{mV}) = 2.52\ \text{V}$$

PRACTICE PROBLEM 10-5 In Fig. 10-6, change the load resistance of stage two from 10 kΩ to 6.8 kΩ and calculate the final output voltage.

10-4 Swamped Amplifier

The voltage gain of a CE amplifier changes with the quiescent current, temperature variations, and transistor replacement because these quantities change r'_e and β.

AC Emitter Feedback

One way to stabilize the voltage gain is to leave some of the emitter resistance unbypassed, as shown in Fig. 10-7a, producing **ac emitter feedback.** When ac

Figure 10–7 (*a*) Swamped amplifier; (*b*) ac equivalent circuit.

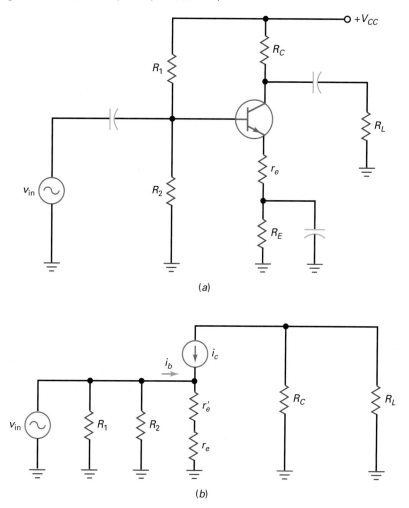

(a)

(b)

emitter current flows through the unbypassed emitter resistance r_e, an ac voltage appears across r_e. This produces *negative feedback* (described in Chap. 8). The ac voltage across r_e opposes changes in voltage gain. The unbypassed resistance r_e is called a **feedback resistor** because it has an ac voltage across it that opposes changes in voltage gain.

For instance, suppose the ac collector current increases because of a temperature increase. This will produce a larger output voltage, but it will also produce a larger ac voltage across r_e. Since v_{be} equals the difference between v_{in} and v_e, the increase in v_e will decrease v_{be}. This decreases the ac collector current. Since this opposes the original increase in ac collector current, we have negative feedback.

Voltage Gain

Figure 10-7*b* shows the ac equivalent circuit with the T model of the transistor. Clearly, the ac emitter current must flow through r_e' and r_e. With Ohm's law, we can write:

$$v_{in} = i_e(r_e + r_e') = v_b$$

In the collector circuit, the current source pumps an ac current i_c through the ac collector resistance. Therefore, the ac output voltage equals:

$$v_{out} = i_c r_c$$

Now, we can divide v_{out} by v_{in} to get:

$$A_V = \frac{v_{out}}{v_{in}} = \frac{i_c r_c}{i_e(r_e + r'_e)} = \frac{v_c}{v_b}$$

Since $i_c \approx i_e$, we can simplify the equation to get:

$$A_V = \frac{r_c}{r_e + r'_e} \qquad (10\text{-}6)$$

When r_e is much greater than r'_e, the foregoing equation simplifies to:

$$A_V = \frac{r_c}{r_e} \qquad (10\text{-}7)$$

This says that the voltage gain equals the ac collector resistance divided by the feedback resistance. Since r'_e no longer appears in the equation for voltage gain, it no longer has an effect on the voltage gain.

The foregoing is an example of **swamping,** making one quantity much larger than a second quantity to eliminate changes in the second. In Eq. (10-6), a large value of r_e swamps out the variations in r'_e. The result is a stable voltage gain, one that does not change with temperature variation or transistor replacement.

Input Impedance of the Base

The negative feedback not only stabilizes the voltage gain, it also increases the input impedance of the base. In Fig. 10-7b, the input impedance of the base is:

$$z_{in(base)} = \frac{v_{in}}{i_b}$$

Applying Ohm's law to the emitter diode of Fig. 10-7b, we can write:

$$v_{in} = i_e(r_e + r'_e)$$

Substitute this equation into the preceding one to get:

$$z_{in(base)} = \frac{v_{in}}{i_b} = \frac{i_e(r_e + r'_e)}{i_b}$$

Since $i_e \approx i_c$, the foregoing equation becomes:

$$z_{in(base)} = \beta(r_e + r'_e) \qquad (10\text{-}8)$$

In a **swamped amplifier,** this simplifies to:

$$z_{in(base)} = \beta r_e \qquad (10\text{-}9)$$

This says that the input impedance of the base equals the current gain times the feedback resistance.

Less Distortion with Large Signals

The nonlinearity of the emitter-diode curve is the source of large-signal distortion. By swamping the emitter diode, we reduce the effect it has on voltage gain. In turn, this reduces the distortion that occurs for large-signal operation.

Put it this way: Without the feedback resistor, the voltage gain is:

$$A_V = \frac{r_c}{r'_e}$$

Since r'_e is current-sensitive, its value changes when a large signal is present. This means that the voltage gain changes during the cycle of a large signal. In other words, changes in r'_e are the cause of distortion with large signals.

With the feedback resistor, however, the swamped voltage gain is:

$$A_V = \frac{r_c}{r_e}$$

Since r'_e is no longer present, the distortion of large signals has been eliminated. A swamped amplifier therefore has three advantages: It stabilizes voltage gain, increases the input impedance of the base, and reduces the distortion of large signals.

Example 10-6

What is the output voltage across the load resistor of MultiSim Fig. 10-8 if $\beta = 200$? Ignore r'_e in the calculations.

IIII **MultiSim** **Figure 10-8**   Single-stage example.

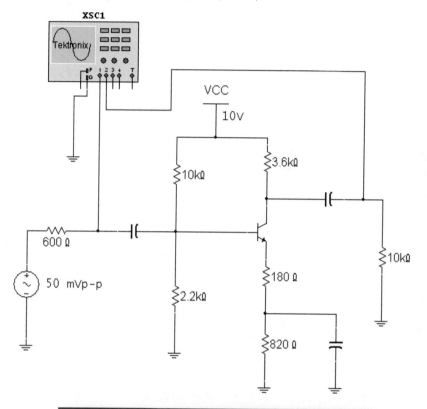

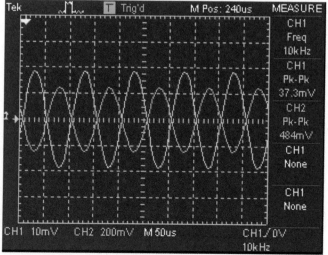

SOLUTION The input impedance of the base is:

$$z_{in(base)} = \beta r_e = (200)(180\ \Omega) = 36\ k\Omega$$

The input impedance of the stage is:

$$z_{in(stage)} = 10\ k\Omega \parallel 2.2\ k\Omega \parallel 36\ k\Omega = 1.71\ k\Omega$$

The ac input voltage to the base is:

$$v_{in} = \frac{1.71\ k\Omega}{600\ \Omega + 1.71\ k\Omega}\ 50\ mV = 37\ mV$$

The voltage gain is:

$$A_V = \frac{r_c}{r_e} = \frac{2.65\ k\Omega}{180\ \Omega} = 14.7$$

The output voltage is:

$$v_{out} = (14.7)(37\ mV) = 544\ mV$$

PRACTICE PROBLEM 10–6 Using Fig. 10-8, change the β value to 300 and solve for the output voltage across the 10 kΩ load.

Example 10-7

Repeat the preceding example, but this time include r'_e in the calculations.

SOLUTION The input impedance of the base is:

$$z_{in(base)} = \beta(r_e + r'_e) = (200)(180\ \Omega + 22.7\ \Omega) = 40.5\ k\Omega$$

The input impedance of the stage is:

$$z_{in(stage)} = 10\ k\Omega \parallel 2.2\ k\Omega \parallel 40.5\ k\Omega = 1.72\ k\Omega$$

The ac input voltage to the base is:

$$v_{in} = \frac{1.72\ k\Omega}{600\ \Omega + 1.72\ k\Omega}\ 50\ mV = 37\ mV$$

The voltage gain is:

$$A_V = \frac{r_c}{r_e + r'_e} = \frac{2.65\ k\Omega}{180\ \Omega + 22.7\ \Omega} = 13.1$$

The output voltage is:

$$v_{out} = (13.1)(37\ mV) = 485\ mV$$

Comparing the results with and without r'_e in the calculations, we can see that it has little effect on the final answer. This is to be expected in a swamped amplifier. When you are troubleshooting, you can assume that the amplifier is swamped when a feedback resistor is used in the emitter. If you need more accuracy, you can include r'_e.

PRACTICE PROBLEM 10–7 Compare the calculated v_{out} value to the measured value using MultiSim.

Example 10-8

What is the output voltage in Fig. 10-9 if $\beta = 200$? Ignore r'_e in the calculations.

SOLUTION   In Example 10-6, we calculated $z_{in(base)} = 36$ kΩ and $z_{in(stage)} = 1.71$ kΩ. The first stage has these values because its circuit values are the same as those of Example 10-6. The ac input voltage to the first base is:

$$v_{in} = \frac{1.71 \text{ k}\Omega}{600 \text{ }\Omega + 1.71 \text{ k}\Omega} \, 1 \text{ mV} = 0.74 \text{ mV}$$

The input impedance of the second stage is the same as the first stage: $z_{in(stage)} = 1.71$ kΩ. Therefore, the ac collector resistance of the first stage is:

$$r_c = 3.6 \text{ k}\Omega \, \| \, 1.71 \text{ k}\Omega = 1.16 \text{ k}\Omega$$

and the voltage gain of the first stage is:

$$A_{V_1} = \frac{1.16 \text{ k}\Omega}{180 \text{ }\Omega} = 6.44$$

The amplified and inverted ac voltage at the first collector and the second base is:

$$v_c = (6.44)(0.74 \text{ mV}) = 4.77 \text{ mV}$$

The second stage has an ac collector resistance of 2.65 kΩ, calculated in Example 10-6. Therefore, it has a voltage gain of:

$$A_{V_2} = \frac{2.65 \text{ k}\Omega}{180 \text{ }\Omega} = 14.7$$

The final output voltage equals:

$$v_{out} = (14.7)(4.77 \text{ mV}) = 70 \text{ mV}$$

Another way to calculate the output voltage is to use the overall voltage gain:

$$A_V = (A_{V_1})(A_{V_2}) = (6.44)(14.7) = 95$$

Then:

$$v_{out} = A_V v_{in} = (95)(0.74 \text{ mV}) = 70 \text{ mV}$$

Figure 10-9   Two-stage amplifier example.

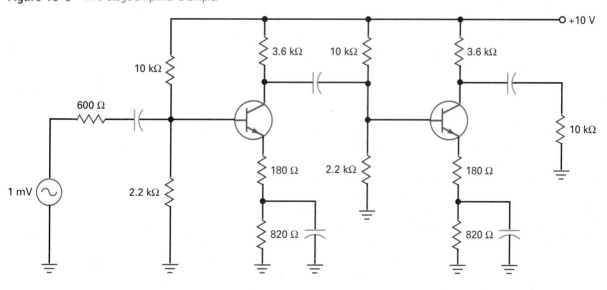

10-5 Two-Stage Feedback

A swamped amplifier is an example of single-stage feedback. It works reasonably well to stabilize the voltage gain, increase the input impedance, and reduce distortion. **Two-stage feedback** works even better.

Basic Idea

Figure 10-10 shows a two-stage feedback amplifier. The first stage has an unbypassed emitter resistance of r_e. The second stage is a CE stage, with the emitter at ac ground to produce maximum gain in this stage. The output signal is coupled back through a feedback resistance r_f to the first emitter. Because of the voltage divider, the ac voltage between the first emitter and ground is:

$$v_e = \frac{r_e}{r_f + r_e} v_{\text{out}}$$

Here is the basic idea of how the two-stage feedback works: Assume that an increase in temperature causes the output voltage to increase. Since part of the output voltage is fed back to the first emitter, v_e increases. This decreases v_{be} in the first stage, decreases v_c in the first stage, and decreases v_{out}. On the other hand, if the output voltage tries to decrease, v_{be} increases and v_{out} increases.

In either case, any attempted change in the output voltage is fed back and the amplified change opposes the original change. The overall effect is that the output voltage will change by a much smaller amount than it would without the negative feedback.

Voltage Gain

In a well-designed two-stage feedback amplifier, the voltage gain is given by this derivation:

$$A_V = \frac{r_f}{r_e} + 1 \tag{10-10}$$

Figure 10-10 Two-stage feedback amplifier.

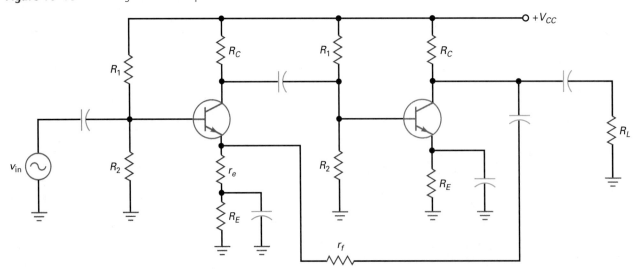

In most designs, the first term in this equation is much greater than 1, so the equation simplifies to:

$$A_V = \frac{r_f}{r_e}$$

When we discuss op amps, we will analyze negative feedback in detail. At that time, you will see what is meant by a *well-designed feedback amplifier.*

What's important about Eq. (10-10) is this: The voltage gain depends only on external resistances r_f and r_e. Since these resistances are fixed in value, the voltage gain is fixed.

Example 10-9

A variable resistor is used in Fig. 10-11. It can be varied from 0 to 10 kΩ. What is the minimum voltage gain of the two-stage amplifier? The maximum?

Figure 10-11 Example of two-stage feedback.

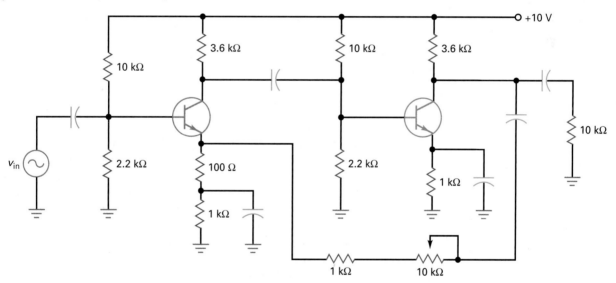

SOLUTION The feedback resistance r_f is the sum of 1 kΩ and the adjustable resistance. The minimum voltage gain occurs when the variable resistor is zeroed:

$$A_V = \frac{r_f}{r_e} = \frac{1 \text{ k}\Omega}{100 \text{ }\Omega} = 10$$

The maximum voltage gain when the variable resistance is 10 kΩ is:

$$A_V = \frac{r_f}{r_e} = \frac{11 \text{ k}\Omega}{100 \text{ }\Omega} = 110$$

PRACTICE PROBLEM 10-9 In Fig. 10-11, what value of resistance would the variable resistor need to be for a voltage gain of 50?

Example 10-10

How could the circuit of Fig. 10-11 be modified to be used as a portable microphone preamplifier?

SOLUTION   The 10 V dc power supply could be replaced with a 9 V battery and on/off switch. Connect a properly sized microphone jack to the preamplifier's input coupling capacitor and ground. The microphone should ideally be a low-impedance dynamic style. If an electret microphone is used, it will be necessary to power it from the 9 V battery through a series resistor. For good low-frequency response, the coupling and bypass capacitors need to have low capacitive reactance. A value of 47 μF for each coupling capacitor and 100 μF for each bypass capacitor can be used. The 10 kΩ output load can be changed to a 10 kΩ potentiometer to vary the output level. If more voltage gain is needed, change the 10 kΩ feedback potentiometer to a higher value. The output should be able to drive the line/CD/aux/tape inputs of a home stereo amplifier. Check your system's specifications for the proper input needed. Placing all components in a small metal box and using shielded cables will reduce external noise and interference.

10-6 Troubleshooting

When a single- or double-stage amplifier is not working, a troubleshooter can start by measuring dc voltages, including the dc power supply voltages. These voltages are estimated mentally as discussed earlier, and then the voltages are measured to see whether they are approximately correct. If the dc voltages are distinctly different from the estimated voltages, the possible troubles include open resistors (burned out), shorted resistors (solder bridges across them), incorrect wiring, shorted capacitors, and transistor failures. A short across a coupling or bypass capacitor will change the dc equivalent circuit, which means radically different dc voltages.

If all dc voltages measure OK, the troubleshooting is continued by considering what can go wrong in the ac equivalent circuit. If there is generator voltage but there is no ac base voltage, something may be open between the generator and the base. Perhaps a connecting wire is not in place, or maybe the input coupling capacitor is open. Similarly, if there is no final output voltage but there is an ac collector voltage, the output coupling capacitor may be open, or a connection may be missing.

Normally, there is no ac voltage between the emitter and ground when the emitter is at ac ground. When an amplifier is not working properly, one of the things a troubleshooter checks with an oscilloscope is the emitter voltage. If there is any ac voltage at a bypass emitter, it means that the bypass capacitor is not working.

For instance, an open bypass capacitor means that the emitter is no longer at ac ground. Because of this, the ac emitter current flows through R_E instead of through the bypass capacitor. This produces an ac emitter voltage which you can see with an oscilloscope. So, if you see an ac emitter voltage comparable in size to the ac base voltage, check the emitter-bypass capacitor. It may be defective or not properly connected.

Under normal conditions, the supply line is an ac ground point because of the filter capacitor in the power supply. If the filter capacitor is defective, the ripple becomes huge. This unwanted ripple gets to the base through the voltage divider. Then it is amplified the same as the generator signal. This amplified ripple will produce a hum of 60 or 120 Hz when the amplifier is connected to a loudspeaker. So, if you ever hear excessive hum coming out of a loudspeaker, one of the prime suspects is an open filter capacitor in the power supply.

When the amplifier consists of more than one or two stages, it is best to isolate the defective stage first by using signal tracing or signal injection techniques. As an example, if the amplifier consists of four stages, split the amplifier in half by either measuring or injecting a signal at the output of the second stage. By doing this, you should be able to determine if the trouble is before or after this circuit point. Depending on the result of the first step, move your next troubleshooting point to the midpoint of the defective half. This split-half method of troubleshooting can quickly isolate a defective stage.

Example 10-11

The CE amplifier of Fig. 10-12 has an ac load voltage of zero. If the dc collector voltage is 6 V and the ac collector voltage is 70 mV, what is the trouble?

SOLUTION   Since the dc and ac collector voltages are normal, there are only two components that can be the trouble: C_2 or R_L. If you ask four what-if questions about these components, you can find the trouble. The four what-ifs are:

> What if C_2 is shorted?
> What if C_2 is open?
> What if R_L is shorted?
> What if R_L is open?

The answers are:

> A shorted C_2 decreases the dc collector voltage significantly.
> An open C_2 breaks the ac path but does not change the dc or ac collector voltages.
> A shorted R_L kills the ac collector voltage.
> An open R_L increases the ac collector voltage significantly.

Figure 10-12   Troubleshooting example.

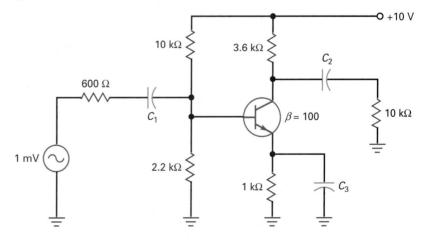

The trouble is an open C_2. When you first learn how to troubleshoot, you may have to ask yourself what-if questions to isolate the trouble. After you gain experience, the whole process becomes automatic. An experienced troubleshooter would have found this trouble almost instantly.

Example 10-12

The CE amplifier of Fig. 10-12 has an ac emitter voltage of 0.75 mV and an ac collector voltage of 2 mV. What is the trouble?

SOLUTION   Because troubleshooting is an art, you have to ask what-if questions that make sense to *you* and in any order that helps you to find the trouble. If you haven't figured out this trouble yet, start to ask what-if questions about each component and see whether you can find the trouble. Then read what comes next.

No matter which component you select, your what-if questions will not produce the symptoms given here until you start asking *these* what-if questions:

What if C_3 is shorted?
What if C_3 is open?

A shorted C_3 cannot produce the symptoms, but an open C_3 does. Why? Because with an open C_3, the input impedance of the base is much higher and the ac base voltage increases from 0.625 to 0.75 mV. Since the emitter is no longer ac grounded, almost all this 0.75 mV appears at the emitter. Since the amplifier has a swamped voltage gain of 2.65, the ac collector voltage is approximately 2 mV.

PRACTICE PROBLEM 10-12   In the CE amplifier of Fig. 10-12, what would happen to the dc and ac transistor voltages if the transistor's BE diode opened?

Summary

SEC. 10-1   VOLTAGE GAIN

The voltage gain of a CE amplifier equals the ac collector resistance divided by the ac resistance of the emitter diode.

SEC. 10-2   THE LOADING EFFECT OF INPUT IMPEDANCE

The input impedance of the stage includes the biasing resistors and the input impedance of the base. When the source is not stiff compared to this input impedance, the input voltage is less than the source voltage.

SEC. 10-3   MULTISTAGE AMPLIFIERS

The overall voltage gain equals the product of the individual voltage gains. The input impedance of the second stage is the load resistance on the first stage.

Two CE stages produce an amplified in-phase signal.

SEC. 10-4   SWAMPED AMPLIFIER

By leaving some of the emitter resistance unbypassed, we get negative feedback. This stabilizes the voltage gain, increases the input impedance, and reduces large-signal distortion.

SEC. 10–5 TWO–STAGE FEEDBACK

We can feed back the output voltage of the second collector to the first emitter through a voltage divider. This produces negative feedback, which stabilizes the voltage gain of the two-stage amplifier.

SEC. 10–6 TROUBLESHOOTING

With single- or double-stage amplifiers, start with dc measurements. If they do not isolate the trouble, you continue with ac measurements until you have found the trouble.

Definition

(10-2)   AC collector resistance:

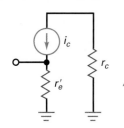

$r_c = R_C \| R_L$

Derivations

(10-3)   CE voltage gain:

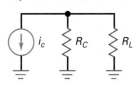

$A_V = \dfrac{r_c}{r_e'}$

(10-4)   Loading effect:

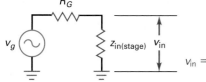

$v_{in} = \dfrac{z_{in(stage)}}{R_G + z_{in(stage)}} \, v_g$

(10-5)   Two-stage voltage gain:

$A_V = (A_{V_1})(A_{V_2})$

(10-6)   Single-stage feedback:

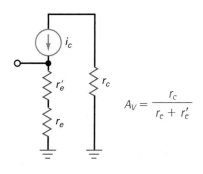

$A_V = \dfrac{r_c}{r_e + r_e'}$

(10-7)   Swamped amplifier:

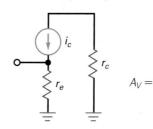

$A_V = \dfrac{r_c}{r_e}$

(10-8)   Input impedance:

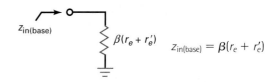

$z_{in(base)} = \beta(r_e + r_e')$

(10-9)   Swamped input impedance:

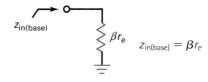

$z_{in(base)} = \beta r_e$

(10-10)   Two-stage feedback gain:

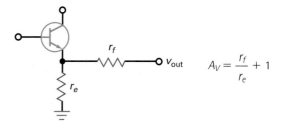

$A_V = \dfrac{r_f}{r_e} + 1$

Student Assignments

1. The emitter is at ac ground in a
   a. CB stage   c. CE stage
   b. CC stage   d. None of these

2. The output voltage of an emitter-bypassed CE stage is usually
   a. Constant
   b. Dependent on r'_e
   c. Small
   d. Less than one

3. The voltage gain equals the output voltage divided by the
   a. Input voltage
   b. AC emitter resistance
   c. AC collector resistance
   d. Generator voltage

4. The input impedance of the base decreases when
   a. β increases
   b. Supply voltage increases
   c. β decreases
   d. AC collector resistance increases

5. Voltage gain is directly proportional to
   a. β
   b. r'_e
   c. DC collector voltage
   d. AC collector resistance

6. Compared to the ac resistance of the emitter diode, the feedback resistance of a swamped amplifier should be
   a. Small   c. Large
   b. Equal   d. Zero

7. Compared to a CE stage, a swamped amplifier has an input impedance that is
   a. Smaller   c. Larger
   b. Equal   d. Zero

8. To reduce the distortion of an amplified signal, you can increase the
   a. Collector resistance
   b. Emitter feedback resistance
   c. Generator resistance
   d. Load resistance

9. The emitter of a swamped amplifier
   a. Is grounded
   b. Has no dc voltage
   c. Has an ac voltage
   d. Has no ac voltage

10. A swamped amplifier uses
    a. Base bias
    b. Positive feedback
    c. Negative feedback
    d. A grounded emitter

11. In a swamped amplifier, the effects of the emitter diode become
    a. Important to voltage gain
    b. Critical to input impedance
    c. Significant to the analysis
    d. Unimportant

12. The feedback resistor
    a. Increases voltage gain
    b. Reduces distortion
    c. Decreases collector resistance
    d. Decreases input impedance

13. The feedback resistor
    a. Stabilizes voltage gain
    b. Increases distortion
    c. Increases collector resistance
    d. Decreases input impedance

14. The ac collector resistance of the first stage includes the
    a. Load resistance
    b. Input impedance of first stage
    c. Emitter resistance of first stage
    d. Input impedance of second stage

15. If the emitter-bypass capacitor opens, the ac output voltage will
    a. Decrease   c. Remain the same
    b. Increase   d. Equal zero

16. If the emitter-bypass capacitor shorts, the base dc voltage will
    a. Decrease   c. Remain the same
    b. Increase   d. Equal zero

17. If the collector resistor is shorted, the ac output voltage will
    a. Decrease   c. Remain the same
    b. Increase   d. Equal zero

18. If the load resistance is open, the ac output voltage will
    a. Decrease
    b. Increase
    c. Remain the same
    d. Equal zero

19. If any capacitor is open, the ac output voltage will
    a. Decrease   c. Remain the same
    b. Increase   d. Equal zero

20. If the input-coupling capacitor is open, the ac input voltage will
    a. Decrease   c. Remain the same
    b. Increase   d. Equal zero

21. If the bypass capacitor is open, the ac input voltage at the base will
    a. Decrease   c. Remain the same
    b. Increase   d. Equal zero

22. If the output-coupling capacitor is open, the ac input voltage will
    a. Decrease   c. Remain the same
    b. Increase   d. Equal zero

23. If the emitter resistor is open, the ac input voltage at the base will
    a. Decrease   c. Remain the same
    b. Increase   d. Equal zero

24. If the collector resistor is open, the ac input voltage at the base will
    a. Decrease
    b. Increase
    c. Remain the same
    d. Equal approximately zero

25. If the emitter-bypass capacitor is shorted, the ac input voltage at the base will
    a. Decrease   c. Remain the same
    b. Increase   d. Equal zero

26. If the input impedance of the second stage decreases, the voltage gain of the first stage will
    a. Decrease   c. Remain the same
    b. Increase   d. Equal zero

27. If the BE diode of the second stage opens, the voltage gain of the first stage will
    a. Decrease   c. Remain the same
    b. Increase   d. Equal zero

28. If the load resistance of the second stage opens, the voltage gain of the first stage will
    a. Decrease
    b. Increase
    c. Remain the same
    d. Equal zero

Problems

Figure 10-13

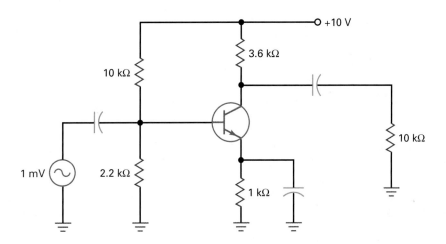

Figure 10-14

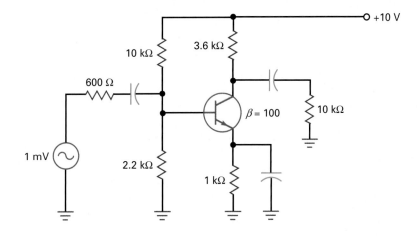

Figure 10-15

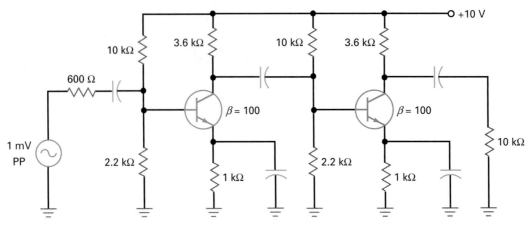

SEC. 10-4 SWAMPED AMPLIFIER

10-10 |||| **MultiSim** The generator voltage of Fig. 10-16 is reduced by half. What is the output voltage? Ignore r'_e.

10-11 |||| **MultiSim** If the generator resistance of Fig. 10-16 is 50 Ω, what is the output voltage?

10-12 |||| **MultiSim** The load resistance of Fig. 10-16 is reduced to 3.6 kΩ. What is the voltage gain?

10-13 |||| **MultiSim** The supply voltage triples in Fig. 10-16. What is the voltage gain?

SEC. 10-5 TWO-STAGE FEEDBACK

10-14 A feedback amplifier like Fig. 10-10 has $r_f = 5$ kΩ and $r_e = 50$ Ω. What is the voltage gain?

10-15 In a feedback amplifier like Fig. 10-11, $r_e = 125$ Ω. If you want a voltage gain of 100, what value should r_f be?

SEC. 10-6 TROUBLESHOOTING

10-16 In Fig. 10-15, the emitter bypass capacitor is open in the first stage. What happens to the dc voltages of the first stage? To the ac input voltage of the second stage? To the final output voltage?

Figure 10-16

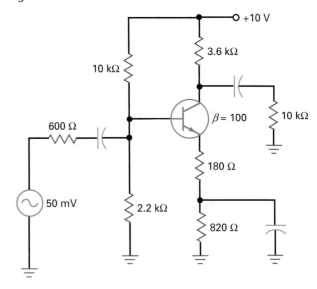

10-17 There is no ac load voltage in Fig. 10-15. The ac input voltage to the second stage is approximately 20 mV. Name some of the possible troubles.

Critical Thinking

10-18 All resistances are doubled in Fig. 10-13. What is the voltage gain?

10-19 If all resistances are doubled in Fig. 10-14, what is the output voltage?

10-20 In Fig. 10-15, all resistances are doubled. What is the output voltage?

10-21 If the load resistor of Fig. 10-15 is disconnected, what is the Thevenin resistance of the second stage?

Troubleshooting

Refer to Fig. 10-17 for the following problems.

10-22 Find Troubles 1 to 4.

10-23 Find Troubles 5 to 8.

10-24 Find Troubles 9 to 12.

Figure 10-17 Troubleshooting.

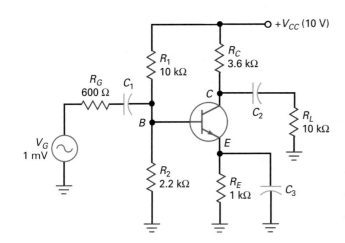

	V_B	V_E	V_C	v_b	v_e	v_c
OK	1.8	1.1	6	0.6 mV	0	73 mV
T1	1.8	1.1	6	0	0	0
T2	1.83	1.13	10	0.75 mV	0	0
T3	1.1	0.4	10	0	0	0
T4	0	0	10	0.8 mV	0	0
T5	1.8	1.1	6	0.6 mV	0	98 mV
T6	3.4	2.7	2.8	0	0	0
T7	1.8	1.1	6	0.75 mV	0.75 mV	1.93 mV
T8	1.1	0.4	0.5	0	0	0
T9	0	0	0	0.75 mV	0	0
T10	1.83	0	10	0.75 mV	0	0
T11	2.1	2.1	2.1	0	0	0
T12	1.8	1.1	6	0	0	0

Job Interview Questions

1. Draw a VDB amplifier. Now, tell me how it works. Include voltage gain and input impedance in your discussion.
2. Draw a swamped amplifier. What is its voltage gain and input impedance? Why does it stabilize voltage gain?
3. In a multistage amplifier, what effect does the input impedance of a stage have on the preceding stage? What effect does a change in β have?
4. What are three improvements that negative feedback makes in an amplifier?
5. You want a circuit like Fig. 10-12 to operate down to 0 Hz. What changes would you have to make?
6. What effect does a swamping resistor have on the voltage gain?
7. What characteristics are desirable in an audio amplifier, and why?
8. What is a swamping resistor, and what does it do?
9. If no value of β is given, what is a reasonable value for a technician to assume?
10. Explain the usefulness of capacitors in multistage voltage amplifiers.
11. What is a swamping resistor? Name three improvements it makes.

Self-Test Answers

1.	c	5.	d	9.	c	13.	a	17.	d	21.	b	25.	a
2.	b	6.	c	10.	c	14.	d	18.	b	22.	c	26.	a
3.	a	7.	c	11.	d	15.	a	19.	a	23.	b	27.	b
4.	c	8.	b	12.	b	16.	a	20.	d	24.	a	28.	c

Practice Problem Answers

10-1 $A_V = 104$

10-2 $V_{out} = 277$ mV

10-3 $V_{out} = 226$ mV

10-4 $V_{out} = 167$ mV

10-5 $V_{out} = 2.24$ V

10-6 $V_{out} = 547$ mV

10-7 Calculated value approximately equal to MultiSim

10-9 $r_f = 4.9$ kΩ

10-12 V_B would increase slightly. $V_E = 0$ V and $V_C = 10$ V. AC measurements would show a slight increase of V_{in} at the base, along with no emitter or collector ac values.

When the load resistance is small compared to the collector resistance, the voltage gain of a CE stage becomes small and the amplifier may become overloaded. One way to prevent overloading is to use a common-collector (CC) amplifier or emitter follower. This type of amplifier has a large input impedance and can drive small load resistances. In addition to emitter followers, this chapter discusses Darlington amplifiers, improved voltage regulation, and common-base (CB) amplifiers.

Objectives

After studying this chapter, you should be able to:

- Draw a diagram of an emitter follower and describe its advantages.
- Analyze an emitter follower for dc and ac operation.
- Describe the purpose of cascading CE and CC amplifiers.
- State the advantages of a Darlington transistor.
- Draw a schematic for a zener follower and discuss how it increases the load current out of a zener regulator.
- Analyze a common-base amplifier for dc and ac operation.
- Compare the characteristics of CE, CC, and CB amplifiers.

Chapter Outline

Vocabulary

buffer

common-base (CB) amplifier

common-collector (CC) amplifier

complementary Darlington

Darlington connection

Darlington pair

Darlington transistor

direct coupled

emitter follower

zener follower

11-1 CC Amplifier

The **emitter follower** is also called a **common-collector (CC) amplifier**. The input signal is coupled to the base, and the output signal is taken from the emitter.

Basic Idea

Figure 11-1a shows an emitter follower. Because the collector is at ac ground, the circuit is a CC amplifier. The input voltage is coupled to the base. This sets up an ac emitter current and produces an ac voltage across the emitter resistor. This ac voltage is then coupled to the load resistor.

Figure 11-1b shows the total voltage between the base and ground. It has a dc component and an ac component. As you can see, the ac input voltage rides on the quiescent base voltage V_{BQ}. Similarly, Fig. 11-1c shows the total voltage between the emitter and ground. This time, the ac input voltage is centered on a quiescent emitter voltage V_{EQ}.

The ac emitter voltage is coupled to the load resistor. This output voltage is shown in Fig. 11-1d, a pure ac voltage. This output voltage is in phase and is

Figure 11-1 Emitter follower and waveforms.

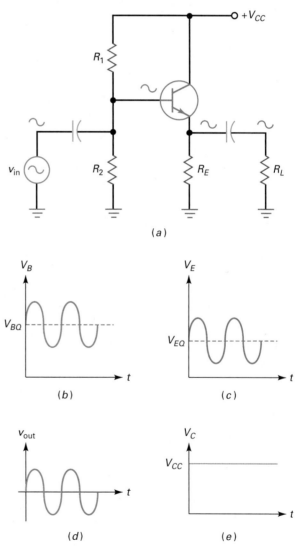

y

b

Chapter 11

approximately equal to the input voltage. The reason the circuit is called an *emitter follower* is because the output voltage follows the input voltage.

Since there is no collector resistor, the total voltage between the collector and ground equals the supply voltage. If you look at the collector voltage with an oscilloscope, you will see a constant dc voltage like Fig. 11-1e. There is no ac signal on the collector because it is an ac ground point.

Negative Feedback

Like a swamped amplifier, the emitter follower uses negative feedback. But with the emitter follower, the negative feedback is massive because the feedback resistance equals all of the emitter resistance. As a result, the voltage gain is ultrastable, the distortion is almost nonexistent, and the input impedance of the base is very high. The trade-off is the voltage gain, which has a maximum value of 1.

AC Emitter Resistance

In Fig. 11-1a, the ac signal coming out of the emitter sees R_E in parallel with R_L. Let us define the ac emitter resistance as follows:

$$r_e = R_E \parallel R_L \tag{11-1}$$

This is the external ac emitter resistance, which is different from the internal ac emitter resistance r'_e.

Voltage Gain

Figure 11-2a shows the ac equivalent with the T model. Using Ohm's law, we can write these two equations:

$$v_{out} = i_e r_e$$

$$v_{in} = i_e(r_e + r'_e)$$

Divide the first equation by the second, and you get the voltage gain of the emitter follower:

$$A_V = \frac{r_e}{r_e + r'_e} \tag{11-2}$$

Usually, a designer makes r_e much greater than r'_e, so that the voltage gain equals 1 (approximately). This is the value to use for all preliminary analysis and troubleshooting.

Figure 11-2 AC equivalent circuits for emitter follower.

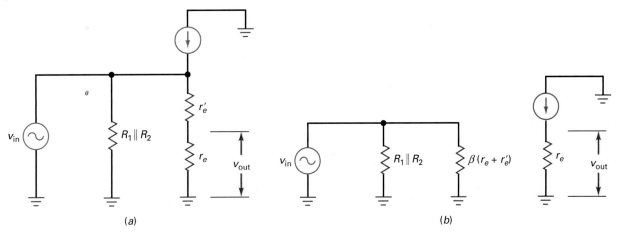

(a)   (b)

Why is an emitter follower called an *amplifier* if its voltage gain is only 1? Because it has a current gain of β. The stages near the end of a system need to produce more current because the final load is usually a low impedance. The emitter follower can produce the large output currents needed by low-impedance loads. In short, although it is not a voltage amplifier, the emitter follower is a current or power amplifier.

Input Impedance of the Base

Figure 11-2b shows the ac equivalent circuit with the π model of the transistor. As far as the input impedance of the base is concerned, the action is the same as that of a swamped amplifier. The current gain transforms the total emitter resistance up by a factor of β. The derivation is therefore identical to that of a swamped amplifier:

$$z_{\text{in(base)}} = \beta(r_e + r_e')$$
(11-3)

For troubleshooting, you can assume that r_e is much greater than r_e', which means that the input impedance is approximately βr_e.

The step-up in impedance is the major advantage of an emitter follower. Small load resistances that would overload a CE amplifier can be used with an emitter follower because it steps up the impedance and prevents overloading.

Input Impedance of the Stage

When the ac source is not stiff, some of the ac signal will be lost across the internal resistance. If you want to calculate the effect of the internal resistance, you will need to use the input impedance of the stage, given by:

$$z_{\text{in(stage)}} = R_1 \| R_2 \| \beta(r_e + r_e')$$
(11-4)

With the input impedance and the source resistance, you can use the voltage divider to calculate the input voltage reaching the base. The calculations are the same as shown in earlier chapters.

GOOD TO KNOW

In Fig. 11-3, the biasing resistors R_1 and R_2 lower z_{in} to a value that is not much different from that of a swamped CE amplifier. This disadvantage is overcome in most emitter follower designs by simply not using the biasing resistors R_1 and R_2. Instead, the emitter follower is dc-biased by the stage driving the emitter follower.

Example 11-1   IIII MultiSim

What is the input impedance of the base in Fig. 11-3 if $\beta = 200$? What is the input impedance of the stage?

Figure 11-3 Example.

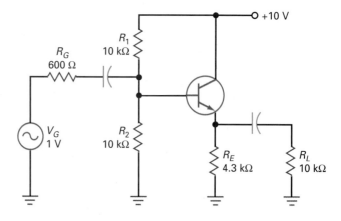

SOLUTION   Because each resistance in the voltage divider is 10 kΩ, the dc base voltage is half the supply voltage, or 5 V. The dc emitter voltage is 0.7 V less, or 4.3 V. The dc emitter voltage is 4.3 V divided by 4.3 kΩ, or 1 mA. Therefore, the ac resistance of the emitter diode is:

$$r_e' = \frac{25\ \text{mV}}{1\ \text{mA}} = 25\ \Omega$$

The external ac emitter resistance is the parallel equivalent of R_E and R_L, which is:

$$r_e = 4.3\ \text{k}\Omega \parallel 10\ \text{k}\Omega = 3\ \text{k}\Omega$$

Since the transistor has an ac current gain of 200, the input impedance of the base is:

$$z_{\text{in(base)}} = 200(3\ \text{k}\Omega + 25\ \Omega) = 605\ \text{k}\Omega$$

The input impedance of the base appears in parallel with the two biasing resistors. The input impedance of the stage is:

$$z_{\text{in(stage)}} = 10\ \text{k}\Omega \parallel 10\ \text{k}\Omega \parallel 605\ \text{k}\Omega = 4.96\ \text{k}\Omega$$

Because the 605 kΩ is much larger than 5 kΩ, troubleshooters usually approximate the input impedance of the stage as the parallel of the biasing resistors only:

$$z_{\text{in(stage)}} = 10\ \text{k}\Omega \parallel 10\ \text{k}\Omega = 5\ \text{k}\Omega$$

PRACTICE PROBLEM 11-1   Find the input impedance of the base and the stage, using Fig. 11-3, if β changes to 100.

Example 11-2

||| MultiSim

Assuming a β of 200, what is the ac input voltage to the emitter follower of Fig. 11-3?

SOLUTION   Figure 11-4 shows the ac equivalent circuit. The ac base voltage appears across z_{in}. Because the input impedance of the stage is large compared to the generator resistance, most of the generator voltage appears at the base. With the voltage-divider theorem:

$$v_{\text{in}} = \frac{5\ \text{k}\Omega}{5\ \text{k}\Omega + 600\ \Omega}\ 1\ \text{V} = 0.893\ \text{V}$$

Figure 11-4   Example.

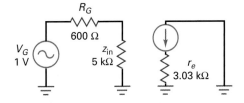

PRACTICE PROBLEM 11-2   If the β value is 100, find the ac input voltage of Fig. 11-3.

Example 11-3

What is the voltage gain of the emitter follower in Fig. 11-5? If $\beta = 150$, what is the ac load voltage?

Figure 11-5 Example.

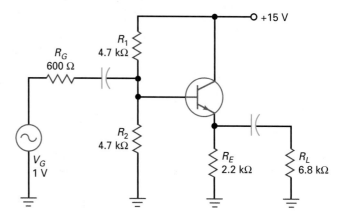

SOLUTION The dc base voltage is half the supply voltage:

$$V_B = 7.5 \text{ V}$$

The dc emitter current is:

$$I_E = \frac{6.8 \text{ V}}{2.2 \text{ k}\Omega} = 3.09 \text{ mA}$$

and the ac resistance of the emitter diode is:

$$r_e' = \frac{25 \text{ mV}}{3.09 \text{ mA}} = 8.09 \text{ }\Omega$$

The external ac emitter resistance is:

$$r_e = 2.2 \text{ k}\Omega \parallel 6.8 \text{ k}\Omega = 1.66 \text{ k}\Omega$$

The voltage gain equals:

$$A_V = \frac{1.66 \text{ k}\Omega}{1.66 \text{ k}\Omega + 8.09 \text{ }\Omega} = 0.995$$

The input impedance of the base is:

$$z_{in(base)} = 150(1.66 \text{ k}\Omega + 8.09 \text{ }\Omega) = 250 \text{ k}\Omega$$

This is much larger than the biasing resistors. Therefore, to a close approximation, the input impedance of the emitter follower is:

$$z_{in(stage)} = 4.7 \text{ k}\Omega \parallel 4.7 \text{ k}\Omega = 2.35 \text{ k}\Omega$$

The ac input voltage is:

$$v_{in} = \frac{2.35 \text{ k}\Omega}{600 \text{ }\Omega + 2.35 \text{ k}\Omega} 1 \text{ V} = 0.797 \text{ V}$$

The ac output voltage is:

$$v_{out} = 0.995(0.797 \text{ V}) = 0.793 \text{ V}$$

PRACTICE PROBLEM 11-3 Repeat Example 11-3 using an R_G value of 50 Ω.

11-2 Output Impedance

The output impedance of an amplifier is the same as its Thevenin impedance. One of the advantages of an emitter follower is its low output impedance.

As discussed in earlier electronics courses, maximum power transfer occurs when the load impedance is *matched* (made equal) to the source (Thevenin) impedance. Sometimes, when maximum load power is wanted, a designer can match the load impedance to the output impedance of an emitter follower. For instance, the low impedance of a speaker can be matched to the output impedance of an emitter follower to deliver maximum power to the speaker.

Basic Idea

Figure 11-6a shows an ac generator driving an amplifier. If the source is not stiff, some of the ac voltage is dropped across the internal resistance R_G. In this case, we need to analyze the voltage divider shown in Fig. 11-6b to get the input voltage v_{in}.

A similar idea can be used with the output side of the amplifier. In Fig. 11-6c, we can apply the Thevenin theorem at the load terminals. Looking back into the amplifier, we see an output impedance z_{out}. In the Thevenin equivalent circuit, this output impedance forms a voltage divider with the load resistance, as shown in Fig. 11-6d. If z_{out} is much smaller than R_L, the output source is stiff and v_{out} equals v_{th}.

CE Amplifiers

Figure 11-7a shows the ac equivalent circuit for the output side of a CE amplifier. When we apply Thevenin's theorem, we get Fig. 11-7b. In other words, the output

Figure 11-6 Input and output impedances.

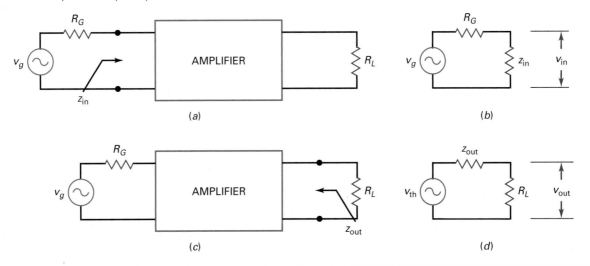

Figure 11-7 Output impedance of CE stage.

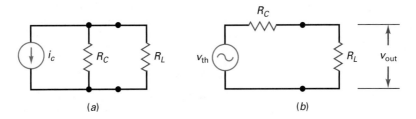

Figure 11-8 Output impedance of emitter follower.

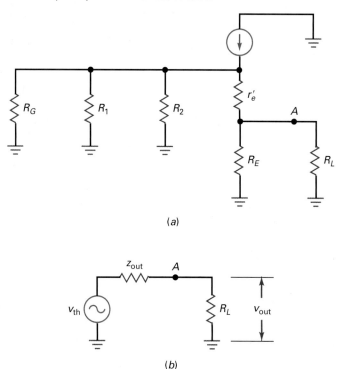

(a)

(b)

impedance facing the load resistance is R_C. Since the voltage gain of a CE amplifier depends on R_C, a designer cannot make R_C too small without losing voltage gain. Stated another way, it is very difficult to get a small output impedance with a CE amplifier. Because of this, CE amplifiers are not suited to driving small load resistances.

Emitter Follower

Figure 11-8a shows the ac equivalent circuit for an emitter follower. When we apply Thevenin's theorem to point A, we get Fig. 11-8b. The output impedance z_{out} is much smaller than you can get with a CE amplifier. It equals:

$$z_{out} = R_E \parallel \left(r_e' + \frac{R_G \parallel R_1 \parallel R_2}{\beta} \right) \qquad (11\text{-}5)$$

The impedance of the base circuit is $R_G \parallel R_1 \parallel R_2$. The current gain of the transistor steps this impedance down by a factor of β. The effect is similar to what we get with a swamped amplifier, except that we are moving from the base back to the emitter. Therefore, we get a reduction of impedance rather than an increase. The stepped-down impedance of $(R_G \parallel R_1 \parallel R_2)/\beta$ is in series with r_e' as indicated by Eq. (11-5).

Ideal Action

In some designs the biasing resistances and the ac resistance of the emitter diode become negligible. In this case, the output impedance of an emitter follower can be approximated by:

$$z_{out} = \frac{R_G}{\beta} \qquad (11\text{-}6)$$

This brings out the key idea of an emitter follower: It steps the impedance of the ac source down by a factor of β. As a result, the emitter follower allows us to build stiff ac sources. Instead of using a stiff ac source that maximizes the load voltage, a designer may prefer to maximize the load power. In this case, instead of designing for:

$$z_{out} \ll R_L \qquad \text{(Stiff voltage source)}$$

the designer will select values to get:

$$z_{out} = R_L \qquad \text{(Maximum power transfer)}$$

In this way, the emitter follower can deliver maximum power to a low-impedance load such as a stereo speaker. By basically removing the effect of R_L on the output voltage, the circuit is acting as a buffer between the input and output.

Equation (11-6) is an ideal formula. You can use it to get an approximate value for the output impedance of an emitter follower. With discrete circuits, the equation usually gives only an estimate of the output impedance. Nevertheless, it is adequate for troubleshooting and preliminary analysis. When necessary, you can use Eq. (11-5) to get an accurate value for the output impedance.

Example 11-4

Estimate the output impedance of the emitter follower of Fig. 11-9a.

SOLUTION   Ideally, the output impedance equals the generator resistance divided by the current gain of the transistor:

$$z_{out} = \frac{600\ \Omega}{300} = 2\ \Omega$$

Figure 11-9b shows the equivalent output circuit. The output impedance is much smaller than the load resistance, so that most of the signal appears across the load resistor. As you can see, the output source of Fig. 11-9b is almost stiff because the ratio of load to source resistance is 50.

PRACTICE PROBLEM 11-4   Using Fig. 11-9, change the source resistance to 1 kΩ and solve for the approximate z_{out} value.

Example 11-5

Calculate the output impedance in Fig. 11-9a using Eq. (11-5).

SOLUTION   The quiescent base voltage is approximately:

$$V_{BQ} = 15\ \text{V}$$

Ignoring V_{BE}, the quiescent emitter current is approximately:

$$I_{EQ} = \frac{15\ \text{V}}{100\ \Omega} = 150\ \text{mA}$$

The ac resistance of the emitter diode is:

$$r'_e = \frac{25\ \text{mV}}{150\ \text{mA}} = 0.167\ \Omega$$

The impedance seen looking back from the base is:

$$R_G \parallel R_1 \parallel R_2 = 600\ \Omega \parallel 10\ \text{k}\Omega \parallel 10\ \text{k}\Omega = 536\ \Omega$$

The current gain steps this down to:

$$\frac{R_G \parallel R_1 \parallel R_2}{\beta} = \frac{536\ \Omega}{300} = 1.78\ \Omega$$

Figure 11-9 Example.

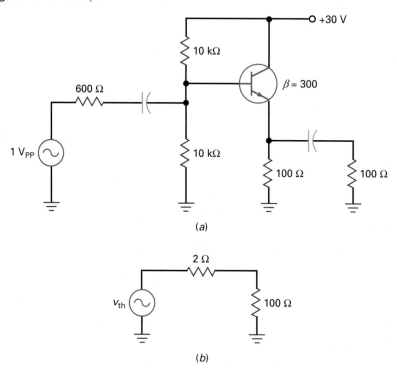

(a)

(b)

This is in series with r_e', so that the impedance looking back into the emitter is:

$$r_e' + \frac{R_G \| R_1 \| R_2}{\beta} = 0.167\ \Omega + 1.78\ \Omega = 1.95\ \Omega$$

This is in parallel with the dc emitter resistance, so that the output impedance is:

$$z_{out} = R_E \| \left(r_e' + \frac{R_G \| R_1 \| R_2}{\beta} \right) = 100\ \Omega \| 1.95\ \Omega = 1.91\ \Omega$$

This accurate answer is extremely close to the ideal answer of 2 Ω. This result is typical of many designs. For all troubleshooting and preliminary analysis, you can use the ideal method to estimate the output impedance.

PRACTICE PROBLEM 11-5 Repeat Example 11-5 using an R_G value of 1 kΩ.

11-3 Cascading CE and CC

To illustrate the buffering action of a CC amplifier, suppose we have a load resistance of 270 Ω. If we try to couple the output of a CE amplifier directly into this load resistance, we may overload the amplifier. One way to avoid this overload is by using an emitter follower between the CE amplifier and the load resistance. The signal can be coupled capacitively (this means through coupling capacitors), or it may be **direct coupled** as shown in Fig. 11-10.

As you can see, the base of the second transistor is connected directly to the collector of the first transistor. Because of this, the dc collector voltage of the first transistor is used to bias the second transistor. If the dc current gain of the second transistor is 100, the dc resistance looking into the base of the second transistor is $R_{in} = 100\ (270\ \Omega) = 27\ \text{k}\Omega$.

Figure 11–10 Direct coupled output stage.

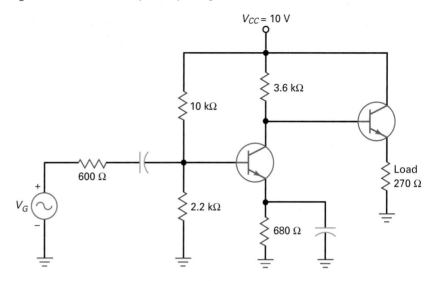

Because 27 kΩ is large compared to the 3.6 kΩ, the dc collector voltage of the first stage is only slightly disturbed.

In Fig. 11-10, the amplified voltage out of the first stage drives the emitter follower and appears across the final load resistance of 270 Ω. Without the emitter follower, the 270 Ω would overload the first stage. But with the emitter follower, its impedance effect is increased by a factor of β. Instead of appearing like 270 Ω, it now looks like 27 kΩ in both the dc and the ac equivalent circuits.

This demonstrates how an emitter follower can act as a **buffer** between a high-output impedance and a low-resistance load.

Example 11-6 ||| MultiSim

What is the voltage gain of the CE stage in Fig. 11-10 for a β of 100?

SOLUTION The dc base voltage of the CE stage is 1.8 V and the dc emitter voltage is 1.1 V. The dc emitter current is $I_E = \dfrac{1.1 \text{ V}}{680 \text{ Ω}} = 1.61$ mA and the ac resistance of the emitter diode is $r'_e = \dfrac{25 \text{ mV}}{1.61 \text{ mA}} = 15.5$ Ω. Next, we need to calculate the input impedance of the emitter follower. Since there are no biasing resistors, the input impedance equals the input impedance looking into the base: $z_{in} = (100)(270 \text{ Ω}) = 27$ kΩ. The ac collector resistance of the CE amplifier is $r_c = 3.6 \text{ kΩ} \parallel 27 \text{ kΩ} = 3.18$ kΩ and the voltage gain of this stage is

$$A_v = \frac{3.18 \text{ kΩ}}{15.5 \text{ Ω}} = 205$$

PRACTICE PROBLEM 11–6 Using Fig. 11-10, find the voltage gain of the CE stage for a β of 300.

Example 11-7 ||| MultiSim

Suppose the emitter follower is removed in Fig. 11-10 and a capacitor is used to couple the ac signal to the 270 Ω load resistor. What happens to the voltage gain of the CE amplifier?

SOLUTION The value of r'_e remains the same for the CE stage: 15.5 Ω. But the ac collector resistance is much lower. To begin with, the ac collector resistance is the parallel resistance of 3.6 kΩ and 270 Ω: $r_c = 3.6 \text{ kΩ} \parallel 270 \text{ Ω} = 251 \text{ Ω}$.

Because this is much lower, the voltage gain decreases to $A_v = \dfrac{251 \text{ Ω}}{15.5 \text{ Ω}} = 16.2$.

PRACTICE PROBLEM 11-7 Repeat Example 11-7 using a load resistance of 100 Ω.

This shows you the effects of overloading a CE amplifier. The load resistance should be much greater than the dc collector resistance to get maximum voltage gain. We have just the opposite; the load resistance (270 Ω) is much smaller than the dc collector resistance (3.6 kΩ).

11-4 Darlington Connections

A **Darlington connection** is a connection of two transistors whose overall current gain equals the product of the individual current gains. Since its current gain is much higher, a Darlington connection can have a very high input impedance and can produce very large output currents. Darlington connections are often used with voltage regulators, power amplifiers, and high current switching applications.

Darlington Pair

Figure 11-11a shows a **Darlington pair**. Since the emitter current of Q_1 is the base current for Q_2, the Darlington pair has an overall current gain of:

$$\beta = \beta_1 \beta_2 \qquad (11\text{-}7)$$

For instance, if each transistor has a current gain of 200, the overall current gain is:

$$\beta = (200)(200) = 40,000$$

Semiconductor manufacturers can put a Darlington pair inside a single case like Fig. 11-11b. This device, known as a **Darlington transistor,** acts like a single transistor with a very high current gain. For instance, the 2N6725 is a Darlington transistor with a current gain of 25,000 at 200 mA. As another example, the TIP102 is a power Darlington with a current gain of 1000 at 3 A.

This is shown in the data sheet of Fig. 11-12. Notice that this device uses a TO-220 case style and has built-in base-emitter shunt resistors, along with an

---

Figure 11-11 (a) Darlington pair; (b) Darlington transistor; (c) complementary Darlington.

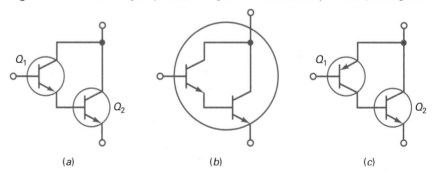

(a)   (b)   (c)

TIP100/101/102

Monolithic Construction With Built In Base-Emitter Shunt Resistors

- High DC Current Gain : h_{FE}=1000 @ V_{CE}=4V, I_C=3A (Min.)
- Collector-Emitter Sustaining Voltage
- Low Collector-Emitter Saturation Voltage
- Industrial Use
- Complementary to TIP105/106/107

TO-220

1.Base   2.Collector   3.Emitter

NPN Epitaxial Silicon Darlington Transistor

Absolute Maximum Ratings T_C=25°C unless otherwise noted

Equivalent Circuit

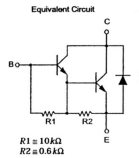

$R1 \cong 10k\Omega$
$R2 \cong 0.6k\Omega$

Symbol	Parameter		Value	Units
V_{CBO}	Collector-Base Voltage	: TIP100	60	V
		: TIP101	80	V
		:TIP102	100	V
V_{CEO}	Collector-Emitter Voltage	: TIP100	60	V
		: TIP101	80	V
		: TIP102	100	V
V_{EBO}	Emitter-Base Voltage		5	V
I_C	Collector Current (DC)		8	A
I_{CP}	Collector Current (Pulse)		15	A
I_B	Base Current (DC)		1	A
P_C	Collector Dissipation (T_a=25°C)		2	W
	Collector Dissipation (T_C=25°C)		80	W
T_J	Junction Temperature		150	°C
T_{STG}	Storage Temperature		- 65 ~ 150	°C

Electrical Characteristics T_C=25°C unless otherwise noted

Symbol	Parameter		Test Condition	Min.	Max.	Units
V_{CEO}(sus)	Collector-Emitter Sustaining Voltage					
		: TIP100	I_C = 30mA, I_B = 0	60		V
		: TIP101		80		V
		: TIP102		100		V
I_{CEO}	Collector Cut-off Current					
		: TIP100	V_{CE} = 30V, I_B = 0		50	μA
		: TIP101	V_{CE} = 40V, I_B = 0		50	μA
		: TIP102	V_{CE} = 50V, I_B = 0		50	μA
I_{CBO}	Collector Cut-off Current					
		: TIP100	V_{CE} = 60V, I_E = 0		50	μA
		: TIP101	V_{CE} = 80V, I_E = 0		50	μA
		: TIP102	V_{CE} = 100V, I_E = 0		50	μA
I_{EBO}	Emitter Cut-off Current		V_{EB} = 5V, I_C = 0		2	mA
h_{FE}	DC Current Gain		V_{CE} = 4V, I_C = 3A	1000	20000	
			V_{CE} = 4V, I_C = 8A	200		
V_{CE}(sat)	Collector-Emitter Saturation Voltage		I_C = 3A, I_B = 6mA		2	V
			I_C = 8A, I_B = 80mA		2.5	V
V_{BE}(on)	Base-Emitter ON Voltage		V_{CE} = 4V, I_C = 8A		2.8	V
C_{ob}	Output Capacitance		V_{CB} = 10V, I_E = 0, f = 0.1MHz		200	pF

Rev. A1, June 2001

internal diode. These internal components must be taken into consideration when testing this device with an ohmmeter.

The analysis of a circuit using a Darlington transistor is almost identical to the emitter follower analysis. With the Darlington transistor, since there are two transistors, there are two V_{BE} drops. The base current of Q_2 is the same as the emitter current of Q_1. Also, the input impedance at the base of Q_1 can be found by $z_{\text{in(base)}} \cong \beta_1\beta_2 r_e$ or stated as:

$$z_{\text{in(base)}} \cong \beta r_e \tag{11-8}$$

Example 11-8 ⫼ MultiSim

If each transistor of Fig. 11-13 has a beta value of 100, what is the overall current gain, base current of Q_1, and input impedance at the base of Q_1?

Figure 11-13 Example.

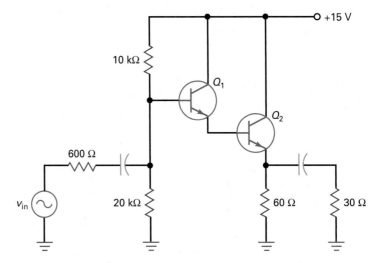

SOLUTION   The overall current gain is found by:

$$\beta = \beta_1\beta_2 = (100)(100) = 10{,}000$$

The dc emitter current of Q_2 is:

$$I_{E2} = \frac{10\text{ V} - 1.4\text{ V}}{60\ \Omega} = 143\text{ mA}$$

The emitter current of Q_1 is equal to the base current of Q_2. This is found by:

$$I_{E1} = I_{B2} \cong \frac{I_{E2}}{\beta_2} = \frac{143\text{ mA}}{100} = 1.43\text{ mA}$$

The base current of Q_1 is:

$$I_{B1} = \frac{I_{E1}}{\beta_1} = \frac{1.43\text{ mA}}{100} = 14.3\ \mu\text{A}$$

To find the input impedance at the base of Q_1, first solve for r_e. The ac emitter resistance is:

$$r_e = 60\ \Omega \parallel 30\ \Omega = 20\ \Omega$$

The input impedance of the Q_1 base is: $z_{\text{in(base)}} = (10{,}000)(20\ \Omega) = 200\text{ k}\Omega$

Complementary Darlington

Figure 11-11c shows another Darlington connection called a **complementary Darlington,** a connection of *npn* and *pnp* transistors. The collector current of Q_1 is the base current of Q_2. If the *pnp* transistor has a current gain of β_1 and the *npn* output transistor has a current gain of β_2 the complementary Darlington acts like a single *pnp* transistor with a current gain of $\beta_1\beta_2$.

Npn and *pnp* Darlington transistors can be manufactured to be complements to each other. As an example, the TIP105/106/107 *pnp* Darlington series is complementary to the TIP/101/102 *npn* series.

11-5 Voltage Regulation

Besides being used in buffer circuits and impedance matching amplifiers, the emitter follower is widely used in voltage regulators. In conjunction with a zener diode, the emitter follower can produce regulated output voltages with much larger output currents.

Zener Follower

Figure 11-14a shows a **zener follower,** a circuit that combines a zener regulator and an emitter follower. Here is how it works: The zener voltage is the input to the base of the emitter follower. The dc output voltage of the emitter follower is:

$$V_{\text{out}} = V_Z - V_{BE} \tag{11-9}$$

This output voltage is fixed so that it is equal to the zener voltage minus the V_{BE} drop of the transistor. If the supply voltage changes, the zener voltage remains approximately constant, and so does the output voltage. In other words, the circuit acts like a voltage regulator because the output voltage is always one V_{BE} drop less than the zener voltage.

Figure 11-14   (a) Zener follower; (b) ac equivalent circuit.

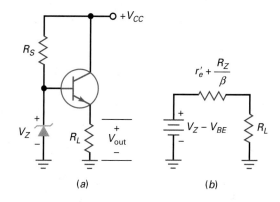

(a)   (b)

The zener follower has two advantages over an ordinary zener regulator: First, the zener diode of Fig. 11-14a has to produce a load current of only

$$I_B = \frac{I_{out}}{\beta_{dc}} \qquad (11\text{-}10)$$

Since this base current is much smaller than the output current, we can use a much smaller zener diode.

For instance, if you are trying to supply several amperes to a load resistor, an ordinary zener regulator requires a zener diode capable of handling several amperes. On the other hand, with the improved regulator of Fig. 11-14a, the zener diode needs to handle only tens of milliamperes.

The second advantage of a zener follower is its low output impedance. In an ordinary zener regulator, the load resistor sees an output impedance of approximately R_Z, the zener impedance. But in the zener follower, the output impedance is:

$$z_{out} = r_e' + \frac{R_Z}{\beta_{dc}} \qquad (11\text{-}11)$$

Figure 11-14b shows the equivalent output circuit. Because z_{out} is usually very small compared to R_L, an emitter follower can hold the dc output voltage almost constant because the source looks stiff.

In summary, the zener follower provides the regulation of a zener diode with the increased current-handling capability of an emitter follower.

Two-Transistor Regulator

Figure 11-15 shows another voltage regulator. The dc input voltage V_{in} comes from an unregulated power supply such as a bridge rectifier with a capacitor-input filter. Typically, V_{in} has a peak-to-peak ripple of about 10 percent of the dc voltage. The final output voltage V_{out} has almost no ripple and is almost constant in value, even though the input voltage and load current may vary over a large range.

How does it work? Any attempted change in output voltage produces an amplified feedback voltage that opposes the original change. For instance, suppose the output voltage increases. Then, the voltage appearing at the base of Q_1 will increase. Since Q_1 and R_2 form a CE amplifier, the collector voltage of Q_1 will decrease because of the voltage gain.

Since the collector voltage of Q_1 has decreased, the base voltage of Q_2 decreases. Because Q_2 is an emitter follower, the output voltage will decrease. In other words, we have negative feedback. The original increase in output voltage produces an opposing decrease in output voltage. The overall effect is that the output voltage will increase only slightly, much less than it would without the negative feedback.

Figure 11-15 Transistor voltage regulator.

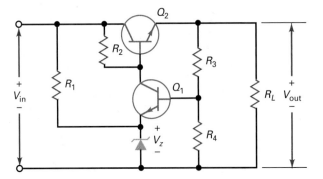

Conversely, if the output voltage tries to decrease, less voltage appears at the Q_1 base, more voltage appears at the Q_1 collector, and more voltage appears at the Q_2 emitter. Again, we have a returning voltage that opposes the original change in output voltage. Therefore, the output voltage will decrease only a little, far less than without the negative feedback.

Because of the zener diode, the Q_1 emitter voltage equals V_Z. The Q_1 base voltage is one V_{BE} drop higher. Therefore, the voltage across R_4 is:

$$V_4 = V_Z + V_{BE}$$

With Ohm's law, the current through R_4 is:

$$I_4 = \frac{V_Z + V_{BE}}{R_4}$$

Since this current flows through R_3 in series with R_4, the output voltage is:

$$V_{out} = I_4(R_3 + R_4)$$

After expanding:

$$V_{out} = \frac{R_3 + R_4}{R_4}(V_Z + V_{BE}) \tag{11-12}$$

Example 11-9 |||| MultiSim

Figure 11-16 shows a zener follower as it is usually drawn on a schematic diagram. What is the output voltage? What is the zener current if $\beta_{dc} = 100$?

Figure 11-16   Example.

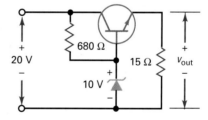

SOLUTION   The output voltage is approximately:

$$V_{out} = 10\text{ V} - 0.7\text{ V} = 9.3\text{ V}$$

With a load resistance of 15 Ω, the load current is:

$$I_{out} = \frac{9.3\text{ V}}{15\,\Omega} = 0.62\text{ A}$$

The base current is:

$$I_B = \frac{0.62\text{ A}}{100} = 6.2\text{ mA}$$

The current through the series resistor is:

$$I_S = \frac{20\text{ V} - 10\text{ V}}{680\,\Omega} = 14.7\text{ mA}$$

The zener current is:

$$I_Z = 14.7 \text{ mA} - 6.2 \text{ mA} = 8.5 \text{ mA}$$

PRACTICE PROBLEM 11-9 Repeat Example 11-9 using an 8.2 V zener diode and an input voltage of 15 V.

Example 11-10

‖‖ MultiSim

What is the output voltage in Fig. 11-17?

Figure 11-17 Example.

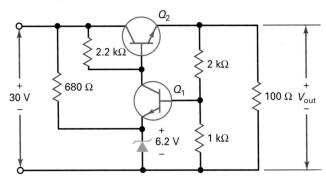

SOLUTION With Eq. (11-12):

$$V_{out} = \frac{2 \text{ k}\Omega + 1 \text{ k}\Omega}{1 \text{ k}\Omega} (6.2 \text{ V} + 0.7 \text{ V}) = 20.7 \text{ V}$$

You can also solve the problem as follows: The current through the 1-kΩ resistor is:

$$I_4 = \frac{6.2 \text{ V} + 0.7 \text{ V}}{1 \text{ k}\Omega} = 6.9 \text{ mA}$$

This current flows through a total resistance of 3 kΩ, which means that the output voltage is:

$$V_{out} = (6.9 \text{ mA})(3 \text{ k}\Omega) = 20.7 \text{ V}$$

PRACTICE PROBLEM 11-10 Using Fig. 11-17, change the zener value to 5.6 V and find the new V_{out} value.

11-6 The Common-Base Amplifier

Figure 11-18a shows a **common-base (CB) amplifier** using a dual polarity or split power supply. Since the base is grounded, this circuit is also called a grounded-based amplifier. The Q point is set by emitter bias as shown by the dc

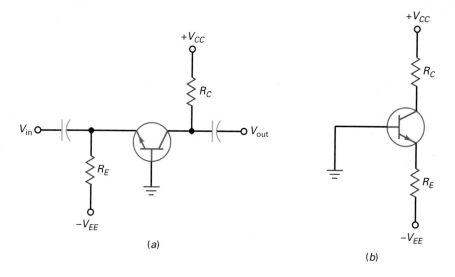

(*a*)

(*b*)

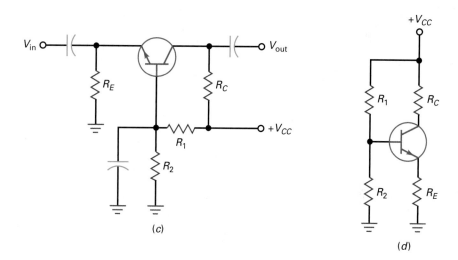

(*c*)

(*d*)

equivalent circuit shown in Fig. 11-18*b*. Therefore, the dc emitter current is found by:

$$I_E = \frac{V_{EE} - V_{BE}}{R_E} \tag{11-13}$$

Figure 11-18*c* shows a voltage-divider bias CB amplifier using a single power supply source. Notice the bypass capacitor across R_2. This places the base at ac ground. By drawing the dc equivalent circuit, as shown in Fig. 11-18*d*, you should recognize the voltage-divider bias configuration.

In either amplifier, the base is at ac ground. The input signal drives the emitter, and the output signal is taken from the collector. Figure 11-19*a* shows the ac equivalent circuit of a CB amplifier during the positive half-cycle of input voltage. In this circuit, the ac collector voltage, or v_{out}, equals:

$$v_{\text{out}} \cong i_c r_c$$

Figure 11-19 AC equivalent circuit.

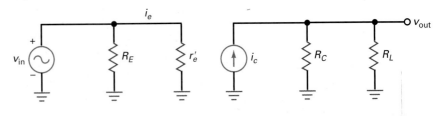

This is in phase with the ac input voltage v_e. Since the input voltage equals:

$$v_{in} = i_e r'_e$$

The voltage gain is:

$$A_V = \frac{v_{out}}{v_{in}} = \frac{i_c r_c}{i_e r'_e}$$

because $i_c \cong i_e$, the equation simplifies to:

$$A_v = \frac{r_c}{r'_e} \tag{11-14}$$

Notice that the voltage gain has the same magnitude as it would in an unswamped CE amplifier. The only difference is the phase of the output voltage. Whereas the output signal of a CE amplifier is 180° out of phase with the input signal, the output voltage of the CB amplifier is in phase with the input signal.

Ideally, the collector current source of Fig. 11-19 has an infinite internal impedance. Therefore, the output impedance of a CB amplifier is:

$$z_{out} \cong R_c \tag{11-15}$$

One of the major differences between the CB amplifier and other amplifier configurations is its low input impedance. Looking into the emitter of Fig. 11-19, we have an input impedance of:

$$z_{in(emitter)} = \frac{v_e}{i_e} = \frac{i_e r'_e}{i_e} \quad \text{or} \quad z_{in(emitter)} = r'_e$$

The input impedance of the circuit is:

$$z_{in} = R_E \| r'_e$$

Since R_E is normally much larger than r'_e, the circuit input impedance is approximately:

$$z_{in} \cong r'_e \tag{11-16}$$

As an example, if $I_E = 1$ mA, the input impedance of a CB amplifier is only 25 Ω. Unless the input ac source is very small, most of the signal will be lost across the source resistance.

The input impedance of a CB amplifier is normally so low that it overloads most signal sources. Because of this, a discrete CB amplifier is not used too often at low frequencies. It is mainly used in high-frequency applications (above 10 MHz) where low source impedances are common. Also, at high frequencies, the base separates the input and output resulting in fewer oscillations at these frequencies.

An emitter follower circuit was used in applications where a high impedance source needed to drive a low impedance load. Just the opposite, a common-base circuit can be used to couple a low impedance source to a high impedance load.

Example 11-11

IIII MultiSim

What is the output voltage of Fig. 11-20?

Figure 11-20 Example.

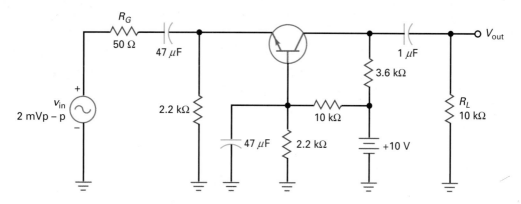

SOLUTION The circuit needs to have its Q point determined.

$$V_B = \frac{2.2 \text{ k}\Omega}{10 \text{ k}\Omega + 2.2 \text{ k}\Omega} (+10 \text{ V}) = 1.8 \text{ V}$$

$$V_E = V_B - 0.7 \text{ V} = 1.8 \text{ V} - 0.7 \text{ V} = 1.1 \text{ V}$$

$$I_E = \frac{V_E}{R_E} = \frac{1.1 \text{ V}}{2.2 \text{ k}\Omega} = 500 \text{ }\mu\text{A}$$

Therefore, $r'_e = \dfrac{25 \text{ mV}}{500 \text{ }\mu\text{A}} = 50 \text{ }\Omega$

Now, solving for the ac circuit values:

$$z_{\text{in}} = R_E \| r'_e = 2.2 \text{ k}\Omega \| 50 \text{ }\Omega \cong 50 \text{ }\Omega$$

$$z_{\text{out}} = R_C = 3.6 \text{ k}\Omega$$

$$A_V = \frac{r_c}{r'_e} = \frac{3.6 \text{ k}\Omega \| 10 \text{ k}\Omega}{50 \text{ }\Omega} = \frac{2.65 \text{ k}\Omega}{50 \text{ }\Omega} = 53$$

$$v_{\text{in(base)}} = \frac{r'_e}{R_G} (v_{\text{in}}) = \frac{50 \text{ }\Omega}{50 \text{ }\Omega + 50 \text{ }\Omega} (2 \text{ mV}_{\text{pp}}) = 1 \text{ mV}_{\text{pp}}$$

$$v_{\text{out}} = (A_v)(v_{\text{in(base)}}) = (53)(1 \text{ mV}_{\text{pp}}) = 53 \text{ mV}_{\text{pp}}$$

PRACTICE PROBLEM 11-11 In Fig. 11-20, change V_{CC} to 20 V and find v_{out}.

A summary of the four common transistor amplifier configurations is shown in Summary Table 11-1. It is important to be able to recognize the amplifier configuration, know its basic characteristics, and understand its common applications.

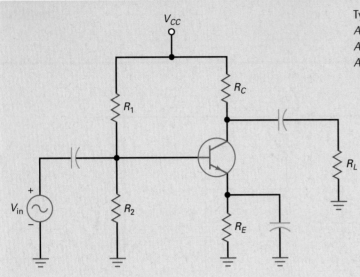

Type: CE
A_v: Medium-High
A_i: β
A_p: High

θ: 180°
z_{in}: Medium
z_{out}: Medium
Applications: General purpose amplifier, with voltage and current gain

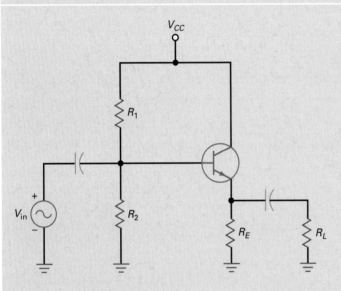

Type: CC
A_v: ≈ 1
A_i: β
A_p: Medium

θ: 0°
z_{in}: High
z_{out}: Low
Applications: Buffer, impedance matching, high current driver

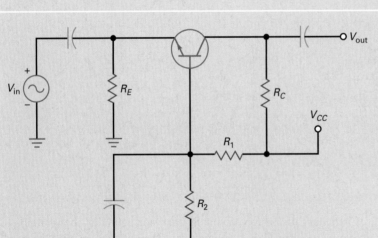

Type: CB
A_v: Medium-high
A_i: ≈ 1
A_p: Medium

θ: 0°
z_{in}: Low
z_{out}: High
Applications: High-frequency amplifier, low to high impedance matching

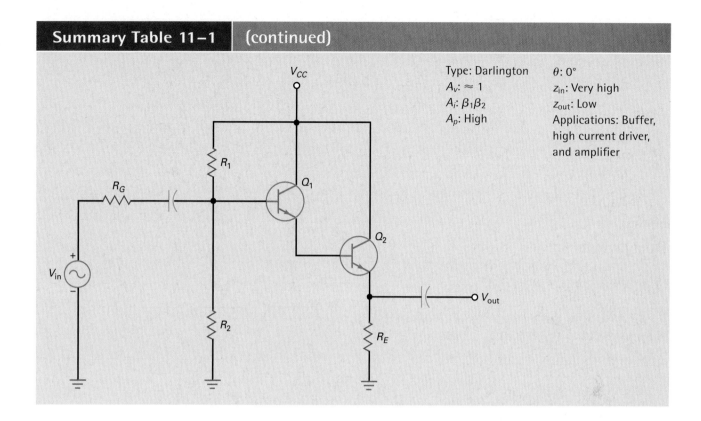

Type: Darlington
A_v: ≈ 1
A_i: $\beta_1\beta_2$
A_p: High

θ: 0°
z_{in}: Very high
z_{out}: Low
Applications: Buffer, high current driver, and amplifier

Summary

SEC. 11–1 CC AMPLIFIER

A CC amplifier, better known as an emitter follower, has its collector at ac ground. The input signal drives the base and the output signal comes from the emitter. Because it is heavily swamped, an emitter follower has stable voltage gain, high input impedance, and low distortion.

SEC. 11–2 OUTPUT IMPEDANCE

The output impedance of an amplifier is the same as its Thevenin impedance. An emitter follower has a low output impedance. The current gain of a transistor transforms the source impedance driving the base to a much lower value when seen from the emitter.

SEC. 11–3 CASCADING CE AND CC

When a low resistance load is connected to the output of a CE amplifier, it may

become overloaded resulting in a very small voltage gain. A CC amplifier placed between the CE output and load will significantly reduce this effect. In this way, the CC amplifier is acting as a buffer.

SEC. 11–4 DARLINGTON CONNECTIONS

Two transistors can be connected as a Darlington pair. The emitter of the first is connected to the base of the second. This produces an overall current gain equal to the product of the individual current gains.

SEC. 11–5 VOLTAGE REGULATION

By combining a zener diode and an emitter follower, we get a zener follower. This circuit produces regulated output voltage with large load currents. The advantage is that the zener current is

much smaller than the load current. By adding a stage of voltage gain, a larger regulated output voltage can be produced.

SEC. 11–6 COMMON–BASE AMPLIFIER

The CB amplifier configuration has its base at ac ground. The input signal drives the emitter and the output signal comes from the collector. Even though this circuit has no current gain, it can produce a significant voltage gain. The CB amplifier has a low input impedance and high output impedance, and is used in high-frequency applications.

Definitions

(11-1) AC emitter resistance:

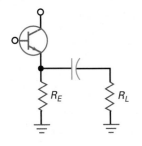

$$r_e = R_E \parallel R_L$$

Derivations

(11-2) Emitter-follower voltage gain:

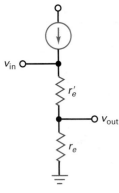

$$A_V = \frac{r_e}{r_e + r_e'}$$

(11-3) Emitter-follower input impedance of base:

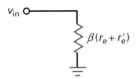

$$z_{in(base)} = \beta(r_e + r_e')$$

(11-5) Emitter-follower output impedance:

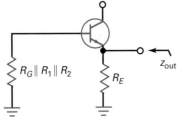

$$z_{out} = R_E \parallel \left(r_e' + \frac{R_G \parallel R_1 \parallel R_2}{\beta} \right)$$

(11-7) Darlington current gain:

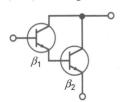

$$\beta = \beta_1 \beta_2$$

(11-9) Zener follower:

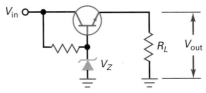

$$V_{out} = V_Z - V_{BE}$$

(11-12) Voltage regulator:

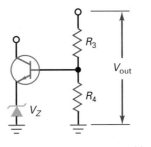

$$V_{out} = \frac{R_3 + R_4}{R_4}(V_Z + V_{BE})$$

(11-14) Common-base voltage gain:

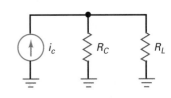

$$A_V = \frac{r_c}{r_e'}$$

(11-16) Common-base input impedance:

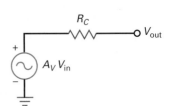

$$z_{in} \cong r_e'$$

Student Assignments

1. An emitter follower has a voltage gain that is
   a. Much less than one
   b. Approximately equal to one
   c. Greater than one
   d. Zero

2. The total ac emitter resistance of an emitter follower equals
   a. r_e'
   b. r_e
   c. $r_e + r_e'$
   d. R_E

3. The input impedance of the base of an emitter follower is usually
   a. Low
   b. High
   c. Shorted to ground
   d. Open

4. The dc current gain of an emitter follower is
   a. 0
   b. ≈ 1
   c. β_{dc}
   d. Dependant on r_e'

5. The ac base voltage of an emitter follower is across the
   a. Emitter diode
   b. DC emitter resistor
   c. Load resistor
   d. Emitter diode and external ac emitter resistance

6. The output voltage of an emitter follower is across the
   a. Emitter diode
   b. DC collector resistor
   c. Load resistor
   d. Emitter diode and external ac emitter resistance

7. If $\beta = 200$ and $r_e = 150\ \Omega$, the input impedance of the base is
   a. 30 kΩ
   b. 600 Ω
   c. 3 kΩ
   d. 5 kΩ

8. The input voltage to an emitter follower is usually
   a. Less than the generator voltage
   b. Equal to the generator voltage
   c. Greater than the generator voltage
   d. Equal to the supply voltage

9. The ac emitter current is closest to
   a. V_G divided by r_e
   b. v_{in} divided by r_e'
   c. V_G divided by r_e'
   d. v_{in} divided by r_e

10. The output voltage of an emitter follower is approximately
    a. 0
    b. V_G
    c. v_{in}
    d. V_{CC}

11. The output voltage of an emitter follower is
    a. In phase with v_{in}
    b. Much greater than v_{in}
    c. 180° out of phase
    d. Generally much less than v_{in}

12. An emitter-follower buffer is generally used when
    a. $R_G \ll R_L$
    b. $R_G = R_L$
    c. $R_L \ll R_G$
    d. R_L is very large

13. For maximum power transfer, a CC amplifier is designed so
    a. $R_G \ll z_{in}$
    b. $z_{out} \gg R_L$
    c. $z_{out} \ll R_L$
    d. $z_{out} = R_L$

14. If a CE stage is directly coupled to an emitter follower
    a. Low and high frequencies will be passed
    b. Only high frequencies will be passed
    c. High-frequency signals will be blocked
    d. Low-frequency signals will be blocked

15. If the load resistance of an emitter follower is very large, the external ac emitter resistance equals
    a. Generator resistance
    b. Impedance of the base
    c. DC emitter resistance
    d. DC collector resistance

16. If an emitter follower has $r_e' = 10\ \Omega$ and $r_e = 90\ \Omega$, the voltage gain is approximately
    a. 0
    b. 0.5
    c. 0.9
    d. 1

17. An emitter follower circuit always makes the source resistance
    a. β times smaller
    b. β times larger
    c. Equal to the load
    d. Zero

18. A Darlington transistor has
    a. A very low input impedance
    b. Three transistors
    c. A very high current gain
    d. One V_{BE} drop

19. The amplifier configuration that produces a 180° phase shift is the
    a. CB
    b. CC
    c. CE
    d. All of the above

20. If the generator voltage is 5 mV in an emitter follower, the output voltage across the load is closest to
    a. 5 mV
    b. 150 mV
    c. 0.25 V
    d. 0.5 V

21. If the load resistor of Fig. 11-1a is shorted, which of the following are different from their normal values:
    a. Only ac voltages
    b. Only dc voltages
    c. Both dc and ac voltages
    d. Neither dc nor ac voltages

22. If R_1 is open in an emitter follower, which of these is true?

a. DC base voltage is V_{CC}

b. DC collector voltage is zero

c. Output voltage is normal

d. DC base voltage is zero

23. Usually, the distortion in an emitter follower is

a. Very low

b. Very high

c. Large

d. Not acceptable

24. The distortion in an emitter follower is

a. Seldom low

b. Often high

c. Always low

d. High when clipping occurs

25. If a CE stage is direct coupled to an emitter follower, how many coupling capacitors are there between the two stages?

a. 0

b. 1

c. 2

d. 3

26. A Darlington transistor has a β of 8000. If $R_E = 1\ k\Omega$ and $R_L = 100\ \Omega$, the input impedance of the base is closest to

a. 8 kΩ c. 800 kΩ

b. 80 kΩ d. 8 MΩ

27. The ac emitter resistance of an emitter follower

a. Equals the dc emitter resistance

b. Is larger than the load resistance

c. Is β times smaller than the load resistance

d. Is usually less than the load resistance

28. A common-base amplifier has a voltage gain that is

a. Much less than one

b. Approximately equal to one

c. Greater than one

d. Zero

29. An application of a common-base amplifier is when

a. $R_{source} \gg R_L$

b. $R_{source} \ll R_L$

c. A high current gain is required

d. High frequencies need to be blocked

30. A common-base amplifier can be used when

a. Matching low to high impedances

b. A voltage gain without a current gain is required

c. A high-frequency amplifier is needed

d. All of the above

31. The zener current in a zener follower is

a. Equal to the output current

b. Smaller than the output current

c. Larger than the output current

d. Prone to thermal runaway

32. In the two-transistor voltage regulator, the output voltage

a. Is regulated

b. Has much smaller ripple than the input voltage

c. Is larger than the zener voltage

d. All of the above

Problems

SEC. 11-1   CC AMPLIFIER

11-1 In Fig. 11-21, what is the input impedance of the base if $\beta = 200$? The input impedance of the stage?

11-2 If $\beta = 150$ in Fig. 11-21, what is the ac input voltage to the emitter follower?

11-3 What is the voltage gain in Fig. 11-21? If $\beta = 175$, what is the ac load voltage?

11-4 What is the input voltage in Fig. 11-21 if β varies over a range of 50 to 300?

11-5 All resistors are doubled in Fig. 11-21. What happens to the input impedance of the stage if $\beta = 150$? To the input voltage?

11-6 What is the input impedance of the base if $\beta = 200$ in Fig. 11-22? The input impedance of the stage?

Figure 11-21

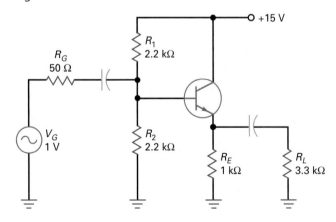

Figure 11-22

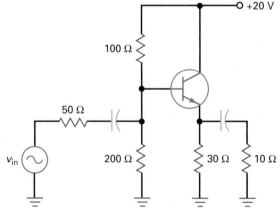

11-7 In Fig. 11-22, what is the ac input voltage to the emitter follower if $\beta = 150$ and $v_{in} = 1$ V?

11-8 What is the voltage gain in Fig. 11-22? If $\beta = 175$, what is the ac load voltage?

SEC. 11-2 OUTPUT IMPEDANCE

11-9 What is the output impedance in Fig. 11-21 if $\beta = 200$?

11-10 What is the output impedance in Fig. 11-22 if $\beta = 100$?

SEC. 11-3 CASCADING CE AND CC

11-11 What is the voltage gain of the CE stage in Fig. 11-23 if the second transistor has a dc and ac current gain of 200?

11-12 If both transistors in Fig. 11-23 have a dc and ac current gain of 150, what is the output voltage when $V_G = 10$ mV?

11-13 If both transistors have a dc and ac current gain of 200 in Fig. 11-23, what is the voltage gain of the CE stage if the load resistance drops to 125 Ω?

11-14 In Fig. 11-23, what would happen to the voltage gain of the CE amplifier if the emitter follower stage were removed and a capacitor were used to couple the ac signal to the 150 Ω load?

SEC. 11-4 DARLINGTON CONNECTIONS

11-15 If the Darlington pair of Fig. 11-24 has an overall current gain of 5000, what is the input impedance of the Q_1 base?

Figure 11-23

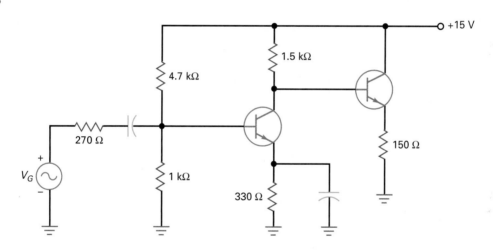

Figure 11-24

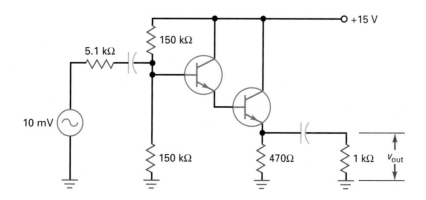

Figure 11–25

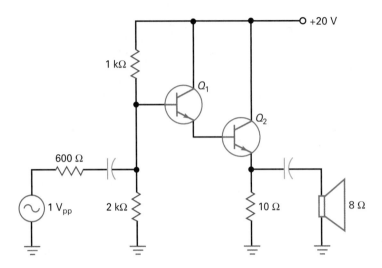

11-16 In Fig. 11-24, what is the ac input voltage to the Q_1 base if the Darlington pair has an overall current gain of 7000?

11-17 Both transistors have a β of 150 in Fig. 11-25. What is the input impedance of the first base?

11-18 In Fig. 11-25, what is the ac input voltage to the Q_1 base if the Darlington pair has an overall current gain of 2000?

SEC. 11–5   VOLTAGE REGULATION

11-19 The transistor of Fig. 11-26 has a current gain of 150. If the 1N958 has a zener voltage of 7.5 V, what is the output voltage? The zener current?

11-20 If the input voltage of Fig. 11-26 changes to 25 V, what is the output voltage? The zener current?

11-21 The potentiometer of Fig. 11-27 can vary from 0 to 1 kΩ. What is the output voltage when the wiper is at the center?

Figure 11-26

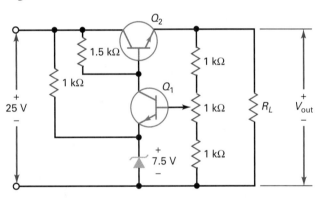

Figure 11-27

11-22 What is the output voltage in Fig. 11-27 if the wiper is all the way up? If it is all the way down?

SEC. 11–6   COMMON–BASE AMPLIFIER

11-23 In Fig. 11-28, what is the Q point emitter current?

11-24 What is the approximate voltage gain of Fig. 11-28?

11-25 In Fig. 11-28, what is the input impedance looking into the emitter? What is the input impedance of the stage?

11-26 In Fig. 11-28, with an input of 2 mV from the generator, what is the value of v_{out}?

11-27 In Fig. 11-28, if the V_{CC} supply voltage were increased to 15 V, what would v_{out} equal?

Figure 11-28

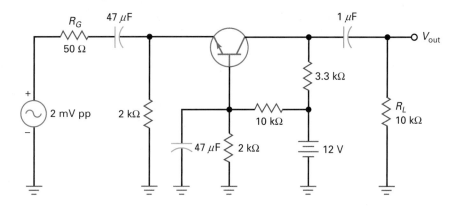

Critical Thinking

11-28 In Fig. 11-26, what is the power dissipation of the transistor if the current gain is 100 and the zener voltage is 7.5 V?

11-29 In Fig. 11-28, the transistor has a β_{dc} of 150. Calculate the following dc quantities: V_B, V_E, V_C, I_E, I_C, and I_B.

11-30 If an input signal with a peak-to-peak value of 5 mV drives the circuit of Fig. 11-29a, what are the two ac output voltages? What do you think is the purpose of this circuit?

11-31 Figure 11-29b shows a circuit in which the control voltage can be 0 V or +5 V. If the audio input voltage is 10 mV, what is the audio output voltage when the control voltage is 0 V? When the control voltage is +5 V? What do you think this circuit is supposed to do?

11-32 In Fig. 11-26, what would the output voltage be if the zener diode opened?

11-33 In Fig. 11-26, if the 33 Ω load shorts, what is the transistor's power dissipation?

11-34 In Fig. 11-27, what is the power dissipation of Q_2 when the wiper is at the center and the load resistance is 100 Ω?

11-35 Using Fig. 11-24, if both transistors have a β of 100, what is the approximate output impedance of the amplifier?

11-36 In Fig. 11-23, if the input voltage from the generator were 100 mV pp and the emitter-bypass capacitor opened, what would the output voltage across the load be?

11-37 In Fig. 11-28, what would be the output voltage if the base-bypass capacitor shorted?

Figure 11-29

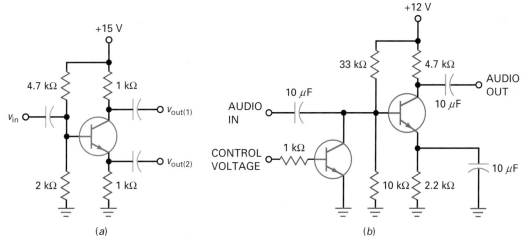

(a)　　　　　　　　　　　　　(b)

Troubleshooting

Use Fig. 11-30 for the remaining problems. The table labeled "Ac Millivolts" contains the measurements of the ac voltages expressed in millivolts. For this exercise, all resistors are OK. The troubles are limited to open capacitors, open connecting wires, and open transistors.

11-38 Find Troubles T1 to T3.

11-39 Find Troubles T4 to T7.

Figure 11-30

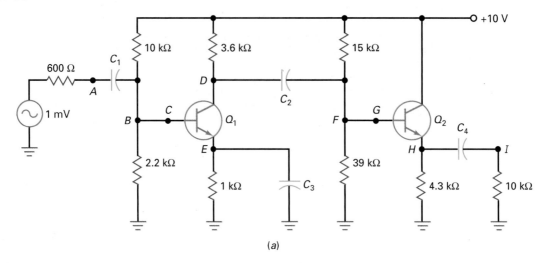

(a)

Ac Millivolts

Trouble	V_A	V_B	V_C	V_D	V_E	V_F	V_G	V_H	V_I
OK	0.6	0.6	0.6	70	0	70	70	70	70
T1	0.6	0.6	0.6	70	0	70	70	70	0
T2	0.6	0.6	0.6	70	0	70	0	0	0
T3	1	0	0	0	0	0	0	0	0
T4	0.75	0.75	0.75	2	0.75	2	2	2	2
T5	0.75	0.75	0	0	0	0	0	0	0
T6	0.6	0.6	0.6	95	0	0	0	0	0
T7	0.6	0.6	0.6	70	0	70	70	0	0

(b)

Job Interview Questions

1. Draw the schematic diagram of an emitter follower. Tell me why this circuit is widely used in power amplifiers and voltage regulators.
2. Tell me all that you know about the output impedance of an emitter follower.
3. Draw a Darlington pair and explain why the overall current gain is the product of the individual current gains.
4. Draw a zener follower and explain why it regulates the output voltage against changes in the input voltage.
5. What is the voltage gain of an emitter follower? This being the case, in what applications would such a circuit be useful?
6. Explain why a Darlington pair has a higher power gain than a single transistor.

7. Why are "follower" circuits so important in acoustic circuits?
8. What is the approximate ac voltage gain for a CC amplifier?
9. What is another name for a common-collector amplifier?
10. What is the relationship between an ac signal phase (output to input) and a common-collector amplifier?
11. If a technician measures unity voltage gain (output voltage divided by input voltage) from a CC amplifier, what is the problem?
12. The Darlington amplifier is used in the final power amplifier (FPA) in most higher-quality audio amplifiers because it increases the power gain. How does a Darlington amplifier increase the power gain?

Self–Test Answers

1.	b	12.	c	23.	a
2.	c	13.	d	24.	d
3.	b	14.	a	25.	a
4.	c	15.	c	26.	c
5.	d	16.	c	27.	d
6.	c	17.	a	28.	c
7.	a	18.	c	29.	b
8.	a	19.	c	30.	d
9.	d	20.	a	31.	b
10.	c	21.	a	32.	d
11.	a	22.	d		

Practice Problem Answers

11-1 $z_{in(base)} = 303 \text{ k}\Omega$;
$z_{in(stage)} = 4.92 \text{ k}\Omega$

11-2 $v_{in} \approx 0.893 \text{ V}$

11-3 $v_{in} = 0.979 \text{ V}$;
$v_{out} = 0.974 \text{ V}$

11-4 $z_{out} = 3.33 \ \Omega$

11-5 $z_{out} = 2.86 \ \Omega$

11-6 $A_v = 222$

11-7 $A_v = 6.28$

11-8 $\beta = 5625$;
$I_{B1} = 14.3 \ \mu\text{A}$;
$z_{in(base)} = 112.5 \text{ k}\Omega$

11-9 $V_{out} = 7.5 \text{ V}$;
$I_z = 5 \text{ mA}$

11-10 $V_{out} = 18.9 \text{ V}$

11-11 $V_{out} = 76.9 \text{ mVpp}$

12 Power Amplifiers

In a stereo, radio, or television, the input signal is small. After several stages of voltage gain, however, the signal becomes large and uses the entire load line. In these later stages of a system, the collector currents are much larger because the load impedances are much smaller. Stereo amplifier speakers, for example, may have an impedance of 8 Ω or less.

As indicated in Chap. 6, small-signal transistors have a power rating of less than 1 W, whereas power transistors have a power rating of more than 1 W. Small-signal transistors are typically used at the front end of systems where the signal power is low, and power transistors are used near the end of systems because the signal power and current are high.

Objectives

After studying this chapter, you should be able to:

- Show how the dc load line, ac load line, and Q point are determined for CE and CC power amplifiers.

- Calculate the maximum peak-to-peak (MPP) unclipped ac voltage that is possible with CE and CC power amplifiers.

- Describe the characteristics of amplifiers, including classes of operation, types of coupling, and frequency ranges.

- Draw a schematic of class B/AB push-pull amplifier and explain its operation.

- Determine the efficiency of transistor power amplifiers.

- Discuss the factors that limit the power rating of a transistor and what can be done to improve the power rating.

Chapter Outline

Vocabulary

ac output compliance
ac load line
audio amplifier
bandwidth (BW)
capacitive coupling
class A operation
class AB operation
class B operation
class C operation
compensating diodes

crossover distortion
current drain
direct coupling
driver stage
duty cycle
efficiency
harmonics
large-signal operation
narrowband amplifier
power amplifier

power gain
preamp
push-pull circuit
radio-frequency (RF) amplifier
thermal runaway
transformer coupling
tuned RF amplifier
wideband amplifier

12-1 Amplifier Terms

There are different ways to describe amplifiers. For instance, we can describe them by their class of operation, by their interstage coupling, or by their frequency range.

Classes of Operation

Class A operation of an amplifier means that the transistor operates in the active region at all times. This implies that collector current flows for 360° of the ac cycle, as shown in Fig. 12-1a. With a class A amplifier, the designer usually tries to locate the Q point somewhere near the middle of the load line. This way, the signal can swing over the maximum possible range without saturating or cutting off the transistor, which would distort the signal.

Class B operation is different. It means that collector current flows for only half the cycle (180°), as shown in Fig. 12-1b. To have this kind of operation, a designer locates the Q point at cutoff. Then, only the positive half of ac base voltage can produce collector current. This reduces the wasted heat in power transistors.

Class C operation means that collector current flows for less than 180° of the ac cycle, as shown in Fig. 12-1c. With class C operation, only part of the positive half cycle of ac base voltage produces collector current. As a result, we get brief pulses of collector current like those of Fig. 12-1c.

Types of Coupling

Figure 12-2a shows **capacitive coupling.** The coupling capacitor transmits the amplified ac voltage to the next stage. Figure 12-2b illustrates **transformer coupling.** Here the ac voltage is coupled through a transformer to the next stage. Capacitive coupling and transformer coupling are both examples of ac coupling, which blocks the dc voltage.

Direct coupling is different. In Fig. 12-2c, there is a direct connection between the collector of the first transistor and the base of the second transistor.

Figure 12-1 Collector current: (a) class A; (b) class B; (c) class C.

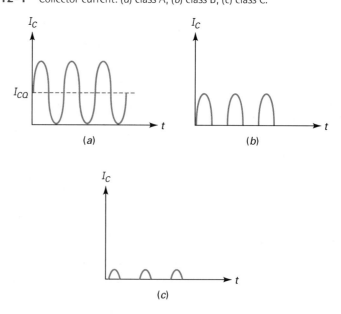

Figure 12-2 Types of coupling: (*a*) capacitive; (*b*) transformer; (*c*) direct.

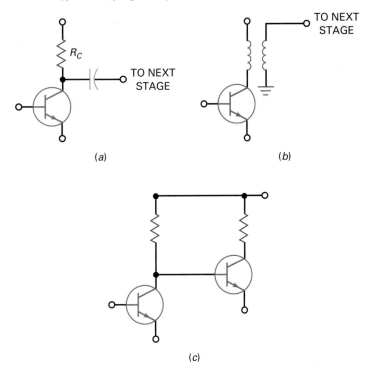

(*a*)　　　　　　　　(*b*)

(*c*)

Because of this, both the dc and the ac voltages are coupled. Since there is no lower frequency limit, a direct-coupled amplifier is sometimes called a *dc amplifier.*

Ranges of Frequency

Another way to describe amplifiers is by stating their frequency range. For instance, an **audio amplifier** refers to an amplifier that operates in the range of 20 Hz to 20 kHz. On the other hand, a **radio-frequency (RF) amplifier** is one that amplifies frequencies above 20 kHz, usually much higher. For instance, the RF amplifiers in AM radios amplify frequencies between 535 and 1605 kHz, and the RF amplifiers in FM radios amplify frequencies between 88 and 108 MHz.

Amplifiers are also classified as **narrowband** or **wideband.** A narrowband amplifier works over a small frequency range like 450 to 460 kHz. A wideband amplifier operates over a large frequency range like 0 to 1 MHz.

Narrowband amplifiers are usually **tuned RF amplifiers,** which means that their ac load is a high-Q resonant tank tuned to a radio station or television channel. Wideband amplifiers are usually untuned; that is, their ac load is resistive.

Figure 12-3*a* is an example of a tuned RF amplifier. The *LC* tank is resonant at some frequency. If the tank has a high Q, the bandwidth is narrow. The output is capacitively coupled to the next stage.

Figure 12-3*b* is another example of a tuned RF amplifier. This time, the narrowband output signal is transformer-coupled to the next stage.

Signal Levels

We have already described *small-signal operation,* in which the peak-to-peak swing in collector current is less than 10 percent of quiescent collector current. In **large-signal operation,** a peak-to-peak signal uses all or most of the load line. In a stereo system, the small signal from a radio tuner, tape player, or compact disc

Figure 12-3 Tuned RF amplifiers: (*a*) capacitive coupling; (*b*) transformer coupling.

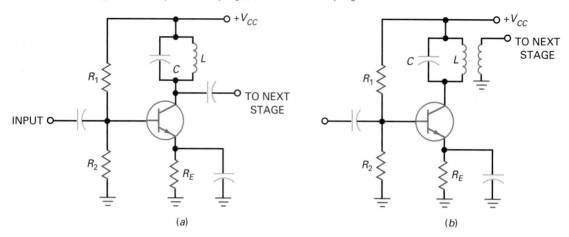

(a) (b)

player is used as the input to a **preamp,** an amplifier that produces a larger output suitable for driving tone and volume controls. The signal is then used as the input to a **power amplifier,** which produces output power ranging from a few hundred milliwatts up to hundreds of watts.

In the remainder of this chapter, we will discuss power amplifiers and related topics like the ac load line, power gain, and efficiency.

12-2 Two Load Lines

Every amplifier has a dc equivalent circuit and an ac equivalent circuit. Because of this, it has two load lines: a dc load line and an ac load line. For small-signal operation, the location of the Q point is not critical. But with large-signal amplifiers, the Q point has to be at the middle of the ac load line to get the maximum possible output swing.

DC Load Line

Figure 12-4*a* is a voltage-divider-based (VDB) amplifier. One way to move the Q point is by varying the value of R_2. For very large values of R_2, the transistor goes into saturation and its current is given by:

$$I_{C(\text{sat})} = \frac{V_{CC}}{R_C + R_E} \tag{12-1}$$

Very small values of R_2 will drive the transistor into cutoff, and its voltage is given by:

$$V_{CE(\text{cutoff})} = V_{CC} \tag{12-2}$$

Figure 12-4*b* shows the dc load line with the Q point.

AC Load Line

Figure 12-4*c* is the ac equivalent circuit for the VDB amplifier. With the emitter at ac ground, R_E has no effect on the ac operation. Furthermore, the ac collector resistance is less than the dc collector resistance. Therefore, when an ac signal comes in, the instantaneous operating point moves along the **ac load line** of Fig. 12-4*d*. In other words, the peak-to-peak sinusoidal current and voltage are determined by the ac load line.

Figure 12-4 (*a*) VDB amplifier; (*b*) dc load line; (*c*) ac equivalent circuit; (*d*) ac load line.

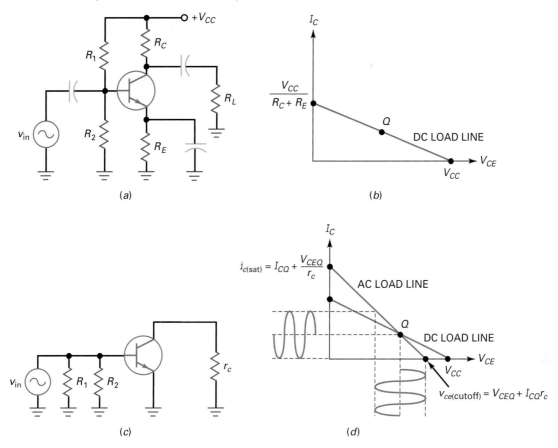

(a)

(b)

(c)

(d)

As shown in Fig. 12-4*d*, the saturation and cutoff points on the ac load line differ from those on the dc load line. Because the ac collector and emitter resistance are lower than the respective dc resistance, the ac load line is much steeper. It's important to note that the ac and dc load lines intersect at the Q point. This happens when the ac input voltage is crossing zero.

Here's how to determine the ends of the ac load line. Writing a collector voltage loop gives us:

$$v_{ce} + i_c r_c = 0$$

or

$$i_c = -\frac{v_{ce}}{r_c} \qquad (12\text{-}3)$$

The ac collector current is given by:

$$i_c = \Delta I_C = I_C - I_{CQ}$$

and the ac collector voltage is:

$$v_{ce} = \Delta V_{CE} = V_{CE} - V_{CEQ}$$

When substituting these expressions into Eq. (12-3) and rearranging, we arrive at:

$$I_C = I_{CQ} + \frac{V_{CEQ}}{r_c} - \frac{V_{CE}}{r_c} \qquad (12\text{-}4)$$

This is the equation of the ac load line. When the transistor goes into saturation, V_{CE} is zero, and Eq. (12-4) gives us:

$$i_{c(sat)} = I_{CQ} + \frac{V_{CEQ}}{r_c} \qquad (12\text{-}5)$$

where $i_{c(sat)}$ = ac saturation current
I_{CQ} = dc collector current
V_{CEQ} = dc collector-emitter voltage
r_c = ac resistance seen by the collector

When the transistor goes into cutoff, I_c equals zero. Since

$$v_{ce(cutoff)} = V_{CEQ} + \Delta V_{CE}$$

and

$$\Delta V_{CE} = (\Delta I_C)(r_c)$$

we can substitute to get:

$$\Delta V_{CE} = (I_{CQ} - 0A)(r_c)$$

resulting in:

$$v_{ce(cutoff)} = V_{CEQ} + I_{CQ}r_c \qquad (12\text{-}6)$$

Because the ac load line has a higher slope than the dc load line, the maximum peak-to-peak (MPP) output is always less than the supply voltage. As a formula:

$$\text{MPP} < V_{CC} \qquad (12\text{-}7)$$

For instance, if the supply voltage is 10 V, the maximum peak-to-peak sinusoidal output is less than 10 V.

Clipping of Large Signals

When the Q point is at the center of the dc load line (Fig. 12-4d), the ac signal cannot use all of the ac load line without clipping. For instance, if the ac signal increases, we will get the cutoff clipping shown in Fig. 12-5a.

If the Q point is moved higher as shown in Fig. 12-5b, a large signal will drive the transistor into saturation. In this case, we get saturation clipping. Both cutoff and saturation clipping are undesirable because they distort the signal. When a distorted signal like this drives a loudspeaker, it sounds terrible.

A well-designed large-signal amplifier has the Q point at the middle of the ac load line (Fig. 12-5c). In this case, we get a maximum peak-to-peak unclipped output. This maximum unclipped peak-to-peak ac voltage is also referred to its **ac output compliance.**

Maximum Output

When the Q point is below the center of the ac load line, the maximum peak (MP) output is $I_{CQ}r_c$, as shown in Fig. 12-6a. On the other hand, if the Q point is above the center of the ac load line, the maximum peak output is V_{CEQ}, as shown in Fig. 12-6b.

For any Q point, therefore, the maximum peak output is:

$$\text{MP} = I_{CQ}r_c \quad \text{or} \quad V_{CEQ}, \quad \text{whichever is smaller} \qquad (12\text{-}8)$$

and the maximum peak-to-peak output is twice this amount:

$$\text{MPP} = 2\text{MP} \qquad (12\text{-}9)$$

Equations (12-8) and (12-9) are useful in troubleshooting to determine the largest unclipped output that is possible.

Figure 12-5 (a) Cutoff clipping; (b) saturation clipping; (c) optimum Q point.

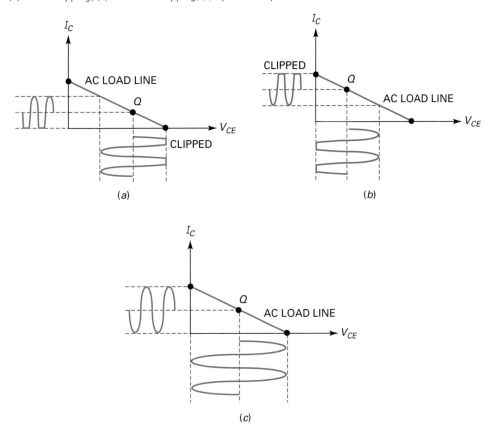

(a)

(b)

(c)

Figure 12-6 Q point at center of ac load line.

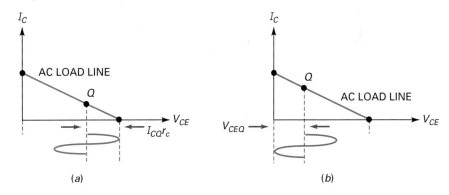

(a)

(b)

When the Q point is at the center of the ac load line:

$$I_{CQ}r_c = V_{CEQ} \tag{12-10}$$

A designer will try to satisfy this condition as closely as possible, given the tolerance of biasing resistors. The circuit's emitter resistance can be adjusted to find the optimum Q point. A formula that can be derived for the optimum emitter resistance is:

$$R_E = \frac{R_C + r_c}{V_{CC}/V_E - 1} \tag{12-11}$$

Example 12-1

|||| MultiSim

What are the values of I_{CQ}, V_{CEQ} and r_c in Fig. 12-7?

Figure 12-7   Example.

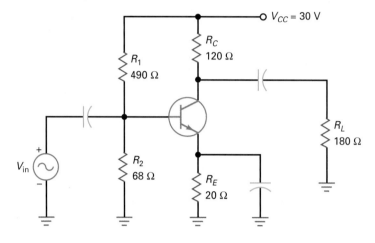

SOLUTION

$$V_B = \frac{68\ \Omega}{68\ \Omega + 490\ \Omega}\ (30\ \text{V}) = 3.7\ \text{V}$$

$$V_E = V_B - 0.7\ \text{V} = 3.7\ \text{V} - 0.7\ \text{V} = 3\ \text{V}$$

$$I_E = \frac{V_E}{R_E} = \frac{3\ \text{V}}{20\ \Omega} = 150\ \text{mA}$$

$$I_{CQ} \cong I_E = 150\ \text{mA}$$

$$V_{CEQ} = V_C - V_E = 12\ \text{V} - 3\ \text{V} = 9\ \text{V}$$

$$r_c = R_C \,\|\, R_L = 120\ \Omega \,\|\, 180\ \Omega\ = 72\ \Omega$$

PRACTICE PROBLEM 12-1   In Fig. 12-7, change R_E from 20 Ω to 30 Ω. Solve for I_{CQ} and V_{CEQ}.

Example 12-2

Determine the ac load line saturation and cutoff points in Fig. 12-7. Also, find the maximum peak-to-peak output voltage.

SOLUTION   From Example 12-1, the transistor's Q point is:

$$I_{CQ} = 150\ \text{mA} \quad \text{and} \quad V_{CEQ} = 9\ \text{V}$$

To find the ac saturation and cutoff points, first determine the ac collector resistance, r_c:

$$r_c = R_C \,\|\, R_L = 120\ \Omega \,\|\, 180\ \Omega = 72\ \Omega$$

Next find the ac load line end points:

$$i_{c(\text{sat})} = I_{CQ} + \frac{V_{CEQ}}{r_c} = 150\ \text{mA} + \frac{9\ \text{V}}{72\ \Omega} = 275\ \text{mA}$$

$$v_{ce(\text{cutoff})} = V_{CEQ} + I_{CQ}r_c = 9\ \text{V} + (150\ \text{mA})(72\ \Omega) = 19.8\ \text{V}$$

Now determine the MPP value. With a supply voltage of 30 V:

MPP < 30 V

MP will be the smaller value of:

$I_{CQ}r_c = (150\ \text{mA})(72\ \Omega) = 10.8\ \text{V}$

or

$V_{CEQ} = 9\ \text{V}$

Therefore, MPP = 2 (9 V) = 18 V

PRACTICE PROBLEM 12-2 Using Example 12-2, change R_E to 30 Ω and find $i_{c(\text{sat})}$, $v_{ce(\text{cutoff})}$, and MPP.

12-3 Class A Operation

The VDB amplifier of Fig. 12-8a is a class A amplifier as long as the output signal is not clipped. With this kind of amplifier, collector current flows throughout the cycle. Stated another way, no clipping of the output signal occurs at any time during the cycle. Now, we discuss a few equations that are useful in the analysis of class A amplifiers.

Power Gain

Besides voltage gain, any amplifier has a **power gain,** defined as:

$$A_p = \frac{p_{\text{out}}}{p_{\text{in}}} \tag{12-12}$$

In words, the power gain equals the ac output power divided by the ac input power.

Figure 12-8 Class A amplifier.

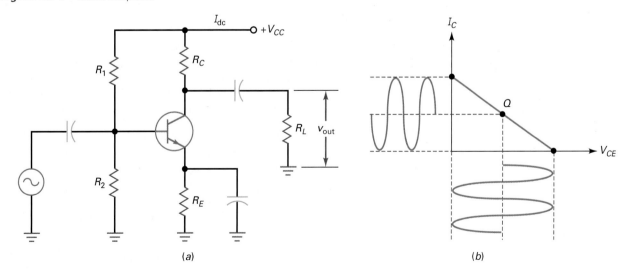

(a)

(b)

For instance, if the amplifier of Fig. 12-8a has an output power of 10 mW and an input power of 10 μW, it has a power gain of:

$$A_p = \frac{10\text{ mW}}{10\ \mu\text{W}} = 1000$$

Output Power

If we measure the output voltage of Fig. 12-8a in rms volts, the output power is given by

$$p_{\text{out}} = \frac{v_{\text{rms}}^2}{R_L} \qquad (12\text{-}13)$$

Usually, we measure the output voltage in peak-to-peak volts with an oscilloscope. In this case, a more convenient equation to use for output power is:

$$p_{\text{out}} = \frac{v_{\text{out}}^2}{8R_L} \qquad (12\text{-}14)$$

The factor of 8 in the denominator occurs because $v_{\text{pp}} = 2\sqrt{2}\ v_{\text{rms}}$. When you square $2\sqrt{2}$, you get 8.

The maximum output power occurs when the amplifier is producing the maximum peak-to-peak output voltage, as shown in Fig. 12-8b. In this case, v_{pp} equals the maximum peak-to-peak output voltage and the maximum output power is:

$$p_{\text{out(max)}} = \frac{\text{MPP}^2}{8R_L} \qquad (12\text{-}15)$$

Transistor Power Dissipation

When no signal drives the amplifier of Fig. 12-8a, the quiescent power dissipation is:

$$P_{DQ} = V_{CEQ}I_{CQ} \qquad (12\text{-}16)$$

This makes sense. It says that the quiescent power dissipation equals the dc voltage times the dc current.

When a signal is present, the power dissipation of a transistor decreases because the transistor converts some of the quiescent power to signal power. For this reason, the quiescent power dissipation is the worst case. Therefore, the power rating of a transistor in a class A amplifier must be greater than P_{DQ}; otherwise, the transistor will be destroyed.

Current Drain

As shown in Fig. 12-8a, the dc voltage source has to supply a dc current I_{dc} to the amplifier. This dc current has two components: the biasing current through the voltage divider and the collector current through the transistor. The dc current is called the **current drain** of the stage. If you have a multistage amplifier, you have to add the individual current drains to get the total current drain.

Efficiency

The dc power supplied to an amplifier by the dc source is:

$$P_{\text{dc}} = V_{CC}I_{\text{dc}} \qquad (12\text{-}17)$$

To compare the design of power amplifiers, we can use the **efficiency,** defined by:

$$\eta = \frac{p_{\text{out}}}{P_{\text{dc}}} \times 100\% \qquad (12\text{-}18)$$

GOOD TO KNOW

Efficiency can also be defined as the amplifier's ability to convert its dc input power to useful ac output power.

This equation says that the efficiency equals the ac output power divided by the dc input power.

The efficiency of any amplifier is between 0 and 100 percent. Efficiency gives us a way to compare two different designs because it indicates how well an amplifier converts the dc input power to ac output power. The higher the efficiency, the better the amplifier is at converting dc power to ac power. This is important in battery-operated equipment because high efficiency means that the batteries last longer.

Since all resistors except the load resistor waste power, the efficiency is less than 100 percent in a class A amplifier. In fact, it can be shown that the maximum efficiency of a class A amplifier with a dc collector resistance and a separate load resistance is 25 percent.

In some applications, the low efficiency of class A is acceptable. For instance, the small-signal stages near the front of a system usually work fine with low efficiency because the dc input power is small. In fact, if the final stage of a system needs to deliver only a few hundred milliwatts, the current drain on the power supply may still be low enough to accept. But when the final stage needs to deliver watts of power, the current drain usually becomes too large with class A operation.

Example 12–3

If the peak-to-peak output voltage is 18 V and the input impedance of the base is 100 Ω, what is the power gain in Fig. 12-9a?

Figure 12–9 Example.

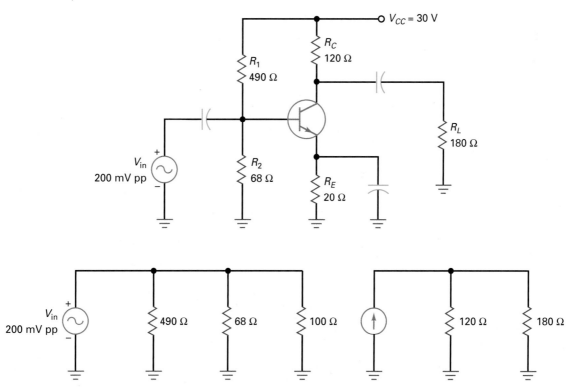

SOLUTION As shown in Fig. 12-9*b*:

$$z_{\text{in(stage)}} = 490 \ \Omega \parallel 68 \ \Omega \parallel 100 \ \Omega = 37.4 \ \Omega$$

The ac input power is:

$$P_{\text{in}} = \frac{(200 \text{ mV})^2}{8 \ (37.4)} = 133.7 \ \mu\text{W}$$

The ac output power is:

$$P_{\text{out}} = \frac{(18 \text{ V})^2}{8 \ (180 \ \Omega)} = 225 \text{ mW}$$

The power gain is:

$$A_p = \frac{225 \text{ mW}}{133.7 \ \mu\text{W}} = 1683$$

PRACTICE PROBLEM 12-3 In Fig. 12-9*a*, if R_L is 120 Ω and the peak-to-peak output voltage equals 12 V, what is the power gain?

Example 12-4

||| MultiSim

What is the transistor power dissipation and efficiency of Fig. 12-9*a*?

SOLUTION The dc emitter current is:

$$I_E = \frac{3 \text{ V}}{20 \ \Omega} = 150 \text{ mA}$$

The dc collector voltage is:

$$V_C = 30 \text{ V} - (150 \text{ mA})(120 \ \Omega) = 12 \text{ V}$$

and the dc collector-emitter voltage is:

$$V_{CEQ} = 12 \text{ V} - 3 \text{ V} = 9 \text{ V}$$

The transistor power dissipation is:

$$P_{DQ} = V_{CEQ} I_{CQ} = (9 \text{ V})(150 \text{ mA}) = 1.35 \text{ W}$$

To find the stage efficiency:

$$I_{\text{bias}} = \frac{30 \text{ V}}{490 \ \Omega + 68 \ \Omega} = 53.8 \text{ mA}$$

$$I_{\text{dc}} = I_{\text{bias}} + I_{CQ} = 53.8 \text{ mA} + 150 \text{ mA} = 203.8 \text{ mA}$$

The dc input power to the stage is:

$$P_{\text{dc}} = V_{CC} I_{\text{dc}} = (30 \text{ V})(203.8 \text{ mA}) = 6.11 \text{ W}$$

Since the output power (found in Example 12-3) is 225 mW, the efficiency of the stage is:

$$\eta = \frac{225 \text{ mW}}{6.11 \text{ W}} \times 100\% = 3.68\%$$

Example 12-5

Describe the action of Fig. 12-10.

Figure 12-10 Class A power amplifier.

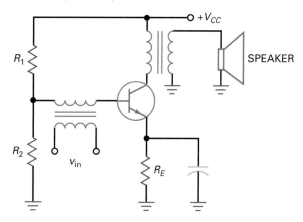

SOLUTION This is a class A power amplifier driving a loudspeaker. The amplifier uses voltage-divider bias, and the ac input signal is transformer-coupled to the base. The transistor produces voltage and power gain to drive the loudspeaker through the output transformer.

A small speaker with an impedance of 3.2 Ω needs only 100 mW in order to operate. A slightly larger speaker with an impedance of 8 Ω needs 300 to 500 mW for proper operation. Therefore, a class A power amplifier like Fig. 12-10 may be adequate if all you need is a few hundred milliwatts of output power. Since the load resistance is also the ac collector resistance, the efficiency of this class A amplifier is higher than that of the class A amplifier discussed earlier. Using the impedance-reflecting ability of the transformer, the speaker load resistance appears $\left(\dfrac{N_P}{N_S}\right)^2$ times larger at the collector. If the transformer's turns ratio were 10:1, a 3.2 Ω speaker would appear as 320 Ω at the collector.

The class A amplifier discussed earlier had a separate collector resistance R_C and a separate load resistance R_L. The best you can do in this case is to match the impedances, $R_L = R_C$, to get a maximum efficiency of 25 percent. When the load resistance becomes the ac collector resistor, as shown in Fig. 12-10, it receives twice as much output power, and the maximum efficiency increases to 50 percent.

PRACTICE PROBLEM 12-5 In Fig. 12-10, what resistance would an 8 Ω speaker appear to the collector as, if the transformer's turns ratio were 5 : 1?

Emitter-Follower Power Amplifier

When the emitter follower is used as class A power amplifier at the end of a system, a designer will usually locate the Q point at the center of the ac load line to get maximum peak-to-peak (MPP) output.

Figure 12-11 DC and ac load lines.

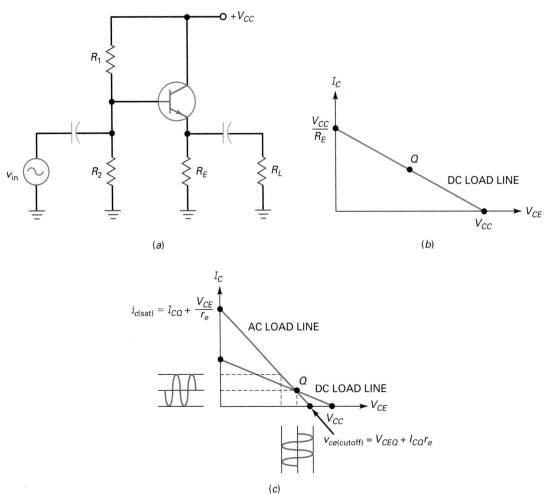

(a)

(b)

(c)

In Fig. 12-11a, large values of R_2 will saturate the transistor, producing a saturation current of:

$$I_{C(\text{sat})} = \frac{V_{CC}}{R_E} \tag{12-19}$$

Small values of R_2 will drive the transistor into cutoff, producing a cutoff voltage of:

$$V_{CE(\text{cutoff})} = V_{CC} \tag{12-20}$$

Fig. 12-11b shows the dc load line with the Q point.

In Fig. 12-11a, the ac emitter resistance is less than the dc emitter resistance. Therefore, when an ac signal comes in, the instantaneous operating point moves along the ac load line of Fig. 12-11c. The peak-to-peak sinusoidal current and voltage are determined by the ac load line.

As shown in Fig. 12-11c, the ac load line end points are found by:

$$i_{c(\text{sat})} = I_{CQ} + \frac{V_{CE}}{r_e} \tag{12-21}$$

and

$$V_{CE(\text{cutoff})} = V_{CE} + I_{CQ}\, r_e \tag{12-22}$$

Figure 12-12 Maximum peak excursions.

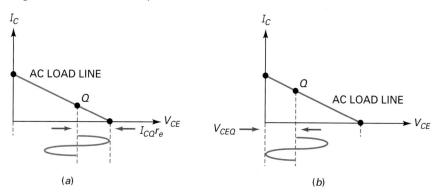

(a)   (b)

Because the ac load line has a higher slope than the dc load line, the maximum peak-to-peak output is always less than the supply voltage. As with the class A CE amplifier, MPP < V_{CC}.

When the Q point is below the center of the ac load line, the maximum peak (MP) output is $I_{CQ}r_e$, as shown in Fig. 12-12a. On the other hand, if the Q point is above the center of the load line, the maximum peak output is V_{CEQ}, as shown in Fig. 12-12b.

As you can see, determining the MPP value for an emitter-follower amplifier is essentially the same as for a CE amplifier. The difference is the need to use the emitter ac resistance, r_e, instead of the collector ac resistance, r_c. To increase the output power level, the emitter follower may also be connected in a Darlington configuration.

Example 12-6

IIII MultiSim

What are the values of I_{CQ}, V_{CEQ}, and r_e in Fig. 12-13?

Figure 12-13 Emitter-follower power amplifier.

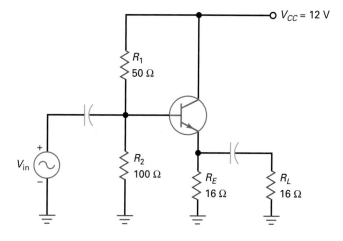

SOLUTION

$$I_{CQ} = \frac{8\text{ V} - 0.7\text{ }v}{16\text{ }\Omega} = 456\text{ mA}$$

$$V_{CEQ} = 12\text{ V} - 7.3\text{ V} = 4.7\text{ V}$$

and

$$r_e = 16\text{ }\Omega \parallel 16\text{ }\Omega = 8\text{ }\Omega$$

PRACTICE PROBLEM 12-6   In Fig. 12-13, change R_1 to 100 Ω and find I_{CQ}, V_{CEQ}, and r_e.

Example 12-7

Determine the ac saturation and cutoff points in Fig. 12-13. Also, find the circuit's MPP output voltage.

SOLUTION   From Example 12-6, the dc Q point is:

$$I_{CQ} = 456\text{ mA}\quad\text{and}\quad V_{CEQ} = 4.7\text{ V}$$

The ac load line saturation and cutoff points are found by:

$$r_e = R_C \parallel R_L = 16\text{ }\Omega \parallel 16\text{ }\Omega = 8\text{ }\Omega$$

$$i_{c(\text{sat})} = I_{CQ} + \frac{V_{CE}}{r_e} = 456\text{ mA} + \frac{4.7\text{ V}}{8\text{ }\Omega} = 1.04\text{ A}$$

$$v_{ce(\text{cutoff})} = V_{CEQ} + I_{CQ}r_e = 4.7\text{ V} + (456\text{ mA})(8\text{ }\Omega) = 8.35\text{ V}$$

MPP is found by determining the smaller value of:

$$\text{MP} = I_{CQ}r_e = (456\text{ mA})(8\text{ }\Omega) = 3.65\text{ V}$$

or

$$\text{MP} = V_{CEQ} = 4.7\text{ V}$$

Therefore, MPP = 2 (3.65 V) = 7.3 V_{pp}.

PRACTICE PROBLEM 12-7   In Fig. 12-13, if $R_1 = 100\text{ }\Omega$, solve for its MPP value.

12-4 Class B Operation

Class A is the common way to run a transistor in linear circuits because it leads to the simplest and most stable biasing circuits. But class A is not the most efficient way to operate a transistor. In some applications, like battery-powered systems, current drain and stage efficiency become important considerations in the design. This section introduces the basic idea of class B operation.

Push–Pull Circuit

Figure 12-14 shows a basic class B amplifier. When a transistor operates as class B, it clips off half a cycle. To avoid the resulting distortion, we can use two

Figure 12-14 Class B push-pull amplifier.

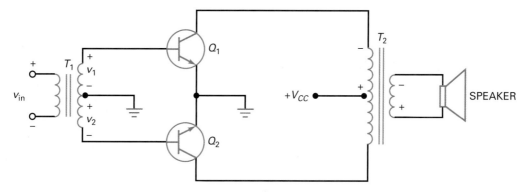

transistors in a push-pull arrangement like that of Fig. 12-14. **Push-pull** means that one transistor conducts for half a cycle while the other is off, and vice versa.

Here is how the circuit works: On the positive half cycle of input voltage, the secondary winding of T_1 has voltage v_1 and v_2, as shown. Therefore, the upper transistor conducts and the lower one cuts off. The collector current through Q_1 flows through the upper half of the output primary winding. This produces an amplified and inverted voltage, which is transformer-coupled to the loudspeaker.

On the next half cycle of input voltage, the polarities reverse. Now, the lower transistor turns on and the upper transistor turns off. The lower transistor amplifies the signal, and the alternate half cycle appears across the loudspeaker.

Since each transistor amplifies one-half of the input cycle, the loudspeaker receives a complete cycle of the amplified signal.

Advantages and Disadvantages

Since there is no bias in Fig. 12-14, each transistor is at cutoff when there is no input signal, an advantage because there is no current drain when the signal is zero.

Another advantage is improved efficiency where there is an input signal. The maximum efficiency of a class B push-pull amplifier is 78.5 percent, so a class B push-pull power amplifier is more commonly used for an output stage than a class A power amplifier.

The main disadvantage of the amplifier shown in Fig. 12-14 is the use of transformers. Audio transformers are bulky and expensive. Although widely used at one time, a transformer-coupled amplifier like Fig. 12-14 is no longer popular. Newer designs have eliminated the need for transformers in most applications.

12-5 Class B Push–Pull Emitter Follower

Class B operation means that the collector current flows for only 180° of the ac cycle. For this to occur, the Q point is located at cutoff on both the dc and the ac load lines. The advantage of class B amplifiers is lower current drain and higher stage efficiency.

Push–Pull Circuit

Figure 12-15*a* shows one way to connect a class B push-pull emitter follower. Here, we have an *npn* emitter follower and a *pnp* emitter follower connected in a push-pull arrangement.

Let's begin the analysis with the dc equivalent circuit of Fig. 12-15*b*. The designer selects biasing resistors to set the Q point at cutoff. This biases the

Figure 12-15 Class B push-pull emitter follower: (*a*) complete circuit; (*b*) dc equivalent circuit.

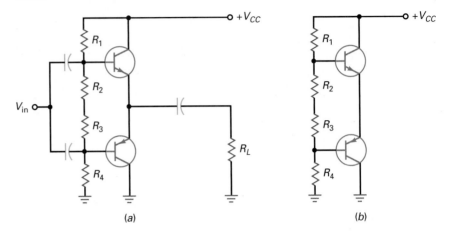

(a)   (b)

emitter diode of each transistor between 0.6 and 0.7 V, so that it is on the verge of conduction. Ideally:

$$I_{CQ} = 0$$

Because the biasing resistors are equal, each emitter diode is biased with the same value of voltage. As a result, half the supply voltage is dropped across each transistor's collector-emitter terminals. That is:

$$V_{CEQ} = \frac{V_{CC}}{2} \qquad (12\text{-}23)$$

DC Load Line

Since there is no dc resistance in the collector or emitter circuits of Fig. 12-15*b*, the dc saturation current is infinite. This means that the dc load line is vertical, as shown in Fig. 12-16*a*. If you think that this is a dangerous situation, you are right. The most difficult thing about designing a class B amplifier is setting up a stable *Q* point at cutoff. Any significant decrease in V_{BE} with temperature can move the *Q* point up the dc load line to dangerously high currents. For the moment, assume that the *Q* point is rock-solid at cutoff, as shown in Fig. 12-16*a*.

AC Load Line

Figure 12-16*a* shows the ac load line. When either transistor is conducting, its operating point moves up along the ac load line. The voltage swing of the

Figure 12-16 (*a*) DC and ac load lines; (*b*) ac equivalent circuit.

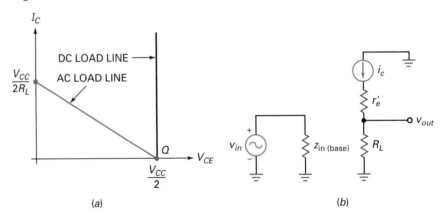

(a)   (b)

conducting transistor can go all the way from cutoff to saturation. On the alternate half cycle, the other transistor does the same thing. This means that the maximum peak-to-peak output is:

$$\text{MPP} = V_{CC} \tag{12-24}$$

AC Analysis

Figure 12-16b shows the ac equivalent of the conducting transistor. This is almost identical to the class A emitter follower. Ignoring r_e', the voltage gain is:

$$A_v \approx 1 \tag{12-25}$$

and the input impedance of the base is:

$$z_{\text{in(base)}} \approx \beta R_L \tag{12-26}$$

Overall Action

On the positive half cycle of input voltage, the upper transistor of Fig. 12-15a conducts and the lower one cuts off. The upper transistor acts like an ordinary emitter follower, so that the output voltage approximately equals the input voltage.

On the negative half cycle of input voltage, the upper transistor cuts off and the lower transistor conducts. The lower transistor acts like an ordinary emitter follower and produces a load voltage approximately equal to the input voltage. The upper transistor handles the positive half cycle of input voltage, and the lower transistor takes care of the negative half cycle. During either half cycle, the source sees a high input impedance looking into either base.

Crossover Distortion

Figure 12-17a shows the ac equivalent circuit of a class B push-pull emitter follower. Suppose that no bias is applied to the emitter diodes. Then, the incoming ac

Figure 12–17   (a) AC equivalent circuit; (b) crossover distortion; (c) Q point is slightly above cutoff.

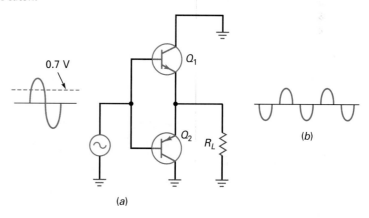

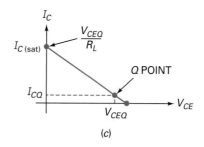

voltage has to rise to about 0.7 V to overcome the barrier potential of the emitter diodes. Because of this, no current flows through Q_1 when the signal is less than 0.7 V.

The action is similar on the other half cycle. No current flows through Q_2 until the ac input voltage is more negative than -0.7 V. For this reason, if no bias is applied to the emitter diodes, the output of a class B push-pull emitter follower looks like Fig. 12-17b.

Because of clipping between half cycles, the output is distorted. Since the clipping occurs between the time one transistor cuts off and the other one comes on, we call it **crossover distortion.** To eliminate crossover distortion, we need to apply a slight forward bias to each emitter diode. This means locating the Q point slightly above cutoff, as shown in Fig. 12-17c. As a guide, an I_{CQ} from 1 to 5 percent of $I_{C(\text{sat})}$ is enough to eliminate crossover distortion.

Class AB

In Fig. 12-17c, the slight forward bias implies that the conduction angle will be slightly greater than 180° because the transistor will conduct for a bit more than half a cycle. Strictly speaking, we no longer have class B operation. Because of this, the operation is sometimes referred to as **class AB,** defined as a conduction angle between 180° and 360°. But it is barely class AB. For this reason, many people still refer to the circuit as a *class B push-pull amplifier* because the operation is class B to a close approximation.

Power Formulas

The formulas shown in Table 12-1 apply to all classes of operation including class B push-pull operation.

When using these formulas to analyze a class B/AB push-pull emitter follower, remember that the class B/AB push-pull amplifier has the ac load line and waveforms of Fig. 12-18a. Each transistor supplies half of a cycle.

Transistor Power Dissipation

Ideally, the transistor power dissipation is zero when there is no input signal because both transistors are cut off. If there is a slight forward bias to prevent crossover distortion, the quiescent power dissipation in each transistor is still very small.

Table 12–1	Amplifier Power Formulas
Equation	**Value**
$A_p = \dfrac{p_{\text{out}}}{p_{\text{in}}}$	Power gain
$p_{\text{out}} = \dfrac{v_{\text{out}}^2}{8R_L}$	AC output power
$p_{\text{out(max)}} = \dfrac{\text{MPP}^2}{8R_L}$	Maximum ac output power
$P_{\text{dc}} = V_{CC}I_{\text{dc}}$	DC input power
$\eta = \dfrac{p_{\text{out}}}{P_{\text{dc}}} \times 100\%$	Efficiency

Figure 12-18 *(a)* Class B load line; *(b)* transistor power dissipation.

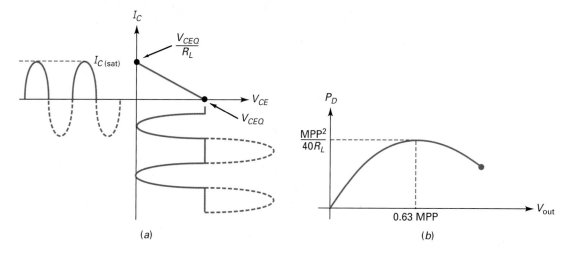

(a)

(b)

When an input signal is present, the transistor power dissipation becomes significant. The transistor power dissipation depends on how much of the ac load line is used. The maximum transistor power dissipation of each transistor is:

$$P_{D(max)} = \frac{MPP^2}{40R_L} \tag{12-27}$$

Figure 12-18b shows how the transistor power dissipation varies according to the peak-to-peak output voltage. As indicated, P_D reaches a maximum when the peak-to-peak output is 63 percent of MPP. Since this is the worst case, each transistor in a class B/AB push-pull amplifier must have a power rating of at least $MPP^2/40R_L$.

Example 12-8

The adjustable resistor of Fig. 12-19 sets both emitter diodes on the verge of conduction. What is the maximum transistor power dissipation? The maximum output power?

Figure 12-19 Example.

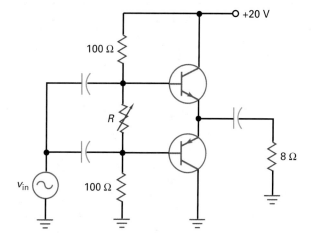

SOLUTION The maximum peak-to-peak output is:

$$MPP = V_{CC} = 20 \text{ V}$$

With Eq. (12-27):

$$P_{D(max)} = \frac{MPP^2}{40R_L} = \frac{(20 \text{ V})^2}{40(8 \text{ }\Omega)} = 1.25 \text{ W}$$

The maximum output power is:

$$P_{out(max)} = \frac{MPP^2}{8R_L} = \frac{(20 \text{ V})^2}{8(8 \text{ }\Omega)} = 6.25 \text{ W}$$

PRACTICE PROBLEM 12–8 In Fig. 12-19, change V_{CC} to +30 V and calculate $P_{D(max)}$ and $P_{out(max)}$.

Example 12-9

If the adjustable resistance is 15 Ω, what is the efficiency in the preceding example?

SOLUTION The dc current though the biasing resistors is:

$$I_{bias} \approx \frac{20 \text{ V}}{215 \text{ }\Omega} = 0.093 \text{ A}$$

Next, we need to calculate the dc current through the upper transistor. Here is how to do it: As shown in Fig. 12-18a, the saturation current is:

$$I_{C(sat)} = \frac{V_{CEQ}}{R_L} = \frac{10 \text{ V}}{8 \text{ }\Omega} = 1.25 \text{ A}$$

The collector current in the conducting transistor is a half-wave signal with a peak of $I_{C(sat)}$. Therefore, it has an average value of:

$$I_{av} = \frac{I_{C(sat)}}{\pi} = \frac{1.25 \text{ A}}{\pi} = 0.398 \text{ A}$$

The total current drain is:

$$I_{dc} = 0.093 \text{ A} + 0.398 \text{ A} = 0.491 \text{ A}$$

The dc input power is:

$$P_{dc} = (20 \text{ V})(0.491 \text{ A}) = 9.82 \text{ W}$$

The efficiency of the stage is:

$$\eta = \frac{P_{out}}{P_{dc}} \times 100\% = \frac{6.25 \text{ W}}{9.82 \text{ W}} \times 100\% = 63.6\%$$

PRACTICE PROBLEM 12–9 Repeat Example 12-9 using +30 V for V_{CC}.

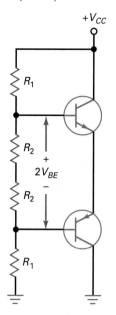

Figure 12-20 Voltage-divider bias of class B push-pull amplifier.

$+V_{CC}$

R_1

R_2

$+$
$2V_{BE}$
$-$

R_2

R_1

12-6 Biasing Class B/AB Amplifiers

As mentioned earlier, the hardest thing about designing a class B/AB amplifier is setting up a stable Q point near cutoff. This section discusses the problem and its solution.

Voltage-Divider Bias

Figure 12-20 shows voltage-divider bias for a class B/AB push-pull circuit. The two transistors have to be complementary; that is, they must have similar V_{BE} curves, maximum ratings, and so forth. For instance, the 2N3904 and 2N3906 are complementary, the first being an *npn* transistor and the second being a *pnp*. They have similar V_{BE} curves, maximum ratings, and so on. Complementary pairs like these are available for almost any class B/AB push-pull design.

To avoid crossover distortion in Fig. 12-20, we set the Q point slightly above cutoff, with the correct V_{BE} somewhere between 0.6 and 0.7 V. But here is the major problem: The collector current is very sensitive to changes in V_{BE}. Data sheets indicate that an increase of 60 mV in V_{BE} produces 10 times as much collector current. Because of this, an adjustable resistor is needed to set the correct Q point.

But an adjustable resistor does not solve the temperature problem. Even though the Q point may be perfect at room temperature, it will change when the temperature changes. As discussed earlier, V_{BE} decreases approximately 2 mV per degree rise. As the temperature increases in Fig. 12-20, the fixed voltage on each emitter diode forces the collector current to increase rapidly. If the temperature increases 30°, the collector current increases by a factor of 10 because the fixed bias is 60 mV too high. Therefore, the Q point is very unstable with voltage-divider bias.

The ultimate danger in Fig. 12-20 is **thermal runaway.** When the temperature increases, the collector current increases. As the collector current increases, the junction temperature increases even more, further reducing the correct V_{BE}. This escalating situation means that the collector current may "run away" by rising until excessive power destroys the transistor.

Whether or not thermal runaway takes place depends on the thermal properties of the transistor, how it is cooled, and the type of heat sink used. More often than not, voltage-divider bias like Fig. 12-20 will produce thermal runaway, which destroys the transistors.

Diode Bias

One way to avoid thermal runaway is with diode bias, shown in Fig. 12-21. The idea is to use **compensating diodes** to produce the bias voltage for the emitter diodes. For this scheme to work, the diode curves must match the V_{BE} curves of the transistors. Then, any increase in temperature reduces the bias voltage developed by the compensating diodes by just the right amount.

For instance, assume that a bias voltage of 0.65 V sets up 2 mA of collector current. If the temperature rises 30°C, the voltage across each compensating diode drops 60 mV. Since the required V_{BE} also decreases by 60 mV, the collector current remains fixed at 2 mA.

For diode bias to be immune to changes in temperature, the diode curves must match the V_{BE} curves over a wide temperature range. This is not easily done with discrete circuits because of the tolerance of components. But diode bias is easy to implement with integrated circuits because the diodes and transistors are on the same chip, which means that they have almost identical curves.

GOOD TO KNOW

In actual designs, the compensating diodes are mounted on the case of the power transistors so that, as the transistors heat up so do the diodes. The diodes are usually mounted to the power transistors with a nonconductive adhesive that has good thermal transfer characteristics.

Figure 12-21 Diode bias of class B push-pull amplifier.

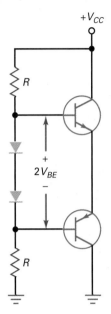

With diode bias, the bias current through the compensating diodes of Fig. 12-21 is:

$$I_{\text{bias}} = \frac{V_{CC} - 2V_{BE}}{2R} \tag{12-28}$$

When the compensating diodes match the V_{BE} curves of the transistors, I_{CQ} has the same value as I_{bias}. (For details, see Sec. 17-7.) As mentioned earlier, I_{CQ} should be between 1 and 5 percent of $I_{C(\text{sat})}$ to avoid crossover distortion.

Example 12-10   ‖‖ MultiSim

What is the quiescent collector current in Fig. 12-22? The maximum efficiency of the amplifier?

Figure 12-22 Example.

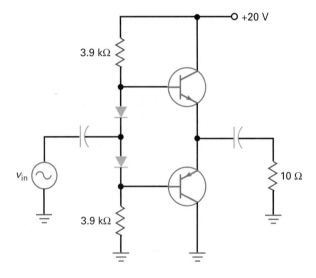

SOLUTION The bias current through the compensating diodes is:

$$I_{\text{bias}} = \frac{20 \text{ V} - 1.4 \text{ V}}{2(3.9 \text{ k}\Omega)} = 2.38 \text{ mA}$$

This is the value of the quiescent collector current, assuming that the compensating diodes match the emitter diodes.

The collector saturation current is:

$$I_{C(\text{sat})} = \frac{V_{CEQ}}{R_L} = \frac{10 \text{ V}}{10 \text{ }\Omega} = 1 \text{ A}$$

The average value of the half-wave collector current is:

$$I_{\text{av}} = \frac{I_{C(\text{sat})}}{\pi} = \frac{1 \text{ A}}{\pi} = 0.318 \text{ A}$$

The total current drain is:

$$I_{\text{dc}} = 2.38 \text{ mA} + 0.318 \text{ A} = 0.32 \text{ A}$$

The dc input power is:

$$P_{dc} = (20 \text{ V})(0.32 \text{ A}) = 6.4 \text{ W}$$

The maximum ac output power is:

$$P_{out(max)} = \frac{MPP^2}{8R_L} = \frac{(20 \text{ V})^2}{8(10 \text{ }\Omega)} = 5 \text{ W}$$

The efficiency of the stage is:

$$\eta = \frac{P_{out}}{P_{dc}} \times 100\% = \frac{5 \text{ W}}{6.4 \text{ W}} \times 100\% = 78.1\%$$

PRACTICE PROBLEM 12-10 Repeat Example 12-10 using $+30$ V for V_{CC}.

12-7 Class B/AB Driver

In the earlier discussion of the class B/AB push-pull emitter follower, the ac signal was capacitively coupled into the bases. This is not the preferred way to drive a class B/AB push-pull amplifier.

CE Driver

The stage that precedes the output stage is called a **driver.** Rather than capacitively couple into the output push-pull stage, we can use the direct-coupled CE driver shown in Fig. 12-23a. Transistor Q_1 is a current source that sets up the dc biasing current through the diodes. By adjusting R_2, we can control the dc emitter current through R_4. This means that Q_1 sources the biasing current through the compensating diodes.

When an ac signal drives the base of Q_1, it acts like a swamped amplifier. The amplified and inverted ac signal at the Q_1 collector drives the bases of Q_2 and Q_3. On the positive half cycle, Q_2 conducts and Q_3 cuts off. On the negative half cycle, Q_2 cuts off and Q_3 conducts. Because the output coupling capacitor is an ac short, the ac signal is coupled to the load resistance.

Figure 12-23b shows the ac equivalent circuit of the CE driver. The diodes are replaced by their ac emitter resistances. In any practical circuit, r_e' is at least 100 times smaller than R_3. Therefore, the ac equivalent circuit simplifies to Fig. 12-23c.

Now, we can see that the driver stage is a swamped amplifier whose amplified and inverted output drives both bases of the output transistors with the same signal. Often, the input impedance of the output transistors is very high, and we can approximate the voltage gain of the driver by:

$$A_V = \frac{R_3}{R_4}$$

In short, the driver stage is a swamped voltage amplifier that produces a large signal for the output push-pull amplifier.

Figure 12-23 (*a*) Direct-coupled CE driver; (*b*) ac equivalent circuit; (*c*) simplified ac equivalent circuit.

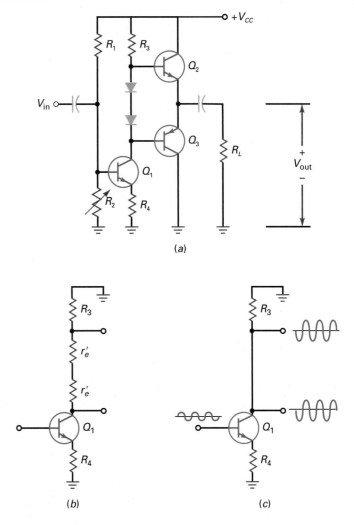

(a)

(b)   (c)

Two-Stage Negative Feedback

Figure 12-24 is another example of using a large-signal CE stage to drive a class B/AB push-pull emitter follower. The input signal is amplified and inverted by the Q_1 driver. The push-pull stage then provides the current gain needed to drive the low-impedance loudspeaker. Notice that the CE driver has its emitter connected to ground. As a result, this driver has more voltage gain than the driver of Fig. 12-23*a*.

The resistance R_2 does two useful things: First, since it is connected to a dc voltage of $+V_{CC}/2$, this resistance provides the dc bias for Q_1. Second, R_2 produces negative feedback for the ac signal. Here's why: A positive-going signal on the base of Q_1 produces a negative-going signal on the Q_1 collector. The output of the emitter follower is therefore negative-going. When fed back through R_2 to the Q_1 base, this returning signal opposes the original input signal. This is negative feedback, which stabilizes the bias and the voltage gain of the overall amplifier.

Integrated circuit (IC) audio power amplifiers are often used in low- to medium-power applications. These amplifiers, such as a LM380 IC, contain class AB biased output transistors and will be discussed in Chap. 18.

Figure 12-24   Two-stage negative feedback to CE driver.

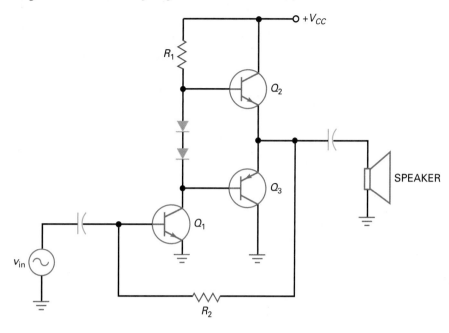

12-8 Class C Operation

With class B, we need to use a push-pull arrangement. That's why almost all class B amplifiers are push-pull amplifiers. With class C, we need to use a resonant circuit for the load. This is why almost all class C amplifiers are tuned amplifiers.

Resonant Frequency

With class C operation, the collector current flows for less than half a cycle. A parallel resonant circuit can filter the pulses of collector current and produce a pure sine wave of output voltage. The main application for class C is with tuned RF amplifiers. The maximum efficiency of a tuned class C amplifier is 100 percent.

Figure 12-25a shows a tuned RF amplifier. The ac input voltage drives the base, and an amplified output voltage appears at the collector. The amplified and inverted signal is then capacitively coupled to the load resistance. Because of the parallel resonant circuit, the output voltage is maximum at the resonant frequency, given by:

$$f_r = \frac{1}{2\pi\sqrt{LC}} \qquad (12\text{-}29)$$

On either side of the resonant frequency f_r, the voltage gain drops off as shown in Fig. 12-25b. For this reason, a tuned class C amplifier is always intended to amplify a narrow band of frequencies. This makes it ideal for amplifying radio and television signals because each station or channel is assigned a narrow band of frequencies on both sides of a center frequency.

The class C amplifier is unbiased, as shown in the dc equivalent circuit of Fig. 12-25c. The resistance R_S in the collector circuit is the series resistance of the inductor.

Load Lines

Figure 12-25d shows the two load lines. The dc load line is approximately vertical because the winding resistance R_S of an RF inductor is very small. The dc load

GOOD TO KNOW

Most class C amplifiers are designed so that the peak value of input voltage is just sufficient to drive the transistor into saturation.

Figure 12-25 (*a*) Tuned class C amplifier; (*b*) voltage gain versus frequency; (*c*) dc equivalent circuit is unbiased; (*d*) two load lines; (*e*) ac equivalent circuit.

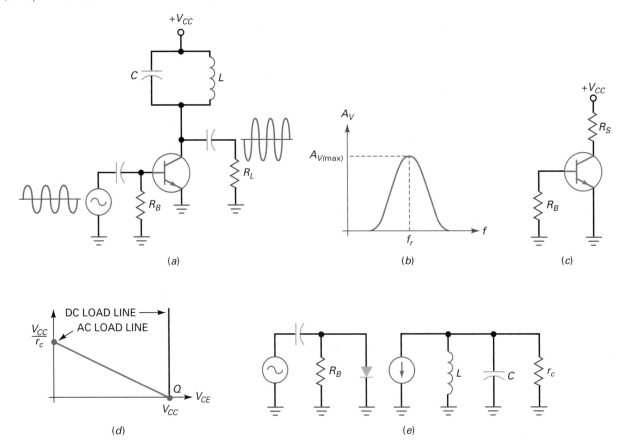

line is not important because the transistor is unbiased. What is important is the ac load line. As indicated, the Q point is at the lower end of the ac load line. When an ac signal is present, the instantaneous operating point moves up the ac load line toward the saturation point. The maximum pulse of collector current is given by the saturation current V_{CC}/r_c.

DC Clamping of Input Signal

Figure 12-25*e* is the ac equivalent circuit. The input signal drives the emitter diode, and the amplified current pulses drive the resonant tank circuit. In a tuned class C amplifier the input capacitor is part of a negative dc clamper. For this reason, the signal appearing across the emitter diode is negatively clamped.

Figure 12-26*a* illustrates the negative clamping. Only the positive peaks of the input signal can turn on the emitter diode. For this reason, the collector current flows in brief pulses like those of Fig. 12-26*b*.

Filtering the Harmonics

Chapter 5 briefly discussed the concept of harmonics. The basic idea is this: A nonsinusoidal waveform like Fig. 12-26*b* is rich in **harmonics,** multiples of the input frequency. In other words, the pulses of Fig. 12-26*b* are equivalent to a group of sine waves with frequencies of f, $2f$, $3f$, . . . , nf.

The resonant tank circuit of Fig. 12-26*c* has a high impedance only at the fundamental frequency f. This produces a large voltage gain at the fundamental frequency. On the other hand, the tank circuit has a very low impedance to the

Figure 12-26 (a) Input signal is negatively clamped at base; (b) collector current flows in pulses; (c) ac collector circuit; (d) collector voltage waveform.

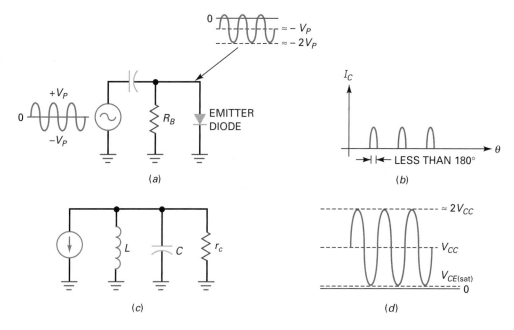

higher harmonics, producing very little voltage gain. This is why the voltage across the resonant tank looks almost like the pure sine wave of Fig. 12-26d. Since all higher harmonics are filtered, only the fundamental frequency appears across the tank circuit.

Troubleshooting

Since the tuned class C amplifier has a negatively clamped input signal, you can use a high-impedance dc voltmeter to measure the voltage across the emitter diode. If the circuit is working correctly, you should read a negative voltage approximately equal to the peak of the input signal.

The voltmeter test just described is useful when an oscilloscope is not handy. If you have an oscilloscope, however, an even better test is to look across the emitter diode. You should see a negatively clamped waveform when the circuit is working properly.

Example 12-11

IIII MultiSim

Describe the action of Fig. 12-27.

SOLUTION   The circuit has a resonant frequency of:

$$f_r = \frac{1}{2\pi\sqrt{(2\ \mu H)(470\ pF)}} = 5.19\ MHz$$

If the input signal has this frequency, the tuned class C circuit will amplify the input signal.

In Fig. 12-27, the input signal has a peak-to-peak value of 10 V. The signal is negatively clamped at the base of the transistor with a positive peak of

Figure 12-27 Example.

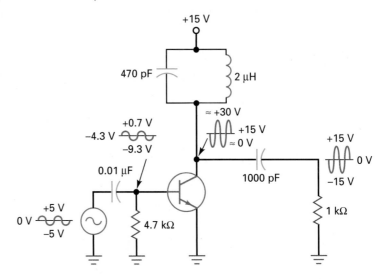

+0.7 V and a negative peak of −9.3 V. The average base voltage is −4.3 V, which can be measured with a high-impedance dc voltmeter.

The collector signal is inverted because of the CE connection. The dc or average voltage of the collector waveform is +15 V, the supply voltage. Therefore, the peak-to-peak collector voltage is 30 V. This voltage is capacitively coupled to the load resistance. The final output voltage has a positive peak of +15 V and a negative peak of −15 V.

PRACTICE PROBLEM 12–11 Using Fig. 12-27, change the 470 pF capacitor to 560 pF and V_{CC} to +12 V. Solve the circuit for f_r and V_{out} peak-to-peak.

12-9 Class C Formulas

A tuned class C amplifier is usually a narrowband amplifier. The input signal in a class C circuit is amplified to get large output power with an efficiency approaching 100 percent.

Bandwidth

As discussed in basic courses, the **bandwidth (BW)** of a resonant circuit is defined as:

$$BW = f_2 - f_1 \tag{12-30}$$

where f_1 = lower half-power frequency
f_2 = upper half-power frequency

The half-power frequencies are identical to the frequencies at which the voltage gain equals 0.707 times the maximum gain, as shown in Fig. 12-28. The smaller BW is, the narrower the bandwidth of the amplifier.

Figure 12-28 Bandwidth.

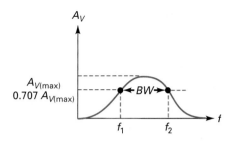

With Eq. (12-30), it is possible to derive this new relation for bandwidth:

$$BW = \frac{f_r}{Q} \tag{12-31}$$

where Q is the quality factor of the circuit. Equation (12-31) says that the bandwidth is inversely proportional to Q. The higher the Q of the circuit, the smaller the bandwidth.

Class C amplifiers almost always have a circuit Q that is greater than 10. This means that the bandwidth is less than 10 percent of the resonant frequency. For this reason, class C amplifiers are narrowband amplifiers. The output of a narrowband amplifier is a large sinusoidal voltage at resonance with a rapid drop-off above and below resonance.

Current Dip at Resonance

When a tank circuit is resonant, the ac load impedance seen by the collector current source is maximum and purely resistive. Therefore, the collector current is minimum at resonance. Above and below resonance, the ac load impedance decreases and the collector current increases.

One way to tune a resonant tank is to look for a decrease in the dc current supplied to the circuit, as shown in Fig. 12-29. The basic idea is to measure the current I_{dc} from the power supply while tuning the circuit (varying either L or C). When the tank is resonant at the input frequency, the ammeter reading will dip to a minimum value. This indicates that the circuit is correctly tuned because the tank has a maximum impedance at this point.

AC Collector Resistance

Any inductor has a series resistance R_S, as indicated in Fig. 12-30a. The Q of the inductor is defined as:

$$Q_L = \frac{X_L}{R_S} \tag{12-32}$$

Figure 12-29 Current dip at resonance.

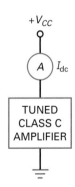

Figure 12-30 (a) Series equivalent resistance for inductor; (b) parallel equivalent resistance for inductor.

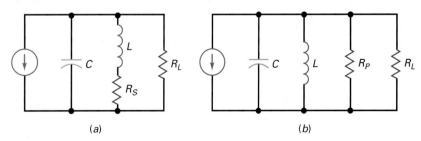

where Q_L = quality factor of coil
X_L = inductive reactance
R_S = coil resistance

Remember that this is the Q of the coil only. The overall circuit has a lower Q because it includes the effect of load resistance as well as coil resistance.

As discussed in basic ac courses, the series resistance of the inductor can be replaced by a parallel resistance R_P, as shown in Fig. 12-30b. When Q is greater than 10, this equivalent resistance is given by:

$$R_P = Q_L X_L \tag{12-33}$$

In Fig. 12-30b, X_L cancels X_C at resonance, leaving only R_P in parallel with R_L. Therefore, the ac resistance seen by the collector at resonance is:

$$r_c = R_P \| R_L \tag{12-34}$$

The *Q of the overall circuit* is given by:

$$Q = \frac{r_c}{X_L} \tag{12-35}$$

This circuit Q is lower than Q_L, the coil Q. In practical class C amplifiers, the Q of the coil is typically 50 or more, and the Q of the circuit is 10 or more. Since the overall Q is 10 or more, the operation is narrowband.

Duty Cycle

The brief turn-on of the emitter diode at each positive peak produces narrow pulses of collector current, as shown in Fig. 12-31a. With pulses like these, it is convenient to define the **duty cycle** as:

$$D = \frac{W}{T} \tag{12-36}$$

where D = duty cycle
W = width of pulse
T = period of pulses

For instance, if an oscilloscope displays a pulse width of 0.2 μs and a period of 1.6 μs, the duty cycle is:

$$D = \frac{0.2\,\mu s}{1.6\,\mu s} = 0.125$$

The smaller the duty cycle, the narrower the pulses compared to the period. The typical class C amplifier has a small duty cycle. In fact, the efficiency of a class C amplifier increases as the duty cycle decreases.

Conduction Angle

An equivalent way to state the duty cycle is by using the conduction angle ϕ, shown in Fig. 12-31b:

$$D = \frac{\phi}{360°} \tag{12-37}$$

Figure 12-31   Duty cycle.

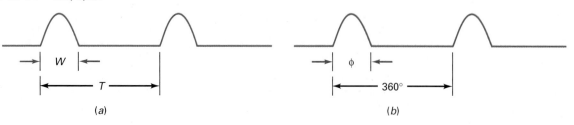

(a)   (b)

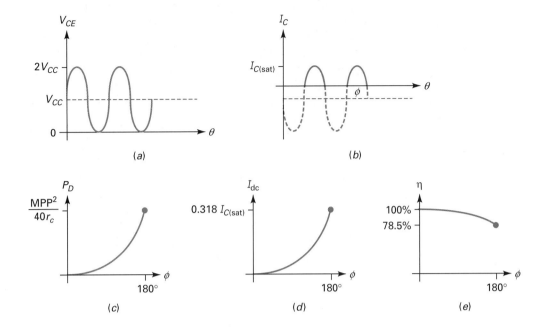

For instance, if the conduction angle is 18°, the duty cycle is:

$$D = \frac{18°}{360°} = 0.05$$

Transistor Power Dissipation

Figure 12-32a shows the ideal collector-emitter voltage in a class C transistor amplifier. In Fig. 12-32a, the maximum output is given by:

$$\mathbf{MPP = 2V_{CC}} \tag{12-38}$$

Since the maximum voltage is approximately $2V_{CC}$, the transistor must have a V_{CEO} rating greater than $2V_{CC}$.

Figure 12-32b shows the collector current for a class C amplifier. Typically, the conduction angle ϕ is much less than 180°. Notice that the collector current reaches a maximum value of $I_{C(sat)}$. The transistor must have a peak current rating greater than this. The dotted parts of the cycle represent the off time of the transistor.

The power dissipation of the transistor depends on the conduction angle. As shown in Fig. 12-32c, the power dissipation increases with the conduction angle up to 180°. The maximum power dissipation of the transistor can be derived with calculus:

$$\mathbf{P_D} = \frac{\mathbf{MPP^2}}{\mathbf{40r_c}} \tag{12-39}$$

Equation (12-39) represents the worst case. A transistor operating as class C must have a power rating greater than this or it will be destroyed. Under normal drive conditions, the conduction angle will be much less than 180° and the transistor power dissipation will be less than $MPP^2/40r_c$.

Stage Efficiency

The dc collector current depends on the conduction angle. For a conduction angle of 180° (a half-wave signal), the average or dc collector current is $I_{C(sat)}/\pi$. For smaller conduction angles, the dc collector current is less than this, as shown in

Fig. 12-32*d*. The dc collector current is the only current drain in a class C amplifier because it has no biasing resistors.

In a class C amplifier, most of the dc input power is converted into ac load power because the transistor and coil losses are small. For this reason, a class C amplifier has high stage efficiency.

Figure 12-32*e* shows how the optimum stage efficiency varies with conduction angle. When the angle is 180°, the stage efficiency is 78.5 percent, the theoretical maximum for a class B amplifier. When the conduction angle decreases, the stage efficiency increases. As indicated, class C has a maximum efficiency of 100 percent, approached at very small conduction angles.

Example 12-12

If Q_L is 100 in Fig. 12-33, what is the bandwidth of the amplifier?

Figure 12-33 Example.

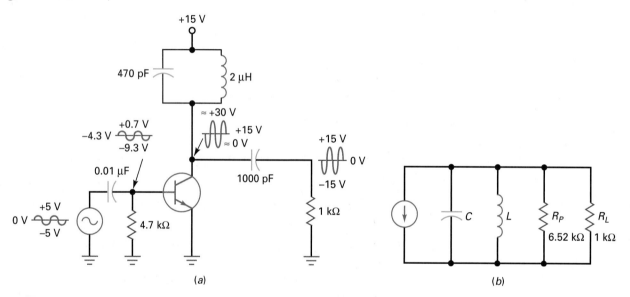

(a) (b)

SOLUTION   At the resonant frequency (found in Example 12-11):

$$X_L = 2\pi fL = 2\pi(5.19 \text{ MHz})(2 \text{ μH}) = 65.2 \text{ Ω}$$

With Eq. (12-33), the equivalent parallel resistance of the coil is:

$$R_P = Q_L X_L = (100)(65.2 \text{ Ω}) = 6.52 \text{ kΩ}$$

This resistance is in parallel with the load resistance, as shown in Fig. 12-33*b*. Therefore, the ac collector resistance is:

$$r_c = 6.52 \text{ kΩ} \| 1 \text{ kΩ} = 867 \text{ Ω}$$

With Eq. (12-35), the Q of the overall circuit is:

$$Q = \frac{r_c}{X_L} = \frac{867 \text{ Ω}}{65.2 \text{ Ω}} = 13.3$$

Since the resonant frequency is 5.19 MHz, the bandwidth is:

$$BW = \frac{5.19 \text{ MHz}}{13.3} = 390 \text{ kHz}$$

Example 12-13

In Fig. 12-33a, what is the worst-case power dissipation?

SOLUTION   The maximum peak-to-peak output is:

$$MPP = 2V_{CC} = 2(15\ V) = 30\ V\ pp$$

Equation (12-39) gives us the worst-case power dissipation of the transistor:

$$P_D = \frac{MPP^2}{40r_c} = \frac{(30\ V)^2}{40(867\ \Omega)} = 26\ mW$$

PRACTICE PROBLEM 12-13   In Fig. 12-33, if V_{CC} is $+12$ V, what is the worst case power dissipation?

Summary Table 12-1 illustrates the characteristics of class A, B/AB, and C amplifiers.

12-10 Transistor Power Rating

The temperature at the collector junction places a limit on the allowable power dissipation P_D. Depending on the transistor type, a junction temperature in the range of 150 to 200°C will destroy the transistor. Data sheets specify this maximum junction temperature as $T_{J(max)}$. For instance, the data sheet of a 2N3904 gives a $T_{J(max)}$ of 150°C; the data sheet of a 2N3719 specifies a $T_{J(max)}$ of 200°C.

Ambient Temperature

The heat produced at the junction passes through the transistor case (metal or plastic housing) and radiates to the surrounding air. The temperature of this air, known as the *ambient temperature,* is around 25°C, but it can get much higher on hot days. Also, the ambient temperature may be much higher inside a piece of electronic equipment.

Derating Factor

Data sheets often specify the $P_{D(max)}$ of a transistor at an ambient temperature of 25°C. For instance, the 2N1936 has a $P_{D(max)}$ of 4 W for an ambient temperature of 25°C. This means that a 2N1936 used in a class A amplifier can have a quiescent power dissipation as high as 4 W. As long as the ambient temperature is 25°C or less, the transistor is within its specified power rating.

What do you do if the ambient temperature is greater than 25°C? You have to derate (reduce) the power rating. Data sheets sometimes include a *derating curve* like the one in Fig. 12-34. As you can see, the power rating decreases when the ambient temperature increases. For instance, at an ambient temperature of 100°C, the power rating is 2 W.

Some data sheets do not give a derating curve like the one in Fig. 12-34. Instead, they list a derating factor D. For instance, the derating factor of a 2N1936

GOOD TO KNOW

With integrated circuits, a maximum junction temperature cannot be specified because there are so many transistors. Therefore, ICs have a maximum device temperature or case temperature instead. For example, the μA741 op amp IC has a power rating of 500 mW if it is in a metal package, 310 mW if it is in a dual-inline package, and 570 mW if it is in a flatpack.

Figure 12-34 Power rating versus ambient temperature.

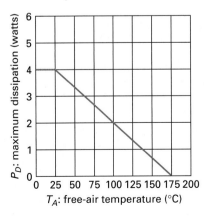

T_A: free-air temperature (°C)

Circuit	Characteristics	Where used
A 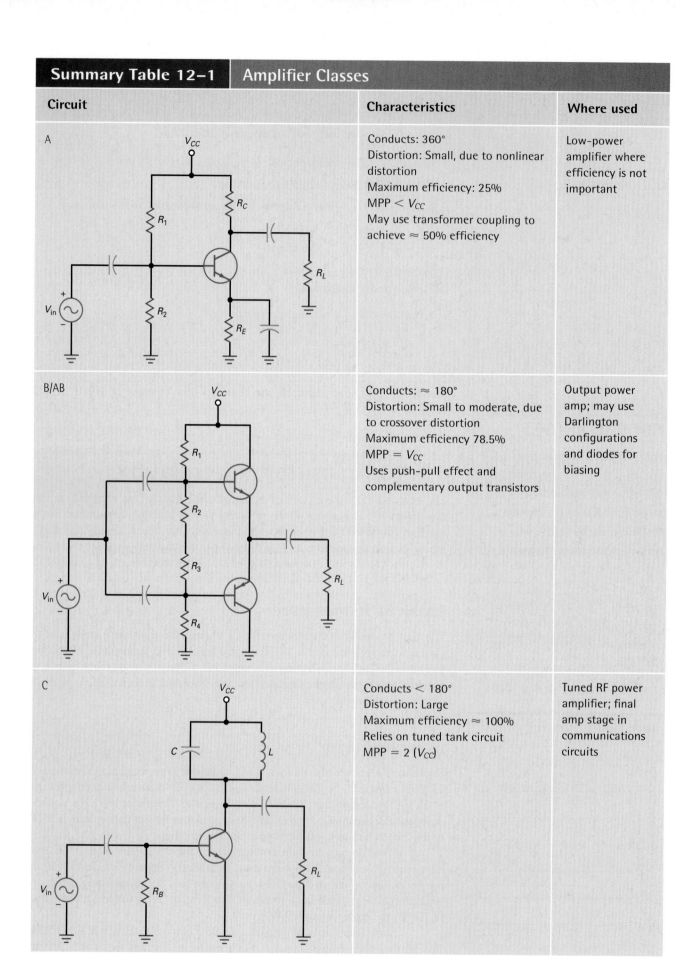	Conducts: 360° Distortion: Small, due to nonlinear distortion Maximum efficiency: 25% MPP $<$ V_{CC} May use transformer coupling to achieve $\approx$ 50% efficiency	Low-power amplifier where efficiency is not important
B/AB	Conducts: $\approx$ 180° Distortion: Small to moderate, due to crossover distortion Maximum efficiency 78.5% MPP = V_{CC} Uses push-pull effect and complementary output transistors	Output power amp; may use Darlington configurations and diodes for biasing
C	Conducts $<$ 180° Distortion: Large Maximum efficiency $\approx$ 100% Relies on tuned tank circuit MPP = 2 (V_{CC})	Tuned RF power amplifier; final amp stage in communications circuits

is 26.7 mW/°C. This means that you have to subtract 26.7 mW for each degree the ambient temperature is above 25°C. In symbols:

$$\Delta P = D(T_A - 25°C) \tag{12-40}$$

where ΔP = decrease in power rating
D = derating factor
T_A = ambient temperature

As an example, if the ambient temperature rises to 75°C, you have to reduce the power rating by:

$$\Delta P = 26.7 \text{ mW}(75 - 25) = 1.34 \text{ W}$$

Since the power rating is 4 W at 25°C, the new power rating is:

$$P_{D(max)} = 4 \text{ W} - 1.34 \text{ W} = 2.66 \text{ W}$$

This agrees with the derating curve of Fig. 12-34.

Whether you get the reduced power rating from a derating curve like the one in Fig. 12-34 or from a formula like the one in Eq. (12-40), the important thing to be aware of is the reduction in power rating as the ambient temperature increases. Just because a circuit works well at 25°C doesn't mean it will perform well over a large temperature range. When you design circuits, therefore, you must take the operating temperature range into account by derating all transistors for the highest expected ambient temperature.

Heat Sinks

One way to increase the power rating of a transistor is to get rid of the heat faster. This is why heat sinks are used. If we increase the surface area of the transistor case, we allow the heat to escape more easily into the surrounding air. Look at Fig. 12-35a. When this type of heat sink is pushed on to the transistor case, heat radiates more quickly because of the increased surface area of the fins.

Figure 12-35b shows the power-tab transistor. The metal tab provides a path out of the transistor for heat. This metal tab can be fastened to the chassis of electronics equipment. Because the chassis is a massive heat sink, heat can easily escape from the transistor to the chassis.

Large power transistors like Fig. 12-35c have the collector connected directly to the case to let heat escape as easily as possible. The transistor case is then fastened to the chassis. To prevent the collector from shorting to the chassis ground, a thin insulating washer and a thermal conductive paste are used between the transistor case and the chassis. The important idea here is that heat can leave

Figure 12-35 (a) Push-on heat sink; (b) power-tab transistor; (c) power transistor with collector connected to case.

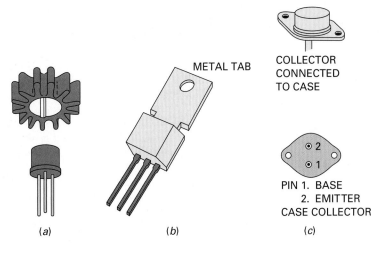

METAL TAB

COLLECTOR
CONNECTED
TO CASE

PIN 1. BASE
2. EMITTER
CASE COLLECTOR

(a) (b) (c)

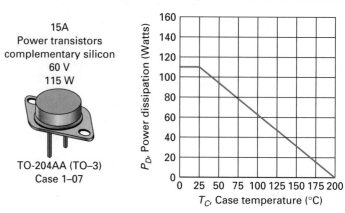

15A
Power transistors
complementary silicon
60 V
115 W

TO-204AA (TO–3)
Case 1–07

the transistor more rapidly, which means that the transistor has a higher power rating at the same ambient temperature.

Case Temperature

When heat flows out of a transistor, it passes through the case of the transistor and into the heat sink, which then radiates the heat into the surrounding air. The temperature of the transistor case T_C will be slightly higher than the temperature of the heat sink T_S which in turn is slightly higher than the ambient temperature T_A.

The data sheets of large power transistors give derating curves for the case temperature rather than the ambient temperature. For instance, Fig. 12-36 shows the derating curve of a 2N3055. The power rating is 115 W at a case temperature of 25°C; then it decreases linearly with temperature until it reaches zero for a case temperature of 200°C.

Sometimes you get a derating factor instead of a derating curve. In this case, you can use the following equation to calculate the reduction in power rating:

$$\Delta P = D(T_C - 25°C) \tag{12-41}$$

where ΔP = decrease in power rating
D = derating factor
T_C = case temperature

To use the derating curve of a large power transistor, you need to know what the case temperature will be in the worst case. Then you can derate the transistor to arrive at its maximum power rating.

Example 12-14

The circuit of Fig. 12-37 is to operate over an ambient temperature range of 0 to 50°C. What is the maximum power rating of the transistor for the worst-case temperature?

SOLUTION   The worst-case temperature is the highest one because you have to derate the power rating given on a data sheet. If you look at the data sheet of a 2N3904 in Fig. 6-15, you will see the maximum power rating is listed as:

$P_D = 625$ mW at 25°C ambient

and the derating factor is given as:

$D = 5$ mW/°C

Figure 12-37 Example.

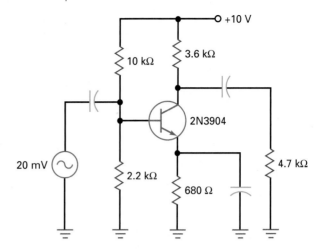

With Eq. (12-40), we can calculate:

$$\Delta P = (5\ \text{mW})(50 - 25) = 125\ \text{mW}$$

Therefore, the maximum power rating at 50°C is:

$$P_{D(\text{max})} = 625\ \text{mW} - 125\ \text{mW} = 500\ \text{mW}$$

PRACTICE PROBLEM 12-14 In Example 12-14, what is the transistor's power rating when the ambient temperature is 65°?

Summary

SEC. 12-1 AMPLIFIER TERMS

The classes of operation are A, B, and C. The types of coupling are capacitive, transformer, and direct. Frequency terms include audio, RF, narrowband, and wideband. Some types of audio amplifiers are preamps and power amplifiers.

SEC. 12-2 TWO LOAD LINES

Every amplifier has a dc load line and an ac load line. To get maximum peak-to-peak output, the Q point should be in the center of the ac load line.

SEC. 12-3 CLASS A OPERATION

The power gain equals the ac output power divided by the ac input power. The power rating of a transistor must be greater than the quiescent power dissipation. The efficiency of an amplifier

stage equals the ac output power divided by the dc input power, times 100 percent. The maximum efficiency of class A with a collector and load resistor is 25%. If the load resistor is the collector resistor or uses a transformer, the maximum efficiency increases to 50 percent.

SEC. 12-4 CLASS B OPERATION

Most class B amplifiers use a push-pull connection of two transistors. While one transistor conducts, the other is cut off, and vice versa. Each transistor amplifies one-half of the ac cycle. The maximum efficiency of class B is 78.5 percent.

SEC. 12-5 CLASS B PUSH-PULL EMITTER FOLLOWER

Class B is more efficient than class A. In a class B push-pull emitter follower,

complementary *npn* and *pnp* transistors are used. The *npn* transistor conducts on one half-cycle, and the *pnp* transistor on the other.

SEC. 12-6 BIASING CLASS B/AB AMPLIFIERS

To avoid crossover distortion, the transistors of a class B push-pull emitter follower have a small quiescent current. This is referred to as a class AB. With voltage divider bias, the Q point is unstable and may result in thermal runaway. Diode bias is preferred because it can produce a stable Q point over a large temperature range.

SEC. 12-7 CLASS B/AB DRIVER

Rather than capacitive couple the signal into the output stage, we can use a direct-coupled driver stage. The collector current out of the driver sets up the quiescent current through the complementary diodes.

SEC. 12-8 CLASS C OPERATION

Most class C amplifiers are tuned RF amplifiers. The input signal is negatively clamped, which produces narrow pulses of collector current. The tank circuit is tuned to the fundamental frequency, so that all higher harmonics are filtered out.

SEC. 12-9 CLASS C FORMULAS

The bandwidth of a class C amplifier is inversely proportional to the Q of the circuit. The ac collector resistance includes the parallel equivalent resistance of the inductor and the load resistance.

SEC. 12-10 TRANSISTOR POWER RATING

The power rating of a transistor decreases as the temperature increases. The data sheet of a transistor either lists a derating factor or shows a graph of the power rating versus temperature. Heat sinks can remove the heat more rapidly, producing a higher power rating.

Definitions

(12-12)   Power gain:

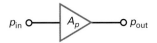

$$A_p = \frac{p_{out}}{p_{in}}$$

(12-18)   Efficiency:

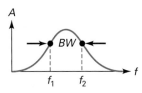

$$\eta = \frac{p_{out}}{P_{dc}} \times 100\%$$

(12-30)   Bandwidth:

$$BW = f_2 - f_1$$

(12-32) Q of inductor:

$$Q_L = \frac{X_L}{R_S}$$

(12-33)   Equivalent parallel R:

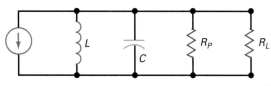

$$R_P = Q_L X_L$$

(12-34)   AC collector resistance:

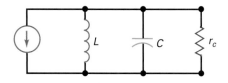

$$r_c = R_P \| R_L$$

(12-35) Q of amplifier:

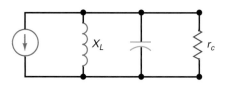

$$Q = \frac{r_c}{X_L}$$

(12-36)   Duty cycle:

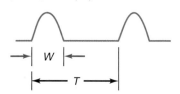

$$D = \frac{W}{T}$$

Derivations

(12-1)   Saturation current:

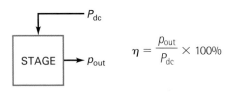

$$I_{C(sat)} = \frac{V_{CC}}{R_C + R_E}$$

(12-2)   Cutoff voltage:

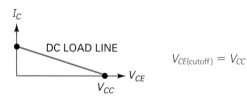

$$V_{CE(cutoff)} = V_{CC}$$

(12-7)   Limit on output:

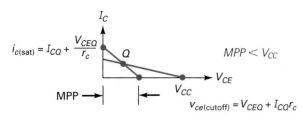

$$i_{c(\text{sat})} = I_{CQ} + \frac{V_{CEQ}}{r_c}$$

$$MPP < V_{CC}$$

$$v_{ce(\text{cutoff})} = V_{CEQ} + I_{CQ}r_c$$

(12-8)   Maximum peak:

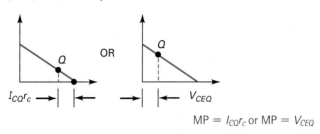

$$MP = I_{CQ}r_c \text{ or } MP = V_{CEQ}$$

(12-9)   Maximum peak-to-peak output:

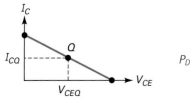

$$MPP = 2MP$$

(12-14)   Output power:

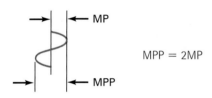

$$p_{\text{out}} = \frac{v_{\text{out}}^2}{8R_L}$$

(12-15)   Maximum output:

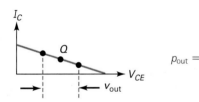

$$p_{\text{out(max)}} = \frac{MPP^2}{8R_L}$$

(12-16)   Transistor power:

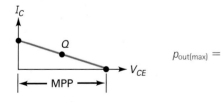

$$P_{DQ} = V_{CEQ}I_{CQ}$$

(12-17)   DC input power:

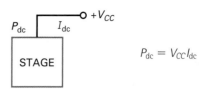

$$P_{\text{dc}} = V_{CC}I_{\text{dc}}$$

(12-24)   Class B maximum output:

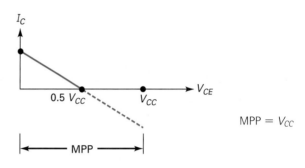

$$MPP = V_{CC}$$

(12-27)   Class B transistor output:

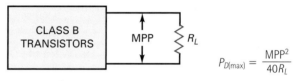

$$P_{D(\text{max})} = \frac{MPP^2}{40R_L}$$

(12-28)   Class B bias:

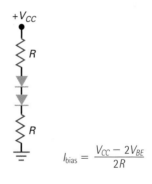

$$I_{\text{bias}} = \frac{V_{CC} - 2V_{BE}}{2R}$$

(12-29)   Resonant frequency:

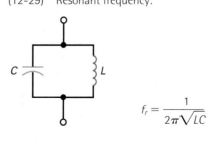

$$f_r = \frac{1}{2\pi\sqrt{LC}}$$

(12-31)   Bandwidth:

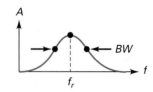

$$BW = \frac{f_r}{Q}$$

(12-39)   Power dissipation:

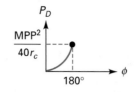

$$P_D = \frac{MPP^2}{40r_c}$$

(12-38)   Maximum output:

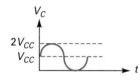

$$MPP = 2V_{CC}$$

Student Assignments

1. **For class B operation, the collector current flows for**
   a. The whole cycle
   b. Half the cycle
   c. Less than half a cycle
   d. Less than a quarter of a cycle

2. **Transformer coupling is an example of**
   a. Direct coupling
   b. AC coupling
   c. DC coupling
   d. Impedance coupling

3. **An audio amplifier operates in the frequency range of**
   a. 0 to 20 Hz
   b. 20 Hz to 2 kHz
   c. 20 to 20 kHz
   d. Above 20 kHz

4. **A tuned RF amplifier is**
   a. Narrowband
   b. Wideband
   c. Direct-coupled
   d. A dc amplifier

5. **The first stage of a preamp is**
   a. A tuned RF stage
   b. Large signal
   c. Small signal
   d. A dc amplifier

6. **For maximum peak-to-peak output voltage, the Q point should be**
   a. Near saturation
   b. Near cutoff
   c. At the center of the dc load line
   d. At the center of the ac load line

7. **An amplifier has two load lines because**
   a. It has ac and dc collector resistances
   b. It has two equivalent circuits
   c. DC acts one way and ac acts another
   d. All of the above

8. **When the Q point is at the center of the ac load line, the maximum peak-to-peak output voltage equals**
   a. V_{CEQ}
   b. $2V_{CEQ}$
   c. I_{CQ}
   d. $2I_{CQ}$

9. **Push-pull is almost always used with**
   a. Class A
   b. Class B
   c. Class C
   d. All of the above

10. **One advantage of a class B push-pull amplifier is**
    a. No quiescent current drain
    b. Maximum efficiency of 78.5 percent
    c. Greater efficiency than class A
    d. All of the above

11. **Class C amplifiers are almost always**
    a. Transformer-coupled between stages
    b. Operated at audio frequencies
    c. Tuned RF amplifiers
    d. Wideband

12. **The input signal of a class C amplifier**
    a. Is negatively clamped at the base
    b. Is amplified and inverted
    c. Produces brief pulses of collector current
    d. All of the above

13. **The collector current of a class C amplifier**
    a. Is an amplified version of the input voltage
    b. Has harmonics
    c. Is negatively clamped
    d. Flows for half a cycle

14. **The bandwidth of a class C amplifier decreases when the**
    a. Resonant frequency increases
    b. Q increases
    c. X_L decreases
    d. Load resistance decreases

15. **The transistor dissipation in a class C amplifier decreases when the**
    a. Resonant frequency increases
    b. coil Q increases
    c. Load resistance decreases
    d. Capacitance increases

16. **The power rating of a transistor can be increased by**
    a. Raising the temperature
    b. Using a heat sink
    c. Using a derating curve
    d. Operating with no input signal

17. The ac load line is the same as the dc load line when the ac collector resistance equals the
    a. DC emitter resistance
    b. AC emitter resistance
    c. DC collector resistance
    d. Supply voltage divided by collector current

18. If $R_C = 100\ \Omega$ and $R_L = 180\ \Omega$, the ac load resistance equals
    a. 64 Ω c. 90 Ω
    b. 100 Ω d. 180 Ω

19. The quiescent collector current is the same as the
    a. DC collector current
    b. AC collector current
    c. Total collector current
    d. Voltage-divider current

20. The ac load line usually
    a. Equals the dc load line
    b. Has less slope than the dc load line
    c. Is steeper than the dc load line
    d. Is horizontal

21. For a Q point closer to cutoff than saturation on a CE dc load line, clipping is more likely to occur on the
    a. Positive peak of input voltage
    b. Negative peak of input voltage
    c. Negative peak of output voltage
    d. Negative peak of emitter voltage

22. In a class A amplifier, the collector current flows for
    a. Less than half the cycle
    b. Half the cycle
    c. Less than the whole cycle
    d. The entire cycle

23. With class A, the output signal should be
    a. Unclipped
    b. Clipped on positive voltage peak
    c. Clipped on negative voltage peak
    d. Clipped on negative current peak

24. The instantaneous operating point swings along the
    a. AC load line
    b. DC load line
    c. Both load lines
    d. Neither load line

25. The current drain of an amplifier is the
    a. Total ac current from the generator
    b. Total dc current from the supply
    c. Current gain from base to collector
    d. Current gain from collector to base

26. The power gain of an amplifier
    a. Is the same as the voltage gain
    b. Is smaller than the voltage gain
    c. Equals output power divided by input power
    d. Equals load power

27. Heat sinks reduce the
    a. Transistor power
    b. Ambient temperature
    c. Junction temperature
    d. Collector current

28. When the ambient temperature increases, the maximum transistor power rating
    a. Decreases
    b. Increases
    c. Remains the same
    d. None of the above

29. If the load power is 300 mW and the dc power is 1.5 W, the efficiency is
    a. 0   c. 3 percent
    b. 2 percent d. 20 percent

30. The ac load line of an emitter follower is usually
    a. The same as the dc load line
    b. Vertical
    c. More horizontal than the dc load line
    d. Steeper than the dc load line

31. If an emitter follower has $V_{CEO} = 6$ V, $I_{CQ} = 200$ mA, and $r_e = 10\ \Omega$, the maximum peak-to-peak unclipped output is
    a. 2 V
    b. 4 V
    c. 6 V
    d. 8 V

32. The ac resistance of compensating diodes
    a. Must be included
    b. Is very high
    c. Is usually small enough to ignore
    d. Compensates for temperature changes

33. If the Q point is at the middle of the dc load line, clipping will first occur on the
    a. Left voltage swing
    b. Upward current swing
    c. Positive half-cycle of input
    d. Negative half-cycle of input

34. The maximum efficiency of a class B push-pull amplifier is
    a. 25 percent
    b. 50 percent
    c. 78.5 percent
    d. 100 percent

35. A small quiescent current is necessary with a class AB push-pull amplifier to avoid
    a. Crossover distortion
    b. Destroying the compensating diodes
    c. Excessive current drain
    d. Loading the driver stage

Problems

SEC. 12-2 TWO LOAD LINES

12-1 What is the dc collector resistance in Fig. 12-38? What is the dc saturation current?

12-2 In Fig. 12-38, what is the ac collector resistance? What is the ac saturation current?

12-3 What is the maximum peak-to-peak output in Fig. 12-38?

12-4 All resistances are doubled in Fig. 12-38. What is the ac collector resistance?

12-5 All resistances are tripled in Fig. 12-38. What is the maximum peak-to-peak output?

Figure 12-38

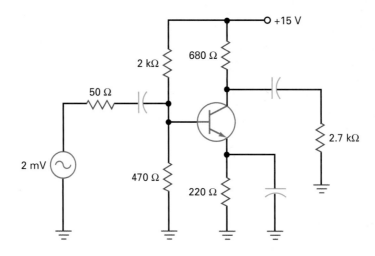

Figure 12-39

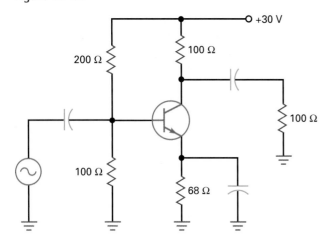

Figure 12-40

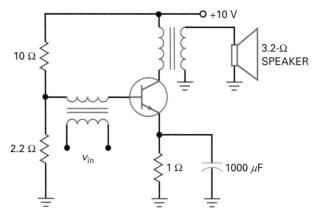

12-6 What is the dc collector resistance in Fig. 12-39? What is the dc saturation current?

12-7 In Fig. 12-39, what is the ac collector resistance? What is the ac saturation current?

12-8 What is the maximum peak-to-peak output in Fig. 12-39?

12-9 All resistances are doubled in Fig. 12-39. What is the ac collector resistance?

12-10 All resistances are tripled in Fig. 12-39. What is the maximum peak-to-peak output?

SEC. 12-3 CLASS A OPERATION

12-11 An amplifier has an input power of 4 mW and output power of 2 W. What is the power gain?

12-12 If an amplifier has a peak-to-peak output voltage of 15 V across a load resistance of 1 kΩ, what is the power gain if the input power is 400 μW?

12-13 What is the current drain in Fig. 12-38?

12-14 What is the dc power supplied to the amplifier of Fig. 12-38?

12-15 The input signal of Fig. 12-38 is increased until maximum peak-to-peak output voltage is across the load resistor. What is the efficiency?

12-16 What is the quiescent power dissipation in Fig. 12-38?

12-17 What is the current drain in Fig. 12-39?

12-18 What is the dc power supplied to the amplifier of Fig. 12-39?

12-19 The input signal of Fig. 12-39 is increased until maximum peak-to-peak output voltage is across the load resistor. What is the efficiency?

12-20 What is the quiescent power dissipation in Fig. 12-39?

12-21 If $V_{BE} = 0.7$ V in Fig. 12-40, what is the dc emitter current?

12-22 The speaker of Fig. 12-40 is equivalent to a load resistance of 3.2 Ω. If the voltage across the speaker is 5 V pp, what is the output power? What is the efficiency of the stage?

SEC. 12-6 BIASING CLASS B/AB AMPLIFIERS

12-23 The ac load line of a class B push-pull emitter follower has a cutoff voltage of 12 V. What is the maximum peak-to-peak voltage?

Figure 12-41

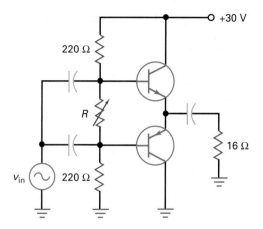

12-24 What is the maximum power dissipation of each transistor of Fig. 12-41?

12-25 What is the maximum output power in Fig. 12-41?

12-26 What is the quiescent collector current in Fig. 12-42?

12-27 In Fig. 12-42, what is the maximum efficiency of the amplifier?

Figure 12-42

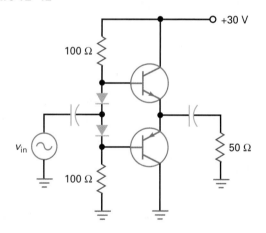

12-28 If the biasing resistors of Fig. 12-42 are changed to 1 kΩ, what is the quiescent collector current? The efficiency of the amplifier?

SEC. 12-7 CLASS B/AB DRIVERS

12-29 What is the maximum output power in Fig. 12-43?

12-30 In Fig. 12-43, what is the voltage gain of the first stage if $\beta = 200$?

12-31 If Q_3 and Q_4 have current gains of 200 in Fig. 12-43, what is the voltage gain of the second stage?

12-32 What is the quiescent collector current in Fig. 12-43?

12-33 What is the overall voltage gain for the three-stage amplifier in Fig. 12-43?

SEC. 12-8 CLASS C OPERATION

12-34 IIII MultiSim If the input voltage equals 5 V rms in Fig. 12-44, what is the peak-to-peak input voltage? If the dc voltage between the base and ground is measured, what will the voltmeter indicate?

12-35 IIII MultiSim What is the resonant frequency in Fig. 12-44?

12-36 IIII MultiSim If the inductance is doubled in Fig. 12-44, what is the resonant frequency?

12-37 IIII MultiSim What is the resonance in Fig. 12-44 if the capacitance is changed to 100 pF?

SEC. 12-9 CLASS C FORMULAS

12-38 If the class C amplifier of Fig. 12-44 has an output power of 11 mW and an input power of 50 μW, what is the power gain?

12-39 What is the output power in Fig. 12-44 if the output voltage is 50 V pp?

12-40 What is the maximum ac output power in Fig. 12-44?

12-41 If the current drain in Fig. 12-44 is 0.5 mA, what is the dc input power?

12-42 What is the efficiency of Fig. 12-44 if the current drain is 0.4 mA and the output voltage is 30 V pp?

Figure 12-43

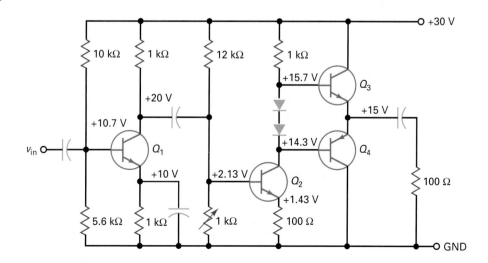

Figure 12-44

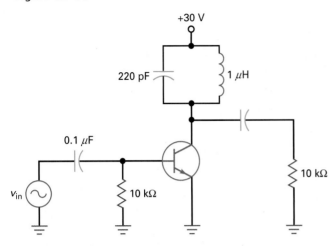

+30 V

220 pF   1 μH

0.1 μF

v_{in}

10 kΩ

10 kΩ

12-43 If the Q of the inductor is 125 in Fig. 12-44, what is the bandwidth of the amplifier?

12-44 What is the worst-case transistor power dissipation in Fig. 12-44 ($Q = 125$)?

SEC. 12-10 TRANSISTOR POWER RATING

12-45 A 2N3904 is used in Fig. 12-44. If the circuit has to operate over an ambient temperature range of 0 to 100°C, what is the maximum power rating of the transistor in the worst case?

12-46 A transistor has the derating curve shown in Fig. 12-34. What is the maximum power rating for an ambient temperature of 100°C?

12-47 The data sheet of a 2N3055 lists a power rating of 115 W for a case temperature of 25°C. If the derating factor is 0.657 W/°C, what is $P_{D(max)}$ when the case temperature is 90°C?

Critical Thinking

12-48 The output of an amplifier is a square-wave output even though the input is a sine wave. What is the explanation?

12-49 A power transistor like the one in Fig. 12-36 is used in an amplifier. Somebody tells you that since the case is grounded, you can safely touch the case. What do you think about this?

12-50 You are in a bookstore and you read the following in an electronics book: "Some power amplifiers can have an efficiency of 125 percent." Would you buy the book? Explain your answer.

12-51 Normally, the ac load line is more vertical than the dc load line. A couple of classmates say that they are willing to bet that they can draw a circuit whose ac load line is less vertical than the dc load line. Would you take the bet? Explain.

12-52 Draw the dc and ac load lines for Fig. 12-38.

Up-Down Analysis

In Fig. 12-45, P_L is the output power in the load resistor, and P_S is the dc input power from the supply.

12-53 Predict the response of the dependent variables to a slight increase in V_{CC}. Use the table to check your predictions.

12-54 Repeat Prob. 12-53 for a slight increase in R_1.

12-55 Repeat Prob. 12-53 for a slight increase in R_2.

12-56 Repeat Prob. 12-53 for a slight increase in R_E.

12-57 Repeat Prob. 12-53 for a slight increase in R_C.

12-58 Repeat Prob. 12-53 for a slight increase in V_G.

12-59 Repeat Prob. 12-53 for a slight increase in R_G.

12-60 Repeat Prob. 12-53 for a slight increase in R_L.

12-61 Repeat Prob. 12-53 for a slight increase in β.

Figure 12-45

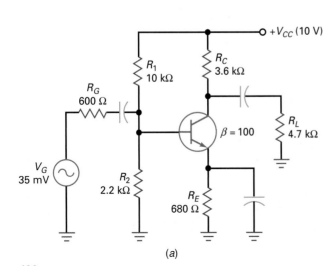

$+V_{CC}$ (10 V)

R_1 10 kΩ

R_C 3.6 kΩ

R_G 600 Ω

$\beta = 100$

R_L 4.7 kΩ

V_G 35 mV

R_2 2.2 kΩ

R_E 680 Ω

(a)

UP-Down Analysis

Slight increase	P_L	P_D	P_S	MPP	η
V_{CC}					
R_1					
R_2					
R_E					
R_C					
V_G					
R_G					
R_L					
β					

(b)

Job Interview Questions

1. Tell me about the three classes of amplifier operation. Illustrate the classes by drawing collector current waveforms.
2. Draw brief schematics showing the three types of coupling used between amplifier stages.
3. Draw a VDB amplifier. Then, draw its dc load line and ac load line. Assuming that the Q point is centered on the ac load lines, what is the ac saturation current? The ac cutoff voltage? The maximum peak-to-peak output?
4. Draw the circuit of a two-stage amplifier and tell me how to calculate the total current drain on the supply.
5. Draw a class C tuned amplifier. Tell me how to calculate the resonant frequency, and tell me what happens to the ac signal at the base. Explain how it is possible that the brief pulses of collector current produce a sine wave of voltage across the resonant tank circuit.
6. What is the most common application of a class C amplifier? Could this type of amplifier be used for an audio application? If not, why not?
7. Explain the purpose of heat sinks. Also, why do we put an insulating washer between the transistor and the heat sink?
8. What is meant by the duty cycle? How is it related to the power supplied by the source?
9. Define Q.
10. Which class of amplifier operation is most efficient? Why?
11. You have ordered a replacement transistor and heat sink. In the box with the heat sink is a package containing a white substance. What is it?
12. Comparing a class A amplifier to a class C amplifier, which has the greater fidelity? Why?
13. What type of amplifier is used when only a small range of frequencies is to be amplified?
14. What other types of amplifiers are you familiar with?

Self-Test Answers

1.	b	13.	b	25.	b
2.	b	14.	b	26.	c
3.	c	15.	b	27.	c
4.	a	16.	b	28.	a
5.	c	17.	c	29.	d
6.	d	18.	a	30.	d
7.	d	19.	a	31.	b
8.	b	20.	c	32.	c
9.	b	21.	b	33.	d
10.	d	22.	d	34.	c
11.	c	23.	a	35.	a
12.	d	24.	a		

Practice Problem Answers

12-1 $I_{CQ} = 100$ mA; $V_{CEQ} = 15$ V

12-2 $i_{c(sat)} = 350$ mA; $V_{CE(cutoff)} = 21$ V; MPP = 12 V

12-3 $Ap = 1122$

12-5 $R = 200\ \Omega$

12-6 $I_{CQ} = 331$ mA; $V_{CEQ} = 6.7$ V; $r_e = 8\ \Omega$

12-7 MPP = 5.3 V

12-8 $P_{D(max)} = 2.8$ W; $P_{out(max)} = 14$ W

12-9 Efficiency = 63%

12-10 Efficiency = 78%

12-11 $f_r = 4.76$ MHz; $V_{out} = 24$ V pp

12-13 $P_D = 16.6$ mW

12-14 $P_{D(max)} = 425$ mW

13 JFETs

The bipolar junction transistor (BJT) relies on two types of charge: free electrons and holes. This is why it is called *bipolar:* the prefix *bi* stands for "two." This chapter discusses another kind of transistor called the **field-effect transistor (FET).** This type of device is *unipolar* because its operation depends on only one type of charge, either free electrons or holes. In other words, an FET has majority carriers but not minority carriers.

For most linear applications, the BJT is the preferred device. But there are some linear applications in which the FET is better suited because of its high input impedance and other properties. Furthermore, the FET is the preferred device for most switching applications. Why? Because there are no minority carriers in an FET. As a result, it can switch off faster since no stored charge has to be removed from the junction area.

There are two kinds of unipolar transistors: JFETs and MOSFETs. This chapter discusses the *junction field-effect transistor (JFET)* and its applications. In Chapter 14, we discuss the *metal-oxide semiconductor FET (MOSFET)* and its applications.

© Eyewire/Getty Images

Objectives

After studying this chapter, you should be able to:

- Describe the basic construction of a JFET.
- Draw diagrams that show common biasing arrangements.
- Identify and describe the significant regions of JFET drain curves and transconductance curves.
- Calculate the proportional pinchoff voltage and determine which region a JFET is operating in.
- Determine the dc operating point using ideal and graphical solutions.
- Determine transconductance and use it to calculate gain in JFET amplifiers.
- Describe several JFET applications including switches, variable resistances, and choppers.
- Test JFETs for proper operation.

Chapter Outline

Vocabulary

automatic gain control (AGC)

channel

chopper

common-source (CS) amplifier

current source bias

drain

field effect

field-effect transistor (FET)

gate

gate bias

gate-source cutoff voltage

ohmic region

pinchoff voltage

self-bias

series switch

shunt switch

source

source follower

transconductance

transconductance curve

voltage-controlled device

voltage-divider bias

13-1 Basic Ideas

Figure 13-1a shows a piece of *n*-type semiconductor. The lower end is called the **source,** and the upper end is called the **drain.** The supply voltage V_{DD} forces free electrons to flow from the source to the drain. To produce a JFET, a manufacturer diffuses two areas of *p*-type semiconductor into the *n*-type semiconductor, as shown in Fig. 13-1b. These *p* regions are connected internally to get a single external **gate** *lead.*

Field Effect

Figure 13-2 shows the normal biasing voltages for a JFET. The drain supply voltage is positive, and the gate supply voltage is negative. The term **field effect** is related to the depletion layers around each *p* region. These depletion layers exist because free electrons diffuse from the *n* regions into the *p* regions. The recombination of free electrons and holes creates the depletion layers shown by the colored areas.

Reverse Bias of Gate

In Fig. 13-2, the *p*-type gate and the *n*-type source form the gate-source diode. With a JFET, we always *reverse-bias* the gate-source diode. Because of reverse bias, the gate current I_G is approximately zero, which is equivalent to saying that the JFET has an almost infinite input resistance.

Figure 13-1   (*a*) Part of JFET; (*b*) single-gate JFET.

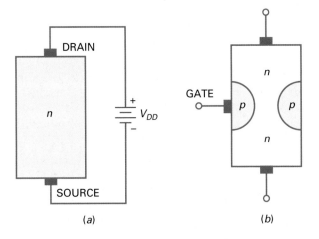

Figure 13-2   Normal biasing of JFET.

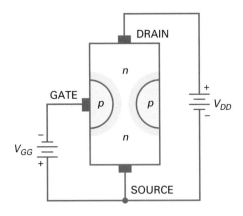

A typical JFET has an input resistance in the hundreds of megohms. This is the big advantage that a JFET has over a bipolar transistor. It is the reason that JFETs excel in applications in which a high input impedance is required. One of the most important applications of the JFET is the *source follower,* a circuit like the emitter follower, except that the input impedance is in the hundreds of megohms for lower frequencies.

Gate Voltage Controls Drain Current

In Fig. 13-2, electrons flowing from the source to the drain must pass through the narrow **channel** between the depletion layers. When the gate voltage becomes more negative, the depletion layers expand and the conducting channel becomes narrower. The more negative the gate voltage, the smaller the current between the source and the drain.

The JFET is a **voltage-controlled device** because an input voltage controls an output current. In a JFET, the gate-to-source voltage V_{GS} determines how much current flows between the source and the drain. When V_{GS} is zero, maximum drain current flows through the JFET. This is why a JFET is referred to as a normally on device. On the other hand, if V_{GS} is negative enough, the depletion layers touch and the drain current is cut off.

Schematic Symbol

The JFET of Fig. 13-2 is an *n-channel JFET* because the channel between the source and the drain is an *n*-type semiconductor. Figure 13-3*a* shows the schematic symbol for an *n*-channel JFET. In many low-frequency applications, the source and the drain are interchangeable because you can use either end as the source and the other end as the drain.

The source and drain terminals are not interchangeable at high frequencies. Almost always, the manufacturer minimizes the internal capacitance on the drain side of the JFET. In other words, the capacitance between the gate and the drain is smaller than the capacitance between the gate and the source. You will learn more about internal capacitances and their effect on circuit action in a later chapter.

Figure 13-3*b* shows an alternative symbol for an *n*-channel JFET. This symbol with its offset gate is preferred by many engineers and technicians. The offset gate points to the source end of the device, a definite advantage in complicated multistage circuits.

There is also a *p*-channel JFET. The schematic symbol for a *p*-channel JFET, shown in Fig. 13-3*c*, is similar to that for the *n*-channel JFET, except that the gate arrow points in the opposite direction. The action of a *p*-channel JFET is complementary; that is, all voltages and currents are reversed. To reverse bias a *p*-channel JFET, the gate is made positive in respect to the source. Therefore, V_{GS} is made positive.

GOOD TO KNOW

The depletion layers are actually wider near the top of the *p*-type materials and narrower at the bottom. The reason for the change in the width can be understood by realizing that the drain current I_D will produce a voltage drop along the length of the channel. With respect to the source, a more positive voltage is present as you move up the channel toward the drain end. Since the width of a depletion layer is proportional to the amount of reverse-bias voltage, the depletion layer of the *pn* junction must be wider at the top, where the amount of reverse-bias voltage is greater.

Figure 13-3 (*a*) Schematic symbol; (*b*) offset-gate symbol; (*c*) *p*-channel symbol.

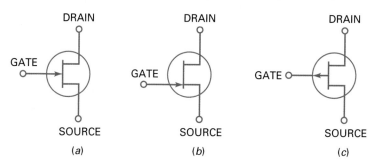

(a)   *(b)*   *(c)*

Example 13-1

A 2N5486 JFET has a gate current of 1 nA when the reverse gate voltage is 20 V. What is the input resistance of this JFET?

SOLUTION Use Ohm's law to calculate:

$$R_{in} = \frac{20\ V}{1\ nA} = 20,000\ M\Omega$$

PRACTICE PROBLEM 13-1 In Example 13-1, calculate the input resistance if the JFET's gate current is 2 nA.

13-2 Drain Curves

Figure 13-4a shows a JFET with normal biasing voltages. In this circuit, the gate-source voltage V_{GS} equals the gate supply voltage V_{GG}, and the drain-source voltage V_{DS} equals the drain supply voltage V_{DD}.

Maximum Drain Current

If we short the gate to the source, as shown in Fig. 13-4b, we will get maximum drain current because $V_{GS} = 0$. Figure 13-4c shows the graph of drain current I_D versus drain-source voltage V_{DS} for this shorted-gate condition. Notice how the drain current increases rapidly and then becomes almost horizontal when V_{DS} is greater than V_P.

Figure 13-4 (a) Normal bias; (b) zero gate voltage; (c) shorted gate drain current.

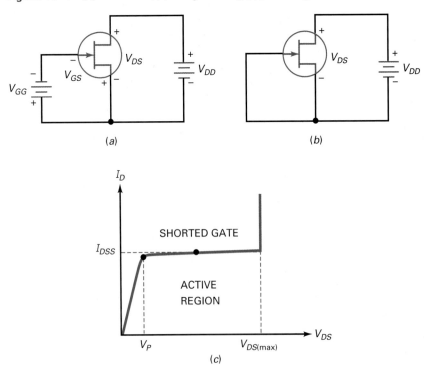

Why does the drain current become almost constant? When V_{DS} increases, the depletion layers expand. When $V_{DS} = V_P$, the depletion layers are almost touching. The narrow conducting channel therefore pinches off or prevents a further increase in current. This is why the current has an upper limit of I_{DSS}.

The active region of a JFET is between V_P and $V_{DS(max)}$. The minimum voltage V_P is called the **pinchoff voltage,** and the maximum voltage $V_{DS(max)}$ is the *breakdown voltage.* Between pinchoff and breakdown, the JFET acts like a current source of approximately I_{DSS} when $V_{GS} = 0$.

I_{DSS} stands for the current drain to source with a shorted gate. This is the maximum drain current a JFET can produce. The data sheet of any JFET lists the value of I_{DSS}. This is one of the most important JFET quantities, and you should always look for it first because it is the upper limit on the JFET current.

The Ohmic Region

In Fig. 13-5, the pinchoff voltage separates two major operating regions of the JFET. The almost-horizontal region is the active region. The almost-vertical part of the drain curve below pinchoff is called the **ohmic region.**

When operated in the ohmic region, a JFET is equivalent to a resistor with a value of approximately:

$$R_{DS} = \frac{V_P}{I_{DSS}} \tag{13-1}$$

R_{DS} is called the *ohmic resistance of the JFET.* In Fig. 13-5, $V_P = 4$ V and $I_{DSS} = 10$ mA. Therefore, the ohmic resistance is:

$$R_{DS} = \frac{4\,\text{V}}{10\,\text{mA}} = 400\;\Omega$$

If the JFET is operating anywhere in the ohmic region, it has an ohmic resistance of 400 Ω.

Gate Cutoff Voltage

Figure 13-5 shows the drain curves for a JFET with an I_{DSS} of 10 mA. The top curve is always for $V_{GS} = 0$, the shorted-gate condition. In this example, the pinchoff voltage is 4 V and the breakdown voltage is 30 V. The next curve down is for $V_{GS} = -1$ V, the next for $V_{GS} = -2$ V, and so on. As you can see, the more negative the gate-source voltage, the smaller the drain current.

Figure 13-5 Drain curves.

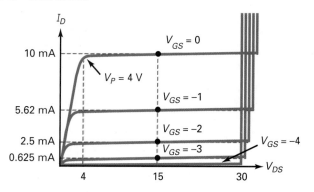

The bottom curve is important. Notice that a V_{GS} of -4 V reduces the drain current to almost zero. This voltage is called the **gate-source cutoff voltage** and is symbolized by $V_{GS(off)}$ on data sheets. At this cutoff voltage the depletion layers touch. In effect, the conducting channel disappears. This is why the drain current is approximately zero.

In Fig. 13-5, notice that

$$V_{GS(off)} = -4 \text{ V} \qquad \text{and} \qquad V_P = 4 \text{ V}$$

This is not a coincidence. The two voltages always have the same magnitude because they are the values where the depletion layers touch or almost touch. Data sheets may list either quantity, and you are expected to know that the other has the same magnitude. As an equation:

$$V_{GS(off)} = -V_P \qquad\qquad (13\text{-}2)$$

Example 13-2

An MPF4857 has $V_P = 6$ V and $I_{DSS} = 100$ mA. What is the ohmic resistance? The gate-source cutoff voltage?

SOLUTION   The ohmic resistance is:

$$R_{DS} = \frac{6 \text{ V}}{100 \text{ mA}} = 60 \ \Omega$$

Since the pinchoff voltage is 6 V, the gate-source cutoff voltage is:

$$V_{GS(off)} = -6 \text{ V}$$

PRACTICE PROBLEM 13-2   A 2N5484 has a $V_{GS(off)} = -3.0$ V and $I_{DSS} = 5$ mA. Find its ohmic resistance and V_p values.

13-3 The Transconductance Curve

The **transconductance curve** of a JFET is a graph of I_D versus V_{GS}. By reading the values of I_D and V_{GS} of each drain curve in Fig. 13-5, we can plot the curve of Fig. 13-6a. Notice that the curve is nonlinear because the current increases faster when V_{GS} approaches zero.

Any JFET has a transconductance curve like Fig. 13-6b. The end points on the curve are $V_{GS(off)}$ and I_{DSS}. The equation for this graph is:

$$I_D = I_{DSS}\left(1 - \frac{V_{GS}}{V_{GS(off)}}\right)^2 \qquad\qquad (13\text{-}3)$$

Because of the squared quantity in this equation, JFETs are often called *square-law devices*. The squaring of the quantity produces the nonlinear curve of Fig. 13-6b.

Figure 13-6c shows a *normalized transconductance curve. Normalized* means that we are graphing ratios like I_D/I_{DSS} and $V_{GS}/V_{GS(off)}$.

Figure 13-6 Transconductance curve.

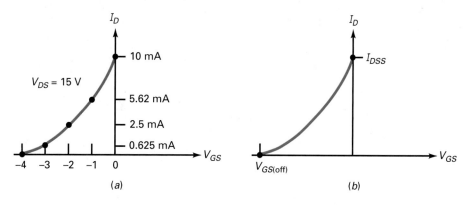

(a)

(b)

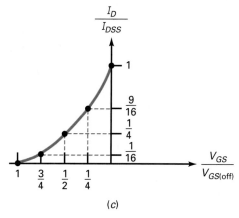

(c)

In Fig. 13-6c, the half-cutoff point

$$\frac{V_{GS}}{V_{GS(\text{off})}} = \frac{1}{2}$$

produces a normalized current of:

$$\frac{I_D}{I_{DSS}} = \frac{1}{4}$$

In words: When the gate voltage is half the cutoff voltage, the drain current is one quarter of maximum.

Example 13-3

A 2N5668 has $V_{GS(\text{off})} = -4$ V and $I_{DSS} = 5$ mA. What are the gate voltage and drain current at the half cutoff point?

SOLUTION   At the half cutoff point:

$$V_{GS} = \frac{-4\,V}{2} = -2\,V$$

and the drain current is:

$$I_D = \frac{5\,mA}{4} = 1.25\,mA$$

Example 13-4

A 2N5459 has $V_{GS(off)} = -8$ V and $I_{DSS} = 16$ mA. What is the drain current at the half cutoff point?

SOLUTION The drain current is one quarter of the maximum, or:

$$I_D = 4 \text{ mA}$$

The gate-source voltage that produces this current is -4 V, half of the cutoff voltage.

PRACTICE PROBLEM 13-4 Repeat Example 13-4 using a JFET with $V_{GS(off)} = -6$ V and $I_{DSS} = 12$ mA.

13-4 Biasing in the Ohmic Region

The JFET can be biased in the ohmic or in the active region. When biased in the ohmic region, the JFET is equivalent to a resistance. When biased in the active region, the JFET is equivalent to a current source. In this section, we discuss gate bias, the method used to bias a JFET in the ohmic region.

Gate Bias

Figure 13-7a shows **gate bias.** A negative gate voltage of $-V_{GG}$ is applied to the gate through biasing resistor R_G. This sets up a drain current that is less than I_{DSS}. When the drain current flows through R_D, it sets up a drain voltage of:

$$V_D = V_{DD} - I_D R_D \tag{13-4}$$

Gate bias is the worst way to bias a JFET in the active region because the Q point is too unstable.

For example, a 2N5459 has the following spreads between minimum and maximum: I_{DSS} varies from 4 to 16 mA, and $V_{GS(off)}$ varies from -2 to -8 V. Figure 13-7b shows the minimum and maximum transconductance curves. If a gate bias of -1 V is used with this JFET, we get the minimum and maximum Q points shown. Q_1 has a drain current of 12.3 mA, and Q_2 has a drain current of only 1 mA.

Hard Saturation

Although not suitable for active-region biasing, gate bias is perfect for ohmic-region biasing because stability of the Q point does not matter. Figure 13-7c shows how to bias a JFET in the ohmic region. The upper end of the dc load line has a drain saturation current of:

$$I_{D(sat)} = \frac{V_{DD}}{R_D}$$

To ensure that a JFET is biased in the ohmic region, all we need to do is use $V_{GS} = 0$ and:

$$I_{D(sat)} \ll I_{DSS} \tag{13-5}$$

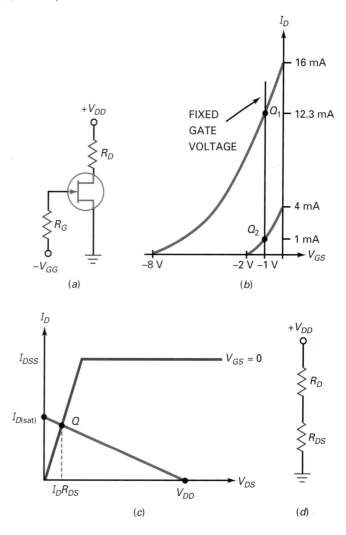

The symbol $\ll$ means "much less than." This equation says that the drain saturation current must be much less than the maximum drain current. For instance, if a JFET has $I_{DSS} = 10$ mA, hard saturation will occur if $V_{GS} = 0$ and $I_{D(\text{sat})} = 1$ mA.

When a JFET is biased in the ohmic region, we can replace it by a resistance of R_{DS}, as shown in Fig. 13-7*d*. With this equivalent circuit, we can calculate the drain voltage. When R_{DS} is much smaller than R_D, the drain voltage is close to zero.

Example 13-5

What is the drain voltage in Fig. 13-8*a*?

SOLUTION Since $V_P = 4$ V, $V_{GS(\text{off})} = -4$ V. Before point *A* in time, the input voltage is -10 V and the JFET is cut off. In this case, the drain voltage is:

$$V_D = 10 \text{ V}$$

Figure 13-8 Example.

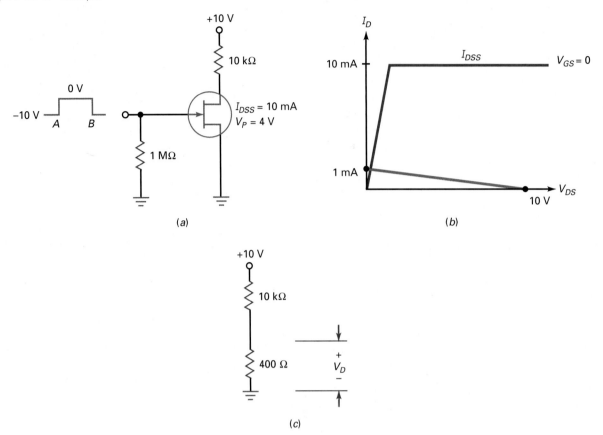

(a)

(b)

(c)

Between points A and B, the input voltage is 0 V. The upper end of the dc load line has a saturation current of:

$$I_{D(\text{sat})} = \frac{10\text{ V}}{10\text{ k}\Omega} = 1\text{ mA}$$

Figure 13-8b shows the dc load line. Since $I_{D(\text{sat})}$ is much less than I_{DSS}, the JFET is in hard saturation.

The ohmic resistance is:

$$R_{DS} = \frac{4\text{ V}}{10\text{ mA}} = 400\ \Omega$$

In the equivalent circuit of Fig. 13-8c, the drain voltage is:

$$V_D = \frac{400\ \Omega}{10\text{ k}\Omega + 400\ \Omega}\ 10\text{ V} = 0.385\text{ V}$$

PRACTICE PROBLEM 13-5 Using Fig. 13-8a, find R_{DS} and V_D if $V_p = 3$ V.

13-5 Biasing in the Active Region

JFET amplifiers need to have a Q point in the active region. Because of the large spread in JFET parameters, we cannot use gate bias. Instead, we need to use other biasing methods. Some of these methods are similar to those used with bipolar junction transistors.

The choice of analysis technique depends on the level of accuracy needed. For example, when doing preliminary analysis and troubleshooting of biasing circuits, it is often desirable to use ideal values and circuit approximations. In JFET circuits, this means that we will often ignore V_{GS} values. Usually, the ideal answers will have an error of less that 10 percent. When closer analysis is called for, we can use graphical solutions to determine a circuit's Q point. If you are designing JFET circuits or need even greater accuracy, you should use a circuit simulator like MultiSim (EWB).

Self-Bias

Figure 13-9a shows **self-bias.** Since drain current flows through the source resistor R_S, a voltage exists between the source and ground, given by:

$$V_S = I_D R_S \tag{13-6}$$

Since V_G is zero,

$$V_{GS} = -I_D R_S \tag{13-7}$$

This says that the gate-source voltage equals the negative of the voltage across the source resistor. Basically, the circuit creates its own bias by using the voltage developed across R_S to reverse bias the gate.

Figure 13-9b shows the effect of different source resistors. There is a medium value of R_S at which the gate-source voltage is half of the cutoff voltage. An approximation for this medium resistance is:

$$R_S \approx R_{DS} \tag{13-8}$$

This equation says that the source resistance should equal the ohmic resistance of the JFET. When this condition is satisfied, the V_{GS} is roughly half the cutoff voltage and the drain current is roughly one-quarter of I_{DSS}.

Figure 13-9 Self-bias.

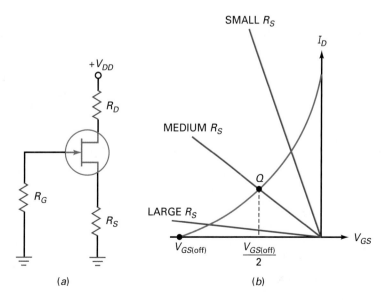

(a)　　　　　　(b)

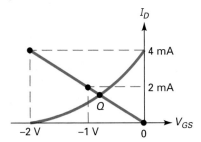

Figure 13-10 Self-bias Q point.

When a JFET's transconductance curves are known, we can analyze a self-bias circuit using graphical methods. Suppose a self-bias JFET has the transconductance curve shown in Fig. 13-10. The maximum drain current is 4 mA, and the gate voltage has to be between 0 and −2 V. By graphing Eq. (13-7), we can find out where it intersects the transconductance curve and determine the values of V_{GS} and I_D. Since Eq. (13-7) is a linear equation, all we have to do is plot two points and draw a line through them.

Suppose the source resistance is 500 Ω. Then Eq. (13-7) becomes:

$$V_{GS} = -I_D \, (500 \; \Omega)$$

Since any two points can be used, we choose the two convenient points corresponding to $I_D = -(0)(500 \; \Omega) = 0$, therefore, the coordinates for the first point are (0, 0), which is the origin. To get the second point, find V_{GS} for $I_D = I_{DSS}$. In this case, $I_D = 4$ mA and $V_{GS} = -(4 \; \text{mA})(500 \; \Omega) = -2$ V, therefore, the coordinates of the second point are at (4 mA, −2 V).

We now have two points on the graph of Eq. (13-7). The two points are (0, 0) and (4 mA, −2 V). By plotting these two points as shown in Fig. 13-10, we can draw a straight line through the two points as shown. This line will, of course, intersect the transconductance curve. This intersection point is the operating point of the self-biased JFET. As you can see, the drain current is slightly less than 2 mA, and the gate-source voltage is slightly less than −1 V.

In summary, here is a process for finding the Q point of any self-biased JFET, provided you have the transconductance curve. If the curve is not available, you can use the $V_{GS(\text{off})}$ and I_{DSS} rated values, along with the square law equation (13-3), to develop one:

1. Multiply I_{DSS} by R_S to get V_{GS} for the second point.
2. Plot the second point (I_{DSS}, V_{GS}).
3. Draw a line through the origin and the second point.
4. Read the coordinates of the intersection point.

The Q point with self-bias is not extremely stable. Because of this, self-bias is used only with small-signal amplifiers. This is why you may see self-biased JFET circuits near the front end of communication receivers where the signal is small.

Example 13-6

In Fig. 13-11a, what is a medium source resistance using the rule discussed earlier? Estimate the drain voltage with this source resistance.

SOLUTION As discussed earlier, self-bias works fine if you use a source resistance equal to the ohmic resistance of the JFET:

$$R_{DS} = \frac{4 \; \text{V}}{10 \; \text{mA}} = 400 \; \Omega$$

Figure 13-11b shows a source resistance of 400 Ω. In this case, the drain current is around one-quarter of 10 mA, or 2.5 mA, and the drain voltage is roughly:

$$V_D = 30 \; \text{V} - (2.5 \; \text{mA})(2 \; \text{k}\Omega) = 25 \; \text{V}$$

Figure 13-11 Example.

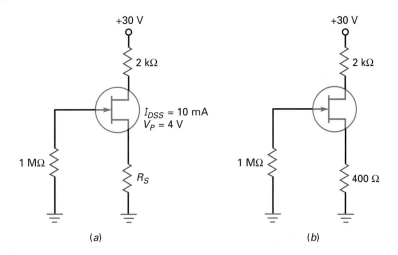

(a)   (b)

PRACTICE PROBLEM 13-6 Repeat Example 13-6 using a JFET with $I_{DSS} = 8$ mA. Determine R_S and V_D.

Example 13-7

Using the MultiSim circuit of Fig. 13-12a, along with the minimum and maximum transconductance curves for a 2N5486 JFET shown in Fig. 13-12b, determine the range of V_{GS} and I_D Q point values. Also, what would be the optimum source resistor for this JFET?

SOLUTION First, multiply I_{DSS} by R_S to get V_{GS}:

$$V_{GS} = -(20 \text{ mA})(270 \ \Omega) = -5.4 \text{ V}$$

Second, plot the second point (I_{DSS}, V_{GS}):

$$(20 \text{ mA}, -5.4 \text{ V})$$

Now draw a line through the origin (0, 0) and the second point. Then read the coordinates of the intersection points for the minimum and maximum Q point values.

$$Q \text{ point (min)} \quad V_{GS} = -0.8 \text{ V} \quad I_D = 2.8 \text{ mA}$$
$$Q \text{ point (max)} \quad V_{GS} = -2.1 \text{ V} \quad I_D = 8.0 \text{ mA}$$

Note that the MultiSim measured values of Fig. 13-12a are between the minimum and maximum values. The optimum source resistor can be found by:

$$R_S = \frac{V_{GS(\text{off})}}{I_{DSS}} \quad \text{or} \quad R_S = \frac{V_P}{I_{DSS}}$$

using minimum values:

$$R_S = \frac{2 \text{ V}}{8 \text{ mA}} = 250 \ \Omega$$

using maximum values:

$$R_S = \frac{6 \text{ V}}{20 \text{ mA}} = 300 \ \Omega$$

Notice the value of R_S in Fig. 13-12a is an approximate midpoint value between $R_{S(\text{min})}$ and $R_{S(\text{max})}$.

JFETs   **437**

Figure 13-12 (a) Self-bias example; (b) transconductance curves.

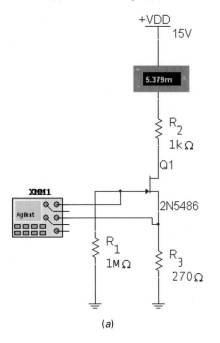

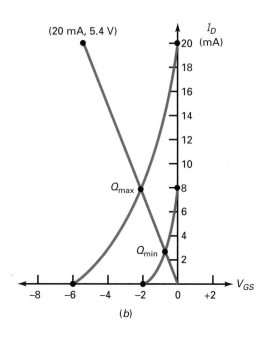

(a)

(b)

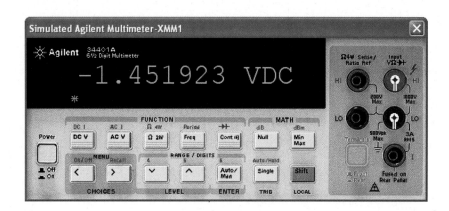

PRACTICE PROBLEM 13-7 In Fig. 13-12a, change R_S to 390 Ω and find the Q point values.

Voltage–Divider Bias

Figure 13-13a shows **voltage-divider bias.** The voltage divider produces a gate voltage that is a fraction of the supply voltage. By subtracting the gate-source voltage, we get the voltage across the source resistor:

$$V_S = V_G - V_{GS}$$

(13-9)

Figure 13-13 Voltage-divider bias.

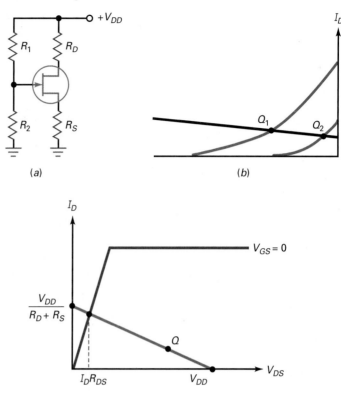

(a)

(b)

(c)

Since V_{GS} is a negative, the source voltage will be slightly larger than the gate voltage. When you divide this source voltage by the source resistance, you get the drain current:

$$I_D = \frac{V_G - V_{GS}}{R_S} \approx \frac{V_G}{R_S} \tag{13-10}$$

When the gate voltage is large, it can swamp out the variations in V_{GS} from one JFET to the next. Ideally, the drain current equals the gate voltage divided by the source resistance. As a result, the drain current is almost constant for any JFET, as shown in Fig. 13-13b.

Figure 13-13c shows the dc load line. For an amplifier, the Q point has to be in the active region. This means that V_{DS} must be greater than $I_D R_{DS}$ (ohmic region) and less than V_{DD} (cutoff). When a large supply voltage is available, voltage-divider bias can set up a stable Q point.

When more accuracy is needed in determining the Q point for a voltage-divider bias circuit, a graphical method can be used. This is especially true when the minimum and maximum V_{GS} values for a JFET vary several volts from each other. In Fig. 13-13a, the voltage applied to the gate is

$$V_G = \frac{R_2}{R_1 + R_2}(V_{DD}) \tag{13-11}$$

Using transconductance curves, as in Fig. 13-14, plot the V_G value on the horizontal, or x-axis, of the graph. This becomes one point on our bias line. To get the second point, use Eq. (13-10) with $V_{GS} = 0$ V to determine I_D. This second point, where $I_D = V_G/R_S$, is plotted on the vertical, or y-axis, of the transconductance curve. Next, draw a line between these two points and extend the line so it intersects the transconductance curves. Finally, read the coordinates of the intersection points.

Figure 13-14 VDB *Q* point.

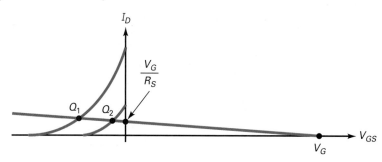

Example 13-8

Draw the dc load line and the *Q* point for Fig. 13-15*a* using ideal methods.

SOLUTION The 3:1 voltage divider produces a gate voltage of 10 V. Ideally, the voltage across the source resistor is:

$$V_S = 10 \text{ V}$$

The drain current is:

$$I_D = \frac{10 \text{ V}}{2 \text{ k}\Omega} = 5 \text{ mA}$$

and the drain voltage is:

$$V_D = 30 \text{ V} - (5 \text{ mA})(1 \text{ k}\Omega) = 25 \text{ V}$$

The drain-source voltage is:

$$V_{DS} = 25 \text{ V} - 10 \text{ V} = 15 \text{ V}$$

The dc saturation current is:

$$I_{D(\text{sat})} = \frac{30 \text{ V}}{3 \text{ k}\Omega} = 10 \text{ mA}$$

and the cutoff voltage is:

$$V_{DS(\text{cutoff})} = 30 \text{ V}$$

Figure 13-15 Example.

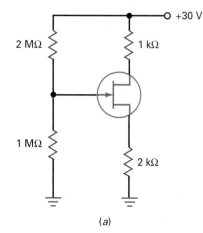

(a)

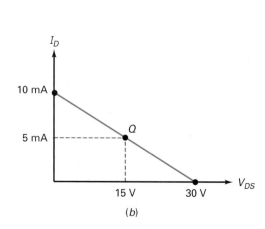

(b)

Figure 13-15*b* shows the dc load line and the *Q* point.

PRACTICE PROBLEM 13–8 In Fig. 13-15, change V_{DD} to 24 V. Solve for I_D and V_{DS} using ideal methods.

Example 13–9

Again using Fig. 13-15*a*, solve for the minimum and maximum *Q* point values using the graphical method and transconductance curves for a 2N5486 JFET shown in Fig. 13-16*a*. How does this compare to the measured values using MultiSim?

Figure 13–16 (*a*) Transconductance; (*b*) MultiSim measurements.

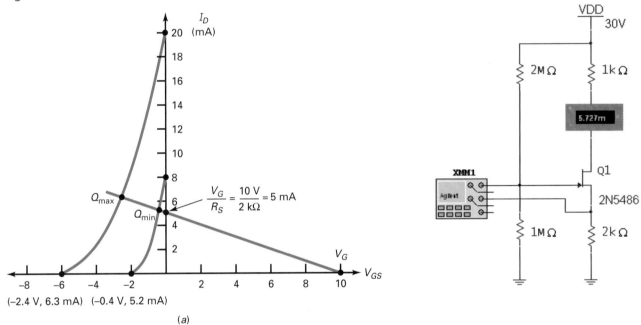

$$\frac{V_G}{R_S} = \frac{10 \text{ V}}{2 \text{ k}\Omega} = 5 \text{ mA}$$

(−2.4 V, 6.3 mA) (−0.4 V, 5.2 mA)

(*a*)

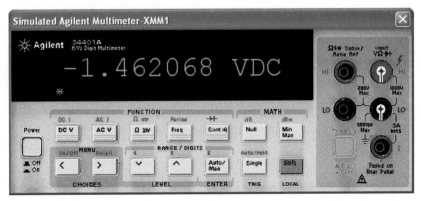

(*b*)

SOLUTION First, the value of V_G is found by:

$$V_G = \frac{1\ M\Omega}{2\ M\Omega + 1\ M\Omega}(30\ V) = 10\ V$$

This value is plotted on the *x*-axis.
Next, find the second point:

$$I_D = \frac{V_G}{R_S} = \frac{10\ V}{2\ K} = 5\ mA$$

This value is plotted on the *y*-axis.
By drawing a line between these two points extending through the minimum and maximum transconductance curves, we find:

$$V_{GS(min)} = -0.4\ V \qquad I_{D(min)} = 5.2\ mA$$

and

$$V_{GS(max)} = -2.4\ V \qquad I_{D(max)} = 6.3\ mA$$

Fig. 13-16*b* shows that the measured MultiSim values fall between the calculated minimum and maximum values.

PRACTICE PROBLEM 13-9 Using Fig. 13-15*a*, find the maximum I_D value using graphical methods when $V_{DD} = 24$ V.

Two-Supply Source Bias

Figure 13-17 shows two-supply source bias. The drain current is given by:

$$I_D = \frac{V_{SS} - V_{GS}}{R_S} \approx \frac{V_{SS}}{R_S} \tag{13-12}$$

Again, the idea is to swamp out the variations in V_{GS} by making V_{SS} much larger than V_{GS}. Ideally, the drain current equals the source supply voltage divided by the source resistance. In this case, the drain current is almost constant in spite of JFET replacement and temperature change.

Current-Source Bias

When the drain supply voltage is not large, there may not be enough gate voltage to swamp out the variations in V_{GS}. In this case, a designer may prefer to use the **current-source bias** of Fig. 13-18*a*. In this circuit, the bipolar junction transistor pumps a fixed current through the JFET. The drain current is given by:

$$I_D = \frac{V_{EE} - V_{BE}}{R_E} \tag{13-13}$$

Figure 13-18*b* illustrates how effective current-source bias is. Both Q points have the same current. Although V_{GS} is different for each Q point, V_{GS} no longer has an effect on the value of drain current.

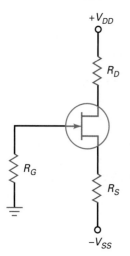

Figure 13-17 Two-supply source bias.

Figure 13-18 Current-source bias.

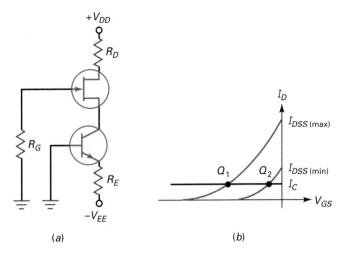

(a)   (b)

Example 13-10

What is the drain current in Fig. 13-19a? The voltage between the drain and ground?

SOLUTION Ideally, 15 V appears across the source resistor, producing a drain current of:

$$I_D = \frac{15\text{ V}}{3\text{ k}\Omega} = 5\text{ mA}$$

The drain voltage is:

$$V_D = 15\text{ V} - (5\text{ mA})(1\text{ k}\Omega) = 10\text{ V}$$

Figure 13-19 Example.

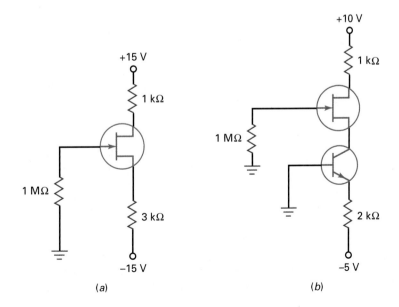

(a)   (b)

Figure 13-19 (continued)

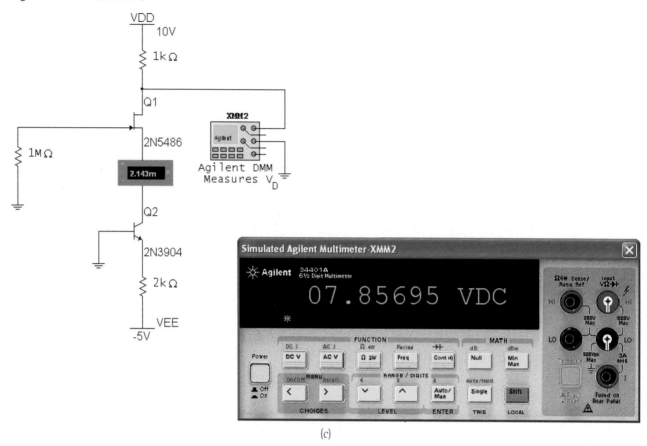

(c)

Example 13-11

|||| MultiSim

In Fig. 13-19*b*, what is the drain current? The drain voltage?

SOLUTION The bipolar junction transistor sets up a drain current of:

$$I_D = \frac{5\text{ V} - 0.7\text{ V}}{2\text{ k}\Omega} = 2.15\text{ mA}$$

The drain voltage is:

$$V_D = 10\text{ V} - (2.15\text{ mA})(1\text{ k}\Omega) = 7.85\text{ V}$$

Fig. 13-19*c* shows how close the MultiSim measured values are to the calculated values.

PRACTICE PROBLEM 13-11 Repeat Example 13-11 with $R_E = 1\text{ k}\Omega$.

Summary Table 13-1 shows the most popular types of JFET bias circuits. The graphical operating points on the transconductance curves should clearly demonstrate the advantage of one biasing technique over another.

Gate bias

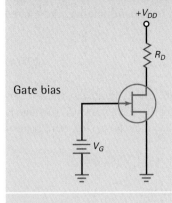
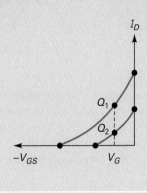

$$I_D = I_{DSS}\left(1 - \frac{V_{GS}}{V_{GS(off)}}\right)^2$$

$$V_{GS} = V_G$$

$$V_D = V_{DD} - I_D R_D$$

Self-bias

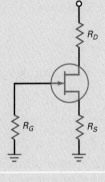
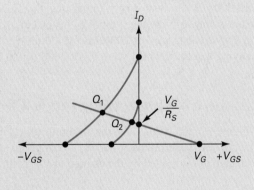

$$V_{GS} = -I_D(R_S)$$

$$\text{Second point} = (I_{DSS})(R_S)$$

VDB

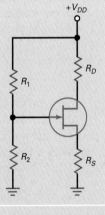
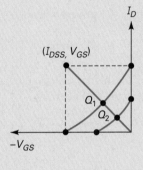

$$V_G = \frac{R_2}{R_1 + R_2}(V_{DD})$$

$$I_D = \frac{V_G}{R_S}$$

$$V_{DS} = V_D - V_S$$

Current-source bias

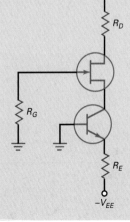
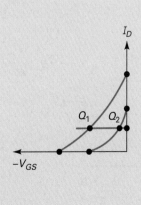

$$I_D = \frac{V_{EE} - V_{BE}}{R_E}$$

$$V_D = V_{DD} - I_D R_D$$

13-6 Transconductance

To analyze JFET amplifiers, we need to discuss **transconductance,** designated g_m and defined as:

$$g_m = \frac{i_d}{v_{gs}} \qquad (13\text{-}14)$$

This says that transconductance equals the ac drain current divided by the ac gate-source voltage. Transconductance tells us how effective the gate-source voltage is in controlling the drain current. The higher the transconductance, the more control the gate voltage has over the drain current.

For instance, if $i_d = 0.2$ mA pp when $v_{gs} = 0.1$ V pp, then:

$$g_m = \frac{0.2 \text{ mA}}{0.1 \text{ V}} = 2(10^{-3}) \text{ mho} = 2000 \ \mu\text{mho}$$

On the other hand, if $i_d = 1$ mA pp when $v_{gs} = 0.1$ V pp, then:

$$g_m = \frac{1 \text{ mA}}{0.1 \text{ V}} = 10,000 \ \mu\text{mho}$$

In the second case, the higher transconductance means that the gate is more effective in controlling the drain current.

Siemen

The unit *mho* is the ratio of current to voltage. An equivalent and modern unit for the mho is the *siemen (S),* so the foregoing answers can be written as 2000 μS and 10,000 μS. On data sheets, either quantity (mho or siemen) may be used. Data sheets may also use the symbol g_{fs} instead of g_m. As an example, the data sheet of a 2N5451 lists a g_{fs} of 2000 μS for a drain current of 1 mA. This is identical to saying that the 2N5451 has a g_m of 2000 μmho for a drain current of 1 mA.

Slope of Transconductance Curve

Figure 13-20a brings out the meaning of g_m in terms of the transconductance curve. Between points A and B, a change in V_{GS} produces a change in I_D. The change in I_D divided by the change in V_{GS} is the value of g_m between A and B. If

Figure 13-20 (a) Transconductance; (b) ac equivalent circuit; (c) variation of g_m.

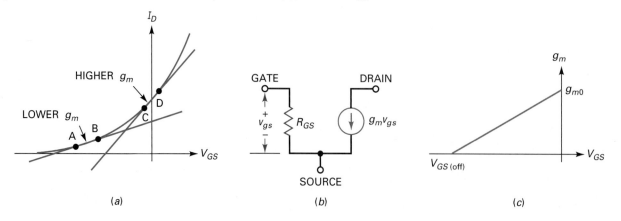

we select another pair of points farther up the curve at C and D, we get a bigger change in I_D for the same change in V_{GS}. Therefore, g_m has a larger value higher up the curve. Stated another way, g_m is the slope of the transconductance curve. The steeper the curve is at the Q point, the higher the transconductance.

Figure 13-20b shows an ac equivalent circuit for a JFET. A very high resistance R_{GS} is between the gate and the source. The drain of a JFET acts like a current source with a value of $g_m v_{gs}$. Given the values of g_m and v_{gs}, we can calculate the ac drain current.

Transconductance and Gate–Source Cutoff Voltage

The quantity $V_{GS(off)}$ is difficult to measure accurately. On the other hand, I_{DSS} and g_{m0} are easy to measure with high accuracy. For this reason, $V_{GS(off)}$ is often calculated with the following equation:

$$V_{GS(off)} = \frac{-2I_{DSS}}{g_{m0}} \qquad (13\text{-}15)$$

In this equation, g_{m0} is the value of transconductance when $V_{GS} = 0$. Typically, a manufacturer will use the foregoing equation to calculate the value of $V_{GS(off)}$ for use on data sheets.

The quantity g_{m0} is the maximum value of g_m for a JFET because it occurs when $V_{GS} = 0$. When V_{GS} becomes negative, g_m decreases. Here is the equation for calculating g_m for any value of V_{GS}:

$$g_m = g_{m0}\left(1 - \frac{V_{GS}}{V_{GS(off)}}\right) \qquad (13\text{-}16)$$

Notice that g_m decreases linearly when V_{GS} becomes more negative, as shown in Fig. 13-20c. Changing the value of g_m is useful in *automatic gain control,* which is discussed later.

Example 13-12

A 2N5457 has $I_{DSS} = 5$ mA and $g_{m0} = 5000$ μS. What is the value of $V_{GS(off)}$? What does g_m equal when $V_{GS} = -1$ V?

SOLUTION With Eq. (13-15):

$$V_{GS(off)} = \frac{-2(5 \text{ mA})}{5000\,\mu S} = -2 \text{ V}$$

Next, use Eq. (13-16) to get:

$$g_m = (5000\ \mu S)\left(1 - \frac{1\,V}{2\,V}\right) = 2500\ \mu S$$

PRACTICE PROBLEM 13-12 Repeat Example 13-12 using $I_{DSS} = 8$ mA and $V_{GS} = -2$ V.

13-7 JFET Amplifiers

Figure 13-21a shows a **common-source (CS) amplifier.** The coupling and bypass capacitors are ac shorts. Because of this, the signal is coupled directly into the gate. Since the source is bypassed to ground, all of the ac input voltage appears between the gate and the source. This produces an ac drain current. Since the ac drain current flows through the drain resistor, we get an amplified and inverted ac output voltage. This output signal is then coupled to the load resistor.

Voltage Gain of CS Amplifier

Figure 13-21b shows the ac equivalent circuit. The ac drain resistance r_d is defined as:

$$r_d = R_D \| R_L$$

The voltage gain is:

$$A_v = \frac{v_{\text{out}}}{v_{\text{in}}} = \frac{g_m v_{\text{in}} r_d}{v_{\text{in}}}$$

which simplifies to:

$$A_v = g_m r_d \tag{13-17}$$

This says that the voltage gain of a CS amplifier equals the transconductance times the ac drain resistance.

Source Follower

Figure 13-22 shows a **source follower.** The input signal drives the gate, and the output signal is coupled from the source to the load resistor. Like the emitter follower, the source follower has a voltage gain less than 1. The main advantage of the source

Figure 13-21 (a) CS amplifier; (b) ac equivalent circuit.

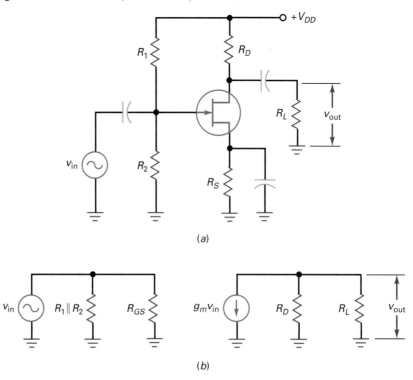

(a)

(b)

Figure 13-22 Source follower.

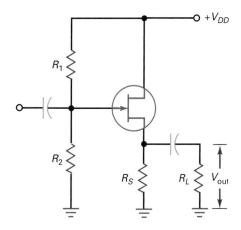

follower is its very high input resistance. Often, you will see a source follower used at the front end of a system, followed by bipolar stages of voltage gain.

In Fig. 13-22, the ac source resistance is defined as:

$$r_s = R_S \parallel R_L$$

It is possible to derive this equation for the voltage gain of a source follower:

$$A_v = \frac{g_m r_s}{1 + g_m r_s} \qquad (13\text{-}18)$$

Because the denominator is always greater than the numerator, the voltage gain is always less than 1.

Example 13–13

IIII **MultiSim**

If $g_m = 5000 \ \mu S$ in Fig. 13-23, what is the output voltage?

SOLUTION   The ac drain resistance is:

$$r_d = 3.6 \ k\Omega \parallel 10 \ k\Omega = 2.65 \ k\Omega$$

The voltage gain is:

$$A_v = (5000 \ \mu S)(2.65 \ k\Omega) = 13.3$$

Figure 13-23   Example of CS amplifier.

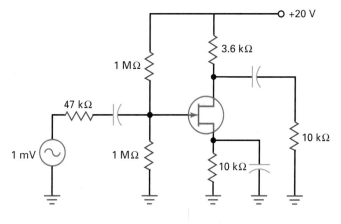

The output voltage is:

$$v_{out} = 13.3(1 \text{ mV pp}) = 13.3 \text{ mV pp}$$

PRACTICE PROBLEM 13–13   Using Fig. 13-23, what is the output voltage if $g_m = 2000 \text{ μS}$?

Example 13–14

|||| MultiSim

If $g_m = 2500 \text{ μS}$ in Fig. 13-24, what is the output voltage of the source follower?

Figure 13-24   Example of source follower.

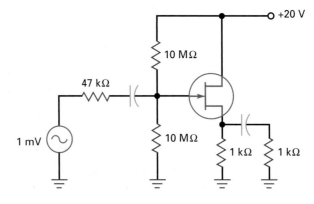

SOLUTION   The ac source resistance is:

$$r_s = 1 \text{ kΩ} \| 1 \text{ kΩ} = 500 \text{ Ω}$$

With Eq. (13-18), the voltage gain is:

$$A_v = \frac{(2500 \text{ μS})(500 \text{ Ω})}{1 + (2500 \text{ μS})(500 \text{ Ω})} = 0.556$$

Because the input impedance of the stage is 5 MΩ, the input signal to the gate is approximately 1 mV. Therefore, the output voltage is:

$$v_{out} = 0.556(1 \text{ mV}) = 0.556 \text{ mV}$$

PRACTICE PROBLEM 13–14   What is the output voltage of Fig. 13-24 if $g_m = 5000 \text{ μS}$?

Example 13–15

|||| MultiSim

Figure 13-25 includes a variable resistor of 1 kΩ. If this is adjusted to 780 Ω, what is the voltage gain?

SOLUTION   The total dc source resistance is:

$$R_S = 780 \text{ Ω} + 220 \text{ Ω} = 1 \text{ kΩ}$$

The ac source resistance is:

$$r_s = 1 \text{ kΩ} \| 3 \text{ kΩ} = 750 \text{ Ω}$$

Figure 13-25 Example.

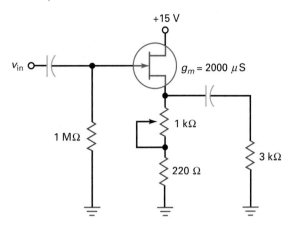

The voltage gain is:

$$A_v = \frac{(2000\ \mu S)(750\ \Omega)}{1 + (2000\ \mu S)(750\ \Omega)} = 0.6$$

PRACTICE PROBLEM 13-15 Using Fig. 13-25, what is the maximum voltage gain possible when adjusting the variable resistor?

Example 13-16

In Fig. 13-26, what is the drain current? The voltage gain?

SOLUTION The 3:1 voltage divider produces a dc gate voltage of 10 V. Ideally, the drain current is:

$$I_D = \frac{10\ V}{2.2\ k\Omega} = 4.55\ mA$$

The ac source resistance is:

$$r_s = 2.2\ k\Omega \,\|\, 3.3\ k\Omega = 1.32\ k\Omega$$

Figure 13-26 Example.

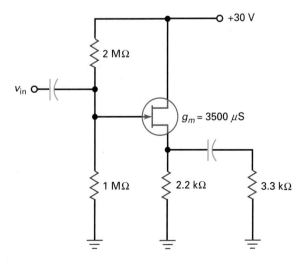

The voltage gain is:

$$A_v = \frac{(3500\ \mu S)(1.32\ k\Omega)}{1 + (3500\ \mu S)(1.32\ k\Omega)} = 0.822$$

PRACTICE PROBLEM 13–16   In Fig. 13-26, what would the voltage gain change to if the 3.3 kΩ resistor opened?

Summary Table 13-2 shows the common-source and source follower amplifier configurations and equations.

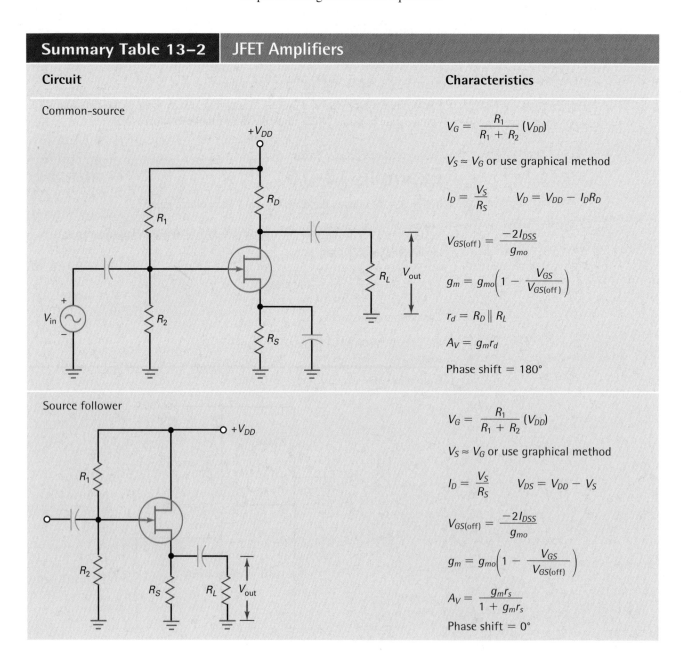

Summary Table 13–2	JFET Amplifiers

Circuit — **Characteristics**

Common-source

$$V_G = \frac{R_1}{R_1 + R_2}(V_{DD})$$

$V_S \approx V_G$ or use graphical method

$$I_D = \frac{V_S}{R_S} \qquad V_D = V_{DD} - I_D R_D$$

$$V_{GS(off)} = \frac{-2I_{DSS}}{g_{mo}}$$

$$g_m = g_{mo}\left(1 - \frac{V_{GS}}{V_{GS(off)}}\right)$$

$$r_d = R_D \parallel R_L$$

$$A_V = g_m r_d$$

Phase shift = 180°

Source follower

$$V_G = \frac{R_1}{R_1 + R_2}(V_{DD})$$

$V_S \approx V_G$ or use graphical method

$$I_D = \frac{V_S}{R_S} \qquad V_{DS} = V_{DD} - V_S$$

$$V_{GS(off)} = \frac{-2I_{DSS}}{g_{mo}}$$

$$g_m = g_{mo}\left(1 - \frac{V_{GS}}{V_{GS(off)}}\right)$$

$$A_V = \frac{g_m r_s}{1 + g_m r_s}$$

Phase shift = 0°

13-8 The JFET Analog Switch

Besides the source follower, another major application of the JFET is *analog switching*. In this application, the JFET acts as a switch that either transmits or blocks a small ac signal. To get this type of action, the gate-source voltage V_{GS} has only two values: either zero or a value that is greater than $V_{GS(\text{off})}$. In this way, the JFET operates either in the ohmic region or in the cutoff region.

Shunt Switch

Figure 13-27a shows a JFET **shunt switch.** The JFET is either conducting or cut off, depending on whether V_{GS} is high or low. When V_{GS} is high (0 V), the JFET operates in the ohmic region. When V_{GS} is low, the JFET is cut off. Because of this, we can use Fig. 13-27b as an equivalent circuit.

For normal operation, the ac input voltage must be a small signal, typically less than 100 mV. A small signal ensures that the JFET remains in the ohmic region when the ac signal reaches its positive peak. Also, R_D is much greater than R_{DS} to ensure hard saturation:

$$R_D \gg R_{DS}$$

When V_{GS} is high, the JFET operates in the ohmic region and the switch of Fig. 13-27b is closed. Since R_{DS} is much smaller than R_D, v_{out} is much smaller than v_{in}. When V_{GS} is low, the JFET cuts off and the switch of Fig. 13-27b opens. In this case, $v_{\text{out}} = v_{\text{in}}$. Therefore, the JFET shunt switch either transmits the ac signal or blocks it.

Series Switch

Figure 13-27c shows a JFET **series switch,** and Fig. 13-27d is its equivalent circuit. When V_{GS} is high, the switch is closed and the JFET is equivalent to a resistance of R_{DS}. In this case, the output approximately equals the input. When V_{GS} is low, the JFET is open and v_{out} is approximately zero.

Figure 13-27 JFET analog switches: (a) Shunt type; (b) shunt equivalent circuit; (c) series type; (d) series equivalent circuit.

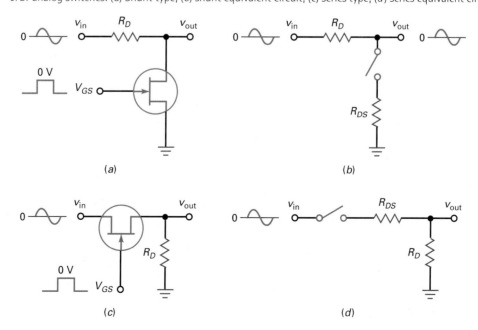

(a) (b)

(c) (d)

Figure 13-28 Chopper.

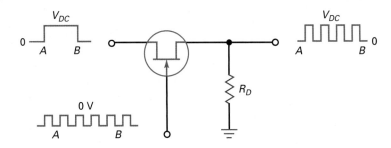

The on-off ratio of a switch is defined as the maximum output voltage divided by the minimum output voltage:

$$\textbf{On-off ratio} = \frac{v_{\text{out(max)}}}{v_{\text{out(min)}}} \tag{13-19}$$

When a high on-off ratio is important, the JFET series switch is a better choice because its on-off ratio is higher than that of the JFET shunt switch.

Chopper

Figure 13-28 shows a JFET **chopper.** The gate voltage is a continuous square wave that continuously switches the JFET on and off. The input voltage is a rectangular pulse with a value of V_{DC}. Because of the square wave on the gate, the output is *chopped* (switched on and off), as shown.

A JFET chopper can use either a shunt or a series switch. Basically, the circuit converts a dc input voltage to a square-wave output. The peak value of the chopped output is V_{DC}. As will be described later, a JFET chopper can be used to build a *dc amplifier,* a circuit that can amplify frequencies all the way down to zero frequencies.

Example 13-17

A JFET shunt switch has $R_D = 10 \text{ k}\Omega$, $I_{DSS} = 10 \text{ mA}$, and $V_{GS(\text{off})} = -2 \text{ V}$. If $v_{\text{in}} = 10 \text{ mV pp}$, what are output voltages? What is the on-off ratio?

SOLUTION   The ohmic resistance is:

$$R_{DS} = \frac{2 \text{ V}}{10 \text{ mA}} = 200 \ \Omega$$

Figure 13-29a shows the equivalent circuit when the JFET is conducting. The output voltage is:

$$v_{\text{out}} = \frac{200 \ \Omega}{10.2 \text{ k}\Omega} (10 \text{ mV pp}) = 0.196 \text{ mV pp}$$

When the JFET is off:

$$v_{\text{out}} = 10 \text{ mV pp}$$

The on-off ratio is:

$$\text{On-off ratio} = \frac{10 \text{ mV pp}}{0.196 \text{ mV pp}} = 51$$

Figure 13-29 Examples.

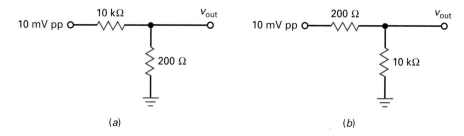

(a)   (b)

PRACTICE PROBLEM 13–17 Repeat Example 13-17 using a $V_{GS(\text{off})}$ value of -4 V.

Example 13–18

A JFET series switch has the same data as in the preceding example. What are output voltages? If the JFET has a resistance of 10 MΩ when off, what is the on-off ratio?

SOLUTION Figure 13-29b shows the equivalent circuit when the JFET is conducting. The output voltage is:

$$v_{\text{out}} = \frac{10 \text{ k}\Omega}{10.2 \text{ k}\Omega} (10 \text{ mV pp}) = 9.8 \text{ mV pp}$$

When the JFET is off:

$$v_{\text{out}} = \frac{10 \text{ k}\Omega}{10 \text{ M}\Omega} (10 \text{ mV pp}) = 10 \text{ } \mu\text{V pp}$$

The on-off ratio of the switch is:

$$\text{On-off ratio} = \frac{9.8 \text{ mV pp}}{10 \text{ } \mu\text{V pp}} = 980$$

Compare this to the preceding example, and you can see that a series switch has a better on-off ratio.

PRACTICE PROBLEM 13–18 Repeat Example 13-18 using a $V_{GS(\text{off})}$ value of -4 V.

Example 13–19   ▌▌▌▌ MultiSim

The square wave on the gate of Fig. 13-30 has a frequency of 20 kHz. What is the frequency of the chopped output? If the MPF4858 has an R_{DS} of 50 Ω, what is the peak value of the chopped output?

Figure 13-30 Example of chopper.

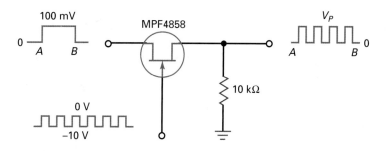

SOLUTION The output frequency is the same as the chopping or gate frequency:

$$f_{out} = 20 \text{ kHz}$$

Since 50 Ω is much smaller than 10 kΩ, almost all the input voltage reaches the output:

$$V_{peak} = \frac{10 \text{ k}\Omega}{10 \text{ k}\Omega + 50 \text{ }\Omega} (100 \text{ mV}) = 99.5 \text{ mV}$$

PRACTICE PROBLEM 13-19 Using Fig. 13-30 and a R_{DS} value of 100 Ω, determine the peak value of the chopped output.

13-9 Other JFET Applications

A JFET cannot compete with a bipolar transistor for most amplifier applications. But its unusual properties make it a better choice in special applications. In this section, we discuss those applications where a JFET has a clear-cut advantage over the bipolar transistor.

Multiplexing

Multiplex means "many into one." Figure 13-31 shows an *analog multiplexer,* a circuit that steers one or more of the input signals to the output line. Each JFET acts like a series switch. The control signals (V_1, V_2, and V_3) turn the JFETs on and off. When a control signal is high, its input signal is transmitted to the output.

For instance, if V_1 is high and the others are low, the output is a sine wave. If V_2 is high and the others are low, the output is a triangular wave. When V_3 is the high input, the output is a square wave. Normally, only one of the control signals is high; this ensures that only one of the input signals is transmitted to the output.

Figure 13-31 Multiplexer.

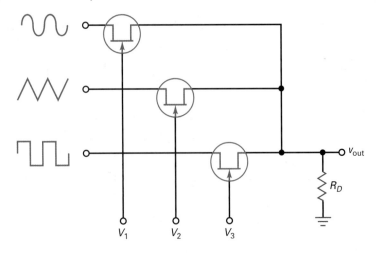

Figure 13-32 Chopper amplifier.

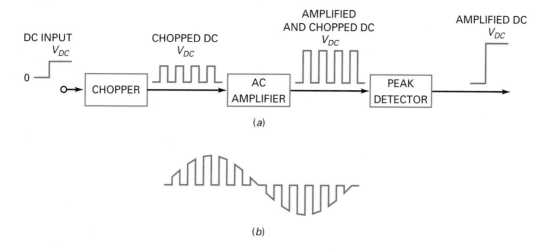

(a)

(b)

Chopper Amplifiers

We can build a direct-coupled amplifier by leaving out the coupling and bypass capacitors and connecting the output of each stage directly to the input of the next stage. In this way, dc voltages are coupled, as are ac voltages. Circuits that can amplify dc signals are called *dc amplifiers*. The major disadvantage of direct coupling is *drift,* a slow shift in the final dc output voltage produced by minor changes in supply voltage, transistor parameters, and temperature variations.

Figure 13-32a shows one way to overcome the drift problem of direct coupling. Instead of using direct coupling, we use a JFET chopper to convert the input dc voltage to a square wave. The peak value of this square wave equals V_{DC}. Because the square wave is an ac signal, we can use a conventional ac amplifier, one with coupling and bypass capacitors. The amplified output can then be peak-detected to recover an amplified dc signal.

A chopper amplifier can amplify low-frequency signals as well as dc signals. If the input is a low-frequency signal, it gets chopped into the ac waveform of Fig. 13-32b. This chopped signal can be amplified by an ac amplifier. The amplified signal can then be peak-detected to recover the original input signal.

Buffer Amplifier

Figure 13-33 shows a buffer amplifier, a stage that isolates the preceding stage from the following stage. Ideally, a buffer should have a high input impedance. If it does, almost all the Thevenin voltage from stage A appears at the buffer input. The buffer should also have a low output impedance. This ensures that all its output voltage reaches the input of stage B.

The source follower is an excellent buffer amplifier because of its high input impedance (well into the megohms at low frequencies) and its low output impedance (typically a few hundred ohms). The high input impedance means

Figure 13-33 Buffer amplifiers isolates stages A and B.

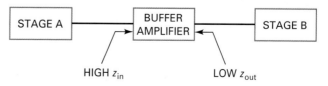

light loading of stage A. The low output impedance means that the buffer can drive heavy loads (small load resistances).

Low–Noise Amplifier

Noise is any unwanted disturbance superimposed on a useful signal. Noise interferes with the information contained in the signal. For instance, the noise in television receivers produces small white or black spots on the picture. Severe noise can wipe out the picture altogether. Similarly, the noise in radio receivers produces crackling and hissing, which sometimes completely masks the signal. Noise is independent of the signal because it exists even when the signal is off.

The JFET is an outstanding low-noise device because it produces much less noise than a bipolar junction transistor. Low noise is very important at the front end of receivers because the later stages amplify front-end noise along with the signal. If we use a JFET amplifier at the front end, we get less amplified noise at the final output.

Other circuits near the front end of receivers include *frequency mixers* and *oscillators.* A frequency mixer is a circuit that converts a higher frequency to a lower one. An oscillator is a circuit that generates an ac signal. JFETs are often used for VHF/UHF amplifiers, mixers, and oscillators. *VHF* stands for "very high frequencies" (30 to 300 MHz), and *UHF*, for "ultra high frequencies") (300 to 3000 MHz).

Voltage-Controlled Resistance

When a JFET operates in the ohmic region, it usually has $V_{GS} = 0$ to ensure hard saturation. But there is an exception. It is possible to operate a JFET in the ohmic region with V_{GS} values between 0 and $V_{GS(off)}$. In this case, the JFET can act like a *voltage-controlled resistance.*

Figure 13-34 shows the drain curves of a 2N5951 near the origin with V_{DS} less than 100 mV. In this region, the small-signal resistance r_{ds} is defined as the drain voltage divided by the drain current:

$$r_{ds} = \frac{V_{DS}}{I_D} \tag{13-20}$$

In Fig. 13-34, you can see that r_{ds} depends on which V_{GS} curve is used. For $V_{GS} = 0$, r_{ds} is minimum and equals R_{DS}. As V_{GS} becomes more negative, r_{ds} increases and becomes greater than R_{DS}.

For instance, when $V_{GS} = 0$ in Fig. 13-34, we can calculate:

$$r_{ds} = \frac{100 \text{ mV}}{0.8 \text{ mA}} = 125 \text{ }\Omega$$

When $V_{GS} = -2$ V:

$$r_{ds} = \frac{100 \text{ mV}}{0.4 \text{ mA}} = 250 \text{ }\Omega$$

When $V_{GS} = -4$ V:

$$r_{ds} = \frac{100 \text{ mV}}{0.1 \text{ mA}} = 1 \text{ k}\Omega$$

This means that a JFET acts like a voltage-controlled resistance in the ohmic region.

Recall that a JFET is a symmetrical device at low frequencies since either end can act like the source or the drain. This is why the drain curves of Fig. 13-34 extend on both sides of the origin. This means that a JFET can be used as a voltage-controlled resistance for small ac signals, typically those with a

Figure 13-34 Small-signal r_{ds} is voltage-controlled.

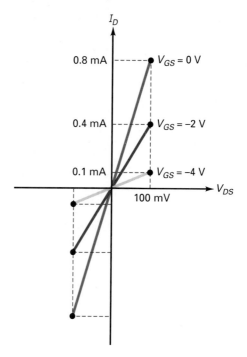

peak-to-peak value of less than 200 mV. When it is used in this way, the JFET does not need a dc drain voltage from the supply because the small ac signal supplies the drain voltage.

Figure 13-35a shows a shunt circuit where the JFET is used as a voltage-controlled resistance. This circuit is identical to the JFET shunt switch discussed earlier. The difference is that the control voltage V_{GS} does not swing from 0 to a

Figure 13-35 Example of voltage-controlled resistance.

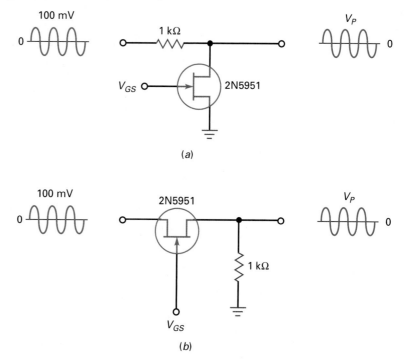

large negative value. Instead, V_{GS} can vary continuously; that is, it can have any value between 0 and $V_{GS(off)}$. In this way, V_{GS} controls the resistance of the JFET, which then changes the peak output voltage.

Figure 13-35*b* is a series circuit with the JFET used as a voltage-controlled resistance. The basic idea is the same. When you change V_{GS}, you change the ac resistance of the JFET, which changes the peak output voltage.

As calculated earlier, when $V_{GS} = 0$ V, the 2N5951 has a small-signal resistance of:

$$r_{ds} = 125 \ \Omega$$

In Fig. 13-35*a*, this means that the voltage divider produces a peak output voltage of:

$$V_p = \frac{125 \ \Omega}{1.125 \ \text{k}\Omega} (100 \ \text{mV}) = 11.1 \ \text{mV}$$

If V_{GS} is changed to -2 V, r_{ds} increases to 250 Ω, and the peak output increases to:

$$V_p = \frac{250 \ \Omega}{1.25 \ \text{k}\Omega} (100 \ \text{mV}) = 20 \ \text{mV}$$

When V_{GS} is changed to -4 V, r_{ds} increases to 1 kΩ, and the peak output increases to:

$$V_p = \frac{1 \ \text{k}\Omega}{2 \ \text{k}\Omega} (100 \ \text{mV}) = 50 \ \text{mV}$$

Automatic Gain Control

When a receiver is tuned from a weak to a strong station, the loudspeaker will blare (become loud) unless the volume is immediately decreased. The volume may also change because of fading, a decrease in signal caused by a change in the path between the transmitter and receiver. To prevent unwanted changes in the volume, most modern receivers use **automatic gain control (AGC).**

Figure 13-36 illustrates the basic idea of AGC. An input signal v_{in} passes through a JFET used as a voltage-controlled resistance. The signal is amplified to get the output voltage v_{out}. The output signal is fed back to a negative peak detector. The output of this peak detector then supplies the V_{GS} for the JFET.

If the input signal suddenly increases by a large amount, the output voltage will increase. This means that a larger negative voltage comes out of the peak detector. Since V_{GS} is more negative, the JFET has a higher ohmic resistance, which reduces the signal to the amplifier and decreases the output signal.

On the other hand, if the input signal fades, the output voltage decreases and the negative peak detector produces a smaller output. Since V_{GS} is less negative,

Figure 13-36 Automatic gain control.

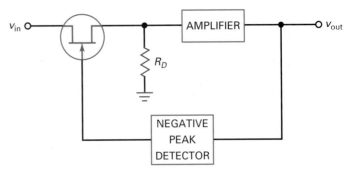

the JFET transmits more signal voltage to the amplifier, which raises the final output. Therefore, the effect of any sudden change in the input signal is offset or at least reduced by the AGC action.

Another AGC Example

As shown earlier, the g_m of a JFET decreases when V_{GS} becomes more negative. The equation is:

$$g_m = g_{m0}\left(1 - \frac{V_{GS}}{V_{GS(\text{off})}}\right)$$

This is a linear equation. When it is graphed, it results in Fig. 13-37a. For a JFET, g_m reaches a maximum value when $V_{GS} = 0$. As V_{GS} becomes more negative, the value of g_m decreases. Since a CS amplifier has a voltage gain of:

$$A_v = g_m r_d$$

we can control the voltage gain by controlling the value of g_m.

Figure 13-37b shows how it's done. A JFET amplifier is near the front end of a receiver. It has a voltage gain of $g_m r_d$. Subsequent stages amplify the JFET output. This amplified output goes into a negative peak detector that produces voltage V_{AGC}. This negative voltage is applied to the gate of the CS amplifier.

When the receiver is tuned from a weak to a strong station, a larger signal is peak-detected and V_{AGC} becomes more negative. This reduces the gain of the JFET amplifier. Conversely, if the signal fades, less AGC voltage is applied to the gate and the JFET stage produces a larger output signal.

The overall effect of AGC is this: The final output signal changes, but not nearly as much as it would without AGC. For instance, in some AGC systems an increase of 100 percent in the input signal results in an increase of less than 1 percent in the final output signal.

Figure 13-37 AGC used with receiver.

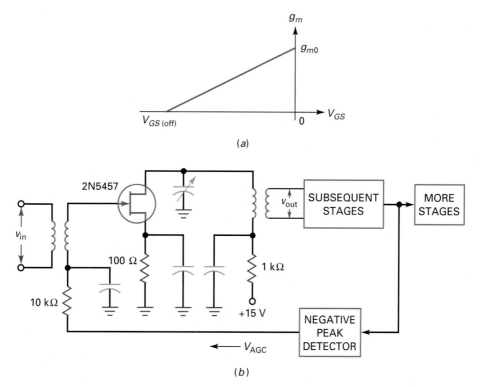

(a)

(b)

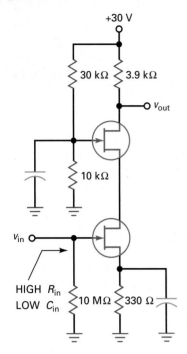

Figure 13-38 Cascode amplifier.

Cascode Amplifier

Figure 13-38 is an example of a cascode amplifier. It can be shown that the overall voltage gain of this two-FET connection is:

$$A_v = g_m r_d$$

This is the same voltage gain as for a CS amplifier.

The advantage of the circuit is its low input capacitance, which is important with VHF and UHF signals. At these higher frequencies, the input capacitance becomes a limiting factor on the voltage gain. With a cascode amplifier, the low input capacitance allows the circuit to amplify higher frequencies than are possible with only a CS amplifier. Chapter 16 will mathematically analyze the effect of capacitance on high-frequency operation.

Current Sourcing

Suppose you have a load that requires a constant current. One solution is to use a shorted-gate JFET to supply the constant current. Figure 13-39a shows the basic idea. If the Q point is in the active region, as shown in Fig. 13-39b, the load current equals I_{DSS}. If the load can tolerate the change in I_{DSS} when JFETs are replaced, the circuit is an excellent solution.

Figure 13-39 JFET used as current source.

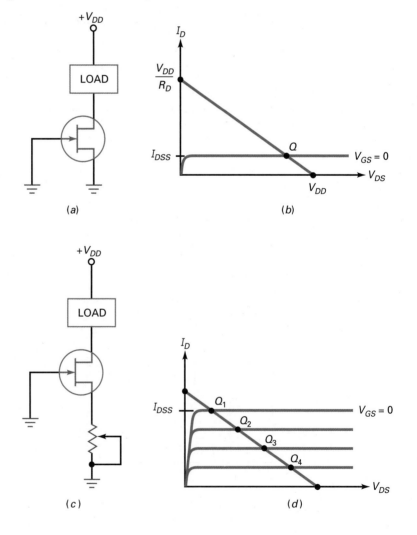

On the other hand, if the constant load current must have a specific value, we can use an adjustable source resistor, as shown in Fig. 13-39c. The self-bias will produce negative values of V_{GS}. By adjusting the resistor, we can set up different Q points, as shown in Fig. 13-39d.

Using JFETs like this is a simple way to produce a fixed load current, one that is constant even though the load resistance changes. In later chapters, we will discuss other ways to produce fixed load currents using op amps.

Current Limiting

Instead of sourcing current, a JFET can limit current. Figure 13-40a shows how. In this application, the JFET operates in the ohmic region rather than the active region. To ensure operation in the ohmic region, the designer selects values to get the dc load line of Fig. 13-40b. The normal Q point is in the ohmic region, and the normal load current is approximately V_{DD}/R_D.

If the load becomes shorted, the dc load line becomes vertical. In this case, the Q point changes to the new position shown in Fig. 13-40b. With this Q point, the current is limited to I_{DSS}. The point to remember is that a shorted load usually produces an excessive current. But with the JFET in series with the load, the current is limited to a safe value.

Conclusion

Look at Summary Table 13-3. Some of the terms are new and will be discussed in later chapters. The JFET buffer has the advantage of high input impedance and low output impedance. This is why the JFET is the natural choice at the front end of voltmeters, oscilloscopes, and other, similar equipment where you need high input resistances (10 MΩ or more). As a guide, the input resistance at the gate of a JFET is more than 100 MΩ.

When a JFET is used as a small-signal amplifier, its output voltage is linearly related to the input voltage because only a small part of the transconductance curve is used. Near the front end of television and radio receivers, the signals are small. Therefore, JFETs are often used as RF amplifiers.

But with larger signals, more of the transconductance curve is used, resulting in square-law distortion. This nonlinear distortion is unwanted in an amplifier. But in a frequency mixer, square-law distortion has a big advantage. This is why the JFET is preferred to the bipolar junction transistor for FM and television mixer applications.

As indicated in Summary Table 13-3, JFETs are also useful in AGC amplifiers, cascode amplifiers, choppers, voltage-controlled resistors, audio amplifiers, and oscillators.

Figure 13-40 JFET limits current if load shorts.

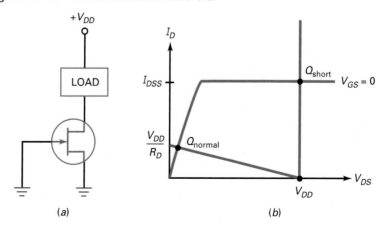

(a)   (b)

Application	Main Advantage	Uses
Buffer	High z_{in}, low z_{out}	General-purpose measuring equipment, receivers
RF amplifier	Low noise	FM tuners, communication equipment
RF mixer	Low distortion	FM and television receivers, communication equipment
AGC amplifier	Ease of gain control	Receivers, signal generators
Cascode amplifier	Low input capacitance	Measuring instruments, test equipment
Chopper amplifier	No drift	DC amplifiers, guidance control systems
Variable resistor	Voltage-controlled	Op amps, organ tone controls
Audio amplifier	Small coupling capacitors	Hearing aids, inductive transducers
RF oscillator	Minimum frequency drift	Frequency standards, receivers

13-10 Reading Data Sheets

JFET data sheets are similar to bipolar junction data sheets. You will find maximum ratings, dc characteristics, ac characteristics, mechanical data, and so on. As usual, a good place to start is with the maximum ratings because these are the limits on the JFET currents, voltages, and other quantities.

Breakdown Ratings

As shown in Fig. 13-41, the data sheet of the MPF102 gives these maximum ratings:

$$V_{DS} \quad 25 \text{ V}$$
$$V_{GS} \quad -25 \text{ V}$$
$$P_D \quad 350 \text{ mW}$$

As usual, a conservative design includes a safety factor for all of these maximum ratings.

As discussed earlier, the derating factor tells you how much to reduce the power rating of a device. The derating factor of an MPF102 is given as 2.8 mW/°C. This means that you have to reduce the power rating by 2.8 mW for each degree above 25°C.

I_{DSS} and $V_{GS(off)}$

Two of the most important pieces of information on the data sheet of a depletion-mode device are the maximum drain current and the gate-source cutoff voltage. These values are given on the data sheet of the MPF102:

Symbol	Minimum	Maximum
$V_{GS(off)}$	—	−8 V
I_{DSS}	2 mA	20 mA

Notice the 10 : 1 spread in I_{DSS}. This large spread is one of the reasons for using ideal approximations with a preliminary analysis of JFET circuits. Another reason

Figure 13–41 MPF102 data sheet.

JFET VHF Amplifier
N–Channel – Depletion

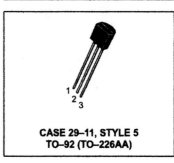

MPF102

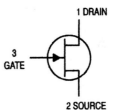

CASE 29–11, STYLE 5
TO–92 (TO–226AA)

1 DRAIN

3 GATE

2 SOURCE

MAXIMUM RATINGS

Rating	Symbol	Value	Unit
Drain–Source Voltage	V_{DS}	25	Vdc
Drain–Gate Voltage	V_{DG}	25	Vdc
Gate–Source Voltage	V_{GS}	−25	Vdc
Gate Current	I_G	10	mAdc
Total Device Dissipation @ T_A = 25°C Derate above 25°C	P_D	350 2.8	mW mW/°C
Junction Temperature Range	T_J	125	°C
Storage Temperature Range	T_{stg}	−65 to +150	°C

ELECTRICAL CHARACTERISTICS (T_A = 25°C unless otherwise noted)

Characteristic	Symbol	Min	Max	Unit
OFF CHARACTERISTICS				
Gate–Source Breakdown Voltage (I_G = −10 μAdc, V_{DS} = 0)	$V_{(BR)GSS}$	−25	–	Vdc
Gate Reverse Current (V_{GS} = −15 Vdc, V_{DS} = 0) (V_{GS} = −15 Vdc, V_{DS} = 0, T_A = 100°C)	I_{GSS}	– –	−2.0 −2.0	nAdc μAdc
Gate–Source Cutoff Voltage (V_{DS} = 15 Vdc, I_D = 2.0 nAdc)	$V_{GS(off)}$	–	−8.0	Vdc
Gate–Source Voltage (V_{DS} = 15 Vdc, I_D = 0.2 mAdc)	V_{GS}	−0.5	−7.5	Vdc
ON CHARACTERISTICS				
Zero–Gate–Voltage Drain Current[1] (V_{DS} = 15 Vdc, V_{GS} = 0 Vdc)	I_{DSS}	2.0	20	mAdc
SMALL–SIGNAL CHARACTERISTICS				
Forward Transfer Admittance[1] (V_{DS} = 15 Vdc, V_{GS} = 0, f = 1.0 kHz) (V_{DS} = 15 Vdc, V_{GS} = 0, f = 100 MHz)	$\lvert y_{fs} \rvert$	2000 1600	7500 –	μmhos
Input Admittance (V_{DS} = 15 Vdc, V_{GS} = 0, f = 100 MHz)	Re(y_{is})	–	800	μmhos
Output Conductance (V_{DS} = 15 Vdc, V_{GS} = 0, f = 100 MHz)	Re(y_{os})	–	200	μmhos
Input Capacitance (V_{DS} = 15 Vdc, V_{GS} = 0, f = 1.0 MHz)	C_{iss}	–	7.0	pF
Reverse Transfer Capacitance (V_{DS} = 15 Vdc, V_{GS} = 0, f = 1.0 MHz)	C_{rss}	–	3.0	pF

1. Pulse Test; Pulse Width ≤ 630 ms, Duty Cycle ≤ 10%.

Table 13–1	JFET Sampler				
Device	$V_{GS(off)}$, V	I_{DSS}, mA	g_{m0}, μS	R_{DS}, Ω	Application
J202	−4	4.5	2,250	888	Audio
2N5668	−4	5	2,500	800	RF
MPF3822	−6	10	3,333	600	Audio
2N5459	−8	16	4,000	500	Audio
MPF102	−8	20	5,000	400	RF
J309	−4	30	15,000	133	RF
BF246B	−14	140	20,000	100	Switching
MPF4857	−6	100	33,000	60	Switching
MPF4858	−4	80	40,000	50	Switching

for using ideal approximations is this: Data sheets often omit values, so you really have no idea what some values may be. In the case of the MPF102, the minimum value of $V_{GS(off)}$ is not listed on the data sheet.

Another important static characteristic of a JFET is I_{GSS}, which is the gate current when the gate-source junction is reverse biased. This current value allows us to determine the dc input resistance of the JFET. As shown on the data sheet, a MPF102 has an I_{GSS} value of 2 nAdc when $V_{GS} = -15$ V. Under these conditions, the gate-source resistance is $R = 15$ V/2 nA $= 7500$ MΩ.

Table of JFETs

Table 13-1 shows a sample of different JFETs. The data are sorted in ascending order for g_{m0}. The data sheets for these JFETs show that some are optimized for use at audio frequencies and others for use at RF frequencies. The last three JFETs are optimized for switching applications.

JFETs are small-signal devices because their power dissipation is usually a watt or less. In audio applications, JFETs are often used as source followers. In RF applications, they are used as VHF/UHF amplifiers, mixers, and oscillators. In switching applications, they are typically used as analog switches.

13–11 JFET Testing

The data sheet for the MPF102 shows a maximum gate current I_G of 10 mA. This is the maximum forward gate-to-source or gate-to-drain current the JFET can handle. This can occur if the gate to channel *pn* junction becomes forward biased. If you are testing a JFET using an ohmmeter or digital multimeter on the diode test range, be sure that your meter doesn't cause excessive gate current. Many analog VOMs will provide approximately 100 mA in the R × 1 range. The R × 100 range generally results in a current of 1–2 mA. Most DMMs output a constant 1–2 mA of current when in the diode test range. This should allow safe testing of a JFET's gate-to-source and gate-to-drain *pn* junctions. To check the JFET's drain-to-source channel resistance, connect the gate lead to the source lead. Otherwise, you will get erratic measurements due to the electric field produced in the channel.

If you have a semiconductor curve tracer available, the JFET can be tested to display its drain curves. A simple test circuit using Multisim, shown in Fig. 13-42a, can also be used to display one drain curve at a time. By using the x-y display capability of most oscilloscopes, a drain curve similar to Fig. 13-42b can be displayed. By varying the reverse-bias voltage of V_1, you can determine the approximate I_{DSS} and $V_{GS(off)}$ values.

For example, as shown in Fig. 13-42a, the oscilloscope's y-input is connected across a 10 Ω source resistor. With the oscilloscope's vertical input set to 50 mV/division, this results in a vertical drain current measurement of

$$I_D = \frac{50 \text{ mV/div.}}{10 \text{ }\Omega} = 5 \text{ mA/}div$$

With V_1 adjusted to 0 V, the resulting I_D value (I_{DSS}) is approximately 12 mA. $V_{GS(off)}$ can be found by increasing V_1 until I_D is zero.

Figure 13–42   (a) JFET test circuit; (b) drain curve.

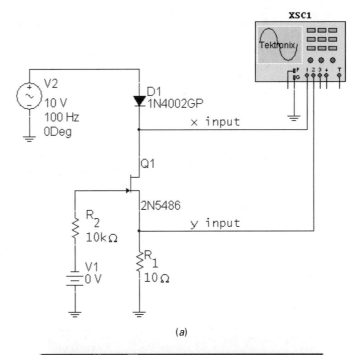

(a)

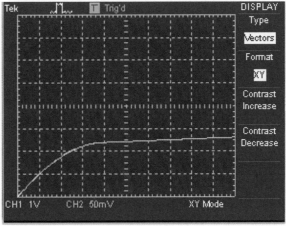

(b)

Summary

SEC. 13-1   BASIC IDEAS

The junction FET, abbreviated *JFET*, has a source, gate, and drain. The JFET has two diodes: the gate-source diode and the gate-drain diode. For normal operation, the gate-source diode is reverse biased. Then, the gate voltage controls the drain current.

SEC. 13-2   DRAIN CURVES

Maximum drain current occurs when the gate-source voltage is zero. The pinchoff voltage separates the ohmic and active regions for $V_{GS} = 0$. The gate-source cutoff voltage has the same magnitude as the pinchoff voltage. $V_{GS(off)}$ turns the JFET off.

SEC. 13-3   THE TRANSCONDUC-
TANCE CURVE

This is a graph of drain current versus gate-source voltage. The drain current increases more rapidly as V_{GS} approaches zero. Because the equation for drain current contains a squared quantity, JFETs are referred to as *square-law devices*. The normalized transconductance curve shows that I_D equals one-quarter of maximum when V_{GS} equals half of cutoff.

SEC. 13-4   BIASING IN THE
OHMIC REGION

Gate bias is used to bias a JFET in the ohmic region. When it operates in the ohmic region, a JFET is equivalent to a small resistance of R_{DS}. To ensure operation in the ohmic region, the JFET is driven into hard saturation by using $V_{GS} = 0$ and $I_{D(sat)} \ll I_{DSS}$.

SEC. 13-5   BIASING IN THE
ACTIVE REGION

When the gate voltage is much larger than V_{GS}, voltage-divider bias can set up a stable Q point in the active region. When positive and negative supply voltages are available, two-supply source bias can be used to swamp out the variations in V_{GS} and set up a stable Q point. When supply voltages are not large, current-source bias can be used to get a stable Q point. Self-bias is used only with small-signal amplifiers because the Q point is less stable than with the other biasing methods.

SEC. 13-6   TRANSCONDUCTANCE

Transconductance g_m tells us how effective the gate voltage is in controlling the drain current. The quantity g_m is the slope of the transconductance curve, which increases as V_{GS} approaches zero. Data sheets may list g_{fs} and siemens, which are equivalent to g_m and mhos.

SEC. 13-7   JFET AMPLIFIERS

A CS amplifier has a voltage gain of $g_m r_d$ and produces an inverted output signal. One of the most important uses of a JFET amplifier is the source follower, which is often used at the front end of systems because of its high input resistance.

SEC. 13-8   THE JFET ANALOG
SWITCH

In this application, the JFET acts like a switch that either transmits or blocks a small ac signal. To get this type of action, the JFET is biased into hard saturation or cutoff, depending on whether V_{GS} is high or low. JFET shunt and series switches are used. The series type has a higher on-off ratio.

SEC. 13-9   OTHER JFET
APPLICATIONS

JFETs are used in multiplexers (ohmic), chopper amplifiers (ohmic), buffer amplifiers (active), voltage-controlled resistors (ohmic), AGC circuits (ohmic), cascode amplifiers (active), current sources (active), and current limiters (ohmic and active).

SEC. 13-10   READING DATA
SHEETS

JFETs are mainly small-signal devices because most JFETs have a power rating of less than 1 W. When reading data sheets, start with the maximum ratings. Sometimes data sheets omit the minimum $V_{GS(off)}$ or other parameters. The large spread in JFET parameters justifies using ideal approximations for preliminary analysis and troubleshooting.

SEC. 13-11   JFET TESTING

JFETs can be tested using an ohmmeter or DMM on the diode test range. Care must be taken not to exceed the JFET's current limits. Curve tracers and circuits can be used to display a JFET's dynamic characteristics.

Definitions

(13-1)   Ohmic resistance at pinchoff:

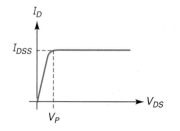

$$R_{DS} = \frac{V_P}{I_{DSS}}$$

(13-5)   Hard saturation:

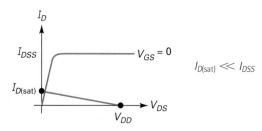

$$I_{D(sat)} \ll I_{DSS}$$

(13-13) Transconductance:

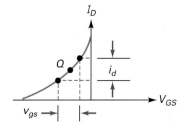

$$g_m = \frac{i_d}{v_{gs}}$$

(13-19) Ohmic resistance near origin:

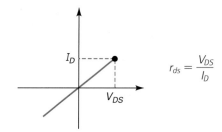

$$r_{ds} = \frac{V_{DS}}{I_D}$$

Derivations

(13-2) Gate-source cutoff voltage:

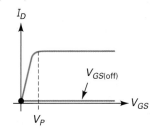

$$V_{GS(off)} = -V_P$$

(13-3) Drain current:

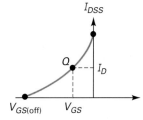

$$I_D = I_{DSS}\left(1 - \frac{V_{GS}}{V_{GS(off)}}\right)^2$$

(13-7) Self-bias:

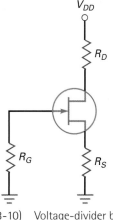

$$V_{GS} = -I_D R_S$$

(13-10) Voltage-divider bias:

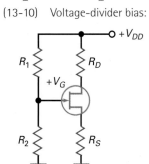

$$I_D = \frac{V_G - V_{GS}}{R_S} \approx \frac{V_G}{R_S}$$

(13-12) Source bias:

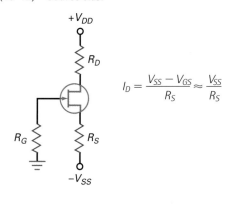

$$I_D = \frac{V_{SS} - V_{GS}}{R_S} \approx \frac{V_{SS}}{R_S}$$

(13-13) Current-source bias:

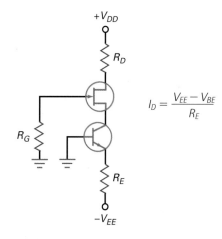

$$I_D = \frac{V_{EE} - V_{BE}}{R_E}$$

(13-15) Gate cutoff voltage:

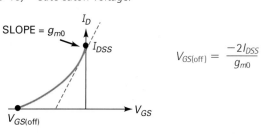

$$V_{GS(off)} = \frac{-2I_{DSS}}{g_{m0}}$$

(13-16)   Transconductance:

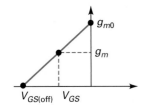

$$g_m = g_{m0}\left(1 - \frac{V_{GS}}{V_{GS(off)}}\right)$$

(13-18)   Source follower:

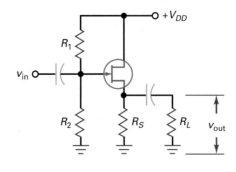

$$A_v = \frac{g_m r_s}{1 + g_m r_s}$$

(13-17)   CS voltage gain:

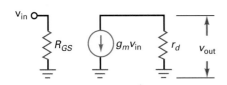

$$A_v = g_m r_d$$

Student Assignments

1. **A JFET**
   a. Is a voltage-controlled device
   b. Is a current-controlled device
   c. Has a low input resistance
   d. Has a very large voltage gain

2. **A unipolar transistor uses**
   a. Both free electrons and holes
   b. Only free electrons
   c. Only holes
   d. Either one or the other, but not both

3. **The input impedance of a JFET**
   a. Approaches zero
   b. Approaches one
   c. Approaches infinity
   d. Is impossible to predict

4. **The gate controls**
   a. The width of the channel
   b. The drain current
   c. The gate voltage
   d. All the above

5. **The gate-source diode of a JFET should be**
   a. Forward biased
   b. Reverse biased
   c. Either forward or reverse biased
   d. None of the above

6. **Compared to a bipolar junction transistor, the JFET has a much higher**
   a. Voltage gain
   b. Input resistance

   c. Supply voltage
   d. Current

7. **The pinchoff voltage has the same magnitude as the**
   a. Gate voltage
   b. Drain-source voltage
   c. Gate-source voltage
   d. Gate-source cutoff voltage

8. **When the drain saturation current is less than I_{DSS}, a JFET acts like a**
   a. Bipolar junction transistor
   b. Current source
   c. Resistor
   d. Battery

9. **R_{DS} equals pinchoff voltage divided by**
   a. Drain current
   b. Gate current
   c. Ideal drain current
   d. Drain current for zero gate voltage

10. **The transconductance curve is**
    a. Linear
    b. Similar to the graph of a resistor
    c. Nonlinear
    d. Like a single drain curve

11. **The transconductance increases when the drain current approaches**
    a. 0
    b. $I_{D(sat)}$
    c. I_{DSS}
    d. I_S

12. **A CS amplifier has a voltage gain of**
    a. $g_m r_d$
    b. $g_m r_s$
    c. $g_m r_s/(1 + g_m r_s)$
    d. $g_m r_d/(1 + g_m r_d)$

13. **A source follower has a voltage gain of**
    a. $g_m r_d$ c. $g_m r_s/(1 + g_m r_s)$
    b. $g_m r_s$ d. $g_m r_d/(1 + g_m r_d)$

14. **When the input signal is large, a source follower has**
    a. A voltage gain of less than 1
    b. Some distortion
    c. A high input resistance
    d. All of these

15. **The input signal used with a JFET analog switch should be**
    a. Small   c. A square wave
    b. Large   d. Chopped

16. **A cascode amplifier has the advantage of**
    a. Large voltage gain
    b. Low input capacitance
    c. Low input impedance
    d. Higher g_m

17. **VHF covers frequencies from**
    a. 300 kHz to 3 MHz
    b. 3 to 30 MHz
    c. 30 to 300 MHz
    d. 300 MHz to 3 GHz

18. **When a JFET is cut off, the depletion layers are**

   a. Far apart
   b. Close together
   c. Touching
   d. Conducting

19. **When the gate voltage becomes more negative in an *n*-channel JFET, the channel between the depletion layers**

   a. Shrinks
   b. Expands
   c. Conducts
   d. Stops conducting

20. **If a JFET has $I_{DSS} = 8$ mA and $V_P = 4$ V, then R_{DS} equals**

   a. 200 Ω
   b. 320 Ω
   c. 500 Ω
   d. 5 kΩ

21. **The easiest way to bias a JFET in the ohmic region is with**

   a. Voltage-divider bias
   b. Self-bias
   c. Gate bias
   d. Source bias

22. **Self-bias produces**

   a. Positive feedback
   b. Negative feedback
   c. Forward feedback
   d. Reverse feedback

23. **To get a negative gate-source voltage in a self-biased JFET circuit, you must have a**

   a. Voltage divider
   b. Source resistor
   c. Ground
   d. Negative gate supply voltage

24. **Transconductance is measured in**

   a. Ohms
   b. Amperes
   c. Volts
   d. Mhos or siemens

25. **Transconductance indicates how effectively the input voltage controls the**

   a. Voltage gain
   b. Input resistance
   c. Supply voltage
   d. Output current

Problems

SEC. 13-1   BASIC IDEAS

13-1 A 2N5458 has a gate current of 1 nA when the reverse voltage is -15 V. What is the input resistance of the gate?

13-2 A 2N5640 has a gate current of 1 μA when the reverse voltage is -20 V and the ambient temperature is 100°C. What is the input resistance of the gate?

SEC. 13-2   DRAIN CURVES

13-3 A JFET has $I_{DSS} = 20$ mA and $V_P = 4$ V. What is the maximum drain current? The gate-source cutoff voltage? The value of R_{DS}?

13-4 A 2N5555 has $I_{DSS} = 16$ mA and $V_{GS(off)} = -2$ V. What is the pinchoff voltage for this JFET? What is the drain-source resistance R_{DS}?

13-5 A 2N5457 has $I_{DSS} = 1$ to 5 mA and $V_{GS(off)} = -0.5$ to -6 V. What are the minimum and maximum values of R_{DS}?

SEC. 13-3   THE TRANSCONDUCTANCE CURVE

13-6 A 2N5462 has $I_{DSS} = 16$ mA and $V_{GS(off)} = -6$ V. What are the gate voltage and drain current at the half cutoff point?

13-7 A 2N5670 has $I_{DSS} = 10$ mA and $V_{GS(off)} = -4$ V. What are the gate voltage and drain current at the half cutoff point?

13-8 If a 2N5486 has $I_{DSS} = 14$ mA and $V_{GS(off)} = -4$ V, what is the drain current when $V_{GS} = -1$ V? When $V_{GS} = -3$ V?

SEC. 13-4   BIASING IN THE OHMIC REGION

13-9 What is the drain saturation current in Fig. 13-43*a*? The drain voltage?

13-10 If the 10-kΩ resistor of Fig. 13-43*a* is increased to 20 kΩ, what is the drain voltage?

13-11 What is the drain voltage in Fig. 13-43*b*?

13-12 If the 20-kΩ resistor of Fig. 13-43*b* is decreased to 10 kΩ, what is the drain saturation current? The drain voltage?

SEC. 13-5   BIASING IN THE ACTIVE REGION

For Problems 13-13 through 13-20, use preliminary analysis.

13-13 What is the ideal drain voltage in Fig. 13-44*a*?

13-14 Draw the dc load line and *Q* point for Fig. 13-44*a*.

13-15 What is the ideal drain voltage in Fig. 13-44*b*?

13-16 If the 18 kΩ of Fig. 13-44*b* is changed to 30 kΩ, what is the drain voltage?

13-17 In Fig. 13-45*a*, what is the drain current? The drain voltage?

Figure 13-43

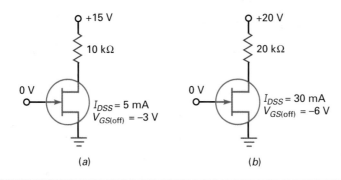

(a)   (b)

Figure 13-44

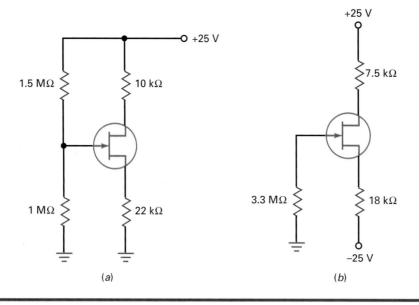

(a)   (b)

Figure 13-45

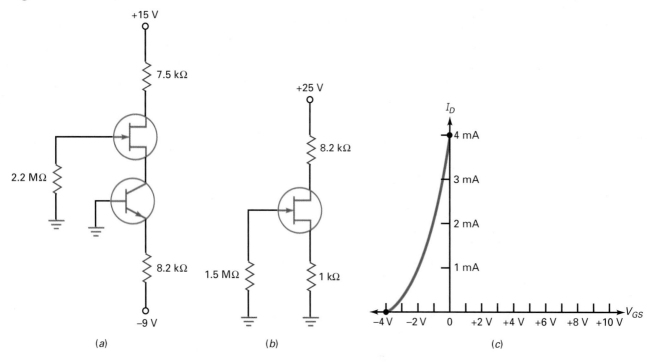

(a)   (b)   (c)

13-18 If the 7.5 kΩ of Fig. 13-45a is changed to 4.7 kΩ, what is the drain current? The drain voltage?

13-19 In Fig. 13-45b, the drain current is 1.5 mA. What does V_{GS} equal? What does V_{DS} equal?

13-20 The voltage across the 1KΩ of Fig. 13-45b is 1.5 V. What is the voltage between the drain and ground?

For Problems 13-21 through 13-24, use Fig. 13-45c and graphical methods to find your answers.

13-21 In Fig. 13-44a, find V_{GS} and I_D using the transconductance curve of Fig. 13-45c.

13-22 In Fig. 13-45a, find V_{GS} and V_D using the transconductance curve of Fig. 13-45c.

13-23 In Fig. 13-45b, find V_{GS} and I_D using the transconductance curve of Fig. 13-45c.

13-24 Change R_S in Fig. 13-45b from 1 kΩ to 2 kΩ. Use the curve of Fig. 13-45c to find V_{GS}, I_D, and V_{DS}.

SEC. 13-6   TRANSCONDUCTANCE

13-25 A 2N4416 has I_{DSS} = 10 mA and g_{m0} = 4000 μS. What is its gate-source cutoff voltage? What is the value of g_m for V_{GS} = −1 V?

13-26 A 2N3370 has I_{DSS} = 2.5 mA and g_{m0} = 1500 μS. What is the value of g_m for V_{GS} = −1 V?

13-27 The JFET of Fig. 13-46a has g_{m0} = 6000 μS. If I_{DSS} = 12 mA, what is the approximate value of I_D for V_{GS} of −2 V? Find the g_m for this I_D.

SEC. 13-7   JFET AMPLIFIERS

13-28 If g_m = 3000 μS in Fig. 13-46a, what is the ac output voltage?

13-29 The JFET amplifier of Fig. 13-46a has the transconductance curve of Fig. 13-46b. What is the approximate ac output voltage?

13-30 If the source follower of Fig. 13-47a has g_m = 2000 μS, what is the ac output voltage?

Figure 13-46

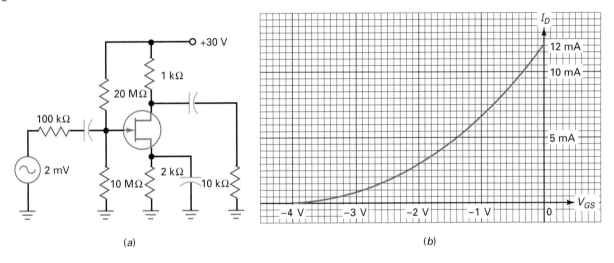

(a)   (b)

Figure 13-47

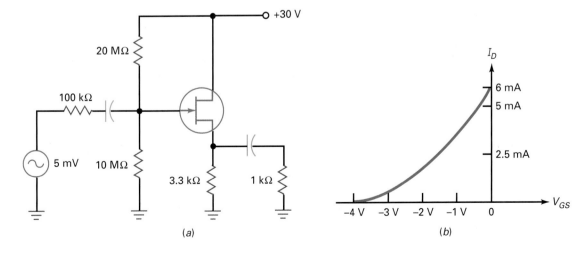

(a)   (b)

Figure 13-48

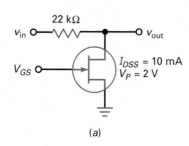

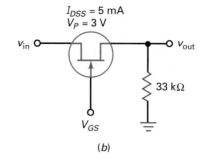

(a)   (b)

13-31 The source follower of Fig. 13-47a has the transconductance curve of Fig. 13-47b. What is the ac output voltage?

SEC. 13-8   THE JFET ANALOG SWITCH

13-32 The input voltage of Fig. 13-48a is 50 mV pp. What is the output voltage when $V_{GS} = 0$ V? When $V_{GS} = -10$ V? The on-off ratio?

13-33 The input voltage of Fig. 13-48b is 25 mV pp. What is the output voltage when $V_{GS} = 0$ V? When $V_{GS} = -10$ V? The on-off ratio?

Critical Thinking

13-34 If a JFET has the drain curves of Fig. 13-49a, what does I_{DSS} equal? What is the maximum V_{DS} in the ohmic region? Over what voltage range of V_{DS} does the JFET act as a current source?

13-35 Write the transconductance equation for the JFET whose curve is shown in Fig. 13-49b. How much drain current is there when $V_{GS} = -4$ V? When $V_{GS} = -2$ V?

13-36 If a JFET has a square-law curve like Fig. 13-49c, how much drain current is there when $V_{GS} = -1$ V?

13-37 What is the dc drain voltage in Fig. 13-50? The ac output voltage if $g_m = 2000$ μS?

13-38 Figure 13-51 shows a JFET dc voltmeter. The zero adjust is set just before a reading is taken. The calibrate adjust is set periodically to give full-scale deflection when $v_{in} = -2.5$ V. A calibrate adjustment like this takes care of variations from one FET to another and FET aging effects.

  a. The current through the 510 Ω equals 4 mA. How much dc voltage is there from the source to ground?

  b. If no current flows through the ammeter, what voltage does the wiper tap off the zero adjust?

  c. If an input voltage of 2.5 V produces a deflection of 1 mA, how much deflection does 1.25 V produce?

13-39 In Fig. 13-52a, the JFET has an I_{DSS} of 16 mA and an R_{DS} of 200 Ω. If the load has a resistance of 10 kΩ, what are the load current and the voltage across the JFET? If the load is accidentally shorted, what are the load current and the voltage across the JFET?

13-40 Figure 13-52b shows part of an AGC amplifier. A dc voltage is fed back from an output stage to an earlier stage such as the one shown here. Figure 13-46b is the transconductance curve. What is the voltage gain for each of these?

  a. $V_{AGC} = 0$
  b. $V_{AGC} = -1$ V
  c. $V_{AGC} = -2$ V
  d. $V_{AGC} = -3$ V
  e. $V_{AGC} = -3.5$ V

Figure 13-49

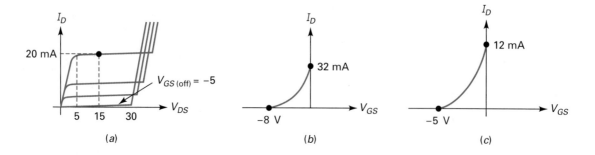

(a)   (b)   (c)

Figure 13-50

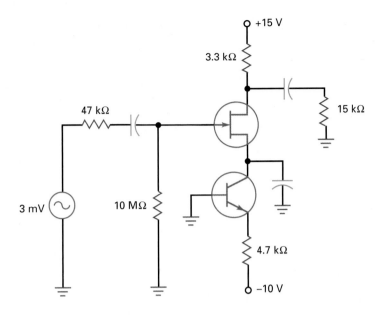

Figure 13-51

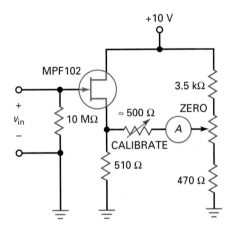

Figure 13-52

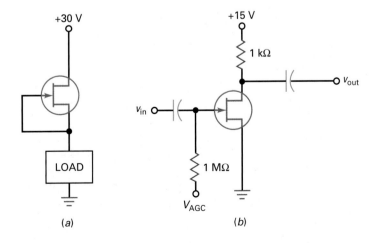

Troubleshooting

||| MultiSim Use Fig. 13-53 and the troubleshooting table to solve the remaining problems.

13-41 Find the trouble T1.

13-42 Find the trouble T2.

13-43 Find the trouble T3.

13-44 Find the trouble T4.

13-45 Find the trouble T5.

13-46 Find the trouble T6.

13-47 Find the trouble T7.

13-48 Find the trouble T8.

Figure 13-53 Troubleshooting.

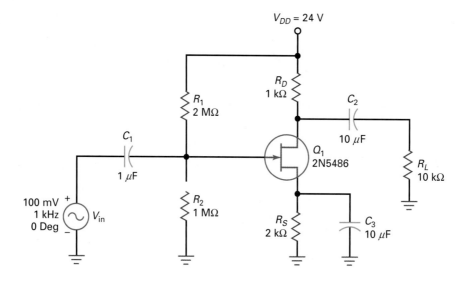

Trouble	V_{GS}	I_D	V_{DS}	V_g	V_s	V_d	V_{out}
OK	−1.6 V	4.8 mA	9.6 V	100 mV	0	357 mV	357 mV
T1	−2.75 V	1.38 mA	19.9 V	100 mV	0	200 mV	200 mV
T2	0.6 V	7.58 mA	1.25 V	100 mV	0	29 mV	29 mV
T3	0.56 V	0	0	100 mV	0	0	0
T4	−8 V	0	8 V	100 mV	0	0	0
T5	8 V	0	24 V	100 mV	0	0	0
T6	−1.6 V	4.8 mA	9.6 V	100 mV	87 mV	40 mV	40 mV
T7	−1.6 V	4.8 mA	9.6 V	100 mV	0	397 mV	0
T8	0	7.5 mA	1.5 V	1 mV	0	0	0

Job Interview Questions

1. Tell me how a JFET works, including the pinchoff and gate-source cutoff voltage in your explanation.
2. Draw the drain curves and the transconductance curve for a JFET.
3. Compare the JFET and the bipolar junction transistor. Your comments should include advantages and disadvantages of each.
4. How can you tell whether an FET is operating in the ohmic region or the active region?
5. Draw a JFET source follower and explain how it works.
6. Draw a JFET shunt switch and a JFET series switch. Explain how each works.
7. How can the JFET be used as a static electricity switch?
8. What input quantity controls the output current in a BJT? A JFET? If the quantities are different, explain.
9. A JFET is a device that controls current flow by placing a voltage on the gate. Explain this.
10. What is the advantage of a cascode amplifier?
11. Tell me why JFETs are sometimes found as the first amplifying device at the front end of radio receivers.

Self-Test Answers

1.	a	10.	c	18.	c
2.	d	11.	c	19.	a
3.	c	12.	a	20.	c
4.	d	13.	c	21.	c
5.	b	14.	d	22.	b
6.	b	15.	a	23.	b
7.	d	16.	b	24.	d
8.	c	17.	c	25.	d
9.	d				

Practice Problem Answers

13-1 $R_{in} = 10,000 \ M\Omega$

13-2 $R_{DS} = 600 \ \Omega$;
$V_p = 3.0 \ V$

13-4 $I_D = 3 \ mA$;
$V_{GS} = -3 \ V$

13-5 $R_{DS} = 300 \ \Omega$;
$V_D = 0.291 \ V$

13-6 $R_S = 500 \ \Omega$;
$V_D = 26 \ V$

13-7 $V_{GS(min)} = -0.85$;
$I_{D(min)} = 2.2 \ mA$;

$V_{GS(max)} = -2.5 \ V$;
$I_{D(max)} = 6.4 \ mA$

13-8 $I_D = 4 \ mA$;
$V_{DS} = 12 \ V$

13-9 $I_{D(max)} = 5.6 \ mA$

13-11 $I_D = 4.3 \ mA$;
$V_D = 5.7 \ V$

13-12 $V_{GS(off)} = -3.2 \ V$;
$g_m = 1,875 \ \mu S$

13-13 $V_{out} = 5.3 \ mV_{pp}$

13-14 $V_{out} = 0.714 \ mV$

13-15 $A_v = 0.634$

13-16 $A_v = 0.885$

13-17 $R_{DS} = 400 \ \Omega$;
on-off ratio = 26

13-18 $V_{out(on)} = 9.6 \ mV$;
$V_{out(off)} = 10 \ \mu V$ on-off ratio = 960

13-19 $V_{peak} = 99.0 \ mV$

14 MOSFETs

The **metal-oxide semiconductor FET,** or **MOSFET,** has a source, gate, and drain. The MOSFET differs from the JFET, however, in that the gate is insulated from the channel. Because of this, the gate current is even smaller than it is in a JFET. The MOSFET is sometimes called an IGFET, which stands for insulated-gate FET.

There are two kinds of MOSFETs, the depletion-mode type and the enhancement-mode type. The enhancement-mode MOSFET is widely used in both discrete and integrated circuits. In discrete circuits, the main use is in power switching, which means turning large currents on and off. In integrated circuits, the main use is in digital switching, the basic process behind modern computers. Although their use has declined, depletion-mode MOSFETs are still found in high-frequency front-end communications circuits as RF amplifiers.

Objectives

After studying this chapter, you should be able to:

- Explain the characteristics and operation of both depletion-mode and enhancement-mode MOSFETs.

- Sketch the characteristic curves for D-MOSFETs and E-MOSFETs.

- Describe how E-MOSFETs are used as digital switches.

- Draw a schematic of a typical CMOS digital switching circuit and explain its operation.

- Compare power FETs with power bipolar junction transistors (BJTs).

- Name and describe several power FET applications.

- Analyze the dc and ac operation of both D-MOSFET and E-MOSFET amplifier circuits.

Chapter Outline

Vocabulary

active-load resistors

analog

complementary MOS (CMOS)

dc-to-ac converter

dc-to-dc converter

depletion-mode MOSFET

digital

drain-feedback bias

enhancement-mode MOSFET

interface

metal-oxide semiconductor FET (MOSFET)

power FET

substrate

threshold voltage

uninterruptible power supply (UPS)

vertical MOS (VMOS)

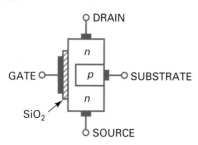

Figure 14-1   Depletion-mode MOSFET.

14-1 The Depletion-Mode MOSFET

Figure 14-1 shows a **depletion-mode MOSFET,** a piece of n material with an insulated gate on the left and a p region on the right. The p region is called the **substrate.** Electrons flowing from source to drain must pass through the narrow channel between the gate and the p substrate.

A thin layer of silicon dioxide (SiO_2) is deposited on the left side of the channel. Silicon dioxide is the same as glass, which is an insulator. In a MOSFET, the gate is metallic. Because the metallic gate is insulated from the channel, negligible gate current flows even when the gate voltage is positive.

Figure 14-2a shows a depletion-mode MOSFET with a negative gate voltage. The V_{DD} supply forces free electrons to flow from source to drain. These electrons flow through the narrow channel on the left of the p substrate. As with a JFET, the gate voltage controls the width of the channel. The more negative the gate voltage, the smaller the drain current. When the gate voltage is negative enough, the drain current is cut off. Therefore, the operation of a depletion-mode MOSFET is similar to that of a JFET when V_{GS} is negative.

Since the gate is insulated, we can also use a positive input voltage, as shown in Fig. 14-2b. The positive gate voltage increases the number of free electrons flowing through the channel. The more positive the gate voltage, the greater the conduction from source to drain.

14-2 D-MOSFET Curves

Figure 14-3a shows the set of drain curves for a typical n-channel, depletion-mode MOSFET. Notice that the curves above $V_{GS} = 0$ are positive and the curves below $V_{GS} = 0$ are negative. As with a JFET, the bottom curve is for $V_{GS} = V_{GS(\text{off})}$ and the drain current will be approximately zero. As shown, when $V_{GS} = 0$ V, the drain current will equal I_{DSS}. This demonstrates that the depletion-mode MOSFET, or D-MOSFET, is a *normally on* device. When V_{GS} is made negative, the drain current will be reduced. In contrast to an n-channel JFET, the n-channel D-MOSFET can have V_{GS} made positive and still function properly. This is because there is no pn junction to become forward biased. When V_{GS} becomes positive, I_D will increase following the square-law equation

$$I_D = I_{DSS}\left(1 - \frac{V_{GS}}{V_{GS(\text{off})}}\right)^2$$

(14-1)

Figure 14-2   (a) D-MOSFET with negative gate; (b) D-MOSFET with positive gate.

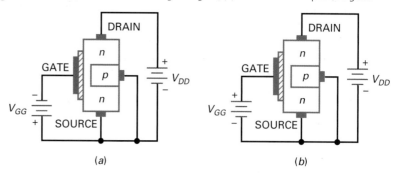

(a)   (b)

Figure 14-3 An *n*-channel, depletion-mode MOSFETs: (*a*) Drain curves; (*b*) transconductance curve.

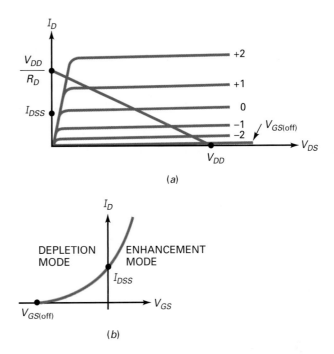

(*a*)

(*b*)

When V_{GS} is negative, the D-MOSFET is operating in the depletion mode. When V_{GS} is positive, the D-MOSFET is operating in the enhancement mode. Like the JFET, the D-MOSFET curves display an ohmic region, a current-source region, and a cutoff region.

Figure 14-3*b* is the transconductance curve for a D-MOSFET. Again, I_{DSS} is the drain current with the gate shorted to the source. I_{DSS} is no longer the maximum possible drain current. The parabolic transconductance curve follows the same square-law relation that exists with a JFET. As a result, the analysis of a depletion-mode MOSFET is almost identical to that of a JFET circuit. The major difference is enabling V_{GS} to be either negative or positive.

There is also a *p*-channel D-MOSFET. It consists of a drain-to-source *p*-channel, along with a *n*-type substrate. Once again, the gate is insulated from the channel. The action of a *p*-channel MOSFET is complementary to the *n*-channel MOSFET. The schematic symbols for both *n*-channel and *p*-channel D-MOSFETs are shown in Fig. 14-4.

Figure 14-4 D-MOSFET schematic symbols: (*a*) *n*-channel; (*b*) *p*-channel.

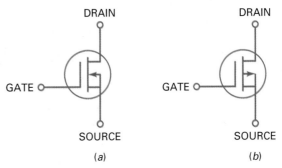

(*a*)

(*b*)

Example 14-1

A D-MOSFET has the values $V_{GS(off)} = -3$ V and $I_{DSS} = 6$ mA. What will the drain current equal when V_{GS} equals -1 V, -2 V, 0 V, $+1$ V, and $+2$ V?

SOLUTION   Following the square-law equation (14-1), when

$$V_{GS} = -1 \text{ V} \qquad I_D = 2.67 \text{ mA}$$

$$V_{GS} = -2 \text{ V} \qquad I_D = 0.667 \text{ mA}$$

$$V_{GS} = 0 \text{ V} \qquad I_D = 6 \text{ mA}$$

$$V_{GS} = +1 \text{ V} \qquad I_D = 10.7 \text{ mA}$$

$$V_{GS} = +2 \text{ V} \qquad I_D = 16.7 \text{ mA}$$

PRACTICE PROBLEM 14-1   Repeat example 14-1 using the values $V_{GS(off)} = -4$ V and $I_{DSS} = 4$ mA.

14-3 Depletion–Mode MOSFET Amplifiers

Figure 14-5   Zero-bias.

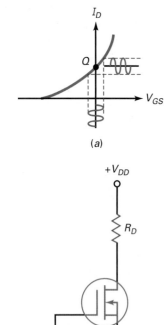

(a)

(b)

A depletion-mode MOSFET is unique because it can operate with a positive or a negative gate voltage. Because of this, we can set its Q point at $V_{GS} = 0$ V, as shown in Fig. 14-5a. When the input signal goes positive, it increases I_D above I_{DSS}. When the input signal goes negative, it decreases I_D below I_{DSS}. Because there is no pn junction to forward bias, the input resistance of the MOSFET remains very high. Being able to use zero V_{GS} allows us to build the very simple bias circuit of Fig. 14-5b. Because I_G is zero, $V_{GS} = 0$ V and $I_D = I_{DSS}$. The drain voltage is:

$$V_{DS} = V_{DD} - I_{DSS} R_D \qquad (14\text{-}2)$$

Due to the fact that a D-MOSFET is a normally on device, it is also possible to use self-bias by adding a source resistor. The operation becomes the same as a self-biased JFET circuit.

Example 14-2

The D-MOSFET amplifier shown in Fig. 14-6 has $V_{GS(off)} = -2$ V, $I_{DSS} = 4$ mA, and $g_{mo} = 2000$ μS. What is the circuit's output voltage?

SOLUTION   With the source grounded, $V_{GS} = 0$ V and $I_D = 4$ mA.

$$V_{DS} = 15 \text{ V} - (4 \text{ mA})(2 \text{ k}\Omega) = 7 \text{ V}$$

Figure 14-6 D-MOSFET amplifier.

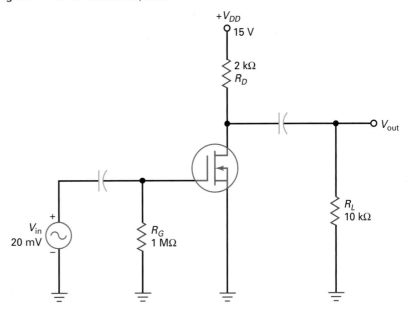

Since $V_{GS} = 0$ V, $gm = g_{mo} = 2000$ µS.
The amplifier's voltage gain is found by:

$$A_V = g_m r_d$$

The ac drain resistance is equal to:

$$r_d = R_D \parallel R_L = 2 \text{ k}\Omega \parallel 10 \text{ k}\Omega = 1.67 \text{ k}\Omega$$

and A_V is:

$$A_V = (2000 \text{ µS})(1.67 \text{ k}\Omega) = 3.34$$

Therefore,

$$V_{\text{out}} = (V_{\text{in}})(A_V) = (20 \text{ mV})(3.34) = 66.8 \text{ mV}$$

PRACTICE PROBLEM 14–2   In Fig. 14-6, if the MOSFET's g_{mo} value is 3000 µS, what is the value of V_{out}?

As shown by Example 14-2, the D-MOSFET has a relatively low voltage gain. One of the major advantages of this device is its extremely high input resistance. This allows us to use this device when circuit loading could be a problem. Also, MOSFETs have excellent low-noise properties. This is a definite advantage for any stage near the front end of a system where the signal is weak. This is very common in many types of electronic communications circuits.

Some D-MOSFETs, as shown in Fig. 14-7, are dual-gate devices. One gate can serve as the input signal point, while the other gate can be connected to an automatic gain control dc voltage. This allows the voltage gain of the MOSFET to be controlled and varied depending on the input signal strength.

Figure 14-7 Dual-gate MOSFET.

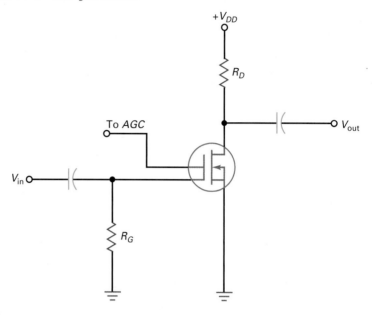

14-4 The Enhancement-Mode MOSFET

The depletion-mode MOSFET was part of the evolution toward the **enhancement-mode MOSFET,** abbreviated *E-MOSFET.* Without the E-MOSFET, the personal computers that are now so widespread would not exist.

The Basic Idea

Figure 14-8a shows an E-MOSFET. The *p* substrate now extends all the way to the silicon dioxide. As you can see, there no longer is an *n* channel between the source and the drain. How does an E-MOSFET work? Figure 14-8b shows normal biasing polarities. When the gate voltage is zero, the current between source and drain is zero. For this reason, an E-MOSFET is *normally off* when the gate voltage is zero.

The only way to get current is with a positive gate voltage. When the gate is positive, it attracts free electrons into the *p* region. The free electrons recombine with the holes next to the silicon dioxide. When the gate voltage is positive enough, all the holes touching the silicon dioxide are filled and free electrons begin to flow from the source to the drain. The effect is the same as creating a thin layer of *n*-type material next to the silicon dioxide. This thin conducting layer is

Figure 14-8 Enhancement-mode MOSFET: (*a*) Unbiased; (*b*) biased.

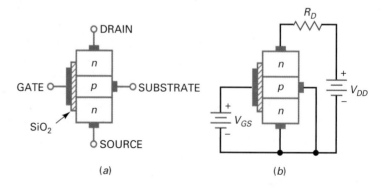

Figure 14-9 EMOS graphs: (a) Drain curves; (b) transconductance curve.

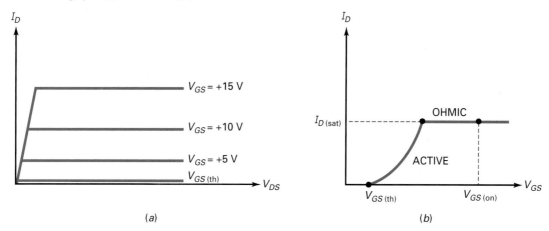

(a)

(b)

called the *n-type inversion layer.* When it exists, free electrons can flow easily from the source to the drain.

The minimum V_{GS} that creates the *n*-type inversion layer is called the **threshold voltage,** symbolized $V_{GS(th)}$. When V_{GS} is less than $V_{GS(th)}$, the drain current is zero. When V_{GS} is greater than $V_{GS(th)}$, an *n*-type inversion layer connects the source to the drain and the drain current can flow. Typical values of $V_{GS(th)}$ for small-signal devices are from 1 to 3 V.

The JFET is referred to as a *depletion-mode device* because its conductivity depends on the action of depletion layers. The E-MOSFET is classified as an *enhancement-mode device* because a gate voltage greater than the threshold voltage enhances its conductivity. With zero gate voltage, a JFET is *on,* whereas an E-MOSFET is *off.* Therefore, the E-MOSFET is considered to be a normally off device.

Drain Curves

A small-signal E-MOSFET has a power rating of 1 W or less. Figure 14-9a shows a set of drain curves for a typical small-signal E-MOSFET. The lowest curve is the $V_{GS(th)}$ curve. When V_{GS} is less than $V_{GS(th)}$, the drain current is approximately zero. When V_{GS} is greater than $V_{GS(th)}$, the device turns on and the drain current is controlled by the gate voltage.

The almost-vertical part of the graph is the ohmic region, and the almost-horizontal parts are the active region. When biased in the ohmic region, the E-MOSFET is equivalent to a resistor. When biased in the active region, it is equivalent to a current source. Although the E-MOSFET can operate in the active region, the main use is the ohmic region.

Figure 14-9b shows a typical transconductance curve. There is no drain current until $V_{GS} = V_{GS(th)}$. The drain current then increases rapidly until it reaches the saturation current $I_{D(sat)}$. Beyond this point, the device is biased in the ohmic region. Therefore, I_D cannot increase, even though V_{GS} increases. To ensure hard saturation, a gate voltage of $V_{GS(on)}$ well above $V_{GS(th)}$ is used, as shown in Fig. 14-9b.

Schematic Symbol

When $V_{GS} = 0$, the E-MOSFET is off because there is no conducting channel between source and drain. The schematic symbol of Fig. 14-10a has a broken channel line to indicate this normally off condition. As you know, a gate voltage greater than the threshold voltage creates an *n*-type inversion layer that connects

GOOD TO KNOW

With the E-MOSFET, V_{GS} has to be greater than $V_{GS(th)}$ to get any drain current at all. Therefore, when E-MOSFETs are biased, self-bias, current-source bias, and zero bias cannot be used because these forms of bias depend on the depletion mode of operation. This leaves gate bias, voltage-divider bias, and source bias as the means for biasing E-MOSFETs.

Figure 14-10 EMOS Schematic symbols: (a) N-channel device; (b) p-channel device.

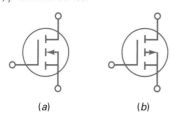

(a)

(b)

the source to the drain. The arrow points to this inversion layer, which acts like an *n* channel when the device is conducting.

There is also a *p*-channel E-MOSFET. The schematic symbol is similar, except that the arrow points outward, as shown in Fig. 14-10*b*.

Maximum Gate-Source Voltage

MOSFETs have a thin layer of silicon dioxide, an insulator that prevents gate current for positive as well as negative gate voltages. This insulating layer is kept as thin as possible to give the gate more control over the drain current. Because the insulating layer is so thin, it is easily destroyed by excessive gate-source voltage.

For instance, a 2N7000 has a $V_{GS(max)}$ rating of ± 20 V. If the gate-source voltage becomes more positive than $+20$ V or more negative than -20 V, the thin insulating layer will be destroyed.

Aside from directly applying an excessive V_{GS}, you can destroy the thin insulating layer in more subtle ways. If you remove or insert a MOSFET into a circuit while the power is on, transient voltages caused by inductive kickback may exceed the $V_{GS(max)}$ rating. Even picking up a MOSFET may deposit enough static charge to exceed the $V_{GS(max)}$ rating. This is the reason why MOSFETs are often shipped with a wire ring around the leads, or wrapped in tin foil, or inserted into conductive foam.

Some MOSFETs are protected by a built-in zener diode in parallel with the gate and the source. The zener voltage is less than the $V_{GS(max)}$ rating. Therefore, the zener diode breaks down before any damage to the thin insulating layer occurs. The disadvantage of these internal zener diodes is that they reduce the MOSFET's high input resistance. The trade-off is worth it in some applications because expensive MOSFETs are easily destroyed without zener protection.

In conclusion, MOSFET devices are delicate and can be easily destroyed. You have to handle them carefully. Furthermore, you should never connect or disconnect them while the power is on. Finally, before you pick up a MOSFET device, you should ground your body by touching the chassis of the equipment you are working on.

14-5 The Ohmic Region

Although the E-MOSFET can be biased in the active region, this is seldom done because it is primarily a switching device. The typical input voltage is either low or high. Low voltage is 0 V, and high voltage is $V_{GS(on)}$, a value specified on data sheets.

Drain-Source on Resistance

When an E-MOSFET is biased in the ohmic region, it is equivalent to a resistance of $R_{DS(on)}$. Almost all data sheets will list the value of this resistance at a specific drain current and gate-source voltage.

Figure 14-11 illustrates the idea. There is a Q_{test} point in the ohmic region of the $V_{GS} = V_{GS(on)}$ curve. The manufacturer measures $I_{D(on)}$ and $V_{DS(on)}$ at this Q_{test} point. From this, the manufacturer calculates the value of $R_{DS(on)}$ using this definition:

$$R_{DS(on)} = \frac{V_{DS(on)}}{I_{D(on)}} \qquad (14\text{-}3)$$

Figure 14-11 Measuring $R_{DS(on)}$.

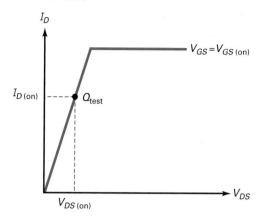

For instance, at the test point, a VN2406L has $V_{DS(on)} = 1$ V and $I_{D(on)} = 100$ mA. With Eq. (14-3):

$$R_{DS(on)} = \frac{1 \text{ V}}{100 \text{ mA}} = 10 \ \Omega$$

Fig. 14-12 shows the data sheet of a 2N7000 n-channel E-MOSFET. Notice that this E-MOSFET can also be packaged as a surface-mount device. Also make note of the internal diode between the drain and source leads. Minimum, typical, and maximum values are listed for this device. These device specifications often have a wide range of values.

Table of E-MOSFETs

Table 14-1 is a sample of small-signal E-MOSFETs. The typical $V_{GS(th)}$ values are 1.5 to 3 V. The $R_{DS(on)}$ values are 0.3 to 28 Ω, which means that the E-MOSFET has a low resistance when biased in the ohmic region. When biased at cutoff, it has a very high resistance, approximately an open circuit. Therefore, E-MOSFETs have excellent on-off ratios.

Table 14-1	Small-Signal EMOS Sampler					
Device	$V_{GS(th)}$, **V**	$V_{GS(on)}$, **V**	$I_{D(on)}$	$R_{DS(on)}$, Ω	$I_{D(max)}$	$P_{D(max)}$
VN2406L	1.5	2.5	100 mA	10	200 mA	350 mW
BS107	1.75	2.6	20 mA	28	250 mA	350 mW
2N7000	2	4.5	75 mA	6	200 mA	350 mW
VN10LM	2.5	5	200 mA	7.5	300 mA	1 W
MPF930	2.5	10	1 A	0.9	2 A	1 W
IRFD120	3	10	600 mA	0.3	1.3 A	1 W

Figure 14–12 The 2N7000 data sheet.

SEMICONDUCTOR ™

2N7000 / 2N7002 / NDS7002A
N-Channel Enhancement Mode Field Effect Transistor

General Description

These N-Channel enhancement mode field effect transistors are produced using Fairchild's proprietary, high cell density, DMOS technology. These products have been designed to minimize on-state resistance while provide rugged, reliable, and fast switching performance. They can be used in most applications requiring up to 400mA DC and can deliver pulsed currents up to 2A. These products are particularly suited for low voltage, low current applications such as small servo motor control, power MOSFET gate drivers, and other switching applications.

Features

- High density cell design for low $R_{DS(ON)}$.
- Voltage controlled small signal switch.
- Rugged and reliable.
- High saturation current capability.

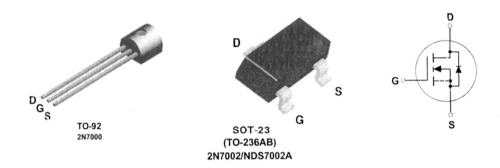

TO-92
2N7000

SOT-23
(TO-236AB)
2N7002/NDS7002A

Absolute Maximum Ratings T_A = 25°C unless otherwise noted

Symbol	Parameter	2N7000	2N7002	NDS7002A	Units
V_{DSS}	Drain-Source Voltage	60			V
V_{DGR}	Drain-Gate Voltage ($R_{GS} \leq$ 1 MΩ)	60			V
V_{GSS}	Gate-Source Voltage - Continuous	$\pm$20			V
	- Non Repetitive (tp < 50µs)	$\pm$40			
I_D	Maximum Drain Current - Continuous	200	115	280	mA
	- Pulsed	500	800	1500	
P_D	Maximum Power Dissipation	400	200	300	mW
	Derated above 25°C	3.2	1.6	2.4	mW/°C
T_J, T_{STG}	Operating and Storage Temperature Range	-55 to 150		-65 to 150	°C
T_L	Maximum Lead Temperature for Soldering Purposes, 1/16" from Case for 10 Seconds	300			°C
THERMAL CHARACTERISTICS					
$R_{\theta JA}$	Thermal Resistance, Junction-to-Ambient	312.5	625	417	°C/W

2N7000.SAM Rev. A1

Figure 14–12 (continued)

Electrical Characteristics $T_A = 25°C$ unless otherwise noted

Symbol	Parameter	Conditions		Type	Min	Typ	Max	Units
OFF CHARACTERISTICS								
BV_{DSS}	Drain-Source Breakdown Voltage	$V_{GS} = 0$ V, $I_D = 10$ μA		All	60			V
I_{DSS}	Zero Gate Voltage Drain Current	$V_{DS} = 48$ V, $V_{GS} = 0$ V		2N7000			1	μA
			$T_J = 125°C$				1	mA
		$V_{DS} = 60$ V, $V_{GS} = 0$ V		2N7002 NDS7002A			1	μA
			$T_J = 125°C$				0.5	mA
I_{GSSF}	Gate - Body Leakage, Forward	$V_{GS} = 15$ V, $V_{DS} = 0$ V		2N7000			10	nA
		$V_{GS} = 20$ V, $V_{DS} = 0$ V		2N7002 NDS7002A			100	nA
I_{GSSR}	Gate - Body Leakage, Reverse	$V_{GS} = -15$ V, $V_{DS} = 0$ V		2N7000			-10	nA
		$V_{GS} = -20$ V, $V_{DS} = 0$ V		2N7002 NDS7002A			-100	nA
ON CHARACTERISTICS (Note 1)								
$V_{GS(th)}$	Gate Threshold Voltage	$V_{DS} = V_{GS}$, $I_D = 1$ mA		2N7000	0.8	2.1	3	V
		$V_{DS} = V_{GS}$, $I_D = 250$ μA		2N7002 NDS7002A	1	2.1	2.5	
$R_{DS(ON)}$	Static Drain-Source On-Resistance	$V_{GS} = 10$ V, $I_D = 500$ mA		2N7000		1.2	5	Ω
			$T_J = 125°C$			1.9	9	
		$V_{GS} = 4.5$ V, $I_D = 75$ mA				1.8	5.3	
		$V_{GS} = 10$ V, $I_D = 500$ mA		2N7002		1.2	7.5	
			$T_J = 100°C$			1.7	13.5	
		$V_{GS} = 5.0$ V, $I_D = 50$ mA				1.7	7.5	
			$T_J = 100C$			2.4	13.5	
		$V_{GS} = 10$ V, $I_D = 500$ mA		NDS7002A		1.2	2	
			$T_J = 125°C$			2	3.5	
		$V_{GS} = 5.0$ V, $I_D = 50$ mA				1.7	3	
			$T_J = 125°C$			2.8	5	
$V_{DS(ON)}$	Drain-Source On-Voltage	$V_{GS} = 10$ V, $I_D = 500$ mA		2N7000		0.6	2.5	V
		$V_{GS} = 4.5$ V, $I_D = 75$ mA				0.14	0.4	
		$V_{GS} = 10$ V, $I_D = 500$ mA		2N7002		0.6	3.75	
		$V_{GS} = 5.0$ V, $I_D = 50$ mA				0.09	1.5	
		$V_{GS} = 10$ V, $I_D = 500$ mA		NDS7002A		0.6	1	
		$V_{GS} = 5.0$ V, $I_D = 50$ mA				0.09	0.15	

Electrical Characteristics $T_A = 25°C$ unless otherwise noted

Symbol	Parameter	Conditions	Type	Min	Typ	Max	Units
ON CHARACTERISTICS Continued (Note 1)							
$I_{D(ON)}$	On-State Drain Current	$V_{GS} = 4.5$ V, $V_{DS} = 10$ V	2N7000	75	600		mA
		$V_{GS} = 10$ V, $V_{DS} \geq 2 V_{DS(on)}$	2N7002	500	2700		
		$V_{GS} = 10$ V, $V_{DS} \geq 2 V_{DS(on)}$	NDS7002A	500	2700		
g_{FS}	Forward Transconductance	$V_{DS} = 10$ V, $I_D = 200$ mA	2N7000	100	320		mS
		$V_{DS} \geq 2 V_{DS(on)}$, $I_D = 200$ mA	2N7002	80	320		
		$V_{DS} \geq 2 V_{DS(on)}$, $I_D = 200$ mA	NDS7002A	80	320		

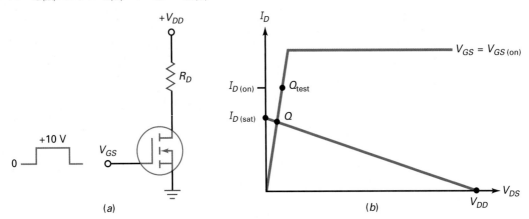

(a)

(b)

Biasing in Ohmic Region

In Fig. 14-13a, the drain saturation current in this circuit is:

$$I_{D(\text{sat})} = \frac{V_{DD}}{R_D} \tag{14-4}$$

and the drain cutoff voltage is V_{DD}. Figure 14-13b shows the dc load line between a saturation current of $I_{D(\text{sat})}$ and a cutoff voltage of V_{DD}.

When $V_{GS} = 0$, the Q point is at the lower end of the dc load line. When $V_{GS} = V_{GS(\text{on})}$, the Q point is at the upper end of the load line. When the Q point is below the Q_{test} point, as shown in Fig. 14-13b, the device is biased in the ohmic region. Stated another way, an E-MOSFET is biased in the ohmic region when this condition is satisfied:

$$I_{D(\text{sat})} < I_{D(\text{on})} \qquad \text{when} \qquad V_{GS} = V_{GS(\text{on})} \tag{14-5}$$

Equation (14-5) is important. It tells us whether an E-MOSFET is operating in the active region or the ohmic region. Given an EMOS circuit, we can calculate the $I_{D(\text{sat})}$. If $I_{D(\text{sat})}$ is less than $I_{D(\text{on})}$ when $V_{GS} = V_{GS(\text{on})}$, we will know that the device is biased in the ohmic region and is equivalent to a small resistance.

Example 14-3

What is the output voltage in Fig. 14-14a?

SOLUTION   For the 2N7000, the most important values in Table 14-1 are:

$$V_{GS(\text{on})} = 4.5 \text{ V}$$

$$I_{D(\text{on})} = 75 \text{ mA}$$

$$R_{DS(\text{on})} = 6 \text{ }\Omega$$

Since the input voltage swings from 0 to 4.5 V, the 2N7000 is being switched on and off.

The drain saturation current in Fig. 14-14a is:

$$I_{D(\text{sat})} = \frac{20 \text{ V}}{1 \text{ k}\Omega} = 20 \text{ mA}$$

Figure 14-14 Switching between cutoff and saturation.

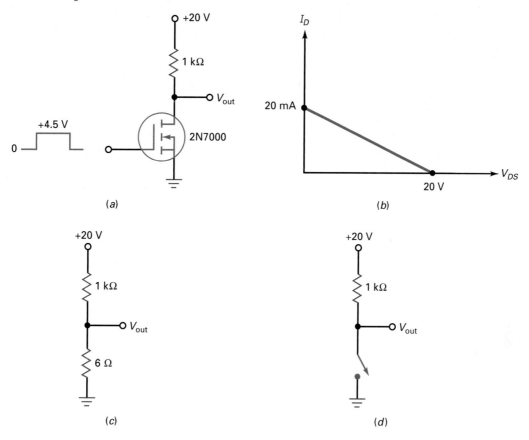

Figure 14-14*b* is the dc load line. Since 20 mA is less than 75 mA, the value of $I_{D(on)}$, the 2N7000 is biased in the ohmic region when the gate voltage is high.

Figure 14-14*c* is the equivalent circuit for a high-input gate voltage. Since the E-MOSFET has a resistance of 6 Ω, the output voltage is:

$$V_{out} = \frac{6\ \Omega}{1\ k\Omega + 6\ \Omega}\ (20\ V) = 0.12\ V$$

On the other hand, when V_{GS} is low, the E-MOSFET is open (Fig. 14-14*d*), and the output voltage is pulled up to the supply voltage:

$$V_{out} = 20\ V$$

PRACTICE PROBLEM 14–3 Using Fig. 14-14*a*, replace the 2N7000 with a VN2406L E-MOSFET and find the output voltage value.

Example 14-4

|||| MultiSim

What is the LED current in Fig. 14-15?

Figure 14-15 Turning an LED on and off.

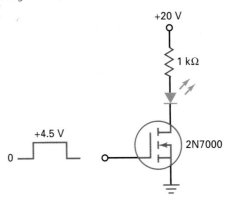

SOLUTION When V_{GS} is low, the LED is off. When V_{GS} is high, the action is similar to that in the preceding example because the 2N7000 goes into hard saturation. If we ignore the LED voltage drop, the LED current is:

$$I_D \approx 20 \text{ mA}$$

If we allow 2 V for the LED drop:

$$I_D = \frac{20 \text{ V} - 2 \text{ V}}{1 \text{ k}\Omega} = 18 \text{ mA}$$

PRACTICE PROBLEM 14-4 Repeat Example 14-4 using a VN2406L E-MOSFET and a 560 Ω drain resistor.

Example 14-5

What does the circuit of Fig. 14-16*a* do if a coil current of 30 mA or more closes the relay contacts?

SOLUTION The E-MOSFET is being used to turn a relay on and off. Since the relay coil has a resistance of 500 Ω, the saturation current is:

$$I_{D(\text{sat})} = \frac{24 \text{ V}}{500 \text{ }\Omega} = 48 \text{ mA}$$

Because this is less than the $I_{D(\text{on})}$ of the VN2406L, the device has a resistance of only 10 Ω (see Table 14-1).

Figure 14-16*b* shows the equivalent circuit for high V_{GS}. The current through the relay coil is approximately 48 mA, more than enough to close the relay. When the relay is closed, the contact circuit looks like Fig. 14-16*c*. Therefore, the final load current is 8 A (120 V divided by 15 Ω).

Figure 14–16 Low-input current signal controls large output current.

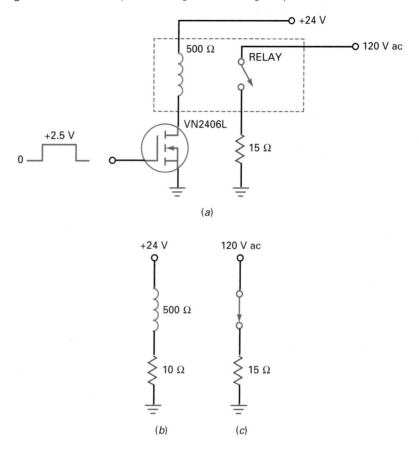

(a)

(b)   (c)

In Figure 14-16*a*, an input voltage of only +2.5 V and almost zero input current control a load voltage of 120 V ac and a load current of 8 A. A circuit like this is useful with remote control. The input voltage could be a signal that has been transmitted a long distance through copper wire, fiber-optic cable, or outer space.

14–6 Digital Switching

Why has the E-MOSFET revolutionized the computer industry? Because of its threshold voltage, it is ideal for use as a switching device. When the gate voltage is well above the threshold voltage, the device switches from cutoff to saturation. This off-on action is the key to building computers. When you study computer circuits, you will see how a typical computer uses millions of E-MOSFETs as off-on switches to process data. (*Data* include numbers, text, graphics, and all other information that can be coded as binary numbers.)

Analog, Digital, and Switching Circuits

The word **analog** means "continuous," like a sine wave. When we speak of an analog signal, we are talking about signals that continuously change in voltage like the one in Fig. 14-17*a*. The signal does not have to be sinusoidal. As long as

Figure 14-17 (*a*) Analog signal; (*b*) digital signal.

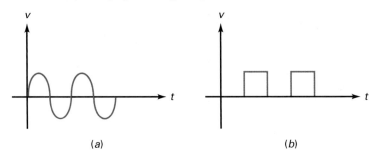

GOOD TO KNOW

Most physical quantities are analog in nature, and these are the quantities that are most often the inputs and outputs being monitored and controlled by a system. Some examples of analog inputs and outputs are temperature, pressure, velocity, position, fluid level, and flow rate. To take advantage of digital techniques when dealing with analog inputs, the physical quantities are converted to a digital format. A circuit that does this is called an *analog-to-digital (A/D) converter.*

there are no sudden jumps between two distinct voltage levels, the signal is referred to as an *analog signal.*

The word **digital** refers to a discontinuous signal. This means that the signal jumps between two distinct voltage levels like the waveform of Fig. 14-17*b*. Digital signals like these are the kind of signals inside computers. These signals are computer codes that represent numbers, letters, and other symbols.

The word *switching* is a broader word than *digital.* Switching circuits include digital circuits as a subset. In other words, switching circuits can also refer to circuits that turn on motors, lamps, heaters, and other heavy-current devices.

Passive-Load Switching

Figure 14-18 shows an E-MOSFET with a passive load. The word *passive* refers to ordinary resistors like R_D. In this circuit, v_{in} is either low or high. When v_{in} is low, the MOSFET is cut off, and v_{out} equals the supply voltage V_{DD}. When v_{in} is high, the MOSFET saturates and v_{out} drops to a low value. For the circuit to work properly, the drain saturation current $I_{D(sat)}$ has to be less than $I_{D(on)}$ when the input voltage is equal to or greater than $V_{GS(on)}$. This is equivalent to saying that the resistance in the ohmic region has to be much smaller than the passive drain resistance. In symbols:

$$R_{DS(on)} \ll R_D$$

A circuit like Fig. 14-18 is the simplest computer circuit that can be built. It is called an *inverter* because the output voltage is the opposite of the input voltage. When the input voltage is low, the output voltage is high. When the input voltage is high, the output voltage is low. Great accuracy is not necessary when analyzing switching circuits. All that matters is that the input and output voltages can be easily recognized as low or high.

Figure 14-18 Passive load.

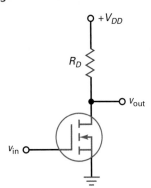

Active-Load Switching

Integrated circuits (ICs) consist of thousands of microscopically small transistors, either bipolar or MOS. The earliest integrated circuits used passive load resistors like the one of Fig. 14-18. But a passive load resistance presents a major problem. It is physically much larger than a MOSFET. Because of this, integrated circuits with passive-load resistors were too big until somebody invented **active-load resistors.** This greatly reduced the size of integrated circuits and led to the personal computers that we have today.

The key idea was to get rid of passive-load resistors. Figure 14-19*a* shows the invention: *active-load switching.* The lower MOSFET acts like a switch, but the upper MOSFET acts like large resistance. Notice that the upper

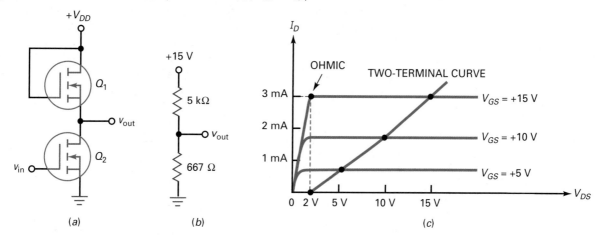

Figure 14-19 (a) Active load; (b) equivalent circuit; (c) $V_{GS} = V_{DS}$ produces a two-terminal curve.

MOSFET has its gate connected to its drain. Because of this, it becomes a *two-terminal device* with an active resistance of:

$$R_D = \frac{V_{DS(\text{active})}}{I_{D(\text{active})}} \tag{14-6}$$

where $V_{DS(\text{active})}$ and $I_{D(\text{active})}$ are voltages and currents in the active region.

For the circuit to work properly, the R_D of the upper MOSFET has to be large compared to the $R_{DS(\text{on})}$ of the lower MOSFET. For instance, if the upper MOSFET acts like an R_D of 5 kΩ and the lower one like an $R_{DS(\text{on})}$ of 667 Ω, as shown in Fig. 14-19b, then the output voltage will be low.

Figure 14-19c shows how to calculate the R_D of the upper MOSFET. Because $V_{GS} = V_{DS}$, each operating point of this MOSFET has to fall along the two-terminal curve shown in Fig. 14-19c. If you check each plotted point on this two-terminal curve, you will see that $V_{GS} = V_{DS}$.

The two-terminal curve of Fig. 14-19c means that the upper MOSFET acts like a resistance of R_D. The value of R_D will change slightly for the different points. For instance, at the highest point shown in Fig. 14-19c, the two-terminal curve has $I_D = 3$ mA and $V_{DS} = 15$ V. With Eq. (14-6), we can calculate:

$$R_D = \frac{15 \text{ V}}{3 \text{ mA}} = 5 \text{ k}\Omega$$

The next point down has these approximate values: $I_D = 1.6$ mA and $V_{DS} = 10$ V. Therefore:

$$R_D = \frac{10 \text{ V}}{1.6 \text{ mA}} = 6.25 \text{ k}\Omega$$

By a similar calculation, the lowest point where $V_{DS} = 5$ V and $I_D = 0.7$ mA has $R_D = 7.2$ kΩ.

If the lower MOSFET has the same set of drain curves as the upper one, then the lower MOSFET has an $R_{DS(\text{on})}$ of:

$$R_{DS(\text{on})} = \frac{2 \text{ V}}{3 \text{ mA}} = 667 \text{ }\Omega$$

This is the value shown in Fig. 14-19b.

As already indicated, exact values don't matter with digital switching circuits as long as the voltages can be easily distinguished as low or high. Therefore, the exact value of R_D does not matter. It can be 5, 6.25, or 7.2 kΩ. Any of these values is large enough to produce a low output voltage in Fig. 14-19b.

Conclusion

Active-load resistors are necessary with digital ICs because a small physical size is important with digital ICs. The designer makes sure that the R_D of upper MOSFET is large compared to the $R_{D(on)}$ of the lower MOSFET. When you see a circuit like Fig. 14-19a, all you have to remember is the basic idea: The circuit acts like a resistance of R_D in series with a switch. As a result, the output voltage is either high or low.

Example 14-6
IIII MultiSim

What is the output voltage in Fig. 14-20a when the input is low? When it is high?

Figure 14-20 Examples.

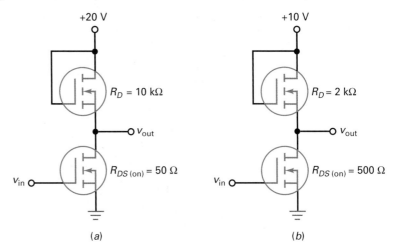

(a) (b)

SOLUTION When the input voltage is low, the lower MOSFET is open and the output voltage is pulled up to the supply voltage:

$$v_{out} = 20 \text{ V}$$

When the input voltage is high, the lower MOSFET has a resistance of 50 Ω. In this case, the output voltage is pulled down toward ground:

$$v_{out} = \frac{50 \ \Omega}{10 \text{ k}\Omega + 50 \ \Omega} (20 \text{ V}) = 100 \text{ mV}$$

PRACTICE PROBLEM 14-6 Repeat Example 14-6 using a $R_{D(on)}$ value of 100 Ω.

Example 14-7

What is the output voltage in Fig. 14-20b?

SOLUTION When the input voltage is low:

$$v_{out} = 10 \text{ V}$$

When the input voltage is high:

$$v_{out} = \frac{500\ \Omega}{2.5\ k\Omega}\ (10\ V) = 2\ V$$

If you compare this to the preceding example, you can see that the on-off ratio is not as good. But with digital circuits, a high on-off ratio is not important. In this example, the output voltage is either 2 or 10 V. These voltages are easily distinguishable as low or high.

PRACTICE PROBLEM 14-7 Using Fig. 14-20b, how high can $R_{DS(on)}$ be and have a V_{out} value below 1 V when V_{in} is high?

14-7 CMOS

With active-load switching, the current drain with a low output is approximately equal to $I_{D(sat)}$. This may create a problem with battery-operated equipment. One way to reduce the current drain of a digital circuit is with **complementary MOS (CMOS)**. In this approach, the IC designer combines n-channel and p-channel MOSFETs.

Figure 14-21a shows the idea. Q_1 is a p-channel MOSFET and Q_2 is an n-channel MOSFET. These two devices are complementary; that is, they have equal and opposite values of $V_{GS(th)}$, $V_{GS(on)}$, $I_{D(on)}$, and so on. The circuit is similar to a class B amplifier because one MOSFET conducts while the other is off.

Basic Action

When a CMOS circuit like Fig. 14-21a is used in a switching application, the input voltage is either high $(+V_{DD})$ or low (0 V). When the input voltage is high, Q_1 is off and Q_2 is on. In this case, the shorted Q_2 pulls the output voltage down to ground. On the other hand, when the input voltage is low, Q_1 is on and Q_2 is off. Now, the shorted Q_1 pulls the output voltage up to $+V_{DD}$. Since the output voltage is inverted, the circuit is called a *CMOS inverter*.

Figure 14-21b shows how the output voltage varies with the input voltage. When the input voltage is zero, the output voltage is high. When the input voltage is high, the output voltage is low. Between the two extremes, there is

Figure 14-21 CMOS inverter: (a) Circuit; (b) input-output graph.

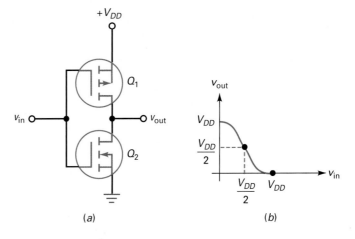

(a)

(b)

a crossover point where the input voltage equals $V_{DD}/2$. At this point, both MOSFETs have equal resistances and the output voltage equals $V_{DD}/2$.

Power Consumption

The main advantage of CMOS is its extremely low power consumption. Because both MOSFETs are in series in Fig. 14-21a, the quiescent current drain is determined by the nonconducting device. Since its resistance is in the megohms, the *quiescent* (idling) power consumption approaches zero.

The power consumption increases when the input signal switches from low to high, and vice versa. The reason is this: At the midway point in a transition from low to high, or vice versa, both MOSFETs are on. This means that the drain current temporarily increases. Since the transition is very rapid, only a brief pulse of current occurs. The product of the drain supply voltage and the brief pulse of current means that the average *dynamic* power consumption is greater than the quiescent power consumption. In other words, a CMOS device dissipates more average power when it has transitions than when it is quiescent.

Since the pulses of current are very short, however, the average power dissipation is very low even when CMOS devices are switching states. In fact, the average power consumption is so small that CMOS circuits are often used for battery-powered applications such as calculators, digital watches, and hearing aids.

Example 14–8

The MOSFETs of Fig. 14-22a have $R_{DS(on)} = 100\ \Omega$ and $R_{DS(off)} = 1\ M\Omega$. What does the output waveform look like?

SOLUTION The input signal is a rectangular pulse that switches from 0 to $+15$ V at point A and from $+15$ V to 0 at point B. Before point A in time, Q_1 is on and Q_2 is off. Since Q_1 has a resistance of 100 Ω compared to a resistance of 1 MΩ for Q_2, the output voltage is pulled up to $+15$ V.

Between points A and B, the input voltage is $+15$ V. This cuts off Q_1 and turns on Q_2. In this case, the low resistance of Q_2 pulls the output voltage down to approximately zero. Figure 14-22b shows the output waveform.

Figure 14–22 Example.

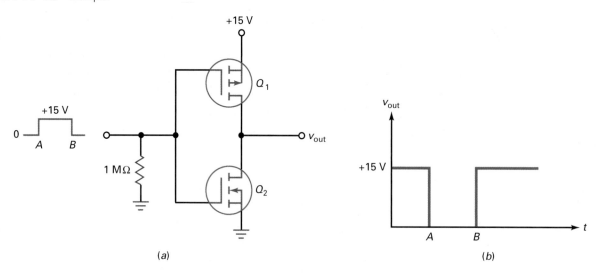

(a)

(b)

PRACTICE PROBLEM 14-8 Repeat Example 14-8 with $V_{in} = +10$ V pulse between A and B.

14-8 Power FETs

In earlier discussions, we emphasized small-signal E-MOSFETs, that is, low-power MOSFETs. Although some discrete low-power E-MOSFETs are commercially available (see Table 14-1), the major use of low-power EMOS is with digital integrated circuits.

High-power EMOS is different. With high-power EMOS, the E-MOSFET is a discrete device widely used in applications that control motors, lamps, disk drives, printers, power supplies, and so on. In these applications, the E-MOSFET is called a **power FET.**

Discrete Devices

Manufacturers are producing different devices such as VMOS, TMOS, hexFET, trench MOSFET, and waveFET. All these power FETs use different channel geometries to increase their maximum ratings. These devices have current ratings from 1 A to more than 200 A, and power ratings from 1 W to more than 500 W.

Fig. 14-23a shows the structure of an enhancement-type MOSFET in an integrated circuit. The source is on the left, the gate in the middle, and the drain on the right. Free electrons flow horizontally from the source to the drain when V_{GS} is greater than V_{GS}(th). This structure limits the maximum current because free electrons must flow along the narrow inversion layer, symbolized by the dashed line. Because the channel is so narrow, conventional MOS devices have small drain currents and low power ratings.

Fig. 14-23b shows the structure of a **vertical MOS (VMOS)** device. It has two sources at the top, which are usually connected, and the substrate acts like the drain. When V_{GS} is greater than V_{GS}(th), free electronics flow vertically downward from the two sources to the drain. Because the conducting channel is much wider along both sides of the V groove, the current can be much larger. This enables the VMOS device to act as a power FET.

Figure 14-23 MOS structures: (a) Conventional MOSFET structure; (b) VMOS structure.

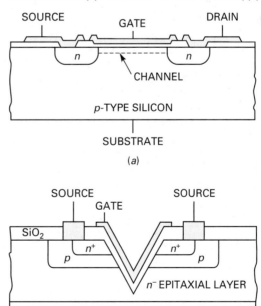

Table 14-2	Power FET Sampler				
Device	$V_{GS(on)}$, V	$I_{D(on)}$, A	$R_{DS(on)}$, Ω	$I_{D(max)}$, A	$P_{D(max)}$, W
MTP4N80E	10	2	1.95	4	125
MTV10N100E	10	5	1.07	10	250
MTW24N40E	10	12	0.13	24	250
MTW45N10E	10	22.5	0.035	45	180
MTE125N20E	10	62.5	0.012	125	460

Table 14-2 is a sample of commercially available power FETs. Notice that $V_{GS(on)}$ is 10 V for all these devices. Because they are physically larger devices, they require higher $V_{GS(on)}$ to ensure operation in the ohmic region. As you can see, the power ratings of these devices are substantial, capable of handling heavy-duty applications like automotive controls, lighting, and heating.

The analysis of a power FET circuit is the same as for small-signal devices. When driven by a $V_{GS(on)}$ of 10 V, a power FET has a small resistance of $R_{DS(on)}$ in the ohmic region. As before, an $I_{D(sat)}$ less than $I_{D(on)}$ when $V_{GS} = V_{GS(on)}$ guarantees that the device is biased in the ohmic region and acts like a small resistance.

Lack of Thermal Runaway

As discussed in Chap. 12, bipolar junction transistors may be destroyed by *thermal runaway*. The problem with bipolar transistors is the negative temperature coefficient of V_{BE}. When the internal temperature increases, V_{BE} decreases. This increases the collector current, forcing the temperature higher. But a higher temperature reduces V_{BE} even more. If not properly heat-sinked, the bipolar transistor will go into thermal runaway and be destroyed.

One major advantage of power FETs over bipolar transistors is the lack of thermal runaway. The $R_{DS(on)}$ of a MOSFET has a positive temperature coefficient. When the internal temperature increases, $R_{DS(on)}$ increases and reduces the drain current, which lowers the temperature. As a result, power FETs are inherently temperature-stable and cannot go into thermal runaway.

Power FETs in Parallel

Bipolar junction transistors cannot be connected in parallel because their V_{BE} drops do not match closely enough. If you try to connect them in parallel, *current hogging* occurs. This means that the transistor with the lower V_{BE} takes more collector current than the others.

Power FETs in parallel do not suffer from the problem of current hogging. If one of the power FETs tries to hog the current, its internal temperature will increase. This increases its $R_{DS(on)}$, which reduces its drain current. The overall effect is for all the power FETs to have equal drain currents.

Faster Turnoff

As mentioned earlier, the minority carriers of bipolar transistors are stored in the junction area during forward bias. When you try to switch off a bipolar transistor, the stored charges flow for a while, preventing a fast turnoff. Since a power FET does not have minority carriers, it can switch a large current off faster than a bipolar transistor can. Typically, a power FET can switch off amperes of current

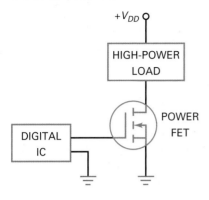

in tens of nanoseconds. This is 10 to 100 times faster than with a comparable bipolar junction transistor.

Power FET as an Interface

Digital ICs are low-power devices because they can supply only small load currents. If we want to use the output of a digital IC to drive a high-current load, we can use a power FET as an **interface** (a device B that allows device A to communicate with or control device C).

Figure 14-24 shows how a digital IC can control a high-power load. The output of the digital IC drives the gate of the power FET. When the digital output is high, the power FET is like a closed switch. When the digital output is low, the power FET is like an open switch. Interfacing digital ICs (small-signal EMOS and CMOS) to high-power loads is one of the important applications of power FETs.

Figure 14-25 is an example of a digital IC controlling a high-power load. When the CMOS output is high, the power FET acts like a closed switch. The motor winding then has approximately 12 V across it, and the motor shaft turns. When the CMOS output is low, the power FET is open and the motor stops turning.

DC-to-AC Converters

When there is a sudden power failure, computers will stop operating and valuable data may be lost. One solution is to use an **uninterruptible power supply (UPS).** A UPS contains a battery and a dc-to-ac converter. The basic idea is this: When there is a power failure, the battery voltage is converted to an ac voltage to drive the computer.

Figure 14-26 shows a **dc-to-ac converter,** the basic idea behind a UPS. When the power fails, other circuits (op amps, discussed later) are activated and

Figure 14-25 Using a power FET to control a motor.

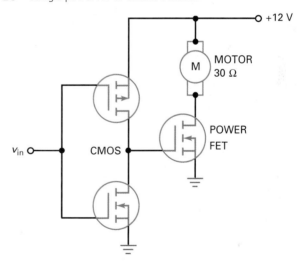

Figure 14-26 A rudimentary dc-to-ac converter.

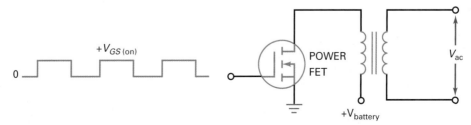

Figure 14–27 A rudimentary dc-to-dc converter.

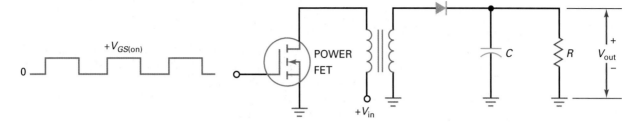

generate a square wave to drive the gate. The square-wave input switches the power FET on and off. Since a square wave will appear across the transformer windings, the secondary winding can supply the ac voltage needed to keep the computer running. A commercial UPS is more complicated than this, but the basic idea of converting dc to ac is the same.

DC–to–DC Converters

Figure 14-27 is a **dc-to-dc converter,** a circuit that converts an input dc voltage to an output dc voltage that is either higher or lower. The power FET switches on and off, producing a square wave across the secondary winding. The half-wave rectifier and capacitor-input filter then produce the dc output voltage V_{out}. By using different turns ratios, we can get a dc output voltage that is higher or lower than the input voltage V_{in}. For lower ripple, a full-wave or bridge rectifier can be used. The dc-to-dc converter is one of the important sections of a switching or switchmode power supply. This application will be examined in Chap. 24.

Example 14–9

What is the current through the motor winding of Fig. 14-28?

SOLUTION Table 14-2 gives $V_{GS(on)} = 10$ V, $I_{D(on)} = 2$ A, and $R_{DS(on)}$ of 1.95 Ω for an MTP4N80E. In Fig. 14-28, the saturation current is:

$$I_{D(sat)} = \frac{30 \text{ V}}{30 \text{ Ω}} = 1 \text{ A}$$

Figure 14–28 Example of controlling a motor.

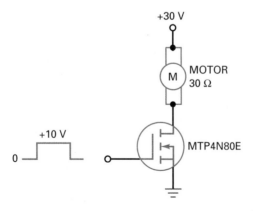

Since this is less than 2 A, the power FET is equivalent to a resistance of 1.95 Ω. Ideally, the current through the motor winding is 1 A. If we include the 1.95 Ω in the calculations, the current is:

$$I_D = \frac{30\ \text{V}}{30\ \Omega + 1.95\ \Omega} = 0.939\ \text{A}$$

PRACTICE PROBLEM 14–9 Repeat Example 14-9 using a MTW24N40E found in Table 14-2.

Example 14-10

During the day, the photodiode of Fig. 14-29 is conducting heavily and the gate voltage is low. At night, the photodiode is off, and the gate voltage rises to +10 V. Therefore, the circuit turns the lamp on automatically at night. What is the current through the lamp?

Figure 14-29 Automatic light control.

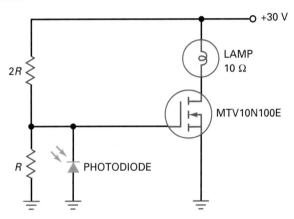

SOLUTION Table 14-2 gives $V_{GS(\text{on})} = 10$ V, $I_{D(\text{on})} = 5$ A, and $R_{DS(\text{on})}$ of 1.07 Ω for an MTV10N100E. In Fig. 14-29, the saturation current is:

$$I_{D(\text{sat})} = \frac{30\ \text{V}}{10\ \Omega} = 3\ \text{A}$$

Since this is less than 5 A, the power FET is equivalent to a resistance of 1.07 Ω, and the lamp current is:

$$I_D = \frac{30\ \text{V}}{10\ \Omega + 1.07\ \Omega} = 2.71\ \text{A}$$

PRACTICE PROBLEM 14–10 Find the lamp current of Fig. 14-29 using a MTP4N80E found in Table 14-2.

Example 14-11

The circuit of Fig. 14-30 automatically fills a swimming pool when the water level is low. When the water level is below the two metal probes, the gate voltage is pulled up to +10 V, the power FET conducts, and the water valve opens to put water in the pool.

When the water level eventually rises above the metal probes, the resistance between the probes becomes very low because water is a good conductor. In this case, the gate voltage goes low, the power FET opens, and the spring-loaded water valve closes.

Figure 14-30 Automatic pool filler.

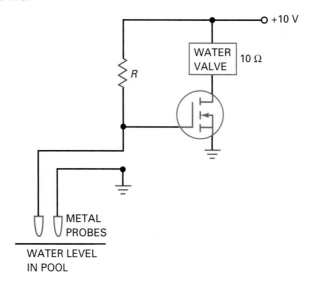

What is the current through the water valve of Fig. 14-30 if the power FET operates in the ohmic region with an $R_{DS(on)}$ of 0.5 Ω?

SOLUTION   The valve current is:

$$I_D = \frac{10\ V}{10\ \Omega + 0.5\ \Omega} = 0.952\ A$$

Example 14-12

What does the circuit of Fig. 14-31a do? What is the *RC* time constant? What is the lamp power at full brightness?

SOLUTION   When the manual switch is closed, the large capacitor charges slowly toward 10 V. As the gate voltage increases above $V_{GS(th)}$, the power FET begins to conduct. Since the gate voltage is changing slowly, the operating point of the power FET has to pass slowly through the active region of Fig. 14-31b. Because of this, the lamp gets gradually

Figure 14-31   Soft turn-on of a lamp.

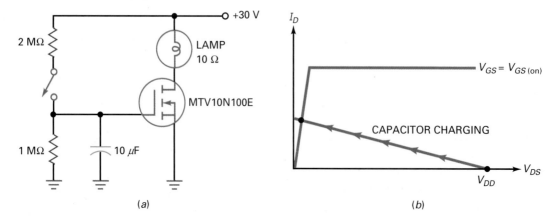

(a)

(b)

brighter. When the operating point of the power FET finally reaches the ohmic region, the lamp brightness is maximum. The overall effect is a *soft turn-on* of the lamp.

The Thevenin resistance facing the capacitor is:

$$R_{TH} = 2 \text{ M}\Omega \parallel 1 \text{ M}\Omega = 667 \text{ k}\Omega$$

The *RC* time constant is:

$$RC = (667 \text{ k}\Omega)(10 \text{ } \mu\text{F}) = 6.67 \text{ s}$$

With Table 14-2, the $R_{DS(on)}$ of the MTV10N100E is 1.07 Ω. The lamp current is:

$$I_D = \frac{30 \text{ V}}{10 \text{ }\Omega + 1.07 \text{ }\Omega} = 2.71 \text{ A}$$

and the lamp power is:

$$P = (2.71 \text{ A})^2(10 \text{ }\Omega) = 73.4 \text{ W}$$

14-9 E-MOSFET Amplifiers

As mentioned in previous sections, the E-MOSFET finds its use primarily as a switch. Applications do exist for this device to be used as an amplifier, however. These applications include front-end high-frequency RF amplifiers used in communications equipment and power E-MOSFETs used in class AB power amplifiers.

With E-MOSFETs, V_{GS} has to be greater than $V_{GS(th)}$ for drain current to flow. This eliminates self-bias, current-source bias, and zero bias because all these will have depletion-mode operation. This leaves gate bias and voltage-divider bias. Both of these biasing arrangements will work with E-MOSFETs because they can achieve enhancement mode operation.

Fig. 14-32 shows the drain curves and the transconductance curve for an *n*-channel E-MOSFET. The parabolic transfer curve is similar to that of D-MOSFET with some important differences. The E-MOSFET operates only in the enhancement mode. Also, the drain current doesn't start until $V_{GS} = V_{GS(th)}$. Again, this demonstrates that the E-MOSFET is a voltage-controlled normally off device. Because the drain current is zero when $V_{GS} = 0$, the standard transconductance formula will not work with the E-MOSFET. The drain current can be found by:

$$I_D = k[V_{GS} - V_{GS(th)}]^2 \tag{14-7}$$

where k is a constant value for the E-MOSFET found by:

$$k = \frac{I_{D(on)}}{[V_{GS(on)} - V_{GS(th)}]^2} \tag{14-8}$$

Figure 14–32 An *n*-channel E-MOSFET: (*a*) Drain curves; (*b*) transconductance curve.

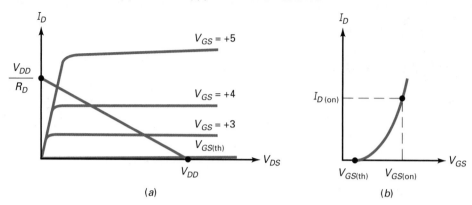

(*a*)

(*b*)

The data sheet for a 2N7000 *n*-channel enhancement-mode FET is shown in Fig. 14-12. Again, the important values needed are $I_{D(on)}$, $V_{GS(on)}$, and $V_{GS(th)}$. The specifications for the 2N7000 show a large variance in values. Typical values will be used in the following calculations. $I_{D(on)}$ is shown to be 600 mA when $V_{GS} = 4.5$ V. Therefore, use 4.5 V for the $V_{GS(on)}$ values. Also shown, $V_{GS(th)}$ has a typical value of 2.1 V when $V_{DS} = V_{GS}$ and $I_D = 1$ mA.

Example 14–13

Using the 2N7000 data sheet and typical values, find the constant k value and I_D at V_{GS} values of 3 V and 4.5 V.

SOLUTION Using these specified values and Eq. (14-8) k is found by:

$$k = \frac{600 \text{ mA}}{[4.5 \text{ V} - 2.1 \text{ V}]^2}$$

$$k = 104 \times 10^{-3} \text{ A/V}^2$$

With the constant value of k known, you then can solve for I_D at various V_{GS} values. For example, if $V_{GS} = 3$ V I_D is:

$$I_D = (104 \times 10^{-3} \text{ A/V}^2)[3 \text{ V} - 2.1 \text{ V}]^2$$

$$I_D = 84.4 \text{ mA}$$

and when $V_{GS} = 4.5$ V I_D is:

$$I_D = (104 \times 10^{-3} \text{ A/V}^2)[4.5 \text{ V} - 2.1 \text{ V}]^2$$

$$I_D = 600 \text{ mA}$$

PRACTICE PROBLEM 14–13 Using the 2N7000 data sheet and the listed minimum values of $I_{D(on)}$ and $V_{GS(th)}$, find the constant k value and I_D when $V_{GS} = 3$ V.

Fig. 14-33*a* shows another biasing method for E-MOSFETs called **drain-feedback bias.** This biasing method is similar to collector-feedback bias used with bipolar junction transistors. When the MOSFET is conducting, it has a

Figure 14-33 Drain-feedback bias: (a) Biasing method; (b) Q point.

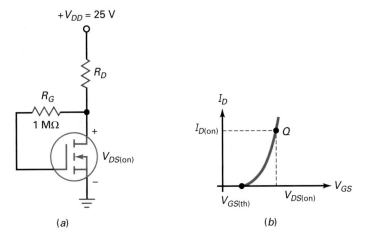

(a)

(b)

drain current of $I_{D(on)}$ and a drain voltage of $V_{DS(on)}$. Because there is virtually no gate current, $V_{GS} = V_{DS(on)}$. As with collector-feedback, drain-feedback bias tends to compensate for changes in FET characteristics. For example, if $I_{D(on)}$ tries to increase for some reason, $V_{DS(on)}$ decreases. This reduces V_{GS} and partially offsets the original increase in $I_{D(on)}$.

Figure 14-33b shows the Q point on the transconductance curve. The Q point has the coordinates of $I_{D(on)}$ and $V_{DS(on)}$. Data sheets for E-MOSFETs often give a value of $I_{D(on)}$ for $V_{GS} = V_{DS(on)}$. When designing this circuit, select a value of R_D that produces the specified value of V_{DS}. This can be found by:

$$R_D = \frac{V_{DD} - V_{DS(on)}}{I_{D(on)}}$$

(14-9)

Example 14–14

The data sheet for the E-MOSFET shown in Fig. 14-33a specifies $I_{D(on)} = 3$ mA and $V_{DS(on)} = 10$ V. If $V_{DD} = 25$ V, select a value of R_D that allows the MOSFET to operate at the specified Q point.

SOLUTION Find the value of R_D using Eq. (14-9):

$$R_D = \frac{25\text{ V} - 10\text{ V}}{3\text{ mA}}$$

$$R_D = 5\text{ k}\Omega$$

PRACTICE PROBLEM 14–14 Using Fig. 14-33a, change V_{DD} to $+22$ V and solve for R_D.

The forward transconductance value, g_{FS}, is listed on most MOSFET data sheets. For the 2N7000, a minimum and typical value is given when $I_D = 200$ mA. The minimum value is 100 mS and the typical value is 320 mS. The transconductance value will vary, depending on the circuit's Q point, following the relationship of $I_D = k [V_{GS} - V_{GS(th)}]^2$ and $g_m = \dfrac{\Delta I_D}{\Delta V_{GS}}$. From these equations, it can be determined that:

$$g_m = 2\,k\,[V_{GS} - V_{GS(th)}] \qquad\qquad\qquad (14\text{-}10)$$

Example 14-15

For the circuit of Fig. 14-34, find V_{GS}, I_D, g_m, and V_{out}. The MOSFET specifications are $k = 104 \times 10^{-3}$ A/V^2, $I_{D(on)} = 600$ mA, and $V_{GS(th)} = 2.1$ V.

Figure 14-34   E-MOSFET amplifier.

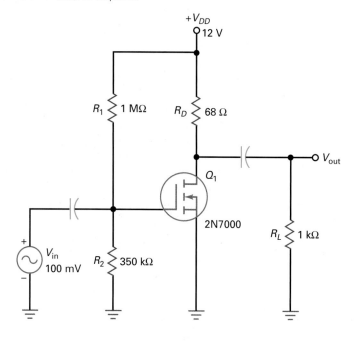

SOLUTION   First, find the value of V_{GS} by:

$$V_{GS} = V_G$$

$$V_{GS} = \frac{350\text{ k}\Omega}{350\text{ k}\Omega + 1\text{ M}\Omega}(12\text{ V}) = 3.11\text{ V}$$

Next, solve for I_D:

$$I_D = (104 \times 10^{-3}\text{ A/V}^2)\,[3.11\text{ V} - 2.1\text{ V}]^2 = 106\text{ mA}$$

The transconductance value, g_m is found by:

$$g_m = 2\,k\,[3.11\text{ V} - 2.1\text{ V}] = 210\text{ mS}$$

The voltage gain of this common-source amplifier is same as other FET devices:

$$A_V = g_m r_d$$

where $r_d = R_D \parallel R_L = 68\ \Omega \parallel 1\text{ k}\Omega = 63.7\ \Omega$.
Therefore,

$$A_V = (210\text{ mS})(63.7\ \Omega) = 13.4$$

and

$$V_{out} = (A_V)(V_{in}) = (13.4)(100\text{ mV}) = 1.34\text{ V}$$

PRACTICE PROBLEM 14-15   Repeat Example 14-15 with $R_2 = 330$ kΩ.

Circuit	Characteristics
D-MOSFET 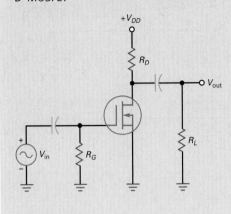	• Normally on device. • Biasing methods used: Zero-bias, gate-bias, self-bias, and voltage-divider bias $I_D = I_{DSS}\left(1 - \dfrac{V_{GS}}{V_{GS(off)}}\right)^2$ $V_{DS} = V_D - V_S$ $g_m = g_{mo}\left(1 - \dfrac{V_{GS}}{V_{GS(off)}}\right)$ $A_V = g_m r_d \quad Z_{in} \approx R_G \quad Z_{out} \approx R_D$
E-MOSFET 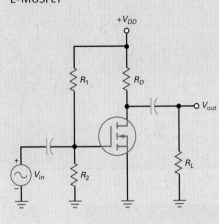	• Normally off device • Biasing methods used: Gate-bias, voltage-divider bias, Drain-feedback bias $I_D = k\,[V_{GS} - V_{GS(th)}]^2$ $k = \dfrac{I_{D(on)}}{[V_{GS(on)} - V_{GS(th)}]^2}$ $g_m = 2\,k\,[V_{GS} - V_{GS(th)}]$ $A_V = g_m r_d \quad Z_{in} \approx R_1 \| R_2$ $Z_{out} \approx R_D$

Summary Table 14-1 shows a D-MOSFET and E-MOSFET amplifier along with their basic characteristics and equations.

14–10 MOSFET Testing

MOSFET devices require special care when being tested for proper operation. As stated previously, the thin layer of silicon dioxide between the gate and channel can be easily destroyed when V_{GS} exceeds $V_{GS(max)}$. Because of the insulated gate, along with the channel construction, testing MOSFET devices with an ohmmeter or DMM is not very effective. A good way to test these devices is with a semi-conductor curve tracer. If a curve tracer is not available, special test circuits can be constructed. Fig. 14-35a shows a circuit capable of testing both depletion-mode and enhancement-mode MOSFETs. By changing the voltage level and polarity of V_1, the device can be tested in either depletion or enhancement modes of operation. The drain curve shown in Fig. 14-35b shows the approximate drain current of 275 mA when $V_{GS} = 4.52$ V. The y-axis is set to display 50 mA/div.

Figure 14-35 MOSFET test circuit.

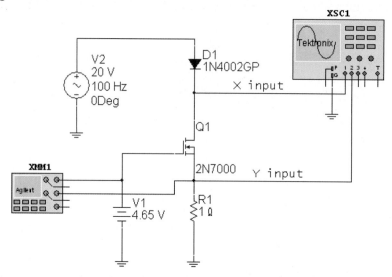

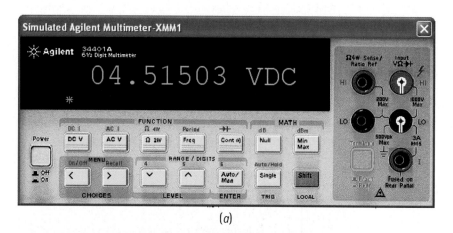

(a)

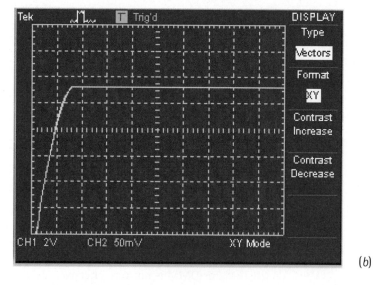

(b)

An alternative to the above testing methods is to simply use component substitution. By measuring in-circuit voltage values, it is often possible to deduct that the MOSFET is defective. Replacing the device with a known good component should lead you to a final conclusion.

Summary

SEC. 14-1 THE DEPLETION-MODE MOSFET

The depletion-mode MOSFET, abbreviated *D-MOSFET*, has a source, gate, and drain. The gate is insulated from the channel. Because of this, the input resistance is very high. The D-MOSFET has limited use, mainly in RF circuits.

SEC. 14-2 D-MOSFET CURVES

The drain curves for a D-MOSFET are similar to those of a JFET when the MOS device is operating in the depletion mode. Unlike JFETs, D-MOSFETs can also operate in the enhancement mode. When operating in the enhancement mode, the drain current is greater than I_{DSS}.

SEC. 14-3 DEPLETION-MODE MOSFET AMPLIFIERS

D-MOSFETs are mainly used as RF amplifiers. D-MOSFETs have good high-frequency response, generate low levels of electrical noise, and maintain high input impedance values when V_{GS} is negative or positive. Dual-gate D-MOSFETs can be used with automatic gain control (AGC) circuits.

SEC. 14-4 THE ENHANCEMENT-MODE MOSFET

The E-MOSFET is normally off. When the gate voltage equals the threshold voltage, an *n*-type inversion layer connects the source to the drain. When the gate voltage is much greater than the threshold voltage, the device conducts heavily. Because of the thin insulating layer, MOSFETs are easily destroyed unless you take precautions in handling them.

SEC. 14-5 THE OHMIC REGION

Since the E-MOSFET is primarily a switching device, it usually operates between cutoff and saturation. When it is biased in the ohmic region, it acts like a small resistance. If $I_{D(sat)}$ is less than $I_{D(on)}$ when $V_{GS} = V_{GS(on)}$, the E-MOSFET is operating in the ohmic region.

SEC. 14-6 DIGITAL SWITCHING

Analog means that the signal changes continuously, that is, with no sudden jumps. *Digital* means that the signal jumps between two distinct voltage levels. Switching includes high-power circuits as well as small-signal digital circuits. Active-load switching means that one of the MOSFETs acts like a large resistor and the other like a switch.

SEC. 14-7 CMOS

CMOS uses two complementary MOSFETs, in which one conducts and the other shuts off. The CMOS inverter is a basic digital circuit. CMOS devices have the advantage of very low power consumption.

SEC. 14-8 POWER FETS

Discrete E-MOSFETs can be manufactured to switch very large currents. Known as *power FETS*, these devices are useful in automotive controls, disk drives, converters, printers, heating, lighting, motors, and other heavy-duty applications.

SEC. 14-9 E-MOSFET AMPLIFIERS

Besides their main use as power switches, E-MOSFETs find applications as amplifiers. The normally off characteristics of E-MOSFETs dictate that V_{GS} be greater than $V_{GS(th)}$ when used as an amplifier. Drain-feedback bias is similar to collector-feedback bias.

SEC. 14-10 MOSFET TESTING

It is difficult to safely test MOSFET devices using an ohmmeter. If a semiconductor curve tracer is not available, MOSFETs can be tested in test circuits or by simple substitution.

Definitions

(14-1) D-MOSFET drain current:

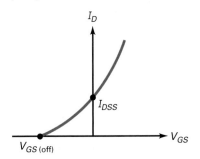

$$I_D = I_{DSS}\left(1 - \frac{V_{GS}}{V_{GS(off)}}\right)^2$$

(14-3) On resistance:

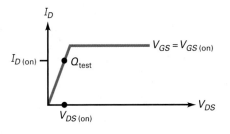

$$R_{DS(on)} = \frac{V_{DS(on)}}{I_{D(on)}}$$

(14-6)   Two-terminal resistance:

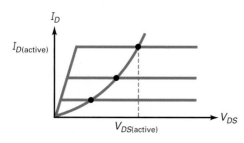

$$R_D = \frac{V_{DS(active)}}{I_{D(active)}}$$

(14-8)   E-MOSFET constant k:

$$k = \frac{I_{D(on)}}{[V_{GS(on)} - V_{GS(th)}]^2}$$

(14-10)   E-MOSFET g_m:

$$g_m = 2\,k\,[V_{GS} - V_{GS(th)}]$$

Derivations

(14-2)   D-MOSFET zero-bias:

$$V_{DS} = V_{DD} - I_{DSS}\,R_D$$

(14-4)   Saturation current:

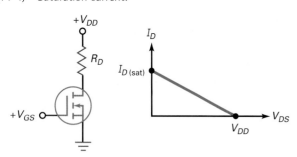

$$I_{D(sat)} = \frac{V_{DD}}{R_D}$$

(14-5)   Ohmic region:

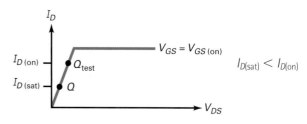

$$I_{D(sat)} < I_{D(on)}$$

(14-7)   E-MOSFET drain current:

$$I_D = k\,[V_{GS} - V_{GS(th)}]^2$$

(14-9) R_D for drain-feedback bias:

$$R_D = \frac{V_{DD} - V_{DS(on)}}{I_{D(on)}}$$

Student Assignments

1. **A D-MOSFET can operate in the**
   a. Depletion-mode only
   b. Enhancement-mode only
   c. Depletion-mode or enhancement-mode
   d. Low-impedance mode

2. **When an *n*-channel D-MOSFET has $I_D > I_{DSS}$, it**
   a. Will be destroyed
   b. Is operating in the depletion mode
   c. Is forward biased
   d. Is operating in the enhancement mode

3. **The voltage gain of a D-MOSFET amplifier is dependent on**
   a. R_D
   b. R_L
   c. g_m
   d. All of the above

4. **Which of the following devices revolutionized the computer industry?**
   a. JFET
   b. D-MOSFET
   c. E-MOSFET
   d. Power FET

5. **The voltage that turns on an EMOS device is the**
   a. Gate-source cutoff voltage
   b. Pinchoff voltage
   c. Threshold voltage
   d. Knee voltage

6. **Which of these may appear on the data sheet of an enhancement-mode MOSFET?**
   a. $V_{GS(th)}$
   b. $I_{D(on)}$
   c. $V_{GS(on)}$
   d. All the above

7. **The $V_{GS(on)}$ of an *n*-channel E-MOSFET is**
   a. Less than the threshold voltage
   b. Equal to the gate-source cutoff voltage
   c. Greater than $V_{DS(on)}$
   d. Greater than $V_{GS(th)}$

8. **An ordinary resistor is an example of**
   a. A three-terminal device
   b. An active load
   c. A passive load
   d. A switching device

9. **An E-MOSFET with its gate connected to its drain is an example of**
   a. A three-terminal device
   b. An active load
   c. A passive load
   d. A switching device

10. An E-MOSFET that operates at cutoff or in the ohmic region is an example of
    a. A current source
    b. An active load
    c. A passive load
    d. A switching device

11. VMOS devices generally
    a. Switch off faster than BJTs
    b. Carry low values of current
    c. Have a negative temperature coefficient
    d. Are used as CMOS inverters

12. A D-MOSFET is considered to be a
    a. Normally off device
    b. Normally on device
    c. Current controlled device
    d. High-power switch

13. CMOS stands for
    a. Common MOS
    b. Active-load switching
    c. p-channel and n-channel devices
    d. Complementary MOS

14. $V_{GS(on)}$ is always
    a. Less than $V_{GS(th)}$
    b. Equal to $V_{DS(on)}$
    c. Greater than $V_{GS(th)}$
    d. Negative

15. With active-load switching, the upper E-MOSFET is a
    a. Two-terminal device
    b. Three-terminal device
    c. Switch
    d. Small resistance

16. CMOS devices use
    a. Bipolar transistors
    b. Complementary E-MOSFETs
    c. Class A operation
    d. DMOS devices

17. The main advantage of CMOS is its
    a. High power rating
    b. Small-signal operation
    c. Switching capability
    d. Low power consumption

18. Power FETs are
    a. Integrated circuits
    b. Small-signal devices
    c. Used mostly with analog signals
    d. Used to switch large currents

19. When the internal temperature increases in a power FET, the
    a. Threshold voltage increases
    b. Gate current decreases
    c. Drain current decreases
    d. Saturation current increases

20. Most small-signal E-MOSFETs are found in
    a. Heavy-current applications
    b. Discrete circuits
    c. Disk drives
    d. Integrated circuits

21. Most power FETS are
    a. Used in high-current applications
    b. Digital computers
    c. RF stages
    d. Integrated circuits

22. An n-channel E-MOSFET conducts when it has
    a. $V_{GS} > V_P$
    b. An n-type inversion layer
    c. $V_{DS} > 0$
    d. Depletion layers

23. With CMOS, the upper MOSFET is
    a. A passive load
    b. An active load
    c. Nonconducting
    d. Complementary

24. The high output of a CMOS inverter is
    a. $V_{DD}/2$
    b. V_{GS}
    c. V_{DS}
    d. V_{DD}

25. The $R_{DS(on)}$ of a power FET
    a. Is always large
    b. Has a negative temperature coefficient
    c. Has a positive temperature coefficient
    d. Is an active load

Problems

SEC. 14–2 D-MOSFET CURVES

14-1 An n-channel D-MOSFET has the specifications $V_{GS(off)} = -2$ V and $I_{DSS} = 4$ mA. Given V_{GS} values of -0.5 V, -1.0 V, -1.5 V, $+0.5$ V, $+1.0$ V, and $+1.5$ V, determine I_D in the depletion mode only.

14-2 Given the same values as in the previous problem, calculate I_D for the enhancement mode only.

14-3 A p-channel D-MOSFET has $V_{GS(off)} = +3$ V and $I_{DSS} = 12$ mA. Given V_{GS} values of -1.0 V, -2.0 V, 0 V, $+1.5$ V, and $+2.5$ V, determine I_D in the depletion mode only.

SEC. 14–3 DEPLETION–MODE MOSFET AMPLIFIERS

14-4 The D-MOSFET in Fig. 14-36 has $V_{GS(off)} = -3$ V and $I_{DSS} = 12$ mA. Determine the circuit's drain current and V_{DS} values.

14-5 In Fig. 14-36, what are the values of r_d, A_v, and V_{out} using a g_{mo} of 4000 μS?

14-6 Using Fig. 14-36 find r_d, A_v, and V_{out} if $R_D = 680$ Ω and $R_L = 10$ kΩ.

14-7 What is the approximate input impedance of Fig. 14-36?

SEC. 14–5 THE OHMIC REGION

14-8 Calculate $R_{DS(on)}$ for each of these E-MOSFET values:

    a. $V_{DS(on)} = 0.1$ V and $I_{D(on)} = 10$ mA
    b. $V_{DS(on)} = 0.25$ V and $I_{D(on)} = 45$ mA
    c. $V_{DS(on)} = 0.75$ V and $I_{D(on)} = 100$ mA
    d. $V_{DS(on)} = 0.15$ V and $I_{D(on)} = 200$ mA

Figure 14-36

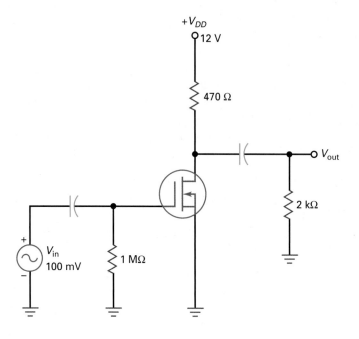

Figure 14-37

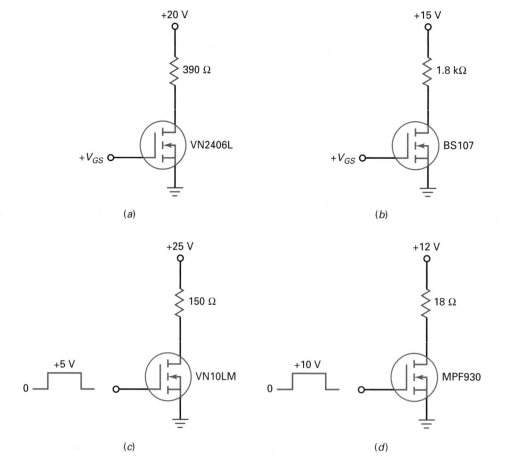

14-9 An E-MOSFET has $R_{DS(on)} = 2\ \Omega$ when $V_{GS(on)} = 3$ V and $I_{D(on)} = 500$ mA. If it is biased in the ohmic region, what is the voltage across it for each of these drain currents:

   a. $I_{D(sat)} = 25$ mA   c. $I_{D(sat)} = 100$ mA

   b. $I_{D(sat)} = 50$ mA   d. $I_{D(sat)} = 200$ mA

14-10 |||| **MultiSim** What is the voltage across the E-MOSFET in Fig. 14-37a if $V_{GS} = 2.5$ V? (Use Table 14-1.)

14-11 |||| **MultiSim** Calculate the drain voltage in Fig. 14-37b for a gate voltage of +3 V. Assume that $R_{DS(on)}$ is approximately the same as the value given in Table 14-1.

14-12 If V_{GS} is high in Fig. 14-37c, what is the voltage across the load resistor of Fig. 14-37c?

14-13 Calculate the voltage across the E-MOSFET of Fig. 14-37d for a high input voltage.

14-14 What is the LED current in Fig. 14-38a when $V_{GS} = 5$ V?

14-15 The relay of Fig. 14-38b closes when $V_{GS} = 2.6$ V. What is the MOSFET current when the gate voltage is high? The current through the final load resistor?

SEC. 14-6 DIGITAL SWITCHING

14-16 An E-MOSFET has these values: $I_{D(active)} = 1$ mA and $V_{DS(active)} = 10$ V. What does its drain resistance equal in the active region?

14-17 What is the output voltage in Fig. 14-39a when the input is low? When it is high?

14-18 In Fig. 14-39b, the input voltage is low. What is the output voltage? If the input goes high, what is the output voltage?

14-19 A square wave drives the gate of Fig. 14-39a. If the square wave has a peak-to-peak value large enough to drive the lower MOSFET into the ohmic region, what is the output waveform?

Figure 14-38

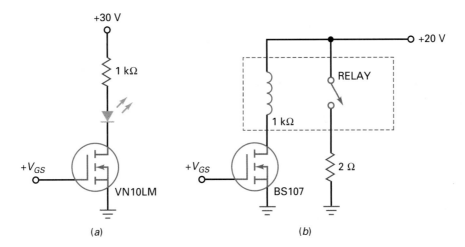

(a)   (b)

Figure 14-39

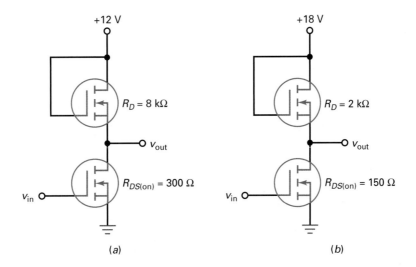

(a)   (b)

MOSFETs

Figure 14-40

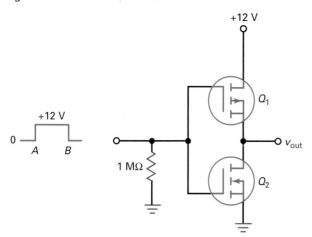

SEC. 14-7 CMOS

14-20 The MOSFETs of Fig. 14-40 have $R_{DS(on)} = 250\ \Omega$ and $R_{DS(off)} = 5\ M\Omega$. What is the output waveform?

14-21 The upper E-MOSFET of Fig. 14-40 has these values: $I_{D(on)} = 1\ mA$, $V_{DS(on)} = 1\ V$, $I_{D(off)} = 1\ \mu A$, and $V_{DS(off)} = 10\ V$. What is the output voltage when the input voltage is low? When it is high?

14-22 A square wave with a peak value of 12 V and a frequency of 1 kHz is the input in Fig. 14-40. Describe the output waveform.

14-23 During the transition from low to high in Fig. 14-40, the input voltage is 6 V for an instant. At this time, both MOSFETs have active resistances of $R_D = 5\ k\Omega$. What is the current drain at this instant?

SEC. 14-8 POWER FETS

14-24 What is the current through the motor winding of Fig. 14-41 when the gate voltage is low? When it is high?

Figure 14-42

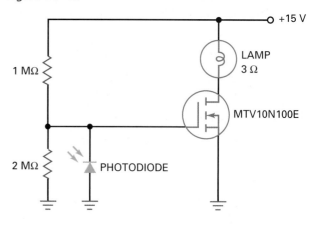

14-25 The motor winding of Fig. 14-41 is replaced by another with a resistance of 6 Ω. What is the current through the winding when the gate voltage is high?

14-26 What is the current through the lamp of Fig. 14-42 when the gate voltage is low? When it is +10 V?

14-27 The lamp of Fig. 14-42 is replaced by another with a resistance of 5 Ω. What is the lamp power when it is dark?

14-28 What is the current through the water valve of Fig. 14-43 when the gate voltage is high? When it is low?

14-29 The supply voltage of Fig. 14-43 is changed to 12 V and the water valve is replaced by another with a resistance of 18 Ω. What is the current through the water valve when the probes are underwater? When the probes are above the water?

Figure 14-41

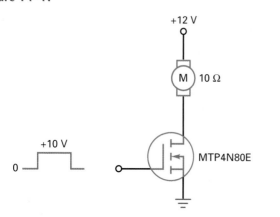

Figure 14-43

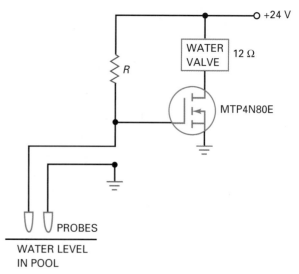

Figure 14-44

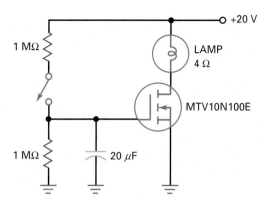

14-30 What is the *RC* time constant in Fig. 14-44? The lamp power at full brightness?

14-31 The two resistances in the gate circuit are doubled in Fig. 14-44. What is the *RC* time constant? If the lamp is changed to one with a resistance of 6 Ω, what is the lamp current at full brightness?

SEC. 14–9 E–MOSFET AMPLIFIERS

14-32 Find the constant *k* value and I_D of Fig. 14-45 using the minimum values of $I_{D(on)}$, $V_{GS(on)}$, and $V_{GS(th)}$ for the 2N7000.

14-33 Determine the g_m, A_V, and V_{out} values for Fig. 14-45 using the minimum rated specifications.

14-34 In Fig. 14-45, change R_D to 50 Ω. Find the constant *k* value and I_D using the typical values of $I_{D(on)}$, $V_{GS(on)}$, and $V_{GS(th)}$ for the 2N7000.

14-35 Determine the g_m, A_V, and V_{out} values for Fig. 14-45 using the typical rated specifications, V_{DD} at +12 V and $R_D = 15$ Ω.

Figure 14-45

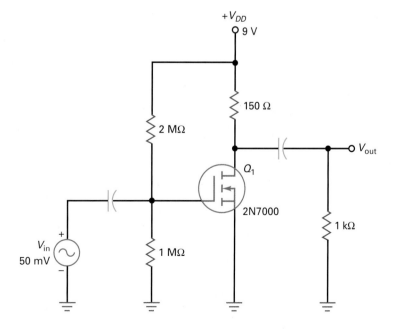

Critical Thinking

14-36 In Fig. 14-37c, the gate input voltage is a square wave with a frequency of 1 kHz and a peak voltage of +5 V. What is the average power dissipation in the load resistor?

14-37 The gate input voltage of Fig. 14-37d is a series of rectangular pulses with a duty cycle of 25 percent. This means that the gate voltage is high for 25 percent of the cycle and low the rest of the time. What is the average power dissipation in the load resistor?

14-38 The CMOS inverter of Fig. 14-40 uses MOSFETs with $R_{DS(on)} = 100\ \Omega$ and $R_{DS(off)} = 10\ M\Omega$. What is the quiescent power consumption of the circuit? When a square wave is the input, the average current through Q_1 is 50 μA. What is the power consumption?

14-39 If the gate voltage is 3 V in Fig. 14-42, what is the photodiode current?

14-40 The data sheet of an MTP16N25E shows a normalized graph of $R_{DS(off)}$ versus temperature. The normalized value increases linearly from 1 to 2.25 as the junction temperature increases from 25 to 125°C. If $R_{DS(on)} = 0.17\ \Omega$ at 25°C, what does it equal at 100°C?

14-41 In Fig. 14-27, $V_{in} = 12$ V. If the transformer has a turns ratio of 4:1 and the output ripple is very small, what is the dc output voltage V_{out}?

Job Interview Questions

1. Draw an E-MOSFET showing the *p* and *n* regions. Then, explain the off-on action.
2. Describe how active-load switching works. Use circuit diagrams in your explanation.
3. Draw a CMOS inverter and explain the circuit action.
4. Draw any circuit that shows a power FET controlling a large load current. Explain the off-on action. Include $R_{DS(on)}$ in your discussion.
5. Some people say that MOS technology revolutionized the world of electronics. Why?
6. List and compare the advantages and disadvantages of BJT and FET amplifiers.
7. Explain what happens when drain current starts to increase through a power FET.
8. Why must an E-MOSFET be handled with care?
9. Why is a thin metal wire connected around all the leads of a MOSFET during shipment?
10. What are some precautionary measures that are taken while working with MOS devices?
11. Why would a designer generally select a MOSFET over a BJT for a power-switching function in a switching power supply?

Self-Test Answers

1.	c	10.	d	18.	d
2.	d	11.	a	19.	c
3.	d	12.	b	20.	d
4.	c	13.	d	21.	a
5.	c	14.	c	22.	b
6.	d	15.	a	23.	d
7.	d	16.	b	24.	d
8.	c	17.	d	25.	c
9.	b				

Practice Problem Answers

14-1

V_{GS}	I_D
-1 V	2.25 mA
-2 V	1 mA
0 V	4 mA
$+1$ V	6.25 mA
$+2$ V	9 mA

14-2 $V_{out} = 105.6$ mV

14-3 $V_{out(off)} = 20$ V; $V_{out(on)} = 0.198$ V

14-4 $I_{LED} = 32$ mA

14-6 $V_{out} = 20$ V and 198 mV

14-7 $R_{DS(on)} \cong 222\ \Omega$

14-8 If $V_{in} > V_{GS(th)}$; $V_{out} = +15$ V pulse

14-9 $I_D = 0.996$ A

14-10 $I_L = 2.5$ A

14-13 $k = 5.48 \times 10^{-3}$ A/V^2; $I_D = 26$ mA

14-14 $R_D = 4\ k\Omega$

14-15 $V_{GS} = 2.98$ V; $I_D = 80$ m A; $g_m = 183$ mS; $A_V = 11.7$; $V_{out} = 1.17$ V

chapter 15 Thyristors

The word **thyristor** comes from the Greek and means "door," as in opening a door and letting something pass through it. A thyristor is a semiconductor device that uses internal feedback to produce switching action. The most important thyristors are the silicon controlled rectifier (SCR) and the triac. Like power FETs, the SCR and the triac can switch large currents on and off. Because of this, they can be used for overvoltage protection, motor controls, heaters, lighting systems, and other heavy-current loads. Insulated-gate bipolar transistors (IGBTs) are not included in the thyristor family, but are covered in this chapter as an important power-switching device.

Objectives

After studying this chapter, you should be able to:

- Describe the four-layer diode, how it is turned on, and how it is turned off.

- Explain the characteristics of SCRs.

- Demonstrate how to test SCRs.

- Calculate the firing and conduction angles of RC phase control circuits.

- Explain the characteristics of triacs and diacs.

- Compare the switching control of IGBTs to power MOSFETs.

- Describe the major characteristics of the photo-SCR and silicon controlled switch.

- Explain the operation of UJT and PUT circuits.

Chapter Outline

Vocabulary

breakover

conduction angle

diac

firing angle

four-layer diode

gate trigger current I_{GT}

gate trigger voltage V_{GT}

holding current

Insulated-gate bipolar transistor (IGBT)

low-current drop-out

programmable unijunction transistor (PUT)

sawtooth generator

Schockley diode

SCR

silicon unilateral switch (SUS)

thyristor

triac

unijunction transistor (UJT)

15-1 The Four-Layer Diode

Thyristor operation can be explained in terms of the equivalent circuit shown in Fig. 15-1a. The upper transistor Q_1 is a *pnp* device, and the lower transistor Q_2 is an *npn* device. The collector of Q_1 drives the base of Q_2. Similarly, the collector of Q_2 drives the base of Q_1.

Positive Feedback

The unusual connection of Fig. 15-1a uses *positive feedback.* Any change in the base current of Q_2 is amplified and fed back through Q_1 to magnify the original change. This positive feedback continues changing the base current of Q_2 until both transistors go into either saturation or cutoff.

For instance, if the base current of Q_2 increases, the collector current of Q_2 increases. This increases the base current of Q_1 and the collector current of Q_1. More collector current in Q_1 will further increase the base current of Q_2. This amplify-and-feedback action continues until both transistors are driven into saturation. In this case, the overall circuit acts like a closed switch (Fig. 15-1b).

On the other hand, if something causes the base current of Q_2 to decrease, the collector current of Q_2 decreases, the base current of Q_1 decreases, the collector current of Q_1 decreases, and the base current of Q_2 decreases further. This action continues until both transistors are driven into cutoff. Then, the circuit acts like an open switch (Fig. 15-1c).

The circuit of Fig. 15-1a is stable in either of two states: *open* or *closed.* It will remain in either state indefinitely until acted on by an outside force. If the circuit is open, it stays open until something increases the base current of Q_2. If the circuit is closed, it stays closed until something decreases the base current of Q_2. Because the circuit can remain in either state indefinitely, it is called a *latch.*

Closing a Latch

Figure 15-2a shows a latch connected to a load resistor with a supply voltage of V_{CC}. Assume that the latch is open, as shown in Fig. 15-2b. Because there is no current through the load resistor, the voltage across the latch equals the supply voltage. So, the operating point is at the lower end of the dc load line (Fig. 15-2d).

The only way to close the latch of Fig. 15-2b is by **breakover.** This means using a large enough supply voltage V_{CC} to break down the Q_1 collector diode. Since the collector current of Q_1 increases the base current of Q_2, the positive

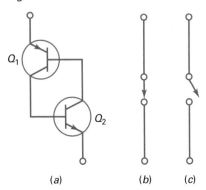

Figure 15-1 Transistor latch.

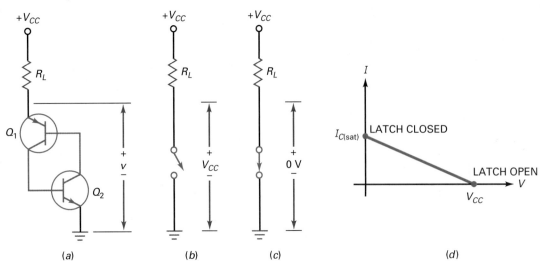

Figure 15-2 Latching circuit.

feedback will start. This drives both transistors into saturation, as previously described. When saturated, both transistors ideally look like short circuits, and the latch is closed (Fig. 15-2c). Ideally, the latch has zero voltage across it when it is closed and the operating point is at the upper end of the load line (Fig. 15-2d).

In Fig. 15-2a, breakover can also occur if Q_2 breaks down first. Although breakover starts with the breakdown of either collector diode, it ends with both transistors in the saturated state. This is why the term *breakover* is used instead of *breakdown* to describe this kind of latch closing.

Opening a Latch

How do we open the latch of Fig. 15-2a? By reducing the V_{CC} supply to zero. This forces the transistors to switch from saturation to cutoff. We call this type of opening **low-current drop-out** because it depends on reducing the latch current to a value low enough to bring the transistors out of saturation.

The Schockley Diode

Figure 15-3a was originally called a **Schockley diode** after the inventor. Several other names are also used for this device: **four-layer diode,** *pnpn diode,* and **silicon unilateral switch (SUS).** The device lets current flow in only one direction.

The easiest way to understand how it works is to visualize it separated into two halves, as shown in Fig. 15-3b. The left half is a *pnp transistor,* and the right half is an *npn transistor.* Therefore, the four-layer diode is equivalent to the latch of Fig. 15-3c.

Figure 15-3d shows the schematic symbol of a four-layer diode. The only way to close a four-layer diode is by breakover. The only way to open it is by low-current drop-out, which means reducing the current to less than the **holding current** (given on data sheets). The holding current is the low value of current where the transistors switch from saturation to cutoff.

After a four-layer diode breaks over, the voltage across it ideally drops to zero. In reality, there is some voltage across the latched diode. Figure 15-3e shows current versus voltage for a 1N5158 that is latched on. As you can see, the voltage across the device increases when the current increases: 1 V at 0.2 A, 1.5 V at 0.95 A, 2 V at 1.8 A, and so on.

Breakover Characteristic

Figure 15-4 shows the graph of current versus voltage of a four-layer diode. The device has two operating regions: cutoff and saturation. The dashed line is the

Figure 15-3 Four-layer diode.

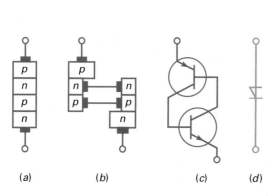

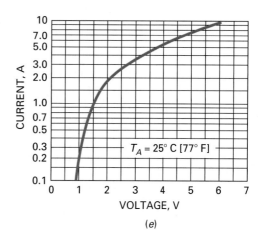

(a)   (b)   (c)   (d)   (e)

Figure 15-4 Breakover characteristic.

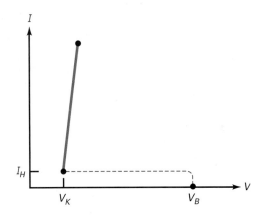

transition path between cutoff and saturation. It is dashed to indicate that the device switches rapidly between the off and on states.

When the device is at cutoff, it has zero current. If the voltage across diode tries to exceed V_B, the device breaks over and moves rapidly along the dashed line to the saturation region. When the diode is in saturation, it is operating on the upper line. As long as the current through it is greater than the holding current I_H, the diode remains latched in the on state. If the current becomes less than I_H, the device switches into cutoff.

The ideal approximation of a four-layer diode is an open switch when cut off and a closed switch when saturated. The second approximation includes the knee voltage V_K, approximately 0.7 V in Fig. 15-4. For higher approximations, use computer simulation software or refer to the data sheet of the four-layer diode.

Example 15-1

Figure 15-5 Example.

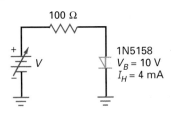

The diode of Fig. 15-5 has a breakover voltage of 10 V. If the input voltage of Fig. 15-5 is increased to +15 V, what is the diode current?

SOLUTION   Since an input voltage of 15 V is more than the break-over voltage of 10 V, the diode breaks over. Ideally, the diode is like a closed switch, so the current is:

$$I = \frac{15 \text{ V}}{100 \text{ }\Omega} = 150 \text{ mA}$$

To a second approximation:

$$I = \frac{15 \text{ V} - 0.7 \text{ V}}{100 \text{ }\Omega} = 143 \text{ mA}$$

For a more accurate answer, look at Fig. 15-3e and you will see that the voltage is 0.9 V when the current is around 150 mA. Therefore, an improved answer is:

$$I = \frac{15 \text{ V} - 0.9 \text{ V}}{100 \text{ }\Omega} = 141 \text{ mA}$$

PRACTICE PROBLEM 15-1   In Fig. 15-5, determine the diode current if the input voltage V is 12 V, to a second approximation.

Example 15-2

The diode of Fig. 15-5 has a holding current of 4 mA. The input voltage is increased to 15 V to latch the diode, and then decreased to open the diode. What is the input voltage that opens the diode?

SOLUTION   The diode opens when the current is slightly less than the holding current, given as 4 mA. At this small current, the diode voltage is approximately equal to the knee voltage, 0.7 V. Since 4 mA flows through 100 Ω, the input voltage is:

$$V_{in} = 0.7 \text{ V} + (4 \text{ mA})(100 \text{ }\Omega) = 1.1 \text{ V}$$

So, the input voltage has to be reduced from 15 V to slightly less than 1.1 V to open the diode.

PRACTICE PROBLEM 15–2   Repeat Example 15-2 using a diode with a holding current of 10 mA.

Example 15-3

Figure 15-6a shows a **sawtooth generator.** The capacitor charges toward the supply voltage, as shown in Fig. 15-6b. When the capacitor voltage reaches +10 V, the diode breaks over. This discharges the capacitor, producing the *flyback* (sudden voltage drop) of the output waveform. When the voltage is ideally zero, the diode opens and the capacitor begins to charge again. In this way, we get the ideal sawtooth shown in Fig. 15-6b.

What is the *RC* time constant for capacitor charging? What is the frequency of the sawtooth wave if its period is approximately 20 percent of the time constant?

Figure 15-6   Sawtooth generator.

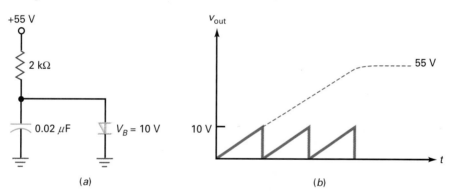

SOLUTION   The *RC* time constant is:

$$RC = (2 \text{ k}\Omega)(0.02 \text{ }\mu\text{F}) = 40 \text{ }\mu\text{s}$$

The period is approximately 20 percent of the time constant. So:

$$T = 0.2(40 \text{ }\mu\text{s}) = 8 \text{ }\mu\text{s}$$

The frequency is:

$$f = \frac{1}{8 \text{ }\mu\text{s}} = 125 \text{ kHz}$$

PRACTICE PROBLEM 15–3   Using Fig. 15-6, change the resistor value to 1 kΩ and solve for the sawtooth frequency.

15-2 The Silicon Controlled Rectifier

The **SCR** is the most widely used thyristor. It can switch very large currents on and off. Because of this, it is used to control motors, ovens, air conditioners, and induction heaters.

Triggering the Latch

By adding an input terminal to the base of Q_2, as shown in Fig. 15-7a, we can create a second way to close the latch. Here is the theory of operation: When the latch is open, as shown in Fig. 15-7b, the operating point is at the lower end of the dc load line (Fig. 15-7d). To close the latch, we can couple a *trigger* (sharp pulse) into the base of Q_2, as shown in Fig. 15-7a. The trigger momentarily increases the base current of Q_2. This starts the positive feedback, which drives both transistors into saturation.

When saturated, both transistors ideally look like short circuits, and the latch is closed (Fig. 15-7c). Ideally, the latch has zero voltage across it when it is closed, and the operating point is at the upper end of the load line (Fig. 15-7d).

Gate Triggering

Figure 15-8a shows the structure of the SCR. The input is called the *gate,* the top is the *anode,* and the bottom is the *cathode.* The SCR is far more useful than a four-layer diode because the gate triggering is easier than breakover triggering.

Figure 15-7 Transistor latch with trigger input.

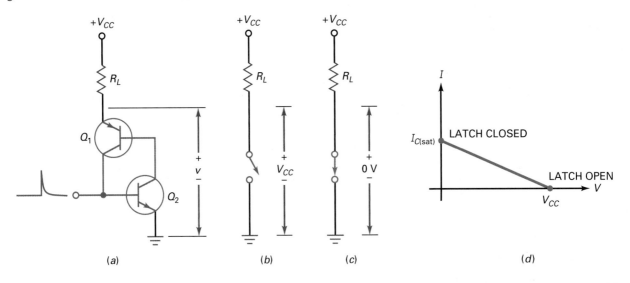

Figure 15-8 Silicon controlled rectifier (SCR).

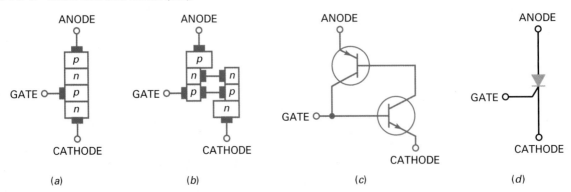

Figure 15-9 Typical SCRs.

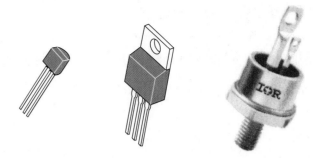

Again, we can visualize the four doped regions separated into two transistors, as shown in Fig. 15-8b. Therefore, the SCR is equivalent to a latch with a trigger input (Fig. 15-8c). Schematic diagrams use the symbol of Fig. 15-8d. Whenever you see this symbol, remember that it is equivalent to a latch with a trigger input. Typical SCRs are shown in Fig. 15-9.

Since the gate of an SCR is connected to the base of an internal transistor, it takes at least 0.7 V to trigger an SCR. Data sheets list this voltage as the **gate trigger voltage V_{GT}**. Rather than specify the input resistance of the gate, a manufacturer gives the minimum input current needed to turn on the SCR. Data sheets list this current as the **gate trigger current I_{GT}**.

Fig. 15-10 shows a data sheet for the 2N6504 series of SCRs. For this series, it shows typical trigger voltage and current values of:

$$V_{GT} = 1.0 \text{ V}$$

$$I_{GT} = 9.0 \text{ mA}$$

This means that the source driving the gate of a typical 2N6504 series SCR has to supply 9.0 mA at 1.0 V to latch the SCR.

Also, the breakover voltage or blocking voltage is specified as its peak repetitive off state forward voltage, V_{DRM}, and its peak repetitive off state reverse voltage, V_{RRM}. Depending on which SCR of the series is used, the breakover voltage ranges from 50 V to 800 V.

Required Input Voltage

An SCR like the one shown in Fig. 15-11 has a gate voltage V_G. When this voltage is more than V_{GT}, the SCR will turn on and the output voltage will drop from $+V_{CC}$ to a low value. Sometimes, a gate resistor is used as shown here. This resistor limits the gate current to a safe value. The input voltage needed to trigger an SCR has to be more than:

$$V_{\text{in}} = V_{GT} + I_{GT}R_G \tag{15-1}$$

In this equation, V_{GT} and I_{GT} are the gate trigger voltage and current for the device. For instance, the data sheet of a 2N4441 gives $V_{GT} = 0.75$ V and $I_{GT} = 10$ mA. When you have the value of R_G, the calculation of V_{in} is straightforward. If a gate resistor is not used, R_G is the Thevenin resistance of the circuit driving the gate. Unless Eq. (15-1) is satisfied, the SCR cannot turn on.

Resetting the SCR

After the SCR has turned on, it stays on even though you reduce the gate supply, V_{in}, to zero. In this case, the output remains low indefinitely. To reset the SCR, you must reduce the anode to cathode current to a value less than its holding current, I_H. This can be done by reducing V_{CC} to a low value. The data sheet for the

Figure 15–10   SCR data sheet.

2N6504 Series

Preferred Device

Silicon Controlled Rectifiers

Reverse Blocking Thyristors

Designed primarily for half-wave ac control applications, such as motor controls, heating controls and power supply crowbar circuits.

Features

- Glass Passivated Junctions with Center Gate Fire for Greater Parameter Uniformity and Stability
- Small, Rugged, Thermowatt Constructed for Low Thermal Resistance, High Heat Dissipation and Durability
- Blocking Voltage to 800 Volts
- 300 A Surge Current Capability
- Pb–Free Packages are Available*

2N6504 Series

Voltage Current Characteristic of SCR

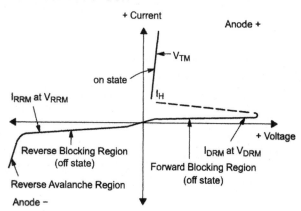

Symbol	Parameter
V_{DRM}	Peak Repetitive Off State Forward Voltage
I_{DRM}	Peak Forward Blocking Current
V_{RRM}	Peak Repetitive Off State Reverse Voltage
I_{RRM}	Peak Reverse Blocking Current
V_{TM}	Peak On State Voltage
I_H	Holding Current

ON Semiconductor®

http://onsemi.com

SCRs
25 AMPERES RMS
50 thru 800 VOLTS

MARKING DIAGRAM

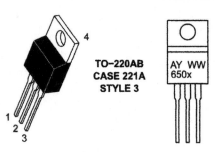

TO–220AB
CASE 221A
STYLE 3

AY WW
650x

x = 4, 5, 7, 8 or 9
A = Assembly Location
Y = Year
WW = Work Week

PIN ASSIGNMENT	
1	Cathode
2	Anode
3	Gate
4	Anode

ORDERING INFORMATION

See detailed ordering and shipping information in the package dimensions section on page 3 of this data sheet.

Preferred devices are recommended choices for future use and best overall value.

*For additional information on our Pb–Free strategy and soldering details, please download the ON Semiconductor Soldering and Mounting Techniques Reference Manual, SOLDERRM/D.

Publication Order Number:
2N6504/D

1

Figure 15–10 (continued)

2N6504 Series

MAXIMUM RATINGS (T_J = 25°C unless otherwise noted)

Rating	Symbol	Value	Unit
*Peak Repetitive Off–State Voltage (Note 1) (Gate Open, Sine Wave 50 to 60 Hz, T_J = 25 to 125°C) 2N6504 2N6505 2N6507 2N6508 2N6509	V_{DRM}, V_{RRM}	 50 100 400 600 800	V
On-State Current RMS (180° Conduction Angles; T_C = 85°C)	$I_{T(RMS)}$	25	A
Average On-State Current (180° Conduction Angles; T_C = 85°C)	$I_{T(AV)}$	16	A
Peak Non-repetitive Surge Current (1/2 Cycle, Sine Wave 60 Hz, T_J = 100°C)	I_{TSM}	250	A
Forward Peak Gate Power (Pulse Width ≤ 1.0 μs, T_C = 85°C)	P_{GM}	20	W
Forward Average Gate Power (t = 8.3 ms, T_C = 85°C)	$P_{G(AV)}$	0.5	W
Forward Peak Gate Current (Pulse Width ≤ 1.0 μs, T_C = 85°C)	I_{GM}	2.0	A
Operating Junction Temperature Range	T_J	–40 to +125	°C
Storage Temperature Range	T_{stg}	–40 to +150	°C

Maximum ratings are those values beyond which device damage can occur. Maximum ratings applied to the device are individual stress limit values (not normal operating conditions) and are not valid simultaneously. If these limits are exceeded, device functional operation is not implied, damage may occur and reliability may be affected.
1. V_{DRM} and V_{RRM} for all types can be applied on a continuous basis. Ratings apply for zero or negative gate voltage; however, positive gate voltage shall not be applied concurrent with negative potential on the anode. Blocking voltages shall not be tested with a constant current source such that the voltage ratings of the devices are exceeded.

THERMAL CHARACTERISTICS

Characteristic	Symbol	Max	Unit
*Thermal Resistance, Junction–to–Case	$R_{\theta JC}$	1.5	°C/W
*Maximum Lead Temperature for Soldering Purposes 1/8 in from Case for 10 Seconds	T_L	260	°C

ELECTRICAL CHARACTERISTICS (T_C = 25°C unless otherwise noted.)

Characteristic	Symbol	Min	Typ	Max	Unit
OFF CHARACTERISTICS					
*Peak Repetitive Forward or Reverse Blocking Current (V_{AK} = Rated V_{DRM} or V_{RRM}, Gate Open) T_J = 25°C T_J = 125°C	I_{DRM}, I_{RRM}	– –	– –	10 2.0	μA mA
ON CHARACTERISTICS					
*Forward On–State Voltage (Note 2) (I_{TM} = 50 A)	V_{TM}	–	–	1.8	V
*Gate Trigger Current (Continuous dc) T_C = 25°C (V_{AK} = 12 Vdc, R_L = 100 Ω) T_C = –40°C	I_{GT}	– –	9.0 –	30 75	mA
*Gate Trigger Voltage (Continuous dc) (V_{AK} = 12 Vdc, R_L = 100 Ω, T_C = –40°C)	V_{GT}	–	1.0	1.5	V
Gate Non-Trigger Voltage (V_{AK} = 12 Vdc, R_L = 100 Ω, T_J = 125°C)	V_{GD}	0.2	–	–	V
*Holding Current T_C = 25°C (V_{AK} = 12 Vdc, Initiating Current = 200 mA, Gate Open) T_C = –40°C	I_H	– –	18 –	40 80	mA
*Turn-On Time (I_{TM} = 25 A, I_{GT} = 50 mAdc)	t_{gt}	–	1.5	2.0	μs
Turn-Off Time (V_{DRM} = rated voltage) (I_{TM} = 25 A, I_R = 25 A) (I_{TM} = 25 A, I_R = 25 A, T_J = 125°C)	t_q	 – –	 15 35	 – –	μs
DYNAMIC CHARACTERISTICS					
Critical Rate of Rise of Off-State Voltage (Gate Open, Rated V_{DRM}, Exponential Waveform)	dv/dt	–	50	–	V/μs

*Indicates JEDEC Registered Data.
2. Pulse Test: Pulse Width ≤ 300 μs, Duty Cycle ≤ 2%.

Figure 15-11   Basic SCR circuit.

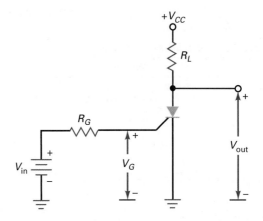

2N6504 lists a typical holding current value of 18 mA. SCRs with lower and higher power ratings generally have lower and higher respective holding current values. Since the holding current flows through the load resistor in Fig. 15-11, the supply voltage for turnoff has to be less than:

$$V_{CC} = 0.7 \text{ V} + I_H R_L \tag{15-2}$$

Besides reducing V_{CC}, other methods can be used to reset the SCR. Two common methods are current interruption and forced commutation. By either opening the series switch, as shown in Fig. 15-12a, or closing the parallel switch in Fig. 15-12b, the anode-to-cathode current will drop down below its holding current value and the SCR will switch to its off state.

Another method used to reset the SCR is forced commutation, as shown in Fig. 15-12c. When the switch is depressed, a negative V_{AK} voltage is momentarily applied. This reduces the forward anode-to-cathode current below I_H and turns off the SCR. In actual circuits, the switch can be replaced with a BJT or FET device.

Figure 15-12   Resetting the SCR.

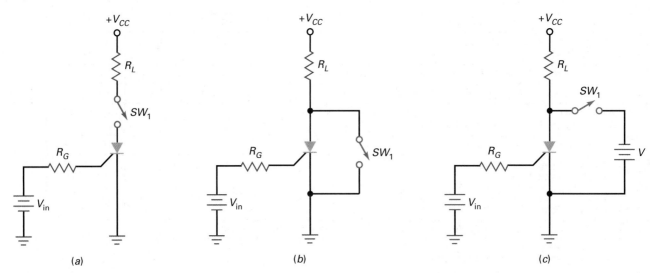

(a)   (b)   (c)

Figure 15-13 Power FET versus SCR.

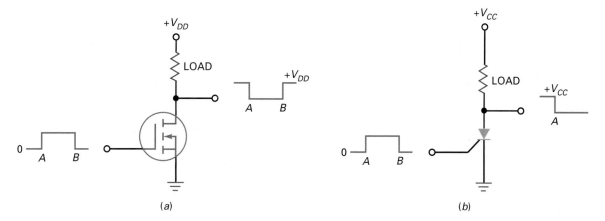

(*a*)   (*b*)

Power FET versus SCR

Although both the power FET and the SCR can switch large currents on and off, the two devices are fundamentally different. The key difference is the way they turn off. The gate voltage of a power FET can turn the device on and off. This is not the case with an SCR. The gate voltage can only turn it on.

Figure 15-13 illustrates the difference. In Fig. 15-13*a*, when the input voltage to the power FET goes high, the output voltage goes low. When the input voltage goes low, the output voltage goes high. In other words, a rectangular input pulse produces an inverted rectangular output pulse.

In Fig. 15-13*b*, when the input voltage to the SCR goes high, the output voltage goes low. But when the input voltage goes low, the output voltage stays low. With an SCR, a rectangular input pulse produces a negative-going output step. The SCR does not reset.

Because the two devices have to be reset in different ways, their applications tend to be different. Power FETs respond like push-button switches, whereas SCRs respond like single-pole single-throw switches. Since it is easier to control the power FET, you will see it used more often as an interface between digital ICs and heavy loads. In applications in which latching is important, you will see the SCR used.

Example 15-4 |||| MultiSim

In Fig. 15-14, the SCR has a trigger voltage of 0.75 V and a trigger current of 7 mA. What is the input voltage that turns the SCR on? If the holding current is 6 mA, what is the supply voltage that turns it off?

SOLUTION   With Eq. (15-1), the minimum input voltage needed to trigger is:

$$V_{in} = 0.75 \text{ V} + (7 \text{ mA})(1 \text{ k}\Omega) = 7.75 \text{ V}$$

With Eq. (15-2), the supply voltage that turns off the SCR is:

$$V_{CC} = 0.7 \text{ V} + (6 \text{ mA})(100 \text{ }\Omega) = 1.3 \text{ V}$$

Figure 15-14 Example.

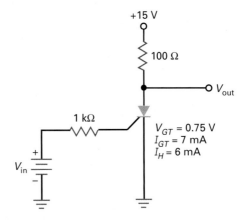

PRACTICE PROBLEM 15-4 In Fig. 15-14, determine the input voltage needed to trigger the SCR on and the supply voltage that turns off the SCR, using the typical rated values for a 2N6504 SCR.

Example 15–5

What does the circuit of Fig. 15-15a do? What is the peak output voltage? What is the frequency of the sawtooth wave if its period is approximately 20 percent of the time constant?

Figure 15-15 Example.

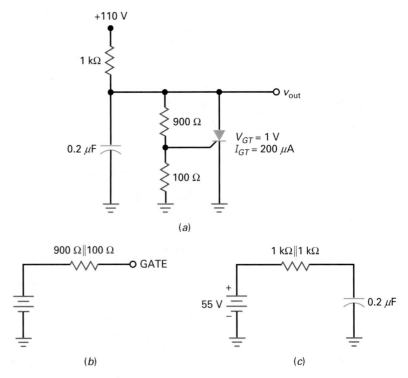

SOLUTION As the capacitor voltage increases, the SCR eventually *fires* (turns on) and rapidly discharges the capacitor. When the SCR opens, the capacitor begins charging again. Therefore, the output voltage is a sawtooth wave similar to the one in Fig. 15-6b, discussed in Example 15-3.

Figure 15-15b shows the Thevenin circuit facing the gate. The Thevenin resistance is:

$$R_{TH} = 900 \ \Omega \parallel 100 \ \Omega = 90 \ \Omega$$

With Eq. (15-1), the input voltage needed to trigger is:

$$V_{\text{in}} = 1 \ V + (200 \ \mu A)(90 \ \Omega) \approx 1 \ V$$

Because of the 10:1 voltage divider, the gate voltage is one-tenth of the output voltage. Therefore, the output voltage at the SCR firing point is:

$$V_{\text{peak}} = 10(1 \ V) = 10 \ V$$

Figure 15-15c shows the Thevenin circuit facing the capacitor when the SCR is off. From this, you can see that the capacitor will try charging to a final voltage of +55 V with a time constant of:

$$RC = (500 \ \Omega)(0.2 \ \mu F) = 100 \ \mu s$$

Since the period of the sawtooth is approximately 20 percent of this:

$$T = 0.2(100 \ \mu s) = 20 \ \mu s$$

The frequency is:

$$f = \frac{1}{20 \ \mu s} = 50 \ \text{kHz}$$

Testing SCRs

Thyristors, like SCRs, handle large amounts of current and must block high-voltage values. Because of this, they may fail under these conditions. Common failures are A-K opens, A-K shorts, and no gate control. Fig. 15-16a shows a circuit that can test the operation of SCRs. Before SW_1 is pushed, I_{AK} should be zero and V_{AK} should be approximately equal to V_A. When SW_1 is momentarily pushed, I_{AK} should rise to a level near V_A/R_L and V_{AK} should drop to approximately 1 V. V_A and R_L must be selected to provide the necessary current and power levels. When SW_1 is released, the SCR should remain in the on state. The anode supply voltage, V_A, can then be reduced until the SCR drops out of conduction. By observing the anode current value just before the SCR turns off, you can determine the SCR's holding current.

Another method to test SCRs is by using an ohmmeter. The ohmmeter must be able to provide the necessary gate voltage and current to turn on the SCR and, just as important, provide the holding current required to keep the SCR in the on state. Many analog VOMs are capable of outputting approximately 1.5 V and 100 mA in the R × 1 range. In Fig. 15-16b, the ohmmeter is placed across the anode-cathode leads. With either polarity connection, the result should be a very high resistance. With the positive test lead connected to the anode and negative test lead connected to the cathode, connect a jumper wire from the anode to the gate. The SCR should turn on and show a low resistance reading. When the gate lead is disconnected, the SCR should remain in the on state. Momentarily disconnecting the anode test lead will turn the SCR off.

Figure 15-16 Testing SCRs: (*a*) Test circuit; (*b*) Ohmmeter.

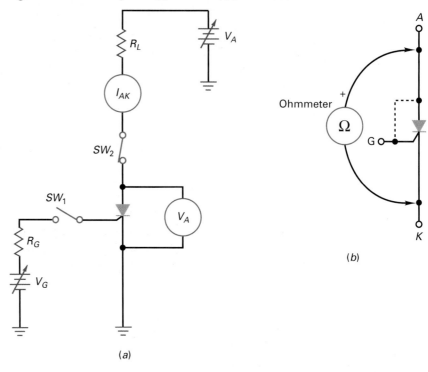

(*a*)

(*b*)

15-3 The SCR Crowbar

If anything happens inside a power supply to cause its output voltage to go excessively high, the results can be devastating. Why? Because some loads such as expensive digital ICs cannot withstand too much supply voltage without being destroyed. One of the most important applications of the SCR is to protect delicate and expensive loads against overvoltages from a power supply.

Prototype

Figure 15-17 shows a power supply of V_{CC} applied to a protected load. Under normal conditions, V_{CC} is less than the breakdown voltage of the zener diode. In this case, there is no voltage across R, and the SCR remains open. The load receives a voltage of V_{CC}, and all is well.

Now, assume that the supply voltage increases for any reason whatsoever. When V_{CC} is too large, the zener diode breaks down and a voltage appears across R. If this voltage is greater than the gate trigger voltage of the SCR, the SCR fires and becomes a closed latch. The action is similar to throwing a *crowbar*

Figure 15-17 SCR crowbar.

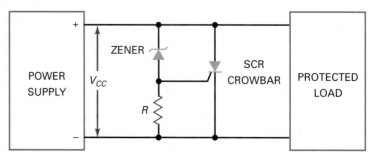

Figure 15-18 Adding transistor gain to crowbar.

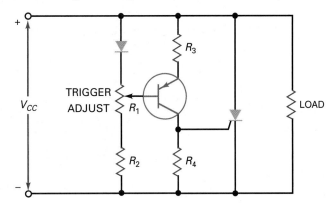

across the load terminals. Because the SCR turn-on is very fast (1 μs for a 2N4441), the load is quickly protected against the damaging effects of a large overvoltage. The overvoltage that fires the SCR is:

$$V_{CC} = V_Z + V_{GT} \qquad (15\text{-}3)$$

Crowbarring, though a drastic form of protection, is necessary with many digital ICs because they can't take much overvoltage. Rather than destroy expensive ICs, therefore, we can use an SCR crowbar to short the load terminals at the first sign of overvoltage. With an SCR crowbar, a fuse or *current limiting* (discussed later) is needed to prevent damage to the power supply.

Adding Voltage Gain

The crowbar of Fig. 15-17 is a *prototype,* a basic circuit that can be modified and improved. It is adequate for many applications as is. But it suffers from a *soft turn-on* because the knee at zener breakdown is curved rather than sharp. When we take into account the tolerance in zener voltages, the soft turn-on can result in the supply voltage becoming dangerously high before the SCR fires.

One way to overcome the soft turn-on is by adding some voltage gain, as shown in Fig. 15-18. Normally, the transistor is off. But when the output voltage increases, the transistor eventually turns on and produces a large voltage across R_4. Since the transistor provides a swamped voltage gain of approximately R_4/R_3, a small overvoltage can trigger the SCR.

Notice that an ordinary diode is being used, not a zener diode. This diode temperature-compensates the transistor's base-emitter diode. The *trigger adjust* allows us to set the *trip point* of the circuit, typically around 10 to 15 percent above the normal voltage.

IC Voltage Gain

Figure 15-19 shows an even better solution. The triangular box is an IC amplifier called a *comparator* (discussed in later chapters). This amplifier has a noninverting (+) input and an inverting (−) input. When the noninverting input is greater than the inverting input, the output is positive. When the inverting input is greater than the noninverting input, the output is negative.

The amplifier has a very large voltage gain, typically 100,000 or more. Because of its large voltage gain, the circuit can detect the slightest overvoltage. The zener diode produces 10 V, which goes to the minus input of the amplifier. When the supply voltage is 20 V (normal output), the trigger adjust is set to produce slightly less than 10 V on the positive input. Since the negative input is greater than the positive, the amplifier output is negative and the SCR is open.

Figure 15-19 Adding IC amplifier to crowbar.

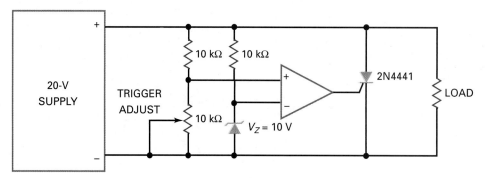

Figure 15-20 IC crowbar.

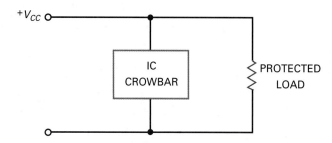

If the supply voltage rises above 20 V, the positive input to the amplifier becomes greater than 10 V. Then, the amplifier output becomes positive and the SCR fires. This rapidly shuts down the supply by crowbarring the load terminals.

Integrated Crowbar

The simplest solution is to use an IC crowbar, as shown in Fig. 15-20. This is an integrated circuit with a built-in zener diode, transistors, and an SCR. The RCA SK9345 series of IC crowbars is an example of what is commercially available. The SK9345 protects power supplies of +5 V, the SK9346 protects +12 V, and the SK9347 protects +15 V.

If an SK9345 is used in Fig. 15-20, it will protect the load with a supply voltage of +5 V. The data sheet of an SK9345 indicates that it fires at +6.6 V with a tolerance of ±0.2 V. This means that it fires between 6.4 and 6.8 V. Since 7 V is the maximum rating of many digital ICs, the SK9345 protects the load under all operating conditions.

Example 15-6

IIII MultiSim

Calculate the supply voltage that turns on the crowbar of Fig. 15-21.

SOLUTION The 1N752 has a breakdown voltage of 5.6 V, and the 2N4441 has a gate trigger voltage of 0.75 V. With Eq. (15-3):

$$V_{CC} = V_Z + V_{GT} = 5.6\ V + 0.75\ V = 6.35\ V$$

When the supply voltage increases to this level, the SCR fires.

Figure 15-21 Example.

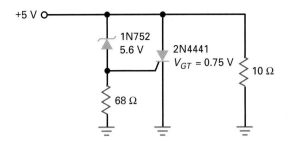

The prototype crowbar is all right if the application is not too critical about the exact supply voltage at which the SCR turns on. For instance, the 1N752 has a tolerance of ± 10 percent, which means that the breakdown voltage may vary from 5.04 to 6.16 V. Furthermore, the trigger voltage of a 2N4441 has a worst-case maximum of 1.5 V. So, the overvoltage can be as high as:

$$V_{CC} = 6.16 \text{ V} + 1.5 \text{ V} = 7.66 \text{ V}$$

Since many digital ICs have a maximum rating of 7 V, the simple crowbar of Fig. 15-21 cannot be used to protect them.

PRACTICE PROBLEM 15-6 Repeat Example 15-6 using a 1N4733A zener diode. This diode has a zener voltage of 5.1 V $\pm$ 5%.

15-4 SCR Phase Control

Table 15-1 shows some commercially available SCRs. The gate trigger voltages vary from 0.8 to 2 V, and the gate trigger currents range from 200 μA to 50 mA. Also notice that anode currents vary from 1.5 to 70 A. Devices like these can control heavy industrial loads by using phase control.

RC Circuit Controls Phase Angle

Figure 15-22*a* shows ac line voltage being applied to an SCR circuit that controls the current through a heavy load. In this circuit, variable resistor R_1 and capacitor

Table 15-1	SCR Sampler			
Device	V_{GT}, **V**	I_{GT}	I_{max}, **A**	V_{max}, **V**
TCR22-2	0.8	200 μA	1.5	50
T106B1	0.8	200 μA	4	200
S4020L	1.5	15 mA	10	400
S6025L	1.5	39 mA	25	600
S1070W	2	50 mA	70	100

Figure 15–22 SCR phase control.

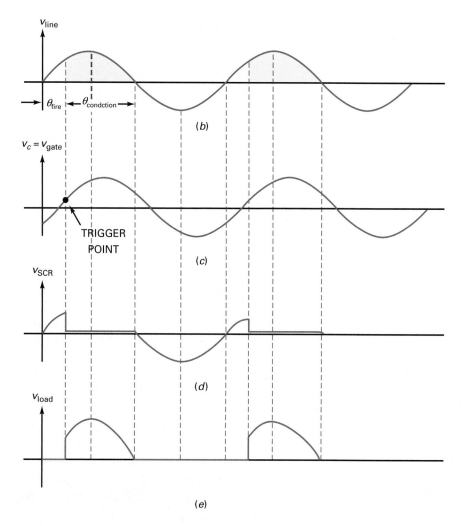

(a)

(b)

(c)

(d)

(e)

C shift the phase angle of the gate signal. When R_1 is zero, the gate voltage is in phase with the line voltage, and the SCR acts like a half-wave rectifier. R_2 limits the gate current to a safe level.

When R_1 increases, however, the ac gate voltage lags the line voltage by an angle between 0 and 90°, as shown in Figs. 15-22b and c. Before the trigger point shown in Fig. 15-22c, the SCR is off and the load current is zero. At the trigger point, the capacitor voltage is large enough to trigger the SCR. When this happens, almost all of the line voltage appears across the load and the load current becomes high. Ideally, the SCR remains latched until the line voltage reverses polarity. This is shown in Figs. 15-22d and e.

The angle at which the SCR fires is called the **firing angle,** shown as θ_{fire} in Fig. 15-22b. The angle between the start and end of conduction is called the **conduction angle,** shown as $\theta_{\text{conduction}}$. The RC phase controller of Fig. 15-22a can change the firing angle between 0 and 90°, which means that the conduction angle changes from 180 to 90°.

The shaded portions of Fig. 15-22b show when the SCR is conducting. Because R_1 is variable, the phase angle of the gate voltage can be changed. This allows us to control the shaded portions of the line voltage. Stated another way: We can control the average current through the load. This is useful for changing the speed of a motor, the brightness of a lamp, or the temperature of an induction furnace.

By using circuit analysis techniques studied in basic electricity courses, we can determine the approximate phase shifted voltage across the capacitor. This gives us the approximate firing angle and conduction angle of the circuit. To determine the voltage across the capacitor, use the following steps:

First, find the capacitive reactance of C by:

$$X_C = \frac{1}{2\pi fc}$$

The impedance and phase angle of the RC phase-shift circuit is:

$$Z_T = \sqrt{R^2 + X_C^2} \tag{15-4}$$

$$\theta_Z = \angle - \arctan \frac{X_C}{R} \tag{15-5}$$

Using the input voltage as our reference point, the current through C is:

$$I_C \angle \theta = \frac{V_{\text{in}} \angle 0°}{Z_T \angle - \arctan \dfrac{X_C}{R}}$$

Now, the voltage value and phase of the capacitor can be found by:

$$V_C = (I_C \angle \theta)(X_C \angle -90°)$$

The amount of delayed phase shift will be the approximate firing angle of the circuit. The conduction angle is found by subtracting the firing angle from 180°.

Example 15-7

IIII MultiSim

Using Fig. 15-22a, find the approximate firing angle and conduction angle when $R_1 = 26$ kΩ and $C = 0.1$ μF.

SOLUTION The approximate firing angle can be found by solving for the voltage value and its phase shift across the capacitor. This is found by:

$$X_C = \frac{1}{2\pi fc} = \frac{1}{(2\pi)(60 \text{ Hz})(0.1 \text{ } \mu\text{F})} = 26.5 \text{ k}\Omega$$

Because capacitive reactance is at an angle of $-90°$, $X_C = 26.5$ k$\Omega \angle -90°$. Next, find the total RC impedance Z_T and its angle by:

$$Z_T = \sqrt{R^2 + X_C^2} = \sqrt{(26 \text{ k}\Omega)^2 + (26.5 \text{ k}\Omega)^2} = 37.1 \text{ k}\Omega$$

$$\theta_Z = \angle - \arctan \frac{X_C}{R} = \angle - \arctan \frac{26.5 \text{ k}\Omega}{26 \text{ k}\Omega} = -45.5°$$

Therefore, $Z_T = 37.1 \text{ k}\Omega \angle -45.5°$.

Using the ac input as our reference, the current through C is:

$$I_C = \frac{V_{in} \angle 0°}{Z_T \angle \theta} = \frac{120 \text{ V}_{ac} \angle 0°}{37.1 \text{ k}\Omega \angle -45.5°} = 3.23 \text{ mA} \angle 45.5°$$

Now, the voltage across C can be found by:

$$V_C = (I_C \angle \theta)(X_C \angle -90°) = (3.23 \text{ mA} \angle 45.5°)(26.5 \text{ k}\Omega \angle -90°)$$

$$V_C = 85.7 \text{ V}_{ac} \angle -44.5°$$

With the voltage phase shift across the capacitor of $-44.5°$, the firing angle of the circuit is approximately $-45.5°$. After the SCR fires, it will remain on until its current drops below I_H. This will occur approximately when the ac input is zero volts.

Therefore, the conduction angle is:

$$\text{conduction } \theta = 180° - 44.5° = 135.5°$$

PRACTICE PROBLEM 15–7   Using Fig. 15-22*a*, find the approximate firing angle and conduction angle when $R = 50 \text{ k}\Omega$.

The *RC* phase controller of Fig. 15-22*a* is a basic way of controlling the average current through the load. The controllable range of current is limited because the phase angle can change from only 0 to 90°. With op amps and more sophisticated *RC* circuits, we can change the phase angle from 0 to 180°. This allows us to vary the average current all the way from zero to maximum.

Figure 15-23   (*a*) *RC* snubber protects SCR against rapid voltage rise; (*b*) inductor protects SCR against rapid current rise.

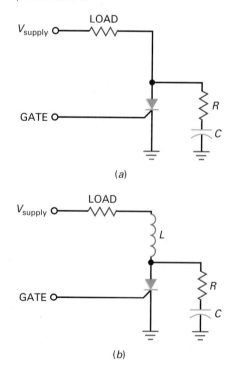

(a)

(b)

Critical Rate of Rise

When ac voltage is used to supply the anode of an SCR, it is possible to get false triggering. Because of capacitances inside an SCR, rapidly changing supply voltages may trigger the SCR. To avoid false triggering of an SCR, the rate of voltage change must not exceed the *critical rate of voltage rise* specified on the data sheet. For instance, the 2N6504 has a critical rate of voltage rise of 50 V/μs. To avoid false triggering, the anode voltage must not rise faster than 50 V/μs.

Switching transients are the main cause of exceeding the critical rate of voltage rise. One way to reduce the effects of switching transients is with an *RC snubber,* shown in Fig. 15-23a. If a high-speed switching transient does appear on the supply voltage, its rate of rise is reduced at the anode because of the *RC* time constant.

Large SCRs also have a *critical rate of current rise.* For instance, the C701 has a critical rate of current rise of 150 A/μs. If the anode current tries to rise faster than this, the SCR will be destroyed. Including an inductor in series with the load (Fig. 15-23b) reduces the rate of current rise to a safe level.

15-5 Bidirectional Thyristors

The two devices discussed so far, the four-layer diode and the SCR, are unidirectional because current can flow in only one direction. The **diac** and **triac** are *bidirectional thyristors.* These devices can conduct in either direction. The diac is sometimes called a *silicon bidirectional switch (SBS).*

Diac

The diac can latch current in either direction. The equivalent circuit of a diac is two four-layer diodes in parallel, as shown in Fig. 15-24a, ideally the same as the latches in Fig. 15-24b. The diac is nonconducting until the voltage across it exceeds the breakover voltage in either direction.

For instance, if *v* has the polarity indicated in Fig. 15-24a, the left diode conducts when *v* exceeds the breakover voltage. In this case, the left latch closes, as shown in Fig. 15-24c. When *v* has the opposite polarity, the right latch closes. Figure 15-24d shows the schematic symbol for a diac.

Triac

The triac acts like two SCRs in reverse parallel (Fig. 15-25a), equivalent to the two latches of Fig. 15-25b. Because of this, the triac can control current in both directions. If *v* has the polarity shown in Fig. 15-25a, a positive trigger will close

Figure 15-24 Diac.

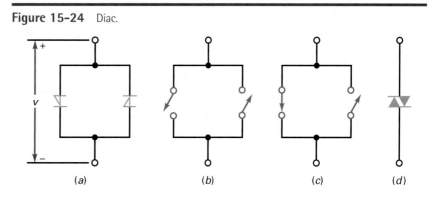

| (a) | (b) | (c) | (d) |

Figure 15–25 Triac.

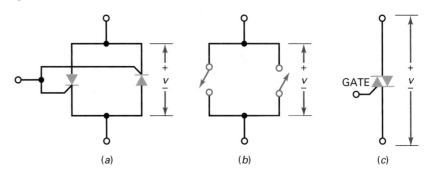

(a)   (b)   (c)

the left latch. When v has opposite polarity, a negative trigger will close the right latch. Figure 15-25c is the schematic symbol for a triac.

Fig. 15-26 shows the data sheet for a FKPF8N80 triac. As the name triac implies, it is a bidirectional (ac) triode thyristor. Notice, at the end of the data sheet, the quadrant definitions or modes of triac operation. The triac normally operates in quadrants I and III during typical ac applications. Since this device is most sensitive in quadrant I, a diac is often used in conjunction with the triac to provide symmetrical ac conduction.

Table 15-2 shows some commercially available triacs. Because of their internal structure, triacs have higher gate trigger voltages and currents than comparable SCRs. As you can see, the gate trigger voltages of Table 15-2 are from 2 to 2.5 V and the gate trigger currents are from 10 to 50 mA. The maximum anode currents are from 1 to 15 A.

Phase Control

Figure 15-27a shows an *RC* circuit that varies the phase angle of the gate voltage to a triac. The circuit can control the current through a heavy load. Figure 15-27b and c shows the line voltage and lagging gate voltage. When the capacitor voltage is large enough to supply the trigger current, the triac conducts. Once on, the triac continues to conduct until the line voltage returns to zero. Figs. 15-27d and 15-27e show the respective voltages across the triac and the load.

Although triacs can handle large currents, they are not in the same class as SCRs, which have much higher current ratings. Nevertheless, when conduction on both half cycles is important, triacs are useful devices especially in industrial applications.

Table 15-2	Triac Sampler			
Device	V_{GT}, **V**	I_{GT}, **mA**	I_{max}, **A**	V_{max}, **V**
Q201E3	2	10	1	200
Q4004L4	2.5	25	4	400
Q5010R5	2.5	50	10	500
Q6015R5	2.5	50	15	600

Figure 15-26 Triac data sheet.

FKPF8N80

Application Explanation

- Switching mode power supply, light dimmer, electric flasher unit, hair drier
- TV sets, stereo, refrigerator, washing machine
- Electric blanket, solenoid driver, small motor control
- Photo copier, electric tool

TO-220F

1 2 3

1: T_1
2: T_2
3: Gate

Bi-Directional Triode Thyristor Planar Silicon

Absolute Maximum Ratings T_C=25°C unless otherwise noted

Symbol	Parameter	Rating	Units
V_{DRM}	Repetitive Peak Off-State Voltage (Note1)	800	V

Symbol	Parameter	Conditions		Rating	Units
$I_{T(RMS)}$	RMS On-State Current	Commercial frequency, sine full wave 360° conduction, T_C=91°C		8	A
I_{TSM}	Surge On-State Current	Sinewave 1 full cycle, peak value, non-repetitive	50Hz	80	A
			60Hz	88	A
I^2t	I^2t for Fusing	Value corresponding to 1 cycle of halfwave, surge on-state current, tp=10ms		32	A^2s
di/dt	Critical Rate of Rise of On-State Current	I_G = 2x I_{GT}, tr ≤ 100ns		50	A/μs
P_{GM}	Peak Gate Power Dissipation			5	W
$P_{G(AV)}$	Average Gate Power Dissipation			0.5	W
V_{GM}	Peak Gate Voltage			10	V
I_{GM}	Peak Gate Current			2	A
T_J	Junction Temperature			- 40 ~ 125	°C
T_{STG}	Storage Temperature			- 40 ~ 125	°C
V_{iso}	Isolation Voltage	Ta=25°C, AC 1 minute, T_1 T_2 G terminal to case		1500	V

Thermal Characteristic

Symbol	Parameter	Test Condition	Min.	Typ.	Max.	Units
$R_{th(J-C)}$	Thermal Resistance	Junction to case (Note 4)	-	-	3.6	°C/W

Rev. B1, April 2004

Figure 15–26 (continued)

Electrical Characteristics T_C=25°C unless otherwise noted

Symbol	Parameter		Test Condition		Min.	Typ.	Max.	Units
I_{DRM}	Repetieive Peak Off-State Current		V_{DRM} applied		-	-	20	μA
V_{TM}	On-State Voltage		T_C=25°C, I_{TM}=12A Instantaneous measurement		-	-	1.5	V
V_{GT}	Gate Trigger Voltage (Note 2)	I	V_D=12V, R_L=20Ω	T2(+), Gate (+)	-	-	1.5	V
		II		T2(+), Gate (-)	-	-	1.5	V
		III		T2(-), Gate (-)	-	-	1.5	V
I_{GT}	Gate Trigger Current (Note 2)	I	V_D=12V, R_L=20Ω	T2(+), Gate (+)	-	-	30	mA
		II		T2(+), Gate (-)	-	-	30	mA
		III		T2(-), Gate (-)	-	-	30	mA
V_{GD}	Gate Non-Trigger Voltage		T_J=125°C, V_D=1/2V_{DRM}		0.2	-	-	V
I_H	Holding Current		V_D = 12V, I_{TM} = 1A				50	mA
I_L	Latching Current	I, III	V_D = 12V, I_G = 1.2I_{GT}				50	mA
		II					70	mA
dv/dt	Critical Rate of Rise of Off-State Voltag		V_{DRM} = Rated, T_j = 125°C, Exponential Rise			300		V/μs
$(dv/dt)_C$	Critical-Rate of Rise of Off-State Commutating Voltage (Note 3)				10	-	-	V/μs

Notes:
1. Gate Open
2. Measurement using the gate trigger characteristics measurement circuit
3. The critical-rate of rise of the off-state commutating voltage is shown in the table below
4. The contact thermal resistance $R_{TH(c-f)}$ in case of greasing is 0.5 °C/W

V_{DRM} (V)	Test Condition	Commutating voltage and current waveforms (inductive load)
FKPF8N80	1. Junction Temperature T_J=125°C 2. Rate of decay of on-state commutating current $(di/dt)_C$ = - 4.5A/ms 3. Peak off-state voltage V_D = 400V	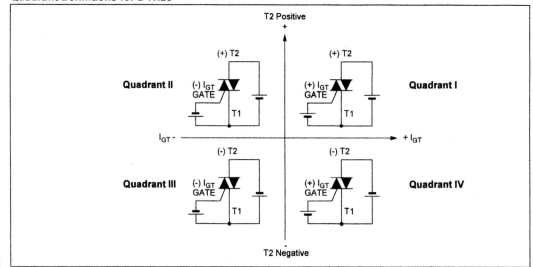

Quadrant Definitions for a Triac

T2 Positive
+

(+) T2   (+) T2

Quadrant II   (-) I_{GT} GATE   (+) I_{GT} GATE   **Quadrant I**

T1   T1

I_{GT} -   + I_{GT}

(-) T2   (-) T2

Quadrant III   (-) I_{GT} GATE   (+) I_{GT} GATE   **Quadrant IV**

T1   T1

-
T2 Negative

Rev. B1, April 2004

Figure 15–27 Triac phase control.

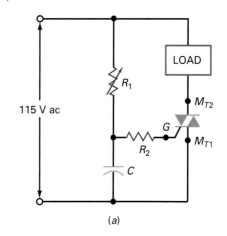

(a)

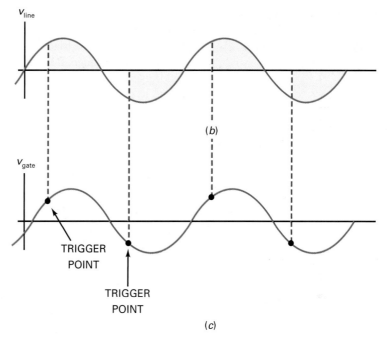

(b)

TRIGGER
POINT

TRIGGER
POINT

(c)

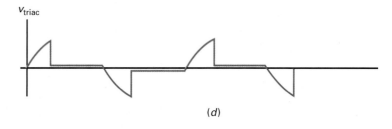

(d)

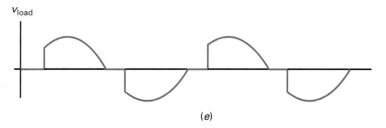

(e)

Figure 15-28 Triac crowbar.

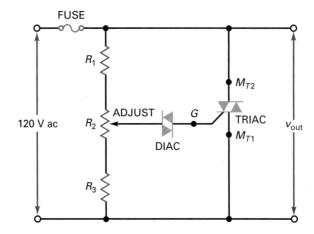

Triac Crowbar

Figure 15-28 shows a triac crowbar that can be used to protect equipment against excessive line voltage. If the line voltage becomes too high, the diac breaks over and triggers the triac. When the triac fires, it blows the fuse. A potentiometer R_2 allows us to set the trigger point.

Example 15-8   ‖‖‖ MultiSim

In Fig. 15-29, the switch is closed. If the triac has fired, what is the approximate current through the 22-Ω resistor?

Figure 15-29 Example.

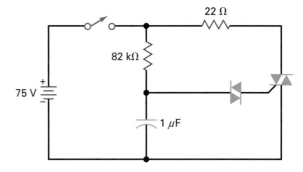

SOLUTION Ideally, the triac has zero volts across it when conducting. Therefore, the current through the 22-Ω resistor is:

$$I = \frac{75 \text{ V}}{22 \text{ } \Omega} = 3.41 \text{ A}$$

If the triac has 1 or 2 V across it, the current is still close to 3.41 A because the large supply voltage swamps the effect of the triac on voltage.

PRACTICE PROBLEM 15-8 Using Fig. 15-29, change V_{in} to 120 V and calculate the approximate current through the 22 Ω resistor.

Example 15-9

In Fig. 15-29, the switch is closed. The MPT32 is a diac with a breakover voltage of 32 V. If the triac has a trigger voltage of 1 V and a trigger current of 10 mA, what is the capacitor voltage that triggers the triac?

SOLUTION As the capacitor charges, the voltage across the diac increases. When the diac voltage is slightly less than 32 V, the diac is on the verge of breakover. Since the triac has a trigger voltage of 1 V, the capacitor voltage needs to be:

$$V_{in} = 32 \text{ V} + 1 \text{ V} = 33 \text{ V}$$

At this input voltage, the diac breaks over and triggers the triac.

PRACTICE PROBLEM 15-9 Repeat Example 15-9 using a diac with a 24 V breakover value.

15-6 IGBTs

Basic Construction

Power MOSFETs and BJTs can both be used in high-power switching applications. The MOSFET has the advantage of greater switching speed, and the BJT has lower conduction losses. By combining the low conduction loss of a BJT with the switching speed of a power MOSFET, we can begin to approach an ideal switch.

This hybrid device exists and is called an **insulated-gate bipolar transistor (IGBT).** The IGBT has essentially evolved from power MOSFET technology. Its structure and operation closely resembles a power MOSFET. Fig. 15-30 shows the basic structure of an *n*-channel IGBT. Its structure resembles an *n*-channel power MOSFET constructed on a *p*-type substrate. As shown, it has gate, emitter, and collector leads.

Figure 15-30 Basic IGBT structure.

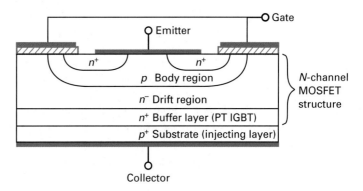

Figure 15-31 IGBTs: (*a*) and (*b*) Schematic symbols; (*c*) simplified equivalent circuit.

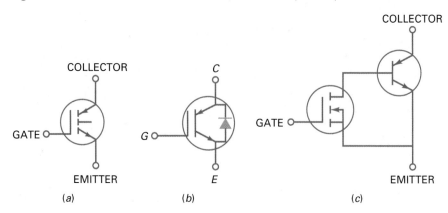

(*a*) (*b*) (*c*)

Two versions of this device are referred to as punch-through (PT) and nonpunch-through (NPT) IGBTs. Fig. 15-30 shows the structure of a PT IGBT. The PT IGBT has an $n+$ buffer layer between its $p+$ and $n-$ regions, and the NPT device has no $n+$ buffer layer.

NPT versions have higher conduction $V_{CE(on)}$ values than PT versions and a positive temperature coefficient. The positive temperature coefficient makes the NPT suited for paralleling. PT versions, with the extra $n+$ layer, have the advantage of higher switching speeds. They also have a negative temperature coefficient.

IGBT Control

Figures 15-31*a* and *b* show two common schematic symbols for an *n*-channel IGBT. Also, Fig. 15-31*c* shows a simplified equivalent circuit for this device. As you can see, the IGBT is essentially a power MOSFET on the input side and a BJT on the output side. The input control is a voltage between the gate and emitter leads. The output is a current between the collector and emitter leads.

The IGBT is a normally off high-input impedance device. When the input voltage, V_{GE}, is large enough, collector current will begin to flow. This minimum voltage value is the gate threshold voltage, $V_{GE(th)}$. Fig. 15-32 shows the data sheet for a FGL60N100BNTD IGBT using NPT-Trench technology. The typical $V_{GE(th)}$ for this device is listed as 5.0 V when $I_C = 60$ mA. The maximum continuous collector current is shown to be 60 A. Another important on characteristic is its collector to emitter saturation voltage, $V_{CE(sat)}$. The typical $V_{CE(sat)}$ value, shown on the data sheet, is 1.5 V at a collector current of 10 A and 2.5 V at a collector current of 60 A.

IGBT Advantages

Conduction losses of IGBTs are related to the forward voltage drop of the device, and the MOSFET conduction loss is based on its $R_{DS(on)}$ values. For low-voltage applications, power MOSFETs can have extremely low $R_{D(on)}$ resistances. In high-voltage applications, however, MOSFETs have increased $R_{DS(on)}$ values resulting in increased conduction losses. The IGBT does not have this characteristic. IGBTs also have a much higher collector-emitter breakdown voltage as compared to the V_{DSS} maximum value of MOSFETs. As shown in the data sheet of Fig. 15-32, the V_{CES} value is 1000 V. This is important in applications using higher-voltage inductive loads. As compared to BJTs, IGBTs have a much higher input impedance and have much simpler gate drive requirements. Although the IGBT cannot match the switching speed of the MOSFET, new IGBT families are being developed for high-frequency applications. IGBTs are, therefore, effective solutions for high-voltage and current applications at moderate frequencies.

Figure 15–32 IGBT data sheet.

SEMICONDUCTOR®

IGBT

FGL60N100BNTD

NPT-Trench IGBT

General Description

Trench insulated gate bipolar transistors (IGBTs) with NPT technology show outstanding performance in conduction and switching characteristics as well as enhanced avalanche ruggedness. These devices are well suited for Induction Heating (I-H) applications

Features

• High Speed Switching
• Low Saturation Voltage : $V_{CE(sat)}$ = 2.5 V @ I_C = 60A
• High Input Impedance
• Built-in Fast Recovery Diode

Application

Micro- Wave Oven, I-H Cooker, I-H Jar, Induction Heater, Home Appliance

TO-264

G   C   E

Absolute Maximum Ratings T_C = 25°C unless otherwise noted

Symbol	Description		FGL60N100BNTD	Units
V_{CES}	Collector-Emitter Voltage		1000	V
V_{GES}	Gate-Emitter Voltage		± 25	V
I_C	Collector Current	@ T_C = 25°C	60	A
	Collector Current	@ T_C = 100°C	42	A
$I_{CM (1)}$	Pulsed Collector Current		120	A
I_F	Diode Continuous Forward Current	@ T_C = 100°C	15	A
P_D	Maximum Power Dissipation	@ T_C = 25°C	180	W
	Maximum Power Dissipation	@ T_C = 100°C	72	W
T_J	Operating Junction Temperature		-55 to +150	°C
T_{stg}	Storage Temperature Range		-55 to +150	°C
T_L	Maximum Lead Temp. for soldering Purposes, 1/8" from case for 5 seconds		300	°C

Notes :
(1) Repetitive rating : Pulse width limited by max. junction temperature

Thermal Characteristics

Symbol	Parameter	Typ.	Max.	Units
$R_{\theta JC}$(IGBT)	Thermal Resistance, Junction-to-Case	--	0.69	°C/W
$R_{\theta JC}$(DIODE)	Thermal Resistance, Junction-to-Case	--	2.08	°C/W
$R_{\theta JA}$	Thermal Resistance, Junction-to-Ambient	--	25	°C/W

FGL60N100BNTD Rev. A

Figure 15–32 (continued)

Electrical Characteristics of IGBT $T_C = 25°C$ unless otherwise noted

Symbol	Parameter	Test Conditions	Min.	Typ.	Max.	Units
Off Characteristics						
BV_{CES}	Collector Emitter Breakdown Voltage	$V_{GE} = 0V$, $I_C = 1mA$	1000	--	--	V
I_{CES}	Collector Cut-Off Current	$V_{CE} = 1000V$, $V_{GE} = 0V$	--	--	1.0	mA
I_{GES}	G-E Leakage Current	$V_{GE} = \pm 25$, $V_{CE} = 0V$	--	--	± 500	nA
On Characteristics						
$V_{GE(th)}$	G-E Threshold Voltage	$I_C = 60mA$, $V_{CE} = V_{GE}$	4.0	5.0	7.0	V
$V_{CE(sat)}$	Collector to Emitter Saturation Voltage	$I_C = 10A$, $V_{GE} = 15V$	--	1.5	1.8	V
		$I_C = 60A$, $V_{GE} = 15V$	--	2.5	2.9	V
Dynamic Characteristics						
C_{ies}	Input Capacitance	$V_{CE}=10V$, $V_{GE} = 0V$, $f = 1MHz$	--	6000	--	pF
C_{oes}	Output Capacitance		--	260	--	pF
C_{res}	Reverse Transfer Capacitance		--	200	--	pF
Switching Characteristics						
$t_{d(on)}$	Turn-On Delay Time	$V_{CC} = 600 V$, $I_C = 60A$, $R_G = 51\Omega$, $V_{GE}=15V$, Resistive Load, $T_C = 25°C$	--	140	--	ns
t_r	Rise Time		--	320	--	ns
$t_{d(off)}$	Turn-Off Delay Time		--	630	--	ns
t_f	Fall Time		--	130	250	ns
Q_g	Total Gate Charge	$V_{CE} = 600 V$, $I_C = 60A$, $V_{GE} = 15V$., $T_C = 25°C$	--	275	350	nC
Q_{ge}	Gate-Emitter Charge		--	45	--	nC
Q_{gc}	Gate-Collector Charge		--	95	--	nC

Electrical Characteristics of DIODE $T_C = 25°C$ unless otherwise noted

Symbol	Parameter	Test Conditions	Min.	Typ.	Max.	Units
V_{FM}	Diode Forward Voltage	$I_F = 15A$	--	1.2	1.7	V
		$I_F = 60A$	--	1.8	2.1	V
t_{rr}	Diode Reverse Recovery Time	$I_F = 60A$ di/dt = 20 A/us		1.2	1.5	us
I_R	Instantaneous Reverse Current	$V_{RRM} = 1000V$	--	0.05	2	uA

Figure 15-33   Photo-SCR.

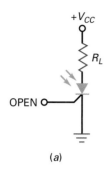

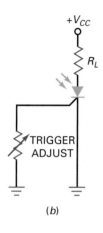

(a)

(b)

15-7 Other Thyristors

SCRs, triacs, and IGBTs are important thyristors. But there are others worth looking at briefly. Some of these thyristors, like the photo-SCR, are still used in special applications. Others, like the UJT, were popular at one time but have been mostly replaced by op amps and timer ICs.

Photo-SCR

Figure 15-33a shows a *photo-SCR,* also known as a *light-activated SCR.* The arrows represent incoming light that passes through a window and hits the depletion layers. When the light is strong enough, valence electrons are dislodged from their orbits and become free electrons. The flow of free electrons starts the positive feedback, and the photo-SCR closes.

After a light trigger has closed the photo-SCR, it remains closed, even though the light disappears. For maximum sensitivity to light, the gate is left open, as shown in Fig. 15-33a. To get an adjustable trip point, we can include the trigger adjust shown in Fig. 15-33b. The resistance between the gate and ground diverts some of the light-produced electrons and reduces the sensitivity of the circuit to the incoming light.

Gate-Controlled Switch

As mentioned earlier, low-current drop-out is the normal way to open an SCR. But the *gate-controlled switch* is designed for easy opening with a reverse-biased trigger. A gate-controlled switch is closed by a positive trigger and opened by a negative trigger.

Figure 15-34 shows a gate-controlled circuit. Each positive trigger closes the gate-controlled switch, and each negative trigger opens it. Because of this, we get the square-wave output shown. The gate-controlled switch has been used in counters, digital circuits, and other applications in which a negative trigger is available.

Silicon Controlled Switch

Figure 15-35a shows the doped regions of a *silicon controlled switch.* Now an external lead is connected to each doped region. Visualize the device separated into two halves (Fig. 15-35b). Therefore, it's equivalent to a latch with access to both bases (Fig. 15-35c). A forward-bias trigger on either base will close the silicon controlled switch. Likewise, a reverse-bias trigger on either base will open the device.

Figure 15-34   Gate-controlled switch.

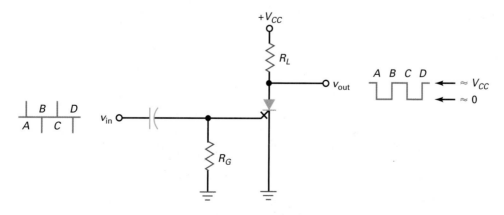

Figure 15-35 Silicon controlled switch.

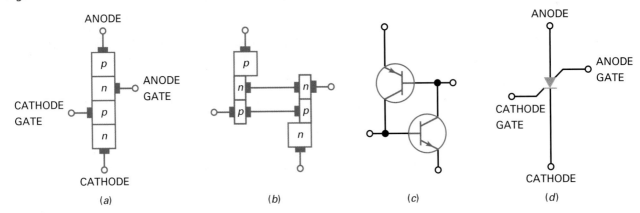

Figure 15-35d shows the schematic symbol for a silicon controlled switch. The lower gate is called the *cathode gate,* and the upper gate is the *anode gate.* The silicon controlled switch is a low-power device compared to the SCR. It handles currents in milliamperes rather than amperes.

Unijunction Transistor and *PUT*

The **unijunction transistor (UJT)** has two doped regions, as shown in Fig. 15-36a. When the input voltage is zero, the device is nonconducting. When we increase the input voltage above the *standoff voltage* (given on a data sheet), the resistance between the *p* region and the lower *n* region becomes very small, as shown in Fig. 15-36b. Figure 15-36c is the schematic symbol for a UJT.

The UJT can be used to form a pulse-generating circuit called a UJT relaxation oscillator, as shown in Fig. 15-37. In this circuit, the capacitor charges toward V_{BB}. When the capacitor voltage reaches a value equal to the standoff voltage, the UJT turns on. The internal lower base (lower *n* region) resistance quickly drops in value allowing the capacitor to discharge. The capacitor discharge continues until low-current dropout occurs. When this happens, the UJT turns off and the capacitor begins to once again charge toward V_{BB}. The charging *RC* time constant is normally significantly larger than the discharge time constant.

The sharp pulse waveform developed across the external resistor at B_1 can be used as a trigger source for controlling the conduction angle of SCR and triac circuits. The waveform developed across the capacitor can be used in applications where a sawtooth generator is needed.

The **programmable unijunction transistor (PUT)** is a four-layer *pnpn* device, which is used to produce trigger pulses and waveforms similar to UJT circuits. The schematic symbol is shown in Fig. 15-38a.

Figure 15-36 Unijunction transistor.

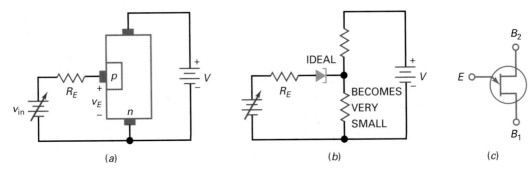

Figure 15-37 UJT relaxation oscillator.

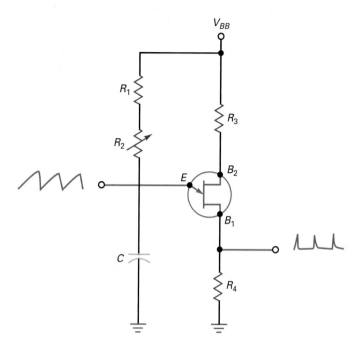

Figure 15-38 PUT: (*a*) symbol; (*b*) structure; (*c*) PUT circuit.

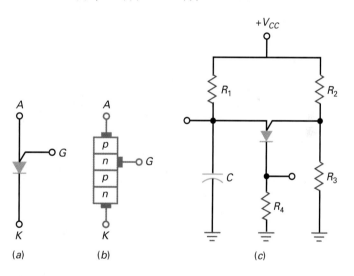

Its basic construction, shown in Fig. 15-38*b*, is very different from a UJT, more closely resembling an SCR. The gate lead is connected to the *n* layer next to the anode. This *pn* junction is used to control the on and off states of the device. The cathode terminal is connected to a voltage point lower than the gate, typically at a ground point. When the anode voltage becomes approximately 0.7 V higher than the gate voltage, the PUT turns on. The device will remain in the on state until its anode current falls below the rated holding current, normally given as its valley current, I_V. When this happens, the device returns to its off state.

The PUT is considered to be programmable because the gate voltage can be determined by an external voltage divider. This is shown in Fig. 15-38*c*. The external resistors, R_2 and R_3, establish the gate voltage, V_G. By changing these resistor values, the voltage on the gate can be modified or "programmed," thus changing the required anode voltage for firing. When the capacitor charges up through R_1, it must reach a voltage value approximately 0.7 V higher than V_G. At that point, the PUT turns and the capacitor discharges. As with the UJT, sawtooth and trigger-pulse waveforms can be developed for controlling thyristors.

UJTs and PUTs were popular at one time for building oscillators, timers, and other circuits. But, as mentioned earlier, op amps and timer ICs (such as the 555), along with microcontrollers, have replaced these devices in many of their applications.

15-8 Troubleshooting

When you troubleshoot a circuit to find faulty resistors, diodes, transistors, and so on, you are troubleshooting at the *component level*. The troubleshooting problems of earlier chapters gave you troubleshooting practice at the component level. Troubleshooting at this level is an excellent foundation for troubleshooting at higher levels because it teaches you how to think logically, using Ohm's law as your guide.

Now, we want to practice troubleshooting at the *system level*. This means thinking in terms of *functional blocks*, which are the smaller jobs being done by the different parts of the overall circuit. To get the idea of this higher level of troubleshooting, look at the troubleshooting section at the end of this chapter (Fig. 15-48).

Here you see a block diagram of a power supply with an SCR crowbar. The power supply has been drawn in terms of its functional blocks. If you measure the voltages at the different points, you can often isolate the trouble to a particular block. Then you can continue troubleshooting at the component level, if necessary.

Often, a manufacturer's instruction manual includes block diagrams of the equipment in which the function of each block is specified. For instance, a television receiver can be drawn in terms of its functional blocks. Once you know what the input and output signals of each block are supposed to be, you can troubleshoot the television receiver to isolate the defective block. After you isolate the defective block, you can either replace the entire block or continue troubleshooting at the component level.

Summary

SEC. 15-1 THE FOUR-LAYER DIODE

A thyristor is a semiconductor device that uses internal positive feedback to produce latching action. The four-layer diode, also called a Schockley diode, is the simplest thyristor. Breakover closes it, and low-current drop-out opens it.

SEC. 15-2 THE SILICON CONTROLLED RECTIFIER

The silicon controlled rectifier (SCR) is the most widely used thyristor. It can switch very large currents on and off. To turn it on, we need to apply a minimum gate trigger voltage and current. To turn it off,

we need to reduce the anode voltage to almost zero.

SEC. 15-3 THE SCR CROWBAR

One important application of the SCR is to protect delicate and expensive loads against supply overvoltages. With an SCR crowbar, a fuse or current-limiting circuit is needed to prevent excessive current from damaging the power supply.

SEC. 15–4   SCR PHASE CONTROL

An *RC* circuit can vary the lag angle of gate voltage from 0 to 90°. This allows us to control the average load current. By using more advanced phase control circuits, we can vary the phase angle from 0 to 180° and have greater control over the average load current.

SEC. 15–5   BIDIRECTIONAL THYRISTORS

The diac can latch current in either direction. It is open until the voltage across it exceeds the breakover voltage. The triac is a gate-controlled device

similar to an SCR. With a phase controller, a triac gives us full-wave control of the average load current.

SEC. 15–6   IGBTs

The IGBT is a hybrid device composed of a power MOSFET on the input side and a BJT on the output side. This combination produces a device with simple input gate drive requirements and low conduction losses on the output. IGBTs have an advantage over power MOSFETs in high-voltage, high-current switching applications.

SEC. 15–7   OTHER THYRISTORS

The photo-SCR latches when the incoming light is strong enough. The

gate-controlled switch is designed to close with a positive trigger and open with a negative trigger. The silicon controlled switch has two input trigger gates, either of which can close or open the device. The unijunction transistor has been used to build oscillators and timing circuits.

SEC. 15–8   TROUBLESHOOTING

When you troubleshoot a circuit to find defective resistors, diodes, transistors, and so on, you are troubleshooting at the component level. When you are troubleshooting to find a defective functional block, you are troubleshooting at the system level.

Derivations

(15-1)   SCR turn-on:

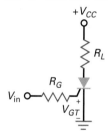

$$V_{in} = V_{GT} + I_{GT}R_G$$

(15-2)   SCR reset:

$$V_{CC} = 0.7\ \text{V} + I_HR_L$$

(15-3)   Overvoltage:

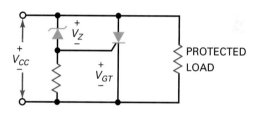

$$V_{CC} = V_Z + V_{GT}$$

(15-4)   RC phase control impedance:

$$Z_T = \sqrt{R^2 + X_C^2}$$

(15-5)   RC phase control angle:

$$\theta_Z = -\arctan\frac{X_C}{R}$$

Student Assignments

1. **A thyristor can be used as**
   a. A resistor   c. A switch
   b. An amplifier   d. A power source

2. **Positive feedback means that the returning signal**
   a. Opposes the original change
   b. Aids the original change
   c. Is equivalent to negative feedback
   d. Is amplified

3. **A latch always uses**
   a. Transistors
   b. Negative feedback
   c. Current
   d. Positive feedback

4. **To turn on a four-layer diode, you need**
   a. A positive trigger
   b. Low-current drop-out

   c. Breakover
   d. Reverse-bias triggering

5. **The minimum input current that can turn on a thyristor is called the**
   a. Holding current
   b. Trigger current
   c. Breakover current
   d. Low-current drop-out

6. The only way to stop a four-layer diode that is conducting is by
   a. A positive trigger
   b. Low-current drop-out
   c. Breakover
   d. Reverse-bias triggering

7. The minimum anode current that keeps a thyristor turned on is called the
   a. Holding current
   b. Trigger current
   c. Breakover current
   d. Low-current drop-out

8. A silicon controlled rectifier has
   a. Two external leads
   b. Three external leads
   c. Four external leads
   d. Three doped regions

9. An SCR is usually turned on by
   a. Breakover
   b. A gate trigger
   c. Breakdown
   d. Holding current

10. SCRs are
    a. Low-power devices
    b. Four-layer diodes
    c. High-current devices
    d. Bidirectional

11. The usual way to protect a load from excessive supply voltage is with a
    a. Crowbar
    b. Zener diode
    c. Four-layer diode
    d. Thyristor

12. An *RC* snubber protects an SCR against
    a. Supply overvoltages
    b. False triggering
    c. Breakover
    d. Crowbarring

13. When a crowbar is used with a power supply, the supply needs to have a fuse or
    a. Adequate trigger current
    b. Holding current
    c. Filtering
    d. Current limiting

14. The photo-SCR responds to
    a. Current
    b. Voltage
    c. Humidity
    d. Light

15. The diac is a
    a. Transistor
    b. Unidirectional device
    c. Three-layer device
    d. Bidirectional device

16. The triac is equivalent to
    a. A four-layer diode
    b. Two diacs in parallel
    c. A thyristor with a gate lead
    d. Two SCRs in parallel

17. The unijunction transistor acts as a
    a. Four-layer diode   c. Triac
    b. Diac   d. Latch

18. Any thyristor can be turned on with
    a. Breakover
    b. Forward-bias triggering
    c. Low-current drop-out
    d. Reverse-bias triggering

19. A Schockley diode is the same as
    a. a four-layer diode   c. a diac
    b. an SCR   d. a triac

20. The trigger voltage of an SCR is closest to
    a. 0   c. 4 V
    b. 0.7 V   d. Breakover voltage

21. Any thyristor can be turned off with
    a. Breakover
    b. Forward-bias triggering
    c. Low-current drop-out
    d. Reverse-bias triggering

22. Exceeding the critical rate of rise produces
    a. Excessive power dissipation
    b. False triggering
    c. Low-current drop-out
    d. Reverse-bias triggering

23. A four-layer diode is sometimes called a
    a. Unijunction transistor
    b. Diac
    c. *pnpn* diode
    d. Switch

24. A latch is based on
    a. Negative feedback
    b. Positive feedback
    c. The four-layer diode
    d. SCR action

25. An SCR can switch to the on state if
    a. Its forward breakover voltage is exceeded
    b. I_{GT} is applied
    c. The critical rate of voltage rise is exceeded
    d. All of the above

26. To properly test an SCR using an ohmmeter
    a. The ohmmeter must supply the SCR's breakover voltage
    b. The ohmmeter cannot supply more than 0.7 V
    c. The ohmmeter must supply the SCR's reverse breakover voltage
    d. The ohmmeter must supply the SCR's holding current

27. The maximum firing angle with a single *RC* phase control circuit is
    a. 45°
    b. 90°
    c. 180°
    d. 360°

28. A triac is generally considered most sensitive in
    a. Quadrant I
    b. Quadrant II
    c. Quadrant III
    d. Quadrant IV

29. An IGBT is essentially a
    a. BJT on the input and MOSFET on the output
    b. MOSFET on the input and MOSFET on the output
    c. MOSFET on the input and BJT on the output
    d. BJT on the input and BJT on the output

30. The maximum on-state output voltage of an IGBT is
    a. $V_{GS(on)}$
    b. $V_{CE(Sat)}$
    c. $R_{DS(on)}$
    d. V_{CES}

31. A PUT is considered programmable by using
    a. External gate resistors
    b. Applying preset cathode voltage levels
    c. An external capacitor
    d. Doped *pn* junctions

Problems

SEC. 15-1 THE FOUR-LAYER DIODE

15-1 The 1N5160 of Fig. 15-39a is conducting. If we allow 0.7 V across the diode at the drop-out point, what is the value of V when the diode opens?

15-2 The capacitor of Fig. 15-39b charges from 0.7 to 12 V, causing the four-layer diode to break over. What is the current through the 5-kΩ resistor just before the diode breaks over? The current through the 5-kΩ resistor when diode is conducting?

15-3 What is the charging time constant in Fig. 15-39b? The period of the sawtooth equals the time constant. What does the frequency equal?

15-4 If the breakover voltage of Fig. 15-39a changes to 20 V and the holding current changes to 3 mA, what is the voltage V that turns on the diode? What is the voltage that turns it off?

15-5 If the supply voltage is changed to 50 V in Fig. 15-39b, what is the maximum voltage across the capacitor? What is the time constant if the resistance is doubled and the capacitance is tripled?

SEC. 15-2 THE SILICON CONTROLLED RECTIFIER

15-6 The SCR of Fig. 15-40 has $V_{GT} = 1.0$ V, $I_{GT} = 2$ mA, and $I_H = 12$ mA. What is the output voltage when the SCR is off? What is the input voltage that triggers the SCR? If V_{CC} is decreased until the SCR opens, what is the value of V_{CC}?

15-7 All resistances are doubled in Fig. 15-40. If the gate trigger current of the SCR is 1.5 mA, what is the input voltage that triggers the SCR?

15-8 What is the peak output voltage in Fig. 15-41 if R is adjusted to 500 Ω?

15-9 If the SCR of Fig. 15-40 has a gate trigger voltage of 1.5 V, a gate trigger current of 15 mA, and a holding current of 10 mA, what is the input voltage that triggers the SCR? The supply voltage that resets the SCR?

15-10 If the resistance is tripled in Fig. 15-40, what is the input voltage that triggers the SCR if $V_{GT} = 2$ V and $I_{GT} = 8$ mA?

Figure 15-39

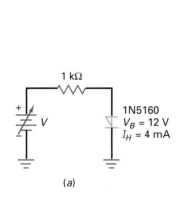

(a)

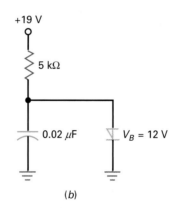

(b)

Figure 15-40

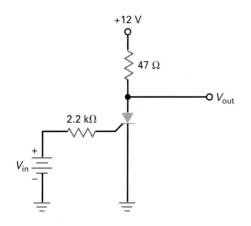

Figure 15-41

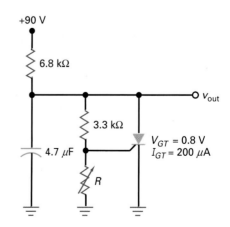

15-11 In Fig. 15-41, R is adjusted to 750 Ω. What is the charging time constant for the capacitor? What is the Thevenin resistance facing the gate?

15-12 The resistor R_2 in Fig. 15-42 is set to 4.6 kΩ. What are the approximate firing and conduction angles for this circuit? How much ac voltage is across C?

15-13 Using Fig. 15-42, when adjusting R_2, what are the minimum and maximum firing angle values?

15-14 What are the minimum and maximum conduction angles of the SCR in Fig. 15-42?

Figure 15-42

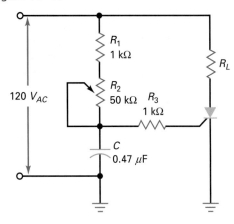

SEC. 15-3 THE SCR CROWBAR

15-15 Calculate the supply voltage that triggers the crowbar of Fig. 15-43.

15-16 If the zener diode of Fig. 15-43 has a tolerance of ± 10 percent and the trigger voltage can be as high as 1.5 V, what is the maximum supply voltage where crowbarring takes place?

15-17 If the zener voltage in Fig. 15-43 is changed from 10 to 12 V, what is the voltage that triggers the SCR?

15-18 The zener diode of Fig. 15-43 is replaced by a 1N759. What is the supply voltage that triggers the SCR crowbar?

Figure 15-43

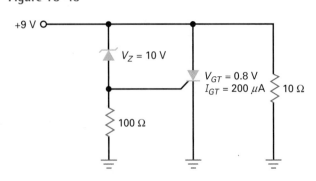

SEC. 15-5 BIDIRECTIONAL THYRISTORS

15-19 The diac of Fig. 15-44 has a breakover voltage of 20 V, and the triac has a V_{GT} of 2.5 V. What is the capacitor voltage that turns on the triac?

15-20 What is the load current in Fig. 15-44 when the triac is conducting?

15-21 All resistances are doubled in Fig. 15-44, and the capacitance is tripled. If the diac has a breakover voltage of 28 V and the triac has a gate trigger voltage of 2.5 V, what is the capacitor voltage that fires the triac?

Figure 15-44

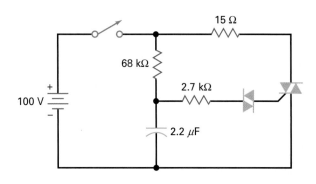

SEC. 15-7 OTHER THYRISTORS

15-22 In Fig. 15-45, what is the anode and gate voltage values when the PUT fires?

15-23 What will be the ideal peak voltage across R_4 in Fig. 15-45, when the PUT fires?

15-24 In Fig. 15-45, what will the voltage waveform across the capacitor look like? What will be the minimum and maximum voltage values of this waveform?

Figure 15-45

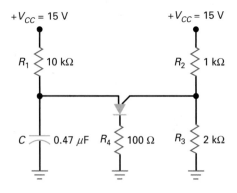

Critical Thinking

15-25 Figure 15-46*a* shows an overvoltage indicator. What is the voltage that turns on the lamp?

15-26 What is the peak output voltage in Fig. 15-46*b*?

15-27 If the period of the sawtooth is 20 percent of the time constant, what is the minimum frequency in Fig. 15-46*b*? What is the maximum frequency?

15-28 The circuit of Fig. 15-47 is in a dark room. What is the output voltage? When a bright light is turned on, the thyristor fires. What is the approximate output voltage? What is the current through the 100 Ω?

Figure 15-46

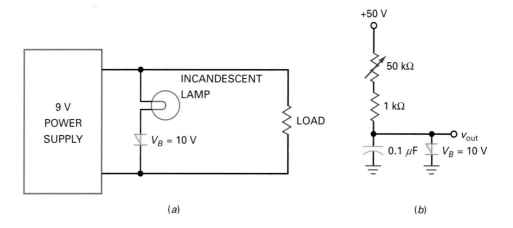

(a)   (b)

Figure 15-47

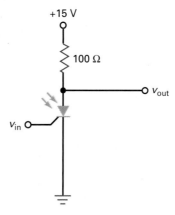

Troubleshooting

Use Fig. 15-48 for the remaining problems. This power supply has a bridge rectifier working into a capacitor-input filter. Therefore, the filtered dc voltage is approximately equal to the peak secondary voltage. All listed values are in volts, unless otherwise indicated. Also, the measured voltages at points *A*, *B*, and *C* are given as rms values. The measured voltages at points *D*, *E*, and *F* are given as dc voltages. In this exercise, you are troubleshooting at the system level; that is, you are to locate the most suspicious block for further testing. For instance, if the voltage is OK at point *B* but incorrect at point *C*, your answer should be *transformer*.

15-29 Find Troubles 1 to 4.

15-30 Find Troubles 4 to 8.

Figure 15-48 Troubleshooting measurements.

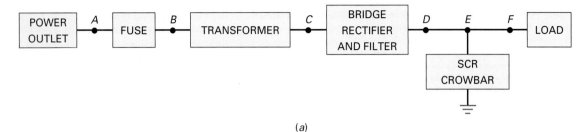

(a)

Troubleshooting

Trouble	V_A	V_B	V_C	V_D	V_E	V_F	R_L	SCR
OK	115	115	12.7	18	18	18	100 Ω	Off
T1	115	115	12.7	18	0	0	100 Ω	Off
T2	0	0	0	0	0	0	100 Ω	Off
T3	115	115	0	0	0	0	100 Ω	Off
T4	115	0	0	0	0	0	0	Off
T5	130	130	14.4	20.5	20.5	20.5	100 Ω	Off
T6	115	115	12.7	0	0	0	100 Ω	Off
T7	115	115	12.7	18	18	0	100 Ω	Off
T8	115	0	0	0	0	0	100 Ω	Off

(b)

Job Interview Questions

1. Draw a two-transistor latch. Then, explain how the positive feedback can drive the transistors into saturation and into cutoff.
2. Draw a basic SCR crowbar. What is the theory of operation behind this circuit? In other words, tell me all the details of how it works.
3. Draw a phase-controlled SCR circuit. Include the waveforms for ac line voltage and gate voltage. Then explain the theory of operation.
4. In thyristor circuits, what is the purpose of snubber networks?
5. How might one employ an SCR in an alarm circuit? Why would this device be preferable to one using a transistor trigger? Draw a simple schematic.
6. Where in the field of electronics would a technician find thyristors in use?
7. Compare a power BJT, a power FET, and an SCR for use in high-power amplification.
8. Explain the differences in operation between the Schockley diode and an SCR.
9. Compare a power MOSFET and an IGBT used for high-power switching.

Self-Test Answers

1.	c	6.	b	11.	a	16.	d	21.	c	26.	d	30.	b
2.	b	7.	a	12.	b	17.	d	22.	b	27.	b	31.	a
3.	d	8.	b	13.	d	18.	a	23.	c	28.	a		
4.	c	9.	b	14.	d	19.	a	24.	b	29.	c		
5.	b	10.	c	15.	d	20.	b	25.	d				

Practice Problem Answers

15-1 $I_D = 113$ mA

15-2 $V_{in} = 1.7$ V

15-3 $F = 250$ kHz

15-4 $V_{in} = 10$ V; $V_{CC} = 2.5$ V

15-6 $V_{CC} = 6.86$ V (worst case)

15-7 $\theta_{firing} = 62°$; $\theta_{conduction} = 118°$

15-8 $I_R = 5.45$ A

15-9 $V_{in} = 25$ V

16

Frequency Effects

Earlier chapters discussed amplifiers operating in their normal frequency range. Now, we want to discuss how an amplifier responds when the input frequency is outside this normal range. With ac amplifiers, the voltage gain decreases when the input frequency is too low or too high. On the other hand, dc amplifiers have voltage gain all the way down to zero frequency. It is only at higher frequencies that the voltage gain of a dc amplifier falls off. We can use decibels to describe the decrease in voltage gain and a Bode plot to graph the response of an amplifier.

Objectives

After studying this chapter, you should be able to:

- Calculate decibel power gain and decibel voltage gain and state the implications of the impedance-matched condition.

- Sketch Bode plots for both magnitude and phase.

- Use Miller's theorem to calculate the equivalent input and output capacitances in a given circuit.

- Describe the risetime-bandwidth relationship.

- Explain how coupling capacitors and emitter-bypass capacitors produce the low-cutoff frequencies in BJT stages.

- Explain how the collector or drain-bypass capacitors and the input Miller capacitance produce the high-cutoff frequencies in BJT and FET stages.

Chapter Outline

Vocabulary

Bode plot
cutoff frequencies
dc amplifier
decibel power gain
decibels
decibel voltage gain
dominant capacitor

feedback capacitor
frequency response
half-power frequencies
internal capacitances
inverting amplifier
lag circuit
logarithmic scale

midband of an amplifier
Miller effect
risetime T_R
stray-wiring capacitance
unity-gain frequency

16-1 Frequency Response of an Amplifier

The **frequency response** of an amplifier is the graph of its gain versus the frequency. In this section, we will discuss the frequency response of ac and dc amplifiers. Earlier, we discussed a CE amplifier with coupling and bypass capacitors. This is an example of an *ac amplifier,* one designed to amplify ac signals. It is also possible to design a *dc amplifier,* one that can amplify dc signals as well as ac signals.

Response of an AC Amplifier

Figure 16-1a shows the *frequency response* of an ac amplifier. In the middle range of frequencies, the voltage gain is maximum. This middle range of frequencies is where the amplifier is normally operated. At low frequencies, the voltage gain decreases because the coupling and bypass capacitors no longer act like short circuits. Instead, their capacitive reactances are large enough to drop some of the ac signal voltage. The result is a loss of voltage gain as we approach zero hertz (0 Hz).

At high frequencies, the voltage gain decreases for other reasons. To begin with, a transistor has **internal capacitances** across its junctions, as shown in Fig. 16-1b. These capacitances provide bypass paths for the ac signal. As the frequency increases, the capacitive reactances become low enough to prevent normal transistor action. The result is a loss of voltage gain.

Stray-wiring capacitance is another reason for a loss of voltage gain at high frequencies. Figure 16-1c illustrates the idea. Any connecting wire in a transistor circuit acts like one plate of a capacitor, and the chassis ground acts like the other plate. The stray-wiring capacitance that exists between this wire and ground is unwanted. At higher frequencies, its low capacitive reactance prevents the ac current from reaching the load resistor. This is equivalent to saying that the voltage gain drops off.

Figure 16-1 (a) Frequency response of ac amplifier; (b) internal capacitance of transistor; (c) connecting wire forms capacitance with chassis.

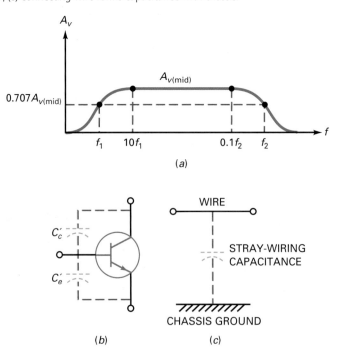

Cutoff Frequencies

The frequencies at which the voltage gain equals 0.707 of its maximum value are called the **cutoff frequencies.** In Fig. 16-1a, f_1 is the lower cutoff frequency and f_2 is the upper cutoff frequency. The cutoff frequencies are also referred to as the **half-power frequencies** because the load power is half of its maximum value at these frequencies.

Why is the output power half of maximum at the cutoff frequencies? When the voltage gain is 0.707 of the maximum value, the output voltage is 0.707 of the maximum value. Recall that power equals the square of voltage divided by resistance. When you square 0.707, you get 0.5. This is why the load power is half of its maximum value at the cutoff frequencies.

Midband

We will define the **midband of an amplifier** as the band of frequencies between $10f_1$ and $0.1f_2$. In the midband, the voltage gain of the amplifier is approximately maximum, designated by $A_{v(\text{mid})}$. Three important characteristics of any ac amplifier are its $A_{v(\text{mid})}$, f_1, and f_2. Given these values, we know how much voltage gain there is in the midband and where the voltage gain is down to $0.707A_{v(\text{mid})}$.

Outside the Midband

Although an amplifier normally operates in the midband, there are times when we want to know what the voltage gain is outside of the midband. Here is an approximation for calculating the voltage gain of an ac amplifier:

$$A_v = \frac{A_{v(\text{mid})}}{\sqrt{1 + (f_1/f)^2}\sqrt{1 + (f/f_2)^2}} \tag{16-1}$$

Given $A_{v(\text{mid})}$, f_1, and f_2, we can calculate the voltage gain at any frequency f. This equation assumes that one dominant capacitor is producing the lower cutoff frequency, and one dominant capacitor is producing the upper cutoff frequency. A **dominant capacitor** is one that is more important than all others in determining the cutoff frequency.

Equation (16-1) is not as formidable as it first appears. There are only three frequency ranges to analyze: the midband, below midband, and above midband. In the midband, $f_1/f \approx 0$ and $f/f_2 \approx 0$. Therefore, both radicals in Eq. (16-1) are approximately 1, and Eq. (16-1) simplifies to:

$$\textbf{Midband: } A_v = A_{v(\text{mid})} \tag{16-2}$$

Below the midband, $f/f_2 \approx 0$. As a result, the second radical equals 1 and Eq. (16-1) simplifies to:

$$\textbf{Below midband: } A_v = \frac{A_{v(\text{mid})}}{\sqrt{1 + (f_1/f)^2}} \tag{16-3}$$

Above midband, $f_1/f \approx 0$. As a result, the first radical equals 1 and Eq. (16-1) simplifies to:

$$\textbf{Above midband: } A_v = \frac{A_{v(\text{mid})}}{\sqrt{1 + (f/f_2)^2}} \tag{16-4}$$

Response of a DC Amplifier

As mentioned in Chap. 12, a designer can use direct coupling between amplifier stages. This allows the circuit to amplify all the way down to zero hertz (0 Hz). This type of amplifier is called a **dc amplifier.**

Figure 16-2a shows the frequency response of a dc amplifier. Since there is no lower cutoff frequency, the two important characteristics of a dc amplifier

Figure 16-2 Frequency response of dc amplifier.

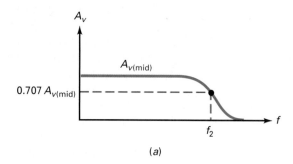

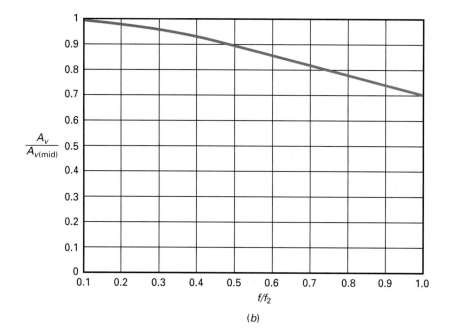

(b)

are $A_{v(mid)}$ and f_2. Given these two values on a data sheet, we have the voltage gain of the amplifier in the midband and its upper cutoff frequency.

The dc amplifier is more widely used than the ac amplifier because most amplifiers are now being designed with op amps instead of with discrete transistors. An *op amp* is a dc amplifier that has high voltage gain, high input impedance, and low output impedance. A wide variety of op amps are commercially available as integrated circuits (ICs).

Most dc amplifiers are designed with one dominant capacitance that produces the cutoff frequency f_2. Because of this, we can use the following formula to calculate the voltage gain of typical dc amplifiers:

$$A_v = \frac{A_{v(mid)}}{\sqrt{1 + (f/f_2)^2}} \tag{16-5}$$

For instance, when $f = 0.1f_2$:

$$A_v = \frac{A_{v(mid)}}{\sqrt{1 + (0.1)^2}} = 0.995\, A_{v(mid)}$$

This says that the voltage gain is within a half percent of maximum when the input frequency is one-tenth of the upper cutoff frequency. In other words, the voltage gain is approximately 100 percent of maximum.

Table 16-1	Between Midband and Cutoff	
f/f_2	$A_v/A_{v\text{(mid)}}$	Percent (approx.)
0.1	0.995	100
0.2	0.981	98
0.3	0.958	96
0.4	0.928	93
0.5	0.894	89
0.6	0.857	86
0.7	0.819	82
0.8	0.781	78
0.9	0.743	74
1	0.707	70

Between Midband and Cutoff

With Eq. (16-5), we can calculate the voltage gain in the region between midband and cutoff. Table 16-1 shows the normalized values of frequency and voltage gain. When $f/f_2 = 0.1$, $A_v/A_{v\text{(mid)}} = 0.995$. When f/f_2 increases, the normalized voltage gain decreases until it reaches 0.707 at the cutoff frequency. As an approximation, we can say that the voltage gain is 100 percent of maximum when $f/f_2 = 0.1$. Then, it decreases to 98 percent, 96 percent, and so on, until it is approximately 70 percent at the cutoff frequency. Figure 16-2b shows the graph of $A_v/A_{v\text{(mid)}}$ versus f/f_2.

Example 16-1

Figure 16-3a shows an ac amplifier with a midband voltage gain of 200. If the cutoff frequencies are $f_1 = 20$ Hz and $f_2 = 20$ kHz, what does the frequency response look like? What is the voltage gain if the input frequency is 5 Hz? If it is 200 kHz?

SOLUTION   In the midband, the voltage gain is 200. At either cutoff frequency, it equals:

$$A_v = 0.707(200) = 141$$

Figure 16-3b shows the frequency response.

With Eq. (16-3), we can calculate the voltage gain for an input frequency of 5 Hz:

$$A_v = \frac{200}{\sqrt{1 + (20/5)^2}} = \frac{200}{\sqrt{1 + (4)^2}} = \frac{200}{\sqrt{17}} = 48.5$$

Figure 16-3 AC amplifier and its frequency response.

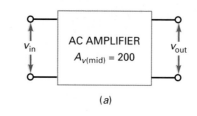

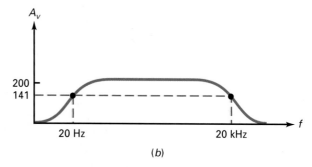

(a)

(b)

In a similar way, we can use Eq. (16-4) to calculate the voltage gain for an input frequency of 200 kHz:

$$A_v = \frac{200}{\sqrt{1 + (200/20)^2}} = 19.9$$

PRACTICE PROBLEM 16–1 Repeat Example 16-1 using an ac amplifier with a midband voltage gain of 100.

Example 16–2

Figure 16-4a shows a 741C, an op amp with a midband voltage gain of 100,000. If $f_2 = 10$ Hz, what does the frequency response look like?

SOLUTION At the cutoff frequency of 10 Hz, the voltage gain is 0.707 of its midband value:

$$A_v = 0.707(100,000) = 70,700$$

Figure 16-4b shows the frequency response. Notice that the voltage gain is 100,000 at a frequency of zero hertz (0 Hz). As the input frequency approaches

Figure 16-4 The 741C and its frequency response.

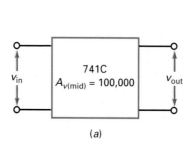

(a)

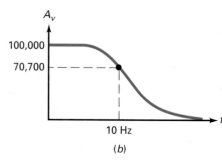

(b)

10 Hz, the voltage gain decreases until it equals approximately 70 percent of maximum.

PRACTICE PROBLEM 16-2   Repeat Example 16-2 with $A_{v(mid)} = 200,000$.

Example 16-3

In the preceding example, what is the voltage gain for each of the following input frequencies: 100 Hz, 1 kHz, 10 kHz, 100 kHz, and 1 MHz?

SOLUTION   Since the cutoff frequency is 10 Hz, an input frequency of:

$$f = 100 \text{ Hz, } 1 \text{ kHz, } 10 \text{ kHz, } \dots$$

gives a ratio f/f_2 of:

$$f/f_2 = 10, 100, 1000, \dots$$

Therefore, we can use Eq. (16-5) as follows to calculate the voltage gains:

$$f = 100 \text{ Hz: } A_v = \frac{100,000}{\sqrt{1 + (10)^2}} \approx 10,000$$

$$f = 1 \text{ kHz: } A_v = \frac{100,000}{\sqrt{1 + (100)^2}} = 1000$$

$$f = 10 \text{ kHz: } A_v = \frac{100,000}{\sqrt{1 + (1000)^2}} = 100$$

$$f = 100 \text{ kHz: } A_v = \frac{100,000}{\sqrt{1 + (10,000)^2}} = 10$$

$$f = 1 \text{ MHz: } A_v = \frac{100,000}{\sqrt{1 + (100,000)^2}} = 1$$

Each time the frequency increases by a *decade* (a factor of 10), the voltage gain decreases by a factor of 10.

PRACTICE PROBLEM 16-3   Repeat Example 16-3 with $A_{v(mid)} = 200,000$.

16-2 Decibel Power Gain

We are about to discuss **decibels,** a useful method for describing frequency response. But before we do, we need to review some ideas from basic mathematics.

Review of Logarithms

Suppose we are given this equation:

$$x = 10^y \tag{16-6}$$

This equation can be solved for y in terms of x to get:

$$y = \log_{10} x$$

This says that y is the logarithm (or exponent) of 10 that gives x. Usually, the 10 is omitted, and the equation is written as:

$$y = \log x \tag{16-7}$$

With a calculator that has the common log function, you can quickly find the y value for any x value. For instance, here is how to calculate the value of y for $x = 10$, 100, and 1000:

$$y = \log 10 = 1$$

$$y = \log 100 = 2$$

$$y = \log 1000 = 3$$

As you can see, each time x increases by a factor of 10, y increases by 1.

You can also calculate y values, given decimal values of x. For instance, here are the values of y for $x = 0.1$, 0.01, and 0.001:

$$y = \log 0.1 = -1$$

$$y = \log 0.01 = -2$$

$$y = \log 0.001 = -3$$

Each time x decreases by a factor of 10, y decreases by 1.

Definition of $A_{p(\text{dB})}$

In Chap. 12, power gain A_p was defined as the output power divided by the input power:

$$A_p = \frac{p_{\text{out}}}{p_{\text{in}}}$$

Decibel power gain is defined as:

$$A_{p(\text{dB})} = 10 \log A_p \tag{16-8}$$

Since A_p is the ratio of output power to input power, A_p has no units or dimensions. When you take the logarithm of A_p, you get a quantity that has no units or dimensions. But to make sure that $A_{p(\text{dB})}$ is never confused with A_p, we attach the unit *decibel* (abbreviated *dB*) to all answers for $A_{p(\text{dB})}$.

For instance, if an amplifier has a power gain of 100, it has a decibel power gain of:

$$A_{p(\text{dB})} = 10 \log 100 = 20 \text{ dB}$$

As another example, if $A_p = 100,000,000$, then:

$$A_{p(\text{dB})} = 10 \log 100,000,000 = 80 \text{ dB}$$

In both of these examples, the log equals the number of zeros: 100 has two zeros, and 100,000,000 has eight zeros. You can use the zero count to find the logarithm whenever the number is a multiple of 10. Then, you can multiply by 10 to get the decibel answer. For instance, a power gain of 1000 has three zeros; multiply by 10 to get 30 dB. A power gain of 100,000 has five zeros; multiply by 10 to get 50 dB. This shortcut is useful for finding decibel equivalents and checking answers.

Decibel power gain is often used on data sheets to specify the power gain of devices. One reason for using decibel power gain is that logarithms compress numbers. For instance, if an amplifier has a power gain that varies from 100 to 100,000,000, the decibel power gain varies from 20 to 80 dB. As you can see, decibel power gain is a more compact notation than ordinary power gain.

Table 16-2	Properties of Power Gain
Factor	**Decibel, dB**
×2	+3
×0.5	−3
×10	+10
×0.1	−10

Two Useful Properties

Decibel power gain has two useful properties:

1. Each time the ordinary power gain increases (decreases) by a factor of 2, the decibel power gain increases (decreases) by 3 dB.
2. Each time the ordinary power gain increases (decreases) by a factor of 10, the decibel power gain increases (decreases) by 10 dB.

Table 16-2 shows these properties in compact form. The following examples will demonstrate these properties.

Example 16–4

Calculate the decibel power gain for the following values: $A_p = 1, 2, 4$, and 8.

SOLUTION With a calculator, we get the following answers:

$$A_{p(\text{dB})} = 10 \log 1 = 0 \text{ dB}$$
$$A_{p(\text{dB})} = 10 \log 2 = 3 \text{ dB}$$
$$A_{p(\text{dB})} = 10 \log 4 = 6 \text{ dB}$$
$$A_{p(\text{dB})} = 10 \log 8 = 9 \text{ dB}$$

Each time A_p increases by a factor of 2, the decibel power gain increases by 3 dB. This property is always true. Whenever you double the power gain, the decibel power gain increases by 3 dB.

PRACTICE PROBLEM 16–4 Find $A_{p(\text{dB})}$ for power gains of 10, 20, and 40.

Example 16–5

Calculate the decibel power gain for each of these values: $A_p = 1, 0.5, 0.25$, and 0.125.

SOLUTION

$$A_{p(\text{dB})} = 10 \log 1 = 0 \text{ dB}$$
$$A_{p(\text{dB})} = 10 \log 0.5 = -3 \text{ dB}$$

$$A_{p(\mathrm{dB})} = 10 \log 0.25 = -6 \text{ dB}$$

$$A_{p(\mathrm{dB})} = 10 \log 0.125 = -9 \text{ dB}$$

Each time A_p decreases by a factor of 2, the decibel power gain decreases by 3 dB.

PRACTICE PROBLEM 16-5 Repeat Example 16-5 for power gains of 4, 2, 1, and 0.5.

Example 16-6

Calculate the decibel power gain for the following values: $A_p = 1, 10, 100,$ and 1000.

SOLUTION

$$A_{p(\mathrm{dB})} = 10 \log 1 = 0 \text{ dB}$$

$$A_{p(\mathrm{dB})} = 10 \log 10 = 10 \text{ dB}$$

$$A_{p(\mathrm{dB})} = 10 \log 100 = 20 \text{ dB}$$

$$A_{p(\mathrm{dB})} = 10 \log 1000 = 30 \text{ dB}$$

Each time A_p increases by a factor of 10, the decibel power gain increases by 10 dB.

PRACTICE PROBLEM 16-6 Calculate the decibel power gain for A_p values of 5, 50, 500, and 5000.

Example 16-7

Calculate the decibel power gain for each of these values: $A_p = 1, 0.1, 0.01,$ and 0.001.

SOLUTION

$$A_{p(\mathrm{dB})} = 10 \log 1 = 0 \text{ dB}$$

$$A_{p(\mathrm{dB})} = 10 \log 0.1 = -10 \text{ dB}$$

$$A_{p(\mathrm{dB})} = 10 \log 0.01 = -20 \text{ dB}$$

$$A_{p(\mathrm{dB})} = 10 \log 0.001 = -30 \text{ dB}$$

Each time the A_p decreases by a factor of 10, the decibel power gain decreases by 10 dB.

PRACTICE PROBLEM 16-7 Calculate the decibel power gain for A_p values of 20, 2, 0.2, 0.02.

16-3 Decibel Voltage Gain

Voltage measurements are more common than power measurements. For this reason, decibels are even more useful with voltage gain.

Definition

As defined in earlier chapters, voltage gain is the output voltage divided by the input voltage:

$$A_v = \frac{v_{\text{out}}}{v_{\text{in}}}$$

Decibel voltage gain is defined as:

$$A_{v(\text{dB})} = 20 \log A_v \qquad\qquad (16\text{-}9)$$

The reason for using 20 instead of 10 in this definition is because power is proportional to the square of voltage. As will be discussed in the next section, this definition produces an important derivation for impedance-matched systems.

If an amplifier has a voltage gain of 100,000, it has a decibel voltage gain of:

$$A_{v(\text{dB})} = 20 \log 100{,}000 = 100 \text{ dB}$$

We can use a shortcut whenever the number is a multiple of 10. Count the number of zeros and multiply by 20 to get the decibel equivalent. In the foregoing calculation, count five zeros and multiply by 20 to get the decibel voltage gain of 100 dB.

As another example, if an amplifier has a voltage gain that varies from 100 to 100,000,000, then its decibel voltage gain varies from 40 to 160 dB.

Basic Rules for Voltage Gain

Here are the useful properties for decibel voltage gain:

1. Each time the voltage gain increases (decreases) by a factor of 2, the decibel voltage gain increases (decreases) by 6 dB.
2. Each time the voltage gain increases (decreases) by a factor of 10, the decibel voltage gain increases (decreases) by 20 dB.

Table 16-3 summarizes these properties.

Table 16–3	Properties of Voltage Gain
Factor	**Decibel, dB**
×2	+6
×0.5	−6
×10	+20
×0.1	−20

Figure 16-5 Two stages of voltage gain.

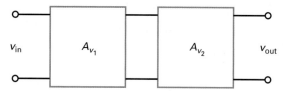

Cascaded Stages

In Fig. 16-5, the total voltage gain of the two-stage amplifier is the product of the individual voltage gains:

$$A_v = (A_{v_1})(A_{v_2}) \tag{16-10}$$

For instance, if the first stage has a voltage gain of 100 and the second stage has a voltage gain of 50, the total voltage gain is:

$$A_v = (100)(50) = 5000$$

Something unusual happens in Eq. (16-10) when we use the decibel voltage gain instead of the ordinary voltage gain:

$$A_{v(dB)} = 20 \log A_v = 20 \log (A_{v_1})(A_{v_2}) = 20 \ \log A_{v_1} + 20 \log A_{v_2}$$

This can be written as:

$$A_{v(dB)} = A_{v_1(dB)} + A_{v_2(dB)} \tag{16-11}$$

This equation says that the total decibel voltage gain of two cascaded stages equals the sum of the individual decibel voltage gains. The same idea applies to any number of stages. This additive property of decibel gain is one reason for its popularity.

Example 16-8

What is the total voltage gain in Fig. 16-6a? Express this in decibels. Next, calculate the decibel voltage gain of each stage and the total decibel voltage gain using Eq. (16-11).

SOLUTION   With Eq. (16-10), the total voltage gain is:

$$A_v = (100)(200) = 20,000$$

In decibels, this is:

$$A_{v(dB)} = 20 \log 20,000 = 86 \text{ dB}$$

You can use a calculator to get 86 dB, or you can use the following shortcut: The number 20,000 is the same as 2 times 10,000. The number 10,000 has four zeros, which means that the decibel equivalent is 80 dB. Because of the factor of 2, the final answer is 6 dB higher, or 86 dB.

Next, we can calculate the decibel voltage gain of each stage as follows:

$$A_{v_1(dB)} = 20 \log 100 = 40 \text{ dB}$$

$$A_{v_2(dB)} = 20 \log 200 = 46 \text{ dB}$$

Figure 16-6 Voltage gains and decibel equivalents.

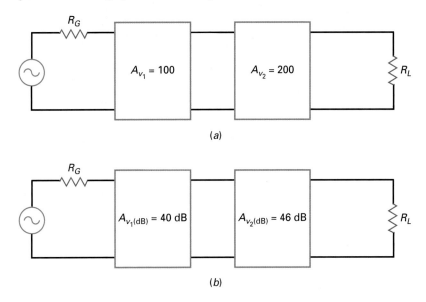

(a)

(b)

Figure 16-6b shows these decibel voltage gains. With Eq. (16-11), the total decibel voltage gain is:

$$A_{v(\text{dB})} = 40 \text{ dB} + 46 \text{ dB} = 86 \text{ dB}$$

As you can see, adding the decibel voltage gain of each stage gives us the same answer calculated earlier.

PRACTICE PROBLEM 16–8   Repeat Example 16-8 with stage voltage gains of 50 and 200.

16–4 Impedance Matching

Figure 16-7a shows an amplifier stage with a generator resistance of R_G, an input resistance of R_{in}, an output resistance of R_{out}, and a load resistance of R_L. Up to now, most of our discussions have used different impedances.

In many communication systems (microwave, television, and telephone), all impedances are matched; that is, $R_G = R_{\text{in}} = R_{\text{out}} = R_L$. Figure 16-7b illustrates the idea. As indicated, all impedances equal R. The impedance R is 50 Ω in microwave systems, 75 Ω (coaxial cable) or 300 Ω (twin-lead) in television systems, and 600 Ω in telephone systems. Impedance matching is used in these systems because it produces maximum power transfer.

In Fig. 16-7b, the input power is:

$$p_{\text{in}} = \frac{V_{\text{in}}^2}{R}$$

and the output power is:

$$p_{\text{out}} = \frac{V_{\text{out}}^2}{R}$$

Figure 16-7 Impedance matching.

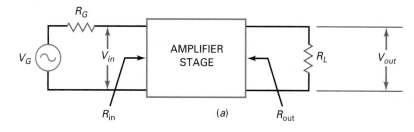

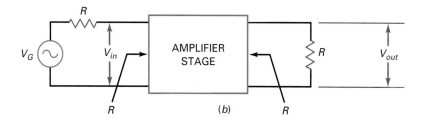

The power gain is:

$$A_p = \frac{p_{out}}{p_{in}} = \frac{V_{out}^2/R}{V_{in}^2/R} = \frac{V_{out}^2}{V_{in}^2} = \left(\frac{V_{out}}{V_{in}}\right)^2$$

or

$$A_p = A_v^2 \qquad\qquad\qquad\qquad \textbf{(16-12)}$$

This says that the power gain equals the square of the voltage gain in any impedance-matched system.

In terms of decibels:

$$A_{p(dB)} = 10 \log A_p = 10 \log A_v^2 = 20 \log A_v$$

or

$$A_{p(\textbf{dB})} = A_{v(\textbf{dB})} \qquad\qquad\qquad\qquad \textbf{(16-13)}$$

This says that the decibel power gain equals the decibel voltage gain. Equation (16-13) is true for any impedance-matched system. If a data sheet states that the gain of a system is 40 dB, then both decibel power gain and voltage gain equal 40 dB.

Converting Decibel to Ordinary Gain

When a data sheet specifies the decibel power gain or voltage gain, you can convert the decibel gain to ordinary gain with the following equations:

$$A_p = \text{antilog } \frac{A_{p(\textbf{dB})}}{10} \qquad\qquad\qquad\qquad \textbf{(16-14)}$$

and

$$A_v = \text{antilog } \frac{A_{v(\textbf{dB})}}{20} \qquad\qquad\qquad\qquad \textbf{(16-15)}$$

The antilog is the inverse logarithm. These conversions are easily done on a scientific calculator that has a log function and an inverse key.

Example 16-9

Figure 16-8 shows impedance-matched stages with $R = 50\ \Omega$. What is the total decibel gain? What is the total power gain? The total voltage gain?

SOLUTION The total decibel voltage gain is:

$$A_{v(dB)} = 23\ \text{dB} + 36\ \text{dB} + 31\ \text{dB} = 90\ \text{dB}$$

The total decibel power gain also equals 90 dB because the stages are impedance-matched.
 With Eq. (16-14), the total power gain is:

$$A_p = \text{antilog}\ \frac{90\ \text{dB}}{10} = 1{,}000{,}000{,}000$$

and the total voltage gain is:

$$A_v = \text{antilog}\ \frac{90\ \text{dB}}{20} = 31{,}623$$

Figure 16-8 Impedance matching in a 50-Ω system.

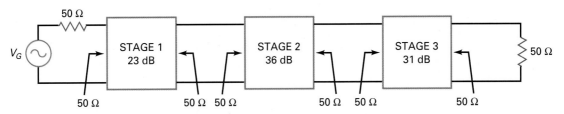

PRACTICE PROBLEM 16-9 Repeat Example 16-9 with stage gains of 10 dB, -6 dB, and 26 dB.

Example 16-10

In the preceding example, what is the ordinary voltage gain of each stage?

SOLUTION The first stage has a voltage gain of:

$$A_{v_1} = \text{antilog}\ \frac{23\ \text{dB}}{20} = 14.1$$

The second stage has a voltage gain of:

$$A_{v_2} = \text{antilog}\ \frac{36\ \text{dB}}{20} = 63.1$$

The third stage has a voltage gain of:

$$A_{v_3} = \text{antilog}\ \frac{31\ \text{dB}}{20} = 35.5$$

PRACTICE PROBLEM 16-10 Repeat Example 16-10 with stage gains of 10 dB, -6 dB, and 26 dB.

16-5 Decibels above a Reference

In this section, we will discuss two more ways to use decibels. Besides applying decibels to power and voltage gains, we can use *decibels above a reference*. The reference levels used in this section are the milliwatt and the volt.

The Milliwatt Reference

Decibels are sometimes used to indicate the power level above 1 mW. In this case, the label *dBm* is used instead of dB. The *m* at the end of dBm reminds us of the milliwatt reference. The dBm equation is:

$$P_{dBm} = 10 \log \frac{P}{1 \text{ mW}} \tag{16-16}$$

where P_{dBm} is the power expressed in dBm. For instance, if the power is 2 W, then:

$$P_{dBm} = 10 \log \frac{2 \text{ W}}{1 \text{ mW}} = 10 \log 2000 = 33 \text{ dBm}$$

Using dBm is a way of comparing the power to 1 mW. If a data sheet says that the output of a power amplifier is 33 dBm, it is saying that the output power is 2 W. Table 16-4 shows some dBm values.

You can convert any dBm value to its equivalent power by using this equation:

$$P = \text{antilog} \frac{P_{dBm}}{10} \tag{16-17}$$

where P is the power in milliwatts.

The Volt Reference

Decibels can also be used to indicate the voltage level above 1 V. In this case, the label *dBV* is used. The dBV equation is:

$$V_{dBV} = 20 \log \frac{V}{1 \text{ V}}$$

Table 16-4	Power in dBm
Power	P_{dBm}
1 μW	−30
10 μW	−20
100 μW	−10
1 mW	0
10 mW	10
100 mW	20
1 W	30

Table 16-5	Voltage in dBV
Voltage	V_{dBV}
10 μV	−100
100 μV	−80
1 mV	−60
10 mV	−40
100 mV	−20
1 V	0
10 V	+20
100 V	+40

Since the denominator equals 1, we can simplify the equation to:

$$V_{dBV} = 20 \log V \tag{16-18}$$

where V is dimensionless. For instance, if the voltage is 25 V, then:

$$V_{dBV} = 20 \log 25 = 28 \text{ dBV}$$

Using dBV is a way of comparing the voltage to 1 V. If a data sheet says that the output of a voltage amplifier is 28 dBV, it is saying that the output voltage is 25 V. If the output level or sensitivity of a microphone is specified as −40 dBV, its output voltage is 10 mV. Table 16-5 shows some dBV values.

You can convert any dBV value to its equivalent voltage using this equation:

$$V = \text{antilog} \frac{V_{dBV}}{20} \tag{16-19}$$

where V is the voltage in volts.

Example 16–11

A data sheet says that the output of an amplifier is 24 dBm. What is the output power?

SOLUTION   With a calculator and Eq. (16-17):

$$P = \text{antilog} \frac{24 \text{ dBm}}{10} = 251 \text{ mW}$$

PRACTICE PROBLEM 16-11   What is the power output of an amplifier rated at 50 dBm?

Example 16-12

If a data sheet says that the output of an amplifier is -34 dBV, what is the output voltage?

SOLUTION With Eq. (16-18):

$$V = \text{antilog } \frac{-34\,\text{dBV}}{20} = 20\,\text{mV}$$

PRACTICE PROBLEM 16-12 Given a microphone rating of -54.5 dBV, what is the output voltage?

16-6 Bode Plots

Figure 16-9 shows the frequency response of an ac amplifier. Although it contains some information such as the midband voltage gain and the cutoff frequencies, it is an incomplete picture of the amplifier's behavior. This is where the **Bode plot** comes in. Because this type of graph uses decibels, it can give us more information about the amplifier's response outside the midband.

Octaves

The middle C on a piano has a frequency of 256 Hz. The next-higher C is an octave higher, and it has a frequency of 512 Hz. The next-higher C has a frequency of 1024 Hz, and so on. In music, the word *octave* refers to a doubling of the frequency. Every time you go up one octave, you have doubled the frequency.

In electronics, an octave has a similar meaning for ratios like f_1/f and f/f_2. For instance, if $f_1 = 100$ Hz and $f = 50$ Hz, the f_1/f ratio is:

$$\frac{f_1}{f} = \frac{100\,\text{Hz}}{50\,\text{Hz}} = 2$$

We can describe this by saying that f is one octave below f_1. As another example, suppose $f = 400$ kHz and $f_2 = 200$ kHz. Then:

$$\frac{f}{f_2} = \frac{400\,\text{kHz}}{200\,\text{kHz}} = 2$$

This means that f is one octave above f_2.

Figure 16-9 Frequency response of an ac amplifier.

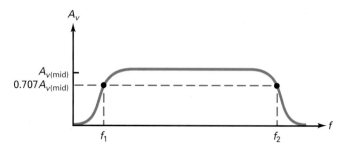

Decades

A *decade* has a similar meaning for ratios like f_1/f and f/f_2, except that a factor of 10 is used instead of a factor of 2. For instance, if $f_1 = 500$ Hz and $f = 50$ Hz, the f_1/f ratio is

$$\frac{f_1}{f} = \frac{500 \text{ Hz}}{50 \text{ Hz}} = 10$$

We can describe this by saying that f is one decade below f_1. As another example, suppose $f = 2$ MHz and $f_2 = 200$ kHz. Then:

$$\frac{f}{f_2} = \frac{2 \text{ MHz}}{200 \text{ kHz}} = 10$$

This means that f is one decade above f_2.

Linear and Logarithmic Scales

Ordinary graph paper has a *linear scale* on both axes. This means that the spaces between the numbers are the same for all numbers, as shown in Fig. 16-10*a*. With a linear scale, you start at 0 and proceed in uniform steps toward higher numbers. All the graphs discussed up to now have used linear scales.

Sometimes we may prefer to use a **logarithmic scale** because it compresses very large values and allows us to see over many decades. Figure 16-10*b* shows a logarithmic scale. Notice that the numbering begins with 1. The space between 1 and 2 is much larger than the space between 9 and 10. By compressing the scale logarithmically as shown here, we can take advantage of certain properties of logarithms and decibels.

Both ordinary graph paper and semilogarithmic paper are available. Semilogarithmic graph paper has a linear scale on the vertical axis and a logarithmic scale on the horizontal axis. People use semilogarithmic paper when they want to graph a quantity like voltage gain over many decades of frequency.

Graph of Decibel Voltage Gain

Figure 16-11*a* shows the frequency response of a typical ac amplifier. The graph is similar to Fig. 16-9, but this time we are looking at the decibel voltage gain versus frequency as it would appear on semilogarithmic paper. A graph like this is called a *Bode plot*. The vertical axis uses a linear scale, and the horizontal axis uses a logarithmic scale.

As shown, the decibel voltage gain is maximum in the midband. At each cutoff frequency, the decibel voltage gain is down slightly from the maximum value. Below f_1, the decibel voltage gain decreases 20 dB per decade. Above f_2, the decibel voltage gain decreases 20 dB per decade. Decreases of 20 dB per decade occur in an amplifier where there is one dominant capacitor producing the lower cutoff frequency and one dominant bypass capacitor producing the upper cutoff frequency, as discussed in Sec. 16-1.

Figure 16–10 Linear and logarithmic scales.

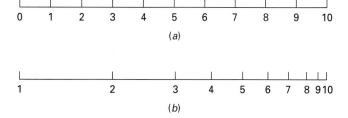

(*a*)

(*b*)

Figure 16-11 (a) Bode plot; (b) ideal Bode plot.

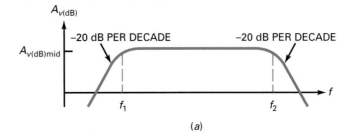

(a)

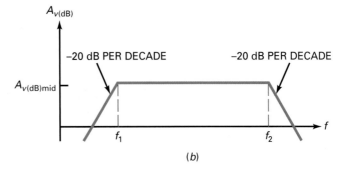

(b)

At the cutoff frequencies, f_1 and f_2, the voltage gain is 0.707 of the midband value. In terms of decibels:

$$A_{v(\text{dB})} = 20 \log 0.707 = -3 \text{ dB}$$

We can describe the frequency response of Fig. 16-11a in this way: In the midband, the voltage gain is maximum. Between the midband and each cutoff frequency, the voltage gain gradually decreases until it is down 3 dB at the cutoff frequency. Then, the voltage gain rolls off (decreases) at a rate of 20 dB per decade.

Ideal Bode Plot

Figure 16-11b shows the frequency response in *ideal* form. Many people prefer using the ideal Bode plot because it is easy to draw and gives approximately the same information. Anyone looking at this ideal graph knows that the decibel voltage gain is down 3 dB at the cutoff frequencies. The ideal Bode plot contains all the original information when this correction of 3 dB is mentally included.

Ideal Bode plots are approximations that allow us to draw the frequency response of an amplifier quickly and easily. They let us concentrate on the main issues rather than being caught in the details of exact calculations. For instance, an ideal Bode plot like Fig. 16-12 gives us a quick visual summary of an amplifier's frequency response. We can see the midband voltage gain (40 dB), the cutoff frequencies (1 kHz and 100 kHz), and roll-off rate (20 dB per decade). Also notice that the voltage gain equals 0 dB (unity or 1) at $f = 10$ Hz and $f = 10$ MHz. Ideal graphs like these are very popular in industry.

Incidentally, many technicians and engineers use the term *corner frequency* instead of *cutoff frequency*. This is because the ideal Bode plot has a sharp corner at each cutoff frequency. Another term often used is *break frequency*. This is because the graph breaks at each cutoff frequency and then decreases at a rate of 20 dB per decade.

Figure 16-12 Ideal Bode plot of an ac amplifier.

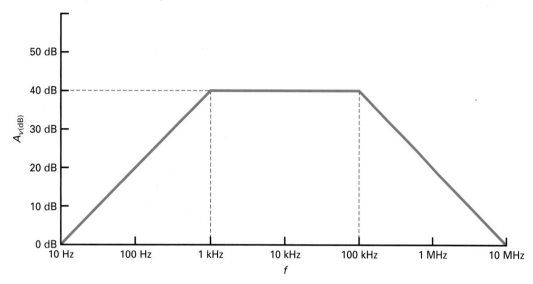

Example 16-13

The data sheet for a 741C op amp gives a midband voltage gain of 100,000, a cutoff frequency of 10 Hz, and roll-off rate of 20 dB per decade. Draw the ideal Bode plot. What is the ordinary voltage gain at 1 MHz?

SOLUTION   As mentioned in Sec. 16-1, op amps are dc amplifiers, so they have only an upper cutoff frequency. For a 741C, $f_2 = 10$ Hz. The midband voltage gain in decibels is:

$$A_{v(dB)} = 20 \log 100{,}000 = 100 \text{ dB}$$

The ideal Bode plot has a midband voltage gain of 100 dB up to 10 Hz. Then, it decreases 20 dB per decade.

Figure 16-13 shows the ideal Bode plot. After breaking at 10 Hz, the response rolls off 20 dB per decade until it equals 0 dB at 1 MHz. The ordinary voltage is unity (1) at this frequency. Data sheets often list the **unity-gain frequency** (symbolized f_{unity}) because it immediately tells you the frequency limitation of the op amp. The device can provide voltage gain up to unity-gain frequency but not beyond it.

Figure 16-13 Ideal Bode plot of a dc amplifier.

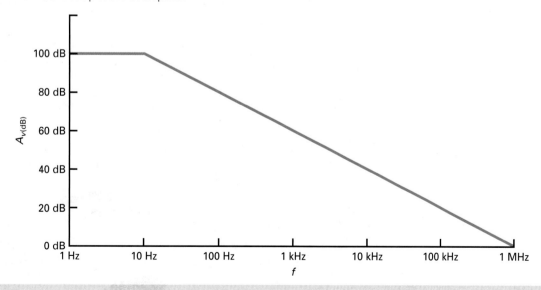

16-7 More Bode Plots

Ideal Bode plots are useful approximations for preliminary analysis. But sometimes, we need more accurate answers. For instance, the voltage gain of an op amp gradually decreases between the midband and the cutoff frequency. Let us look at this transition region more closely.

Between Midband and Cutoff

In Sec. 16-1, we introduced the following equation for the voltage gain of an amplifier above midband:

$$A_v = \frac{A_{v(mid)}}{\sqrt{1 + (f/f_2)^2}} \tag{16-20}$$

With this equation, we can calculate the voltage gain in the transition region between midband and cutoff. For instance, here are the calculations for $f/f_2 = 0.1$, 0.2, and 0.3:

$$A_v = \frac{A_{v(mid)}}{\sqrt{1 + (0.1)^2}} = 0.995\, A_{v(mid)}$$

$$A_v = \frac{A_{v(mid)}}{\sqrt{1 + (0.2)^2}} = 0.981\, A_{v(mid)}$$

$$A_v = \frac{A_{v(mid)}}{\sqrt{1 + (0.3)^2}} = 0.958\, A_{v(mid)}$$

Continuing like this, we can calculate the remaining values shown in Table 16-6.

Table 16-6 includes the dB values for $A_v/A_{v(mid)}$. The decibel entries are calculated as follows:

$$(A_v/A_{v(mid)})_{dB} = 20 \log 0.995 = -0.04 \text{ dB}$$

$$(A_v/A_{v(mid)})_{dB} = 20 \log 0.981 = -0.17 \text{ dB}$$

$$(A_v/A_{v(mid)})_{dB} = 20 \log 0.958 = -0.37 \text{ dB}$$

Table 16-6	Between Midband and Cutoff	
f/f_2	$A_v/A_{v(mid)}$	$A_v/A_{v(mid)dB},$ dB
0.1	0.995	−0.04
0.2	0.981	−0.17
0.3	0.958	−0.37
0.4	0.928	−0.65
0.5	0.894	−0.97
0.6	0.857	−1.3
0.7	0.819	−1.7
0.8	0.781	−2.2
0.9	0.743	−2.6
1	0.707	−3

and so on. We seldom need the values of Table 16-6. But occasionally, we may want to refer to this table for an accurate value of voltage gain in the region between midband and cutoff.

Lag Circuit

Most op amps include an *RC* lag circuit that rolls off the voltage gain at a rate of 20 dB per decade. This prevents *oscillations,* unwanted signals that can appear under certain conditions. Later chapters will explain oscillations and how the internal lag circuit of an op amp prevents these unwanted signals.

Figure 16-14 shows a circuit with bypass capacitor. As discussed in Sec. 9-2, *R* represents the Thevenized resistance facing the capacitor. This circuit is often called a **lag circuit** because the output voltage lags the input voltage at higher frequencies. Stated another way: If the input voltage has a phase angle of 0°, the output voltage has a phase angle between 0° and −90°.

At low frequencies, the capacitive reactance approaches infinity, and the output voltage equals the input voltage. As the frequency increases, the capacitive reactance decreases, which decreases the output voltage. Recall from basic courses in electricity that the output voltage for this circuit is:

$$V_{\text{out}} = \frac{X_C}{\sqrt{R^2 + X_C^2}}\, V_{\text{in}}$$

If we rearrange the foregoing equation, the voltage gain of Fig. 16-14 is:

$$A_v = \frac{X_C}{\sqrt{R^2 + X_C^2}} \tag{16-21}$$

Because the circuit has only passive devices, the voltage gain is always less than or equal to 1.

The cutoff frequency of a lag circuit is where the voltage gain is 0.707. The equation for cutoff frequency is:

$$f_2 = \frac{1}{2\pi RC} \tag{16-22}$$

At this frequency, $X_C = R$ and the voltage gain is 0.707.

Bode Plot of Voltage Gain

By substituting $X_C = 1/2\pi fC$ into Eq. (16-21) and rearranging, we can derive this equation:

$$A_v = \frac{1}{\sqrt{1 + (f/f_2)^2}} \tag{16-23}$$

This equation is similar to Eq. (16-20), where $A_{v(\text{mid})}$ equals 1. For example, when $f/f_2 = 0.1, 0.2,$ and 0.3, we get:

$$A_v = \frac{1}{\sqrt{1 + (0.1)^2}} = 0.995$$

$$A_v = \frac{1}{\sqrt{1 + (0.2)^2}} = 0.981$$

$$A_v = \frac{1}{\sqrt{1 + (0.3)^2}} = 0.958$$

Continuing like this and converting to decibels, we get the values shown in Table 16-7.

Figure 16-14 An *RC* bypass circuit.

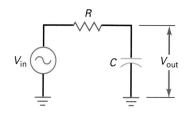

Table 16-7	Response of Lag Circuit	
f/f_2	A_v	$A_{v(dB)}$, dB
0.1	0.995	−0.04
1	0.707	−3
10	0.1	−20
100	0.01	−40
1000	0.001	−60

Figure 16-15 Ideal Bode plot of a lag circuit.

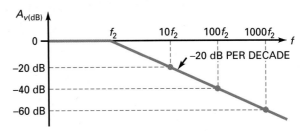

Figure 16-15 shows the ideal Bode plot of a lag circuit. In the midband, the decibel voltage gain is 0 dB. The response breaks at f_2 and then rolls off at a rate of 20 dB per decade.

6 dB per Octave

Above the cutoff frequency, the decibel voltage gain of a lag circuit decreases 20 dB per decade. This is equivalent to 6 dB per octave, which is easily proved as follows: When $f/f_2 = 10$, 20, and 40, the voltage gain is:

$$A_v = \frac{1}{\sqrt{1 + (10)^2}} = 0.1$$

$$A_v = \frac{1}{\sqrt{1 + (20)^2}} = 0.05$$

$$A_v = \frac{1}{\sqrt{1 + (40)^2}} = 0.025$$

The corresponding decibel voltage gains are:

$$A_{v(dB)} = 20 \log 0.1 = -20 \text{ dB}$$

$$A_{v(dB)} = 20 \log 0.05 = -26 \text{ dB}$$

$$A_{v(dB)} = 20 \log 0.025 = -32 \text{ dB}$$

In other words, you can describe the frequency response of a lag circuit above the cutoff frequency in either of two ways: You can say that the decibel voltage gain decreases at a rate of 20 dB per decade, or you can say that it decreases at a rate of 6 dB per octave.

Figure 16-16 Phasor diagram of lag circuit.

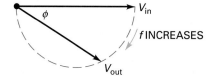

Phase Angle

The charging and discharging of a capacitor produce a lag in the output voltage of an *RC* bypass circuit. In other words, the output voltage will lag the input voltage by a phase angle ϕ. Figure 16-16 shows how ϕ varies with frequency. At zero hertz (0 Hz), the phase angle is 0°. As the frequency increases, the phase angle of the output voltage changes gradually from 0 to −90°. At very high frequencies, $\phi = -90°$.

When necessary, we can calculate the phase angle with this equation from basic courses:

$$\phi = -\arctan \frac{R}{X_C} \qquad (16\text{-}24)$$

By substituting $X_C = 1/2\pi fC$ into Eq. (16-24) and rearranging, we can derive this equation:

$$\phi = -\arctan \frac{f}{f_2} \qquad (16\text{-}25)$$

With a calculator that has the tangent function and an inverse key, we can easily calculate the phase angle for any value of f/f_2. Table 16-8 shows a few values for ϕ. For example, when $f/f_2 = 0.1$, 1, and 10, the phase angles are:

$$\phi = -\arctan 0.1 = -5.71°$$

$$\phi = -\arctan 1 = -45°$$

$$\phi = -\arctan 10 = -84.3°$$

Bode Plot of Phase Angle

Figure 16-17 shows how the phase angle of a lag circuit varies with the frequency. At very low frequencies, the phase angle is zero. When $f = 0.1f_2$, the phase angle is approximately −6°. When $f = f_2$, the phase angle equals −45°. When $f = 10f_2$, the phase angle is approximately −84°. Further increases in frequency produce little change because the limiting value is −90°. As you can see, the phase angle of a lag circuit is between 0 and −90°.

A graph like Fig. 16-17*a* is a Bode plot of the phase angle. Knowing that the phase angle is −6° at $0.1f_2$ and 84° at $10f_2$ is of little value except to

Table 16-8	Response of Lag Circuit
f/f_2	ϕ
0.1	−5.71°
1	−45°
10	−84.3°
100	−89.4°
1000	−89.9°

Figure 16-17   Bode plots of phase angle.

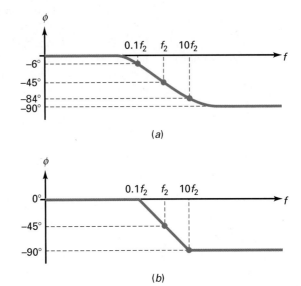

(a)

(b)

indicate how close the phase angle is to its limiting value. The ideal Bode plot of Fig. 16-17b is more useful for preliminary analysis. This is the one to remember because it emphasizes these ideas:

1. When $f = 0.1f_2$, the phase angle is approximately zero.

2. When $f = f_2$, the phase angle is $-45°$.

3. When $f = 10f_2$, the phase angle is approximately $-90°$.

Another way to summarize the Bode plot of the phase angle is this: At the cutoff frequency, the phase angle equals $-45°$. A decade below the cutoff frequency, the phase angle is approximately $0°$. A decade above the cutoff frequency, the phase angle is approximately $-90°$.

Example 16-14   IIII MultiSim

Draw the ideal Bode plot for the lag circuit of Fig. 16-18a.

Figure 16-18   A lag circuit and its Bode plot.

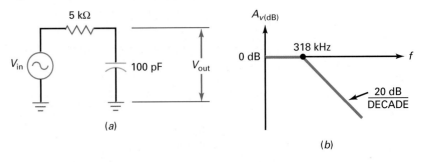

(a)

(b)

SOLUTION   With Eq. (16-22), we can calculate the cutoff frequency:

$$f_2 = \frac{1}{2\pi(5 \text{ k}\Omega)(100 \text{ pF})} = 318 \text{ kHz}$$

Figure 16-18b shows the ideal Bode plot. The voltage gain is 0 dB at low frequencies. The frequency response breaks at 318 kHz and then rolls off at a rate of 20 dB/decade.

PRACTICE PROBLEM 16-14   Using Fig. 16-18, change *R* to 10 kΩ and calculate the cutoff frequency.

Example 16-15

In Fig. 16-19a, the dc amplifier stage has a midband voltage gain of 100. If the Thevenin resistance facing the bypass capacitor is 2 kΩ, what is the ideal Bode plot? Ignore all capacitances inside the amplifier stage.

Figure 16-19   (a) DC amplifier and bypass capacitor; (b) ideal Bode plot; (c) Bode plot with second break frequency.

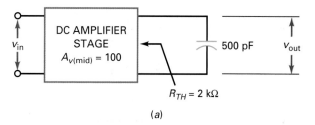

(a)

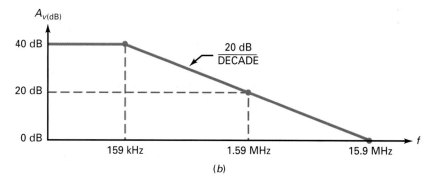

(b)

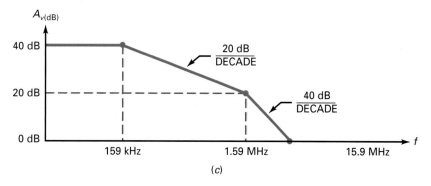

(c)

SOLUTION The Thevenin resistance and the bypass capacitor are a lag circuit with a cutoff frequency of:

$$f_2 = \frac{1}{2\pi(2\ \text{k}\Omega)(500\ \text{pF})} = 159\ \text{kHz}$$

The amplifier has a midband voltage gain of 100, which is equivalent to 40 dB.

Figure 16-19b shows the ideal Bode plot. The decibel voltage gain is 40 dB from zero to the cutoff frequency of 159 kHz. The response then rolls off at a rate of 20 dB per decade until it reaches an f_{unity} of 15.9 MHz.

PRACTICE PROBLEM 16–15 Repeat Example 16-15 using a Thevenin resistance of 1 kΩ.

Example 16–16

Suppose the amplifier stage of Fig. 16-19a has an internal lag circuit with a cutoff frequency of 1.59 MHz. What effect will this have on the ideal Bode plot?

SOLUTION Figure 16-19c shows the frequency response. The response breaks at 159 kHz, the cutoff frequency produced by the external 500-pF capacitor. The voltage gain rolls off at 20 dB per decade until the frequency is 1.59 MHz. At this point, the response breaks again because this is the cutoff frequency of internal lag circuit. The gain then rolls off at a rate of 40 dB per decade.

16-8 The Miller Effect

Figure 16-20a shows an **inverting amplifier** with a voltage gain of A_v. Recall that an inverting amplifier produces an output voltage that is 180° out of phase with the input voltage.

Figure 16–20 (a) Inverting amplifier; (b) Miller effect produces large input capacitor.

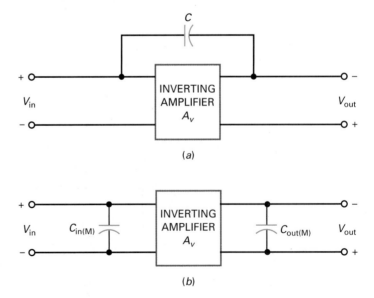

Feedback Capacitor

In Fig. 16-20a, the capacitor between the input and output terminals is called a **feedback capacitor** because the amplified output signal is being fed back to the input. A circuit like this is difficult to analyze because the feedback capacitor affects the input and output circuits simultaneously.

Converting the Feedback Capacitor

Fortunately, there is a shortcut called *Miller's theorem* that converts the capacitor into two separate capacitors, as shown in Fig. 16-20b. This equivalent circuit is easier to work with because the feedback capacitor has been split into two new capacitances, $C_{in(M)}$ and $C_{out(M)}$. With complex algebra, it is possible to derive the following equations:

$$C_{in(M)} = C(A_v + 1) \tag{16-26}$$

$$C_{out(M)} = C\left(\frac{A_v + 1}{A_v}\right) \tag{16-27}$$

Miller's theorem converts the feedback capacitor into two equivalent capacitors, one for the input side and the other for the output side. This makes two simple problems out of one big one. Equations (16-26) and (16-27) are valid for any inverting amplifier such as a CE amplifier, a swamped CE amplifier, or an inverting op amp. In these equations, A_v is the midband voltage gain.

Usually, A_v is much greater than 1, and $C_{out(M)}$ is approximately equal to the feedback capacitance. The striking thing about Miller's theorem is the effect it has on the input capacitance $C_{in(M)}$. It's as though the feedback capacitance has been amplified to get a new capacitance that is $A_v + 1$ times larger. This phenomenon, known as the **Miller effect,** has useful applications because it creates artificial or virtual capacitors that are much larger than the feedback capacitor.

Compensating an Op Amp

As discussed in Sec. 16-7, most op amps are *internally compensated,* which means that they include one dominant bypass capacitor that rolls off the voltage gain at a rate of 20 dB per decade. The Miller effect is used to produce this dominant bypass capacitor.

Figure 16–21   Miller effect produces an input lag circuit.

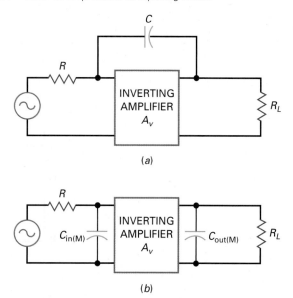

(a)

(b)

Here is the basic idea: One of the amplifier stages in an op amp has a feedback capacitor as shown in Fig. 16-21a. With Miller's theorem, we can convert this feedback capacitor into the two equivalent capacitors shown in Fig. 16-21b. Now, there are two lag circuits, one on the input side and one on the output side. Because of the Miller effect, the bypass capacitor on the input side is much larger than the bypass capacitor on the output side. As a result, the input lag circuit is dominant; that is, it determines the cutoff frequency of the stage. The output bypass capacitor usually has no effect until the input frequency is several decades higher.

In a typical op amp, the input lag circuit of Fig. 16-21b produces a dominant cutoff frequency. The voltage gain breaks at this cutoff frequency and then rolls off at a rate of 20 dB per decade until the input frequency reaches the unity-gain frequency.

Example 16–17

The amplifier of Fig. 16-22a has a voltage gain of 100,000. Draw the ideal Bode plot.

Figure 16–22 Amplifier with feedback capacitor and its Bode plot.

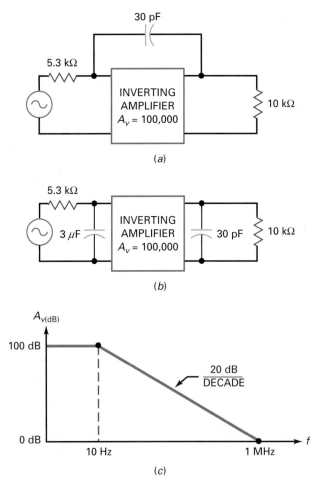

SOLUTION Start by converting the feedback capacitor to its Miller components. Since the voltage gain is much greater than 1:

$$C_{in(M)} = 100,000(30 \text{ pF}) = 3 \text{ } \mu\text{F}$$

$$C_{out(M)} = 30 \text{ pF}$$

Figure 16-22b shows the input and output Miller capacitances. The dominant lag circuit on the input side has a cutoff frequency of:

$$f_2 = \frac{1}{2\pi RC} = \frac{1}{2\pi(5.3 \text{ k}\Omega)(3 \text{ } \mu\text{F})} = 10 \text{ Hz}$$

Since a voltage gain of 100,000 is equivalent to 100 dB, we can draw the ideal Bode plot shown in Fig. 16-22c.

PRACTICE PROBLEM 16–17 Using Fig. 16-22a, determine $C_{in(M)}$ and $C_{out(M)}$ if the voltage gain is 10,000.

16–9 Risetime–Bandwidth Relationship

Sine-wave testing of an amplifier means that we use a sinusoidal input voltage and measure the sinusoidal output voltage. To find the upper cutoff frequency, we have to vary the input frequency until the voltage gain drops 3 dB from the mid-band value. Sine-wave testing is one approach. But there is a faster and simpler way to test an amplifier by using a square wave instead of a sine wave.

Risetime

The capacitor is initially uncharged in Fig. 16-23a. If we close the switch, the capacitor voltage will rise exponentially toward the supply voltage V. The **risetime** T_R is the time it takes the capacitor voltage to go from $0.1V$ (called the *10 percent point*) to $0.9V$ (called the *90 percent point*). If it takes 10 μs for the exponential waveform to go from the 10 percent point to the 90 percent point, the waveform has a risetime of:

$$T_R = 10 \text{ } \mu\text{s}$$

Instead of using a switch to apply the sudden step in voltage, we can use a square-wave generator. For instance, Fig. 16-23b shows the leading edge of a square wave driving the same RC circuit as before. The risetime is still the time it takes for the voltage to go from the 10 percent point to the 90 percent point.

Figure 16-23c shows how several cycles will look. Although the input voltage changes almost instantly from one voltage level to another, the output voltage takes much longer to make its transitions because of the bypass capacitor. The output voltage cannot suddenly step, because the capacitor has to charge and discharge through the resistance.

Relationship between T_R and RC

By analyzing the exponential charge of a capacitor, it is possible to derive this equation for the risetime:

$$T_R = 2.2RC \tag{16-28}$$

Figure 16-23 (*a*) Risetime; (*b*) voltage step produces output exponential; (*c*) square-wave testing.

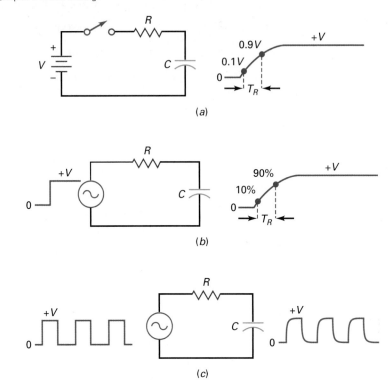

(a)

(b)

(c)

This says that the risetime is slightly more than two RC time constants. For instance, if R equals 10 kΩ and C is 50 pF, then:

$$RC = (10 \text{ k}\Omega)(50 \text{ pF}) = 0.5 \ \mu s$$

The risetime of the output waveform equals:

$$T_R = 2.2RC = 2.2(0.5 \ \mu s) = 1.1 \ \mu s$$

Data sheets often specify the risetime because it is useful to know the response to a voltage step when analyzing switching circuits.

An Important Relationship

As mentioned earlier, a dc amplifier typically has one dominant lag circuit that rolls off the voltage gain at a rate of 20 dB per decade until f_{unity} is reached. The cutoff frequency of this lag circuit is given by:

$$f_2 = \frac{1}{2\pi RC}$$

which can be solved for RC to get:

$$RC = \frac{1}{2\pi f_2}$$

When we substitute this into Eq. (16-28) and simplify, we get this widely used equation:

$$f_2 = \frac{0.35}{T_R} \tag{16-29}$$

This is an important result because it converts risetime to cutoff frequency. It means that we can test an amplifier with a square wave to find the cutoff frequency. Since square-wave testing is much faster than sine-wave testing, many engineers and technicians use Eq. (16-29) to find the upper cutoff frequency of an amplifier.

Equation (16-29) is called the *risetime-bandwidth relationship*. In a dc amplifier, the word *bandwidth* refers to all the frequencies from zero up to the cutoff frequency. Often, bandwidth is used as a synonym for *cutoff frequency*. If the data sheet for a dc amplifier gives a bandwidth of 100 kHz, it means that the upper cutoff frequency equals 100 kHz.

Example 16-18

What is the upper cutoff frequency for the circuit shown in Fig. 16-24a?

SOLUTION In Fig. 16-24a, the risetime is 1 μs. With Eq. (16-29):

$$f_2 = \frac{0.35}{1 \ \mu s} = 350 \ \text{kHz}$$

Therefore, the circuit of Fig. 16-24a has an upper cutoff frequency of 350 kHz. An equivalent statement is that the circuit has a bandwidth of 350 kHz.

Figure 16-24b illustrates the meaning of sine-wave testing. If we change the input voltage from a square wave to a sine wave, we will get a sine-wave output. By increasing the input frequency, we can eventually find the cutoff frequency of 350 kHz. In other words, we would get the same result with sine-wave testing, except that it is slower than square-wave testing.

Figure 16-24 Risetime and cutoff frequency are related.

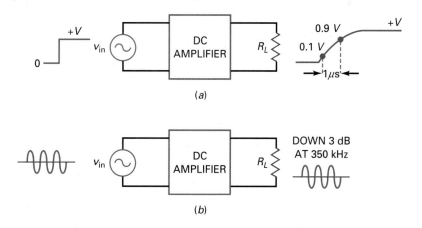

(a)

(b)

PRACTICE PROBLEM 16-18 An *RC* circuit has $R = 2$ kΩ and $C = 100$ pF. Determine the risetime of the output waveform and its upper cutoff frequency.

16-10 Frequency Analysis of BJT Stages

A wide variety of op amps is now commercially available with unity-gain frequencies from 1 to over 200 MHz. Because of this, most amplifiers are now built using op amps. Since op amps are the heart of analog systems, the analysis of discrete amplifier stages is less important than it once was. The next section briefly discusses the low- and high-cutoff frequencies of a voltage-divider biased CE stage. We will look at the effects of individual components on the circuit's frequency response, starting with the low-frequency cutoff point.

Input Coupling Capacitor

When an ac signal is coupled into an amplifier stage, the equivalent looks like Fig. 16-25a. Facing the capacitor is the generator resistance and the input resistance of the stage. This coupling circuit has a cutoff frequency of:

$$f_1 = \frac{1}{2\pi RC} \tag{16-30}$$

where R is the sum of R_G and R_{in}. Figure 16-25b shows the frequency response.

Output Coupling Capacitor

Figure 16-26a shows the output side of a BJT stage. After applying Thevenin's theorem, we get the equivalent circuit of Fig. 16-26b. Equation (16-30) can be used to calculate the cutoff frequency, where R is the sum of R_C and R_L.

Emitter Bypass Capacitor

Figure 16-27a shows a CE amplifier. Figure 16-27b shows the effect that the emitter bypass capacitor has on the output voltage. Facing the emitter bypass capacitor

---

Figure 16-25 Coupling circuit and its frequency response.

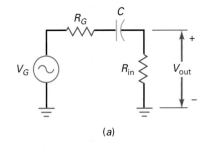

(a)

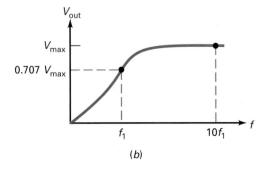

(b)

Figure 16-26 Output coupling capacitor.

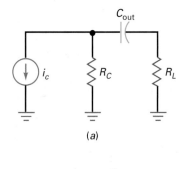

(a)

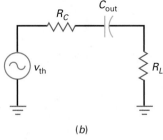

(b)

Figure 16-27 Effect of the emitter bypass capacitor.

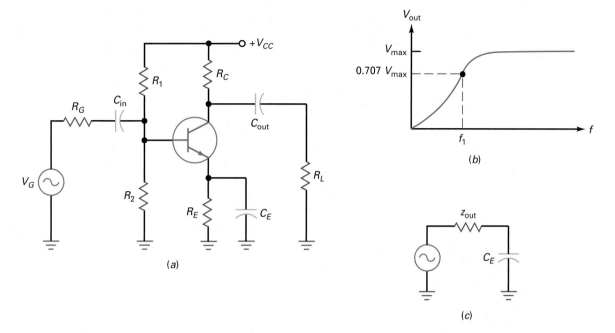

(a)

(b)

(c)

is the Thevenin circuit of Fig. 16-27c. The cutoff frequency is given by:

$$f_1 = \frac{1}{2\pi z_{out}C_E} \tag{16-31}$$

The output impedance z_{out} was discussed in Chap. 11 and is given by Eqs. (11-5) and (11-6).

The input coupling, output coupling, and emitter bypass capacitors each produce a cutoff frequency. Usually, one of these is dominant. When the frequency decreases, the gain breaks at this dominant cutoff frequency. Then, it rolls off at a rate of 20 dB per decade until it breaks again at the next cutoff frequency. It then rolls off at 40 dB per decade until it breaks a third time. With further decreases in frequency, the voltage gain rolls off at 60 dB per decade.

Example 16-19

Using the circuit values shown in Fig. 16-28a, calculate the low-cutoff frequency for each coupling and bypass capacitor. Compare the results to a measurement using a Bode plot. (Use 150 for the dc and ac beta values.)

SOLUTION   In Fig. 16-28a, we will analyze each coupling capacitor and each bypass capacitor separately. When analyzing each capacitor, treat the other two capacitors as ac shorts.

From past dc calculations of this circuit, $r'_e = 22.7\ \Omega$. The Thevenin resistance facing the input coupling capacitor is:

$$R = R_G + R_1 \| R_2 \| R_{in(base)}$$

where

$$R_{in(base)} = (\beta)(r'_e) = (150)(22.7\ \Omega) = 3.41\ k\Omega$$

Therefore,

$$R = 600\ \Omega + (10\ k\Omega \, \| \, 2.2\ k\Omega \, \| \, 3.41\ k\Omega)$$

$$R = 600\ \Omega + 1.18\ k\Omega = 1.78\ k\Omega$$

Figure 16-28 (*a*) CE amplifier using MultiSim; (*b*) low-frequency response; (*c*) high-frequency response.

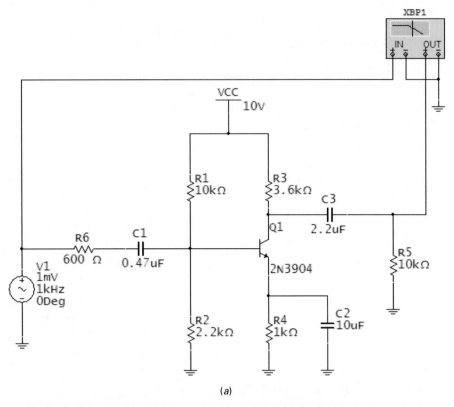

(*a*)

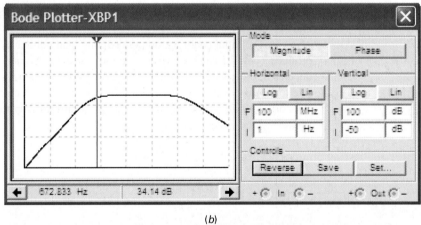

(*b*)

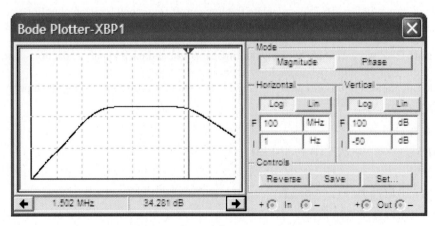

(*c*)

Using Eq. (16-30), the input coupling circuit has a cutoff frequency of:

$$f_1 = \frac{1}{2\pi RC} = \frac{1}{(2\pi)(1.78 \text{ k}\Omega)(0.47 \text{ } \mu\text{F})} = 190 \text{ Hz}$$

Next, the Thevenin resistance facing the output coupling capacitor is:

$$R = R_C + R_L = 3.6 \text{ k}\Omega + 10 \text{ k}\Omega = 13.6 \text{ k}\Omega$$

The output coupling circuit will have a cutoff frequency of:

$$f_1 = \frac{1}{2\pi RC} = \frac{1}{(2\pi)(13.6 \text{ k}\Omega)(2.2 \text{ } \mu\text{F})} = 5.32 \text{ Hz}$$

Now, the Thevenin resistance facing the emitter-bypass capacitor is found by:

$$Z_{\text{out}} = 1 \text{ k}\Omega \| 22.7 \text{ } \Omega + \frac{10 \text{ k}\Omega \| 2.2 \text{ k}\Omega \| 600 \text{ } \Omega}{150}$$

$$Z_{\text{out}} = 1 \text{ k}\Omega \| (22.7 \text{ } \Omega + 3.0 \text{ } \Omega)$$

$$Z_{\text{out}} = 1 \text{ k}\Omega \| 25.7 \text{ } \Omega = 25.1 \text{ } \Omega$$

Therefore, the cutoff frequency for the bypass circuit is:

$$f_1 = \frac{1}{2\pi Z_{\text{out}} C_E} = \frac{1}{(2\pi)(25.1 \text{ } \Omega)(10 \text{ } \mu\text{F})} = 635 \text{ Hz}$$

The results show that:

$f_1 = 190 \text{ Hz}$ input-coupling capacitor

$f_1 = 5.32 \text{ Hz}$ output-coupling capacitor

$f_1 = 635 \text{ Hz}$ emitter-bypass capacitor

As you can see by the results, the emitter-bypass circuit becomes the dominant lower frequency cutoff value.

The measured midpoint voltage gain, $A_{v(\text{mid})}$, in the Bode plot of Fig. 16-28b, is 37.1 dB. The Bode plot shows an approximate 3 dB drop at a frequency of 673 Hz. This is close to our calculation.

PRACTICE PROBLEM 16–19 Using Fig. 16-28a, change the input coupling capacitor to 10 μF and the emitter bypass capacitor to 100 μF. Determine the new dominant cutoff frequency.

Collector Bypass Circuit

The high-frequency response of an amplifier involves a significant amount of detail and requires accurate values to get good results. We will use some detail in our discussion, but more accurate results can be obtained with circuit simulation software.

Figure 16-29a shows a CE stage with stray-wiring capacitance C_{stray}. Just to the left is C_c', a quantity usually specified on the data sheet of a transistor. This is the internal capacitance between the collector and the base. Although C_c' and C_{stray} are very small, they will have an effect when the input frequency is high enough.

Figure 16-29b is the ac equivalent circuit, and Fig. 16-29c is the Thevenin equivalent circuit. The cutoff frequency of this lag circuit is:

$$f_2 = \frac{1}{2\pi RC} \tag{16-32}$$

Figure 16-29 Internal and stray-wiring capacitance produce upper cutoff frequency.

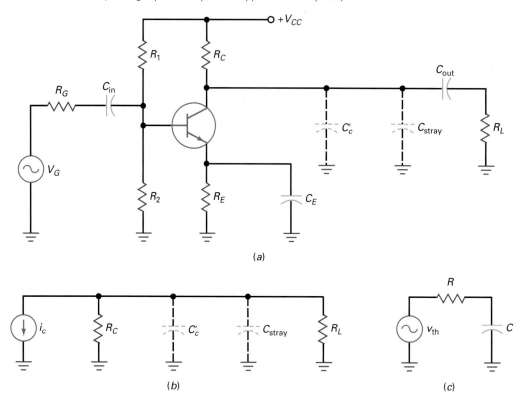

(a)

(b)   (c)

where $R = R_C \| R_L$ and $C = C_c' + C_{stray}$. It is important to keep wires as short as possible in high-frequency work because the stray-wiring capacitance degrades bandwidth by lowering the cutoff frequency.

Base Bypass Circuit

The transistor has two internal capacitances, C_c' and C_e', as shown in Fig. 16-30. Since C_c' is a feedback capacitor, it can be converted into two components. The input Miller component then appears in parallel with C_e'. The cutoff frequency of this base bypass circuit is given by Eq. (16-32), where R is the Thevenin resistance facing the capacitance. The capacitance is the sum of C_e' and the input Miller component.

The collector bypass capacitor and Miller input capacitance each produce a cutoff frequency. Normally, one of these is dominant. When the frequency increases, the gain breaks at this dominant cutoff frequency. Then, it rolls off at a rate

Figure 16-30 High-frequency analysis includes internal transistor capacitances.

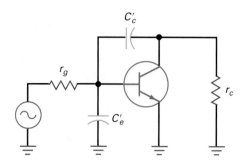

Figure 16–31 The 2N3904 data: (*a*) internal capacitance; (*b*) changes with reverse voltage.

Small signal characteristics

f_T	Current gain—bandwidth product	I_C = 10 mA, V_{CE} = 20 V, f = 100 MHz	300		MHz
C_{obo}	Output capacitance	V_{CB} = 5.0 V, I_E = 0, f = 1.0 MHz		4.0	pF
C_{ibo}	Input capacitance	V_{EB} = 0.5 V, I_C = 0, f = 1.0 MHz		8.0	pF
NF	Noise figure	I_C = 100 μA, V_{CE} = 5.0 V, R_S = 1.0 kΩ, f = 10 Hz to 15.7 kHz		5.0	dB

(*a*)

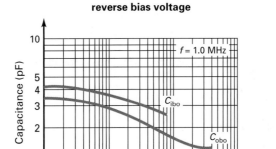

(*b*)

of 20 dB per decade until it breaks again at the second cutoff frequency. With further decreases in frequency, the voltage gain rolls off at 40 dB per decade.

On data sheets, C_C' may be listed as C_{bc}, C_{ob}, or C_{obo}. This value is specified at a particular transistor operating condition. As an example, the C_{obo} value for a 2N3904 is specified as 4.0 pF when V_{CB} = 5.0 V, I_E = 0, and frequency is 1 MHz. C_e' is often listed as C_{be}, C_{ib}, or C_{ibo} on data sheets. The data sheet for a 2N3904 specifies a C_{ibo} value of 4 pF when V_{CB} = 5.0 V, I_E = 0, and frequency is 1 MHz. These values are shown in Fig. 16-31*a* under Small Signal Characteristics.

Each of these internal capacitance values will vary, depending on the circuit condition. Fig. 16-31*b* shows how C_{obo} changes as the amount of reverse bias V_{CB} changes. Also, C_{be} is dependent on the transistor's operating point. When not given on a data sheet, C_{be} can be approximated by:

$$C_{be} \cong \frac{1}{2\pi f_T r_e'} \tag{16-33}$$

where f_T is the current gain-bandwidth product normally listed on the data sheet. The value r_g, shown in Fig. 16-30, is equal to:

$$r_g = R_G \| R_1 \| R_2 \tag{16-34}$$

and r_c is found by:

$$r_c = R_C \| R_L \tag{16-35}$$

Frequency Effects

Example 16–20

Using the circuit values shown in Fig. 16-28a, calculate the high-frequency cutoff values for the base bypass circuit and the collector bypass circuit. Use 150 for beta and 10 pF for the stray output capacitance. Compare the results to a Bode plot using simulation software.

SOLUTION   First determine the values of transistor input and output capacitance.

In our previous dc calculations of this circuit, we determined that $V_B = 1.8$ V and $V_C = 6.04$ V. This results in a collector-to-base reverse voltage of approximately 4.2 V. Using the graph of Fig. 16-31b, the value of C_{obo} or C'_c at this reverse voltage is 2.1 pF. The value of C'_e can be found using Eq. (16-33) as:

$$C'_e = \frac{1}{(2\pi)(300 \text{ MHz})(22.7 \text{ } \Omega)} = 23.4 \text{ pF}$$

Since the voltage gain for this amplifier circuit is:

$$A_v = \frac{r_c}{r'_e} = \frac{2.65 \text{ k}\Omega}{22.7 \text{ } \Omega} = 117$$

The input Miller capacitance is found by:

$$C_{\text{in(M)}} = C'_C (A_v + 1) = 2.1 \text{ pF } (117 + 1) = 248 \text{ pF}$$

Therefore, the base bypass capacitance equals:

$$C = C'_e + C_{\text{in(M)}} = 23.4 \text{ pF} + 248 \text{ pF} = 271 \text{ pF}$$

The resistance value facing this capacitance is:

$$R = r_g \, \| \, R_{\text{in(base)}} = 450 \text{ } \Omega \, \| \, (150)(22.7 \text{ } \Omega) = 397 \text{ } \Omega$$

Now, using Eq. (16-32), the base bypass circuit cutoff frequency is:

$$f_2 = \frac{1}{(2\pi)(397 \text{ } \Omega)(271 \text{ pF})} = 1.48 \text{ MHz}$$

The collector bypass circuit cutoff frequency is found by first determining the total output bypass capacitance:

$$C = C'_C + C_{\text{stray}}$$

Using Eq. (16-27), the output Miller capacitance is found by:

$$C_{\text{out(M)}} = C_C \left(\frac{A_v + 1}{A_v} \right) = 2.1 \text{ pF} \left(\frac{117 + 1}{117} \right) \cong 2.1 \text{ pF}$$

The total output bypass capacitance is:

$$C = 2.1 \text{ pF} + 10 \text{ pF} = 12.1 \text{ pF}$$

The resistance facing this capacitance is:

$$R = R_C \, \| \, R_L = 3.6 \text{ k}\Omega \, \| \, 10 \text{ k}\Omega = 2.65 \text{ k}\Omega$$

Therefore, the collector bypass circuit cutoff frequency is:

$$f_2 = \frac{1}{(2\pi)(2.65 \text{ k}\Omega)(12.1 \text{ pF})} = 4.96 \text{ MHz}$$

The dominant cutoff frequency is determined by the lower of the two cutoff frequencies. In Fig. 16-28a, the Bode plot using MultiSim shows a high-frequency cutoff of approximately 1.5 MHz.

PRACTICE PROBLEM 16-20 If the stray capacitance in Example 16-20 is 40 pF, determine the collector bypass cutoff frequency.

16-11 Frequency Analysis of FET Stages

The frequency response analysis of FET circuits is very similar to that of BJT circuits. In most cases, the FET will have an input coupling circuit and an output coupling circuit, one of which will determine the low-frequency cutoff point. The gate and drain will have an unwanted bypass circuit mainly as the result of the FET's internal capacitances. Along with stray wiring capacitance, this will determine the high-frequency cutoff point.

Low-Frequency Analysis

Fig. 16-32 shows an E-MOSFET common-source amplifier circuit using voltage-divider bias. Because of the very high input resistance of the MOSFET, the resistance R facing the input coupling capacitor is:

$$R = R_G + R_1 \| R_2 \qquad\qquad (16\text{-}36)$$

and the input coupling cutoff frequency is found by:

$$f_1 = \frac{1}{2\pi RC}$$

The output resistance facing the output coupling capacitor is:

$$R = R_D + R_L$$

and the output coupling cutoff frequency is found by:

$$f_1 = \frac{1}{2\pi RC}$$

As you can see, the low-frequency analysis of the FET circuit is very similar to the BJT circuit. Because of the very high input resistance of the FET, larger voltage-divider-resistor values can be used. This results in being able to use a much smaller input-coupling capacitor.

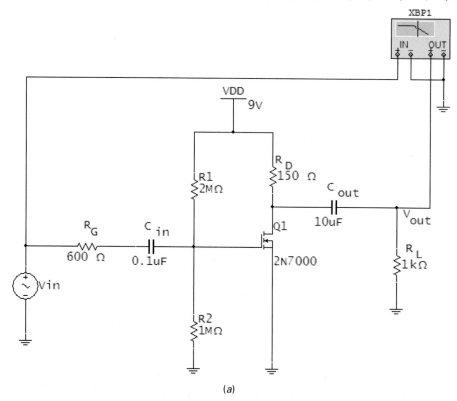

(a)

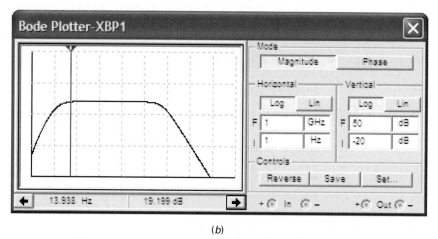

(b)

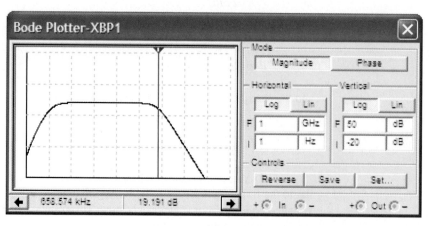

(c)

Example 16-21

Using the circuit shown in Fig. 16-32, determine the input-coupling circuit and output-coupling circuit low-frequency cutoff points. Compare the calculated values to a Bode plot using MultiSim.

SOLUTION The Thevenin resistance facing the input-coupling capacitor is:

$$R = 600\ \Omega + 2\ \text{M}\Omega \| 1\ \text{M}\Omega = 667\ \text{k}\Omega$$

and the input-coupling cutoff frequency is:

$$f_1 = \frac{1}{(2\pi)(667\ \text{k}\Omega)(0.1\ \mu\text{F})} = 2.39\ \text{Hz}$$

Next, the Thevenin resistance facing the output coupling capacitor is found by:

$$R = 150\ \Omega + 1\ \text{k}\Omega = 1.15\ \text{k}\Omega$$

and the output coupling cutoff frequency is:

$$f_1 = \frac{1}{(2\pi)(1.15\ \text{k}\Omega)(10\ \mu\text{F})} = 13.8\ \text{Hz}$$

Therefore, the dominant low-frequency cutoff value is 13.8 Hz. The midpoint voltage gain of this circuit is 22.2 dB. The Bode plot in 16-32b shows a 3 dB loss at approximately 14 Hz. This is very close to the calculated value.

High–Frequency Analysis

Like the high-frequency analysis of a BJT circuit, determining the high-frequency cutoff point of a FET involves a significant amount of detail and requires the use of accurate values. As with the BJT, FETs have internal capacitances C_{gs}, C_{gd}, and C_{ds} as shown in Fig. 16-33a. These capacitance values are not important at low frequencies, but become significant at high frequencies.

Because these capacitances are difficult to measure, manufacturers measure and list the FET capacitances under short-circuit conditions. For example, C_{iss} is the input capacitance with an ac short across the output. When doing this, C_{gd} becomes in parallel with C_{gs} (Fig. 16-33b) so C_{iss} is found by:

$$C_{iss} = C_{ds} + C_{gd}$$

Figure 16-33 Measuring FET capacitances.

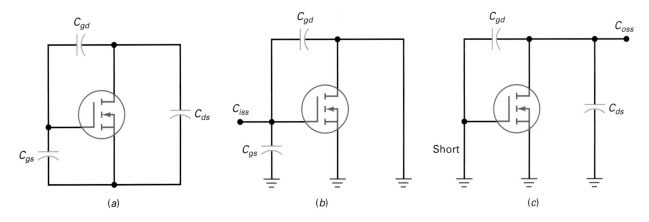

(a)　　　　　(b)　　　　　(c)

Data sheets often list C_{oss}, the capacitance looking back into the FET with a short across the input terminals (Fig. 16-33c), as:

$$C_{oss} = C_{ds} + C_{gd}$$

Data sheets also commonly list the feedback capacitance C_{rss}. The feedback capacitance is equal to:

$$C_{rss} = C_{gd}$$

By using these equations, we can determine that:

$$C_{gd} = C_{rss} \tag{16-37}$$

$$C_{gs} = C_{iss} - C_{rss} \tag{16-38}$$

$$C_{ds} = C_{oss} - C_{rss} \tag{16-39}$$

The gate-to-drain capacitance C_{gd} is used to determine in input Miller capacitance $C_{in(M)}$ and the output Miller capacitance $C_{out(M)}$. These values are found by:

$$C_{in(M)} = C_{gd}(A_v + 1) \tag{16-40}$$

and

$$C_{out(M)} = C_{gd}\left(\frac{A_v + 1}{A_v}\right) \tag{16-41}$$

where $A_v = g_m r_d$ for the common-source amplifier.

Example 16–22 |||| MultiSim

In the MOSFET amplifier circuit of Fig. 16-32, the 2N7000 has these capacitances given on a data sheet:

$$C_{iss} = 60 \text{ pF}$$

$$C_{oss} = 25 \text{ pF}$$

$$C_{rss} = 5.0 \text{ pF}$$

If $g_m = 97$ mS, what are the high-frequency cutoff values for the gate and drain circuits? Compare the calculations to a Bode plot.

SOLUTION   Using the given data sheet capacitance values, we can determine the FET's internal capacitances by:

$$C_{gd} = C_{rss} = 5.0 \text{ pF}$$

$$C_{gs} = C_{iss} - C_{rss} = 60 \text{ pF} - 5 \text{ pF} = 55 \text{ pF}$$

$$C_{ds} = C_{oss} - C_{rss} = 25 \text{ pF} - 5 \text{ pF} = 20 \text{ pF}$$

To determine the Miller input capacitance, we must first find the voltage gain of the amplifier. This is found by:

$$A_v = g_m r_d = (93 \text{ mS})(150 \text{ } \Omega \parallel 1 \text{ k}\Omega) = 12.1$$

Therefore, $C_{in(M)}$ is:

$$C_{in(M)} = C_{gd}(A_v + 1) = 5.0 \text{ pF}(12.1 + 1) = 65.5 \text{ pF}$$

The gate bypass capacitance is found by:

$$C = C_{gs} + C_{in(M)} = 55 \text{ pF} + 65.5 \text{ pF} = 120.5 \text{ pF}$$

The resistance facing C is:

$$R = R_G \parallel R_1 \parallel R_2 = 600\ \Omega \parallel 2\ \text{M}\Omega \parallel 1\ \text{M}\Omega \cong 600\ \Omega$$

The gate-bypass cutoff frequency is:

$$f_2 = \frac{1}{(2\pi)(600\ \Omega)(120.5\ \text{pF})} = 2.2\ \text{MHz}$$

Next, the drain-bypass capacitance is found by:

$$C = C_{ds} + C_{\text{out(M)}}$$

$$C = 20\ \text{pF} + 5.0\ \text{pF}\left(\frac{12.1 + 1}{12.1}\right) = 25.4\ \text{pF}$$

The resistance r_d facing this capacitance is:

$$r_d = R_D \parallel R_L = 150\ \Omega \parallel 1\ \text{k}\Omega = 130\ \Omega$$

The drain-bypass cutoff frequency is therefore:

$$f_2 = \frac{1}{(2\pi)(130\ \Omega)(25.4\ \text{pF})} = 48\ \text{MHz}$$

As shown in Fig. 16-32c, the high-frequency cutoff frequency measured using MultiSim is approximately 638 kHz. As you can see, this measurement differs significantly from our calculations. This somewhat inaccurate result demonstrates the difficulty of choosing the correct internal capacitance values of the device, which are critical to the calculations.

PRACTICE PROBLEM 16–22 Given that $C_{iss} = 25$ pF, $C_{oss} = 10$ pF, and $C_{rss} = 5$ pF, determine the values of C_{gd}, C_{gs}, and C_{ds}.

Summary Table 16-1 shows some of the equations used for frequency analysis of a common-emitter BJT amplifier stage and a common-source FET amplifier stage.

Conclusion

We have examined some of the issues involved in the frequency analysis of discrete BJT and FET amplifier stages. If done manually, the analysis can be tedious and time-consuming. The discussion was deliberately kept somewhat brief here because the frequency analysis of discrete amplifiers is now mainly done on a computer. We hope, you can see how some of the individual components shape the frequency response.

If you need to analyze a discrete amplifier stage, use MultiSim or an equivalent circuit simulator. MultiSim loads all the parameters of the BJT or FET, quantities like C'_C, C'_e, C_{rss}, and C_{oss}, as well as the midband quantities like β, r'_e, and g_m. In other words, MultiSim contains built-in data sheets of devices. For instance, when you select a 2N3904, MultiSim will load all the parameters (including at high frequencies) for a 2N3904. This is a tremendous time saver.

Furthermore, you can use the Bode plotter in MultiSim to see the frequency response. With a Bode plotter, you can measure the midband voltage gain and the cutoff frequencies. In short, using MultiSim or other circuit simulation software is the fastest and most accurate way to analyze the frequency response of a discrete BJT or FET amplifier.

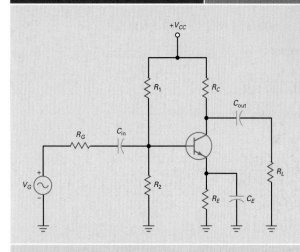 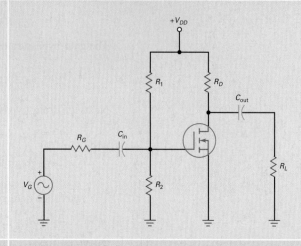

Low-frequency analysis

Base input:
$R = R_G + R_1 \parallel R_2 \parallel R_{in(base)}$

$$f_1 = \frac{1}{2\pi(R)(C_{in})}$$

Collector output:
$R = R_C + R_L$

$$f_1 = \frac{1}{2\pi(R)(C_{out})}$$

Emitter bypass:

$$Z_{out} = R_E \parallel r'_e + \frac{R_1 \parallel R_2 \parallel R_G}{\beta}$$

$$f_1 = \frac{1}{2\pi(R)(C_E)}$$

High-frequency analysis

Base bypass:
$R = R_G \parallel R_1 \parallel R_2 \parallel R_{in(base)}$

$C_{in(M)} = C'_C (A_v + 1)$

$C = C'_e + C_{in(M)}$

$$f_2 = \frac{1}{2\pi(R)(C)}$$

Collector bypass:
$R = R_C \parallel R_L$

$C_{out(M)} = C'_C \left(\dfrac{A_v + 1}{A_v} \right)$

$C = C_{out(M)} + C_{stray}$

$$f_2 = \frac{1}{2\pi(R)(C)}$$

Low-frequency analysis

Gate input:
$R = R_G + R_1 \parallel R_2$

$$f_1 = \frac{1}{2\pi(R)(C_{in})}$$

Drain output
$R = R_D + R_L$

$$f_1 = \frac{1}{2\pi(R)(C_{out})}$$

High-frequency analysis

Gate bypass:
$R = R_G \parallel R_1 \parallel R_2$

$C_{in(M)} = C_{gd} (A_v + 1)$

$C = C_{gs} + C_{in(M)}$

$$f_2 = \frac{1}{2\pi(R)(C)}$$

Drain bypass:
$R = R_D \parallel R_L$

$C_{out(M)} = C_{gd} \left(\dfrac{A_v + 1}{A_v} \right)$

$C = C_{ds} + C_{out(M)} + C_{stray}$

$$f_2 = \frac{1}{2\pi(R)(C)}$$

16-12 Frequency Effects of Surface-Mount Circuits

Stray capacitance and inductance become serious considerations for discrete and IC devices that are operating above 100 kHz. With conventional feed-through components, there are three sources of stray effects:

1. The geometry and internal structure of the device.

2. The printed-circuit layout, including the orientation of the devices and the conductive tracks.

3. The external leads on the device.

Using SM components virtually eliminates item 3 from the list, thus increasing the amount of control design engineers have over stray effects among components on a circuit board.

Summary

SEC. 16-1 FREQUENCY RESPONSE OF AN AMPLIFIER

The frequency response is the graph of voltage gain versus input frequency. An ac amplifier has a lower and an upper cutoff frequency. A dc amplifier has only an upper cutoff frequency. Coupling and bypass capacitors produce the lower cutoff frequency. Internal transistor capacitances and stray-wiring capacitances produce the upper cutoff frequency.

SEC. 16-2 DECIBEL POWER GAIN

Decibel power gain is defined as 10 times the common logarithm of the power gain. When the power gain increases by a factor of 2, the decibel power gain increases by 3 dB. When the power gain increases by a factor of 10, the decibel power gain increases by 10 dB.

SEC. 16-3 DECIBEL VOLTAGE GAIN

Decibel voltage gain is defined as 20 times the common logarithm of the voltage gain. When the voltage gain increases by a factor of 2, the decibel voltage gain

increases by 6 dB. When the voltage gain increases by a factor of 10, the decibel voltage gain increases by 20 dB. The total decibel voltage gain of cascaded stages equals the sum of the individual decibel voltage gains.

SEC. 16-4 IMPEDANCE MATCHING

In many systems, all impedances are matched because this produces maximum power transfer. In an impedance-matched system, the decibel power gain and the decibel voltage gain are equal.

SEC. 16-5 DECIBELS ABOVE A REFERENCE

Besides using decibels with power and voltage gains, we can use decibels above a reference. Two popular references are the milliwatt and the volt. Decibels with the 1 milliwatt reference are labeled dBm, and decibels with the 1 volt reference are labeled dBV.

SEC. 16-6 BODE PLOTS

An octave refers to a factor of 2 change of frequency. A decade refers to a factor of 10 change in frequency. A graph of

decibel voltage gain versus frequency is called a Bode plot. Ideal Bode plots are approximations that allow us to draw the frequency response quickly and easily.

SEC. 16-7 MORE BODE PLOTS

In a lag circuit, the voltage gain breaks at the upper cutoff frequency and then rolls off at a rate of 20 dB per decade, equivalent to 6 dB per octave. We can also draw a Bode plot of phase angle versus frequency. With a lag circuit, the phase angle is between 0 and $-90°$.

SEC. 16-8 THE MILLER EFFECT

A feedback capacitor from the output to the input of an inverting amplifier is equivalent to two capacitors. One capacitor is across the input terminals, and the other is across the output terminals. The Miller effect refers to the input capacitance being $A_v + 1$ times the feedback capacitance.

16-9 RISETIME-BANDWIDTH RELATIONSHIP

When a voltage step is used as the input to a dc amplifier, the risetime of the output is the time between the 10 and

90 percent points. The upper cutoff frequency equals 0.35 divided by the risetime. This gives us a quick and easy way to measure the bandwidth of a dc amplifier.

16-10 FREQUENCY ANALYSIS OF BJT STAGES

The input coupling capacitor, output coupling capacitor, and emitter bypass capacitor produce the low cutoff frequencies. The collector bypass capacitor and the input Miller capacitance produce the high cutoff frequencies. Frequency analysis of bipolar and FET stages is typically done with MultiSim or an equivalent circuit simulator.

16-11 FREQUENCY ANALYSIS OF FET STAGES

The input and output coupling capacitors of a FET stage produce the low cutoff frequencies (like a BJT stage). The drain bypass capacitances, along with the gate capacitance and input Miller capacitance, produce the high cutoff frequencies. Frequency analysis of BJT and FET stages are typically done with MultiSim or an equivalent circuit simulator.

Definitions

(16-8) Decibel power gain:

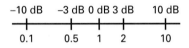

$$A_p(\text{dB}) = 10 \log A_p$$

(16-9) Decibel voltage gain:

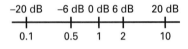

$$A_{V(\text{dB})} = 20 \log A_V$$

(16-16) Decibels referenced to 1 mW:

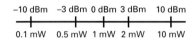

$$P_{\text{dBm}} = 10 \log \frac{P}{1 \text{ mW}}$$

(16-18) Decibels referenced to 1 V:

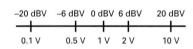

$$V_{\text{dBV}} = 20 \log V$$

Derivations

(16-3) Below the midband:

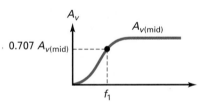

$$A_V = \frac{A_{V(\text{mid})}}{\sqrt{1 + (f_1/f)^2}}$$

(16-4) Above the midband:

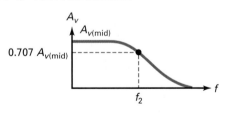

$$A_V = \frac{A_{V(\text{mid})}}{\sqrt{1 + (f/f_2)^2}}$$

(16-10) Total voltage gain:

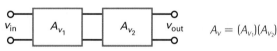

$$A_V = (A_{V_1})(A_{V_2})$$

(16-11) Total decibel voltage gain:

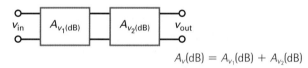

$$A_V(\text{dB}) = A_{V_1}(\text{dB}) + A_{V_2}(\text{dB})$$

(16-13) Impedance-matched system:

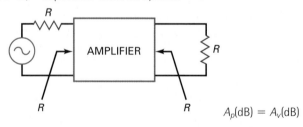

$$A_p(\text{dB}) = A_V(\text{dB})$$

(16-22)   Cutoff frequency:

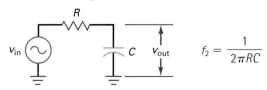

$$f_2 = \frac{1}{2\pi RC}$$

(16-26)   Miller effect: $C_{in(M)} = C(A_v + 1)$

and

(16-27) $$C_{out(M)} = C\left(\frac{A_v + 1}{A_v}\right)$$

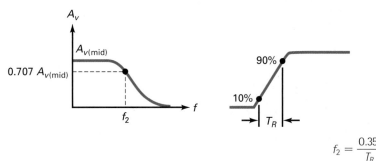

(16-33)   BJT base-emitter capacitance:

$$C_{be} \cong \frac{1}{2\pi f_T r_e'}$$

(16-37)   FET internal capacitance:

$$C_{gd} = C_{rss}$$

(16-38)   FET internal capacitance:

$$C_{gs} = C_{iss} - C_{rss}$$

(16-39)   FET internal capacitance:

$$C_{ds} = C_{oss} - C_{rss}$$

(16-29)   Risetime-bandwidth:

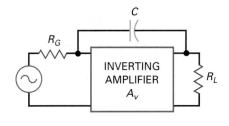

$$f_2 = \frac{0.35}{T_R}$$

Student Assignments

1. **Frequency response is a graph of voltage gain versus**
   a. Frequency
   b. Power gain
   c. Input voltage
   d. Output voltage

2. **At low frequencies, the coupling capacitors produce a decrease in**
   a. Input resistance
   b. Voltage gain
   c. Generator resistance
   d. Generator voltage

3. **The stray-wiring capacitance has an effect on the**
   a. Lower cutoff frequency
   b. Midband voltage gain
   c. Upper cutoff frequency
   d. Input resistance

4. **At the lower or upper cutoff frequency, the voltage gain is**
   a. $0.35A_{v(mid)}$
   b. $0.5A_{v(mid)}$
   c. $0.707A_{v(mid)}$
   d. $0.995A_{v(mid)}$

5. **If the power gain doubles, the decibel power gain increases by**
   a. A factor of 2
   b. 3 dB
   c. 6 dB
   d. 10 dB

6. **If the voltage gain doubles, the decibel voltage gain increases by**
   a. A factor of 2
   b. 3 dB
   c. 6 dB
   d. 10 dB

7. If the voltage gain is 10, the decibel voltage gain is
   a. 6 dB
   b. 20 dB
   c. 40 dB
   d. 60 dB

8. If the voltage gain is 100, the decibel voltage gain is
   a. 6 dB
   b. 20 dB
   c. 40 dB
   d. 60 dB

9. If the voltage gain is 2000, the decibel voltage gain is
   a. 40 dB
   b. 46 dB
   c. 66 dB
   d. 86 dB

10. Two stages have decibel voltage gains of 20 and 40 dB. The total ordinary voltage gain is
    a. 1
    b. 10
    c. 100
    d. 1000

11. Two stages have voltage gains of 100 and 200. The total decibel voltage gain is
    a. 46 dB
    b. 66 dB
    c. 86 dB
    d. 106 dB

12. One frequency is 8 times another frequency. How many octaves apart are the two frequencies?
    a. 1   c. 3
    b. 2   d. 4

13. If $f = 1$ MHz, and $f_2 = 10$ Hz, the ratio f/f_2 represents how many decades?
    a. 2   c. 4
    b. 3   d. 5

14. Semilogarithmic paper means that
    a. One axis is linear, and the other is logarithmic
    b. One axis is linear, and the other is semilogarithmic
    c. Both axes are semilogarithmic
    d. Neither axis is linear

15. If you want to improve the high-frequency response of an amplifier, which of these approaches would you try?
    a. Decrease the coupling capacitances
    b. Increase the emitter bypass capacitance
    c. Shorten leads as much as possible
    d. Increase the generator resistance

16. The voltage gain of an amplifier decreases 20 dB per decade above 20 kHz. If the midband voltage gain is 86 dB, what is the ordinary voltage gain at 20 MHz?
    a. 20   c. 2000
    b. 200   d. 20,000

17. In a BJT amplifier circuit, C'_e is the same as
    a. C_{be}
    b. C_{ib}
    c. C_{ibo}
    d. Any of the above

18. In a BJT amplifier circuit, increasing the value of C_{in} and C_{out} will
    a. Decrease A_v at low frequencies
    b. Increase A_v at low frequencies
    c. Decrease A_v at high frequencies
    d. Increase A_v at high frequencies

19. Input coupling capacitors in FET circuits
    a. Are normally larger than in BJT circuits
    b. Determine the high-frequency cutoff value
    c. Are normally smaller than in BJT circuits
    d. Are treated as ac opens

20. On FET data sheets, C_{oss} is
    a. Equal to $C_{ds} + C_{gd}$
    b. Equal to $C_{gs} - C_{rss}$
    c. Equal to C_{gd}
    d. Equal to $C_{iss} - C_{rss}$

Problems

SEC. 16–1   FREQUENCY RESPONSE OF AN AMPLIFIER

16-1 An amplifier has a midband voltage gain of 1000. If cutoff frequencies are $f_1 = 100$ Hz and $f_2 = 100$ kHz, what does the frequency response look like? What is the voltage gain if the input frequency is 20 Hz? If it is 300 kHz?

16-2 Suppose an op amp has a midband voltage gain of 500,000. If the upper cutoff frequency is 15 Hz, what does the frequency response look like?

16-3 A dc amplifier has a midband voltage gain of 200. If the upper cutoff frequency is 10 kHz, what is the voltage gain for each of these input frequencies: 100 kHz, 200 kHz, 500 kHz, and 1 MHz?

SEC. 16–2   DECIBEL POWER GAIN

16-4 Calculate the decibel power gain for $A_p = 5, 10, 20,$ and 40.

16-5 Calculate the decibel power gain for $A_p = 0.4, 0.2, 0.1,$ and 0.05.

16-6 Calculate the decibel power gain for $A_p = 2, 20, 200,$ and 2000.

16-7 Calculate the decibel power gain for $A_p = 0.4, 0.04,$ and 0.004.

SEC. 16–3   DECIBEL VOLTAGE GAIN

16-8 What is the total voltage gain in Fig. 16-34a? Convert the answer to decibels.

Figure 16-34

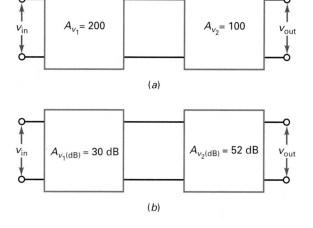

(a)

(b)

16-9 Convert each stage gain in Fig. 16-34a to decibels.

16-10 What is the total decibel voltage gain in Fig. 16-34b? Convert this to ordinary voltage gain.

16-11 What is the ordinary voltage gain of each stage in Fig. 16-34b?

16-12 What is the decibel voltage gain of an amplifier if it has an ordinary voltage gain of 100,000?

16-13 The data sheet of an LM380, an audio power amplifier, gives a decibel voltage gain of 34 dB. Convert this to ordinary voltage gain.

16-14 A two-stage amplifier has these stage gains: $A_{v_1} = 25.8$ and $A_{v_2} = 117$. What is the decibel voltage gain of each stage? The total decibel voltage gain?

SEC. 16-4 IMPEDANCE MATCHING

16-15 If Fig. 16-35 is an impedance-matched system, what is the total decibel voltage gain? The decibel voltage gain of each stage?

16-16 If the stages of Fig. 16-35 are impedance-matched, what is the load voltage? The load power?

SEC. 16-5 DECIBELS ABOVE A REFERENCE

16-17 If the output power of a preamplifier is 20 dBm, how much power is this in milliwatts?

16-18 How much output voltage does a microphone have when its output is −45 dBV?

16-19 Convert the following powers to dBm: 25 mW, 93.5 mW, and 4.87 W.

16-20 Convert the following voltages to dBV: 1 μV, 34.8 mV, 12.9 V, and 345 V.

SEC. 16-6 BODE PLOTS

16-21 The data sheet of an op amp gives a midband voltage gain of 200,000, a cutoff frequency of 10 Hz, and a roll-off rate of 20 dB per decade. Draw the ideal Bode plot. What is the ordinary voltage gain at 1 MHz?

16-22 The LF351 is an op amp with a voltage gain of 316,000, a cutoff frequency of 40 Hz, and a roll-off rate of 20 dB per decade. Draw the ideal Bode plot.

SEC. 16-7 MORE BODE PLOTS

16-23 **IIII MultiSim** Draw the ideal Bode plot for the lag circuit of Fig. 16-36a.

16-24 **IIII MultiSim** Draw the ideal Bode plot for the lag circuit of Fig. 16-36b.

16-25 What is the ideal Bode plot for the stage of Fig. 16-37?

Figure 16-35

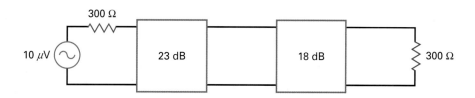

Figure 16-36

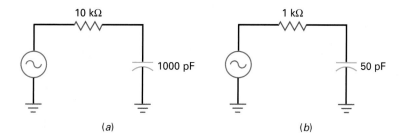

(a)   (b)

Figure 16-37

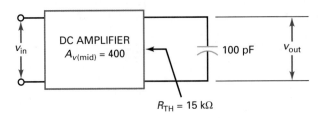

SEC. 16-8 THE MILLER EFFECT

16-26 What is the input Miller capacitance in Fig. 16-38 if $C = 5$ pF and $A_v = 200,000$?

16-27 Draw the ideal Bode plot for the input lag circuit of Fig. 16-38 with $A_v = 250,000$ and $C = 15$ pF.

16-28 If the feedback capacitor of Fig. 16-38 is 50 pF, what is the input Miller capacitance when $A_v = 200,000$?

16-29 Draw the ideal Bode plot for Fig. 16-38 with a feedback capacitance of 100 pF and a voltage gain of 150,000.

SEC. 16-9 RISETIME–BANDWIDTH RELATIONSHIP

16-30 An amplifier has the step response shown in Fig. 16-39*a*. What is its upper cutoff frequency?

16-31 What is the bandwidth of an amplifier if the risetime is 0.25 μs?

16-32 The upper cutoff frequency of an amplifier is 100 kHz. If it is square-wave tested, what would the risetime of the amplifier output be?

16-33 In Fig. 16-40, what is the low-cutoff frequency for the base coupling circuit?

Figure 16-38

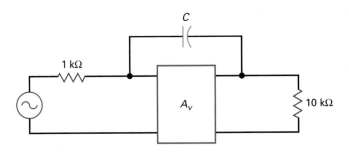

Figure 16-39

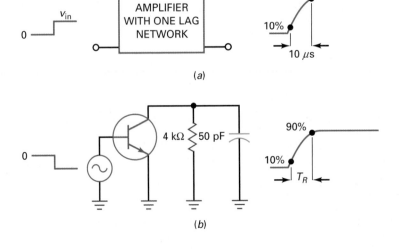

Figure 16-40

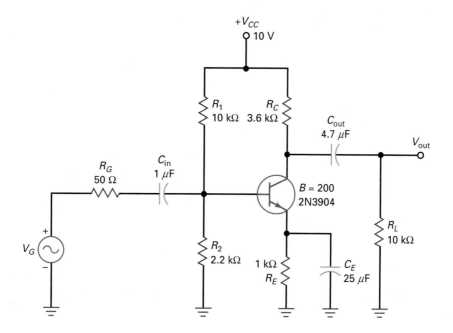

16-34 In Fig. 16-40, what is the low-cutoff frequency for the collector coupling circuit?

16-35 In Fig. 16-40, what is the low-cutoff frequency for the emitter bypass circuit?

16-36 In Fig. 16-40, C'_C is given as 2 pF, C'_e = 10 pF, and C_{stray} is 5 pF. Determine the high-frequency cutoff values for both base-input and collector-output circuits.

16-37 The circuit of Fig. 16-41 uses an E-MOSFET with these specifications: g_m = 16.5 mS, C_{iss} = 30 pF, C_{oss} = 20 pF,

and C_{rss} = 5.0 pF. Determine the FET's internal capacitance values for C_{gd}, C_{gs}, and C_{ds}.

16-38 In Fig. 16-41, what is the dominant low-cutoff frequency?

16-39 In Fig. 16-41, determine the high-frequency cutoff values for both gate input and drain output circuits.

Figure 16-41

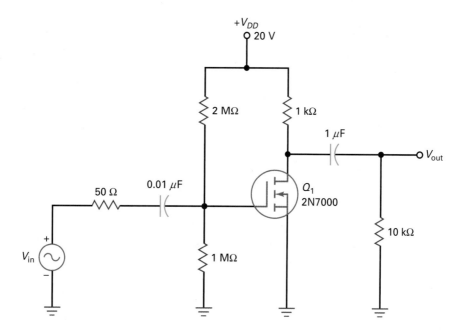

Critical Thinking

16-40 In Fig. 16-42*a*, what is the decibel voltage gain when *f* = 20 kHz? When *f* = 44.4 kHz?

16-41 In Fig. 16-42*b*, what is the decibel voltage gain when *f* = 100 kHz?

16-42 The amplifier of Fig. 16-39*a* has a midband voltage gain of 100. If the input voltage is a step of 20 mV, what is the output voltage at the 10 percent point? The 90 percent point?

16-43 Figure 16-39*b* is an equivalent circuit. What is the risetime of the output voltage?

16-44 You have two data sheets for amplifiers. The first shows a cutoff frequency of 1 MHz. The second gives a risetime of 1 μs. Which amplifier has the greater bandwidth?

Figure 16-42

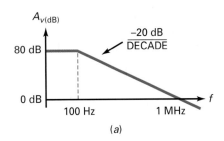

(a)

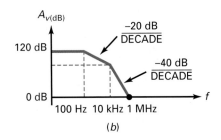

(b)

Job Interview Questions

1. This morning I breadboarded an amplifier stage and used a lot of wire. The upper cutoff frequency tested much lower than it should be. Do you have any suggestions?
2. On my lab bench is a dc amplifier, an oscilloscope, and a function generator that can produce sine, square, or triangular waves. Tell me how to find the bandwidth of the amplifier.
3. Without using your calculator, I want you to convert a voltage gain of 250 to its decibel equivalent.
4. I would like you to draw an inverting amplifier with a feedback capacitor of 50 pF and a voltage gain of 10,000. Next, I want you to draw the ideal Bode plot for the input lag circuit.
5. Assume that the front panel of your oscilloscope notes that its vertical amplifier has a risetime of 7 ns. What does this say about the bandwidth of the instrument?

6. How would you measure the bandwidth of a dc amplifier?
7. Why does the decibel voltage gain use a factor of 20 but the power gain uses a factor of 10?
8. Why is impedance matching important in some systems?
9. What is the difference between dB and dBm?
10. Why is a dc amplifier called a dc amplifier?
11. A radio station engineer needs to test the voltage gain over several decades. What type of graph paper would be most useful in this situation?
12. Have you ever heard of MultiSim (EWB)? If so, what is it?

Self-Test Answers

1.	a	8.	c	15.	c	
2.	b	9.	c	16.	a	
3.	c	10.	d	17.	d	
4.	c	11.	c	18.	b	
5.	b	12.	c	19.	c	
6.	c	13.	d	20.	a	
7.	b	14.	a			

Practice Problem Answers

16-1 $A_{v(mid)} = 70.7$; A_v at 5 Hz = 24.3; A_v at 200 kHz = 9.95

16-2 A_v at 10 Hz = 141

16-3 20,000 at 100 Hz; 2000 at 1 kHz; 200 at 10 kHz; 20 at 100 kHz; 2.0 at 1 MHz

16-4 10 A_p = 10 dB; 20 A_p = 13 dB; 40 A_p = 16 dB

16-5 4 A_p = 6 dB; 2 A_p = 3 dB; 1 A_p = 0 dB; 0.5 A_p = −3 dB

16-6 5 A_p = 7 dB; 50 A_p = 17 dB; 500 A_p = 27 dB; 5000 A_p = 37 dB

16-7 20 A_p = 13 dB; 2 A_p = 3 dB; 0.2 A_p = −7 dB; 0.02 A_p = −17 dB

16-8 50 A_v = 34 dB; 200 A_v = 46 dB; A_{vT} = 10,000; $A_{v(dB)}$ = 80 dB

16-9 $A_{v(dB)}$ = 30 dB; A_p = 1,000; A_v = 31.6

16-10 A_{v1} = 3.16; A_{v2} = 0.5; A_{v3} = 20

16-11 P = 1,000 W

16-12 V_{out} = 1.88 mV

16-14 f_2 = 159 kHz

16-15 f_2 = 318 kHz; f_{unity} = 31.8 MHz

16-17 $C_{in(M)}$ = 0.3 µF; $C_{out(M)}$ = 30 pF

16-18 T_R = 440 ns; f_2 = 795 kHz

16-19 f_1 = 63 Hz

16-20 f_2 = 1.43 MHz

16-22 C_{gd} = 5 pF; C_{gs} = 20 pF; C_{ds} = 5 pF

17

Differential Amplifiers

The term **operational amplifier (op amp)** refers to an amplifier that performs a mathematical operation. Historically, the first op amps were used in analog computers, where they did addition, subtraction, multiplication, and so on. At one time, op amps were built as discrete circuits. Now, most op amps are integrated circuits (ICs).

The typical op amp is a dc amplifier with very high voltage gain, very high input impedance, and very low output impedance. The unity-gain frequency is from 1 to more than 20 MHz, depending on the part number. An IC op amp is a complete functional block with external pins. By connecting these pins to supply voltages and a few components, we can quickly build all kinds of useful circuits.

The input circuit used in most op amps is the differential amplifier. This amplifier configuration establishes many of the IC's input characteristics. The differential amplifier may also be configured in a discrete form to be used in communications, instrumentation, and industrial control circuits. This chapter will focus on the differential amplifier used in ICs.

Objectives

After studying this chapter, you should be able to:

- Perform a dc analysis of a differential amplifier.
- Perform an ac analysis of a differential amplifier.
- Define input bias current, input offset current, and input offset voltage.
- Explain common-mode gain and common-mode rejection ratio.
- Describe how integrated circuits are manufactured.
- Apply Thevenin's theorem to a loaded differential amplifier.

Chapter Outline

Vocabulary

active load resistor

common-mode rejection ratio (CMRR)

common-mode signal

compensating diode

current mirror

differential amplifier (diff amp)

differential input

differential output

hybrid IC

input bias current

input offset current

input offset voltage

integrated circuit (IC)

inverting input

monolithic IC

noninverting input

operational amplifier (op amp)

single-ended

tail current

17-1 The Differential Amplifier

Transistors, diodes, and resistors are the only practical components in typical ICs. Capacitors may also be used, but they are small, usually less than 50 pF. For this reason, IC designers cannot use coupling and bypass capacitors the way a discrete circuit designer can. Instead, the IC designer has to use direct coupling between stages and also needs to eliminate the emitter bypass capacitor without losing too much voltage gain.

The **differential amplifier (diff amp)** is the key. The design of this circuit is extremely clever because it eliminates the need for an emitter bypass capacitor. For this and other reasons, the diff amp is used as the input stage of almost every IC op amp.

Differential Input and Output

Figure 17-1 shows a diff amp. It is two CE stages in parallel with a common emitter resistor. Although it has two input voltages (v_1 and v_2) and two collector voltages (v_{c1} and v_{c2}), the overall circuit is considered to be one stage. Because there are no coupling or bypass capacitors, there is no lower cutoff frequency.

The ac output voltage v_{out} is defined as the voltage between the collectors with the polarity shown in Fig. 17-1:

$$v_{out} = v_{c2} - v_{c1} \tag{17-1}$$

This voltage is called a **differential output** because it combines the two ac collector voltages into one voltage that equals the difference of the collector voltages. *Note:* We will use lowercase letters for v_{out}, v_{c1}, and v_{c2} because they are ac voltages that include zero hertz (0 Hz) as a special case.

Ideally, the circuit has identical transistors and equal collector resistors. With perfect symmetry, v_{out} is zero when the two input voltages are equal. When v_1 is greater than v_2, the output voltage has the polarity shown in Fig. 17-1. When v_2 is greater than v_1, the output voltage is inverted and has the opposite polarity.

The diff amp of Fig. 17-1 has two separate inputs. Input v_1 is called the **noninverting input** because v_{out} is in phase with v_1. On the other hand, v_2 is called the **inverting input** because v_{out} is 180° out of phase with v_2. In some applications, only the noninverting input is used and the inverting input is grounded. In

Figure 17-1 Differential input and differential output.

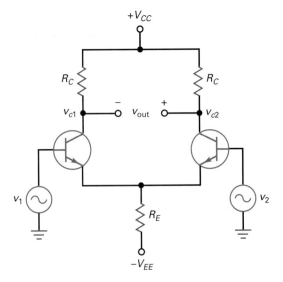

other applications, only the inverting input is active and the noninverting input is grounded.

When both the noninverting and inverting input voltages are present, the total input is called a **differential input** because the output voltage equals the voltage gain times the difference of the two input voltages. The equation for the output voltage is:

$$v_{\text{out}} = A_v(v_1 - v_2) \tag{17-2}$$

where A_v is the voltage gain. We will derive the equation for voltage gain in Sec. 17-3.

Single-Ended Output

A differential output like that of Fig. 17-1 requires a floating load because neither end of the load can be grounded. This is inconvenient in many applications since loads are often **single-ended;** that is, one end is grounded.

Figure 17-2a shows a widely used form of the diff amp. This has many applications because it can drive single-ended loads like CE stages, emitter followers, and other circuits. As you can see, the ac output signal is taken from the collector on the right side. The collector resistor on the left has been removed because it serves no useful purpose.

Because the input is differential, the ac output voltage is still given by $A_v(v_1 - v_2)$. With a single-ended output, however, the voltage gain is half as much as with a differential output. We get half as much voltage gain with a single-ended output because the output is coming from only one of the collectors.

Incidentally, Fig. 17-2b shows the block-diagram symbol for a diff amp with a differential input and a single-ended output. The same symbol is used for

Figure 17-2 (a) Differential input and single-ended output; (b) block diagram symbol.

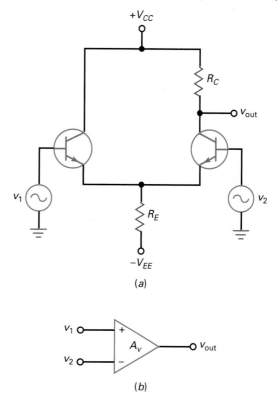

(a)

(b)

Figure 17-3 *(a)* Noninverting input and differential output; *(b)* noninverting input and single-ended output.

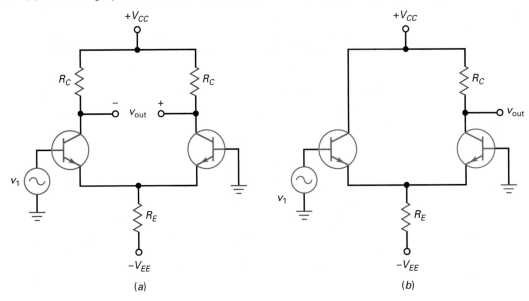

an op amp. The plus sign $(+)$ represents the noninverting input, and the minus sign $(-)$ is the inverting input.

Noninverting–Input Configurations

Often, only one of the inputs is active and the other is grounded, as shown in Fig. 17-3a. This configuration has a noninverting input and a differential output. Since $v_2 = 0$, Eq. (17-2) gives:

$$v_{out} = A_v \, (v_1) \qquad\qquad\qquad (17\text{-}3)$$

Figure 17-3b shows another configuration for the diff amp. This one has a noninverting input and a single-ended output. Since v_{out} is the ac output voltage, Eq. (17-3) is still valid, but the voltage gain A_v will be half as much because the output is taken from only one side of the diff amp.

Inverting–Input Configurations

In some applications, v_2 is the active input and v_1 is the grounded input, as shown in Fig. 17-4a. In this case, Eq. (17-2) simplifies to:

$$v_{out} = -A_v \, (v_2) \qquad\qquad\qquad (17\text{-}4)$$

The minus sign in Eq. (17-4) indicates phase inversion.

Figure 17-4b shows the final configuration that we will discuss. Here we are using the inverting input with a single-ended output. In this case, the ac output voltage is still given by Eq. (17-4).

Conclusion

Table 17-1 summarizes the four basic configurations of a diff amp. The general case has a differential input and differential output. The remaining cases are subsets of the general case. For instance, to get single-ended input operation, one of the inputs is used and the other is grounded. When the input is single-ended, either the noninverting input v_1 or the inverting input v_2 may be used.

Figure 17-4 (*a*) Inverting input and differential output; (*b*) inverting input and single-ended output.

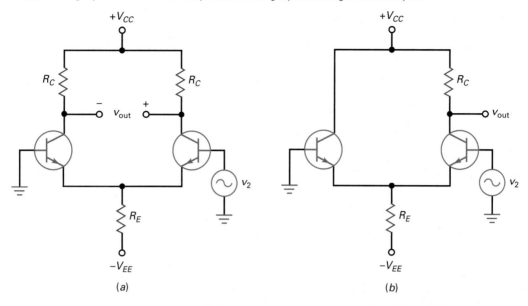

(a)   (b)

Table 17-1	Diff-Amp Configurations		
Input	**Output**	v_{in}	v_{out}
Differential	Differential	$v_1 - v_2$	$v_{c2} - v_{c1}$
Differential	Single-ended	$v_1 - v_2$	v_{c2}
Single-ended	Differential	v_1 or v_2	$v_{c2} - v_{c1}$
Single-ended	Single-ended	v_1 or v_2	v_{c2}

17-2 DC Analysis of a Diff Amp

Figure 17-5*a* shows the dc equivalent circuit for a diff amp. Throughout this discussion, we will assume identical transistors and equal collector resistors. Also, both bases are grounded in this preliminary analysis.

The bias used here should look familiar. It is almost identical to the two-supply emitter bias (TSEB) discussed in Chap. 8. If you recall, most of the negative supply voltage in a TSEB circuit appears across the emitter resistor. This sets up a fixed emitter current.

Ideal Analysis

A diff amp is sometimes called a *long-tail pair* because the two transistors share a common resistor R_E. The current through this common resistor is called the **tail current.** If we ignore the V_{BE} drops across the emitter diodes of Fig. 17-5*a*, then the top of the emitter resistor is ideally a dc ground point. In this case, all of V_{EE} appears across R_E and the tail current is:

$$I_T = \frac{V_{EE}}{R_E} \qquad (17\text{-}5)$$

Figure 17–5 (a) Ideal dc analysis; (b) second approximation.

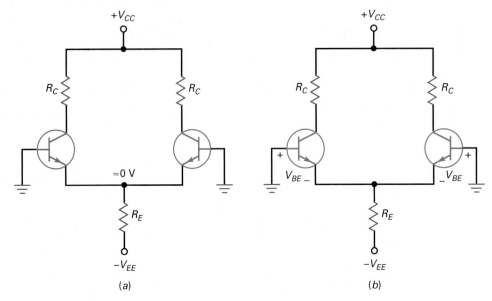

(a)   (b)

This equation is fine for troubleshooting and preliminary analysis because it quickly gets to the point, which is that almost all the emitter supply voltage appears across the emitter resistor.

When the two halves of Fig. 17-5a are perfectly matched, the tail current will split equally. Therefore, each transistor has an emitter current of:

$$I_E = \frac{I_T}{2} \tag{17-6}$$

The dc voltage on either collector is given by this familiar equation:

$$V_C = V_{CC} - I_C R_C \tag{17-7}$$

Second Approximation

We can improve the dc analysis by including the V_{BE} drop across each emitter diode. In Fig. 17-5b, the voltage at the top of the emitter resistor is one V_{BE} drop below ground. Therefore, the tail current is:

$$I_T = \frac{V_{EE} - V_{BE}}{R_E} \tag{17-8}$$

where $V_{BE} = 0.7$ V for silicon transistors.

Effect of Base Resistors on Tail Current

In Fig. 17-5b, both bases are grounded for simplicity. When base resistors are used, they have a negligible effect on the tail current in a well-designed diff amp. Here is the reason: When base resistors are included in the analysis, the equation for tail current becomes:

$$I_T = \frac{V_{EE} - V_{BE}}{R_E + R_B/2\beta_{dc}}$$

In any practical design, $R_B/2\beta_{dc}$ is less than 1 percent of R_E. This is why we prefer using either Eq. (17-5) or Eq. (17-8) to calculate tail current.

Although base resistors have a negligible effect on the tail current, they can produce input error voltages when the two halves of the diff amp are not perfectly symmetrical. We will discuss these input error voltages in a later section.

Example 17-1

What are the ideal currents and voltages in Fig. 17-6a?

SOLUTION With Eq. (17-5), the tail current is:

$$I_T = \frac{15\text{ V}}{7.5\text{ k}\Omega} = 2\text{ mA}$$

Each emitter current is half of the tail current:

$$I_E = \frac{2\text{ mA}}{2} = 1\text{ mA}$$

Figure 17-6 Example.

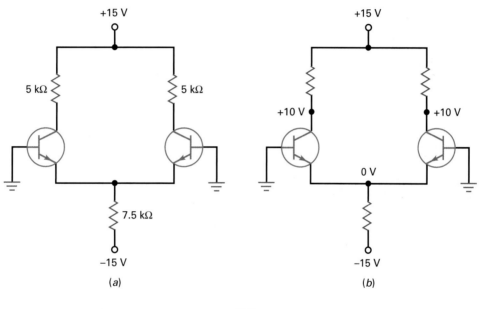

(a)

(b)

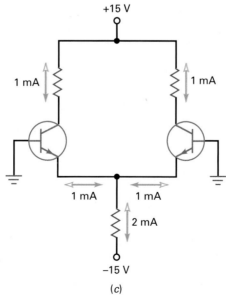

(c)

Each collector has a quiescent voltage of approximately:

$$V_C = 15 \text{ V} - (1 \text{ mA})(5 \text{ k}\Omega) = 10 \text{ V}$$

Figure 17-6b shows the dc voltages, and Fig. 17-6c shows the currents. (*Note:* The standard arrowhead indicates conventional flow, and the triangular arrowhead indicates electron flow.)

PRACTICE PROBLEM 17–1 In Fig. 17-6a, change R_E to 5 kΩ and find the ideal currents and voltages.

Example 17-2

|||| MultiSim

Recalculate the currents and voltages for Fig. 17-6a using the second approximation.

SOLUTION The tail current is:

$$I_T = \frac{15 \text{ V} - 0.7 \text{ V}}{7.5 \text{ k}\Omega} = 1.91 \text{ mA}$$

Each emitter current is half of the tail current:

$$I_E = \frac{1.91 \text{ mA}}{2} = 0.955 \text{ mA}$$

and each collector has a quiescent voltage of:

$$V_C = 15 \text{ V} - (0.955 \text{ mA})(5 \text{ k}\Omega) = 10.2 \text{ V}$$

As you can see, the answers change only slightly when the second approximation is used. In fact, if the same circuit is built and tested with MultiSim (EWB), the following answers result with 2N3904 transistors:

$$I_T = 1.912 \text{ mA}$$

$$I_E = 0.956 \text{ mA}$$

$$I_C = 0.950 \text{ mA}$$

$$V_C = 10.25 \text{ V}$$

These answers are almost the same as the second approximation and not much different from the ideal answers. The point is that ideal analysis is adequate for many situations. If you need more accuracy, use either the second approximation or MultiSim analysis.

PRACTICE PROBLEM 17–2 Repeat Example 17-2 using a 5 kΩ emitter resistor.

Example 17-3

|||| MultiSim

What are the currents and voltages in the single-ended output circuit of Fig. 17-7a?

SOLUTION Ideally, the tail current is:

$$I_T = \frac{12 \text{ V}}{5 \text{ k}\Omega} = 2.4 \text{ mA}$$

Each emitter current is half of the tail current:

$$I_E = \frac{2.4 \text{ mA}}{2} = 1.2 \text{ mA}$$

The collector on the right has a quiescent voltage of approximately:

$$V_C = 12 \text{ V} - (1.2 \text{ mA})(3 \text{ k}\Omega) = 8.4 \text{ V}$$

and the one on the left has 12 V.

Figure 17-7 Example.

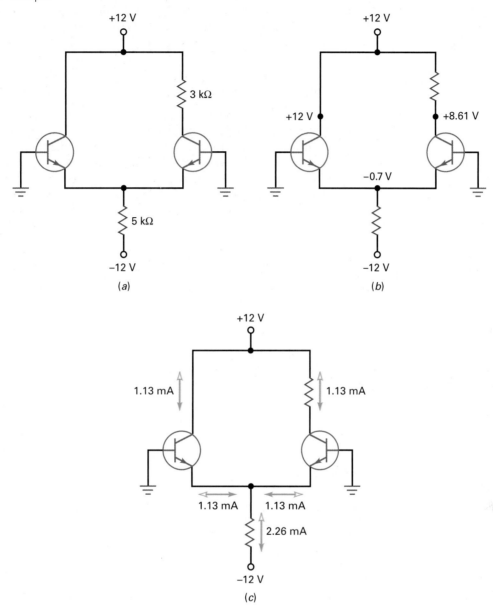

With the second approximation, we can calculate:

$$I_T = \frac{12\ \text{V} - 0.7\ \text{V}}{5\ \text{k}\Omega} = 2.26\ \text{mA}$$

$$I_E = \frac{2.26\ \text{mA}}{2} = 1.13\ \text{mA}$$

$$V_C = 12\ \text{V} - (1.13\ \text{mA})(3\ \text{k}\Omega) = 8.61\ \text{V}$$

Figure 17-7b shows the dc voltages, and Fig. 17-7c shows the currents for the second approximation.

PRACTICE PROBLEM 17-3   In Fig. 17-7a, change R_E to 3 kΩ. Determine the currents and voltages with the second approximation.

17-3 AC Analysis of a Diff Amp

In this section, we will derive the equation for the voltage gain of a diff amp. We will start with the simplest configuration, the noninverting input and single-ended output. After deriving its voltage gain, we will extend the results to the other configurations.

Theory of Operation

Figure 17-8a shows a noninverting input and single-ended output. With a large R_E, the tail current is almost constant when a small ac signal is present. Because of this, the two halves of a diff amp respond in a complementary manner to the noninverting input. In other words, an increase in the emitter current of Q_1 produces a decrease in the emitter current of Q_2. Conversely, a decrease in the emitter current of Q_1 produces an increase in the emitter current of Q_2.

In Fig. 17-8a, the left transistor Q_1 acts like an emitter follower that produces an ac voltage across the emitter resistor. This ac voltage is half of the input voltage v_1. On the positive half cycle of input voltage, the Q_1 emitter current increases, the Q_2 emitter current decreases, and the Q_2 collector voltage increases.

Figure 17-8 (a) Noninverting input and single-ended output; (b) ac equivalent circuit; (c) simplified ac equivalent circuit.

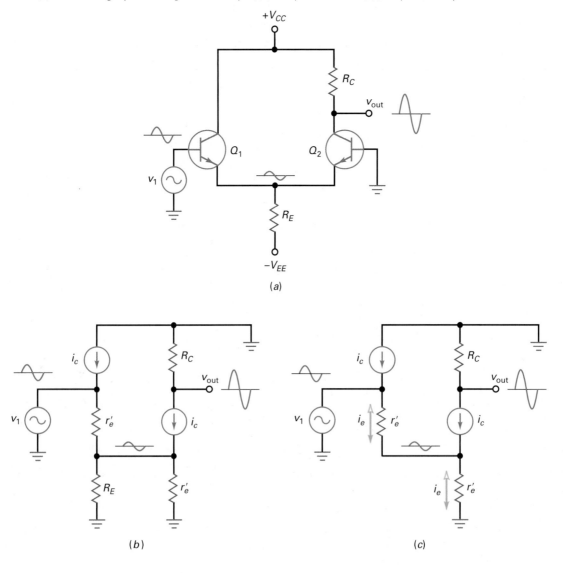

Similarly, on the negative half cycle of input voltage, the Q_1 emitter current decreases, the Q_2 emitter current increases, and the Q_2 collector voltage decreases. This is why the amplified output sine wave is in phase with the noninverting input.

Single-Ended Output Gain

Figure 17-8b shows the ac equivalent circuit. Notice that each transistor has an r_e'. Also, the biasing resistor R_E is in parallel with the r_e' of the right transistor. In any practical design, R_E is much greater than r_e'. Because of this, we can ignore R_E in a preliminary analysis.

Figure 17-8c shows the simplified equivalent circuit. Notice that the input voltage v_1 is across the first r_e' in series with the second r_e'. Since the two resistances are equal, the voltage across each r_e' is half of the input voltage. This is why the ac voltage across the tail resistor of Fig. 17-8a is half of the input voltage.

In Fig. 17-8c, the ac output voltage is:

$$v_{\text{out}} = i_c R_C$$

and the ac input voltage is:

$$v_{\text{in}} = i_e r_e' + i_e r_e' = 2i_e r_e'$$

Dividing v_{out} by v_{in} gives the voltage gain:

Single-ended output: $A_v = \dfrac{R_C}{2r_e'}$ (17-9)

A final point: In Fig. 17-8a, a quiescent dc voltage V_C exists at the output terminal. This voltage is not part of the ac signal. The ac voltage v_{out} is any change from the quiescent voltage. In an op amp, the quiescent dc voltage is removed in a later stage because it is unimportant.

Differential-Output Gain

Figure 17-9 shows the ac equivalent circuit for a noninverting input and differential output. The analysis is almost identical to the previous example, except that the output voltage is twice as much since there are two collector resistors:

$$v_{\text{out}} = v_{c2} - v_{c1} = i_c R_C - (-i_c R_C) = 2i_c R_C$$

(*Note:* The second minus sign appears because the v_{c1} signal is 180° out of phase with v_{c2}, as shown in Fig. 17-9.)

Figure 17-9 Noninverting input and differential output.

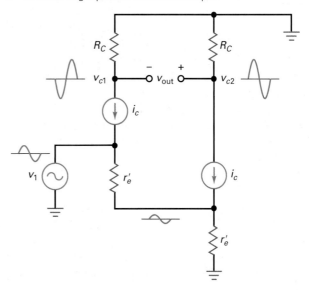

Figure 17-10 (*a*) Inverting input with single-ended output; (*b*) *pnp* version.

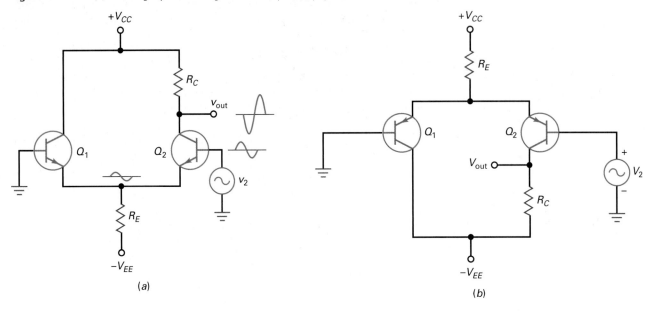

(*a*)

(*b*)

The ac input voltage is still equal to:

$$v_{in} = 2i_e r'_e$$

Dividing the output voltage by the input voltage gives the voltage gain:

Differential output: $A_v = \dfrac{R_C}{r'_e}$ (17-10)

This is easy to remember because it is the same as the voltage gain for a CE stage.

Inverting–Input Configurations

Figure 17-10*a* shows an inverting input and single-ended output. The ac analysis is almost identical to the noninverting analysis. In this circuit, the inverting input v_2 produces an amplified and inverted ac voltage at the final output. The r'_e of each transistor is still part of a voltage divider in the ac equivalent circuit. This is why the ac voltage across R_E is half of the inverting input voltage. If a differential output is used, the voltage gain is twice as much as previously discussed.

The diff amp in Fig. 17-10*b* is an upside-down *pnp* version of Fig. 17-10*a*. As discussed in Chap. 8, *pnp* transistors are often used in transistor circuits using positive power supplies. These *pnp* transistors are drawn in an upside-down configuration. As with the *npn* versions, the inputs and outputs may be either differential or single ended.

Differential–Input Configurations

The differential-input configurations have both inputs active at the same time. The ac analysis can be simplified by using the superposition theorem as follows: Since we know how a diff amp behaves with noninverting and inverting inputs, we can combine the two results to get the equations for differential-input configurations.

Table 17-2	Diff-Amp Voltage Gains		
Input	**Output**	A_v	v_{out}
Differential	Differential	R_C/r_e'	$A_v(v_1 - v_2)$
Differential	Single-ended	$R_C/2r_e'$	$A_v(v_1 - v_2)$
Single-ended	Differential	R_C/r_e'	$A_v v_1$ or $-A_v v_2$
Single-ended	Single-ended	$R_C/2r_e'$	$A_v v_1$ or $-A_v v_2$

The output voltage for a noninverting input is:

$$A_v(v_1)$$

and the output voltage for an inverting input is:

$$v_{out} = -A_v(v_2)$$

By combining the two results, we get the equation for a differential input:

$$v_{out} = A_v(v_1 - v_2)$$

Table of Voltage Gains

Table 17-2 summarizes the voltage gains for the diff-amp configurations. As you can see, the voltage gain is maximum with a differential output. The voltage gain is cut in half when a single-ended output is used. Also, when a single-ended output is used, the input may be noninverting or inverting.

Input Impedance

In a CE stage, the input impedance of the base is:

$$z_{in} = \beta r_e'$$

In a diff amp, the input impedance of either base is twice as high:

$$\mathbf{z_{in} = 2\beta r_e'} \qquad \textbf{(17-11)}$$

The input impedance of a diff amp is twice as high because there are two ac emitter resistances r_e' in the ac equivalent circuit instead of one. Equation (17-11) is valid for all configurations because any ac input signal sees two ac emitter resistances in the path between the base and ground.

Example 17-4

IIII MultiSim

In Fig. 17-11, what is the ac output voltage? If $\beta = 300$, what is the input impedance of the diff amp?

SOLUTION   We analyzed the dc equivalent circuit in Example 17-1. Ideally, 15 V is across the emitter resistor, producing a tail current of 2 mA, which means that the dc emitter current in each transistor is:

$$I_E = 1 \text{ mA}$$

Figure 17-11 Example.

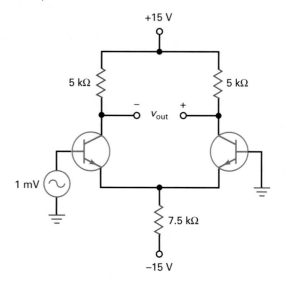

Now, we can calculate the ac emitter resistance:

$$r_e' = \frac{25 \text{ mV}}{1 \text{ mA}} = 25 \ \Omega$$

The voltage gain is:

$$A_v = \frac{5 \text{ k}\Omega}{25 \text{ V}} = 200$$

The ac output voltage is:

$$v_{out} = 200(1 \text{ mV}) = 200 \text{ mV}$$

and the input impedance of the diff amp is:

$$z_{in(base)} = 2(300)(25 \ \Omega) = 15 \text{ k}\Omega$$

PRACTICE PROBLEM 17-4   Repeat Example 17-4 with R_E changed to 5 kΩ.

Example 17-5

Repeat the preceding example using the second approximation to calculate the quiescent emitter current.

SOLUTION   In Example 17-2, we calculated a dc emitter current of:

$$I_E = 0.955 \text{ mA}$$

The ac emitter resistance is:

$$r_e' = \frac{25 \text{ mV}}{0.955 \text{ mA}} = 26.2 \ \Omega$$

Since the circuit has a differential output, the voltage gain is:

$$A_v = \frac{5 \text{ k}\Omega}{26.2 \text{ }\Omega} = 191$$

The ac output voltage is:

$$v_{\text{out}} = 191(1 \text{ mV}) = 191 \text{ mV}$$

and the input impedance of the diff amp is:

$$z_{\text{in(base)}} = 2(300)(26.2 \text{ }\Omega) = 15.7 \text{ k}\Omega$$

If the circuit is simulated with MultiSim, the following answers result for 2N3904 transistors:

$$v_{\text{out}} = 172 \text{ mV}$$

$$z_{\text{in(base)}} = 13.4 \text{ k}\Omega$$

The MultiSim output voltage and the input impedance are both slightly lower than our calculated values. When using specific part numbers for transistors, MultiSim loads in all kinds of higher-order transistor parameters that produce almost exact answers. This is why you must use a computer if you need high accuracy. Otherwise, rest satisfied with approximate methods of analysis.

Example 17-6

Repeat Example 17-4 for $v_2 = 1$ mV and $v_1 = 0$.

SOLUTION   Instead of driving the noninverting input, we are driving the inverting input. Ideally, the output voltage has the same magnitude, 200 mV, but it is inverted. The input impedance is approximately 15 kΩ.

Example 17-7

What is the ac output voltage in Fig. 17-12? If $\beta = 300$, what is the input impedance of the diff amp?

SOLUTION   Ideally, 15 V is across the emitter resistor, so that the tail current is:

$$I_T = \frac{15 \text{ V}}{1 \text{ M}\Omega} = 15 \text{ }\mu\text{A}$$

Since the emitter current in each transistor is half of the tail current:

$$r'_e = \frac{25 \text{ mV}}{7.5 \text{ }\mu\text{A}} = 3.33 \text{ k}\Omega$$

The voltage gain for the single-ended output is:

$$A_v = \frac{1 \text{ M}\Omega}{2(3.33 \text{ k}\Omega)} = 150$$

The ac output voltage is:

$$v_{\text{out}} = 150(7 \text{ mV}) = 1.05 \text{ V}$$

Figure 17-12 Example.

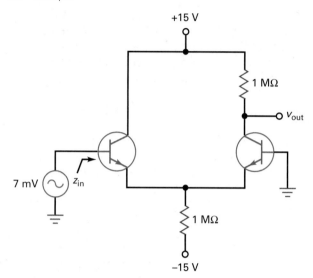

and the input impedance of the base is:

$$z_{in} = 2(300)(3.33 \text{ k}\Omega) = 2 \text{ M}\Omega$$

PRACTICE PROBLEM 17-7 Repeat Example 17-7 with R_E changed to 500 kΩ.

17-4 Input Characteristics of an Op Amp

Assuming perfect symmetry in a diff amp is a good approximation for many applications. But in precision applications, we can no longer treat the two halves of a diff amp as identical. There are three characteristics on the data sheet of every op amp that a designer uses when more accurate answers are needed. They are the input bias current, the input offset current, and the input offset voltage.

Input Bias Current

In an integrated op amp, the β_{dc} of each transistor in the first stage is slightly different, which means that the base currents in Fig. 17-13 are slightly different. The **input bias current** is defined as the average of the dc base currents:

$$I_{in(bias)} = \frac{I_{B1} + I_{B2}}{2} \tag{17-12}$$

For instance, if $I_{B1} = 90$ nA and $I_{B2} = 70$ nA, the input bias current is:

$$I_{in(bias)} = \frac{90 \text{ nA} + 70 \text{ nA}}{2} = 80 \text{ nA}$$

With bipolar op amps, the input bias current is typically in nanoamperes. When op amps use JFETs in the input diff amp, the input bias current is in picoamperes.

Figure 17-13 Different base currents.

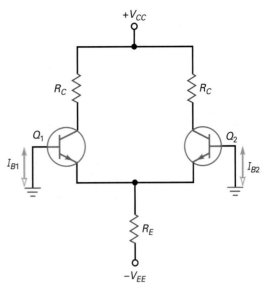

The input bias current will flow through the resistances between the bases and ground. These resistances may be discrete resistances, or they may be the Thevenin resistances of the input sources.

Input Offset Current

The **input offset current** is defined as the difference of the dc base currents:

$$I_{\text{in(off)}} = I_{B1} - I_{B2} \tag{17-13}$$

This difference in the base currents indicates how closely the transistors are matched. If the transistors are identical, the input offset current is zero because both base currents will be equal. But almost always, the two transistors are slightly different and the two base currents are not equal.

As an example, suppose $I_{B1} = 90$ nA and $I_{B2} = 70$ nA. Then:

$$I_{\text{in(off)}} = 90 \text{ nA} - 70 \text{ nA} = 20 \text{ nA}$$

The Q_1 transistor has 20 nA more base current than the Q_2 transistor. This can cause a problem when large base resistances are used.

Base Currents and Offsets

By rearranging Eqs. (17-12) and (17-13), we can derive these two equations for the base currents:

$$I_{B1} = I_{\text{in(bias)}} + \frac{I_{\text{in(off)}}}{2} \tag{17-13a}$$

$$I_{B2} = I_{\text{in(bias)}} - \frac{I_{\text{in(off)}}}{2} \tag{17-13b}$$

Data sheets always list $I_{\text{in(bias)}}$ and $I_{\text{in(off)}}$, but not I_{B1} and I_{B2}. With these equations, we can calculate the base currents. These equations assume that I_{B1} is greater than I_{B2}. If I_{B2} is greater than I_{B1}, transpose the equations.

Effect of Base Current

Some diff amps are operated with a base resistance on only one side, as shown in Fig. 17-14a. Because of the base current direction, the base current through R_B produces a noninverting dc input voltage of:

$$V_1 = -I_{B1}R_B$$

Differential Amplifiers

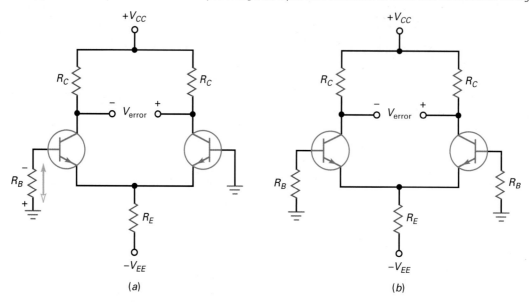

(a)   (b)

(*Note:* Capital letters are used here and elsewhere for dc error voltages like V_1. For simplicity, we will treat V_1 as an absolute value. This voltage has the same effect as a genuine input signal. When this false signal is amplified, an unwanted dc voltage V_{error} appears across the output, as shown in Fig. 17-14*a*.)

For instance, if a data sheet gives $I_{\text{in(bias)}} = 80$ nA and $I_{\text{in(off)}} = 20$ nA, Eqs. (17-13a) and (17-13b) give:

$$I_{B1} = 80 \text{ nA} + \frac{20 \text{ nA}}{2} = 90 \text{ nA}$$

$$I_{B2} = 80 \text{ nA} - \frac{20 \text{ nA}}{2} = 70 \text{ nA}$$

If $R_B = 1$ kΩ, the noninverting input has an error voltage of:

$$V_1 = (90 \text{ nA})(1 \text{ k}\Omega) = 90 \ \mu\text{V}$$

Effect of Input Offset Current

One way to reduce the output error voltage is by using an equal base resistance on the other side of the diff amp, as shown in Fig. 17-14*b*. In this case, we have a differential dc input of:

$$V_{\text{in}} = I_{B1}R_B - I_{B2}R_B = (I_{B1} - I_{B2})R_B$$

or

$$V_{\text{in}} = I_{\text{in(off)}}R_B \tag{17-14}$$

Since $I_{\text{in(off)}}$ is usually less than 25 percent of $I_{\text{in(bias)}}$, the input error voltage is much less when equal base resistors are used. For this reason, designers often include an equal base resistance on the opposite side of a diff amp, as shown in Fig. 17-14*b*.

For instance, if $I_{\text{in(bias)}} = 80$ nA and $I_{\text{in(off)}} = 20$ nA, then a base resistance of 1 kΩ produces an input error voltage of:

$$V_{\text{in}} = (20 \text{ nA})(1 \text{ k}\Omega) = 20 \ \mu\text{V}$$

Figure 17-15 (*a*) Different collector resistors produce error when bases are grounded; (*b*) different base-emitter curves added to error; (*c*) input offset voltage is equivalent to an unwanted input voltage.

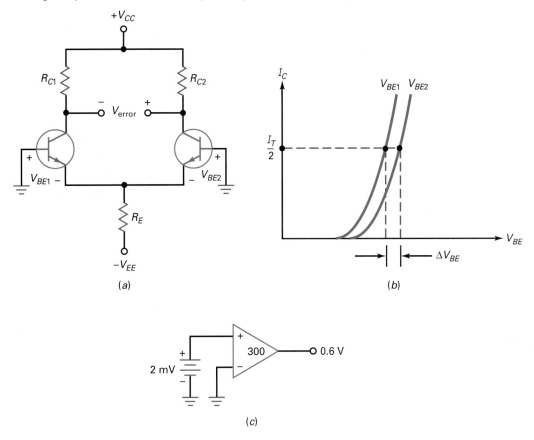

(a)

(b)

(c)

Input Offset Voltage

When a diff amp is integrated as the first stage of an op amp, the two halves are almost but not quite identical. To begin with, the two collector resistances may be different, as shown in Fig. 17-15*a*. Because of this, an error voltage appears across the output.

Another source of error is the different V_{BE} curves for each transistor. For instance, suppose that the two base-emitter curves have the same current, as shown in Fig. 17-15*b*. Because the curves are slightly different, there is a difference between the two V_{BE} values. This difference adds to the error voltage. Besides R_C and V_{BE}, other transistor parameters may differ slightly on each half of the diff amp.

The **input offset voltage** is defined as the input voltage that would produce the same output error voltage in a perfect diff amp. As an equation:

$$V_{\text{in(off)}} = \frac{V_{\text{error}}}{A_v} \tag{17-15}$$

In this equation, V_{error} does not include the effects of input bias and offset current because both bases are grounded when V_{error} is measured.

For instance, if a diff amp has an output error voltage of 0.6 V and a voltage gain of 300, the input offset voltage is:

$$V_{\text{in(off)}} = \frac{0.6 \text{ V}}{300} = 2 \text{ mV}$$

Figure 17-15*c* illustrates the idea. An input offset voltage of 2 mV is driving a diff amp with a voltage gain of 300 to produce an error voltage of 0.6 V.

Figure 17-16 Output of diff amp includes desired signal and error voltage.

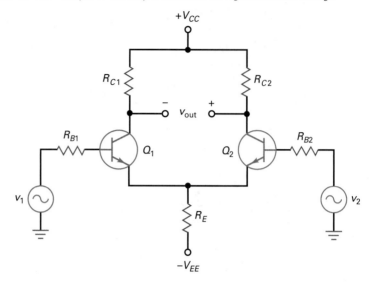

Combined Effects

In Fig. 17-16, the output voltage is the superposition of all input effects. To begin with, there is the ideal ac input:

$$v_{in} = v_1 - v_2$$

This is what we want. It is the voltage coming from the two input sources. It is amplified to produce the desired ac output:

$$v_{out} = A_v(v_1 - v_2)$$

Then, there are the three unwanted dc error inputs. With Eqs. (17-13a) and (17-13b), we can derive these formulas:

$$V_{1err} = (R_{B1} - R_{B2})I_{in(bias)} \tag{17-16}$$

$$V_{2err} = (R_{B1} + R_{B2})\frac{I_{in(off)}}{2} \tag{17-17}$$

$$V_{3err} = V_{in(off)} \tag{17-18}$$

The advantage of these formulas is that they use $I_{in(bias)}$ and $I_{in(off)}$, quantities on the data sheet. The three dc errors are amplified to produce the output error voltage:

$$V_{error} = A_v(V_{1err} + V_{2err} + V_{3err}) \tag{17-19}$$

In many cases, V_{error} can be ignored. This will depend on the application. For instance, if we are building an ac amplifier, V_{error} may not be important. It is only when we are building some kind of precision dc amplifier that V_{error} needs to be taken into account.

Equal Base Resistances

When the bias and offset errors cannot be ignored, here are the remedies. As already mentioned, one of the first things a designer can do is to use equal base resistances: $R_{B1} = R_{B2} = R_B$. This brings the two halves of the diff amp into a closer alignment because Eqs. (17-16) through (17-19) become:

$$V_{1err} = 0$$

$$V_{2err} = R_B I_{in(off)}$$

$$V_{3err} = V_{in(off)}$$

Table 17–3	Sources of Output Error Voltage	
Description	**Cause**	**Solution**
Input bias current	Voltage across a single R_B	Use equal R_B on other side
Input offset current	Unequal current gains	Data sheet nulling methods
Input offset voltage	Unequal R_C and V_{BE}	Data sheet nulling methods

If further compensation is necessary, the best approach is to use the *nulling circuits* suggested on the data sheets. Manufacturers optimize the design of these nulling circuits, which should be used if output error voltage is a problem. We will discuss nulling circuits in a later chapter.

Conclusion

Table 17-3 summarizes the sources of output error voltage. In many applications, the output error voltage is either small enough to ignore or not important in the particular application. In precision applications, in which the dc output is important, some form of nulling is used to eliminate the effects of input bias and offset. Designers usually null the output with methods suggested on the manufacturer's data sheet.

Example 17-8

The diff amp of Fig. 17-17 has $A_v = 200$, $I_{in(bias)} = 3 \ \mu A$, $I_{in(off)} = 0.5 \ \mu A$, and $V_{in(off)} = 1$ mV. What is the output error voltage? If a matching base resistor is used, what is the output error voltage?

Figure 17-17 Example.

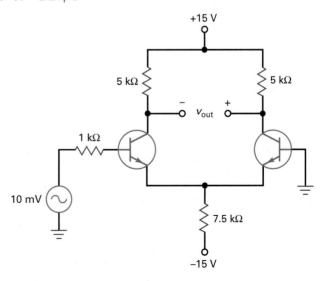

With Eqs. (17-16) to (17-18):

$$V_{1\text{err}} = (R_{B1} - R_{B2})I_{\text{in(bias)}} = (1\ \text{k}\Omega)(3\ \mu\text{A}) = 3\ \text{mV}$$

$$V_{2\text{err}} = (R_{B1} + R_{B2})\frac{I_{\text{in(off)}}}{2} = (1\ \text{k}\Omega)(0.25\ \mu\text{A}) = 0.25\ \text{mV}$$

$$V_{3\text{err}} = V_{\text{in(off)}} = 1\ \text{mV}$$

The output error voltage is:

$$V_{\text{error}} = 200(3\ \text{mV} + 0.25\ \text{mV} + 1\ \text{mV}) = 850\ \text{mV}$$

When a matching base resistance of 1 kΩ is used on the inverting side,

$$V_{1\text{err}} = 0$$

$$V_{2\text{err}} = R_B I_{\text{in(off)}} = (1\ \text{k}\Omega)(0.5\ \mu\text{A}) = 0.5\ \text{mV}$$

$$V_{3\text{err}} = V_{\text{in(off)}} = 1\ \text{mV}$$

The output error voltage is:

$$V_{\text{error}} = 200(0.5\ \text{mV} + 1\ \text{mV}) = 300\ \text{mV}$$

PRACTICE PROBLEM 17–8 In Fig. 17-17, what is the output error voltage if the diff amp has a voltage gain of 150?

Example 17-9

The diff amp of Fig. 17-18 has $A_v = 300$, $I_{\text{in(bias)}} = 80\ \text{nA}$, $I_{\text{in(off)}} = 20\ \text{nA}$, and $V_{\text{in(off)}} = 5\ \text{mV}$. What is the output error voltage?

Figure 17-18 Example.

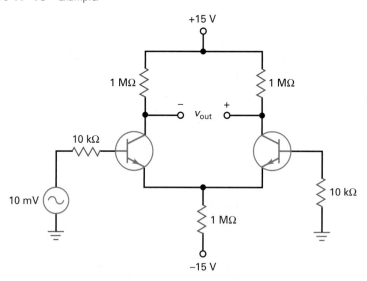

SOLUTION The circuit uses equal base resistors. With the equations shown above:

$$V_{1\text{err}} = 0$$

$$V_{2\text{err}} = (10\ \text{k}\Omega)(20\ \text{nA}) = 0.2\ \text{mV}$$

$$V_{3\text{err}} = 5\ \text{mV}$$

The total output error voltage is:

$$V_{\text{error}} = 300(0.2 \text{ mV} + 5 \text{ mV}) = 1.56 \text{ V}$$

PRACTICE PROBLEM 17-9   Repeat Example 17-9 using $I_{\text{in(off)}} = 10$ nA.

17-5 Common-Mode Gain

Figure 17-19*a* shows a differential input and single-ended output. The same input voltage, $v_{\text{in(CM)}}$ is being applied to each base. This voltage is called a **common-mode signal.** If the diff amp is perfectly symmetrical, there is no ac output voltage with a common-mode input signal because $v_1 = v_2$. When a diff amp is not perfectly symmetrical, there will be a small ac output voltage.

In Fig. 17-19*a*, equal voltages are applied to the noninverting and inverting inputs. Nobody would deliberately use a diff amp this way because the output voltage is ideally zero. The reason for discussing this type of input is because most static, interference, and other kinds of undesirable pickup are common-mode signals.

Here is how a common-mode signal appears: The connecting wires on the input bases act like small antennas. If the diff amp is operating in an environment with a lot of electromagnetic interference, each base acts like a small antenna that picks up an unwanted signal voltage. One of the reasons the diff amp is so popular is because it discriminates against these common-mode signals. In other words, a diff amp does not amplify common-mode signals.

Here is an easy way to find the voltage gain for a common-mode signal: We can redraw the circuit, as shown in Fig. 17-19*b*. Since equal voltages $v_{\text{in(CM)}}$ drive both inputs simultaneously, there is almost no current through the wire between the emitters. Therefore, we can remove the connecting wire, as shown in Fig. 17-20.

Figure 17-19   (*a*) Common-mode input signal; (*b*) equivalent circuit.

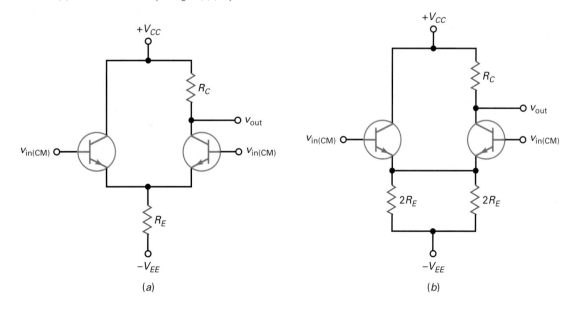

(*a*)   (*b*)

Figure 17-20 Right side acts like swamped amplifier with common-mode input.

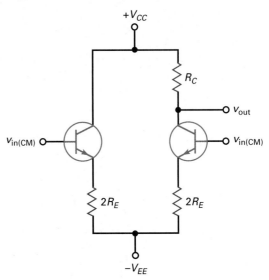

With a common-mode signal, the right side of the circuit is equivalent to a heavily swamped CE amplifier. Since R_E is always much greater than r'_e, the swamped voltage gain is approximately:

$$A_{v(CM)} = \frac{R_C}{2R_E} \tag{17-20}$$

With typical values of R_C and R_E, the common-mode voltage gain is usually less than 1.

Common-Mode Rejection Ratio

The **common-mode rejection ratio (CMRR)** is defined as the voltage gain divided by common-mode voltage gain. In symbols:

$$CMRR = \frac{A_v}{A_{v(CM)}} \tag{17-21}$$

For instance, if $A_v = 200$ and $A_{v(CM)} = 0.5$, CMRR = 400.

The higher the CMRR, the better. A high CMRR means that the diff amp is amplifying the wanted signal and discriminating against the common-mode signal.

Data sheets usually specify CMRR in decibels, using the following formula for the decibel conversion:

$$CMRR_{dB} = 20 \log CMRR \tag{17-22}$$

As an example, if CMRR = 400:

$$CMRR_{dB} = 20 \log 400 = 52 \text{ dB}$$

Example 17-10

In Fig. 17-21, what is the common-mode voltage gain? The output voltage?

SOLUTION With Eq. (17-20):

$$A_{v(CM)} = \frac{1 \text{ M}\Omega}{2 \text{ M}\Omega} = 0.5$$

Figure 17–21 Example.

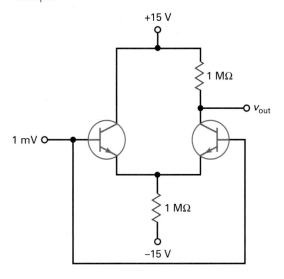

The output voltage is:

$$v_{\text{out}} = 0.5(1 \text{ mV}) = 0.5 \text{ mV}$$

As you can see, the diff amp attenuates (weakens) the common-mode signal rather than amplifying it.

PRACTICE PROBLEM 17-10 Repeat Example 17-10 with R_E changed to 2 MΩ.

Example 17–11

In Fig. 17-22, $A_v = 150$, $A_{v(\text{CM})} = 0.5$, and $v_{\text{in}} = 1$ mV. If the base leads are picking up a common-mode signal of 1 mV, what is the output voltage?

SOLUTION   The input has two components, the desired signal and a common-mode signal. Both are equal in amplitude. The desired component is amplified to get an output of:

$$v_{\text{out1}} = 150(1 \text{ mV}) = 150 \text{ mV}$$

The common-mode signal is attenuated to get an output of:

$$v_{\text{out2}} = 0.5(1 \text{ mV}) = 0.5 \text{ mV}$$

The total output is the sum of these two components:

$$v_{\text{out}} = v_{\text{out1}} + v_{\text{out2}}$$

The output contains both components, but the desired component is 300 times greater than the unwanted component.

This example shows why the diff amp is useful as the input stage of an op amp. It attenuates the common-mode signal. This is a distinct advantage over

Figure 17-22 Example.

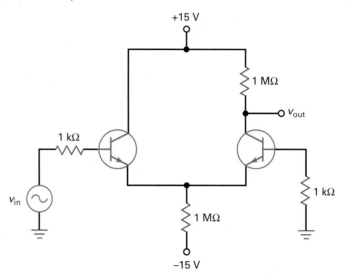

the ordinary CE amplifier, which amplifies a stray pickup signal the same way it amplifies the desired signal.

PRACTICE PROBLEM 17–11 In Fig. 17-22, change A_v to 200 and find the output voltage.

Example 17-12

A 741 is an op amp with $A_v = 200,000$ and $CMRR_{dB} = 90$ dB. What is the common-mode voltage gain? If the desired and common-mode signal each has a value of 1 μV, what is the output voltage?

SOLUTION

$$CMRR = \text{antilog} \frac{90\,dB}{20} = 31,600$$

Rearranging Eq. (17-21):

$$A_{v(CM)} = \frac{A_v}{CMRR} = \frac{200,000}{31,600} = 6.32$$

The desired output component is:

$$v_{out1} = 200,000(1\ \mu V) = 0.2\ V$$

The common-mode output is:

$$v_{out2} = 6.32(1\ \mu V) = 6.32\ \mu V$$

As you can see, the desired output is much larger than the common-mode output.

PRACTICE PROBLEM 17–12 Repeat Example 17-12 using an op amp gain of 100,000.

Figure 17-23 (a) P crystal; (b) wafer; (c) epitaxial layer; (d) insulating layer.

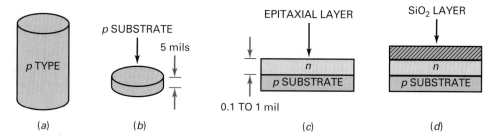

(a) (b) (c) (d)

17-6 Integrated Circuits

The invention of the **integrated circuit (IC)** in 1959 was a major breakthrough because the components are no longer discrete; they are *integrated*. This means that they are produced and connected during the manufacturing process on a single *chip*, a small piece of semiconductor material. Because the components are microscopically small, a manufacturer can place thousands of these integrated components in the space occupied by a single discrete transistor.

What follows is a brief description of how an IC is made. Current manufacturing processes are much more complicated, but the simplified discussion will give you the basic idea behind the making of a bipolar IC.

Basic Idea

First, the manufacturer produces a *p* crystal several inches long (Fig. 17-23*a*). This is sliced into many thin *wafers,* as in Fig. 17-23*b*. One side of the wafer is lapped and polished to get rid of surface imperfections. This wafer is called the *p* substrate. It will be used as a chassis for the integrated components. Next, the wafers are put into a furnace. A gas mixture of silicon atoms and pentavalent atoms passes over the wafers. This forms a thin layer of *n*-type semiconductor on the heated surface of the substrate (see Fig. 17-23*c*). We call this thin layer an *epitaxial layer.* As shown in Fig. 17-23*c*, the epitaxial layer is about 0.1 to 1 mil thick.

To prevent contamination of the epitaxial layer, pure oxygen is blown over the surface. The oxygen atoms combine with the silicon atoms to form a layer of silicon dioxide (SiO_2) on the surface, as shown in Fig. 17-23*d*. This glasslike layer of SiO_2 seals off the surface and prevents further chemical reactions. Sealing off the surface like this is known as *passivation.*

The wafer is then cut into the rectangular areas shown in Fig. 17-24. Each of these areas will be a separate chip after the wafer is cut. But before the wafer is cut, the manufacturer produces hundreds of circuits on the wafer, one on each chip area of Fig. 17-24. This simultaneous mass production is the reason for the low cost of ICs.

Here is how an integrated transistor is formed: Part of the SiO_2 is etched off, exposing the epitaxial layer (see Fig. 17-25*a*). The wafer is then put into a furnace, and trivalent atoms are diffused into the epitaxial layer. The concentration of trivalent atoms is enough to change the exposed epitaxial layer from *n* material to *p* material. Therefore, we get an island of *n* material under the SiO_2 layer (Fig. 17-25*b*). Oxygen is again blown over the surface to form the complete SiO_2 layer shown in Fig. 17-25*c*.

A hole is now etched in the center of the SiO_2 layer. This exposes the *n* epitaxial layer (Fig. 17-25*d*). The hole in the SiO_2 layer is called a *window.* We are now looking down at what will be the collector of the transistor.

Figure 17-24 Cutting wafer into chips.

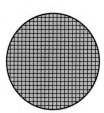

Differential Amplifiers

645

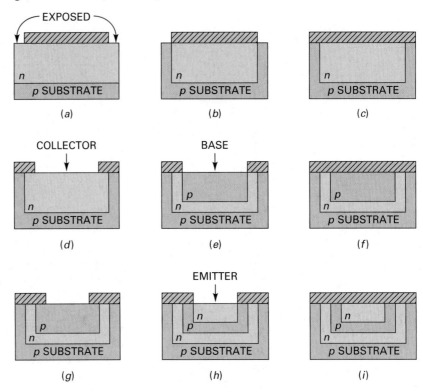

Figure 17-25 Steps in making a transistor.

To get the base, we pass trivalent atoms through this window; these impurities diffuse into the epitaxial layer and form an island of p-type material (Fig. 17-25e). Then, the SiO_2 layer is re-formed by passing oxygen over the wafer (Fig. 17-25f).

To form the emitter, we etch a window in the SiO_2 layer and expose the p island (Fig. 17-25g). By diffusing pentavalent atoms into the p island, we can form the small n island shown in Fig. 17-25h.

We then passivate the structure by blowing oxygen over the wafer (Fig. 17-25i). By etching windows in the SiO_2 layer, we can deposit metal to make electrical contact with the emitter, base, and collector. This gives us the integrated transistor of Fig. 17-26a.

To get a diode, we follow the same steps up to the point at which the p island has been formed and sealed off (Fig. 17-25f). Then, we etch windows to expose the p and n islands. By depositing metal through these windows, we make electrical contact with the cathode and anode of the integrated diode (Fig. 17-26b). By etching two windows above the p island of Fig. 17-25f, we can make metallic contact with this p island; this gives us an integrated resistor (Fig. 17-26c).

Figure 17-26 Integrated components: (a) Transistor; (b) diode; (c) resistor.

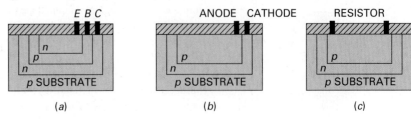

Figure 17-27 Simple IC.

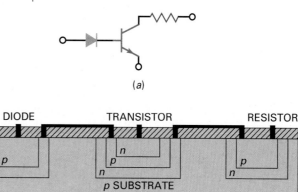

(a)

DIODE   TRANSISTOR   RESISTOR

p SUBSTRATE

(b)

Transistors, diodes, and resistors are easy to fabricate on a chip. For this reason, almost all ICs use these components. It is not practical to integrate inductors and large capacitors on the surface of a chip.

A Simple Example

To give you an idea of how a circuit is produced, look at the simple three-component circuit of Fig. 17-27a. To fabricate this circuit, we would simultaneously produce hundreds of circuits like this on a wafer. Each chip area would resemble Fig. 17-27b. The diode and resistor would be formed at the point mentioned earlier. At a later step, the emitter of the transistor would be formed. Then we would etch windows and deposit metal to connect the diode, transistor, and resistor, as shown in Fig. 17-27b.

Regardless of how complicated a circuit may be, producing it is mainly a process of etching windows, forming p and n islands, and connecting the integrated components. The p substrate isolates the integrated components from each other. In Fig. 17-27b, there are depletion layers between the p substrate and the three n islands that touch it. Because the depletion layers have essentially no current carriers, the integrated components are insulated from one another. This kind of insulation is known as *depletion-layer isolation.*

Types of ICs

The integrated circuits we have described are called **monolithic ICs.** The word *monolithic* is from the Greek and means "one stone." The word is appropriate because the components are part of one chip. Monolithic ICs are the most common type of IC. Since their invention, manufacturers have been producing monolithic ICs to carry out all kinds of functions.

Commercially available types can be used as amplifiers, voltage regulators, crowbars, AM receivers, television circuits, and computer circuits. But the monolithic IC has power limitations. Since most monolithic ICs are about the size of a discrete small-signal transistor, they are used in low-power applications.

When higher power is needed, thin-film and thick-film ICs may be used. These devices are larger than monolithic ICs but smaller than discrete circuits. With a thin- or thick-film IC, the passive components like resistors and capacitors are integrated, but the transistors and diodes are connected as discrete components to form a complete circuit. Therefore, commercially available thin- and thick-film circuits are combinations of integrated and discrete components.

Another IC used in high-power applications is the **hybrid IC.** Hybrid ICs combine two or more monolithic ICs in one package, or they combine monolithic ICs with thin- or thick-film circuits. Hybrid ICs are widely used for high-power audio-amplifier applications from 5 to more than 50 W.

Levels of Integration

Figure 17-27*b* is an example of *small-scale integration (SSI);* only a few components have been integrated to form a complete circuit. SSI refers to ICs with fewer than 12 integrated components. Most SSI chips use integrated resistors, diodes, and bipolar transistors.

Medium-scale integration (MSI) refers to ICs that have from 12 to 100 integrated components per chip. Either bipolar transistors or MOS transistors (enhancement-mode MOSFETS) can be used as the integrated transistors of an IC. Again, most MSI chips use bipolar components.

Large-scale integration (LSI) refers to ICs with more than a hundred components. Since it takes fewer steps to make an integrated MOS transistor, a manufacturer can produce more components on a chip than is possible with bipolar transistors.

Very large scale integration (VLSI) refers to placing thousands (or hundreds of thousands) of components on a single chip. Nearly all modern chips employ VLSI.

Finally, there is *ultra large scale integration (ULSI),* which refers to placing more than 1 million components on a single chip. The Intel Pentium P4 microprocessor uses ULSI technology. Various versions of this microprocessor have been developed, with the Intel P4 Prescott version containing approximately 125 million transistors. Current expectations are to have 1 billion components on a chip by the year 2011. The exponential growth often referred to as Moore's law will be challenged at this time. However, new technologies, such as nanotechnology, will allow the continued growth to occur.

17-7 The Current Mirror

With ICs, there is a way to increase the voltage gain and CMRR of a diff amp. Figure 17-28*a* shows a **compensating diode** in parallel with the emitter diode of a transistor. The current through the resistor is given by:

$$I_R = \frac{V_{CC} - V_{BE}}{R} \qquad (17\text{-}23)$$

Figure 17-28 The current mirror.

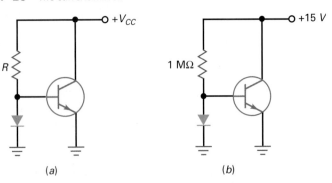

(a)   (b)

If the compensating diode and the emitter diode have identical current-voltage curves, the collector current will equal the current through the resistor:

$$I_C = I_R \tag{17-24}$$

A circuit like Fig. 17-28a is called a **current mirror** because the collector current is a mirror image of the resistor current. With ICs, it is relatively easy to match the characteristics of the compensating diode and the emitter diode because both components are on the same chip. Current mirrors are used as current sources and active loads in the design of IC op amps.

Current Mirror Sources the Tail Current

With a single-ended output, the voltage gain of a diff amp is $R_C/2r_e'$ and the common-mode voltage gain is $R_C/2R_E$. The ratio of the two gains gives:

$$\text{CMRR} = \frac{R_E}{r_e'}$$

The larger we can make R_E, the greater the CMRR.

One way to get a high equivalent R_E is to use a current mirror to produce the tail current, as shown in Fig. 17-29. The current through the compensating diode is:

$$I_R = \frac{V_{CC} + V_{EE} - V_{BE}}{R} \tag{17-25}$$

Because of the current mirror, the tail current has the same value. Since Q_4 acts like a current source, it has a very high output impedance. As a result, the equivalent R_E of the diff amp is in hundreds of megohms and the CMRR is dramatically improved.

Active Load

The voltage gain of a single-ended diff amp is $R_C/2r_e'$. The larger we can make R_C, the greater the voltage gain. Figure 17-30 shows a current mirror used as an **active load resistor.** Since Q_6 is a *pnp* current source Q_2 sees an equivalent R_C that is hundreds of megohms. As a result, the voltage gain is much higher with an active load than with an ordinary resistor. Active loading like this is used in most op amps.

Figure 17-29   Current mirror sources the tail current.

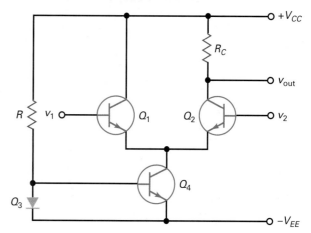

Figure 17-30 Current mirror is an active load.

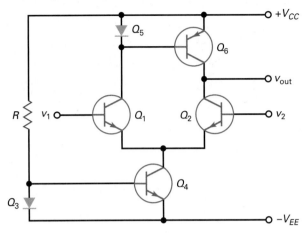

17-8 The Loaded Diff Amp

In the earlier discussions of a diff amp, we did not use a load resistor. When a load resistor is used, the analysis becomes much more complicated, especially with a differential output.

Figure 17-31a shows a differential output with a load resistor between the collectors. There are several ways to calculate the effect that this load resistor has on the output voltage. If you try to solve this with Kirchhoff loop equations, you will have a very difficult problem. But with Thevenin's theorem, the problem unravels very quickly.

Here is how it is done: If we open the load resistor in Fig. 17-31a, the Thevenin voltage is the same as the v_{out} calculated in earlier discussions. Also, looking into the open AB terminals with all sources zeroed, we see a Thevenin resistance of $2R_C$. (*Note:* Because the transistors are current sources, they become open when zeroed.)

Figure 17-31 (*a*) Diff amp with load resistor; (*b*) Thevenin equivalent circuit for differential output; (*c*) Thevenin equivalent circuit for single-ended output.

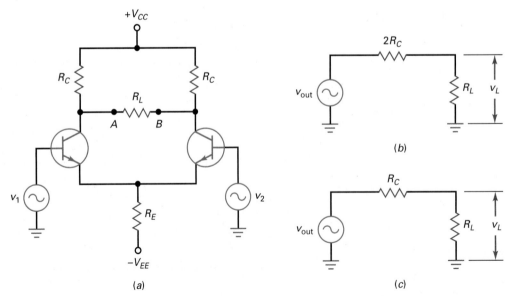

Figure 17-31*b* shows the Thevenin equivalent circuit. The ac output voltage v_{out} is the same output voltage discussed in earlier sections. After calculating v_{out}, finding the load voltage is easy because all we need is Ohm's law. If a diff amp has a single-ended output, the Thevenin equivalent circuit simplifies to Fig. 17-31*c*.

Example 17-13

What is the load voltage in Fig. 17-32*a* when $R_L = 15$ kΩ?

Figure 17-32 Example.

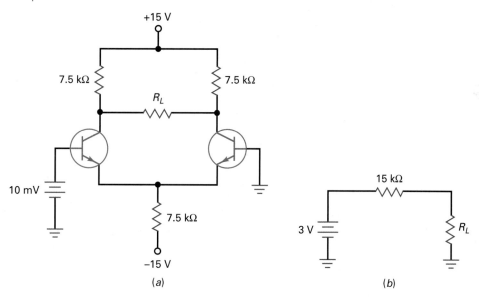

(a)

(b)

SOLUTION   Ideally, the tail current is 2 mA, the emitter current is 1 mA, and $r'_e = 25$ Ω. The unloaded voltage gain is:

$$A_v = \frac{R_C}{r'_e} = \frac{7.5 \text{ k}\Omega}{25 \text{ }\Omega} = 300$$

The Thevenin or unloaded output voltage is:

$$v_{out} = A_v(v_1) = 300(10 \text{ mV}) = 3 \text{ V}$$

The Thevenin resistance is:

$$R_{TH} = 2R_C = 2(7.5 \text{ k}\Omega) = 15 \text{ k}\Omega$$

Figure 17-32*b* is the Thevenin equivalent circuit. With a load resistance of 15 kΩ, the load voltage is:

$$v_L = 0.5(3 \text{ V}) = 1.5 \text{ V}$$

PRACTICE PROBLEM 17-13   In Fig. 17-32*a*, find the load voltage when $R_L = 10$ kΩ.

Differential Amplifiers

Example 17-14

An ammeter is used for the load resistance in Fig. 17-32a. What is the current through the ammeter?

SOLUTION   In Fig. 17-32b, the load resistance is ideally zero and the load current is:

$$i_L = \frac{3 \text{ V}}{15 \text{ k}\Omega} = 0.2 \text{ mA}$$

Without Thevenin's theorem, this would be a very difficult problem to solve.

PRACTICE PROBLEM 17-14   Repeat Example 17-14 with an input voltage of 20 mV.

Summary

SEC. 17-1 THE DIFFERENTIAL AMPLIFIER

A diff amp is the typical input stage of an op amp. It has no coupling or bypass capacitors. Because of this, it has no lower cutoff frequency. The input may be differential, noninverting, or inverting. The output may be single-ended or differential.

SEC. 17-2 DC ANALYSIS OF A DIFF AMP

The diff amp uses two-supply emitter bias to produce the tail current. When a diff amp is perfectly symmetrical, each emitter current is half the tail current. Ideally, the voltage across the emitter resistor equals the negative supply voltage.

SEC. 17-3 AC ANALYSIS OF A DIFF AMP

Because the tail current is ideally constant, an increase in the emitter current of one transistor produces a decrease in the emitter current of the other transistor. With a differential output, the voltage gain is R_C/r'_e. With a

single-ended output, the voltage gain is half as much.

SEC. 17-4 INPUT CHARACTERISTICS OF AN OP AMP

Three important input characteristics of an op amp are the input bias current, input offset current, and input offset voltage. The input bias and offset currents produce unwanted input error voltages when they flow through the base resistors. The input offset voltage is an equivalent input error produced by differences in R_C and V_{BE}.

SEC. 17-5 COMMON-MODE GAIN

Most static, interference, and other kinds of electromagnetic pickup are common-mode signals. The diff amp discriminates against common-mode signals. The CMRR is the voltage gain divided by the common-mode gain. The higher the CMRR, the better.

SEC. 17-6 INTEGRATED CIRCUITS

Monolithic ICs are complete circuit functions on a single chip such as amplifiers, voltage regulators, and

computer circuits. For high-power applications, thin-film, thick-film, and hybrid ICs may be used. SSI refers to fewer than 12 components, MSI to between 12 and 100 components, LSI to more than 100 components, VLSI to more than 1000 components, and ULSI to more than 1 million components.

SEC. 17-7 THE CURRENT MIRROR

The current mirror is used in ICs because it is a convenient way to create current sources and active loads. The advantages of using current mirrors are increases in voltage gain and CMRR.

SEC. 17-8 THE LOADED DIFF AMP

When a load resistance is used with a diff amp, the best approach is to use Thevenin's theorem. Calculate the ac output voltage v_{out} as discussed in earlier sections. This voltage is equal to the Thevenin voltage. Use a Thevenin resistance of $2R_C$ with a differential output and R_C with a single-ended output.

Definitions

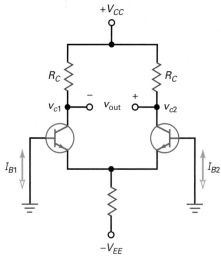

(17-1)   Differential output:

$$v_{out} = v_{c2} - v_{c1}$$

(17-12)   Input bias current:

$$I_{in(bias)} = \frac{I_{B1} + I_{B2}}{2}$$

(17-13)   Input offset current:

$$I_{in(off)} = I_{B1} - I_{B2}$$

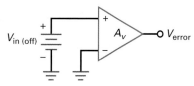

(17-15)   Input offset voltage:

$$V_{in(off)} = \frac{V_{error}}{A_v}$$

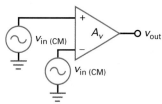

(17-21)   Common-mode rejection ratio:

$$CMRR = \frac{A_v}{A_{v(CM)}}$$

(17-22)   Decibel CMRR:

$$CMRR_{dB} = 20 \log CMRR$$

Derivations

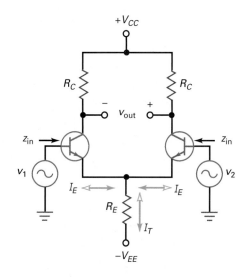

(17-2)   Differential output:

$$v_{out} = A_v(v_1 - v_2)$$

(17-5)   Tail current:

$$I_T = \frac{V_{EE}}{R_E}$$

(17-6)   Emitter current:

$$I_E = \frac{I_T}{2}$$

(17-9)   Single-ended output:

$$A_v = \frac{R_C}{2r'_e}$$

(17-10)   Differential output:

$$A_v = \frac{R_C}{r'_e}$$

(17-11)   Input impedance:

$$z_{in} = 2\beta r'_e$$

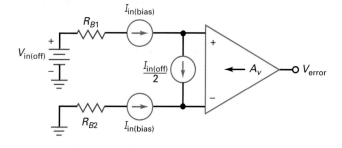

(17-16) First error voltage:

$$V_{1err} = (R_{B1} - R_{B2})I_{in(bias)}$$

(17-17) Second error voltage:

$$V_{2err} = (R_{B1} + R_{B2})\frac{I_{in(off)}}{2}$$

(17-18) Third error voltage:

$$V_{3err} = V_{in(off)}$$

(17-19) Total output error voltage:

$$V_{error} = A_v(V_{1err} + V_{2err} + V_{3err})$$

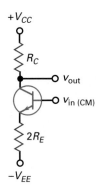

(17-20) Common-mode voltage gain:

$$A_{v(CM)} = \frac{R_C}{2R_E}$$

Student Assignments

1. **Monolithic ICs are**
   a. Forms of discrete circuits
   b. On a single chip
   c. Combinations of thin-film and thick-film circuits
   d. Also called hybrid ICs

2. **The op amp can amplify**
   a. AC signals only
   b. DC signals only
   c. Both ac and dc signals
   d. Neither ac nor dc signals

3. **Components are soldered together in**
   a. Discrete circuits
   b. Integrated circuits
   c. SSI
   d. Monolithic ICs

4. **The tail current of a diff amp is**
   a. Half of either collector current
   b. Equal to either collector current
   c. Two times either collector current
   d. Equal to the difference in base currents

5. **The node voltage at the top of the tail resistor is closest to**
   a. Collector supply voltage
   b. Zero
   c. Emitter supply voltage
   d. Tail current times base resistance

6. **The input offset current equals the**
   a. Difference between the two base currents
   b. Average of the two base currents

   c. Collector current divided by current gain
   d. Difference between the two base-emitter voltages

7. **The tail current equals the**
   a. Difference between the two emitter currents
   b. Sum of the two emitter currents
   c. Collector current divided by current gain
   d. Collector voltage divided by collector resistance

8. **The voltage gain of a diff amp with an unloaded differential output is equal to R_C divided by**
   a. r_e'
   b. $r_e'/2$
   c. $2r_e'$
   d. R_E

9. **The input impedance of a diff amp equals r_e' times**
   a. 0
   b. R_C
   c. R_E
   d. 2β

10. **A dc signal has a frequency of**
    a. 0 Hz
    b. 60 Hz
    c. 0 to more than 1 MHz
    d. 1 MHz

11. **When the two input terminals of a diff amp are grounded,**
    a. The base currents are equal
    b. The collector currents are equal

    c. An output error voltage usually exists
    d. The ac output voltage is zero

12. **One source of output error voltage is**
    a. Input bias current
    b. Difference in collector resistors
    c. Tail current
    d. Common-mode voltage gain

13. **A common-mode signal is applied to**
    a. The noninverting input
    b. The inverting input
    c. Both inputs
    d. The top of the tail resistor

14. **The common-mode voltage gain is**
    a. Smaller than the voltage gain
    b. Equal to the voltage gain
    c. Greater than the voltage gain
    d. None of the above

15. **The input stage of an op amp is usually a**
    a. Diff amp
    b. Class B push-pull amplifier
    c. CE amplifier
    d. Swamped amplifier

16. **The tail of a diff amp acts like a**
    a. Battery
    b. Current source
    c. Transistor
    d. Diode

17. **The common–mode voltage gain of a diff amp is equal to R_C divided by**
   a. r'_e
   b. $r'_e/2$
   c. $2r'_e$
   d. $2R_E$

18. **When the two bases are grounded in a diff amp, the voltage across each emitter diode is**
   a. Zero
   b. 0.7 V
   c. The same
   d. High

19. **The common–mode rejection ratio is**
   a. Very low
   b. Often expressed in decibels

   c. Equal to the voltage gain
   d. Equal to the common–mode voltage gain

20. **The typical input stage of an op amp has a**
   a. Single-ended input and single-ended output
   b. Single-ended input and differential output
   c. Differential input and single-ended output
   d. Differential input and differential output

21. **The input offset current is usually**
   a. Less than the input bias current
   b. Equal to zero
   c. Less than the input offset voltage
   d. Unimportant when a base resistor is used

22. **With both bases grounded, the only offset that produces an error is the**
   a. Input offset current
   b. Input bias current
   c. Input offset voltage
   d. β

23. **The voltage gain of a loaded diff amp is**
   a. Larger than the unloaded voltage gain
   b. Equal to $\dfrac{R_C}{r'_e}$
   c. Smaller than the unloaded voltage gain
   d. Impossible to determine

Problems

SEC. 17–2   DC ANALYSIS OF A DIFF AMP

17-1   What are the ideal currents and voltages in Fig. 17-33?

17-2 |||| **MultiSim** Repeat Prob. 17-1 using the second approximation.

17-3   What are the ideal currents and voltages in Fig. 17-34?

17-4 |||| **MultiSim** Repeat Prob. 17-3 using the second approximation.

Figure 17-33

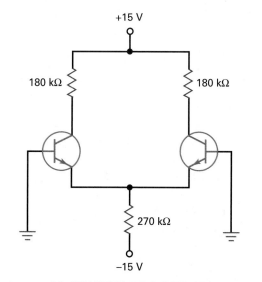

SEC. 17–3   AC ANALYSIS OF A DIFF AMP

17-5   In Fig. 17-35, what is the ac output voltage? If $\beta = 275$, what is the input impedance of the diff amp? Use the ideal approximation to get the tail current.

17-6   Repeat Prob. 17-5 using the second approximation.

17-7   Repeat Prob. 17-5 by grounding the noninverting input and using an input of $v_2 = 1$ mV.

SEC. 17–4   INPUT CHARACTERISTICS OF AN OP AMP

17-8   The diff amp of Fig. 17-36 has $A_v = 360$, $I_{in(bias)} = 600$ nA, $I_{in(off)} = 100$ nA, and $V_{in(off)} = 1$ mV. What is the output error voltage? If a matching base resistor is used, what is the output error voltage?

17-9   The diff amp of Fig. 17-36 has $A_v = 250$, $I_{in(bias)} = 1$ μA, $I_{in(off)} = 200$ nA, and $V_{in(off)} = 5$ mV. What is the output error voltage? If a matching base resistor is used, what is the output error voltage?

SEC. 17–5   COMMON–MODE GAIN

17-10   What is the common-mode voltage gain of Fig. 17-37? If a common-mode voltage of 20 μV exists on both bases, what is the common-mode output voltage?

17-11   In Fig. 17-37, $v_{in} = 2$ mV and $v_{in(CM)} = 5$ mV. What is the ac output voltage?

17-12   A 741C is an op amp with $A_v = 100{,}000$ and a minimum CMRR$_{dB} = 70$ dB. What is the common-mode voltage gain? If a desired and common-mode signal each has a value of 5 μV, what is the output voltage?

17-13   If the supply voltages are reduced to $+10$ V and -10 V, what is the common-mode rejection ratio of Fig. 17-37? Express the answer in decibels.

17-14   The data sheet of an op amp gives $A_v = 150{,}000$ and CMRR $= 85$ dB. What is the common-mode voltage gain?

SEC. 17–8   THE LOADED DIFF AMP

17-15   A load resistance of 27 kΩ is connected across the differential output of Fig. 17-36. What is the load voltage?

17-16   What is the load current in Fig. 17-36 if an ammeter is across the output?

Figure 17-34

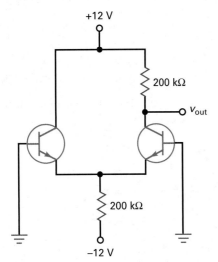

Figure 17-36

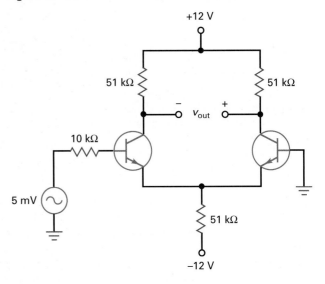

Figure 17-35

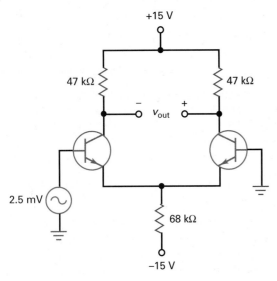

Figure 17-37

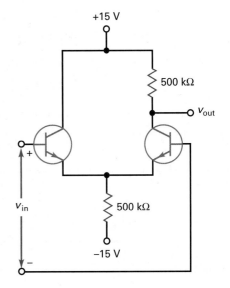

Troubleshooting

17-17 Somebody builds the diff amp of Fig. 17-35 without a ground on the inverting input. What does the output voltage equal? Based on your preceding answer, what does any diff amp or op amp need to work properly?

17-18 In Fig. 17-34, 20 kΩ is mistakenly used for the upper 200 kΩ. What does the output voltage equal?

17-19 In Fig. 17-34, V_{out} is almost zero. The input bias current is 80 nA. Which of the following is the trouble?

a. Upper 200 kΩ shorted

b. Lower 200 kΩ open

c. Left base open

d. Both inputs shorted together

Critical Thinking

17-20 In Fig. 17-34, the transistors are identical with $\beta_{dc} = 200$. What is the output voltage?

17-21 What are base voltages in Fig. 17-34 if each transistor has $\beta_{dc} = 300$?

17-22 In Fig. 17-38, transistors Q_3 and Q_5 are connected to act like compensating diodes for Q_4 and Q_6. What is the tail current? The current through the active load?

Figure 17-38

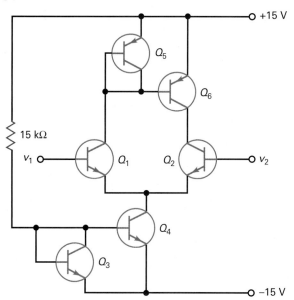

17-23 The 15 kΩ of Fig. 17-38 is changed to get a tail current of 15 μA. What is the new value of resistance?

17-24 At room temperature, the output voltage of Fig. 17-34 has a value of 6.0 V. As the temperature increases, the V_{BE} of each emitter diode decreases. If the left V_{BE} decreases 2 mV per degree and the right V_{BE} decreases 2.1 mV per degree, what is the output voltage at 75°C?

17-25 The dc resistance of each signal source in Fig. 17-39a is zero. What is the r'_e of each transistor? If the ac output voltage is between the collectors, what is the voltage gain?

17-26 If the transistors are identical in Fig. 17-39b, what is the tail current? The voltage between the left collector and ground? Between the right collector and ground?

Figure 17-39

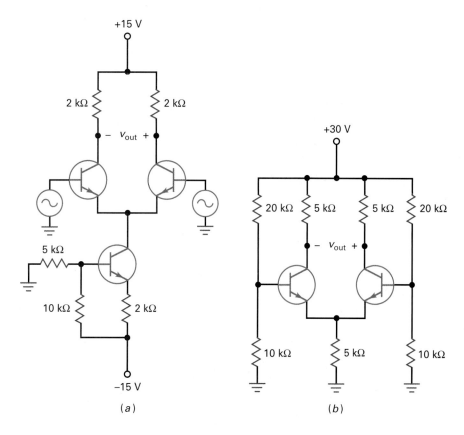

(a)

(b)

Up-Down Circuit Analysis

17-27 In Fig. 17-40, predict the response of each dependent variable in the rows labeled I_{B1} and I_{B2}.

17-28 In Fig. 17-40, predict the response of each dependent variable in the rows labeled R_E and R_C.

17-29 In Fig. 17-40, predict the response of each dependent variable in the row labeled V_{CC}.

Figure 17-40

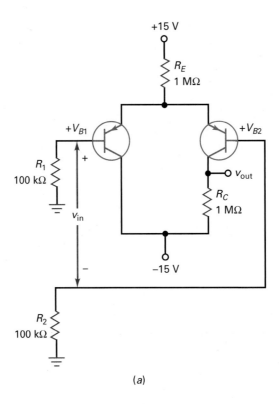

+15 V

R_E
1 MΩ

+V_{B1}

+V_{B2}

R_1
100 kΩ

v_{out}

R_C
1 MΩ

v_{in}

−15 V

R_2
100 kΩ

(a)

Up-Down Circuit Analysis

Increase	V_{B1}	V_{B2}	V_{in}	V_{out}	I_T
I_{B1}					
I_{B2}					
R_E					
R_C					
±V_{CC}					

(b)

Job–Interview Questions

1. Please draw the six configurations of a diff amp and identify the inputs and outputs as noninverting, inverting, single-ended, or differential.
2. Draw a diff amp with a differential input and single-ended output. Tell me how you would calculate the tail current, the emitter currents, and the collector voltages.
3. Draw any diff amp that has a voltage gain of R_C/r'_e. Now, draw any other diff amp that has a voltage gain of $R_C/2r'_e$.
4. Tell me what a common-mode signal is and what advantage a diff amp has when this kind of signal is present at the input.
5. A diff amp has an ammeter connected across its differential output. How would you go about calculating the current through the ammeter?
6. Assume that you have a diff amp circuit with a tail resistor. You have determined that the CMRR of the circuit is not acceptable. How could you improve the CMRR?
7. Explain the concept of a current mirror and why it is used.
8. Should CMRR be a large or a small number? Why?
9. In a diff amp, both emitters are tied together and get current from a common resistor. If you were to replace the common resistor with any type of component, what would you use to improve operation?
10. Why does a diff amp have a higher input impedance than a CE amplifier?
11. What does a current mirror simulate; that is, what is it used as?
12. What are the advantages of using current mirrors?
13. How do you test a 741 op amp with an ohmmeter?

Self–Test Answers

1.	b	9.	d	17.	d
2.	c	10.	a	18.	c
3.	a	11.	c	19.	b
4.	c	12.	b	20.	c
5.	b	13.	c	21.	a
6.	a	14.	a	22.	c
7.	b	15.	a	23.	c
8.	a	16.	b		

Practice Problem Answers

17-1 $I_T = 3$ mA; $I_E = 1.5$ mA; $V_C = 7.5$ V; $V_E = 0$ V

17-2 $I_T = 2.86$ mA; $I_E = 1.42$ mA; $V_C = 7.85$ V; $V_E = -0.7$ V

17-3 $I_T = 3.77$ mA; $I_E = 1.88$ mA; $V_E = 6.35$ V

17-4 $I_E = 1.5$ mA; $r'_e = 1.67$ Ω; $A_v = 300$; $V_{out} = 300$ mV; $z_{in(base)} = 10$ kΩ

17-7 $I_T = 30$ μA; $r'_e = 1.67$ kΩ; $A_v = 300$; $V_{out} = 2.1$ V; $z_{in} = 1$ MΩ

17-9 $V_{error} = 638$ mV

17-10 $A_{v(CM)} = 0.25$; $V_{out} = 0.25$ V

17-11 $V_{out1} = 200$ mV; $V_{out2} = 0.5$ mV; $V_{out} = 200$ mV $+ 0.5$ mV

17-12 $A_{v(CM)} = 3.16$; $V_{out1} = 0.1$ V; $V_{out2} = 3.16$ μV

17-13 $V_L = 1.2$ V

17-14 $I_L = 0.4$ mA

18

Operational Amplifiers

Although some high-power op amps are available, most are low-power devices with a maximum power rating of less than a watt. Some op amps are optimized for their bandwidth, others for low input offsets, others for low noise, and so on. This is why the variety of commercially available op amps is so large. You can find an op amp for almost any analog application.

Op amps are some of the most basic active components in analog systems. For instance, by connecting two external resistors, we can adjust the voltage gain and bandwidth of an op amp to our exact requirements. Furthermore, with other external components, we can build waveform converters, oscillators, active filters, and other interesting circuits.

Objectives

After studying this chapter, you should be able to:

- List the characteristics of ideal op amps and 741 op amps.

- Define slew rate and use it to find the power bandwidth of an op amp.

- Analyze an op-amp inverting amplifier.

- Analyze an op-amp noninverting amplifier.

- Explain how summing amplifiers and voltage followers work.

- List other linear integrated circuits and discuss how they are applied.

Chapter Outline

Vocabulary

BIFET op amp

bootstrapping

closed-loop voltage gain

compensating capacitor

first-order response

gain-bandwidth product (GBW)

inverting amplifier

mixer

noninverting amplifier

nulling circuit

open-loop bandwidth

open-loop voltage gain

output error voltage

power bandwidth

power supply rejection ratio (PSRR)

short-circuit output current

slew rate

summing amplifier

virtual ground

virtual short

voltage-controlled voltage source (VCVS)

voltage follower

voltage step

18-1 Introduction to Op Amps

Figure 18-1 shows a block diagram of an op amp. The input stage is a diff amp, followed by more stages of gain, and a class B push-pull emitter follower. Because a diff amp is the first stage, it determines the input characteristics of the op amp. In most op amps the output is single-ended, as shown. With positive and negative supplies, the single-ended output is designed to have a quiescent value of zero. This way, zero input voltage ideally results in zero output voltage.

Not all op amps are designed like Fig. 18-1. For instance, some do not use a class B push-pull output, and others may have a double-ended output. Also, op amps are not as simple as Fig. 18-1 suggests. The internal design of a monolithic op amp is very complicated, using dozens of transistors as current mirrors, active loads, and other innovations that are not possible in discrete designs. For our needs, Fig. 18-1 captures two important features that apply to typical op amps: the differential input and the single-ended output.

Figure 18-2a is the schematic symbol of an op amp. It has noninverting and inverting inputs and a single-ended output. Ideally, this symbol means that the

Figure 18-1   Block diagram of an op amp.

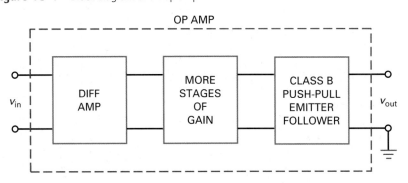

Figure 18-2   (a) Schematic symbol for op amp; (b) equivalent circuit of op amp.

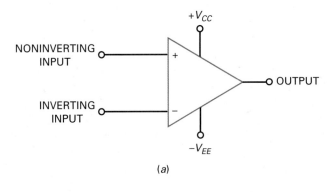

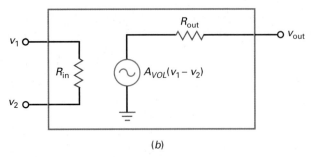

Table 18-1 | Typical Op-Amp Characteristics

Quantity	Symbol	Ideal	LM741C	LF157A
Open-loop voltage gain	A_{VOL}	Infinite	100,000	200,000
Unity-gain frequency	f_{unity}	Infinite	1 MHz	20 MHz
Input resistance	R_{in}	Infinite	2 MΩ	10^{12} Ω
Output resistance	R_{out}	Zero	75 Ω	100 Ω
Input bias current	$I_{in(bias)}$	Zero	80 nA	30 pA
Input offset current	$I_{in(off)}$	Zero	20 nA	3 pA
Input offset voltage	$V_{in(off)}$	Zero	2 mV	1 mV
Common-mode rejection ratio	CMRR	Infinite	90 dB	100 dB

amplifier has infinite voltage gain, infinite input impedance, and zero output impedance. The ideal op amp represents a perfect voltage amplifier and is often referred to as a **voltage-controlled voltage source (VCVS).** We can visualize a VCVS as shown in Fig. 18-2b, where R_{in} is infinite and R_{out} is zero.

Table 18-1 summarizes the characteristics of an ideal op amp. The ideal op amp has infinite voltage gain, infinite unity-gain frequency, infinite input impedance, and infinite CMRR. It also has zero output resistance, zero bias current, and zero offsets. This is what manufacturers would build if they could. What they actually can build approaches these ideal values.

For instance, the LM741C of Table 18-1 is a standard op amp, a classic that has been available since the 1960s. Its characteristics are the minimum of what to expect from a monolithic op amp. The LM741C has a voltage gain of 100,000, a unity-gain frequency of 1 MHz, an input impedance of 2 MΩ, and so on. Because the voltage gain is so high, the input offsets can easily saturate the op amp. This is why practical circuits need external components between the input and output of an op amp to stabilize the voltage gain. For instance, in many applications negative feedback is used to adjust the overall voltage gain to a much lower value in exchange for stable linear operation.

When no feedback path (or loop) is used, the voltage gain is maximum and is called the **open-loop voltage gain,** designated A_{VOL}. In Table 18-1, notice that the A_{VOL} of the LM741C is 100,000. Although not infinite, this open-loop voltage gain is very high. For instance, an input as small as 10 μV produces an output of 1 V. Because the open-loop voltage gain is very high, we can use heavy negative feedback to improve the overall performance of a circuit.

The 741C has a unity-gain frequency of 1 MHz. This means that we can get usable voltage gain almost as high as 1 MHz. The 741C has an input resistance of 2 MΩ, an output resistance of 75 Ω, an input bias current of 80 nA, an input offset current of 20 nA, an input offset voltage of 2 mV, and a CMRR of 90 dB.

When higher input resistance is needed, a designer can use a **BIFET op amp.** This type of op amp incorporates JFETs and bipolar transistors on the same chip. The JFETs are used in the input stage to get smaller input bias and offset currents; the bipolar transistors are used in the later stages to get more voltage gain.

The LF157A is an example of a BIFET op amp. As shown in Table 18-1, the input bias current is only 30 pA, and the input resistance if 10^{12} Ω. The

LF157A has a voltage gain of 200,000 and a unity-gain frequency of 20 MHz. With this device, we can get voltage gain up to 20 MHz.

18-2 The 741 Op Amp

In 1965 Fairchild Semiconductor introduced the μA709, the first widely used monolithic op amp. Although successful, this first-generation op amp had many disadvantages. These led to an improved op amp known as the μA741. Because it is inexpensive and easy to use, the μA741 has been an enormous success. Other 741 designs have appeared from various manufacturers. For instance, Motorola produces the MC1741, National Semiconductor the LM741, and Texas Instruments the SN72741. All these monolithic op amps are equivalent to the μA741 because they have the same specifications on their data sheets. For convenience, most people drop the prefixes and refer to this widely used op amp simply as the 741.

An Industry Standard

The 741 has become an industry standard. As a rule, you try to use it first in your designs. In cases when you cannot meet a design specification with a 741, you upgrade to a better op amp. Because it is a standard, we will use the 741 as a basic device in our discussions. Once you understand the 741, you can branch out to other op amps.

Incidentally, the 741 has different versions numbered 741, 741A, 741C, 741E, and 741N. These differ in their voltage gain, temperature range, noise level, and other characteristics. The 741C (the C stands for "commercial grade") is the least expensive and most widely used. It has an open-loop voltage gain of 100,000, an input impedance of 2 MΩ, and an output impedance of 75 Ω. Fig. 18-3 shows three popular package styles and their respective pin outs.

Figure 18-3   The 741 package style and pin outs: (a) Dual-in-line; (b) ceramic flatpak; and (c) metal can.

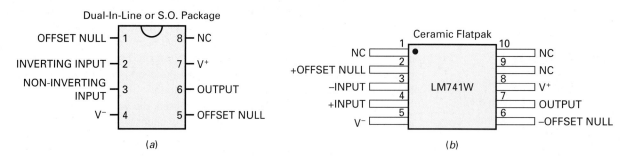

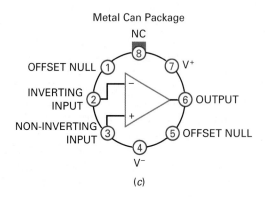

Figure 18-4 Simplified schematic diagram of 741.

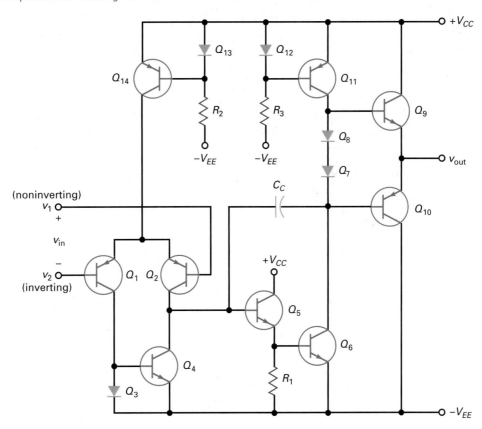

The Input Diff Amp

Figure 18-4 is a simplified schematic diagram of the 741. This circuit is equivalent to the 741 and many later-generation op amps. You do not need to understand every detail about the circuit design, but you should have a general idea of how the circuit works. With that in mind, here is the basic idea behind a 741.

The input stage is a diff amp (Q_1 and Q_2). In the 741, Q_{14} is a current source that replaces the tail resistor. R_2, Q_{13}, and Q_{14} are a current mirror that produces the tail current for Q_1 and Q_2. Instead of using an ordinary resistor as the collector resistor of the diff amp, the 741 uses an active load resistor. This active load Q_4 acts like a current source with an extremely high impedance. Because of this, the voltage gain of the diff amp is much higher than with a passive load resistor.

The amplified signal from the diff amp drives the base of Q_5, an emitter follower. This stage steps up the impedance level to avoid loading down the diff amp. The signal out of Q_5 goes to Q_6. Diodes Q_7 and Q_8 are part of the biasing for the final stage. Q_{11} is an active load resistor for Q_6. Therefore, Q_6 and Q_{11} are like a CE driver stage with a very high voltage gain.

The Final Stage

The amplified signal out of the CE driver stage (Q_6) goes to the final stage, which is a class B push-pull emitter follower (Q_9 and Q_{10}). Because of the split supply (equal positive V_{CC} and negative V_{EE} voltages), the quiescent output is ideally 0 V when the input voltage is zero. Any deviation from 0 V is called the **output error voltage.**

When v_1 is greater than v_2, the input voltage v_{in} produces a positive output voltage v_{out}. When v_2 is greater than v_1, the input voltage v_{in} produces a negative output voltage v_{out}. Ideally, v_{out} can be as positive as $+V_{CC}$ and as negative as $-V_{EE}$ before clipping occurs. The output swing is within 1 to 2 V of each supply voltage because of voltage drops inside the 741.

Active Loading

In Fig. 18-4, we have two examples of *active loading* (using transistors instead of resistors for loads) as discussed in Chap. 17. First, there is the active load Q_4 on the input diff amp. Second, there is the active load Q_{11} in the CE driver stage. Because current sources have high output impedances, active loads produce much higher voltage gain than is possible with resistors. These active loads produce a typical voltage gain of 100,000 for the 741C. Active loading is very popular in integrated circuits (ICs) because it is easier and less expensive to fabricate transistors on a chip than it is to fabricate resistors.

Frequency Compensation

In Fig. 18-4, C_c is a **compensating capacitor.** Because of the Miller effect (discussed in Chap. 16), this small capacitor (typically 30 pF) is multiplied by the voltage gain of Q_5 and Q_6 to get a much larger equivalent capacitance of:

$$C_{in(M)} = (A_v + 1)C_c$$

where A_v is the voltage gain of the Q_5 and Q_6 stages.

The resistance facing this Miller capacitance is the output impedance of the diff amp. Therefore, we have a lag circuit, as described in Chap. 16. This lag circuit produces a cutoff frequency of 10 Hz in a 741C. The open-loop gain of the op amp is down 3 dB at this cutoff frequency. Then, A_{VOL} decreases approximately 20 dB per decade until reaching the unity-gain frequency.

Figure 18-5 shows the ideal Bode plot of open-loop voltage gain versus frequency. The 741C has an open-loop voltage gain of 100,000, equivalent to 100 dB. Since the open-loop cutoff frequency is 10 Hz, the voltage gain breaks at 10 Hz and then rolls off at a rate of 20 dB per decade until it is 0 dB at 1 MHz.

A later chapter discusses *active filters,* circuits that use op amps, resistors, and capacitors to tailor the frequency response for different applications. At that time, we will discuss circuits that produce a first-order response (20 dB per decade rolloff), a second-order response (40 dB per decade rolloff), a third-order response (60 dB per decade rolloff), and so on. An op amp that is internally compensated, such as the 741C, has a **first-order response.**

Figure 18–5 Ideal Bode plot of open-loop voltage gain for 741C.

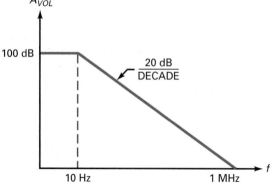

Incidentally, not all op amps are internally compensated. Some require the user to connect an external compensating capacitor to prevent oscillations. The advantage of using external compensation is that a designer has more control over the high-frequency performance. Although an external capacitor is the simplest way to compensate, more complicated circuits can be used that not only provide compensation but also produce a higher f_{unity} than is possible with internal compensation.

Bias and Offsets

As discussed in Chap. 17, a diff amp has input bias and offsets that produce an output error when there is no input signal. In many applications, the output error is small enough to ignore. But when the output error cannot be ignored, a designer can reduce it by using equal base resistors. This eliminates the problem of bias current, but not the offset current or offset voltage.

This is why it is best to eliminate output error by using the **nulling circuit** given on the data sheet. This recommended nulling circuit works with the internal circuitry to eliminate the output error and also to minimize *thermal drift,* a slow change in output voltage caused by the effect of changing temperature on op-amp parameters. Sometimes, the data sheet of an op amp does not include a nulling circuit. In this case, we have to apply a small input voltage to null the output. We will discuss this method later.

Figure 18-6 shows the nulling method suggested on the data sheet of a 741C. The ac source driving the inverting input has a Thevenin resistance of R_B. To neutralize the effect of input bias current (80 nA) flowing through this source resistance, a discrete resistor of equal value is added to the noninverting input, as shown.

To eliminate the effect of an input offset current of 20 nA and an input offset voltage of 2 mV, the data sheet of a 741C recommends using a 10-kΩ potentiometer between pins 1 and 5. By adjusting this potentiometer with no input signal, we can null or zero the output voltage.

Common–Mode Rejection Ratio

For a 741C, CMRR is 90 dB at low frequencies. Given equal signals, one a desired signal and the other a common-mode signal, the desired signal will be 90 dB larger at the output than the common-mode signal. In ordinary numbers, this means that the desired signal will be approximately 30,000 times larger than the common-mode signal. At higher frequencies, reactive effects degrade

Figure 18-6 Compensation and nulling used with 741C.

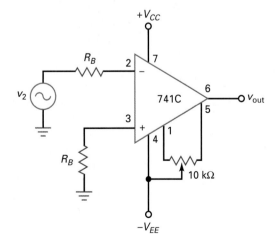

Figure 18-7 Typical 741C graphs for CMRR, MPP, and A_{VOL}.

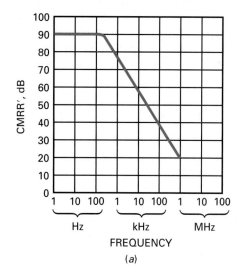

(a)

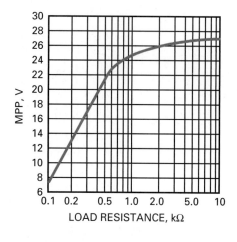

(b)

(c)

CMRR, as shown in Fig. 18-7a. Notice that CMRR is approximately 75 dB at 1 kHz, 56 dB at 10 kHz, and so on.

Maximum Peak-to-Peak Output

As discussed in Chap. 12, the MPP value of an amplifier is the maximum peak-to-peak output that the amplifier can produce. Since the quiescent output of an op amp is ideally zero, the ac output voltage can swing positively or negatively. For load resistances that are much larger than R_{out}, the output voltage can swing almost to the supply voltages. For instance, if $V_{CC} = +15$ V and $V_{EE} = -15$ V, the MPP value with a load resistance of 10 kΩ is ideally 30 V.

With a nonideal op amp, the output cannot swing all the way to the value of the supply voltages because there are small voltage drops in the final stage of the op amp. Furthermore, when the load resistance is not large compared to R_{out}, some of the amplified voltage is dropped across R_{out}, which means that the final output voltage is smaller.

Figure 18-7b shows MPP versus load resistance for a 741C with supply voltages of +15 V and −15 V. Notice that MPP is approximately 27 V for an R_L of 10 kΩ. This means that the output saturates positively at +13.5 V and negatively at -13.5 V. When the load resistance decreases, MPP decreases as shown. For instance, if the load resistance is only 275 Ω, MPP decreases to 16 V, which means that the output saturates positively at $+8$ V and negatively at -8 V.

Short–Circuit Current

In some applications, an op amp may drive a load resistance of approximately zero. In this case, you need to know the value of the **short-circuit output current.** The data sheet of a 741C lists a short-circuit output current of 25 mA. This is the maximum output current the op amp can produce. If you are using small load resistors (less than 75 Ω), don't expect to get a large output voltage because the voltage cannot be greater than the 25 mA times the load resistance.

Frequency Response

Figure 18-7c shows the small-signal frequency response of a 741C. In the midband, the voltage gain is 100,000. The 741C has a cutoff frequency f_c of 10 Hz. As indicated, the voltage gain is 70,700 (down 3 dB) at 10 Hz. Above the cutoff frequency, the voltage gain decreases at a rate of 20 dB per decade (first-order response).

The unity-gain frequency is the frequency at which the voltage gain equals 1. In Fig. 18-7c, f_{unity} is 1 MHz. Data sheets usually specify the value of f_{unity} because it represents the upper limit on the useful gain of an op amp. For instance, the data sheet of a 741C lists an f_{unity} of 1 MHz. This means that the 741C can amplify signals up to 1 MHz. Beyond 1 MHz, the voltage gain is less than 1 and the 741C is useless. If a designer needs a higher f_{unity}, better op amps are available. For instance, the LM318 has an f_{unity} of 15 MHz, which means that it can produce usable voltage gain all the way to 15 MHz.

Slew Rate

The compensating capacitor inside a 741C performs a very important function: It prevents oscillations that would interfere with the desired signal. But there is a disadvantage. The compensating capacitor needs to be charged and discharged. This creates a speed limit on how fast the output of the op amp can change.

Here is the basic idea: Suppose the input voltage to an op amp is a positive **voltage step,** a sudden transition in voltage from one dc level to a higher dc level. If the op amp were perfect, we would get the ideal response shown in Fig. 18-8a. Instead, the output is the positive exponential waveform shown. This occurs because the compensating capacitor must be charged before the output voltage can change to the higher level.

In Fig. 18-8a, the initial slope of the exponential waveform is called the **slew rate,** symbolized S_R. The definition of slew rate is:

$$S_R = \frac{\Delta v_{out}}{\Delta t} \tag{18-1}$$

where the Greek letter Δ (delta) stands for "the change in." In words, the equation says that slew rate equals the change in output voltage divided by the change in time.

Figure 18-8b illustrates the meaning of slew rate. The initial slope equals the vertical change divided by the horizontal change between two points on the early part of the exponential wave. For instance, if the exponential wave increases 0.5 V during the first microsecond, as shown in Fig. 18-8c, the slew rate is:

$$S_R = \frac{0.5 \text{ V}}{1 \ \mu\text{s}} = 0.5 \text{ V}/\mu\text{s}$$

Figure 18-8 (*a*) Ideal and actual responses to an input step voltage; (*b*) illustrating definition of slew rate; (*c*) slew rate equals 0.5 V/μs.

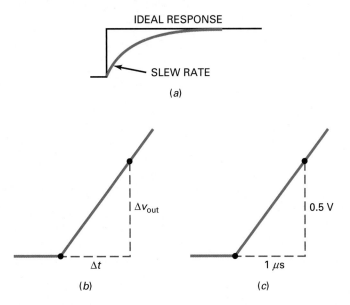

The slew rate represents the fastest response that an op amp can have. For instance, the slew rate of a 741C is 0.5 V/μs. This means that the output of a 741C can change no faster than 0.5 V in a microsecond. In other words, if a 741C is driven by a large step in input voltage, we do not get a sudden step in output voltage. Instead, we get an exponential output wave. The initial part of this output waveform will look like Fig. 18-8*c*.

We can also get slew-rate limiting with a sinusoidal signal. Here is how it occurs: In Fig. 18-9*a*, the op amp can produce the output sine wave shown only if the initial slope of the sine wave is less than the slew rate. For instance, if the output sine wave has an initial slope of 0.1 V/μs, a 741C can produce this sine wave with no trouble at all because its slew rate is 0.5 V/μs. On the other hand, if the sine wave has an initial slope of 1 V/μs, the output is smaller than it should be and it looks triangular instead of sinusoidal, as shown in Fig. 18-9*b*.

The data sheet of an op amp always specifies the slew rate because this quantity limits the large-signal response of an op amp. If the output sine wave is very small or the frequency is very low, slew rate is no problem. But when the signal is large and the frequency is high, slew rate will distort the output signal.

With calculus, it is possible to derive this equation:

$$S_S = 2\pi f V_p$$

Figure 18-9 (*a*) Initial slope of a sine wave; (*b*) distortion occurs if initial slope exceeds slew rate.

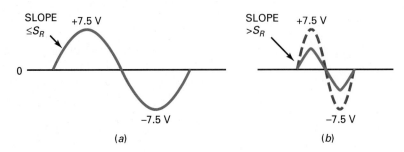

Chapter 18

where S_S is the initial slope of the sine wave, f is its frequency, and V_p is its peak value. To avoid slew-rate distortion of a sine wave, S_S has to be less than or equal to S_R. When the two are equal, we are at the limit, on the verge of slew-rate distortion. In this case:

$$S_R = S_S = 2\pi f V_p$$

Solving for f gives:

$$f_{\max} = \frac{S_R}{2\pi V_p} \qquad\qquad (18\text{-}2)$$

where $f_{\max}$ is the highest frequency that can be amplified without slew-rate distortion. Given the slew rate of an op amp and the peak output voltage desired, we can use Eq. (18-2) to calculate the maximum undistorted frequency. Above this frequency, we will see slew-rate distortion on an oscilloscope.

The frequency $f_{\max}$ is sometimes called the **power bandwidth** or *large-signal bandwidth* of the op amp. Figure 18-10 is a graph of Eq. (18-2) for three slew rates. Since the bottom graph is for a slew rate of 0.5 V/μs, it is useful with a 741C. Since the top graph is for a slew rate of 50 V/μs, it is useful with an LM318 (it has a minimum slew rate of 50 V/μs).

For instance, suppose we are using a 741C. To get an undistorted output peak voltage of 8 V, the frequency can be no higher than 10 kHz (see Fig. 18-10). One way to increase the $f_{\max}$ is to accept less output voltage. By trading off peak value for frequency, we can improve the power bandwidth. As an example, if our application can accept a peak output voltage of 1 V, $f_{\max}$ increases to 80 kHz.

There are two bandwidths to consider when analyzing the operation of an op-amp circuit: the small-signal bandwidth determined by the first-order response of the op amp and the large-signal or power bandwidth determined by the slew rate. More will be said about these two bandwidths later.

Figure 18-10 Graph of power bandwidth versus peak voltage.

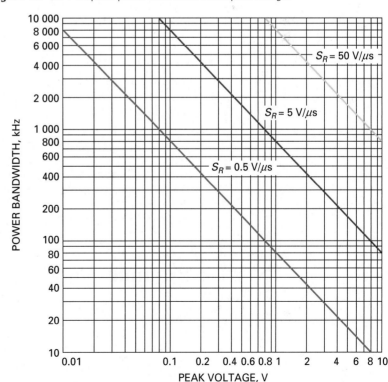

Example 18-1

How much inverting input voltage does it take to drive the 741C of Fig. 18-11a into negative saturation?

Figure 18-11 Example.

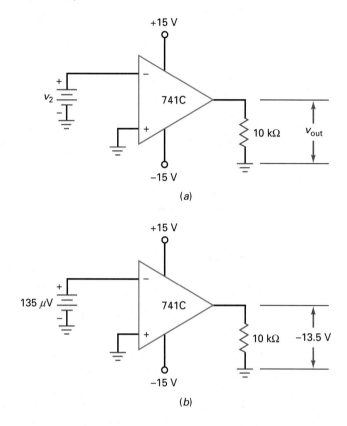

(a)

(b)

SOLUTION Figure 18-7b shows that MPP equals 27 V for a load resistance of 10 kΩ, which translates into an output of −13.5 V for negative saturation. Since the 741C has an open-loop voltage gain of 100,000, the required input voltage is:

$$v_2 = \frac{13.5 \text{ V}}{100,000} = 135 \ \mu V$$

Figure 18-11b summarizes the answer. As you can see, an inverting input of 135 μV produces negative saturation, an output voltage of −13.5 V.

PRACTICE PROBLEM 18-1 Repeat Example 18-1 where $A_{VOL} = 200,000$.

Example 18-2

What is the common-mode rejection ratio of a 741C when the input frequency is 100 kHz?

SOLUTION In Fig. 18-7a, we can read a CMRR of approximately 40 dB at 100 kHz. This is equivalent to 100, which means that the desired signal receives

100 times more amplification than a common-mode signal when the input frequency is 100 kHz.

PRACTICE PROBLEM 18-2 What is the CMRR of a 741C when the input frequency is 10 kHz?

Example 18-3

What is the open-loop voltage gain of a 741C when the input frequency is 1 kHz? 10 kHz? 100 kHz?

SOLUTION In Fig. 18-7c, the voltage gain is 1000 for 1 kHz, 100 for 10 kHz, and 10 for 100 kHz. As you can see, the voltage gain decreases by a factor of 10 each time the frequency increases by a factor of 10.

Example 18-4

The input voltage to an op amp is a large voltage step. The output is an exponential waveform that changes to 0.25 V in 0.1 μs. What is the slew rate of the op amp?

SOLUTION With Eq. (18-1):

$$S_R = \frac{0.25 \text{ V}}{0.1 \text{ }\mu\text{s}} = 2.5 \text{ V/}\mu\text{s}$$

PRACTICE PROBLEM 18-4 If the measured output voltage changes 0.8 V in 0.2 μS, what is the slew rate?

Example 18-5

The LF411A has a slew rate of 15 V/μs. What is the power bandwidth for a peak output voltage of 10 V?

SOLUTION With Eq. (18-2):

$$f_{\max} = \frac{S_R}{2\pi V_p} = \frac{15 \text{ V/}\mu\text{s}}{2\pi(10 \text{ V})} = 239 \text{ kHz}$$

PRACTICE PROBLEM 18-5 Repeat Example 18-5 using a 741C and $V_p = 200$ mV.

Example 18-6

What is the power bandwidth for each of the following?

$S_R = 0.5$ V/μs and $V_p = 8$ V
$S_R = 5$ V/μs and $V_p = 8$ V
$S_R = 50$ V/μs and $V_p = 8$ V

SOLUTION With Fig. 18-10, read each power bandwidth to get these approximate answers: 10 kHz, 100 kHz, and 1 MHz.

PRACTICE PROBLEM 18-6 Repeat Example 18-6 with $V_p = 1$ V.

18-3 The Inverting Amplifier

The **inverting amplifier** is the most basic op-amp circuit. It uses negative feedback to stabilize the overall voltage gain. The reason we need to stabilize the overall voltage gain is because A_{VOL} is too high and unstable to be of any use without some form of feedback. For instance, the 741C has a minimum A_{VOL} of 20,000 and a maximum A_{VOL} of more than 200,000. An unpredictable voltage gain of this magnitude and variation is useless without feedback.

Inverting Negative Feedback

Figure 18-12 shows an inverting amplifier. To keep the drawing simple, the power-supply voltages are not shown. In other words, we are looking at the ac equivalent circuit. An input voltage v_{in} drives the inverting input through resistor R_1. This results in an inverting input voltage of v_2. The input voltage is amplified by the open-loop voltage gain to produce an inverted output voltage. The output voltage is fed back to the input through feedback resistor R_f. This results in negative feedback because the output is 180° out of phase with the input. In other words, any changes in v_2 produced by the input voltage are opposed by the output signal.

     Here is how the negative feedback stabilizes the overall voltage gain: If the open-loop voltage gain A_{VOL} increases for any reason, the output voltage will increase and feed back more voltage to the inverting input. This opposing feedback voltage reduces v_2. Therefore, even though A_{VOL} has increased, v_2 has decreased, and the final output increases much less than it would without the negative feedback. The overall result is a very slight increase in output voltage, so small that it is hardly noticeable. In Chap. 19, we will discuss the mathematical details of negative feedback and you will better understand how small the changes are.

Virtual Ground

When we connect a piece of wire between some point in a circuit and ground, the voltage of the point becomes zero. Furthermore, the wire provides a path for current to flow to ground. A *mechanical ground* (a wire between a point and ground) is ground to both voltage and current.

    A **virtual ground** is different. This type of ground is a widely used shortcut for analyzing an inverting amplifier. With a virtual ground, the analysis of an inverting amplifier and related circuits becomes incredibly easy.

    The concept of a virtual ground is based on an ideal op amp. When an op amp is ideal, it has infinite open-loop voltage gain and infinite input resistance. Because of this, we can deduce the following ideal properties for the inverting amplifier of Fig. 18-13:

1. Since R_{in} is infinite, i_2 is zero.
2. Since A_{VOL} is infinite, v_2 is zero.

Figure 18-12 The inverting amplifier.

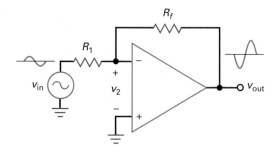

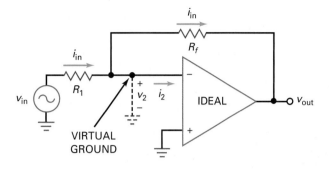

Since i_2 is zero in Fig. 18-13, the current through R_f must equal the input current through R_1, as shown. Furthermore, since v_2 is zero, the virtual ground shown in Fig. 18-13 means that the inverting input acts like a ground for voltage but an open for current!

Virtual ground is very unusual. It is like half of a ground because it is a short for voltage but an open for current. To remind us of this half-ground quality, Fig. 18-13 uses a dashed line between the inverting input and ground. The dashed line means that no current can flow to ground. Although virtual ground is an ideal approximation, it gives very accurate answers when used with heavy negative feedback.

Voltage Gain

In Fig. 18-14, visualize a virtual ground on the inverting input. Then, the right end of R_1 is a voltage ground, so we can write:

$$v_{in} = i_{in} R_1$$

Similarly, the left end of R_f is a voltage ground, so the magnitude of output voltage is:

$$v_{out} = -i_{in} R_f$$

Divide v_{out} by v_{in} to get the voltage gain:

$$A_{v(CL)} = \frac{-R_f}{R_1} \qquad (18\text{-}3)$$

where $A_{v(CL)}$ is the closed-loop voltage gain. This is called the **closed-loop voltage gain** because it is the voltage when there is a feedback path between the output and the input. Because of the negative feedback, the closed-loop voltage gain $A_{v(CL)}$ is always smaller than the open-loop voltage gain A_{VOL}.

Figure 18-14 Inverting amplifier has same current through both resistors.

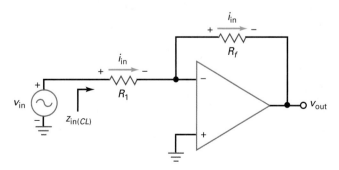

Look at how simple and elegant Eq. (18-3) is. The closed-loop voltage gain equals the ratio of the feedback resistance to the input resistance. For instance, if $R_1 = 1$ kΩ and $R_f = 50$ kΩ, the closed-loop voltage gain is 50. Because of the heavy negative feedback, this closed-loop voltage gain is very stable. If A_{VOL} varies because of temperature change, supply voltage variations, or op-amp replacement, $A_{v(CL)}$ will still be very close to 50. Chapter 19 will discuss gain stability in more detail. The negative sign in the voltage gain equation indicates a 180° phase shift.

Input Impedance

In some applications, a designer may want a specific input impedance. This is one of the advantages of an inverting amplifier; it is easy to set up a desired input impedance. Here is why: Since the right end of R_1 is virtually grounded, the closed-loop input impedance is:

$$z_{in(CL)} = R_1 \qquad\qquad (18\text{-}4)$$

This is the impedance looking into the left end of R_1, as shown in Fig. 18-14. For instance, if an input impedance of 2 kΩ and a closed-loop voltage gain of 50 is needed, a designer can use $R_1 = 2$ kΩ and $R_f = 100$ kΩ.

Bandwidth

The **open-loop bandwidth** or cutoff frequency of an op amp is very low because of the internal compensating capacitor. For a 741C:

$$f_{2(OL)} = 10 \text{ Hz}$$

At this frequency, the open-loop voltage gain breaks and rolls off in a first-order response.

When negative feedback is used, the overall bandwidth increases. Here is the reason: When the input frequency is greater than $f_{2(OL)}$, A_{VOL} decreases 20 dB per decade. When v_{out} tries to decrease, less opposing voltage is fed back to the inverting input. Therefore, v_2 increases and compensates for the decrease in A_{VOL}. Because of this, $A_{v(CL)}$ breaks at a higher frequency than $f_{2(OL)}$. The greater the negative feedback, the higher the closed-loop cutoff frequency. Stated another way: The smaller $A_{v(CL)}$ is, the higher $f_{2(CL)}$ is.

Figure 18-15 illustrates how the closed-loop bandwidth increases with negative feedback. As you can see, the heavier the negative feedback (smaller $A_{v(CL)}$), the greater the closed-loop bandwidth. Here is the equation for closed-loop bandwidth:

$$f_{2(CL)} = \frac{f_{unity}}{A_{v(CL)} + 1} \qquad\qquad \text{(inverting amplifier only)}$$

In most applications, $A_{v(CL)}$ is greater than 10 and the equation simplifies to:

$$f_{2(CL)} = \frac{f_{unity}}{A_{v(CL)}} \qquad\qquad \textbf{(noninverting)} \qquad (18\text{-}5)$$

For instance, when $A_{v(CL)}$ is 10:

$$f_{2(CL)} = \frac{1 \text{ MHz}}{10} = 100 \text{ kHz}$$

which agrees with Fig. 18-14. If $A_{v(CL)}$ is 100:

$$f_{2(CL)} = \frac{1 \text{ MHz}}{100} = 10 \text{ kHz}$$

which also agrees.

Figure 18-15 Lower voltage gain produces more bandwidth.

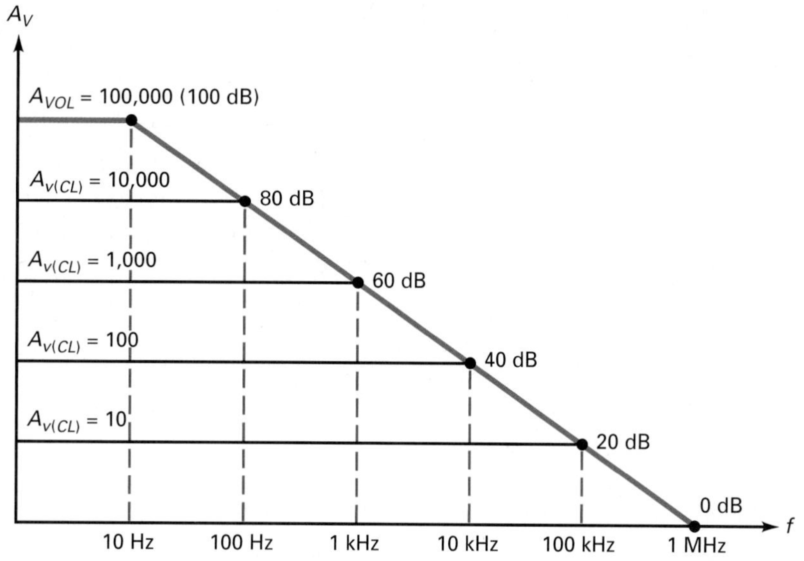

Equation (18-5) can be rearranged into:

$$f_{\text{unity}} = A_{v(CL)}f_{2(CL)} \tag{18-6}$$

Notice that the unity-gain frequency equals the product of gain and bandwidth. For this reason, many data sheets refer to the unity-gain frequency as the **gain-bandwidth product (GBW).**

(*Note:* No consistent symbol is used on data sheets for the open-loop voltage gain. You may see any of the following: A_{OL}, A_v, A_{vo}, and A_{vol}. It is usually clear from the data sheet that all these symbols represent the open-loop voltage gain of the op amp. We will use A_{VOL} in this book.)

Bias and Offsets

Negative feedback reduces the output error caused by input bias current, input offset current, and input offset voltage. Chapter 17 discussed the three input error voltages and the equation for total output error voltage:

$$V_{\text{error}} = A_{VOL}(V_{1\text{err}} + V_{2\text{err}} + V_{3\text{err}})$$

When negative feedback is used, this equation may be written as:

$$V_{\text{error}} \cong \pm A_{v(CL)}(\pm V_{1\text{err}} \pm V_{2\text{err}} \pm V_{3\text{err}}) \tag{18-7}$$

where V_{error} is the total output error voltage. Notice that the Eq. (18-7) includes $\pm$ signs. Data sheets do not include $\pm$ signs because it is implied that errors can be in either direction. For instance, either base current can be larger than the other, and the input offset voltage can be plus or minus.

In mass production, the input errors may add up in the worst possible way. The input errors were discussed in Chap. 17 and are repeated here:

$$V_{1\text{err}} = (R_{B1} - R_{B2})I_{\text{in(bias)}} \tag{18-8}$$

$$V_{2\text{err}} = (R_{B1} + R_{B2})\frac{I_{\text{in(off)}}}{2} \tag{18-9}$$

$$V_{3\text{err}} = V_{\text{in(off)}} \tag{18-10}$$

When $A_{v(CL)}$ is small, the total output error given by Eq. (18-7) may be small enough to ignore. If not, resistor compensation and offset nulling will be necessary.

In an inverting amplifier, R_{B2} is the Thevenin resistance seen when looking back from the inverting input toward the source. This resistance is given by:

$$R_{B2} = R_1 \,\|\, R_f \qquad\qquad (18\text{-}11)$$

If it is necessary to compensate for input bias current, an equal resistance R_{B1} should be connected to the noninverting input. This resistance has no effect on the virtual-ground approximation because no ac signal current flows through it.

Example 18-7

|||| MultiSim

Figure 18-16a is an ac equivalent circuit, so we can ignore the output error caused by input bias and offsets. What are closed-loop voltage gain and bandwidth? What is the output voltage at 1 kHz? At 1 MHz?

Figure 18-16 Example.

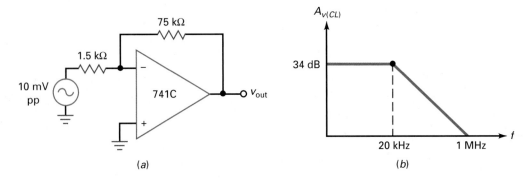

(a)

(b)

SOLUTION With Eq. (18-3), the closed-loop voltage gain is:

$$A_{v(CL)} = \frac{-75\ \text{k}\Omega}{1.5\ \text{k}\Omega} = -50$$

With Eq. (18-5), the closed-loop bandwidth is:

$$f_{2(CL)} = \frac{1\ \text{MHz}}{50} = 20\ \text{kHz}$$

Figure 18-16b shows the ideal Bode plot of the closed-loop voltage gain. The decibel equivalent of 50 is 34 dB. (Shortcut: 50 is half of 100, or down 6 dB from 40 dB.)

The output voltage at 1 kHz is:

$$v_{\text{out}} = (-50)(10\ \text{mV pp}) = -500\ \text{mV pp}$$

Since 1 MHz is the unity-gain frequency, the output voltage at 1 MHz is:

$$v_{\text{out}} = -10\ \text{mV pp}$$

Again, the minus ($-$) output value indicates a 180° phase shift between the input and output.

PRACTICE PROBLEM 18-7 In Fig. 18-16a, what is the output voltage at 100 kHz? [*Hint:* Use Eq. (16-20).]

Example 18-8

What is the output voltage in Fig. 18-17 when v_{in} is zero? Use the typical values given in Table 18-1.

Figure 18-17 Example.

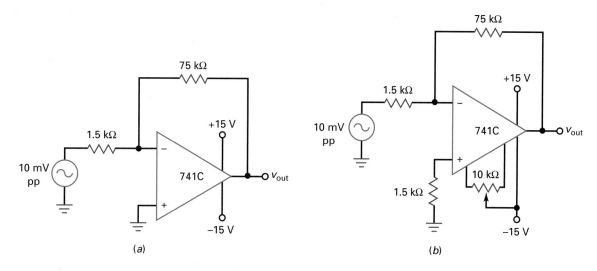

(a)

(b)

SOLUTION Table 18-1 shows these values for a 741C: $I_{in(bias)} = 80$ nA, $I_{in(off)} = 20$ nA, and $V_{in(off)} = 2$ mV. With Eq. (18-11):

$$R_{B2} = R_1 \| R_f = 1.5 \text{ k}\Omega \| 75 \text{ k}\Omega = 1.47 \text{ k}\Omega$$

With Eqs. (18-8) to (18-10), the three input error voltages are:

$$V_{1err} = (R_{B1} - R_{B2})I_{in(bias)} = (-1.47 \text{ k}\Omega)(80 \text{ nA}) = -0.118 \text{ mV}$$

$$V_{2err} = (R_{B1} + R_{B2})\frac{I_{in(off)}}{2} = (1.47 \text{ k}\Omega)(10 \text{ nA}) = 0.0147 \text{ mV}$$

$$V_{3err} = V_{in(off)} = 2 \text{ mV}$$

The closed-loop voltage gain is 50, calculated in the previous example. With Eq. (18-7), adding the errors in the worst possible way gives an output error voltage of:

$$V_{error} = \pm 50(0.118 \text{ mV} + 0.0147 \text{ mV} + 2 \text{ mV}) = \pm 107 \text{ mV}$$

PRACTICE PROBLEM 18-8 Repeat Example 18-8 using an LF157A op amp.

Example 18-9

In the foregoing example, we used typical parameters. The data sheet of a 741C lists the following worst-case parameters: $I_{in(bias)} = 500$ nA, $I_{in(off)} = 200$ nA, and $V_{in(off)} = 6$ mV. Recalculate the output voltage when v_{in} is zero in Fig. 18-17a.

SOLUTION With Eqs. (18-8) to (18-10), the three input error voltages are:

$$V_{1err} = (R_{B1} - R_{B2})I_{in(bias)} = (-1.47 \text{ k}\Omega)(500 \text{ nA}) = -0.735 \text{ mV}$$

$$V_{2err} = (R_{B1} + R_{B2})\frac{I_{in(off)}}{2} = (1.47 \text{ k}\Omega)(100 \text{ nA}) = 0.147 \text{ mV}$$

$$V_{3err} = V_{in(off)} = 6 \text{ mV}$$

Adding the errors in the worst possible way gives an output error voltage of:

$$V_{\text{error}} = \pm 50(0.735 \text{ mV} + 0.147 \text{ mV} + 6 \text{ mV}) = \pm 344 \text{ mV}$$

In Example 18-7, the desired output voltage was 500 mV pp. Can we ignore the large output error voltage? It depends on the application. For instance, suppose we only need to amplify audio signals with frequencies between 20 Hz and 20 kHz. Then, we can capacitively couple the output into the load resistor or next stage. This will block the dc output error voltage but transmit the ac signal. In this case, the output error is irrelevant.

On the other hand, if we want to amplify signals with frequencies from 0 to 20 kHz, then we need to use a better op amp (lower bias and offsets), or modify the circuit as shown in Fig. 18-17b. Here, we have added a compensating resistor to the noninverting input to eliminate the effect of input bias current. Also, we are using a 10-kΩ potentiometer to null the effects of input offset current and input offset voltage.

18-4 The Noninverting Amplifier

The **noninverting amplifier** is another basic op-amp circuit. It uses negative feedback to stabilize the overall voltage gain. With this type of amplifier, the negative feedback also increases the input impedance and decreases the output impedance.

Basic Circuit

Figure 18-18 shows the ac equivalent circuit of a noninverting amplifier. An input voltage v_{in} drives the noninverting input. This input voltage is amplified to produce the in-phase output voltage shown. Part of output voltage is fed back to the input through a voltage divider. The voltage across R_1 is the feedback voltage applied to the inverting input. This feedback voltage is almost equal to the input voltage. Because of the high open-loop voltage gain, the difference between v_1 and v_2 is very small. Since the feedback voltage opposes the input voltage, we have negative feedback.

Here is how the negative feedback stabilizes the overall voltage gain: If the open-loop voltage gain A_{VOL} increases for any reason, the output voltage will increase and feed back more voltage to the inverting input. This opposing feedback voltage reduces the net input voltage $v_1 - v_2$. Therefore, even though A_{VOL} increases, $v_1 - v_2$ decreases, and the final output increases much less than it

Figure 18-18 The noninverting amplifier.

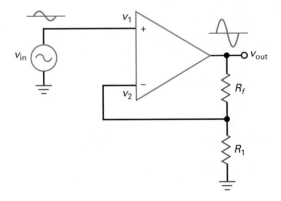

Figure 18-19 A virtual short exists between the two op-amp inputs.

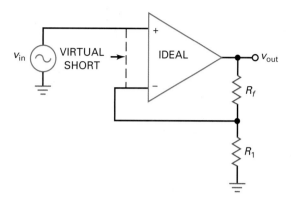

would without the negative feedback. The overall result is only a very slight increase in output voltage.

Virtual Short

When we connect a piece of wire between two points in a circuit, the voltage of both points with respect to ground is equal. Furthermore, the wire provides a path for current to flow between the two points. A *mechanical short* (a wire between two points) is a short for both voltage and current.

A **virtual short** is different. This type of short can be used for analyzing noninverting amplifiers. With a virtual short, we can quickly and easily analyze noninverting amplifiers and related circuits.

The virtual short uses these two properties of an ideal op amp:

1. Since R_{in} is infinite, both input currents are zero.
2. Since A_{VOL} is infinite, $v_1 - v_2$ is zero.

Figure 18-19 shows a virtual short between the input terminals of the op amp. The virtual short is a short for voltage but an open for current. As a reminder, the dashed line means that no current can flow through it. Although the virtual short is an ideal approximation, it gives very accurate answers when used with heavy negative feedback.

Here is how we will use the virtual short: Whenever we analyze a noninverting amplifier or a similar circuit, we can visualize a virtual short between the input terminals of the op amp. As long as the op amp is operating in the linear region (not positively or negatively saturated), the open-loop voltage gain approaches infinity and a virtual short exists between the two input terminals.

One more point: Because of the virtual short, the inverting input voltage follows the noninverting input voltage. If the noninverting input voltage increases or decreases, the inverting input voltage immediately increases or decreases to the same value. This follow-the-leader action is called **bootstrapping** (as in "pulling yourself up by your bootstraps"). The noninverting input pulls the inverting input up or down to an equal value. Described another way, the inverting input is bootstrapped to the noninverting input.

Voltage Gain

In Fig. 18-20, visualize a virtual short between the input terminals of the op amp. Then, the virtual short means that the input voltage appears across R_1, as shown. So, we can write:

$$v_{in} = i_1 R_1$$

Figure 18-20 Input voltage appears across R_1 and same current flows through resistors.

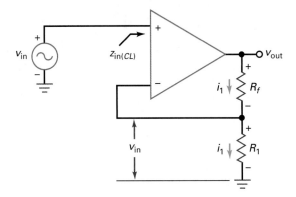

Since no current can flow through a virtual short, the same i_1 current must flow through R_f, which means that the output voltage is given by:

$$v_{\text{out}} = i_1(R_f + R_1)$$

Divide v_{out} by v_{in} to get the voltage gain:

$$A_{v(CL)} = \frac{R_f + R_1}{R_1}$$

or

$$A_{v(CL)} = \frac{R_f}{R_1} + 1 \tag{18-12}$$

This is easy to remember because it is the same as the equation for an inverting amplifier, except that we add 1 to the ratio of resistances. Also note that the output is in phase with the input. Therefore, no $(-)$ sign is used in the voltage gain equation.

Other Quantities

The closed-loop input impedance approaches infinity. In the next chapter, we will mathematically analyze the effect of negative feedback and will show that negative feedback increases the input impedance. Since the open-loop input impedance is already very high (2 MΩ for a 741C), the closed-loop input impedance will be even higher.

The effect of negative feedback on bandwidth is the same as with an inverting amplifier:

$$f_{2(CL)} = \frac{f_{\text{unity}}}{A_{v(CL)}}$$

Again, we can trade off voltage gain for bandwidth. The smaller the closed-loop voltage gain, the greater the bandwidth.

The input error voltages caused by input bias current, input offset current, and input offset voltage are analyzed the same way as with an inverting amplifier. After calculating each input error, we can multiply by the closed-loop voltage gain to get the total output error.

R_{B2} is the Thevenin resistance seen when looking from the inverting input toward the voltage divider. This resistance is the same as for an inverting amplifier:

$$R_{B2} = R_1 \, \| \, R_f$$

If it is necessary to compensate for input bias current, an equal resistance R_{B1} should be connected to the noninverting input. This resistance has no effect on the virtual-short approximation because no ac signal current flows through it.

Figure 18-21 Output error voltage reduces MPP.

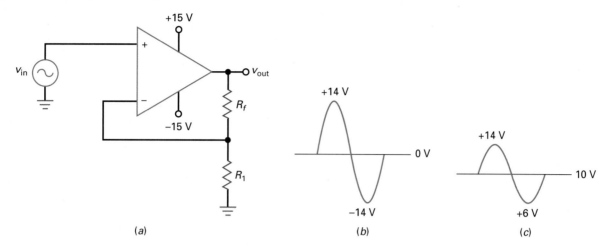

(a)　　　　(b)　　　　(c)

Output Error Voltage Reduces MPP

As previously discussed, if we are amplifying ac signals, we can capacitively couple the output signal to the load. In this case, we can ignore the output error voltage unless it is excessively large. If the output error voltage is large, it will significantly reduce the MPP, the maximum unclipped peak-to-peak output.

For instance, if there is no output error voltage, the noninverting amplifier of Fig. 18-21a can swing to within approximately a volt or two of either supply voltage. For simplicity, assume that the output signal can swing from +14 to −14 V, giving an MPP of 28 V, as shown in Fig. 18-21b. Now, suppose the output error voltage is +10 V, as shown in Fig. 18-21c. With this large output error voltage, the maximum unclipped peak-to-peak swing is from +14 to +6 V, an MPP of only 8 V. This may still be all right if the application does not require a large output signal. Here is the point to remember: The greater the output error voltage, the smaller the MPP value.

Example 18-10

||||| MultiSim

In Fig. 18-22a, what is closed-loop voltage gain and bandwidth? What is the output voltage at 250 kHz?

SOLUTION With Eq. (18-12):

$$A_{v(CL)} = \frac{3.9 \text{ k}\Omega}{100 \ \Omega} + 1 = 40$$

Dividing the unity-gain frequency by the closed-loop voltage gain gives:

$$f_{2(CL)} = \frac{1 \text{ MHz}}{40} = 25 \text{ kHz}$$

Figure 18-22b shows the ideal Bode plot of closed-loop voltage gain. The decibel equivalent of 40 is 32 dB. (Shortcut: $40 = 10 \times 2 \times 2$ or 20 dB + 6 dB + 6 dB = 32 dB.) Since the $A_{v(CL)}$ breaks at 25 kHz, it is down 20 dB at 250 kHz. This means that $A_{v(CL)} = 12$ dB at 250 kHz, which is equivalent to an

Figure 18-22 Example.

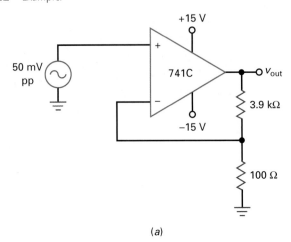

(a)

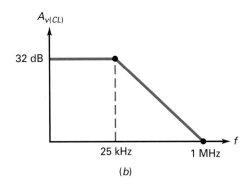

(b)

ordinary voltage gain of 4. Therefore, the output voltage at 250 kHz is:

$$v_{\text{out}} = 4 \ (50 \text{ mV pp}) = 200 \text{ mV pp}$$

PRACTICE PROBLEM 18-10 In Fig. 18-22, change the 3.9 kΩ resistor to 4.9 kΩ. Solve for $A_{v(CL)}$ and v_{out} at 200 kHz.

Example 18-11

For convenience, we repeat the worst-case parameters of a 741C: $I_{\text{in(bias)}} = 500$ nA, $I_{\text{in(off)}} = 200$ nA, and $V_{\text{in(off)}} = 6$ mV. What is the output error voltage in Fig. 18-22a?

SOLUTION R_{B2} is the parallel equivalent of 3.9 kΩ and 100 Ω, which is approximately 100 Ω. With Eqs. (18-8) to (18-10), the three input error voltages are:

$$V_{1\text{err}} = (R_{B1} - R_{B2})I_{\text{in(bias)}} = (-100 \ \Omega)(500 \text{ nA}) = -0.05 \text{ mV}$$

$$V_{2\text{err}} = (R_{B1} + R_{B2})\frac{I_{\text{in(off)}}}{2} = (100 \ \Omega)(100 \text{ nA}) = 0.01 \text{ mV}$$

$$V_{3\text{err}} = V_{\text{in(off)}} = 6 \text{ mV}$$

Adding the errors in the worst possible way gives an output error voltage of:

$$V_{\text{error}} = \pm 40(0.05 \text{ mV} + 0.01 \text{ mV} + 6 \text{ mV}) = \pm 242 \text{ mV}$$

If this output error voltage is a problem, we can use a 10-kΩ potentiometer, as previously described, to null the output.

18-5 Two Op-Amp Applications

Op-amp applications are so broad and varied that it is impossible to discuss them comprehensively in this chapter. Besides, we need to understand negative feedback better before looking at some of the more advanced applications. For now, let us take a look at two practical circuits.

The Summing Amplifier

Whenever we need to combine two or more analog signals into a single output, the **summing amplifier** of Fig. 18-23a is a natural choice. For simplicity, the circuit shows only two inputs, but we can have as many inputs as needed for the application. A circuit like this amplifies each input signal. The gain for each *channel* or input is given by the ratio of the feedback resistance to the appropriate input resistance. For instance, the closed-loop voltage gains of Fig. 18-23a are:

$$A_{v1(CL)} = \frac{-R_f}{R_1} \quad \text{and} \quad A_{v2(CL)} = \frac{-R_f}{R_2}$$

The summing circuit combines all the amplified input signals into a single output, given by:

$$v_{out} = A_{v1(CL)}v_1 + A_{v2(CL)}v_2 \qquad (18\text{-}13)$$

It is easy to prove Eq. (18-13). Since the inverting input is a virtual ground, the total input current is:

$$i_{in} = i_1 + i_2 = \frac{v_1}{R_1} + \frac{v_2}{R_2}$$

Figure 18-23 Summing amplifier.

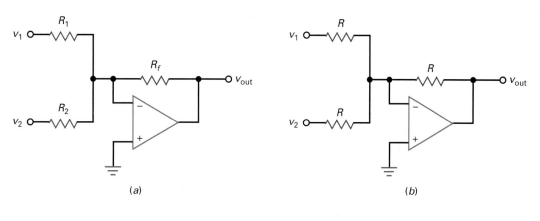

(a)   (b)

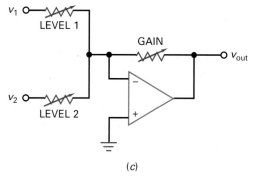

(c)

Because of the virtual ground, all this current flows through the feedback resistor, producing an output voltage with a magnitude of:

$$v_{\text{out}} = (i_1 + i_2)R_f = -\left(\frac{R_f}{R_1}v_1 + \frac{R_f}{R_2}v_2\right)$$

Here you see that each input voltage is multiplied by its channel gain and added to produce the total output. The same result applies to any number of inputs.

In some applications, all resistances are equal, as shown in Fig. 18-23b. In this case, each channel has a closed-loop voltage gain of unity (1) and the output is given by:

$$v_{\text{out}} = -(v_1 + v_2 + \ldots + v_n)$$

This is a convenient way of combining input signals and maintaining their relative sizes. The combined output signal can then be processed by more circuits.

Figure 18-23c is a **mixer,** a convenient way to combine audio signals in a high-fidelity audio system. The adjustable resistors allow us to set the level of each input, and the gain control allows us to adjust the combined output volume. By decreasing LEVEL 1, we can make the v_1 signal louder at the output. By decreasing LEVEL 2, we can make the v_2 signal louder. By increasing GAIN, we can make both signals louder.

A final point: If a summing circuit needs to be compensated by adding an equal resistance to the noninverting input, the resistance to use is the Thevenin resistance looking from the inverting input back to the sources. This resistance is given by the parallel equivalent of all resistances connected to the virtual ground:

$$R_{B2} = R_1 \parallel R_2 \parallel R_f \parallel \ldots \parallel R_n \tag{18-14}$$

Voltage Follower

In Chap. 11, we discussed the emitter follower and saw how useful it was for increasing the input impedance while producing an output signal that was almost equal to the input. The **voltage follower** is the equivalent of an emitter follower, except that it works much better.

Figure 18-24a shows the ac equivalent circuit for a voltage follower. Although it appears deceptively simple, the circuit is very close to ideal because the negative feedback is maximum. As you can see, the feedback resistance is zero. Therefore, all the output voltage is fed back to the inverting input. Because of the virtual short between the op-amp inputs, the output voltage equals the input voltage:

$$v_{\text{out}} = v_{\text{in}}$$

which means that the closed-loop voltage gain is:

$$A_{v(CL)} = 1 \tag{18-15}$$

We can get the same result by calculating the closed-loop voltage gain with Eq. (18-12). Since $R_f = 0$ and $R_1 = \infty$:

$$A_{v(CL)} = \frac{R_f}{R_1} + 1 = 1$$

Therefore, the voltage follower is a perfect follower circuit because it produces an output voltage that is exactly equal to the input voltage (or close enough to satisfy almost any application).

Furthermore, the maximum negative feedback produces a closed-loop input impedance that is much higher than the open-loop input impedance (2 MΩ for a 741C). Also, a maximum negative feedback produces a closed-loop output impedance that is much lower than the open-loop output impedance (75 Ω for a 741C). Therefore, we have an almost perfect method for converting a high-impedance source to a low-impedance source.

Figure 18-24 (*a*) Voltage follower has unity gain and maximum bandwidth; (*b*) voltage follower allows high-impedance source to drive low-impedance load with no loss of voltage.

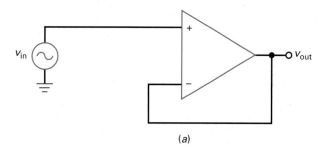

(*a*)

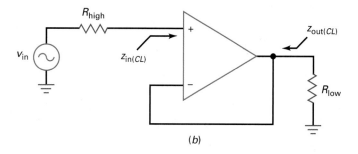

(*b*)

Figure 18-24*b* illustrates the idea. The input ac source has a high output impedance R_{high}. The load has a low impedance R_{low}. Because of the maximum negative feedback in a voltage follower, the closed-loop input impedance $z_{in(CL)}$ is incredibly high and the closed-loop output impedance $z_{out(CL)}$ is incredibly low. As a result, all the input source voltage appears across the load resistor.

The crucial point to understand is this: The voltage follower is the ideal interface to use between a high-impedance source and a low-impedance load. Basically, it transforms the high-impedance voltage source into a low-impedance voltage source. You will see the voltage follower used a great deal in practice.

Since $A_{v(CL)} = 1$ in a voltage follower, the closed-loop bandwidth is maximum and equal to:

$$f_{2(CL)} = f_{unity} \qquad (18\text{-}16)$$

Another advantage is the low output offset error because the input errors are not amplified. Since $A_{v(CL)} = 1$, the total output error voltage equals the worst-case sum of the input errors.

Example 18-12

IIII MultiSim

Three audio signals drive the summing amplifier of Fig. 18-25. What is the ac output voltage?

SOLUTION The channels have closed-loop voltage gains of:

$$A_{v1(CL)} = \frac{-100 \text{ k}\Omega}{20 \text{ k}\Omega} = -5$$

$$A_{v2(CL)} = \frac{-100 \text{ k}\Omega}{10 \text{ k}\Omega} = -10$$

$$A_{v3(CL)} = \frac{-100 \text{ k}\Omega}{50 \text{ k}\Omega} = -2$$

Figure 18-25 Example.

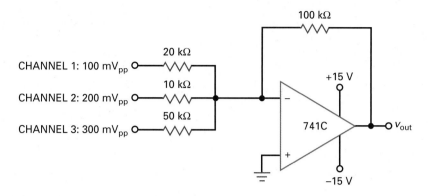

The output voltage is:

$$v_{out} = (-5)(100 \text{ mV}_{pp}) + (-10)(200 \text{ mV}_{pp}) + (-2)(300 \text{ mV}_{pp}) = -3.1 \text{ V}_{pp}$$

Again, the negative sign indicates a 180° phase shift.

If it is necessary to compensate for input bias by adding an equal R_B to the noninverting input, the resistance to use is:

$$R_{B2} = 20 \text{ k}\Omega \parallel 10 \text{ k}\Omega \parallel 50 \text{ k}\Omega \parallel 100 \text{ k}\Omega = 5.56 \text{ k}\Omega$$

The nearest standard value of 5.6 kΩ would be fine. A nulling circuit would take care of the remaining input errors.

PRACTICE PROBLEM 18-12 Using Fig. 18-25, the input channel voltages are changed from peak-to-peak values to positive dc values. What is the output dc voltage?

Example 18-13   IIII MultiSim

An ac voltage source of 10 mV$_{pp}$ with an internal resistance of 100 kΩ drives the voltage follower of Fig. 18-26a. The load resistance is 1 Ω. What is the output voltage? The bandwidth?

SOLUTION The closed-loop voltage gain is unity. Therefore:

$$v_{out} = 10 \text{ mV}_{pp}$$

and the bandwidth is:

$$f_{2(CL)} = 1 \text{ MHz}$$

This example echoes the idea discussed earlier. The voltage follower is an easy way to transform a high-impedance source into a low-impedance source. It does what the emitter follower does, only far better.

PRACTICE PROBLEM 18-13 Repeat Example 18-13 using an LF157A op amp.

Example 18-14

When the voltage follower of Fig. 18-26a is built with MultiSim, the output voltage across the 1-Ω load is 9.99 mV. Show how to calculate the closed-loop output impedance.

Figure 18-26 Example.

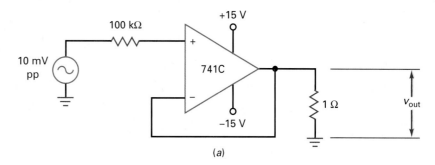

(a)

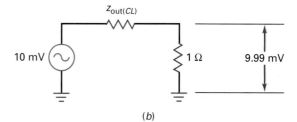

(b)

SOLUTION

$$v_{out} = 9.99 \text{ mV}$$

The closed-loop output impedance is the same as the Thevenin resistance facing the load resistor. In Fig. 18-26b, the load current is:

$$i_{out} = \frac{9.99 \text{ mV}}{1 \text{ }\Omega} = 9.99 \text{ mA}$$

This load current flows through $z_{out(CL)}$. Since the voltage across $z_{out(CL)}$ is 0.01 mV:

$$z_{out(CL)} = \frac{0.01 \text{ mV}}{9.99 \text{ mA}} = 0.001 \text{ }\Omega$$

Let the significance of this sink in. In Fig. 18-26a, the voltage source with an internal impedance of 100 kΩ has been converted to a voltage source with an internal impedance of only 0.001 Ω. Small output impedances like this mean that we are approaching the ideal voltage source first discussed in Chap. 1.

PRACTICE PROBLEM 18-14 In Fig. 18-26a, if the loaded output voltage is 9.95 mV, calculate the closed-loop output impedance.

Summary Table 18-1 shows the basic op-amp circuits we have discussed to this point.

18-6 Linear ICs

Op amps represent about a third of all linear ICs. With op amps we can build a wide variety of useful circuits. Although the op amp is the most important linear IC, other linear ICs such as audio amplifiers, video amplifiers, and voltage regulators are also widely used.

Inverting amp

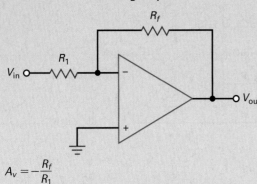

$$A_v = -\frac{R_f}{R_1}$$

Summing amp

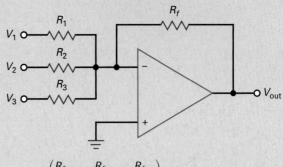

$$V_{out} = -\left(\frac{R_f}{R_1}V_1 + \frac{R_f}{R_2}V_2 + \frac{R_f}{R_3}V_3\right)$$

Noninverting amp

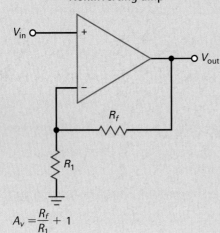

$$A_v = \frac{R_f}{R_1} + 1$$

Voltage follower

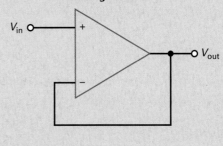

$$A_v = 1$$

Table of Op Amps

In Table 18-2, the prefix *LF* indicates a BIFET op amp. For instance, LF351 is the first entry in the table. This BIFET op amp has a maximum input offset voltage of 10 mV, a maximum input bias current of 0.2 nA, and a maximum input offset current of 0.1 nA. It can deliver a short-circuit current of 10 mA. It has a unity-gain frequency of 4 MHz, a slew rate of 13 V/μs, an open-loop voltage gain of 88 dB, and a common-mode rejection ratio of 70 dB.

The table contains two more quantities not previously discussed. First, there is the **power supply rejection ratio (PSRR).** This quantity is defined as:

$$\textbf{PSRR} = \frac{\Delta V_{in(off)}}{\Delta V_S} \tag{18-17}$$

In words, the equation says that the power-supply rejection ratio equals the change in the input offset voltage divided by the change in the supply voltages. In making this measurement, the manufacturer varies both supplies simultaneously and symmetrically. If $V_{CC} = +15$ V, $V_{EE} = -15$ V, and $\Delta V_S = +1$ V, then V_{CC} becomes +16 V and V_{EE} becomes −16 V.

Here is what Eq. (18-17) means: Because of the imbalance in the input diff amp plus other internal effects, a change in the supply voltage will produce an output error voltage. Dividing this output error voltage by the closed-loop voltage gain gives the change in the input offset voltage. For instance, the LF351 of

Table 18–2 Typical Parameters of Selected Op Amps at 25°C

Number	$V_{in(off)}$ max, mV	$I_{in(bias)}$ max, nA	$I_{in(off)}$ max, nA	I_{out} max, mA	f_{unity} typ, MHz	S_R typ, V/μs	A_{VOL} typ, dB	CMRR min, dB	PSRR min, dB	Drift typ, μV/°C	Description of Op Amps
LF351	10	0.2	0.1	10	4	13	88	70	−76	10	BIFET
LF353	10	0.2	0.1	10	4	13	88	70	−76	10	Dual BIFET
LF356	5	0.2	0.05	20	5	12	94	85	−85	5	BIFET, wideband
LF411A	0.5	200	100	20	4	15	88	80	−80	10	Low offset BIFET
LM12	7	300	100	10 A⁺	0.7	9	94	75	−80	50	High-power, 80 W out
LM301A	7.5	250	50	10	1+	0.5+	108	70	−70	30	External compensation
LM307	7.5	250	50	10	1	0.5	108	70	−70	30	Improved 709, internal comp
LM308	7.5	7	1	5	0.3	0.15	108	80	−80	30	Precision
LM318	10	500	200	10	15	70	86	70	−65	–	High speed, high slew rate
LM324	4	10	2	5	0.1	0.05	94	80	−90	10	Low-power quad
LM348	6	500	200	25	1	0.5	100	70	−70	–	Quad 741
LM675	10	2 μA*	500	3 A⁺	5.5	8	90	70	−70	25	High-power, 25 W out
LM741C	6	500	200	25	1	0.5	100	70	−70	–	Original classic
LM747C	6	500	200	25	1	0.5	100	70	−70	–	Dual 741
LM833	5	1 μA*	200	10	15	7	90	80	−80	2	Low noise
LM1458	6	500	200	20	1	0.5	104	70	−77	–	Dual
OP-07A	0.025	2	1	10	0.6	0.17	110	110	−100	0.6	Precision
OP-21A	0.1	100	4	–	0.6	0.25	120	100	−104	1	Low-power precision
OP-42E	0.75	0.2	0.04	25	10	58	114	88	−86	10	High-speed BIFET
OP-64E	0.75	300	100	20	200	200	100	110	−105	–	Very high speed and bandwidth
TL072	10	0.2	0.05	10	3	13	88	70	−70	10	Low-noise BIFET dual
TL074	10	0.2	0.05	10	3	13	88	70	−70	10	Low-noise BIFET quad
TL082	3	0.2	0.01	10	3	13	94	80	−80	10	Low-noise BIFET dual
TL084	3	0.2	0.01	10	3	13	94	80	−80	10	Low-noise BIFET quad

*For the LM675 and LM833, this value is commonly expressed in microamperes.
⁺For the LM12 and LM675, this value is commonly expressed in amperes.

Table 18-2 has a PSRR in decibels of -76 dB. When we convert this to an ordinary number we get:

$$PSRR = antilog \frac{-76\,dB}{20} = 0.000158$$

or, as it is sometimes written:

$$PSRR = 158\;\mu V/V$$

This tells us that a change of 1 V in the supply voltage will produce a change in the input offset voltage of 158 μV. Therefore, we have one more source of input error that joins the three input errors discussed earlier.

The last parameter shown for the LF351 is the *drift* of 10 μV/°C. This is defined as the temperature coefficient of the input offset voltage. It tells us how much the input offset voltage increases with temperature. A drift of 10 μV/°C means that the input offset voltage increases 10 μV for each degree increase in degrees Celsius. If the internal temperature of the op amp increases by 50°C, the input offset voltage of an LF351 increases by 500 μV.

The op amps in Table 18-2 were selected to show you the variety of commercially available devices. For example, the LF411A is a low-offset BIFET with an input offset voltage of only 0.5 mV. Most op amps are low-power devices, but not all. The LM675 is a high-power op amp. It has a short-circuit current of 3 A and can deliver 25 W to a load resistor. Even more powerful is the LM12. It has a short-circuit current of 10 A and can produce a load power of 80 W. Several LM12s can be operated in parallel for even greater power output. Applications include heavy-duty voltage regulators, high-quality audio amplifiers, and servo-control systems.

When you need a high slew rate, an LM318 can slew at a rate of 70 V/μs. And then there is the OP-64E, which has a slew rate of 200 V/μs. High slew rate and bandwidth usually go together. As you can see, LM318 has an f_{unity} of 15 MHz, and the OP-64E has an f_{unity} of 200 MHz.

Many of the op amps are available as dual and quad op amps. This means that there are either two or four op amps in the same package. For instance, the LM747C is a dual 741C. The LM348 is a quad 741. The single and dual op amps fit in a package with 8 pins, and the quad op amp comes in packages with 14 pins.

Not all op amps need two supply voltages. For the instance, the LM324 has four internally compensated op amps. Although it can operate with two supplies like most op amps, it was specifically designed for a single power supply, a definite advantage in many applications. Another convenience of the LM324 is that it can work with a single power supply as low as $+5$ V, the standard voltage for many digital systems.

Internal compensation is convenient and safe because an internally compensated op amp will not break into oscillations under any condition. The price paid for this safety is a loss of design control. This is why some op amps offer external compensation. For instance, LM301A is compensated by connecting an external 30-pF capacitor. But the designer has the option of overcompensating with a larger capacitor or undercompensating with smaller capacitor. Overcompensation can improve low-frequency operation, whereas undercompensation can increase the bandwidth and slew rate. This is why a plus sign ($+$) has been added to the f_{unity} and S_R of the LM301A in Table 18-2.

All op amps have imperfections, as we have seen. Precision op amps try to minimize these imperfections. For instance, the OP-07A is a precision op amp with the following worst-case parameters: input offset voltage is only 0.025 mV, CMRR is at least 110 dB, PSRR is at least 100 dB, and drift is only 0.6 μV/°C. Precision op amps are necessary for stringent applications such as measurement and control.

In subsequent chapters, we will discuss more applications of op amps. At that time, you will see how op amps can be used in a wide variety of linear circuits, nonlinear circuits, oscillators, voltage regulators, and active filters.

Audio Amplifiers

Preamplifiers (preamps) are audio amplifiers with less than 50 mW of output power. Preamps are optimized for low noise because they are used at the front end of audio systems, where they amplify weak signals from optical sensors, magnetic tape heads, microphones, and so on.

An example of an IC preamp is the LM381, a low-noise dual preamp. Each amplifier is completely independent of the other. The LM381 has a voltage gain of 112 dB and a 10-V power bandwidth of 75 kHz. It operates from a positive supply of 9 to 40 V. Its input impedance is 100 kΩ, and its output impedance is 150 Ω. The LM381's input stage is a diff amp, which allows differential or single-ended input.

Medium-level audio amplifiers have output powers from 50 to 500 mW. These are useful near the output end of portable electronic devices such as cell phones and CD players. An example is the LM4818 audio power amplifier, which has an output power of 350 mW.

Audio power amplifiers deliver more than 500 mW of output power. They are used in high-fidelity amplifiers, intercoms, AM-FM radios, and other applications. The LM380 is an example. It has a voltage gain of 34 dB, a bandwidth of 100 kHz, and an output power of 2 W. As another example, the LM4756 power amp has an internally set voltage gain of 30 dB and can deliver 7W/channel. Fig. 18-27 shows the package style and pin out for this IC. Notice the dual offset pin arrangement.

Figure 18-28 shows a simplified schematic diagram of the LM380. The input diff amp uses *pnp* inputs. The signal can be directly coupled, which is an advantage with transducers. The diff amp drives a current-mirror load (Q_5 and Q_6). The output of the current mirror goes to an emitter follower (Q_7) and CE driver (Q_8). The output stage is a class B push-pull emitter follower (Q_{13} and Q_{14}). There is an internal compensating capacitor of 10 pF that rolls off the decibel voltage gain at a rate of 20 dB per decade. This capacitor produces a slew rate of approximately 5 V/μs.

Figure 18-27 The LM4756 package style and pin out.

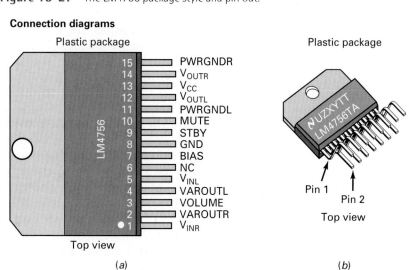

Connection diagrams

Plastic package

Pin	Signal
15	PWRGNDR
14	V_{OUTR}
13	V_{CC}
12	V_{OUTL}
11	PWRGNDL
10	MUTE
9	STBY
8	GND
7	BIAS
6	NC
5	V_{INL}
4	VAROUTL
3	VOLUME
2	VAROUTR
1	V_{INR}

Top view

Plastic package

Pin 1   Pin 2

Top view

(a)   (b)

Figure 18-28 Simplified schematic diagram of LM380.

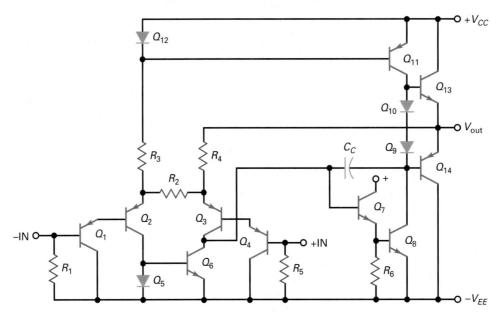

Video Amplifiers

A video or wideband amplifier has a flat response (constant decibel voltage gain) over a very broad range of frequencies. Typical bandwidths are well into the megahertz region. Video amps are not necessarily dc amps, but they often do have a response that extends down to zero frequency. They are used in applications in which the range of input frequencies is very large. For instance, many oscilloscopes handle frequencies from 0 to over 100 MHz; instruments like these use video amps to increase the signal strength before applying it to the cathode-ray tube. As another example, the LM7171 is a very high speed amplifier with a wide unity-gain bandwidth of 200 MHz and a slew rate of 4100 V/μS. This amplifier finds applications in video cameras, copiers and scanners, and HDTV amplifiers.

IC video amps have voltage gains and bandwidths that you can adjust by connecting different external resistors. For instance, the VLA702 has a decibel voltage gain of 40 dB and a cutoff frequency of 5 MHz; by changing external components, you can get useful gain to 30 MHz. The MC1553 has a decibel voltage gain of 52 dB and a bandwidth of 20 MHz; these are adjustable by changing external components. The LM733 has a very wide bandwidth; it can be set up to give 20 dB gain and a bandwidth of 120 MHz.

RF and IF Amplifiers

A radio-frequency (RF) amplifier is usually the first stage in an AM, FM, or television receiver. Intermediate-frequency (IF) amplifiers typically are the middle stages. ICs like the LM703 include RF and IF amplifiers on the same chip. The amplifiers are tuned (resonant) so that they amplify only a narrow band of frequencies. This allows the receiver to tune a desired signal from a particular radio or television station. As mentioned earlier, it is impractical to integrate inductors and large capacitors on a chip. For this reason, you have to connect external inductors and capacitors to the chip to get tuned amplifiers. Another example of RF ICs is the MBC13720. This low-noise amplifier is designed to operate in the 400 MHz to 2.4 GHz range, which is where many broadband wireless applications are found.

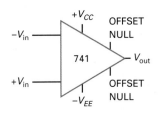

The SM version of the LM741 op amp.

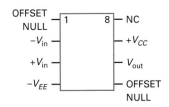

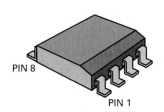

Voltage Regulators

Chapter 4 discussed rectifiers and power supplies. After filtering, we have a dc voltage with ripple. This dc voltage is proportional to the line voltage; that is, it will change 10 percent if the line voltage changes 10 percent. In most applications, a 10 percent change in dc voltage is too much, and voltage regulation is necessary. Typical of IC voltage regulators is the LM340 series. Chips of this type can hold the output dc voltage to within 0.01 percent for normal changes in line voltage and load resistance. Other features include positive or negative output, adjustable output voltage, and short-circuit protection.

18-7 Op Amps as Surface–Mount Devices

Operational amplifiers and similar kinds of analog circuits are frequently available in surface-mount (SM) packages as well as in the more tradition dual-in-line IC forms. Because the pinout for most op amp tends to be relatively simple, the small outline package (SOP) is the preferred SM style.

For example, the LM741 op amp—the mainstay of school electronics labs for many years—is now available in the latest SOP package (at left). In this instance, the pinout of the surface-mount device (SMD) is the same as the pinout for the more familiar dual-in-line version.

The LM2900, a quad op amp, is an example of a more complex op-amp SMD package. This device is provided in a feed-through, 14-pin DIP and a 14-pin SOT (below). Conveniently, the pinouts are identical for the two packages.

A typical quad op-amp circuit provided in a 14-pin SOT package.

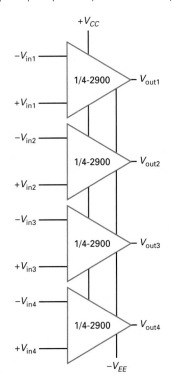

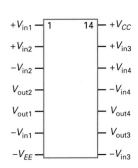

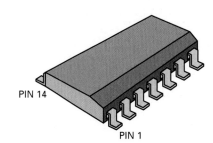

Summary

SEC. 18-1 INTRODUCTION TO OP AMPS

A typical op amp has a noninverting input, an inverting input, and a single-ended output. An ideal op amp has infinite open-loop voltage gain, infinite input resistance, and zero output impedance. It is a perfect amplifier, a voltage-controlled voltage source (VCVS).

SEC. 18-2 THE 741 OP AMP

The 741 is a standard op amp that is widely used. It includes an internal compensating capacitor to prevent oscillations. With a large load resistance, the output signal can swing to within 1 or 2 V of either supply. With small load resistances, MPP is limited by the short-circuit current. The slew rate is the maximum speed at which the output voltage can change when driven by a step input. The power bandwidth is directly proportional to slew rate and inversely proportional to the peak output voltage.

SEC. 18-3 THE INVERTING AMPLIFIER

The inverting amplifier is the most basic op-amp circuit. It uses negative feedback to stabilize the closed-loop voltage gain. The inverting input is a virtual ground because it is a short for voltage but an open for current. The closed-loop voltage gain equals the feedback resistance divided by the input resistance. The closed-loop bandwidth equals the unity-gain frequency divided by the closed-loop voltage gain.

SEC. 18-4 THE NONINVERTING AMPLIFIER

The noninverting amplifier is another basic op-amp circuit. It uses negative feedback to stabilize the closed-loop voltage gain. A virtual short is between the noninverting input and the inverting input. The closed-loop voltage gain equals $R_f/R_1 + 1$. The closed-loop bandwidth equals the unity-gain frequency divided by the closed-loop voltage gain.

SEC. 18-5 TWO OP-AMP APPLICATIONS

The summing amplifier has two or more inputs and one output. Each input is amplified by its channel gain. The output is the sum of the amplified inputs. If all channel gains equal unity, the output equals the sum of the inputs. In a mixer, a summing amplifier can amplify and combine audio signals. A voltage follower has a closed-loop voltage gain of unity and a bandwidth of f_{unity}. The circuit is useful as an interface between a high-impedance source and a low-impedance load.

SEC. 18-6 LINEAR ICS

Op amps represent about a third of all linear ICs. A wide variety of op amps exists for almost any application. Some have very low input offsets, other have high bandwidths and slew rates, and others have low drifts. Dual and quad op amps are available. Even high-power op amps exist that can produce large load power. Other linear ICs include audio and video amplifiers, RF and IF amplifiers, and voltage regulators.

Definitions

(18-1) Slew rate:

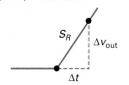

$$S_R = \frac{\Delta v_{out}}{\Delta t}$$

(18-17) Power-supply rejection ratio:

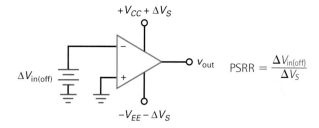

$$PSRR = \frac{\Delta V_{in(off)}}{\Delta V_S}$$

Derivations

(18-2) Power bandwidth:

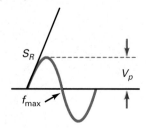

$$f_{max} = \frac{S_R}{2\pi V_p}$$

(18-3) Closed-loop voltage gain:

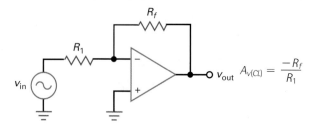

$$A_{V(CL)} = \frac{-R_f}{R_1}$$

(18-4)   Closed-loop input impedance:

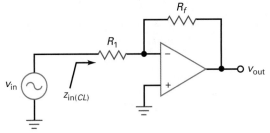

$$z_{in(CL)} = R_1$$

(18-5)   Closed-loop bandwidth:

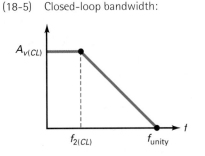

$$f_{2(CL)} = \frac{f_{unity}}{A_{v(CL)}}$$

(18-11)   Compensating resistor:

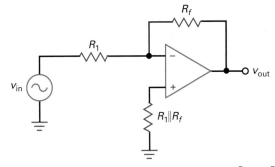

$$R_{B1} = R_1 \parallel R_f$$

(18-12)   Noninverting amplifier:

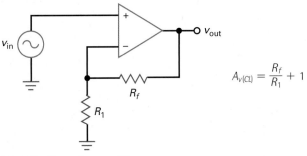

$$A_{v(CL)} = \frac{R_f}{R_1} + 1$$

(18-13)   Summing amplifier:

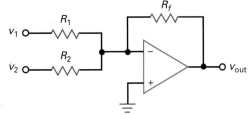

$$v_{out} = A_{v1(CL)}v_1 + A_{v2(CL)}v_2$$

(18-15)   Voltage follower:

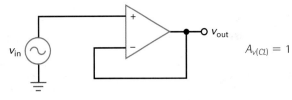

$$A_{v(CL)} = 1$$

(18-16)   Follower bandwidth:

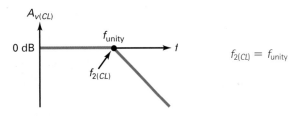

$$f_{2(CL)} = f_{unity}$$

Student Assignments

1. What usually controls the open-loop cutoff frequency of an op amp?
   a. Stray-wiring capacitance
   b. Base-emitter capacitance
   c. Collector-base capacitance
   d. Compensating capacitance

2. A compensating capacitor prevents
   a. Voltage gain
   b. Oscillations
   c. Input offset current
   d. Power bandwidth

3. At the unity–gain frequency, the open-loop voltage gain is
   a. 1
   b. $A_{v(mid)}$
   c. Zero
   d. Very large

4. The cutoff frequency of an op amp equals the unity–gain frequency divided by
   a. The cutoff frequency
   b. Closed-loop voltage gain
   c. Unity
   d. Common-mode voltage gain

5. If the cutoff frequency is 20 Hz and the midband open-loop voltage gain is 1,000,000, the unity–gain frequency is
   a. 20 Hz
   b. 1 MHz
   c. 2 MHz
   d. 20 MHz

6. If the unity–gain frequency is 5 MHz and the midband open-loop voltage gain is 100,000, the cutoff frequency is
   a. 50 Hz
   b. 1 MHz
   c. 1.5 MHz
   d. 15 MHz

7. The initial slope of a sine wave is directly proportional to
   a. Slew rate
   b. Frequency
   c. Voltage gain
   d. Capacitance

8. When the initial slope of a sine wave is greater than the slew rate,
   a. Distortion occurs
   b. Linear operation occurs
   c. Voltage gain is maximum
   d. The op amp works best

9. The power bandwidth increases when
   a. Frequency decreases
   b. Peak value decreases
   c. Initial slope decreases
   d. Voltage gain increases

10. A 741C contains
    a. Discrete resistors
    b. Inductors
    c. Active-load resistors
    d. A large coupling capacitor

11. A 741C cannot work without
    a. Discrete resistors
    b. Passive loading
    c. DC return paths on the two bases
    d. A small coupling capacitor

12. The input impedance of a BIFET op amp is
    a. Low
    b. Medium
    c. High
    d. Extremely high

13. An LF157A is a
    a. Diff amp
    b. Source follower
    c. Bipolar op amp
    d. BIFET op amp

14. If the two supply voltages are ±12 V, the MPP value of an op amp is closest to
    a. 0   c. −12 V
    b. +12 V   d. 24 V

15. The open-loop cutoff frequency of a 741C is controlled by
    a. A coupling capacitor
    b. The output short circuit current
    c. The power bandwidth
    d. A compensating capacitor

16. The 741C has a unity-gain frequency of
    a. 10 Hz   c. 1 MHz
    b. 20 kHz   d. 15 MHz

17. The unity-gain frequency equals the product of closed-loop voltage gain and the
    a. Compensating capacitance
    b. Tail current
    c. Closed-loop cutoff frequency
    d. Load resistance

18. If f_{unity} is 10 MHz and midband open-loop voltage gain is 200,000, then the open-loop cutoff frequency of the op amp is
    a. 10 Hz   c. 50 Hz
    b. 20 Hz   d. 100 Hz

19. The initial slope of a sine wave increases when
    a. Frequency decreases
    b. Peak value increases
    c. C_c increases
    d. Slew rate decreases

20. If the frequency of the input signal is greater than the power bandwidth,
    a. Slew-rate distortion occurs
    b. A normal output signal occurs
    c. Output offset voltage increases
    d. Distortion may occur

21. An op amp has an open base resistor. The output voltage will be
    a. Zero
    b. Slightly different from zero
    c. Maximum positive or negative
    d. An amplified sine wave

22. An op amp has a voltage gain of 200,000. If the output voltage is 1 V, the input voltage is
    a. 2 μV   c. 10 mV
    b. 5 μV   d. 1 V

23. A 741C has supply voltages of ±15 V. If the load resistance is large, the MPP value is approximately
    a. 0   c. 27 V
    b. +15 V   d. 30 V

24. Above the cutoff frequency, the voltage gain of a 741C decreases approximately
    a. 10 dB per decade
    b. 20 dB per octave

    c. 10 dB per octave
    d. 20 dB per decade

25. The voltage gain of an op amp is unity at the
    a. Cutoff frequency
    b. Unity-gain frequency
    c. Generator frequency
    d. Power bandwidth

26. When slew-rate distortion of a sine wave occurs, the output
    a. Is larger
    b. Appears triangular
    c. Is normal
    d. Has no offset

27. A 741C has
    a. A voltage gain of 100,000
    b. An input impedance of 2 MΩ
    c. An output impedance of 75 Ω
    d. All of the above

28. The closed-loop voltage gain of an inverting amplifier equals
    a. The ratio of the input resistance to the feedback resistance
    b. The open-loop voltage gain
    c. The feedback resistance divided by the input resistance
    d. The input resistance

29. The noninverting amplifier has a
    a. Large closed-loop voltage gain
    b. Small open-loop voltage gain
    c. Large closed-loop input impedance
    d. Large closed-loop output impedance

30. The voltage follower has a
    a. Closed-loop voltage gain of unity
    b. Small open-loop voltage gain
    c. Closed-loop bandwidth of zero
    d. Large closed-loop output impedance

31. A summing amplifier can have
    a. No more than two input signals
    b. Two or more input signals
    c. A closed-loop input impedance of infinity
    d. A small open-loop voltage gain

Problems

SEC. 18-2   THE 741 OP AMP

18-1 Assume that negative saturation occurs at 1 V less than the supply voltage with an 741C. How much inverting input voltage does it take to drive the op amp of Fig. 18-29 into negative saturation?

Figure 18-29

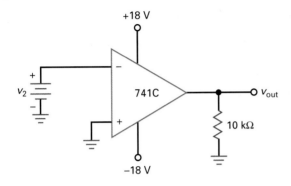

18-2 What is the common-mode rejection ratio of an LF157A at low frequencies? Convert this decibel value to an ordinary number.

18-3 What is the open-loop voltage gain of an LF157A when the input frequency is 1 kHz? 10 kHz? 100 kHz? (Assume a first-order response, that is, 20 dB per decade rolloff.)

18-4 The input voltage to an op amp is a large voltage step. The output is an exponential waveform that changes 2.0 V in 0.4 μs. What is the slew rate of the op amp?

18-5 An LM318 has a slew rate of 70 V/μs. What is the power bandwidth for a peak output voltage of 7 V?

18-6 Use Eq. (18-2) to calculate the power bandwidth for each of the following:

a. $S_R = 0.5$ V/μs and $V_p = 1$ V
b. $S_R = 3$ V/μs and $V_p = 5$ V
c. $S_R = 15$ V/μs and $V_p = 10$ V

SEC. 18-3   THE INVERTING AMPLIFIER

18-7 IIII **MultiSim** What are closed-loop voltage gain and bandwidth in Fig. 18-30? What is the output voltage at 1 kHz? At 10 MHz? Draw the ideal Bode plot of closed-loop voltage gain.

Figure 18-30

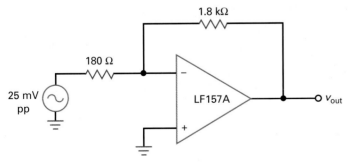

Figure 18-31

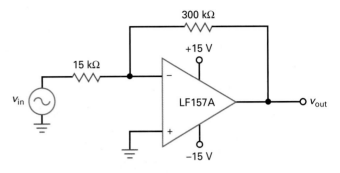

18-8 What is the output voltage in Fig. 18-31 when v_{in} is zero? Use the typical values of Table 18-1.

18-9 The data sheet of an LF157A lists the following worst-case parameters: $I_{in(bias)} = 50$ pA, $I_{in(off)} = 10$ pA, and $V_{in(off)} = 2$ mV. Recalculate the output voltage when v_{in} is zero in Fig. 18-31.

SEC. 18-4   THE NONINVERTING AMPLIFIER

18-10 IIII **MultiSim** In Fig. 18-32, what are the closed-loop voltage gain and bandwidth? The ac output voltage at 100 kHz?

18-11 What is the output voltage when v_{in} is reduced to zero in Fig. 18-32? Use the worst-case parameters given in Prob. 18-9.

Figure 18-32

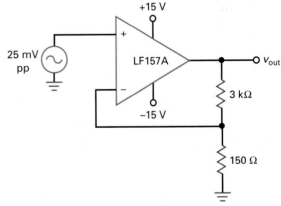

18-12 ▌▌▌▌ MultiSim In Fig. 18-33*a*, what is the ac output voltage? If a compensating resistor needs to be added to the noninverting input, what size should it be?

18-13 What is the output voltage in Fig. 18-33*b*? The bandwidth?

Figure 18-33

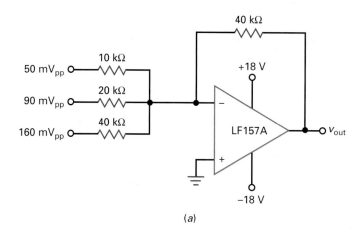

(*a*)

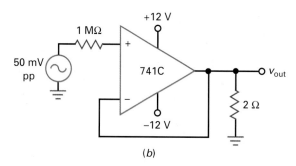

(*b*)

Critical Thinking

18-14 The adjustable resistor of Fig. 18-34 can be varied from 0 to 100 kΩ. Calculate the minimum and maximum closed-loop voltage gain and bandwidth.

18-15 Calculate the minimum and maximum closed-loop voltage gain and bandwidth in Fig. 18-35.

18-16 In Fig. 18-33*b*, the ac output voltage is 49.98 mV. What is the closed-loop output impedance?

18-17 What is the initial slope of a sine wave with a frequency of 15 kHz and a peak value of 2 V? What happens to the initial slope if the frequency increases to 30 kHz?

Figure 18-34

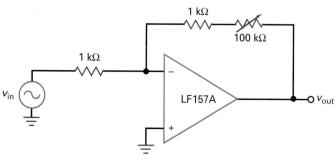

Figure 18-35

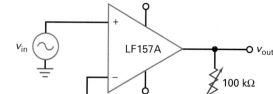

18-18 Which op amp in Table 18-3 has the following:

    a. Minimum input offset voltage
    b. Minimum input offset current
    c. Maximum output-current capability
    d. Maximum bandwidth
    e. Minimum drift

18-19 What is the CMRR of a 741C at 100 kHz? The MPP value when the load resistance is 500 Ω? The open-loop voltage gain at 1 kHz?

18-20 If the feedback resistor in Fig. 18-33*a* is changed to a 100-kΩ variable resistor, what is the maximum output voltage? The minimum?

18-21 In Fig. 18-36, what is the closed-loop voltage gain for each switch position?

Figure 18-36

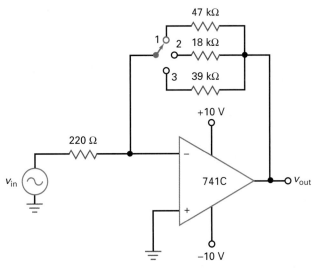

18-22 What is the closed-loop voltage gain for each switch position of Fig. 18-37? The bandwidth?

18-23 In wiring the circuit of Fig. 18-37, a technician leaves the ground off the 6-kΩ resistor. What is the closed-loop voltage gain in each switch position?

Figure 18-37

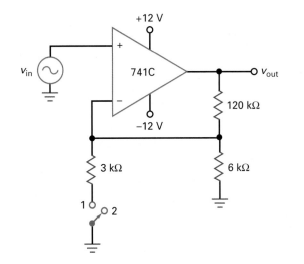

18-24 If the 120-kΩ resistor opens in Fig. 18-37, what is the output voltage most likely to do?

18-25 What is the closed-loop voltage gain for each switch position of Fig. 18-38? The bandwidth?

Figure 18-38

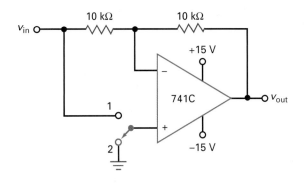

18-26 If the input resistor of Fig. 18-38 opens, what is the closed-loop voltage gain for each switch position?

18-27 If the feedback resistor opens in Fig. 18-38, what is the output voltage most likely to do?

18-28 The worst-case parameters for a 741C are $I_{in(bias)} = 500$ nA, $I_{in(off)} = 200$ nA, and $V_{in(off)} = 6$ mV. What is the total output error voltage in Fig. 18-39?

18-29 In Fig. 18-39, the input signal has a frequency of 1 kHz. What is the ac output voltage?

18-30 If the capacitor is shorted in Fig. 18-39, what is the total output error voltage? Use the worst-case parameters given in Prob. 18-28.

Figure 18-39

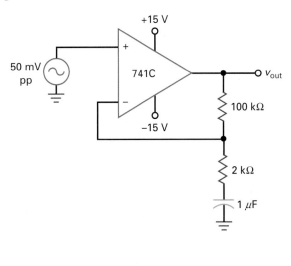

Up-Down Circuit Analysis

Use Fig. 18-40 for the remaining problems. A circuit like this is impractical for mass production because it has no feedback. The input offset error voltages are most likely to drive the op amp into positive or negative saturation. But assume that we have hand-selected a 741C to get a zero output error voltage for this theoretical exercise.

18-31 Predict the responses for each input base current.

18-32 Predict the responses for supply-voltage variations.

18-33 Predict the responses for slew-rate changes.

18-34 Predict the responses for peak-voltage changes.

Figure 18-40

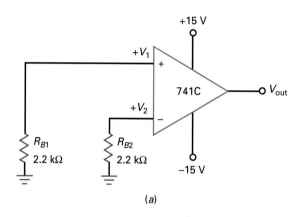

(a)

Up-Down Circuit Analysis

Increase	V_1	V_2	V_{in}	V_{out}	MPP	f_{max}
I_{B1}						
I_{B2}						
$\pm V_{CC}$						
S_R						
V_P						

(b)

Job Interview Questions

1. What is an ideal op amp? Compare the properties of a 741C to those of an ideal op amp.
2. Draw an op amp with an input voltage step. What is slew rate, and why is it important?
3. Draw an inverting amplifier using an op amp with component values. Now, tell me where the virtual ground is. What are the properties of a virtual ground? What is the closed-loop voltage gain, input impedance, and bandwidth?
4. Draw a noninverting amplifier using an op amp with component values. Now, tell me where the virtual short is. What are the properties of a virtual short? What is the closed-loop voltage gain and bandwidth?
5. Draw a summing amplifier and tell me the theory of operation.
6. Draw a voltage follower. What are the closed-loop voltage gain and bandwidth? Describe the closed-loop input and output impedances. What good is this circuit if its voltage gain is so low?
7. What are the input and output impedances of a typical op amp? What advantage do these values have?
8. How does the frequency of the input signal to an op amp affect voltage gain?
9. The LM318 is a much faster op amp than the LM741C. In what applications might the 318 be preferred to the 741C? What are some possible disadvantages of using the 318?
10. With zero input voltage to an ideal op amp, why is there exactly zero output voltage?
11. Name a few linear ICs besides the op amp.
12. What condition is needed for an LM741 to produce maximum voltage gain?
13. Draw an inverting op amp and derive the formula for voltage gain.
14. Draw a noninverting op amp and derive the formula for voltage gain.
15. Why is a 741C thought of as a dc or low-frequency amplifier?

Self–Test Answers

1.	d	6.	a	11.	c	16.	c	21.	c	26.	b	30.	a
2.	b	7.	b	12.	d	17.	c	22.	b	27.	d	31.	b
3.	a	8.	a	13.	d	18.	c	23.	c	28.	c		
4.	b	9.	b	14.	d	19.	b	24.	d	29.	c		
5.	d	10.	c	15.	d	20.	a	25.	b				

Practice Problem Answers

18-1 $V_2 = 67.5\ \mu V$

18-2 CMRR = 60 dB

18-4 $S_R = 4\ V/\mu S$

18-5 $f_{max} = 398$ kHz

18-6 $f_{max} = 80$ kHz, 800 kHz, 8 MHz

18-7 $V_{out} = 98$ mV

18-8 $V_{out} = 50$ mV

18-10 $A_{v(CL)} = 50;\ V_{out} = 250\ mV_{pp}$

18-12 $V_{out} = -3.1$ Vdc

18-13 $V_{out} = 10$ mV; $f_{2(CL)} = 20$ MHz

18-14 $z_{out} = 0.005\ \Omega$

19 Negative Feedback

In August 1927, a young engineer named Harold Black took a ferry from Staten Island, New York, to work. To pass the time on that summer morning, he jotted down some equations about a new idea. During the next few months, he polished the idea and then applied for a patent. But as so often happens with a truly new idea, it was ridiculed. The patent office rejected his application and classified it as another one of those "perpetual-motion follies." But only for a while. Black's idea was negative feedback.

Objectives

After studying this chapter, you should be able to:

- Define four types of negative feedback.

- Discuss the effect of VCVS negative feedback on voltage gain, input impedance, output impedance, and harmonic distortion.

- Explain the operation of a transresistance amplifier.

- Explain the operation of a transconductance amplifier.

- Describer how ICIS negative feedback can be used to realize a nearly ideal current amplifier.

- Discuss the relationship between bandwidth and negative feedback.

Chapter Outline

Vocabulary

current amplifier

current-controlled current source (ICIS)

current-controlled voltage source (ICVS)

current-to-voltage converter

feedback attenuation factor

feedback fraction *B*

gain-bandwidth product (GBP)

harmonic distortion

loop gain

negative feedback

transconductance amplifier

transresistance amplifier

voltage-controlled current source (VCIS)

voltage-controlled voltage source (VCVS)

voltage-to-current converter

19-1 Four Types of Negative Feedback

Black invented only one type of **negative feedback,** the kind that stabilizes the voltage gain, increases the input impedance, and decreases the output impedance. With the advent of transistors and op amps, three more kinds of negative feedback became available.

Basic Ideas

The input to a negative-feedback amplifier can be either a voltage or a current. Also, the output signal can be either a voltage or a current. This implies that four types of negative feedback exist. As shown in Table 19-1, the first type has an input voltage and an output voltage. The circuit that uses this type of negative feedback is called a **voltage-controlled voltage source (VCVS).** A VCVS is an ideal voltage amplifier because it has a stabilized voltage gain, infinite input impedance, and zero output impedance as shown.

In the second type of negative feedback, an input current controls an output voltage. The circuit using this type of feedback is called a **current-controlled voltage source (ICVS).** Because an input current controls an output voltage, an ICVS is sometimes called a **transresistance amplifier.** The word *resistance* is used because the ratio of v_{out}/i_{in} has the unit of ohms. The prefix *trans* refers to taking the ratio of an output quantity to an input quantity.

The third type of negative feedback has an input voltage controlling an output current. The circuit using this type of negative feedback is called a **voltage-controlled current source (VCIS).** Because an input voltage controls an output current, a VCIS is sometimes called a **transconductance amplifier.** The word *conductance* is used because the ratio of i_{out}/v_{in} has the unit of siemens (mhos).

In the fourth type of negative feedback, an input current is amplified to get a larger output current. The circuit with this type of negative feedback is called a **current-controlled current source (ICIS).** An ICIS is an ideal current amplifier because it has a stabilized current gain, zero input impedance, and infinite output impedance.

Converters

Referring to VCVS and ICIS circuits as amplifiers makes sense because the first is a voltage amplifier and the second is a current amplifier. But the use of the word *amplifier* with transconductance and transresistance amplifiers may seem a bit odd at first, because the input and output quantities are different. Because of this, many engineers and technicians prefer to think of these circuits as converters. For instance, the VCIS is also called a **voltage-to-current converter.** You put volts in, and you get amperes out. Similarly, the ICVS is also called a **current-to-voltage converter.** Current goes in, and voltage comes out.

Table 19-1		Ideal Negative Feedback						
Input	**Output**	**Circuit**	z_{in}	z_{out}	**Converts**	**Ratio**	**Symbol**	**Type of amplifier**
V	V	VCVS	∞	0	—	v_{out}/v_{in}	A_v	Voltage amplifier
I	V	ICVS	0	0	*i* to *v*	v_{out}/i_{in}	r_m	Transresistance amplifier
V	I	VCIS	∞	∞	*v* to *i*	i_{out}/v_{in}	g_m	Transconductance amplifier
I	I	ICIS	0	∞	—	i_{out}/i_{in}	A_i	Current amplifier

Figure 19-1 (*a*) Voltage-controlled voltage source; (*b*) current-controlled voltage source.

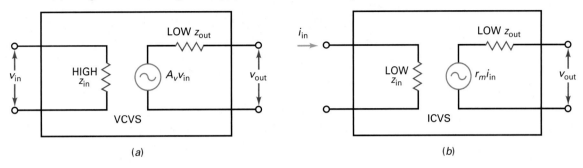

(*a*)   (*b*)

Figure 19-2 (*a*) Voltage-controlled current source; (*b*) current-controlled current source.

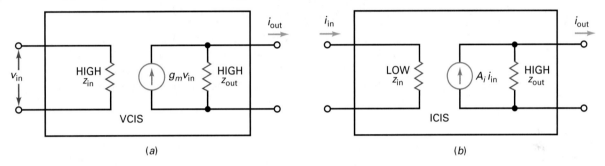

(*a*)   (*b*)

Diagrams

Figure 19-1*a* shows the VCVS, a voltage amplifier. With practical circuits, the input impedance is not infinite, but it is very high. Likewise, the output impedance is not zero, but it is very low. The voltage gain of the VCVS is symbolized A_v. Since z_{out} approaches zero, the output side of a VCVS is a stiff voltage source to any practical load resistance.

Figure 19-1*b* shows an ICVS, a transresistance amplifier (current-to-voltage converter). It has a very low input impedance and a very low output impedance. The conversion factor of the ICVS is called *transresistance,* symbolized r_m and expressed in ohms. For instance, if $r_m = 1$ kΩ, an input current of 1 mA will produce a constant voltage of 1 V across the load. Because z_{out} approaches zero, the output side of a ICVS is a stiff voltage source for practical load resistances.

Figure 19-2*a* shows a VCIS, a transconductance amplifier (voltage-to-current converter). It has a very high input impedance and a very high output impedance. The conversion factor of the VCIS is called *transconductance,* symbolized g_m and expressed in siemens (mhos). For instance, if $g_m = 1$ mS, an input voltage of 1 V will pump a current of 1 mA through the load. Because z_{out} approaches infinity, the output side of a VCIS is a stiff current source for any practical load resistance.

Figure 19-2*b* shows an ICIS, a current amplifier. It has very low input impedance and very high output impedance. The current gain of the ICIS is symbolized A_i. Since z_{out} approaches infinity, the output side of a VCVS is a stiff current source to any practical load resistance.

19-2 VCVS Voltage Gain

In Chap. 18, we analyzed the noninverting amplifier, a widely used *implementation* (circuit realization) of a VCVS. In this section, we want to reexamine the noninverting amplifier and delve more deeply into its voltage gain.

Figure 19-3 VCVS amplifier.

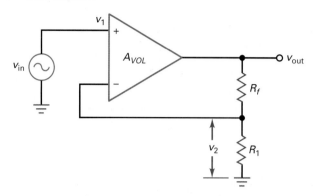

Exact Closed–Loop Voltage Gain

Figure 19-3 shows a noninverting amplifier. The op amp has an open-loop voltage gain of A_{VOL}, typically 100,000 or more. Because of the voltage divider, part of the output voltage is fed back to the inverting input. The **feedback fraction B** of any VCVS circuit is defined as feedback voltage divided by the output voltage. In Fig. 19-3:

$$B = \frac{v_2}{v_{\text{out}}} \tag{19-1}$$

The feedback fraction is also called the **feedback attenuation factor** because it indicates how much the output voltage is attenuated before the feedback signal reaches the inverting input.

With some algebra, we can derive the following exact equation for the closed-loop voltage gain:

$$A_{v(CL)} = \frac{A_{VOL}}{1 + A_{VOL}\,B} \tag{19-2}$$

or with the notation of Table 19-1, where $A_v = A_{v(CL)}$:

$$A_v = \frac{A_{VOL}}{1 + A_{VOL}\,B} \tag{19-3}$$

This is the exact equation for the closed-loop voltage gain of any VCVS amplifier.

Loop Gain

The second term in the denominator, $A_{VOL}B$, is called the **loop gain** because it is the voltage gain of the forward and feedback paths. The loop gain is a very important value in the design of a negative-feedback amplifier. In any practical design, the loop gain is made very large. The larger the loop gain, the better, because it stabilizes the voltage gain and has an enhancing or curative effect on quantities such as gain stability, distortion, offsets, input impedance, and output impedance.

Ideal Closed–Loop Voltage Gain

For a VCVS to work well, the loop gain $A_{VOL}B$ must be much greater than unity. When the designer satisfies this condition, Eq. (19-3) becomes:

$$A_v = \frac{A_{VOL}}{1 + A_{VOL}\,B} \cong \frac{A_{VOL}}{A_{VOL}\,B}$$

or

$$A_v \cong \frac{1}{B} \tag{19-4}$$

This ideal equation gives almost exact answers when $A_{VOL}B \gg 1$. The exact closed-loop voltage gain is slightly less than this ideal closed-loop voltage gain. If necessary, we can calculate the percent error between the ideal and exact values with:

$$\% \text{ Error} = \frac{100\%}{1 + A_{VOL}B} \qquad (19\text{-}5)$$

For instance, if $1 + A_{VOL}B$ is 1000 (60 dB), the error is only 0.1 percent. This means that the exact answer is only 0.1 percent less than the ideal answer.

Using the Ideal Equation

Equation (19-4) can be used to calculate the ideal closed-loop voltage gain of any VCVS amplifier. All you have to do is calculate the feedback fraction with Eq. (19-1) and take the reciprocal. For instance, in Fig. 19-3, the feedback fraction is:

$$B = \frac{v_2}{v_{out}} = \frac{R_1}{R_1 + R_f} \qquad (19\text{-}6)$$

Taking the reciprocal gives:

$$A_v \cong \frac{1}{B} = \frac{R_1 + R_f}{R_1} = \frac{R_f}{R_1} + 1$$

Except for replacing $A_{v(CL)}$ with A_v, this is the same formula derived in Chap. 18 with a virtual short between the input terminals of the op amp.

Example 19-1

In Fig. 19-4, calculate the feedback fraction, the ideal closed-loop voltage gain, the percent error, and the exact closed-loop voltage gain. Use a typical A_{VOL} of 100,000 for the 741C.

Figure 19-4 Example.

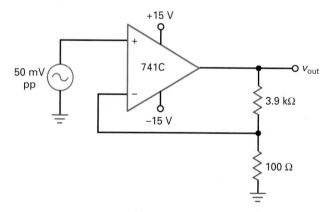

SOLUTION With Eq. (19-6), the feedback fraction is:

$$B = \frac{100\ \Omega}{100\ \Omega + 3.9\ \text{k}\Omega} = 0.025$$

With Eq. (19-4), the ideal closed-loop voltage gain is:

$$A_v = \frac{1}{0.025} = 40$$

With Eq. (19-5), the percent error is:

$$\% \text{ Error} = \frac{100\%}{1 + A_{VOL}B} = \frac{100\%}{1 + (100,000)(0.025)} = 0.04\%$$

We can calculate the exact closed-loop voltage gain in either of two ways: We can reduce the ideal answer by 0.04 percent, or we can use the exact formula, Eq. (19-3). Here are the calculations for both approaches:

$$A_v = 40 - (0.04\%)(40) = 40 - (0.0004)(40) = 39.984$$

This unrounded-off answer allows us to see how close the ideal answer (40) is to the exact answer. We can get the same exact answer with Eq. (19-3):

$$A_v = \frac{A_{VOL}}{1 + A_{VOL}B} = \frac{100,000}{1 + (100,000)(0.025)} = 39.984$$

In conclusion, this example has demonstrated the accuracy of the ideal equation for closed-loop voltage gain. Except for the most stringent analysis, we can always use the ideal equation. In those rare cases when we need to know how much error exists, we can fall back on Eq. (19-5) to calculate the percent error.

This example also validates the use of a virtual short between the input terminals of an op amp. In more complicated circuits, the virtual short allows us to analyze the effect of feedback with logical methods based on Ohm's law rather than having to derive more equations.

PRACTICE PROBLEM 19-1   In Fig. 19-4, change the feedback resistor from 3.9 kΩ to 4.9 kΩ. Calculate the feedback fraction, the ideal-closed-loop voltage gain, the percent error, and the exact closed-loop gain.

19-3 Other VCVS Equations

Negative feedback has a curative effect on the flaws or shortcomings of an amplifier, whether it is made up of ICs or discrete components. For instance, the open-loop voltage gain may have wide variations from one op amp to the next. Negative feedback *stabilizes* the voltage gain; that is, it almost eliminates the internal op-amp variations and makes the closed-loop voltage gain dependent primarily on external resistances. Since these resistances can be precision resistors with very low temperature coefficients, the closed-loop voltage gain becomes ultrastable.

Similarly, negative feedback in a VCVS amplifier increases the input impedance, decreases the output impedance, and reduces any nonlinear distortion of the amplified signal. In this section, we will find out just how much improvement occurs with negative feedback.

Gain Stability

The gain stability depends on having a very low percent error between the ideal and the exact closed-loop voltage gains. The smaller the percent error, the better the stability. The *worst-case error* of closed-loop voltage gain occurs when the open-loop voltage gain is minimum. As an equation:

$$\% \textbf{ Maximum error} = \frac{100\%}{1 + A_{VOL(\text{min})}B} \tag{19-7}$$

GOOD TO KNOW

Basically, any op-amp circuit that does not use negative feedback is considered too unstable to be useful.

where $A_{VOL(min)}$ is the minimum or worst-case open-loop voltage gain shown on a data sheet. With a 741C, $A_{VOL(min)} = 20{,}000$.

For instance, if $1 + A_{VOL(min)}B$ equals 500:

$$\% \text{ Maximum error} = \frac{100\%}{500} = 0.2\%$$

In mass production the closed-loop voltage gain of any VCVS amplifier with the foregoing numbers will be within 0.2 percent of the ideal value.

Closed–Loop Input Impedance

Figure 19-5a shows a noninverting amplifier. Here is the exact equation for the closed-loop input impedance of this VCVS amplifier:

$$z_{in(CL)} = (1 + A_{VOL}B)R_{in} \parallel R_{CM} \tag{19-8}$$

where R_{in} = the open-loop input resistance of the op amp

R_{CM} = the common-mode input resistance of the op amp

A word or two about the resistances that appear in this equation: First, R_{in} is the input resistance shown on a data sheet. In a discrete bipolar diff amp, it equals $2\beta r'_e$, discussed in Chap. 17. We also discussed R_{in}, and Table 18-1 listed an input resistance of 2 MΩ for a 741C.

Figure 19-5 (a) VCVS amplifier; (b) nonlinear distortion; (c) fundamental and harmonics.

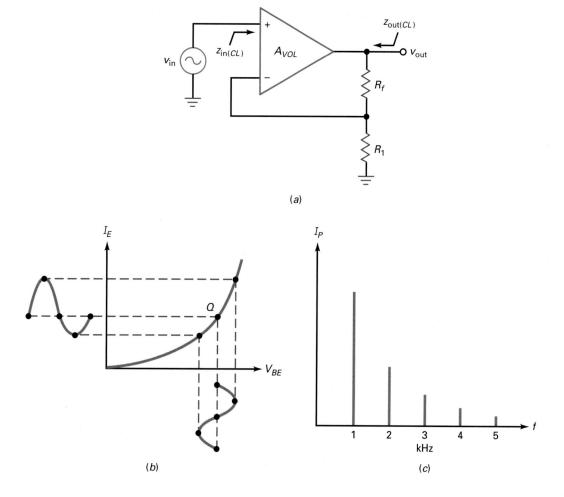

(a)

(b)

(c)

Second, R_{CM} is the equivalent tail resistance of the input diff-amp stage. In a discrete bipolar diff amp, R_{CM} equals R_E. In op amps, a current mirror is used in place of R_E. Because of this, the R_{CM} of an op amp has an extremely high value. For instance, a 741C has an R_{CM} that is greater than 100 MΩ.

Often, R_{CM} is ignored because it is large, and Eq. (19-8) is approximated by:

$$z_{in(CL)} \cong (1 + A_{VOL}B)R_{in} \tag{19-9}$$

Since $1 + A_{VOL}B$ is much greater than unity in a practical VCVS amplifier, the closed-input impedance is extremely high. In a voltage follower, B is 1 and $z_{in(CL)}$ would approach infinity, except for the parallel effect of R_{CM} in Eq. (19-8). In other words, the ultimate limit on the closed-loop input impedance is:

$$z_{in(CL)} = R_{CM}$$

The main point to get is this: The exact value of closed-loop input impedance is not important. What is important is that it is very large, usually much larger than R_{in} but less than the ultimate limit of R_{CM}.

Closed–Loop Output Impedance

In Fig. 19-5a, the closed-loop output impedance is the overall output impedance looking back into the VCVS amplifier. The exact equation for this closed-loop output impedance is:

$$z_{out(CL)} = \frac{R_{out}}{1 + A_{VOL}B} \tag{19-10}$$

where R_{out} is the open-loop output resistance of the op amp shown on a data sheet. We discussed R_{out}, and Table 18-1 listed an output resistance of 75 Ω for a 741C.

Since $1 + A_{VOL}B$ is much greater than unity in a practical VCVS amplifier, the closed-loop output impedance is less than 1 Ω and may even approach zero in a voltage follower. For a voltage follower, the closed-loop impedance is so low that the resistance of the connecting wires may become the limiting factor.

Again, the main point is not the exact value of closed-loop output impedance but, rather, the fact that VCVS negative feedback reduces it to values much smaller than 1 Ω. For this reason, the output side of a VCVS amplifier approaches an ideal voltage source.

Nonlinear Distortion

One more improvement worth mentioning is the effect of negative feedback on distortion. In the later stages of an amplifier, *nonlinear distortion* will occur with large signals because the input/output response of the amplifying devices becomes nonlinear. For instance, the nonlinear graph of the base-emitter diode distorts a large signal by elongating the positive half cycle and compressing the negative half cycle, as shown in Fig. 19-5b.

Nonlinear distortion produces *harmonics* of the input signal. For instance, if a sinusoidal voltage signal has a frequency of 1 kHz, the distorted output current will contain sinusoidal signals with frequencies of 1, 2, 3 kHz, and so forth, as shown in the *spectrum diagram* of Fig. 19-5c. The fundamental frequency is 1 kHz, and all others are harmonics. The rms value of all the harmonics measured together tells us how much distortion has occurred. This is why nonlinear distortion is often called **harmonic distortion.**

We can measure harmonic distortion with an instrument called a *distortion analyzer.* This instrument measures the total harmonic voltage and divides it by the fundamental voltage to get the *percent of total harmonic distortion,* defined as:

$$THD = \frac{\textbf{Total harmonic voltage}}{\textbf{Fundamental voltage}} \times 100\% \qquad (19\text{-}11)$$

For instance, if the total harmonic voltage is 0.1 V rms and the fundamental voltage is 1 V, then $THD = 10$ percent.

Negative feedback reduces harmonic distortion. The exact equation for closed-loop harmonic distortion is:

$$THD_{CL} = \frac{THD_{OL}}{1 + A_{VOL}B} \qquad (19\text{-}12)$$

where THD_{OL} = open-loop harmonic distortion

THD_{CL} = closed-loop harmonic distortion

Once again, the quantity $1 + A_{VOL}B$ has a curative effect. When it is large, it reduces the harmonic distortion to negligible levels. In stereo amplifiers, this means that we hear high-fidelity music instead of distorted sounds.

Discrete Negative Feedback Amplifier

The idea of a voltage amplifier (VCVS), whose voltage gain is controlled by external resistors, was briefly described in Chap. 10, "Voltage Amplifiers." The discrete two-stage feedback amplifier, shown in Fig. 10-10, is essentially a noninverting voltage amplifier using negative feedback.

Looking back at this circuit, the two CE stages produce an open-loop voltage gain equal to:

$$A_{VOL} = (A_{v1})(A_{v2})$$

The output voltage drives a voltage divider formed by r_f and r_e. Because the bottom of r_e is at ac ground, the feedback fraction is approximately:

$$B \cong \frac{r_e}{r_e + r_f}$$

This ignores the loading effect of the input transistor's emitter.

The input V_{in} drives the base of the first transistor, while the feedback voltage drives the emitter. An error voltage appears across the base-emitter diode. The mathematical analysis is similar to that given earlier. The closed-loop voltage gain is approximately $\frac{1}{B}$, the input impedance is $(1 + A_{VOL}B)R_{in}$, the output impedance is $\frac{R_{out}}{(1 + A_{VOL}B)}$, and the distortion is $\frac{THD_{OL}}{(1 + A_{VOL}B)}$. It is very common to find the use of negative feedback in a variety of discrete amplifier configurations.

Example 19-2

In Fig. 19-6, the 741C has an R_{in} of 2 MΩ and an R_{CM} of 200 MΩ. What is the closed-loop input impedance? Use a typical A_{VOL} of 100,000 for the 741C.

SOLUTION   In Example 19-1, we calculated $B = 0.025$. Therefore:

$$1 + A_{VOL}B = 1 + (100,000)(0.025) \cong 2500$$

With Eq. (19-9):

$$z_{in(CL)} \cong (1 + A_{VOL}B)R_{in} = (2500)(2\ \text{MΩ}) = 5000\ \text{MΩ}$$

Figure 19-6   Example.

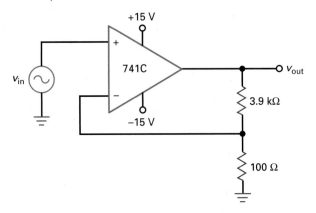

Whenever you get an answer over 100 MΩ, Eq. (19-8) should be used. With Eq. (19-8):

$$z_{in(CL)} = (5000 \text{ M}\Omega) \parallel 200 \text{ M}\Omega = 192 \text{ M}\Omega$$

This high input impedance means that a VCVS approaches an ideal voltage amplifier.

PRACTICE PROBLEM 19-2   In Fig. 19-6, change the 3.9 kΩ resistor to 4.9 kΩ and solve for $z_{in(CL)}$.

Example 19-3

Use the data and results of the preceding example to calculate the closed-loop output impedance in Fig. 19-6. Use an A_{VOL} of 100,000 and R_{out} of 75 Ω.

SOLUTION   With Eq. (19-10):

$$z_{out(CL)} = \frac{75 \ \Omega}{2500} = 0.03 \ \Omega$$

This low output impedance means that a VCVS approaches an ideal voltage amplifier.

PRACTICE PROBLEM 19-3   Repeat Example 19-3 with $A_{VOL} = 200,000$ and $B = 0.025$.

Example 19-4

Suppose the amplifier has an open-loop total harmonic distortion of 7.5 percent. What is the closed-loop total harmonic distortion?

SOLUTION   With Eq. (19-12):

$$THD_{(CL)} = \frac{7.5\%}{2500} = 0.003\%$$

PRACTICE PROBLEM 19-4   Repeat Example 19-4 with the 3.9 kΩ resistor changed to 4.9 kΩ.

19-4 The ICVS Amplifier

Figure 19-7 shows a transresistance amplifier. It has an input current and an output voltage. The ICVS amplifier is an almost perfect *current-to-voltage converter* because it has zero input impedance and zero output impedance.

Output Voltage

The exact equation for output voltage is:

$$v_{\text{out}} = -\left(i_{\text{in}}R_f \frac{A_{VOL}}{1 + A_{VOL}}\right) \tag{19-13}$$

Because A_{VOL} is much greater than unity, the equation simplifies to:

$$v_{\text{out}} = -(i_{\text{in}}R_f) \tag{19-14}$$

where R_f is the transresistance.

An easy way to derive and remember Eq. (19-14) is to use the concept of a virtual ground. Remember, the inverting input is a virtual ground to voltage, not current. When you visualize a virtual ground on the inverting input, you can see that all of the input current must flow through the feedback resistor. Since the left end of this resistor is grounded, the magnitude of the output voltage is given by:

$$v_{\text{out}} = -(i_{\text{in}}R_f)$$

The circuit is a current-to-voltage converter. We can select different values of R_f to get different conversion factors (transresistances). For instance, if $R_f = 1 \text{ k}\Omega$, then an input of 1 mA produces an output of 1 V. If $R_f = 10 \text{ k}\Omega$, the same input current produces an output of 10 V. The current direction shown in Fig. 19-8 is conventional current flow.

Figure 19-7 ICVS amplifier.

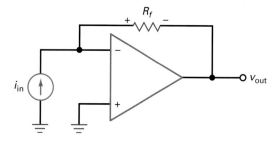

Figure 19-8 Inverting amplifier.

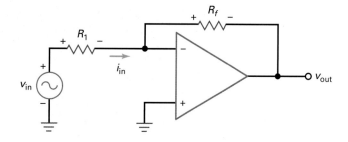

Noninverting Input and Output Impedances

In Figure 19-7, the exact equations for closed-loop input and output impedances are:

$$z_{in(CL)} = \frac{R_f}{1 + A_{VOL}}$$ (19-15)

$$z_{out(CL)} = \frac{R_{out}}{1 + A_{VOL}}$$ (19-16)

In both equations, the large denominator will reduce the impedance to a very low value.

The Inverting Amplifier

In Chap. 18, we discussed the inverting amplifier of Fig. 19-8. Recall that it has a closed-loop voltage gain of:

$$A_v = \frac{-R_f}{R_1}$$ (19-17)

This type of amplifier uses ICVS negative feedback. Because of the virtual ground on the inverting input, the input current equals:

$$i_{in} = \frac{v_{in}}{R_1}$$

Example 19-5　　　　　|||| MultiSim

In Fig. 19-9, what is the output voltage if the input frequency is 1 kHz?

Figure 19-9 Example.

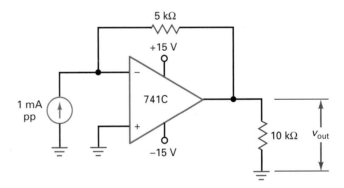

SOLUTION Visualize the input current of 1 mA pp flowing through the 5-kΩ resistor. With either Ohm's law or Eq. (19-14):

$$v_{out} = -(1 \text{ mA pp})(5 \text{ k}\Omega) = -5 \text{ V pp}$$

Again, the negative sign indicates a 180° phase shift. The output voltage is an ac voltage with a peak-to-peak value of 5 V and a frequency of 1 kHz.

PRACTICE PROBLEM 19-5 In Fig. 19-9, change the feedback resistor to 2 kΩ and calculate v_{out}.

Example 19-6

What are the closed-loop input and output impedances in Fig. 19-9? Use typical 741C parameters.

SOLUTION With Eq. (19-15):

$$z_{in(CL)} = \frac{5\,k\Omega}{1 + 100{,}000} \cong \frac{5\,k\Omega}{100{,}000} = 0.05\,\Omega$$

With Eq. (19-16):

$$z_{out(CL)} = \frac{75\,\Omega}{1 + 100{,}000} \cong \frac{75\,\Omega}{100{,}000} = 0.00075\,\Omega$$

PRACTICE PROBLEM 19-6 Repeat Example 19-6 with $A_{VOL} = 200{,}000$.

19-5 The VCIS Amplifier

With a VCIS amplifier, an input voltage controls an output current. Because of the heavy negative feedback in this kind of amplifier, the input voltage is converted to a precise value of output current.

Figure 19-10 shows a transconductance amplifier. It is similar to a VCVS amplifier, except that R_L is the load resistor as well as the feedback resistor. In other words, the active output is not the voltage across $R_1 + R_L$; rather, it is the current through R_L. This output current is stabilized; that is, a specific value of input voltage produces a precise value of output current.

In Fig. 19-10, the exact equation for output current is:

$$i_{out} = \frac{v_{in}}{R_1 + (R_1 + R_L)/A_{VOL}} \qquad (19\text{-}18)$$

Figure 19-10 VCIS amplifier.

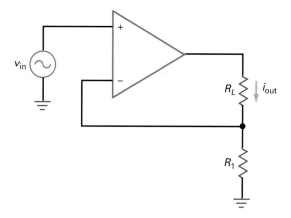

In a practical circuit, the second term in the denominator is much smaller than the first and the equation simplifies to:

$$i_{\text{out}} = \frac{v_{\text{in}}}{R_1} \tag{19-19}$$

This is sometimes written as:

$$i_{\text{out}} = g_m v_{\text{in}}$$

where $g_m = 1/R_1$.

Here is an easy way to derive and remember Eq. (19-19): When you visualize a virtual short between the input terminals of Fig. 19-10, the inverting input is bootstrapped to the noninverting input. Therefore, all the input voltage appears across R_1. The current through this resistor is:

$$i_1 = \frac{v_{\text{in}}}{R_1}$$

In Fig. 19-10, the only path for this current is through R_L. This is why Eq. (19-19) gives the value of output current.

The circuit is a *voltage-to-current converter*. We can select different values of R_1 to get different conversion factors (transconductances). For instance, if $R_1 = 1\ k\Omega$, an input voltage of 1 V produces an output current of 1 mA. If $R_1 = 100\ \Omega$, the same input voltage produces an output current of 10 mA.

Since the input side of Fig. 19-10 is the same as the input side of a VCVS amplifier, the approximate equation for the closed-loop input impedance of a VCIS amplifier is:

$$z_{\text{in}(CL)} = (1 + A_{VOL}B)R_{\text{in}} \tag{19-20}$$

where R_{in} is the input resistance of the op amp. The stabilized output current sees a closed-loop output impedance of:

$$z_{\text{out}(CL)} = (1 + A_{VOL})R_1 \tag{19-21}$$

In both equations, a large A_{VOL} increases both impedances toward infinity, exactly what we want for a VCIS amplifier. The circuit is an almost perfect voltage-to-current converter because it has very high input and output impedances.

The transconductance amplifier of Fig. 19-10 operates with a floating load resistor. This is not always convenient because many loads are single-ended. In this case, you may see the following linear ICs used as transconductance amplifiers: LM3080, LM13600, and LM13700. These monolithic transconductance amplifiers can drive a single-ended load resistance.

Example 19-7

IIII MultiSim

What is the load current in Fig. 19-11? The load power? What happens if the load resistance changes to 4 Ω?

SOLUTION   Visualize a virtual short across the input terminals of the op amp. With the inverting input bootstrapped to the noninverting input, all the input voltage is across the 1-Ω resistor. With Ohm's law or Eq. (19-19), we can calculate an output current of:

$$i_{\text{out}} = \frac{2\ \text{V rms}}{1\ \Omega} = 2\ \text{A rms}$$

This 2 A flows through the load resistance of 2 Ω, producing a load power:

$$P_L = (2\ \text{A})^2 (2\ \Omega) = 8\ \text{W}$$

Figure 19-11   Example.

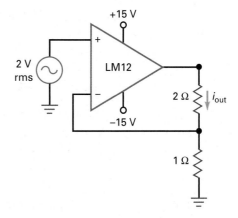

If the load resistance is changed to 4 Ω, the output current is still 2 A rms, but the load power increases to:

$$P_L = (2 \text{ A})^2(4 \text{ Ω}) = 16 \text{ W}$$

As long as the op amp does not saturate, we can change the load resistance to any value and still have a stabilized output current of 2 A rms.

PRACTICE PROBLEM 19-7   In Fig. 19-11, change the input voltage to 3 V rms and solve for i_{out} and P_L.

19-6 The ICIS Amplifier

An ICIS circuit amplifies the input current. Because of the heavy negative feedback, the ICIS amplifier tends to act like a perfect **current amplifier.** It has a very low input impedance and a very high output impedance.

Figure 19-12 shows an inverting current amplifier. The closed-loop current gain is stabilized and given by:

$$A_i = \frac{A_{VOL}(R_1 + R_2)}{R_L + A_{VOL}R_1} \tag{19-22}$$

Figure 19-12   ICIS amplifier.

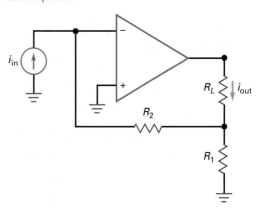

Usually, the second term in the denominator is much larger than the first and the equation simplifies to:

$$A_i \cong \frac{R_2}{R_1} + 1 \qquad (19\text{-}23)$$

The equation for the closed-loop input impedance of an ICIS amplifier is:

$$z_{\text{in}(CL)} = \frac{R_2}{1 + A_{VOL}B} \qquad (19\text{-}24)$$

where the feedback fraction is given by:

$$B = \frac{R_1}{R_1 + R_2} \qquad (19\text{-}25)$$

The stabilized output current sees a closed-loop output impedance of:

$$z_{\text{out}(CL)} = (1 + A_{VOL})R_1 \qquad (19\text{-}26)$$

A large A_{VOL} produces a very small input impedance and a very large output impedance. Because of this, the ICIS circuit is an almost perfect current amplifier.

Example 19-8   IIII MultiSim

What is the load current in Fig. 19-13? The load power? If the load resistance is changed to 2 Ω, what are the load current and power?

Figure 19-13   Example.

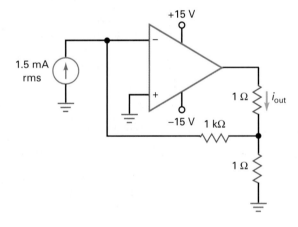

SOLUTION   With Eq. (19-23), the current gain is:

$$A_i = \frac{1\,k\Omega}{1\,\Omega} + 1 \cong 1000$$

The load current is:

$$i_{\text{out}} = (1000)(1.5\ \text{mA rms}) = 1.5\ \text{A rms}$$

The load power is:

$$P_L = (1.5\ \text{A})^2 (1\ \Omega) = 2.25\ \text{W}$$

If the load resistance is increased to 2 Ω, the load current is still 1.5 A rms, but the load power increases to:

$$P_L = (1.5 \text{ A})^2(2 \text{ }\Omega) = 4.5 \text{ W}$$

PRACTICE PROBLEM 19-8 Using Fig. 19-13, change i_{in} to 2 mA. Calculate i_{out} and P_L.

19-7 Bandwidth

Negative feedback increases the bandwidth of an amplifier because the roll-off in open-loop voltage gain means that less voltage is fed back, which produces more input voltage as a compensation. Because of this, the closed-loop cutoff frequency is higher than the open-loop cutoff frequency.

Gain–Bandwidth Product Is Constant

We discussed VCVS bandwidth in Chap. 18. Recall that the closed-loop cutoff bandwidth is given by:

$$f_{2(CL)} = \frac{f_{unity}}{A_{v(CL)}} \tag{19-27}$$

We can also derive two more VCVS equations for closed-loop bandwidth:

$$f_{2(CL)} = (1 + A_{VOL}B)f_{2(OL)} \tag{19-28}$$

$$f_{2(CL)} = \frac{A_{VOL}}{A_{v(CL)}} f_{2(OL)} \tag{19-29}$$

where $A_{v(CL)}$ is the same as A_v.

You can use any of these equations to calculate the closed-loop bandwidth of a VCVS amplifier. The one to use depends on the given data. For instance, if you know the values of f_{unity} and $A_{v(CL)}$, then Eq. (19-27) is the one to use. If you have the values of A_{VOL}, B, and $f_{2(OL)}$, use Eq. (19-28). Sometimes, you know the values of A_{VOL}, $A_{v(CL)}$, and $f_{2(OL)}$. In this case, Eq. (19-29) is useful.

Gain–Bandwidth Product Is Constant

Equation (19-27) can be rewritten as:

$$A_{v(CL)}f_{2(CL)} = f_{unity}$$

The left side of this equation is the product of gain and bandwidth, and is called the **gain-bandwidth product (GBP).** The right side of the equation is a constant for a given op amp. In words, the equation says that the *gain-bandwidth product is a constant.* Because GBP is a constant for a given op amp, a designer has to trade off gain for bandwidth. The less gain used, the more bandwidth results. Conversely, if the designer wants more gain, he or she has to settle for less bandwidth.

The only way to improve matters is to use an op amp with a higher GBP, equivalent to a higher f_{unity}. If an op amp does not have enough GBP for an application, a designer can select a better op amp, one with a greater GBP. For instance, a 741C has a GBP of 1 MHz. If this is too low for a given application, we can use an LM318 which has a GBP of 15 MHz. This way, we would get 15 times as much bandwidth for the same closed-loop voltage gain.

Bandwidth and Slew-Rate Distortion

Although negative feedback reduces the nonlinear distortion of the later stages of an amplifier, it has absolutely no effect on slew-rate distortion. Therefore, after you calculate the closed-loop bandwidth, you can calculate the power bandwidth with Eq. (18-2). For an undistorted output over the entire closed-loop bandwidth, the closed-loop cutoff frequency must be less than the power bandwidth:

$$f_{2(CL)} < f_{max} \tag{19-30}$$

This means that the peak value of the output should be less than:

$$V_{p(max)} = \frac{S_R}{2\pi f_{2(CL)}} \tag{19-31}$$

Here is why negative feedback has no effect on slew-rate distortion: In Chap. 18, we discussed how the compensating capacitor of an op amp produces a large-input Miller capacitance. For a 741C, this large capacitance loads down the input diff amp, as shown in Fig. 19-14a. When slew-rate distortion occurs, v_{in} is high enough to saturate one transistor and cut off the other. Since the op amp is no longer operating in the linear region, the curative effect of negative feedback is temporarily suspended.

Figure 19-14b shows what happens when Q_1 is saturated and Q_2 is cut off. Since the 3000-pF capacitor must charge through a 1-MΩ resistor, we get the slew shown in the figure. After the capacitor charges, Q_1 comes out of saturation, Q_2 comes out of cutoff, and the curative effect of negative feedback reappears.

Table of Negative Feedback

Summary Table 19-1 displays the four ideal prototypes of negative feedback. These prototypes are basic circuits that can be modified to get more advanced circuits. For instance, by using a voltage source and an input resistor of R_1, the ICVS prototype becomes the widely used inverting amplifier discussed in Chap. 18. As

---

Figure 19-14   (a) Input diff amp of 741C; (b) capacitor charging causes slew.

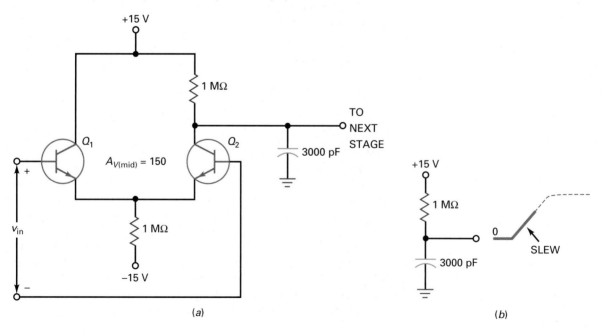

Type	Stabilized	Equation	$z_{in(CL)}$	$z_{out(CL)}$	$f_{2(CL)}$	$f_{2(CL)}$	$f_{2(CL)}$
VCVS	A_v	$\dfrac{R_f}{R_1} + 1$	$(1 + A_{VOL}B)R_{in}$	$\dfrac{R_{out}}{(1 + A_{VOL}B)}$	$(1 + A_{VOL}B)f_{2(OL)}$	$\dfrac{A_{VOL}}{A_{v(CL)}}f_{2(OL)}$	$\dfrac{f_{unity}}{A_{v(CL)}}$
ICVS	$\dfrac{v_{out}}{i_{in}}$	$v_{out} = -(i_{in}R_f)$	$\dfrac{R_f}{1 + A_{VOL}}$	$\dfrac{R_{out}}{1 + A_{VOL}}$	$(1 + A_{VOL})f_{2(OL)}$	–	–
VCIS	$\dfrac{i_{out}}{v_{in}}$	$i_{out} = \dfrac{v_{in}}{R_1}$	$(1 + A_{VOL}B)R_{in}$	$(1 + A_{VOL})R_1$	$(1 + A_{VOL})f_{2(OL)}$	–	–
ICIS	A_i	$\dfrac{R_2}{R_1} + 1$	$\dfrac{R_2}{(1 + A_{VOL}B)}$	$(1 + A_{VOL})R_1$	$(1 + A_{VOL}B)f_{2(OL)}$	–	–

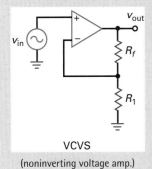

VCVS

(noninverting voltage amp.)

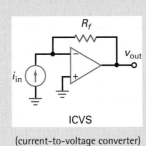

ICVS

(current-to-voltage converter)

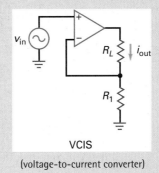

VCIS

(voltage-to-current converter)

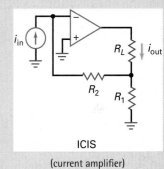

ICIS

(current amplifier)

another example, we can add coupling capacitors to the VCVS prototype to get an ac amplifier. In the next few chapters, we will modify these basic prototypes to get a wide variety of useful circuits.

Example 19-9

If the VCVS amplifier of Summary Table 19-1 uses an LF411A with $(1 + A_{VOL}B) = 1000$ and $f_{2(OL)} = 160$ Hz, what is the closed-loop bandwidth?

SOLUTION   With Eq. (19-28):

$$f_{2(CL)} = (1 + A_{VOL}B)f_{2(OL)} = (1000)(160 \text{ Hz}) = 160 \text{ kHz}$$

PRACTICE PROBLEM 19-9   Repeat Example 19-9 with $f_{2(OL)} = 100$ Hz.

Example 19-10

If a VCVS amplifier of Summary Table 19-1 uses an LM308 with $A_{VOL} = 250{,}000$ and $f_{2(OL)} = 1.2$ Hz, what is the closed-loop bandwidth for an $A_{v(CL)} = 50$?

SOLUTION With Eq. (19-29):

$$f_{2(CL)} = \frac{A_{VOL}}{A_{v(CL)}} f_{2(OL)} = \frac{250{,}000}{50} (1.2 \text{ Hz}) = 6 \text{ kHz}$$

PRACTICE PROBLEM 19-10 Repeat Example 19-10 using $A_{VOL} = 200{,}000$ and $f_{2(OL)} = 2$ Hz.

Example 19-11

If the ICVS amplifier of Summary Table 19-1 uses an LM12 with $A_{VOL} = 50{,}000$ and $f_{2(OL)} = 14$ Hz, what is the closed-loop bandwidth?

SOLUTION With the equation given in Summary Table 19-1:

$$f_{2(CL)} = (1 + A_{VOL})f_{2(OL)} = (1 + 50{,}000)(14 \text{ Hz}) = 700 \text{ kHz}$$

PRACTICE PROBLEM 19-11 In Example 19-11, if $A_{VOL} = 75{,}000$ and $f_{2(OL)} = 750$ kHz, find the open-loop bandwidth.

Example 19-12

If the ICIS amplifier of Summary Table 19-1 uses an OP-07A with $f_{2(OL)} = 20$ Hz and if $(1 + A_{VOL}B) = 2500$, what is the closed-loop bandwidth?

SOLUTION With the equation given in Summary Table 19-1:

$$f_{2(CL)} = (1 + A_{VOL}B)f_{2(OL)} = (2500)(20 \text{ Hz}) = 50 \text{ kHz}$$

PRACTICE PROBLEM 19-12 Repeat Example 19-12 with $f_{2(OL)} = 50$ Hz.

Example 19-13

A VCVS amplifier uses an LM741C with $f_{unity} = 1$ MHz and $S_R = 0.5$ V/μs. If $A_{v(CL)} = 10$, what is the closed-loop bandwidth? The largest undistorted peak output voltage at $f_{2(CL)}$?

SOLUTION With Eq. (19-27):

$$f_{2(CL)} = \frac{f_{unity}}{A_{v(CL)}} = \frac{1 \text{ MHz}}{10} = 100 \text{ kHz}$$

With Eq. (19-31):

$$V_{p(max)} = \frac{S_R}{2\pi f_{2(CL)}} = \frac{0.5 \text{ V/}\mu\text{s}}{2\pi(100 \text{ kHz})} = 0.795 \text{ V}$$

PRACTICE PROBLEM 19-13 Calculate the closed-loop bandwidth and $V_{p(max)}$ in Example 19-13 with $A_{v(CL)} = 100$.

Summary

SEC. 19-1 FOUR TYPES OF NEGATIVE FEEDBACK

There are four ideal types of negative feedback: VCVS, ICVS, VCIS, and ICIS. Two types (VCVS and VCIS) are controlled by an input voltage, and the other two types (ICVS and ICIS) are controlled by an input current. The output sides of VCVS and ICVS act like a voltage source, and the output sides of VCIS and ICIS act like a current source.

SEC. 19-2 VCVS VOLTAGE GAIN

The loop gain is the voltage gain of the forward and feedback paths. In any practical design, the loop gain is very large. As a result, the closed-loop voltage gain is ultrastable because it no longer depends on the characteristics of the amplifier. Instead, it depends almost entirely on the characteristics of external resistors.

SEC. 19-3 OTHER VCVS EQUATIONS

VCVS negative feedback has a curative effect on the flaws of an amplifier because it stabilizes the voltage gain, increases the input impedance, decreases the output impedance, and decreases harmonic distortion.

SEC. 19-4 THE ICVS AMPLIFIER

This is a transresistance amplifier, equivalent to a current-to-voltage converter. Because of the virtual ground, it ideally has zero input impedance. The input current produces a precise value of output voltage.

SEC. 19-5 THE VCIS AMPLIFIER

This is a transconductance amplifier, equivalent to a voltage-to-current converter. It ideally has infinite input impedance. The input voltage produces a precise value of output current. The output impedance approaches infinity.

SEC. 19-6 THE ICIS AMPLIFIER

Because of the heavy negative feedback, the ICIS amplifier approaches the perfect current amplifier, one with zero input impedance and infinite output impedance.

SEC. 19-7 BANDWIDTH

Negative feedback increases the bandwidth of an amplifier because the roll-off in open-loop voltage gain means that less voltage is fed back, which produces more input voltage as a compensation. Because of this, the closed-loop cutoff frequency is higher than the open-loop cutoff frequency.

Definitions

(19-1)   Feedback fraction:

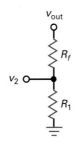

$$B = \frac{v_2}{v_{out}}$$

(19-11)   Total harmonic distortion:

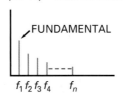

$$THD = \frac{\text{Total harmonic voltage}}{\text{Fundamental voltage}} \times 100\%$$

Derivations

(19-4)   VCVS voltage gain:

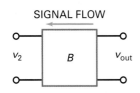

$$A_v \cong \frac{1}{B}$$

(19-5)   VCVS percent error:

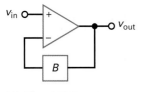

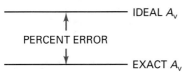

$$\% \text{ Error} = \frac{100\%}{1 + A_{VOL}B}$$

(19-6)   VCVS feedback fraction:

$$B = \frac{v_2}{v_{out}} = \frac{R_1}{R_1 + R_f}$$

(19-9)   VCVS input impedance:

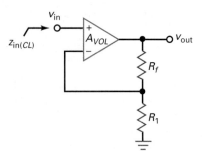

$$z_{in(CL)} \cong (1 + A_{VOL}B)R_{in}$$

(19-10)   VCVS output impedance:

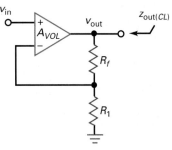

$$z_{out(CL)} = \frac{R_{out}}{1 + A_{VOL}B}$$

(19-12)   Closed-loop distortion:

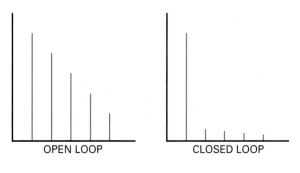

OPEN LOOP   CLOSED LOOP

$$THD_{CL} = \frac{THD_{OL}}{1 + A_{VOL}B}$$

(19-14)   ICVS output voltage:

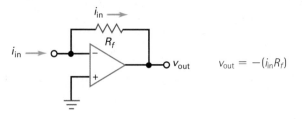

$$v_{out} = -(i_{in}R_f)$$

(19-15)   ICVS input impedance:

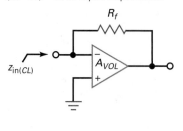

$$z_{in(CL)} = \frac{R_f}{1 + A_{VOL}}$$

(19-16)   ICVS output impedance:

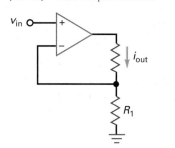

$$z_{out(CL)} = \frac{R_{out}}{1 + A_{VOL}}$$

(19-19)   VCIS output current:

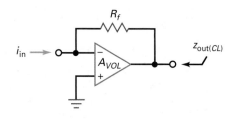

$$i_{out} = \frac{v_{in}}{R_1}$$

(19-23)   ICIS current gain:

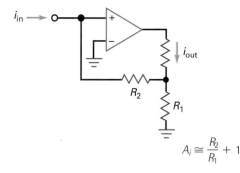

$$A_i \cong \frac{R_2}{R_1} + 1$$

(19-27)   Closed-loop bandwidth:

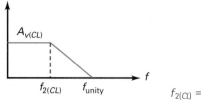

$$f_{2(CL)} = \frac{f_{unity}}{A_{v(CL)}}$$

Student Assignments

1. With negative feedback, the returning signal
   a. Aids the input signal
   b. Opposes the input signal
   c. Is proportional to output current
   d. Is proportional to differential voltage gain

2. How many types of negative feedback are there?
   a. One   c. Three
   b. Two   d. Four

3. A VCVS amplifier approximates an ideal
   a. Voltage amplifier
   b. Current-to-voltage converter
   c. Voltage-to-current converter
   d. Current amplifier

4. The voltage between the input terminals of an ideal op amp is
   a. Zero
   b. Very small
   c. Very large
   d. Equal to the input voltage

5. When an op amp is not saturated, the voltages at the noninverting and inverting inputs are
   a. Almost equal
   b. Much different
   c. Equal to the output voltage
   d. Equal to ± 15 V

6. The feedback fraction B
   a. Is always less than 1
   b. Is usually greater than 1
   c. May equal 1
   d. May not equal 1

7. An ICVS amplifier has no output voltage. A possible trouble is
   a. No negative supply voltage
   b. Shorted feedback resistor
   c. No feedback voltage
   d. Open load resistor

8. In a VCVS amplifier, any decrease in open-loop voltage gain produces an increase in
   a. Output voltage
   b. Error voltage
   c. Feedback voltage
   d. Input voltage

9. The open-loop voltage gain equals the
   a. Gain with negative feedback
   b. Differential voltage gain of the op amp
   c. Gain when B is 1
   d. Gain at f_{unity}

10. The loop gain $A_{VOL}B$
    a. Is usually much smaller than 1
    b. Is usually much greater than 1
    c. May not equal 1
    d. Is between 0 and 1

11. The closed-loop input impedance with an ICVS amplifier is
    a. Usually larger than the open-loop input impedance
    b. Equal to the open-loop input impedance
    c. Sometimes less than the open-loop impedance
    d. Ideally zero

12. With an ICVS amplifier, the circuit approximates an ideal
    a. Voltage amplifier
    b. Current-to-voltage converter
    c. Voltage-to-current converter
    d. Current amplifier

13. Negative feedback reduces
    a. The feedback fraction
    b. Distortion
    c. The input offset voltage
    d. The open-loop gain

14. A voltage follower has a voltage gain of
    a. Much less than 1
    b. 1
    c. More than 1
    d. A_{VOL}

15. The voltage between the input terminals of a real op amp is
    a. Zero
    b. Very small
    c. Very large
    d. Equal to the input voltage

16. The transresistance of an amplifier is the ratio of its
    a. Output current to input voltage
    b. Input voltage to output current
    c. Output voltage to input voltage
    d. Output voltage to input current

17. Current cannot flow to ground through
    a. A mechanical ground
    b. An ac ground
    c. A virtual ground
    d. An ordinary ground

18. In a current-to-voltage converter, the input current flows
    a. Through the input impedance of the op amp
    b. Through the feedback resistor
    c. To ground
    d. Through the load resistor

19. The input impedance of a current-to-voltage converter is
    a. Small
    b. Large
    c. Ideally zero
    d. Ideally infinite

20. The open-loop bandwidth equals
    a. f_{unity}
    b. $f_{2(OL)}$
    c. $f_{unity}/A_{v(CL)}$
    d. f_{max}

21. The closed-loop bandwidth equals
    a. f_{unity}
    b. $f_{2(OL)}$
    c. $f_{unity}/A_{v(CL)}$
    d. f_{max}

22. For a given op amp, which of these is constant?
    a. $f_{2(OL)}$
    b. Feedback voltage
    c. $A_{v(CL)}$
    d. $A_{v(CL)}f_{(CL)}$

23. Negative feedback does not improve
    a. Stability of voltage gain
    b. Nonlinear distortion in later stages
    c. Output offset voltage
    d. Power bandwidth

24. An ICVS amplifier is saturated. A possible trouble is

a. No supply voltages

b. Open feedback resistor

c. No input voltage

d. Open load resistor

25. A VCVS amplifier has no output voltage. A possible trouble is

a. Shorted load resistor

b. Open feedback resistor

c. Excessive input voltage

d. Open load resistor

26. An ICIS amplifier is saturated. A possible trouble is

a. Shorted load resistor

b. R_2 is open

c. No input voltage

d. Open load resistor

27. An ICVS amplifier has no output voltage. A possible trouble is

a. No positive supply voltage

b. Open feedback resistor

c. No feedback voltage

d. Shorted load resistor

28. The closed–loop input impedance in a VCVS amplifier is

a. Usually larger than the open-loop input impedance

b. Equal to the open-loop input impedance

c. Sometimes less than the open-loop input impedance

d. Ideally zero

Problems

In the following problems, refer to Table 18-2 as needed for the parameters of the op amps

SEC. 19-2 VCVS VOLTAGE GAIN

19-1 In Fig. 19-15, calculate the feedback fraction, the ideal closed-loop voltage gain, the percent error, and the exact voltage gain.

Figure 19-15

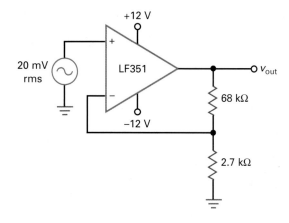

19-2 If the 68-kΩ resistor of Fig. 19-15 is changed to 39 kΩ, what is the feedback fraction? The closed-loop voltage gain.

19-3 In Fig. 19-15, the 2.7-kΩ resistor is changed to 4.7 kΩ. What is the feedback fraction? The closed-loop voltage gain?

19-4 If the LF351 of Fig. 19-15 is replaced by an LM308, what is the feedback fraction, the ideal closed-loop voltage gain, the percent error, and the exact voltage gain?

SEC. 19-3 OTHER VCVS EQUATIONS

19-5 In Fig. 19-16, the op amp has an R_{in} of 3 MΩ and an R_{CM} of 500 MΩ. What is the closed-loop input impedance? Use an A_{VOL} of 200,000 for the op amp.

Figure 19-16

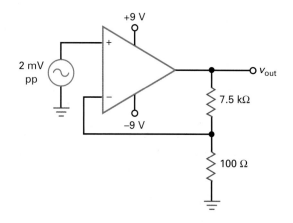

19-6 What is the closed-loop output impedance in Fig. 19-16? Use an A_{VOL} of 75,000 and an R_{out} of 50 Ω.

19-7 Suppose the amplifier of Fig. 19-16 has an open-loop total harmonic distortion of 10 percent. What is the closed-loop total harmonic distortion?

SEC. 19-4 THE ICVS AMPLIFIER

19-8 ▌▌▌ MultiSim In Fig. 19-17, the frequency is 1 kHz. What is the output voltage?

Figure 19-17

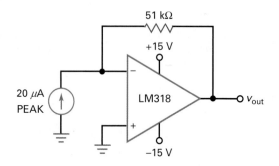

19-9 ||||| MultiSim What is the output voltage in Fig. 19-17 if the feedback resistor is changed from 51 to 33 kΩ?

19-10 In Fig. 19-17, the input current is changed to 10.0 μA rms. What is the peak-to-peak output voltage?

SEC. 19-5 THE VCIS AMPLIFIER

19-11 ||||| MultiSim What is the output current in Fig. 19-18? The load power?

Figure 19-18

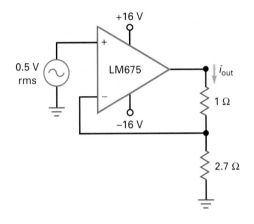

19-12 If the load resistor is changed from 1 to 3 Ω in Fig. 19-18, what is the output current? The load power?

19-13 ||||| MultiSim If the 2.7-Ω resistor is changed to 4.7 Ω in Fig. 19-18, what are the output current and load power?

SEC. 19-6 THE ICIS AMPLIFIER

19-14 ||||| MultiSim What is the current gain in Fig. 19-19? The load power?

19-15 ||||| MultiSim If the load resistor is changed from 1 to 2 Ω in Fig. 19-19, what is the output current? The load power?

Figure 19-19

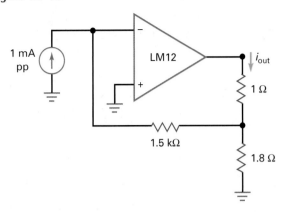

19-16 If the 1.8-Ω resistor is changed to 7.5 Ω in Fig. 19-19, what are the current gain and load power?

SEC. 19-7 BANDWIDTH

19-17 A VCVS amplifier uses an LM324 with $(1 + A_{VOL}B)$ = 1000 and $f_{2(OL)}$ = 2 Hz. What is the closed-loop bandwidth?

19-18 If a VCVS amplifier uses an LM833 with A_{VOL} = 316,000 and $f_{2(OL)}$ = 4.5 Hz, what is the closed-loop bandwidth for $A_{v(CL)}$ = 75?

19-19 An ICVS amplifier uses an LM318 with A_{VOL} = 20,000 and $f_{2(OL)}$ = 750 Hz. What is the closed-loop bandwidth?

19-20 An ICIS amplifier uses a TL072 with $f_{2(OL)}$ = 120 Hz. If $(1 + A_{VOL}B)$ = 5000, what is the closed-loop bandwidth?

19-21 A VCVS amplifier uses an LM741C with f_{unity} = 1 MHz and S_R = 0.5 V/μs. If $A_{v(CL)}$ = 10, what is the closed-loop bandwidth? The largest undistorted peak output voltage at $f_{2(CL)}$?

Critical Thinking

19-22 Figure 19-20 is a current-to-voltage converter that can be used to measure current. What does the voltmeter read when the input current is 4 μA?

19-23 What is the output voltage in Fig. 19-21?

19-24 In Fig. 19-22, what is the voltage gain of the amplifier for each position of the switch?

19-25 In Fig. 19-22, what is the output voltage for each position of the switch if the input voltage is 10 mV?

19-26 A 741C with A_{VOL} = 100,000, R_{in} = 2 MΩ, and R_{out} = 75 Ω is used in Fig. 19-22. What are the closed-loop input and output impedances for each switch position?

Figure 19-20

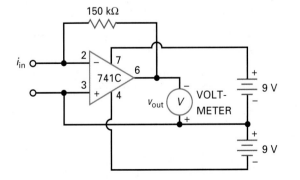

Negative Feedback

Figure 19-21

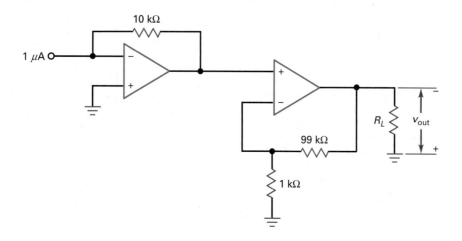

Figure 19-22

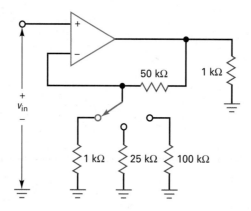

19-27 A 741C with $A_{VOL} = 100,000$, $I_{in(bias)} = 80$ nA, $I_{in(offset)} = 20$ nA, $V_{in(offset)} = 1$ mV and $R_f = 100$ kΩ is used in Fig. 19-22. What is the output offset voltage for each position of the switch?

19-28 What does the output voltage equal in Fig. 19-23a for each position of the switch?

19-29 The photodiode of Fig. 19-23b produces a current of 2 μA. What is the output voltage?

19-30 If the unknown resistor of Fig. 19-23c has a value of 3.3 kΩ, what is the output voltage?

19-31 If the output voltage is 2 V in Fig. 19-23c, what is the value of the unknown resistor?

19-32 The feedback resistor of Fig. 19-24 has a resistance that is controlled by sound waves. If the feedback resistance varies sinusoidally between 9 and 11 kΩ, what is the output voltage?

19-33 Temperature controls the feedback resistance of Fig. 19-24. If the feedback resistance varies from 1 to 10 kΩ, what is the range of output voltage?

19-34 Figure 19-25 shows a sensitive dc voltmeter that uses a BIFET op amp. Assume that the output voltage has been nulled with the zero adjustment. What is the input voltage that produces full-scale deflection for each switch position?

Figure 19-23

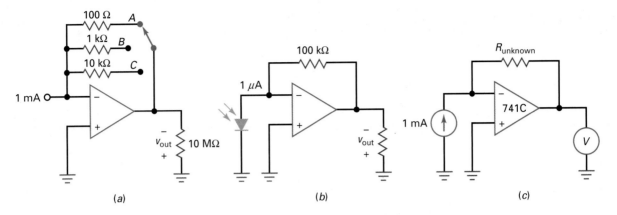

(a)   (b)   (c)

Figure 19-24

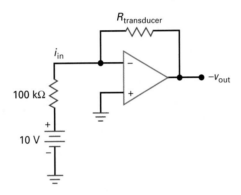

Figure 19-25

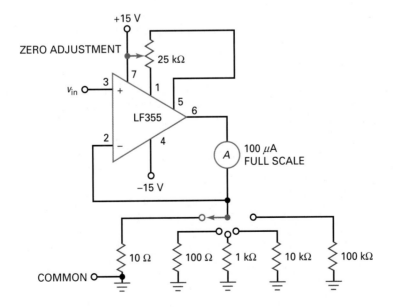

Troubleshooting

||||| **MultiSim** Use Fig. 19-26 for the remaining problems. Any resistor R_2 through R_4 may be open or shorted. Also, connecting wires *AB*, *CD*, or *FG* may be open.

19-35 Find Troubles 1 to 3.

19-36 Find Troubles 4 to 6.

19-37 Find Troubles 7 to 9.

Figure 19-26

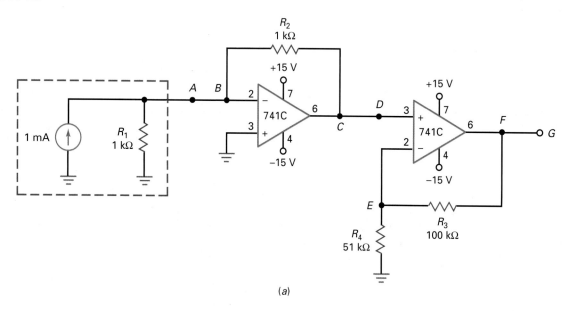

(a)

Troubleshooting

Trouble	V_A	V_B	V_C	V_D	V_E	V_F	V_G	R_4
OK	0	0	−1	−1	−1	−3	−3	OK
T1	0	0	−1	0	0	0	0	OK
T2	0	0	0	0	0	0	0	OK
T3	0	0	−1	−1	0	−13.5	−13.5	0
T4	0	0	−13.5	−13.5	−4.5	−13.5	−13.5	OK
T5	0	0	−1	−1	−1	−3	0	OK
T6	0	0	−1	−1	0	−13.5	−13.5	OK
T7	+1	−4.5	0	0	0	0	0	OK
T8	0	0	−1	−1	−1	−1	−1	OK
T9	0	0	−1	−1	−1	−1	−1	∞

(b)

Job Interview Questions

1. Draw the equivalent circuit for VCVS negative feedback. Write the equations for closed-loop voltage gain, input and output impedances, and backwidth.
2. Draw the equivalent circuit for ICVS negative feedback. How is this related to the inverting amplifier?
3. What is the difference between the closed-loop bandwidth and the power bandwidth?
4. What are the four kinds of negative feedback? Briefly describe what the circuits do.
5. What effect does negative feedback have on an amplifier's bandwidth?
6. Is the closed-loop cutoff frequency higher or lower than the open-loop cutoff frequency?
7. Why does any circuit use negative feedback?
8. What effect does positive feedback have on an amplifier?
9. What is feedback attenuation (also called *feedback attenuation factor*)?
10. What is negative feedback, and why is it used?
11. Why might you provide negative feedback to an amplifier stage when doing so will reduce the overall voltage gain?
12. What type of amplifiers are the BJT and the FET?

Self-Test Answers

1.	b	11.	d	20.	b
2.	d	12.	b	21.	c
3.	a	13.	b	22.	d
4.	a	14.	b	23.	d
5.	a	15.	b	24.	b
6.	c	16.	d	25.	a
7.	b	17.	c	26.	b
8.	b	18.	b	27.	d
9.	b	19.	c	28.	a
10.	b				

Practice Problem Answers

19-1 $B = 0.020$; $A_{v(ideal)} = 50$; % error = 0.05%; $A_{v(exact)} = 49.975$

19-2 $z_{in(CL)} = 191\ M\Omega$

19-3 $z_{out(CL)} = 0.015\ \Omega$

19-4 $THD_{(CL)} = 0.004\%$

19-5 $v_{out} = 2\ V_{pp}$

19-6 $z_{in(CL)} = 0.025\ \Omega$; $z_{out(CL)} = 0.000375\ \Omega$

19-7 $i_{out} = 3\ A\ rms$; $P_L = 18\ W$

19-8 $i_{out} = 2\ A\ rms$; $P_L = 4\ W$

19-9 $f_{2(CL)} = 100\ kHz$

19-10 $f_{2(CL)} = 8\ kHz$

19-11 $f_{2(CL)} = 10\ Hz$

19-12 $f_{2(CL)} = 125\ kHz$

19-13 $f_{2(CL)} = 10\ kHz$; $V_{p(max)} = 7.96\ Hz$

chapter 20 Linear Op-Amp Circuits

The output of a **linear op-amp circuit** has the same shape as the input signal. If the input is sinusoidal, the output is sinusoidal. At no time during the cycle does the op amp go into saturation. This chapter discusses a variety of linear op-amp circuits including inverting amplifiers, noninverting amplifiers, differential amplifiers, instrumentation amplifiers, current boosters, controlled current sources, and automatic gain control circuits.

Objectives

After studying this chapter, you should be able to:

- Describe several applications for inverting amplifiers.
- Describe several applications for noninverting amplifiers.
- Calculate the voltage gain of inverting and noninverting amplifiers.
- Explain the operation and characteristics of differential amplifiers and instrumentation amplifiers.
- Calculate the output voltage of binary weighted and R/2R D/A converters.
- Discuss current boosters and voltage-controlled current sources.
- Draw a circuit showing how an op amp can be operated from a single power supply.

Chapter Outline

Vocabulary

automatic gain control (AGC)

averager

buffer

current booster

differential amplifier

differential input voltage

differential voltage gain

digital-to-analog (D/A) converter

floating load

guard driving

input transducer

instrumentation amplifier

laser trimming

linear op-amp circuit

output transducer

R/2R ladder D/A converter

rail-to-rail op amp

sign changer

squelch circuit

thermistor

voltage reference

20-1 Inverting–Amplifier Circuits

In this chapter and succeeding chapters, we will be discussing many different types of op-amp circuits. Instead of providing a summary page showing all of the circuits, small summary boxes will be given containing the important formulas for circuit understanding. Also, where needed, the feedback resistor, R_f, will be labeled as R, R_2, or other designations.

The inverting amplifier is one of the most basic circuits. Chapters 18 and 19 discussed the prototype for this amplifier. One advantage of this amplifier is that its voltage gain equals the ratio of the feedback resistance to the input resistance. Let us look at a few applications.

High–Impedance Probe

Figure 20-1 shows a high-impedance probe that can be used with a digital multimeter. Because of the virtual ground in the first stage, the probe has an input impedance of 100 MΩ at low frequencies. The first stage is an inverting amplifier with a voltage gain of 0.1. The second stage is an inverting amplifier with a voltage gain of either 1 or 10.

The circuit of Fig. 20-1 gives you the basic idea of the 10:1 probe. It has a very high input impedance, and an overall voltage gain of either 0.1 or 1. In the X10 position of the switch, the output signal is attenuated by a factor of 10. In the X1 position, there is no attenuation of the output signal. The basic circuit shown here can be improved by adding more components to increase the bandwidth.

AC–Coupled Amplifier

In some applications, you do not need a response that extends down to zero frequency because only ac signals drive the input. Figure 20-2 shows an ac-coupled amplifier and its equations. The voltage gain is shown as:

$$A_v = \frac{-R_f}{R_1}$$

For the values given in Fig. 20-2, the closed-loop voltage gain is:

$$A_v = \frac{-100 \text{ k}\Omega}{10 \text{ k}\Omega} = -10$$

Figure 20-1   High-impedance probe.

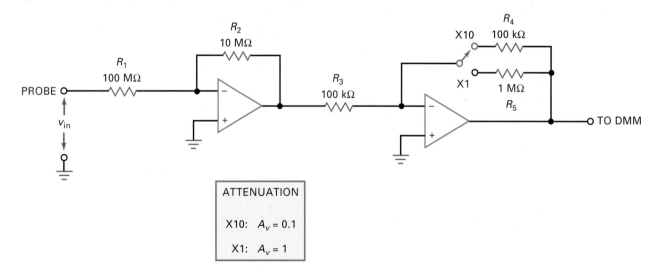

Figure 20-2 AC-coupled inverting amplifier.

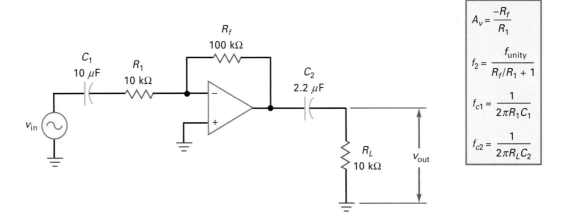

If f_{unity} is 1 MHz, the bandwidth is:

$$f_{2(CL)} = \frac{1 \text{ MHz}}{10 + 1} = 90.9 \text{ kHz}$$

The input coupling capacitor C_1 and the input resistor R_1 produce one of the lower cutoff frequencies f_{c1}. For the values shown:

$$f_{c1} = \frac{1}{2\pi(10 \text{ k}\Omega)(10 \text{ } \mu\text{F})} = 1.59 \text{ Hz}$$

Similarly, the output coupling capacitor C_2 and the load resistance R_L produce the cutoff frequency f_{c2}:

$$f_{c2} = \frac{1}{2\pi(10 \text{ k}\Omega)(2.2 \text{ } \mu\text{F})} = 7.23 \text{ Hz}$$

Adjustable-Bandwidth Circuit

Sometimes we would like to change the closed-loop bandwidth of an inverting voltage amplifier without changing the closed-loop voltage gain. Figure 20-3 shows one way to do it. When R is varied, the bandwidth will change but the voltage gain will remain constant.

Figure 20-3 Adjustable bandwidth circuit.

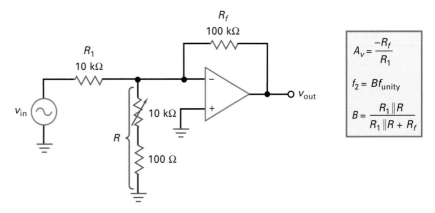

With the equations and values given in Fig. 20-3, the closed-loop voltage gain is

$$A_v = \frac{-100\ k\Omega}{10\ k\Omega} = -10$$

The minimum feedback fraction is:

$$B_{\min} \cong \frac{10\ k\Omega \parallel 100\ \Omega}{100\ k\Omega} \cong 0.001$$

The maximum feedback fraction is:

$$B_{\max} \cong \frac{10\ k\Omega \parallel 10.1\ k\Omega}{100\ k\Omega} \cong 0.05$$

If $f_{\text{unity}} = 1$ MHz, the minimum and maximum bandwidths are:

$$f_{2(CL)\min} = (0.001)(1\ \text{MHz}) = 1\ \text{kHz}$$
$$f_{2(CL)\max} = (0.05)(1\ \text{MHz}) = 50\ \text{kHz}$$

In summary, when R varies from 100 Ω to 10 kΩ, the voltage gain remains constant but the bandwidth varies from 1 to 50 kHz.

20-2 Noninverting-Amplifier Circuits

The noninverting amplifier is another basic op-amp circuit. Advantages include stable voltage gain, high input impedance, and low output impedance. Here are some applications.

AC-Coupled Amplifier

Figure 20-4 shows an ac-coupled noninverting amplifier and its analysis equations. C_1 and C_2 are coupling capacitors. C_3 is a bypass capacitor. Using a bypass capacitor has the advantage of minimizing the output offset voltage. Here's why: In the midband of the amplifier, the bypass capacitor has a very low impedance.

Figure 20-4   AC-coupled noninverting amplifier.

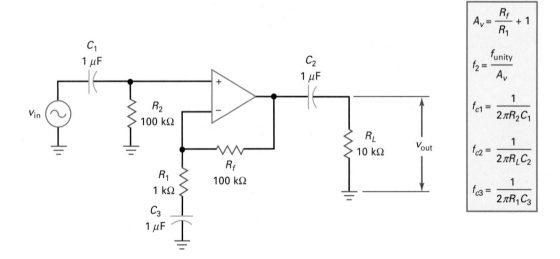

$$A_v = \frac{R_f}{R_1} + 1$$

$$f_2 = \frac{f_{\text{unity}}}{A_v}$$

$$f_{c1} = \frac{1}{2\pi R_2 C_1}$$

$$f_{c2} = \frac{1}{2\pi R_L C_2}$$

$$f_{c3} = \frac{1}{2\pi R_1 C_3}$$

Therefore, the bottom of R_1 is at ac ground. In the midband, the feedback fraction is:

$$B = \frac{R_1}{R_1 + R_f} \qquad (20\text{-}1)$$

In this case, the circuit amplifies the input voltage as previously described.

When the frequency is zero, the bypass capacitor C_3 is open and the feedback fraction B increases to unity because:

$$B = \frac{\infty}{\infty + 1} = 1$$

This equation is valid if we define ∞ as an extremely large value, which is what the impedance equals at zero frequency. With B equal to 1, the closed-loop voltage gain is unity. This reduces the output offset voltage to a minimum.

With values given in Fig. 20-4, we can calculate the midband voltage gain as:

$$A_v = \frac{100 \text{ k}\Omega}{1 \text{ k}\Omega} + 1 = 101$$

If f_{unity} is 15 MHz, the bandwidth is:

$$f_{2(CL)} = \frac{15 \text{ MHz}}{101} = 149 \text{ kHz}$$

The input coupling capacitor produces a cutoff frequency of:

$$f_{c1} = \frac{1}{2\pi (100 \text{ k}\Omega)(1 \text{ } \mu\text{F})} = 1.59 \text{ Hz}$$

Similarly, the output coupling capacitor, C_2 and the load resistance R_L produce a cutoff frequency f_{c2}:

$$f_{c2} = \frac{1}{2\pi (10 \text{ k}\Omega)(1 \text{ } \mu\text{F})} = 15.9 \text{ Hz}$$

The bypass capacitor produces a cutoff frequency of:

$$f_{c3} = \frac{1}{2\pi (1 \text{ k}\Omega)(1 \text{ } \mu\text{F})} = 159 \text{ Hz}$$

Audio Distribution Amplifier

Figure 20-5 shows an ac-coupled noninverting amplifier driving three voltage followers. This is one way to distribute an audio signal to several different outputs. The closed-loop voltage gain and bandwidth of the first stage are given by the familiar equations shown in Fig. 20-5. For the values shown, the closed-loop voltage gain is 40. If f_{unity} is 1 MHz, the closed-loop bandwidth is 25 kHz.

Incidentally, an op amp like an LM348 is convenient to use in a circuit like Fig. 20-5 because the LM348 is a quad 741—four 741s in a 14-pin package. One of the op amps can be the first stage, and the others can be the voltage followers.

JFET–Switched Voltage Gain

Some applications require a change in closed-loop voltage gain. Figure 20-6 shows a noninverting amplifier whose voltage gain is controlled by a JFET that acts like a switch. The input voltage to the JFET is a two-state voltage, either zero or $V_{GS(\text{off})}$. When the control voltage is low, it equals $V_{GS(\text{off})}$ and the JFET is open.

Figure 20-5 Distribution amplifier.

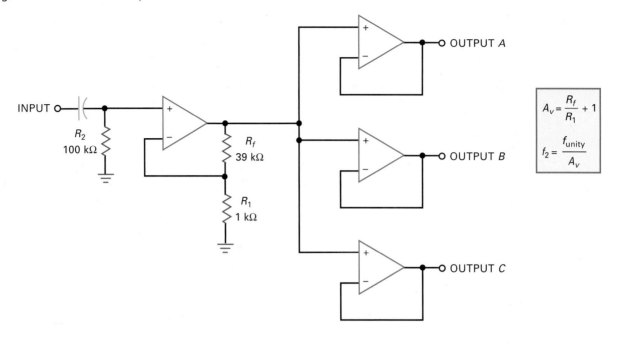

$$A_v = \frac{R_f}{R_1} + 1$$

$$f_2 = \frac{f_{unity}}{A_v}$$

Figure 20-6 JFET switch controls voltage gain.

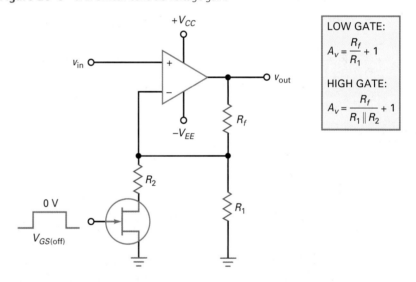

LOW GATE:

$$A_v = \frac{R_f}{R_1} + 1$$

HIGH GATE:

$$A_v = \frac{R_f}{R_1 \parallel R_2} + 1$$

In this case, R_2 is ungrounded and the voltage gain is given by the usual equation for a noninverting amplifier (the top equation in Fig. 20-6).

When the control voltage is high, it equals 0 V and the JFET switch is closed. This puts R_2 in parallel with R_1, and the closed-loop voltage gain decreases to:

$$A_v = \frac{R_f}{R_1 \parallel R_2} + 1 \tag{20-2}$$

In most designs, R_2 is made much larger than $r_{ds(on)}$ to prevent the JFET resistance from affecting the closed-loop voltage gain. Sometimes, you may see several resistors and JFET switches in parallel with R_1 to provide a selection of different voltage gains.

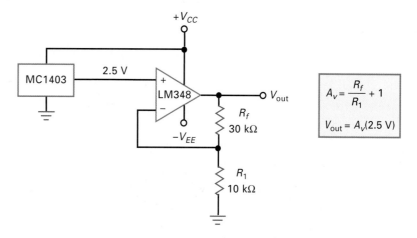

Figure 20-7 Voltage reference.

$$A_v = \frac{R_f}{R_1} + 1$$

$$V_{out} = A_v(2.5\ V)$$

Voltage Reference

The MC1403 is a special-function IC called a **voltage reference,** a circuit that produces an extremely accurate and stable value of output voltage. For any positive supply voltage between 4.5 to 40 V, it produces an output voltage of 2.5 V with a tolerance of ± 1 percent. The temperature coefficient is only 10 ppm/°C. The abbreviation *ppm* stands for "part per million" (1 ppm is equivalent to 0.0001 percent). Therefore, 10 ppm/°C produces a change of only 2.5 mV for a 100°C change in temperature (10×0.0001 percent $\times 100 \times 2.5\ V$). The point is that the output voltage is ultra-stable and equal to 2.5 V over a large temperature range.

The only problem is that 2.5 V may be too low a voltage reference for many applications. For instance, suppose we want a voltage reference of 10 V. Then, one solution is to use an MC1403 and a noninverting amplifier as shown in Fig. 20-7. With the circuit values shown, the voltage gain is:

$$A_v = \frac{30\ k\Omega}{10\ k\Omega} + 1 = 4$$

and the output voltage is:

$$V_{out} = 4(2.5\ V) = 10\ V$$

Because the closed-loop voltage gain of the noninverting amplifier is only 4, the output voltage will be a stable voltage reference of 10 V.

Example 20-1

One application for Fig. 20-6 is in a **squelch circuit.** This kind of circuit is used in communication receivers to reduce listener fatigue by having a low voltage gain when no signal is being received. This way, the user does not have to listen to static when there is no communication signal. When a signal comes in, the voltage gain is switched to high.

If $R_1 = 100\ k\Omega$, $R_f = 100\ k\Omega$, and $R_2 = 1\ k\Omega$ in Fig. 20-6, what is the voltage gain when the JFET is on? What is the voltage gain when the JFET is off? Explain how the circuit can be used as part of a squelch circuit.

SOLUTION With the equations given in Fig. 20-6, the maximum voltage gain is:

$$A_v = \frac{100 \text{ k}\Omega}{100 \text{ k}\Omega \| 1 \text{ k}\Omega} + 1 = 102$$

The minimum voltage gain is:

$$A_v = \frac{100 \text{ k}\Omega}{100 \text{ k}\Omega} + 1 = 2$$

When a communication signal is being received, we can use a peak detector and other circuits to produce a high gate voltage for the JFET in Fig. 20-6. This produces maximum voltage gain while the signal is being received. On the other hand, when no signal is being received, the output of the peak detector is low and the JFET is cut off, producing minimum voltage gain.

20-3 Inverter/Noninverter Circuits

In this section, we will discuss circuits in which the input signal drives both inputs of the op amp simultaneously. When an input signal drives both inputs, we get both inverting and noninverting amplification at the same time. This produces some interesting results because the output is the superposition of two amplified signals.

The total voltage gain with an input signal driving both sides of the op amp equals the voltage gain of the inverting channel plus the voltage gain of the noninverting channel:

$$A_v = A_{v(\text{inv})} + A_{v(\text{non})} \tag{20-3}$$

We will use this equation to analyze the circuits of this section.

Switchable Inverter/Noninverter

Figure 20-8 shows an op amp that can function as either an inverter or a noninverter. With the switch in the lower position, the noninverting input is grounded and the circuit is an inverting amplifier. Since the feedback and input resistances are equal, the inverting amplifier has a closed-loop voltage gain of:

$$A_v = \frac{-R}{R} = -1$$

Figure 20-8 Reversible voltage gain.

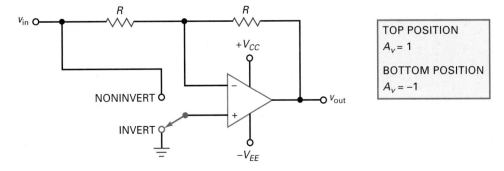

When the switch is moved to the upper position, the input signal drives both the inverting and the noninverting inputs simultaneously. The voltage gain of the inverting channel is still:

$$A_{v(\text{inv})} = -1$$

The voltage gain of the noninverting channel is:

$$A_{v(\text{non})} = \frac{R}{R} + 1 = 2$$

The total voltage gain is the superposition or algebraic sum of the two gains:

$$A_v = A_{v(\text{inv})} + A_{v(\text{non})} = -1 + 2 = 1$$

The circuit is a switchable inverter/noninverter. It has a voltage gain of either 1 or -1, depending on the position of the switch. In other words, the circuit produces an output voltage with the same magnitude as the input voltage, but the phase can be switched between $0°$ and $-180°$.

JFET-Controlled Switchable Inverter

Figure 20-9 is a modification of Fig. 20-8. The JFET acts like a voltage-controlled resistance r_{ds}, discussed in Sec. 13-9. The JFET has either a very low or a very high resistance, depending on the gate voltage.

When the gate voltage is low, it equals $V_{GS(\text{off})}$ and the JFET is open. Therefore, the input signal drives both inputs. In this case:

$$A_{v(\text{non})} = 2$$
$$A_{v(\text{inv})} = -1$$

and

$$A_v = A_{v(\text{inv})} + A_{v(\text{non})} = 1$$

The circuit acts like a noninverting voltage amplifier with a closed-loop voltage gain of 1.

When the gate voltage is high, it equals 0 V and the JFET has a very low resistance. Therefore, the noninverting input is approximately grounded. In this case, the circuit acts like an inverting voltage amplifier with a closed-loop voltage gain of -1. For proper operation, R should be at least 100 times greater than the r_{ds} of the JFET.

In summary, the circuit has a voltage gain that can be either 1 or -1, depending on whether the control voltage to the JFET is low or high.

Figure 20-9 JFET-controlled reversible gain.

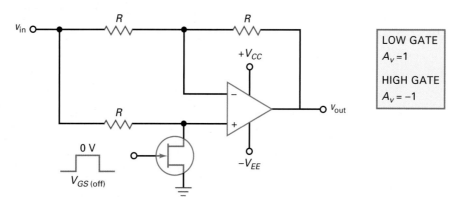

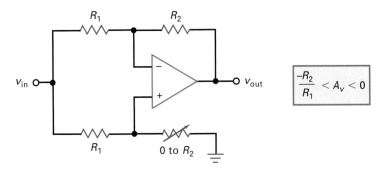

Figure 20-10 Inverter with adjustable gain.

Inverter with Adjustable Gain

When the variable resistor of Fig. 20-10 is zero, the noninverting input is grounded and the circuit becomes an inverting amplifier with a voltage gain of $-R_2/R_1$. When the variable resistor is increased to R_2, equal voltages drive the noninverting and inverting inputs of the op amp (common-mode input). Because of the common-mode rejection, the output voltage is approximately zero. Therefore, the circuit of Fig. 20-10 has a voltage gain that is continuously variable from $-R_2/R_1$ to 0.

Sign Changer

The circuit of Fig. 20-11 is called a **sign changer,** a rather unusual circuit because its voltage gain can be varied from -1 to 1. Here is the theory of operation: When the wiper is all the way to the right, the noninverting input is grounded and the circuit has a voltage gain of:

$$A_v = -1$$

When the wiper is all the way to the left, the input signal drives the noninverting input as well as the inverting input. In this case, the total voltage gain is the superposition of the inverting and noninverting voltage gains:

$$A_{v(non)} = 2$$
$$A_{v(inv)} = -1$$
$$A_v = A_{v(inv)} + A_{v(non)} = 1$$

In summary, when the wiper is moved from right to left, the voltage gain changes continuously from -1 to 1. At the crossover point (wiper at center), a common-mode signal drives the op amp and the output is ideally zero.

Figure 20-11 Reversible and adjustable gain of ± 1.

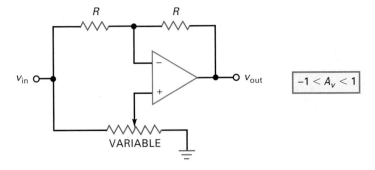

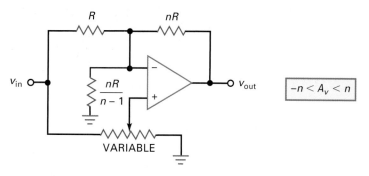

Figure 20-12 Reversible and adjustable gain of ±*n*.

$$-n < A_v < n$$

Adjustable and Reversible Gain

Figure 20-12 shows another unusual circuit. It allows us to adjust the voltage gain between $-n$ and n. The theory of operation is similar to that of the sign changer. When the wiper is all the way to the right, the noninverting input is grounded and the circuit becomes an inverting amplifier with a closed-loop voltage gain of:

$$A_v = \frac{-nR}{R} = -n$$

When the wiper is all the way to the left, it can be shown that:

$$A_{v(\text{inv})} = -n$$
$$A_{v(\text{non})} = 2n$$
$$A_v = A_{v(\text{non})} + A_{v(\text{inv})} = n$$

These results can be derived by applying Thevenin's theorem to the circuit and simplifying with algebra.

Circuits like those in Figs. 20-11 and 20-12 are unusual because they have no simple discrete counterparts. They are good examples of circuits that would be difficult to implement with discrete components but are easy to build with op amps.

Phase Shifter

Figure 20-13 shows a circuit that can ideally produce a phase shift of 0° to $-180°$. The noninverting channel has an RC lag circuit, and the inverting channel has two equal resistors with a value of R'. Therefore, the voltage gain of the inverting channel is always unity. But the voltage gain of the noninverting channel depends on the cutoff frequency of RC lag circuit.

When the input frequency is much lower than the cutoff frequency ($f \ll f_c$), the capacitor appears open and:

$$A_{v(\text{non})} = 2$$
$$A_{v(\text{inv})} = -1$$
$$A_v = A_{v(\text{non})} + A_{v(\text{inv})} = 1$$

This means that the output signal has the same magnitude as the input signal, and the phase shift is 0°, well below the cutoff frequency of the lag network.

When the input frequency is much greater than the cutoff frequency ($f \gg f_c$), the capacitor appears shorted. In this case, the noninverting channel has

Figure 20-13 Phase shifter.

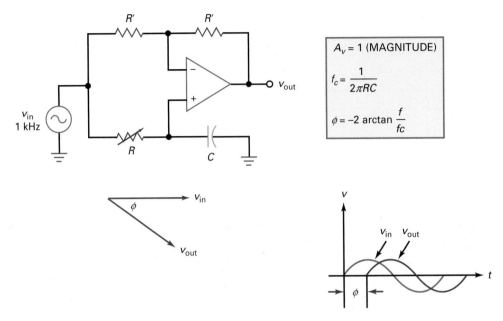

a voltage gain of zero. The overall gain therefore equals the gain of inverting channel, which is -1, equivalent to a phase shift of $-180°$.

To calculate the phase shift between the two extremes, we need to calculate the cutoff frequency using the equation given in Fig. 20-13. For instance, if $C = 0.022\ \mu\text{F}$ and variable resistor of Fig. 20-13 is set to $1\ \text{k}\Omega$, the cutoff frequency is:

$$f_c = \frac{1}{2\pi(1\ \text{k}\Omega)(0.022\ \mu\text{F})} = 7.23\ \text{kHz}$$

With a source frequency of 1 kHz, the phase shift is:

$$\phi = -2\ \arctan \frac{1\ \text{kHz}}{7.23\ \text{kHz}} = -15.7°$$

If the variable resistor is increased to $10\ \text{k}\Omega$, the cutoff frequency decreases to 723 Hz and the phase shift increases to:

$$\phi = -2\ \arctan \frac{1\ \text{kHz}}{723\ \text{Hz}} = -108°$$

If the variable resistor is increased to $100\ \text{k}\Omega$, the cutoff frequency decreases to 72.3 Hz and the phase shift increases to:

$$\phi = -2\ \arctan \frac{1\ \text{kHz}}{72.3\ \text{Hz}} = -172°$$

In summary, the phase shifter produces an output voltage with the same magnitude as the input voltage, but with a phase angle that can be varied continuously between $0°$ and $-180°$.

Example 20–2

When we need to vary the amplitude of an out-of-phase signal, we can use a circuit like the one in Fig. 20-10. If $R_1 = 1.2$ kΩ and $R_2 = 91$ kΩ, what are the values of the maximum and minimum voltage gain?

SOLUTION With the equation given in Fig. 20-10, the maximum voltage gain is:

$$A_v = \frac{-91 \text{ k}\Omega}{1.2 \text{ k}\Omega} = -75.8$$

The minimum voltage gain is zero.

PRACTICE PROBLEM 20–2 In Example 20-2, what value should R_2 be changed to for a maximum gain of -50?

Example 20–3

If $R = 1.5$ kΩ and $nR = 7.5$ kΩ in Fig. 20-12, what is the maximum positive voltage gain? What is the value of the other fixed resistance?

SOLUTION The value of n is:

$$n = \frac{7.5 \text{ k}\Omega}{1.5 \text{ k}\Omega} = 5$$

The maximum positive voltage gain is 5. The other fixed resistor has a value of:

$$\frac{nR}{n-1} = \frac{5(1.5 \text{ k}\Omega)}{5-1} = 1.875 \text{ k}\Omega$$

With a circuit like this, we have to use a precision resistor to get a nonstandard value like 1.875 kΩ.

PRACTICE PROBLEM 20–3 Using Fig. 20-12, if $R = 1$ kΩ, what is the maximum positive voltage gain and value of the other fixed resistance?

20-4 Differential Amplifiers

This section will discuss how to build a **differential amplifier** using an op amp. One of the most important characteristics of a differential amplifier is its CMRR because the typical input signal is a small differential voltage and a large common-mode voltage.

Basic Differential Amplifier

Figure 20-14 shows an op amp connected as a differential amplifier. The resistor R_1' has the same nominal value as R_1 but differs slightly in value because of tolerances.

Figure 20-14 Differential amplifier.

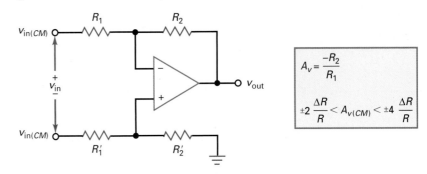

$$A_v = \frac{-R_2}{R_1}$$

$$\pm 2\,\frac{\Delta R}{R} < A_{v(CM)} < \pm 4\,\frac{\Delta R}{R}$$

For instance, if the resistors are 1 kΩ $\pm$ 1 percent, R_1 may be as high as 1010 Ω and R_1' may be as low as 990 Ω, and vice versa. Similarly, R_2 and R_2' are nominally equal but may differ slightly because of tolerances.

In Fig. 20-14, the desired input voltage v_{in} is called the **differential input voltage** to distinguish it from the common-mode input voltage $v_{in(CM)}$. A circuit like Fig. 20-14 amplifies the differential input voltage v_{in} to get an output voltage of v_{out}. Using the superposition theorem, it can be shown that:

$$v_{out} = A_v v_{in}$$

where

$$A_v = \frac{-R_2}{R_1} \tag{20-4}$$

This voltage gain is called the **differential voltage gain** to distinguish it from the common-mode voltage gain $A_{v(CM)}$. By using precision resistors, we can build a differential amplifier with a precise voltage gain.

A differential amplifier is often used in applications in which the differential input signal v_{in} is a small dc voltage (millivolts) and the common-mode input signal is a large dc voltage (volts). As a result, the CMRR of the circuit becomes a critical parameter. For instance, if the differential input signal is 7.5 mV and the common-mode signal is 7.5 V, the differential input signal is 60 dB less than the common-mode input signal. Unless the circuit has a very high CMRR, the common-mode output signal will be objectionably large.

CMRR of the Op Amp

In Fig. 20-14, two factors determine the overall CMRR of the circuit. First, there is the CMRR of the op amp itself. For a 741C, the minimum CMRR is 70 dB at low frequencies. If the differential input signal is 60 dB less than the common-mode input signal, the differential output signal will be only 10 dB greater than the common-mode output signal. This means that the desired signal is only 3.16 times greater than the undesired signal. Therefore, a 741C would be useless in an application such as this.

The solution is to use a precision op amp like an OP-07A. It has a minimum CMRR of 110 dB. This will significantly improve the operation. If the differential input signal is 60 dB less than the common-mode input signal, the differential output signal will be 50 dB greater than the common-mode output signal. This would be fine if the CMRR of the op amp were the only source of error.

CMRR of External Resistors

There is a second source of common-mode error: the tolerance of the resistors in Fig. 20-14. When the resistors are perfectly matched:

$$R_1 = R_1'$$
$$R_2 = R_2'$$

In this case, the common-mode input voltage of Fig. 20-14 produces zero voltage across the op-amp input terminals.

On the other hand, when the resistors have a tolerance of ± 1 percent, the common-mode input voltage of Fig. 20-14 will produce a common-mode output voltage because the mismatch in the resistances produces a differential input voltage to the op amp.

As discussed in Sec. 20-3, the overall voltage gain when the same signal drives both sides of an op amp is given by:

$$A_{v(CM)} = A_{v(\text{inv})} + A_{v(\text{non})} \tag{20-5}$$

In Fig. 20-14, the inverting voltage gain is:

$$A_{v(\text{inv})} = \frac{-R_2}{R_1} \tag{20-6}$$

and the noninverting voltage gain is:

$$A_{v(\text{non})} = \left(\frac{R_2}{R_1} + 1\right)\left(\frac{R_2'}{R_1' + R_2'}\right) \tag{20-7}$$

where the second factor is the decrease in the noninverting input signal caused by the voltage divider on the noninverting side.

With Eqs. (20-5) to (20-7), we can derive these useful formulas:

$$A_{v(CM)} = \pm 2\frac{\Delta R}{R} \quad \text{for } R_1 = R_2 \tag{20-8}$$

$$A_{v(CM)} = \pm 4\frac{\Delta R}{R} \quad \text{for } R_1 \ll R_2 \tag{20-9}$$

or

$$\pm 2\frac{\Delta R}{R} < A_{v(CM)} < \pm 4\frac{\Delta R}{R} \tag{20-10}$$

In these equations, $\Delta R/R$ is the tolerance of the resistors converted to the decimal equivalent.

For instance, if the resistors have a tolerance of ± 1 percent, Eq. (20-8) gives:

$$A_{v(CM)} = \pm 2(1\%) = \pm 2(0.01) = \pm 0.02$$

Equation (20-9) gives:

$$A_{v(CM)} = \pm 4(1\%) = \pm 4(0.01) = \pm 0.04$$

Inequality (20-10) gives:

$$\pm 0.02 < A_{v(CM)} < \pm 0.04$$

This says that the common-mode voltage gain is between ± 0.02 and ± 0.04. When necessary, we can calculate the exact value of $A_{v(CM)}$ with Eqs. (20-5) to (20-7).

Calculating CMRR

Here is an example of how to calculate the CMRR: In a circuit like the one in Fig. 20-14, resistors with a tolerance of ± 0.1 percent are commonly used. When $R_1 = R_2$, Eq. (20-4) gives a differential voltage gain of:

$$A_v = -1$$

and Eq. (20-8) gives a common-mode voltage gain of:

$$A_{v(CM)} = \pm 2(0.1\%) = \pm 2(0.001) = \pm 0.002$$

The CMRR has a magnitude of:

$$\text{CMRR} = \frac{|A_v|}{|A_{v(CM)}|} = \frac{1}{0.002} = 500$$

which is equivalent to 54 dB. (*Note:* The vertical bars around A_v and $A_{v(CM)}$ indicate absolute values.)

Buffered Inputs

The source resistances driving the differential amplifier of Fig. 20-14 effectively become part of R_1 and R_1', which changes the voltage gain and may degrade the CMRR. This is a very serious disadvantage. The solution is to increase the input impedance of the circuit.

Figure 20-15 shows one way to do it. The first stage (the preamp) consists of two voltage followers that **buffer** (isolate) the inputs, as shown in Fig. 20-15. This can increase the input impedance to well over 100 MΩ. The voltage gain of the first stage is unity for both the differential and the common-mode input signal. Therefore, the second stage (the differential amplifier) still has to provide all the CMRR for the circuit.

Wheatstone Bridge

As previously mentioned, the differential input signal is often a small dc voltage. The reason it is small is because it is usually the output of a Wheatstone bridge like that in Fig. 20-16a. A Wheatstone bridge is balanced when the ratio of resistances on the left side equals the ratio of resistances on the right side:

$$\frac{R_1}{R_2} = \frac{R_3}{R_4} \tag{20-11}$$

When this condition is satisfied, the voltage across R_2 equals the voltage across R_4 and the output voltage of the bridge is zero.

Figure 20-15 Differential input with buffered inputs.

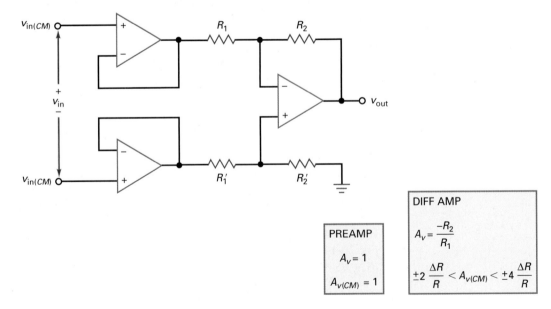

Figure 20-16 (*a*) Wheatstone bridge; (*b*) slightly unbalanced bridge.

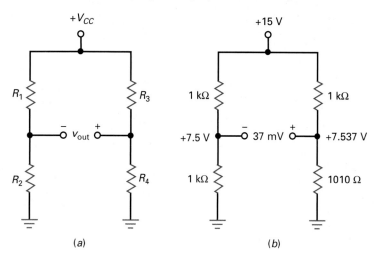

(*a*)　　　　　　　　(*b*)

The Wheatstone bridge can detect small changes in one of the resistors. For instance, suppose we have a bridge with three resistors of 1 kΩ and a fourth resistor of 1010 Ω, as shown in Fig. 20-16*b*. The voltage across R_2 is:

$$v_2 = \frac{1\ \text{k}\Omega}{2\ \text{k}\Omega}\ (15\ \text{V}) = 7.5\ \text{V}$$

and the voltage across R_4 is approximately:

$$v_4 = \frac{1010\ \Omega}{2010\ \Omega}\ (15\ \text{V}) = 7.537\ \text{V}$$

The output voltage of the bridge is approximately:

$$v_{\text{out}} = v_4 - v_2 = 7.537\ \text{V} - 7.5\ \text{V} = 37\ \text{mV}$$

Transducers

Resistance R_4 may be an **input transducer,** a device that converts a nonelectrical quantity into an electrical quantity. For instance, a photoresistor converts a change in light intensity into a change in resistance and a **thermistor** converts a change in temperature into a change in resistance.

There is also the **output transducer,** a device that converts an electrical quantity into a nonelectrical quantity. For instance, an LED converts current into light and a loudspeaker converts ac voltage into sound waves.

A wide variety of transducers are commercially available for quantities such as temperature, sound, light, humidity, velocity, acceleration, force, radioactivity, strain, and pressure, to mention a few. These transducers can be used with a Wheatstone bridge to measure nonelectrical quantities. Because the output of a Wheatstone bridge is a small dc voltage with a large common-mode voltage, we need to use dc amplifiers that have very high CMRRs.

A Typical Application

Figure 20-17 shows a typical application. Three of the bridge resistors have a value of:

$$R = 1\ \text{k}\Omega$$

The transducer has a resistance of:

$$R + \Delta R = 1010\ \Omega$$

Figure 20-17 Bridge with transducer drives instrumentation amplifier.

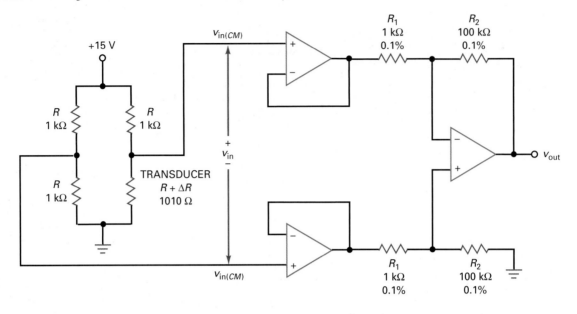

The common-mode signal is:

$$v_{\text{in(CM)}} = 0.5V_{CC} = 0.5(15\ V) = 7.5\ V$$

This is the voltage across each of the lower bridge resistors when $\Delta R = 0$.

When a bridge transducer is acted on by an outside quantity such as light, temperature, or pressure, its resistance will change. Figure 20-17 shows a transducer resistance of 1010 Ω, which implies that $\Delta R = 10\ \Omega$. It is possible to derive this equation for the input voltage in Fig. 20-17:

$$v_{\text{in}} = \frac{\Delta R}{4R + 2\Delta R}\ V_{CC} \tag{20-12}$$

In a typical application, $2\Delta R \ll 4R$ and the equation simplifies to:

$$v_{\text{in}} \cong \frac{\Delta R}{4R}\ V_{CC} \tag{20-13}$$

For the values shown in Fig. 20-17:

$$v_{\text{in}} \cong \frac{10\ \Omega}{4\ k\Omega}\ (15\ V) = 37.5\ mV$$

Since the differential amplifier has a voltage gain of -100, the differential output voltage is:

$$v_{\text{out}} = -100(37.5\ mV) = -3.75\ V$$

As far as the common-mode signal is concerned, Eq. (20-9) gives:

$$A_{v(CM)} = \pm 4(0.1\%) = \pm 4(0.001) = \pm 0.004$$

for the tolerance of ± 0.1 percent shown in Fig. 20-17. Therefore, the common-mode output voltage is:

$$v_{out(CM)} = \pm 0.004(7.5\text{ V}) = \pm 0.03\text{ V}$$

The magnitude of CMRR is:

$$\text{CMRR} = \frac{100}{0.004} = 25000$$

which is equivalent to 88 dB.

That gives you the basic idea of how a differential amplifier is used with a Wheatstone bridge. A circuit like Fig. 20-17 is adequate for some applications but can be improved, as will be discussed in the following section.

20-5 Instrumentation Amplifiers

This section discusses the **instrumentation amplifier,** a differential amplifier optimized for its dc performance. An instrumentation amplifier has a large voltage gain, a high CMRR, low input offsets, low temperature drift, and high input impedance.

Basic Instrumentation Amplifier

Figure 20-18 shows the classic design used for most instrumentation amplifiers. The output op amp is a differential amplifier with the voltage gain of unity. The resistors used in this output stage are usually matched to within ± 0.1 percent or better. This means that the CMRR of the output stage is at least 54 dB.

Figure 20-18 Standard three op-amp instrumentation amplifier.

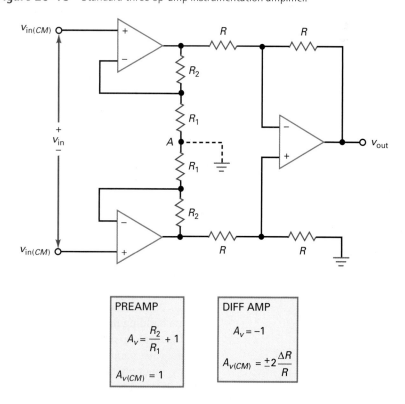

Precision resistors are commercially available from less than $1\,\Omega$ to more than $10\,M\Omega$, with tolerances of ±0.01 to ±1 percent. If we use matched resistors that are within ±0.01 percent of each other, the CMRR of the output stage can be as high as 74 dB. Also, temperature drift of precision resistors can be as low as 1 ppm/°C.

The first stage consists of two input op amps that act like a preamplifier. The design of the first stage is extremely clever. What makes it so ingenious is the action of point A, the junction between the two R_1 resistors. Point A acts like a virtual ground for a differential input signal and like a floating point for the common-mode signal. Because of this action, the differential signal is amplified but the common-mode signal is not.

Point *A*

The key to understanding how the first stage works is to understand what point A does. With the superposition theorem, we can calculate the effect of each input with the other zeroed. For instance, assume that the differential input signal is zero. Then only the common-mode signal is active. Since the common-mode signal applies the same positive voltage to each noninverting input, equal voltages appear at the op-amp outputs. Because of this, the same voltage appears everywhere along the branch that contains R_1 and R_2. Therefore, point A is floating and each input op amp acts like a voltage follower. As a result, the first stage has a common-mode gain of:

$$A_{v(CM)} = 1$$

Unlike the second stage, where the R resistors have to be closely matched to minimize the common-mode gain, in the first stage the tolerance of the resistors has no effect on the common-mode gain. This is because the entire branch containing these resistors is floating at a voltage of $v_{in(CM)}$ above ground. So, the resistor values do not matter. This is another advantage of the three op-amp design of Fig. 20-18.

The second step in applying the superposition theorem is to reduce the common-mode input to zero and to calculate the effect of the differential input signal. Since the differential input signal drives the noninverting inputs with equal and opposite input voltages, one op-amp output will be positive and the other will be negative. With equal and opposite voltages across the branch containing the R_1 and R_2 resistors, point A will have a voltage of zero with respect to ground.

In other words, point A is a virtual ground for the differential signal. For this reason, each input op amp is a noninverting amplifier and the first stage has a differential voltage gain of:

$$A_v = \frac{R_2}{R_1} + 1 \tag{20-14}$$

Since the second stage has a gain of unity, the differential voltage gain of the instrumentation amplifier is given by Eq. (20-14).

Because the first stage has a common-mode gain of unity, the overall common-mode gain equals the common-mode gain of the second stage:

$$A_{v(CM)} = \pm2\,\frac{\Delta R}{R} \tag{20-15}$$

To have high CMRR and low offsets, precision op amps must be used when building the instrumentation amplifier of Fig. 20-18. A typical op amp used in the three op-amp approach of Fig. 20-18 is the OP-07A. It has the following worst-case parameters: Input offset voltage is 0.025 mV, input bias current is 2 nA, input offset current is 1 nA, A_{OL} is 110 dB, CMRR is 110 dB, and temperature drift is 0.6 μV/°C.

A final point about Fig. 20-18: Since point A is a virtual ground rather than a mechanical ground, the R_1 resistors in the first stage do not have to be separate resistors. We can use a single resistor R_G that equals $2R_1$ without changing the operation of the first stage. The only difference is that the differential voltage gain is written as:

$$A_v = \frac{2R_2}{R_G} + 1 \tag{20-16}$$

The factor of 2 appears because $R_G = 2R_1$.

Guard Driving

Because the differential signal out of a bridge is small, a shielded cable is often used to isolate the signal-carrying wires from electromagnetic interference. But this creates a problem. Any leakage current between the inner wires and the shield will add to the low input bias and offset currents. Besides the leakage current, the shielded cable adds capacitance to the circuit, which slows down the response of the circuit to a change in transducer resistance. To minimize the effects of leakage current and cable capacitance, the shield should be bootstrapped to the common-mode potential. This technique is known as **guard driving.**

Figure 20-19a shows one way to bootstrap the shield to the common-mode voltage. A new branch containing the resistors labeled R_3 is added to the output of the first stage. This voltage divider picks off the common-mode voltage and feeds it to a voltage follower. The guard voltage is fed back to the shield, as shown. Sometimes, separate cables are used for each input. In this case, the guard voltage is connected to both shields, as shown in Fig. 20-19b.

Figure 20-19 Guard driving to reduce leakage currents and capacitance of shield cable.

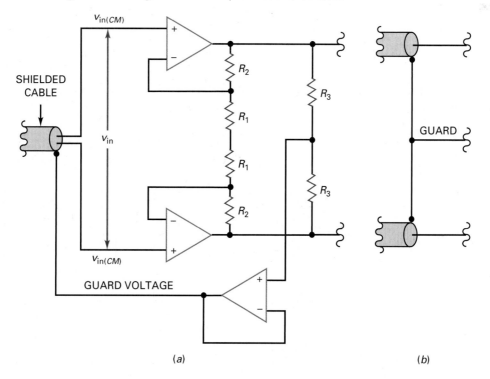

(a)   (b)

Integrated Instrumentation Amplifiers

The classic design of Fig. 20-18 can be integrated on a chip with all the components shown in Fig. 20-18, except R_G. This external resistance is used to control the voltage gain of the instrumentation amplifier. For instance, the AD620 is a monolithic instrumentation amplifier. The data sheet gives this equation for its voltage gain:

$$A_v = \frac{49.4 \text{ k}\Omega}{R_G} + 1 \tag{20-17}$$

The quantity 49.4 kΩ is the sum of the two R_2 resistors. The IC manufacturer uses **laser trimming** to get a precise value of 49.4 kΩ. The word *trim* refers to a fine adjustment rather than a coarse adjustment. Laser trimming means burning off resistor areas on a semiconductor chip with a laser to get an extremely precise value of resistance.

Figure 20-20a shows the AD620 with an R_G of 499 Ω. This is a precision resistor with a tolerance of ±0.1 percent. The voltage gain is:

$$A_v = \frac{49.4 \text{ k}\Omega}{499} + 1 = 100$$

The *pinout* (pin numbers) of the AD620 is similar to that of a 741C since pins 2 and 3 are for the input signals, pins 4 and 7 are for the supply voltages, and pin 6 is the output. Pin 5 is shown grounded, the usual case for the AD620. But this pin does not have to be grounded. If necessary for interfacing with another circuit, we can offset the output signal by applying a dc voltage to pin 5.

If guard driving is used, the circuit can be modified as shown in Fig. 20-20b. The common-mode voltage drives a voltage follower, whose output is connected to the shield of the cable. A similar modification is used if separate cables are used for the inputs.

In summary, monolithic instrumentation amplifiers typically have a voltage gain between 1 and 1000 that can be set with one external resistor, a CMRR greater than 100 dB, an input impedance greater than 100 MΩ, an input offset voltage less than 0.1 mV, a drift of less than 0.5 μV/°C, and other outstanding parameters.

Figure 20-20 (a) A monolithic instrumentation amplifier; (b) guard driving with an AD620.

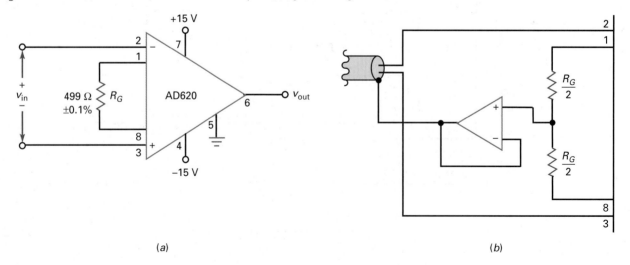

(a)

(b)

Example 20-4

In Fig. 20-18, $R_1 = 1$ kΩ, $R_2 = 100$ kΩ, and $R = 10$ kΩ. What is the differential voltage gain of the instrumentation amplifier? What is the common-mode voltage gain if the resistor tolerances in the second stage are ± 0.01 percent? If $v_{in} = 10$ mV and $v_{in(CM)} = 10$ V, what are the values of the differential and common-mode output signals?

SOLUTION With the equations given in Fig. 20-18, the voltage gain of the preamp is:

$$A_v = \frac{100 \text{ k}\Omega}{1 \text{ k}\Omega} + 1 = 101$$

Since the voltage gain of the second stage is -1, the voltage gain of the instrumentation amplifier is -101.

The common-mode voltage gain of the second stage is:

$$A_{v(CM)} = \pm 2(0.01\%) = \pm 2(0.0001) = \pm 0.0002$$

Since the first stage has a common-mode voltage gain of 1, the common-mode voltage gain of the instrumentation amplifier is ± 0.0002.

A differential input signal of 10 mV will produce an output signal of:

$$v_{out} = -101(10 \text{ mV}) = -1.01 \text{ V}$$

A common-mode signal of 10 V will produce an output signal of:

$$v_{out(CM)} = \pm 0.0002(10 \text{ V}) = \pm 2 \text{ mV}$$

Even though the common-mode input signal is 1000 times greater than the differential input, the CMRR of the instrumentation amplifier produces a common-mode output signal that is approximately 500 times smaller than the differential output signal.

PRACTICE PROBLEM 20-4 Repeat Example 20-4 with $R_2 = 50$ kΩ and $\pm 0.1\%$ second stage resistor tolerance.

20-6 Summing Amplifier Circuits

We discussed the basic summing amplifier in Chap. 18. Now, let us look at some variations of this circuit.

The Subtracter

Figure 20-21 shows a circuit that subtracts two input voltages to produce an output voltage equal to the difference of v_1 and v_2. Here is how it works: Input v_1 drives an inverter with a voltage gain of unity. The output of the first stage is $-v_1$. This voltage is one of the inputs to the second-stage summing circuit. The other input is v_2. Since the gain of each channel is unity, the final output voltage equals v_1 minus v_2.

Figure 20-21   Subtracter.

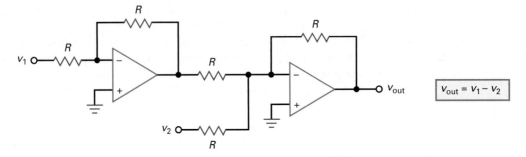

$$v_{out} = v_1 - v_2$$

Summing on Both Inputs

Sometimes you may see a circuit like Fig. 20-22. It is nothing more than a summing circuit that has inverting and noninverting inputs. The inverting side of the amplifier has two input channels, and the noninverting side has two input channels. The total gain is the superposition of the channel gains.

The gain of each inverting channel is the ratio of the feedback resistor R_f to input channel resistance, either R_1 or R_2. The gain of each noninverting channel is:

$$\frac{R_f}{R_1 \parallel R_2} + 1$$

reduced by the voltage-divider factor of the channel, either:

$$\frac{R_4 \parallel R_5}{R_3 + R_4 \parallel R_5}$$

or

$$\frac{R_3 \parallel R_5}{R_4 + R_3 \parallel R_5}$$

Figure 20-22 gives the equations for the gain of each channel. After getting each channel gain, we can calculate total output voltage.

Figure 20-22   Summing amplifier using both sides of op amp.

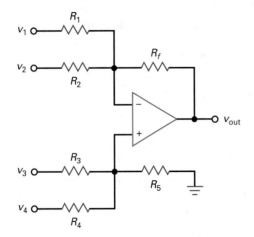

$$v_{out} = A_{v1}v_1 + A_{v2}v_2 + A_{v3}v_3 + A_{v4}v_4$$

$$A_{v1} = \frac{-R_f}{R_1}$$

$$A_{v2} = \frac{-R_f}{R_2}$$

$$A_{v3} = \left(\frac{R_f}{R_1 \parallel R_2} + 1\right)\left(\frac{R_4 \parallel R_5}{R_3 + R_4 \parallel R_5}\right)$$

$$A_{v4} = \left(\frac{R_f}{R_1 \parallel R_2} + 1\right)\left(\frac{R_3 \parallel R_5}{R_4 + R_3 \parallel R_5}\right)$$

Example 20-5

In Fig. 20-22, $R_1 = 1$ kΩ, $R_2 = 2$ kΩ, $R_3 = 3$ kΩ, $R_4 = 4$ kΩ, $R_5 = 5$ kΩ, and $R_f = 6$ kΩ. What is the voltage gain of each channel?

SOLUTION With the equations given in Fig. 20-22, the voltage gains are:

$$A_{v1} = \frac{-6\,\text{k}\Omega}{1\,\text{k}\Omega} = -6$$

$$A_{v2} = \frac{-6\,\text{k}\Omega}{2\,\text{k}\Omega} = -3$$

$$A_{v3} = \left(\frac{6\,\text{k}\Omega}{1\,\text{k}\Omega \,\|\, 2\,\text{k}\Omega} + 1\right) \frac{4\,\text{k}\Omega \,\|\, 5\,\text{k}\Omega}{3\,\text{k}\Omega + 4\,\text{k}\Omega \,\|\, 5\,\text{k}\Omega} = 4.26$$

$$A_{v4} = \left(\frac{6\,\text{k}\Omega}{1\,\text{k}\Omega \,\|\, 2\,\text{k}\Omega} + 1\right) \frac{3\,\text{k}\Omega \,\|\, 5\,\text{k}\Omega}{4\,\text{k}\Omega + 3\,\text{k}\Omega \,\|\, 5\,\text{k}\Omega} = 3.19$$

PRACTICE PROBLEM 20-5 Repeat Example 20-5 using 1 kΩ for R_f.

The Averager

Figure 20-23 is an **averager,** a circuit whose output equals the average of the input voltages. Each channel has a voltage gain of:

$$A_v = \frac{R}{3R} = \frac{1}{3}$$

When all amplified outputs are added, we get an output that is the average of all input voltages.

The circuit shown in Fig. 20-23 has three inputs. Any number of inputs can be used, as long as each channel input resistance is changed to nR, where n is the number of channels.

D/A Converter

In digital electronics, a **digital-to-analog (D/A) converter** takes a binary represented value and converts it into a voltage or current. This voltage or current will be proportional to the input binary value. Two methods of D/A conversion are

Figure 20-23 Averaging circuit.

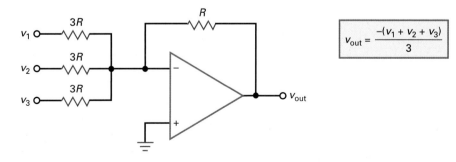

$$v_{\text{out}} = \frac{-(v_1 + v_2 + v_3)}{3}$$

Figure 20-24 Binary-weighted D/A converter changes digital input to analog voltage.

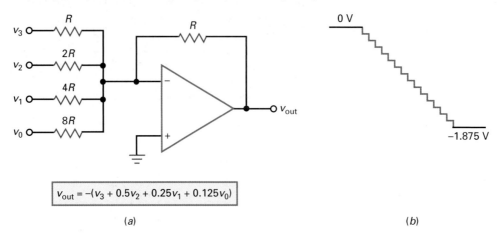

$$V_{out} = -(v_3 + 0.5v_2 + 0.25v_1 + 0.125v_0)$$

(a)

(b)

often used, the binary-weighted D/A converter and the R/2R ladder D/A converter.

The binary-weighted D/A converter is shown in Fig. 20-24. This circuit produces an output voltage equal to the weighted sum of the inputs. The *weight* is the same as the gain of the channel. For instance, in Fig. 20-24a the channel gains are:

$$A_{v3} = -1$$
$$A_{v2} = -0.5$$
$$A_{v1} = -0.25$$
$$A_{v0} = -0.125$$

The input voltages are digital or two-state, which means that they have a value of either 1 or 0. With 4 inputs, there are 16 possible input combinations of $v_3v_2v_1v_0$: 0000, 0001, 0010, 0011, 0100, 0101, 0110, 0111, 1000, 1001, 1010, 1011, 1100, 1101, 1110, and 1111.

When all inputs are zero (0000), the output is:

$$v_{out} = 0$$

When $v_3v_2v_1v_0$ is 0001, the output is:

$$v_{out} = -(0.125) = -0.125$$

When $v_3v_2v_1v_0$ is 0010, the output is:

$$v_{out} = -(0.25) = -0.25$$

and so on. When the inputs are all 1s (1111), the output is maximum and equals:

$$v_{out} = -(1 + 0.5 + 0.25 + 0.125) = -1.875$$

If the D/A converter of Fig. 20-24 is driven by a circuit that produces the 0000 to 1111 sequence of numbers given earlier, it will produce these output voltages: 0, −0.125, −0.25, −0.375, −0.5, −0.625, −0.75, −0.875, −1, −1.125, −1.25, −1.375, −1.5, −1.625, −1.75, and −1.875. When viewed on an oscilloscope, the output voltage of the D/A converter will look like the negative-going staircase shown in Fig. 20-24b.

The staircase voltage demonstrates that the D/A converter does not produce a continuous range of output values. Therefore, strictly speaking its output is not truly analog. Low-pass filter circuits can be connected to the output to provide a smoother transition between output steps.

A 4-input D/A converter has 16 possible outputs, an 8-input A/D converter has 256 possible outputs, and a 16-input D/A converter has 65,536 possible outputs. This means that the negative-going staircase voltage of Fig. 20-24b can have 256 steps with an 8-input converter and 65,536 steps with a 16-input converter. A negative-going staircase voltage like this is used in a digital multimeter along with other circuits to measure the voltage numerically.

The binary-weighted D/A converter can be used in applications where the number of inputs are limited and where high precision is not required. When a higher number of inputs is used, a higher number of different resistor values is required. The accuracy and stability of the D/A converter depends on the absolute accuracy of the resistors and their ability to track each other with temperature variations. Because the input resistors all have different values, identical tracking characteristics is difficult to obtain. Loading problems can also exist with this type of D/A converter because each input has a different input impedance value.

The **R/2R ladder D/A converter,** shown in Fig. 20-25, overcomes the limitations of the binary-weighted D/A converter and is the method most often used in integrated-circuit D/A converters. Because only two resistor values are required, this method lends itself to ICs with 8 bit or higher binary inputs and provides a higher degree of accuracy. For simplicity, Fig. 20-25 is shown as a 4-bit D/A converter. The switches, $D_0 - D_3$, would normally be some type of active switch. The switches connect the four inputs to either ground (logic 0) or $+V_{ref}$ (logic 1). The ladder network converts the possible binary input values from 0000 through 1111 to one of 16 unique output voltage levels. In the D/A converter shown in Fig. 20-25, D_0 is considered to be the least significant input bit (LSB) while D_3 is the most significant bit (MSB).

To determine the D/A converter's output voltage, you must first change the binary input value to its decimal equivalent value, BIN. This can be done by:

$$\textbf{BIN} = (D_0 \times 2^0) + (D_1 \times 2^1) + (D_2 \times 2^2) + (D_3 \times 2^3) \qquad \textbf{(20-18)}$$

Then, the output voltage will be found by:

$$V_{\textbf{out}} = -\left(\frac{\textbf{BIN}}{2^N} \times 2V_{\textbf{ref}}\right) \qquad \textbf{(20-19)}$$

where N equals the number of inputs.

For more detail of this circuit's operation, the D/A converter can be Thevenized. This analysis can be found in Appendix D.

Figure 20-25   R/2R ladder D/A converter.

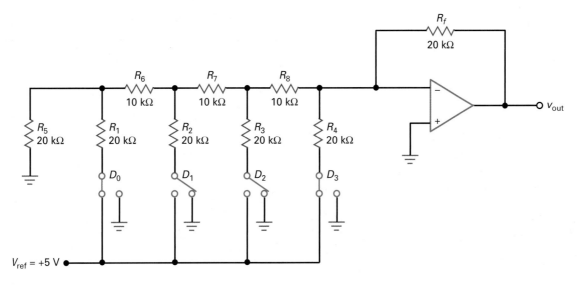

Example 20-6

In Fig. 20-25, $D_0 = 1$, $D_1 = 0$, $D_2 = 0$, and $D_3 = 1$. Using a V_{ref} value of $+5$ V, determine the decimal equivalent of the binary input (BIN) and the output voltage of the converter.

SOLUTION   Using Eq. 20-18, the decimal equivalent can be found by:

$$BIN = (1 \times 2^0) + (0 \times 2^1) + (0 \times 2^2) + (1 \times 2^3) = 9$$

The output voltage of the converter is found by using Eq. 20-19 as:

$$V_{out} = -\left(\frac{9}{2^4}\right) \times 2\ (5\ V)$$

$$V_{out} = -\left(\frac{9}{16}\right)(10\ V) = -5.625\ V$$

PRACTICE PROBLEM 20-6   Using Fig. 20-25, what is the largest and smallest output voltage possible with at least one input being a logic 1.

20-7 Current Boosters

The short-circuit output current of an op amp is typically 25 mA or less. One way to get more output current is to use a power op amp like the LM675 or LM12. These op amps have short-circuit output currents of 3 and 10 A. Another way to get more short-circuit output current is to use a **current booster,** a power transistor or other device that has a current gain and a higher current rating than the op amp.

Unidirectional Booster

Figure 20-26 shows one way to increase the maximum load current. The output of an op amp drives an emitter follower. The closed-loop voltage gain is:

$$A_v = \frac{R_2}{R_1} + 1 \tag{20-20}$$

Figure 20-26   Unidirectional current booster increases short-circuit output current.

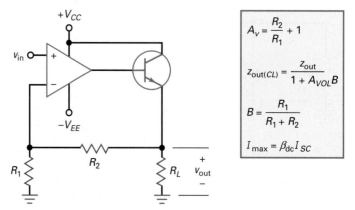

$$A_v = \frac{R_2}{R_1} + 1$$

$$z_{out(CL)} = \frac{z_{out}}{1 + A_{VOL}B}$$

$$B = \frac{R_1}{R_1 + R_2}$$

$$I_{max} = \beta_{dc} I_{SC}$$

In this circuit, the op amp no longer has to supply the load current. Instead, it only has to supply base current to the emitter follower. Because of the current gain of the transistor, the maximum load current is increased to:

$$I_{max} = \beta_{dc}I_{SC} \tag{20-21}$$

where I_{SC} is the short-circuit output current of the op amp. This means that an op amp like a 741C can have a maximum output current of 25 mA increased by a factor of β_{dc}. For instance, a BU806 is an *npn* power transistor with $\beta_{dc} = 100$. If it is used with a 741C, the short-circuit output current increases to:

$$I_{max} = 100(25 \text{ mA}) = 2.5 \text{ A}$$

The circuit can drive low-impedance loads because the negative feedback reduces the output impedance of the emitter follower by a factor of $1 + A_{VOL}B$. Since the emitter follower already has a low output impedance, the closed-loop output impedance will be very small.

Bidirectional Current

The disadvantage of the current booster shown in Fig. 20-26 is its *unidirectional load current*. Figure 20-27 shows one way to get a *bidirectional load current*. An inverting amplifier drives a class B push-pull emitter follower. In this circuit, the closed-loop voltage gain is:

$$A_v = \frac{-R_2}{R_1} \tag{20-22}$$

When the input voltage is positive, the lower transistor is conducting and the load voltage is negative. When the input voltage is negative, the upper transistor is conducting and the output voltage is positive. In either case, the maximum output current is increased by the current gain of the conducting transistor. Since the class B push-pull emitter follower is inside the feedback loop, the closed-loop output impedance is very small.

Rail-to-Rail Op Amps

Current boosters are sometimes used in the final stage of an op amp. For instance, the MC33206 is a **rail-to-rail op amp** that has a current-boosted output of 80 mA. *Rail-to-rail* refers to the supply lines of an op amp because they look like rails on a schematic diagram. *Rail-to-rail operation* means that the input and output voltages can swing all the way to the positive or negative supply voltages.

Figure 20-27 Bidirectional current booster.

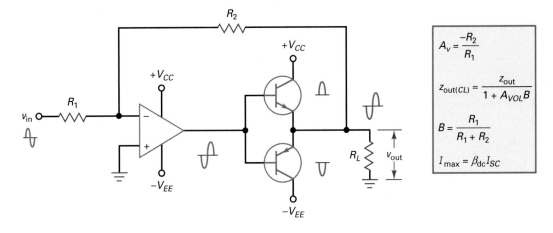

For instance, the 741C does not have a rail-to-rail output because the output is always 1 to 2 V less than either supply voltage. On the other hand, the MC33206 does have a rail-to-rail output because its output voltage can swing to within 50 mV of either supply voltage, close enough to qualify as rail-to-rail. Rail-to-rail op amps allow a designer to make full use of the available supply voltage range.

Example 20-7

In Fig. 20-27, $R_1 = 1\ k\Omega$, and $R_2 = 51\ k\Omega$. If a 741C is used for the op amp, what is the voltage gain of the circuit? What is the closed-loop output impedance? What is the shorted-load current of the circuit if each transistor has a current gain of 125?

SOLUTION   With the equations given in Fig. 20-26, the voltage gain is:

$$A_v = \frac{-51\ k\Omega}{1\ k\Omega} = -51$$

The feedback fraction is:

$$B = \frac{1\ k\Omega}{1\ k\Omega + 51\ k\Omega} = 0.0192$$

Since the 741C has a typical voltage gain of 100,000 and an open-loop output impedance of 75 Ω, the closed-loop output impedance is:

$$z_{out(CL)} = \frac{75\ \Omega}{1 + (100,000)(0.0192)} = 0.039\ \Omega$$

Since the 741C has a shorted-load current of 25 mA, the boosted value of the shorted-load current is:

$$I_{max} = 125(25\ mA) = 3.13\ A$$

PRACTICE PROBLEM 20-7   Using Fig. 20-27, change R_2 to 27 kΩ. Determine the new voltage gain, $z_{out(CL)}$ and I_{max}, when each transistor has a current gain of 100.

20-8 Voltage-Controlled Current Sources

This section discusses circuits that allow an input voltage to control an output current. The load may be floating or grounded. All the circuits are variations of the VCIS prototype discussed in Chap. 19, which means that they are voltage-controlled current sources also known as voltage-to-current converters.

Floating Load

Figure 20-28 shows the VCIS prototype. The load may be a resistor, a relay, or a motor. Because of the virtual short between the input terminals, the inverting

Figure 20-28 Unidirectional VCIS with floating load.

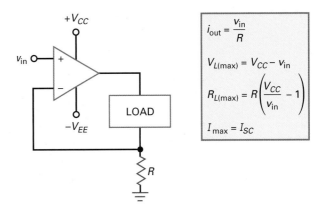

$$i_{out} = \frac{v_{in}}{R}$$

$$V_{L(max)} = V_{CC} - v_{in}$$

$$R_{L(max)} = R\left(\frac{V_{CC}}{v_{in}} - 1\right)$$

$$I_{max} = I_{SC}$$

input is bootstrapped to within microvolts of the noninverting input. Since voltage v_{in} appears across R, the load current is:

$$i_{out} = \frac{v_{in}}{R} \tag{20-23}$$

Since the load resistance does not appear in this equation, the current is independent of the load resistance. Stated another way, the load appears to be driven by a very stiff current source. As an example, if v_{in} is 1 V and R is 1 kΩ, i_{out} is 1 mA.

If the load resistance is too large in Fig. 20-28, the op amp goes into saturation and the circuit no longer acts like a stiff current source. If a rail-to-rail op amp is used, the output can swing all the way to $+V_{CC}$. Therefore, the maximum load voltage is:

$$V_{L(max)} = V_{CC} - v_{in} \tag{20-24}$$

For example, if V_{CC} is 15 V and v_{in} is 1 V, $V_{L(max)}$ is 14 V. If the op amp does not have a rail-to-rail output, we can subtract 1 to 2 V from $V_{L(max)}$.

Since the load current equals v_{in}/R, we can derive this equation for the maximum load resistance that can be used without saturating the op amp:

$$R_{L(max)} = R\left(\frac{V_{CC}}{v_{in}} - 1\right) \tag{20-25}$$

As an example, if R is 1 kΩ, V_{CC} is 15 V, and v_{in} is 1 V, then $R_{L(max)} = 14$ kΩ.

Another limitation on a voltage-controlled current source is the short-circuit output current of the op amp. For instance, a 741C has a short-circuit output current of 25 mA. Short-circuit output currents for various op amps were discussed in Chap. 18 and listed in Table 18-2. As an equation, the short-circuit current out of the controlled current source in Fig. 20-28 is:

$$I_{max} = I_{SC} \tag{20-26}$$

where I_{SC} is the short-circuit output current of the op amp.

Grounded Load

If a **floating load** is all right and the short-circuit current is adequate, a circuit like Fig. 20-28 works well. But if the load needs to be grounded or more short-circuit current is needed, we can modify the basic circuit as shown in Fig. 20-29. Since the collector and emitter currents of the transistor are almost equal, the current through R is approximately equal to the load current. Because of the virtual short

Figure 20-29 Unidirectional VCIS with single-ended load.

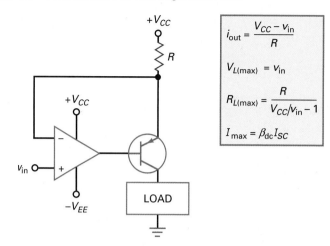

between the op-amp inputs, the inverting input voltage approximately equals v_{in}. Therefore, the voltage across R equals V_{CC} minus v_{in} and the current through R is given by:

$$i_{out} = \frac{V_{CC} - v_{in}}{R} \qquad (20\text{-}27)$$

Figure 20-29 shows the equations for maximum load voltage, maximum load resistance, and short-circuit output current. Notice that the circuit uses a current booster on the output side. This increases the short-circuit output current to:

$$I_{max} = \beta_{dc} I_{SC} \qquad (20\text{-}28)$$

Output Current Directly Proportional to Input Voltage

In Fig. 20-29, the load current decreases when the input voltage increases. Figure 20-30 shows a circuit in which the load current is directly proportional to the input voltage. Because of the virtual short on the input terminals of the first op amp, the emitter current in Q_1 is v_{in}/R. Since the Q_1 collector current is approximately the same as the emitter current, the voltage across the collector R is v_{in} and the voltage at node A is:

$$V_A = V_{CC} - v_{in}$$

This is the noninverting input to the second op amp.

Because of the virtual short between the input terminals of the second op amp, the voltage at node B is:

$$V_B = V_A$$

The voltage across the final R is:

$$V_R = V_{CC} - V_B = V_{CC} - (V_{CC} - v_{in}) = v_{in}$$

Therefore, the output current is approximately:

$$i_{out} = \frac{v_{in}}{R} \qquad (20\text{-}29)$$

Figure 20-30 shows the equations for analyzing this circuit. Again, a current booster increases the short-circuit output current by a factor β_{dc}.

Figure 20-30 Another unidirectional VCIS with single-ended load.

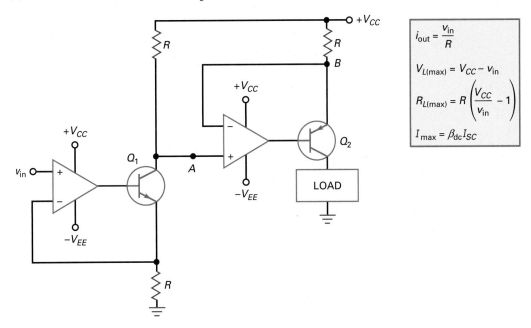

$$i_{out} = \frac{v_{in}}{R}$$

$$V_{L(max)} = V_{CC} - v_{in}$$

$$R_{L(max)} = R\left(\frac{V_{CC}}{v_{in}} - 1\right)$$

$$I_{max} = \beta_{dc} I_{SC}$$

Howland Current Source

The current source of Fig. 20-30 produces a unidirectional load current. When a bidirectional current is needed, the Howland current source of Fig. 20-31 may be used. For a preliminary understanding of how it works, consider the special case of $R_L = 0$. When the load is shorted, the noninverting input is grounded, the inverting input is at virtual ground, and the output voltage is:

$$v_{out} = -v_{in}$$

On the lower side of the circuit, the output voltage will appear across R in series with the shorted load. The current through R is:

$$i_{out} = \frac{-v_{in}}{R} \tag{20-30}$$

Figure 20-31 Howland current source is a bidirectional VCIS.

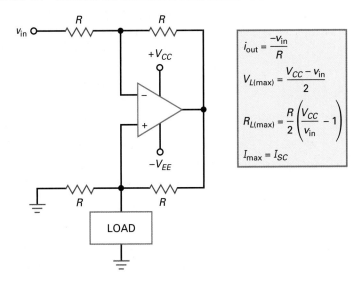

$$i_{out} = \frac{-v_{in}}{R}$$

$$V_{L(max)} = \frac{V_{CC} - v_{in}}{2}$$

$$R_{L(max)} = \frac{R}{2}\left(\frac{V_{CC}}{v_{in}} - 1\right)$$

$$I_{max} = I_{SC}$$

When the load is shorted, all this current flows through the load. The minus sign means that the load voltage is inverted.

When the load resistance is greater than zero, the analysis is much more complicated because the noninverting input is no longer grounded and the inverting input is no longer a virtual ground. Instead, the noninverting input voltage equals the voltage across the load resistor. After writing and solving several equations, we can show that Eq. (20-30) is still valid for any load resistance, provided the op amp does not go into saturation. Since R_L does not appear in the equation, the circuit acts like a stiff current source.

Figure 20-31 shows the analysis equations. For instance, if $V_{CC} = 15$, $v_{in} = 3$ V, and $R = 1$ kΩ, the maximum load resistance that can be used without saturating the op amp is:

$$R_{L(max)} = \frac{1\ k\Omega}{2}\left(\frac{15\ V}{3\ V} - 1\right) = 2\ k\Omega$$

Example 20-8

|||| MultiSim

If the current source of Fig. 20-28 has $R = 10$ kΩ, $v_{in} = 1$ V, and $V_{CC} = 15$ V, what is the output current? What is the maximum load resistance that can be used with this circuit if v_{in} can be as large as 10 V?

SOLUTION   With the equations of Fig. 20-28, the output current is:

$$i_{out} = \frac{1\ V}{10\ k\Omega} = 0.1\ mA$$

The maximum load resistance is:

$$R_{L(max)} = (10\ k\Omega)\left(\frac{15\ V}{10\ V} - 1\right) = 5\ k\Omega$$

PRACTICE PROBLEM 20–8   Change R to 2 kΩ and repeat Example 20-8.

Example 20-9

The Howland current source of Fig. 20-31 has $R = 15$ kΩ, $v_{in} = 3$ V, and $V_{CC} = 15$ V. What is the output current? What is the largest load resistance that can be used with this circuit if the maximum input voltage is 9 V?

SOLUTION   With the equations of Fig. 20-31:

$$i_{out} = \frac{-3\ V}{15\ k\Omega} = -0.2\ mA$$

The maximum load resistance is:

$$R_{L(max)} = \frac{15\ k\Omega}{2}\left(\frac{15\ V}{12\ V} - 1\right) = 1.88\ k\Omega$$

PRACTICE PROBLEM 20–9   Repeat Example 20-9 with $R = 10$ kΩ.

20-9 Automatic Gain Control

AGC stands for **automatic gain control.** In many applications like radio and television, we want the voltage gain to change automatically when the input signal changes. Specifically, when the input signal increases, we want the voltage gain to decrease. In this way, the output voltage of an amplifier will be approximately constant. One reason for wanting AGC in a radio or television is to keep the volume from changing abruptly when we tune in different stations.

Audio AGC

Figure 20-32 shows an audio AGC circuit. Q_1 is a JFET used as a voltage-controlled resistance. For small-signal operation with drain voltages near zero, the JFET operates in the ohmic region and has a resistance of r_{ds} to ac signals (discussed in Sec. 13-9). The r_{ds} of a JFET can be controlled by the gate voltage. The more negative the gate voltage is, the larger r_{ds} becomes. With a JFET like the 2N4861, r_{ds} can vary from 100 Ω to more than 10 MΩ.

R_3 and Q_1 act like a voltage divider whose output varies between $0.001v_{in}$ and v_{in}. Therefore, the noninverting input voltage is between $0.001v_{in}$ and v_{in}, a 60-dB range. The output voltage of the noninverting amplifier is $(R_2/R_1 + 1)$ times this input voltage.

In Fig. 20-32, the output voltage is coupled to the base of Q_2. For a peak-to-peak output less than 1.4 V, Q_2 is cut off because there is no bias on it. With Q_2 off, capacitor C_2 is uncharged and the gate of Q_1 is at $-V_{EE}$, enough negative voltage to cut off the JFET. This means that maximum input voltage reaches the noninverting input. In other words, an output voltage of less than 1.4 V pp implies that the circuit acts like a noninverting voltage amplifier with a maximum input signal.

When the output peak-to-peak voltage is greater than 1.4 V, Q_2 conducts and charges capacitor C_2. This increases the gate voltage and decreases r_{ds}. With a smaller r_{ds}, the output of the R_3 and Q_1 voltage divider decreases and there is less input voltage to the noninverting input. In other words, the overall voltage gain of the circuit decreases when the peak-to-peak output voltage is greater than 1.4 V.

Figure 20-32 JFET used as a voltage-controlled resistance in AGC circuit.

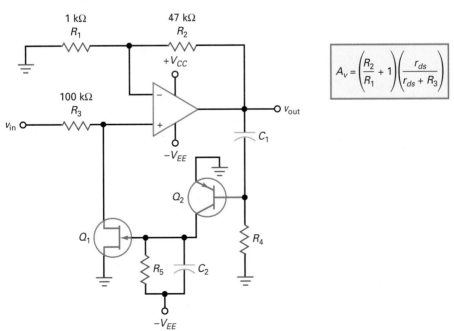

$$A_v = \left(\frac{R_2}{R_1} + 1\right)\left(\frac{r_{ds}}{r_{ds} + R_3}\right)$$

Figure 20-33   AGC circuit used with small input signals.

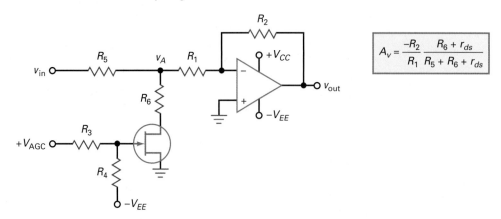

The larger the output voltage, the smaller the voltage gain. This way, the output voltage increases only slightly for large increases in the input signal. One reason for using AGC is to reduce sudden increases in signal level and prevent overdriving a loudspeaker. If you are listening to a radio, you do not want an unexpected increase in the signal level to bombard your ears. In summary, even though the input voltage of Fig. 20-32 varies over a 60-dB range, the peak-to-peak output is only slightly more than 1.4 V.

Low-Level Video AGC

The signal out of a television camera has frequencies from 0 to well over 4 MHz. Frequencies in this range are called *video frequencies*. Figure 20-33 shows a standard technique for video AGC that has been used for frequencies up to 10 MHz. In this circuit the JFET acts like a voltage-controlled resistance. When the AGC voltage is zero, the JFET is cut off by the negative bias and its r_{ds} is maximum. As the AGC voltage increases, the r_{ds} of the JFET decreases.

The input voltage to the inverting amplifier comes from the voltage divider formed by R_5, R_6, and r_{ds}. This voltage is given by:

$$v_A = \frac{R_6 + r_{ds}}{R_5 + R_6 + r_{ds}} v_{in}$$

The voltage gain of the inverting amplifier is:

$$A_v = \frac{-R_2}{R_1}$$

In this circuit the JFET is a voltage-controlled resistance. The more positive the AGC voltage, the smaller the value of r_{ds} and the lower the input voltage to the inverting amplifier. This means that the AGC voltage controls the overall voltage gain of the circuit.

With a wideband op amp, the circuit works well for input signals up to approximately 100 mV. Beyond this level, the JFET resistance becomes a function of the signal level in addition to the AGC voltage. This is undesirable because only the AGC voltage should control the overall voltage gain.

High-Level Video AGC

For high-level video signals, we can replace the JFET by an LED-photoresistor combination like Fig. 20-34. The resistance R_7 of the photoresistor decreases as the amount of light increases. Therefore, the larger the AGC voltage, the lower the

Figure 20-34 AGC circuit used with large input signals.

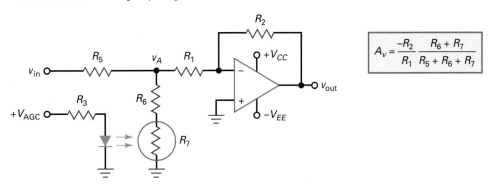

value of R_7. As before, the input voltage divider controls the amount of voltage driving the inverting voltage amplifier. This voltage is given by:

$$v_A = \frac{R_6 + R_7}{R_5 + R_6 + R_7}\, v_{in}$$

The circuit can handle high-level input voltages up to 10 V because the photocell resistance is unaffected by larger voltages and is a function only of V_{AGC}. Also, there is almost total isolation between the AGC voltage and the input voltage v_{in}.

Example 20-10

If r_{ds} varies from 50 Ω to 120 kΩ in Fig. 20-32, what is the maximum voltage gain? What is the minimum voltage gain?

SOLUTION   Using the values and equations of Fig. 20-32, the maximum voltage gain is:

$$A_v = \left(\frac{47\text{ k}\Omega}{1\text{ k}\Omega} + 1\right) \frac{120\text{ k}\Omega}{120\text{ k}\Omega + 100\text{ k}\Omega} = 26.2$$

The minimum voltage gain is:

$$A_v = \left(\frac{47\text{ k}\Omega}{1\text{ k}\Omega} + 1\right) \frac{50\ \Omega}{50\ \Omega + 100\text{ k}\Omega} = 0.024$$

PRACTICE PROBLEM 20-10   In Example 20-10, what value should r_{ds} drop to for a voltage gain of 1?

20-10 Single-Supply Operation

Using dual supplies is the typical way to power op amps. But this is not necessary or even desirable in some applications. This section discusses the inverting and noninverting amplifiers running off a single positive supply.

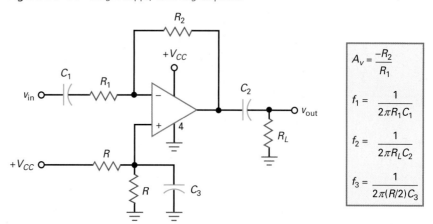

Figure 20-35 Single-supply inverting amplifier.

$$A_v = \frac{-R_2}{R_1}$$

$$f_1 = \frac{1}{2\pi R_1 C_1}$$

$$f_2 = \frac{1}{2\pi R_L C_2}$$

$$f_3 = \frac{1}{2\pi (R/2) C_3}$$

Inverting Amplifier

Figure 20-35 shows a single-supply inverting voltage amplifier that can be used with ac signals. The V_{EE} supply (pin 4) is grounded, and a voltage divider applies half the V_{CC} supply to the noninverting input. Because the two inputs are virtually shorted, the inverting input has a quiescent voltage of approximately $+0.5V_{CC}$.

In the dc equivalent circuit, all capacitors are open and the circuit is a voltage follower that produces a dc output voltage of $+0.5V_{CC}$. Input offsets are minimized because the voltage gain is unity.

In the ac equivalent circuit, all capacitors are shorted and the circuit is an inverting amplifier with a voltage gain of $-R_2/R_1$. Figure 20-35 shows the analysis equations. With these, we calculate the three lower cutoff frequencies.

A bypass capacitor is used on the noninverting input, as shown in Fig. 20-35. This reduces the power-supply ripple and noise appearing at the noninverting input. To be effective, the cutoff frequency of this bypass circuit should be much lower than the ripple frequency out of the power supply. You can calculate the cutoff frequency of this bypass circuit with the equation given in Fig. 20-35.

Noninverting Amplifier

In Fig. 20-36, only a positive supply is being used. To get maximum output swing, you need to bias the noninverting input at half the supply voltage, which is conveniently done with an equal-resistor voltage divider. This produces a dc input of $+0.5V_{CC}$ at the noninverting input. Because of the negative feedback, the inverting input is bootstrapped to the same value.

In the dc equivalent circuit, all capacitors are open and the circuit has a voltage gain of unity, which minimizes the output offset voltage. The dc output voltage of the op amp is $+0.5V_{CC}$, but this is blocked from the final load by the output coupling capacitor.

In the ac equivalent circuit, all capacitors are shorted. When an ac signal drives the circuit, an amplified output signal appears across R_L. If a rail-to-rail op amp is used, the maximum peak-to-peak unclipped output is V_{CC}. Figure 20-36 gives the equations for calculating the cutoff frequencies.

Single–Supply Op Amps

Although we can use ordinary op amps with a single supply, as shown in Figs. 20-35 and 20-36, there are some op amps that are optimized for single-supply operation. For instance, the LM324 is a quad op amp that eliminates the

Figure 20-36 Single-supply noninverting amplifier.

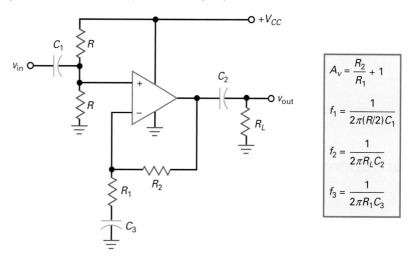

$$A_v = \frac{R_2}{R_1} + 1$$

$$f_1 = \frac{1}{2\pi(R/2)C_1}$$

$$f_2 = \frac{1}{2\pi R_L C_2}$$

$$f_3 = \frac{1}{2\pi R_1 C_3}$$

need for dual supplies. It contains four internally compensated op amps in a single package, each with an open-loop voltage gain of 100 dB, input biasing current of 45 nA, input offset current of 5 nA, and input offset voltage of 2 mV. It runs off a single positive supply voltage that can have any value between 3 and 32 V. Because of this, the LM324 is convenient to use as an interface with digital circuits that run off a single positive supply of +5 V.

Summary

SEC. 20–1 INVERTING-AMPLIFIER CIRCUITS

Inverting-amplifier circuits discussed in this section included a high-impedance probe (X10 and X1), an ac-coupled amplifier, and an adjustable-bandwidth circuit.

SEC. 20–2 NONINVERTING-AMPLIFIER CIRCUITS

Noninverting-amplifier circuits discussed in this section included an ac-coupled amplifier, an audio distribution amplifier, a JFET-switched amplifier, and a voltage reference.

SEC. 20–3 INVERTER/NONINVERTER CIRCUITS

The circuits discussed in this section are the switchable inverter/noninverter, the JFET-controlled switchable inverter, the

sign changer, the adjustable and reversible gain circuit, and the phase shifter.

SEC. 20–4 DIFFERENTIAL AMPLIFIERS

Two factors determine the overall CMRR of a differential amplifier: the CMRR of each op amp and the CMRR of the matched resistors. The input signal is usually a small differential voltage and a large common-mode voltage coming from a Wheatstone bridge.

SEC. 20–5 INSTRUMENTATION AMPLIFIERS

An instrumentation amplifier is a differential amplifier optimized for large voltage gain, high CMRR, low input offsets, low-temperature drift, and high input impedance. Instrumentation amplifiers can be built with the classic

three op-amp circuit, using precision op amps, or with an integrated instrumentation amplifier.

SEC. 20–6 SUMMING AMPLIFIER CIRCUITS

The topics discussed in this section were the subtracter, summing on both inputs, the averager, and the D/A converter. The D/A converter is used in digital multimeters to measure voltages, currents, and resistances.

SEC. 20–7 CURRENT BOOSTERS

When the short-circuit output current of an op amp is too low, one solution is to use a current booster on the output side of the circuit. Typically, the current booster is a transistor whose base current is supplied by the op amp. Because of the transistor current gain, the short-circuit output current is increased by the β factor.

SEC. 20-8 VOLTAGE-CONTROLLED CURRENT SOURCES

We can build current sources that are controlled by an input voltage. The loads may be floating or grounded. The load currents may be unidirectional or bidirectional. The Howland current source is a bidirectional voltage-controlled current source.

SEC. 20-9 AUTOMATIC GAIN CONTROL

In many applications we want the voltage gain of a system to change automatically as needed to maintain an almost constant output voltage. In radio and television receivers, AGC prevents sudden and large changes in the volume of the sound out of the speakers.

SEC. 20-10 SINGLE-SUPPLY OPERATION

Although op amps normally use dual supplies, there are applications for which only a single supply is preferred. When ac-coupled amplifiers are needed, single-supply amplifiers are easily implemented by biasing the nonsignal side of the op amp to half the positive supply voltage. Some op amps are optimized for single-supply operation.

Derivations

(20-3) Gain for inverter/noninverter circuits:

$$A_v = A_{v(inv)} + A_{v(non)}$$

See Figs. 20-8 to 20-13. The total voltage gain is the superposition of the inverting and noninverting voltage gains. We use it when the input signal is being applied to both inputs.

(20-5) Common-mode voltage gain:

$$A_{v(CM)} = A_{v(inv)} + A_{v(non)}$$

See Fig. 20-14, 20-15, and 20-18. This is similar to Eq. (20-3) because it is the superposition of gains.

(20-7) Overall noninverter gain:

$$A_{v(non)} = \left(\frac{R_2}{R_1} + 1\right)\left(\frac{R_2'}{R_1' + R_2'}\right)$$

See Fig. 20-14. This is the voltage gain of the noninverting side reduced by the voltage-divider factor.

(20-8) Common-mode gain for $R_1 = R_2$:

$$A_{v(CM)} = \pm 2\frac{\Delta R}{R}$$

See Figs. 20-15 and 20-18. This is the common-mode gain caused by the resistor tolerances when the resistors of the differential amplifier are equal and matched.

(20-11) Wheatstone bridge:

$$\frac{R_1}{R_2} = \frac{R_3}{R_4}$$

See Fig. 20-16a. This is the equation for balance in a Wheatstone bridge.

(20-13) Unbalanced Wheatstone bridge:

$$v_{in} \cong \frac{\Delta R}{4R} V_{CC}$$

See Fig. 20-17. This equation is valid for small changes in the resistance of the transducer.

(20-16) Instrumentation amplifier:

$$A_v = \frac{2R_2}{R_G} + 1$$

See Figs. 20-18 and 20-20. This is the voltage gain of the first stage of the classic three op-amp instrumentation amplifier.

(20-18) Binary-to-decimal equivalent:

$$BIN = (D_0 \times 2^0) + (D_1 \times 2^1) + (D_2 \times 2^2) + (D_3 \times 2^3)$$

(20-19) R/2R Ladder output voltage:

$$V_{out} = -\left(\frac{BIN}{2^N} \times 2V_{ref}\right)$$

(20-21) Current booster:

$$I_{max} = \beta_{dc} I_{SC}$$

See Figs. 20-26 to 20-30. The short-circuit current of an op amp is increased by the current gain of a transistor between the op amp and the load.

(20-23) Voltage-controlled current sources:

$$i_{out} = \frac{v_{in}}{R}$$

See Figs. 20-28 to 20-31. In voltage-controlled current sources, the input voltage is converted to a stiff output current.

Student Assignments

1. In a linear op-amp circuit, the
   a. Signals are always sine waves
   b. Op amp does not go into saturation
   c. Input impedance is ideally infinite
   d. Gain-bandwidth product is constant

2. In an ac amplifier using an op amp with coupling and bypass capacitors, the output offset voltage is
   a. Zero
   b. Minimum
   c. Maximum
   d. Unchanged

3. To use an op amp, you need at least
   a. One supply voltage
   b. Two supply voltages
   c. One coupling capacitor
   d. One bypass capacitor

4. In a controlled current source with op amps, the circuit acts like a
   a. Voltage amplifier
   b. Current-to-voltage converter
   c. Voltage-to-current converter
   d. Current amplifier

5. An instrumentation amplifier has a high
   a. Output impedance
   b. Power gain
   c. CMRR
   d. Supply voltage

6. A current booster on the output of an op amp will increase the short-circuit current by
   a. $A_{v(CL)}$
   b. β_{dc}
   c. f_{unity}
   d. A_v

7. Given a voltage reference of +2.5 V, we can get a voltage reference of +15 V by using
   a. An inverting amplifier
   b. A noninverting amplifier
   c. A differential amplifier
   d. An instrumentation amplifier

8. In a differential amplifier, the CMRR is limited mostly by the
   a. CMRR of the op amp
   b. Gain-bandwidth product
   c. Supply voltages
   d. Tolerance of the resistors

9. The input signal for an instrumentation amplifier usually comes from
   a. An inverting amplifier
   b. A resistor
   c. A differential amplifier
   d. A Wheatstone bridge

10. In the classic three op-amp instrumentation amplifier, the differential voltage gain is usually produced by the
    a. First stage
    b. Second stage
    c. Mismatched resistors
    d. Output op amp

11. Guard driving reduces the
    a. CMRR of an instrumentation amplifier
    b. Leakage current in the shielded cable
    c. Voltage gain of the first stage
    d. Common-mode input voltage

12. In an averaging circuit, the input resistances are
    a. Equal to the feedback resistance
    b. Less than the feedback resistance
    c. Greater than the feedback resistance
    d. Unequal

13. A D/A converter is an application of the
    a. Adjustable bandwidth circuit
    b. Noninverting amplifier
    c. Voltage-to-current converter
    d. Summing amplifier

14. In a voltage-controlled current source,
    a. A current booster is never used
    b. The load is always floated
    c. A stiff current source drives the load
    d. The load current equals I_{SC}

15. The Howland current source produces a
    a. Unidirectional floating load current
    b. Bidirectional single-ended load current
    c. Unidirectional single-ended load current
    d. Bidirectional floating load current

16. The purpose of AGC is to
    a. Increase the voltage gain when the input signal increases
    b. Convert voltage to current
    c. Keep the output voltage almost constant
    d. Reduce the CMRR of the circuit

17. 1 ppm is equivalent to
    a. 0.1 percent
    b. 0.01 percent
    c. 0.001 percent
    d. 0.0001 percent

18. An input transducer converts
    a. Voltage to current
    b. Current to voltage
    c. An electrical quantity to a nonelectrical quantity
    d. A nonelectrical quantity to an electrical quantity

19. A thermistor converts
    a. Light to resistance
    b. Temperature to resistance
    c. Voltage to sound
    d. Current to voltage

20. When we trim a resistor, we are
    a. Making a fine adjustment
    b. Reducing its value
    c. Increasing its value
    d. Making a coarse adjustment

21. A D/A converter with four inputs has
    a. Two output values
    b. Four output values
    c. Eight output values
    d. Sixteen output values

22. An op amp with a rail-to-rail output
    a. Has a current-boosted output
    b. Can swing all the way to either supply voltage
    c. Has a high output impedance
    d. Cannot be less than 0 V

23. When a JFET is used in an AGC circuit, it acts like a
    a. Switch
    b. Voltage-controlled current source
    c. Voltage-controlled resistance
    d. Capacitance

24. If an op amp has only a positive supply voltage, its output cannot
    a. Be negative
    b. Be zero
    c. Equal the supply voltage
    d. Be ac-coupled

Problems

SEC. 20-1 INVERTING–AMPLIFIER CIRCUITS

20-1 In the probe of Fig. 20-1, $R_1 = 10$ MΩ, $R_2 = 20$ MΩ, $R_3 = 15$ kΩ, $R_4 = 15$ kΩ, and $R_5 = 75$ kΩ. What is the attenuation of the probe in each switch position?

20-2 In the ac-coupled inverting amplifier of Fig. 20-2, $R_1 = 1.5$ kΩ, $R_f = 75$ kΩ, $R_L = 15$ kΩ, $C_1 = 1$ μF, $C_2 = 4.7$ μF, and $f_{unity} = 1$ MHz. What is the voltage gain in the midband of the amplifier? What are the upper and lower cutoff frequencies?

20-3 In the adjustable-bandwidth circuit of Fig. 20-3, $R_1 = 10$ kΩ and $R_f = 180$ kΩ. If the 100-Ω resistor is changed to 130 Ω and the variable resistor to 25 kΩ, what is the voltage gain? What are the minimum and maximum bandwidth if $f_{unity} = 1$ MHz?

20-4 What is the output voltage in Fig. 20-37? What are the minimum and maximum bandwidth? (Use $f_{unity} = 1$ MHz.)

SEC. 20-2 NONINVERTING–AMPLIFIER CIRCUITS

20-5 In Fig. 20-4, $R_1 = 2$ kΩ, $R_f = 82$ kΩ, $R_L = 25$ kΩ, $C_1 = 2.2$ μF, $C_2 = 4.7$ μF, and $f_{unity} = 3$ MHz. What is the voltage gain in the midband of the amplifier? What are the upper and lower cutoff frequencies?

20-6 What is the voltage gain in the midband of Fig. 20-38? What are the upper and lower cutoff frequencies?

20-7 **||| MultiSim** In the distribution amplifier of Fig. 20-5, $R_1 = 2$ kΩ, $R_f = 100$ kΩ, and $v_{in} = 10$ mV. What is the output voltage for A, B, and C?

20-8 The JFET-switched amplifier of Fig. 20-6 has these values: $R_1 = 91$ kΩ, $R_f = 12$ kΩ, and $R_2 = 1$ kΩ. If $v_{in} = 2$ mV, what is the output voltage when the gate is low? When it is high?

20-9 If $V_{GS(off)} = -5$ V, what are the minimum and maximum output voltage in Fig. 20-39?

20-10 The voltage reference of Fig. 20-7 is modified to get $R_1 = 10$ kΩ and $R_f = 10$ kΩ. What is the new output reference voltage?

SEC. 20-3 INVERTER/NONINVERTER CIRCUITS

20-11 In the adjustable inverter of Fig. 20-10, $R_1 = 1$ kΩ and $R_2 = 10$ kΩ. What is the maximum positive gain? The maximum negative gain?

20-12 What is the voltage gain in Fig. 20-11 when the wiper is at the ground end? When it is 10 percent away from ground?

20-13 Precision resistors are used in Fig. 20-12. If $R = 5$ kΩ, $nR = 75$ kΩ, and $nR/(n-1)R = 5.36$ kΩ, what are the maximum positive and negative gains?

20-14 In the phase shifter of Fig. 20-13, $R' = 10$ kΩ, $R = 22$ kΩ, and $C = 0.02$ μF. What is the phase shift when the input frequency is 100 Hz? 1 kHz? 10 kHz?

Figure 20-37

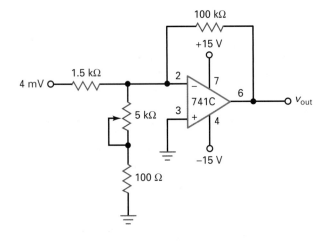

Figure 20-38

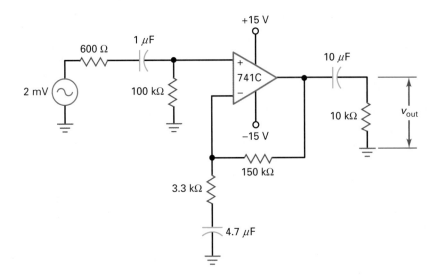

Figure 20-39

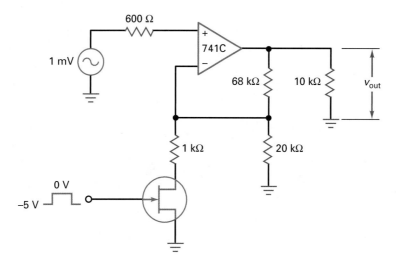

SEC. 20-4 DIFFERENTIAL AMPLIFIERS

20-15 The differential amplifier of Fig. 20-14 has $R_1 = 1.5\ k\Omega$ and $R_2 = 30\ k\Omega$. What is the differential voltage gain? The common-mode gain? (Resistor tolerance = ±0.1 percent.)

20-16 In Fig. 20-15, $R_1 = 1\ k\Omega$ and $R_2 = 20\ k\Omega$. What is the differential voltage gain? The common-mode gain? (Resistor tolerance = ±1 percent.)

20-17 In the Wheatstone bridge of Fig. 20-16, $R_1 = 10\ k\Omega$, $R_2 = 20\ k\Omega$, $R_3 = 20\ k\Omega$, and $R_4 = 10\ k\Omega$. Is the bridge balanced?

20-18 In the typical application of Fig. 20-17, transducer resistance changes to 985 Ω. What is the final output voltage?

SEC. 20-5 INSTRUMENTATION AMPLIFIERS

20-19 In the instrumentation amplifier of Fig. 20-18, $R_1 = 1\ k\Omega$ and $R_2 = 99\ k\Omega$. What is the output voltage if $v_{in} = 2\ mV$? If three OP-07A op amps are used and $R = 10\ k\Omega$ ± 0.5 percent, what is the CMRR of the instrumentation amplifier?

20-20 In Fig. 20-19, $v_{in(CM)} = 5\ V$. If $R_3 = 10\ k\Omega$, what does the guard voltage equal?

20-21 The value of R_G is changed to 1008 Ω in Fig. 20-20. What is the differential output voltage if the differential input voltage is 20 mV?

SEC. 20-6 SUMMING AMPLIFIER CIRCUITS

20-22 What does the output voltage equal in Fig. 20-21 if $R = 10\ k\Omega$, $v_1 = -50\ mV$ and $v_2 = -30\ mV$?

20-23 ▍▍▍ **MultiSim** In the summing circuit of Fig. 20-22, $R_1 = 10\ k\Omega$, $R_2 = 20\ k\Omega$, $R_3 = 15\ k\Omega$, $R_4 = 15\ k\Omega$, $R_5 = 30\ k\Omega$, and $R_f = 75\ k\Omega$. What is the output voltage if $v_0 = 1\ mV$, $v_1 = 2\ mV$, $v_2 = 3\ mV$, and $v_3 = 4\ mV$?

20-24 The averaging circuit of Fig. 20-23 has $R = 10\ k\Omega$. What is the output if $v_1 = 1.5\ V$, $v_2 = 2.5\ V$, and $v_3 = 4.0\ V$?

20-25 The D/A converter of Fig. 20-24 has an input of $v_0 = 5\ V$, $v_1 = 0$, $v_2 = 5\ V$, and $v_3 = 0$. What is the output voltage?

20-26 In Fig. 20-25, if the number of binary inputs is expanded to eight and D_7 to D_0 equals 10100101, determine the decimal equivalent input value, BIN.

20-27 In Fig. 20-25, if the binary inputs were expanded so D_7 to D_0 equaled 01100110, what would be the output voltage?

20-28 In Fig. 20-25, using an input reference voltage of 2.5 V, determine the smallest incremental output voltage step.

SEC. 20-7 CURRENT BOOSTERS

20-29 The noninverting amplifier of Fig. 20-40 has a current-boosted output. What is the voltage gain of the circuit? If the transistor has a current gain of 100, what is the short-circuit output current?

20-30 What is the voltage gain in Fig. 20-41? If the transistors have a current gain of 125, what is the short-circuit output current?

Figure 20-40

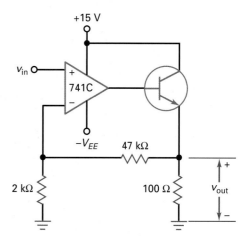

Figure 20-41

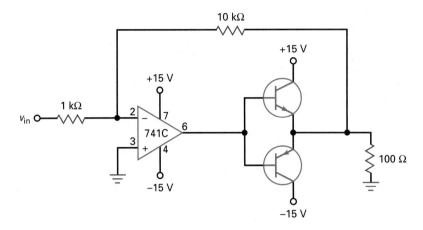

SEC. 20-8 VOLTAGE-CONTROLLED CURRENT SOURCES

20-31 What is the load current in Fig. 20-42a? The maximum load resistance that can be used without saturating the op amp?

20-32 Calculate the output current in Fig. 20-42b. Also, work out the maximum value of load resistance.

20-33 If $R = 10$ kΩ and $v_{cc} = 15$ V in the voltage-controlled current source of Fig. 20-30, what is the output current when the input voltage is 3 V? The maximum load resistance?

20-34 The Howland current source of Fig. 20-31 has $R = 2$ kΩ and $R_L = 500$ Ω. What is the output current when the input voltage is 6 V? What is the maximum load resistance that can be used with this circuit if the input voltage is never greater than 7.5 V? (Use supply voltages of ± 15 V.)

SEC. 20-9 AUTOMATIC GAIN CONTROL

20-35 In the AGC circuit of Fig. 20-32, $R_1 = 10$ kΩ, $R_2 = 100$ kΩ, $R_3 = 100$ kΩ, and $R_4 = 10$ kΩ. If r_{ds} can vary from 200 Ω to 1 MΩ, what is the minimum voltage gain of the circuit? The maximum?

20-36 In the low-level AGC circuit of Fig. 20-33, $R_1 = 5.1$ kΩ, $R_2 = 51$ kΩ, $R_5 = 68$ kΩ, and $R_6 = 1$ kΩ. If r_{ds} can vary from 120 Ω to 5 MΩ, what is the minimum voltage gain of the circuit? The maximum?

20-37 In the high-level AGC circuit of Fig. 20-34, $R_1 = 10$ kΩ, $R_2 = 10$ kΩ, $R_5 = 75$ kΩ, and $R_6 = 1.2$ kΩ. If R_7 can vary from 180 Ω to 10 MΩ, what is the minimum voltage gain of the circuit? The maximum?

Figure 20-42

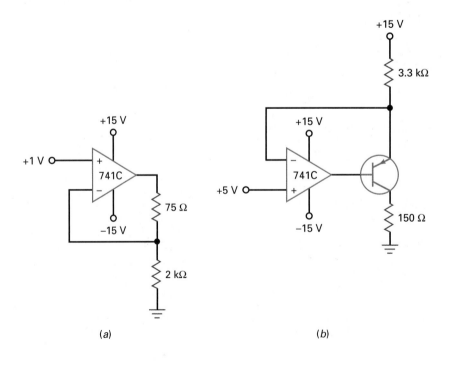

(a)   (b)

20-38 What is the voltage gain in the single-supply inverting amplifier of Fig. 20-43? The three lower cutoff frequencies?

20-39 In the single-supply noninverting amplifier of Fig. 20-36, $R = 68$ kΩ, $R_1 = 1.5$ kΩ, $R_2 = 15$ kΩ, $R_L = 15$ kΩ, Ω, $C_1 = 1$ μF, $C_2 = 2.2$ μF, and $C_3 = 3.3$ μF. What is the voltage gain? The three lower cutoff frequencies?

Figure 20-43

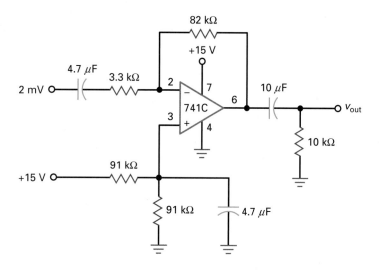

Critical Thinking

20-40 When switching between the positions of Fig. 20-8, there is a brief period of time when the switch is temporarily open. What is the output voltage at this time? Can you suggest how to prevent this from happening?

20-41 An inverting amplifier has $R_1 = 1$ kΩ and $R_f = 100$ kΩ. If these resistances have tolerances of ± 1 percent, what is the maximum possible voltage gain? The minimum?

20-42 What is the voltage gain in the midband of the circuit shown in Fig. 20-44?

20-43 The transistors of Fig. 20-41 have $\beta_{dc} = 50$. If the input voltage is 0.5 V, what is the base current in the conducting transistor?

Figure 20-44

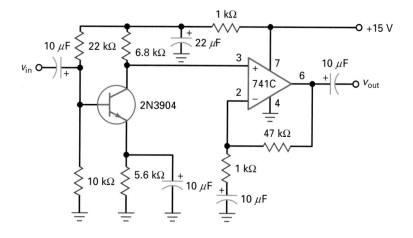

Troubleshooting

MultiSim Use Fig. 20-45 for the remaining problems. Any resistor may be open or shorted. Also, connecting wires CD, EF, JA, or KB may be open. Voltage values are in millivolts unless otherwise indicated.

20-44 Find Troubles T1 to T3.

20-45 Find Troubles T4 to T6.

20-46 Find Troubles T7 to T10.

Figure 20-45

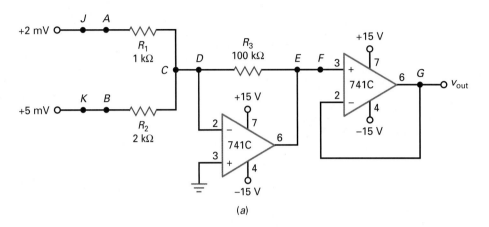

(a)

Troubleshooting

Trouble	V_A	V_B	V_C	V_D	V_E	V_F	V_G
OK	2	5	0	0	450	450	450
T1	2	5	0	0	450	0	0
T2	2	5	0	0	200	200	200
T3	2	5	2	2	−13.5 V	−13.5 V	−13.5 V
T4	2	0	0	0	200	200	200
T5	2	5	3	0	0	0	0
T6	0	5	0	0	250	250	250
T7	2	5	3	3	−13.5 V	−13.5 V	−13.5 V
T8	2	5	0	0	250	250	250
T9	2	5	0	0	0	0	0
T10	2	5	5	5	−13.5 V	−13.5 V	−13.5 V

(b)

Job Interview Questions

1. Draw the schematic diagram of an ac-coupled inverting amplifier with a voltage gain of 100. Discuss the theory of operation.

2. Draw the schematic diagram of a differential amplifier built with an op amp. What are the factors that determine the CMRR?

3. Draw the schematic diagram of the classic three op-amp instrumentation amplifier. Tell me what the first stage does to the differential and common-mode signals.
4. Why does the instrumentation amplifier have more than one stage?
5. You have designed a simple op-amp circuit for a particular application. During your initial testing, you find that the op amp is very hot to the touch. Assuming that the circuit has been correctly breadboarded, what is the most likely problem and what can you do to correct it?
6. Explain how an inverting amplifier is used in a high-impedance probe (X10 and X1).
7. In Fig. 20-1, why does the probe see a high impedance? Explain how the voltage gain is calculated in each switch position.

8. What can be said about the analog output of a D/A converter when it is compared with the digital input?
9. You want to construct a portable op-amp circuit that runs off a single 9-V battery using a 741C. What is one way you can do this? How would you have to modify this circuit if a dc response is required?
10. How could you increase the output current of an op amp?
11. Why is no resistor or diode biasing required in the circuit of Fig. 20-27?
12. When working with op amps, one often hears the term *rail*, as in a *rail-to-rail amplifier*. To what does that term refer?
13. Can a 741 be operated with a single supply voltage? If so, discuss what would be required for an inverting amplifier.

Self–Test Answers

1.	b	9.	d	17.	d
2.	b	10.	a	18.	d
3.	a	11.	b	19.	b
4.	c	12.	c	20.	a
5.	c	13.	d	21.	d
6.	b	14.	c	22.	b
7.	b	15.	b	23.	c
8.	d	16.	c	24.	a

Practice Problem Answers

20-2 $R_2 = 60 \text{ k}\Omega$

20-3 $N = 7.5; nR = 1.154 \text{ k}\Omega$

20-4 $A_v = 51; A_{v(CM)} = 0.002;$
$V_{out} = -510 \text{ mV};$
$V_{out(CM)} = \pm 20 \text{ mV}$

20-5 $A_{v1} = -1; A_{v2} = -0.5;$
$A_{v3} = -1.06; A_{v4} = -0.798$

20-6 largest $V_{out} = -9.375 \text{ V};$
smallest $V_{out} = -0.625 \text{ V}$

20-7 $A_v = -27; z_{out(CL)} = 0.021 \ \Omega;$
$I_{max} = 2.5 \text{ A}$

20-8 $i_{out} = 0.5 \text{ mA}; R_{L(max)} = 1 \text{ k}\Omega$

20-9 $i_{out} = -0.3 \text{ mA}; R_{L(max)} = 1.25 \text{ k}\Omega$

20-10 $r_{ds} = 2.13 \text{ k}\Omega$

21

Active Filters

Almost all communication systems use filters. A filter passes one band of frequencies while rejecting another. A filter can be either passive or active. **Passive filters** are built with resistors, capacitors, and inductors. They are generally used above 1 MHz, have no power gain, and are relatively difficult to tune. **Active filters** are built with resistors, capacitors, and op amps. They are useful below 1 MHz, have power gain, and are relatively easy to tune. Filters can separate desired signals from undesired signals, block interfering signals, enhance speech and video, and alter signals in other ways.

Objectives

After studying this chapter, you should be able to:

- Discuss the five basic filter responses.
- Describe the difference between passive and active filters.
- Differentiate between brick wall responses and approximate responses.
- Explain filter terminology, including passband, stopband, cutoff, Q, ripple, and order.
- Determine the order of passive and active filters.
- Discuss the reasons why filter stages are sometimes cascaded, and describe the results.

Chapter Outline

Vocabulary

active filters

all-pass filter

attenuation

bandpass filter

bandstop filter

Bessel approximation

biquadratic bandpass/lowpass filter

Butterworth approximation

Chebyshev approximation

damping factor

delay equalizer

edge frequency

elliptic approximation

frequency scaling factor (FSF)

geometric average

high-pass filter

inverse Chebyshev approximation

linear phase shift

low-pass filter

monotonic

multiple-feedback (MFB)

narrowband filter

order of a filter

passband

passive filters

pole frequency (f_p)

poles

predistortion

Sallen-Key equal-component filter

Sallen-Key low-pass filter

Sallen-Key second-order notch filter

state-variable filter

stopband

transition

wideband filter

783

21-1 Ideal Responses

This chapter is a comprehensive look at a variety of passive and active filter circuits. Basic filter terminology and first-order stages are covered through Sec. 21-4. Sections 21-5 and beyond contain more detailed circuit analysis of higher-order filters.

The *frequency response of a filter* is the graph of its voltage gain versus frequency. There are five types of filters: *low-pass, high-pass, bandpass, bandstop,* and *all-pass.* This section discusses the ideal frequency response of each. The next section describes the approximations for these ideal responses.

Low-Pass Filter

Figure 21-1 shows the ideal frequency response of a **low-pass filter.** It is sometimes called a *brick wall response* because the right edge of the rectangle looks like a brick wall. A low-pass filter passes all frequencies from zero to the cutoff frequency and blocks all frequencies above the cutoff frequency.

With a low-pass filter, the frequencies between zero and the cutoff frequency are called the **passband.** The frequencies above the cutoff frequency are called the **stopband.** The roll-off region between the passband and the stopband is called the **transition.** An ideal low-pass filter has zero *attenuation* (signal loss) in the passband, infinite attenuation in the stopband, and a vertical transition.

One more point: The ideal low-pass filter has zero phase shift for all frequencies in the passband. Zero phase shift is important when the input signal is nonsinusoidal. When a filter has zero phase shift, the shape of the nonsinusoidal signal is preserved as it passes through the ideal filter. For instance, if the input signal is a square wave, it has a fundamental frequency and harmonics. If the fundamental frequency and all significant harmonics (approximately the first 10) are inside the passband, the square wave will have approximately the same shape at the output.

High-Pass Filter

Figure 21-2 shows the ideal frequency response of a **high-pass filter.** A high-pass filter blocks all frequencies from zero up to the cutoff frequency and passes all frequencies above the cutoff frequency.

With a high-pass filter, the frequencies between zero and the cutoff frequency are the stopband. The frequencies above the cutoff frequency are the passband. An ideal high-pass filter has infinite attenuation in the stopband, zero attenuation in the passband, and a vertical transition.

Figure 21-1 Ideal low-pass response.

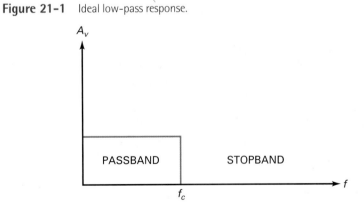

Figure 21-2 Ideal high-pass response.

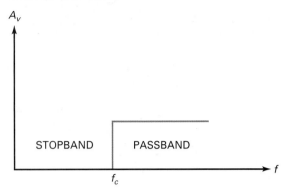

Bandpass Filter

A **bandpass filter** is useful when you want to tune in a radio or television signal. It is also useful in telephone communications equipment for separating the different phone conversations that are being simultaneously transmitted over the same communication path.

Figure 21-3 shows the ideal frequency response of a bandpass filter. A brick wall response like this blocks all frequencies from zero up to the lower cutoff frequency. Then, it passes all the frequencies between the lower and upper cutoff frequencies. Finally, it blocks all frequencies above the upper cutoff frequency.

With a bandpass filter, the passband is all the frequencies between the lower and upper cutoff frequencies. The frequencies below the lower cutoff frequency and above the upper cutoff frequency are the stopband. An ideal bandpass filter has zero attenuation in the passband, infinite attenuation in the stopband, and two vertical transitions.

The *bandwidth (BW)* of a bandpass filter is the difference between its upper and lower 3-dB cutoff frequencies:

$$BW = f_2 - f_1 \tag{21-1}$$

For instance, if the cutoff frequencies are 450 and 460 kHz, the bandwidth is:

$$BW = 460 \text{ kHz} - 450 \text{ kHz} = 10 \text{ kHz}$$

As another example, if the cutoff frequencies are 300 and 3300 Hz, the bandwidth is:

$$BW = 3300 \text{ Hz} - 300 \text{ Hz} = 3000 \text{ Hz}$$

Figure 21-3 Ideal bandpass response.

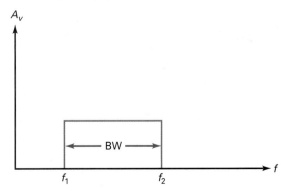

The center frequency is symbolized by f_0 and is given by the **geometric average** of the two cutoff frequencies:

$$f_0 = \sqrt{f_1 f_2} \tag{21-2}$$

For instance, telephone companies use a bandpass filter with cutoff frequencies of 300 and 3300 Hz to separate phone conversations. The center frequency of these filters is:

$$f_0 = \sqrt{(300 \text{ Hz})(3300 \text{ Hz})} = 995 \text{ Hz}$$

To avoid interference between different phone conversations, the bandpass filters have responses that approach the brick wall response shown in Fig. 21-3.

The Q of a bandpass filter is defined as the center frequency divided by the bandwidth:

$$Q = \frac{f_0}{\text{BW}} \tag{21-3}$$

For instance, if $f_0 = 200$ kHz and BW $= 40$ kHz, then $Q = 5$.

When the Q is greater than 10, the center frequency can be approximated by the *arithmetic average* of the cutoff frequencies:

$$f_0 \cong \frac{f_1 + f_2}{2}$$

For instance, in a radio receiver the cutoff frequencies of the bandpass filter (IF stage) are 450 and 460 kHz. The center frequency is approximately:

$$f_0 \cong \frac{450 \text{ kHz} + 460 \text{ kHz}}{2} = 455 \text{ kHz}$$

If Q is less than 1, the bandpass filter is called a **wideband filter.** If Q is greater than 1, the filter is called a **narrowband filter.** For example, a filter with cutoff frequencies of 95 and 105 kHz has a bandwidth of 10 kHz. This is a narrowband because Q is approximately 10. A filter with cutoff frequencies of 300 and 3300 Hz has a center frequency of approximately 1000 Hz and a bandwidth of 3000 Hz. This is wideband because Q is approximately 0.333.

Bandstop Filter

Figure 21-4 shows the ideal frequency response of a **bandstop filter.** This type of filter passes all frequencies from zero up to the lower cutoff frequency. Then, it blocks all the frequencies between the lower and upper cutoff frequencies. Finally, it passes all frequencies above the upper cutoff frequency.

Figure 21-4 Ideal bandstop response.

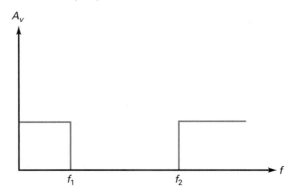

Chapter 21

Figure 21–5 Ideal all-pass response.

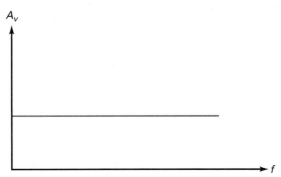

With a bandstop filter, the stopband is all the frequencies between the lower and upper cutoff frequencies. The frequencies below the lower cutoff frequency and above the upper cutoff frequency are the passband. An ideal bandstop filter has infinite attenuation in the stopband, no attenuation in the passband, and two vertical transitions.

The definitions for bandwidth, narrowband, and center frequency are the same as before. In other words, with a bandstop filter, we use Eqs. (21-1) through (21-3) to calculate BW, f_0, and Q. Incidentally, the bandstop filter is sometimes called a *notch filter* because it notches out or removes all frequencies in the stopband.

All–Pass Filter

Figure 21-5 shows the frequency response of an ideal **all-pass filter.** It has a passband and no stopband. Because of this, it passes all frequencies between zero and infinite frequency. It may seem rather unusual to call it a filter since it has zero attenuation for all frequencies. The reason it is called a filter is because of the effect it has on the *phase* of signals passing through it. The all-pass filter is useful when we want to produce a certain amount of phase shift for the signal being filtered without changing its amplitude.

The *phase response of a filter* is defined as the graph of phase shift versus frequency. As mentioned earlier, the ideal low-pass filter has a phase response of $0°$ at all frequencies. Because of this, a nonsinusoidal input signal has the same shape after passing through an ideal low-pass filter, provided its fundamental frequency and all significant harmonics are in the passband.

The phase response of an all-pass filter is different from that of the ideal low-pass filter. With the all-pass filter, each distinct frequency can be shifted by a certain amount as it passes through the filter. For instance, the phase shifter discussed in Sec. 20-3 was a noninverting op-amp circuit with zero attenuation at all frequencies but an output phase angle between 0 and $-180°$. The phase shifter is a simple example of an all-pass filter. In later sections, we will discuss more complicated all-pass filters that can produce larger phase shifts.

21–2 Approximate Responses

The ideal responses discussed in the preceding section are impossible to realize with practical circuits, but there are five standard approximations used as compromises for the ideal responses. Each of these approximations offers an advantage that the others do not have. The approximation chosen by a designer will depend on what is acceptable in an application.

Attenuation

Attenuation refers to a loss of signal. With a constant input voltage, attenuation is defined as the output voltage at any frequency divided by the output voltage in the midband:

$$\text{Attenuation} = \frac{v_{out}}{v_{out(mid)}} \qquad \text{(21-3a)}$$

For instance, if the output voltage is 1 V at some frequency and the output voltage in the midband is 2 V, then:

$$\text{Attenuation} = \frac{1 \text{ V}}{2 \text{ V}} = 0.5$$

Attenuation is normally expressed in decibels using this equation:

$$\textbf{Decibel attenuation} = \textbf{20 log attenuation} \qquad \text{(21-3b)}$$

For an attenuation of 0.5, the decibel attenuation is:

$$\text{Decibel attenuation} = -20 \log 0.5 = 6 \text{ dB}$$

Because of the minus sign, decibel attenuation always is a positive number. Decibel attenuation uses the midband output voltage as a reference. Basically, we are comparing the output voltage at any frequency to the output voltage in the midband of the filter. Because attenuation is almost always expressed in decibels, we will use the term *attenuation* to mean decibel attenuation.

For instance, an attenuation of 3 dB means that the output voltage is 0.707 of its midband value. An attenuation of 6 dB means that the output voltage is 0.5 of its midband value. An attenuation of 12 dB means that the output voltage is 0.25 of its midband value. An attenuation of 20 dB means that the output voltage is 0.1 of its midband value.

Passband and Stopband Attenuation

In filter analysis and design, the low-pass filter is a *prototype*, a basic circuit that can be modified to get other circuits. Typically, any filter problem is converted into an equivalent low-pass filter problem and solved as a low-pass filter problem; the solution is converted back to the original filter type. For this reason, our discussion will focus on the low-pass filter and extend the discussion to other filters.

Zero attenuation in the passband, infinite attenuation in the stopband, and a vertical transition are unrealistic. To build a practical low-pass filter, the three regions are approximated as shown in Fig. 21-6. The passband is the set of frequencies between 0 and f_c. The stopband is all the frequencies above f_s. The transition region is between f_c and f_s.

As shown in Fig. 21-6, the passband no longer has zero attenuation. Instead, we are allowing for an attenuation between 0 and A_p. For instance, in some applications the passband can have $A_p = 0.5$ dB. This means that we are compromising the ideal response by allowing up to 0.5 dB of signal loss anywhere in the passband.

Similarly, the stopband no longer has infinite attenuation. Instead, we are allowing the stopband attenuation to be anywhere from A_s to infinity. For instance, in some applications, $A_s = 60$ dB may be adequate. This means that we are accepting an attenuation of 60 dB or more anywhere in the stopband.

Figure 21-6 Realistic low-pass response.

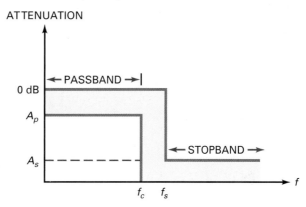

In Fig. 21-6, the transition region is no longer vertical. Instead, we are accepting a nonvertical roll-off. The roll-off rate will be determined by the values of f_c, f_s, A_p, and A_s. For instance, if $f_c = 1$ kHz, $f_s = 2$ kHz, $A_p = 0.5$ dB, and $A_s = 60$ dB, the required roll-off is approximately 60 dB per octave.

The five approximations we are about to discuss are a trade-off between the characteristics of the passband, stopband, and transition region. The approximations may optimize the flatness of the passband, or the roll-off rate, or the phase shift.

A final point: The highest frequency in the passband of a low-pass filter is called the *cutoff frequency* (f_c). This frequency is also referred to as the **edge frequency** because it is on the edge of the passband. In some filters, the attenuation at the edge frequency is less than 3 dB. For this reason, we will use f_{3dB} for the frequency when the attenuation is down 3 dB and f_c for the edge frequency, which may have a different attenuation.

Order of Filter

The **order of a passive filter** (symbolized by n) equals the number of inductors and capacitors in the filter. If a passive filter has two inductors and two capacitors, $n = 4$. If a passive filter has five inductors and five capacitors, $n = 10$. Therefore, the order tells us how complicated the filter is. The higher the order, the more complicated the filter.

The **order of an active filter** depends on the number of RC circuits (called **poles**) it contains. If an active filter contains eight RC circuits, $n = 8$. Counting the individual RC circuits in an active filter is usually difficult. Therefore, we will use a simpler method to determine the order of an active filter:

$$n \cong \text{\# capacitors} \qquad\qquad (21\text{-}4)$$

where the symbol # stands for "the number of." For instance, if an active filter contains 12 capacitors, it has an order of 12.

Bear in mind that Eq. (21-4) is a guideline. Since we are counting capacitors rather than RC circuits, exceptions may occur. Aside from the occasional exception, Eq. (21-4) gives us a quick and easy way to determine the order or number of poles in an active filter.

Butterworth Approximation

The **Butterworth approximation** is sometimes called the *maximally flat approximation* because the passband attenuation is zero through most of the passband

Figure 21-7 Butterworth low-pass response.

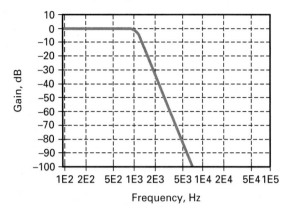

and decreases gradually to A_p at the edge of the passband. Well above the edge frequency, the response rolls off at a rate of approximately $20n$ dB per decade, where n is the order of the filter:

$$\textbf{Roll-off} = \textbf{20}\textbf{\textit{n}} \qquad \textbf{dB/decade} \tag{21-4a}$$

An equivalent roll-off in terms of octaves is:

$$\textbf{Roll-off} = \textbf{6}\textbf{\textit{n}} \qquad \textbf{dB/octave} \tag{21-4b}$$

For instance, a first-order Butterworth filter rolls off at a rate of 20 dB decade, or 6 dB per octave; a fourth-order filter rolls off at a rate of 80 dB per decade, or 24 dB per octave; a ninth-order filter rolls off at a rate of 180 dB per decade, or 54 dB per octave; and so on.

Figure 21-7 shows the response of a Butterworth low-pass filter with the following specifications: $n = 6$, $A_p = 2.5$ dB, and $f_c = 1$ kHz. These specifications tell us that it is a sixth-order or 6-pole filter with passband attenuation of 2.5 dB and an edge frequency of 1 kHz. The numbers along the frequency axis of Fig. 21-7 are abbreviated as follows: $2E3 = 2 \times 10^3 = 2000$. (*Note: E* stands for "exponent.")

Notice how flat the response is in the passband. The major advantage of a Butterworth filter is the flatness of the passband response. The major disadvantage is the relatively slow roll-off rate compared with the other approximations.

Chebyshev Approximation

In some applications, a flat passband response is not important. In this case, a **Chebyshev approximation** may be preferred because it rolls off faster in the transition region than a Butterworth filter. The price paid for this faster roll-off is that ripples appear in the passband of the frequency response.

Figure 21-8a shows the response of a Chebyshev low-pass filter with the following specifications: $n = 6$, $A_p = 2.5$ dB, and $f_c = 1$ kHz. These are the same specifications as those of the preceding Butterworth filter. When we compare Fig. 21-7 with Fig. 21-8a, we can see that a Chebyshev filter of the same order has a faster roll-off in the transition region. Because of this, the attenuation with a Chebyshev filter is always greater than the attenuation of a Butterworth filter of the same order.

The number of ripples in the passband of a Chebyshev low-pass filter equals half of the filter order:

$$\textbf{\# Ripples} = \frac{\textbf{\textit{n}}}{\textbf{2}} \tag{21-5}$$

Figure 21-8 (a) Chebyshev low-pass response; (b) magnified view of passband ripples.

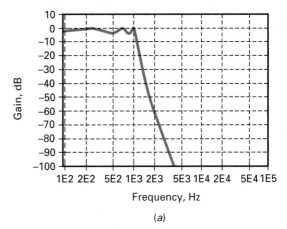

(a)

(b)

If a filter has an order of 10, it will have 5 ripples in the passband; if a filter has an order of 15, it will have 7.5 ripples. Figure 21-8b shows a magnified view of a Chebyshev response for an order of 20. It has 10 ripples in the passband.

In Fig. 21-8b, the ripples have the same peak-to-peak value. This is why the Chebyshev approximation is sometimes called the *equal-ripple approximation*. Typically, a designer will choose a ripple depth between 0.1 and 3 dB, depending on the needs of the application.

Inverse Chebyshev Approximation

In applications in which a flat passband response is required, as well as a fast roll-off, a designer may use the **inverse Chebyshev approximation.** It has a flat passband response and a rippled stopband response. The roll-off rate in the transition region is comparable to the roll-off rate of a Chebyshev filter.

Figure 21-9 shows the response of an inverse Chebyshev low-pass filter with the following specifications: $n = 6$, $A_p = 2.5$ dB, and $f_c = 1$ kHz. When we compare Fig. 21-9 with Figs. 21-7 and 21-8a, we can see that the inverse Chebyshev filter has a flat passband, a fast roll-off, and a rippled stopband.

Monotonic means that the stopband has no ripples. With the approximations discussed so far, the Butterworth and Chebyshev filters have monotonic stopbands. The inverse Chebyshev has a rippled stopband.

Figure 21-9 Inverse Chebyshev low-pass response.

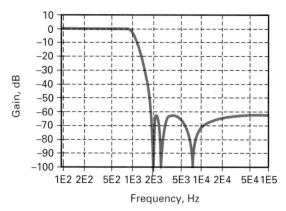

When specifying an inverse Chebyshev filter, the minimum acceptable attenuation throughout the stopband must be given because the stopband has ripples that may reach this value. For instance, in Fig. 21-9, the inverse Chebyshev filter has a stopband attenuation of 60 dB. As you can see, the ripples do approach this level at different frequencies in the stopband.

The unusual stopband response of Fig. 21-9 occurs because the inverse Chebyshev filter has components that notch the response at certain frequencies in the stopband. In other words, there are frequencies in the stopband at which the attenuation approaches infinity.

Elliptic Approximation

Some applications need the fastest possible roll-off in the transition region. If a rippled passband and a rippled stopband are acceptable, a designer may choose the **elliptic approximation**. Also known as the *Cauer filter,* this filter optimizes the transition region at the expense of the passband and stopband.

Figure 21-10 shows the response of an elliptic low-pass filter with the same specifications as before: $n = 6$, $A_p = 2.5$ dB, and $f_c = 1$ kHz. Notice that the elliptic filter has a rippled passband, a very fast roll-off, and a rippled stopband. After the response breaks at the edge frequency, the initial roll-off is very rapid, slows down slightly in the middle of the transition, and then becomes very rapid toward the end of the transition. Given a set of specifications for any complicated filter, the elliptic approximation will always produce the most efficient design; that is, it will have the lowest order.

Figure 21-10 Elliptic low-pass response.

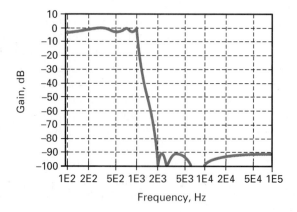

For instance, suppose we are given the following specifications: $A_p = 0.5$ dB, $f_c = 1$ kHz, $A_s = 60$ dB, and $f_s = 1.5$ kHz. Here are the required orders or number of poles for each of the approximations: Butterworth (20), Chebyshev (9), inverse Chebyshev (9), and elliptic (6). In other words, the elliptic filter requires the fewest capacitors, which translates to the simplest circuit.

Bessel Approximation

The **Bessel approximation** has a flat passband and a monotonic stopband similar to those of the Butterworth approximation. For the same filter order, however, the roll-off in the transition region is much less with a Bessel filter than with a Butterworth filter.

Figure 21-11a shows the response of a Bessel low-pass filter with the same specifications as before: $n = 6$, $A_p = 2.5$ dB, and $f_c = 1$ kHz. Notice that the Bessel filter has a flat passband, a relatively slow roll-off, and a monotonic stopband. Given a set of specifications for a complicated filter, the Bessel approximation will always produce the least roll-off of all the approximations. Stated another way: It has the highest order or greatest circuit complexity of all approximations.

Why is the order of a Bessel filter the highest for the same specifications? Because the Butterworth, Chebyshev, inverse Chebyshev, and elliptic approximations are optimized for frequency response only. With these approximations, no

---

Figure 21-11 (a) Bessel low-pass frequency response; (b) Bessel low-pass phase response.

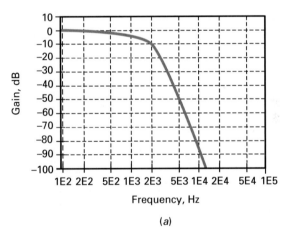

(a)

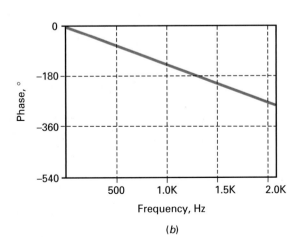

(b)

attempt is made to control the phase of the output signal. On the other hand, the Bessel approximation is optimized to produce a **linear phase shift** with frequency. In other words, the Bessel filter trades off some of the roll-off rate to get a linear phase shift.

Why bother with a linear phase shift? Recall the earlier discussion of the ideal low-pass filter. One of its ideal properties was a phase shift of $0°$. This was desirable because it meant that the shape of a nonsinusoidal signal would be preserved as it passed through the filter. With a Bessel filter, we cannot get a phase shift of $0°$, but we can get a linear phase response. This is a phase response in which the phase shift increases linearly with frequency.

Figure 21-11b shows the phase response of a Bessel filter with $n = 6$, $A_p = 2.5$ dB, and $f_c = 1$ kHz. As you can see, the phase response is linear. The phase shift is approximately $14°$ at 100 Hz, $28°$ at 200 Hz, $42°$ at 300 Hz, and so on. This linearity exists through the entire passband and somewhat beyond. At higher frequencies, the phase response becomes nonlinear, but that is not what matters. What counts is the linear phase response to all frequencies in the passband.

The linear phase shift for all frequencies in the passband means that the fundamental frequency and harmonics of a nonsinusoidal input signal will shift linearly in phase as they pass through the filter. Because of this, the shape of the output signal will be the same as the shape of the input signal.

The major advantage of the Bessel filter is that it produces the least distortion of nonsinusoidal signals. One easy way to measure this type of distortion is by the step response of the filter. This means applying a voltage step to the input and looking at the output with an oscilloscope. The Bessel filter has the best step response of all the filters.

Figure 21-12a to c shows the different step responses for a low-pass filter with $A_p = 3$ dB, $f_c = 1$ kHz, and $n = 10$. Notice how the step response of a Butterworth filter (Fig. 21-12a) overshoots the final level, rings a couple of times, and then settles on the final value of 1 V. A step response like this might be acceptable in some applications, but it is not ideal. The step response of a Chebyshev filter (Fig. 21-12b) is worse. It overshoots and rings many times before settling on its final value. A step response like this is far from ideal and not acceptable in some applications. The step response of the inverse Chebyshev filter is similar to that of the Butterworth because both responses are maximally flat in the passband. The step response of the elliptic filter is similar to that of the Chebyshev because both responses have rippled passbands.

Figure 21-12c shows the step response of a Bessel filter. This is almost an ideal reproduction of an input voltage step. The only deviation from a perfect step is the risetime. The Bessel step response has no noticeable overshoot or ringing. Since digital data consist of positive and negative steps, a clean step response like that shown in Fig. 21-12c is preferred to the distortion of Fig. 21-12a and b. For this reason, the Bessel filter may be used in some data communication systems.

A linear phase response implies a *constant time delay,* which means that all frequencies in the passband are delayed by the same amount of time as they pass through the filter. The amount of time it takes for a signal to pass through a filter depends on the order of the filter. With all filters except the Bessel filter, this amount of time changes with the frequency. With the Bessel filter, the time delay is constant at all frequencies in the passband.

As an illustration, Fig. 21-13a shows the time delay for an elliptic filter with $A_p = 3$ dB, $f_c = 1$ kHz, and $n = 10$. Notice how the time delay changes with frequency. Figure 21-13b shows the time delay of a Bessel filter with the same specifications. Notice how the time delay is constant through the passband and beyond. This is why the Bessel filter is sometimes referred to as a *maximally flat delay filter.* Constant time delay implies linear phase shift, and vice versa.

Figure 21–12 Step responses: (a) Butterworth and inverse Chebyshev; (b) Chebyshev and elliptic; (c) Bessel.

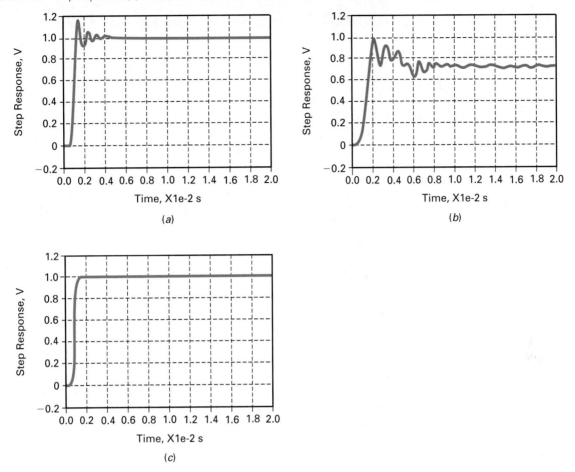

Figure 21–13 Time delays: (a) Elliptic; (b) Bessel.

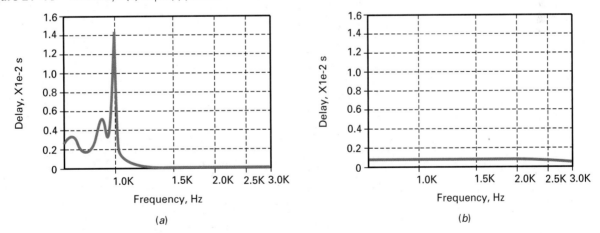

Roll-off of Different Approximations

The Butterworth roll-off rate is neatly summarized by Eqs. (21-4a) and (21-4b):

$$\text{Roll off} = 20n \quad \text{dB/decade}$$

$$\text{Roll off} = 6n \quad \text{dB/octave}$$

The Chebyshev, inverse Chebyshev, and elliptic approximations have faster roll-offs in the transition region, but the Bessel has a slower roll-off.

Table 21–1	Attenuation for Sixth–Order Approximations	
Type	**f_c, dB**	**$2f_c$, dB**
Bessel	3	14
Butterworth	3	36
Chebyshev	3	63
Inverse Chebyshev	3	63
Elliptic	3	93

The transition roll-off rates of non-Butterworth filters cannot be summarized with simple equations because the roll-offs are nonlinear and depend on the filter order, the ripple depth, and the other factors. Although we cannot write equations for these nonlinear roll-offs, we can compare the different roll-off rates in the transition region as follows.

Table 21-1 shows the attenuation for $n = 6$ and $A_p = 3$ dB. The filters have been ranked by their attenuations 1 octave above the edge frequency. The Bessel filter has the slowest roll-off, the Butterworth filter is next, and so on. All filters with rippled passbands or stopbands have transition roll-off rates that are faster than those of the Bessel and Butterworth filters, which have no ripples in their frequency response.

Other Types of Filters

Most of the preceding discussion applies to the high-pass, bandpass, and bandstop filters. The approximations for a high-pass filter are the same as those for a low-pass filter, except that the responses are rotated horizontally around the edge frequency. For instance, Fig. 21-14 shows the Butterworth response for a high-pass filter with $n = 6$, $A_p = 2.5$ dB, and $f_c = 1$ kHz. This is a mirror image of the low-pass response discussed earlier. The Chebyshev, inverse Chebyshev, elliptic, and Bessel high-pass responses are likewise mirror images of their low-pass counterparts.

Figure 21–14 Butterworth high-pass response.

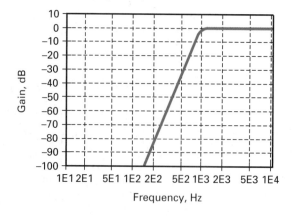

The bandpass responses are different. Here are the specifications used for the following examples: $n = 12$, $A_p = 3$ dB, $f_0 = 1$ kHz, and BW = 3 kHz. Figure 21-15a shows the Butterworth response. As expected, the passband is maximally flat and the stopband is monotonic. The Chebyshev response of Fig. 21-15b shows a rippled passband and a monotonic stopband. There are six passband ripples, half the order, which agrees with Eq. (21-5). Figure 21-15c is the response for an inverse Chebyshev filter. Here we see the flat passband and a rippled stopband. Figure 21-15d shows the elliptic response with its rippled passband and rippled stopband. Finally, Fig. 21-15e shows the Bessel response.

Figure 21–15 Bandpass responses: (a) Butterworth; (b) Chebyshev; (c) inverse Chebyshev; (d) elliptic; (e) Bessel.

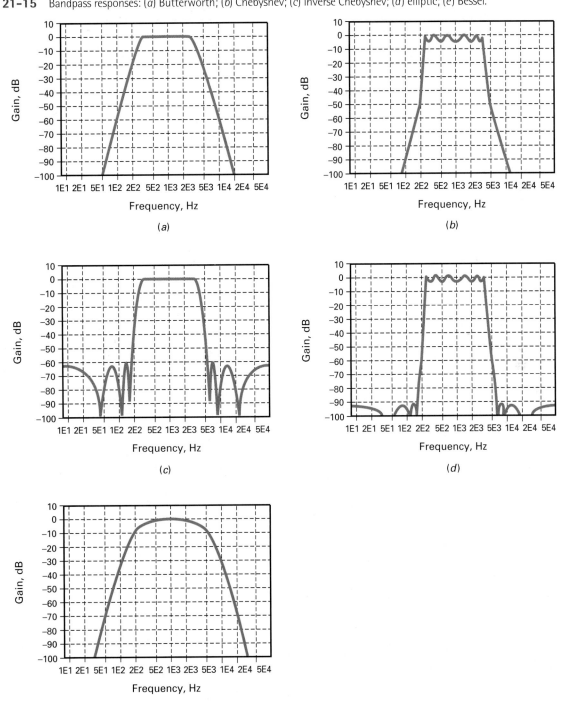

The bandstop responses are the opposite of the bandpass responses. Here are the bandstop responses for $n = 12$: $A_p = 3\text{B}$, $f_0 = 1$ kHz, and BW = 3 kHz. Figure 21-16a shows the Butterworth response. As expected, the passband is maximally flat and the stopband is monotonic. The Chebyshev response of Fig. 21-16b shows a rippled passband and a monotonic stopband. Figure 21-16c is the response for an inverse Chebyshev filter. Here we see a flat passband and a rippled stopband. Figure 21-16d shows the elliptic response with its rippled passband and rippled stopband. Finally, Fig. 21-16e shows the bandstop response for a Bessel filter.

Figure 21-16 Bandstop responses: (a) Butterworth; (b) Chebyshev; (c) inverse Chebyshev; (d) elliptic; (e) Bessel.

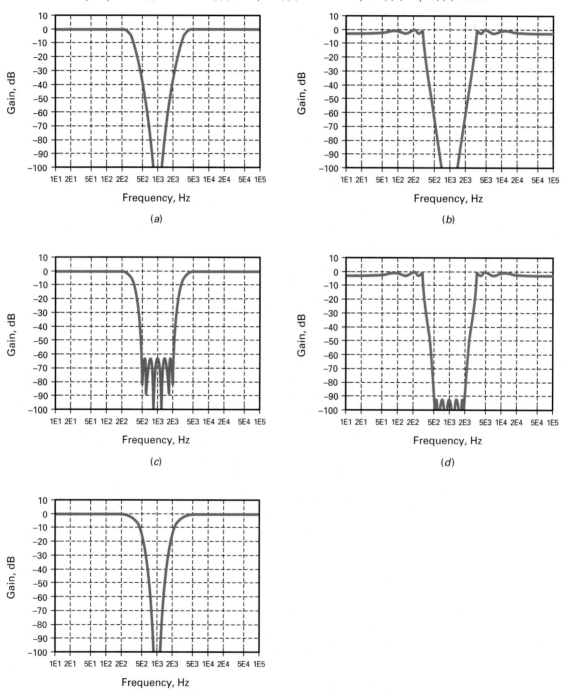

Table 21-2	Filter Approximations			
Type	Passband	Stopband	Roll-off	Step response
Butterworth	Flat	Monotonic	Good	Good
Chebyshev	Rippled	Monotonic	Very good	Poor
Inverse Chebyshev	Flat	Rippled	Very good	Good
Elliptic	Rippled	Rippled	Best	Poor
Bessel	Flat	Monotonic	Poor	Best

Conclusion

Table 21-2 summarizes the five approximations used in designing filters. Each has its advantages and disadvantages. When a flat passband is needed, the Butterworth and inverse Chebyshev filters are the logical candidates. The required roll-off, order, and other design considerations will then determine which of the two will be used.

If a rippled passband is acceptable, the Chebyshev and elliptic filters are the best candidates. Again, the required roll-off, order, and other design considerations will then determine the final choice.

When the step response is important, the Bessel filter is the logical candidate if it can meet the attenuation requirements. The Bessel approximation is the only one shown in the table that preserves the shape of a nonsinusoidal signal. This is critical in data communications because digital signals consist of positive and negative steps.

In applications in which a Bessel filter cannot provide sufficient attenuation, we can cascade an all-pass filter with a non-Bessel filter. When properly designed, the all-pass filter can linearize the overall phase response to get an almost perfect step response. A later section discusses this in more detail.

Op-amp circuits with resistors and capacitors can implement any of the five approximations. As we will see, many different circuits are available that offer a trade-off between complexity of design, sensitivity of components, and ease of tuning. For instance, some second-order circuits use only one op amp and a few components. But these simple circuits have cutoff frequencies that are heavily dependent on component tolerance and drift. Other second-order circuits may use three or more op amps, but these complex circuits are much less dependent on component tolerance and drift.

21-3 Passive Filters

Before discussing active-filter circuits, there are two more ideas that we need to explore. A second-order low-pass LC filter has a resonant frequency and a Q—similar to a series or parallel resonant circuit. By keeping the resonant frequency constant but varying the Q, we can get ripples to appear in the passband of higher-order filters. This section will describe the concept because it explains a great deal about the operation of active filters.

Figure 21-17 *Second-order LC filter.*

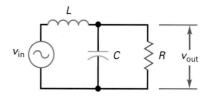

Resonant Frequency and *Q*

Figure 21-17 shows a low-pass *LC* filter. It has an order of 2 because it contains two reactive components, an inductor and a capacitor. A second-order *LC* filter has a resonant frequency and a *Q* defined as follows:

$$f_0 = \frac{1}{2\pi\sqrt{LC}} \tag{21-6}$$

$$Q = \frac{R}{X_L} \tag{21-7}$$

where X_L is calculated at the resonant frequency.

For instance, the filter of Fig. 21-18*a* has a resonant frequency and *Q* of:

$$f_0 = \frac{1}{2\pi\sqrt{(9.55 \text{ mH})(2.65 \text{ }\mu\text{F})}} = 1 \text{ kHz}$$

$$Q = \frac{600 \text{ }\Omega}{2\pi(1 \text{ kHz})(9.55 \text{ mH})} = 10$$

Figure 21-18*b* shows the frequency response. Notice how the response peaks at 1 kHz, the resonant frequency of the filter. Notice also how the voltage gain increases 20 dB at 1 kHz. The higher *Q* is, the greater the increase in voltage gain at the resonant frequency.

The filter of Fig. 21-18*c* has a resonant frequency and a *Q* of:

$$f_0 = \frac{1}{2\pi\sqrt{(47.7 \text{ mH})(531 \text{ nF})}} = 1 \text{ kHz}$$

$$Q = \frac{600 \text{ }\Omega}{2\pi(1 \text{ kHz})(47.7 \text{ mH})} = 2$$

In Fig. 21-18*c*, the inductance has been increased by a factor 5 and the capacitance has been decreased by a factor of 5 from the values of Fig. 21-18*a*. Because the *LC* product is the same, the resonant frequency is still 1 kHz.

On the other hand, *Q* has decreased by a factor of 5 since it is inversely proportional to inductance. Figure 21-18*d* shows the frequency response. Notice how the response again peaks at 1 kHz, but the increase in voltage gain is only 6 dB, a result of the lower *Q*.

If we continue to decrease *Q*, the resonant peak will disappear. For instance, the filter of Fig. 21-18*e* has:

$$f_0 = \frac{1}{2\pi\sqrt{(135 \text{ mH})(187 \text{ nF})}} = 1 \text{ kHz}$$

$$Q = \frac{600 \text{ }\Omega}{2\pi(1 \text{ kHz})(135 \text{ mH})} = 0.707$$

Figure 21-18*f* shows the frequency response, which is a Butterworth response. With *Q* of 0.707, the resonant peak disappears and the passband becomes maximally flat. Any second-order filter with a *Q* of 0.707 always has a Butterworth response.

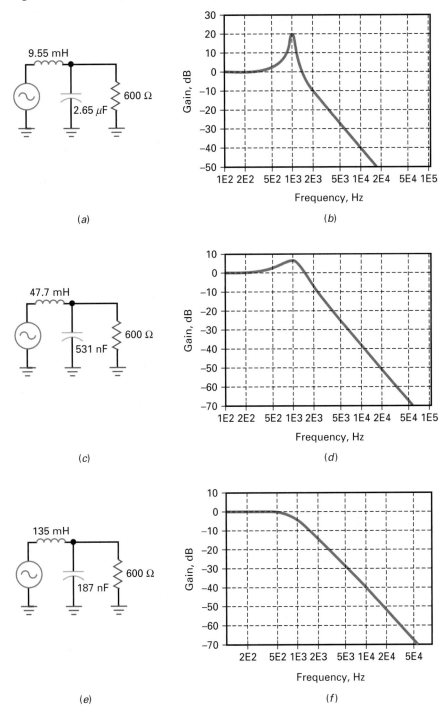

Figure 21-18 Examples.

Damping Factor

Another way of explaining the peaking action at resonance is to use the **damping factor,** defined as:

$$\alpha = \frac{1}{Q} \tag{21-8}$$

For $Q = 10$, the damping factor is:

$$\alpha = \frac{1}{10} = 0.1$$

Figure 21-19 Effect of Q on second-order response.

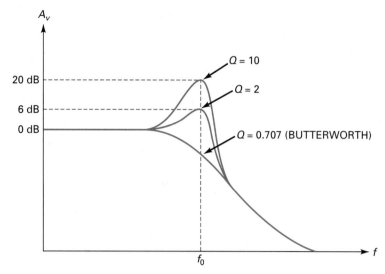

Similarly, a Q of 2 gives $\alpha = 0.5$, and a Q of 0.707 gives $\alpha = 1.414$.

Figure 21-18b has a low damping factor, only 0.1. In Fig. 21-18d, the damping factor increases to 0.5 and the resonant peak decreases. In Fig. 21-18f, the damping factor increases to 1.414 and the resonant peak disappears. As the word implies, *damping* means "reducing" or "diminishing." The higher the damping factor, the smaller the peak.

Butterworth and Chebyshev Responses

Figure 21-19 summarizes the effect of Q on a second-order filter. As indicated in Fig. 21-19, a Q of 0.707 produces the Butterworth or maximally flat response. A Q of 2 produces a ripple depth of 6 dB, and a Q of 10 produces a ripple depth of 20 dB. In terms of damping, the Butterworth response is *critically damped,* whereas the rippled responses are *underdamped.* A Bessel response (not shown) is *overdamped* because its Q equals 0.577.

Higher-Order *LC* Filters

Higher-order filters are usually built by cascading second-order stages. For example, Fig. 21-20 shows a Chebyshev filter with an edge frequency of 1 kHz and a ripple depth of 1 dB. The filter consists of three second-order stages, which means that the overall filter has an order of 6. Since $n = 6$, the filter has three passband ripples.

Notice how each stage has its own resonant frequency and Q. The staggered resonant frequencies produce the three ripples in the passband. The staggered Qs maintain a ripple depth of 1 dB by producing peaks at frequencies at which other stages have rolled off. For instance, the second stage has a resonant frequency of 747 Hz. At this frequency, the first stage has rolled off because its cutoff frequency is 353 Hz. The second stage compensates for the roll-off in the

Figure 21-20 Staggered resonant frequencies and Qs in higher-order filter.

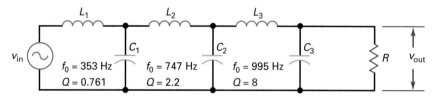

first stage by producing a resonant peak at 747 Hz. Similarly, the third stage has a cutoff frequency of 995 Hz. At this frequency, the first and second stages have rolled off, but the third stage compensates for their roll-offs by producing a high-Q peak at 995 Hz.

The idea of staggering the resonant frequencies and Qs of second-order stages applies to active filters as well as to passive filters. In other words, to build a high-order active filter, we can cascade second-order stages whose resonant frequencies and Qs are staggered in precisely the right way to get the desired overall response.

21-4 First-Order Stages

First-order or 1-pole active-filter stages have only one capacitor. Because of this, they can produce only a low-pass or a high-pass response. Bandpass and bandstop filters can be implemented only when n is greater than 1.

Low-Pass Stage

Figure 21-21a shows the simplest way to build a first-order low-pass active filter. It is nothing more than an RC lag circuit and a voltage follower. The voltage gain is:

$$A_v = 1$$

Figure 21-21   First-order low-pass stages: (a) Noninverting unity gain; (b) noninverting with voltage gain; (c) inverting with voltage gain.

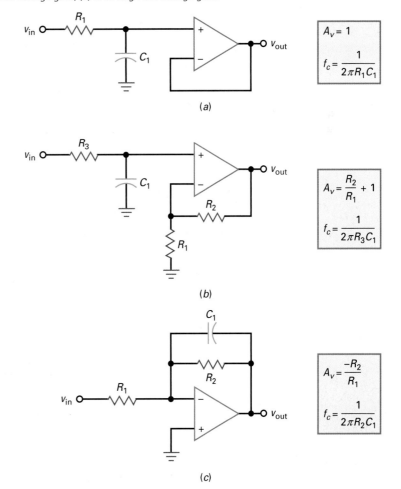

(a)

(b)

(c)

The 3-dB cutoff frequency is given by:

$$f_c = \frac{1}{2\pi R_1 C_1} \tag{21-9}$$

When the frequency increases above the cutoff frequency, the capacitive reactance decreases and reduces the noninverting input voltage. Since the $R_1 C_1$ lag circuit is outside the feedback loop, the output voltage rolls off. As the frequency approaches infinity, the capacitor becomes a short and there is zero input voltage.

Figure 21-21b shows another noninverting first-order low-pass filter. Although it has two additional resistors, it has the advantage of voltage gain. The voltage gain well below the cutoff frequency is given by:

$$A_v = \frac{R_2}{R_1} + 1 \tag{21-10}$$

The cutoff frequency is given by:

$$f_c = \frac{1}{2\pi R_3 C_1} \tag{21-11}$$

Above the cutoff frequency, the lag circuit reduces the noninverting input voltage. Since the $R_3 C_1$ lag circuit is outside the feedback loop, the output voltage rolls off at a rate of 20 dB per decade.

Figure 21-21c shows an inverting first-order low-pass filter and its equations. At low frequencies, the capacitor appears to be open and the circuit acts like an inverting amplifier with a voltage gain of:

$$A_v = \frac{-R_2}{R_1} \tag{21-12}$$

As the frequency increases, the capacitive reactance decreases and reduces the impedance of the feedback branch. This implies less voltage gain. As the frequency approaches infinity, the capacitor becomes a short and there is no voltage gain. As shown in Fig. 21-21c, the cutoff frequency is given by:

$$f_c = \frac{1}{2\pi R_2 C_1} \tag{21-13}$$

There is no other way to implement a first-order low-pass filter. In other words, the circuits shown in Fig. 21-21 are the only three configurations available for an active-filter low-pass stage.

A final point about all first-order stages. They can implement only a Butterworth response. The reason is that a first-order stage has no resonant frequency. Therefore, it cannot produce the peaking that produces a rippled passband. This means that all first-order stages are maximally flat in the passband and monotonic in the stopband, and they roll off at a rate of 20 dB per decade.

High-Pass Stage

Figure 21-22a shows the simplest way to build a first-order high-pass active filter. The voltage gain is:

$$A_v = 1$$

The 3-dB cutoff frequency is given by:

$$f_c = \frac{1}{2\pi R_1 C_1} \tag{21-14}$$

When the frequency decreases below the cutoff frequency, the capacitive reactance increases and reduces the noninverting input voltage. Since the $R_1 C_1$ circuit is

Figure 21-22 First-order high-pass stages: (*a*) Noninverting unity gain; (*b*) noninverting with voltage gain; (*c*) inverting with voltage gain.

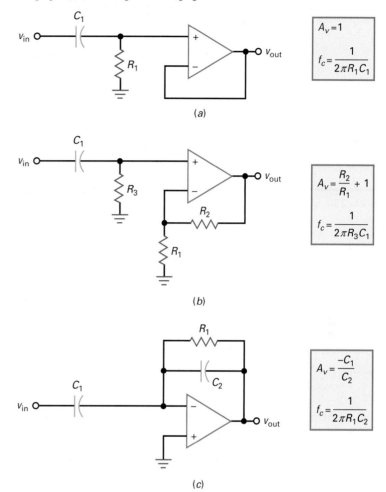

(a)

(b)

(c)

outside the feedback loop, the output voltage rolls off. As the frequency approaches zero, the capacitor becomes an open and there is zero input voltage.

Figure 21-22*b* shows another noninverting first-order high-pass filter. The voltage gain well above the cutoff frequency is given by:

$$A_v = \frac{R_2}{R_1} + 1 \tag{21-15}$$

The 3-dB cutoff frequency is given by:

$$f_c = \frac{1}{2\pi R_3 C_1} \tag{21-16}$$

Well below the cutoff frequency, the *RC* circuit reduces the noninverting input voltage. Since the $R_3 C_1$ lag circuit is outside the feedback loop, the output voltage rolls off at a rate of 20 dB per decade.

Figure 21-22*c* shows another first-order high-pass filter and its equations. At high frequencies, the circuit acts like an inverting amplifier with a voltage gain of:

$$A_v = \frac{-X_{C2}}{X_{C1}} = \frac{-C_1}{C_2} \tag{21-17}$$

As the frequency decreases, the capacitive reactances increase and eventually reduce the input signal and the feedback. This implies less voltage gain. As the frequency approaches zero, the capacitors become open and there is no input signal. As shown in Fig. 21-22c, the 3-dB cutoff frequency is given by:

$$f_c = \frac{1}{2\pi R_1 C_2} \tag{21-18}$$

Example 21–1

What is the voltage gain in Fig. 21-23a? What is the cutoff frequency? What is the frequency response?

Figure 21–23 Example.

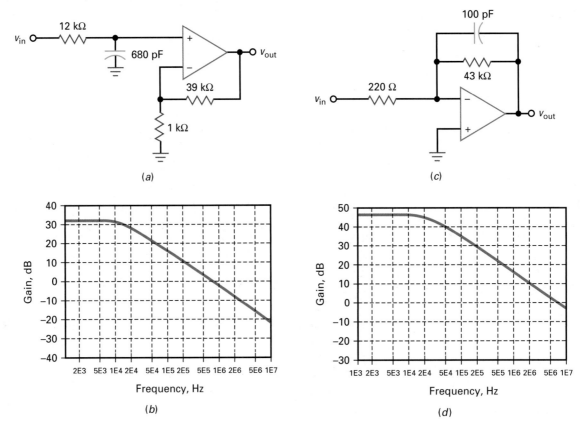

(a)

(b)

(c)

(d)

SOLUTION   This is a noninverting first-order low-pass filter. With Eqs. (21-10) and (21-11), the voltage gain and cutoff frequencies are:

$$A_v = \frac{39\ k\Omega}{1\ k\Omega} + 1 = 40$$

$$f_c = \frac{1}{2\pi(12\ k\Omega)(680\ pF)} = 19.5\ kHz$$

Figure 21-23b shows the frequency response. The voltage gain is 32 dB in the passband. The response breaks at 19.5 kHz and then rolls off at a rate of 20 dB per decade.

PRACTICE PROBLEM 21–1   Using Fig. 21-23a, change the 12 kΩ resistor to 6.8 kΩ. Find the new cutoff frequency.

Example 21-2

What is the voltage gain in Fig. 21-23c? What is the cutoff frequency? What is the frequency response?

SOLUTION This is an inverting first-order low-pass filter. With Eqs. (21-12) and (21-13), the voltage gain and cutoff frequencies are:

$$A_v = \frac{-43 \text{ k}\Omega}{220 \text{ }\Omega} = -195$$

$$f_c = \frac{1}{2\pi(43 \text{ k}\Omega)(100 \text{ pF})} = 37 \text{ kHz}$$

Figure 21-23d shows the frequency response. The voltage gain is 45.8 dB in the passband. The response breaks at 37 kHz and then rolls off at a rate of 20 dB per decade.

PRACTICE PROBLEM 21-2 In Fig. 21-23c, change the 100 pF capacitor to 220 pF. What is the new cutoff frequency?

21-5 VCVS Unity-Gain Second-Order Low-Pass Filters

Second-order or 2-pole stages are the most common because they are easy to build and analyze. Higher-order filters are usually made by cascading second-order stages. Each second-order stage has a resonant frequency and a Q to determine how much peaking occurs.

This section discusses the **Sallen-Key low-pass filters** (named after the inventors). These filters are also called *VCVS filters* because the op amp is used as a voltage-controlled voltage source. VCVS low-pass circuits can implement three of the basic approximations: Butterworth, Chebyshev, and Bessel.

Circuit Implementation

Figure 21-24 shows a Sallen-Key second-order low-pass filter. Notice that the two resistors have the same value, but the two capacitors are different. There is a lag

Figure 21-24 Second-order VCVS stage for Butterworth and Bessel.

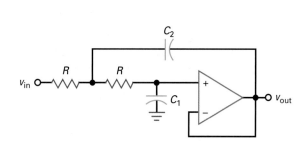

$A_v = 1$

$$Q = 0.5\sqrt{\frac{C_2}{C_1}}$$

$$f_p = \frac{1}{2\pi R\sqrt{C_1 C_2}}$$

Butterworth:
$Q = 0.707$
$K_c = 1$

Bessel:
$Q = 0.577$
$K_c = 0.786$

circuit on the noninverting input, but this time there is a feedback path through a second capacitor C_2. At low frequencies, both capacitors appear to be open and the circuit has a unity gain because the op amp is connected as a voltage follower.

As the frequency increases, the impedance of C_1 decreases and the noninverting input voltage decreases. At the same time, capacitor C_2 is feeding back a signal that is in phase with the input signal. Since the feedback signal adds to the source signal, the feedback is *positive*. As a result, the decrease in the noninverting input voltage caused by C_1 is not as large as it would be without the positive feedback.

The larger C_2 is with respect to C_1, the more the positive feedback; this is equivalent to increasing the Q of the circuit. If C_2 is large enough to make Q greater than 0.707, peaking appears in the frequency response.

Pole Frequency

As shown in Fig. 21-24:

$$Q = 0.5\sqrt{\frac{C_2}{C_1}} \tag{21-19}$$

and

$$f_p = \frac{1}{2\pi R\sqrt{C_1 C_2}} \tag{21-20}$$

The **pole frequency** (f_p) is a special frequency used in the design of active filters. The mathematics behind the pole frequency is too complicated to go into here because it involves an advanced topic called the *s plane*. Advanced courses analyze and design filters using the *s plane*. (*Note: s* is a complex number given by $\sigma + j\omega$.)

For our needs, it is enough to understand how to calculate the pole frequency. In more complicated circuits, the pole frequency is given by:

$$f_p = \frac{1}{2\pi\sqrt{R_1 R_2 C_1 C_2}}$$

In a Sallen-Key unity-gain filter, $R_1 = R_2$ and the equation simplifies to Eq. (21-20).

Butterworth and Bessel Responses

When analyzing a circuit like the one shown in Fig. 21-24, we start by calculating Q and f_p. If $Q = 0.707$, we have a Butterworth response and a K_c value of 1. If $Q = 0.577$, we have a Bessel response and a K_c value of 0.786. Next, we can calculate the cutoff frequency with:

$$f_c = K_c f_p \tag{21-21}$$

With Butterworth and Bessel filters, the cutoff frequency is always the frequency at which the attenuation is 3 dB.

Peaked Response

Figure 21-25 shows how to analyze the circuit when Q is greater than 0.707. After calculating the Q and the pole frequency of the circuit, we can calculate three other frequencies with these equations:

$$f_0 = K_0 f_p \tag{21-22}$$

$$f_c = K_c f_p \tag{21-23}$$

$$f_{3dB} = K_3 f_p \tag{21-24}$$

The first of these frequencies is the resonant frequency where peaking appears. The second is the edge frequency, and the third is the 3-dB frequency.

Figure 21-25 Second-order VCVS stage for $Q > 0.707$.

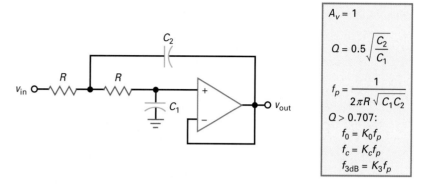

$$A_v = 1$$

$$Q = 0.5\sqrt{\frac{C_2}{C_1}}$$

$$f_p = \frac{1}{2\pi R\sqrt{C_1 C_2}}$$

$Q > 0.707$:
$$f_0 = K_0 f_p$$
$$f_c = K_c f_p$$
$$f_{3dB} = K_3 f_p$$

Table 21-3 shows the K and A_p values versus Q. The Bessel and Butterworth values appear first. Because these responses have no noticeable resonant frequency, the K_0 and A_p values do not apply. When Q is greater than 0.707, a noticeable resonant frequency appears and all K and A_p values are present. By

Table 21-3		K Values and Ripple Depth of Second-Order Stages		
Q	**K_0**	**K_c**	**K_3**	**A_p(dB)**
0.577	—	0.786	1	—
0.707	—	1	1	—
0.75	0.333	0.471	1.057	0.054
0.8	0.467	0.661	1.115	0.213
0.9	0.620	0.874	1.206	0.688
1	0.708	1.000	1.272	1.25
2	0.935	1.322	1.485	6.3
3	0.972	1.374	1.523	9.66
4	0.984	1.391	1.537	12.1
5	0.990	1.400	1.543	14
6	0.992	1.402	1.546	15.6
7	0.994	1.404	1.548	16.9
8	0.995	1.406	1.549	18
9	0.997	1.408	1.550	19
10	0.998	1.410	1.551	20
100	1.000	1.414	1.554	40

Figure 21-26 (*a*) *K* values versus *Q*; (*b*) peaking versus *Q*.

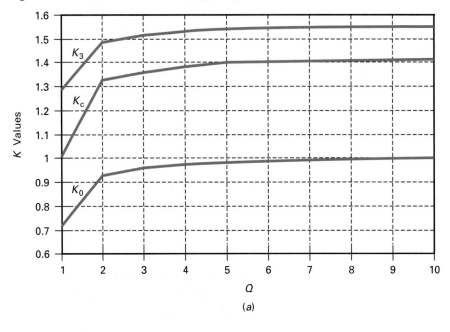

(*a*)

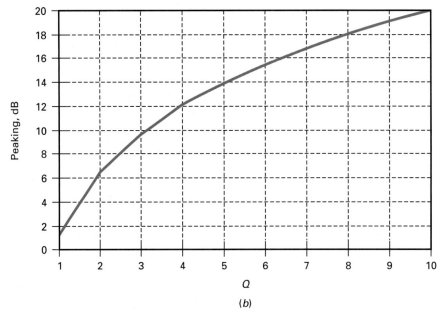

(*b*)

plotting the values of Table 21-3, we get Fig. 21-26*a* and *b*. We can use the table for integral values of *Q* and the graphs for intermediate values of *Q*. For instance, if we calculate a *Q* of 5, we can read the following approximate values from either Table 21-3 or Fig. 21-26: $K_0 = 0.99$, $K_c = 1.4$, $K_3 = 1.54$, and $A_p = 14$ dB.

In Fig. 21-26*a*, notice how the *K* values level off as *Q* approaches 10. For *Q* greater than 10, we will use these approximations:

$$K_0 = 1 \tag{21-25}$$

$$K_c = 1.414 \tag{21-26}$$

$$K_3 = 1.55 \tag{21-27}$$

$$A_p = 20 \log Q \tag{21-28}$$

The values shown in Table 21-3 and Fig. 21-26 apply to all second-order low-pass stages.

Gain-Bandwidth Product of Op Amps

In all our discussions about active filters, we will assume that the op amps have enough *gain-bandwidth product (GBW)* not to affect filter performance. Limited GBW increases the Q of a stage. With high cutoff frequencies, a designer must be aware of limited GBW because it may change the performance of a filter.

One way to correct for limited GBW is by means of **predistortion.** This refers to decreasing the design value of Q as needed to compensate for limited GBW. For instance, if a stage should have a Q of 10 and a limited GBW increases it to 11, a designer can predistort by designing the stage with a Q of 9.1. The limited GBW will increase 9.1 to 10. Designers try to avoid predistortion because low-Q and high-Q stages sometimes interact adversely. The best approach is to use a better op amp, one with a higher GBW (same as f_{unity}).

Example 21–3 IIII MultiSim

What are the pole frequency and Q of the filter shown in Fig. 21-27? What is the cutoff frequency?

Figure 21-27 Butterworth unity-gain example.

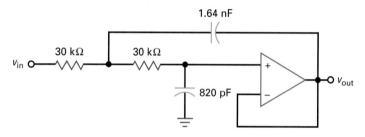

SOLUTION The Q and pole frequency are:

$$Q = 0.5 \sqrt{\frac{C_2}{C_1}} = 0.5 \sqrt{\frac{1.64 \text{ nF}}{820 \text{ pF}}} = 0.707$$

$$f_p = \frac{1}{2\pi R \sqrt{C_1 C_2}} = \frac{1}{2\pi (30 \text{ k}\Omega) \sqrt{(820 \text{ pF})(1.64 \text{ nF})}} = 4.58 \text{ kHz}$$

The Q value of 0.707 tells us that this is a Butterworth response, so the cutoff frequency is the same as the pole frequency:

$$f_c = f_p = 4.58 \text{ kHz}$$

The response of the filter breaks at 4.58 kHz and rolls off at a rate of 40 dB per decade because $n = 2$.

PRACTICE PROBLEM 21–3 Repeat Example 21-3 with the resistor values changed to 10 kΩ.

Example 21–4

In Fig. 21-28, what are the pole frequency and Q? What is the cutoff frequency?

Figure 21–28 Bessel unity-gain example.

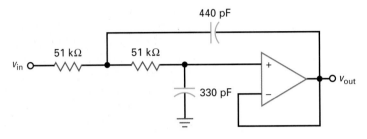

SOLUTION The Q and pole frequency are:

$$Q = 0.5 \sqrt{\frac{C_2}{C_1}} = 0.5 \sqrt{\frac{440 \text{ pF}}{330 \text{ pF}}} = 0.577$$

$$f_p = \frac{1}{2\pi R \sqrt{C_1 C_2}} = \frac{1}{2\pi(51 \text{ k}\Omega)\sqrt{(330 \text{ pF})(440 \text{ pF})}} = 8.19 \text{ kHz}$$

The Q value of 0.577 tells us that this is a Bessel response. With Eq. (21-21), the cutoff frequency is given by:

$$f_c = K_c f_p = 0.786(8.19 \text{ kHz}) = 6.44 \text{ kHz}$$

PRACTICE PROBLEM 21–4 In Example 21-4, if the value of C_1 changed to 680 pF, what value should C_2 be to maintain a Q of 0.577?

Example 21–5

What are the pole frequency and Q in Fig. 21-29? What are the cutoff and 3-dB frequencies?

Figure 21–29 Unity-gain example with $Q > 0.707$.

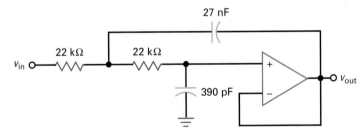

SOLUTION The Q and pole frequency are:

$$Q = 0.5 \sqrt{\frac{C_2}{C_1}} = 0.5 \sqrt{\frac{27 \text{ nF}}{390 \text{ pF}}} = 4.16$$

$$f_p = \frac{1}{2\pi R \sqrt{C_1 C_2}} = \frac{1}{2\pi(22 \text{ k}\Omega)\sqrt{(390 \text{ pF})(27 \text{ pF})}} = 2.23 \text{ kHz}$$

Referring to Fig. 21-26, we can read the following approximate K and A_p values:

$$K_0 = 0.99$$

$$K_c = 1.38$$

$$K_3 = 1.54$$

$$A_p = 12.5 \text{ dB}$$

The cutoff or edge frequency is:

$$f_c = K_c f_p = 1.38(2.23 \text{ kHz}) = 3.08 \text{ kHz}$$

and the 3-dB frequency is:

$$f_{3dB} = K_3 f_p = 1.54(2.23 \text{ kHz}) = 3.43 \text{ kHz}$$

PRACTICE PROBLEM 21-5 In Fig. 21-29, change the 27 nF capacitor to 14 nF and repeat Example 21.5.

21-6 Higher-Order Filters

The standard approach in building higher-order filters is to cascade first- and second-order stages. When the order is even, we need to cascade only second-order stages. When the order is odd, we need to cascade second-order stages and a single first-order stage. For instance, if we want to build a sixth-order filter, we can cascade three second-order stages. If we want to build a fifth-order filter, we can cascade two second-order stages and one first-order stage.

Butterworth Filters

When filter stages are cascaded, we can add the decibel attenuation of each stage to get the total attenuation. For instance, Fig. 21-30a shows two cascaded second-order stages. If each has a Q of 0.707 and a pole frequency of 1 kHz, then each stage has a Butterworth response with an attenuation of 3 dB at 1 kHz. Although

Figure 21-30 (a) Cascading two stages; (b) equal stages produce a droop at the cutoff frequency; (c) low-Q and high-Q stages compensate to produce Butterworth response.

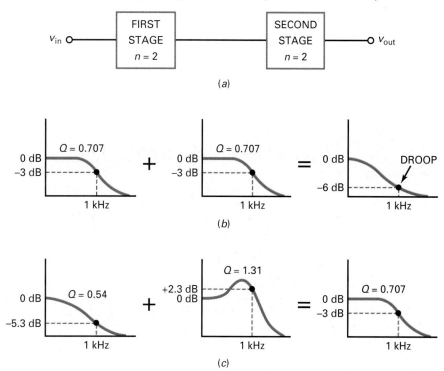

Table 21–4	Staggered Qs for Butterworth Low-Pass Filters				
Order	Stage 1	Stage 2	Stage 3	Stage 4	Stage 5
2	0.707				
4	0.54	1.31			
6	0.52	1.93	0.707		
8	0.51	2.56	0.6	0.9	
10	0.51	3.2	0.56	1.1	0.707

each stage has a Butterworth response, the overall response is not a Butterworth response because it droops at the pole frequency, as shown in Fig. 21-30b. Since each stage has an attenuation of 3 dB at the cutoff frequency of 1 kHz, the overall attenuation is 6 dB at 1 kHz.

To get a Butterworth response, the pole frequencies are still 1 kHz, but the Qs of the stages have to be staggered above and below 0.707. Figure 21-30c shows how to get a Butterworth response for the overall filter. The first stage has $Q = 0.54$, and the second stage has $Q = 1.31$. The peaking in the second stage offsets the droop in the first stage to get an attenuation of 3 dB at 1 kHz. Furthermore, it can be shown that the passband response is maximally flat with these Q values.

Table 21-4 shows the staggered Q values of the stages used in higher-order Butterworth filters. All the stages have the same pole frequency, but each stage has a different Q. For instance, the fourth-order filter described by Fig. 21-30c uses Q values of 0.54 and 1.31, the same values shown in Table 21-4. To build a tenth-order Butterworth filter, we would need five stages with Q values of 0.51, 3.2, 0.56, 1.1, and 0.707.

Bessel Filters

With higher-order Bessel filters, we need to stagger both the Qs and the pole frequencies of the stages. Table 21-5 shows the Q and f_p for each stage in a filter with a cutoff frequency of 1000 Hz. For instance, a fourth-order Bessel filter needs a

Table 21–5	Staggered Qs and Pole Frequencies for Bessel Low-Pass Filters (f_c = 1000 Hz)									
Order	Q_1	f_{p1}	Q_2	f_{p2}	Q_3	f_{p3}	Q_4	f_{p4}	Q_5	f_{p5}
2	0.577	1274								
4	0.52	1432	0.81	1606						
6	0.51	1607	1.02	1908	0.61	1692				
8	0.51	1781	1.23	2192	0.71	1956	0.56	1835		
10	0.50	1946	1.42	2455	0.81	2207	0.62	2066	0.54	1984

Order	A_p, dB	Q_1	f_{p1}	Q_2	f_{p2}	Q_3	f_{p3}	Q_4	f_{p4}
2	1	0.96	1050						
	2	1.13	907						
	3	1.3	841						
4	1	0.78	529	3.56	993				
	2	0.93	471	4.59	964				
	3	1.08	443	5.58	950				
6	1	0.76	353	8	995	2.2	747		
	2	0.9	316	10.7	983	2.84	730		
	3	1.04	298	12.8	977	3.46	722		
8	1	0.75	265	14.2	997	4.27	851	1.96	584
	2	0.89	238	18.7	990	5.58	842	2.53	572
	3	1.03	224	22.9	987	6.83	839	3.08	566

first stage with $Q = 0.52$ and $f_p = 1432$ Hz, and a second stage with $Q = 0.81$ and $f_p = 1606$ Hz.

If the frequency is different from 1000 Hz, the pole frequencies in Table 21-5 are scaled in direct proportion by a **frequency scaling factor (FSF)** of:

$$\text{FSF} = \frac{f_c}{1 \text{ kHz}}$$

For instance, if a sixth-order Bessel filter has a cutoff frequency of 7.5 kHz, we would multiply each pole frequency in Table 21-5 by 7.5.

Chebyshev Filters

With Chebyshev filters, we have to stagger Q and f_p. Furthermore, we have to include the ripple depth. Table 21-6 shows the Q and f_p for each stage of a Chebyshev filter. As an example, a sixth-order Chebyshev filter with a ripple depth of 2 dB needs a first stage with $Q = 0.9$ and $f_p = 316$ Hz. The second stage must have $Q = 10.7$ and $f_p = 983$ Hz, and a third stage needs $Q = 2.84$ and $f_p = 730$ Hz.

Filter Design

The foregoing discussion gives you the basic idea behind the design of higher-order filters. So far, we have discussed only the simplest circuit implementation, which is the Sallen-Key unity-gain second-order stage. By cascading Sallen-Key unity-gain stages with staggered Qs and pole frequencies, we can implement higher-order filters for the Butterworth, Bessel, and Chebyshev approximations.

The tables shown earlier indicate how the Qs and pole frequencies need to be staggered in different designs. Larger, comprehensive tables are available in filter handbooks. The design of active filters is very complicated, especially when filters need to be designed with orders up to 20 and trade-offs are made between circuit complexity, component sensitivity, and ease of tuning.

Which brings us to an important point: All serious filter design is done on computers because the calculations are too difficult and time-consuming to attempt by hand. An active-filter computer program stores all the equations, tables, and circuits needed to implement the five approximations discussed earlier (Butterworth, Chebyshev, inverse Chebyshev, elliptic, and Bessel). The circuits used to build filters range from a simple one op-amp stage to complex five op-amp stages.

21-7 VCVS Equal-Component Low-Pass Filters

Figure 21-31 shows another Sallen-Key second-order low-pass filter. This time, both resistors and both capacitors have the same value. This is why the circuit is called a **Sallen-Key equal-component filter.** The circuit has a midband voltage gain of:

$$A_v = \frac{R_2}{R_1} + 1 \tag{21-29}$$

The operation of the circuit is similar to that of Sallen-Key unity-gain filter, except for the effect of the voltage gain. Since the voltage gain can produce more positive feedback through the feedback capacitor, the Q of the stage becomes a function of voltage gain and is given by:

$$Q = \frac{1}{3 - A_v} \tag{21-30}$$

Because A_v can be no smaller than unity, the minimum Q is 0.5. When A_v increases from 1 to 3, Q varies from 0.5 to infinity. Therefore, the allowable range of A_v is between 1 and 3. If we try to run the circuit with A_v greater than 3, it will break into oscillations because the positive feedback is too large. In fact, it is dangerous to use a voltage gain that even approaches 3 because component tolerance and drift may cause the voltage gain to exceed 3. A later example will bring this point out more clearly.

After we calculate A_v, Q, and f_p with the equations shown in Fig. 21-31, the rest of the analysis is the same as before because a Butterworth filter has $Q = 0.707$ and $K_c = 1$. A Bessel filter has $Q = 0.577$ and $K_c = 0.786$. For other Qs, we can get the approximate K and A_p values by interpolating from Table 21-3 or by using Fig. 21-26.

Figure 21-31 VCVS equal-component stage.

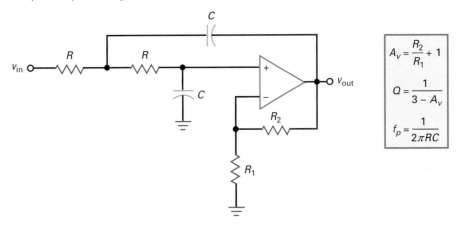

Example 21-6

What are the pole frequency and Q of the filter shown in Fig. 21-32? What is the cutoff frequency?

Figure 21-32 Butterworth equal-component example.

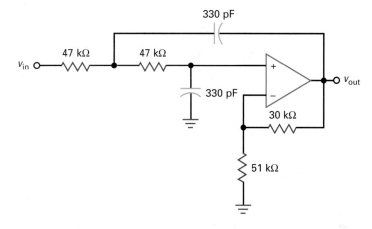

SOLUTION The A_v, Q, and f_p are:

$$A_v = \frac{30 \text{ k}\Omega}{51 \text{ k}\Omega} + 1 = 1.59$$

$$Q = \frac{1}{3 - A_v} = \frac{1}{3 - 1.59} = 0.709$$

$$f_p = \frac{1}{2\pi RC} = \frac{1}{2\pi(47 \text{ k}\Omega)(330 \text{ pF})} = 10.3 \text{ kHz}$$

It takes a Q of 0.77 to produce a ripple of 0.1 dB. Therefore, a Q of 0.709 produces a ripple of less than 0.003 dB. For all practical purposes, the calculated Q of 0.709 means that we have a Butterworth response to a very close approximation.

The cutoff frequency of a Butterworth filter is equal to the pole frequency of 10.3 kHz.

PRACTICE PROBLEM 21-6 In Example 21-6, change the 47 kΩ resistors to 22 kΩ and solve for A_v, Q, and f_p.

Example 21-7

In Fig. 21-33, what are the pole frequency and Q? What is the cutoff frequency?

SOLUTION The A_v, Q, and f_p are:

$$A_v = \frac{15 \text{ k}\Omega}{56 \text{ k}\Omega} + 1 = 1.27$$

$$Q = \frac{1}{3 - A_v} = \frac{1}{3 - 1.27} = 0.578$$

$$f_p = \frac{1}{2\pi RC} = \frac{1}{2\pi(82 \text{ k}\Omega)(100 \text{ pF})} = 19.4 \text{ kHz}$$

Figure 21-33   Bessel equal-component example.

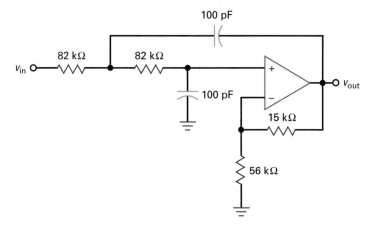

This is the Q of a Bessel second-order response. Therefore, $K_c = 0.786$ and the cutoff frequency is:

$$f_c = 0.786 f_p = 0.786(19.4 \text{ kHz}) = 15.2 \text{ kHz}$$

PRACTICE PROBLEM 21-7   Repeat Example 21-7 with the capacitors equal to 330 pF and the R value set to 100 kΩ.

Example 21-8

What are the pole frequency and Q in Fig. 21-34? What are the resonant, cutoff, and 3-dB frequencies? What is the ripple depth in decibels?

SOLUTION   The A_v, Q, and f_p are:

$$A_v = \frac{39 \text{ k}\Omega}{20 \text{ k}\Omega} + 1 = 2.95$$

$$Q = \frac{1}{3 - A_v} = \frac{1}{3 - 2.95} = 20$$

$$f_p = \frac{1}{2\pi RC} = \frac{1}{2\pi(56 \text{ k}\Omega)(220 \text{ pF})} = 12.9 \text{ kHz}$$

Figure 21-34   Equal-component example with $Q > 0.707$.

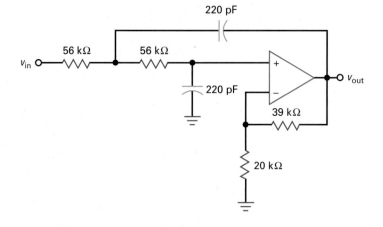

Figure 21-26 has Qs only between 1 and 10. In this case, we need to use Eqs. (21-25) to (21-28) to get the K and Q values:

$$K_0 = 1$$

$$K_c = 1.414$$

$$K_3 = 1.55$$

$$A_p = 20 \log Q = 20 \log 20 = 26 \text{ dB}$$

The resonant frequency is:

$$f_0 = K_0 f_p = 12.9 \text{ kHz}$$

The cutoff or edge frequency is:

$$f_c = K_c f_p = 1.414 \, (12.9 \text{ kHz}) = 18.2 \text{ kHz}$$

and the 3-dB frequency is:

$$f_{3dB} = K_3 f_p = 1.55(12.9 \text{ kHz}) = 20 \text{ kHz}$$

The circuit produces a 26-dB peak in the response at 12.9 kHz, rolls off to 0 dB at the cutoff frequency, and is down 3 dB at the 20 kHz.

A Sallen-Key circuit like this is impractical because the Q is too high. Since the voltage gain is 2.95, any error in the values of R_1 and R_2 can cause large increases in Q. For instance, if the tolerance of the resistors is ± 1 percent, the voltage gain can be as high as:

$$A_v = \frac{1.01(39 \text{ k}\Omega)}{0.99(20 \text{ k}\Omega)} + 1 = 2.989$$

This voltage gain produces a Q of:

$$Q = \frac{1}{3 - A_v} = \frac{1}{3 - 2.989} = 90.9$$

The Q has changed from a design value of 20 to an approximate value of 90.9, which means that the frequency response is radically different from the intended response.

Even though the Sallen-Key equal-component filter is simple compared to other filters, it has the disadvantage of component sensitivity when high Qs are used. This is why more complicated circuits are typically used for high-Q stages. The added complexity reduces the component sensitivity.

21-8 VCVS High–Pass Filters

Figure 21-35a shows the Sallen-Key unity-gain high-pass filter and its equations. Notice that the positions of resistors and capacitors have been reversed. Also notice that Q depends on the ratio of resistances rather than capacitances. The calculations are similar to those discussed for low-pass filters, except that we have to divide the pole frequency by the K value. To calculate the cutoff frequency of a high-pass filter, we use:

$$f_c = \frac{f_p}{K_c} \tag{21-31}$$

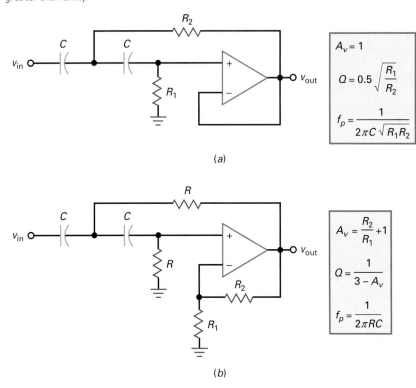

Figure 21-35 Second-order VCVS high-pass stages: (*a*) Unity gain; (*b*) voltage gain greater than unity.

(*a*)

(*b*)

Similarly, we divide the pole frequency by K_0 or K_3 for the other frequencies. For instance, if the pole frequency is 2.5 kHz and we read $K_c = 1.3$ in Fig. 21-26, the cutoff frequency for the high-pass filter is:

$$f_c = \frac{2.5 \text{ kHz}}{1.3} = 1.92 \text{ kHz}$$

Figure 21-35*b* shows the Sallen-Key equal-component high-pass filter and its equations. All the equations are the same as for a low-pass filter. The positions of the resistors and capacitors have been reversed. The following examples show you how to analyze high-pass filters.

Example 21-9

What are the pole frequency and Q of the filter shown in Fig. 21-36? What is the cutoff frequency?

SOLUTION   The Q and pole frequency are:

$$Q = 0.5 \sqrt{\frac{R_1}{R_2}} = 0.5 \sqrt{\frac{24 \text{ k}\Omega}{12 \text{ k}\Omega}} = 0.707$$

$$f_p = \frac{1}{2\pi C \sqrt{R_1 R_2}} = \frac{1}{2\pi(4.7 \text{ nF}) \sqrt{(24 \text{ k}\Omega)(12 \text{ k}\Omega)}} = 2 \text{ kHz}$$

Figure 21–36 High-pass Butterworth example.

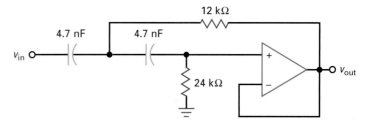

Since $Q = 0.707$, the filter has a Butterworth second-order response and:

$$f_c = f_p = 2 \text{ kHz}$$

The filter has a high-pass response with a break at 2 kHz, and it rolls off at 40 dB per decade below 2 kHz.

PRACTICE PROBLEM 21–9 In Fig. 21-36, double the two resistor values. Find the circuit's $Q, f_p,$ and f_c values.

Example 21–10

What are the pole frequency and Q in Fig. 21-37? What are the resonant, cutoff, and 3-dB frequencies? What is the ripple depth or peaking in decibels?

Figure 21–37 High-pass example with $Q > 1$.

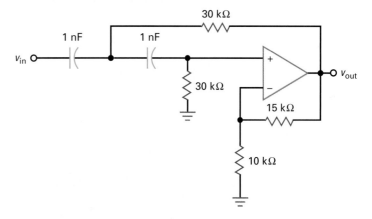

SOLUTION The A_v, Q, and f_p are:

$$A_v = \frac{15 \text{ k}\Omega}{10 \text{ k}\Omega} + 1 = 2.5$$

$$Q = \frac{1}{3 - A_v} = \frac{1}{3 - 2.5} = 2$$

$$f_p = \frac{1}{2\pi RC} = \frac{1}{2\pi(30 \text{ k}\Omega)(1 \text{ nF})} = 5.31 \text{ kHz}$$

In Fig. 21-26, a Q of 2 gives the following approximate values:

$$K_0 = 0.94$$
$$K_c = 1.32$$
$$K_3 = 1.48$$
$$A_p = 20 \log Q = 20 \log 2 = 6.3 \text{ dB}$$

The resonant frequency is:

$$f_0 = \frac{f_p}{K_0} = \frac{5.31 \text{ kHz}}{0.94} = 5.65 \text{ kHz}$$

The cutoff frequency is:

$$f_c = \frac{f_p}{K_c} = \frac{5.31 \text{ kHz}}{1.32} = 4.02 \text{ kHz}$$

The 3-dB frequency is:

$$f_{3dB} = \frac{f_p}{K_3} = \frac{5.31 \text{ kHz}}{1.48} = 3.59 \text{ kHz}$$

The circuit produces a 6.3-dB peak in the response at 5.65 kHz, rolls off to 0 dB at the cutoff frequency of 4.02 kHz, and is down 3 dB at 3.59 kHz.

PRACTICE PROBLEM 21-10 Repeat Example 21-10 with the 15 kΩ resistor changed to 17.5 kΩ.

21-9 MFB Bandpass Filters

A bandpass filter has a center frequency and a bandwidth. Recall the basic equations for a bandpass response:

$$\text{BW} = f_2 - f_1$$
$$f_0 = \sqrt{f_1 f_2}$$
$$Q = \frac{f_0}{\text{BW}}$$

When Q is less than 1, the filter has a wideband response. In this case, a bandpass filter is usually built by cascading a low-pass stage with a high-pass stage. When Q is greater than 1, the filter has a narrowband response and a different approach is used.

Wideband Filters

Suppose we want to build a bandpass filter with a lower cutoff frequency of 300 Hz and an upper cutoff frequency of 3.3 kHz. The center frequency of the filter is:

$$f_0 = \sqrt{f_1 f_2} = \sqrt{(300 \text{ Hz})(3.3 \text{ kHz})} = 995 \text{ Hz}$$

The bandwidth is:

$$\text{BW} = f_2 - f_1 = 3.3 \text{ kHz} - 300 \text{ Hz} = 3 \text{ kHz}$$

Q is:

$$Q = \frac{f_0}{\text{BW}} = \frac{995 \text{ Hz}}{3 \text{ kHz}} = 0.332$$

Since Q is less than 1, we can use cascaded low-pass and high-pass stages, as shown in Fig. 21-38. The high-pass filter has a cutoff frequency of 300 Hz, and the low-pass filter has a cutoff frequency of 3.3 kHz. When the two decibel responses are added, we get a bandpass response with cutoff frequencies of 300 Hz and 3.3 kHz.

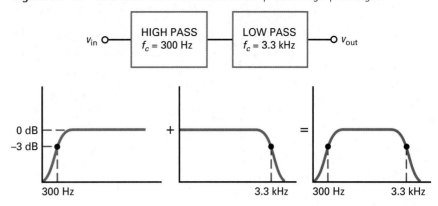

When Q is greater than 1, the cutoff frequencies are much closer than shown in Fig. 21-38. Because of this, the sum of the passband attenuations is greater than 3 dB at the cutoff frequencies. This is why we use another approach for narrowband filters.

Narrowband Filters

When Q is greater than 1, we can use the **multiple-feedback (MFB)** filter shown in Fig. 21-39. First, notice that the input signal goes to the inverting input rather than the noninverting input. Second, notice that the circuit has two feedback paths, one through a capacitor and another through a resistor.

At low frequencies the capacitors appear to be open. Therefore, the input signal cannot reach the op amp, and the output is zero. At high frequencies the capacitors appear to be shorted. In this case, the voltage gain is zero because the feedback capacitor has zero impedance. Between the low and high extremes in frequency, there is a band of frequencies where the circuit acts like an inverting amplifier.

The voltage gain at the center frequency is given by:

$$A_v = \frac{-R_2}{2R_1} \tag{21-32}$$

This is almost identical to the voltage gain of an inverting amplifier: except for the factor of 2 in the denominator. The Q of the circuit is given by:

$$Q = 0.5 \sqrt{\frac{R_2}{R_1}} \tag{21-33}$$

which is equivalent to:

$$Q = 0.707 \sqrt{-A_v} \tag{21-34}$$

Figure 21–39 Multiple-feedback bandpass stage.

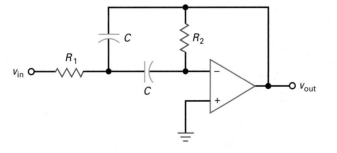

For instance, if $A_v = -100$:

$$Q = 0.707 \sqrt{100} = 7.07$$

Equation (21-34) tells us that the greater the voltage gain, the higher the Q.
The center frequency is given by:

$$f_0 = \frac{1}{2\pi\sqrt{R_1 R_2 C_1 C_2}} \tag{21-35}$$

Since $C_1 = C_2$ in Fig. 21-39, the equation simplifies to:

$$f_0 = \frac{1}{2\pi C \sqrt{R_1 R_2}} \tag{21-36}$$

Increasing the Input Impedance

Equation (21-33) tells us that Q is proportional to the square root of R_2/R_1. To get high Qs, we need to use a high ratio of R_2/R_1. For instance, to get a Q of 5, R_2/R_1 must equal 100. To avoid problems with input offset and bias current, R_2 is usually kept under 100 kΩ, which means that R_1 has to be less than 1 kΩ. For Qs greater than 5, R_1 must be even smaller. This means that the input impedance of Fig. 21-39 may be too low at higher Qs.

Figure 21-40a shows an MFB bandpass filter that increases the input impedance. The circuit is identical to the earlier MFB circuit, except for the new resistor R_3. Notice that R_1 and R_3 form a voltage divider. By applying Thevenin's theorem, the circuit simplifies to Fig. 21-40b. This configuration is the same as that shown in Fig. 21-39, but some of the equations are different. To begin with, the voltage gain is still given by Eq. (21-32). But the Q and center frequency become:

$$Q = 0.5 \sqrt{\frac{R_2}{R_1 \| R_3}} \tag{21-37}$$

$$f_0 = \frac{1}{2\pi C \sqrt{(R_1 \| R_3) R_2}} \tag{21.38}$$

Figure 21-40 Increasing input impedance of MFB stage.

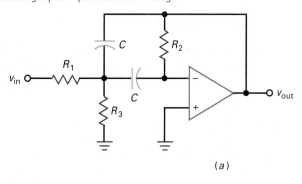

$$A_v = \frac{-R_2}{2R_1}$$

$$Q = 0.5 \sqrt{\frac{R_2}{R_1 \| R_3}}$$

$$f_0 = \frac{1}{2\pi C \sqrt{(R_1 \| R_3) R_2}}$$

(a)

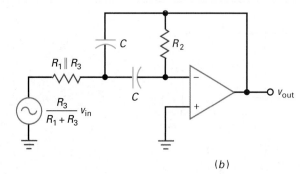

(b)

Figure 21–41 MFB stage with variable center frequency and constant bandwidth.

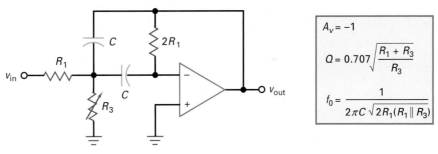

The circuit has the advantage of higher input impedance because R_1 can be made higher for a given Q.

Tunable Center Frequency with Constant Bandwidth

Having a voltage gain greater than 1 is not necessary in many applications because voltage gain is usually available in another stage. If unity voltage gain is acceptable, then we can use a clever circuit that varies the center frequency while holding the bandwidth constant.

Figure 21-41 shows a modified MFB circuit in which $R_2 = 2R_1$ and R_3 is variable. With this circuit, the analysis equations are:

$$A_v = -1 \tag{21-39}$$

$$Q = 0.707\sqrt{\frac{R_1 + R_3}{R_3}} \tag{21-40}$$

$$f_0 = \frac{1}{2\pi C \sqrt{2R_1(R_1 \parallel R_3)}} \tag{21-41}$$

Since BW $= f_0/Q$, we can derive this equation for bandwidth:

$$\text{BW} = \frac{1}{2\pi R_1 C} \tag{21-42}$$

Equation (21-41) says that varying R_3 will vary f_0, but Eq. (21-42) shows that bandwidth is independent of R_3. Therefore, we can have a constant bandwidth while varying the center frequency.

Variable resistor R_3 in Fig. 21-41 is often a JFET used as a voltage-controlled resistance (discussed in Sec. 13-9). Since the gate voltage changes the resistance of the JFET, the center frequency of the circuit can be tuned electronically.

Example 21–11

The gate voltage of Fig. 21-42 can vary the JFET resistance from 15 to 80 Ω. What is the bandwidth? What are the minimum and maximum center frequencies?

SOLUTION Equation (21-42) gives the bandwidth:

$$\text{BW} = \frac{1}{2\pi\,R_1 C} = \frac{1}{2\pi(18\text{ k}\Omega)(8.2\text{ nF})} = 1.08\text{ kHz}$$

Figure 21-42 Tuning an MFB filter with a voltage-controlled resistance.

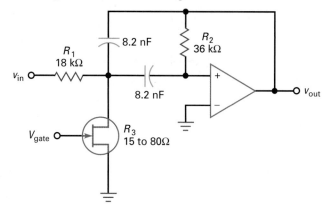

With Eq. (21-41), the minimum center frequency is:

$$f_0 = \frac{1}{2\pi C \sqrt{2R_1(R_1 \parallel R_3)}}$$

$$= \frac{1}{2\pi(8.2 \text{ nF})\sqrt{2(18 \text{ k}\Omega)(18 \text{ k}\Omega \parallel 80 \text{ }\Omega)}}$$

$$= 11.4 \text{ kHz}$$

The maximum frequency is:

$$f_0 = \frac{1}{2\pi(8.2 \text{ nF})\sqrt{2(18 \text{ k}\Omega)(18 \text{ k}\Omega \parallel 15 \text{ }\Omega)}} = 26.4 \text{ kHz}$$

PRACTICE PROBLEM 21–11 Using Fig. 21-42, change R_1 to 10 kΩ and repeat Example 21-11.

21-10 Bandstop Filters

There are many circuit implementations for bandstop filters. They use from one to four op amps in each second-order stage. In many applications, a bandstop filter needs to block only a single frequency. For instance, the ac power lines may induce a hum of 60 Hz in sensitive circuits; this may interfere with a desired signal. In this case, we can use a bandstop filter to notch out the unwanted hum signal.

Figure 21-43 shows a **Sallen-Key second-order notch filter** and its analysis equations. At low frequencies all capacitors are open. As a result, all the input signal reaches the noninverting input. The circuit has a passband voltage gain of:

$$A_v = \frac{R_2}{R_1} + 1 \tag{21-43}$$

At very high frequencies, the capacitors are shorted. Again, all the input signal reaches the noninverting input.

Figure 21-43 Sallen-Key second-order notch filter.

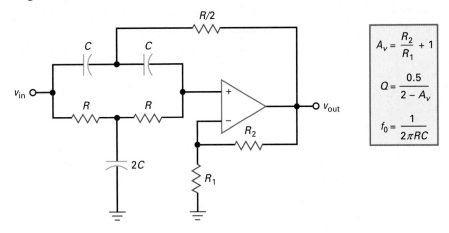

$$A_v = \frac{R_2}{R_1} + 1$$

$$Q = \frac{0.5}{2 - A_v}$$

$$f_0 = \frac{1}{2\pi RC}$$

Between the low and high extremes in frequency, there is a center frequency given by:

$$f_0 = \frac{1}{2\pi RC} \tag{21-44}$$

At this frequency, the feedback signal returns with the correct amplitude and phase to attenuate the signal on the noninverting input. Because of this, the output voltage drops to a very low value.

The Q of the circuit is given by:

$$Q = \frac{0.5}{2 - A_v} \tag{21-45}$$

The voltage gain of a Sallen-Key notch filter must be less than 2 to avoid oscillations. Because of the tolerance of the R_1 and R_2 resistors, the circuit Q should be much less than 10. At higher Qs, the tolerance of these resistors may produce a voltage gain greater than 2, which would produce oscillations.

Example 21-12

||| MultiSim

What are the voltage gain, center frequency, and Q for the bandstop filter shown in Fig. 21-43 if $R = 22$ kΩ, $C = 120$ nF, $R_1 = 13$ kΩ, and $R_2 = 10$ kΩ?

SOLUTION With Eqs. (21-43) to (21-45):

$$A_v = \frac{10\,\text{k}\Omega}{13\,\text{k}\Omega} + 1 = 1.77$$

$$f_0 = \frac{1}{2\pi(22\text{ k}\Omega)(120\text{ nF})} = 60.3\text{ Hz}$$

$$Q = \frac{0.5}{2 - A_v} = \frac{0.5}{2 - 1.77} = 2.17$$

Figure 21-44a shows the response. Notice how sharp the notch is for a second-order filter.

By increasing the order of the filter, we can broaden the notch. For instance, Fig. 21-44b shows the frequency response for a notch filter with $n = 20$. The broader notch reduces component sensitivity and guarantees that the 60-Hz hum will be heavily attenuated.

Figure 21-44 (*a*) Second-order notch filter at 60 Hz; (*b*) notch filter with *n* = 20.

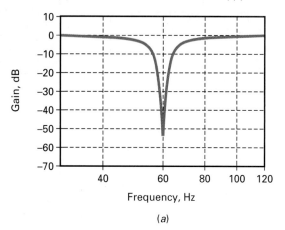

(*a*)

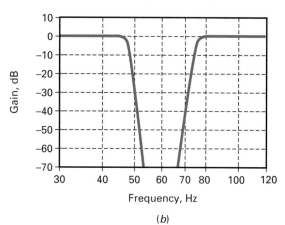

(*b*)

PRACTICE PROBLEM 21–12 In. Fig. 21-43, change R_2 to obtain a Q value of 3. Also, change the C value for a center frequency of 120 Hz.

21-11 The All-Pass Filter

Section 21-1 discussed the basic idea of the *all-pass filter*. Although the term *all-pass filter* is widely used in industry, a more descriptive name would be the *phase filter* because the filter shifts the phase of the output signal without changing the magnitude. Another descriptive title would be the *time-delay filter*, since time delay is related to a phase shift.

First-Order All-Pass Stage

The all-pass filter has a constant voltage gain for all frequencies. This type of filter is useful when we want to produce a certain amount of phase shift for a signal without changing the amplitude.

Figure 21-45*a* shows a *first-order all-pass lag filter*. It is first order because it has only one capacitor. This is the phase shifter we discussed in Chap. 20. Recall that it shifts the phase of the output signal between 0 and $-180°$. The center frequency of an all-pass filter is where the phase shift is half of maximum. For a first-order lag filter, the center frequency has a phase shift of $-90°$.

Figure 21-45*b* shows a *first-order all-pass lead filter*. In this case, the circuit shifts the phase of the output signal between 180 and 0°. This means that the output signal can lead the input signal by up to $+180°$. For a first-order lead filter, the phase shift is $+90°$ at the center frequency.

Second-Order All-Pass Filter

A second-order all-pass filter has at least one op amp, two capacitors, and several resistors that can shift the phase between 0 and $\pm360°$. Furthermore, it is possible to adjust the Q of a second-order all-pass filter to change the shape of the phase response between 0 and $\pm360°$. The center frequency of a second-order filter is where the phase shift equals $\pm180°$.

Figure 21-46 shows a *second-order MFB all-pass lag filter*. It has one op amp, four resistors, and two capacitors, which is the simplest configuration. More

Figure 21-45 First-order all-pass stages: (*a*) Lagging output phase; (*b*) leading output phase.

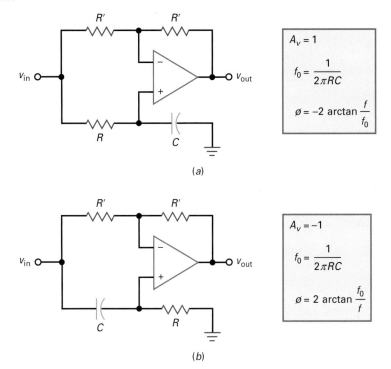

(a)

(b)

Figure 21-46 Second-order all-pass stage.

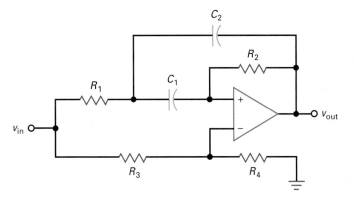

complex configurations use two or more op amps, two capacitors, and several resistors. With a second-order all-pass filter, we can set the center frequency and Q of the circuit.

Figure 21-47 shows the phase response of a second-order all-pass lag filter with $Q = 0.707$. Notice how the output phase increases from $0°$ to $-360°$. By increasing the Q to 2, we can get the phase response as shown in Fig. 21-47*b*. The higher Q does not change the center frequency, but the phase change is faster near the center frequency. A Q of 10 produces the even steeper phase response of Fig. 21-47*c*.

Linear Phase Shift

To prevent distortion of digital signals (rectangular pulses), a filter must have a linear phase shift for the fundamental and all significant harmonics. An equivalent requirement is a constant time delay for all frequencies in the passband. The

Figure 21-47 Second-order phase responses: (*a*) *Q* = 0.707; (*b*) *Q* = 2; (*c*) *Q* = 10.

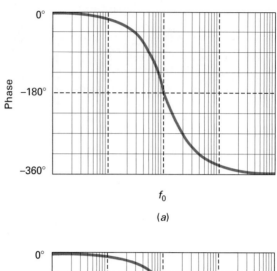

(*a*)

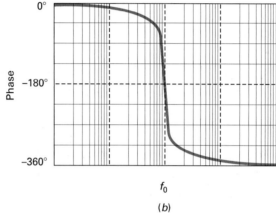

(*b*)

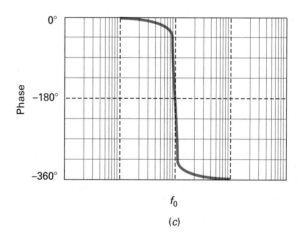

(*c*)

Bessel approximation produces an almost linear phase shift and constant time delay. But in some applications, the slow roll-off rate of the Bessel approximation may not be adequate. Sometimes, the only solution is to use one of the other approximations to get the required roll-off rate, and then use an all-pass filter to correct the phase shift as needed to get an overall linear phase shift.

Bessel Responses

For instance, suppose we need a low-pass filter with A_p = 3 dB, f_c = 1 kHz, A_s = 60 dB, and f_s = 2 kHz and with a linear phase shift for all frequencies in the

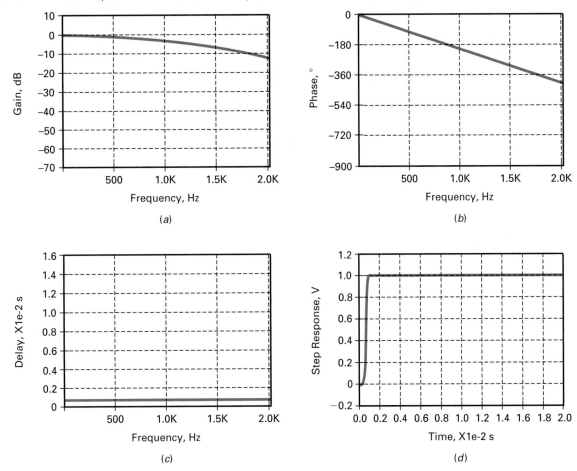

(*a*)

(*b*)

(*c*)

(*d*)

passband. If a tenth-order Bessel filter is used, it would produce the frequency response of Fig. 21-48*a*, the phase response of Fig. 21-48*b*, the time-delay response of Fig. 21-48*c*, and the step response of Fig. 21-48*d*.

First off, notice how slow the roll-off is in Fig. 21-48*a*. The cutoff frequency is 1 kHz. An octave higher, the attenuation is only 12 dB, which does not meet the required specification of $A_s = 60$ dB and $f_s = 2$ kHz. But look at how linear the phase response of Fig. 21-48*b* is. This is the kind of phase response that is almost perfect for digital signals. Linear phase shift and constant time delay are synonymous. This is why the time delay is constant in Fig. 21-48*c*. Finally, look at how sharp the step response of Fig. 21-48*d* is. It may not be perfect, but it is close.

Butterworth Responses

To meet the specifications, we can do the following: We can cascade a tenth-order Butterworth filter and an all-pass filter. The Butterworth filter will produce the required roll-off rate, and the all-pass filter will produce a phase response that complements the Butterworth phase response to get a linear phase response.

A tenth-order Butterworth filter will produce the frequency response of Fig. 21-49*a*, the phase response of Fig. 21-49*b*, the time-delay response of Fig. 21-49*c*, and the step response of Fig. 21-49*d*. As we can see, the attenuation is 60 dB at 2 kHz (Fig. 21-49*a*), which meets the specifications of $A_s = 60$ dB and

Figure 21-49 Butterworth responses for n 5 10: (*a*) Gain; (*b*) phase; (*c*) time delay; (*d*) step response.

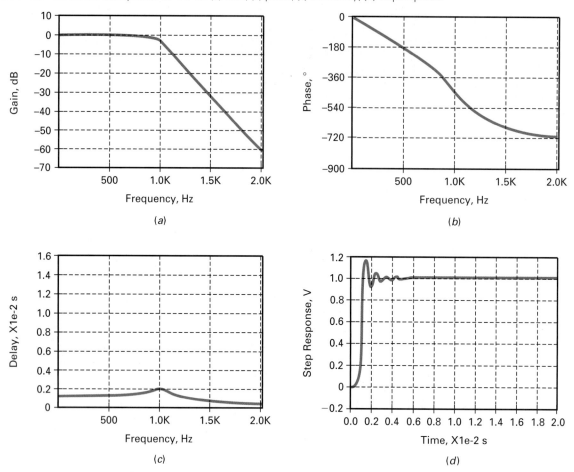

(*a*)

(*b*)

(*c*)

(*d*)

$f_s = 2$ kHz. But look at how nonlinear the phase response of Fig. 21-49*b* is. This kind of phase response will distort digital signals. Likewise, look at the peaked time delay of Fig. 21-49*c*. Finally, look at the overshoot in the step response of Fig. 21-49*d*.

Delay Equalizers

One of the main uses of all-pass filters is to correct the overall phase response by adding the necessary phase shift at each frequency to linearize the overall phase response. When this is done, the time delay becomes constant and the overshoot disappears. When used to compensate for the time delay of another filter, the all-pass filter is sometimes called a **delay equalizer.** A delay equalizer has a time delay that looks like an inverted image of the original time delay. For instance, to compensate for the time delay of Fig. 21-49*c*, the delay equalizer needs to have an upside-down version of Fig. 21-49*c*. Since the total time delay is the sum of the two delays, the total time delay will be flat or constant.

The problem of designing a delay equalizer is extremely complicated. Because of the difficult calculations that are required, only computers can find the component values in a reasonable amount of time. To synthesize an all-pass filter, the computer has to cascade several second-order all-pass stages and then stagger the center frequencies and *Q*s as needed to get the final design.

Example 21-13

In Fig. 21-45b, $R = 1$ kΩ and $C = 100$ nF. What is the phase shift of the output voltage when $f = 1$ kHz?

SOLUTION   Figure 21-45b gives the equation for the cutoff frequency:

$$f_0 = \frac{1}{2\pi(1 \text{ k}\Omega)(100 \text{ nF})} = 1.59 \text{ kHz}$$

The phase shift is:

$$\phi = 2 \arctan \frac{1.59 \text{ kHz}}{1 \text{ kHz}} = 116°$$

21-12 Biquadratic and State-Variable Filters

All second-order filters discussed up to now have used only one op amp. These single op-amp stages are adequate for many applications. In the most stringent applications, more complicated second-order stages are used.

Biquadratic Filter

Figure 21-50 shows a **second-order biquadratic bandpass/lowpass filter.** It has three op amps, two equal capacitors, and six resistors. Resistors R_2 and R_1 set the voltage gain. Resistors R_3 and R_3' have the same nominal value, as do R_4 and R_4'. The circuit equations are shown in Fig. 21-50.

The biquadratic filter is also referred to as a *TT (Tow-Thomas) filter.* This type of filter can be tuned by varying R_3. This has no effect on the voltage gain, which is an advantage. The biquadratic filter of Fig. 21-50 also has a low-pass output. In some applications, getting bandpass and low-pass responses simultaneously is an advantage.

Here is another advantage of the biquadratic filter: As shown in Fig. 21-50, the bandwidth of a biquadratic filter is given by:

$$\text{BW} = \frac{1}{2\pi R_2 C}$$

With the biquadratic filter of Fig. 21-50, we can independently vary the voltage gain with R_1, the bandwidth with R_2, and the center frequency with R_3. Having voltage gain, center frequency, and bandwidth all independently tunable is a major advantage and one of the reasons for the popularity of biquadratic filters (also called *biquads*).

By adding a fourth op amp and more components, we can also build biquadratic high-pass, bandstop, and all-pass filters. When component tolerance is a problem, biquadratic filters are often used because they have less sensitivity to changes in the component values than do the Sallen-Key and MFB filters.

Figure 21–50 Biquadratic stage.

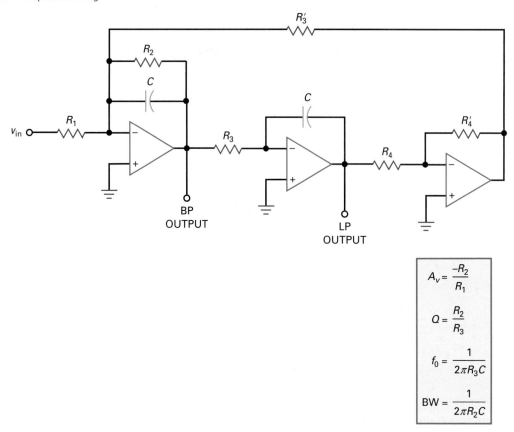

$$A_v = \frac{-R_2}{R_1}$$

$$Q = \frac{R_2}{R_3}$$

$$f_0 = \frac{1}{2\pi R_3 C}$$

$$BW = \frac{1}{2\pi R_2 C}$$

State–Variable Filter

The **state-variable filter** is also called a *KHN filter* after the inventors (Kerwin, Huelsman, and Newcomb). Two configurations are available: inverting and noninverting. Figure 21-51 shows a second-order state-variable filter. It has three simultaneous outputs: low-pass, high-pass, and bandpass. This may be an advantage in some applications.

By adding a fourth op amp and a few more components, the Q of the circuit becomes independent of the voltage gain and the center frequency. This means that the Q is constant when the center frequency is varied. A constant Q means that bandwidth is a fixed percentage of the center frequency. For instance, if $Q = 10$, bandwidth will be 10 percent of f_0. This is desirable in some applications where the center frequency is varied.

Like the biquad, the state-variable filter uses more parts than do the VCVS and MFB filters. But the additional op amps and other components make it more suitable for higher-order filters and critical applications. Furthermore, the biquad and state-variable filters exhibit less component sensitivity, which results in a filter that is easier to produce and requires less adjustment.

Conclusion

Summary Table 21-1 presents the four basic filter circuits used to implement the different approximations. As indicated, the Sallen-Key filters fall into the general class of VCVS filters, the multiple-feedback filters are abbreviated *MFB,* the biquadratic filters may be referred to as *TT filters,* and the state-variable filters are known as *KHN filters.* The complexity of the VCVS and MFB filters is low

Figure 21–51 State-variable stage.

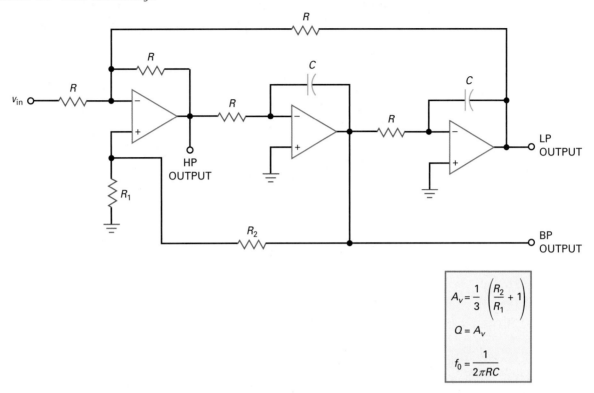

$$A_v = \frac{1}{3}\left(\frac{R_2}{R_1} + 1\right)$$

$$Q = A_v$$

$$f_0 = \frac{1}{2\pi RC}$$

Summary Table 21–1	Basic Filter Circuits				
Type	**Other names**	**Complexity**	**Sensitivity**	**Tuning**	**Advantages**
Sallen-Key	VCVS	Low	High	Difficult	Simplicity, noninverting
Multiple-feedback	MFB	Low	High	Difficult	Simplicity, inverting
Biquadratic	TT	High	Low	Easy	Stability, extra outputs, constant BW
State-variable	KHN	High	Low	Easy	Stability, extra outputs, constant Q

because they use only one op amp, whereas the complexity of the TT and KHN filters is high because they may use three to five op amps in a second-order stage.

The VCVS and MFB filters have a high sensitivity to component tolerance, whereas the TT and KHN filters have a much lower component sensitivity. VCVS and MFB filters may be somewhat difficult to tune because of interaction between voltage gain, cutoff and center frequencies, and Q. The TT filter is easier to tune because its voltage gain, center frequency, and bandwidth are independently tunable. The KHN has independently tunable voltage gain, center frequency, and Q. Finally, the VCVS and MFB filters offer simplicity, and the TT and KHN filters offer stability and additional outputs. When the center frequency of a bandpass filter is varied, the TT filter has a constant bandwidth and the KHN filter has a constant Q.

Although any of the five basic approximations (Butterworth, Chebyshev, inverse Chebyshev, elliptic, and Bessel) can be implemented with op-amp circuits, the more complicated approximations (inverse Chebyshev and elliptic)

Summary Table 21-2		Approximations and Circuits		
Type	**Passband**	**Stopband**	**Usable stages**	
Butterworth	Flat	Monotonic	VCVS, MFB, TT, KHN	
Chebyshev	Rippled	Monotonic	VCVS, MFB, TT, KHN	
Inverse Chebyshev	Flat	Rippled	KHN	
Elliptic	Rippled	Rippled	KHN	
Bessel	Flat	Monotonic	VCVS, MFB, TT, KHN	

cannot be implemented with VCVS or MFB circuits. Summary Table 21-2 shows the five approximations and the types of stages that can be used with them. As we can see, the rippled stopband responses of the inverse Chebyshev and the elliptic approximations require a complex filter like the KHN (state-variable) for implementation.

This chapter discussed four of the most basic filter circuits, shown in Summary Table 21-1. These basic circuits are quite popular and widely used. But we should be aware of the fact that many more circuits are available in computer programs that do filter design. These include the following second-order stages: Akerberg-Mossberg, Bach, Berha-Herpy, Boctor, Dliyannis-Friend, Fliege, Mikhael-Bhattacharyya, Scultety, and the twin-T. All the active-filter circuits used today have advantages and disadvantages that allow a designer to choose the best compromise for an application.

Summary

SEC. 21-1 IDEAL RESPONSES

There are five basic types of responses: low-pass, high-pass, bandpass, bandstop, and all-pass. The first four have a passband and a stopband. Ideally, the attenuation should be zero in the passband and infinite in the stopband with a brick wall transition.

SEC. 21-2 APPROXIMATE RESPONSES

The passband is identified by its low attenuation and its edge frequency. The stopband is identified by its high attenuation and edge frequency. The order of a filter is the number of reactive components. With active filters, it is usually the number of capacitors. The five approximations are the Butterworth (maximally flat passband), the Chebyshev

(rippled passband), the inverse Chebyshev (flat passband and rippled stopband), the elliptic (rippled passband and stopband), and the Bessel (maximally flat time delay).

SEC. 21-3 PASSIVE FILTERS

A low-pass LC filter has a resonant frequency f_0 and a Q. The response is maximally flat when $Q = 0.707$. As Q increases, a peak appears in the response, centered on the resonant frequency. The Chebyshev response occurs with Q greater than 0.707, and the Bessel with $Q = 0.577$. The higher the Q, the faster the roll-off in the transition region.

SEC. 21-4 FIRST-ORDER STAGES

First-order stages have a single capacitor and one or more resistors. All first-order

stages produce a Butterworth response because peaking is possible only in second-order stages. A first-order stage can produce either a low-pass or a high-pass response.

SEC. 21-5 VCVS UNITY-GAIN SECOND-ORDER LOW-PASS FILTERS

Second-order stages are the most common stage because they are easy to implement and analyze. The Q of the stage produces different K values. The pole frequency of a low-pass stage can be multiplied by its K values to get the resonant frequency if there is a peak, a cutoff frequency, and a 3-dB frequency.

SEC. 21-6 HIGHER-ORDER FILTERS

Higher-order filters are usually made by cascading second-order stages or a first-order stage when the total order is odd. When filter stages are cascaded, we add the decibel gains of the stages to get the total decibel gain. To get the Butterworth response for a higher-order filter, we have to stagger the Qs of the stages. To get the Chebyshev and other responses, we have to stagger the pole frequencies and the Qs.

SEC. 21-7 VCVS EQUAL-COMPONENT LOW-PASS FILTERS

The Sallen-Key equal-component filters control the Q by setting the voltage gain. The voltage gain must be less than 3 to avoid oscillations. Higher Qs are difficult to get with this circuit because the component tolerance becomes very important in determining the voltage gain and Q.

SEC. 21-8 VCVS HIGH-PASS FILTERS

VCVS high-pass filters have the same configuration as low-pass filters, except that the resistors and capacitors are interchanged. Again, the Q values determine the K values. We have to divide pole frequency by the K values to get the resonant frequency, cutoff frequency, and 3-dB frequency.

SEC. 21-9 MFB BANDPASS FILTERS

Low-pass and high-pass filters may be cascaded to get a bandpass filter, provided that Q is less than 1. When Q is greater than 1, we have a narrowband filter rather than a wideband filter.

SEC. 21-10 BANDSTOP FILTERS

Bandstop filters can be used to notch out a specific frequency such as the 60-Hz hum induced in circuits by ac power lines. With a Sallen-Key notch filter, the voltage gain controls the Q of the circuit. The voltage gain must be less than 2 to avoid oscillations.

SEC. 21-11 THE ALL-PASS FILTER

Somewhat of a misnomer, the all-pass filter does more than pass all frequencies with no attenuation. This type of filter is designed to control the phase of the output signal. Especially important is the use of an all-pass filter as a phase or time-delay equalizer. With one of the other filters producing the desired frequency response and an all-pass filter producing the desired phase response, the overall filter has a linear phase response, equivalent to a maximally flat time delay.

SEC. 21-12 BIQUADRATIC AND STATE-VARIABLE FILTERS

The biquadratic or TT filters use three or four op amps. Although more complex, the biquadratic filter offers lower component sensitivity and easier tuning. This type of filter also has simultaneous low-pass and bandpass outputs, or high-pass and bandstop outputs. The state-variable or KHN filters also use three or more op amps. When a fourth op amp is used, it offers easy tuning because voltage gain, center frequency, and Q are all independently tunable.

Definitions

(21-1) Bandwidth:

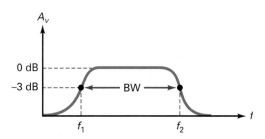

(21-4) Order of a filter:

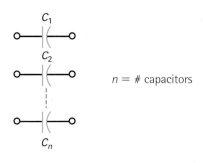

$n = \#$ capacitors

(21-5) Number of ripples:

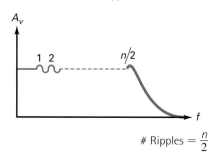

$$\# \text{Ripples} = \frac{n}{2}$$

Derivations

(21-2)   Center frequency:

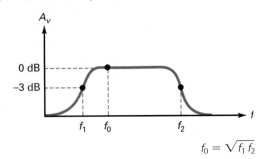

$$f_0 = \sqrt{f_1 f_2}$$

(21-22) to (21-24)   Center, cutoff, and 3-dB frequencies:

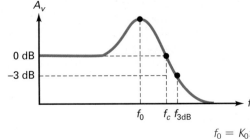

$$f_0 = K_0 f_p$$
$$f_c = K_c f_p$$
$$f_{3dB} = K_3 f_p$$

(21-3)   Q of stage:

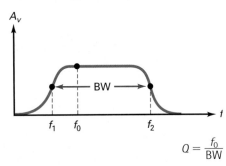

$$Q = \frac{f_0}{BW}$$

Student Assignments

1. **The region between the passband and the stopband is called the**
   a. Attenuation
   b. Center
   c. Transition
   d. Ripple

2. **The center frequency of a bandpass filter is always equal to the**
   a. Bandwidth
   b. Geometric average of the cutoff frequencies
   c. Bandwidth divided by Q
   d. 3-dB frequency

3. **The Q of a narrowband filter is always**
   a. Small
   b. Equal to BW divided by f_0
   c. Less than 1
   d. Greater than 1

4. **A bandstop filter is sometimes called a**
   a. Snubber
   b. Phase shifter
   c. Notch filter
   d. Time-delay circuit

5. **The all-pass filter has**
   a. No passband
   b. One stopband
   c. The same gain at all frequencies
   d. A fast roll-off above cutoff

6. **The approximation with a maximally flat passband is the**
   a. Chebyshev
   b. Inverse Chebyshev
   c. Elliptic
   d. Cauer

7. **The approximation with a rippled passband is the**
   a. Butterworth
   b. Inverse Chebyshev
   c. Elliptic
   d. Bessel

8. **The approximation that distorts digital signals the least is the**
   a. Butterworth
   b. Chebyshev
   c. Elliptic
   d. Bessel

9. **If a filter has six second-order stages and one first-order stage, the order is**
   a. 2   c. 7
   b. 6   d. 13

10. **If a Butterworth filter has nine second-order stages, its roll-off rate is**
    a. 20 dB per decade
    b. 40 dB per decade
    c. 180 dB per decade
    d. 360 dB per decade

11. **If $n = 10$, the approximation with the fastest roll-off in the transition region is**
    a. Butterworth
    b. Chebyshev
    c. Inverse Chebyshev
    d. Elliptic

12. **The elliptic approximation has a**
    a. Slow roll-off rate compared to the Cauer approximation
    b. Rippled stopband
    c. Maximally flat passband
    d. Monotonic stopband

13. Linear phase shift is equivalent to a
   a. Q of 0.707
   b. Maximally flat stopband
   c. Constant time delay
   d. Rippled passband

14. The filter with the slowest roll-off rate is the
   a. Butterworth
   b. Chebyshev
   c. Elliptic
   d. Bessel

15. A first-order active-filter stage has
   a. One capacitor
   b. Two op amps
   c. Three resistors
   d. A high Q

16. A first-order stage cannot have a
   a. Butterworth response
   b. Chebyshev response
   c. Maximally flat passband
   d. Roll-off rate of 20 dB per decade

17. Sallen-Key filters are also called
   a. VCVS filters
   b. MFB filters
   c. Biquadratic filters
   d. State-variable filters

18. To build a tenth-order filter, we should cascade
   a. 10 first-order stages
   b. 5 second-order stages
   c. 3 third-order stages
   d. 2 fourth-order stages

19. To get a Butterworth response with an eighth-order filter, the stages need to have
   a. Equal Qs
   b. Unequal center frequencies
   c. Inductors
   d. Staggered Qs

20. To get a Chebyshev response with a twelfth-order filter, the stages need to have
   a. Equal Qs
   b. Equal center frequencies
   c. Staggered bandwidths
   d. Staggered pole frequencies and Qs

21. The Q of a Sallen-Key equal-component second-order stage depends on the
   a. Voltage gain
   b. Center frequency
   c. Bandwidth
   d. GBW of the op amp

22. With Sallen-Key high-pass filters, the pole frequency must be
   a. Added to the K values
   b. Subtracted from the K values
   c. Multiplied by the K values
   d. Divided by the K values

23. If bandwidth increases,
   a. The center frequency decreases
   b. Q decreases
   c. The roll-off rate increases
   d. Ripples appear in the stopband

24. When Q is greater than 1, a bandpass filter should be built with
   a. Low-pass and high-pass stages
   b. MFB stages
   c. Notch stages
   d. All-pass stages

25. The all-pass filter is used when
   a. High roll-off rates are needed
   b. Phase shift is important
   c. A maximally flat passband is needed
   d. A rippled stopband is important

26. A second-order all-pass filter can vary the output phase from
   a. 90 to $-90°$
   b. 0 to $-180°$
   c. 0 to $-360°$
   d. 0 to $-720°$

27. The all-pass filter is sometimes called a
   a. Tow-Thomas filter
   b. Delay equalizer
   c. KHN filter
   d. State-variable filter

28. The biquadratic filter
   a. Has low component sensitivity
   b. Uses three or more op amps
   c. Is also called a Tow-Thomas filter
   d. All of the above

29. The state-variable filter
   a. Has a low-pass, high-pass, and bandpass output
   b. Is difficult to tune
   c. Has high component sensitivity
   d. Uses fewer than three op amps

30. If GBW is limited, the Q of the stage will
   a. Remain the same
   b. Double
   c. Decrease
   d. Increase

31. To correct for limited GBW, a designer may use
   a. A constant time delay
   b. Predistortion
   c. Linear phase shift
   d. A rippled passband

Problems

SEC. 21-1 IDEAL RESPONSES

21-1 A bandpass filter has lower and upper cutoff frequencies of 445 and 7800 Hz, respectively. What are the bandwidth, center frequency, and Q? Is this a wideband or narrowband filter?

21-2 If a bandpass filter has cutoff frequencies of 20 and 22.5 kHz, what are the bandwidth, center frequency, and Q? Is this a wideband or narrowband filter?

21-3 Identify the following filters as narrowband or wideband:
   a. $f_1 = 2.3$ kHz and $f_2 = 4.5$ kHz
   b. $f_1 = 47$ kHz and $f_2 = 75$ kHz
   c. $f_1 = 2$ Hz and $f_2 = 5$ Hz
   d. $f_1 = 80$ Hz and $f_2 = 160$ Hz

SEC. 21-2 APPROXIMATE RESPONSES

21-4 An active filter contains 7 capacitors. What is the order of the filter?

21-5 If a Butterworth filter contains 10 capacitors, what is its roll-off rate?

21-6 A Chebyshev filter has 14 capacitors. How many ripples does its passband have?

SEC. 21-3 PASSIVE FILTERS

21-7 The filter of Fig. 21-17 has $L = 20$ mH, $C = 5$ μF, and $R = 600$ Ω. What is the resonant frequency? What is the Q?

21-8 If the inductance is reduced by a factor of 2 in Prob. 21-7, what is the resonant frequency? What is the Q?

SEC. 21-4 FIRST-ORDER STAGES

21-9 In Fig. 21-21a, $R_1 = 15$ kΩ and $C_1 = 270$ nF. What is the cutoff frequency?

21-10 **MultiSim** In Fig. 21-21b, $R_1 = 7.5$ kΩ, $R_2 = 33$ kΩ, $R_3 = 20$ kΩ, and $C_1 = 680$ pF. What is the cutoff frequency? What is the voltage gain in the passband?

21-11 **MultiSim** In Fig. 21-21c, $R_1 = 2.2$ kΩ, $R_2 = 47$ kΩ, and $C_1 = 330$ pF. What is the cutoff frequency? What is the voltage gain in the passband?

21-12 In Fig. 21-22a, $R_1 = 10$ kΩ and $C_1 = 15$ nF. What is the cutoff frequency?

21-13 In Fig. 21-22b, $R_1 = 12$ kΩ, $R_2 = 24$ kΩ, $R_3 = 20$ kΩ, and $C_1 = 220$ pF. What is the cutoff frequency? What is the voltage gain in the passband?

21-14 In Fig. 21-22c, $R_1 = 8.2$ kΩ, $C_1 = 560$ pF, and $C_2 = 680$ pF. What is the cutoff frequency? What is the voltage gain in the passband?

SEC. 21-5 VCVS UNITY-GAIN SECOND-ORDER LOW-PASS FILTERS

21-15 **MultiSim** In Fig. 21-24, $R = 75$ kΩ, $C_1 = 100$ pF, and $C_2 = 200$ pF. What are the pole frequency and Q? What are the cutoff and 3-dB frequencies?

21-16 In Fig. 21-25, $R = 51$ kΩ, $C_1 = 100$ pF, and $C_2 = 680$ pF. What are the pole frequency and Q? What are the cutoff and 3-dB frequencies?

SEC. 21-7 VCVS EQUAL-COMPONENT LOW-PASS FILTERS

21-17 In Fig. 21-31, $R_1 = 51$ kΩ, $R_2 = 30$ kΩ, $R = 33$ kΩ, and $C = 220$ pF. What are the pole frequency and Q? What are the cutoff and 3-dB frequencies?

21-18 In Fig. 21-31, $R_1 = 33$ kΩ, $R_2 = 33$ kΩ, $R = 75$ kΩ, and $C = 100$ pF. What are the pole frequency and Q? What are the cutoff and 3-dB frequencies?

21-19 In Fig. 21-31, $R_1 = 75$ kΩ, $R_2 = 56$ kΩ, $R = 68$ kΩ, and $C = 120$ pF. What are the pole frequency and Q? What are the cutoff and 3-dB frequencies?

SEC. 21-8 VCVS HIGH-PASS FILTERS

21-20 In Fig. 21-35a, $R_1 = 56$ kΩ, $R_2 = 10$ kΩ, and $C = 680$ pF. What are the pole frequency and Q? What are the cutoff and 3-dB frequencies?

21-21 **MultiSim** In Fig. 21-35a, $R_1 = 91$ kΩ, $R_2 = 15$ kΩ, and $C = 220$ nF. What are the pole frequency and Q? What are the cutoff and 3-dB frequencies?

SEC. 21-9 MFB BANDPASS FILTERS

21-22 In Fig. 21-39, $R_1 = 2$ kΩ, $R_2 = 56$ kΩ, and $C = 270$ pF. What are the voltage gain, Q, and center frequency?

21-23 In Fig. 21-40, $R_1 = 3.6$ kΩ, $R_2 = 7.5$ kΩ, $R_3 = 27$ Ω, and $C = 22$ nF. What are the voltage gain, Q, and center frequency?

21-24 In Fig. 21-41, $R_1 = 28$ kΩ, $R_3 = 1.8$ kΩ, and $C = 1.8$ nF. What are the voltage gain, Q, and center frequency?

SEC. 21-10 BANDSTOP FILTERS

21-25 **MultiSim** What are the voltage gain, center frequency, and Q for the bandstop filter shown in Fig. 21-43 if $R = 56$ kΩ, $C = 180$ nF, $R_1 = 20$ kΩ, and $R_2 = 10$ kΩ? What is the bandwidth?

SEC. 21-11 THE ALL-PASS FILTER

21-26 In Fig. 21-45a, $R = 3.3$ kΩ and $C = 220$ nF. What is the center frequency? The phase shift 1 octave above the center frequency?

21-27 **MultiSim** In Fig. 21-45b, $R = 47$ kΩ and $C = 6.8$ nF. What is the center frequency? The phase shift 1 octave below the center frequency?

SEC. 21-12 BIQUADRATIC AND STATE-VARIABLE FILTERS

21-28 In Fig. 21-50, $R_1 = 24$ kΩ, $R_2 = 100$ kΩ, $R_3 = 10$ kΩ, $R_4 = 15$ kΩ, and $C = 3.3$ nF. What are the voltage gain, Q, center frequency, and bandwidth?

21-29 In Prob. 21-28, R_3 is varied from 10 kΩ to 2 kΩ. What are the maximum center frequency and the maximum Q? What are the minimum and maximum bandwidths?

21-30 In Fig. 21-51, $R = 6.8$ kΩ, $C = 5.6$ nF, $R_1 = 6.8$ kΩ, and $R_2 = 100$ kΩ. What are the voltage gain, Q, and center frequency?

Critical Thinking

21-31 A bandpass filter has a center frequency of 50 kHz and a Q of 20. What are the cutoff frequencies?

21-32 A bandpass filter has an upper cutoff frequency of 84.7 kHz and a bandwidth of 12.3 kHz. What is the lower cutoff frequency?

21-33 You are testing a Butterworth filter with the following specifications: $n = 10$, $A_p = 3$ dB, and $f_c = 2$ kHz. What is the attenuation at each of the following frequencies: 4, 8, and 20 kHz?

21-34 A Sallen-Key unity-gain low-pass filter has a cutoff frequency of 5 kHz. If $n = 2$ and $R = 10$ kΩ, what do C_1 and C_2 equal for a Butterworth response?

21-35 A Chebyshev Sallen-Key unity-gain low-pass filter has a cutoff frequency of 7.5 kHz. The ripple depth is 12 dB. If $n = 2$ and $R = 25$ kΩ, what do C_1 and C_2 equal?

Job Interview Questions

1. Draw the four brick wall responses. Identify the passband, stopband, and cutoff frequencies of each.
2. Describe the five approximations used in filter design. Use sketches as needed to show what happens in the passbands and stopbands.
3. In digital systems, filters need a linear phase response or maximally flat time delay. What does this mean, and why is it important?
4. Tell me what you can about how a tenth-order low-pass Chebyshev filter is implemented. Your discussion should include the center frequencies and Qs of the stages.
5. To get a fast roll-off and a linear-phase response, somebody has cascaded a Butterworth filter with an all-pass filter. Tell me what each of these filters does.
6. What are the distinguishing features of the response in the passband? In the stopband?
7. What is an all-pass filter?
8. What does the frequency response of a filter measure or indicate?
9. What is the roll-off rate (per decade and per octave) for an active filter?
10. What is an MFB filter and where is it used?
11. Which type of filter is used for delay equalization?

Self-Test Answers

1.	c	12.	b	23.	b
2.	b	13.	c	24.	b
3.	d	14.	d	25.	b
4.	c	15.	a	26.	c
5.	c	16.	b	27.	b
6.	b	17.	a	28.	d
7.	c	18.	b	29.	a
8.	d	19.	d	30.	d
9.	d	20.	d	31.	b
10.	d	21.	a		
11.	d	22.	d		

Practice Problem Answers

21-1 $f_c = 34.4$ kHz

21-2 $f_c = 16.8$ kHz

21-3 $Q = 0.707$; $f_p = 13.7$ kHz; $f_c = 13.7$ kHz

21-4 $C_2 = 904$ pF

21-5 $Q = 3$; $f_p = 3.1$ kHz; $K_0 = 0.96$; $K_C = 1.35$; $K_3 = 1.52$; $A_p = 9.8$ dB; $f_c = 4.19$ kHz; $f_{3dB} = 4.71$ kHz

21-6 $A_v = 1.59$; $Q = 0.709$; $f_p = 21.9$ kHz

21-7 $A_v = 1.27$; $Q = 0.578$; $f_p = 4.82$ kHz; $f_c = 3.79$ kHz

21-9 $Q = 0.707$; $f_p = 998$ Hz; $f_c = 998$ Hz

21-10 $A_v = 2.75$; $Q = 4$; $f_p = 5.31$ kHz; $K_0 = 0.98$; $K_C = 1.38$; $K_3 = 1.53$; $A_p = 12$ dB; $f_0 = 5.42$ kHz; $f_C = 3.85$ kHz; $f_{3dB} = 3.47$ kHz

21-11 BW $= 1.94$ kHz; $f_{0(min)} = 15$ kHz; $f_{0(max)} = 35.5$ kHz

21-12 $R_2 = 12$ kHz; $C = 60$ nF

22 Nonlinear Op-Amp Circuit:

Monolithic op amps are inexpensive, versatile, and reliable. They can be used not only for linear circuits like voltage amplifiers, current sources, and active filters but also for **nonlinear circuits** such as comparators, waveshapers, and active-diode circuits. The output of a nonlinear op-amp circuit usually has a different shape from the input signal because the op amp saturates during part of the input cycle. Because of this, we have to analyze two different modes of operation to see what happens during an entire cycle.

Objectives

After studying this chapter, you should be able to:

- Explain how a comparator works and describe the importance of the reference point.

- Discuss comparators that have positive feedback and calculate the trip points and hysteresis for these circuits.

- Identify and discuss waveform conversion circuits.

- Identify and discuss waveform generation circuits.

- Describe how several active diode circuits work.

- Explain integrators and differentiators.

- Explain the circuit operation of a Class-D amplifier.

Chapter Outline

Vocabulary

active half-wave rectifier
active peak detector
active positive clamper
active positive clipper
Class-D amplifier
comparator
differentiator
hysteresis
integrator

Lissajous pattern
nonlinear circuits
open-collector comparator
oscillators
pullup resistor
pulse-width-modulated (PWM)
relaxation oscillator
Schmitt trigger

speed-up capacitor
thermal noise
threshold
transfer characteristic
trip point
window comparator
zero-crossing detector

22-1 Comparators with Zero Reference

Often we want to compare one voltage with another to see which is larger. In this situation, a **comparator** may be the perfect solution. A comparator is similar to an op amp because it has two input voltages (noninverting and inverting) and one output voltage. It differs from a linear op-amp circuit because it has a two-state output, either a low or a high voltage. Because of this, comparators are often used to interface with analog and digital circuits.

Basic Idea

The simplest way to build a comparator is to connect an op amp without feedback resistors, as shown in Fig. 22-1a. Because of the high open-loop voltage gain, a positive input voltage produces positive saturation, and a negative input voltage produces negative saturation.

The comparator of Fig. 22-1a is called a **zero-crossing detector** because the output voltage ideally switches from low to high or vice versa whenever the input voltage crosses zero. Figure 22-1b shows the input-output response of a zero-crossing detector. The minimum input voltage that produces saturation is:

$$v_{in(min)} = \frac{\pm V_{sat}}{A_{VOL}} \qquad (22\text{-}1)$$

If $V_{sat} = 14$ V, the output swing of the comparator is from approximately -14 to $+14$ V. If the open-loop voltage gain is 100,000, the input voltage needed to produce saturation is:

$$v_{in(min)} = \frac{\pm 14 \text{ V}}{100,000} = \pm 0.14 \text{ mV}$$

Figure 22-1 (a) Comparator; (b) input/output response; (c) 741C response.

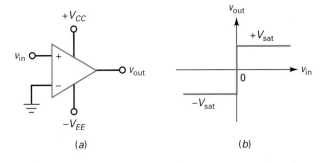

(a)   (b)

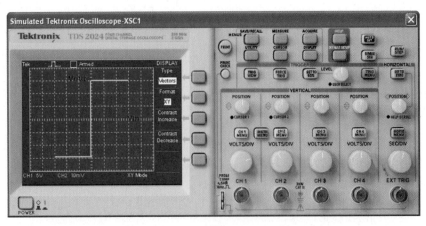

(c)

This means that an input voltage more positive than +0.014 mV drives the comparator into positive saturation, and an input voltage more negative than −0.014 mV drives it into negative saturation.

Input voltages used with comparators are usually much greater than ±0.014 mV. This is why the output voltage is a two-state output, either $+V_{sat}$ or $-V_{sat}$. By looking at the output voltage, we can instantly tell whether the input voltage is: greater than or less than zero.

Lissajous Pattern

A **Lissajous pattern** appears on an oscilloscope when harmonically related signals are applied to the horizontal and vertical inputs. One convenient way to display the input/output response of any circuit is with a Lissajous pattern in which the two harmonically related signals are the input and output voltages of the circuit.

For instance, Fig. 22-1c shows the input/output response for a 741C with supplies of ±15 V. Channel 1 (the vertical axis) has a sensitivity of 5 V/Div. As we can see, the output voltage is either −14 or +14 V, depending on whether the comparator is in negative or positive saturation.

Channel 2 (the horizontal axis) has sensitivity of 10 mV/Div. In Fig. 22-1c, the transition appears to be vertical. This means that the slightest positive input voltage produces positive saturation, and the slightest negative input produces negative saturation.

Inverting Comparator

Sometimes, we may prefer to use an inverting comparator like Fig. 22-2a. The noninverting input is grounded. The input signal drives the inverting input of the comparator. In this case, a slightly positive input voltage produces a maximum

Figure 22-2 (a) Inverting comparator with clamping diodes; (b) input/output response.

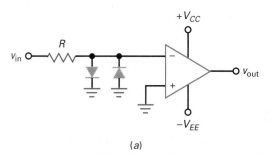

(a)

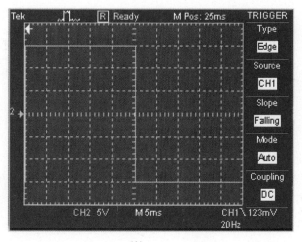

(b)

negative output, as shown in Fig. 22-2b. On the other hand, a slightly negative input voltage produces a maximum positive output.

Diode Clamps

Section 4-10 discussed the use of *diode clamps* to protect sensitive circuits. Figure 22-2a is a practical example. Here we see two diode clamps protecting the comparator against excessively large input voltages. For instance, the LF311 is an IC comparator with an absolute maximum input rating of ±15 V. If the input voltage exceeds these limits, the LF311 will be destroyed.

With some comparators, the maximum input voltage rating may be as little as ±5 V, whereas with others it may be more than ±30 V. In any case, we can protect a comparator against destructively large input voltages by using the diode clamps shown in Fig. 22-2a. These diodes have no effect on the operation of the circuit as long as the magnitude of the input voltage is less than 0.7 V. When the magnitude of the input voltage is greater than 0.7 V, one of the diodes will turn on and clamp the magnitude of the inverting input voltage to approximately 0.7 V.

Some ICs are optimized for use as comparators. These IC comparators often have diode clamps built into their input stages. When using one of these comparators, we have to add an external resistor in series with the input terminal. This series resistor will limit the internal diode currents to a safe level.

Converting Sine Waves to Square Waves

The **trip point** (also called the **threshold** or *reference*) of a comparator is the input voltage that causes the output voltage to switch states (from low to high or from high to low). In the noninverting and inverting comparators discussed earlier, the trip point is zero because this is the value of input voltage where the output switches states. Since a zero-crossing detector has a *two-state output,* any periodic input signal that crosses zero threshold will produce a rectangular output waveform.

For instance, if a sine wave is the input to a noninverting comparator with a threshold of 0 V, the output will be the square wave shown in Fig. 22-3a. As we can see, the output of a zero-crossing detector switches states each time the input voltage crosses the zero threshold.

Figure 22-3b shows the input sine wave and the output square wave for an inverting comparator with a threshold of 0 V. With this zero-crossing detector, the output square wave is 180° out of phase with the input sine wave.

Figure 22-3 Comparator converts sine waves to square waves: (*a*) Noninverting; (*b*) inverting.

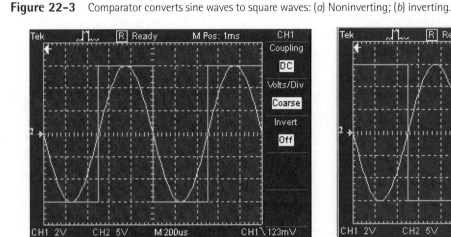
(a)

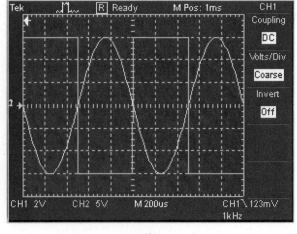
(b)

Figure 22–4 Narrow linear region of typical comparator.

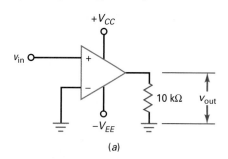

(a)

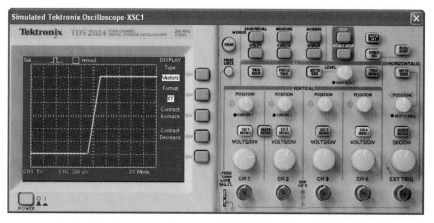

(b)

Linear Region

Figure 22-4a shows a zero-crossing detector. If this comparator had an infinite open-loop gain, the transition between negative and positive saturation would be vertical. In Fig. 22-1c, the transition appears to be vertical because the sensitivity of the 2 channel is 10 mV/Div.

When the sensitivity of the 2 channel is changed to 200 μV/Div, we can see that the transition is not vertical, as shown in Fig. 22-4b. It takes approximately ± 100 μV to get positive or negative saturation. This is typical for a comparator. The narrow input region between approximately -100 to $+100$ μV is called the *linear region of the comparator*. During a zero crossing, a changing input signal usually passes through the linear region so quickly that we see only a sudden jump between negative and positive saturation, or vice versa.

Interfacing Analog and Digital Circuits

Comparators usually interface at their outputs with digital circuits such as CMOS, EMOS, or TTL (stands for *transistor-transistor logic,* a family of digital circuits).

Figure 22-5a shows how a zero-crossing detector can interface with an EMOS circuit. Whenever the input voltage is greater than zero, the output of the comparator is high. This turns on the power FET and produces a large load current.

Figure 22-5b shows a zero-crossing detector interfacing with a CMOS inverter. The idea is basically the same. A comparator input greater than zero produces a high input to the CMOS inverter.

Most EMOS devices can handle input voltages greater than ± 15 V, and most CMOS devices can handle input voltages up to ± 15 V. Therefore, we can interface the output of a typical comparator without any level shifting or clamping. TTL logic, on the other hand, operates with lower input voltages. Because of this,

Figure 22-5 Comparator interfaces with (*a*) power FET; (*b*) CMOS.

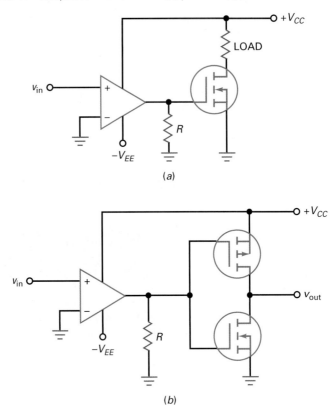

(*a*)

(*b*)

interfacing a comparator with TTL requires a different approach (to be discussed in the next section).

Clamping Diodes and Compensating Resistors

When a current-limiting resistor is used with clamping diodes, a compensating resistor of equal size may be used on the other input of the comparator, as shown in Fig. 22-6. This is still a zero-crossing detector, except that it now has a compensating resistor to eliminate the effect of input bias current.

As before, the diodes are normally off and have no effect on the operation of the circuit. It is only when the input tries to exceed ±0.7 V that one of the clamping diodes turns on and protects the comparator against excessive input voltage.

Bounded Output

The output swing of a zero-crossing detector may be too large in some applications. If so, we can *bound the output* by using back-to-back zener diodes, as

Figure 22-6 Using a compensating resistor to minimize the effect of $I_{in(bias)}$.

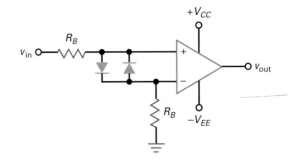

Figure 22-7 Bounded outputs: (a) Zener diodes; (b) rectifier diode.

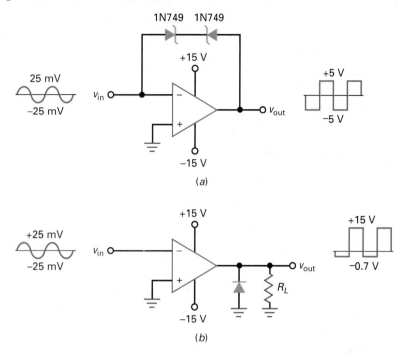

shown in Fig. 22-7a. In this circuit, the inverting comparator has a bounded output because one of the diodes will be conducting in the forward direction and the other will be operating in the breakdown region.

For instance, a 1N749 has a zener voltage of 4.3 V. Therefore, the voltage across the two diodes will be approximately ±5 V. If the input voltage is a sine wave with a peak value of 25 mV, then the output voltage will be an inverted square wave with a peak voltage of 5 V.

Figure 22-7b shows another example of a bounded output. This time, the output diode will clip off the negative half cycles of the output voltage. Given an input sine wave with a peak of 25 mV, the output is bounded between −0.7 and +15 V as shown.

A third approach to bounding the output is to connect zener diodes across the output. For instance, if we connect the back-to-back zener diodes of Fig. 22-7a across the output, the output will be bounded at ±5 V.

Example 22-1

IIII MultiSim

What does the circuit of Fig. 22-8 do?

Figure 22-8 Comparing voltages of different polarities.

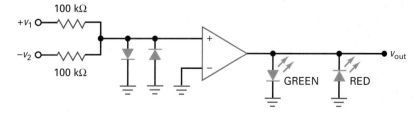

SOLUTION   This circuit compares two voltages of opposite polarity to determine which is greater. If the magnitude of v_1 is greater than the magnitude of v_2, the noninverting input is positive, the comparator output is positive, and the green LED is on. On the other hand, if the magnitude of v_1 is less than the magnitude of v_2, the noninverting input is negative, the comparator output is negative, and the red LED is on.

Example 22-2

What does the circuit of Fig. 22-9 do?

Figure 22-9   Bounded comparator with strobe.

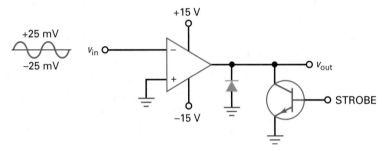

SOLUTION   To begin with, the output diode clips off the negative half cycles. Figure 22-9 also contains a signal called a *strobe*. When the strobe is positive, the transistor saturates and pulls the output voltage down to approximately zero. When the strobe is zero, the transistor is cut off and the comparator output can swing positively. Therefore, the comparator output can swing from -0.7 to $+15$ V when the strobe is low. When the strobe is high, the output is disabled. In this circuit, the strobe is a signal used to turn the output off at certain times or under certain conditions.

Example 22-3

What does the circuit of Fig. 22-10 do?

Figure 22-10   Generating a 60-Hz clock.

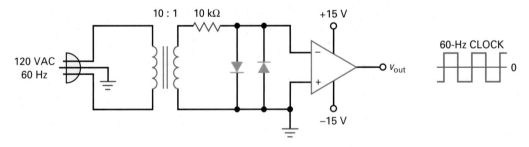

SOLUTION   This is one way to create a *60-Hz clock,* a square-wave signal used as the basic timing mechanism for inexpensive digital clocks. The transformer steps the line voltage down to 12 V ac. The diode clamps then bound the input to ±0.7 V. The inverting comparator produces an output square wave with a frequency of 60 Hz. The output signal is called a *clock* because its frequency can be used to get seconds, minutes, and hours.

A digital circuit called a *frequency divider* can divide the 60 Hz by 60 to get a square wave with a period of 1 s. Another divide-by-60 circuit can divide this signal to get a square wave with a period of 1 min. A final divide-by-60 circuit produces a square wave with a period of 1 hr. Using the three square waves (1 s, 1 min, 1 hr) with other digital circuits and seven-segment LED indicators, we can display the time of day numerically.

22-2 Comparators with Nonzero References

In some applications a threshold voltage different from zero may be preferred. By biasing either input, we can change the threshold voltage as needed.

Moving the Trip Point

In Fig. 22-11a, a voltage divider produces the following reference voltage for the inverting input:

$$v_{ref} = \frac{R_2}{R_1 + R_2} V_{CC} \tag{22-2}$$

When v_{in} is greater than v_{ref}, the differential input voltage is positive and the output voltage is high. When v_{in} is less than v_{ref}, the differential input voltage is negative and the output voltage is low.

A bypass capacitor is typically used on the inverting input, as shown in Fig. 22-11a. This reduces the amount of power-supply ripple and other noise appearing at the inverting input. To be effective, the cutoff frequency of this bypass circuit should be much lower than the ripple frequency of the power supply. The cutoff frequency is given by:

$$f_c = \frac{1}{2\pi (R_1 \parallel R_2)C_{BY}} \tag{22-3}$$

Figure 22-11b shows the **transfer characteristic** (input/output response). The trip point is now equal to v_{ref}. When v_{in} is greater than v_{ref}, the output of the comparator goes into positive saturation. When v_{in} is less than v_{ref}, the output goes into negative saturation.

Figure 22-11 (a) Positive threshold; (b) positive input/output response; (c) negative threshold; (d) negative input/output response.

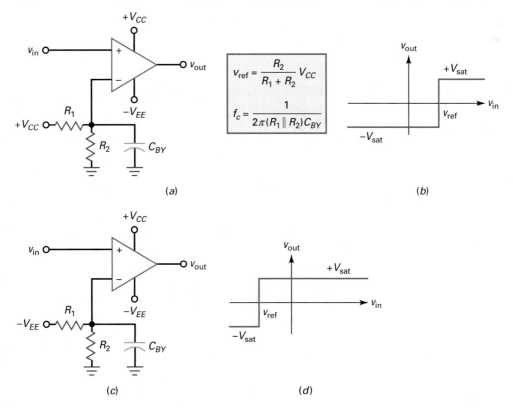

(a)

(b)

(c)

(d)

Figure 22-12 (*a*) Single-supply comparator; (*b*) input/output response.

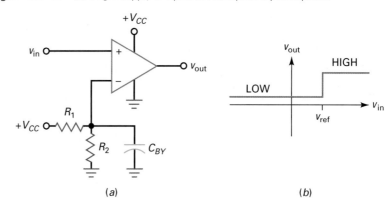

(a)   (b)

A comparator like this is sometimes called a *limit detector* because a positive output indicates that the input voltage exceeds a specific limit. With different values of R_1 and R_2, we can set the limit anywhere between 0 and V_{CC}. If a negative limit is preferred, connect $-V_{EE}$ to the voltage divider, as shown in Fig. 22-11*c*. Now a negative reference voltage is applied to the inverting input. When v_{in} is more positive than v_{ref}, the differential input voltage is positive and the output is high, as shown in Fig. 22-11*d*. When v_{in} is more negative than v_{ref}, the output is low.

Single-Supply Comparator

A typical op amp like the 741C can run on a single positive supply by grounding the $-V_{EE}$ pin, as shown in Fig. 22-12*a*. The output voltage has only one polarity, either a low or a high positive voltage. For instance, with V_{CC} equal to $+15$ V, the output swing is from approximately $+1.5$ V (low state) to around $+13.5$ V (high state).

When v_{in} is greater than v_{ref}, the output is high, as shown in Fig. 22-12*b*. When v_{in} is less than v_{ref}, the output is low. In either case, the output has a positive polarity. For many digital applications, this kind of positive output is preferred.

IC Comparators

An op amp like a 741C can be used as a comparator, but it has speed limitations because of its slew rate. With a 741C, the output can change no faster than 0.5 V/μs. Because of this, a 741C takes more than 50 μs to switch output states with supplies of ±15 V. One solution to the slew-rate problem is to use a faster op amp like an LM318. Since it has a slew rate of 70 V/μs, it can switch from $-V_{sat}$ to $+V_{sat}$ in approximately 0.3 μs.

Another solution is to eliminate the compensating capacitor found in a typical op amp. Since a comparator is always used as a nonlinear circuit, a compensating capacitor is unnecessary. A manufacturer can delete the compensating capacitor and significantly increase the slew rate. When an IC has been optimized for use as a comparator, the device is listed in a separate section of the manufacturer's data book. This is why you will find a section on op amps and another section on comparators in the typical data book.

Open-Collector Devices

Figure 22-13*a* is a simplified schematic diagram for an **open-collector comparator.** Notice that it runs off a single positive supply. The input stage is a diff amp (Q_1 and Q_2). A current source Q_6 supplies the tail current. The diff amp drives an active-load Q_4. The output stage is a single transistor Q_5 with an open collector. This open collector allows the user to control the output swing of the comparator.

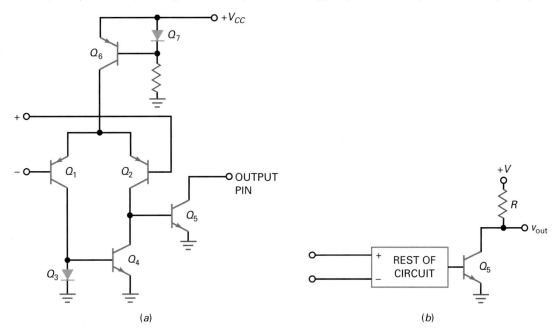

(a)

(b)

The typical op amp discussed in Chap. 18 has an output stage that can be described as an *active-pullup stage* because it contains two devices in a class B push-pull connection. With the active pullup, the upper device turns on and pulls the output up to the high output state. On the other hand, an open-collector output stage of Fig. 22-13a needs external components to be connected to it.

For the output stage to work properly, the user has to connect the open collector to an external resistor and supply voltage, as shown in Fig. 22-13b. The resistor is called a **pullup resistor** because it pulls the output voltage up to the supply voltage when Q_5 is cut off. When Q_5 is saturated, the output voltage is low. Since the output stage is a transistor switch, the comparator produces a two-state output.

With no compensating capacitor in the circuit, the output in Fig. 22-13a can slew very rapidly because only small stray capacitances remain in the circuit. The main limitation on the switching speed is the amount of capacitance across Q_5. This output capacitance is the sum of the internal collector capacitance and the external stray wiring capacitance.

The output time constant is the product of the pullup resistance and the output capacitance. For this reason, the smaller the pullup resistance in Fig. 22-13b, the faster the output voltage can change. Typically, R is from a couple of hundred to a couple of thousand ohms.

Examples of IC comparators are the LM311, LM339, and NE529. They all have an open-collector output stage, which means that you have to connect the output pin to a pullup resistor and a positive supply voltage. Because of their high slew rates, these IC comparators can switch output states in a microsecond or less.

The LM339 is a *quad comparator*—four comparators in a single IC package. It can run off a single supply or off dual supplies. Because it is inexpensive and easy to use, the LM339 is a popular comparator for general-purpose applications.

Not all IC comparators have an open-collector output stage. Some, like the LM360, LM361, and LM760, have an active-collector output stage. The active pullup produces faster switching. These high-speed IC comparators require dual supplies.

Figure 22-14 (a) LM339 comparator; (b) input/output response.

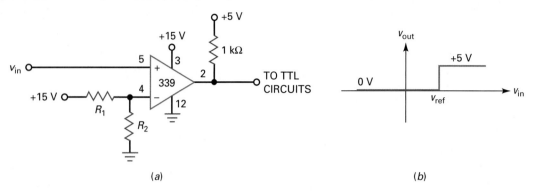

(a)　　　　　　　　　(b)

Driving TTL

The LM339 is an open-collector device. Figure 22-14a shows how an LM339 can be connected to interface with TTL devices. A positive supply of $+15$ V is used for the comparator, but the open collector of the LM339 is connected to a supply of $+5$ V through a pullup resistor of 1 kΩ. Because of this, the output swings between 0 and $+5$ V, as shown in Fig. 22-14b. This output signal is ideal for TTL devices because they are designed to work with supplies of $+5$ V.

Example 22–4

In Fig. 22-15a, the input voltage is a sine wave with a peak value of 10 V. What is the trip point of the circuit? What is the cutoff frequency of the bypass circuit? What does the output waveform look like?

SOLUTION Since $+15$ V is applied to a 3:1 voltage divider, the reference voltage is:

$$v_{\text{ref}} = +5 \text{ V}$$

This is the trip point of the comparator. When the sine wave crosses through this level, the output voltage switches states.

With Eq. (22-3), the cutoff frequency of the bypass circuit is:

$$f_c = \frac{1}{2\pi(200 \text{ k}\Omega \parallel 100 \text{ k}\Omega)(10 \text{ }\mu\text{F})}$$

$$= 0.239 \text{ Hz}$$

This low cutoff frequency means that any 60-Hz ripple on the reference supply voltage will be heavily attenuated.

Figure 22-15b shows the input sine wave. It has a peak value of 10 V. The rectangular output has a peak value of approximately 15 V. Notice how the output voltage switches states when the input sine wave crosses the trip point of $+5$ V.

Figure 22-15 Calculating duty cycle.

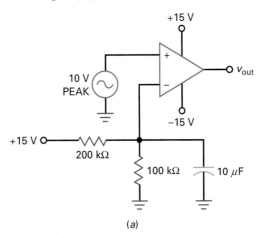

(a)

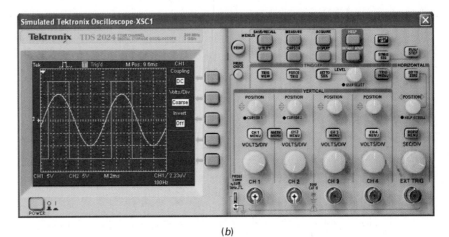

(b)

PRACTICE PROBLEM 22-4 Using Fig. 22-15a, change the 200 kΩ resistor to 100 kΩ and the 10 μF capacitor to 4.7 μF. Solve for the circuit's trip point and cutoff frequency.

Example 22-5

What is the duty cycle of the output waveform in Fig. 22-15b?

SOLUTION In Chap. 11, we defined the *duty cycle* as the pulse width divided by the period. Equation (11-22) gave this equivalent definition: Duty cycle equals the conduction angle divided by 360°.

In Fig. 22-15b, the sine wave has a peak value of 10 V. Therefore, the input voltage is given by:

$$v_{in} = 10 \sin \theta$$

The rectangular output switches states when the input voltage crosses +5 V. At this point, the foregoing equation becomes:

$$5 = 10 \sin \theta$$

Now, we can solve for the angle θ where switching occurs:

$$\sin \theta = 0.5$$

or

$$\theta = \arcsin 0.5 = 30° \text{ and } 150°$$

The first solution, $\theta = 30°$, is where the output switches from low to high. The second solution, $\theta = 150°$, is where the output switches from high to low. The duty cycle is:

$$D = \frac{\text{Conduction angle}}{360°} = \frac{150° - 30°}{360°} = 0.333$$

The duty cycle in Fig. 22-15b can be expressed as 33.3 percent.

22-3 Comparators with Hysteresis

If the input to a comparator contains a large amount of noise, the output will be erratic when v_{in} is near the trip point. One way to reduce the effect of noise is by using a comparator with positive feedback. The positive feedback produces two separate trip points that prevent a noisy input from producing false transitions.

Noise

Noise is any kind of unwanted signal that is not derived from or harmonically related to the input signal. Electric motors, neon signs, power lines, car ignitions, lightning, and so on, produce electromagnetic fields that can induce noise voltages into electronic circuits. Power-supply ripple is also classified as noise since it is not related to the input signal. By using regulated power supplies and shielding, we usually can reduce the ripple and induced noise to an acceptable level.

Thermal noise, on the other hand, is caused by the random motion of free electrons inside a resistor (see Fig. 22-16a). The energy for this electron motion comes from the thermal energy of the surrounding air. The higher the ambient temperature, the more active the electrons.

The motion of billions of free electrons inside a resistor is pure chaos. At some instants, more electrons move up than down, producing a small negative voltage across the resistor. At other instants, more electrons move down than up, producing a positive voltage. If this type of noise were amplified and viewed on an oscilloscope, it would resemble Fig. 22-16b. Like any voltage, noise has an rms or effective value. As an approximation, the highest noise peaks are about four times the rms value.

The randomness of the electron motion inside a resistor produces a distribution of noise at virtually all frequencies. The rms value of this noise increases with temperature, bandwidth, and resistance. For our purposes, we need to be aware of how noise may affect the output of a comparator.

Figure 22-16 Thermal noise: (*a*) Random electron motion in resistor; (*b*) noise on oscilloscope.

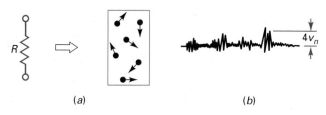

(*a*)   (*b*)

Figure 22-17   Noise produces false triggering of comparator.

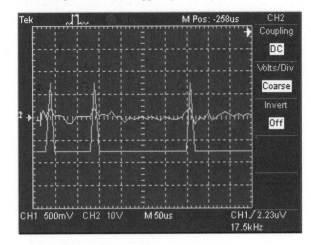

Noise Triggering

As discussed in Sec. 22-1, the high open-loop gain of a comparator means that an input of only 100 μV may be enough to switch the output from one state to another. If the input contains noise with a peak of 100 μV or more, the comparator will detect the zero crossings produced by the noise.

Figure 22-17 shows the output of a comparator with no input signal, except for noise. When the noise peaks are large enough, they produce unwanted changes in the comparator output. For instance, the noise peaks at A, B, and C are producing unwanted transitions from low to high. When an input signal is present, the noise is superimposed on the input signal and produces erratic triggering.

Schmitt Trigger

The standard solution for a noisy input is to use a comparator like the one shown in Fig. 22-18a. The input voltage is applied to the inverting input. Because the feedback voltage is aiding the input voltage, the feedback is *positive*. A comparator using positive feedback like this is usually called a **Schmitt trigger.**

When the comparator is positively saturated, a positive voltage is fed back to the noninverting input. This positive feedback voltage holds the output in the high state. Similarly, when the output voltage is negatively saturated, a negative

Figure 22-18   (a) Inverting Schmitt trigger; (b) input/output response has hysteresis.

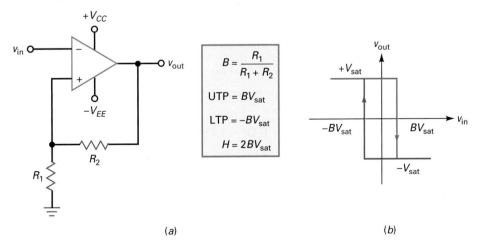

$$B = \frac{R_1}{R_1 + R_2}$$

$$UTP = BV_{sat}$$

$$LTP = -BV_{sat}$$

$$H = 2BV_{sat}$$

(a)

(b)

voltage is fed back to the noninverting input, holding the output in the low state. In either case, the positive feedback reinforces the existing output state.

The feedback fraction is:

$$B = \frac{R_1}{R_1 + R_2} \tag{22-4}$$

When the output is positively saturated, the reference voltage applied to the noninverting input is:

$$v_{\text{ref}} = +BV_{\text{sat}} \tag{22-5a}$$

When the output is negatively saturated, the reference voltage is:

$$v_{\text{ref}} = -BV_{\text{sat}} \tag{22-5b}$$

The output voltage will remain in a given state until the input voltage exceeds the reference voltage for that state. For instance, if the output is positively saturated, the reference voltage is $+BV_{\text{sat}}$. The input voltage must be increased to slightly more than $+BV_{\text{sat}}$ to switch the output voltage from positive to negative, as shown in Fig. 22-18b. Once the output is in the negative state, it will remain there indefinitely until the input voltage becomes more negative than $-BV_{\text{sat}}$. Then, the output switches from negative to positive (Fig. 22-18b).

Hysteresis

The unusual response of Fig. 22-18b has a useful property called **hysteresis.** To understand this concept, put your finger on the upper end of the graph where it says $+V_{\text{sat}}$. Assume that this is the current value of output voltage. Move your finger to the right along the horizontal line. Along this horizontal line, the input voltage is changing but the output voltage is still equal to $+V_{\text{sat}}$. When you reach the upper right corner, v_{in} equals $+BV_{\text{sat}}$. When v_{in} increases to slightly more than $+BV_{\text{sat}}$, the output voltage goes into the transition region between the high and the low states.

If you move your finger down along the vertical line, you will simulate the transition of the output voltage from high to low. When your finger is on the lower horizontal line, the output voltage is negatively saturated and equal to $-V_{\text{sat}}$.

To switch back to the high output state, move your finger until it reaches the lower left corner. At this point, v_{in} equals $-BV_{\text{sat}}$. When v_{in} becomes slightly more negative than $-BV_{\text{sat}}$, the output voltage goes into the transition from low to high. If you move your finger up along the vertical line, you will simulate the switching of the output voltage from low to high.

In Fig. 22-18b, the trip points are defined as the two input voltages where the output voltage changes states. The *upper trip point (UTP)* has the value:

$$\text{UTP} = BV_{\text{sat}} \tag{22-6}$$

and the *lower trip point (LTP)* has the value:

$$\text{LTP} = -BV_{\text{sat}} \tag{22-7}$$

The difference between these trip points is defined as the hysteresis (also called the *deadband*):

$$H = \text{UTP} - \text{LTP} \tag{22-8}$$

With Eqs. (22-6) and (22-7), this becomes:

$$H = BV_{\text{sat}} - (-BV_{\text{sat}})$$

which equals:

$$H = 2BV_{\text{sat}} \tag{22-9}$$

Figure 22-19 (a) Noninverting Schmitt trigger; (b) input/output response.

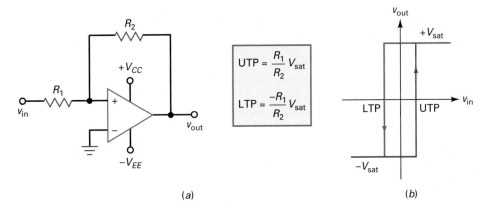

(a) (b)

Positive feedback causes the hysteresis of Fig. 22-18b. If there were no positive feedback, B would equal zero and the hysteresis would disappear, because both trip points would equal zero.

Hysteresis is desirable in a Schmitt trigger because it prevents noise from causing false triggering. If the peak-to-peak noise voltage is less than the hysteresis, the noise cannot produce false triggering. For instance, if UTP $= +1$ V and LTP $= -1$ V, then $H = 2$ V. In this case, the Schmitt trigger is immune to false triggering as long as the peak-to-peak noise voltage is less than 2 V.

Noninverting Circuit

Figure 22-19a shows a *noninverting Schmitt trigger.* The input/output response has a hysteresis loop, as shown in Fig. 22-19b. Here is how the circuit works: If the output is positively saturated in Fig. 22-19a, the feedback voltage to the noninverting input is positive, which reinforces the positive saturation. Similarly, if the output is negatively saturated, the feedback voltage to the noninverting input is negative, which reinforces the negative saturation.

Assume that the output is negatively saturated. The feedback voltage will hold the output in negative saturation until the input voltage becomes slightly more positive than UTP. When this happens, the output switches from negative to positive saturation. Once in positive saturation, the output stays there until the input voltage becomes slightly less than LTP. Then, the output can change back to the negative state.

The equations for the trip points of a noninverting Schmitt trigger are given by:

$$\mathbf{UTP} = \frac{R_1}{R_2} V_{\text{sat}} \tag{22-10}$$

$$\mathbf{LTP} = \frac{-R_1}{R_2} V_{\text{sat}} \tag{22-11}$$

The ratio of R_1 to R_2 determines how much hysteresis the Schmitt trigger has. A designer can create enough hysteresis to prevent unwanted noise triggers.

Speed-Up Capacitor

Besides suppressing the effects of noise, positive feedback speeds up the switching of output states. When the output voltage begins to change, this change is fed back to the noninverting input and amplified, forcing the output to change faster.

Figure 22-20 Speed-up capacitor compensates for stray capacitance.

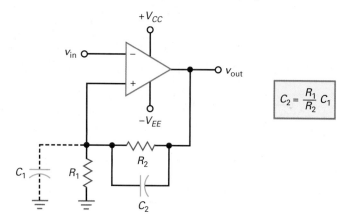

$$C_2 = \frac{R_1}{R_2} C_1$$

Sometimes a capacitor C_2 is connected in parallel with R_2, as shown in Fig. 22-20a. Known as a **speed-up capacitor,** it helps to cancel the bypass circuit formed by the stray capacitance across R_1. This stray capacitance C_1 has to be charged before the noninverting input voltage can change. The speed-up capacitor supplies this charge.

To neutralize the stray capacitance, the minimum speed-up capacitance must be at least:

$$C_2 = \frac{R_1}{R_2} C_1 \tag{22-12}$$

As long as C_2 is equal to or greater than the value given by Eq. (22-12), the output will switch states at maximum speed. Since a designer often has to estimate the stray capacitance C_1, he or she usually makes C_2 at least two times larger than the value given by Eq. (22-12). In typical circuits C_2 is from 10 to 100 pF.

Example 22-6

If $V_{\text{sat}} = 13.5$ V, what are the trip points and hysteresis in Fig. 22-21?

Figure 22-21 Example.

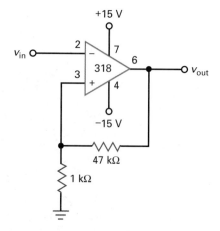

SOLUTION With Eq. (22-4), the feedback fraction is:

$$B = \frac{1 \text{ k}\Omega}{48 \text{ k}\Omega} = 0.0208$$

With Eqs. (22-6) and (22-7), the trip points are:

$$\text{UTP} = 0.0208(13.5 \text{ V}) = 0.281 \text{ V}$$
$$\text{LTP} = -0.0208(13.5 \text{ V}) = -0.281 \text{ V}$$

With Eq. (22-9), the hysteresis is:

$$H = 2(0.0208 \text{ V})(13.5 \text{ V}) = 0.562 \text{ V}$$

This means that the Schmitt trigger of Fig. 22-21 can withstand a peak-to-peak noise voltage up to 0.562 V without false triggering.

PRACTICE PROBLEM 22-6 Repeat Example 22-6 with the 47 kΩ resistor changed to 22 kΩ.

22-4 Window Comparator

An ordinary comparator indicates when the input voltage exceeds a certain limit or threshold. A **window comparator** (also called a *double-ended limit detector*) detects when the input voltage is between two limits called the *window*. To create a window comparator, we will use two comparators with different thresholds.

Low Output between Limits

Figure 22-22a shows a window comparator that can produce a low output voltage when the input voltage is between a lower and an upper limit. The circuit has an

Figure 22-22 (a) Inverting window comparator; (b) output is low when input is in window.

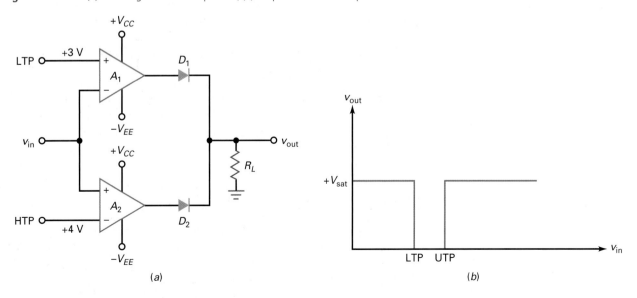

(a) (b)

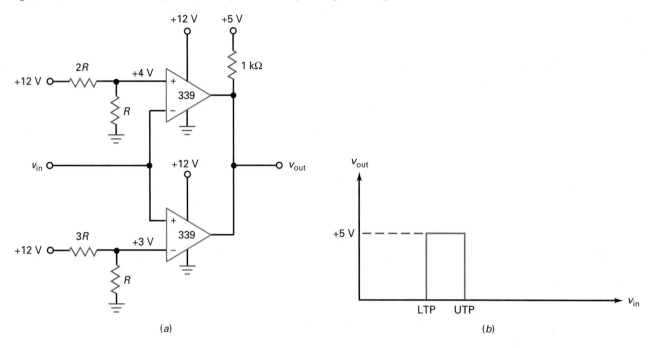

(*a*)

(*b*)

LTP and a UTP. The reference voltages can be derived from voltage dividers, zener diodes, or other circuits. Figure 22-22*b* shows the input/output response of the window comparator. When v_{in} is less than LTP or greater than UTP, the output is high. When v_{in} is between LTP and UTP, the output is low.

Here is the theory of operation: For this discussion, assume the following positive trip points: LTP = 3 V and UTP = 4 V. When $v_{in} < 3$ V, comparator A_1 has a positive output and A_2 has a negative output. Diode D_1 is on and D_2 is off. Therefore, the output voltage is high. Similarly, when $v_{in} > 4$ V, comparator A_1 has a negative output and A_2 has a positive output. Diode D_1 is off, D_2 is on, and the output voltage is high. When 3 V $< v_{in} < 4$ V, A_1 has a negative output, A_2 has a negative output, D_1 is off, D_2 is off, and the output voltage is low.

High Output between Limits

Figure 22-23*a* shows another window comparator. The circuit uses an LM339, which is a quad comparator that needs external pullup resistors. When used with a pullup supply of +5 V, the output can drive TTL circuits. Figure 22-23*b* shows the input/output response. As we can see, the output voltage is high when the input voltage is between the two limits.

For this discussion, we are assuming the same reference voltages as in the preceding example. When the input voltage is less than 3 V, the lower comparator pulls the output down to zero. When the input voltage is greater than 4 V, the upper comparator pulls the output down to zero. When v_{in} is between 3 and 4 V, the output transistor of each comparator is cut off, so the output is pulled up to +5 V.

22-5 The Integrator

An **integrator** is a circuit that performs a mathematical operation called *integration*. The most popular application of an integrator is in producing a *ramp* of output voltage, which is a linearly increasing or decreasing voltage. The integrator is sometimes called the *Miller integrator,* after the inventor.

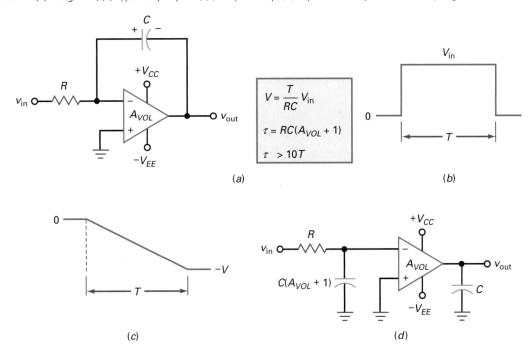

(*a*)   (*b*)

(*c*)   (*d*)

Basic Circuit

Figure 22-24*a* is an op-amp integrator. As you can see, the feedback component is a capacitor instead of a resistor. The usual input to an integrator is a rectangular pulse like the one shown in Fig. 22-24*b*. The width of this pulse is equal to T. When the pulse is low, $v_{in} = 0$. When the pulse is high, $v_{in} = V_{in}$. Visualize this pulse applied to the left end of R. Because of the virtual ground on the inverting input, a high input voltage produces an input current of:

$$I_{in} = \frac{V_{in}}{R}$$

All this input current goes into the capacitor. As a result, the capacitor charges and its voltage increases with the polarity shown in Fig. 22-24*a*. The virtual ground implies that the output voltage equals the voltage across the capacitor. For a positive input voltage, the output voltage will increase negatively, as shown in Fig. 22-24*c*.

Since a constant current is flowing into the capacitor, the charge Q increases linearly with time. This means that the capacitor voltage increases linearly, which is equivalent to a negative ramp of output voltage, as shown in Fig. 22-24*c*. At the end of the pulse period in Fig. 22-24*b*, the input voltage returns to zero and the capacitor charging stops. Because the capacitor retains its charge, the output voltage remains constant at a negative voltage of $-V$. The magnitude of this voltage is given by:

$$V = \frac{T}{RC} V_{in} \tag{22-13}$$

A final point: Because of the Miller effect, we can split the feedback capacitor into two equivalent capacitances, as shown in Fig. 22-24*d*. The closed-loop time constant τ for the input bypass circuit is:

$$\tau = RC(A_{VOL} + 1) \tag{22-14}$$

For the integrator to work properly, the closed-loop time constant should be much greater than the width of the input pulse (at least 10 times greater). As a formula:

$$\tau > 10T \tag{22-15}$$

In the typical op-amp integrator, the closed-loop time constant is extremely long, so this condition is easily satisfied.

Eliminating Output Offset

The circuit of Fig. 22-24a needs a slight modification to make it practical. Because a capacitor is open to dc signals, there is no negative feedback at zero frequency. Without negative feedback, the circuit treats any input offset voltage as a valid input voltage. The result is that the capacitor charges and the output goes into positive or negative saturation, where it stays indefinitely.

One way to reduce the effect of input offset voltage is to decrease the voltage gain at zero frequency by inserting a resistor in parallel with the capacitor, as shown in Fig. 22-25a. This resistor should be at least 10 times larger than the input resistor. If the added resistance equals $10R$, the closed-loop voltage gain is 10 and the output offset voltage is reduced to an acceptable level. When a valid input voltage is present, the additional resistor has almost no effect on the charging of a capacitor, so the output voltage is still almost a perfect ramp.

Another way to suppress the effect of input offset voltage is to use a JFET switch, as shown in Fig. 22-25b. The reset voltage on the gate of the JFET is either 0 V or $-V_{CC}$, which is enough to cut off the JFET. Therefore, we can set the JFET to a low resistance when the integrator is idle and to a high resistance when the integrator is active.

The JFET discharges the capacitor in preparation for the next input pulse. Just before the beginning of the next input pulse, the reset voltage is made equal to 0 V. This discharges the capacitor. At the instant the next pulse begins, the reset voltage becomes $-V_{CC}$, which cuts off the JFET. The integrator then produces an output voltage ramp.

Figure 22-25 (a) Resistor across capacitor reduces output offset voltage; (b) JFET used to reset integrator.

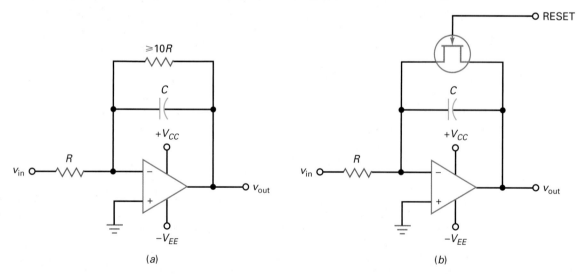

(a)

(b)

Example 22-7

In Fig. 22-26, what is the output voltage at the end of the input pulse? If the 741C has an open-loop voltage gain of 100,000, what is the closed-loop time constant of the integrator?

Figure 22-26 Example.

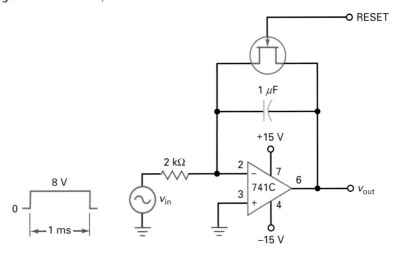

SOLUTION With Eq. (22-13), the magnitude of the negative output voltage at the end of the pulse is:

$$V = \frac{1 \text{ ms}}{(2 \text{ k}\Omega)(1 \text{ } \mu\text{F})} (8 \text{ V}) = 4 \text{ V}$$

With Eq. (22-14), the closed-loop time constant is:

$$\tau = RC(A_{VOL} + 1) = (2 \text{ k}\Omega)(1 \text{ } \mu\text{F})(100{,}001) = 200 \text{ s}$$

Since the pulse width of 1 ms is much smaller than the closed-loop time constant, only the earliest part of an exponential function is involved in the capacitor charging. Because the initial part of an exponential function is almost linear, the output voltage is almost a perfect ramp. Using an integrator to generate linear ramps is how the linear sweep voltages of an oscilloscope are produced.

PRACTICE PROBLEM 22-7 Using Fig. 22-26, change the 2 kΩ resistor to 10 kΩ and repeat Example 22-7.

22-6 Waveform Conversion

With op amps we can convert sine waves to rectangular waves, rectangular waves to triangular waves, and so on. This section is about some basic circuits that convert an input waveform to an output waveform of a different shape.

Figure 22-27 Schmitt trigger always produces rectangular output.

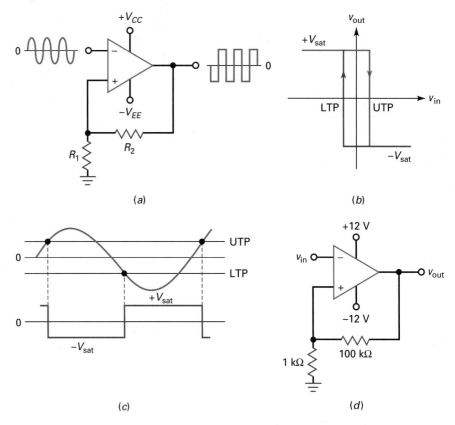

(a)

(b)

(c)

(d)

Sine to Rectangular

Figure 22-27*a* shows a Schmitt trigger, and Fig. 22-27*b* is the graph of output voltage versus input voltage. When the input signal is *periodic* (repeating cycles), the Schmitt trigger produces a rectangular output, as shown. This assumes that the input signal is large enough to pass through both trip points of Fig. 22-27*c*. When the input voltage exceeds UTP on the upward swing of the positive half cycle, the output voltage switches to $-V_{sat}$. One half cycle later, the input voltage becomes more negative than LTP, and the output switches back to $+V_{sat}$.

A Schmitt trigger always produces a rectangular output, regardless of the shape of the input signal. In other words, the input voltage does not have to be sinusoidal. As long as the waveform is periodic and has an amplitude large enough to pass through the trip points, we get a rectangular output from the Schmitt trigger. This rectangular wave has the same frequency as the input signal.

As an example, Fig. 22-27*d* shows a Schmitt trigger with trip points of approximately UTP = +0.1 V and LTP = −0.1 V. If the input voltage is repetitive and has a peak-to-peak value greater than 0.2 V, the output voltage is a rectangular wave with a peak-to-peak value of approximately $2V_{sat}$.

Rectangular to Triangular

In Fig. 22-28*a*, a rectangular wave is the input to an integrator. Since the input voltage has a dc or average value of zero, the dc or average value of the output is also zero. As shown in Fig. 22-28*b*, the ramp is decreasing during the positive half cycle of input voltage and increasing during the negative half cycle. Therefore,

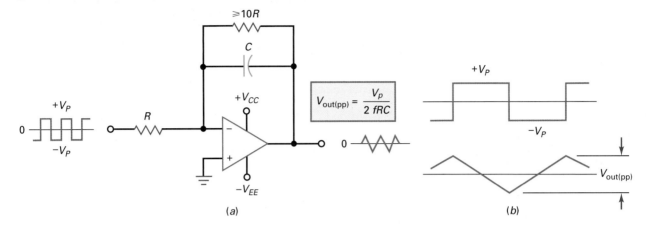

(a)   (b)

the output is a triangular wave with the same frequency as the input. It can be shown that the triangular output waveform has a peak-to-peak value of:

$$V_{out(pp)} = \frac{T}{2RC} V_p \tag{22-16}$$

where T is the period of the signal. An equivalent expression in terms of frequency is:

$$V_{out(pp)} = \frac{V_p}{2fRC} \tag{22-17}$$

where V_p is the peak input voltage and f is the input frequency.

Triangle to Pulse

Figure 22-29*a* shows a circuit that converts a triangular input to a rectangular output. By varying R_2, we can change the width of the output pulses, which is equivalent to varying the duty cycle. In Fig. 22-29*b*, W represents the width of the pulse and T is the period. As previously discussed, the duty cycle D is the width of the pulse divided by the period.

In some applications, we want to vary the duty cycle. The adjustable limit detector of Fig. 22-29*a* is ideal for this purpose. With this circuit, we can move the trip point from zero to a positive level. When the triangular input voltage exceeds

Figure 22-29 Triangular input to limit detector produces rectangular output.

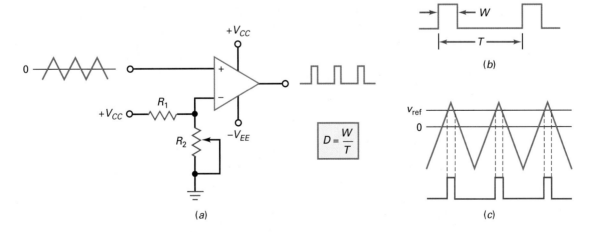

(a)   (c)

the trip point, the output is high, as shown in Fig. 22-29c. Since v_{ref} is adjustable, we can vary the width of the output pulse, which is equivalent to changing the duty cycle. With a circuit like this, we can vary the duty cycle from approximately 0 to 50 percent.

Example 22-8

What is the output voltage in Fig. 22-30 if the input frequency is 1 kHz?

Figure 22-30 Example.

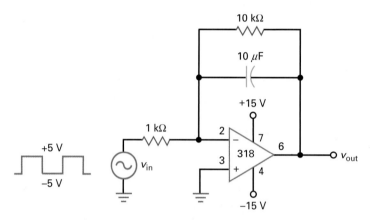

SOLUTION With Eq. (22-17), the output is a triangular wave with a peak-to-peak voltage of:

$$V_{out(pp)} = \frac{5\ V}{2(1\ \text{kHz})(1\ \text{k}\Omega)(10\ \mu\text{F})} = 0.25\ \text{V pp}$$

PRACTICE PROBLEM 22-8 In Fig. 22-30, what value of capacitor is needed to produce an output voltage of 1 V pp?

Example 22-9

|||| MultiSim

A triangular input drives the circuit of Fig. 22-31a. The variable resistance has a maximum value of 10 kΩ. If the triangular input has a frequency of 1 kHz, what is the duty cycle when the wiper is at the middle of its range?

Figure 22-31 Example.

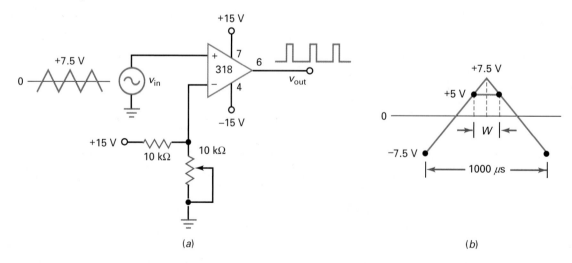

(a)

(b)

SOLUTION When the wiper is at the middle of its range, it has a resistance of 5 kΩ. This means that the reference voltage is:

$$v_{ref} = \frac{5\ k\Omega}{15\ k\Omega}\ 15\ V = 5\ V$$

The period of the signal is:

$$T = \frac{1}{1\ kHz} = 1000\ \mu s$$

Figure 22-31b shows this value. It takes 500 μs for the input voltage to increase from −7.5 to +7.5 V because this is half of the cycle. The trip point of the comparator is +5 V. This means that the output pulse has a width of W, as shown in Fig. 22-31b.

Because of the geometry of Fig. 22-31b, we can set up a proportion between voltage and time as follows:

$$\frac{W/2}{500\ \mu s} = \frac{7.5\ V - 5\ V}{15\ V}$$

Solving for W gives:

$$W = 167\ \mu S$$

The duty cycle is:

$$D = \frac{167\ \mu s}{1000\ \mu s} = 0.167$$

In Fig. 22-31a, moving the wiper down will increase the reference voltage and decrease the output duty cycle. Moving the wiper up will decrease the reference voltage and increase the output duty cycle. For all values given in Fig. 22-31a, the duty cycle can vary from 0 to 50 percent.

PRACTICE PROBLEM 22-9 Repeat Example 22-9 using an input frequency of 2 kHz.

22-7 Waveform Generation

With positive feedback, we can build **oscillators,** circuits that generate or create an output signal with no external input signal. This section discusses some op-amp circuits that can generate nonsinusoidal signals.

Relaxation Oscillator

In Fig. 22-32a, there is no input signal. Nevertheless, the circuit produces a rectangular output signal. This output is a square wave that swings between $-V_{sat}$ and $+V_{sat}$. How is this possible? Assume that the output of Fig. 22-32a is in positive saturation. Because of feedback resistor R, the capacitor will charge exponentially toward $+V_{sat}$, as shown in Fig. 22-32b. But the capacitor voltage never reaches $+V_{sat}$ because the voltage crosses the UTP. When this happens, the output square wave switches to $-V_{sat}$.

With the output now in negative saturation, the capacitor discharges, as shown in Fig. 22-32b. When the capacitor voltage crosses through zero, the capacitor starts charging negatively toward $-V_{sat}$. When the capacitor voltage crosses the LTP, the output square wave switches back to $+V_{sat}$. The cycle then repeats.

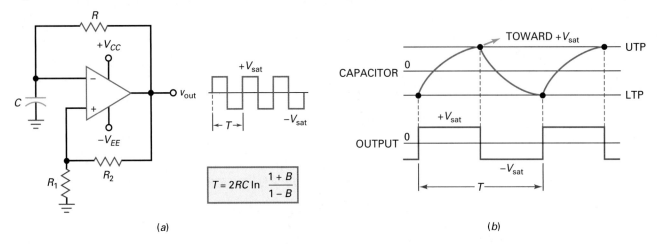

(*a*)   (*b*)

Because of the continuous charging and discharging of the capacitor, the output is a rectangular wave with a duty cycle of 50 percent. By analyzing the exponential charge and discharge of the capacitor, we can derive this formula for the period of the rectangular output:

$$T = 2RC \ln \frac{1 + B}{1 - B} \tag{22-18}$$

where B is the feedback fraction given by:

$$B = \frac{R_1}{R_1 + R_2}$$

Equation (22-18) uses the *natural logarithm,* which is a logarithm to base *e*. A scientific calculator or table of natural logarithms must be used with this equation.

Figure 22-32*a* is called a **relaxation oscillator,** defined as a circuit that generates an output signal whose frequency depends on the charging of a capacitor. If we increase the RC time constant, it takes longer for the capacitor voltage to reach the trip points. Therefore, the frequency is lower. By making R adjustable, we can get a 50 : 1 tuning range.

Generating Triangular Waves

By cascading a relaxation oscillator and an integrator, we get a circuit that produces the triangular output shown in Fig. 22-33. The rectangular wave out of the

Figure 22-33 Relaxation oscillator drives integrator to produce triangular output.

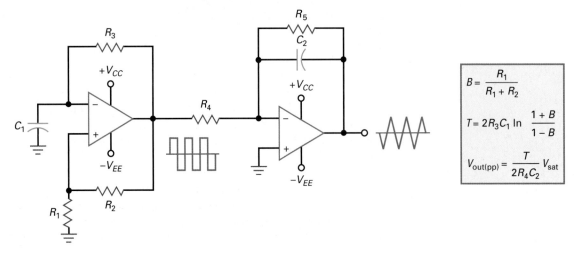

relaxation oscillator drives the integrator, which produces a triangular output waveform. The rectangular wave swings between $+V_{sat}$ and $-V_{sat}$. You can calculate its period with Eq. (22-18). The triangular wave has the same period and frequency. You can calculate its peak-to-peak value with Eq. (22-16).

Example 22-10

||||| MultiSim

What is the frequency of the output signal in Fig. 22-34?

Figure 22-34 Example.

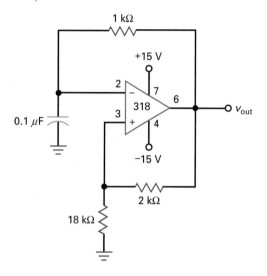

SOLUTION The feedback fraction is:

$$B = \frac{18 \text{ k}\Omega}{20 \text{ k}\Omega} = 0.9$$

With Eq. (22-18):

$$T = 2RC \ln \frac{1+B}{1-B} = 2(1 \text{ k}\Omega)(0.1 \text{ }\mu\text{F}) \ln \frac{1+0.9}{1-0.9} = 589 \text{ }\mu\text{s}$$

The frequency is:

$$f = \frac{1}{589 \text{ }\mu\text{s}} = 1.7 \text{ kHz}$$

The square-wave output voltage has a frequency of 1.7 kHz and a peak-to-peak value of $2V_{sat}$, approximately 27 V for the circuit of Fig. 22-34.

PRACTICE PROBLEM 22-10 In Fig. 22-34, change the 18 kΩ resistor to 10 kΩ and calculate the new output frequency.

Example 22-11

||||| MultiSim

The relaxation oscillator of Example 22-10 is used in Fig. 22-33 to drive the integrator. Assume that the peak voltage out of the relaxation oscillator is 13.5 V. If the integrator has $R_4 = 10$ kΩ and $C_2 = 10$ μF, what is the peak-to-peak value of the triangular output wave?

SOLUTION With the equations shown in Fig. 22-33, we can analyze the circuit. In Example 22-10, we calculated a feedback fraction of 0.9 and a period of 589 μs. Now, we can calculate the peak-to-peak value of the triangular output:

$$V_{out(pp)} = \frac{589 \ \mu s}{2(10 \ k\Omega)(10 \ \mu F)} (13.5 \ V) = 39.8 \ mV \ pp$$

The circuit generates a square wave with a peak-to-peak value of approximately 27 V and a triangular wave with a peak-to-peak value of 39.8 mV.

PRACTICE PROBLEM 22-11 Repeat Example 22-11 with the 18 kΩ resistor, in Fig. 22-34, changed to 10 kΩ.

22-8 Another Triangular Generator

In Fig. 22-35a, the output of a noninverting Schmitt trigger is a rectangular wave that drives an integrator. The output of the integrator is a triangular wave. This triangular wave is fed back and drives the Schmitt trigger. So we have a very interesting circuit. The first stage drives the second, and the second drives the first.

Figure 22-35b is the transfer characteristic of the Schmitt trigger. When the output is low, the input must increase to the UTP to switch the output to high.

Figure 22-35 Schmitt trigger and integrator generate square wave and triangular wave.

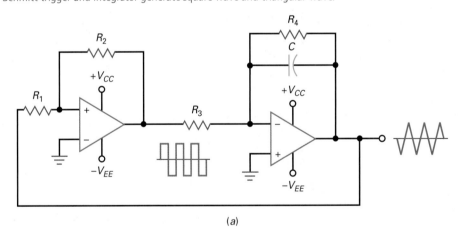

(a)

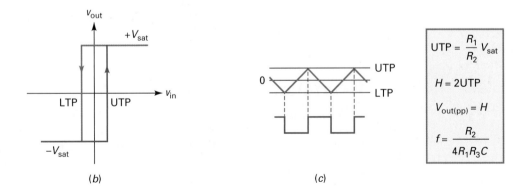

(b)

(c)

$$UTP = \frac{R_1}{R_2} V_{sat}$$

$$H = 2UTP$$

$$V_{out(pp)} = H$$

$$f = \frac{R_2}{4R_1R_3C}$$

Likewise, when the output is high, the input must decrease to the LTP to switch the output to low.

When the Schmitt trigger output is low in Fig. 22-35c, the integrator produces a positive ramp, which increases until it reaches the UTP. At this point, the output of the Schmitt trigger switches to the high state and the triangular wave reverses direction. The negative ramp then decreases until it reaches the LTP, where another Schmitt output change takes place.

In Fig. 22-35c, the peak-to-peak value of the triangular wave equals the difference between the UTP and the LTP. We can derive this equation for the frequency:

$$f = \frac{R_2}{4R_1R_3C} \tag{22-19}$$

Figure 22-35 shows this equation, along with other analysis equations.

Example 22-12

The triangular-wave generator of Fig. 22-35a has these circuit values: $R_1 = 1$ kΩ, $R_2 = 100$ kΩ, $R_3 = 10$ kΩ, $R_4 = 100$ kΩ, and $C = 10$ μF. What is the peak-to-peak output if $V_{sat} = 13$ V? What is the frequency of the triangular wave?

SOLUTION With the equations shown in Fig. 22-35, the UTP has a value of:

$$\text{UTP} = \frac{1 \text{ k}\Omega}{100 \text{ k}\Omega}(13 \text{ V}) = 0.13 \text{ V}$$

The peak-to-peak value of the triangular output equals the hysteresis:

$$V_{out(pp)} = H = 2\text{UTP} = 2(0.13 \text{ V}) = 0.26 \text{ V}$$

The frequency is:

$$f = \frac{100 \text{ k}\Omega}{4(1 \text{ k}\Omega)(10 \text{ k}\Omega)(10 \text{ }\mu\text{F})} = 250 \text{ Hz}$$

PRACTICE PROBLEM 22-12 Using Fig. 22-35, change R_1 to 2 kΩ and C to 1 μF. Calculate $V_{out(pp)}$ and the output frequency.

22-9 Active Diode Circuits

Op amps can enhance the performance of diode circuits. For one thing, an op amp with negative feedback reduces the effect of the knee voltage, allowing us to rectify, peak-detect, clip, and clamp low-level signals (those with amplitudes less than the knee voltage). And because of their buffering action, op amps can eliminate the effects of the source and load on diode circuits.

Half-Wave Rectifier

Figure 22-36 is an **active half-wave rectifier.** When the input signal goes positive, the output goes positive and turns on the diode. The circuit then acts like a voltage follower, and the positive half cycle appears across the load resistor.

Figure 22-36 Active half-wave rectifier.

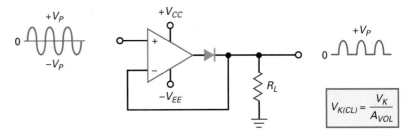

When the input goes negative, the op-amp output goes negative and turns off the diode. Since the diode is open, no voltage appears across the load resistor. The final output is almost a perfect half-wave signal.

There are two distinct *modes* or regions of operation. First, when the input voltage is positive, the diode is conducting and the operation is linear. In this case, the output voltage is fed back to the input, and we have negative feedback. Second, when the input voltage is negative, the diode is nonconducting and the feedback path is open. In this case, the op-amp output is isolated from the load resistor.

The high open-loop voltage gain of the op amp almost eliminates the effect of the knee voltage. For instance, if the knee voltage is 0.7 V and A_{VOL} is 100,000, the input voltage that just turns on the diode is 7 μV.

The closed-loop knee voltage is given by:

$$V_{K(CL)} = \frac{V_K}{A_{VOL}}$$

where $V_K = 0.7$ V for a silicon diode. Because the closed-loop knee voltage is so small, the active half-wave rectifier may be used with low-level signals in the microvolt region.

Active Peak Detector

To peak-detect small signals, we can use an **active peak detector** like the one shown in Fig. 22-37a. Again, the closed-loop knee voltage is in the microvolt region, which means that we can peak-detect low-level signals. When the diode is on, the negative feedback produces a Thevenin output impedance that approaches zero. This means that the charging time constant is very low, so the capacitor can quickly charge to the positive peak value. When the diode is off, the capacitor has to discharge through R_L. Because the discharging time constant R_LC can be made much longer than the period of the input signal, we can get almost perfect peak detection of low-level signals.

There are two distinct regions of operation. First, when the input voltage is positive, the diode is conducting and the operation is linear. In this case, the capacitor charges to the peak of the input voltage. Second, when the input voltage is negative, the diode is nonconducting and the feedback path is open. In this case, the capacitor discharges through the load resistor. As long as the discharging time constant is much greater than the period of the input signal, the output voltage will be approximately equal to the peak value of the input voltage.

If the peak-detected signal has to drive a small load, we can avoid loading effects by using an op-amp buffer. For instance, if we connect point *A* of Fig. 22-37a to point *B* of Fig. 22-37b, the voltage follower isolates the small load resistor from the peak detector. This prevents the small load resistor from discharging the capacitor too quickly.

At a minimum, the R_LC time constant should be at least 10 times longer than the period T of the lowest input frequency. In symbols:

$$R_LC > 10T \tag{22-20}$$

Figure 22-37 (a) Active peak detector; (b) buffer amplifier; (c) peak detector with reset.

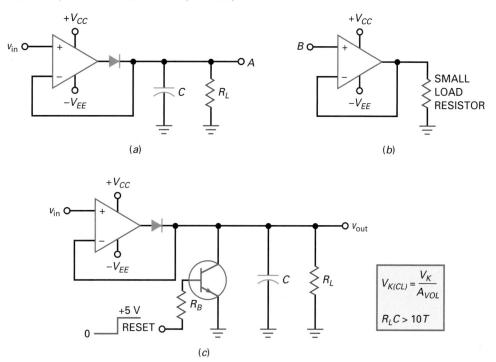

$$V_{K(CL)} = \frac{V_K}{A_{VOL}}$$

$$R_L C > 10T$$

(a)

(b)

(c)

If this condition is satisfied, the output voltage will be within 5 percent of the peak input. For instance, if the lowest frequency is 1 kHz, the period is 1 ms. In this case, the $R_L C$ time constant should be at least 10 ms for an error of less than 5 percent.

Often, a *reset* is included with an active peak detector, as shown in Fig. 22-37c. When the reset input is low, the transistor switch is off. This allows the circuit to work as previously described. When the reset input is high, the transistor switch is closed. This rapidly discharges the capacitor. The reason you may need a reset is because the long discharge time constant means that the capacitor will hold its charge for a long time, even though the input signal is removed. By using a high reset input, we can quickly discharge the capacitor in preparation for another input signal with a different peak value.

Active Positive Clipper

Figure 22-38a is an **active positive clipper.** With the wiper all the way to the left, v_{ref} is zero and the noninverting input is grounded. When v_{in} is positive, the op-amp output is negative and the diode is conducting. The low impedance of the diode produces heavy negative feedback because the feedback resistance approaches zero. For this condition, the output node is at virtual ground for all positive values of v_{in}.

When v_{in} goes negative, the output of the op amp is positive, which turns off the diode and opens the loop. When the loop is open, the virtual ground is lost and v_{out} equals the negative half cycle of input voltage. This is why the negative half cycle appears at the output as shown.

We can adjust the clipping level by moving the wiper to get different values of v_{ref}. In this way, we get the output waveform shown in Fig. 22-38a. The reference level can be varied between 0 and $+V$.

Figure 22-38b shows an active circuit that clips on both half cycles. Notice the back-to-back zener diodes in the feedback loop. Below the zener voltage, the circuit has a closed-loop gain of R_2/R_1. When the output tries to exceed the zener voltage plus one forward diode drop, the zener diode breaks down and

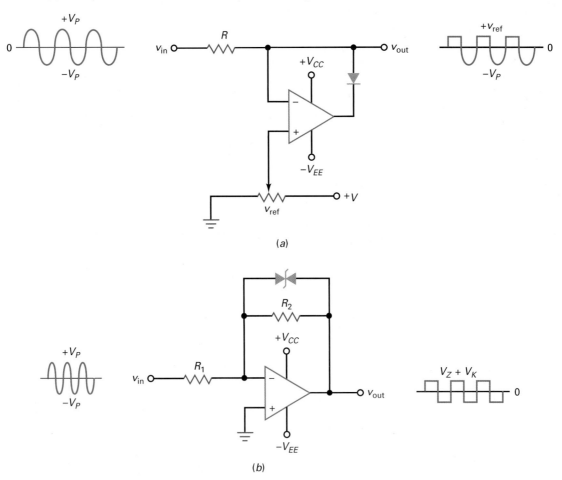

(a)

(b)

the output voltage is $V_Z + V_K$ away from virtual ground. This is why the output is clipped as shown.

Active Positive Clamper

Figure 22-39 is an **active positive clamper.** This circuit adds a dc component to the input signal. As a consequence, the output has the same size and shape as the input signal, except for the dc shift.

Here is the theory of operation: The first negative input half cycle is coupled through the uncharged capacitor and produces a positive op-amp output that

Figure 22-39   Active positive clamper.

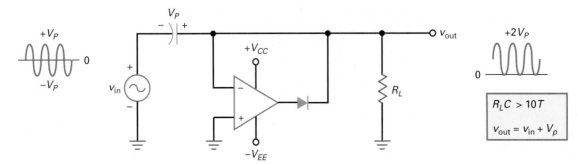

turns on the diode. Because of the virtual ground, the capacitor charges to the peak value of the negative input half cycle with the polarity shown in Fig. 22-39. Just beyond the negative input peak, the diode turns off, the loop opens, and the virtual ground is lost. In this case, the output voltage is the sum of the input voltage and the capacitor voltage:

$$v_{\text{out}} = v_{\text{in}} + V_p \tag{22-21}$$

Since V_p is being added to a sinusoidal input voltage, the final output waveform is shifted positively through V_p, as shown in Fig. 22-39. The positively clamped waveform swings from 0 to $+2V_p$, which means that it has a peak-to-peak value of $2V_p$, the same as the input. Again, the negative feedback reduces the knee voltage by a factor of approximately A_{VOL}, which means that we can build excellent clampers for low-level inputs.

Figure 22-39 shows the op-amp output. During most of the cycle, the op amp operates in negative saturation. Right at the negative input peak, however, the op amp produces a sharp, positive-going pulse that replaces any charge lost by the clamping capacitor between negative input peaks.

22–10 The Differentiator

A **differentiator** is a circuit that performs a calculus operation called *differentiation*. It produces an output voltage proportional to the instantaneous rate of change of the input voltage. Common applications of a differentiator are to detect the leading and trailing edges of a rectangular pulse or to produce a rectangular output from a ramp input.

RC Differentiator

An *RC* circuit like the one shown in Fig. 22-40*a* can be used to differentiate an input signal. The typical input signal is a rectangular pulse, as shown in Fig. 22-40*b*. The output of the circuit is a series of positive and negative spikes. The positive spike occurs at the same instant as the leading edge of the input, and the negative spike occurs at the same instant as the trailing edge. Spikes like these are useful signals because they indicate when the rectangular input signal starts and ends.

To understand how the *RC* differentiator works, look at Fig. 22-40*c*. When the input voltage changes from 0 to $+V$, the capacitor begins to charge

Figure 22–40 (*a*) *RC* differentiator; (*b*) rectangular input produces spiked output; (*c*) charging waveforms; (*d*) example.

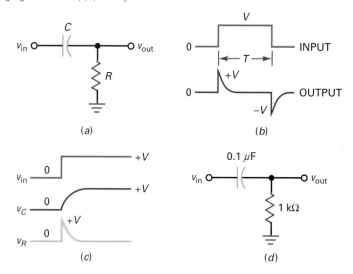

exponentially, as shown. After five time constants, the capacitor voltage is within 1 percent of the final voltage. To satisfy Kirchhoff's voltage law, the voltage across the resistor of Fig. 22-40a is:

$$v_R = v_{in} - v_C$$

Since v_C is initially zero, the output voltage suddenly jumps from 0 to V and then decays exponentially, as shown in Fig. 22-40b. By a similar argument, the trailing edge of a rectangular pulse produces a negative spike. Incidentally, each spike in Fig. 22-40b has a peak value of approximately V, the size of the voltage step.

If an RC differentiator is to produce narrow spikes, the time constant should be at least 10 times smaller than the pulse width T:

$$RC < 10T$$

If the pulse width is 1 ms, the RC time constant should be less than 0.1 ms. Figure 22-40d shows an RC differentiator with a time constant of 0.1 ms. If you drive this circuit with any rectangular pulse that has T greater than 1 ms, the output is a series of sharp positive and negative voltage spikes.

Op–Amp Differentiator

Figure 22-41a shows an op-amp differentiator. Notice the similarity to the op-amp integrator. The difference is that the resistor and capacitor are interchanged. Because of the virtual ground, the capacitor current passes through the feedback resistor, producing a voltage across this resistor. The capacitor current is given by this fundamental relation:

$$i = C\frac{dv}{dt}$$

The quantity dv/dt equals the slope of the input voltage.

One common application of the op-amp differentiator is to produce very narrow spikes, as shown in Fig. 22-41b. The advantage of an op-amp differentiator over a simple RC differentiator is that the spikes are coming from a low-impedance source, which makes driving typical load resistances easier.

Practical Op–Amp Differentiator

The op-amp differentiator of Fig. 22-41a has a tendency to oscillate. To avoid this, a practical op-amp differentiator usually includes some resistance in series with the capacitor, as shown in Fig. 22-42. A typical value for this added resistance is between $0.01R$ and $0.1R$. With this resistor, the closed-loop voltage gain is between 10 and 100. The effect is to limit the closed-loop voltage gain at higher frequencies, where the oscillation problem arises.

Figure 22-41   (a) Op-amp differentiator; (b) rectangular input produces spiked output.

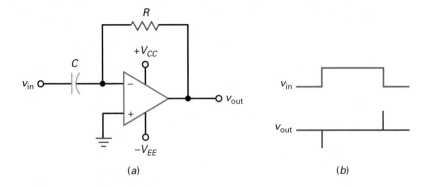

(a)   (b)

Figure 22-42 Resistance added to input to prevent oscillations.

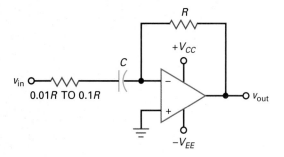

v_{in} —— 0.01R TO 0.1R —— C

$+V_{CC}$

R

$-V_{EE}$

v_{out}

22-11 Class-D Amplifier

The Class-B or Class-AB amplifier has been the main for choice of many designers for audio amplifiers. This linear amplifier configuration has been able to provide the necessary conventional performance and cost requirements. Now, products such as LCD TVs, plasma TVs, and desktop PCs are driving the necessity for greater power output while maintaining or reducing the form-factor, without increasing costs. Portable powered devices, such as PDAs, cell phones, and notebook PCs, are demanding higher circuit efficiencies. Due to very high efficiency and low heat dissipation, the Class-D amplifier is now challenging the Class-AB amplifiers in many applications. The Class-D amplifier demonstrates a practical application of many of the circuits and devices we have been discussing.

Instead of being biased for linear operation, a **Class-D amplifier** uses output transistors operated as switches. This enables each transistor to either be in a cutoff or saturated mode. When in cutoff, its current is zero. When it is saturated, the voltage across it is low. In each mode, its power dissipation is very low. This concept increases the circuit efficiency, therefore, requires less power from the power supply and enables the use of smaller heat sinks for the amplifier.

A basic Class-D amplifier is shown in Fig. 22-43. The amplifier consists of a comparator op amp driving two MOSFETs operating as switches. The comparator has two input signals: one signal is the audio signal V_A, and the other input

Figure 22-43 Basic class-D amplifier.

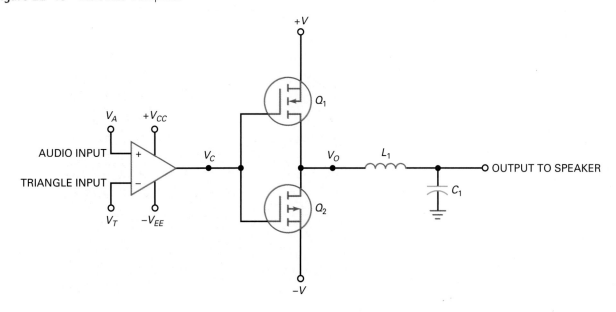

Figure 22-44 Input waveforms.

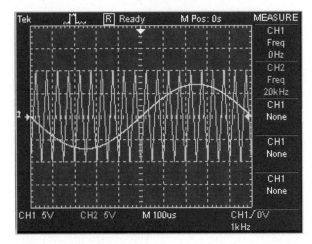

is a triangle wave V_T with a much higher frequency. The voltage value out of the comparator, V_C, will be approximately at either $+V_{CC}$ or $-V_{EE}$. When $V_A > V_T$, $V_C = +V_{CC}$. When $V_A < V_T$, $V_C = -V_{EE}$.

The comparator's positive or negative output voltage drives two complementary common-source MOSFETs. When V_C is positive, Q_1 is switched on and Q_2 is off. When V_C is negative, Q_2 is switched on and Q_1 is off. The output voltage of each transistor will be slightly less than their $+V$ and $-V$ supply values. L_1 and C_1 act as a low-pass filter. When their values are properly chosen, this filter passes the average value of the switching transistors' output to the speaker. If the audio input signal V_A were zero, V_O would be a symmetrical square wave with an average value of zero volts.

To illustrate the operation of this circuit, examine Fig. 22-44. A 1 kHz sine wave is applied to the input at V_A, and a 20 kHz triangle wave is applied to input V_T. In practice, the triangle-wave input frequency would be many times higher than in this illustration. A frequency of 250–300 kHz is often used. The frequency should be as high as possible compared to the cutoff frequency, f_c, of L_1C_1 for minimum output distortion. Also, note that the maximum voltage of V_A is at approximately 70 percent of V_T.

The resulting output V_O of the switching transistors is a **pulse-width-modulated (PWM)** waveform. The duty cycle of the waveform produces an output whose average value follows the audio input signal. This is shown in Fig. 22-45.

Figure 22-45 Output waveform following the input.

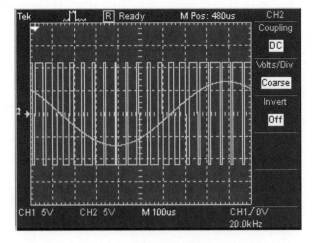

More sophisticated Class-D amplifiers use a MOSFET H-bridge circuit configuration for the switching devices and incorporate active low-pass filters. Resulting efficiencies can reach upwards from 85–90 percent, even at lower power levels. This exceeds the efficiency of the Class-AB amplifier, whose efficiency reaches a theoretical maximum of 78 percent at high-output levels and is much less efficient at lower power levels.

New generation IC Class-D amplifiers, such as the NJU8755, amplify analog input signals and produce PWM digital output signals. This provides for a fusion between digital and analog systems. The NJU8755, configured as a stereo bridge-tied load (BTL) and connected to an analog input signal, is capable of delivering 1.2 W/channel at 5 V into 8 ohms. This type of circuit also employs a standby mode designed to reduce power consumption to minimum levels during silent periods.

Summary

SEC. 22–1 COMPARATORS WITH ZERO REFERENCE

A comparator with a reference voltage of zero is called a zero-crossing detector. Diode clamps are often used to protect the comparator against excessively large input voltages. Comparators usually interface their outputs with digital circuits.

SEC. 22–2 COMPARATORS WITH NONZERO REFERENCES

In some applications a threshold voltage different from zero may be preferred. Comparators with a nonzero reference voltage are sometimes called limit detectors. Although op amps may be used as comparators, IC comparators are optimized for this application by removing the internal compensating capacitor. This increases the switching speed.

SEC. 22–3 COMPARATORS WITH HYSTERESIS

Noise is any kind of unwanted signal that is not derived from or harmonically related to the input signal. Because noise can cause false triggering of a comparator, positive feedback is used to create hysteresis. This prevents noise from producing false triggering. The positive feedback also speeds up the switching between output states.

SEC. 22–4 WINDOW COMPARATOR

A window comparator, also called a double-ended limit detector, detects when the input voltage is between two limits. To create the window, a window comparator uses two comparators with two different trip points.

SEC. 22–5 THE INTEGRATOR

An integrator is useful for converting rectangular pulses into linear ramps. Because of the large input Miller capacitance, only the earliest part of an exponential charge is used. Since this early part is almost linear, the output ramps are almost perfect. Integrators are used to create the time bases of oscilloscopes.

SEC. 22–6 WAVEFORM CONVERSION

We can use a Schmitt trigger to convert a sine wave to a rectangular wave. An integrator can convert a square wave to a triangular wave. With an adjustable resistor, we can control the duty cycle with a limit detector.

SEC. 22–7 WAVEFORM GENERATION

With positive feedback, we can build oscillators, circuits that generate or create an output signal with no external input signal. A relaxation oscillator uses the charging of a capacitor to generate an output signal. By cascading a relaxation oscillator and an integrator, we can produce a triangular output waveform.

SEC. 22–8 ANOTHER TRIANGULAR GENERATOR

The output of a noninverting Schmitt trigger can be used to drive an integrator. If the output of the integrator is used as the input to the Schmitt trigger, we have an oscillator that produces both square waves and triangular waves.

SEC. 22–9 ACTIVE DIODE CIRCUITS

With op amps, we can build active half-wave rectifiers, peak detectors, clippers, and clampers. In all these circuits, the closed-loop knee voltage equals the knee voltage divided by the open-loop voltage gain. Because of this, we can process low-level signals.

SEC. 22–10 THE DIFFERENTIATOR

When a square wave drives an RC differentiator, the output is a series of narrow positive and negative voltage spikes. With an op amp, we can improve the differentiation and get a low output impedance.

SEC. 22–11 CLASS–D AMPLIFIER

The Class-D amplifier uses output transistors operated as switches. Instead of operating in a linear region, these transistors are alternately driven into saturation and cutoff by the output signal of a comparator circuit. The Class-D amplifier is capable of very high circuit efficiencies and is gaining popularity in portable equipment needing audio amplification.

Definitions

(22-8)   Hysteresis:

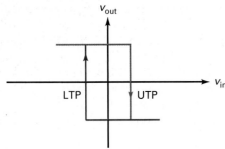

$$H = \text{UTP} - \text{LTP}$$

Derivations

For all derivations not shown here, see appropriate figures in chapter.

(22-9)   Hysteresis:

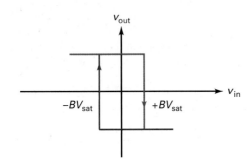

$$H = 2BV_{sat}$$

(22-12)   Speed-up capacitance:

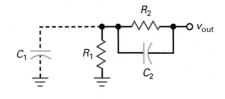

$$C_2 = \frac{R_1}{R_2} C_1$$

Student Assignments

1. **In a nonlinear op-amp circuit, the**
    a. Op amp never saturates
    b. Feedback loop is never opened
    c. Output shape is the same as the input shape
    d. Op amp may saturate

2. **To detect when the input is greater than a particular value, use a**
    a. Comparator
    b. Clamper
    c. Limiter
    d. Relaxation oscillator

3. **The voltage out of a Schmitt trigger is**
    a. A low voltage
    b. A high voltage
    c. Either a low or a high voltage
    d. A sine wave

4. **Hysteresis prevents false triggering associated with**
    a. A sinusoidal input
    b. Noise voltages
    c. Stray capacitances
    d. Trip points

5. **If the input is a rectangular pulse, the output of an integrator is a**
    a. Sine wave
    b. Square wave
    c. Ramp
    d. Rectangular pulse

6. **When a large sine wave drives a Schmitt trigger, the output is a**
    a. Rectangular wave
    b. Triangular wave
    c. Rectified sine wave
    d. Series of ramps

7. If pulse width decreases and the period stays the same, the duty cycle
   a. Decreases
   b. Stays the same
   c. Increases
   d. Is zero

8. The output of a relaxation oscillator is a
   a. Sine wave
   b. Square wave
   c. Ramp
   d. Spike

9. If A_{VOL} = 100,000, the closed-loop knee voltage of a silicon diode is
   a. 1 μV
   b. 3.5 μV
   c. 7 μV
   d. 14 μV

10. The input to a peak detector is a triangular wave with a peak-to-peak value of 8 V and an average value of 0. The output is
    a. 0
    b. 4 V
    c. 8 V
    d. 16 V

11. The input to a positive limiter is a triangular wave with a peak-to-peak value of 8 V and an average value of 0. If the reference level is 2 V, the output has a peak-to-peak value of
    a. 0   c. 6 V
    b. 2 V   d. 8 V

12. The discharging time constant of a peak detector is 100 ms. The lowest frequency you should use is
    a. 10 Hz   c. 1 kHz
    b. 100 Hz   d. 10 kHz

13. A comparator with a trip point of zero is sometimes called a
    a. Threshold detector
    b. Zero-crossing detector
    c. Positive limit detector
    d. Half-wave detector

14. To work properly, many IC comparators need an external
    a. Compensating capacitor
    b. Pullup resistor
    c. Bypass circuit
    d. Output stage

15. A Schmitt trigger uses
    a. Positive feedback
    b. Negative feedback
    c. Compensating capacitors
    d. Pullup resistors

16. A Schmitt trigger
    a. Is a zero-crossing detector
    b. Has two trip points
    c. Produces triangular output waves
    d. Is designed to trigger on noise voltage

17. A relaxation oscillator depends on the charging of a capacitor through a
    a. Resistor
    b. Inductor
    c. Capacitor
    d. Noninverting input

18. A ramp of voltage
    a. Always increases
    b. Is a rectangular pulse
    c. Increases or decreases at a linear rate
    d. Is produced by hysteresis

19. The op-amp integrator uses
    a. Inductors
    b. The Miller effect
    c. Sinusoidal inputs
    d. Hysteresis

20. The trip point of a comparator is the input voltage that causes
    a. The circuit to oscillate
    b. Peak detection of the input signal
    c. The output to switch states
    d. Clamping to occur

21. In an op-amp integrator, the current through the input resistor flows into the
    a. Inverting input
    b. Noninverting input
    c. Bypass capacitor
    d. Feedback capacitor

22. An active half-wave rectifier has a knee voltage of
    a. V_K
    b. 0.7 V
    c. More than 0.7 V
    d. Much less than 0.7 V

23. In an active peak detector, the discharging time constant is
    a. Much longer than the period
    b. Much shorter than the period
    c. Equal to the period
    d. The same as the charging time constant

24. If the reference voltage is zero, the output of an active positive limiter is
    a. Positive
    b. Negative
    c. Either positive or negative
    d. A ramp

25. The output of an active positive clamper is
    a. Positive
    b. Negative
    c. Either positive or negative
    d. A ramp

26. The positive clamper adds
    a. A positive dc voltage to the input
    b. A negative dc voltage to the input
    c. An ac signal to the output
    d. A trip point to the point

27. A window comparator
    a. Has only one usable threshold
    b. Uses hysteresis to speed up response
    c. Clamps the input positively
    d. Detects an input voltage between two limits

28. An RC differentiator circuit produces an output voltage related to the instantaneous rate of change of the input
    a. Current   c. Resistance
    b. Voltage   d. Frequency

29. An op-amp differentiator is used to produce
    a. Output square waves
    b. Output sine waves
    c. Output voltage spikes
    d. Output dc levels

30. Class-D amplifiers are very efficient because
    a. The output transistors are either cutoff or saturated
    b. They do not require a dc voltage source
    c. They use RF tuned stages
    d. They conduct for 360° of the input voltage

Problems

SEC. 22-1 COMPARATORS WITH ZERO REFERENCE

22-1 In Fig. 22-1a, the comparator has an open-loop voltage gain of 106 dB. What is the input voltage that produces positive saturation if the supply voltages are ±20 V?

22-2 If the input voltage is 50 V in Fig. 22-2a, what is the approximation current through the left clamping diode if $R = 10$ kΩ?

22-3 In Fig. 22-7a, each zener diode is a 1N4736A. If the supply voltages are ±15 V, what is the output voltage?

22-4 The dual supplies of Fig. 22-7b are reduced to ±12 V, and the diode is reversed. What is the output voltage?

22-5 If the diode of Fig. 22-9 is reversed and the supplies changed to ±9 V, what is the output when the strobe is high? When it is low?

SEC. 22-2 COMPARATORS WITH NONZERO REFERENCES

22-6 In Fig. 22-11a, the dual supply voltages are ±15 V. If $R_1 = 47$ kΩ and $R_2 = 12$ kΩ, what is the reference voltage? If the bypass capacitance is 0.5 μF, what is the cutoff frequency?

22-7 In Fig. 22-11c, the dual supply voltages are ±12 V. If $R_1 = 15$ kΩ and $R_2 = 7.5$ kΩ, what is the reference voltage? If the bypass capacitance is 1.0 μF, what is the cutoff frequency?

22-8 In Fig. 22-12, $V_{CC} = 9$ V, $R_1 = 22$ kΩ, and $R_2 = 4.7$ kΩ. What is the output duty cycle if the input is a sine wave with a peak of 7.5 V?

22-9 In Fig. 22-46, what is the output duty cycle if the input is a sine wave with a peak of 5 V?

SEC. 22-3 COMPARATORS WITH HYSTERESIS

22-10 In Fig. 22-18a, $R_1 = 2.2$ kΩ, and $R_2 = 18$ kΩ. If $V_{sat} = 14$ V, what are the trip points? What is the hysteresis?

Figure 22-46

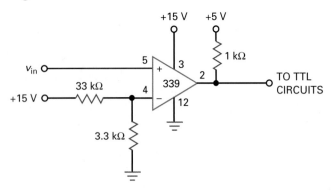

22-11 If $R_1 = 1$ kΩ, $R_2 = 20$ kΩ, and $V_{sat} = 15$ V, what is the maximum peak-to-peak noise the circuit of Fig. 22-19a can withstand without false triggering?

22-12 The Schmitt trigger of Fig. 22-20 has $R_1 = 1$ kΩ and $R_2 = 18$ kΩ. If the stray capacitance across R_1 is 3.0 pF, what size should the speed-up capacitor be?

22-13 If $V_{sat} = 13.5$ V in Fig. 22-47, what are the trip points and hysteresis?

Figure 22-47

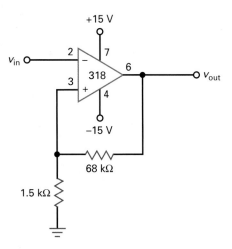

22-14 What are the trip points and hysteresis if, $V_{sat} = 14$ V in Fig. 22-48?

Figure 22-48

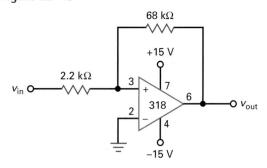

SEC. 22-4 WINDOW COMPARATOR

22-15 In Fig. 22-22a, the LTP and UTP are changed to +3.5 V and +4.75 V. If $V_{sat} = 12$ V and the input is a sine wave with a peak of 10 V, what is the output voltage waveform?

22-16 In Fig. 22-23a, the 2R resistance is changed to 4R, and the 3R resistance is changed to 6R. What are the new reference voltages?

22-17 What is the capacitor charging current in Fig. 22-49 when the input pulse is high?

22-18 In Fig. 22-49, the output voltage is reset just before the pulse begins. What is the output voltage at the end of the pulse?

22-19 The input voltage is changed from 5 to 0.1 V in Fig. 22-49. The capacitance of Fig. 22-49 is changed to each of these values: 0.1, 1, 10, and 100 μF. A reset is done at the beginning of the pulse. What is the output voltage at the end of the pulse for each capacitance?

22-20 What is the output voltage in Fig. 22-50?

22-21 If the capacitance is changed to 0.068 μF in Fig. 22-50, what is the output voltage?

22-22 In Fig. 22-50, what happens to the output voltage if the frequency changes to 5 kHz? To 20 kHz?

22-23 **MultiSim** What is the duty cycle in Fig. 22-51 when the wiper is at the top? What is the duty cycle when the wiper is at the bottom?

22-24 **MultiSim** What is the duty cycle in Fig. 22-51 when the wiper is one-half of the way from the top?

Figure 22-49

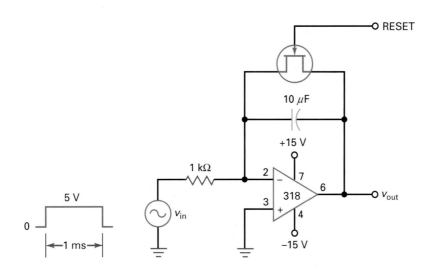

Figure 22-50

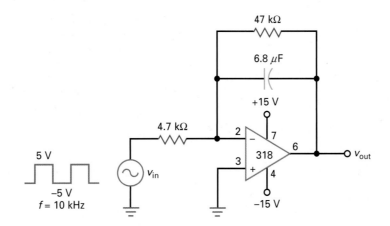

Figure 22-51

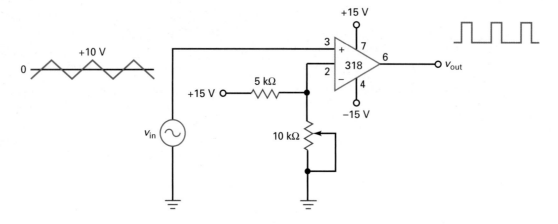

22-25 ‖‖ **MultiSim** What is the frequency of the output signal in Fig. 22-52?

Figure 22-52

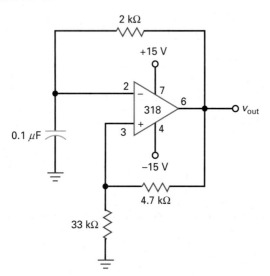

22-26 ‖‖ **MultiSim** If all resistors are doubled in Fig. 22-52, what happens to the frequency?

22-27 The capacitor of Fig. 22-52 is changed to 0.47 μF. What is the new frequency?

SEC. 22-8 ANOTHER TRIANGULAR GENERATOR

22-28 In Fig. 22-35a, $R_1 = 2.2$ kΩ, and $R_2 = 22$ kΩ. If $V_{sat} = 12$ V, what are the trip points of the Schmitt trigger? What is the hysteresis?

22-29 In Fig. 22-35a, $R_3 = 2.2$ kΩ, $R_4 = 22$ kΩ, and $C = 4.7$ μF. If the output of the Schmitt trigger is a square wave with a peak-to-peak value of 28 V and a frequency of 5 kHz, what is the peak-to-peak output of the triangular wave generator?

SEC. 22-9 ACTIVE DIODE CIRCUITS

22-30 In Fig. 22-36, the input sine wave has a peak of 100 mV. What is the output voltage?

22-31 What is the output voltage in Fig. 22-53?

22-32 What is the lowest recommended frequency in Fig. 22-53?

22-33 Suppose the diode of Fig. 22-53 is reversed. What is the output voltage?

22-34 The input voltage of Fig. 22-53 is changed from 75 mV rms to 150 mV pp. What is the output voltage?

22-35 If the peak input voltage is 100 mV in Fig. 22-39, what is the output voltage?

22-36 A positive clamper like Fig. 22-39 has $R_L = 10$ kΩ and $C = 4.7$ μF. What is the lowest recommended frequency for this clamper?

SEC. 22-10 THE DIFFERENTIATOR

22-37 In Fig. 22-40, the input voltage is a square wave with a frequency of 10 kHz. How many positive and negative spikes does the differentiator produce in 1 s?

22-38 In Fig. 22-41, the input voltage is a square wave with frequency of 1 kHz. What is the time between the negative and positive output spike?

Figure 22-53

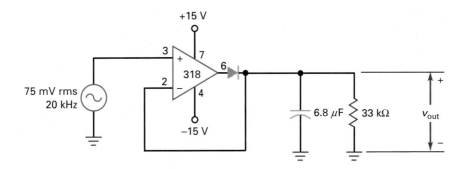

Critical Thinking

22-39 Suggest one or more changes in Fig. 22-46 to get a reference voltage of 1 V.

22-40 The stray capacitance across the output in Fig. 22-46 is 50 pF. What is the risetime of the output waveform when it switches from low to high?

22-41 A bypass capacitor of 47 μF is connected across the 3.3 kΩ of Fig. 22-46. What is the cutoff frequency of the bypass circuit? If the supply ripple is 1 V rms, what is the approximate ripple at the inverting input?

22-42 What is the average current through the 1-kΩ resistor of Fig. 22-14a if the input is a sine wave with a peak of 5 V? Assume $R_1 = 33$ kΩ and $R_2 = 3.3$ kΩ.

22-43 The resistors of Fig. 22-47 have a tolerance of ± 5 percent. What is the minimum hysteresis?

22-44 In Fig. 22-23a, the LTP and UTP are changed to +3.5 V and +4.75 V. If $V_{sat} = 12$ V and the input is a sine wave with a peak of 10 V, what is the output duty cycle?

22-45 We want to produce ramp output voltages in Fig. 22-49 that swing from 0 to +10 V with times of 0.1, 1, and 10 ms. What changes can you make in the circuit to accomplish this? (Many right answers are possible.)

22-46 We want the output frequency of Fig. 22-52 to be 20 kHz. Suggest some changes that will accomplish this.

22-47 The noise voltage at the input of Fig. 22-48 may be as large as 1 V pp. Suggest one or more changes that make the circuit immune to noise voltage.

22-48 Company XYZ is mass-producing relaxation oscillators. The output voltage is supposed to be at least 10 V pp. Suggest some ways to check the output of each unit to see whether it is at least 10 V pp. (There are many right answers here. See how many you can think of. You can use any device or circuit in this and earlier chapters.)

22-49 How can you build a circuit that turns on the lights when it gets dark and turns them off when it gets light? (Use this and earlier chapters to find as many right answers as you can think of.)

22-50 You have some electronics equipment that malfunctions when the line voltage is too low. Suggest one or more ways to set off an audible alarm when the line voltage is less than 105 V rms.

22-51 Radar waves travel at 186,000 mi/s. A transmitter on earth sends a radar wave to the moon, and an echo of this radar wave returns to earth. In Fig. 22-49, 1 kΩ is changed to 1 MΩ. The input rectangular pulse starts at the instant the radar wave is sent to the moon, and the pulse ends at the instant the radar wave arrives back on earth. If the output ramp has decreased from 0 to a final voltage of -1.23 V, how far away is the moon?

Troubleshooting

Use Fig. 22-54 for the remaining problems. Each test point, A through E, will show an oscilloscope display. Based on your knowledge of the circuits and waveforms, you are to locate the most suspicious block for further testing. Familiarize yourself with normal operation by using the OK measurements. When ready to troubleshoot, do the following problems.

22-52 Find Troubles 1 and 2.

22-53 Find Troubles 3 through 5.

22-54 Find Troubles 6 and 7.

22-55 Find Troubles 8 through 10.

Figure 22-54

Troubleshooting					
Trouble	V_A	V_B	V_C	V_D	V_E
OK	K	I	H	J	L
T1	K	N	M	S	P
T2	K	I	H	J	O
T3	M	M	M	S	P
T4	R	I	M	S	P
T5	K	M	M	S	P
T6	K	I	H	S	P
T7	K	I	H	J	J
T8	K	I	Q	S	P
T9	R	I	H	J	S
T10	K	I	H	M	M

WAVEFORMS

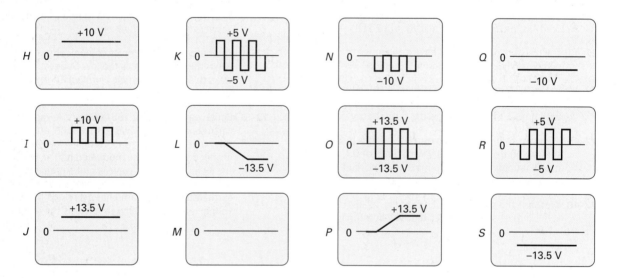

Job Interview Questions

1. Sketch a zero-crossing detector and describe its theory of operation.
2. How can I prevent a noisy input from triggering a comparator? Draw a schematic diagram and some waveforms to support your discussion.
3. Tell me how an integrator works by drawing a schematic diagram and some waveforms.

4. You are going to mass-produce a circuit that is supposed to have a dc output voltage between 3 and 4 V. What kind of a comparator would you use? How would you connect green and red LEDs across the comparator output to indicate pass or fail?
5. What does the term *bounded output* mean? How can the task be easily accomplished?

6. How does a Schmitt trigger differ from a zero-crossing detector?
7. How can we protect the input of a comparator from excessively large input voltages?
8. How does an IC comparator differ from a typical op amp?
9. If a rectangular pulse drives an integrator, what kind of output can we expect?

10. What effect does an active diode circuit have on the knee voltage?
11. What does a relaxation oscillator do? Explain the general idea of how it does this.
12. If a rectangular pulse drives a differentiator, what kind of output can we expect?

Self-Test Answers

1.	d	11.	c	21.	d
2.	a	12.	b	22.	d
3.	c	13.	b	23.	a
4.	b	14.	b	24.	b
5.	c	15.	a	25.	a
6.	a	16.	b	26.	a
7.	a	17.	a	27.	d
8.	b	18.	c	28.	b
9.	c	19.	b	29.	c
10.	b	20.	c	30.	a

Practice Problem Answers

22-4 $V_{ref} = 7.5$ V;
$f_C = 0.508$ Hz

22-6 $B = 0.0435$;
UTP $= 0.587$ V;
LTP $= -0.587$ V;
$H = 1.17$ V

22-7 $V = 0.800$ V;
time constant $= 1000$ sec.

22-8 $C = 2.5$ µF

22-9 $W = 83.3$ µS;
$D = 0.167$

22-10 $T = 479$ µS;
$f = 2.1$ kHz

22-11 $V_{out(pp)} = 32.3$ mV pp

22-12 $V_{out(pp)} = 0.52$ V;
$f = 2.5$ kHz

23 Oscillators

At frequencies under 1 MHz, we can use *RC* oscillators to produce almost perfect sine waves. These low-frequency oscillators use op amps and *RC* resonant circuits to determine the frequency of oscillation. Above 1 MHz, *LC* oscillators are used. These high-frequency oscillators use transistors and *LC* resonant circuits. This chapter also discusses a popular chip called the 555 timer. It is used in many applications to produce time delays, voltage-controlled oscillators, and modulated output signals. The chapter also covers an important communications circuit called the phase-locked loop (PLL) and concludes with the popular XR-2206 function generator IC.

Objectives

After studying this chapter, you should be able to:

- Explain loop gain and phase and how they relate to sinusoidal oscillators.

- Describe the operation of several *RC* sinusoidal oscillators.

- Describe the operation of several *LC* sinusoidal oscillators.

- Explain how crystal-controlled oscillators work.

- Discuss the 555 timer IC, its modes of operation, and how it is used as an oscillator.

- Explain the operation of phase-locked loops.

- Describe the operation of the XR-2206 function generator IC.

Chapter Outline

Vocabulary

Armstrong oscillator

astable

bistable multivibrator

capture range

carrier

Clapp oscillator

Colpitts oscillator

frequency modulation (FM)

frequency-shift keying (FSK)

fundamental frequency

Hartley oscillator

lead-lag circuit

lock range

modulating signal

monostable

mounting capacitance

multivibrator

natural logarithm

notch filter

phase detector

phase-locked loop (PLL)

phase-shift oscillator

Pierce crystal oscillator

piezoelectric effect

pulse-position modulation (PPM)

pulse-width modulation (PWM)

quartz-crystal oscillator

resonant frequency f_r

twin-T oscillator

voltage-controlled oscillator (VCO)

voltage-to-frequency converter

Wien-bridge oscillator

23-1 Theory of Sinusoidal Oscillation

To build a sinusoidal oscillator, we need to use an amplifier with positive feedback. The idea is to use the feedback signal in place of the input signal. If the feedback signal is large enough and has the correct phase, there will be an output signal even though there is no external input signal.

Loop Gain and Phase

Figure 23-1a shows an ac voltage source driving the input terminals of an amplifier. The amplified output voltage is:

$$v_{out} = A_v(v_{in})$$

This voltage drives a feedback circuit that is usually a resonant circuit. Because of this, we get maximum feedback at one frequency. In Fig. 23-1a, the feedback voltage returning to point x is given by:

$$v_f = A_v B(v_{in})$$

where B is the feedback fraction.

If the phase shift through the amplifier and feedback circuit is equivalent to 0°, $A_v B(v_{in})$ is in phase with v_{in}.

Suppose we connect point x to point y and simultaneously remove voltage source v_{in}. Then the feedback voltage $A_v B(v_{in})$ drives the input of the amplifier, as shown in Fig. 23-1b.

What happens to the output voltage? If $A_v B$ is less than 1, $A_v B(v_{in})$ is less than v_{in} and the output signal will die out, as shown in Fig. 23-1c. However, if $A_v B$ is greater than 1, $A_v B(v_{in})$ is greater than v_{in} and the output voltage builds up (Fig. 23-1d). If $A_v B$ equals 1, then $A_v B(v_{in})$ equals v_{in} and the output voltage is a steady sine wave like the one in Fig. 23-1e. In this case, the circuit supplies its own input signal.

In any oscillator the loop gain $A_v B$ is greater than 1 when the power is first turned on. A small starting voltage is applied to the input terminals, and the output voltage builds up, as shown in Fig. 23-1d. After the output voltage reaches a certain level, $A_v B$ automatically decreases to 1, and the peak-to-peak output becomes constant (Fig. 23-1e).

GOOD TO KNOW

In most oscillators, the feedback voltage is a fractional part of the output voltage. When this is the case, the voltage gain A_v must be large enough to ensure that $A_v B = 1$. In other words, the amplifier voltage gain must at least be large enough to overcome the losses in the feedback network. However, if an emitter follower is used as the amplifier, the feedback network must provide a slight amount of gain to ensure that $A_v B = 1$. For example, if the voltage gain A_v of an emitter follower equals 0.9, then B must equal 1/0.9 or 1.11. RF communication circuits sometimes use oscillators that contain an emitter follower for the amplifier.

Figure 23-1 (a) Feedback voltage returns to point x; (b) connecting points x and y; (c) oscillations die out; (d) oscillations increase; (e) oscillations are fixed in amplitude.

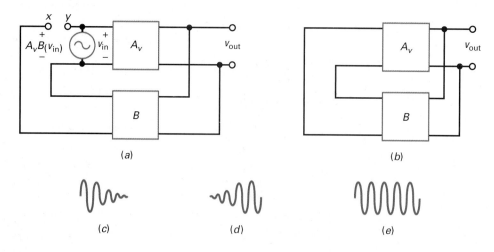

Starting Voltage Is Thermal Noise

Where does the starting voltage come from? As discussed in Chap. 22, every resistor contains some free electrons. Because of the ambient temperature, these free electrons move randomly in different directions and generate a noise voltage across the resistor. The motion is so random that it contains frequencies to over 1000 GHz. You can think of each resistor as a small ac voltage source producing all frequencies.

In Fig. 23-1b, here is what happens: When you first turn on the power, the only signals in the system are the noise voltages generated by the resistors. These noise voltages are amplified and appear at the output terminals. The amplified noise, which contains all frequencies, drives the resonant feedback circuit. By deliberate design, we can make the loop gain greater than 1 and the loop phase shift equal to 0° at the resonant frequency. Above and below the resonant frequency, the phase shift is different from 0°. As a result, oscillations will build up only at the resonant frequency of the feedback circuit.

A_vB Decreases to Unity

There are two ways in which A_vB can decrease to 1. Either A_v can decrease or B can decrease. In some oscillators, the signal is allowed to build up until clipping occurs because of saturation and cutoff. This is equivalent to reducing voltage gain A_v. In other oscillators, the signal builds up and causes B to decrease before clipping occurs. In either case, the product A_vB decreases until it equals 1.

Here are the key ideas behind any feedback oscillator:

1. Initially, loop gain A_vB is greater than 1 at the frequency where the loop phase shift is 0°.
2. After the desired output level is reached, A_vB must decrease to 1 by reducing either A_v or B.

23-2 The Wien-Bridge Oscillator

The **Wien-bridge oscillator** is the standard oscillator circuit for low to moderate frequencies, in the range of 5 Hz to about 1 MHz. It is almost always used in commercial audio generators and is usually preferred for other low-frequency applications.

Lag Circuit

The voltage gain of the bypass circuit of Fig. 23-2a is:

$$\frac{V_{out}}{V_{in}} = \frac{X_C}{\sqrt{R^2 + X_C^2}}$$

and the phase angle is:

$$\phi = -\arctan\frac{R}{X_C}$$

where ϕ is the phase angle between the output and the input.

Notice the minus sign in this equation for phase angle. It means that the output voltage lags the input voltage, as shown in Fig. 23-2b. Because of this, a bypass circuit is also called a *lag circuit*. In Fig. 23-2b, the half circle shows the possible positions of the output phasor voltage. This implies that the output phasor can lag the input phasor by an angle between 0° and −90°.

Figure 23-2 (a) Bypass capacitor; (b) phasor diagram.

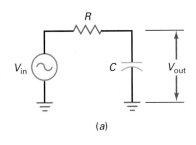

(a)

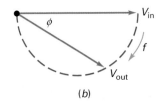

(b)

Figure 23-3 (*a*) Coupling circuit;
(*b*) phasor diagram.

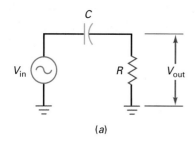

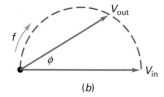

Lead Circuit

Figure 23-3*a* shows a coupling circuit. The voltage gain in this circuit is:

$$\frac{V_{\text{out}}}{V_{\text{in}}} = \frac{R}{\sqrt{R^2 + X_C^2}}$$

and the phase angle is:

$$\phi = \arctan\frac{X_C}{R}$$

Notice that the phase angle is positive. It means that the output voltage leads the input voltage, as shown in Fig. 23-3*b*. Because of this, a coupling circuit is also called a *lead circuit*. In Fig. 23-3*b*, the half circle shows the possible positions of the output phasor voltage. This implies that the output phasor can lead the input phasor by an angle between 0° and +90°.

Coupling and bypass circuits are examples of phase-shifting circuits. These circuits shift the phase of the output signal either positive (leading) or negative (lagging) with respect to the input signal. A sinusoidal oscillator always uses some kind of phase-shifting circuit to produce oscillation at one frequency.

Lead–Lag Circuit

The Wien-bridge oscillator uses a resonant feedback circuit called a **lead-lag circuit** (Fig. 23-4). At very low frequencies, the series capacitor appears open to the input signal, and there is no output signal. At very high frequencies, the shunt capacitor looks shorted, and there is no output. In between these extremes, the output voltage reaches a maximum value (see Fig. 23-5*a*). The frequency where the output is maximum is the **resonant frequency** f_r. At this frequency, the feedback fraction B reaches a maximum value of ⅓.

Figure 23-5*b* shows the phase angle of the output voltage versus input voltage. At very low frequencies, the phase angle is positive (leading). At very high frequencies, the phase angle is negative (lagging). At the resonant frequency, the phase shift is 0°. Figure 23-5*c* shows the phasor diagram of the input and output voltages. The tip of the phasor can lie anywhere on the dashed circle. Because of this, the phase angle may vary from +90° to −90°.

The lead-lag circuit of Fig. 23-4 acts like a resonant circuit. At the resonant frequency f_r, the feedback fraction B reaches a maximum value of ⅓, and the phase angle equals 0°. Above and below the resonant frequency, the feedback fraction is less than ⅓, and the phase angle no longer equals 0°.

Formula for Resonant Frequency

By analyzing Fig. 23-4 with complex numbers, we can derive these two equations:

$$B = \frac{1}{\sqrt{9 - (X_C/R - R/X_C)^2}} \tag{23-1}$$

Figure 23-4 Lead-lag circuit.

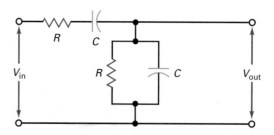

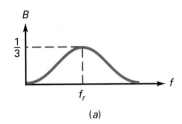

Figure 23-5 (a) Voltage gain; (b) phase response; (c) phasor diagram.

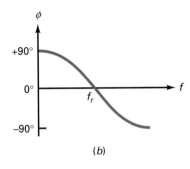

(a)

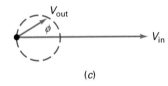

(b)

(c)

and

$$\phi = \text{arctan} \frac{X_C/R - R/X_C}{3} \tag{23-2}$$

Graphing these equations produces Fig. 23-5a and b.

The feedback fraction given by Eq. (23-1) has a maximum value at the resonant frequency. At this frequency, $X_C = R$:

$$\frac{1}{2\pi f_r C} = R$$

Solving for f_r gives:

$$f_r = \frac{1}{2\pi RC} \tag{23-3}$$

How It Works

Figure 23-6a shows a Wien-bridge oscillator. It uses positive and negative feedback because there are two paths for feedback. There is a path for positive feedback from the output through the lead-lag circuit to the noninverting input. There is also a path for negative feedback from the output through the voltage divider to the inverting input.

When the circuit is initially turned on, there is more positive feedback than negative feedback. This allows the oscillations to build up, as previously described. After the output signal reaches a desired level, the negative feedback becomes large enough to reduce loop gain $A_v B$ to 1.

Here is why $A_v B$ decreases to 1: At power-up, the tungsten lamp has a low resistance, and the negative feedback is small. For this reason, the loop gain is greater than 1, and the oscillations can build up at the resonant frequency. As the oscillations build up, the tungsten lamp heats slightly and its resistance increases. In most circuits, the current through the lamp is not enough to make the lamp glow, but it is enough to increase the resistance.

At some high output level, the tungsten lamp has a resistance of exactly R'. At this point, the closed-loop voltage gain from the noninverting input to the

Figure 23-6 Wien-bridge oscillator.

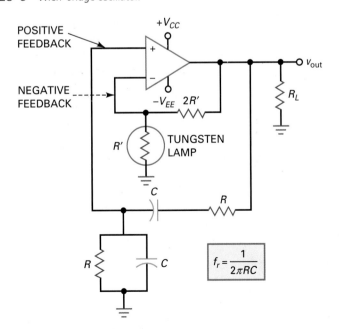

output decreases to:

$$A_{v(CL)} = \frac{2R'}{R'} + 1 = 3$$

Since the lead-lag circuit has a B of $\frac{1}{3}$, the loop gain is:

$$A_{v(CL)}B = 3(\frac{1}{3}) = 1$$

When the power is first turned on, the resistance of the tungsten lamp is less than R'. As a result, the closed-loop voltage gain from the noninverting input to the output is greater than 3 and $A_{v(CL)}B$ is greater than 1.

As the oscillations build up, the peak-to-peak output becomes large enough to increase the resistance of the tungsten lamp. When its resistance equals R', the loop gain $A_{v(CL)}B$ is exactly equal to 1. At this point, the oscillations become stable, and the output voltage has a constant peak-to-peak value.

Initial Conditions

At power-up, the output voltage is zero and the resistance of the tungsten lamp is less than R', as shown in Fig. 23-7. When the output voltage increases, the resistance of the lamp increases, as shown in the graph. When the voltage across the tungsten lamp is V', the tungsten lamp has a resistance of R'. This means that $A_{v(CL)}$ has a value of 3 and the loop gain is 1. When this happens, the output amplitude levels off and becomes constant.

Notch Filter

Figure 23-8 shows another way to draw the Wien-bridge oscillator. The lead-lag circuit is the left side of a bridge, and the voltage divider is the right side. This ac bridge, called a *Wien bridge,* is used in other applications besides oscillators. The *error voltage* is the output of the bridge. When the bridge approaches balance, the error voltage approaches zero.

The Wien bridge acts like a **notch filter,** a circuit with zero output at one particular frequency. For a Wien bridge, the notch frequency equals:

$$f_r = \frac{1}{2\pi RC} \tag{23-4}$$

Because the required error voltage for the op amp is so small, the Wien bridge is almost perfectly balanced, and the oscillation frequency equals f_r to a close approximation.

Figure 23-7 Resistance of tungsten lamp.

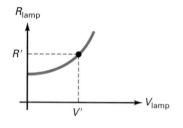

Figure 23-8 Wien-bridge oscillator redrawn.

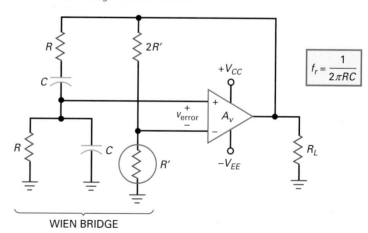

Example 23-1

Calculate the minimum and maximum frequencies in Fig. 23-9. The two variable resistors are *ganged,* which means that they change together and have the same value for any wiper position.

Figure 23-9 Example.

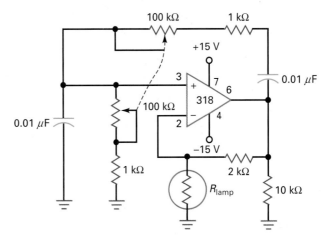

SOLUTION With Eq. (23-4), the minimum frequency of oscillation is:

$$f_r = \frac{1}{2\pi(101\ \text{k}\Omega)(0.01\ \mu\text{F})} = 158\ \text{Hz}$$

The maximum frequency of oscillation is:

$$f_r = \frac{1}{2\pi(1\ \text{k}\Omega)(0.01\ \mu\text{F})} = 15.9\ \text{kHz}$$

PRACTICE PROBLEM 23-1 Using Fig. 23-9, determine the variable resistor value for an output frequency of 1000 Hz.

Example 23-2

Figure 23-10 shows the lamp resistance of Fig. 23-9 versus lamp voltage. If the lamp voltage is expressed in rms volts, what is the output voltage of the oscillator?

SOLUTION In Fig. 23-9, the feedback resistance is 2 kΩ. Therefore, the oscillator output signal becomes constant when the lamp resistance equals 1 kΩ because this produces a closed-loop gain of 3.

In Fig. 23-10, a lamp resistance of 1 kΩ corresponds to a lamp voltage of 2 V rms. The lamp current is:

$$I_{\text{lamp}} = \frac{2\ \text{V}}{1\ \text{k}\Omega} = 2\ \text{mA}$$

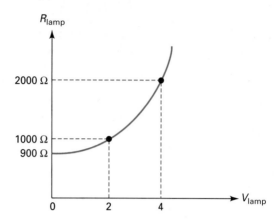

Figure 23-10 Example.

This 2 mA of current flows through the feedback resistance of 2 kΩ, which means that the output voltage of the oscillator is:

$$V_{out} = (2 \text{ mA})(1 \text{ k}\Omega + 2 \text{ k}\Omega) = 6 \text{ V rms}$$

PRACTICE PROBLEM 23-2 Repeat Example 23-2 using a feedback resistance of 3 kΩ.

23-3 Other *RC* Oscillators

Although the Wien-bridge oscillator is the industry standard for frequencies up to 1 MHz, other *RC* oscillators can be used in different applications. This section discusses two other basic designs, called the **twin-T oscillator** and the **phase-shift oscillator.**

Twin-T Filter

Figure 23-11*a* is a twin-T filter. A mathematical analysis of this circuit shows that it acts like a lead-lag circuit with a changing phase angle, as shown in Fig. 23-11*b*. Again, there is a frequency f_r where the phase shift equals 0°. In Fig. 23-11*c*, the voltage gain equals 1 at low and high frequencies. In between, there is a frequency f_r at which the voltage gain drops to 0. The twin-T filter is another example of a notch filter because it can notch out frequencies near f_r. The equation for the resonant frequency of a twin-T filter is the same as for a Wien-bridge oscillator:

$$f_r = \frac{1}{2\pi RC}$$

Twin-T Oscillator

Figure 23-12 shows a twin-T oscillator. The positive feedback to the noninverting input is through a voltage divider. The negative feedback is through the twin-T filter. When power is first turned on, the lamp resistance R_2 is low and the positive feedback is maximum. As the oscillations build up, the lamp resistance increases and the positive feedback decreases. As the feedback decreases, the oscillations level off and become constant. In this way, the lamp stabilizes the level of the output voltage.

Figure 23-11 (*a*) Twin-T filter; (*b*) phase response; (*c*) frequency response.

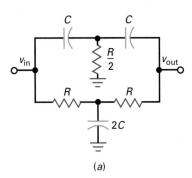

(*a*)

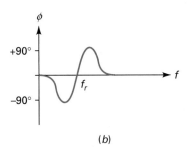

(*b*)

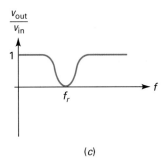

(*c*)

Figure 23-12 Twin-T oscillator.

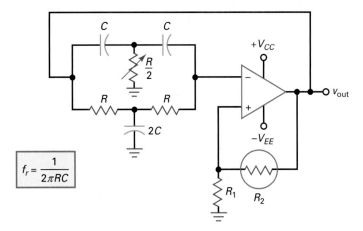

$$f_r = \frac{1}{2\pi RC}$$

In the twin-T filter, resistance $R/2$ is adjusted. This is necessary because the circuit oscillates at a frequency slightly different from the ideal resonant frequency. To ensure that the oscillation frequency is close to the notch frequency, the voltage divider should have R_2 much larger than R_1. As a guide, R_2/R_1 is in the range of 10 to 1000. This forces the oscillator to operate at a frequency near the notch frequency.

Although it is occasionally used, the twin-T oscillator is not a popular circuit because it works well only at one frequency. That is, unlike the Wien-bridge oscillator, it cannot be easily adjusted over a large frequency range.

Phase-Shift Oscillator

Figure 23-13 is a phase-shift oscillator with three lead circuits in the feedback path. As you recall, a lead circuit produces a phase shift between 0° and 90°, depending on the frequency. At some frequency, the total phase shift of the three lead circuits equals 180° (approximately 60° each). Some phase-shift oscillator configurations use four lead circuits to produce the required 180° of phase shift. The amplifier has an additional 180° of phase shift because the signal drives the inverting input. As a result, the phase shift around the loop will be 360°, equivalent to 0°. If A_vB is greater than 1 at this particular frequency, oscillations can start.

Figure 23-14 shows an alternative design. It uses three lag circuits. The operation is similar. The amplifier produces 180° of phase shift, and the lag circuits contribute $-180°$ at some higher frequency to get a loop phase shift of 0°. If A_vB is greater than 1 at this frequency, oscillations can start. The phase-shift oscillator is not a popular circuit. Again, the main problem with the circuit is that it cannot be easily adjusted over a large frequency range.

Figure 23-13 Phase-shift oscillator with three lead circuits.

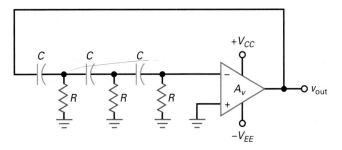

Figure 23-14   Phase-shift oscillator with three lag circuits.

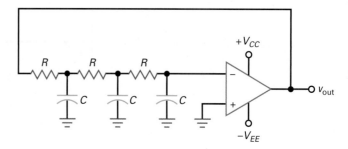

23-4 The Colpitts Oscillator

Although it is superb at low frequencies, the Wien-bridge oscillator is not suited to high frequencies (well above 1 MHz). The main problem is the limited bandwidth (f_{unity}) of the op amp.

LC Oscillators

One way to produce high-frequency oscillations is with an *LC* oscillator, a circuit that can be used for frequencies between 1 and 500 MHz. This frequency range is beyond the f_{unity} of most op amps. This is why a bipolar junction transistor or an FET is typically used for the amplifier. With an amplifier and *LC* tank circuit, we can feed back a signal with the right amplitude and phase to sustain oscillations.

  The analysis and the design of high-frequency oscillators are difficult. Why? Because at higher frequencies, stray capacitances and lead inductances become very important in determining the oscillation frequency, feedback fraction, output power, and other ac quantities. This is why many designers use computer approximations for an initial design and adjust the built-up oscillator as needed to get the desired performance.

CE Connection

Figure 23-15 shows a **Colpitts oscillator.** The voltage-divider bias sets up a quiescent operating point. The RF choke has a very high inductive reactance, so it appears open to the ac signal. The circuit has a low-frequency voltage gain of r_c/r_e', where r_c is the ac collector resistance. Because the RF choke appears open to the ac signal, the ac collector resistance is primarily the ac resistance of the resonant tank circuit. This ac resistance has a maximum value at resonance.

Figure 23-15   Colpitts oscillator.

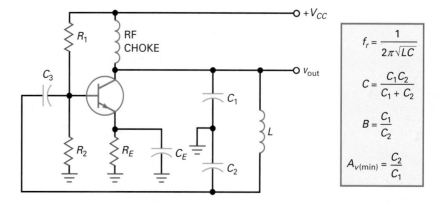

$$f_r = \frac{1}{2\pi\sqrt{LC}}$$

$$C = \frac{C_1 C_2}{C_1 + C_2}$$

$$B = \frac{C_1}{C_2}$$

$$A_{v(\text{min})} = \frac{C_2}{C_1}$$

Figure 23-16 Equivalent circuit of Colpitts oscillator.

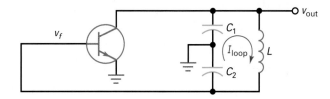

You will encounter many variations of the Colpitts oscillator. One way to recognize a Colpitts oscillator is by the capacitive voltage divider formed by C_1 and C_2. It produces the feedback voltage necessary for oscillations. In other kinds of oscillators, the feedback voltage is produced by transformers, inductive voltage dividers, and so on.

AC Equivalent Circuit

Figure 23-16 is a simplified ac equivalent circuit for the Colpitts oscillator. The circulating or loop current in the tank flows through C_1 in series with C_2. Notice that v_{out} equals the ac voltage across C_1. Also, the feedback voltage v_f appears across C_2. This feedback voltage drives the base and sustains the oscillations developed across the tank circuit, provided there is enough voltage gain at the oscillation frequency. Since the emitter is at ac ground, the circuit is a CE connection.

Resonant Frequency

Most LC oscillators use tank circuits with a Q greater than 10. Because of this, we can calculate the approximate resonant frequency as:

$$f_r = \frac{1}{2\pi\sqrt{LC}} \tag{23-5}$$

This is accurate to better than 1 percent when Q is greater than 10.

The capacitance to use in Eq. (23-5) is the equivalent capacitance through which the circulating current passes. In the Colpitts tank circuit of Fig. 23-16, the circulating current flows through C_1 in series with C_2. Therefore, the equivalent capacitance is:

$$C = \frac{C_1 C_2}{C_1 + C_2} \tag{23-6}$$

For instance, if C_1 and C_2 are 100 pF each, you would use 50 pF in Eq. (23-5).

Starting Condition

The required starting condition for any oscillator is $A_v B > 1$ at the resonant frequency of the tank circuit. This is equivalent to $A_v > 1/B$. In Fig. 23-16, the output voltage appears across C_1 and the feedback voltage appears across C_2. The feedback fraction in this type of oscillator is given by:

$$B = \frac{C_1}{C_2} \tag{23-7}$$

For the oscillator to start, the minimum voltage gain is:

$$A_{v(min)} = \frac{C_2}{C_1} \tag{23-8}$$

What does A_v equal? This depends on the upper cutoff frequencies of the amplifier. There are base and collector bypass circuits in a bipolar amplifier. If the cutoff frequencies of these bypass circuits are greater than the oscillation frequency, A_v is approximately equal to r_c/r_e'. If the cutoff frequencies are lower than the oscillation frequency, the voltage gain is less than r_c/r_e' and there is additional phase shift through the amplifier.

Output Voltage

With light feedback (small B), A_v is only slightly larger than $1/B$, and the operation is approximately class A. When you first turn on the power, the oscillations build up, and the signal swings over more and more of the ac load line. With this increased signal swing, the operation changes from small-signal to large-signal. When this happens, the voltage gain decreases slightly. With light feedback, the value of A_vB can decrease to 1 without excessive clipping.

With heavy feedback (large B), the large feedback signal drives the base of Fig. 23-15 into saturation and cutoff. This charges capacitor C_3, producing negative dc clamping at the base. The negative clamping automatically adjusts the value of A_vB to 1. If the feedback is too heavy, you may lose some of the output voltage because of stray power losses.

When you build an oscillator, you can adjust the feedback to maximize the output voltage. The idea is to use enough feedback to start under all conditions (different transistors, temperature, voltage, and so on), but not so much that you lose output signal. Designing reliable high-frequency oscillators is a challenge. Most designers use computers to model high-frequency oscillators.

Coupling to a Load

The exact frequency of oscillation depends on the Q of the circuit and is given by:

$$f_r = \frac{1}{2\pi \sqrt{LC}} \sqrt{\frac{Q^2}{Q^2 + 1}} \tag{23-9}$$

When Q is greater than 10, this equation simplifies to the ideal value given by Eq. (23-5). If Q is less than 10, the frequency is lower than the ideal value. Furthermore, a low Q may prevent the oscillator from starting because it may reduce the high-frequency voltage gain below the starting value of $1/B$.

Figure 23-17a shows one way to couple the oscillator signal to the load resistance. If the load resistance is large, it will load the resonant circuit only slightly and the Q will be greater than 10. But if the load resistance is small, Q drops under 10 and the oscillations may not start. One solution to a small load resistance is to use a small capacitance C_4, one whose X_C is large compared with the load resistance. This prevents excessive loading of the tank circuit.

Figure 23-17b shows link coupling, another way of coupling the signal to a small load resistance. Link coupling means using only a few turns on the secondary winding of an RF transformer. This light coupling ensures that the load resistance will not lower the Q of the tank circuit to the point at which the oscillator will not start.

Whether capacitive or link coupling is used, the loading effect is kept as small as possible. In this way, the high Q of the tank ensures an undistorted sinusoidal output with a reliable start for the oscillations.

CB Connection

When the feedback signal in an oscillator drives the base, a large Miller capacitance appears across the input. This produces a relatively low cutoff frequency, which means that the voltage gain may be too low at the desired resonant frequency.

To get a higher cutoff frequency, the feedback signal can be applied to the emitter, as shown in Fig. 23-18. Capacitor C_3 ac-grounds the base, and so the transistor acts like a common-base (CB) amplifier. A circuit like this can oscillate

Figure 23-17   (a) Capacitor coupling; (b) link coupling.

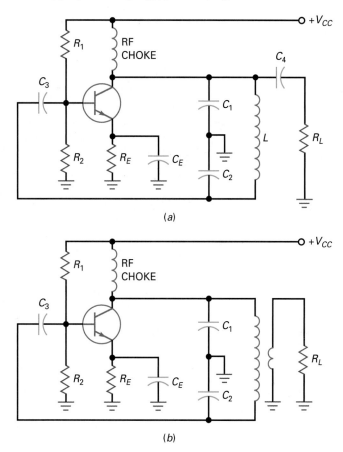

(a)

(b)

Figure 23-18   CB oscillator can oscillate at higher frequencies than CE oscillator.

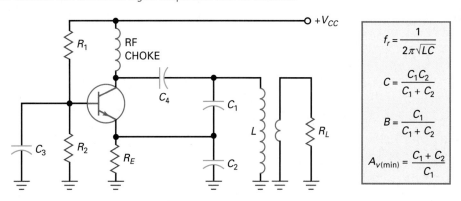

$$f_r = \frac{1}{2\pi\sqrt{LC}}$$

$$C = \frac{C_1 C_2}{C_1 + C_2}$$

$$B = \frac{C_1}{C_1 + C_2}$$

$$A_{v(min)} = \frac{C_1 + C_2}{C_1}$$

at higher frequencies because its high-frequency gain is larger than that of a CE oscillator. With link coupling on the output, the tank is lightly loaded, and the resonant frequency is given by Eq. (23-5).

The feedback fraction is slightly different in a CB oscillator. The output voltage appears across C_1 and C_2 in series, and the feedback voltage appears across C_2. Ideally, the feedback fraction is:

$$B = \frac{C_1}{C_1 + C_2} \tag{23-10}$$

Figure 23-19 JFET oscillator has less loading effect on tank circuit.

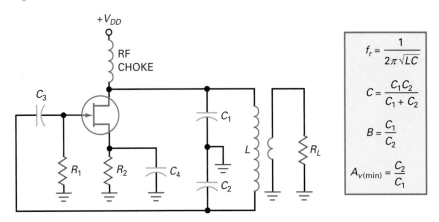

For the oscillations to start, A_v must be greater than $1/B$. As an approximation, this means that:

$$A_{v(min)} = \frac{C_1 + C_2}{C_1} \qquad (23\text{-}11)$$

This is an approximation because it ignores the input impedance of the emitter, which is in parallel with C_2.

FET Colpitts Oscillator

Figure 23-19 is an example of an FET Colpitts oscillator in which the feedback signal is applied to the gate. Since the gate has a high input resistance, the loading effect on the tank circuit is much less than with a bipolar junction transistor. The feedback fraction for the circuit is:

$$B = \frac{C_1}{C_2} \qquad (23\text{-}12)$$

The minimum gain needed to start the FET oscillator is:

$$A_{v(min)} = \frac{C_2}{C_1} \qquad (23\text{-}13)$$

In an FET oscillator, the low-frequency voltage gain is $g_m r_d$. Above the cutoff frequency of the FET amplifier, the voltage gain decreases. In Eq. (23-13), $A_{v(min)}$ is the voltage gain at the oscillation frequency. As a rule, we try to keep the oscillation frequency lower than the cutoff frequency of the FET amplifier. Otherwise, the additional phase shift through the amplifier may prevent the oscillator from starting.

Example 23-3 ⫴ MultiSim

What is the frequency of oscillation in Fig. 23-20? What is the feedback fraction? How much voltage gain does the circuit need to start oscillating?

SOLUTION This is a Colpitts oscillator using the CE connection of a transistor. With Eq. (23-6), the equivalent capacitance is:

$$C = \frac{(0.001 \ \mu F)(0.01 \ \mu F)}{0.001 \ \mu F + 0.01 \ \mu F} = 909 \ pF$$

Figure 23-20 Example.

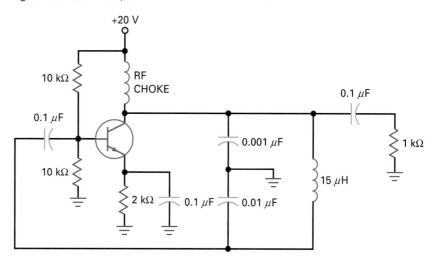

The inductance is 15 μH. With Eq. (23-5), the frequency of oscillation is:

$$f_r = \frac{1}{2\pi \sqrt{(15~\mu\text{H})(909~\text{pF})}} = 1.36~\text{MHz}$$

With Eq. (23-7), the feedback fraction is:

$$B = \frac{0.001~\mu\text{F}}{0.01~\mu\text{F}} = 0.1$$

To start oscillating, the circuit needs a minimum voltage gain of:

$$A_{v(\text{min})} = \frac{0.01~\mu\text{F}}{0.001~\mu\text{F}} = 10$$

PRACTICE PROBLEM 23-3   In Fig. 23-20, what is the approximate value that the 15 μH would need to equal for an output frequency of 1 MHz?

23-5 Other *LC* Oscillators

The Colpitts oscillator is the most widely used *LC* oscillator. The capacitive voltage divider in the resonant circuit is a convenient way to develop the feedback voltage. But other kinds of oscillators can also be used.

Armstrong Oscillator

Figure 23-21 is an example of an **Armstrong oscillator.** In this circuit, the collector drives an *LC* resonant tank. The feedback signal is taken from a small secondary winding and fed back to the base. There is a phase shift of 180° in the transformer, which means that the phase shift around the loop is zero. If we ignore the loading effect of the base, the feedback fraction is:

$$B = \frac{M}{L} \tag{23-14}$$

where *M* is the mutual inductance and *L* is the primary inductance. For the Armstrong oscillator to start, the voltage gain must be greater than 1/*B*.

Figure 23-21   Armstrong oscillator.

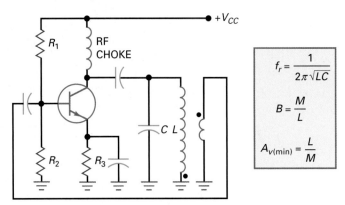

An Armstrong oscillator uses transformer coupling for the feedback signal. This is how you can recognize variations of this basic circuit. The small secondary winding is sometimes called a *tickler coil* because it feeds back the signal that sustains the oscillations. The resonant frequency is given by Eq. (23-5), using the L and C shown in Fig. 23-21. As a rule, you do not see the Armstrong oscillator used much because many designers avoid transformers whenever possible.

Hartley Oscillator

Figure 23-22 is an example of the **Hartley oscillator.** When the LC tank is resonant, the circulating current flows through L_1 in series with L_2. The equivalent L to use in Eq. (23-5) is:

$$L = L_1 + L_2 \tag{23-15}$$

In a Hartley oscillator, the feedback voltage is developed by the inductive voltage divider, L_1 and L_2. Since the output voltage appears across L_1 and the feedback voltage appears across L_2, the feedback fraction is:

$$B = \frac{L_2}{L_1} \tag{23-16}$$

As usual, this ignores the loading effects of the base. For oscillations to start, the voltage gain must be greater than $1/B$.

Often a Hartley oscillator uses a single tapped inductor instead of two separate inductors. Another variation sends the feedback signal to the emitter instead of to the base. Also, you may see an FET used instead of a bipolar junction transistor. The output signal can be either capacitively coupled or link-coupled.

Figure 23-22   Hartley oscillator.

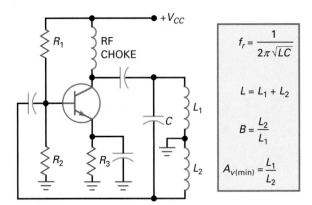

Figure 23-23 Clapp oscillator.

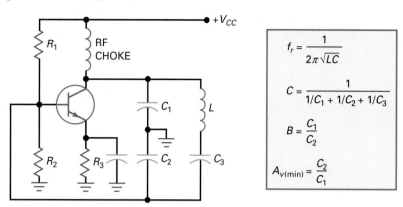

$$f_r = \frac{1}{2\pi\sqrt{LC}}$$

$$C = \frac{1}{1/C_1 + 1/C_2 + 1/C_3}$$

$$B = \frac{C_1}{C_2}$$

$$A_{v(min)} = \frac{C_2}{C_1}$$

Clapp Oscillator

The **Clapp oscillator** of Fig. 23-23 is a refinement of the Colpitts oscillator. The capacitive voltage divider produces the feedback signal as before. An additional capacitor C_3 is in series with the inductor. Since the circulating tank current flows through C_1, C_2, and C_3 in series, the equivalent capacitance used to calculate the resonant frequency is:

$$C = \frac{1}{1/C_1 + 1/C_2 + 1/C_3} \tag{23-17}$$

In a Clapp oscillator, C_3 is much smaller than C_1 and C_2. As a result, C is approximately equal to C_3, and the resonant frequency is given by:

$$f_r \cong \frac{1}{2\pi\sqrt{LC_3}} \tag{23-18}$$

Why is this important? Because C_1 and C_2 are shunted by transistor and stray capacitances. These extra capacitances alter the values of C_1 and C_2 slightly. In a Colpitts oscillator, the resonant frequency therefore depends on the transistor and stray capacitances. But in a Clapp oscillator, the transistor and stray capacitances have no effect on C_3, so the oscillation frequency is more stable and accurate. This is why you occasionally see the Clapp oscillator used.

Crystal Oscillator

When accuracy and stability of the oscillation frequency are important, a **quartz-crystal oscillator** is used. In Fig. 23-24, the feedback signal comes from a

Figure 23-24 Crystal oscillator.

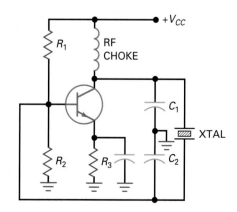

capacitive tap. As will be discussed in the next section, the crystal (abbreviated *XTAL*) acts like a large inductor in series with a small capacitor (similar to the Clapp). Because of this, the resonant frequency is almost totally unaffected by transistor and stray capacitances.

Example 23-4 ||| MultiSim

If 50 pF is added in series with the 15-μH inductor of Fig. 23-20, the circuit becomes a Clapp oscillator. What is the frequency of oscillation?

SOLUTION We can calculate the equivalent capacitance with Eq. (23-17):

$$C = \frac{1}{1/0.001 \ \mu F + 1/0.01 \ \mu F + 1/50 \ pF} \cong 50 \ pF$$

Notice how the term 1/50 pF swamps the other values because the 50 pF is much smaller than the other capacitances. The frequency of oscillation is:

$$f_r = \frac{1}{2\pi \sqrt{(15 \ \mu H)(50 \ pF)}} = 5.81 \ MHz$$

PRACTICE PROBLEM 23-4 Repeat Example 23-4 by replacing the 50 pF capacitor with a 120 pF.

23-6 Quartz Crystals

When the frequency of oscillation needs to be accurate and stable, a crystal oscillator is the natural choice. Electronic wristwatches and other critical timing applications use crystal oscillators because they provide an accurate clock frequency.

Piezoelectric Effect

Some crystals found in nature exhibit the **piezoelectric effect.** When you apply an ac voltage across them, they vibrate at the frequency of the applied voltage. Conversely, if you mechanically force them to vibrate, they generate an ac voltage of the same frequency. The main substances that produce the piezoelectric effect are quartz, Rochelle salts, and tourmaline.

Rochelle salts have the greatest piezoelectric activity. For a given ac voltage, they vibrate more than quartz or tourmaline. Mechanically, they are the weakest because they break easily. Rochelle salts have been used to make microphones, phonograph pickups, headsets, and loudspeakers. Tourmaline shows the least piezoelectric activity but is the strongest of the three. It is also the most expensive. It is occasionally used at very high frequencies.

Quartz is a compromise between the piezoelectric activity of Rochelle salts and the strength of tourmaline. Because it is inexpensive and readily available in nature, quartz is widely used for RF oscillators and filters.

Crystal Slab

The natural shape of a quartz crystal is a hexagonal prism with pyramids at the ends (see Fig. 23-25a). To get a usable crystal out of this, a manufacturer slices a

Figure 23-25 (*a*) Natural quartz crystal; (*b*) slab; (*c*) input current is maximum at resonance.

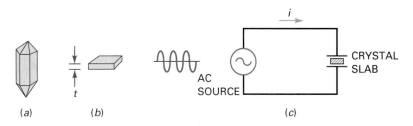

(*a*)　　　(*b*)　　　　　　　　　(*c*)

rectangular slab out of the natural crystal. Figure 23-25*b* shows this slab with thickness *t*. The number of slabs we can get from a natural crystal depends on the size of the slabs and the angle of the cut.

For use in electronic circuits, the slab must be mounted between two metal plates, as shown in Fig. 23-25*c*. In this circuit the amount of crystal vibration depends on the frequency of the applied voltage. By changing the frequency, we can find resonant frequencies at which the crystal vibrations reach a maximum. Since the energy for the vibrations must be supplied by the ac source, the ac current is maximum at each resonant frequency.

Fundamental Frequency and Overtones

Most of the time, the crystal is cut and mounted to vibrate best at one of its resonant frequencies, usually the **fundamental frequency,** or lowest frequency. Higher resonant frequencies, called *overtones,* are almost exact multiples of the fundamental frequency. As an example, a crystal with a fundamental frequency of 1 MHz has a first overtone of approximately 2 MHz, a second overtone of approximately 3 MHz, and so on.

The formula for the fundamental frequency of a crystal is:

$$f = \frac{K}{t} \tag{23-19}$$

where K is a constant and t is the thickness of the crystal. Since the fundamental frequency is inversely proportional to the thickness, there is a limit to the highest fundamental frequency. The thinner the crystal, the more fragile it becomes and the more likely it is to break when vibrating.

Quartz crystals work well up to 10 MHz on the fundamental frequency. To reach higher frequencies, we can use a crystal that vibrates on overtones. In this way, we can reach frequencies up to 100 MHz. Occasionally, the more expensive but stronger tourmaline is used at higher frequencies.

AC Equivalent Circuit

What does the crystal look like to an ac source? When the crystal of Fig. 23-26*a* is not vibrating, it is equivalent to a capacitance C_m because it has two metal plates separated by a dielectric. The capacitance C_m is known as the **mounting capacitance.**

When a crystal is vibrating, it acts like a tuned circuit. Figure 23-26*b* shows the ac equivalent circuit of a crystal vibrating at its fundamental frequency. Typical values are L in henrys, C_s in fractions of a picofarad, R in hundreds of ohms, and C_m in picofarads. For instance, a crystal can have values such as $L = 3$ H, $C_s = 0.05$ pF, $R = 2$ kΩ, and $C_m = 10$ pF.

Crystals have an incredibly high Q. For the values just given, Q is almost 4000. The Q of a crystal can easily be over 10,000. The extremely high Q of a

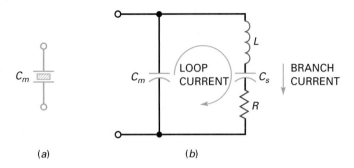

(*a*) (*b*)

crystal means that crystal oscillators have a very stable frequency. We can understand why this is true when we examine Eq. (23-9), the exact equation for resonant frequency:

$$f_r = \frac{1}{2\pi \sqrt{LC}} \sqrt{\frac{Q^2}{Q^2 + 1}}$$

When Q approaches infinity, the resonant frequency approaches the ideal value determined by the values of L and C, which are precise in a crystal. By comparison, the L and C values of a Colpitts oscillator have large tolerances, which means that the frequency is less precise.

Series and Parallel Resonance

The *series resonant frequency* f_s of a crystal is the resonant frequency of the *LCR* branch in Fig. 23-26*b*. At this frequency, the *branch current* reaches a maximum value because L resonates with C_s. The formula for this resonant frequency is:

$$f_s = \frac{1}{2\pi \sqrt{LC_s}} \tag{23-20}$$

The *parallel resonant frequency* f_p of the crystal is the frequency at which the circulating or loop current of Fig. 23-26*b* reaches a maximum value. Since this loop current must flow through the series combination of C_s and C_m, the equivalent parallel capacitance is:

$$C_p = \frac{C_m C_s}{C_m + C_s} \tag{23-21}$$

and the parallel resonant frequency is:

$$f_p = \frac{1}{2\pi \sqrt{LC_p}} \tag{23-22}$$

In any crystal, C_s is much smaller than C_m. Because of this, f_p is only slightly greater than f_s. When we use a crystal in an ac equivalent circuit like Fig. 23-27, the additional circuit capacitances appear in shunt with C_m. Because of this, the oscillation frequency will lie between f_s and f_p.

Crystal Stability

The frequency of any oscillator tends to change slightly with time. This *drift* is produced by temperature, aging, and other causes. In a crystal oscillator, the frequency drift is very small, typically less than 1 part in 10^6 per day. Stability like this is important in electronic wristwatches because they use quartz-crystal oscillators as the basic timing device.

By putting a crystal oscillator in a temperature-controlled oven, we can get a frequency drift of less than 1 part in 10^{10} per day. A clock with this drift will

Figure 23-27 Stray capacitances are in parallel with mounting capacitance.

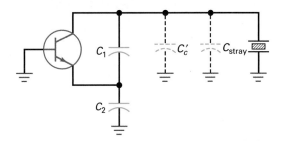

take 300 years to gain or lose 1 s. Stability like this is needed in frequency and time standards.

Crystal Oscillators

Figure 23-28a shows a Colpitts crystal oscillator. The capacitive voltage divider produces the feedback voltage for the base of the transistor. The crystal acts like an inductor that resonates with C_1 and C_2. The oscillation frequency is between the series and parallel resonant frequencies of the crystal.

Figure 23-28b is a variation of the Colpitts crystal oscillator. The feedback signal is applied to the emitter instead of the base. This variation allows the circuit to work at higher resonant frequencies.

Figure 23-28c is an FET Clapp oscillator. The intention is to improve the frequency stability by reducing the effect of stray capacitances. Figure 23-28d is a circuit called a **Pierce crystal oscillator.** Its main advantage is simplicity.

Figure 23-28 Crystal oscillators: (a) Colpitts; (b) variation of Colpitts; (c) Clapp; (d) Pierce.

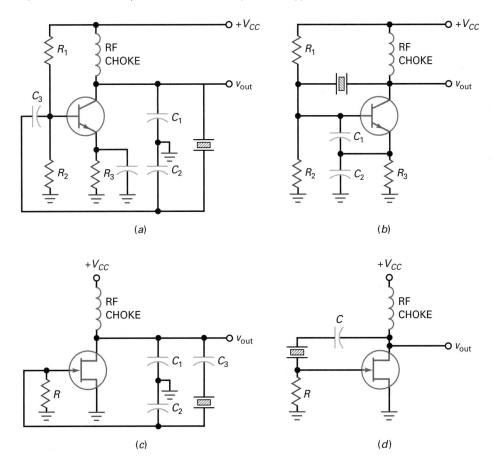

Example 23-5

A crystal has these values: $L = 3$ H, $C_s = 0.05$ pF, $R = 2$ kΩ, and $C_m = 10$ pF. What are the series and parallel resonant frequencies of the crystal?

SOLUTION Equation (23-20) gives the series resonant frequency:

$$f_s = \frac{1}{2\pi \sqrt{(3H)(0.05pF)}} = 411 \text{ kHz}$$

Equation (23-21) gives the equivalent parallel capacitance:

$$C_p = \frac{(10 \text{ pF})(0.05 \text{ pF})}{10 \text{ pF} + 0.05 \text{ pF}} = 0.0498 \text{ pF}$$

Equation (23-22) gives the parallel resonant frequency:

$$f_p = \frac{1}{2\pi \sqrt{(3H)(0.0498 \text{ pF})}} = 412 \text{ kHz}$$

As we can see, the series and parallel resonant frequencies of the crystal are very close in value. If this crystal is used in an oscillator, the frequency of oscillation will be between 411 and 412 kHz.

PRACTICE PROBLEM 23-5 Repeat Example 23-5 with $C_s = 0.1$ pF and $C_m = 15$ pF.

Summary Table 23-1 presents some of the characteristics of *RC* and *LC* oscillators.

23-7 The 555 Timer

The NE555 (also LM555, CA555, and MC1455) is a widely used *IC timer,* a circuit that can run in either of two modes: **monostable** (one stable state) or **astable** (no stable states). In the monostable mode, it can produce accurate time delays from microseconds to hours. In the astable mode, it can produce rectangular waves with a variable duty cycle.

Monostable Operation

Figure 23-29 illustrates monostable operation. Initially, the 555 timer has a low output voltage at which it can remain indefinitely. When the 555 timer receives a

Figure 23-29 The 555 timer used in monostable (one-shot) mode.

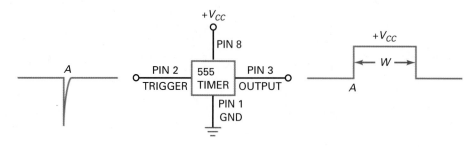

Summary Table 23–1	Oscillators
Type	**Characteristics**
RC oscillators	
Wien-bridge	• Uses lead-lag feedback circuits • Needs ganged Rs for tuning • Low distortion output from 5 Hz to 1 MHz (limited bandwidth) • $f_r = \dfrac{1}{2\pi RC}$
Twin-T	• Uses a notch filter circuit • Works well at one frequency • Difficult to adjust over a wide output frequency • $f_r = \dfrac{1}{2\pi RC}$
Phase-shift	• Uses 3–4 lead or lag circuits • Cannot be adjusted over wide frequency range
LC oscillators	
Colpitts	• Uses a pair of tapped capacitors $\quad C = \dfrac{C_1 C_2}{C_1 + C_2} \qquad f_r = \dfrac{1}{2\pi\sqrt{LC}}$ • Widely used
Armstrong	• Uses a transformer for feedback • Not used frequently • $f_r = \dfrac{1}{2\pi\sqrt{LC}}$
Hartley	• Uses a pair of tapped inductors • $L = L_1 + L_2 \qquad f_r = \dfrac{1}{2\pi\sqrt{LC}}$
Clapp	• Uses tapped capacitors and a capacitor in series with the inductor • Stable and accurate output • $C = \dfrac{1}{\dfrac{1}{C_1} + \dfrac{1}{C_2} + \dfrac{1}{C_3}} \qquad f_r = \dfrac{1}{2\pi\sqrt{LC}}$
Crystal	• Uses a quartz crystal • Very accurate and stable • $f_p = \dfrac{1}{2\pi\sqrt{LC_p}} \qquad f_s = \dfrac{1}{2\pi\sqrt{LC_s}}$

trigger at point *A* in time, the output voltage switches from low to high, as shown. The output remains high for a while and then returns to the low state after a time delay of *W*. The output will remain in the low state until another trigger arrives.

A **multivibrator** is a two-state circuit that has zero, one, or two stable output states. When the 555 timer is used in the monostable mode, it is sometimes called a *monostable multivibrator* because it has only one stable state. It is stable in the low state until it receives a trigger, which causes the output to temporarily change to the high state. The high state, however, is not stable because the output returns to the low state when the pulse ends.

When operating in the monostable mode, the 555 timer is often referred to as a *one-shot multivibrator* because it produces only one output pulse for each input trigger. The duration of this output pulse can be precisely controlled with an external resistor and capacitor.

The 555 timer is an 8-pin IC. Figure 23-29 shows four of the pins. Pin 1 is connected to ground, and pin 8 is connected to the positive supply voltage. The 555 timer will work with any supply voltage between +4.5 and +18 V. The trigger goes into pin 2, and the output comes from pin 3. The other pins, which are not shown here, are connected to external components that determine the pulse width of the output.

Astable Operation

The 555 timer can also be connected to run as an *astable multivibrator.* When used in this way, the 555 timer has no stable states, which means that it cannot remain indefinitely in either state. Stated another way, it oscillates when operated in the astable mode and it produces a rectangular output signal.

Figure 23-30 shows the 555 timer used in the astable mode. As we can see, the output is a series of rectangular pulses. Since no input trigger is needed to get an output, the 555 timer operating in the astable mode is sometimes called a *free-running multivibrator.*

Functional Block Diagram

The schematic diagram of a 555 timer is complicated because it has about two dozen components connected as diodes, current mirrors, and transistors. Figure 23-31 shows a functional diagram of the 555 timer. This diagram captures all the key ideas we need for our discussion of the 555 timer.

As shown in Fig. 23-31, the 555 timer contains a voltage divider, two comparators, an *RS* flip-flop, and an *npn* transistor. Since the voltage divider has equal resistors, the top comparator has a trip point of:

$$\text{UTP} = \frac{2V_{CC}}{3} \tag{23-23}$$

The lower comparator has a trip point of:

$$\text{LTP} = \frac{V_{CC}}{3} \tag{23-24}$$

Figure 23-30 The 555 timer used in astable (free-running) mode.

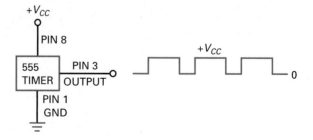

Figure 23-31 Simplified functional block diagram of a 555 timer.

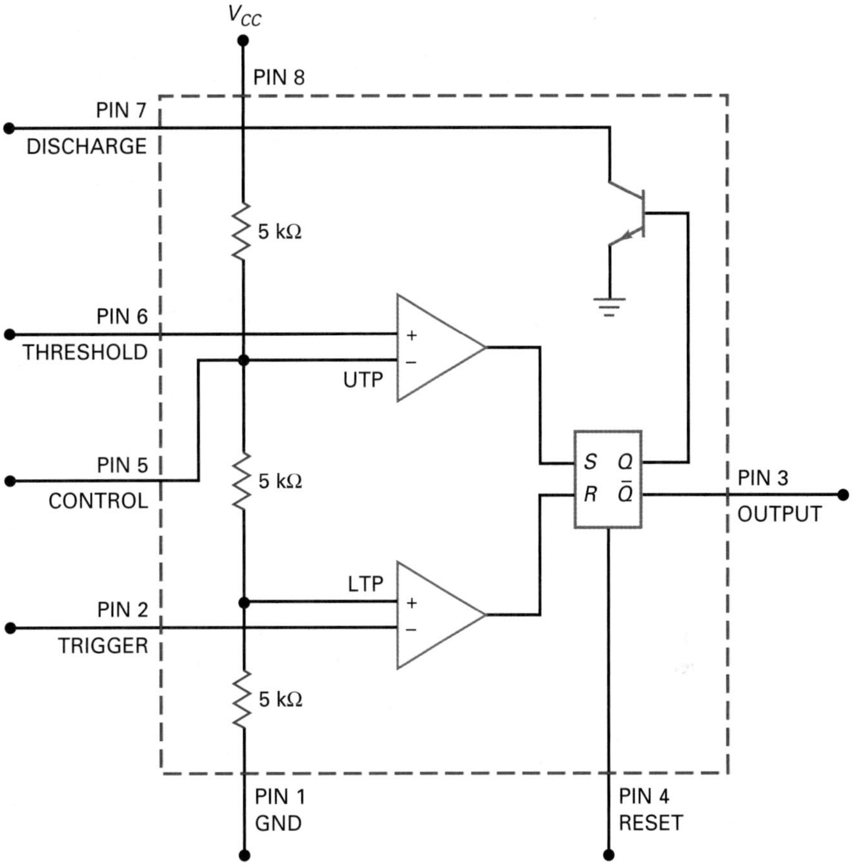

In Fig. 23-31, pin 6 is connected to the upper comparator. The voltage on pin 6 is called the *threshold.* This voltage comes from external components not shown. When the *threshold voltage* is greater than the UTP, the upper comparator has a high output.

Pin 2 is connected to the lower comparator. The voltage on pin 2 is called the *trigger.* This is the trigger voltage that is used for the monostable operation of the 555 timer. When the timer is inactive, the trigger voltage is high. When the trigger voltage falls to less than the LTP, the lower comparator produces a high output.

Pin 4 may be used to reset the output voltage to zero. Pin 5 may be used to control the output frequency when the 555 timer is used in the astable mode. In many applications, these two pins are made inactive as follows: Pin 4 is connected to $+V_{CC}$, and pin 5 is bypassed to ground through a capacitor. Later, we will discuss how pins 4 and 5 are used in some advanced circuits.

RS Flip–Flop

Before we can understand how a 555 timer works with external components, we need to discuss the action of the block that contains, S, R, Q, and $\overline{Q}$. This block is called an *RS flip-flop,* a circuit that has two stable states.

Figure 23-32 shows one way to build an *RS* flip-flop. In a circuit like this, one of the transistors is saturated, and the other is cut off. For instance, if the right transistor is saturated, its collector voltage will be approximately zero. This means that there is no base current in the left transistor. As a result, the left transistor is cut off, producing a high collector voltage. This high collector voltage produces a large base current that keeps the right transistor in saturation.

Figure 23-32 *RS* flip-flop built with transistors.

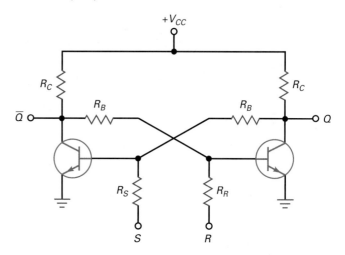

The *RS* flip-flop has two outputs, Q and $\overline{Q}$. These are two-state outputs, either low or high voltages. Furthermore, the two outputs are always in opposite states. When Q is low, $\overline{Q}$ is high. When Q is high, $\overline{Q}$ is low. For this reason, $\overline{Q}$ is called the *complement of Q*. The overbar on $\overline{Q}$ is used to indicate that it is the complement of Q.

We can control the output states with the S and R inputs. If we apply a large positive voltage to the S input, we can drive the left transistor into saturation. This will cut off the right transistor. In this case, Q will be high and $\overline{Q}$ will be low. The high S input can then be removed, because the saturated left transistor will keep the right transistor in cutoff.

Similarly, we can apply a large positive voltage to the R input. This will saturate the right transistor and cut off the left transistor. For this condition, Q is low and $\overline{Q}$ is high. After this transition has occurred, the high R input can be removed because it is no longer needed.

Since the circuit is stable in either of two states, it is sometimes called a **bistable multivibrator.** A bistable multivibrator latches in either of two states. A high S input forces Q into the high state, and a high R input forces Q to return to the low state. The output Q remains in a given state until it is triggered into the opposite state.

Incidentally, the S input is sometimes called the *set input* because it sets the Q output to high. The R input is called the *reset input* because it resets the Q output to low.

Monostable Operation

Figure 23-33 shows the 555 timer connected for monostable operation. The circuit has an external resistor R and a capacitor C. The voltage across the capacitor is used for the threshold voltage to pin 6. When the trigger arrives at pin 2, the circuit produces a rectangular output pulse from pin 3.

Here is the theory of operation. Initially, the Q output of the *RS* flip-flop is high. This saturates the transistor and clamps the capacitor voltage at ground. The circuit will remain in this state until a trigger arrives. Because of the voltage divider, the trip points are the same as previously discussed: UTP $= 2V_{CC}/3$ and LTP $= V_{CC}/3$.

When the trigger input falls to slightly less than $V_{CC}/3$, the lower comparator resets the flip-flop. Since Q has changed to low, the transistor goes into cutoff, allowing the capacitor to charge. At this time, $\overline{Q}$ has changed to high. The

Figure 23-33 555 timer connected for monostable operation.

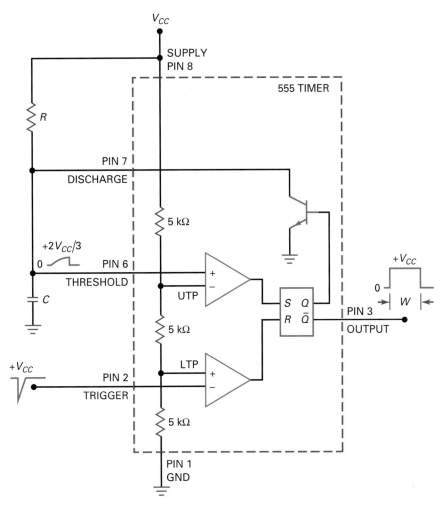

capacitor now charges exponentially as shown. When the capacitor voltage is slightly greater than $2V_{CC}/3$, the upper comparator sets the flip-flop. The high Q turns on the transistor, which discharges the capacitor almost instantly. At the same instant, $\overline{Q}$ returns to the low state and the output pulse ends. $\overline{Q}$ remains low until another input trigger arrives.

The complementary output $\overline{Q}$ comes out of pin 3. The width of the rectangular pulse depends on how long it takes to charge the capacitor through resistance R. The longer the time constant, the longer it takes for the capacitor voltage to reach $2V_{CC}/3$. In one time constant, the capacitor can charge to 63.2 percent of V_{CC}. Since $2V_{CC}/3$ is equivalent to 66.7 percent of V_{CC}, it takes slightly more than one time constant for the capacitor voltage to reach $2V_{CC}/3$. By solving the exponential charging equation, it is possible to derive this formula for the pulse width:

$$W = 1.1RC \qquad (23\text{-}25)$$

Figure 23-34 shows the schematic diagram for the monostable 555 circuit as it usually appears. Only the pins and external components are shown. Notice that pin 4 (reset) is connected to $+V_{CC}$. As discussed earlier, this prevents pin 4 from having any effect on the circuit. In some applications, pin 4 may be temporarily grounded to suspend the operation. When pin 4 is taken high, the operation resumes. A later discussion will describe this type of reset in more detail.

Figure 23-34 Monostable timer circuit.

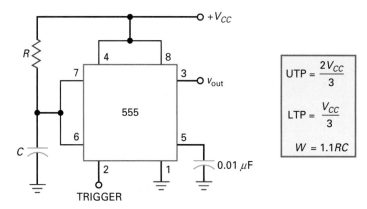

Pin 5 (control) is a special input that can be used to change the UTP, which changes the width of the pulse. Later, we will discuss *pulse-width modulation,* in which an external voltage is applied to pin 5 to change the pulse width. For now, we will bypass pin 5 to ground as shown. By ac-grounding pin 5, we prevent stray electromagnetic noise from interfering with the operation of the 555 timer.

In summary, the monostable 555 timer produces a single pulse whose width is determined by the external R and C used in Fig. 23-34. The pulse begins with the leading edge of the input trigger. A one-shot operation like this has a number of applications in digital and switching circuits.

Example 23-6

||| MultiSim

In Fig. 23-34, $V_{CC} = 12$ V, $R = 33$ kΩ, and $C = 0.47$ μF. What is the minimum trigger voltage that produces an output pulse? What is the maximum capacitor voltage? What is the width of the output pulse?

SOLUTION As shown in Fig. 23-33, the lower comparator has a trip point of LTP. Therefore, the input trigger on pin 2 has to fall from $+V_{CC}$ to slightly less than LTP. With the equations shown in Fig. 23-34:

$$\text{LTP} = \frac{12\,\text{V}}{3} = 4\,\text{V}$$

After a trigger arrives, the capacitor charges from 0 V to a maximum of UTP, which is:

$$\text{UTP} = \frac{2(12\,\text{V})}{3} = 8\,\text{V}$$

The pulse width of the one-shot output is:

$$W = 1.1(33\,\text{kΩ})(0.47\,\text{μF}) = 17.1\,\text{ms}$$

This means that the falling edge of the output pulse occurs 17.1 ms after the trigger arrives. You can think of this 17.1 ms as a time delay, because the falling edge of the output pulse can be used to trigger some other circuit.

PRACTICE PROBLEM 23-6 Using Fig. 23-34, change V_{CC} to 15 V, R to 100 kΩ, and repeat Example 23-6.

Example 23-7

What is the pulse width in Fig. 23-34 if $R = 10 \text{ M}\Omega$ and $C = 470 \text{ }\mu\text{F}$?

SOLUTION

$$W = 1.1(10 \text{ M}\Omega)(470 \text{ }\mu\text{F}) = 5170 \text{ s} = 86.2 \text{ min} = 1.44 \text{ hr}$$

Here we have a pulse width of more than an hour. The falling edge of the pulse occurs after a time delay of 1.44 hr.

23-8 Astable Operation of the 555 Timer

Generating time delays from microseconds to hours is useful in many applications. The 555 timer can also be used as an astable or free-running multivibrator. In this mode, it requires two external resistors and one capacitor to set the frequency of oscillations.

Astable Operation

Figure 23-35 shows the 555 timer connected for astable operation. The trip points are the same as for monostable operation:

$$\text{UTP} = \frac{2V_{CC}}{3}$$

$$\text{LTP} = \frac{V_{CC}}{3}$$

When Q is low, the transistor is cut off and the capacitor is charging through a total resistance of:

$$R = R_1 + R_2$$

Because of this, the charging time constant is $(R_1 + R_2)C$. As the capacitor charges, the threshold voltage (pin 6) increases.

Eventually, the threshold voltage exceeds $+2V_{CC}/3$. Then, the upper comparator sets the flip-flop. With Q high, the transistor saturates and grounds pin 7. The capacitor now discharges through R_2. Therefore, the discharging time constant is R_2C. When the capacitor voltage drops to slightly less than $V_{CC}/3$, the lower comparator resets the flip-flop.

Figure 23-36 shows the waveforms. The timing capacitor has exponentially rising and falling voltages between UTP and LTP. The output is a rectangular wave that swings between 0 and V_{CC}. Since the charging time constant is longer than the discharging time constant, the output is nonsymmetrical. Depending on resistances R_1 and R_2, the duty cycle is between 50 and 100 percent.

By analyzing the equations for charging and discharging, we can derive the following formulas. The pulse width is given by:

$$W = 0.693(R_1 + R_2)C \qquad (23\text{-}26)$$

Figure 23-35 555 timer connected for astable operation.

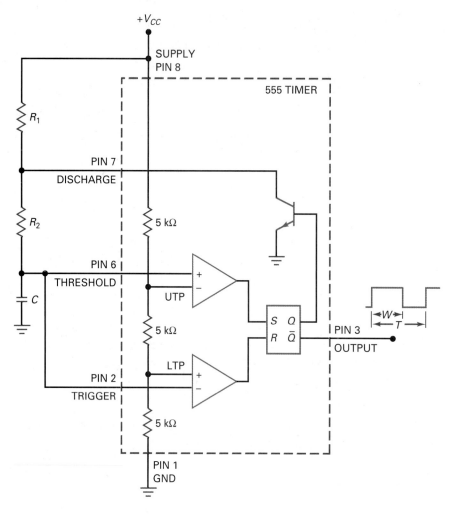

Figure 23-36 Capacitor and output waveforms for astable operation.

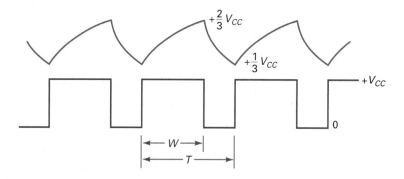

The period of the output equals:

$$T = 0.693(R_1 + 2R_2)C \qquad (23\text{-}27)$$

The reciprocal of the period is the frequency:

$$f = \frac{1.44}{(R_1 + 2R_2)C} \qquad (23\text{-}28)$$

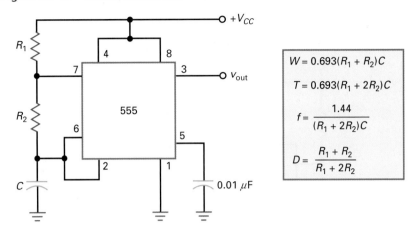

Figure 23-37 Astable multivibrator.

$$W = 0.693(R_1 + R_2)C$$

$$T = 0.693(R_1 + 2R_2)C$$

$$f = \frac{1.44}{(R_1 + 2R_2)C}$$

$$D = \frac{R_1 + R_2}{R_1 + 2R_2}$$

Dividing the pulse width by the period gives the duty cycle:

$$D = \frac{R_1 + R_2}{R_1 + 2R_2} \tag{23-29}$$

When R_1 is much smaller than R_2, the duty cycle approaches 50 percent. Conversely, when R_1 is much greater than R_2, the duty cycle approaches 100 percent.

Figure 23-37 shows the astable 555 timer as it usually appears on a schematic diagram. Again notice how pin 4 (reset) is tied to the supply voltage and how pin 5 (control) is bypassed to ground through a 0.01-μF capacitor.

The circuit of Fig. 23-37 can be modified to enable the duty cycle to become less than 50 percent. By placing a diode in parallel with R_2 (anode connected to pin 7), the capacitor will effectively charge through R_1 and the diode. The capacitor will discharge through R_2. Therefore, the duty cycle becomes:

$$D = \frac{R_1}{R_1 + R_2} \tag{23-30}$$

VCO Operation

Figure 23-38a shows a **voltage-controlled oscillator (VCO),** another application for a 555 timer. The circuit is sometimes called a **voltage-to-frequency converter** because an input voltage can change the output frequency.

Here is how the circuit works: Recall that pin 5 connects to the inverting input of the upper comparator (Fig. 23-31). Normally, pin 5 is bypassed to ground through a capacitor, so that UTP equals $+2V_{CC}/3$. In Fig. 23-38a, however, the voltage from a stiff potentiometer overrides the internal voltage. In other words, UTP equals V_{con}. By adjusting the potentiometer, we can change UTP to a value between 0 and V_{CC}.

Figure 23-38b shows the voltage waveform across the timing capacitor. Notice that the waveform has a minimum value of $+V_{con}/2$ and a maximum value of $+V_{con}$. If we increase V_{con}, it takes the capacitor longer to charge and discharge. Therefore, the frequency decreases. As a result, we can change the frequency of the circuit by varying the control voltage. Incidentally, the control voltage may come from a potentiometer, as shown, or it may be the output of a transistor circuit, an op amp, or some other device.

By analyzing the exponential charging and discharging of the capacitor, we can derive these equations:

$$W = -(R_1 + R_2)C \ln \frac{V_{CC} - V_{con}}{V_{CC} - 0.5V_{con}} \tag{23-31}$$

Figure 23-38 (a) Voltage-controlled oscillator; (b) capacitor voltage waveform.

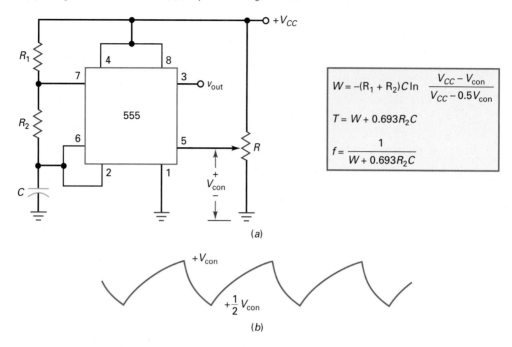

$$W = -(R_1 + R_2)C \ln \frac{V_{CC} - V_{con}}{V_{CC} - 0.5 V_{con}}$$

$$T = W + 0.693R_2C$$

$$f = \frac{1}{W + 0.693R_2C}$$

(a)

(b)

To use this equation, you need to take the **natural logarithm,** which is the logarithm to the base e. If you have a scientific calculator, look for the ln key. The period is given by:

$$T = W + 0.693R_2C \tag{23-32}$$

The frequency is given by:

$$f = \frac{1}{W + 0.693R_2C} \tag{23-33}$$

Example 23-8

‖‖‖ MultiSim

The 555 timer of Fig. 23-37 has $R_1 = 75$ kΩ, $R_2 = 30$ kΩ, and $C = 47$ nF. What is frequency of the output signal? What is the duty cycle?

SOLUTION   With the equations shown in Fig. 23-37:

$$f = \frac{1.44}{(75 \text{ k}\Omega + 60 \text{ k}\Omega)(47 \text{ nF})} = 227 \text{ Hz}$$

$$D = \frac{75 \text{ k}\Omega + 30 \text{ k}\Omega}{75 \text{ k}\Omega + 60 \text{ k}\Omega} = 0.778$$

This is equivalent to 77.8 percent.

PRACTICE PROBLEM 23-8   Repeat Example 23-8 with both R_1 and $R_2 = 75$ kΩ.

Example 23-9

The VCO of Fig. 23-38a has the same R_1, R_2, and C as in Example 23-8. What are the frequency and duty cycle when V_{con} is 11 V? What are the frequency and duty cycle when V_{con} is 1 V?

SOLUTION Using the equations of Fig. 23-38:

$$W = -(75 \text{ k}\Omega + 30 \text{ k}\Omega)(47 \text{ nF}) \ln \frac{12 \text{ V} - 11 \text{ V}}{12 \text{ V} - 5.5 \text{ V}} = 9.24 \text{ ms}$$

$$T = 9.24 \text{ ms} + 0.693(30 \text{ k}\Omega)(47 \text{ nF}) = 10.2 \text{ ms}$$

The duty cycle is:

$$D = \frac{W}{T} = \frac{9.24 \text{ ms}}{10.2 \text{ ms}} = 0.906$$

The frequency is:

$$f = \frac{1}{T} = \frac{1}{10.2 \text{ ms}} = 98 \text{ Hz}$$

When V_{con} is 1 V, the calculations give:

$$W = -(75 \text{ k}\Omega + 30 \text{ k}\Omega)(47 \text{ nF}) \ln \frac{12 \text{ V} - 1 \text{ V}}{12 \text{ V} - 0.5 \text{ V}} = 0.219 \text{ ms}$$

$$T = 0.219 \text{ ms} + 0.693(30 \text{ k}\Omega)(47 \text{ nF}) = 1.2 \text{ ms}$$

$$D = \frac{W}{T} = \frac{0.219 \text{ ms}}{1.2 \text{ ms}} = 0.183$$

$$f = \frac{1}{T} = \frac{1}{1.2 \text{ ms}} = 833 \text{ Hz}$$

PRACTICE PROBLEM 23-9 Repeat Example 23-9 with $V_{CC} = 15$ V and $V_{con} = 10$ V.

23-9 555 Circuits

The output stage of a 555 timer can *source* 200 mA. This means that a high output can produce up to 200 mA of load current (sourcing). Because of this, the 555 timer can drive relatively heavy loads such as relays, lamps, and loudspeakers. The output stage of a 555 timer can also *sink* 200 mA. This means that a low output can allow up to 200 mA to flow to ground (sinking). For instance, when a 555 timer drives a TTL load, the timer sources current when the output is high and it sinks current when the output is low. In this section, we discuss some applications for a 555 timer.

Start and Reset

Figure 23-39 shows a circuit with a number of modifications from the monostable timer shown earlier. To begin with, the trigger input (pin 2) is controlled by a

Figure 23–39 Monostable timer with adjustable pulse width and START and RESET buttons.

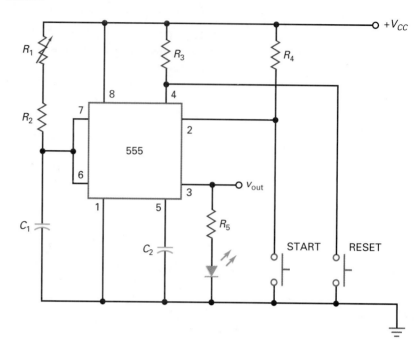

push-button switch (START). Since the switch is normally open, pin 2 is high and the circuit is inactive.

When somebody pushes and releases the START switch, pin 2 is temporarily pulled down to ground. Therefore, the output goes high and the LED turns on. Capacitor C_1 charges positively, as previously described. The charging time constant can be varied with R_1. In this way, we can get time delays of seconds to hours. When the capacitor voltage is slightly greater than $2V_{CC}/3$, the circuit resets and the output goes low. When this happens, the LED turns off.

Notice the RESET switch. It can be used to reset the circuit at any time during the output pulse. Since the switch is normally open, pin 4 is high and has no effect on the operation of the timer. When the RESET switch is closed, however, pin 4 is pulled down to ground and the output is reset to zero. The RESET is included because the user may want to terminate the high output. For instance, if the output pulse width has been set to 5 min, the user can terminate the pulse prematurely by pushing the RESET.

Incidentally, the output signal v_{out} can be used to drive a relay, a power FET, an IGBT, a buzzer, and so on. The LED serves as an indicator of the high output being delivered to some other circuit.

Sirens and Alarms

Figure 23-40 shows how to use an astable 555 timer as a siren or an alarm. Normally, the ALARM switch is closed, which pulls pin 4 down to ground. In this case, the 555 timer is inactive and there is no output. When the ALARM switch is opened, however, the circuit will generate a rectangular output whose frequency is determined by R_1, R_2, and C_1.

The output from pin 3 drives a loudspeaker through a resistance of R_4. The size of this resistance depends on the supply voltage and the impedance of the loudspeaker. The impedance of the branch with R_4 and the speaker should limit the output current to 200 mA or less because this is the maximum current a 555 timer can source.

Figure 23-40 Astable 555 circuit used for siren or alarm.

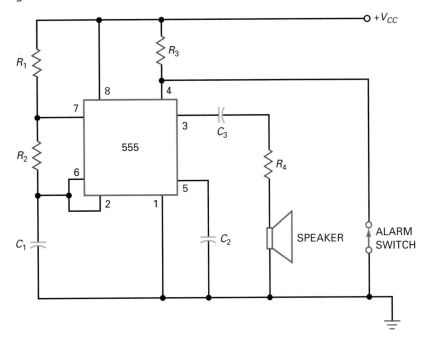

The circuit of Fig. 23-40 can be modified to produce more output power for the speaker. For instance, we can use the output from pin 3 to drive a class B push-pull power amplifier, the output of which then drives the speaker.

Pulse-Width Modulator

Figure 23-41 shows a circuit used for **pulse-width modulation (PWM).** The 555 timer is connected in the monostable mode. The values of R, C, UTP, and V_{CC} determine the width of the output pulse as follows:

$$W = -RC \ln \left(1 - \frac{\text{UTP}}{V_{CC}} \right) \tag{23-34}$$

A low-frequency signal called a **modulating signal** is capacitively coupled into pin 5. This modulating signal is voice or computer data. Since pin 5 controls the value of UTP, v_{mod} is being added to the quiescent UTP. Therefore, the instantaneous UTP is given by:

$$\text{UTP} = \frac{2V_{CC}}{3} + v_{\text{mod}} \tag{23-35}$$

For instance, if $V_{CC} = 12$ V and the modulating signal has a peak value of 1 V, then Eq. (23-31) gives:

$$\text{UTP}_{\text{max}} = 8 \text{ V} + 1 \text{ V} = 9 \text{ V}$$
$$\text{UTP}_{\text{min}} = 8 \text{ V} - 1 \text{ V} = 7 \text{ V}$$

This means that the instantaneous UTP varies sinusoidally between 7 and 9 V.

A train of triggers called the *clock* is the input to pin 2. Each trigger produces an output pulse. Since the period of the triggers is T, the output will be a series of rectangular pulses with a period of T. The modulating signal has no effect on the period T, but it does change the width of each output pulse. At point A, the positive peak of the modulating signal, the output pulse is wide as shown. At point B, the negative peak of the modulating signal, the output pulse is narrow.

PWM is used in communications. It allows a low-frequency modulating signal (voice or data) to change the pulse width of a high-frequency signal called the **carrier.** The modulated carrier can be transmitted over copper wire, over fiber-optic cable, or through space to a receiver. The receiver recovers the modulating signal to drive a speaker (voice) or a computer (data).

Pulse–Position Modulation

With PWM, the pulse width changes, but the period is constant because it is determined by the frequency of the input triggers. Because the period is fixed, the position of each pulse is the same, which means that the leading edge of the pulse always occurs after a fixed interval of time.

Pulse-position modulation (PPM) is different. With this type of modulation, the position (leading edge) of each pulse changes. With PPM, both the width and the period of pulses vary with the modulating signal.

Figure 23-42a shows a *pulse-position modulator.* It is similar to the VCO discussed earlier. Since the modulating signal is coupled into pin 5, the instantaneous UTP is given by Eq. (23-35):

$$\text{UTP} = \frac{2V_{CC}}{3} + v_{\text{mod}}$$

When the modulating signal increases, UTP increases and the pulse width increases. When the modulating signal decreases, UTP decreases and the pulse width decreases. This is why the pulse width varies as shown in Fig. 23-42b.

The pulse width and period equations are:

$$W = -(R_1 + R_2)C \ln \frac{V_{CC} - \text{UTP}}{V_{CC} - 0.5\,\text{UTP}} \tag{23-36}$$

$$T = W + 0.693R_2C \tag{23-37}$$

In Eq. (23-37), the second term is the *space* between pulses:

$$\textbf{Space} = \textbf{0.693}R_2C \tag{23-38}$$

This space is the time between the trailing edge of one pulse and the leading edge of the next pulse. Since V_{con} does not appear in Eq. (23-38), the space between pulses is constant, as shown in Fig. 23-42b.

Since the space is constant, the position of the leading edge of any pulse depends on how wide the preceding pulse is. This is why this type of modulation is called *pulse-position modulation.* Like PWM, PPM is used in communication systems to transfer voice or data.

Ramp Generation

Charging a capacitor through a resistor produces an exponential waveform. If we use a constant current source instead of a resistor to charge a capacitor, the capacitor voltage is ramp. This is the idea behind the circuit of Fig. 23-43a. Here we have replaced the resistor of a monostable circuit with a *pnp* current source that produces a constant charging current of:

$$I_C = \frac{V_{CC} - V_E}{R_E} \tag{23-39}$$

When a trigger starts the monostable 555 timer of Fig. 23-43a, the *pnp* current source forces a constant charging current into the capacitor. Therefore, the voltage across the capacitor is a ramp, as shown in Fig. 23-43b. The slope *S* of the ramp is given by:

$$S = \frac{I_C}{C} \tag{23-40}$$

Since the capacitor voltage reaches a maximum value $2V_{CC}/3$ before discharge occurs, the peak value of the ramp shown in Fig. 23-43b is:

$$V = \frac{2V_{CC}}{3} \qquad (23\text{-}41)$$

and the duration T of the ramp is:

$$T = \frac{2V_{CC}}{3S} \qquad (23\text{-}42)$$

Example 23–10

A pulse-width modulator like Fig. 23-41 has $V_{CC} = 12$ V, $R = 9.1$ kΩ, and $C = 0.01$ μF. The clock has a frequency of 2.5 kHz. If a modulating signal has a peak value of 2 V, what is the period of the output pulses? What is the quiescent pulse width? What are the minimum and maximum pulse widths? What are the minimum and maximum duty cycles?

SOLUTION   The period of the output pulses equals the period of the clock:

$$T = \frac{1}{2.5 \text{ kHz}} = 400 \ \mu s$$

The quiescent pulse width is:

$$W = 1.1RC = 1.1(9.1 \text{ k}\Omega)(0.01 \ \mu\text{F}) = 100 \ \mu s$$

With Eq. (23-35), calculate the minimum and maximum UTP:

$$\text{UTP}_{\min} = 8 \text{ V} - 2 \text{ V} = 6 \text{ V}$$
$$\text{UTP}_{\max} = 8 \text{ V} + 2 \text{ V} = 10 \text{ V}$$

Now, calculate the minimum and maximum pulse widths with Eq. (23-34):

$$W_{\min} = -(9.1 \text{ k}\Omega)(0.01 \ \mu\text{F}) \ln\left(1 - \frac{6 \text{ V}}{12 \text{ V}}\right) = 63.1 \ \mu s$$

$$W_{\max} = -(9.1 \text{ k}\Omega)(0.01 \ \mu\text{F}) \ln\left(1 - \frac{10 \text{ V}}{12 \text{ V}}\right) = 163 \ \mu s$$

Figure 23-41   555 timer connected as pulse-width modulator.

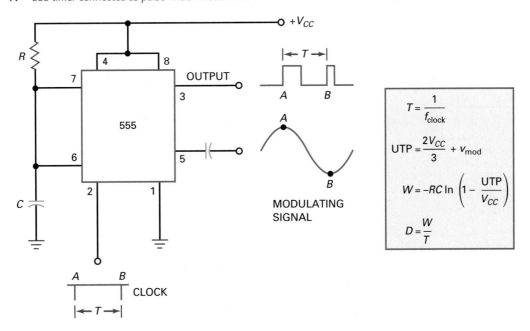

The minimum and maximum duty cycles are:

$$D_{min} = \frac{63.1\,\mu s}{400\,\mu s} = 0.158$$

$$D_{max} = \frac{163\,\mu s}{400\,\mu s} = 0.408$$

PRACTICE PROBLEM 23-10   Following Example 23-10, change V_{CC} to 15 V. Calculate the maximum pulse width and maximum duty cycle.

Example 23-11

A pulse-position modulator like Fig. 23-42 has V_{CC} = 12 V, R_1 = 3.9 kΩ, R_2 = 3 kΩ, and C = 0.01 μF. What are the quiescent width and period of the output pulses? If a modulating signal has a peak value of 1.5 V, what are the minimum and maximum pulse widths. What is the space between pulses?

SOLUTION   With no modulating signal, the quiescent period of the output pulses is that of a 555 timer used as an astable multivibrator. With Eqs. (23-26) and (23-27), we can calculate the quiescent width and period as follows:

$$W = 0.693(3.9\text{ k}\Omega + 3\text{ k}\Omega)(0.01\ \mu F) = 47.8\ \mu s$$
$$T = 0.693(3.9\text{ k}\Omega + 6\text{ k}\Omega)(0.01\ \mu F) = 68.6\ \mu s$$

With Eq. (23-35), calculate the minimum and maximum UTP:

$$UTP_{min} = 8\text{ V} - 1.5\text{ V} = 6.5\text{ V}$$
$$UTP_{max} = 8\text{ V} + 1.5\text{ V} = 9.5\text{ V}$$

Figure 23-42   The 555 timer connected as pulse-position modulator.

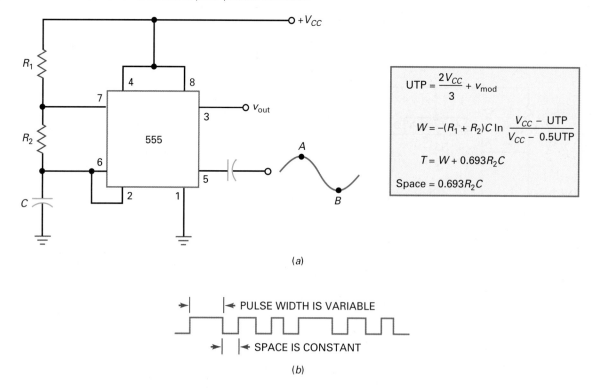

(a)

PULSE WIDTH IS VARIABLE

SPACE IS CONSTANT

(b)

With Eq. (23-36), the minimum and maximum pulse widths are:

$$W_{min} = -(3.9 \text{ k}\Omega + 3 \text{ k}\Omega)(0.01 \ \mu\text{F}) \ln \frac{12 \text{ V} - 6.5 \text{ V}}{12 \text{ V} - 3.25 \text{ V}} = 32 \ \mu\text{s}$$

$$W_{max} = -(3.9 \text{ k}\Omega + 3 \text{ k}\Omega)(0.01 \ \mu\text{F}) \ln \frac{12 \text{ V} - 9.5 \text{ V}}{12 \text{ V} - 4.75 \text{ V}} = 73.5 \ \mu\text{s}$$

With Eq. (23-37), the minimum and maximum periods are:

$$T_{min} = 32 \ \mu\text{s} + 0.693(3 \text{ k}\Omega)(0.01 \ \mu\text{F}) = 52.8 \ \mu\text{s}$$

$$T_{max} = 73.5 \ \mu\text{s} + 0.693(3 \text{ k}\Omega)(0.01 \ \mu\text{F}) = 94.3 \ \mu\text{s}$$

The space between the trailing edge of any pulse and the leading edge of the next pulse is:

$$\text{Space} = 0.693(3 \text{ k}\Omega)(0.01 \ \mu\text{F}) = 20.8 \ \mu\text{s}$$

Example 23–12

The ramp generator of Fig. 23-43 has a constant collector current of 1 mA. If $V_{CC} = 15$ V and $C = 100$ nF, what is the slope of the output ramp? What is its peak value? What is its duration?

SOLUTION   The slope is

$$S = \frac{1 \text{ mA}}{100 \text{ nF}} = 10 \text{ V/ms}$$

Figure 23–43   (a) Bipolar junction transistor and 555 timer produce ramp output; (b) trigger and ramp waveforms.

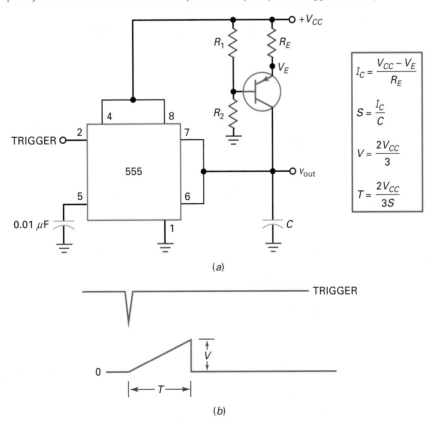

(a)

(b)

The peak value is

$$V = \frac{2(15\text{ V})}{3} = 10\text{ V}$$

The duration of the ramp is:

$$T = \frac{2(15\text{ V})}{3(10\text{ V/ms})} = 1\text{ ms}$$

PRACTICE PROBLEM 23-12   Using Fig. 23-43, with $V_{CC} = 12$ V and $C = 0.2$ μF, repeat Example 23-12.

23-10 The Phase-Locked Loop

A **phase-locked loop (PLL)** contains a phase detector, a dc amplifier, a low-pass filter, and a voltage-controlled oscillator (VCO). When a PLL has an input signal with a frequency of f_{in}, its VCO will produce an output frequency that equals f_{in}.

Phase Detector

Figure 23-44a shows a **phase detector,** the first stage in a PLL. This circuit produces an output voltage proportional to the phase difference between two input signals. For instance, Fig. 23-44b shows two input signals with a phase difference of $\Delta\phi$. The phase detector responds to this phase difference by producing a dc output voltage, which is proportional to $\Delta\phi$, as shown in Fig. 23-44c.

When v_1 leads v_2, as shown in Fig. 23-44b, $\Delta\phi$ is positive. If v_1 were to lag v_2, $\Delta\phi$ would be negative. The typical phase detector produces a linear response between $-90°$ and $+90°$, as shown in Fig. 23-44c. As we can see, the output of the phase detector is zero when $\Delta\phi = 0°$. When $\Delta\phi$ is between $0°$ and $90°$, the output is a positive voltage. When $\Delta\phi$ is between $0°$ and $-90°$, the output is a negative voltage. The key idea here is that the phase detector produces an output voltage that is directly proportional to the phase difference between its two input signals.

The VCO

In Fig. 23-45a, the input voltage v_{in} to the VCO determines the output frequency f_{out}. A typical VCO can be varied over a 10:1 range of frequency. Furthermore, the variation is linear as shown in Fig. 23-45b. When the input voltage to the VCO is zero, the VCO is free-running at a quiescent frequency f_0. When the input voltage is positive, the VCO frequency is greater than f_0. If the input voltage is negative, the VCO frequency is less than f_0.

Block Diagram of a PLL

Figure 23-46 is a block diagram of a PLL. The phase detector produces a dc voltage that is proportional to the phase difference of its two input signals. The output voltage of the phase detector is usually small. This is why the second stage is a dc

Figure 23-44   (a) Phase detector has two input signals and one output signal; (b) equal-frequency sine waves with phase difference; (c) output of phase detector is directly proportional to phase difference.

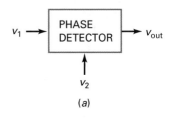

(a)

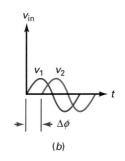

(b)

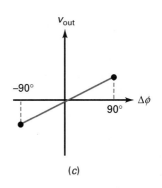

(c)

Figure 23-45 (a) Input voltage controls output frequency of VCO; (b) output frequency is directly proportional to input voltage.

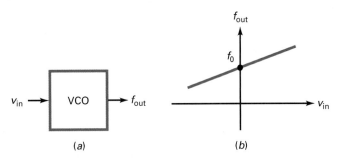

(a) (b)

Figure 23-46 Block diagram of a phase-locked loop.

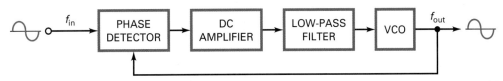

amplifier. The amplified phase difference is filtered before being applied to the VCO. Notice that the VCO output is being fed back to the phase detector.

Input Frequency Equals Free-Running Frequency

To understand PLL action, let us start with the case of input frequency equal to f_0, the free-running frequency of the VCO. In this case, the two input signals to the phase detector have the same frequency and phase. Because of this, the phase difference $\Delta\phi$ is $0°$ and the output of the phase detector is zero. As a result, the input voltage to the VCO is zero, which means that the VCO is free-running with a frequency of f_0. As long as the frequency and phase of the input signal remain the same, the input voltage to the VCO will be zero.

Input Frequency Differs from Free-Running Frequency

Let us assume that the input and free-running VCO frequencies are each 10 kHz. Now, suppose the input frequency increases to 11 kHz. This increase will appear to be an increase in phase because v_1 leads v_2 at the end of the first cycle, as shown in Fig. 23-47a. Since input signal leads the VCO signal, $\Delta\phi$ is positive. In this case, the phase detector of Fig. 23-46 produces a positive output voltage. After being amplified and filtered, this positive voltage increases the VCO frequency.

The VCO frequency will increase until it equals 11 kHz, the frequency of the input signal. When the VCO frequency equals the input frequency, the VCO is *locked on* to the input signal. Even though each of the two input signals to the phase detector has a frequency of 11 kHz, the signals have a different phase, as shown in Fig. 23-47b. This positive phase difference produces the voltage needed to keep the VCO frequency slightly above its free-running frequency.

Figure 23-47 (a) An increase in the frequency of v_1 produces a phase difference; (b) a phase difference exists after the VCO frequency increases.

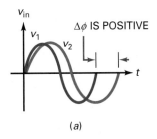

(a)

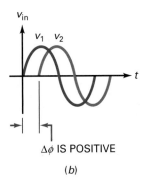

(b)

If the input frequency increases further, the VCO frequency also increases as needed to maintain the lock. For instance, if the input frequency increases to 12 kHz, the VCO frequency increases to 12 kHz. The phase difference between the two input signals will increase as needed to produce the correct control voltage for the VCO.

Lock Range

The **lock range** of a PLL is the range of input frequencies over which the VCO can remain locked on to the input frequency. It is related to the maximum phase difference that can be detected. In our discussion, we are assuming that the phase detector can produce an output voltage for $\Delta\phi$ between $-90°$ and $+90°$. At these limits, the phase detector produces a maximum output voltage, either negative or positive.

If the input frequency is too low or too high, the phase difference is outside the range of $-90°$ and $+90°$. Therefore, the phase detector cannot produce the additional voltage needed for the VCO to remain locked on. At these limits, therefore, the PLL loses its lock on the input signal.

The lock range is usually specified as a percentage of the VCO frequency. For instance, if the VCO frequency is 10 kHz and the lock range is ±20 percent, the PLL can remain locked on any input frequency between 8 and 12 kHz.

Capture Range

The capture range is different. Assume that the input frequency is outside the lock range. Then, the VCO is free-running at 10 kHz. Now, assume that the input frequency changes toward the VCO frequency. At some point, the PLL will be able to lock on to the input frequency. The range of input frequencies within which the PLL can reestablish the lock is called the **capture range.**

The capture range is specified as a percentage of the free-running frequency. If $f_0 = 10$ kHz and the capture range is ±5 percent, the PLL can lock on to an input frequency between 9.5 and 10.5 kHz. Typically, the capture range is less than the lock range because the capture range depends on the cutoff frequency of the low-pass filter. The lower the cutoff frequency, the smaller the capture range.

The cutoff frequency of the low-pass filter is kept low to prevent high-frequency components like noise or other unwanted signals from reaching the VCO. The lower the cutoff frequency of the filter, the cleaner the signal driving the VCO. Therefore, a designer has to trade off capture range against low-pass bandwidth to get a clean signal for the VCO.

Applications

A PLL can be used in two fundamentally different ways. First, it can be used to lock on to the input signal. The output frequency then equals the input frequency. This has the advantage of cleaning up a noisy input signal because the low-pass filter will remove high-frequency noise and other components. Since the output signal comes from the VCO, the final output is stable and almost noise-free.

Second, a PLL can be used as an FM demodulator. The theory of **frequency modulation (FM)** is covered in communication courses, so we will discuss only the basic idea. The *LC* oscillator of Fig. 23-48*a* has a variable capacitance. If a modulating signal controls this capacitance, the oscillator output will be *frequency-modulated,* as shown in Fig. 23-48*b*. Notice how the frequency of

Figure 23-48 (*a*) Variable capacitance changes resonant frequency of LC oscillator; (*b*) sine wave has been frequency-modulated.

(*a*)

$f_{x\,(min)}$ $f_{x\,(max)}$

(*b*)

this FM wave varies from a minimum to a maximum, corresponding to the minimum and maximum peaks of the modulating signal.

If the FM signal is the input to a PLL, the VCO frequency will lock on to the FM signal. Since the VCO frequency varies, $\Delta\phi$ follows the variations in the modulating signal. Therefore, the output of the phase detector will be a low-frequency signal that is a replica of the original modulating signal. When used in this way, the PLL is being used as an *FM demodulator,* a circuit that recovers the modulating signal from the FM wave.

PLLs are available as monolithic ICs. For instance, the NE565 is a PLL that contains a phase detector, a VCO, and a dc amplifier. The user connects external components like a timing resistor and capacitor to set the free-running frequency of the VCO. Another external capacitor sets the cutoff frequency of the low-pass filter. The NE565 can be used for FM demodulation, frequency synthesis, telemetry receivers, modems, tone decoding, and so forth.

23-11 Function Generator ICs

Special function generator ICs have been developed that combine many of the individual circuit capabilities we have been discussing. These ICs are able to provide waveform generation including sine, square, triangle, ramp, and pulse signals. The output waveforms can be made to vary in amplitude and frequency by changing the values of external resistors and capacitors or by applying an external voltage. This external voltage enables the IC to perform useful applications such as voltage-to-frequency (V/F) conversion, AM and FM signal generation, voltage-controlled oscillation (VCO), and frequency-shift keying (FSK).

The XR-2206

An example of a special function generator IC is the XR-2206. This monolithic IC can provide externally controlled frequencies from 0.01 Hz to more than 1.00 MHz. A block diagram of this IC is shown in Fig. 23-49. The diagram shows four main functional blocks, which include a VCO, an analog multiplier and sine-shaper, a unity gain buffer amplifier, and a set of current switches.

The output frequency of the VCO is proportional to an input current, which is determined by a set of external timing resistors. These resistors connect to pins 7 and 8, respectively, and ground. Because there are two timing pins, two discrete output frequencies can be obtained. A high or low input signal on pin 9 controls the current switches. The current switches then select which one of the timing resistors will be used. If the input signal on pin 9 alternately changes from high to low, the output frequency of the VCO will be shifted from one frequency to another. This action is referred to as **frequency-shift keying (FSK)** and is used in electronic communication applications.

The output of the VCO drives the multiplier and sine-shaper block, along with an output switching transistor. The output switching transistor is driven to

Figure 23-49 XR-2206 block diagram.

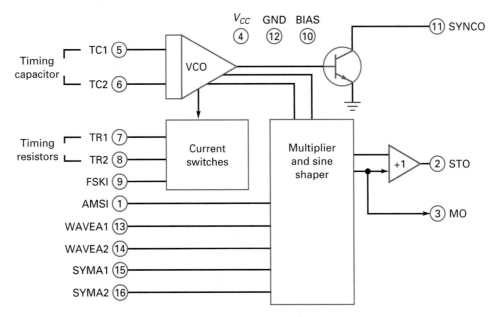

cutoff and saturation, which provides a square wave output signal at pin 11. The output of the multiplier and sine-shaper block is connected to a unity gain buffer amplifier, which determines the IC's output current capability and its output impedance. The output at pin 2 can be either a sine wave or a triangle wave.

Sine- and Triangle-Wave Output

Fig. 23-50a shows the external circuit connections and components for generating sine waves or triangle waves. The frequency of oscillation f_0 is determined by the timing resistor R, connected at either pin 7 or pin 8, and the external capacitor C, connected across pins 5 and 6. The value of oscillation is found by:

$$f_0 = \frac{1}{RC} \tag{23-31}$$

Even though R can be up to 2 MΩ, maximum temperature stability occurs when 4 kΩ < R < 200 kΩ. A graph of R versus oscillation frequency is shown in Fig. 23-50b. Also, the recommended value of C should be from 1000 pF to 100 μF.

In Fig. 23-50a, when the switch S_1 is closed, the output at pin 2 will be a sine wave. The potentiometer R_1 at pin 7 provides the desired frequency tuning. Adjustable resistors R_A and R_B enable the output waveform to be modified for proper waveform symmetry and distortion levels. When S_1 is open, the output at pin 2 changes from a sine wave to a triangle wave. Resistor R_3, connected at pin 3, controls the amplitude of the output waveform. As shown in Fig. 23-50c, the output amplitude is directly proportional to the value of R_3. Notice that the value of the triangle waveform is approximately double the output of a sine waveform for a given R_3 setting.

Figure 23–50 Sine-wave generation: (*a*) Circuit; (*b*) R versus oscillation frequency; (*c*) output amplitude.

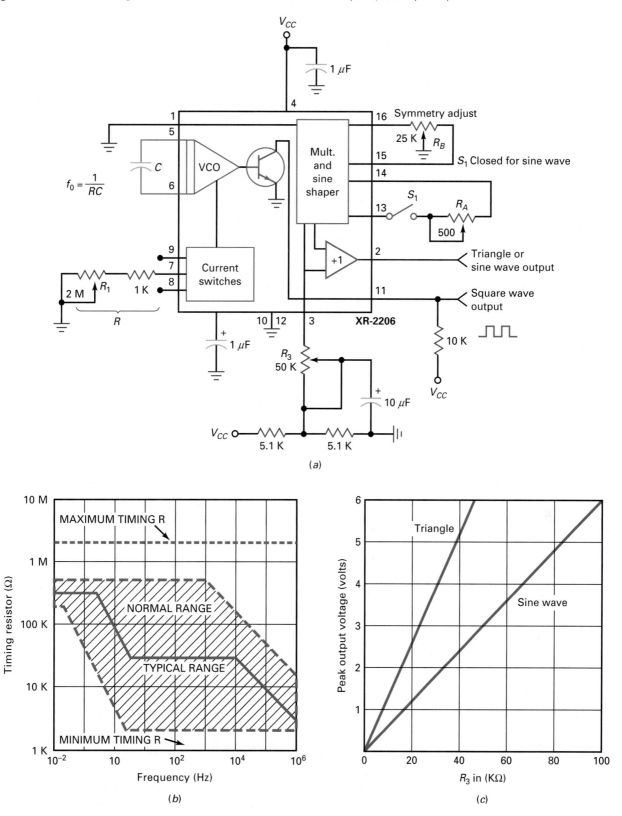

(*a*)

(*b*)

(*c*)

Figure 23-51 Pulse and ramp generation.

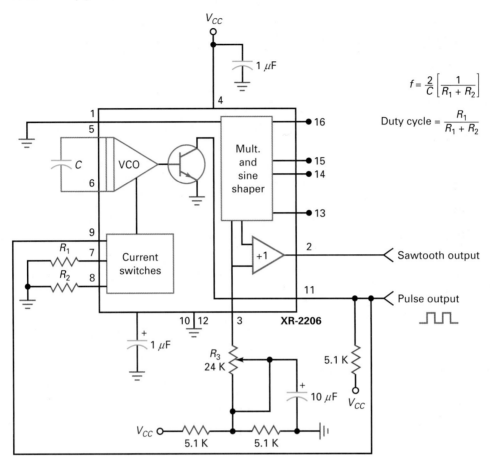

$$f = \frac{2}{C}\left[\frac{1}{R_1 + R_2}\right]$$

$$\text{Duty cycle} = \frac{R_1}{R_1 + R_2}$$

Pulse and Ramp Generation

Fig. 23-51 shows the external connections of the circuit used to create sawtooth (ramp) and pulse outputs. Notice that the square wave output at pin 11 is shorted to the FSK terminal at pin 9. This allows the circuit to automatically frequency-shift between two separate frequencies. This frequency shift occurs when the output at pin 11 changes from a high-level output to a low-level output or from a low-level output to a high-level output. The output frequency is found by:

$$f = \frac{2}{C}\left[\frac{1}{R_1 + R_2}\right] \tag{23-32}$$

and the circuit's duty cycle is found by:

$$D = \frac{R_1}{R_1 + R_2} \tag{23-33}$$

Fig. 23-52 shows a data sheet for the XR-2206. If operated with a single positive supply voltage, the supply can range from 10 V to 26 V. If a split- or dual-supply voltage is used, notice how the values range from ±5 V to ±13 V. Fig. 23-52 also shows recommended R and C values for generating maximum and minimum output frequencies. Also specified is the typical sweep range of 2000:1. As shown in the data sheet, the triangle- and sine-wave output has an output impedance value of 600 Ω. This makes the XR-2206 function generator IC well suited for many electronic communications applications.

Figure 23-52   The XR-2206 data sheet.

XR-2206

DC ELECTRICAL CHARACTERISTICS

Test Conditions: $V_{CC} = 12V$, $T_A = 25°C$, $C = 0.01\mu F$, $R_1 = 100k\Omega$, $R_2 = 10k\Omega$, $R_3 = 25k\Omega$
Unless Otherwise Specified. S_1 open for triangle, closed for sine wave.

Parameters	XR-2206M/P			XR-2206CP/D			Units	Conditions
	Min.	Typ.	Max.	Min.	Typ.	Max.		
General Characteristics								
Single Supply Voltage	**10**		**26**	10		26	V	
Split-Supply Voltage	**±5**		**±13**	±5		±13	V	
Supply Current		12	**17**		14	20	mA	$R_1 \geq 10k\Omega$
Oscillator Section								
Max. Operating Frequency	**0.5**	1		0.5	1		MHz	$C = 1000pF$, $R_1 = 1k\Omega$
Lowest Practical Frequency		0.01			0.01		Hz	$C = 50\mu F$, $R_1 = 2M\Omega$
Frequency Accuracy		±1	**±4**		±2		% of f_o	$f_o = 1/R_1C$
Temperature Stability Frequency		±10	**±50**		±20		ppm/°C	$0°C \leq T_A \leq 70°C$ $R_1 = R_2 = 20k\Omega$
Sine Wave Amplitude Stability[2]		4800			4800		ppm/°C	
Supply Sensitivity		0.01	**0.1**		0.01		%/V	$V_{LOW} = 10V$, $V_{HIGH} = 20V$, $R_1 = R_2 = 20k\Omega$
Sweep Range	1000:1	2000:1			2000:1		$f_H = f_L$	f_H @ $R_1 = 1k\Omega$ f_L @ $R_1 = 2M\Omega$
Sweep Linearity								
10:1 Sweep		2			2		%	$f_L = 1kHz$, $f_H = 10kHz$
1000:1 Sweep		8			8		%	$f_L = 100Hz$, $f_H = 100kHz$
FM Distortion		0.1			0.1		%	±10% Deviation
Recommended Timing Components								
Timing Capacitor: C	**0.001**		100	0.001		100	μF	
Timing Resistors: R_1 & R_2	**1**		2000	1		2000	$k\Omega$	
Triangle Sine Wave Output[1]								
Triangle Amplitude		160			160		mV/$k\Omega$	S_1 Open
Sine Wave Amplitude	**40**	60	80		60		mV/$k\Omega$	S_1 Closed
Max. Output Swing		6			6		Vp-p	
Output Impedance		600			600		Ω	
Triangle Linearity		1			1		%	
Amplitude Stability		0.5			0.5		dB	For 1000:1 Sweep
Sine Wave Distortion								
Without Adjustment		2.5			2.5		%	$R_1 = 30k\Omega$
With Adjustment		0.4	**1.0**		0.5	1.5	%	

Notes
[1] Output amplitude is directly proportional to the resistance, R_3, on Pin 3.
[2] For maximum amplitude stability, R_3 should be a positive temperature coefficient resistor.
Bold face parameters are covered by production test and guaranteed over operating temperature range.

Rev. 1.03

4

Figure 23-52 (*continued*).

DC ELECTRICAL CHARACTERISTICS (CONT'D)

Parameters	XR-2206M/P Min.	XR-2206M/P Typ.	XR-2206M/P Max.	XR-2206CP/D Min.	XR-2206CP/D Typ.	XR-2206CP/D Max.	Units	Conditions
Amplitude Modulation								
Input Impedance	50	100		50	100		kΩ	
Modulation Range		100			100		%	
Carrier Suppression		55			55		dB	
Linearity		2			2		%	For 95% modulation
Square-Wave Output								
Amplitude		12			12		Vp-p	Measured at Pin 11.
Rise Time		250			250		ns	C_L = 10pF
Fall Time		50			50		ns	C_L = 10pF
Saturation Voltage		0.2	**0.4**		0.2	0.6	V	I_L = 2mA
Leakage Current		0.1	**20**		0.1	100	µA	V_{CC} = 26V
FSK Keying Level (Pin 9)	0.8	1.4	**2.4**	0.8	1.4	2.4	V	See section on circuit controls
Reference Bypass Voltage	2.9	3.1	**3.3**	2.5	3	3.5	V	Measured at Pin 10.

Example 23-13

In Fig. 23-50, $R = 10$ kΩ and $C = 0.01$ µF. With S_1 closed, what are the output waveforms and output frequency at pins 2 and 11?

SOLUTION   Because S_1 is closed, the output at pin 2 will be a sine wave and the output at pin 11 will be a square wave. Both output waveforms will have the same frequency. The output frequency is found by:

$$f_0 = \frac{1}{RC} = \frac{1}{(10 \text{ k}\Omega)(0.01 \text{ µF})} = 10 \text{ kHz}$$

PRACTICE PROBLEM 23-13   Repeat Example 23-13 with $R = 20$ kΩ, $C = 0.01$ µF, and S_1 open.

Example 23-14

In Fig. 23-51, $R_1 = 1$ kΩ, $R_2 = 2$ kΩ, and $C = 0.1$ µF. Determine the square-wave output frequency and duty cycle.

SOLUTION   Using Eq. (23-32), the frequency out at pin 11 is:

$$f = \frac{2}{0.1 \text{ µF}} \left[\frac{1}{1 \text{ k}\Omega + 2 \text{ k}\Omega} \right] = 6.67 \text{ kHz}$$

The duty cycle is found using Eq. (23-33) as:

$$D = \frac{1 \text{ k}\Omega}{1 \text{ k}\Omega + 2 \text{ k}\Omega} = 0.333$$

PRACTICE PROBLEM 23-14   Repeat Example 23-14 with R_1 and $R_2 = 2$ kΩ and $C = 0.2$ µF.

Summary

SEC. 23-1 THEORY OF SINUSOIDAL OSCILLATION

To build a sinusoidal oscillator, we need to use an amplifier with positive feedback. For the oscillator to start, the loop gain must be greater than 1 when the phase shift around the loop is 0°.

SEC. 23-2 THE WIEN-BRIDGE OSCILLATOR

This is the standard oscillator for low to moderate frequencies in the range of 5 Hz to 1 MHz. It produces an almost perfect sine wave. A tungsten lamp or other nonlinear resistance is used to decrease the loop gain to 1.

SEC. 23-3 OTHER RC OSCILLATORS

The twin-T oscillator uses an amplifier and RC circuits to produce the required loop gain and phase shift at the resonant frequency. It works well at one frequency but is not suitable for an adjustable frequency oscillator. The phase-shift oscillator also uses an amplifier and RC circuits to produce oscillations. An amplifier can act like a phase-shift oscillator because of the stray lead and lag circuits in each stage.

SEC. 23-4 THE COLPITTS OSCILLATOR

RC oscillators usually do not work well above 1 MHz because of the additional phase shift inside the amplifier. This is why LC oscillators are preferred for frequencies between 1 and 500 MHz. This frequency range is beyond the f_{unity} of most op amps, which is why a bipolar junction transistor or FET is commonly used for the amplifying device. The Colpitts oscillator is one of the most widely used LC oscillators.

SEC. 23-5 OTHER LC OSCILLATORS

The Armstrong oscillator uses a transformer to produce the feedback signal. The Hartley oscillator uses an inductive voltage divider to produce the feedback signal. The Clapp oscillator has a small series capacitor in the inductive branch of the resonant circuit. This reduces the effect that stray capacitances have on the resonant frequency.

SEC. 23-6 QUARTZ CRYSTALS

Some crystals exhibit the piezoelectric effect. Because of this effect, a vibrating crystal acts like an LC resonant circuit with an extremely high Q. Quartz is the most important crystal producing the piezoelectric effect. It is used in crystal oscillators, in which a precise and reliable frequency is needed.

SEC. 23-7 THE 555 TIMER

The 555 timer contains two comparators, an RS flip-flop, and an npn transistor. It has an upper and lower trip point. When used in the monostable mode, the input triggers must fall below LTP to start the action. When the capacitor voltage slightly exceeds UTP, the discharge transistor turns on to discharge the capacitor.

SEC. 23-8 ASTABLE OPERATION OF THE 555 TIMER

When used in the astable mode, the 555 timer produces a rectangular output whose duty cycle can be set between 50 and 100 percent. The capacitor charges between $V_{CC}/3$ and $2V_{CC}/3$. When a control voltage is used, it changes UTP to V_{con}. This control voltage determines the frequency.

SEC. 23-9 555 CIRCUITS

The 555 timer can be used to create time delays, alarms, and ramp outputs. It can also be used to build a pulse-width modulator by applying a modulating signal to the control input and a train of negative-going triggers to the trigger input. The 555 time can also be used to build a pulse-position modulator by applying a modulating signal to the control input when the timer is in the astable mode.

SEC. 23-10 THE PHASE-LOCKED LOOP

A PLL contains a phase detector, a dc amplifier, a low-pass filter, and a VCO. The phase detector produces a control voltage that is proportional to the phase difference between its two input signals. The amplified and filtered control voltage then changes the frequency of the VCO as needed to lock on to the input signal.

SEC. 23-11 FUNCTION GENERATOR ICS

Function generator ICs have the ability to produce sine, square, triangle, pulse, and sawtooth waveforms. By connecting external resistors and capacitors, the output waveforms can be made to vary in frequency and amplitude. Special functions including AM/FM generation, voltage-to-frequency conversion, and frequency-shift keying can also be performed by these ICs.

Definitions

(23-20) Series resonance of crystal:

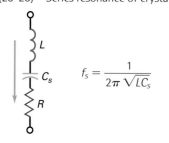

$$f_s = \frac{1}{2\pi \sqrt{LC_s}}$$

(23-22) Parallel resonance crystal:

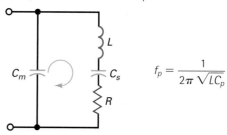

$$f_p = \frac{1}{2\pi \sqrt{LC_p}}$$

Derivations

(23-1) and (23-2)   Feedback factor and phase angle of lead-lag circuit:

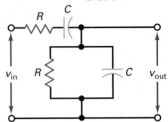

$$B = \frac{1}{\sqrt{9 - (X_C/R - R/X_C)^2}}$$

$$\phi = \arctan \frac{X_C/R - R/X_C}{3}$$

(23-9)   Exact resonant frequency:

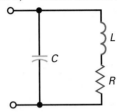

$$f_r = \frac{1}{2\pi \sqrt{LC}} \sqrt{\frac{Q^2}{Q^2 + 1}}$$

(23-19)   Frequency of crystal:

$$f = \frac{K}{t}$$

(23-21)   Equivalent parallel capacitance:

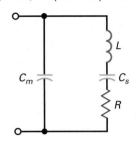

$$C_p = \frac{C_m C_s}{C_m + C_s}$$

(23-23) and (23-24)   Trip points of 555 timer:

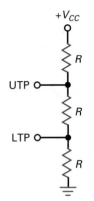

$$UTP = \frac{2V_{CC}}{3}$$

$$LTP = \frac{V_{CC}}{3}$$

Student Assignments

1. **An oscillator always needs an amplifier with**
   a. Positive feedback
   b. Negative feedback
   c. Both types of feedback
   d. An *LC* tank circuit

2. **The voltage that starts an oscillator is caused by**
   a. Ripple from the power supply
   b. Noise voltage in resistors
   c. The input signal from a generator
   d. Positive feedback

3. **The Wien-bridge oscillator is useful**
   a. At low frequencies
   b. At high frequencies
   c. With *LC* tank circuits
   d. At small input signals

4. **A lag circuit has a phase angle that is**
   a. Between 0 and +90°
   b. Greater than 90°
   c. Between 0 and −90°
   d. The same as the input voltage

5. **A coupling circuit is a**
   a. Lag circuit
   b. Lead circuit
   c. Lead-lag circuit
   d. Resonant circuit

6. **A lead circuit has a phase angle that is**
   a. Between 0 and +90°
   b. Greater than 90°
   c. Between 0 and −90°
   d. The same as the input voltage

7. **A Wien-bridge oscillator uses**
   a. Positive feedback
   b. Negative feedback
   c. Both types of feedback
   d. An *LC* tank circuit

8. **Initially, the loop gain of a Wien bridge is**
   a. 0   c. Low
   b. 1   d. High

9. **A Wien bridge is sometimes called a**
   a. Notch filter
   b. Twin-T oscillator
   c. Phase shifter
   d. Wheatstone bridge

10. **To vary the frequency of a Wien bridge, you can vary**
    a. One resistor   c. Three resistors
    b. Two resistors   d. One capacitor

11. **The phase-shift oscillator usually has**
   a. Two lead or lag circuits
   b. Three lead or lag circuits
   c. A lead-lag circuit
   d. A twin-T filter

12. **For oscillations to start in a circuit, the loop gain must be greater than 1 when the phase shift around the loop is**
   a. 90°   c. 270°
   b. 180°   d. 360°

13. **The most widely used *LC* oscillator is the**
   a. Armstrong   c. Colpitts
   b. Clapp   d. Hartley

14. **Heavy feedback in an *LC* oscillator**
   a. Prevents the circuit from starting
   b. Causes saturation and cutoff
   c. Produces maximum output voltage
   d. Means that *B* is small

15. **When *Q* decreases in a Colpitts oscillator, the frequency of oscillation**
   a. Decreases
   b. Remains the same
   c. Increases
   d. Becomes erratic

16. **Link coupling refers to**
   a. Capacitive coupling
   b. Transformer coupling
   c. Resistive coupling
   d. Power coupling

17. **The Hartley oscillator uses**
   a. Negative feedback
   b. Two inductors
   c. A tungsten lamp
   d. A tickler coil

18. **To vary the frequency of an *LC* oscillator, you can vary**
   a. One resistor
   b. Two resistors
   c. Three resistors
   d. One capacitor

19. **Of the following oscillators, the one with the most stable frequency is the**
   a. Armstrong
   b. Clapp
   c. Colpitts
   d. Hartley

20. **The material that has the piezoelectric effect is**
   a. Quartz
   b. Rochelle salts
   c. Tourmaline
   d. All the above

21. **Crystals have a very**
   a. Low Q
   b. High Q
   c. Small inductance
   d. Large resistance

22. **The series and parallel resonant frequencies of a crystal are**
   a. Very close together
   b. Very far apart
   c. Equal
   d. Low frequencies

23. **The kind of oscillator found in an electronic wristwatch is the**
   a. Armstrong
   b. Clapp
   c. Colpitts
   d. Quartz crystal

24. **A monostable 555 timer has the following number of stable states:**
   a. 0   c. 2
   b. 1   d. 3

25. **An astable 555 timer has the following number of stable states:**
   a. 0   c. 2
   b. 1   d. 3

26. **The pulse width from a one-shot multivibrator increases when the**
   a. Supply voltage increases
   b. Timing resistor decreases
   c. UTP decreases
   d. Timing capacitance increases

27. **The output waveform of a 555 timer is**
   a. Sinusoidal
   b. Triangular
   c. Rectangular
   d. Elliptical

28. **The quantity that remains constant in a pulse-width modulator is**
   a. Pulse width
   b. Period
   c. Duty cycle
   d. Space

29. **The quantity that remains constant in a pulse-position modulator is**
   a. Pulse width
   b. Period
   c. Duty cycle
   d. Space

30. **When a PPL is locked on the input frequency, the VCO frequency**
   a. Is less than f_0
   b. Is greater than f_0
   c. Equals f_0
   d. Equals f_{in}

31. **The bandwidth of the low-pass filter in a PPL determines the**
   a. Capture range
   b. Lock range
   c. Free-running frequency
   d. Phase difference

32. **The output frequency of the XR-2206 can be varied with**
   a. An external resistor
   b. An external capacitor
   c. An external voltage
   d. Any of the above

33. **FSK is a method of controlling the output**
   a. Functions
   b. Amplitude
   c. Frequency
   d. Phase

Problems

SEC. 23-2 THE WIEN-BRIDGE OSCILLATOR

23-1 The Wien-bridge oscillator of Fig. 23-53*a* uses a lamp with the characteristics of Fig. 23-53*b*. How much output voltage is there?

23-2 Position *D* in Fig. 23-53*a* is the highest frequency range of the oscillator. We can vary the frequency by using ganged rheostats. What are the minimum and maximum frequencies of oscillation on this range?

Figure 23-53

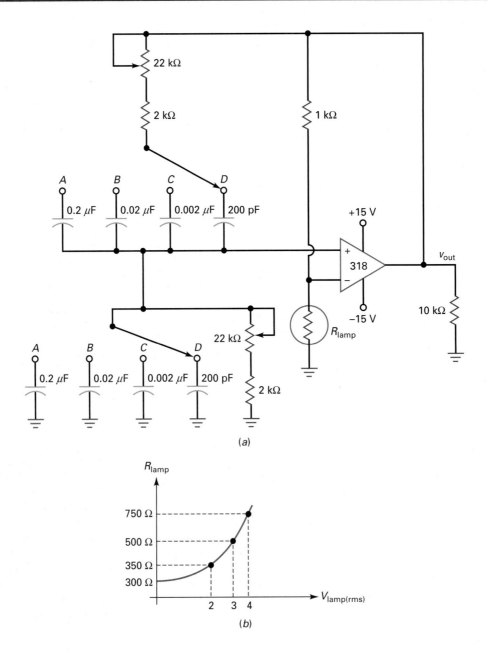

(a)

(b)

23-3 Calculate the minimum and maximum frequency of oscillation for each position of the ganged switch of Fig. 23-53a.

23-4 To change the output voltage of Fig. 23-53a to a value of 6 V rms, what change can you make?

23-5 In Fig. 23-53a, the cutoff frequency of the amplifier with negative feedback is at least 1 decade above the highest frequency of oscillation. What is the cutoff frequency?

SEC. 23-3 OTHER *RC* OSCILLATORS

23-6 The twin-T oscillator of Fig. 23-12 has $R = 10$ kΩ and $C = 0.01$ μF. What is the frequency of oscillation?

23-7 If the values in Prob. 23-6 are doubled, what happens to the frequency of oscillation?

SEC. 23-4 THE COLPITTS OSCILLATOR

23-8 What is the approximate value of the dc emitter current in Fig. 23-54? What is the dc voltage from the collector to the emitter?

23-9 What is the approximate frequency of oscillation in Fig. 23-54? The value of B? For the oscillator to start, what is the minimum value of A_v?

23-10 If the oscillator of Fig. 23-54 is redesigned to get a CB amplifier similar to the one in Fig. 23-18, what is the feedback fraction?

Figure 23-54

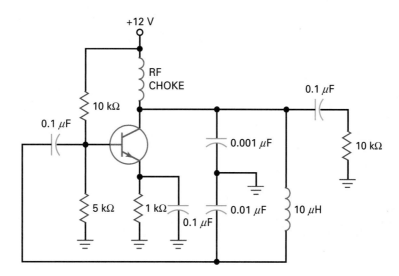

23-11 If the value of L is doubled in Fig. 23-54, what is the frequency of oscillation?

23-12 What can you do to the inductance of Fig. 23-54 to double the frequency of oscillation?

SEC. 23-5   OTHER *LC* OSCILLATORS

23-13 If 47 pF is connected in series with the 10 μH of Fig. 23-54, the circuit becomes a Clapp oscillator. What is the frequency of oscillation?

23-14 A Hartley oscillator like the one in Fig. 23-22 has $L_1 = 1$ μH and $L_2 = 0.2$ μH. What is the feedback fraction? The frequency of oscillation if $C = 1000$ pF? The minimum voltage gain needed to start oscillations?

23-15 An Armstrong oscillator has $M = 0.1$ μH and $L = 3.3$ μH. What is the feedback fraction? What is the minimum voltage gain needed to start the oscillations?

SEC. 23-6   QUARTZ CRYSTALS

23-16 A crystal has a fundamental frequency of 5 MHz. What is the approximate value of the first overtone frequency? The second overtone? The third?

23-17 A crystal has a thickness of t. If you reduce t by 1 percent, what happens to the frequency?

23-18 A crystal has these values: $L = 1$ H, $C_s = 0.01$ pF, $R = 1$ kΩ, and $C_m = 20$ pF. What is the series resonant frequency? The parallel resonant frequency? The Q at each frequency?

SEC. 23-7   THE 555 TIMER

23-19 A 555 timer is connected for monostable operation. If $R = 10$ kΩ and $C = 0.047$ μF, what is the width of the output pulse?

23-20 In Fig. 23-34, $V_{CC} = 10$ V, $R = 2.2$ kΩ, and $C = 0.2$ μF. What is the minimum trigger voltage that produces an output pulse? What is the maximum capacitor voltage? What is the width of the output pulse?

SEC. 23-8   ASTABLE OPERATION OF THE 555 TIMER

23-21 An astable 555 timer has $R_1 = 10$ kΩ, $R_2 = 2$ kΩ, and $C = 0.0022$ μF. What is the frequency?

23-22 The 555 timer of Fig. 23-37 has $R_1 = 20$ kΩ, $R_2 = 10$ kΩ, and $C = 0.047$ μF. What is frequency of the output signal? What is the duty cycle?

SEC. 23-9   555 CIRCUITS

23-23 A pulse-width modulator like the one in Fig. 23-41 has $V_{CC} = 10$ V, $R = 5.1$ kΩ, and $C = 1$ nF. The clock has a frequency of 10 kHz. If a modulating signal has a peak value of 1.5 V, what is the period of the output pulses? What is the quiescent pulse width? What are the minimum and maximum pulse widths? What are the minimum and maximum duty cycles?

23-24 A pulse-position modulator like the one in Fig. 23-42 has $V_{CC} = 10$ V, $R_1 = 1.2$ kΩ, $R_2 = 1.5$ kΩ, and $C = 4.7$ nF. What are the quiescent width and period of the output pulses? If the modulating signal has a peak value of 1.5 V, what are the minimum and maximum pulse widths? What is the space between pulses?

23-25 The ramp generator of Fig. 23-43 has a constant collector current of 0.5 mA. If $V_{CC} = 10$ V and $C = 47$ nF, what is the slope of the output ramp? What is its peak value? What is its duration?

SEC. 23-11   FUNCTION GENERATOR ICS

23-26 In Fig. 23-50, S_1 is closed, $R = 20$ kΩ, $R_3 = 40$ kΩ, and $C = 0.1$ μF. What is the output wave shape, frequency, and amplitude at pin 2?

23-27 In Fig. 23-50, with S_1 open and $R = 10$ kΩ, $R_3 = 40$ kΩ, and $C = 0.01$ μF, what is the output wave shape, frequency, and amplitude at pin 2?

23-28 In Fig. 23-51, $R_1 = 2$ kΩ, $R_2 = 10$ kΩ, and $C = 0.1$ μF. What is the output frequency and duty cycle at pin 11?

Troubleshooting

23-29 Does the output voltage of the Wien-bridge oscillator (Fig. 23-53*a*) increase, decrease, or stay the same for each of these troubles?

a. Lamp open

b. Lamp shorted

c. Upper potentiometer shorted

d. Supply voltages 20 percent low

e. 10 kΩ open

23-30 The Colpitts oscillator of Fig. 23-54 will not start. Name at least three possible troubles.

23-31 You have designed and built an amplifier. It does amplify an input signal, but the output looks fuzzy on an oscilloscope. When you touch the circuit, the fuzz disappears, leaving a perfect signal. What do you think the trouble is, and how would you try to eliminate it?

Critical Thinking

23-32 Design a Wien-bridge oscillator similar to the one in Fig. 23-53*a* that meets these specifications: three decade frequency ranges covering 20 Hz to 20 kHz with an output voltage of 5 V rms.

23-33 Select a value of *L* in Fig. 23-54 to get an oscillation frequency of 2.5 MHz.

23-34 Figure 23-55 shows an op-amp phase-shift oscillator. If $f_{2(CL)}$ = 1 kHz, what is the loop phase shift at 15.9 kHz?

23-35 Design a 555 timer that free-runs at a frequency of 1 kHz and a duty cycle of 75 percent.

Figure 23-55

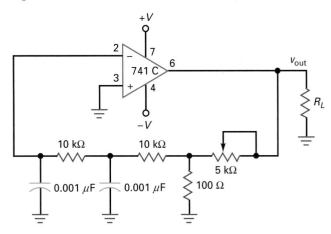

Job Interview Questions

1. How does a sinusoidal oscillator produce an output signal without an input signal?
2. Which oscillator is used for many applications in the range of 5 Hz to 1 MHz? Why is the output sinusoidal rather than clipped?
3. What type of oscillators are used most often for the range of 1 to 500 MHz?
4. To produce oscillations of a precise and reliable frequency, what kind of oscillator is used most often?
5. The 555 timer is used extensively in general applications as a timer. What is the difference between the circuit construction of a monostable and astable multivibrator?
6. Draw a simple block diagram of a PLL and explain the basic idea of how it remains locked on to the incoming frequency.
7. What does *pulse-width modulation* mean? What does *pulse-position modulation* mean? Illustrate your explanation by sketching waveforms.
8. Assume that you are building a three-stage amplifier. When you test it, you discover that it is producing an output signal with no input signal. Explain how this is possible. What are some of the things you can do to eliminate the unwanted signal?
9. How does an oscillator start if there is no input signal?

Self-Test Answers

1.	a	6.	a	11.	b	16.	b	21.	b	26.	d	31.	a
2.	b	7.	c	12.	d	17.	b	22.	a	27.	c	32.	d
3.	a	8.	d	13.	c	18.	d	23.	d	28.	b	33.	c
4.	c	9.	a	14.	b	19.	b	24.	b	29.	d		
5.	b	10.	b	15.	a	20.	d	25.	a	30.	d		

Practice Problem Answers

23-1 $R = 14.9\ \text{k}\Omega$

23-2 $R_{\text{lamp}} = 1.5\ \text{k}\Omega$; $I_{\text{lamp}} = 2\ \text{mA}$; $V_{\text{out}} = 9\ \text{V rms}$

23-3 $L = 28\ \mu\text{H}$

23-4 $C = 106\ \text{pF}$; $f_r = 4\ \text{MHz}$

23-5 $f_S = 291\ \text{kHz}$; $f_p = 292\ \text{kHz}$

23-6 LPT $= 5\ \text{V}$; UTP $= 10\ \text{V}$; $W = 51.7\ \text{ms}$

23-8 $f = 136\ \text{Hz}$; $D = 0.667$ or 66.7%

23-9 $W = 3.42\ \text{ms}$; $T = 4.4\ \text{ms}$; $D = 0.778$; $f = 227\ \text{Hz}$

23-10 $W_{\text{max}} = 146.5\ \mu\text{s}$; $D_{\text{max}} = 0.366$

23-12 $S = 5\ \text{V/ms}$; $V = 8\ \text{V}$; $T = 1.6\ \text{ms}$

23-13 Triangle waveform at pin 2. Square wave at pin 11. Both waveform frequencies are at 500 Hz

23-14 $f = 2.5\ \text{kHz}$; $D = 0.5$

chapter 24

Regulated Power Supplies

With a zener diode, we can build simple voltage regulators. Now, we want to discuss the use of negative feedback to improve voltage regulation. The discussion begins with linear regulators, the kind in which the regulating device is operating in the linear region. We will discuss two types of linear regulators: the shunt type and the series type. This chapter concludes with switching regulators, the type in which the regulating device switches on and off to improve the power efficiency.

Objectives

After studying this chapter, you should be able to:

- Describe how shunt regulators work.
- Describe how series regulators work.
- Explain the operation and characteristics of IC voltage regulators.
- Explain how dc-to-dc converters work.
- State the purposes and functions of current-booster and current-limiting circuits.
- Describe the three basic topologies of switching regulators.

Chapter Outline

Vocabulary

boost regulator

buck-boost regulator

buck regulator

current booster

current limiting

current-sensing resistor

dc-to-dc converter

dropout voltage

electromagnetic interference (EMI)

foldback current limiting

headroom voltage

IC voltage regulator

line regulation

load regulation

outboard transistor

pass transistor

phase splitter

radio-frequency interference (RFI)

short-circuit protection

shunt regulator

switching regulator

thermal shutdown

topology

24-1 Supply Characteristics

The quality of a power supply depends on its load regulation, line regulation, and output resistance. In this section, we will look at these characteristics because they are often used on data sheets to specify power supplies.

Load Regulation

Figure 24-1 shows a bridge rectifier with a capacitor-input filter. Changing the load resistance will change the load voltage. If we reduce the load resistance, we get more ripple and additional voltage drop across the transformer windings and diodes. Because of this, an increase in load current always decreases the load voltage.

Load regulation indicates how much the load voltage changes when the load current changes. The definition for load regulation is:

$$\textbf{Load regulation} = \frac{V_{NL} - V_{FL}}{V_{FL}} \times \textbf{100\%} \qquad (24\text{-}1)$$

where V_{NL} = load voltage with no load current

V_{FL} = load voltage with full load current

With this definition, V_{NL} occurs when the load current is zero, and V_{FL} occurs when the load current is the maximum value for the design.

For instance, suppose that the power supply of Fig. 24-1 has these values:

$V_{NL} = 10.6$ V for $I_L = 0$
$V_{FL} = 9.25$ V for $I_L = 1$ A

Then, Eq. (24-1) gives:

$$\text{Load regulation} = \frac{10.6 \text{ V} - 9.25 \text{ V}}{9.25 \text{ V}} \times 100\% = 14.6\%$$

The smaller the load regulation, the better the power supply. For instance, a well-regulated power supply can have a load regulation of less than

Figure 24-1 Power supply with capacitor-input filter.

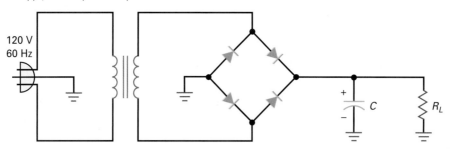

Load regulation $= \dfrac{V_{NL} - V_{FL}}{V_{FL}} \times 100\%$

V_{NL} = Load voltage with no load current

V_{FL} = Load voltage with full load current

Line regulation $= \dfrac{V_{HL} - V_{LL}}{V_{LL}} \times 100\%$

V_{LL} = Load voltage with low line voltage

V_{HL} = Load voltage with high line voltage

1 percent. This means that the load voltage varies less than 1 percent over the full range of load current.

Line Regulation

In Fig. 24-1, the input line voltage has a nominal value of 120 V. The actual voltage coming out of a power outlet may vary from 105 to 125 V rms, depending on the time of day, the locality, and other factors. Since the secondary voltage is directly proportional to the line voltage, the load voltage in Fig. 24-1 will change when line voltage changes.

Another way to specify the quality of a power supply is by its **line regulation,** defined as:

$$\text{Line regulation} = \frac{V_{HL} - V_{LL}}{V_{LL}} \times 100\% \tag{24-2}$$

where V_{HL} = load voltage with high line

V_{LL} = load voltage with low line

For instance, suppose that the power supply of Fig. 24-1 has these measured values:

$V_{LL} = 9.2$ V for line voltage = 105 V rms

$V_{HL} = 11.2$ V for line voltage = 125 V rms

Then, Eq. (24-2) gives:

$$\text{Line regulation} = \frac{11.2 \text{ V} - 9.2 \text{ V}}{9.2 \text{ V}} \times 100\% = 21.7\%$$

As with load regulation, the smaller the line regulation, the better the power supply. For example, a well-regulated power supply can have a line regulation of less than 0.1 percent. This means that the load voltage varies less than 0.1 percent when the line voltage varies from 105 to 125 V rms.

Output Resistance

The Thevenin or output resistance of a power supply determines the load regulation. If a power supply has a low output resistance, its load regulation will also be low. Here is one way to calculate the output resistance:

$$R_{TH} = \frac{V_{NL} - V_{FL}}{I_{FL}} \tag{24-3}$$

GOOD TO KNOW

Equation (24-3) can also be shown as

$$R_{TH} = \frac{V_{NL} - V_{FL}}{V_{FL}} \times R_L$$

For example, here are the values given earlier for Fig. 24-1:

$V_{NL} = 10.6$ V for $I_L = 0$

$V_{FL} = 9.25$ V for $I_L = 1$ A

For this power supply, the output resistance is:

$$R_{TH} = \frac{10.6 \text{ V} - 9.25 \text{ V}}{1 \text{ A}} = 1.35 \ \Omega$$

Figure 24-2 shows a graph of load voltage versus load current. As we can see, the load voltage decreases when the load current increases. The change in load voltage ($V_{NL} - V_{FL}$) divided by the change in current (I_{FL}) equals the output resistance of the power supply. The output resistance is related to the slope of this graph. The more horizontal the graph, the lower the output resistance.

In Fig. 24-2, the maximum load current I_{FL} occurs when the load resistance is minimum. Because of this, an equivalent expression for load regulation is:

$$\text{Load regulation} = \frac{R_{TH}}{R_{L(\text{min})}} \times 100\% \tag{24-4}$$

Regulated Power Supplies

Figure 24-2   Graph of load voltage versus load current.

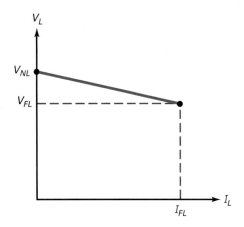

$$R_{TH} = \frac{V_{NL} - V_{FL}}{I_{FL}}$$

V_{NL} = Load voltage with no load current

V_{FL} = Load voltage with full load current

I_{FL} = Full load current

For example, if a power supply has an output resistance of 1.5 Ω and the minimum load resistance is 10 Ω, it has a load regulation of:

$$\text{Load regulation} = \frac{1.5\,\Omega}{10\,\Omega} \times 100\% = 15\%$$

24-2 Shunt Regulators

The line regulation and load regulation of an unregulated power supply are too high for most applications. By using a voltage regulator between the power supply and the load, we can significantly improve the line and load regulation. A linear voltage regulator uses a device operating in the linear region to hold the load voltage constant. There are two fundamental types of linear regulators: the shunt type and the series type. With the shunt type, the regulating device is in parallel with the load.

Zener Regulator

The simplest **shunt regulator** is the zener-diode circuit of Fig. 24-3. As discussed in Chap. 5, the zener diode operates in the breakdown region, producing an output voltage equal to the zener voltage. When the load current changes, the zener current increases or decreases to keep the current through R_S constant. With any shunt regulator, a change in load current is complemented by an opposing change in shunt current. If load current increases by 1 mA, the shunt current decreases by

Figure 24-3   Zener regulator is a shunt regulator.

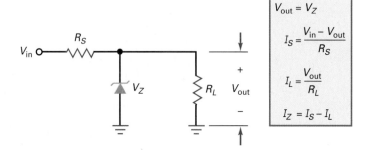

$V_{out} = V_Z$

$$I_S = \frac{V_{in} - V_{out}}{R_S}$$

$$I_L = \frac{V_{out}}{R_L}$$

$$I_Z = I_S - I_L$$

1 mA. Conversely, if the load current decreases by 1 mA, the shunt current increases by 1 mA.

As shown in Fig. 24-3, the equation for the current through the series resistor is:

$$I_S = \frac{V_{in} - V_{out}}{R_S}$$

This series current equals the *input current* to the shunt regulator. When the input voltage is constant, the input current is almost constant when the load current changes. This is how you can recognize any shunt regulator. A change in load current has almost no effect on the input current.

A final point: In Fig. 24-3, the maximum load current with regulation occurs when the zener current is almost zero. Therefore, the maximum load current in Fig. 24-3 equals the input current. This is true for any shunt regulator. The maximum load current with a regulated output voltage is equal to the input current.

Zener Voltage Plus One Diode Drop

At larger load currents, the load regulation of a zener regulator like Fig. 24-3 gets worse (increases) because the changing current through the zener resistance can change the output voltage significantly. One way to improve load regulation at larger load currents is to add a transistor to the circuit, as shown in Fig. 24-4. With this shunt regulator, the load voltage equals:

$$V_{out} = V_Z + V_{BE} \tag{24-5}$$

Here is how the circuit holds the output voltage constant: If the output voltage tries to increase, the increase is coupled through the zener diode to the base of the transistor. The larger base voltage produces more collector current through R_S. The larger voltage drop across R_S will offset most of the attempted increase in output voltage. The only noticeable change will be a slight increase in load voltage.

Conversely, if the output voltage tries to decrease, the voltage fed back to the base reduces the collector current and there is less voltage drop across R_S. Again, the attempted change in output voltage is offset by an opposing change in voltage across the series resistor. This time, the only noticeable change is a slight decrease in the output voltage.

Higher Output Voltage

Figure 24-5 shows another shunt regulator. This circuit has the advantage of being able to use low-temperature-coefficient zener voltages (between 5 and 6 V). The regulated output voltage will have approximately the same temperature coefficient as the zener diode, but the voltage will be higher.

Figure 24-4 Improved shunt regulator.

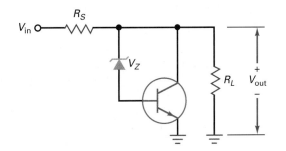

Figure 24-5 Shunt regulator with higher output.

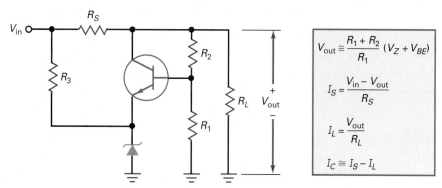

$$V_{out} \cong \frac{R_1 + R_2}{R_1}(V_Z + V_{BE})$$

$$I_S = \frac{V_{in} - V_{out}}{R_S}$$

$$I_L = \frac{V_{out}}{R_L}$$

$$I_C \cong I_S - I_L$$

The negative feedback is similar to that of the preceding regulator. Any attempted change in output voltage is fed back to the transistor, the output of which then almost completely offsets the attempted change in output voltage. The result is an output voltage that changes much less than it would without the negative feedback.

The base voltage is given by:

$$V_B \cong \frac{R_1}{R_1 + R_2} V_{out}$$

This is an approximation because it does not include the loading effect of the base current on the voltage divider. Usually, the base current is small enough to ignore. By solving the foregoing equation for output voltage, we get:

$$V_{out} \cong \frac{R_1 + R_2}{R_1} V_B$$

In Fig. 24-5, the base voltage is the sum of the zener voltage plus one V_{BE} drop:

$$V_B = V_Z + V_{BE}$$

Substituting this into the preceding equation gives:

$$\boldsymbol{V_{out} \cong \frac{R_1 + R_2}{R_1}(V_Z + V_{BE})} \tag{24-6}$$

Figure 24-5 shows the equations for analyzing the circuit. The equation for collector current is an approximation because it does not include the current through the voltage divider (R_1 and R_2). To keep the efficiency of the regulator as high as possible, a designer normally makes R_1 and R_2 much larger than the load resistance. As a result, the current through the voltage divider is usually small enough to neglect in preliminary analysis.

The disadvantage of this regulator is that any changes in V_{BE} will translate into changes in the output voltage. Although useful in simpler applications, the circuit of Fig. 24-5 can be improved.

Improved Regulation

One way to reduce the effect of V_{BE} on output voltage is with the shunt regulator of Fig. 24-6. The zener diode holds the inverting input of the op amp at a constant voltage. The voltage divider consisting of R_1 and R_2 samples the load voltage and returns a feedback voltage to the noninverting input. The output of the op amp drives the base of the shunt transistor. Because of the negative feedback, the output voltage is held almost constant in spite of line and load changes.

Figure 24-6 Shunt regulator with large negative feedback.

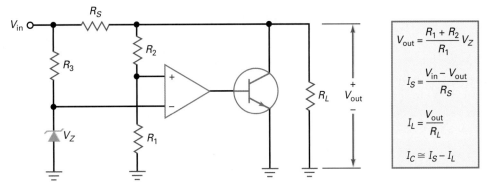

For instance, if the load voltage tries to increase, the feedback signal to the noninverting input increases. The output of the op amp drives the base harder and increases the collector current. The larger collector current through R_S produces a larger voltage across R_S, which offsets most of the attempted increase in load voltage. A similar correction occurs when the load voltage tries to decrease. In short, any attempted change in output voltage is offset by the negative feedback.

In Fig. 24-6, the high voltage gain of the op amp takes V_{BE} out of Eq. (24-6), (a situation similar to the one with active diode circuits discussed in Chap. 22). Because of this, the load voltage is given by:

$$V_{\text{out}} = \frac{R_1 + R_2}{R_1} V_Z \tag{24-7}$$

Short-Circuit Protection

One advantage of shunt regulators is that they have built-in **short-circuit protection**. For instance, if we deliberately place a short circuit across the load terminals in Fig. 24-6, none of the components in the shunt regulator will be damaged. All that happens is that the input current increases to:

$$I_S = \frac{V_{\text{in}}}{R_S}$$

This current is not large enough to damage any of the components in a typical shunt regulator.

Efficiency

One way to compare regulators of different designs is by using *efficiency*, defined as:

$$\textbf{Efficiency} = \frac{P_{\text{out}}}{P_{\text{in}}} \times 100\% \tag{24-8}$$

where P_{out} is the load power ($V_{\text{out}} I_L$) and P_{in} is the input power ($V_{\text{in}} I_{\text{in}}$). The difference between P_{in} and P_{out} is P_{reg}, the power wasted in the regulator components:

$$P_{\text{reg}} = P_{\text{in}} - P_{\text{out}}$$

In the shunt regulators of Figs. 24-4 to 24-6, the power dissipation of R_S and the transistor account for most of the power wasted in the regulator.

Example 24-1

In Fig. 24-4, $V_{in} = 15$ V, $R_S = 10\ \Omega$, $V_Z = 9.1$ V, $V_{BE} = 0.8$ V, and $R_L = 40\ \Omega$. What are the values of output voltage, input current, the load current, and collector current?

SOLUTION   With the equations of Fig. 24-4, we can calculate as follows:

$$V_{out} = V_Z + V_{BE} = 9.1\text{ V} + 0.8\text{ V} = 9.9\text{ V}$$

$$I_S = \frac{V_{in} - V_{out}}{R_S} = \frac{15\text{ V} - 9.9\text{ V}}{10\ \Omega} = 510\text{ mA}$$

$$I_L = \frac{V_{out}}{R_L} = \frac{9.9\text{ V}}{40\ \Omega} = 248\text{ mA}$$

$$I_C \cong I_S - I_L = 510\text{ mA} - 248\text{ mA} = 262\text{ mA}$$

PRACTICE PROBLEM 24–1   Repeat Example 24-1 with $V_{in} = 12$ V and $V_Z = 6.8$ V.

Example 24-2

The shunt regulator of Fig. 24-5 has these circuit values: $V_{in} = 15$ V, $R_S = 10\ \Omega$, $V_Z = 6.2$ V, $V_{BE} = 0.81$ V, and $R_L = 40\ \Omega$. If $R_1 = 750\ \Omega$ and $R_2 = 250\ \Omega$, what are approximate values of output voltage, input current, load current, and collector current?

SOLUTION   With the equations of Fig. 24-5:

$$V_{out} \cong \frac{R_1 + R_2}{R_1}(V_Z + V_{BE})$$

$$= \frac{750\ \Omega + 250\ \Omega}{750\ \Omega}(6.2\text{ V} + 0.81\text{ V}) = 9.35\text{ V}$$

The exact output voltage will be slightly higher than this because of the base current through R_2. The approximate currents are:

$$I_S = \frac{V_{in} - V_{out}}{R_S} = \frac{15\text{ V} - 9.35\text{ V}}{10\ \Omega} = 565\text{ mA}$$

$$I_L = \frac{V_{out}}{R_L} = \frac{9.35\text{ V}}{40\ \Omega} = 234\text{ mA}$$

$$I_C \cong I_S - I_L = 565\text{ mA} - 234\text{ mA} = 331\text{ mA}$$

PRACTICE PROBLEM 24–2   With $V_Z = 7.5$ V, repeat Example 24-2.

Example 24-3

What is the approximate efficiency in the preceding example? How much power does the regulator dissipate?

SOLUTION   The load voltage is approximately 9.35 V, and the load current is approximately 234 mA. The load power is:

$$P_{out} = V_{out}I_L = (9.35 \text{ V})(234 \text{ mA}) = 2.19 \text{ W}$$

In Fig. 24-5, the input current is:

$$I_{in} = I_S + I_3$$

In any well-designed shunt regulator, I_S is much greater than I_3 to keep the efficiency high. Therefore, the input power is:

$$P_{in} = V_{in}I_{in} \cong V_{in}I_S = (15 \text{ V})(565 \text{ mA}) = 8.48 \text{ W}$$

The efficiency of the regulator is:

$$\text{Efficiency} = \frac{P_{out}}{P_{in}} \times 100\% = \frac{2.19 \text{ W}}{8.48 \text{ W}} \times 100\% = 25.8\%$$

This efficiency is low compared to the efficiency of other regulators to be discussed (series regulators and switching regulators). Low efficiency is one of the disadvantages of a shunt regulator. Low efficiency occurs because of the power dissipation in the series resistor and the shunt transistor, which is:

$$P_{reg} = P_{in} - P_{out} \cong 8.48 \text{ W} - 2.19 \text{ W} = 6.29 \text{ W}$$

PRACTICE PROBLEM 24-3   Repeat Example 24-3 with $V_Z = 7.5$ V.

Example 24-4

The shunt regulator of Fig. 24-6 has these circuit values: $V_{in} = 15$ V, $R_S = 10 \ \Omega$, $V_Z = 6.8$ V, and $R_L = 40 \ \Omega$. If $R_1 = 7.5 \ \text{k}\Omega$ and $R_2 = 2.5 \ \text{k}\Omega$, what are approximate values of output voltage, the input current, the load current, and the collector current?

SOLUTION   With the equations of Fig. 24-6:

$$V_{out} \cong \frac{R_1 + R_2}{R_1} V_Z = \frac{7.5 \ \text{k}\Omega + 2.5 \ \text{k}\Omega}{7.5 \ \text{k}\Omega} (6.8 \text{ V}) = 9.07 \text{ V}$$

$$I_S = \frac{V_{in} - V_{out}}{R_S} = \frac{15 \text{ V} - 9.07 \text{ V}}{10 \ \Omega} = 593 \text{ mA}$$

$$I_L = \frac{V_{out}}{R_L} = \frac{9.07 \text{ V}}{40 \ \Omega} = 227 \text{ mA}$$

$$I_C \cong I_S - I_L = 593 \text{ mA} - 227 \text{ mA} = 366 \text{ mA}$$

PRACTICE PROBLEM 24-4   Using Example 24-4, change V_{in} to 12 V and calculate the approximate transistor collector current. What is the approximate power dissipated by R_S?

Example 24-5

Calculate the maximum load currents for Examples 24-1, 24-2, and 24-4.

SOLUTION As discussed earlier, any shunt regulator has a maximum load current approximately equal to the current through R_S. Since we have already calculated I_S in Examples 24-1, 24-2, and 24-4, the maximum load currents are:

$$I_{max} = 510 \text{ mA}$$

$$I_{max} = 565 \text{ mA}$$

$$I_{max} = 593 \text{ mA}$$

Example 24-6

When the shunt regulator of Fig. 24-5 is built and tested, the following values are measured: $V_{NL} = 9.91$ V, $V_{FL} = 9.81$ V, $V_{HL} = 9.94$ V, and $V_{LL} = 9.79$ V. What is the load regulation? What is the line regulation?

SOLUTION

$$\text{Load regulation} = \frac{9.91 \text{ V} - 9.81 \text{ V}}{9.81 \text{ V}} \times 100\% = 1.02\%$$

$$\text{Line regulation} = \frac{9.94 \text{ V} - 9.79 \text{ V}}{9.79 \text{ V}} \times 100\% = 1.53\%$$

PRACTICE PROBLEM 24-6 Repeat Example 24-6 using the following values: $V_{NL} = 9.91$ V, $V_{FL} = 9.70$ V, $V_{HL} = 10.0$ V, and $V_{LL} = 9.68$ V.

24-3 Series Regulators

The disadvantage of a shunt regulator is its low efficiency, caused by large power losses in the series resistor and the shunt transistor. When efficiency is not important, shunt regulators may be used because they have the advantage of simplicity.

Better Efficiency

When efficiency is important, a series regulator or a switching regulator may be used. A switching regulator is the most efficient of all voltage regulators. It has a full-load efficiency from about 75 to more than 95 percent. But a switching regulator is *noisy* because it produces **radio-frequency interference (RFI),** caused by switching a transistor on and off at frequencies from about 10 to more than 100 kHz. Another disadvantage is that a switching regulator is the most complicated regulator to design and build.

On the other hand, the series regulator is *quiet* because its transistor always operates in the linear region. Furthermore, a series regulator is relatively simple to design and build compared to a switching regulator. Finally, a series

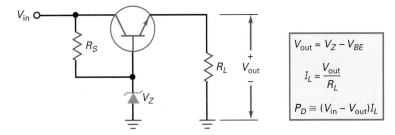

Figure 24–7 Zener follower is a series regulator.

$$V_{out} = V_Z - V_{BE}$$

$$I_L = \frac{V_{out}}{R_L}$$

$$P_D \cong (V_{in} - V_{out})I_L$$

regulator has full-load efficiencies from 50 to 70 percent, good enough for most applications in which the load power is less than 10 W.

Because of the foregoing reasons, the series regulator has emerged as the preferred choice for most applications when the load power is not too high. Its relative simplicity, quiet operation, and acceptable transistor power dissipation make the series regulator the natural choice for many applications. The remainder of this section discusses the series regulator.

The Zener Follower

The simplest series regulator is the zener follower of Fig. 24-7. As discussed in Chap. 11, the zener diode operates in the breakdown region, producing a base voltage equal to the zener voltage. The transistor is connected as an emitter follower. Therefore, the load voltage equals:

$$V_{out} = V_Z - V_{BE} \tag{24-9}$$

If the line voltage or load current changes, the zener voltage and base-emitter voltage will change only slightly. Because of this, the output voltage shows only small changes for large changes in line voltage or load current.

With a series regulator, the load current approximately equals the input current because the current through R_S is usually small enough to ignore in preliminary analysis. The transistor of a series regulator is called a **pass transistor** because all the load current passes through it.

A series regulator is more efficient than a shunt regulator because we have replaced the series resistor by the pass transistor. Now, the only significant power loss is in the transistor. Higher efficiency is one of the main reasons the series regulator is preferred to the shunt regulator when larger load currents are needed.

Recall that the shunt regulator has a constant input current when the load current changes. The series regulator is different because its input current is approximately equal to the load current. When the load current changes in a series regulator, the input current changes by the same amount. This is how you can recognize design variations of shunt and series regulators. In shunt regulators, the input current is constant when the load current changes, whereas in series regulators it changes when the load current changes.

Two-Transistor Regulator

Figure 24-8 shows the two-transistor series regulator discussed in Chap. 11. If V_{out} tries to increase because of an increase in line voltage or an increase in load resistance, more voltage is fed back to the base of Q_1. This produces a larger Q_1 collector current through R_4 and less base voltage at Q_2. The reduced base voltage to the Q_2 emitter follower almost offsets all the attempted increase in output voltage.

Similarly, if the output voltage tries to decrease because of a decrease in line voltage or a decrease in load resistance, there is less feedback voltage at the

Figure 24–8 Discrete series regulator.

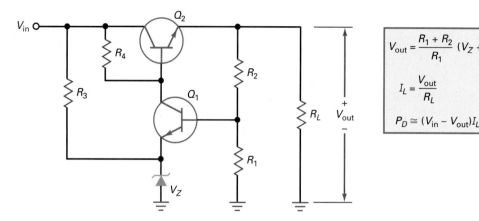

base of Q_1. This produces more base voltage at Q_2, which increases the output voltage and almost completely offsets the attempted decrease in output voltage. The net effect is only a slight decrease in output voltage.

Output Voltage

As discussed in Chap. 11, the output voltage in Fig. 24-8 is given by:

$$V_{\text{out}} = \frac{R_1 + R_2}{R_1} (V_Z + V_{BE}) \qquad (24\text{-}10)$$

With a series regulator like Fig. 24-8, we can use a low zener voltage (5 to 6 V) where the temperature coefficient approaches zero. The output voltage has approximately the same temperature coefficient as the zener voltage.

Headroom, Power Dissipation, and Efficiency

In Fig. 24-8, the **headroom voltage** is defined as the difference between the input and output voltage:

$$\textbf{Headroom voltage} = V_{\text{in}} - V_{\text{out}} \qquad (24\text{-}11)$$

The current through the pass transistor of Fig. 24-8 equals:

$$I_C = I_L + I_2$$

where I_2 is the current through R_2. To keep the efficiency high, a designer will make I_2 much smaller than the full-load value of I_L. Therefore, we can ignore I_2 at larger load currents and write:

$$I_C \cong I_L$$

At high load currents, the power dissipation in the pass transistor is given by the product of headroom voltage and load current:

$$P_D \cong (V_{\text{in}} - V_{\text{out}})I_L \qquad (24\text{-}12)$$

The power dissipation in the pass transistor is very large in some series regulators. In this case, a large heat sink is used. Sometimes, a fan is needed to remove the excess heat inside enclosed equipment.

At full load current, most of the regulator power dissipation is in the pass transistor. Since the current in the pass transistor is approximately equal to the load current, the efficiency is given by:

$$\textbf{Efficiency} \cong \frac{V_{\text{out}}}{V_{\text{in}}} \times 100\% \qquad (24\text{-}13)$$

Figure 24-9 Series regulator with large negative feedback.

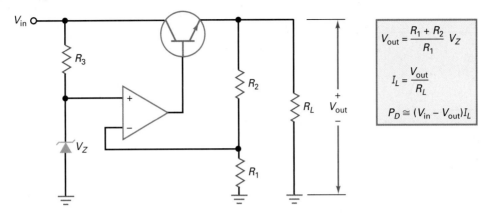

$$V_{out} = \frac{R_1 + R_2}{R_1} V_Z$$

$$I_L = \frac{V_{out}}{R_L}$$

$$P_D \cong (V_{in} - V_{out})I_L$$

With this approximation, the best efficiency occurs when the output voltage is almost as large as the input voltage. It implies that the smaller the headroom voltage, the better the efficiency.

To improve the operation of the series regulator, a Darlington connection is often used for the pass transistor. This allows us to use a low-power transistor to drive a power transistor. The Darlington connection allows us to use larger values of R_1 to R_4 to improve the efficiency.

Improved Regulation

Figure 24-9 shows how we can use an op amp to get better regulation. If the output voltage tries to increase, more voltage is fed back to the inverting input. This reduces the output of the op amp, the base voltage of the pass transistor, and the attempted increase in output voltage. If the output voltage tries to decrease, less voltage is fed back to the op amp, increasing the base voltage of the pass transistor, which almost completely offsets the attempted decrease in output voltage.

The derivation of output voltage is almost the same as for the regulator of Fig. 24-8, except that the high voltage gain of the op amp takes V_{BE} out of the equation. Because of this, the load voltage is given by:

$$V_{out} = \frac{R_1 + R_2}{R_1} V_Z \qquad (24\text{-}14)$$

In Fig. 24-9, the op amp is being used as a noninverting amplifier with a closed-loop voltage gain of:

$$A_{v(CL)} = \frac{R_2}{R_1} + 1 \qquad (24\text{-}15)$$

The input voltage being amplified is the zener voltage. This is why you sometimes see Eq. (24-14) written as:

$$V_{out} = A_{v(CL)} V_Z \qquad (24\text{-}16)$$

For instance, if $A_{v(CL)} = 2$ and $V_Z = 5.6$ V, the output voltage will be 11.2 V.

Current Limiting

Unlike the shunt regulator, the series regulator of Fig. 24-9 has no *short-circuit protection*. If we accidentally short the load terminals, the load current will try to approach infinity, which destroys the pass transistor. It may also destroy one or more diodes in the unregulated power supply that is driving the series regulator. To protect against an accidental short across the load, series regulators usually include some form of **current limiting.**

Figure 24–10 Series regulator with current limiting.

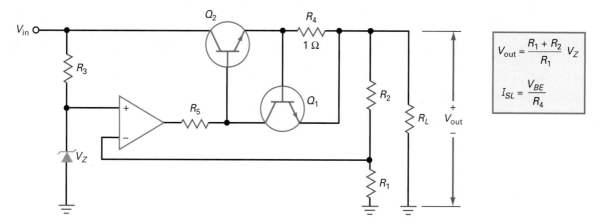

$$V_{out} = \frac{R_1 + R_2}{R_1} V_Z$$

$$I_{SL} = \frac{V_{BE}}{R_4}$$

Figure 24-10 shows one way to limit the load current to safe values. R_4 is a small resistor called a **current-sensing resistor.** For our discussion, we will use an R_4 of 1 Ω. Since the load current has to pass through R_4, the current-sensing resistor produces the base-emitter voltage for Q_1.

When the load current is less than 600 mA, the voltage across R_4 is less than 0.6 V. In this case, Q_1 is cut off and the regulator works as previously described. When the load current is between 600 and 700 mA, the voltage across R_4 is between 0.6 and 0.7 V. This turns on Q_1. The collector current of Q_1 flows through R_5. This decreases the base voltage to Q_2, which reduces the load voltage and the load current.

When the load is shorted, Q_1 conducts heavily and brings the base voltage of Q_2 down to approximately 1.4 V (two V_{BE} drops above ground). The current through the pass transistor is typically limited to 700 mA. It may be slightly more or less than this, depending on the characteristics of the two transistors.

Incidentally, resistor R_5 is added to the circuit because the output impedance of the op amp is very low (75 Ω is typical). Without R_5, the current-sensing transistor does not have enough voltage gain to produce sensitive current limiting. A designer will select a value of R_5 high enough to produce voltage gain in the current-sensing transistor, but not so high that it prevents the op amp from driving the pass transistor. Typical values of R_5 are from a few hundred to a few thousand ohms.

Figure 24-11 summarizes the concept of current limiting. As an approximation, the graph shows 0.6 V as the voltage at which current limiting begins and 0.7 V as the voltage under shorted-load conditions. When the load current is

Figure 24–11 Graph of load voltage versus load current with simple current limiting.

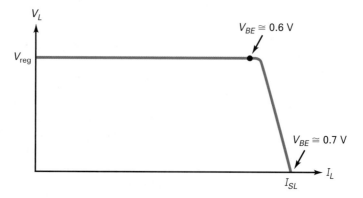

small, the output voltage is regulated and has a value of V_{reg}. When I_L increases, the load voltage remains constant up to a V_{BE} of approximately 0.6 V. Beyond this point, Q_1 turns on and current limiting sets in. Further increases in I_L decrease the load voltage, and regulation is lost. When the load is shorted, the load current is limited to a value of I_{SL}, the load current with *shorted-load terminals*.

When the load terminals are shorted in Fig. 24-10, the load current is given by:

$$I_{SL} = \frac{V_{BE}}{R_4} \tag{24-17}$$

where V_{BE} can be approximated as 0.7 V. For heavy load currents, the V_{BE} of the current-sensing transistor may be somewhat higher. We used an R_4 of 1 Ω in our discussion. By changing the value of R_4, we can get current limiting to begin at any level. For instance, if $R_4 = 10$ Ω, current limiting would start at approximately 60 mA with a shorted-load current of approximately 70 mA.

Foldback Current Limiting

Current limiting is a big improvement because it will protect the pass transistor and rectifier diodes in case the load terminals are accidentally shorted. But it has the disadvantage of a large power dissipation in the pass transistor when the load terminals are shorted. With a short across the load, almost all the input voltage appears across the pass transistor.

To avoid excessive power dissipation in the pass transistor under shorted-load conditions, a designer can add **foldback current limiting** (Fig. 24-12). The voltage across the current-sensing resistor R_4 is fed to a voltage divider (R_6 and R_7) whose output drives the base of Q_1. Over most of the load current range, the base voltage of Q_1 is less than the emitter voltage, and V_{BE} is negative. This keeps Q_1 cut off.

When the load current is high enough, however, the base voltage of Q_1 becomes higher than the emitter voltage. When V_{BE} is between 0.6 and 0.7 V, current limiting starts. Beyond this point, further decreases in load resistance cause the current to fold back (decrease). As a result, the shorted-load current is much smaller than it would be without the foldback limiting.

Figure 24-13 shows how the output voltage varies with load current. The load voltage is constant up to a maximum value of I_{max}. At this point, current limiting starts. When the load resistance decreases further, the current folds back. When a short is across the load terminals, the load current equals I_{SL}. The main advantage of foldback current limiting is the reduced power dissipation in the pass transistor when the load terminals are accidentally shorted.

Figure 24-12 Series regulator with foldback current limiting.

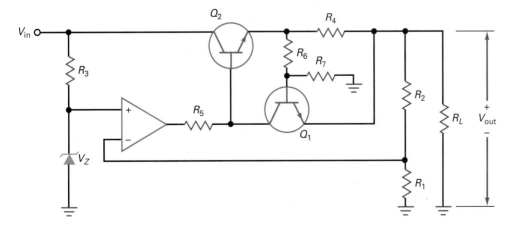

$$V_{\text{out}} = \frac{R_1 + R_2}{R_1} V_Z$$

$$K = \frac{R_7}{R_6 + R_7}$$

$$I_{SL} = \frac{V_{BE}}{K R_4}$$

$$I_{\text{max}} = I_{SL} + \frac{(1 - K) V_{\text{out}}}{K R_4}$$

Figure 24-13 Graph of load voltage versus load current with foldback current limiting.

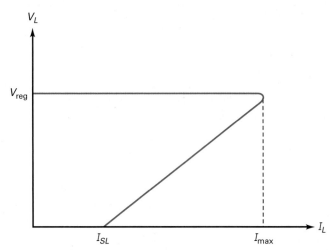

In Fig. 24-13, the power dissipation of the transistor under full-load conditions is:

$$P_D = (V_{in} - V_{reg})I_{max}$$

Under shorted-loaded conditions, the power dissipation is approximately:

$$P_D \cong V_{in}I_{SL}$$

Typically, a designer will use an I_{SL} that is two to three times smaller than I_{max}. By doing this, he or she can keep the power dissipation in the pass transistor down to the level it has under full-load conditions.

Example 24–7 |||| MultiSim

Calculate the approximate output voltage in Fig. 24-14. What is the power dissipation in the pass transistor?

SOLUTION   With the equations of Fig. 24-8:

$$V_{out} = \frac{3\text{ k}\Omega + 1\text{ k}\Omega}{3\text{ k}\Omega}(6.2\text{ V} + 0.7\text{ V}) = 9.2\text{ V}$$

Figure 24-14 Example.

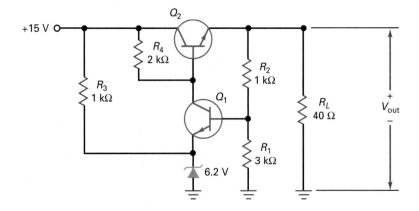

The transistor current is approximately the same as the load current:

$$I_C = \frac{9.2 \text{ V}}{40 \ \Omega} = 230 \text{ mA}$$

The transistor power dissipation is:

$$P_D = (15 \text{ V} - 9.2 \text{ V})(230 \text{ mA}) = 1.33 \text{ W}$$

PRACTICE PROBLEM 24–7   In Fig. 24-14, change the input voltage to $+12$ V and V_Z to 5.6 V. Calculate V_{out} and P_D.

Example 24-8

What is the approximate efficiency in Example 24-7?

SOLUTION   The load voltage is 9.2 V, and the load current is 230 mA. The output power is:

$$P_{\text{out}} = (9.2 \text{ V})(230 \text{ mA}) = 2.12 \text{ W}$$

The input voltage is 15 V, and the input current is approximately 230 mA, the value of the load current. Therefore, the input power is:

$$P_{\text{in}} = (15 \text{ V})(230 \text{ mA}) = 3.45 \text{ W}$$

The efficiency is:

$$\text{Efficiency} = \frac{2.12 \text{ W}}{3.45 \text{ W}} \times 100\% = 61.4\%$$

We can also use Eq. (24-13) to calculate the efficiency of a series regulator:

$$\text{Efficiency} = \frac{V_{\text{out}}}{V_{\text{in}}} \times 100\% = \frac{9.2 \text{ V}}{15 \text{ V}} \times 100\% = 61.3\%$$

This is much better than 25.8 percent, the efficiency of the shunt regulator in Example 24-3. Typically, a series regulator has an efficiency about twice that of a shunt regulator.

PRACTICE PROBLEM 24–8   Repeat Example 24-8 with $V_{\text{in}} = +12$ V and $V_Z = 5.6$ V.

Example 24-9
IIII MultiSim

What is the approximate output voltage in Fig. 24-15? Why is a Darlington transistor used?

SOLUTION   With the equations of Fig. 24-9:

$$V_{\text{out}} = \frac{2.7 \text{ k}\Omega + 2.2 \text{ k}\Omega}{2.7 \text{ k}\Omega} (5.6 \text{ V}) = 10.2 \text{ V}$$

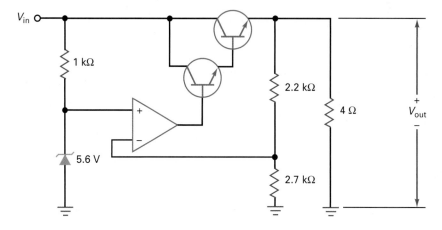

Figure 24-15 Series regulator with Darlington transistor.

The load current is:

$$I_L = \frac{10.2 \text{ V}}{4 \text{ }\Omega} = 2.55 \text{ A}$$

If an ordinary transistor with a current gain of 100 were used for the pass transistor, the required base current would be:

$$I_B = \frac{2.55 \text{ A}}{100} = 25.5 \text{ mA}$$

This is too much output current for a typical op amp. If a Darlington transistor is used, the base current of the pass transistor is reduced to a much lower value. For instance, a Darlington transistor with a current gain of 1000 would require a base current of only 2.55 mA.

PRACTICE PROBLEM 24-9 In Fig. 24-15, determine the output voltage if the zener voltage is changed to 6.2 V.

Example 24-10

When the series regulator of Fig. 24-15 is built and tested, the following values are measured: $V_{NL} = 10.16$ V, $V_{FL} = 10.15$ V, $V_{HL} = 10.16$ V, and $V_{LL} = 10.07$ V. What is the load regulation? What is the line regulation?

SOLUTION

$$\text{Load regulation} = \frac{10.16 \text{ V} - 10.15 \text{ V}}{10.15 \text{ V}} \times 100\% = 0.0985\%$$

$$\text{Line regulation} = \frac{10.16 \text{ V} - 10.07 \text{ V}}{10.07 \text{ V}} \times 100\% = 0.894\%$$

This example shows how effective negative feedback is in reducing the effects of line and load changes. In both cases, the change in the regulated output voltage is less than 1 percent.

Example 24-11

In Fig. 24-16, V_{in} can vary from 17.5 to 22.5 V. What is the maximum zener current? What are the minimum and maximum regulated output voltages? If the regulated output voltage is 12.5 V, what is the load resistance where current limiting starts? What is the approximate shorted-load current?

Figure 24-16 Example.

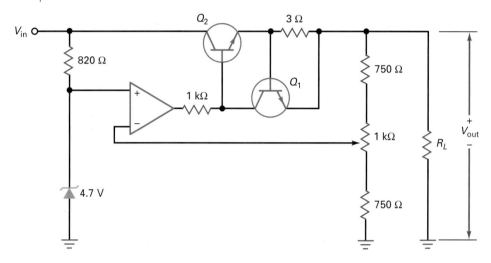

SOLUTION The maximum zener current occurs when the input voltage is 22.5 V:

$$I_Z = \frac{22.5\ \text{V} - 4.7\ \text{V}}{820\ \Omega} = 21.7\ \text{mA}$$

The minimum regulated output voltage occurs when the wiper of the 1-kΩ potentiometer is all the way up. In this case, $R_1 = 1750\ \Omega$, $R_2 = 750\ \Omega$, and the output voltage is:

$$V_{out} = \frac{1750\ \Omega + 750\ \Omega}{1750\ \Omega}(4.7\ \text{V}) = 6.71\ \text{V}$$

The maximum regulated output voltage occurs when the wiper of the 1-kΩ potentiometer is all the way down. In this case, $R_1 = 750\ \Omega$, and $R_2 = 1750\ \Omega$, and the output voltage is:

$$V_{out} = \frac{750\ \Omega + 1750\ \Omega}{750\ \Omega}(4.7\ \text{V}) = 15.7\ \text{V}$$

Current limiting starts when the voltage across the current-limiting resistor is approximately 0.6 V. In this case, the load current is:

$$I_L = \frac{0.6\ \text{V}}{3\ \Omega} = 200\ \text{mA}$$

With an output voltage of 12.5 V, the load resistance where current limiting starts is approximately:

$$R_L = \frac{12.5\ \text{V}}{200\ \text{mA}} = 62.5\ \Omega$$

With a short across the load terminals, the voltage across the current-sensing resistor is approximately 0.7 V and the shorted-load current is:

$$I_{SL} = \frac{0.7\ \text{V}}{3\ \Omega} = 233\ \text{mA}$$

PRACTICE PROBLEM 24-11 Repeat Example 24-11 using a 3.9 V zener and a 2 Ω current-sensing resistor.

24-4 Monolithic Linear Regulators

There is a wide variety of linear **IC voltage regulators** with pin counts from 3 to 14. All are series regulators because the series regulator is more efficient than the shunt regulator. Some IC regulators are used in special applications in which external resistors can set the current limiting, the output voltage, and so on. By far, the most widely used IC regulators are those with only three pins: one for the unregulated input voltage, one for the regulated output voltage, and one for ground.

Available in plastic or metal packages, the three-terminal regulators have become extremely popular because they are inexpensive and easy to use. Aside from two optional bypass capacitors, three-terminal IC voltage regulators require no external components.

Basic Types of IC Regulators

Most IC voltage regulators have one of these types of output voltage: fixed positive, fixed negative, or adjustable. IC regulators with fixed positive or negative outputs are factory-trimmed to get different fixed voltages with magnitudes from about 5 to 24 V. IC regulators with an adjustable output can vary the regulated output voltage from less than 2 to more than 40 V.

IC regulators are also classified as standard, low-power, and low dropout. Standard IC regulators are designed for straightforward and noncritical applications. With heat sinks, a standard IC regulator can have a load current of more than 1 A.

If load currents up to 100 mA are adequate, *low-power IC regulators* are available in TO-92 packages, the same size used for small-signal transistors like the 2N3904. Since these regulators do not require heat sinking, they are convenient and easy to use.

The **dropout voltage** of an IC regulator is defined as the minimum headroom voltage needed for regulation. For instance, standard IC regulators have a dropout voltage of 2 to 3 V. This means that the input voltage has to be at least 2 to 3 V greater than the regulated output voltage for the chip to regulate to specifications. In applications in which 2 to 3 V of headroom is not available, *low dropout IC regulators* can be used. These regulators have typical dropout voltages of 0.15 V for a load current of 100 mA and 0.7 V for a load current of 1 A.

On-Card Regulation versus Single-Point Regulation

With *single-point regulation,* we need to build a power supply with a large voltage regulator and then distribute the regulated voltage to all the different *cards* (printed-circuit boards) in the system. This creates problems. To begin with, the single regulator has to provide a large load current equal to the sum of all the card currents. Second, noise or other **electromagnetic interference (EMI)** can be induced on the connecting wires between the regulated power supply and the cards.

Because IC regulators are inexpensive, electronic systems that have many cards often use *on-card regulation.* This means that each card has its own three-terminal regulator to supply the voltage used by the components on that card. By using on-card regulation, we can deliver an unregulated voltage from a power supply to each card and have a local IC regulator take care of regulating the voltage for its card. This eliminates the problems of the large load current and noise pickup associated with single-point regulation.

Load and Line Regulation Redefined

Up to now, we have used the original definitions for load and line regulation. Manufacturers of fixed IC regulators prefer to specify the change in load voltage for a range of load and line conditions. Here are definitions for load and line regulation used on the data sheets of fixed regulators:

$$\text{Load regulation} = \Delta V_{\text{out}} \text{ for a range of load current}$$
$$\text{Line regulation} = \Delta V_{\text{out}} \text{ for a range of input voltage}$$

For instance, the LM7815 is an IC regulator that produces a fixed positive output voltage of 15 V. The data sheet lists the typical load and line regulation as follows:

$$\text{Load regulation} = 12 \text{ mV for } I_L = 5 \text{ mA to } 1.5 \text{ A}$$
$$\text{Line regulation} = 4 \text{ mV for } V_{\text{in}} = 17.5 \text{ V to } 30 \text{ V}$$

The load regulation will depend on the conditions of measurement. The foregoing load regulation is for $T_J = 25°C$ and $V_{\text{in}} = 23$ V. Similarly, the foregoing line regulation is for $T_J = 25°C$ and $I_L = 500$ mA. In each case, the junction temperature of the device is 25°C.

The LM7800 Series

The LM78XX series (where XX = 05, 06, 08, 10, 12, 15, 18, or 24) is typical of the three-terminal voltage regulators. The 7805 produces an output of +5 V, the 7806 produces +6 V, the 7808 produces +8 V, and so on, up to the 7824, which produces an output of +24 V.

Figure 24-17 shows the functional block diagram for the 78XX series. A built-in reference voltage V_{ref} drives the noninverting input of an amplifier. The voltage regulation is the similar to our earlier discussion. A voltage divider consisting of R_1' and R_2' samples the output voltage and returns a feedback voltage to the inverting input of a high-gain amplifier. The output voltage is given by:

$$V_{\text{out}} = \frac{R_1' + R_2'}{R_1'} V_{\text{ref}}$$

In this equation, the reference voltage is equivalent to the zener voltage in our earlier discussions. The primes attached to R_1' and R_2' indicate that these resistors are inside the IC itself, rather than being external resistors. These resistors are factory-trimmed to get the different output voltages (5 to 24 V) in the 78XX series. The tolerance of the output voltage is ±4 percent.

The LM78XX includes a pass transistor that can handle 1 A of load current, provided that adequate heat sinking is used. Also included are thermal shutdown and current limiting. **Thermal shutdown** means that the chip will shut

Figure 24–17 Functional block diagram of three-terminal IC regulator.

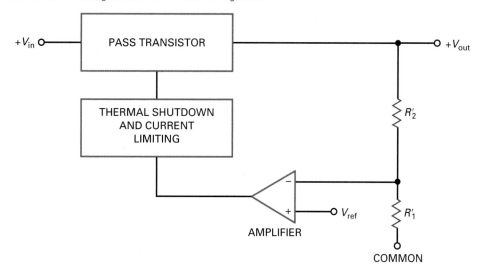

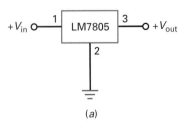

(a)

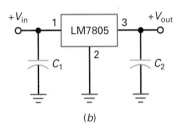

(b)

itself off when the internal temperature becomes too high, around 175°C. This is a precaution against excessive power dissipation, which depends on the ambient temperature, type of heat sinking, and other variables. Because of thermal shutdown and current limiting, devices in the 78XX series are almost indestructible.

Fixed Regulator

Figure 24-18*a* shows an LM7805 connected as a fixed voltage regulator. Pin 1 is the input, pin 3 is the output, and pin 2 is ground. The LM7805 has an output voltage of +5 V and a maximum load current over 1 A. The typical load regulation is 10 mV for a load current between 5 mA and 1.5 A. The typical line regulation is 3 mV for an input voltage of 7 to 25 V. It also has a ripple rejection of 80 dB, which means that it will reduce the input ripple by a factor of 10,000. With an output resistance of approximately 0.01 Ω, the LM7805 is a very stiff voltage source to all loads within its current rating.

When an IC is more than 6 in from the filter capacitor of the unregulated power supply, the inductance of the connecting wire may produce oscillations inside the IC. This is why manufacturers recommend using a bypass capacitor C_1 on pin 1 (Fig. 24-18*b*). To improve the transient response of the regulated output voltage, a bypass capacitor C_2 is sometimes used on pin 3. Typical values for either bypass capacitor are from 0.1 to 1 μF. The data sheet of the 78XX series suggests 0.22 μF for the input capacitor and 0.1 μF for the output capacitor.

Any regulator in the 78XX series has a dropout voltage of 2 to 3 V, depending on the output voltage. This means that the input voltage must be at least 2 to 3 V greater than the output voltage. Otherwise, the chip stops regulating. Also, there is a maximum input voltage because of excessive power dissipation. For instance, the LM7805 will regulate over an input range of approximately 8 to 20 V. The data sheet for the 78XX series gives the minimum and maximum input voltages for the other preset output voltages.

The LM79XX Series

The LM79XX series is a group of negative voltage regulators with preset voltages of −5, −6, −8, −10, −12, −15, −18, or −24 V. For instance, an LM7905 produces a regulated output voltage of −5 V. At the other extreme, an LM7924 produces an output of −24 V. With the LM79XX series, the load-current capability is over 1 A with adequate heat sinking. The LM79XX series is similar to the 78XX series and includes current limiting, thermal shutdown, and excellent ripple rejection.

Regulated Dual Supplies

By combining an LM78XX and an LM79XX, as shown in Fig. 24-19, we can regulate the output of a dual supply. The LM78XX regulates the positive output, and the LM79XX handles the negative output. The input capacitors prevent oscillations,

Figure 24-19 Using the LM78XX and LM79XX for dual outputs.

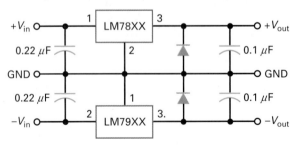

and the output capacitors improve transient response. The manufacturer's data sheet recommends the addition of two diodes to ensure that both regulators can turn on under all operating conditions.

An alternative solution for dual supplies is to use a dual-tracking regulator. This is an IC that contains a positive and a negative regulator in a single IC package. When adjustable, this type of IC can vary the dual supplies with a single variable resistor.

Adjustable Regulators

A number of IC regulators (LM317, LM337, LM338, and LM350) are adjustable. These have maximum load currents from 1.5 to 5 A. For instance, the LM317 is a three-terminal positive voltage regulator that can supply 1.5 A of load current over an adjustable output range of 1.25 to 37 V. The ripple rejection is 80 dB. This means that the input ripple is 10,000 smaller at the output of the IC regulator.

Again, manufacturers redefine the load and line regulation to suit the characteristics of the IC regulator. Here are definitions for load and line regulation used on the data sheets of adjustable regulators:

Load regulation = Percent change in V_{out} for a range in load current

Line regulation = Percent change in V_{out} per volt of input change

For instance, the data sheet of an LM317 lists these typical load and line regulations:

Load regulation = 0.3% for I_L = 10 mA to 1.5 A

Line regulation = 0.02% per volt

Since the output voltage is adjustable between 1.25 and 37 V, it makes sense to specify the load regulation as a percent. For instance, if the regulated voltage is adjusted to 10 V, the foregoing load regulation means that the output voltage will remain within 0.3 percent of 10 V (or 30 mV) when the load current changes from 10 mA to 1.5 A.

The line regulation is 0.02 percent per volt. This means that the output voltage changes only 0.02 percent for each volt of input change. If the regulated output is set at 10 V and the input voltage increases by 3 V, the output voltage will increase by 0.06 percent, equivalent to 60 mV.

Figure 24-20 shows an unregulated supply driving an LM317 circuit. The data sheet of an LM317 gives this formula for output voltage:

$$V_{out} = \frac{R_1 + R_2}{R_1} V_{ref} + I_{ADJ}R_2 \qquad (24\text{-}18)$$

In this equation, V_{ref} has a value of 1.25 V and I_{ADJ} has a typical value of 50 μA. In Fig. 24-20, I_{ADJ} is the current flowing through the middle pin (the one between the input and the output pins). Because this current can change with temperature, load current, and other factors, a designer usually makes the first term in Eq. (24-18)

Figure 24–20 Using an LM317 to regulate output voltage.

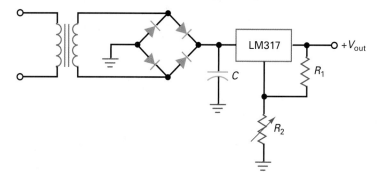

much greater than the second. This is why we can use the following equation for all preliminary analyses of an LM317:

$$V_{out} \cong \frac{R_1 + R_2}{R_1} \, (1.25 \text{ V})$$

(24-19)

Ripple Rejection

The ripple rejection of an IC voltage regulator is high, from about 65 to 80 dB. This is a tremendous advantage because it means that we do not have to use bulky *LC* filters in the power supply to minimize the ripple. All we need is a capacitor-input filter that reduces the peak-to-peak ripple to about 10 percent of the unregulated voltage out of the power supply.

For instance, the LM7805 has a typical ripple rejection of 80 dB. If a bridge rectifier and a capacitor-input filter produce an unregulated output voltage of 10 V with a peak-to-peak ripple of 1 V, we can use an LM7805 to produce a regulated output voltage of 5 V with a peak-to-peak ripple of only 0.1 mV. Eliminating bulky *LC* filters in an unregulated power supply is a bonus that comes with IC voltage regulators.

Regulator Table

Table 24-1 lists some widely used IC regulators. The first group, the LM78XX series, is for fixed positive output voltages from 5 to 24 V. With heat sinking, these

Table 24–1	Typical Parameters of Popular IC Voltage Regulators at 25°C							
Number	V_{out}, V	I_{max}, A	Load Reg, mV	Line Reg, mV	Rip Rej, dB	Dropout, V	R_{out}, mΩ	I_{SL}, A
LM7805	5	1.5	10	3	80	2	8	2.1
LM7806	6	1.5	12	5	75	2	9	0.55
LM7808	8	1.5	12	6	72	2	16	0.45
LM7812	12	1.5	12	4	72	2	18	1.5
LM7815	15	1.5	12	4	70	2	19	1.2
LM7818	18	1.5	12	15	69	2	22	0.20
LM7824	24	1.5	12	18	66	2	28	0.15
LM78L05	5	100 mA	20	18	80	1.7	190	0.14
LM78L12	12	100 mA	30	30	80	1.7	190	0.14
LM2931	3 to 24	100 mA	14	4	80	0.3	200	0.14
LM7905	−5	1.5	10	3	80	2	8	2.1
LM7912	−12	1.5	12	4	72	2	18	1.5
LM7915	−15	1.5	12	4	70	2	19	1.2
LM317	1.2 to 37	1.5	0.3%	0.02%/V	80	2	10	2.2
LM337	−1.2 to −37	1.5	0.3%	0.01%/V	77	2	10	2.2
LM338	1.2 to 32	5	0.3%	0.02%/V	75	2.7	5	8

regulators can produce a load current up to 1.5 A. Load regulation is between 10 and 12 mV. Line regulation is between 3 and 18 mV. Ripple rejection is best at the lowest voltage (80 dB) and worst at the high voltage (66 dB). The dropout voltage is 2 V for the entire series. Output resistance increases from 8 to 28 mΩ between the lowest and highest output voltage.

The LM78L05 and LM78L12 are low-power versions of their standard counterparts, the LM7805 and LM7812. These *low-power IC regulators* are available in TO-92 packages, which do not require heat sinking. As shown in Table 24-1, the LM78L05 and LM78L12 can produce load currents up to 100 mA.

The LM2931 is included as an example of a low-dropout regulator. This adjustable regulator can produce output voltages between 3 and 24 V with a load current up to 100 mA. Notice that the dropout voltage is only 0.3 V, which means that the input voltage need be only 0.3 V greater than the regulated output voltage.

The LM7905, LM7912, and LM7915 are widely used negative regulators. Their parameters are similar to those of their LM78XX counterparts. The LM317 and LM337 are adjustable positive and negative regulators that can deliver load currents up to 1.5 A. Finally, the LM338 is an adjustable positive regulator that can produce a load voltage between 1.2 and 32 V with a load current up to 5 A.

All the regulators listed in Table 24-1 have *thermal shutdown*. This means that the regulator will cut off the pass transistor and shut down the operation if the chip temperature becomes too high. When the device cools off, it will attempt to restart. If whatever caused the excessive temperature has been removed, the regulator will function normally. If not, it will shut down again. Thermal shutdown is an advantage that monolithic regulators offer for safe operation.

Example 24-12

What is the load current in Fig. 24-21? What is the output ripple?

SOLUTION The LM7812 produces a regulated output voltage of +12 V. Therefore, the load current is:

$$I_L = \frac{12\ V}{100\ \Omega} = 120\ mA$$

We can calculate the peak-to-peak input ripple with the equation given in Chap. 4:

$$V_R = \frac{I_L}{fC} = \frac{120\ mA}{(120\ Hz)(1000\ \mu F)} = 1\ V$$

Figure 24-21 Example.

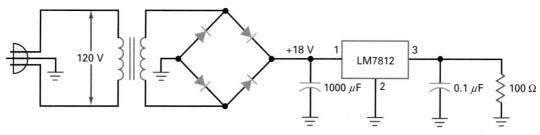

Table 24-1 shows a typical ripple rejection of 72 dB for the LM7812. If we convert 72 dB mentally (60 dB + 12 dB), we get approximately 4000. With a scientific calculator, the exact ripple rejection is:

$$RR = \text{antilog} \frac{72 \text{ dB}}{20} = 3981$$

The peak-to-peak output ripple is approximately:

$$V_R = \frac{1 \text{ V}}{4000} = 0.25 \text{ mV}$$

PRACTICE PROBLEM 24-12   Repeat Example 24-12 using an LM7815 voltage regulator and a 2000 μF capacitor.

Example 24-13

If $R_1 = 2 \text{ k}\Omega$ and $R_2 = 22 \text{ k}\Omega$ in Fig. 24-20, what is the output voltage? If R_2 is increased to 46 kΩ, what is the output voltage?

SOLUTION   With Eq. (24-19):

$$V_{\text{out}} = \frac{2 \text{ k}\Omega + 22 \text{ k}\Omega}{2 \text{ k}\Omega} (1.25 \text{ V}) = 15 \text{ V}$$

When R_2 is increased to 46 kΩ, the output voltage increases to:

$$V_{\text{out}} = \frac{2 \text{ k}\Omega + 46 \text{ k}\Omega}{2 \text{ k}\Omega} (1.25 \text{ V}) = 30 \text{ V}$$

PRACTICE PROBLEM 24-13   In Fig. 24-20, what is the output voltage if $R_1 = 330 \ \Omega$ and $R_2 = 2 \text{ k}\Omega$?

Example 24-14

The LM7805 can regulate to specifications with an input voltage between 7.5 and 20 V. What is the maximum efficiency? What is the minimum efficiency?

SOLUTION   The LM7805 produces an output of 5 V. With Eq. (24-13), the maximum efficiency is:

$$\text{Efficiency} \cong \frac{V_{\text{out}}}{V_{\text{in}}} \times 100\% = \frac{5 \text{ V}}{7.5 \text{ V}} \times 100\% = 67\%$$

This high efficiency is possible only because the headroom voltage is approaching the dropout voltage.

On the other hand, the minimum efficiency occurs when the input voltage is maximum. For this condition, the headroom voltage is maximum and the power dissipation in the pass transistor is maximum. The minimum efficiency is:

$$\text{Efficiency} \cong \frac{5 \text{ V}}{20 \text{ V}} \times 100\% = 25\%$$

Since the unregulated input voltage is usually somewhere between the extremes in input voltage, the efficiency we can expect with an LM7805 is in the range of 40 to 50 percent.

24-5 Current Boosters

Even though the 78XX regulators of Table 24-1 have a maximum load current of 1.5 A, the data sheet shows many parameters measured at 1 A. For instance, a load current of 1 A is used for measuring line regulation, ripple rejection, and output resistance. For this reason, we will establish 1 A as a practical limit on the load current when using a 78XX device.

The Outboard Transistor

One way to get more load current is to use a **current booster**. The idea is similar to what we did to boost the output current of an op amp. Recall that we used the op amp to supply the base current for an external transistor, which produced a much larger output current.

Figure 24-22 shows how we can use an external transistor to boost the output current. The external transistor, called an **outboard transistor,** is a power transistor. R_1 is a current-sensing resistor of 0.7 Ω. Notice that we are using 0.7 Ω instead of 0.6 Ω. We are using 0.7 Ω because a power transistor needs more base voltage than does a small-signal transistor (used in the previous discussion).

When the current is less than 1 A, the voltage across the current-sensing resistor is less than 0.7 V and the transistor is off. When the load current is greater than 1 A, the transistor turns on and supplies almost all the load current above 1 A. Here is why: When the load current increases, the current through the 78XX increases slightly. This produces more voltage across the current-sensing resistor, which makes the outboard transistor conduct more heavily.

Each time we increase the load current, the current through the 78XX device increases slightly, producing more voltage across the current-sensing resistor. In this way, the outboard transistor produces the bulk of any increase in load current above 1 A, with only a small increase in the current through the 78XX.

For large load currents, the base current in the outboard transistor becomes large. The 78XX chip has to supply this base current in addition to its share of load current. When the large base current becomes a problem, a designer may use a Darlington connection for the outboard transistor. In this case, the current-sensing voltage is approximately 1.4 V, which means that R_1 should be increased to approximately 1.4 Ω.

Short-Circuit Protection

Figure 24-23 shows how to add short-circuit protection to the circuit. We are using two current-sensing resistors, one to drive the outboard transistor Q_2 and a second to turn on Q_1 for short-circuit protection. For this discussion, 1 A is where Q_2 conducts, and 10 A is where Q_1 provides short-circuit protection.

Figure 24-22 Outboard transistor increases load current.

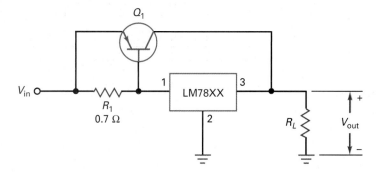

Figure 24-23 Outboard transistor with current limiting.

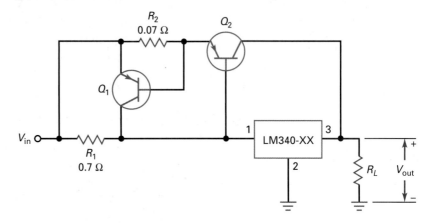

Here is how the circuit works: When the load current is greater than 1 A, the voltage across R_1 is greater than 0.7 V. This turns on the outboard transistor Q_2, which supplies all the load current above 1 A. The outboard current has to pass through R_2. Since R_2 is only 0.07 Ω, the voltage across it is less than 0.7 V as long as the outboard current is less than 10 A.

When the outboard current is 10 A, the voltage across R_2 is:

$$V_2 = (10\ \text{A})(0.07\ \Omega) = 0.7\ \text{V}$$

This means that the current-limiting transistor Q_1 is on the verge of turning on. When the outboard current is greater than 10 A, Q_1 conducts heavily. Since the collector current of Q_1 passes through the 78XX, the device overheats and produces thermal shutdown.

A final point: Using an outboard transistor does not improve the efficiency of a series regulator. With typical headroom voltages, the efficiency is around 40 to 50 percent. To get higher efficiency with large headroom voltages, we have to use a fundamentally different approach to voltage regulation.

24-6 DC-to-DC Converters

Sometimes we want to convert a dc voltage of one value to a dc voltage of another value. For instance, if we have a system with a positive supply of +5 V, we can use a **dc-to-dc converter** to convert this +5 V to an output of +15 V. Then we would have two supply voltages for our system: +5 and +15 V.

DC-to-dc converters are very efficient. Because they switch transistors on and off, transistor power dissipation is greatly reduced. Typical efficiencies are from 65 to 85 percent. This section discusses unregulated dc-to-dc converters. The next section is about regulated dc-to-dc converters that use pulse-width modulation. These dc-to-dc converters are usually called **switching regulators**.

Basic Idea

In a typical unregulated dc-to-dc converter, the input dc voltage is applied to a square-wave oscillator. The peak-to-peak value of the square wave is proportional to the input voltage. The square wave is used to drive the primary winding of a transformer, as shown in Fig. 24-24. The higher the frequency, the smaller the transformer and filter components. If the frequency is too high, however, it is difficult to produce a square wave with vertical transitions. Usually, the frequency of the square wave is between 10 and 100 kHz.

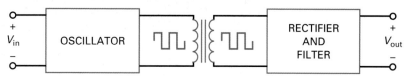

To improve the efficiency, a special kind of transformer is used in more expensive dc-to-dc converters. The transformer has a toroidal core with a rectangular hysteresis loop. This produces a secondary voltage that is a square wave. The secondary voltage can then be rectified and filtered to get a dc output voltage. By selecting a different turns ratio, we can step the secondary voltage up or down. This way, we can build dc-to-dc converters that step the dc input voltage either up or down.

One common dc-to-dc conversion is +5 to ±15 V. In digital systems, +5 V is a standard supply voltage for most ICs. But linear ICs, like op amps, may require ±15 V. In a case like this, you may find a low-power dc-to-dc converter converting an input +5 V dc to dual outputs of ±15 V dc.

One Possible Design

There are many ways to design a dc-to-dc converter, depending on whether bipolar junction or power FETs are used, the switching frequency, whether the input voltage is stepped up or down, and so on. Figure 24-25 shows a design example that uses bipolar junction power transistors. Here is how it works: A relaxation oscillator produces a square wave whose frequency is set by R_3 and C_2. This frequency is in the kilohertz range; a value like 20 kHz would be typical.

The square wave drives a **phase splitter** Q_1, a circuit that produces two equal-magnitude and out-of-phase square waves. These square waves are the

Figure 24-25 An unregulated dc-to-dc converter.

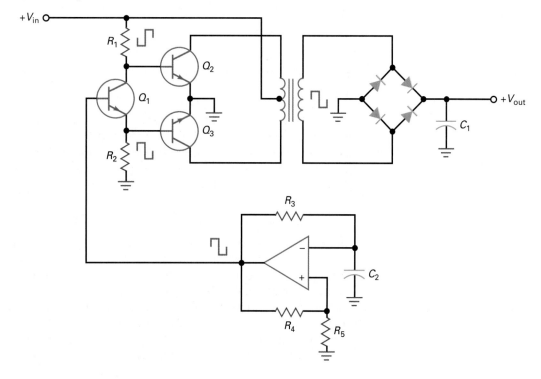

input to class B push-pull switching transistors Q_2 and Q_3. Transistor Q_2 conducts during one half cycle, and Q_3 during the other half cycle. The primary current in the transformer is a square wave. This induces a square wave across the secondary winding, as previously described.

The square wave of voltage out of the secondary winding drives a bridge rectifier and a capacitor-input filter. Because the signal is a rectified square wave in kilohertz, it is easy to filter. The final output is a dc voltage at some level different from the input.

Commercial DC-to-DC Converters

In Fig. 24-25, notice that the output of the dc-to-dc converter is unregulated. This is typical of inexpensive dc-to-dc converters. Unregulated dc-to-dc converters are commercially available with efficiencies of about 65 to more than 85 percent. For instance, inexpensive dc-to-dc converters are available for converting +5 to ±12 V at 375 mA, +5 to +9 V dc at 200 mA, +12 to ±5 V at 250 mA, and so on. All of these converters require a fixed input voltage because they do not include voltage regulation. Also, they use switching frequencies between 10 and 100 kHz. Because of this, they include RFI shielding. Some of the units have an *MTBF* of 200,000 hr. (*Note: MTBF* stands for "mean time between failure.")

24-7 Switching Regulators

A *switching regulator* falls into the general class of dc-to-dc converters because it will convert a dc input voltage to another dc output voltage, either lower or higher. But the switching regulator also includes voltage regulation, typically pulse-width modulation controlling the on-off time of the transistor. By changing the duty cycle, a switching regulator can hold the output voltage constant under varying line and load conditions.

The Pass Transistor

In a series regulator the power dissipation of the pass transistor approximately equals the headroom voltage times the load current:

$$P_D = (V_{\text{in}} - V_{\text{out}})I_L$$

If the headroom voltage equals the output voltage, the efficiency is approximately 50 percent. For instance, if 10 V is the input to a 7805, the load voltage is 5 V and the efficiency is around 50 percent.

Three-terminal series regulators are very popular because they are easy to use and fill most of our needs when the load power is less than about 10 W. When the load power is 10 W and the efficiency is 50 percent, the power dissipation of the pass transistor is also 10 W. This represents a lot of wasted power, as well as heat created inside the equipment. Around load powers of 10 W, heat sinks get very bulky and the temperature of enclosed equipment may rise to objectionable levels.

Switching the Pass Transistor On and Off

The ultimate solution to the problem of low efficiency and high equipment temperature is the switching regulator, briefly described earlier. With this type of regulator, the pass transistor is switched between cutoff and saturation. When the transistor is cut off, the power dissipation is virtually zero. When the transistor is saturated, the power dissipation is still very low because $V_{CE(\text{sat})}$ is much less than the headroom voltage in a series regulator. As mentioned earlier, switching regulators can have efficiencies from about 75 to more than 95 percent. Because of the high efficiency and small size, switching regulators have become widely used.

Topology	Step	Choke	Transformer	Diodes	Transistors	Power, W	Complexity
Buck	Down	Yes	No	1	1	0–150	Low
Boost	Up	Yes	No	1	1	0–150	Low
Buck-boost	Both	Yes	No	1	1	0–150	Low
Flyback	Both	No	Yes	1	1	0–150	Medium
Half-forward	Both	Yes	Yes	1	1	0–150	Medium
Push-pull	Both	Yes	Yes	2	2	100–1000	High
Half bridge	Both	Yes	Yes	4	2	100–500	High
Full bridge	Both	Yes	Yes	4	4	400–2000	Very high

Table 24-2 Switching-Regulator Topologies

Topologies

Topology is a term often used in switching-regulator literature. It is the design technique or fundamental layout of a circuit. Many topologies have evolved for switching regulators because some are better suited to an application than others.

Table 24-2 shows the topologies used for switching regulators. The first three are the most basic. They use the fewest number of parts and can deliver load power up to about 150 W. Because their complexity is low, they are widely used, especially with IC switching regulators.

When transformer isolation is preferred, the flyback and the half-forward topologies can be used for load power up to 150 W. When the load power is from 150 to 2000 W, the push-pull, half bridge, and full bridge topologies are used. Since the last three topologies use more components, the circuit complexity is high.

Buck Regulator

Figure 24-26a shows a **buck regulator**, the most basic topology for switching regulators. A buck regulator always steps the voltage down. A transistor, either a bipolar junction or a power FET, is used as the switching device. A rectangular signal out of the pulse-width modulator closes and opens the switch. A comparator controls the duty cycle of the pulses. For instance, the pulse-width modulator may be a one-shot multivibrator with a comparator driving the control input. As discussed in Chap. 23 with a monostable 555 timer, an increase in control voltage increases the duty cycle.

When the pulse is high, the switch is closed. This reverse-biases the diode, so that all the input current flows through the inductor. This current creates a magnetic field around the inductor. The amount of stored energy in the magnetic field is given by:

$$\text{Energy} = 0.5Li^2 \qquad \text{(24-20)}$$

The current through the inductor also charges the capacitor and supplies current to the load. While the switch is closed, the voltage across the inductor has the plus-minus polarity shown in Fig. 24-26b. As the current through the inductor increases, more energy is stored in the magnetic field.

Figure 24-26 (*a*) Buck regulator; (*b*) polarity with closed switch; (*c*) polarity with open switch; (*d*) choke-input filter passes dc value to output.

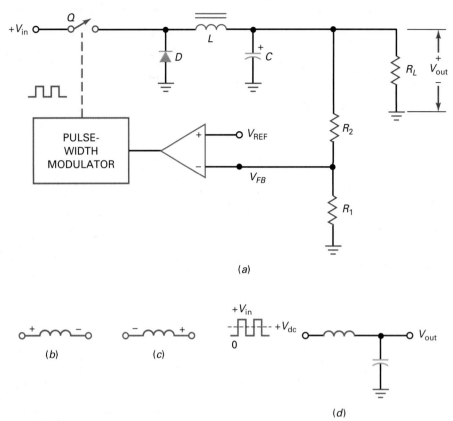

(*a*)

(*b*)   (*c*)

(*d*)

When the pulse goes low, the switch opens. At this instant, the magnetic field around the inductor starts collapsing and induces a reverse voltage across the inductor, as shown in Fig. 24-26*c*. This reverse voltage is called the *inductive kick*. Because of the inductive kick, the diode is forward biased and the current through the inductor continues to flow in the same direction. At this time, the inductor is returning its stored energy to the circuit. In other words, the inductor acts like a source and continues supplying current for the load.

Current flows through the inductor until the inductor returns all its energy to the circuit (discontinuous mode) or until the switch closes again (continuous mode), whichever comes first. In either case, the capacitor will also source load current during part of the time that the switch is open. This way, the ripple across the load is minimized.

The switch is being continuously closed and opened. The frequency of this switching can be from 10 to more than 100 kHz. (Some IC regulators switch at more than 1 MHz.) The current through the inductor is always in the same direction, passing through either the switch or the diode at different times in the cycle.

With a stiff input voltage and an ideal diode, a rectangular voltage waveform appears at the input to the *choke-input filter* (see Fig. 24-26*d*). If you recall from Chap. 4, the output of a choke-input filter equals the dc or average value of the input to the filter. The average value is related to the duty cycle and is given by:

$$V_{out} = DV_{in} \qquad (24\text{-}21)$$

The larger the duty cycle, the larger the dc output voltage.

When the power is first turned on, there is no output voltage and no feedback voltage from the R_1-R_2 voltage divider. Therefore, the comparator output is very large and the duty cycle approaches 100 percent. As the output voltage builds up, however, the feedback voltage V_{FB} reduces the comparator output, which reduces the duty cycle. At some point, the output voltage reaches an equilibrium value at which the feedback voltage produces a duty cycle that gives the same output voltage.

Because of the high gain of the comparator, the virtual short between the input terminals of the comparator means that:

$$V_{FB} \cong V_{REF}$$

From this, we can derive this expression for the output voltage:

$$V_{out} = \frac{R_1 + R_2}{R_1} V_{REF} \qquad (24\text{-}22)$$

After equilibrium sets in, any attempted change in the output voltage, whether caused by line or load changes, will be almost entirely offset by the negative feedback. For instance, if the output voltage tries to increase, the feedback voltage reduces the comparator output. This reduces the duty cycle and the output voltage. The net effect is only a slight increase in output voltage, much less than without the negative feedback.

Similarly, if the output voltage tries to decrease because of a line or load change, the feedback voltage is smaller and the comparator output is larger. This increases the duty cycle and produces a larger output voltage that offsets almost all the attempted decrease in output voltage.

Boost Regulator

Figure 24-27a shows a **boost regulator,** another basic topology for switching regulators. A boost regulator always steps the voltage up. The theory of operation is

---

Figure 24-27 (*a*) Boost regulator; (*b*) kick voltage adds to input when switch is open; (*c*) capacitor-input filter produces output voltage equal to peak input.

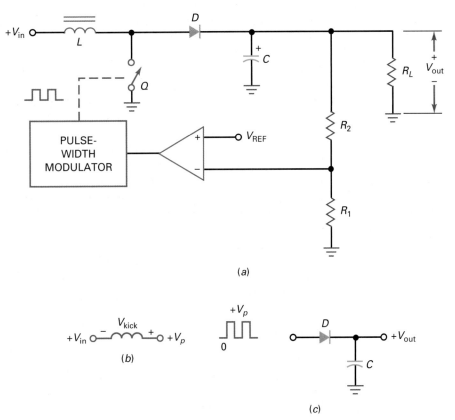

similar to that for a buck regulator in some ways but very different in others. For instance, when the pulse is high, the switch is closed and energy is stored in the magnetic field, as previously described.

When the pulse goes low, the switch opens. Again, the magnetic field around the inductor collapses and induces a reverse voltage across the inductor, as shown in Fig. 24-27b. Notice that the input voltage now adds to the inductive kick. This means that the peak voltage on the right end of the inductor is:

$$V_p = V_{in} + V_{kick} \tag{24-23}$$

The inductive kick depends on how much energy is stored in the magnetic field. Stated another way, V_{kick} is proportional to the duty cycle.

With a stiff input voltage, a rectangular voltage waveform appears at the input to the *capacitor-input filter* of Fig. 24-27c. Therefore, the regulated output voltage approximately equals the peak voltage given by Eq. (24-23). Because V_{kick} is always greater than zero, V_p is always greater than V_{in}. This is why a boost regulator always steps the voltage up.

Aside from using a capacitor-input filter rather than a choke-input filter, the regulation with boost topology is similar to that with buck topology. Because of the high gain of the comparator, the feedback almost equals the reference voltage. Therefore, the regulated output voltage is still given by Eq. (24-22). If the output voltage tries to increase, there is less feedback voltage, less comparator output, a smaller duty cycle, and less inductive kick. This reduces the peak voltage, which offsets the attempted increase in output voltage. If the output voltage tries to decrease, the smaller feedback voltage results in a larger peak voltage, which offsets the attempted decrease in output voltage.

Buck–Boost Regulator

Figure 24-28a shows a **buck-boost regulator,** the third most basic topology for switching regulators. A buck-boost regulator always produces a negative output voltage when driven by a positive input voltage. When the PWM output is high, the switch is closed and energy is stored in the magnetic field. At this time, the voltage across the inductor equals V_{in}, with the polarity shown in Fig. 24-28b.

When the pulse goes low, the switch opens. Again, the magnetic field around the inductor collapses and induces a kick voltage across the inductor, as shown in Fig. 24-28c. The kick voltage is proportional to the energy stored in the magnetic field, which is controlled by the duty cycle. If the duty cycle is low, the kick voltage approaches zero. If the duty cycle is high, the kick voltage can be greater than V_{in}, depending on how much energy is stored in the magnetic field.

In Fig. 24-28d, the magnitude of the peak voltage may be less than or greater than the input voltage. The diode and the capacitor-input filter then produce an output voltage equal to $-V_p$. Since the magnitude of this output voltage can be less than or greater than the input voltage, the topology is called *buck-boost*.

An inverting amplifier is used in Fig. 24-28a to invert the feedback voltage before it reaches the inverting input of the comparator. The voltage regulation then works as previously described. Attempted increases in output voltage reduce the duty cycle, which reduces the peak voltage. Attempted decreases in output voltage increase the duty cycle. Either way, the negative feedback holds the output voltage almost constant.

Monolithic Buck Regulators

Some IC switching regulators have only five external pins. For instance, the LT1074 is a monolithic bipolar switching regulator that uses buck topology. It contains most of the components discussed earlier, such as a reference voltage of 2.21 V, a switching device, an internal oscillator, a pulse-width modulator, and a

Figure 24-28 (*a*) Buck-boost regulator; (*b*) polarity with closed switch; (*c*) polarity with open switch; (*d*) capacitor-input filter produces output equal to negative peak.

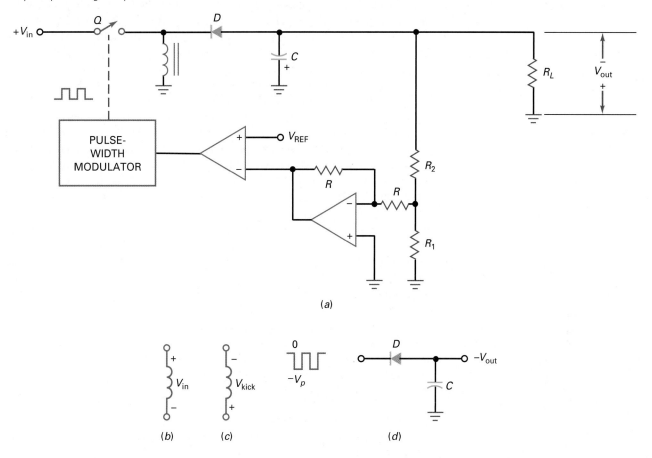

(*a*)

(*b*) (*c*) (*d*)

comparator. It runs at a switching frequency of 100 kHz, can handle input voltages from +8 to +40 V dc, and has efficiency of 75 to 90 percent for load currents from 1 to 5 A.

Figure 24-29 shows an LT1074 connected as a buck regulator. Pin 1 (FB) is for the feedback voltage. Pin 2 (COMP) is for frequency compensation to prevent oscillations at higher frequencies. Pin 3 (GND) is ground. Pin 4 (OUT) is the switched output of the internal switching device. Pin 5 (IN) is for the dc input voltage.

Figure 24-29 Buck regulator using LT1074.

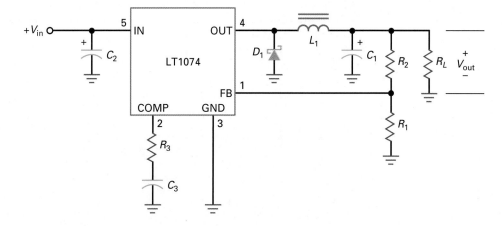

D_1, L_1, C_1, R_1, and R_2 serve the same functions as described in the earlier discussion of a buck regulator. But notice the use of a Schottky diode to improve the efficiency of the regulator. Because the Schottky diode has a lower knee voltage, it wastes less power. The data sheet of an LT1074 recommends adding a capacitor C_2 from 200 to 470 μF across the input for line filtering. Also recommended are a resistor R_3 of 2.7 kΩ and a capacitor C_3 of 0.01 μF to stabilize the feedback loop (prevent oscillations).

The LT1074 is widely used. A look at Fig. 24-29 tells us why. The circuit is incredibly simple, considering that it is a switching regulator, one of the most difficult of all circuits to design and build in discrete form. Fortunately, the IC designers have done all the hard work because the LT1074 includes everything except the components that cannot be integrated (choke and filter capacitors) and those left for the user to select (R_1 and R_2). By selecting values for R_1 and R_2, we can get regulated output voltage from about 2.5 to 38 V. Since the reference voltage of an LT1074 is 2.21 V, the output voltage is given by:

$$V_{out} = \frac{R_1 + R_2}{R_1} (2.21 \text{ V}) \tag{24-24}$$

The headroom voltage should be at least 2 V because the internal switching device consists of a *pnp* transistor driving an *npn* Darlington. The overall switch drop can be as high as 2 V with high currents.

Monolithic Boost Regulators

The MAX631 is a monolithic CMOS switching regulator that uses boost topology to produce a regulated output. This low-power IC switching regulator has a switching frequency of 50 kHz, an input voltage of 2 to 5 V, and an efficiency of about 80 percent. The MAX631 is the ultimate in simplicity because it requires only two external components.

For instance, Fig. 24-30 shows a MAX631 connected as a boost regulator, producing a fixed output voltage of +5 V with an input voltage of +2 to +5 V. The input voltage often comes from a battery because one of the applications for these IC regulators is in portable instruments. The data sheet recommends an inductor of 330 μH and a capacitor of 100 μF.

The MAX631 is an 8-pin device whose unused pins are either grounded or left unconnected. In Fig. 24-30, pin 1 (LBI) can be used for low-battery detection. When grounded, it has no effect. Although typically used as a fixed output regulator, the MAX631 can use an external voltage divider to provide a feedback voltage to pin 7 (FB). When pin 7 is grounded as shown, the output voltage is the factory preset value of +5 V.

Besides the MAX631, there is the MAX632, which produces an output of +12 V, and the MAX633, which produces an output of +15 V. The MAX631 to MAX633 regulators include pin 6, called the *charge pump*, which is a

Figure 24-30 Boost regulator using MAX631.

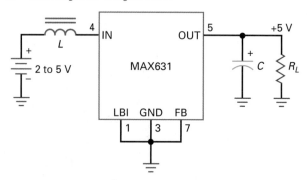

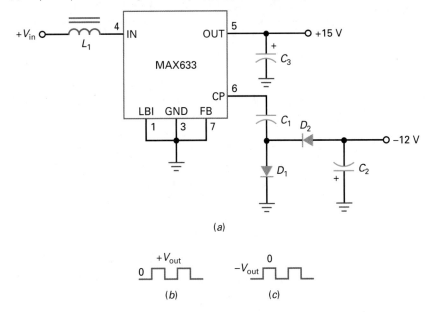

Figure 24-31 (a) Using charge pump of MAX633 to produce negative output voltage; (b) output of pin 6 drives negative clamper; (c) input to negative peak detector.

low-impedance buffer that produces a rectangular output signal. This signal swings from 0 to V_{out} at the oscillator frequency and can be negatively clamped and peak-detected to get a negative output voltage.

For instance, Fig. 24-31a shows how a MAX633 uses its charge pump to get an output of approximately -12 V. C_1 and D_1 are a negative clamper. D_2 and C_2 are a negative peak detector. Here is how the charge pump works: Figure 24-31b shows the ideal voltage waveform coming out of pin 6. Because of the negative clamper, the ideal voltage waveform across D_1 is the negatively clamped waveform of Fig. 24-31c. This waveform drives the negative peak detector to produce an output of approximately -12 V at 20 mA. The magnitude of this voltage is approximately 3 V less than the output voltage because of the two diode drops (D_1 and D_2) and the drop across the output impedance of the buffer (around 30 Ω).

If we use a battery to supply the input voltage to a linear regulator, the output voltage is always smaller. Boost regulators not only have better efficiency than linear regulators, they also can step up the voltage in a battery-powered system. This is very important and explains why monolithic boost regulators are so widely used. The availability of low-cost rechargeable batteries has made the monolithic boost regulator a standard choice for battery-powered systems.

The MAX631 to MAX633 devices have an internal reference voltage of 1.31 V. When these switching regulators are used with an external voltage divider, the following equation gives the regulated output voltage:

$$V_{out} = \frac{R_1 + R_2}{R_1} (1.31 \text{ V}) \tag{24-25}$$

Monolithic Buck–Boost Regulators

The internal design of the LT1074 can support a buck-boost external connection. Figure 24-32 shows the LT1074 connected as a buck-boost regulator. Again, we are using a Schottky diode to improve the efficiency. As previously discussed, energy is stored in the inductor's magnetic field when the internal switch is closed. When the switch opens, the magnetic field collapses and forward biases the diode. The negative kick voltage across the inductor is peak-detected by the capacitor-input filter to produce $-V_{out}$.

Figure 24-32 Using the LT1074 as buck-boost regulator.

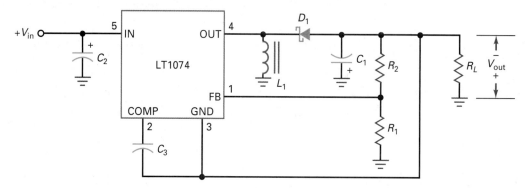

In the earlier discussion of buck-boost topology (Fig. 24-28*a*), we used an inverting amplifier to get a positive feedback voltage because the output sample from the voltage divider was negative. The internal design of the LT1074 takes care of this problem. The data sheet recommends returning the GND pin to the negative output voltage, as shown in Fig. 24-32. This produces the correct error voltage to the comparator that controls the pulse-width modulator.

Example 24-15

In the buck regulator of Fig. 24-29, $R_1 = 2.21$ kΩ and $R_2 = 2.8$ kΩ. What is the output voltage? What is the minimum input voltage that can be used with the output voltage?

SOLUTION   With Eq. (24-24), we can calculate:

$$V_{out} = \frac{R_1 + R_2}{R_1} V_{REF} = \frac{2.21 \text{ k}\Omega + 2.8 \text{ k}\Omega}{2.21 \text{ k}\Omega} (2.21 \text{ V}) = 5.01 \text{ V}$$

Because of the drop across the switching device of an LT1074, the input voltage has to be at least 2 V greater than the output of 5 V, which means minimum input voltage of 7 V. A more comfortable headroom will use an input voltage of 8 V.

PRACTICE PROBLEM 24-15   Repeat Example 24-15, change R_2 to 5.6 kΩ and calculate the new output voltage. With $R_1 = 2.2$ kΩ, what value of R_2 is needed to produce an output of 10 V?

Example 24-16

In the buck-boost regulator of Fig. 24-32, $R_1 = 1$ kΩ and $R_2 = 5.79$ kΩ. What is the output voltage?

SOLUTION   With Eq. (24-24), we can calculate:

$$V_{out} = \frac{R_1 + R_2}{R_1} V_{REF} = \frac{1 \text{ k}\Omega + 5.79 \text{ k}\Omega}{1 \text{ k}\Omega} (2.21 \text{ V}) = 15 \text{ V}$$

PRACTICE PROBLEM 24-16   Using Fig. 24-32, what is the output voltage if $R_1 = 1$ kΩ and $R_2 = 4.7$ kΩ?

Summary Table 24-1 shows a variety of voltage regulators and lists some of their characteristics.

Summary Table 24–1 | Voltage Regulators

Type	Characteristics

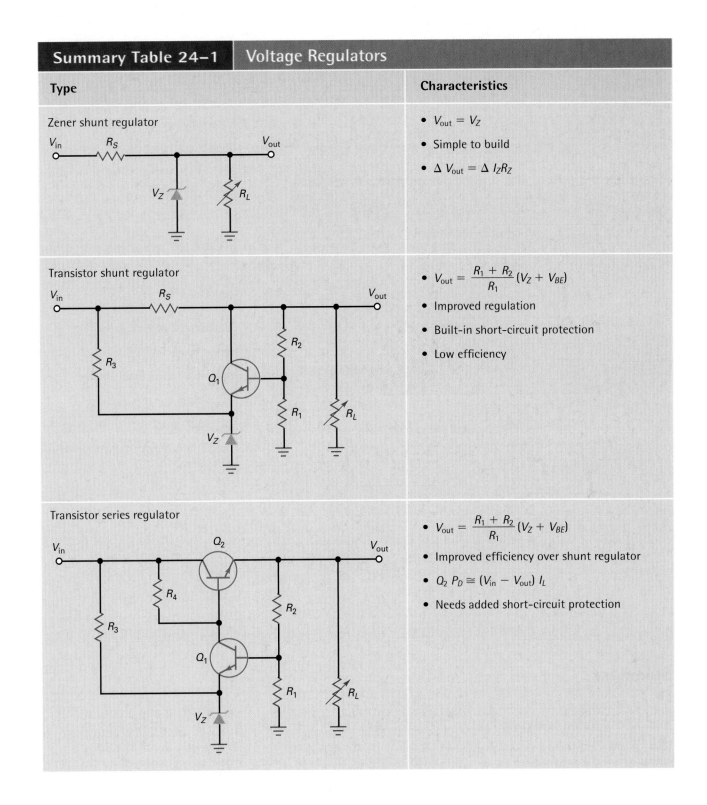

Zener shunt regulator

- $V_{out} = V_Z$
- Simple to build
- $\Delta V_{out} = \Delta I_Z R_Z$

Transistor shunt regulator

- $V_{out} = \dfrac{R_1 + R_2}{R_1}(V_Z + V_{BE})$
- Improved regulation
- Built-in short-circuit protection
- Low efficiency

Transistor series regulator

- $V_{out} = \dfrac{R_1 + R_2}{R_1}(V_Z + V_{BE})$
- Improved efficiency over shunt regulator
- $Q_2\ P_D \cong (V_{in} - V_{out})\ I_L$
- Needs added short-circuit protection

Type | **Characteristics**

IC linear regulator

- Easy to use
- Fixed or adjustable outputs
- $V_{out} = V_{reg}$ or $\dfrac{R_1 + R_2}{R_1}(V_{ref})$
- Essentially a series regulator
- Good ripple rejection
- Built-in short-circuit and temperature protection possible

IC switching regulator

- Uses pulse-wide modulation
- High efficiencies
- Steps up or down input voltage
- May require complex circuitry
- Somewhat noisy
- Popular in computer and consumer electronics

Summary

SEC. 24-1 SUPPLY CHARACTERISTICS

Load regulation indicates how much the output voltage changes when the load current changes. Line regulation indicates how much the load voltage changes when the line voltage changes. The output resistance determines the load regulation.

SEC. 24-2 SHUNT REGULATORS

The zener regulator is the simplest example of a shunt regulator. By adding transistors and an op amp, we can build a shunt regulator that has excellent line and load regulation. The main disadvantage of a shunt regulator is its low efficiency, caused by power losses in the series resistor and shunt transistor.

SEC. 24-3 SERIES REGULATORS

By using a pass transistor instead of a series resistor, we can build series regulators with higher efficiencies than shunt regulators. The zener follower is the simplest example of a series regulator. By adding transistors and an op amp, we can build series regulators with excellent line and load regulation, plus current limiting.

SEC. 24-4 MONOLITHIC LINEAR REGULATORS

IC voltage regulators have one of the following voltages: fixed positive, fixed negative, or adjustable. IC regulators are also classified as standard, low-power, and low-dropout. The LM78XX series is a standard line of fixed regulators with output voltages from 5 to 24 V.

SEC. 24-5 CURRENT BOOSTERS

To increase the regulated load current of an IC regulator such as a 78XX device, we can use an outboard transistor to carry most of the current above 1 A. By adding another transistor, we can have short-circuit protection.

SEC. 24-6 DC-TO-DC CONVERTERS

When we want to convert an input dc voltage to an output dc voltage of another value, a dc-to-dc converter is useful. Unregulated dc-to-dc converters have an oscillator whose output voltage is proportional to the input voltage. Typically, a push-pull arrangement of transistors and a transformer can step this voltage up or down. Then, it is rectified and filtered to get an output voltage different from the input voltage.

SEC. 24-7 SWITCHING REGULATORS

A switching regulator is a dc-to-dc converter that uses pulse-width modulation to regulate the output voltage. By switching the pass transistor on and off, the switching regulator can attain efficiencies from 70 to 95 percent. The basic topologies are the buck (step-down), boost (step-up), and buck-boost (inverting). This type of regulator is very popular in computer and portable electronic systems.

Definitions

(24-8) Efficiency:

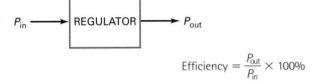

$$\text{Efficiency} = \frac{P_{out}}{P_{in}} \times 100\%$$

(24-11) Headroom:

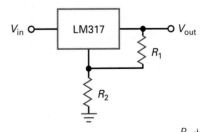

$$\text{Headroom voltage} = V_{in} - V_{out}$$

Derivations

(24-4) Load regulation:

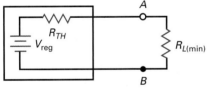

$$\text{Load regulation} = \frac{R_{TH}}{R_{L(min)}} \times 100\%$$

(24-12) Pass dissipation:

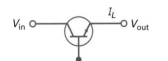

$$P_D \cong (V_{in} - V_{out})I_L$$

(24-13) Efficiency:

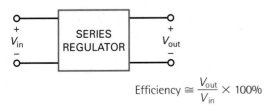

$$\text{Efficiency} \cong \frac{V_{out}}{V_{in}} \times 100\%$$

(24-17) Shorted-load current:

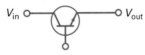

$$I_{SL} = \frac{V_{BE}}{R_4}$$

(24-19) LM317 output voltage:

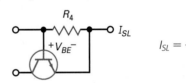

$$V_{out} = \frac{R_1 + R_2}{R_1}(1.25 \text{ V})$$

(24-20) Stored energy in magnetic field:

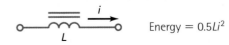

$$\text{Energy} = 0.5Li^2$$

(24-21) Average value of input to filter

$$V_{out} = DV_{in}$$

(24-23) Boost peak voltage:

$$V_p = V_{in} + V_{kick}$$

(24-22) Output of switching regulator:

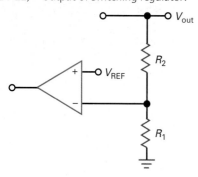

$$V_{out} = \frac{R_1 + R_2}{R_1} V_{REF}$$

Student Assignments

1. **Voltage regulators normally use**
   a. Negative feedback
   b. Positive feedback
   c. No feedback
   d. Phase limiting

2. **During regulation, the power dissipation of the pass transistor equals the collector–emitter voltage times the**
   a. Base current
   b. Load current
   c. Zener current
   d. Foldback current

3. **Without current limiting, a shorted load will probably**
   a. Produce zero load current
   b. Destroy diodes and transistors
   c. Have a load voltage equal to the zener voltage
   d. Have too little load current

4. **A current-sensing resistor is usually**
   a. Zero   c. Large
   b. Small   d. Open

5. **Simple current limiting produces too much heat in the**
   a. Zener diode
   b. Load resistor
   c. Pass transistor
   d. Ambient air

6. **With foldback current limiting, the load voltage approaches zero and the load current approaches**
   a. A small value
   b. Infinity
   c. The zener current
   d. A destructive level

7. **A capacitor may be needed in a discrete voltage regulator to prevent**
   a. Negative feedback
   b. Excessive load current
   c. Oscillations
   d. Current sensing

8. **If the output of a voltage regulator varies from 15 to 14.7 V between the minimum and maximum load current, the load regulation is**
   a. 0
   b. 1 percent
   c. 2 percent
   d. 5 percent

9. **If the output of a voltage regulator varies from 20 to 19.8 V when the line voltage varies over its specified range, the source regulation is**
   a. 0   c. 2 percent
   b. 1 percent   d. 5 percent

10. **The output impedance of a voltage regulator is**
    a. Very small
    b. Very large
    c. Equal to the load voltage divided by the load current
    d. Equal to the input voltage divided by the output current

11. **Compared to the ripple into a voltage regulator, the ripple out of a voltage regulator is**
    a. Equal in value
    b. Much larger
    c. Much smaller
    d. Impossible to determine

12. **A voltage regulator has a ripple rejection of −60 dB. If the input ripple is 1 V, the output ripple is**
    a. −60 mV
    b. 1 mV
    c. 10 mV
    d. 1000 V

13. **Thermal shutdown occurs in an IC regulator if**
    a. Power dissipation is too low
    b. Internal temperature is too high
    c. Current through the device is too low
    d. Any of the above occur

14. If a linear three-terminal IC regulator is more than a few inches from the filter capacitor, you may get oscillations inside the IC unless you use

a. Current limiting

b. A bypass capacitor on the input pin

c. A coupling capacitor on the output pin

d. A regulated input voltage

15. The 78XX series of voltage regulators produces an output voltage that is

a. Positive

b. Negative

c. Either positive or negative

d. Unregulated

16. The LM7812 produces a regulated output voltage of

a. 3 V   c. 12 V

b. 4 V   d. 78 V

17. A current booster is a transistor in

a. Series with the IC regulator

b. Parallel with the IC regulator

c. Either series or parallel

d. Shunt with the load

18. To turn on a current booster, we can drive its base–emitter terminals with the voltage across

a. A load resistor

b. A zener impedance

c. Another transistor

d. A current-sensing resistor

19. A phase splitter produces two output voltages that are

a. Equal in phase

b. Unequal in amplitude

c. Opposite in phase

d. Very small

20. A series regulator is an example of a

a. Linear regulator

b. Switching regulator

c. Shunt regulator

d. DC-to-dc converter

21. To get more output voltage from a buck switching regulator, you have to

a. Decrease the duty cycle

b. Decrease the input voltage

c. Increase the duty cycle

d. Increase the switching frequency

22. An increase of line voltage into a power supply usually produces

a. A decrease in load resistance

b. An increase in load voltage

c. A decrease in efficiency

d. Less power dissipation in the rectifier diodes

23. A power supply with low output impedance has low

a. Load regulation

b. Current limiting

c. Line regulation

d. Efficiency

24. A zener-diode regulator is a

a. Shunt regulator

b. Series regulator

c. Switching regulator

d. Zener follower

25. The input current to a shunt regulator is

a. Variable

b. Constant

c. Equal to load current

d. Used to store energy in a magnetic field

26. An advantage of shunt regulation is

a. Built-in short-circuit protection

b. Low power dissipation in the pass transistor

c. High efficiency

d. Little wasted power

27. The efficiency of a voltage regulator is high when

a. Input power is low

b. Output power is high

c. Little power is wasted

d. Input power is high

28. A shunt regulator is inefficient because

a. It wastes power

b. It uses a series resistor and a shunt transistor

c. The ratio of output to input power is low

d. All of the above

29. A switching regulator is considered

a. Quiet   c. Inefficient

b. Noisy   d. Linear

30. The zener follower is an example of a

a. Boost regulator   c. Buck regulator

b. Shunt regulator   d. Series regulator

31. A series regulator is more efficient than a shunt regulator because

a. It has a series resistor

b. It can boost the voltage

c. The pass transistor replaces the series resistor

d. It switches the pass transistor on and off

32. The efficiency of a linear regulator is high when the

a. Headroom voltage is low

b. Pass transistor has a high power dissipation

c. Zener voltage is low

d. Output voltage is low

33. If the load is shorted, the pass transistor has the least power dissipation when the regulator has

a. Foldback limiting

b. Low efficiency

c. Buck topology

d. A high zener voltage

34. The dropout voltage of standard monolithic linear regulators is closest to

a. 0.3 V   c. 2 V

b. 0.7 V   d. 3.1 V

35. In a buck regulator, the output voltage is filtered with a

a. Choke-input filter

b. Capacitor-input filter

c. Diode

d. Voltage divider

36. The regulator with the highest efficiency is the

a. Shunt regulator

b. Series regulator

c. Switching regulator

d. DC-to-dc converter

37. In a boost regulator, the output voltage is filtered with a

a. Choke-input filter

b. Capacitor-input filter

c. Diode

d. Voltage divider

38. The buck-boost regulator is also

a. A step-down regulator

b. A step-up regulator

c. An inverting regulator

d. All of the above

Problems

SEC. 24-1 SUPPLY CHARACTERISTICS

24-1 A power supply has $V_{NL} = 15$ V and $V_{FL} = 14.5$ V. What is the load regulation?

24-2 A power supply has $V_{HL} = 20$ V and $V_{LL} = 19$ V. What is the line regulation?

24-3 If line voltage changes from 108 to 135 V and load voltage changes from 12 to 12.3 V, what is the line regulation?

24-4 A power supply has an output resistance of 2 Ω. If the minimum load resistance is 50 Ω, what is the load regulation?

SEC. 24-2 SHUNT REGULATORS

24-5 In Fig. 24-4, $V_{in} = 25$ V, $R_S = 22$ Ω, $V_Z = 18$ V, $V_{BE} = 0.75$ V, and $R_L = 100$ Ω. What are the values of output voltage, the input current, the load current, and the collector current?

24-6 The shunt regulator of Fig. 24-5 has these circuit values: $V_{in} = 25$ V, $R_S = 15$ Ω, $V_Z = 5.6$ V, $V_{BE} = 0.77$ V, and $R_L = 80$ Ω. If $R_1 = 330$ Ω and $R_2 = 680$ Ω, what are the approximate values of output voltage, the input current, the load current, and the collector current?

24-7 The shunt regulator of Fig. 24-6 has these circuit values: $V_{in} = 25$ V, $R_S = 8.2$ Ω, $V_Z = 5.6$ V, and $R_L = 50$ Ω. If $R_1 = 2.7$ kΩ and $R_2 = 6.2$ kΩ, what are the approximate values of output voltage, the input current, the load current, and the collector current?

SEC. 24-3 SERIES REGULATORS

24-8 In Fig. 24-8, $V_{in} = 20$ V, $V_Z = 4.7$ V, $R_1 = 2.2$ kΩ, $R_2 = 4.7$ kΩ, $R_3 = 1.5$ kΩ, $R_4 = 2.7$ kΩ, and $R_L = 50$ Ω. What is the output voltage? What is the power dissipation in the pass transistor?

24-9 What is the approximate efficiency in Prob. 24-8?

24-10 In Fig. 24-15, the zener voltage is changed to 6.2 V. What is the approximate output voltage?

24-11 In Fig. 24-16, V_{in} can vary from 20 to 30 V. What is the maximum zener current?

24-12 If the 1-kΩ potentiometer of Fig. 24-16 is changed to 1.5 kΩ, what are the minimum and maximum regulated output voltages?

24-13 If the regulated output voltage is 8 V in Fig. 24-16, what is the load resistance where current limiting starts? What is the approximate shorted-load current?

SEC. 24-4 MONOLITHIC LINEAR REGULATORS

24-14 What is the load current in Fig. 24-33? The headroom voltage? The power dissipation of the LM7815?

24-15 What is the output ripple in Fig. 24-33?

24-16 If $R_1 = 2.7$ kΩ and $R_2 = 20$ kΩ in Fig. 24-20, what is the output voltage?

24-17 The LM7815 is used with an input voltage that can vary from 18 to 25 V. What is the maximum efficiency? The minimum efficiency?

SEC. 24-6 DC-TO-DC CONVERTERS

24-18 A dc-to-dc converter has an input voltage of 5 V and an output voltage of 12 V. If the input current is 1 A and the output current is 0.25 A, what is the efficiency of the dc-to-dc converter?

24-19 A dc-to-dc converter has an input voltage of 12 V and an output voltage of 5 V. If the input current is 2 A and the efficiency is 80 percent, what is the output current?

SEC. 24-7 SWITCHING REGULATORS

24-20 A buck regulator has $V_{REF} = 2.5$ V, $R_1 = 1.5$ kΩ, and $R_2 = 10$ kΩ. What is the output voltage?

24-21 If the duty cycle is 30 percent and the peak value of the pulses to the choke-input filter is 20 V, what is the regulated output voltage?

24-22 A boost regulator has $V_{REF} = 1.25$ V, $R_1 = 1.2$ kΩ, and $R_2 = 15$ kΩ. What is the output voltage?

24-23 A buck-boost regulator has $V_{REF} = 2.1$ V, $R_1 = 2.1$ kΩ, and $R_2 = 12$ kΩ. What is the output voltage?

Figure 24-33 Example.

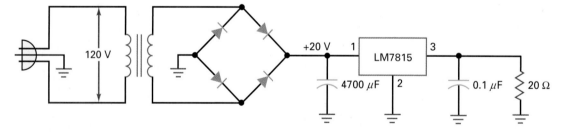

Critical Thinking

24-24 Figure 24-34 shows an LM317 regulator with electronic shutdown. When the shutdown voltage is zero, the transistor is cut off and has no effect on the operation. But when the shutdown voltage is approximately 5 V, the transistor saturates. What is the adjustable range of output voltage when the shutdown voltage is zero?

What does the output voltage equal when the shutdown voltage is 5 V?

24-25 The transistor of Fig. 24-34 is cut off. To get an output voltage of 18 V, what value should the adjustable resistor have?

Figure 24-34

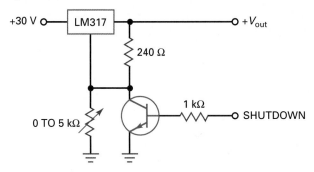

24-26 When a bridge rectifier and a capacitor-input filter drive a voltage regulator, the capacitor voltage during discharge is almost a perfect ramp. Why do we get a ramp instead of the usual exponential wave?

24-27 If the load regulation is 5 percent and the no-load voltage is 12.5 V, what is the full-load voltage?

24-28 If the line regulation is 3 percent and the low-line voltage is 16 V, what is the high-line voltage?

24-29 A power supply has a load regulation of 1 percent and a minimum load resistance of 10 Ω. What is the output resistance of the power supply?

24-30 The shunt regulator of Fig. 24-6 has an input voltage of 35 V, a collector current of 60 mA, a load current of 140 mA. If the series resistance is 100 Ω, what is the load resistance?

24-31 In Fig. 24-10, we want current limiting to start at approximately 250 mA. What value should we use for R_4?

24-32 Figure 24-12 has an output voltage of 10 V. If $V_{BE} = 0.7$ V for the current-limiting transistor, what are the values of shorted-load current and the maximum load current? Use $K = 0.7$ and $R_4 = 1$ Ω.

24-33 In Fig. 24-35, $R_5 = 7.5$ kΩ, $R_6 = 1$ kΩ, $R_7 = 9$ kΩ, and $C_3 = 0.001$ μF. What is the switching frequency of the buck regulator?

24-34 In Fig. 24-16, the wiper is at the middle of its range. What is the output voltage?

Troubleshooting

Use Fig. 24-35 for the remaining problems. In this set of problems, you are troubleshooting a switching regulator. Before you start, look at the OK row in the troubleshooting table to see the normal waveforms with their correct peak voltages. In this exercise, most of the troubles are IC failures rather than resistor failures. When an IC fails, anything can happen. Pins may be internally open, shorted, and so on. No matter what the trouble is inside the IC, the most common symptom is a *stuck output*. This refers to the output voltage being stuck at either positive or negative saturation. If the input signals are OK, an IC with a stuck output has to be replaced. The following problems will give you a chance to work with outputs that are stuck at either +13.5 or −13.5 V.

24-35 Find Trouble 1.

24-36 Find Trouble 2.

24-37 Find Trouble 3.

24-38 Find Trouble 4.

24-39 Find Trouble 5.

24-40 Find Trouble 6.

24-41 Find Trouble 7.

24-42 Find Trouble 8.

24-43 Find Trouble 9.

Figure 24-35

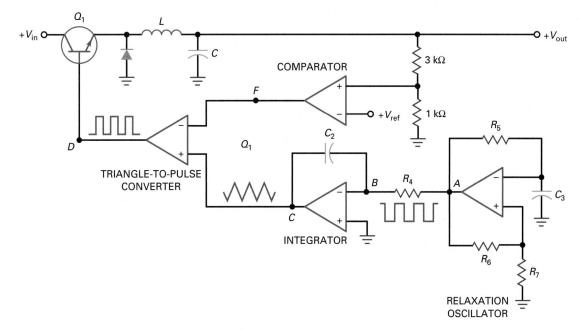

Figure 24–35 (continued)

Troubleshooting

Trouble	V_A	V_B	V_C	V_D	V_E	V_F
OK	N	I	M	J	K	H
T1	P	I	U	T	I	L
T2	T	L	V	O	R	O
T3	N	Q	M	V	I	T
T4	P	N	L	T	Q	L
T5	P	V	L	T	I	L
T6	N	Q	M	O	R	T
T7	P	I	U	I	Q	L
T8	P	I	U	L	Q	V
T9	N	Q	M	O	R	V

Waveforms

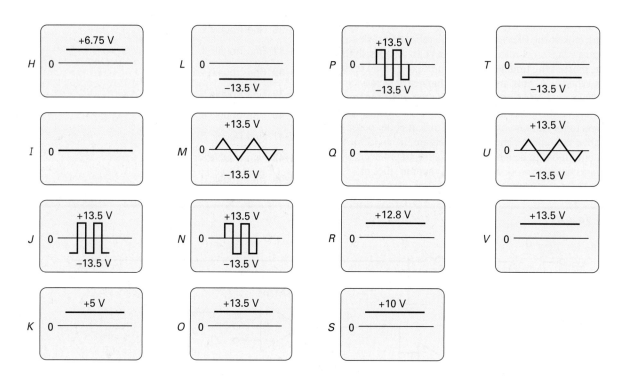

Job Interview Questions

1. Draw any shunt regulator and tell me how it works.
2. Draw any series regulator and tell me how it works.
3. Explain why the efficiency of a series regulator is better than that of a shunt regulator.
4. What are the three basic types of switching regulators? Which one steps the voltage up? Which one produces a negative output from a positive input? Which one steps the voltage down?
5. In a series regulator, what does *headroom voltage* mean? How is the efficiency related to headroom voltage?
6. What is the difference between the LM7806 and the LM7912?

7. Explain what line and load regulation mean. Should they be high or low if you want a quality power supply?
8. How is the Thevenin or output resistance of a power supply related to the load regulation? For a quality power supply, should the output resistance be high or low?
9. What is the difference between simple current limiting and foldback current limiting?

10. What does *thermal shutdown* mean?
11. The manufacturer of a three-terminal regulator recommends using a bypass capacitor on the input if the IC is more than 6 in from the unregulated power supply. What is the purpose of this capacitor?
12. What is the typical dropout voltage for the LM78XX series? What does it mean?

Self-Test Answers

1.	a	14.	b	27.	c
2.	b	15.	a	28.	d
3.	b	16.	c	29.	b
4.	b	17.	b	30.	d
5.	c	18.	d	31.	c
6.	a	19.	c	32.	a
7.	c	20.	a	33.	a
8.	c	21.	c	34.	c
9.	b	22.	b	35.	a
10.	a	23.	a	36.	c
11.	c	24.	a	37.	b
12.	b	25.	b	38.	d
13.	b	26.	a		

Practice Problem Answers

24-1 $V_{out} = 7.6$ V;
$I_S = 440$ mA;
$I_L = 190$ mA;
$I_C = 250$ mA

24-2 $V_{out} = 11.1$ V;
$I_S = 392$ mA;
$I_L = 277$ mA;
$I_C = 115$ mA

24-3 $P_{out} = 3.07$ W;
$P_{in} = 5.88$ W;
% Eff. = 52.2%

24-4 $I_C = 66$ mA;
$P_D = 858$ mW

24-6 Load regulation = 2.16%;
Line regulation = 3.31%

24-7 $V_{out} = 8.4$ V;
$P_D = 756$ mW

24-8 Efficiency = 70%

24-9 $V_{out} = 11.25$ V

24-11 $I_Z = 22.7$ mA;
$V_{out(min)} = 5.57$ V;
$V_{out(max)} = 13$ V;
$R_L = 41.7$ Ω;
$I_{SL} = 350$ mA

24-12 $I_L = 150$ mA;
$V_R = 198$ μV

24-13 $V_{out} = 7.58$ V

24-15 $V_{out} = 7.81$ V;
$R_2 = 7.8$ kΩ

24-16 $V_{out} = 7.47$ V

Appendix A

Data Sheets

1N4001 to 1N4007 (rectifier diodes)

1N957B Series (zener diodes)

1N4728A Series (zener diodes)

2N3903, 2N3904 (general purpose silicon transistors: npn)

2N3906 (general purpose silicon transistors: pnp)

TIP 100/101/102 (silicon Darlington transistor)

MPF102 (JFET n-channel RF amplifier)

2N7000 (MOSFET n-channel enhancement mode)

2N6504 (silicon controlled rectifiers)

FKPF8N80 (triode thyristor)

FGL60N100BNTD (NPT-Trench IGBT)

LM741 (general purpose operational amplifier)

LM118/218/318 (precision high-speed operational amplifiers)

LM555 (timer)

XR-2206 (function generator IC)

LM78XX Series (three terminal voltage regulators)

FAIRCHILD
SEMICONDUCTOR®

1N4001 - 1N4007

Features

- Low forward voltage drop.

- High surge current capability.

DO-41
COLOR BAND DENOTES CATHODE

General Purpose Rectifiers

Absolute Maximum Ratings* $T_A = 25°C$ unless otherwise noted

Symbol	Parameter	Value							Units
		4001	4002	4003	4004	4005	4006	4007	
V_{RRM}	Peak Repetitive Reverse Voltage	50	100	200	400	600	800	1000	V
$I_{F(AV)}$	Average Rectified Forward Current, .375 " lead length @ $T_A = 75°C$	1.0							A
I_{FSM}	Non-repetitive Peak Forward Surge Current 8.3 ms Single Half-Sine-Wave	30							A
T_{stg}	Storage Temperature Range	-55 to +175							°C
T_J	Operating Junction Temperature	-55 to +175							°C

*These ratings are limiting values above which the serviceability of any semiconductor device may be impaired.

Thermal Characteristics

Symbol	Parameter	Value	Units
P_D	Power Dissipation	3.0	W
$R_{\theta JA}$	Thermal Resistance, Junction to Ambient	50	°C/W

Electrical Characteristics $T_A = 25°C$ unless otherwise noted

Symbol	Parameter	Device							Units
		4001	4002	4003	4004	4005	4006	4007	
V_F	Forward Voltage @ 1.0 A	1.1							V
I_{rr}	Maximum Full Load Reverse Current, Full Cycle $T_A = 75°C$	30							μA
I_R	Reverse Current @ rated V_R $T_A = 25°C$ $T_A = 100°C$	5.0 500							μA μA
C_T	Total Capacitance $V_R = 4.0$ V, f = 1.0 MHz	15							pF

1N4001-1N4007, Rev. C1

Typical Characteristics

Forward Current Derating Curve

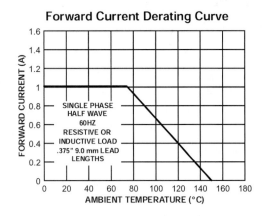

Forward Characteristics

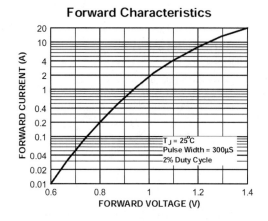

Non-Repetitive Surge Current

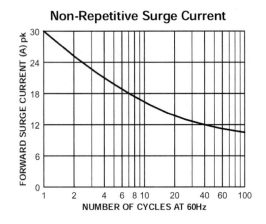

Reverse Characteristics

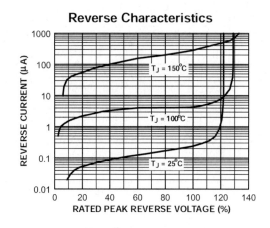

1N4001-1N4007, Rev. C1

SEMICONDUCTOR®

Zeners
1N957B - 1N991B

Absolute Maximum Ratings * T_A = 25°C unless otherwise noted

Tolerance = 5%

Symbol	Parameter	Value	Units
P_D	Power Dissipation @ TL ≤ 75°C, Lead Length = 3/8"	500	mW
	Derate above 75°C	4.0	mW/°C
T_J, T_{STG}	Operating and Storage Temperature Range	-65 to +200	°C

* These ratings are limiting values above which the serviceability of the diode may be impaired.

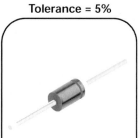

DO-35 Glass case
COLOR BAND DENOTES CATHODE

Electrical Characteristics T_A=25°C unless otherwise noted

Device	V_Z (Volts) (Note 1)				Z_Z (Ω) (Note 2)			I_R @ V_R		I_{ZM} (mA) (Note 3)
	Min.	Typ.	Max.	@ I_Z (mA)	Z_Z @ I_Z	Z_{ZK} @ I_{ZK} Ω	mA	μA	Volts	
1N957B	6.46	6.8	7.14	18.5	4.5	700	1.0	150	5.2	47
1N958B	7.125	7.5	7.875	16.5	5.5	700	0.5	75	5.7	42
1N959B	7.79	8.2	8.61	15	6.5	700	0.5	50	6.2	38
1N960B	8.645	9.1	9.555	14	7.5	700	0.5	25	6.9	35
1N961B	9.5	10	10.5	12.5	8.5	700	0.25	10	7.6	32
1N962B	10.45	11	11.55	11.5	9.5	700	0.25	5	8.4	28
1N963B	11.4	12	12.6	10.5	11.5	700	0.25	5	9.1	26
1N964B	12.35	13	13.65	9.5	13	700	0.25	5	9.9	24
1N965B	14.25	15	15.75	8.5	16	700	0.25	5	11.4	21
1N966B	15.2	16	16.8	7.8	17	700	0.25	5	12.2	19
1N967B	17.1	18	18.9	7.0	21	750	0.25	5	13.7	17
1N968B	19	20	21	6.2	25	750	0.25	5	15.2	15
1N969B	20.9	22	23.1	5.6	29	750	0.25	5	16.7	14
1N970B	22.8	24	25.2	5.2	33	750	0.25	5	18.2	13
1N971B	25.652	27	28.35	4.6	41	750	0.25	5	20.6	11
1N972B	8.5	30	31.5	4.2	49	1000	0.25	5	22.8	10
1N973B	31.35	33	34.65	3.8	58	1000	0.25	5	25.1	9.2
1N974B	34.2	36	37.8	3.4	70	1000	0.25	5	27.4	8.5
1N975B	37.05	39	40.95	3.2	80	1000	0.25	5	29.7	7.8
1N976B	40.85	43	45.15	3.0	93	1500	0.25	5	32.7	7.0
1N977B	44.65	47	49.35	2.7	105	1500	0.25	5	35.8	6.4
1N978B	48.45	51	53.55	2.5	125	1500	0.25	5	38.8	5.9
1N979B	53.2	56	58.8	2.2	150	2000	0.25	5	42.6	5.4
1N980B	58.9	62	65.1	2.0	185	2000	0.25	5	47.1	4.9
1N981B	64.6	68	71.4	1.8	230	2000	0.25	5	51.7	4.5

©2004 Fairchild Semiconductor Corporation

1N957B - 1N991B, Rev. E2

Electrical Characteristics (Continued) T_A=25°C unless otherwise noted

| Device | V_Z (Volts) (Note 1) | | | | Z_Z (Ω) (Note 2) | | | I_R @ V_R | | I_{ZM} (mA) (Note 3) |
| | Min. | Typ. | Max. | @ I_Z | Z_Z @ I_Z | Z_{ZK} @ I_{ZK} | | μA | Volts | |
						Ω	mA			
1N982B	71.25	75	78.75	1.7	270	2000	0.25	5	56.0	4.1
1N983B	77.9	82	86.1	1.5	330	3000	0.25	5	62.2	3.7
1N984B	86.45	91	95.55	1.4	400	3000	0.25	5	69.2	3.3
1N985B	95	100	105	1.3	500	3000	0.25	5	76.0	3.0
1N986B	104.5	110	115.5	1.1	750	4000	0.25	5	83.6	2.7
1N987B	114	120	126	1.0	900	4500	0.25	5	91.2	2.5
1N988B	123.5	130	136.5	0.95	1100	5000	0.25	5	98.8	2.3
1N989B	142.5	150	157.5	0.85	1500	6000	0.25	5	114	2.0
1N990B	152	160	168	0.80	1700	6500	0.25	5	121.6	1.9
1N991B	171	180	189	0.68	2200	7100	0.25	5	136.8	1.7

Notes:
1. Zener Voltage (V_Z) Measurement
   Nominal zener voltage is measured with the device junction in the thermal equilibrium at the lead temperature (T_L) at 30°C ± 1°C and 3/8" lead length.
2. Zener Impedance (Z_Z) Derivation
   Z_{ZT} and Z_{ZK} are measured by dividing the ac voltage drop across the device by the ac current applied. The specified limits are for $I_{Z(ac)}$ = 0.1 $I_{Z(dc)}$ with the ac frequency = 60Hz.
3. Maximum Zener Current Ratings (I_{ZM})
   The maximum current handling capability on a worst case basis is limited by the actual zener voltage at the operation point and the power derating curve.

1N957B - 1N991B, Rev. E2

Appendix A

SEMICONDUCTOR®

January 2005

1N4728A - 1N4764A

Zeners

DO-41 Glass case
COLOR BAND DENOTES CATHODE

Absolute Maximum Ratings * $T_a = 25°C$ unless otherwise noted

Symbol	Parameter	Value	Units
P_D	Power Dissipation @ TL ≤ 50°C, Lead Length = 3/8"	1.0	W
	Derate above 50°C	6.67	mW/°C
T_J, T_{STG}	Operating and Storage Temperature Range	-65 to +200	°C

* These ratings are limiting values above which the serviceability of the diode may be impaired.

Electrical Characteristics $T_a = 25°C$ unless otherwise noted

Device	V_Z (V) @ I_Z (Note 1)			Test Current I_Z (mA)	Max. Zener Impedance			Leakage Current	
	Min.	Typ.	Max.		Z_Z @ I_Z (Ω)	Z_{ZK} @ I_{ZK} (Ω)	I_{ZK} (mA)	I_R (μA)	V_R (V)
1N4728A	3.315	3.3	3.465	76	10	400	1	100	1
1N4729A	3.42	3.6	3.78	69	10	400	1	100	1
1N4730A	3.705	3.9	4.095	64	9	400	1	50	1
1N4731A	4.085	4.3	4.515	58	9	400	1	10	1
1N4732A	4.465	4.7	4.935	53	8	500	1	10	1
1N4733A	4.845	5.1	5.355	49	7	550	1	10	1
1N4734A	5.32	5.6	5.88	45	5	600	1	10	2
1N4735A	5.89	6.2	6.51	41	2	700	1	10	3
1N4736A	6.46	6.8	7.14	37	3.5	700	1	10	4
1N4737A	7.125	7.5	7.875	34	4	700	0.5	10	5
1N4738A	7.79	8.2	8.61	31	4.5	700	0.5	10	6
1N4739A	8.645	9.1	9.555	28	5	700	0.5	10	7
1N4740A	9.5	10	10.5	25	7	700	0.25	10	7.6
1N4741A	10.45	11	11.55	23	8	700	0.25	5	8.4
1N4742A	11.4	12	12.6	21	9	700	0.25	5	9.1
1N4743A	12.35	13	13.65	19	10	700	0.25	5	9.9
1N4744A	14.25	15	15.75	17	14	700	0.25	5	11.4
1N4745A	15.2	16	16.8	15.5	16	700	0.25	5	12.2
1N4746A	17.1	18	18.9	14	20	750	0.25	5	13.7
1N4747A	19	20	21	12.5	22	750	0.25	5	15.2

1N4728A - 1N4764A Rev. G2

www.fairchildsemi.com

Data Sheets

Electrical Characteristics $T_C = 25°C$ unless otherwise noted

Device	V_Z (V) @ I_Z (Note 1)			Test Current I_Z (mA)	Max. Zener Impedance			Leakage Current	
	Min.	Typ.	Max.		Z_Z @ I_Z (Ω)	Z_{ZK} @ I_{ZK} (Ω)	I_{ZK} (mA)	I_R (μA)	V_R (V)
1N4748A	20.9	22	23.1	11.5	23	750	0.25	5	16.7
1N4749A	22.8	24	25.2	10.5	25	750	0.25	5	18.2
1N4750A	25.65	27	28.35	9.5	35	750	0.25	5	20.6
1N4751A	28.5	30	31.5	8.5	40	1000	0.25	5	22.8
1N4752A	31.35	33	34.65	7.5	45	1000	0.25	5	25.1
1N4753A	34.2	36	37.8	7	50	1000	0.25	5	27.4
1N4754A	37.05	39	40.95	6.5	60	1000	0.25	5	29.7
1N4755A	40.85	43	45.15	6	70	1500	0.25	5	32.7
1N4756A	44.65	47	49.35	5.5	80	1500	0.25	5	35.8
1N4757A	48.45	51	53.55	5	95	1500	0.25	5	38.8
1N4758A	53.2	56	58.8	4.5	110	2000	0.25	5	42.6
1N4759A	58.9	62	65.1	4	125	2000	0.25	5	47.1
1N4760A	64.6	68	71.4	3.7	150	2000	0.25	5	51.7
1N4761A	71.25	75	78.75	3.3	175	2000	0.25	5	56
1N4762A	77.9	82	86.1	3	200	3000	0.25	5	62.2
1N4763A	86.45	91	95.55	2.8	250	3000	0.25	5	69.2
1N4764A	95	100	105	2.5	350	3000	0.25	5	76

Notes:

1. Zener Voltage (V_Z)
   The zener voltage is measured with the device junction in the thermal equilibrium at the lead temperature (T_L) at 30°C ± 1°C and 3/8" lead length.

1N4728A - 1N4764A Rev. G2

2N3903, 2N3904

2N3903 is a Preferred Device

General Purpose Transistors

NPN Silicon

Features

- Pb−Free Packages are Available*

MAXIMUM RATINGS

Rating	Symbol	Value	Unit
Collector−Emitter Voltage	V_{CEO}	40	Vdc
Collector−Base Voltage	V_{CBO}	60	Vdc
Emitter−Base Voltage	V_{EBO}	6.0	Vdc
Collector Current − Continuous	I_C	200	mAdc
Total Device Dissipation @ T_A = 25°C Derate above 25°C	P_D	625 5.0	mW mW/°C
Total Device Dissipation @ T_C = 25°C Derate above 25°C	P_D	1.5 12	W mW/°C
Operating and Storage Junction Temperature Range	T_J, T_{stg}	−55 to +150	°C

Maximum ratings are those values beyond which device damage can occur. Maximum ratings applied to the device are individual stress limit values (not normal operating conditions) and are not valid simultaneously. If these limits are exceeded, device functional operation is not implied, damage may occur and reliability may be affected.

THERMAL CHARACTERISTICS (Note 1)

Characteristic	Symbol	Max	Unit
Thermal Resistance, Junction−to−Ambient	$R_{\theta JA}$	200	°C/W
Thermal Resistance, Junction−to−Case	$R_{\theta JC}$	83.3	°C/W

1. Indicates Data in addition to JEDEC Requirements.

ON Semiconductor®

http://onsemi.com

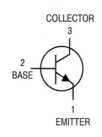

COLLECTOR
3

2
BASE

1
EMITTER

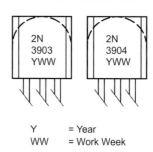

TO−92
CASE 29
STYLE 1

MARKING DIAGRAMS

```
2N
3903
YWW
```

```
2N
3904
YWW
```

Y = Year
WW = Work Week

ORDERING INFORMATION

See detailed ordering and shipping information in the package dimensions section on page 3 of this data sheet.

Preferred devices are recommended choices for future use and best overall value.

*For additional information on our Pb−Free strategy and soldering details, please download the ON Semiconductor Soldering and Mounting Techniques Reference Manual, SOLDERRM/D.

June, 2004 − Rev. 5

Publication Order Number:
2N3903/D

1

2N3903, 2N3904

ELECTRICAL CHARACTERISTICS (T_A = 25°C unless otherwise noted)

Characteristic	Symbol	Min	Max	Unit
OFF CHARACTERISTICS				
Collector–Emitter Breakdown Voltage (Note 2) (I_C = 1.0 mAdc, I_B = 0)	$V_{(BR)CEO}$	40	–	Vdc
Collector–Base Breakdown Voltage (I_C = 10 μAdc, I_E = 0)	$V_{(BR)CBO}$	60	–	Vdc
Emitter–Base Breakdown Voltage (I_E = 10 μAdc, I_C = 0)	$V_{(BR)EBO}$	6.0	–	Vdc
Base Cutoff Current (V_{CE} = 30 Vdc, V_{EB} = 3.0 Vdc)	I_{BL}	–	50	nAdc
Collector Cutoff Current (V_{CE} = 30 Vdc, V_{EB} = 3.0 Vdc)	I_{CEX}	–	50	nAdc
ON CHARACTERISTICS				
DC Current Gain (Note 2)	h_{FE}			–
(I_C = 0.1 mAdc, V_{CE} = 1.0 Vdc) 2N3903		20	–	
2N3904		40	–	
(I_C = 1.0 mAdc, V_{CE} = 1.0 Vdc) 2N3903		35	–	
2N3904		70	–	
(I_C = 10 mAdc, V_{CE} = 1.0 Vdc) 2N3903		50	150	
2N3904		100	300	
(I_C = 50 mAdc, V_{CE} = 1.0 Vdc) 2N3903		30	–	
2N3904		60	–	
(I_C = 100 mAdc, V_{CE} = 1.0 Vdc) 2N3903		15	–	
2N3904		30	–	
Collector–Emitter Saturation Voltage (Note 2)	$V_{CE(sat)}$			Vdc
(I_C = 10 mAdc, I_B = 1.0 mAdc)		–	0.2	
(I_C = 50 mAdc, I_B = 5.0 mAdc		–	0.3	
Base–Emitter Saturation Voltage (Note 2)	$V_{BE(sat)}$			Vdc
(I_C = 10 mAdc, I_B = 1.0 mAdc)		0.65	0.85	
(I_C = 50 mAdc, I_B = 5.0 mAdc)		–	0.95	
SMALL–SIGNAL CHARACTERISTICS				
Current–Gain – Bandwidth Product	f_T			MHz
(I_C = 10 mAdc, V_{CE} = 20 Vdc, f = 100 MHz) 2N3903		250	–	
2N3904		300	–	
Output Capacitance (V_{CB} = 5.0 Vdc, I_E = 0, f = 1.0 MHz)	C_{obo}	–	4.0	pF
Input Capacitance (V_{EB} = 0.5 Vdc, I_C = 0, f = 1.0 MHz)	C_{ibo}	–	8.0	pF
Input Impedance	h_{ie}			kΩ
(I_C = 1.0 mAdc, V_{CE} = 10 Vdc, f = 1.0 kHz) 2N3903		1.0	8.0	
2N3904		1.0	10	
Voltage Feedback Ratio	h_{re}			X 10^{-4}
(I_C = 1.0 mAdc, V_{CE} = 10 Vdc, f = 1.0 kHz) 2N3903		0.1	5.0	
2N3904		0.5	8.0	
Small–Signal Current Gain	h_{fe}			–
(I_C = 1.0 mAdc, V_{CE} = 10 Vdc, f = 1.0 kHz) 2N3903		50	200	
2N3904		100	400	
Output Admittance (I_C = 1.0 mAdc, V_{CE} = 10 Vdc, f = 1.0 kHz)	h_{oe}	1.0	40	μmhos
Noise Figure	NF			dB
(I_C = 100 μAdc, V_{CE} = 5.0 Vdc, R_S = 1.0 kΩ, f = 1.0 kHz) 2N3903		–	6.0	
2N3904		–	5.0	

SWITCHING CHARACTERISTICS

			Symbol	Min	Max	Unit
Delay Time	(V_{CC} = 3.0 Vdc, V_{BE} = 0.5 Vdc,		t_d	–	35	ns
Rise Time	I_C = 10 mAdc, I_{B1} = 1.0 mAdc)		t_r	–	35	ns
Storage Time	(V_{CC} = 3.0 Vdc, I_C = 10 mAdc,	2N3903	t_s	–	175	ns
	I_{B1} = I_{B2} = 1.0 mAdc)	2N3904		–	200	ns
Fall Time			t_f	–	50	ns

2. Pulse Test: Pulse Width ≤ 300 μs; Duty Cycle ≤ 2%.

2N3903, 2N3904

ORDERING INFORMATION

Device	Package	Shipping†
2N3903	TO−92	5,000 Units / Box
2N3903RLRM	TO−92	2,000 / Ammo Pack
2N3904	TO−92	5,000 Units / Box
2N3904G	TO−92 (Pb−Free)	5,000 Units / Box
2N3904RLRA	TO−92	2,000 / Tape & Reel
2N3904RLRAG	TO−92 (Pb−Free)	2,000 / Tape & Reel
2N3904RLRE	TO−92	2,000 / Tape & Reel
2N3904RLRM	TO−92	2,000 / Ammo Pack
2N3904RLRMG	TO−92 (Pb−Free)	2,000 / Ammo Pack
2N3904RLRP	TO−92	2,000 / Ammo Pack
2N3904RLRPG	TO−92 (Pb−Free)	2,000 / Ammo Pack
2N3904RL1	TO−92	2,000 / Tape & Reel
2N3904ZL1	TO−92	2,000 / Ammo Pack

†For information on tape and reel specifications, including part orientation and tape sizes, please refer to our Tape and Reel Packaging Specifications Brochure, BRD8011/D.

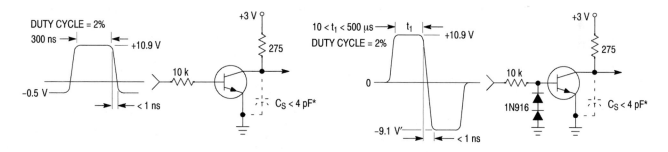

* Total shunt capacitance of test jig and connectors

Figure 1. Delay and Rise Time Equivalent Test Circuit

Figure 2. Storage and Fall Time Equivalent Test Circuit

2N3903, 2N3904

TYPICAL TRANSIENT CHARACTERISTICS

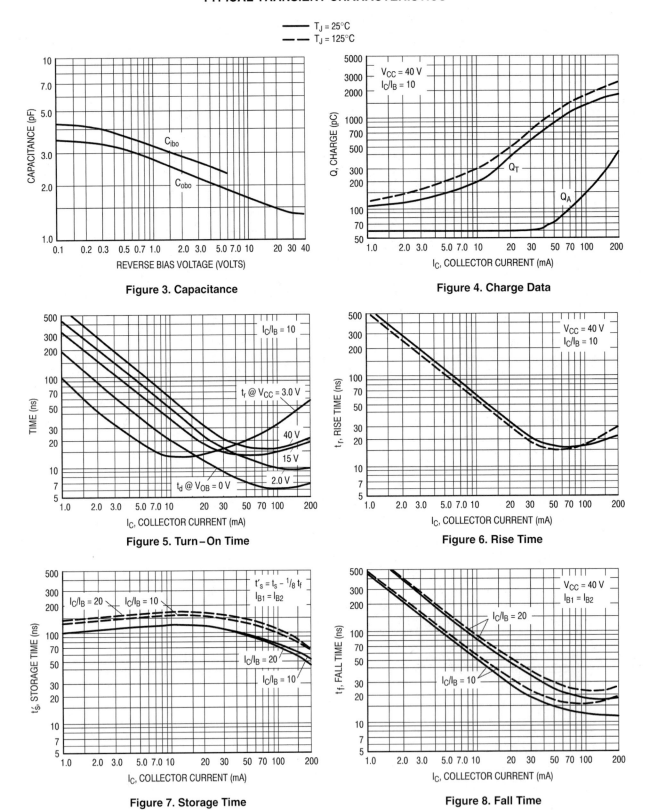

Figure 3. Capacitance

Figure 4. Charge Data

Figure 5. Turn−On Time

Figure 6. Rise Time

Figure 7. Storage Time

Figure 8. Fall Time

TYPICAL AUDIO SMALL-SIGNAL CHARACTERISTICS
NOISE FIGURE VARIATIONS
(V_{CE} = 5.0 Vdc, T_A = 25°C, Bandwidth = 1.0 Hz)

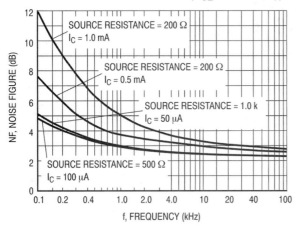

Figure 9.

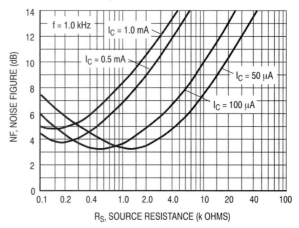

Figure 10.

h PARAMETERS
(V_{CE} = 10 Vdc, f = 1.0 kHz, T_A = 25°C)

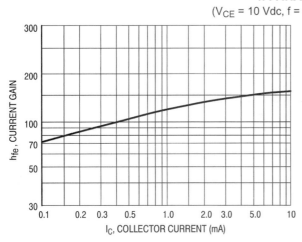

Figure 11. Current Gain

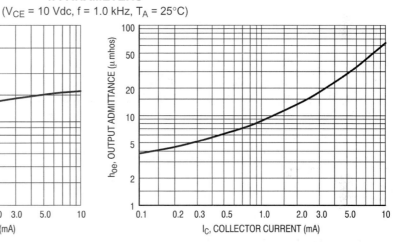

Figure 12. Output Admittance

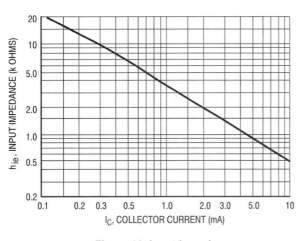

Figure 13. Input Impedance

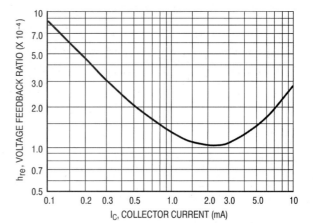

Figure 14. Voltage Feedback Ratio

2N3903, 2N3904

TYPICAL STATIC CHARACTERISTICS

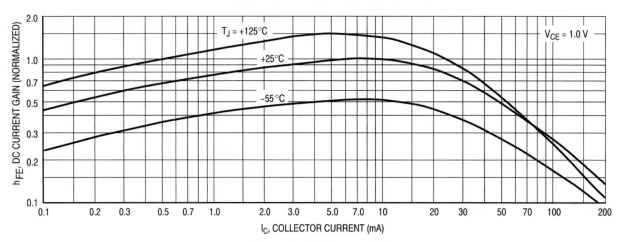

Figure 15. DC Current Gain

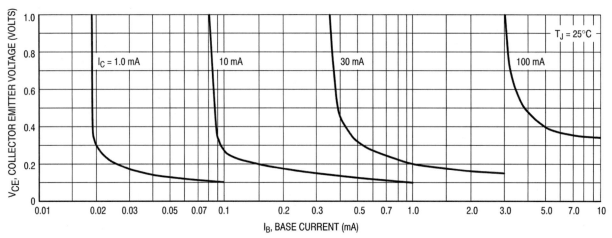

Figure 16. Collector Saturation Region

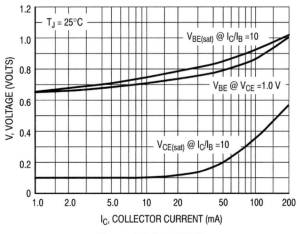

Figure 17. "ON" Voltages

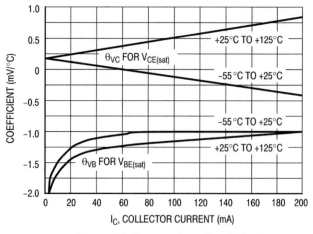

Figure 18. Temperature Coefficients

Appendix A

2N3906

Preferred Device

General Purpose Transistors

PNP Silicon

ON Semiconductor®

http://onsemi.com

Features

- Pb−Free Packages are Available*

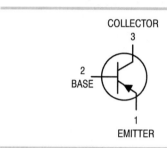

MAXIMUM RATINGS

Rating	Symbol	Value	Unit
Collector − Emitter Voltage	V_{CEO}	40	Vdc
Collector − Base Voltage	V_{CBO}	40	Vdc
Emitter − Base Voltage	V_{EBO}	5.0	Vdc
Collector Current − Continuous	I_C	200	mAdc
Total Device Dissipation @ $T_A = 25°C$ Derate above 25°C	P_D	625 5.0	mW mW/°C
Total Power Dissipation @ $T_A = 60°C$	P_D	250	mW
Total Device Dissipation @ $T_C = 25°C$ Derate above 25°C	P_D	1.5 12	Watts mW/°C
Operating and Storage Junction Temperature Range	T_J, T_{stg}	−55 to +150	°C

Maximum ratings are those values beyond which device damage can occur. Maximum ratings applied to the device are individual stress limit values (not normal operating conditions) and are not valid simultaneously. If these limits are exceeded, device functional operation is not implied, damage may occur and reliability may be affected.

THERMAL CHARACTERISTICS (Note 1)

Characteristic	Symbol	Max	Unit
Thermal Resistance, Junction−to−Ambient	$R_{\theta JA}$	200	°C/W
Thermal Resistance, Junction−to−Case	$R_{\theta JC}$	83.3	°C/W

1. Indicates Data in addition to JEDEC Requirements.

*For additional information on our Pb−Free strategy and soldering details, please download the ON Semiconductor Soldering and Mounting Techniques Reference Manual, SOLDERRM/D.

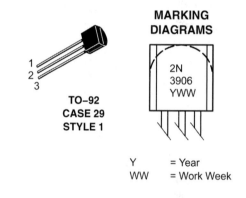

MARKING DIAGRAMS

TO−92
CASE 29
STYLE 1

2N
3906
YWW

Y = Year
WW = Work Week

ORDERING INFORMATION

See detailed ordering and shipping information in the package dimensions section on page 2 of this data sheet.

Preferred devices are recommended choices for future use and best overall value.

June, 2004 − Rev. 2

1

Publication Order Number:
2N3906/D

2N3906

ELECTRICAL CHARACTERISTICS (T_A = 25°C unless otherwise noted)

Characteristic	Symbol	Min	Max	Unit
OFF CHARACTERISTICS				
Collector–Emitter Breakdown Voltage (Note 2) (I_C = 1.0 mAdc, I_B = 0)	$V_{(BR)CEO}$	40	–	Vdc
Collector–Base Breakdown Voltage (I_C = 10 μAdc, I_E = 0)	$V_{(BR)CBO}$	40	–	Vdc
Emitter–Base Breakdown Voltage (I_E = 10 μAdc, I_C = 0)	$V_{(BR)EBO}$	5.0	–	Vdc
Base Cutoff Current (V_{CE} = 30 Vdc, V_{EB} = 3.0 Vdc)	I_{BL}	–	50	nAdc
Collector Cutoff Current (V_{CE} = 30 Vdc, V_{EB} = 3.0 Vdc)	I_{CEX}	–	50	nAdc
ON CHARACTERISTICS (Note 2)				
DC Current Gain	h_{FE}			–
(I_C = 0.1 mAdc, V_{CE} = 1.0 Vdc)		60	–	
(I_C = 1.0 mAdc, V_{CE} = 1.0 Vdc)		80	–	
(I_C = 10 mAdc, V_{CE} = 1.0 Vdc)		100	300	
(I_C = 50 mAdc, V_{CE} = 1.0 Vdc)		60	–	
(I_C = 100 mAdc, V_{CE} = 1.0 Vdc)		30	–	
Collector–Emitter Saturation Voltage	$V_{CE(sat)}$			Vdc
(I_C = 10 mAdc, I_B = 1.0 mAdc)		–	0.25	
(I_C = 50 mAdc, I_B = 5.0 mAdc)		–	0.4	
Base–Emitter Saturation Voltage	$V_{BE(sat)}$			Vdc
(I_C = 10 mAdc, I_B = 1.0 mAdc)		0.65	0.85	
(I_C = 50 mAdc, I_B = 5.0 mAdc)		–	0.95	
SMALL–SIGNAL CHARACTERISTICS				
Current–Gain–Bandwidth Product (I_C = 10 mAdc, V_{CE} = 20 Vdc, f = 100 MHz)	f_T	250	–	MHz
Output Capacitance (V_{CB} = 5.0 Vdc, I_E = 0, f = 1.0 MHz)	C_{obo}	–	4.5	pF
Input Capacitance (V_{EB} = 0.5 Vdc, I_C = 0, f = 1.0 MHz)	C_{ibo}	–	10	pF
Input Impedance (I_C = 1.0 mAdc, V_{CE} = 10 Vdc, f = 1.0 kHz)	h_{ie}	2.0	12	kΩ
Voltage Feedback Ratio (I_C = 1.0 mAdc, V_{CE} = 10 Vdc, f = 1.0 kHz)	h_{re}	0.1	10	X 10^{-4}
Small–Signal Current Gain (I_C = 1.0 mAdc, V_{CE} = 10 Vdc, f = 1.0 kHz)	h_{fe}	100	400	–
Output Admittance (I_C = 1.0 mAdc, V_{CE} = 10 Vdc, f = 1.0 kHz)	h_{oe}	3.0	60	μmhos
Noise Figure (I_C = 100 μAdc, V_{CE} = 5.0 Vdc, R_S = 1.0 kΩ, f = 1.0 kHz)	NF	–	4.0	dB

SWITCHING CHARACTERISTICS

		Symbol	Min	Max	Unit
Delay Time	(V_{CC} = 3.0 Vdc, V_{BE} = 0.5 Vdc,	t_d	–	35	ns
Rise Time	I_C = 10 mAdc, I_{B1} = 1.0 mAdc)	t_r	–	35	ns
Storage Time	(V_{CC} = 3.0 Vdc, I_C = 10 mAdc, I_{B1} = I_{B2} = 1.0 mAdc)	t_s	–	225	ns
Fall Time	(V_{CC} = 3.0 Vdc, I_C = 10 mAdc, I_{B1} = I_{B2} = 1.0 mAdc)	t_f	–	75	ns

2. Pulse Test: Pulse Width ≤ 300 μs; Duty Cycle ≤ 2%.

ORDERING INFORMATION

Device	Package	Shipping†
2N3906	TO–92	5,000 Units / Box
2N3906G	TO–92 (Pb–Free)	5,000 Units / Box
2N3906RL1	TO–92	5,000 Units / Box
2N3906RLRA	TO–92	2,000 / Tape & Reel
2N3906RLRAG	TO–92 (Pb–Free)	2,000 / Tape & Reel
2N3906RLRM	TO–92	2,000 / Ammo Pack
2N3906RLRMG	TO–92 (Pb–Free)	2,000 / Ammo Pack
2N3906RLRP	TO–92	2,000 / Tape & Reel
2N3906ZL1	TO–92	2,000 / Ammo Pack

†For information on tape and reel specifications, including part orientation and tape sizes, please refer to our Tape and Reel Packaging Specifications Brochure, BRD8011/D.

SEMICONDUCTOR®

TIP100/101/102

Monolithic Construction With Built In Base-Emitter Shunt Resistors

- High DC Current Gain : h_{FE}=1000 @ V_{CE}=4V, I_C=3A (Min.)
- Collector-Emitter Sustaining Voltage
- Low Collector-Emitter Saturation Voltage
- Industrial Use
- Complementary to TIP105/106/107

TO-220

1.Base   2.Collector   3.Emitter

NPN Epitaxial Silicon Darlington Transistor

Absolute Maximum Ratings T_C=25°C unless otherwise noted

Symbol	Parameter		Value	Units
V_{CBO}	Collector-Base Voltage	: TIP100	60	V
		: TIP101	80	V
		:TIP102	100	V
V_{CEO}	Collector-Emitter Voltage	: TIP100	60	V
		: TIP101	80	V
		: TIP102	100	V
V_{EBO}	Emitter-Base Voltage		5	V
I_C	Collector Current (DC)		8	A
I_{CP}	Collector Current (Pulse)		15	A
I_B	Base Current (DC)		1	A
P_C	Collector Dissipation (T_a=25°C)		2	W
	Collector Dissipation (T_C=25°C)		80	W
T_J	Junction Temperature		150	°C
T_{STG}	Storage Temperature		- 65 ~ 150	°C

Equivalent Circuit

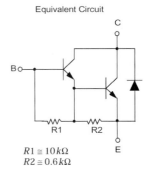

$R1 \cong 10\,k\Omega$
$R2 \cong 0.6\,k\Omega$

Electrical Characteristics T_C=25°C unless otherwise noted

Symbol	Parameter		Test Condition	Min.	Max.	Units
V_{CEO}(sus)	Collector-Emitter Sustaining Voltage					
		: TIP100	I_C = 30mA, I_B = 0	60		V
		: TIP101		80		V
		: TIP102		100		V
I_{CEO}	Collector Cut-off Current					
		: TIP100	V_{CE} = 30V, I_B = 0		50	μA
		: TIP101	V_{CE} = 40V, I_B = 0		50	μA
		: TIP102	V_{CE} = 50V, I_B = 0		50	μA
I_{CBO}	Collector Cut-off Current					
		: TIP100	V_{CE} = 60V, I_E = 0		50	μA
		: TIP101	V_{CE} = 80V, I_E = 0		50	μA
		: TIP102	V_{CE} = 100V, I_E = 0		50	μA
I_{EBO}	Emitter Cut-off Current		V_{EB} = 5V, I_C = 0		2	mA
h_{FE}	DC Current Gain		V_{CE} = 4V, I_C = 3A	1000	20000	
			V_{CE} = 4V, I_C = 8A	200		
V_{CE}(sat)	Collector-Emitter Saturation Voltage		I_C = 3A, I_B = 6mA		2	V
			I_C = 8A, I_B = 80mA		2.5	V
V_{BE}(on)	Base-Emitter ON Voltage		V_{CE} = 4V, I_C = 8A		2.8	V
C_{ob}	Output Capacitance		V_{CB} = 10V, I_E = 0, f = 0.1MHz		200	pF

Rev. A1, June 2001

Typical Characteristics

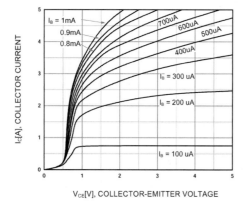

Figure 1. Static Characteristic

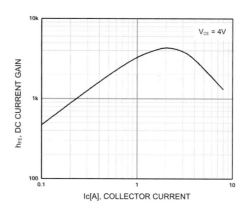

Figure 2. DC current Gain

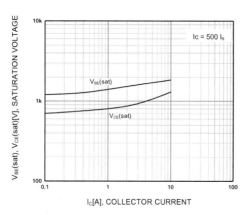

Figure 3. Collector-Emitter Saturation Voltage
Base-Emitter Saturation Voltage

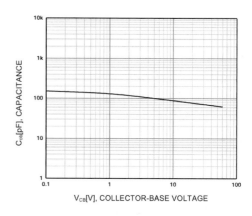

Figure 4. Collector Output Capacitance

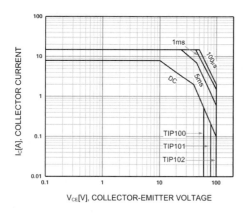

Figure 5. Safe Operating Area

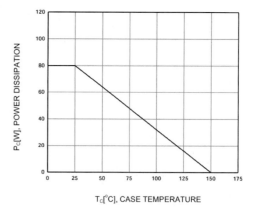

Figure 6. Power Derating

Rev. A1, June 2001

Appendix A

JFET VHF Amplifier
N–Channel – Depletion

MPF102

CASE 29–11, STYLE 5
TO–92 (TO–226AA)

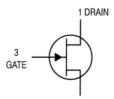

MAXIMUM RATINGS

Rating	Symbol	Value	Unit
Drain–Source Voltage	V_{DS}	25	Vdc
Drain–Gate Voltage	V_{DG}	25	Vdc
Gate–Source Voltage	V_{GS}	−25	Vdc
Gate Current	I_G	10	mAdc
Total Device Dissipation @ T_A = 25°C Derate above 25°C	P_D	350 2.8	mW mW/°C
Junction Temperature Range	T_J	125	°C
Storage Temperature Range	T_{stg}	−65 to +150	°C

ELECTRICAL CHARACTERISTICS (T_A = 25°C unless otherwise noted)

Characteristic	Symbol	Min	Max	Unit
OFF CHARACTERISTICS				
Gate–Source Breakdown Voltage (I_G = −10 µAdc, V_{DS} = 0)	$V_{(BR)GSS}$	−25	–	Vdc
Gate Reverse Current (V_{GS} = −15 Vdc, V_{DS} = 0) (V_{GS} = −15 Vdc, V_{DS} = 0, T_A = 100°C)	I_{GSS}	– –	−2.0 −2.0	nAdc µAdc
Gate–Source Cutoff Voltage (V_{DS} = 15 Vdc, I_D = 2.0 nAdc)	$V_{GS(off)}$	–	−8.0	Vdc
Gate–Source Voltage (V_{DS} = 15 Vdc, I_D = 0.2 mAdc)	V_{GS}	−0.5	−7.5	Vdc
ON CHARACTERISTICS				
Zero–Gate–Voltage Drain Current[1] (V_{DS} = 15 Vdc, V_{GS} = 0 Vdc)	I_{DSS}	2.0	20	mAdc
SMALL–SIGNAL CHARACTERISTICS				
Forward Transfer Admittance[1] (V_{DS} = 15 Vdc, V_{GS} = 0, f = 1.0 kHz) (V_{DS} = 15 Vdc, V_{GS} = 0, f = 100 MHz)	$\|y_{fs}\|$	 2000 1600	 7500 –	µmhos
Input Admittance (V_{DS} = 15 Vdc, V_{GS} = 0, f = 100 MHz)	$Re(y_{is})$	–	800	µmhos
Output Conductance (V_{DS} = 15 Vdc, V_{GS} = 0, f = 100 MHz)	$Re(y_{os})$	–	200	µmhos
Input Capacitance (V_{DS} = 15 Vdc, V_{GS} = 0, f = 1.0 MHz)	C_{iss}	–	7.0	pF
Reverse Transfer Capacitance (V_{DS} = 15 Vdc, V_{GS} = 0, f = 1.0 MHz)	C_{rss}	–	3.0	pF

1. Pulse Test; Pulse Width ≤ 630 ms, Duty Cycle ≤ 10%.

© Semiconductor Components Industries, LLC, 2001
November, 2001 – Rev. 2

1

Publication Order Number:
MPF102/D

COMMON SOURCE CHARACTERISTICS
ADMITTANCE PARAMETERS
(V_{DS} = 15 Vdc, $T_{channel}$ = 25°C)

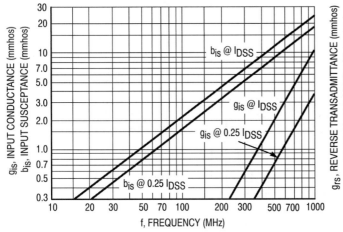

Figure 1. Input Admittance (y_{is})

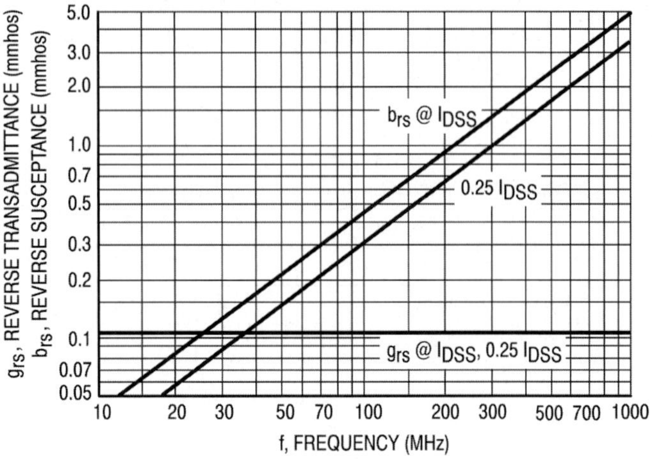

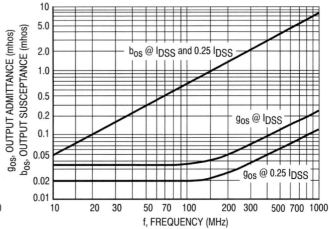

Figure 2. Reverse Transfer Admittance (y_{rs})

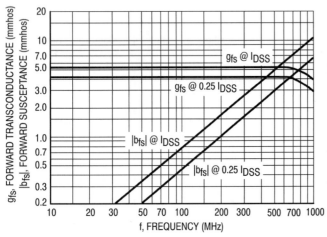

Figure 3. Forward Transadmittance (y_{fs})

Figure 4. Output Admittance (y_{os})

November 1995

2N7000 / 2N7002 / NDS7002A
N-Channel Enhancement Mode Field Effect Transistor

General Description

These N-Channel enhancement mode field effect transistors are produced using Fairchild's proprietary, high cell density, DMOS technology. These products have been designed to minimize on-state resistance while provide rugged, reliable, and fast switching performance. They can be used in most applications requiring up to 400mA DC and can deliver pulsed currents up to 2A. These products are particularly suited for low voltage, low current applications such as small servo motor control, power MOSFET gate drivers, and other switching applications.

Features

- High density cell design for low $R_{DS(ON)}$.
- Voltage controlled small signal switch.
- Rugged and reliable.
- High saturation current capability.

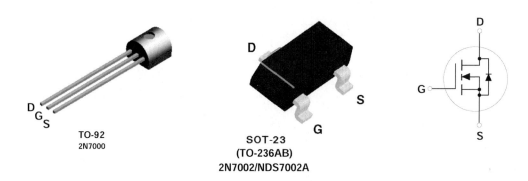

TO-92
2N7000

SOT-23
(TO-236AB)
2N7002/NDS7002A

Absolute Maximum Ratings T_A = 25°C unless otherwise noted

Symbol	Parameter	2N7000	2N7002	NDS7002A	Units
V_{DSS}	Drain-Source Voltage		60		V
V_{DGR}	Drain-Gate Voltage ($R_{GS} \leq 1$ MΩ)		60		V
V_{GSS}	Gate-Source Voltage - Continuous		±20		V
	- Non Repetitive (tp < 50µs)		±40		
I_D	Maximum Drain Current - Continuous	200	115	280	mA
	- Pulsed	500	800	1500	
P_D	Maximum Power Dissipation	400	200	300	mW
	Derated above 25°C	3.2	1.6	2.4	mW/°C
T_J, T_{STG}	Operating and Storage Temperature Range	-55 to 150		-65 to 150	°C
T_L	Maximum Lead Temperature for Soldering Purposes, 1/16" from Case for 10 Seconds		300		°C
THERMAL CHARACTERISTICS					
$R_{\theta JA}$	Thermal Resistance, Junction-to-Ambient	312.5	625	417	°C/W

2N7000.SAM Rev. A1

Electrical Characteristics T_A = 25°C unless otherwise noted

Symbol	Parameter	Conditions		Type	Min	Typ	Max	Units
OFF CHARACTERISTICS								
BV_{DSS}	Drain-Source Breakdown Voltage	V_{GS} = 0 V, I_D = 10 µA		All	60			V
I_{DSS}	Zero Gate Voltage Drain Current	V_{DS} = 48 V, V_{GS} = 0 V		2N7000			1	µA
			T_J=125°C				1	mA
		V_{DS} = 60 V, V_{GS} = 0 V		2N7002 NDS7002A			1	µA
			T_J=125°C				0.5	mA
I_{GSSF}	Gate - Body Leakage, Forward	V_{GS} = 15 V, V_{DS} = 0 V		2N7000			10	nA
		V_{GS} = 20 V, V_{DS} = 0 V		2N7002 NDS7002A			100	nA
I_{GSSR}	Gate - Body Leakage, Reverse	V_{GS} = -15 V, V_{DS} = 0 V		2N7000			-10	nA
		V_{GS} = -20 V, V_{DS} = 0 V		2N7002 NDS7002A			-100	nA
ON CHARACTERISTICS (Note 1)								
$V_{GS(th)}$	Gate Threshold Voltage	V_{DS} = V_{GS}, I_D = 1 mA		2N7000	0.8	2.1	3	V
		V_{DS} = V_{GS}, I_D = 250 µA		2N7002 NDS7002A	1	2.1	2.5	
$R_{DS(ON)}$	Static Drain-Source On-Resistance	V_{GS} = 10 V, I_D = 500 mA		2N7000		1.2	5	Ω
			T_J =125°C			1.9	9	
		V_{GS} = 4.5 V, I_D = 75 mA				1.8	5.3	
		V_{GS} = 10 V, I_D = 500 mA		2N7002		1.2	7.5	
			T_J =100°C			1.7	13.5	
		V_{GS} = 5.0 V, I_D = 50 mA				1.7	7.5	
			T_J =100C			2.4	13.5	
		V_{GS} = 10 V, I_D = 500 mA		NDS7002A		1.2	2	
			T_J =125°C			2	3.5	
		V_{GS} = 5.0 V, I_D = 50 mA				1.7	3	
			T_J =125°C			2.8	5	
$V_{DS(ON)}$	Drain-Source On-Voltage	V_{GS} = 10 V, I_D = 500 mA		2N7000		0.6	2.5	V
		V_{GS} = 4.5 V, I_D = 75 mA				0.14	0.4	
		V_{GS} = 10 V, I_D = 500mA		2N7002		0.6	3.75	
		V_{GS} = 5.0 V, I_D = 50 mA				0.09	1.5	
		V_{GS} = 10 V, I_D = 500mA		NDS7002A		0.6	1	
		V_{GS} = 5.0 V, I_D = 50 mA				0.09	0.15	

Electrical Characteristics T_A = 25°C unless otherwise noted

Symbol	Parameter	Conditions	Type	Min	Typ	Max	Units
ON CHARACTERISTICS Continued (Note 1)							
$I_{D(ON)}$	On-State Drain Current	V_{GS} = 4.5 V, V_{DS} = 10 V	2N7000	75	600		mA
		V_{GS} = 10 V, $V_{DS} \geq$ 2 $V_{DS(on)}$	2N7002	500	2700		
		V_{GS} = 10 V, $V_{DS} \geq$ 2 $V_{DS(on)}$	NDS7002A	500	2700		
g_{FS}	Forward Transconductance	V_{DS} = 10 V, I_D = 200 mA	2N7000	100	320		mS
		$V_{DS} \geq$ 2 $V_{DS(on)}$, I_D = 200 mA	2N7002	80	320		
		$V_{DS} \geq$ 2 $V_{DS(on)}$, I_D = 200 mA	NDS7002A	80	320		
DYNAMIC CHARACTERISTICS							
C_{iss}	Input Capacitance	V_{DS} = 25 V, V_{GS} = 0 V, f = 1.0 MHz	All		20	50	pF
C_{oss}	Output Capacitance		All		11	25	pF
C_{rss}	Reverse Transfer Capacitance		All		4	5	pF
t_{on}	Turn-On Time	V_{DD} = 15 V, R_L = 25 Ω, I_D = 500 mA, V_{GS} = 10 V, R_{GEN} = 25	2N7000			10	ns
		V_{DD} = 30 V, R_L = 150 Ω, I_D = 200 mA, V_{GS} = 10 V, R_{GEN} = 25 Ω	2N700 NDS7002A			20	
t_{off}	Turn-Off Time	V_{DD} = 15 V, R_L = 25 Ω, I_D = 500 mA, V_{GS} = 10 V, R_{GEN} = 25	2N7000			10	ns
		V_{DD} = 30 V, R_L = 150 Ω, I_D = 200 mA, V_{GS} = 10 V, R_{GEN} = 25 Ω	2N700 NDS7002A			20	
DRAIN-SOURCE DIODE CHARACTERISTICS AND MAXIMUM RATINGS							
I_S	Maximum Continuous Drain-Source Diode Forward Current		2N7002			115	mA
			NDS7002A			280	
I_{SM}	Maximum Pulsed Drain-Source Diode Forward Current		2N7002			0.8	A
			NDS7002A			1.5	
V_{SD}	Drain-Source Diode Forward Voltage	V_{GS} = 0 V, I_S = 115 mA (Note 1)	2N7002		0.88	1.5	V
		V_{GS} = 0 V, I_S = 400 mA (Note 1)	NDS7002A		0.88	1.2	

Note:
1. Pulse Test: Pulse Width ≤ 300µs, Duty Cycle ≤ 2.0%.

2N7000.SAM Rev. A1

Typical Electrical Characteristics

2N7000 / 2N7002 / NDS7002A

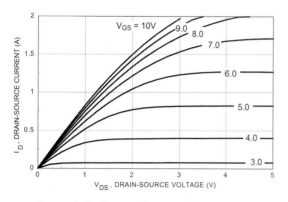

Figure 1. On-Region Characteristics

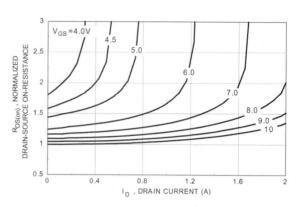

Figure 2. On-Resistance Variation with Gate
Voltage and Drain Current

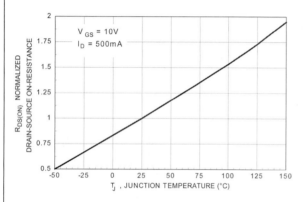

Figure 3. On-Resistance Variation
with Temperature

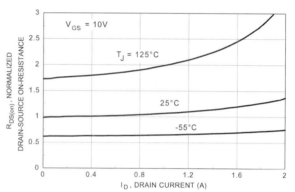

Figure 4. On-Resistance Variation with Drain
Current and Temperature

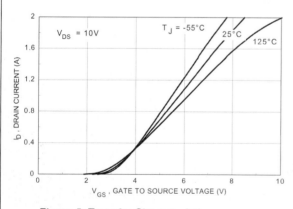

Figure 5. Transfer Characteristics

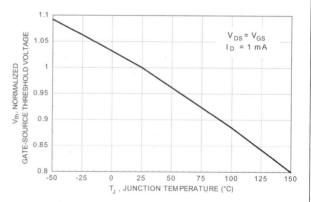

Figure 6. Gate Threshold Variation with
Temperature

2N7000.SAM Rev. A1

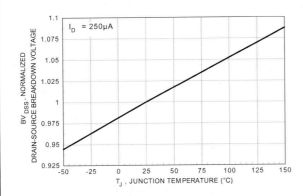

Figure 7. Breakdown Voltage Variation
with Temperature

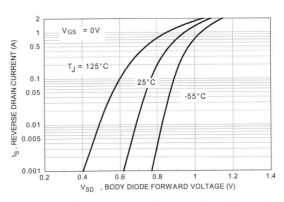

Figure 8. Body Diode Forward Voltage Variation with

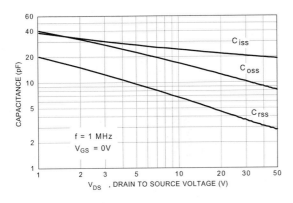

Figure 9. Capacitance Characteristics

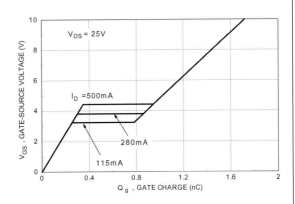

Figure 10. Gate Charge Characteristics

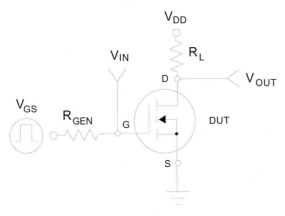

Figure 11.

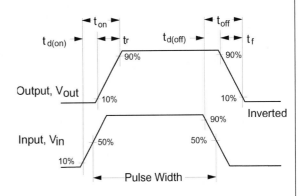

Figure 12. Switching Waveforms

2N7000.SAM Rev. A1

2N6504 Series

Preferred Device

Silicon Controlled Rectifiers

Reverse Blocking Thyristors

Designed primarily for half-wave ac control applications, such as motor controls, heating controls and power supply crowbar circuits.

Features

- Glass Passivated Junctions with Center Gate Fire for Greater Parameter Uniformity and Stability
- Small, Rugged, Thermowatt Constructed for Low Thermal Resistance, High Heat Dissipation and Durability
- Blocking Voltage to 800 Volts
- 300 A Surge Current Capability
- Pb−Free Packages are Available*

ON Semiconductor®

http://onsemi.com

SCRs
25 AMPERES RMS
50 thru 800 VOLTS

MARKING DIAGRAM

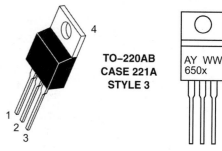

TO−220AB
CASE 221A
STYLE 3

AY WW
650x

x = 4, 5, 7, 8 or 9
A = Assembly Location
Y = Year
WW = Work Week

PIN ASSIGNMENT	
1	Cathode
2	Anode
3	Gate
4	Anode

ORDERING INFORMATION

See detailed ordering and shipping information in the package dimensions section on page 6 of this data sheet.

Preferred devices are recommended choices for future use and best overall value.

*For additional information on our Pb−Free strategy and soldering details, please download the ON Semiconductor Soldering and Mounting Techniques Reference Manual, SOLDERRM/D.

December, 2004 – Rev. 5

1

Publication Order Number:
2N6504/D

Appendix A

MAXIMUM RATINGS (T_J = 25°C unless otherwise noted)

Rating	Symbol	Value	Unit
*Peak Repetitive Off−State Voltage (Note 1) (Gate Open, Sine Wave 50 to 60 Hz, T_J = 25 to 125°C) 2N6504 2N6505 2N6507 2N6508 2N6509	V_{DRM}, V_{RRM}	 50 100 400 600 800	V
On−State Current RMS (180° Conduction Angles; T_C = 85°C)	$I_{T(RMS)}$	25	A
Average On−State Current (180° Conduction Angles; T_C = 85°C)	$I_{T(AV)}$	16	A
Peak Non-repetitive Surge Current (1/2 Cycle, Sine Wave 60 Hz, T_J = 100°C)	I_{TSM}	250	A
Forward Peak Gate Power (Pulse Width ≤ 1.0 μs, T_C = 85°C)	P_{GM}	20	W
Forward Average Gate Power (t = 8.3 ms, T_C = 85°C)	$P_{G(AV)}$	0.5	W
Forward Peak Gate Current (Pulse Width ≤ 1.0 μs, T_C = 85°C)	I_{GM}	2.0	A
Operating Junction Temperature Range	T_J	−40 to +125	°C
Storage Temperature Range	T_{stg}	−40 to +150	°C

Maximum ratings are those values beyond which device damage can occur. Maximum ratings applied to the device are individual stress limit values (not normal operating conditions) and are not valid simultaneously. If these limits are exceeded, device functional operation is not implied, damage may occur and reliability may be affected.

1. V_{DRM} and V_{RRM} for all types can be applied on a continuous basis. Ratings apply for zero or negative gate voltage; however, positive gate voltage shall not be applied concurrent with negative potential on the anode. Blocking voltages shall not be tested with a constant current source such that the voltage ratings of the devices are exceeded.

THERMAL CHARACTERISTICS

Characteristic	Symbol	Max	Unit
*Thermal Resistance, Junction−to−Case	$R_{\theta JC}$	1.5	°C/W
*Maximum Lead Temperature for Soldering Purposes 1/8 in from Case for 10 Seconds	T_L	260	°C

ELECTRICAL CHARACTERISTICS (T_C = 25°C unless otherwise noted.)

Characteristic	Symbol	Min	Typ	Max	Unit
OFF CHARACTERISTICS					
*Peak Repetitive Forward or Reverse Blocking Current (V_{AK} = Rated V_{DRM} or V_{RRM}, Gate Open) T_J = 25°C T_J = 125°C	I_{DRM}, I_{RRM}	– –	– –	10 2.0	μA mA
ON CHARACTERISTICS					
*Forward On−State Voltage (Note 2) (I_{TM} = 50 A)	V_{TM}	–	–	1.8	V
*Gate Trigger Current (Continuous dc) T_C = 25°C (V_{AK} = 12 Vdc, R_L = 100 Ω) T_C = −40°C	I_{GT}	– –	9.0 –	30 75	mA
*Gate Trigger Voltage (Continuous dc) (V_{AK} = 12 Vdc, R_L = 100 Ω, T_C = −40°C)	V_{GT}	–	1.0	1.5	V
Gate Non-Trigger Voltage (V_{AK} = 12 Vdc, R_L = 100 Ω, T_J = 125°C)	V_{GD}	0.2	–	–	V
*Holding Current T_C = 25°C (V_{AK} = 12 Vdc, Initiating Current = 200 mA, Gate Open) T_C = −40°C	I_H	– –	18 –	40 80	mA
*Turn-On Time (I_{TM} = 25 A, I_{GT} = 50 mAdc)	t_{gt}	–	1.5	2.0	μs
Turn-Off Time (V_{DRM} = rated voltage) (I_{TM} = 25 A, I_R = 25 A) (I_{TM} = 25 A, I_R = 25 A, T_J = 125°C)	t_q	 – –	 15 35	 – –	μs
DYNAMIC CHARACTERISTICS					
Critical Rate of Rise of Off-State Voltage (Gate Open, Rated V_{DRM}, Exponential Waveform)	dv/dt	–	50	–	V/μs

*Indicates JEDEC Registered Data.

2. Pulse Test: Pulse Width ≤ 300 μs, Duty Cycle ≤ 2%.

2N6504 Series

Voltage Current Characteristic of SCR

Symbol	Parameter
V_{DRM}	Peak Repetitive Off State Forward Voltage
I_{DRM}	Peak Forward Blocking Current
V_{RRM}	Peak Repetitive Off State Reverse Voltage
I_{RRM}	Peak Reverse Blocking Current
V_{TM}	Peak On State Voltage
I_H	Holding Current

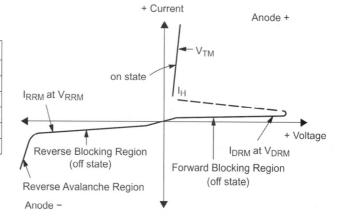

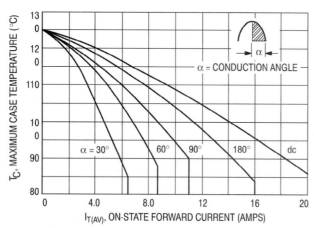

Figure 1. Average Current Derating

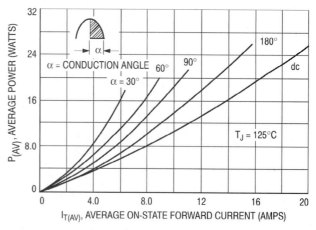

Figure 2. Maximum On–State Power Dissipation

ORDERING INFORMATION

Device	Package	Shipping†
2N6404	TO–220AB	
2N6405	TO–220AB	
2N6405T	TO–220AB	
2N6407	TO–220AB	
2N6407T	TO–220AB	
2N6407TG	TO–220AB (Pb–Free)	
2N6408	TO–220AB	500 Units / Box
2N6408G	TO–220AB (Pb–Free)	
2N6409	TO–220AB	
2N6409G	TO–220AB (Pb–Free)	
2N6409T	TO–220AB	

†For information on tape and reel specification, including part orientation and tape sizes, please refer to our Tape and Reel Packaging Specifications Brochure, BRD8011/D.

FKPF8N80

Application Explanation
- Switching mode power supply, light dimmer, electric flasher unit, hair drier
- TV sets, stereo, refrigerator, washing machine
- Electric blanket, solenoid driver, small motor control
- Photo copier, electric tool

TO-220F

1 2 3

2
o

o 3

1
o

1: T_1
2: T_2
3: Gate

Bi-Directional Triode Thyristor Planar Silicon

Absolute Maximum Ratings T_C=25°C unless otherwise noted

Symbol	Parameter	Rating	Units
V_{DRM}	Repetitive Peak Off-State Voltage (Note1)	800	V

Symbol	Parameter	Conditions		Rating	Units
$I_{T (RMS)}$	RMS On-State Current	Commercial frequency, sine full wave 360° conduction, T_C=91°C		8	A
I_{TSM}	Surge On-State Current	Sinewave 1 full cycle, peak value, non-repetitive	50Hz	80	A
			60Hz	88	A
I^2t	I^2t for Fusing	Value corresponding to 1 cycle of halfwave, surge on-state current, tp=10ms		32	A^2s
di/dt	Critical Rate of Rise of On-State Current	I_G = 2x I_{GT}, tr ≤ 100ns		50	A/μs
P_{GM}	Peak Gate Power Dissipation			5	W
$P_{G (AV)}$	Average Gate Power Dissipation			0.5	W
V_{GM}	Peak Gate Voltage			10	V
I_{GM}	Peak Gate Current			2	A
T_J	Junction Temperature			- 40 ~ 125	°C
T_{STG}	Storage Temperature			- 40 ~ 125	°C
V_{iso}	Isolation Voltage	Ta=25°C, AC 1 minute, T_1 T_2 G terminal to case		1500	V

Thermal Characteristic

Symbol	Parameter	Test Condition	Min.	Typ.	Max.	Units
$R_{th(J-C)}$	Thermal Resistance	Junction to case (Note 4)	-	-	3.6	°C/W

Rev. B1, April 2004

Electrical Characteristics T_C=25°C unless otherwise noted

Symbol	Parameter		Test Condition		Min.	Typ.	Max.	Units
I_{DRM}	Repetieive Peak Off-State Current		V_{DRM} applied		-	-	20	μA
V_{TM}	On-State Voltage		T_C=25°C, I_{TM}=12A Instantaneous measurement		-	-	1.5	V
V_{GT}	Gate Trigger Voltage (Note 2)	I	V_D=12V, R_L=20Ω	T2(+), Gate (+)	-	-	1.5	V
		II		T2(+), Gate (-)	-	-	1.5	V
		III		T2(-), Gate (-)	-	-	1.5	V
I_{GT}	Gate Trigger Current (Note 2)	I	V_D=12V, R_L=20Ω	T2(+), Gate (+)	-	-	30	mA
		II		T2(+), Gate (-)	-	-	30	mA
		III		T2(-), Gate (-)	-	-	30	mA
V_{GD}	Gate Non-Trigger Voltage		T_J=125°C, V_D=1/2V_{DRM}		0.2	-	-	V
I_H	Holding Current		V_D = 12V, I_{TM} = 1A				50	mA
I_L	Latching Current	I, III	V_D = 12V, I_G = 1.2I_{GT}				50	mA
		II					70	mA
dv/dt	Critical Rate of Rise of Off-State Voltag		V_{DRM} = Rated, T_j = 125°C, Exponential Rise			300		V/μs
$(dv/dt)_C$	Critical-Rate of Rise of Off-State Commutating Voltage (Note 3)				10	-	-	V/μs

Notes:
1. Gate Open
2. Measurement using the gate trigger characteristics measurement circuit
3. The critical-rate of rise of the off-state commutating voltage is shown in the table below
4. The contact thermal resistance $R_{TH(c-f)}$ in case of greasing is 0.5 °C/W

V_{DRM} (V)	Test Condition	Commutating voltage and current waveforms (inductive load)
FKPF8N80	1. Junction Temperature T_J=125°C 2. Rate of decay of on-state commutating current $(di/dt)_C$ = - 4.5A/ms 3. Peak off-state voltage V_D = 400V	

Quadrant Definitions for a Triac

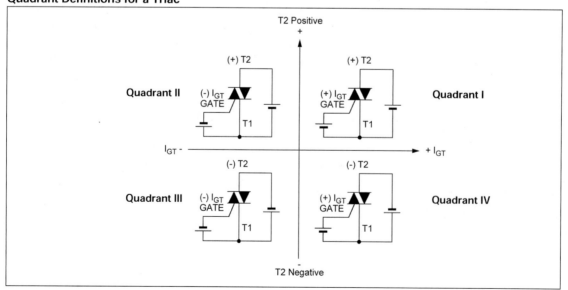

Rev. B1, April 2004

SEMICONDUCTOR®

IGBT

FGL60N100BNTD

NPT-Trench IGBT

General Description

Trench insulated gate bipolar transistors (IGBTs) with NPT technology show outstanding performance in conduction and switching characteristics as well as enhanced avalanche ruggedness. These devices are well suited for Induction Heating (I-H) applications

Features

- High Speed Switching
- Low Saturation Voltage : $V_{CE(sat)}$ = 2.5 V @ I_C = 60A
- High Input Impedance
- Built-in Fast Recovery Diode

Application

Micro- Wave Oven, I-H Cooker, I-H Jar, Induction Heater, Home Appliance

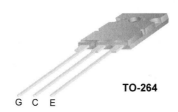

TO-264

G C E

Absolute Maximum Ratings
T_C = 25°C unless otherwise noted

Symbol	Description		FGL60N100BNTD	Units
V_{CES}	Collector-Emitter Voltage		1000	V
V_{GES}	Gate-Emitter Voltage		± 25	V
I_C	Collector Current	@ T_C = 25°C	60	A
	Collector Current	@ T_C = 100°C	42	A
$I_{CM (1)}$	Pulsed Collector Current		120	A
I_F	Diode Continuous Forward Current	@ T_C = 100°C	15	A
P_D	Maximum Power Dissipation	@ T_C = 25°C	180	W
	Maximum Power Dissipation	@ T_C = 100°C	72	W
T_J	Operating Junction Temperature		-55 to +150	°C
T_{stg}	Storage Temperature Range		-55 to +150	°C
T_L	Maximum Lead Temp. for soldering Purposes, 1/8" from case for 5 seconds		300	°C

Notes :
(1) Repetitive rating : Pulse width limited by max. junction temperature

Thermal Characteristics

Symbol	Parameter	Typ.	Max.	Units
$R_{\theta JC}$(IGBT)	Thermal Resistance, Junction-to-Case	--	0.69	°C/W
$R_{\theta JC}$(DIODE)	Thermal Resistance, Junction-to-Case	--	2.08	°C/W
$R_{\theta JA}$	Thermal Resistance, Junction-to-Ambient	--	25	°C/W

FGL60N100BNTD Rev. A

Electrical Characteristics of IGBT T_C = 25°C unless otherwise noted

Symbol	Parameter	Test Conditions	Min.	Typ.	Max.	Units
Off Characteristics						
BV_{CES}	Collector Emitter Breakdown Voltage	V_{GE} = 0V, I_C = 1mA	1000	--	--	V
I_{CES}	Collector Cut-Off Current	V_{CE} = 1000V, V_{GE} = 0V	--	--	1.0	mA
I_{GES}	G-E Leakage Current	V_{GE} = ± 25, V_{CE} = 0V	--	--	± 500	nA
On Characteristics						
$V_{GE(th)}$	G-E Threshold Voltage	I_C = 60mA, V_{CE} = V_{GE}	4.0	5.0	7.0	V
$V_{CE(sat)}$	Collector to Emitter	I_C = 10A, V_{GE} = 15V	--	1.5	1.8	V
	Saturation Voltage	I_C = 60A, V_{GE} = 15V	--	2.5	2.9	V
Dynamic Characteristics						
C_{ies}	Input Capacitance	V_{CE}=10V, V_{GE} = 0V, f = 1MHz	--	6000	--	pF
C_{oes}	Output Capacitance		--	260	--	pF
C_{res}	Reverse Transfer Capacitance		--	200	--	pF
Switching Characteristics						
$t_{d(on)}$	Turn-On Delay Time	V_{CC} = 600 V, I_C = 60A, R_G = 51Ω, V_{GE}=15V, Resistive Load, T_C = 25°C	--	140	--	ns
t_r	Rise Time		--	320	--	ns
$t_{d(off)}$	Turn-Off Delay Time		--	630	--	ns
t_f	Fall Time		--	130	250	ns
Q_g	Total Gate Charge	V_{CE} = 600 V, I_C = 60A, V_{GE} = 15V, , T_C = 25°C	--	275	350	nC
Q_{ge}	Gate-Emitter Charge		--	45	--	nC
Q_{gc}	Gate-Collector Charge		--	95	--	nC

Electrical Characteristics of DIODE T_C = 25°C unless otherwise noted

Symbol	Parameter	Test Conditions	Min.	Typ.	Max.	Units
V_{FM}	Diode Forward Voltage	I_F = 15A	--	1.2	1.7	V
		I_F = 60A	--	1.8	2.1	V
t_{rr}	Diode Reverse Recovery Time	I_F = 60A di/dt = 20 A/us		1.2	1.5	us
I_R	Instantaneous Reverse Current	V_{RRM} = 1000V	--	0.05	2	uA

FGL60N100BNTD Rev. A

Appendix A

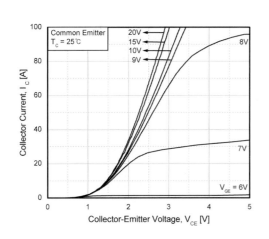

Fig 1. Typical Output Characteristics

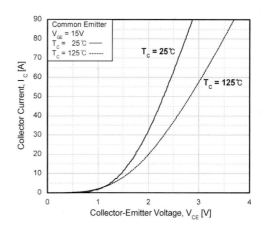

Fig 2. Typical Saturation Voltage Characteristics

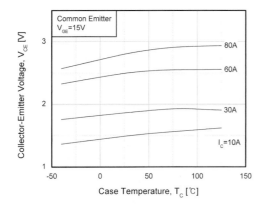

Fig 3. Saturation Voltage vs. Case
Temperature at Varient Current Level

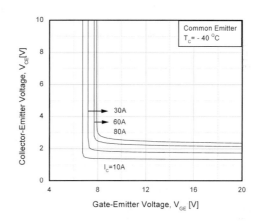

Fig 4. Saturation Voltage vs. V_{GE}

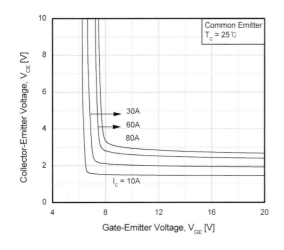

Fig 5. Saturation Voltage vs. V_{GE}

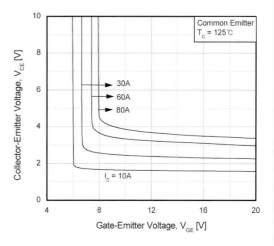

Fig 6. Saturation Voltage vs. V_{GE}

FGL60N100BNTD Rev. A

August 2000

LM741
Operational Amplifier

General Description

The LM741 series are general purpose operational amplifiers which feature improved performance over industry standards like the LM709. They are direct, plug-in replacements for the 709C, LM201, MC1439 and 748 in most applications.

The amplifiers offer many features which make their application nearly foolproof: overload protection on the input and output, no latch-up when the common mode range is exceeded, as well as freedom from oscillations.

The LM741C is identical to the LM741/LM741A except that the LM741C has their performance guaranteed over a 0°C to +70°C temperature range, instead of −55°C to +125°C.

Features

Connection Diagrams

Metal Can Package

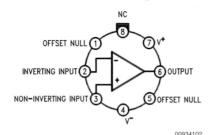

00934102

Note 1: LM741H is available per JM38510/10101

Order Number LM741H, LM741H/883 (Note 1),
LM741AH/883 or LM741CH
See NS Package Number H08C

Dual-In-Line or S.O. Package

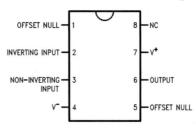

00934103

Order Number LM741J, LM741J/883, LM741CN
See NS Package Number J08A, M08A or N08E

Ceramic Flatpak

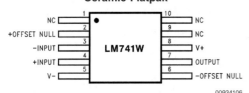

00934106

Order Number LM741W/883
See NS Package Number W10A

Typical Application

Offset Nulling Circuit

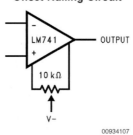

00934107

© 2004 National Semiconductor Corporation   DS009341

www.national.com

Appendix A

Absolute Maximum Ratings (Note 2)

If Military/Aerospace specified devices are required, please contact the National Semiconductor Sales Office/Distributors for availability and specifications.

(Note 7)

	LM741A	LM741	LM741C
Supply Voltage	±22V	±22V	±18V
Power Dissipation (Note 3)	500 mW	500 mW	500 mW
Differential Input Voltage	±30V	±30V	±30V
Input Voltage (Note 4)	±15V	±15V	±15V
Output Short Circuit Duration	Continuous	Continuous	Continuous
Operating Temperature Range	−55°C to +125°C	−55°C to +125°C	0°C to +70°C
Storage Temperature Range	−65°C to +150°C	−65°C to +150°C	−65°C to +150°C
Junction Temperature	150°C	150°C	100°C
Soldering Information			
N-Package (10 seconds)	260°C	260°C	260°C
J- or H-Package (10 seconds)	300°C	300°C	300°C
M-Package			
Vapor Phase (60 seconds)	215°C	215°C	215°C
Infrared (15 seconds)	215°C	215°C	215°C

See AN-450 "Surface Mounting Methods and Their Effect on Product Reliability" for other methods of soldering

surface mount devices.

ESD Tolerance (Note 8)	400V	400V	400V

Electrical Characteristics (Note 5)

Parameter	Conditions	LM741A			LM741			LM741C			Units
		Min	Typ	Max	Min	Typ	Max	Min	Typ	Max	
Input Offset Voltage	$T_A = 25°C$										
	$R_S \leq 10\ k\Omega$					1.0	5.0		2.0	6.0	mV
	$R_S \leq 50\Omega$		0.8	3.0							mV
	$T_{AMIN} \leq T_A \leq T_{AMAX}$										
	$R_S \leq 50\Omega$			4.0							mV
	$R_S \leq 10\ k\Omega$						6.0			7.5	mV
Average Input Offset Voltage Drift				15							µV/°C
Input Offset Voltage Adjustment Range	$T_A = 25°C, V_S = ±20V$	±10				±15			±15		mV
Input Offset Current	$T_A = 25°C$		3.0	30		20	200		20	200	nA
	$T_{AMIN} \leq T_A \leq T_{AMAX}$			70		85	500			300	nA
Average Input Offset Current Drift				0.5							nA/°C
Input Bias Current	$T_A = 25°C$		30	80		80	500		80	500	nA
	$T_{AMIN} \leq T_A \leq T_{AMAX}$			0.210			1.5			0.8	µA
Input Resistance	$T_A = 25°C, V_S = ±20V$	1.0	6.0		0.3	2.0		0.3	2.0		MΩ
	$T_{AMIN} \leq T_A \leq T_{AMAX}, V_S = ±20V$	0.5									MΩ
Input Voltage Range	$T_A = 25°C$							±12	±13		V
	$T_{AMIN} \leq T_A \leq T_{AMAX}$				±12	±13					V

Electrical Characteristics (Note 5) (Continued)

Parameter	Conditions	LM741A			LM741			LM741C			Units
		Min	Typ	Max	Min	Typ	Max	Min	Typ	Max	
Large Signal Voltage Gain	$T_A = 25°C$, $R_L \geq 2$ kΩ										
	$V_S = \pm20V$, $V_O = \pm15V$	50									V/mV
	$V_S = \pm15V$, $V_O = \pm10V$				50	200		20	200		V/mV
	$T_{AMIN} \leq T_A \leq T_{AMAX}$,										
	$R_L \geq 2$ kΩ,										
	$V_S = \pm20V$, $V_O = \pm15V$	32									V/mV
	$V_S = \pm15V$, $V_O = \pm10V$				25			15			V/mV
	$V_S = \pm5V$, $V_O = \pm2V$	10									V/mV
Output Voltage Swing	$V_S = \pm20V$										
	$R_L \geq 10$ kΩ	±16									V
	$R_L \geq 2$ kΩ	±15									V
	$V_S = \pm15V$										
	$R_L \geq 10$ kΩ				±12	±14		±12	±14		V
	$R_L \geq 2$ kΩ				±10	±13		±10	±13		V
Output Short Circuit Current	$T_A = 25°C$	10	25	35		25			25		mA
	$T_{AMIN} \leq T_A \leq T_{AMAX}$	10		40							mA
Common-Mode Rejection Ratio	$T_{AMIN} \leq T_A \leq T_{AMAX}$										
	$R_S \leq 10$ kΩ, $V_{CM} = \pm12V$				70	90		70	90		dB
	$R_S \leq 50\Omega$, $V_{CM} = \pm12V$	80	95								dB
Supply Voltage Rejection Ratio	$T_{AMIN} \leq T_A \leq T_{AMAX}$,										
	$V_S = \pm20V$ to $V_S = \pm5V$										
	$R_S \leq 50\Omega$	86	96								dB
	$R_S \leq 10$ kΩ				77	96		77	96		dB
Transient Response	$T_A = 25°C$, Unity Gain										
Rise Time			0.25	0.8		0.3			0.3		μs
Overshoot			6.0	20		5			5		%
Bandwidth (Note 6)	$T_A = 25°C$	0.437	1.5								MHz
Slew Rate	$T_A = 25°C$, Unity Gain	0.3	0.7			0.5			0.5		V/μs
Supply Current	$T_A = 25°C$					1.7	2.8		1.7	2.8	mA
Power Consumption	$T_A = 25°C$										
	$V_S = \pm20V$		80	150							mW
	$V_S = \pm15V$					50	85		50	85	mW
LM741A	$V_S = \pm20V$										
	$T_A = T_{AMIN}$			165							mW
	$T_A = T_{AMAX}$			135							mW
LM741	$V_S = \pm15V$										
	$T_A = T_{AMIN}$					60	100				mW
	$T_A = T_{AMAX}$					45	75				mW

Note 2: "Absolute Maximum Ratings" indicate limits beyond which damage to the device may occur. Operating Ratings indicate conditions for which the device is functional, but do not guarantee specific performance limits.

Electrical Characteristics (Note 5) (Continued)

Note 3: For operation at elevated temperatures, these devices must be derated based on thermal resistance, and T_j max. (listed under "Absolute Maximum Ratings"). $T_j = T_A + (\theta_{jA} P_D)$.

Thermal Resistance	Cerdip (J)	DIP (N)	HO8 (H)	SO-8 (M)
θ_{jA} (Junction to Ambient)	100°C/W	100°C/W	170°C/W	195°C/W
θ_{jC} (Junction to Case)	N/A	N/A	25°C/W	N/A

Note 4: For supply voltages less than ±15V, the absolute maximum input voltage is equal to the supply voltage.

Note 5: Unless otherwise specified, these specifications apply for V_S = ±15V, −55°C ≤ T_A ≤ +125°C (LM741/LM741A). For the LM741C/LM741E, these specifications are limited to 0°C ≤ T_A ≤ +70°C.

Note 6: Calculated value from: BW (MHz) = 0.35/Rise Time(µs).

Note 7: For military specifications see RETS741X for LM741 and RETS741AX for LM741A.

Note 8: Human body model, 1.5 kΩ in series with 100 pF.

Schematic Diagram

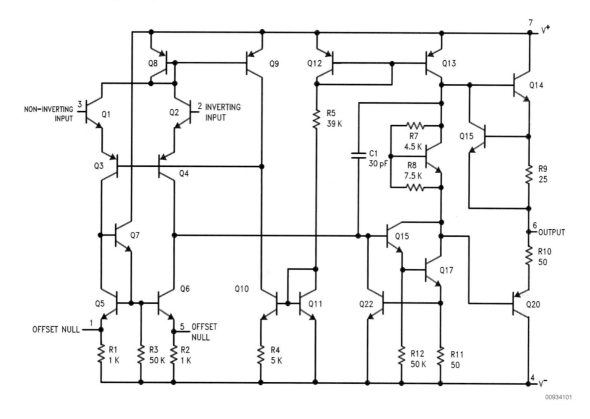

00934101

LM118/LM218/LM318
Operational Amplifiers

General Description

The LM118 series are precision high speed operational amplifiers designed for applications requiring wide bandwidth and high slew rate. They feature a factor of ten increase in speed over general purpose devices without sacrificing DC performance.

The LM118 series has internal unity gain frequency compensation. This considerably simplifies its application since no external components are necessary for operation. However, unlike most internally compensated amplifiers, external frequency compensation may be added for optimum performance. For inverting applications, feedforward compensation will boost the slew rate to over 150V/μs and almost double the bandwidth. Overcompensation can be used with the amplifier for greater stability when maximum bandwidth is not needed. Further, a single capacitor can be added to reduce the 0.1% settling time to under 1 μs.

The high speed and fast settling time of these op amps make them useful in A/D converters, oscillators, active filters, sample and hold circuits, or general purpose amplifiers. These devices are easy to apply and offer an order of magnitude better AC performance than industry standards such as the LM709.

The LM218 is identical to the LM118 except that the LM218 has its performance specified over a −25˚C to +85˚C temperature range. The LM318 is specified from 0˚C to +70˚C.

Features

- 15 MHz small signal bandwidth
- Guaranteed 50V/μs slew rate
- Maximum bias current of 250 nA
- Operates from supplies of ±5V to ±20V
- Internal frequency compensation
- Input and output overload protected
- Pin compatible with general purpose op amps

Fast Voltage Follower

(Note 1)

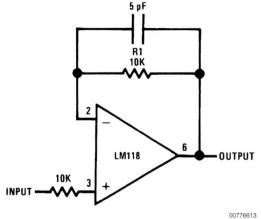

00776613

Note 1: Do not hard-wire as voltage follower (R1 ≥ 5 kΩ)

Absolute Maximum Ratings (Note 7)

If Military/Aerospace specified devices are required, please contact the National Semiconductor Sales Office/ Distributors for availability and specifications.

Supply Voltage	±20V
Power Dissipation (Note 2)	500 mW
Differential Input Current (Note 3)	±10 mA
Input Voltage (Note 4)	±15V
Output Short-Circuit Duration	Continuous
Operating Temperature Range	
LM118	−55˚C to +125˚C
LM218	−25˚C to +85˚C
LM318	0˚C to +70˚C
Storage Temperature Range	−65˚C to +150˚C

Lead Temperature (Soldering, 10 sec.)	
Hermetic Package	300˚C
Plastic Package	260˚C
Soldering Information	
Dual-In-Line Package	
Soldering (10 sec.)	260˚C
Small Outline Package	
Vapor Phase (60 sec.)	215˚C
Infrared (15 sec.)	220˚C

See AN-450 "Surface Mounting Methods and Their Effect on Product Reliability" for other methods of soldering surface mount devices.

ESD Tolerance (Note 8)	2000V

Electrical Characteristics (Note 5)

Parameter	Conditions	LM118/LM218 Min	LM118/LM218 Typ	LM118/LM218 Max	LM318 Min	LM318 Typ	LM318 Max	Units
Input Offset Voltage	$T_A = 25˚C$		2	4		4	10	mV
Input Offset Current	$T_A = 25˚C$		6	50		30	200	nA
Input Bias Current	$T_A = 25˚C$		120	250		150	500	nA
Input Resistance	$T_A = 25˚C$	1	3		0.5	3		MΩ
Supply Current	$T_A = 25˚C$		5	8		5	10	mA
Large Signal Voltage Gain	$T_A = 25˚C, V_S = ±15V$ $V_{OUT} = ±10V, R_L ≥ 2 kΩ$	50	200		25	200		V/mV
Slew Rate	$T_A = 25˚C, V_S = ±15V, A_V = 1$ (Note 6)	50	70		50	70		V/µs
Small Signal Bandwidth	$T_A = 25˚C, V_S = ±15V$		15			15		MHz
Input Offset Voltage			6				15	mV
Input Offset Current			100				300	nA
Input Bias Current			500				750	nA
Supply Current	$T_A = 125˚C$		4.5	7				mA
Large Signal Voltage Gain	$V_S = ±15V, V_{OUT} = ±10V$ $R_L ≥ 2 kΩ$	25			20			V/mV
Output Voltage Swing	$V_S = ±15V, R_L = 2 kΩ$	±12	±13		±12	±13		V
Input Voltage Range	$V_S = ±15V$	±11.5			±11.5			V
Common-Mode Rejection Ratio		80	100		70	100		dB
Supply Voltage Rejection Ratio		70	80		65	80		dB

Note 2: The maximum junction temperature of the LM118 is 150˚C, the LM218 is 110˚C, and the LM318 is 110˚C. For operating at elevated temperatures, devices in the H08 package must be derated based on a thermal resistance of 160˚C/W, junction to ambient, or 20˚C/W, junction to case. The thermal resistance of the dual-in-line package is 100˚C/W, junction to ambient.

Note 3: The inputs are shunted with back-to-back diodes for overvoltage protection. Therefore, excessive current will flow if a differential input voltage in excess of 1V is applied between the inputs unless some limiting resistance is used.

Note 4: For supply voltages less than ±15V, the absolute maximum input voltage is equal to the supply voltage.

Note 5: These specifications apply for ±5V ≤ V_S ≤ ±20V and −55˚C ≤ T_A ≤ +125˚C (LM118), −25˚C ≤ T_A ≤ +85˚C (LM218), and 0˚C ≤ T_A ≤ +70˚C (LM318). Also, power supplies must be bypassed with 0.1 µF disc capacitors.

Note 6: Slew rate is tested with $V_S = ±15V$. The LM118 is in a unity-gain non-inverting configuration. V_{IN} is stepped from −7.5V to +7.5V and vice versa. The slew rates between −5.0V and +5.0V and vice versa are tested and guaranteed to exceed 50V/µs.

Note 7: Refer to RETS118X for LM118H and LM118J military specifications.

Note 8: Human body model, 1.5 kΩ in series with 100 pF.

National Semiconductor

LM555
Timer

General Description

The LM555 is a highly stable device for generating accurate time delays or oscillation. Additional terminals are provided for triggering or resetting if desired. In the time delay mode of operation, the time is precisely controlled by one external resistor and capacitor. For astable operation as an oscillator, the free running frequency and duty cycle are accurately controlled with two external resistors and one capacitor. The circuit may be triggered and reset on falling waveforms, and the output circuit can source or sink up to 200mA or drive TTL circuits.

Features

- Direct replacement for SE555/NE555
- Timing from microseconds through hours
- Operates in both astable and monostable modes
- Adjustable duty cycle
- Output can source or sink 200 mA
- Output and supply TTL compatible
- Temperature stability better than 0.005% per °C
- Normally on and normally off output
- Available in 8-pin MSOP package

Applications

- Precision timing
- Pulse generation
- Sequential timing
- Time delay generation
- Pulse width modulation
- Pulse position modulation
- Linear ramp generator

Schematic Diagram

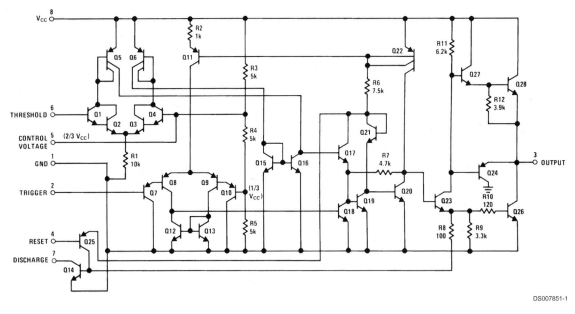

DS007851-1

www.national.com

Connection Diagram

Dual-In-Line, Small Outline
and Molded Mini Small Outline Packages

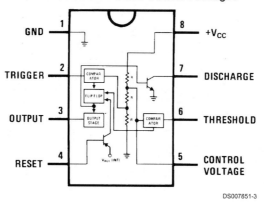

DS007851-3

Top View

Ordering Information

Package	Part Number	Package Marking	Media Transport	NSC Drawing
8-Pin SOIC	LM555CM	LM555CM	Rails	M08A
	LM555CMX	LM555CM	2.5k Units Tape and Reel	
8-Pin MSOP	LM555CMM	Z55	1k Units Tape and Reel	MUA08A
	LM555CMMX	Z55	3.5k Units Tape and Reel	
8-Pin MDIP	LM555CN	LM555CN	Rails	N08E

2

Absolute Maximum Ratings (Note 2)

If Military/Aerospace specified devices are required, please contact the National Semiconductor Sales Office/Distributors for availability and specifications.

Supply Voltage	+18V
Power Dissipation (Note 3)	
LM555CM, LM555CN	1180 mW
LM555CMM	613 mW
Operating Temperature Ranges	
LM555C	0˚C to +70˚C
Storage Temperature Range	−65˚C to +150˚C

Soldering Information

Dual-In-Line Package	
Soldering (10 Seconds)	260˚C
Small Outline Packages	
(SOIC and MSOP)	
Vapor Phase (60 Seconds)	215˚C
Infrared (15 Seconds)	220˚C

See AN-450 "Surface Mounting Methods and Their Effect on Product Reliability" for other methods of soldering surface mount devices.

Electrical Characteristics (Notes 1, 2)

(T_A = 25˚C, V_{CC} = +5V to +15V, unless othewise specified)

Parameter	Conditions	Limits LM555C			Units
		Min	Typ	Max	
Supply Voltage		4.5		16	V
Supply Current	V_{CC} = 5V, R_L = ∞		3	6	
	V_{CC} = 15V, R_L = ∞		10	15	mA
	(Low State) (Note 4)				
Timing Error, Monostable					
Initial Accuracy			1		%
Drift with Temperature	R_A = 1k to 100kΩ,		50		ppm/˚C
	C = 0.1µF, (Note 5)				
Accuracy over Temperature			1.5		%
Drift with Supply			0.1		%/V
Timing Error, Astable					
Initial Accuracy			2.25		%
Drift with Temperature	R_A, R_B = 1k to 100kΩ,		150		ppm/˚C
	C = 0.1µF, (Note 5)				
Accuracy over Temperature			3.0		%
Drift with Supply			0.30		%/V
Threshold Voltage			0.667		x V_{CC}
Trigger Voltage	V_{CC} = 15V		5		V
	V_{CC} = 5V		1.67		V
Trigger Current			0.5	0.9	µA
Reset Voltage		0.4	0.5	1	V
Reset Current			0.1	0.4	mA
Threshold Current	(Note 6)		0.1	0.25	µA
Control Voltage Level	V_{CC} = 15V	9	10	11	V
	V_{CC} = 5V	2.6	3.33	4	
Pin 7 Leakage Output High			1	100	nA
Pin 7 Sat (Note 7)					
Output Low	V_{CC} = 15V, I_7 = 15mA		180		mV
Output Low	V_{CC} = 4.5V, I_7 = 4.5mA		80	200	mV

Electrical Characteristics (Notes 1, 2) (Continued)

(T_A = 25°C, V_{CC} = +5V to +15V, unless othewise specified)

Parameter	Conditions	Limits LM555C			Units
		Min	Typ	Max	
Output Voltage Drop (Low)	V_{CC} = 15V				
	I_{SINK} = 10mA		0.1	0.25	V
	I_{SINK} = 50mA		0.4	0.75	V
	I_{SINK} = 100mA		2	2.5	V
	I_{SINK} = 200mA		2.5		V
	V_{CC} = 5V				
	I_{SINK} = 8mA				V
	I_{SINK} = 5mA		0.25	0.35	V
Output Voltage Drop (High)	I_{SOURCE} = 200mA, V_{CC} = 15V		12.5		V
	I_{SOURCE} = 100mA, V_{CC} = 15V	12.75	13.3		V
	V_{CC} = 5V	2.75	3.3		V
Rise Time of Output			100		ns
Fall Time of Output			100		ns

Note 1: All voltages are measured with respect to the ground pin, unless otherwise specified.

Note 2: Absolute Maximum Ratings indicate limits beyond which damage to the device may occur. Operating Ratings indicate conditions for which the device is functional, but do not guarantee specific performance limits. Electrical Characteristics state DC and AC electrical specifications under particular test conditions which guarantee specific performance limits. This assumes that the device is within the Operating Ratings. Specifications are not guaranteed for parameters where no limit is given, however, the typical value is a good indication of device performance.

Note 3: For operating at elevated temperatures the device must be derated above 25°C based on a +150°C maximum junction temperature and a thermal resistance of 106°C/W (DIP), 170°C/W (SO-8), and 204°C/W (MSOP) junction to ambient.

Note 4: Supply current when output high typically 1 mA less at V_{CC} = 5V.

Note 5: Tested at V_{CC} = 5V and V_{CC} = 15V.

Note 6: This will determine the maximum value of R_A + R_B for 15V operation. The maximum total (R_A + R_B) is 20MΩ.

Note 7: No protection against excessive pin 7 current is necessary providing the package dissipation rating will not be exceeded.

Note 8: Refer to RETS555X drawing of military LM555H and LM555J versions for specifications.

XR-2206
Monolithic
Function Generator

FEATURES

- Low-Sine Wave Distortion, 0.5%, Typical
- Excellent Temperature Stability, 20ppm/°C, Typ.
- Wide Sweep Range, 2000:1, Typical
- Low-Supply Sensitivity, 0.01%V, Typ.
- Linear Amplitude Modulation
- TTL Compatible FSK Controls
- Wide Supply Range, 10V to 26V
- Adjustable Duty Cycle, 1% TO 99%

APPLICATIONS

- Waveform Generation
- Sweep Generation
- AM/FM Generation
- V/F Conversion
- FSK Generation
- Phase-Locked Loops (VCO)

GENERAL DESCRIPTION

The XR-2206 is a monolithic function generator integrated circuit capable of producing high quality sine, square, triangle, ramp, and pulse waveforms of high-stability and accuracy. The output waveforms can be both amplitude and frequency modulated by an external voltage. Frequency of operation can be selected externally over a range of 0.01Hz to more than 1MHz.

The circuit is ideally suited for communications, instrumentation, and function generator applications requiring sinusoidal tone, AM, FM, or FSK generation. It has a typical drift specification of 20ppm/°C. The oscillator frequency can be linearly swept over a 2000:1 frequency range with an external control voltage, while maintaining low distortion.

ORDERING INFORMATION

Part No.	Package	Operating Temperature Range
XR-2206M	16 Lead 300 Mil CDIP	-55°C to +125°C
XR-2206P	16 Lead 300 Mil PDIP	−40°C to +85°C
XR-2206CP	16 Lead 300 Mil PDIP	0°C to +70°C
XR-2206D	16 Lead 300 Mil JEDEC SOIC	0°C to +70°C

Rev. 1.03
©1972

EXAR Corporation, 48720 Kato Road, Fremont, CA 94538 ◆ (510) 668-7000 ◆ (510) 668-7017

Appendix A

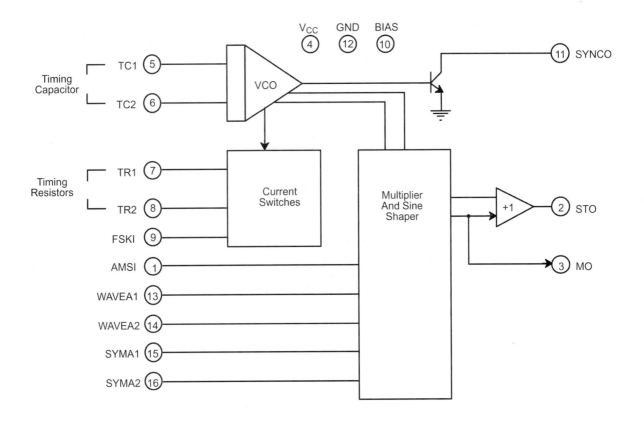

Figure 1.   XR-2206 Block Diagram

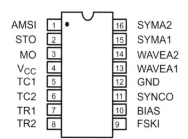

16 Lead PDIP, CDIP (0.300")

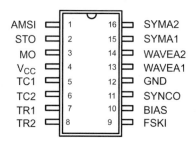

16 Lead SOIC (Jedec, 0.300")

PIN DESCRIPTION

Pin #	Symbol	Type	Description
1	AMSI	I	**Amplitude Modulating Signal Input.**
2	STO	O	**Sine or Triangle Wave Output.**
3	MO	O	**Multiplier Output.**
4	V_{CC}		**Positive Power Supply.**
5	TC1	I	**Timing Capacitor Input.**
6	TC2	I	**Timing Capacitor Input.**
7	TR1	O	**Timing Resistor 1 Output.**
8	TR2	O	**Timing Resistor 2 Output.**
9	FSKI	I	**Frequency Shift Keying Input.**
10	BIAS	O	**Internal Voltage Reference.**
11	SYNCO	O	**Sync Output.** This output is a open collector and needs a pull up resistor to V_{CC}.
12	GND		**Ground pin.**
13	WAVEA1	I	**Wave Form Adjust Input 1.**
14	WAVEA2	I	**Wave Form Adjust Input 2.**
15	SYMA1	I	**Wave Symetry Adjust 1.**
16	SYMA2	I	**Wave Symetry Adjust 2.**

XR-2206

DC ELECTRICAL CHARACTERISTICS

Test Conditions: Test Circuit of *Figure 2* **Vcc = 12V, T_A = 25°C, C = 0.01μF, R_1 = 100kΩ, R_2 = 10kΩ, R_3 = 25kΩ** Unless Otherwise Specified. S_1 open for triangle, closed for sine wave.

Parameters	XR-2206M/P Min.	XR-2206M/P Typ.	XR-2206M/P Max.	XR-2206CP/D Min.	XR-2206CP/D Typ.	XR-2206CP/D Max.	Units	Conditions
General Characteristics								
Single Supply Voltage	**10**		**26**	10		26	V	
Split-Supply Voltage	**±5**		**±13**	±5		±13	V	
Supply Current		12	**17**		14	20	mA	$R_1 \geq 10kΩ$
Oscillator Section								
Max. Operating Frequency	**0.5**	1		0.5	1		MHz	C = 1000pF, R_1 = 1kΩ
Lowest Practical Frequency		0.01			0.01		Hz	C = 50μF, R_1 = 2MΩ
Frequency Accuracy		±1	**±4**		±2		% of f_o	f_o = 1/R_1C
Temperature Stability Frequency		±10	**±50**		±20		ppm/°C	0°C $\leq T_A \leq$ 70°C R_1 = R_2 = 20kΩ
Sine Wave Amplitude Stability[2]		4800			4800		ppm/°C	
Supply Sensitivity		0.01	**0.1**		0.01		%/V	V_{LOW} = 10V, V_{HIGH} = 20V, R_1 = R_2 = 20kΩ
Sweep Range	1000:1	2000:1		2000:1			$f_H = f_L$	f_H @ R_1 = 1kΩ f_L @ R_1 = 2MΩ
Sweep Linearity								
10:1 Sweep		2			2		%	f_L = 1kHz, f_H = 10kHz
1000:1 Sweep		8			8		%	f_L = 100Hz, f_H = 100kHz
FM Distortion		0.1			0.1		%	±10% Deviation
Recommended Timing Components								
Timing Capacitor: C	**0.001**		100	0.001		100	μF	*Figure 5*
Timing Resistors: R_1 & R_2	**1**		2000	1		2000	kΩ	*Figure 3*
Triangle Sine Wave Output[1]								*Figure 3*
Triangle Amplitude		160			160		mV/kΩ	*Figure 2*, S_1 Open
Sine Wave Amplitude	**40**	60	80		60		mV/kΩ	*Figure 2*, S_1 Closed
Max. Output Swing		6			6		Vp-p	
Output Impedance		600			600		Ω	
Triangle Linearity		1			1		%	
Amplitude Stability		0.5			0.5		dB	For 1000:1 Sweep
Sine Wave Distortion								
Without Adjustment		2.5			2.5		%	R_1 = 30kΩ
With Adjustment		0.4	**1.0**		0.5	1.5	%	See *Figure 7* and *Figure 8*

Notes

[1] Output amplitude is directly proportional to the resistance, R_3, on Pin 3. See *Figure 3*.

[2] For maximum amplitude stability, R_3 should be a positive temperature coefficient resistor.

Bold face parameters are covered by production test and guaranteed over operating temperature range.

Rev. 1.03

4

DC ELECTRICAL CHARACTERISTICS (CONT'D)

Parameters	XR-2206M/P			XR-2206CP/D			Units	Conditions
	Min.	Typ.	Max.	Min.	Typ.	Max.		
Amplitude Modulation								
Input Impedance	50	100		50	100		kΩ	
Modulation Range		100			100		%	
Carrier Suppression		55			55		dB	
Linearity		2			2		%	For 95% modulation
Square-Wave Output								
Amplitude		12			12		Vp-p	Measured at Pin 11.
Rise Time		250			250		ns	$C_L = 10pF$
Fall Time		50			50		ns	$C_L = 10pF$
Saturation Voltage		0.2	**0.4**		0.2	0.6	V	$I_L = 2mA$
Leakage Current		0.1	**20**		0.1	100	µA	$V_{CC} = 26V$
FSK Keying Level (Pin 9)	0.8	1.4	**2.4**	0.8	1.4	2.4	V	See section on circuit controls
Reference Bypass Voltage	2.9	3.1	**3.3**	2.5	3	3.5	V	Measured at Pin 10.

Notes
[1] Output amplitude is directly proportional to the resistance, R_3, on Pin 3. See Figure 3.
[2] For maximum amplitude stability, R_3 should be a positive temperature coefficient resistor.
Bold face parameters are covered by production test and guaranteed over operating temperature range.

Specifications are subject to change without notice

ABSOLUTE MAXIMUM RATINGS

Power Supply 26V
Power Dissipation 750mW
Derate Above 25°C . . . . 5mW/°C

Total Timing Current 6mA
Storage Temperature . . . . -65°C to +150°C

SYSTEM DESCRIPTION

The XR-2206 is comprised of four functional blocks; a voltage-controlled oscillator (VCO), an analog multiplier and sine-shaper; a unity gain buffer amplifier; and a set of current switches.

The VCO produces an output frequency proportional to an input current, which is set by a resistor from the timing terminals to ground. With two timing pins, two discrete output frequencies can be independently produced for FSK generation applications by using the FSK input control pin. This input controls the current switches which select one of the timing resistor currents, and routes it to the VCO.

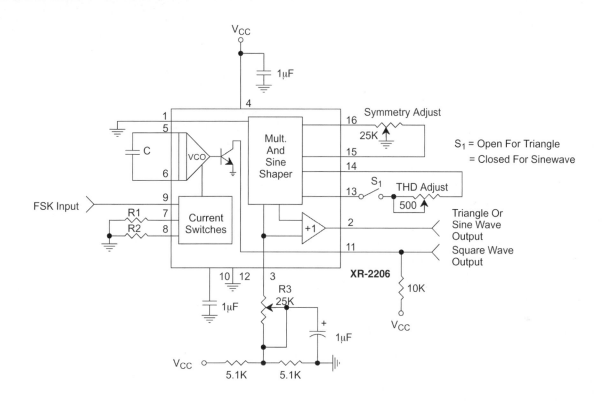

Figure 2. Basic Test Circuit

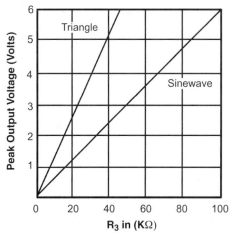

**Figure 3. Output Amplitude
as a Function of the Resistor,
R3, at Pin 3**

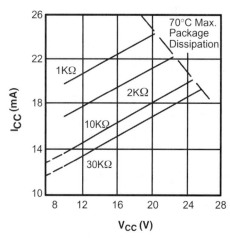

**Figure 4. Supply Current vs
Supply Voltage, Timing, R**

May 2000

LM78XX
Series Voltage Regulators

General Description

The LM78XX series of three terminal regulators is available with several fixed output voltages making them useful in a wide range of applications. One of these is local on card regulation, eliminating the distribution problems associated with single point regulation. The voltages available allow these regulators to be used in logic systems, instrumentation, HiFi, and other solid state electronic equipment. Although designed primarily as fixed voltage regulators these devices can be used with external components to obtain adjustable voltages and currents.

The LM78XX series is available in an aluminum TO-3 package which will allow over 1.0A load current if adequate heat sinking is provided. Current limiting is included to limit the peak output current to a safe value. Safe area protection for the output transistor is provided to limit internal power dissipation. If internal power dissipation becomes too high for the heat sinking provided, the thermal shutdown circuit takes over preventing the IC from overheating.

Considerable effort was expanded to make the LM78XX series of regulators easy to use and minimize the number of external components. It is not necessary to bypass the output, although this does improve transient response. Input bypassing is needed only if the regulator is located far from the filter capacitor of the power supply.

For output voltage other than 5V, 12V and 15V the LM117 series provides an output voltage range from 1.2V to 57V.

Features

- Output current in excess of 1A
- Internal thermal overload protection
- No external components required
- Output transistor safe area protection
- Internal short circuit current limit
- Available in the aluminum TO-3 package

Voltage Range

LM7805C	5V
LM7812C	12V
LM7815C	15V

Connection Diagrams

Metal Can Package
TO-3 (K)
Aluminum

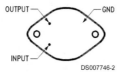

DS007746-2

Bottom View
Order Number LM7805CK,
LM7812CK or LM7815CK
See NS Package Number KC02A

Plastic Package
TO-220 (T)

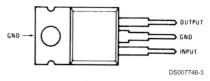

DS007746-3

Top View
Order Number LM7805CT,
LM7812CT or LM7815CT
See NS Package Number T03B

www.national.com

Appendix A

Schematic

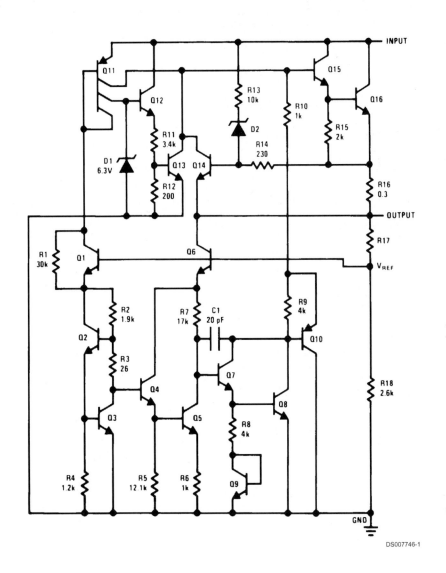

DS007746-1

Absolute Maximum Ratings (Note 3)

If Military/Aerospace specified devices are required, please contact the National Semiconductor Sales Office/Distributors for availability and specifications.

Input Voltage
  (V_O = 5V, 12V and 15V)   35V
Internal Power Dissipation (Note 1)   Internally Limited
Operating Temperature Range (T_A)   0˚C to +70˚C

Maximum Junction Temperature
  (K Package)   150˚C
  (T Package)   150˚C
Storage Temperature Range   −65˚C to +150˚C
Lead Temperature (Soldering, 10 sec.)
  TO-3 Package K   300˚C
  TO-220 Package T   230˚C

Electrical Characteristics LM78XXC (Note 2)

$0˚C \leq T_J \leq 125˚C$ unless otherwise noted.

	Output Voltage		5V			12V			15V			
	Input Voltage (unless otherwise noted)		10V			19V			23V			Units
Symbol	Parameter	Conditions	Min	Typ	Max	Min	Typ	Max	Min	Typ	Max	
V_O	Output Voltage	Tj = 25˚C, 5 mA ≤ I_O ≤ 1A	4.8	5	5.2	11.5	12	12.5	14.4	15	15.6	V
		P_D ≤ 15W, 5 mA ≤ I_O ≤ 1A	4.75		5.25	11.4		12.6	14.25		15.75	V
		V_{MIN} ≤ V_{IN} ≤ V_{MAX}	(7.5 ≤ V_{IN} ≤ 20)			(14.5 ≤ V_{IN} ≤ 27)			(17.5 ≤ V_{IN} ≤ 30)			V
ΔV_O	Line Regulation	I_O = 500 mA Tj = 25˚C		3	50		4	120		4	150	mV
		ΔV_{IN}	(7 ≤ V_{IN} ≤ 25)			14.5 ≤ V_{IN} ≤ 30)			(17.5 ≤ V_{IN} ≤ 30)			V
		0˚C ≤ Tj ≤ +125˚C			50			120			150	mV
		ΔV_{IN}	(8 ≤ V_{IN} ≤ 20)			(15 ≤ V_{IN} ≤ 27)			(18.5 ≤ V_{IN} ≤ 30)			V
		I_O ≤ 1A Tj = 25˚C			50			120			150	mV
		ΔV_{IN}	(7.5 ≤ V_{IN} ≤ 20)			(14.6 ≤ V_{IN} ≤ 27)			(17.7 ≤ V_{IN} ≤ 30)			V
		0˚C ≤ Tj ≤ +125˚C			25			60			75	mV
		ΔV_{IN}	(8 ≤ V_{IN} ≤ 12)			(16 ≤ V_{IN} ≤ 22)			(20 ≤ V_{IN} ≤ 26)			V
ΔV_O	Load Regulation	Tj = 25˚C 5 mA ≤ I_O ≤ 1.5A		10	50		12	120		12	150	mV
		250 mA ≤ I_O ≤ 750 mA			25			60			75	mV
		5 mA ≤ I_O ≤ 1A, 0˚C ≤ Tj ≤ +125˚C			50			120			150	mV
I_Q	Quiescent Current	I_O ≤ 1A Tj = 25˚C			8			8			8	mA
		0˚C ≤ Tj ≤ +125˚C			8.5			8.5			8.5	mA
ΔI_Q	Quiescent Current Change	5 mA ≤ I_O ≤ 1A			0.5			0.5			0.5	mA
		Tj = 25˚C, I_O ≤ 1A			1.0			1.0			1.0	mA
		V_{MIN} ≤ V_{IN} ≤ V_{MAX}	(7.5 ≤ V_{IN} ≤ 20)			(14.8 ≤ V_{IN} ≤ 27)			(17.9 ≤ V_{IN} ≤ 30)			V
		I_O ≤ 500 mA, 0˚C ≤ Tj ≤ +125˚C			1.0			1.0			1.0	mA
		V_{MIN} ≤ V_{IN} ≤ V_{MAX}	(7 ≤ V_{IN} ≤ 25)			(14.5 ≤ V_{IN} ≤ 30)			(17.5 ≤ V_{IN} ≤ 30)			V
V_N	Output Noise Voltage	T_A =25˚C, 10 Hz ≤ f ≤ 100 kHz		40			75			90		µV
$\frac{\Delta V_{IN}}{\Delta V_{OUT}}$	Ripple Rejection	I_O ≤ 1A, Tj = 25˚C or	62	80		55	72		54	70		dB
		f = 120 Hz I_O ≤ 500 mA 0˚C ≤ Tj ≤ +125˚C	62			55			54			dB
		V_{MIN} ≤ V_{IN} ≤ V_{MAX}	(8 ≤ V_{IN} ≤ 18)			(15 ≤ V_{IN} ≤ 25)			(18.5 ≤ V_{IN} ≤ 28.5)			V
R_O	Dropout Voltage	Tj = 25˚C, I_{OUT} = 1A		2.0			2.0			2.0		V
	Output Resistance	f = 1 kHz		8			18			19		mΩ

Electrical Characteristics LM78XXC (Note 2) (Continued)

$0°C \leq T_J \leq 125°C$ unless otherwise noted.

| Symbol | Parameter | Conditions | \
Output Voltage \
5V \
Input Voltage (unless otherwise noted) \
10V | | | 12V \
19V | | | 15V \
23V | | | Units |
|---|---|---|---|---|---|---|---|---|---|---|---|---|
| | | | Min | Typ | Max | Min | Typ | Max | Min | Typ | Max | |
| | Short-Circuit Current | Tj = 25°C | | 2.1 | | | 1.5 | | | 1.2 | | A |
| | Peak Output Current | Tj = 25°C | | 2.4 | | | 2.4 | | | 2.4 | | A |
| | Average TC of V$_{OUT}$ | $0°C \leq Tj \leq +125°C$, I$_O$ = 5 mA | | 0.6 | | | 1.5 | | | 1.8 | | mV/°C |
| V$_{IN}$ | Input Voltage Required to Maintain Line Regulation | Tj = 25°C, I$_O \leq$ 1A | 7.5 | | | 14.6 | | | 17.7 | | | | V |

Note 1: Thermal resistance of the TO-3 package (K, KC) is typically 4°C/W junction to case and 35°C/W case to ambient. Thermal resistance of the TO-220 package (T) is typically 4°C/W junction to case and 50°C/W case to ambient.

Note 2: All characteristics are measured with capacitor across the input of 0.22 µF, and a capacitor across the output of 0.1µF. All characteristics except noise voltage and ripple rejection ratio are measured using pulse techniques (t$_w \leq$ 10 ms, duty cycle $\leq$ 5%). Output voltage changes due to changes in internal temperature must be taken into account separately.

Note 3: Absolute Maximum Ratings indicate limits beyond which damage to the device may occur. For guaranteed specifications and the test conditions, see Electrical Characteristics.

Appendix B

Mathematical Derivations

This appendix contains a few selected derivations. More derivations are available at the Web site that supports this book: www.malvino.com.

Proof of Eq. (9–10)

The starting point for this derivation is the rectangular *pn* junction equation derived by Schockley:

$$I = I_s(\epsilon^{Vq/kT} - 1) \tag{B-1}$$

where I = total diode current

I_s = reverse saturation current

V = total voltage across the depletion layer

q = charge on an electron

k = Boltzmann's constant

T = absolute temperature, $^\circ C + 273$

Equation (B-1) does *not* include the bulk resistance on either side of the junction. For this reason, the equation applies to the total diode only when the voltage across the bulk resistance is negligible.

At room temperature, q/kT equals approximately 40, and Eq. (B-1) becomes:

$$I = I_s(\epsilon^{40V} - 1) \tag{B-2}$$

(Some books are $39V$, but this is a small difference.) To get r_e', we differentiate I with respect to V:

$$\frac{dI}{dV} = 40 I_s \epsilon^{40V}$$

Using Eq. (B-2), we can rewrite this as:

$$\frac{dI}{dV} = 40 \, (I + I_s)$$

Taking the reciprocal gives r_e':

$$r_e' = \frac{dV}{dI} = \frac{1}{40(I + I_s)} = \frac{25 \, \text{mV}}{I + I_s} \tag{B-3}$$

Equation (B-3) includes the effect of reverse saturation current. In a practical linear amplifier, I is much greater than I_s (otherwise, the bias is unstable). For this reason, the practical value of r_e' is

$$r_e' = \frac{25 \, \text{mV}}{I}$$

Since we are talking about the emitter depletion layer, we add the supscript E to get

$$r_e' = \frac{25 \, \text{mV}}{I_E}$$

Proof of Eq. (12–27)

In Fig. 12-18a, the instantaneous power dissipation during the *on* time of the transistor is

$$p = V_{CE}I_C$$
$$= V_{CEQ}(1 - \sin \theta)I_{C(\text{sat})} \sin \theta$$

This is for the half-cycle when the transistor is conducting; during the *off* half-cycle, $p = 0$ ideally.

The average power dissipation equals:

$$p_{\text{av}} = \frac{\text{area}}{\text{period}} = \frac{1}{2\pi} \int_0^{\pi} V_{CEQ}(1 - \sin \theta)I_{C(\text{sat})} \sin \theta \, d\theta$$

After evaluating the definite integral over the half-cycle limits of 0 to π, and dividing by the period 2π, we have the average power over the *entire cycle* for one transistor:

$$p_{\text{av}} = \frac{1}{2\pi} V_{CEQ}I_{C(\text{sat})} \left[-\cos \theta - \frac{\theta}{2} \right]_0^{\pi}$$

$$= 0.068 \, V_{CEQ}I_{C(\text{sat})} \tag{B-4}$$

This is power dissipation in each transistor over the entire cycle, assuming 100 percent swing over the ac load line.

If the signal does not swing over the entire load line, the instantaneous power equals

$$p = V_{CE}I_C = V_{CEQ}(1 - k \sin \theta)I_{C(\text{sat})}k \sin \theta$$

where k is a constant between 0 and 1; k represents the fraction of the load line being used. After integrating:

$$p_{\text{av}} = \frac{1}{2\pi} \int_0^{\pi} p \, d\theta$$

you get:

$$p_{\text{av}} = \frac{V_{CEQ}I_{C(\text{sat})}}{2\pi} \left(2k - \frac{\pi k^2}{2} \right) \tag{B-5}$$

Since p_{av} is a function of k, we can differentiate and set dp_{av}/dk equal to zero to find the maximizing value of k:

$$\frac{dp_{\text{av}}}{dk} = \frac{V_{CEQ}I_{C(\text{sat})}}{2\pi}(2 - k\pi) = 0$$

Solving for k gives:

$$k = \frac{2}{\pi} = 0.636$$

With this value of k, Eq. (B-5) reduces to

$$p_{\text{av}} = 0.107 V_{CEQ}I_{C(\text{sat})} \cong 0.1 V_{CEQ}I_{C(\text{sat})}$$

Since $I_{C(\text{sat})} = V_{CEQ}/R_L$ and $V_{CEQ} = \text{MPP}/2$, the foregoing equation can be written as

$$P_{D(\text{max})} = \frac{\text{MPP}^2}{40R_L}$$

Proof of Eqs. (13–15) and (13–16)

Start with the transconductance equation:

$$I_D = I_{DSS} \left[1 - \frac{V_{GS}}{V_{GS(off)}} \right]^2 \qquad \text{(B-6)}$$

The derivative of this is:

$$\frac{dI_D}{dV_{GS}} = g_m = 2I_{DSS} \left[1 - \frac{V_{GS}}{V_{GS(off)}} \right] \left[-\frac{1}{V_{GS(off)}} \right]$$

or

$$g_m = -\frac{2I_{DSS}}{V_{GS(off)}} \left[1 - \frac{V_{GS}}{V_{GS(off)}} \right] \qquad \text{(B-7)}$$

When $V_{GS} = 0$, we get:

$$g_{m0} = -\frac{2I_{DSS}}{V_{GS(off)}} \qquad \text{(B-8)}$$

or by rearranging:

$$V_{GS(off)} = -\frac{2I_{DSS}}{g_{m0}}$$

This proves Eq. (13-15). By substituting the left member of Eq. (B-8) into Eq. (B-7):

$$g_m = g_{m0} \left[1 - \frac{V_{GS}}{V_{GS(off)}} \right]$$

This is the proof of Eq. (13-16).

Proof of Eq. (18–2)

The equation of a sinusoidal voltage is:

$$v = V_P \sin \omega t$$

The derivative with respect to time is:

$$\frac{dv}{dt} = \omega V_P \cos \omega t$$

The maximum rate of change occurs for $t = 0$. Furthermore, as the frequency increases, we reach the point at which the maximum rate of change just equals the slew rate. At this critical point:

$$S_R = \left(\frac{dv}{dt} \right)_{max} = \omega_{max} V_P = 2\pi f_{max} V_P$$

Solving for f_{max} in terms of S_R, we get:

$$f_{max} = \frac{S_R}{2\pi V_P}$$

Proof of Eq. (19–10)

Here is the derivation for closed-loop output impedance. Begin with:

$$A_{v(CL)} = \frac{A_{VOL}}{1 + A_{VOL}B}$$

Substitute:

$$A_v = A_u \frac{R_L}{r_{\text{out}} + R_L}$$

where A_v is the loaded gain (R_L connected) and A_u is the unloaded gain (R_L disconnected). After substitution for A_v, the closed-loop gain simplifies to:

$$A_{v(CL)} = \frac{A_u}{1 + A_u B + r_{\text{out}}/R_L}$$

When:

$$1 + A_u B = \frac{r_{\text{out}}}{R_L}$$

$A_{v(CL)}$ will drop in half, implying that the load resistance matches the Thevenin output resistance of the feedback amplifier. Solving for R_L gives:

$$R_L = \frac{r_{\text{out}}}{1 + A_u B}$$

This is the value of load resistance that forces the closed-loop voltage gain to drop in half, equivalent to saying it equals the closed-loop output impedance:

$$r_{\text{out}(CL)} = \frac{r_{\text{out}}}{1 + A_u B}$$

In any practical feedback amplifier, r_{out} is much smaller than R_L, so that A_{VOL} approximately equals A_u. This is why you almost always see the following expression for output impedance:

$$r_{\text{out}(CL)} = \frac{r_{\text{out}}}{1 + A_{VOL} B}$$

where $r_{\text{out}(CL)}$ = closed-loop output impedance
r_{out} = open-loop output impedance
$A_{VOL}B$ = open-loop gain

Proof of Eq. (19–23)

Because of the virtual ground in Fig. 19-12, essentially all of the input current flows through R_1. Summing voltages around the circuit gives:

$$-v_{\text{error}} + i_{\text{in}}R_2 - (i_{\text{out}} - i_{\text{in}})R_1 = 0 \qquad \text{(B-9)}$$

With the following substitutions:

$$v_{\text{error}} = \frac{v_{\text{out}}}{A_{VOL}}$$

and

$$v_{\text{out}} = i_{\text{out}}R_L + (i_{\text{out}} - i_{\text{in}})R_1$$

Eq. (B-9) can be rearranged as:

$$\frac{i_{\text{out}}}{i_{\text{in}}} = \frac{A_{VOL}R_2 + (1 + A_{VOL})R_1}{R_L + (1 + A_{VOL})R_1}$$

Since A_{VOL} is usually much greater than 1, this reduces to:

$$\frac{i_{\text{out}}}{i_{\text{in}}} = \frac{A_{VOL}(R_1 + R_2)}{R_L + A_{VOL}R_1}$$

Furthermore, $A_{VOL}R_2$ is usually much greater than R_L, and the foregoing simplifies to

$$\frac{i_{\text{out}}}{i_{\text{in}}} = \frac{R_2}{R_1} + 1$$

Proof of Eq. (22–17)

The change in capacitor voltage is given by:

$$\Delta V = \frac{IT}{C} \tag{B-10}$$

In the positive half-cycle of input voltage (Fig. 22-28a), the capacitor charging current is ideally:

$$I = \frac{V_P}{R}$$

Since T is the rundown time of the output ramp, it represents half of the output period. If f is the frequency of the input square wave, then $T = 1/2f$. Substituting for I and T in Eq. (B-10) gives:

$$\Delta V = \frac{V_P}{2fRC}$$

The input voltage has a peak value of V_P, while the output voltage has a peak-to-peak value of ΔV. Therefore, the equation can be written as

$$v_{\text{out(p-p)}} = \frac{V_P}{2fRC}$$

Proof of Eq. (22–18)

The UTP has a value of $+BV_{\text{sat}}$ and the LTP a value of $-BV_{\text{sat}}$. Start with the basic switching equation that applies to any RC circuit:

$$v = v_i + (v_f - v_i)\,(1 - e^{-t/RC}) \tag{B-11}$$

where $v = $ instantaneous capacitor voltage
$v_i = $ initial capacitor voltage
$v_f = $ target capacitor voltage
$t = $ charging time
$RC = $ time constant

In Fig. 22-32b, the capacitor charge starts with an initial value of $-BV_{\text{sat}}$ and ends with a value of $+BV_{\text{sat}}$. The target voltage for the capacitor voltage is $+V_{\text{sat}}$ and the capacitor charging time is half a period, $T/2$. Substitute into Eq. (B-11) to get:

$$BV_{\text{sat}} = -BV_{\text{sat}} + (V_{\text{sat}} + BV_{\text{sat}})\,(1 - e^{-T/2RC})$$

This simplifies to:

$$\frac{2B}{1 + B} = 1 - e^{-T/2RC}$$

By rearranging and taking the antilog, the foregoing becomes:

$$T = 2RC \ln \frac{1 + B}{1 - B}$$

Proof of Eq. (23-25)

Start with Eq. (B-11), the switching equation for any RC circuit. In Fig. 23-33, the initial capacitor voltage is zero, the target capacitor voltage is $+V_{CC}$, and the final capacitor voltage is $+2V_{CC}/3$. Substitute into Eq. (B-11) to get:

$$\frac{2V_{CC}}{3} = V_{CC}(1 - e^{-W/RC})$$

This simplifies to:

$$e^{-W/RC} = \frac{1}{3}$$

Solving for W gives

$$W = 1.0986RC \cong 1.1RC$$

Proof of Eqs. (23-28) and (23-29)

In Fig. 23-36, the capacitor upward charge takes time W. The capacitor voltage starts at $+V_{CC}/3$ and ends at $+2V_{CC}/3$ with a target voltage of $+V_{CC}$. Substitute into Eq. (B-11) to get:

$$\frac{2V_{CC}}{3} = \frac{V_{CC}}{3} + \left(V_{CC} - \frac{V_{CC}}{3}\right)(1 - e^{-W/RC})$$

This simplifies to:

$$e^{-W/RC} = 0.5$$

or

$$W = 0.693RC = 0.693(R_1 + R_2)C$$

The discharge equation is similar, except that R_2 is used instead of $R_1 + R_2$. In Fig. 23-36, the discharge time is $T - W$, which leads to:

$$T - W = 0.693R_2C$$

Therefore, the period is:

$$T = 0.693(R_1 + R_2)C + 0.693R_2C$$

and the duty cycle is:

$$D = \frac{0.693(R_1 + R_2)C}{0.693(R_1 + R_2)C + 0.693R_2C} \times 100\%$$

or

$$D = \frac{R_1 + R_2}{R_1 + 2R_2} \times 100\%$$

To get the frequency, take the reciprocal of the period T:

$$f = \frac{1}{T} = \frac{1}{0.693(R_1 + R_2)C + 0.693R_2C}$$

or

$$f = \frac{1.44}{(R_1 + 2R_2)C}$$

Appendix C

Introduction

In an effort to help the reader understand the concepts presented in this textbook, key examples and problems will be presented through the use of computer simulation using MultiSim. MultiSim is an interactive circuit simulation package that allows the student to view their circuit in schematic form while measuring the different parameters of the circuit. The ability to create a schematic quickly and then analyze the circuit through simulation makes MultiSim a wonderful tool to help students understand the concepts covered in the study of electronics.

This primer will introduce the reader to the features of MultiSim that directly relate to the study of DC, AC, and semiconductor electronics. The topics covered are:

- Work Area
- Opening a File
- Running a Simulation
- Saving a File
- Components
- Sources
- Measurement Equipment
- Circuit Examples

Work Area

The power of this software lies in its simplicity. With just a few steps, a circuit can be either retrieved from disk or drawn from scratch and simulated. The main screen, as shown in Figure C-1, is divided into three areas: The drop down menu, the tool bars, and the work area.

The drop down menu gives the user access to all the functions of the program. Initially, the user will only utilize a few of the different menu selections. Each of the drop down menu main topics can be accessed by either a mouse click or by pressing the <alt> key and the underlined letter. For example, to access the File menu simply press <ALT><F> at the same time. The File menu will drop down as shown in Figure C-2.

Initially, there are only two selections from the drop down menu that need be mastered: Opening a file and saving a file. The rest of the menu options can be explored as time permits.

The tool bars beneath the drop down menu provide access to all of the menu selections. Typically, a user will access them through the tool bars instead of the drop down menus. The most important icon in the assorted tool bars is the on-off switch. The on-off switch starts and stops the simulation. The push button next to the on-off switch will pause the simulation. Pressing the Pause button while the simulation is running allows the viewing of a waveform or meter reading without the display changing.

Figure C-1 Main Screen

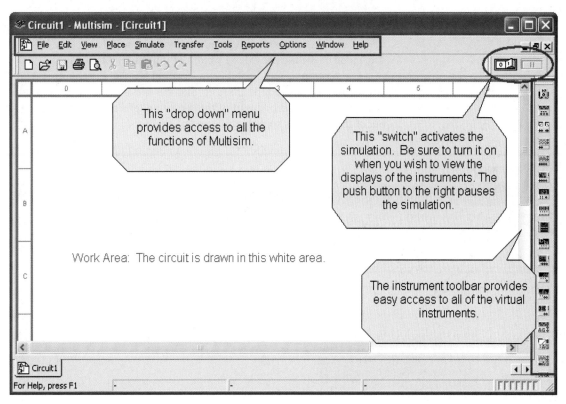

Figure C-2 File Drop Down Menu

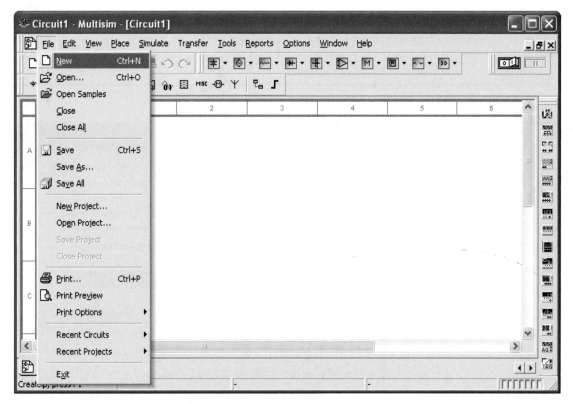

Opening a File

The circuits referenced in this textbook are included on a compact disk located in the back of this textbook. The files are divided into folders, one for each chapter. The name of the file provides a wealth of information to the user.

Example: A typical file name would be "Ch 4 Problems 4-11." This first part of the file name tells the user that the file is located in the folder labeled "Chapter 4." The second part of the file name tells the user that it is question 11 out of the Problems section at the end of chapter 4.

Figure C–3   Opening a File

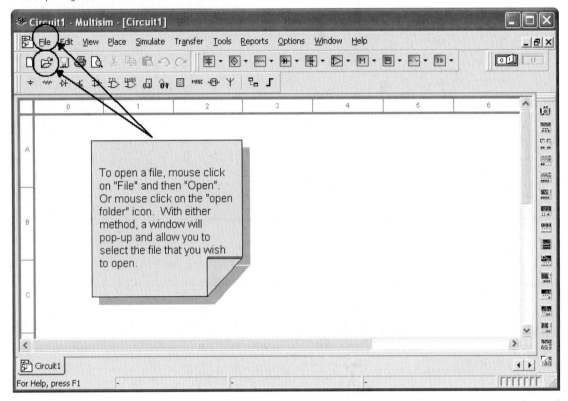

Figure C–4   Open file Dialog Box

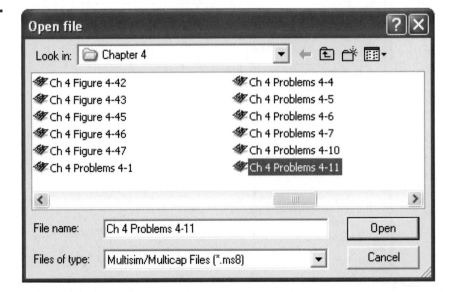

To open a file, either mouse click on the word "File" located on the drop down menu bar and then mouse click on the Open command or mouse click on the open folder icon located on the tool bar as shown in Figure C-3. Both methods will cause the Open File dialog box shown in Figure C-4 to pop open. Navigate to the appropriate chapter folder and retrieve the file needed.

Running a Simulation

The MultiSim files developed for this textbook present the circuit in a standard format. The instrumentation is typically connected to the appropriate places within the circuit. If the instrument's display is not visible, double mouse click on the instrument icon and the display will pop up. There is a help screen referred to as the "Description Box" with helpful information relating to the circuit and the instruments within the circuit. This Description Box is opened by pressing <Control><D> while in MultiSim.

The simulation can be started three ways:

- Selecting "Simulate" from the drop down menu and then selecting "Run"
- Pressing the <F5> key
- Pressing the toggle switch with a mouse click

All three of these ways are illustrated in Figure C-5.

Figure C-5   Starting the Simulation

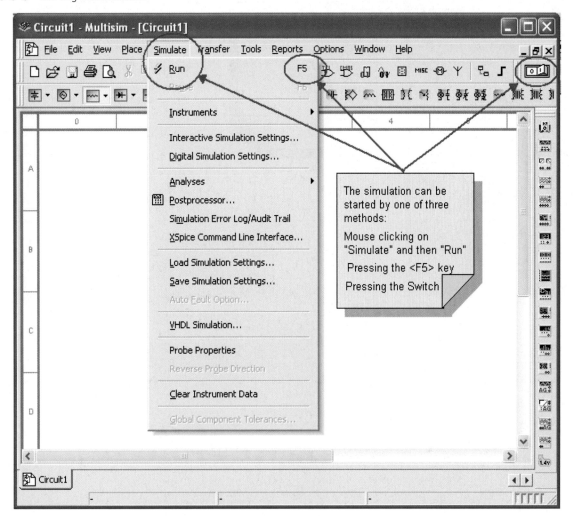

Saving a File

If the file has been modified, it needs to be saved under a new file name. As shown in Figure C-6, select "File" with a mouse click located on the drop down menu bar. Then select "Save As" from the drop down menu. This will cause the "Save As" dialog box to open. Give the file a new name and "Press" the save button with a mouse click. The process is demonstrated in Figure C-7.

Figure C-6 *"Save As . . . " Screen*

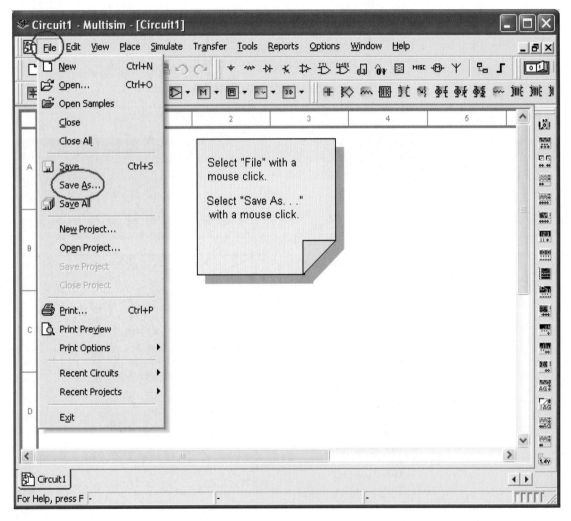

Figure C-7 "Save As . . ." Dialog Box

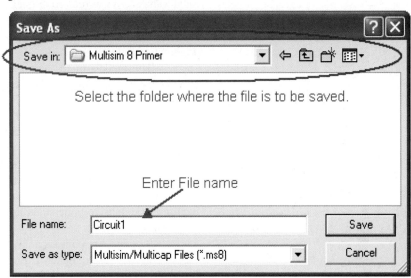

Components

There are two kinds of component models used in MultiSim: Those modeled after actual components and those modeled after "ideal" components. Those modeled after ideal components are referred to as "virtual" components. There is a broad selection of virtual components available, as shown in Figure C-8.

Figure C-8 Virtual Component List

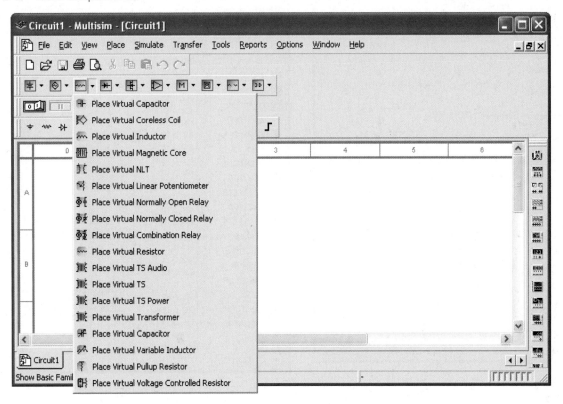

The difference between the two types of components resides in their rated values. The virtual components can have any of their parameters varied, whereas those modeled after actual components are limited to real world values. For example, a virtual resistor can have any value resistance and percent tolerance as shown in Figure C-9.

The models of the actual resistors are limited to standard values with either a 1% or 5% tolerance. The same is true for all other components modeled after real components, as shown in Figure C-10. This is especially important when the semiconductor devices are used in a simulation. Each of the models of actual semiconductors will function in accordance with their data sheets. These components will be listed by their actual device number as identified by the manufacturers. For example, a common diode is the 1N4001. This diode, along with many others, can be found in the semiconductor library of actual components. The parameters of the actual components libraries can also be modified, but that requires an extensive understanding of component modeling and is beyond the scope of this primer.

If actual components are selected for a circuit to be simulated, the measured value may differ slightly from the calculated values as the software will utilize the tolerances to vary the results. If precise results are required, the virtual components can be set to specific values with a zero percent tolerance.

Two of the components require interaction with the user. The switch is probably the most commonly used of these two devices. The movement of the switch is triggered by pressing the key associated with each switch as shown in Figure C-11. The key is selected while in the switch configuration screen, Figure C-12. If two switches are assigned the same key, they both will move when the key is pressed.

Figure C-9 Configuration Screen for a Virtual Resistor

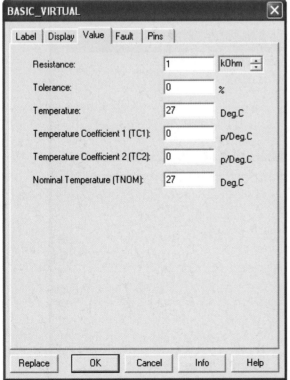

Figure C-10 Component Listing for Resistors

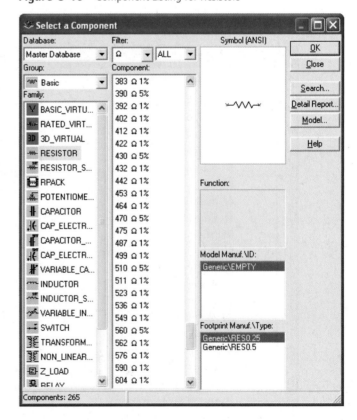

Figure C-11 Switches

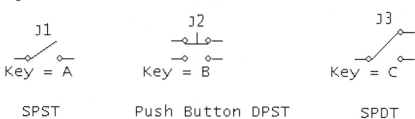

J1

Key = A

SPST

J2

Key = B

Push Button DPST

J3

Key = C

SPDT

Figure C-12 Switch Configuration Screen

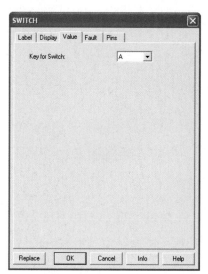

The second component that requires interaction with the user is the potentiometer. The potentiometer will vary its resistance in predetermined steps with each key press. The pressing of the associated letter on the keyboard will increase the resistance and the pressing of the <shift> key and the letter will decrease the resistance. As shown in Figure C-13, the percent of the total resistance is displayed next to the potentiometer. The incremental increase or decrease of resistance is set by the user in the configuration screen. The associated key is also set in the configuration screen as shown in Figure C-14.

Figure C-13 Potentiometer

R1

1KΩ_LIN   50%
Key = A

Figure C-14 Potentiometer Configuration Screen

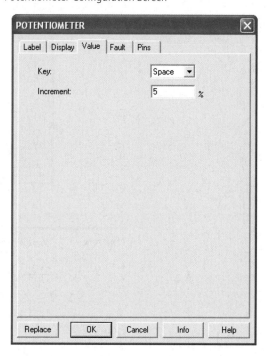

Sources

Figure C-15 DC Source as a Battery

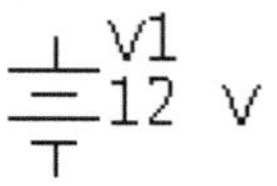

V1
12 V

In the study of DC and AC electronics, the majority of the circuits include either a voltage or current source. There are two main types of voltage sources: DC and AC sources. The DC source can be represented two ways: As a battery in Figure C-15 and as a voltage supply.

The voltage rating is fully adjustable. The default value is 12 VDC. If the component is double clicked, the configuration screen shown in Figure C-16 will pop up and the voltage value can be changed.

The voltage supplies are used in semiconductor circuits to represent either a positive or negative voltage supply. Figure C-17 contains the V_{CC} voltage supply used in transistor circuits. FET circuits will utilize the V_{DD} voltage supply as illustrated in Figure C-18.

Figure C-19 depicts the $+V_{CC}$ and the $-V_{EE}$ voltage sources. These sources are found in operational amplifier circuits. Operational amplifiers typically have two voltage supplies: a negative (V_{EE}) and a positive (V_{CC}) voltage supply, Figure C-20.

Figure C-16 Configuration Screen for the DC Source

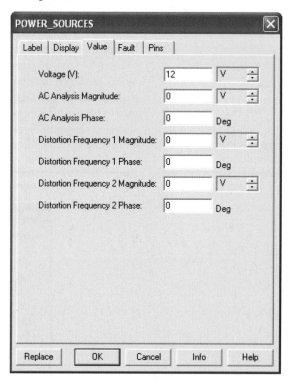

Figure C-17 V_{CC} Voltage Source

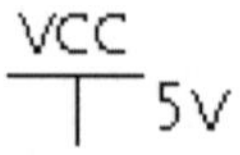

VCC
5V

Figure C-18 V_{DD} Voltage Source

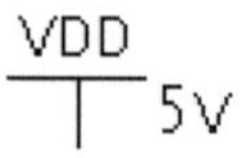

VDD
5V

Figure C-19 V_{CC} and V_{EE} Voltage Sources

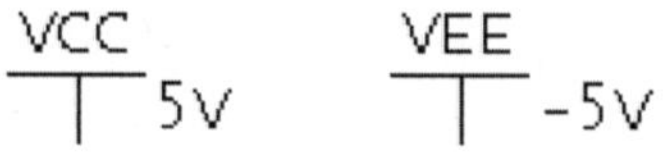

VCC
5V

VEE
-5V

Figure C-20 V_CC and V_EE Op Amp
Example

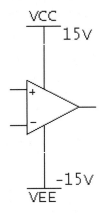

Figure C-21 V_CC Configuration Screen

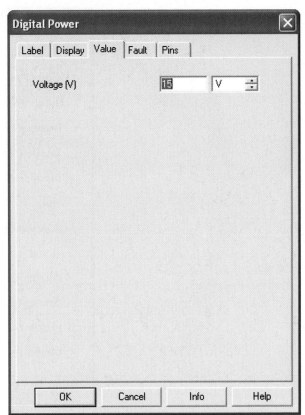

The voltage rating is fully adjustable for all three voltage sources. The default value is +5 VDC for V_{CC} and V_{DD}. The default value for V_{EE} is −5 VDC. If the component is double clicked, the configuration screen shown in Figure C-21 will pop up and the voltage value can be changed.

The AC source can be represented as either a schematic symbol or it can take the form of a function generator. The schematic symbol as shown in Figure C-22 will include information about the AC source. This information will include the device reference number, V_{RMS} value, frequency, and phase shift. These values are fully adjustable. The default values are shown in Figure C-23. If the component is double clicked, the configuration screen will pop up and the value can be changed.

Figure C-22 AC Source

Figure C-23   Configuration Screen for an AC Source

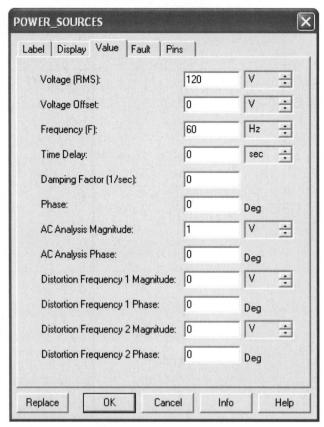

MultiSim provides two function generators: The generic model and the Agilent model. The Agilent model 33120A has the same functionality as the actual Agilent function generator.

The generic function generator icon is shown in Figure C-24, along with the configuration screen. The configuration screen is displayed when the function generator icon is doubled clicked. The generic function generator can produce three types of waveforms: Sinusoidal wave, triangular wave, and a square wave. The frequency, duty cycle, amplitude, and DC offset are all fully adjustable.

The Agilent function generator is controlled via the front panel as shown in Figure C-25. The buttons are "pushed" by a mouse click. The dial can be turned by dragging the mouse over it or by placing the cursor over it and spinning the wheel on the mouse. The latter is by far the preferred method.

Figure C-24   Generic Function Generator and Configuration Screen

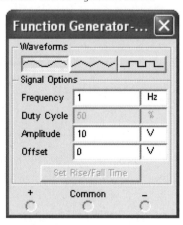

Figure C-25 Agilent Function Generator

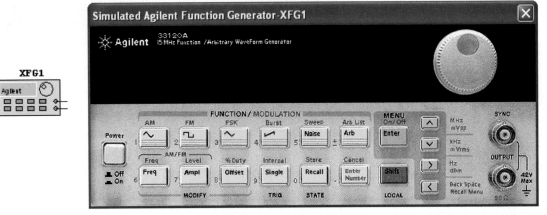

XFG1

There are two types of current sources: DC and AC sources. The DC current source is represented as a circle with a downward pointing arrow in it. The arrow in Figure C-26 represents the direction of current flow. The arrow can be pointed downward for conventional current flow. Electron current flow can be simulated by rotating the symbol 180° and the arrow will point upwards.

The current rating is fully adjustable. The default value is 1 A. If the component is double clicked, the configuration screen in Figure C-27 will pop up and the current value can be changed.

Figure C-26 DC Current Source

Figure C-27 DC Current Source Configuration Screen

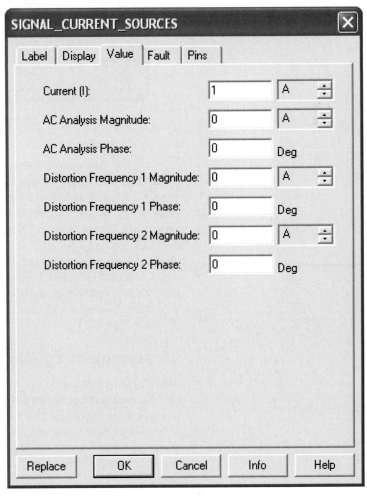

Figure C-28   AC Current Source

Figure C-29   AC Current Source Configuration Screen

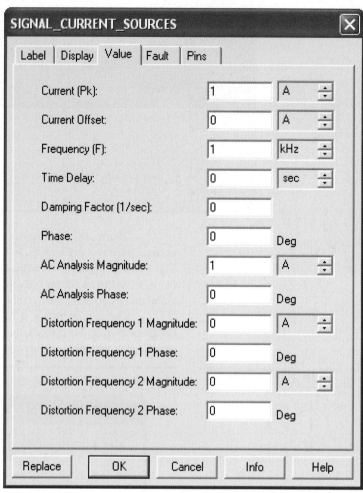

The AC current source is represented as a circle with a downward pointing arrow. There is a sine wave across the arrow. The schematic symbol in Figure C-28 will include information about the AC current source. This information will include the device reference number, I_{Peak} value, frequency, and phase shift. These values are fully adjustable. The default values are shown in Figure C-29. If the component is double clicked, the configuration screen will pop up and the values can be changed.

MultiSim requires a ground to be present in the circuit in order for the simulation to function properly. The circuit must contain a ground and all instrumentation must have a ground connection. The schematic symbol for ground is shown in Figure C-30.

Figure C-30   Ground

Measurement Equipment

MultiSim provides a wide assortment of measurement equipment. In the study of DC, AC and semiconductor electronics, the three main pieces of measurement equipment are the digital Multimeter, the oscilloscope, and the Bode Plotter. The first two pieces of equipment are found in test labs across the world. The Bode Plotter is a fictitious device that automates the task of plotting output voltage verses frequency. This is usually done by taking many measurements and plotting the results in a spreadsheet. The Bode Plotter performs this task for you.

Multimeters

There are two Multimeters to choose from: The Generic Multimeter and the Agilent Multimeter. The Generic Multimeter will measure current, voltage, resistance, and decibels. The meter can be used for both DC and AC measurements. The different functions of the meter are selected by mouse clicking on the icon to the left in Figure C-31. The mouse click will cause the Multimeter display to pop up. The different functions on the display can be selected "pushing" the different buttons via a mouse click.

The Agilent Multimeter icon and meter display are shown in Figure C-32. The display is brought up by mouse clicking on the Agilent Multimeter icon. This Multimeter has the same functionality as the actual Agilent Multimeter. The different functions are accessed by "pushing" the buttons. This is accomplished by mouse clicking on the button. The input jacks on the right side of the meter display correspond to the five inputs on the icon. If something is connected to the icon, the associated jacks on the display will have a white "X" in them to show a connection.

Figure C-31 Generic Multimeter icon and Configuration Screen

Figure C-32 Agilent Multimeter Icon and Meter Display

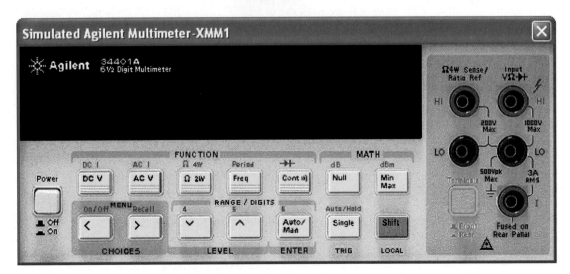

Oscilloscopes

There are three oscilloscopes to choose from: The Generic oscilloscope and the Agilent oscilloscope, and the Tektronix oscilloscope. The Generic oscilloscope shown in Figure C-33 is a dual channel oscilloscope. The oscilloscope display is brought up by mouse clicking on the oscilloscope icon. The settings can be changed by clicking in each box and bringing up the scroll arrows.

The Agilent oscilloscope icon in Figure C-34 has all of the functionality of the Model 54622D two channel oscilloscope. The Agilent oscilloscope is controlled via the front panel as shown in Figure C-35. The buttons are "pushed" by a mouse click. The dials can be turned by dragging the mouse over it or by placing the cursor over it and spinning the wheel on the mouse. The latter is by far the preferred method.

Figure C-33 Generic Oscilloscope Icon and Oscilloscope Display

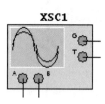

Figure C-34 Agilent Oscilloscope Icon

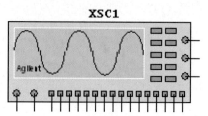

Figure C-35   Agilent Oscilloscope Display

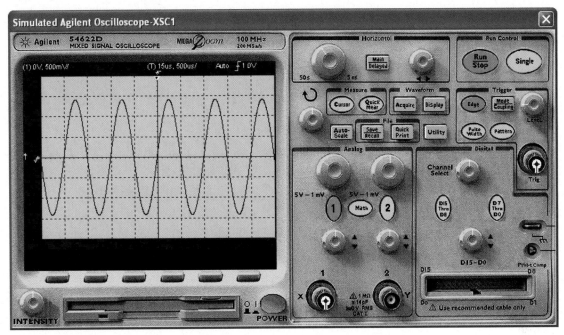

Figure C-36   Tektronix Oscilloscope Icon

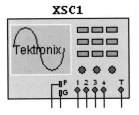

The Tektronix oscilloscope icon shown in Figure C-36 has all of the functionality of the Model TDS2024 four channel Digital Storage oscilloscope. The color of the four channels is the same as the channel selection buttons on the display: Yellow, blue, purple, and green for channels one through four respectively. The Tektronix oscilloscope is controlled via the front panel as seen in Figure C-37. The buttons are "pushed" by a mouse click. The dials can be turned by dragging the mouse over it or by placing the cursor over it and spinning the wheel on the mouse. The latter is by far the preferred method.

Figure C-37   Tektronix Oscilloscope Display

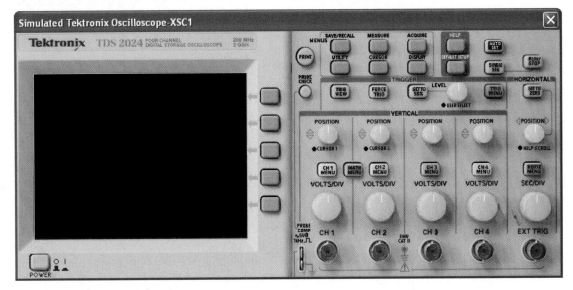

Figure C-38 Voltage and Current Meters

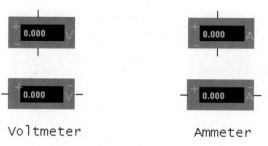

Voltmeter   Ammeter

Figure C-39 Voltmeter Configuration Screen

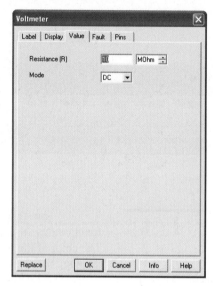

Voltage and Current Meters

When voltage or current need to be measured, MultiSim provides very simple voltmeters and ammeters as shown in Figure C-38. These meters can be placed throughout the circuit. The meters can be rotated to match the polarity needs of the circuit. The default is "DC." If the meters are to be used for AC measurement, then the configuration screen shown in Figure C-39 must be opened and that parameter changed to reflect AC measurement. To open the configuration screen, double click on the meter.

Bode Plotter

The Bode Plotter is used to view the frequency response of a circuit. In the actual lab setting, the circuit would be operated at a base frequency and the output of the circuit measured. The frequency would be incremented by a fixed amount and the measurement repeated. After operating the circuit at a sufficient number of incremental frequencies, the data would be graphed, with independent variable "frequency" on the X axis and dependant variable "amplitude" on the Y axis. This process can be very time consuming. MultiSim provides a simpler method of determining the frequency response of a circuit through the use of the virtual Bode Plotter.

In Figure C-40, the positive terminal of the input is connected to the applied signal source. The positive terminal of the output is connected to the output voltage of the circuit. The other two terminals are connected to ground. The value of the AC source does not matter; the AC source just needs to be in the circuit. The Bode Plotter will provide the input signal.

Figure C–40 Bode Plotter

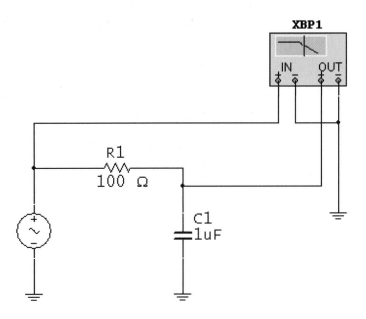

Figure C–41 Bode Plotter Display

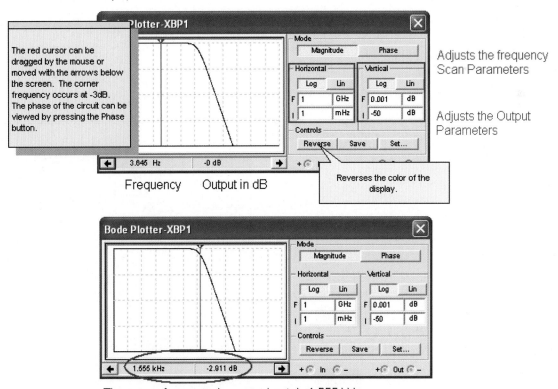

The red cursor can be dragged by the mouse or moved with the arrows below the screen. The corner frequency occurs at -3dB. The phase of the circuit can be viewed by pressing the Phase button.

Frequency   Output in dB

Adjusts the frequency Scan Parameters

Adjusts the Output Parameters

Reverses the color of the display.

The corner frequency is approximately 1.555 kHz

Circuit Examples

Example 1 Voltage Measurement Using a Voltmeter in a Series DC Circuit

A voltmeter in Figure C-42 is placed in parallel with the resistor to measure the voltage across it. The default is set for "DC" measurement. If AC is required, double click on the meter to bring up the configuration screen. All circuits must have a ground. Figure C-43 contains a *Quick Hint* on the use of the Voltmeter.

Example 2 Voltage Measurement Using a Generic Multimeter in a Series DC Circuit

A generic Multimeter is placed in parallel with the resistor to measure the voltage across it. Be sure to double click the generic Multimeter icon to bring up meter display as shown in Figure C-44. Press the appropriate buttons for "Voltage" and then "DC" or "AC" measurement. All circuits must have a ground. Figure C-45 contains a *Quick Hint* on the use of the Generic Multimeter.

Figure C-42   DC Voltage Measurement with a Voltmeter

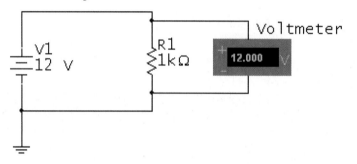

Figure C-43   Voltmeter Quick Hint

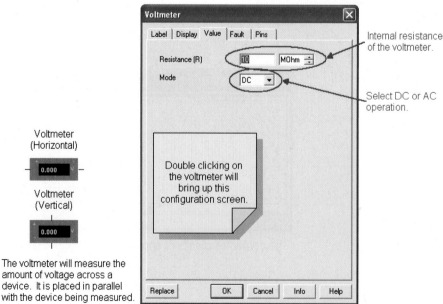

Figure C-44 DC Voltage measurement with a Generic Multimeter

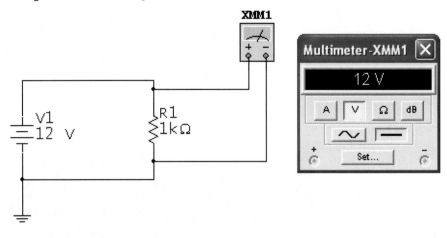

Figure C-45 Generic Multimeter Quick Hint

Generic Multimeter Icon

Double clicking on the Multimeter
brings up this display.

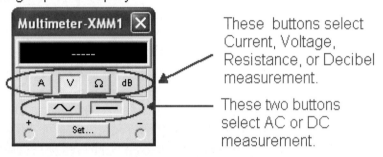

These buttons select
Current, Voltage,
Resistance, or Decibel
measurement.

These two buttons
select AC or DC
measurement.

Simply mouse click on the appropiate
buttons to set up the multimeter for the
type of measurement you wish to
make.

Example 3 Voltage Measurement Using an Agilent Multimeter in a Series DC Circuit

In Figure C-46, an Agilent Multimeter is placed in parallel with the resistor to measure the voltage across it. Be sure to double click the Agilent Multimeter icon to bring up the meter display. Press the appropriate buttons for "Voltage" and then "DC" or "AC" measurement. All circuits must have a ground. Note the two white circles and black x's on the right side of the display to indicate a connection to the meter. Note: This instrument requires that its power button be pressed to "turn on" the meter. Figure C-47 contains a *Quick Hint* on the use of the Agilent Multimeter.

Figure C–46   DC Voltage measurement with an Agilent Multimeter

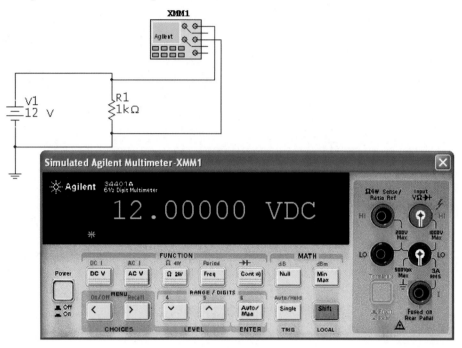

Figure C–47   Agilent Multimeter Quick Hint

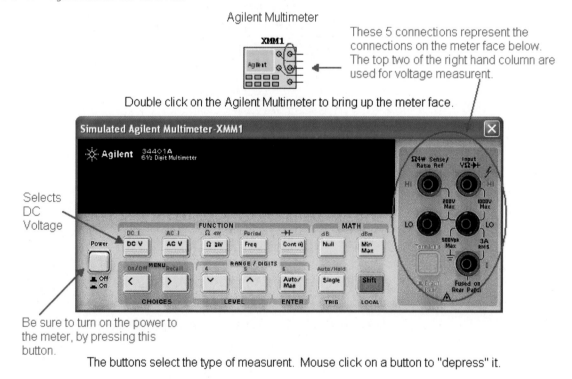

Agilent Multimeter

These 5 connections represent the connections on the meter face below. The top two of the right hand column are used for voltage measurent.

Double click on the Agilent Multimeter to bring up the meter face.

Selects DC Voltage

Be sure to turn on the power to the meter, by pressing this button.

The buttons select the type of measurent. Mouse click on a button to "depress" it.

Example 4 Current Measurement Using an Ammeter in a Series DC Circuit

In Figure C-48, an ammeter is placed in series with the resistor and DC source to measure the current flowing through the circuit. The default is set for "DC" measurement. If AC is required, double click on the meter to bring up the configuration screen. All circuits must have a ground. Figure C-49 contains a *Quick Hint* on the use of the ammeter.

Figure C-48   DC Current measurement with an Ammeter

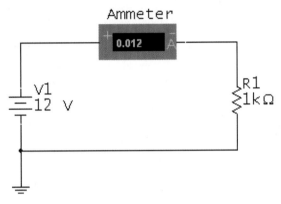

Figure C-49   Ammeter Quick Hint

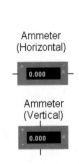

Ammeter
(Horizontal)

Ammeter
(Vertical)

The ammeter will measure the amount of current flowing through it. The meter will display the appropiate prefix, if needed. (m = milli, n = nano, p = pico)

The ammeter is placed in series.

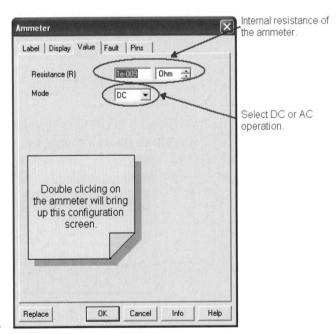

Example 5 Current Measurement Using a Generic Multimeter in a Series DC Circuit

In Figure C-50, a Generic Multimeter is placed in series with the resistor and DC source to measure the current flowing through the circuit. Be sure to double click the generic Multimeter icon to bring up the meter display. The current function is selected by clicking on the "A" on the meter display. Since the source is DC, the DC function of the meter is also selected, as indicated by the depressed button. All circuits must have a ground. Figure C-51 contains a *Quick Hint* on the use of the Generic Multimeter.

Figure C-50   DC Current measurement with a Generic Multimeter

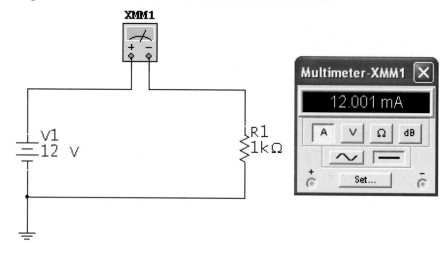

Figure C-51   Generic Multimeter Quick Hint

Generic Multimeter Icon

Double clicking on the Multimeter brings up this display.

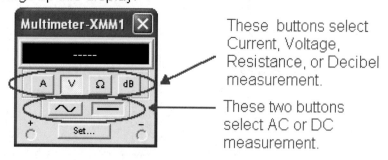

These buttons select Current, Voltage, Resistance, or Decibel measurement.

These two buttons select AC or DC measurement.

Simply mouse click on the appropiate buttons to set up the multimeter for the type of measurement you wish to make.

Example 6 Current Measurement Using an Agilent Multimeter in a Series DC Circuit

In Figure C-52, an Agilent Multimeter is placed in series with the resistor and source to measure the current flowing through the circuit. Be sure to double click the Agilent Multimeter icon to bring up the meter display. Selection of DC current measurement is the second function of the DC voltage measurement button. Be sure to press the "shift" button to access the second function of the voltage button. Note the two white circles and black x's on the right side of the display to indicate a connection to the meter. All circuits must have a ground.

This instrument requires that its power button be pressed to "turn on" the meter. Figure C-53 contains a *Quick Hint* on the use of the Agilent Multimeter for current measurement.

Figure C-52 DC Current measurement with an Agilent Multimeter

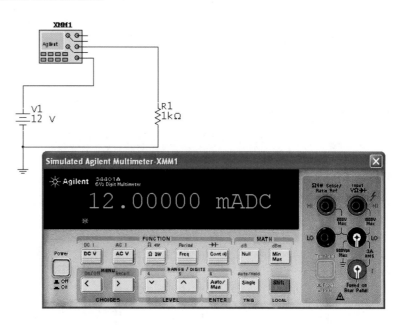

Figure C-53 Agilent Multimeter Quick Hint for Current Measurement

Agilent Multimeter

These 5 connections represent the connections on the meter face below. The bottom two of the right hand column are used for current measurent.

Double click on the Agilent Multimeter to bring up the meter face.

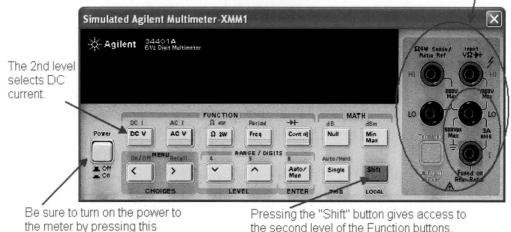

The 2nd level selects DC current.

Be sure to turn on the power to the meter by pressing this button.

Pressing the "Shift" button gives access to the second level of the Function buttons.

The buttons select the type of measurent. Mouse click on a button to "depress" it.

Example 7 Voltage Measurement Using a Generic Oscilloscope in a Series AC Circuit

In Figure C-54, channel 1 of the Generic oscilloscope is connected to the positive side of the resistor. The ground connection of the scope and the circuit must be grounded. The oscilloscope will use this ground as a reference point. This Generic oscilloscope's operation follows that of an actual oscilloscope. The oscilloscope display is brought up by mouse clicking on the oscilloscope icon. The settings can be changed by clicking in each box and bringing up the scroll arrows. Adjust the volts per division on the channel under measurement until the amplitude of the waveform fills the majority of the screen. Adjust the Time-base such that a complete cycle or two are displayed. Figures C-55, C-56 and C-57 contain *Quick Hint* on the use of the Generic oscilloscope.

Figure C-54   Voltage Measurement with a Generic Oscilloscope

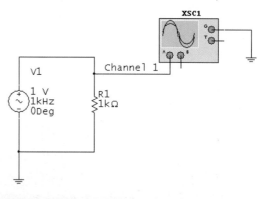

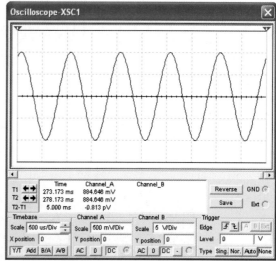

Figure C-55   Generic Oscilloscope Icon Quick Hint

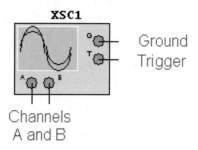

Figure C-56 Generic Oscilloscope
Quick Hint

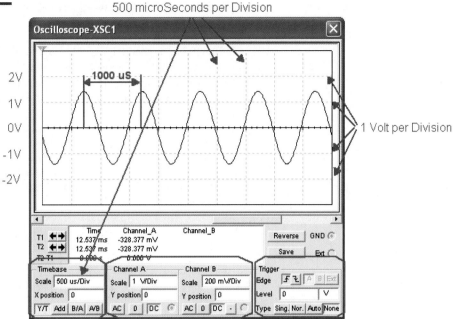

The Time-base controls the units for the "X" axis. The "X" axis represents time and the units are seconds per division.

Channel "A" and "B" control the units for the "Y" axis. The "Y" axis represents Voltage and the units are volts per division.

The Trigger controls when the oscilloscope starts to display the waveform. It can display it on the rising edge or the falling edge. The voltage level at which the oscilloscope is triggered is also selected here.

Figure C-57 Generic Oscilloscope
Quick Hint

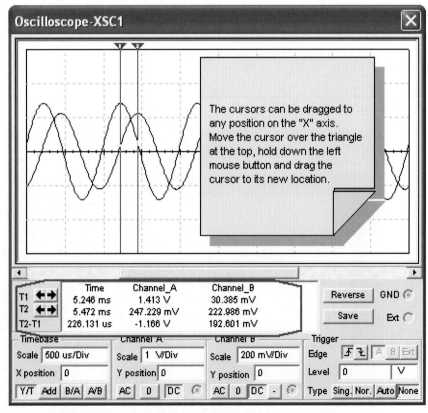

This box and data are from the cursors (1 and 2). The voltage level and time are measured at the point where the waveform passes through the cursor.

Example 8 Voltage Measurement Using the Agilent Oscilloscope in a Series AC Circuit

In Figure C-58, channel 1 of the Agilent oscilloscope is connected to the positive side of the resistor. The ground connection of the scope and the circuit must be grounded. The oscilloscope will use this ground as a reference point. This Agilent oscilloscope's operation follows that of a 2-channel, +16 logic channel, 100-MHz bandwidth Agilent Model 54622D Oscilloscope. The oscilloscope display as shown in Figure C-59, is brought up by mouse clicking on the oscilloscope icon.

Figure C-58   Voltage Measurement with an Agilent Oscilloscope

Figure C-59   Agilent Oscilloscope Display

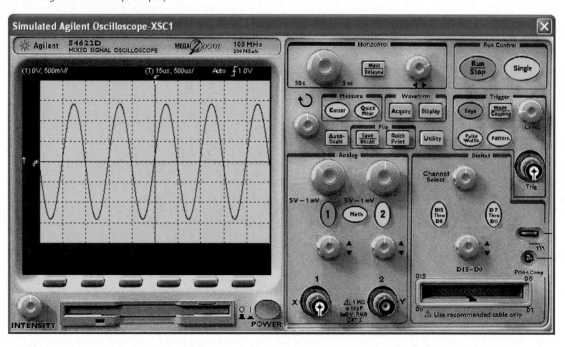

The settings can be changed by placing the mouse over the dials and spinning the mouse wheel or by "pressing" the buttons with a mouse click. Adjust the volts per division on the channel under measurement until the amplitude of the waveform fills the majority of the screen. Adjust the Time-base in the Horizontal section such that a complete cycle or two are displayed. This instrument requires that its power button be pressed to "turn on" the oscilloscope. Figures C-60, C-61 and C-62 contain *Quick Hints* on the use of the Agilent oscilloscope.

Figure C-60   Agilent Oscilloscope Icon Quick Hint

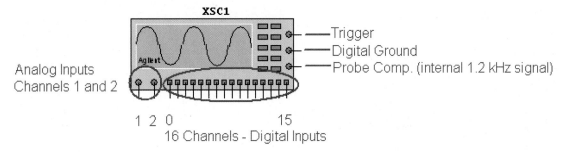

Figure C-61   Agilent Oscilloscope Quick Hint

The Horizontal section controls the units for the "X" axis. The "X" axis represents time and the units are seconds per division.

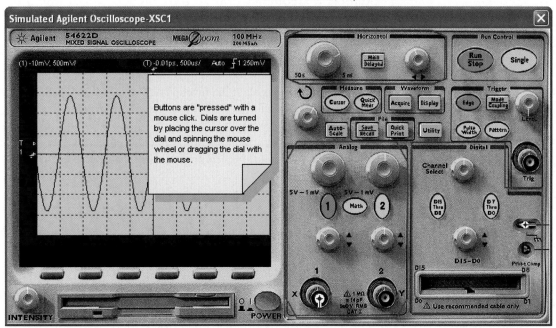

Channel "1" and "2" control the units for the "Y" axis. The "Y" axis represents Voltage and the units are volts per division.

16 Channel Digital Logic Input

Figure C-62 Agilent Oscilloscope Quick Hint

The Measure section includes cursor operation.
The Waveform section allows for waveform storage.
The File section saves, recalls, and prints.

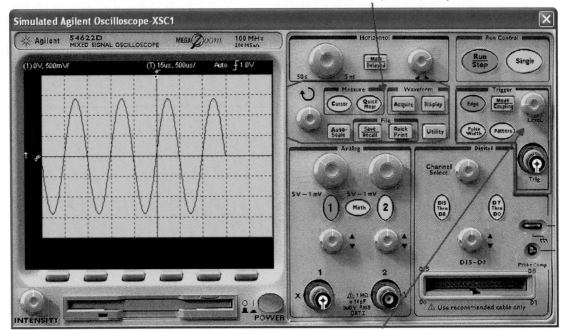

The Trigger controls when the oscilloscope starts to
display the waveform. It can display it on the rising edge
or the falling edge. The voltage level at which the
oscilloscope is triggered is also selected here.

Example 9 Frequency and Voltage Measurement Using the Tektronix Oscilloscope in a Series AC Circuit

In Figure C-63, channel 1 of the Tektronix oscilloscope is connected to the positive side of the resistor. The ground connection of the scope and the circuit must be grounded. The oscilloscope will use this ground as a reference point. The Tektronix oscilloscope has all of the functionality of the Model TDS2024 four channel Digital Storage oscilloscope. The oscilloscope display is brought up by mouse

Figure C-63 Voltage and Frequency measurement with a Tektronix Oscilloscope

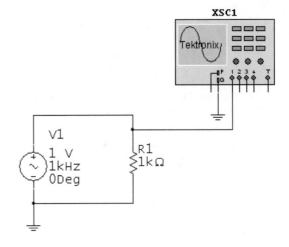

clicking on the oscilloscope icon. The settings can be changed by placing the mouse over the dials and spinning the mouse wheel or by "pressing" the buttons with a mouse click. Adjust the volts per division on the channel under measurement until the amplitude of the waveform fills the majority of the screen. Adjust the Time-base in the Horizontal section such that a complete cycle or two of the waveform is displayed. This instrument requires that its power button be pressed to "turn on" the oscilloscope. The voltage and frequency can be measured by the user or by using the "Measure" function of the oscilloscope.

Using volts per division and the seconds per division settings, the amplitude and frequency of the waveform in Figure C-64 can be determined. The amplitude of the waveform is two divisions above zero volts. (The yellow arrow points to the zero reference point.) The volts per division setting are set to 500 mV per division.

$$V_P = 2 \; divisions \times \frac{500 \; mV}{division}$$

$$V_P = 1 \; V$$

$$V_{PP} = 4 \; divisions \times \frac{500 \; mV}{division}$$

$$V_{PP} = 2 \; V$$

The period of the waveform is measured to be 1 mS. Since frequency is the reciprocal of the period, the frequency can be calculated.

$$F = \frac{1}{T}$$

$$F = \frac{1}{1 \; mS}$$

$$F = 1 \; kHz$$

Figure C-64   Measurement of the Period of the Waveform

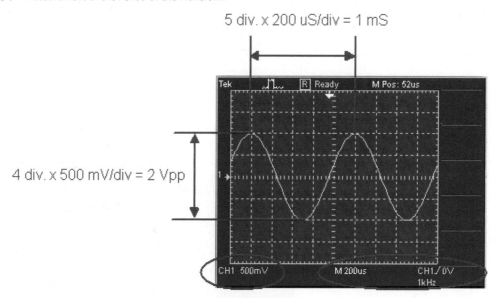

5 div. x 200 uS/div = 1 mS

4 div. x 500 mV/div = 2 Vpp

The oscilloscope displays the volts per division for each channel.

The oscilloscope displays the Time-base in seconds per division and calculates the frequency.

The Tektronix oscilloscope can also perform the voltage and frequency measurements automatically through the use of the "Measure" function. The four steps and the resulting display are shown in Figures C-65 and C-66. To set up the oscilloscope to measure these values automatically:

1. Press the Measure button.
2. Select the channel to be measured.
3. Select what is to be measured: Vpp, Frequency, etc.
4. Return to the Main Screen.
5. Repeat for other channels and or values.

Figures C-67, C-68 and C-69 contain *Quick Hint* on the use of the Tektronix oscilloscope.

Figure C-65 Tektronix Oscilloscope Measurement Function Set-up

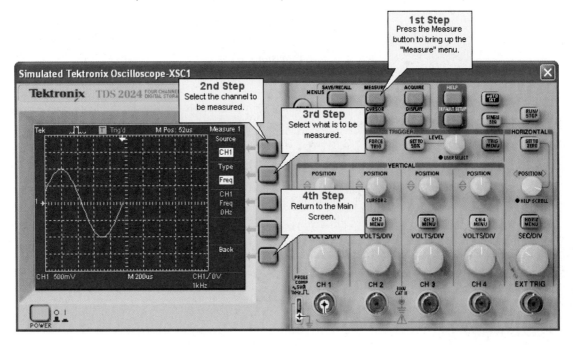

Figure C-66 Tektronix Oscilloscope Measurement Display

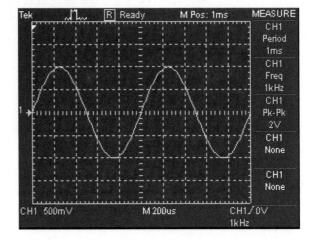

Figure C-67 Tektronix Oscilloscope Icon Quick Hint

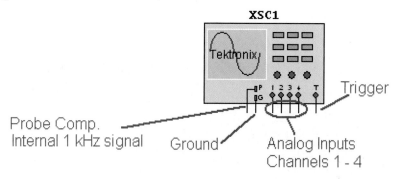

Trigger

Probe Comp.
Internal 1 kHz signal

Ground

Analog Inputs
Channels 1 - 4

Figure C-68 Tektronix Oscilloscope Quick Hint

Hint: The buttons are "pressed" by a mouse click. The dials
can be turned by dragging them with the cursor or by placing
the cursor over the dials and spinning the mouse wheel.

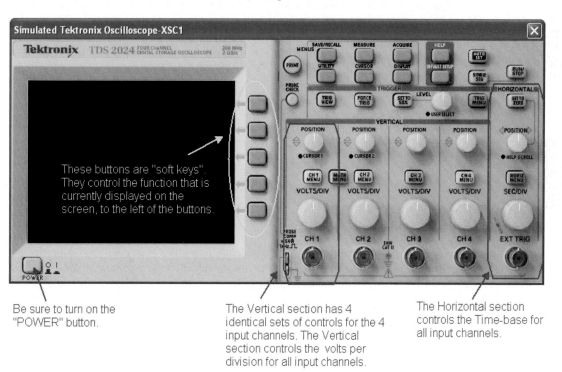

Be sure to turn on the
"POWER" button.

The Vertical section has 4
identical sets of controls for the 4
input channels. The Vertical
section controls the volts per
division for all input channels.

The Horizontal section
controls the Time-base for
all input channels.

Figure C–69   Tektronix Oscilloscope Quick Hint

These buttons bring up the different menus.

This button automatically sets up the oscilloscope.

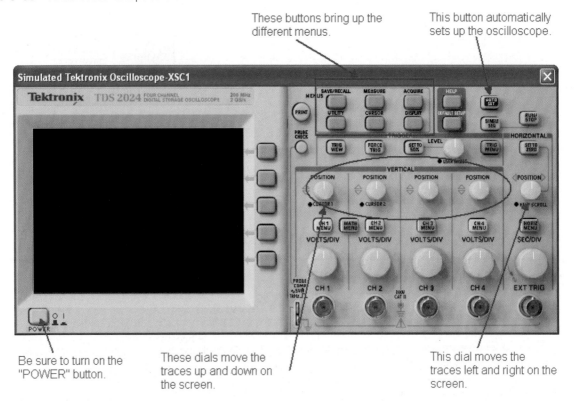

Be sure to turn on the "POWER" button.

These dials move the traces up and down on the screen.

This dial moves the traces left and right on the screen.

Conclusion

The ability to create a schematic quickly and then analyze the circuit through simulation makes MultiSim a wonderful tool to help students understand the concepts covered in the study of DC, AC, and semiconductor electronics.

This primer introduced the reader to the features of MultiSim that directly relate to the topics covered in this textbook. The features covered in this appendix will help the student to utilize the compact disk included with this textbook to its fullest extent.

Appendix D

Thevenizing the R/2R D/A Converter

With the switches $D_0 - D_4$ connected as in Fig. D-1a the binary input is $D_0 = 1$, $D_1 = 0$, $D_2 = 0$, and $D_3 = 0$. First, Thevenize the circuit from point A, looking toward D_0. When doing so, R_5 (20 kΩ) becomes in parallel with R_1 (20 kΩ) and the equivalent equals 10 kΩ. The Thevenized voltage at point A is one-half of V_{ref} and is equal to $+2.5$ V. This equivalent is shown in Fig. D-1b.

Next, Thevenize Fig. D-1b from point B. Notice how R_{TH} (10 kΩ) is in series with R_6 (10 kΩ). This 20 kΩ value is in parallel with R_2 (20 kΩ) and again gives us 10 kΩ. The Thevenized voltage seen from point B is again reduced by half to 1.25 V. This equivalent is shown in Fig. D-1c.

Now, Thevenize Fig. D-1c from point C. Again, R_{TH} (10 kΩ) is in series with R_7 (10 kΩ) and this 20 kΩ value becomes in parallel with R_3 (10 kΩ). V_{TH} is equal to 0.625 V. Notice how the V_{TH} values have been cut in half at each step. The Thevenin equivalent has been reduced to that of Fig. D-1d.

In Fig. D-1d, the op amp's inverting input and the top of R_4 (20 kΩ) is at a virtual ground. The voltage is equal to zero volts at this point. This places the entire 0.625 V of V_{TH} across R_{TH} and R_8 (10 kΩ). This results in an input current I_{in} of:

$$I_{in} = \frac{0.625 \text{ V}}{20 \text{ k}\Omega} = 31.25 \text{ }\mu\text{A}$$

Again, because of the virtual ground, this input current is forced to flow through R_f (20 kΩ) and produces an output voltage of:

$$V_{out} = -(I_{in} R_f) = -(31.25 \text{ }\mu\text{A})(20 \text{ k}\Omega) = -0.625 \text{ V}$$

This output voltage is the smallest output increment above 0 V and is referred to as the circuit's output resolution.

Figure D-1   (a) Original circuit; (b) Thevenized at point A; (c) Thevenized at point B; (d) Thevenized at point C.

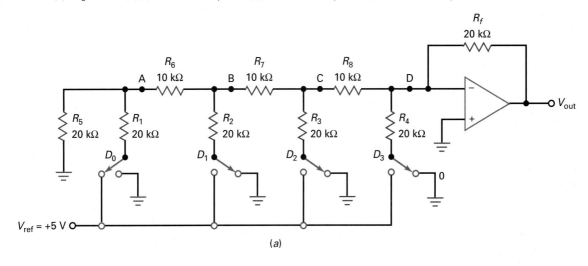

(a)

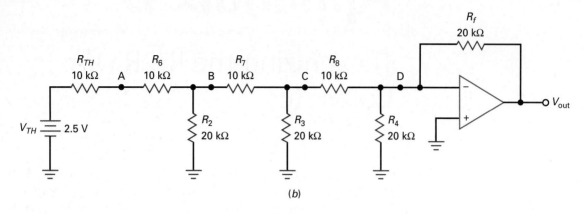

(b)

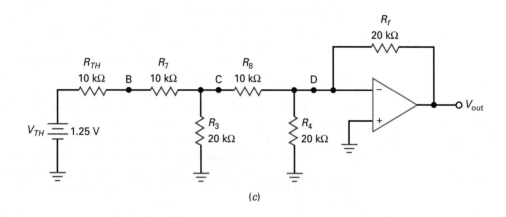

(c)

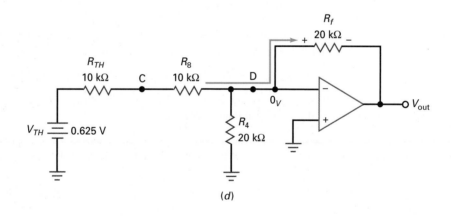

(d)

Appendix E

Summary Table Listing

Glossary

A

absolute value The value of an expression without regard for its sign. Sometimes called the *magnitude*. Given +5 and –5, the absolute value is 5.

acceptor A trivalent atom, one that has three valence electrons. Each trivalent atom produces one hole in a silicon crystal.

ac collector resistance The total ac load resistance found at the circuit's collector. This is often the parallel combination of R_C and R_L. This value is important when determining the voltage gain of a common-base or common-emitter amplifier.

ac compliance A large-signal amplifier with its Q point at the middle of the ac load line allowing a maximum peak-to-peak unclipped output.

ac current gain With a transistor, the ratio of ac collector current to ac base current.

ac cutoff The lower end of the ac load line. At this point, the transistor goes into cutoff and clips the ac signal.

Ac emitter feedback The ac signal developed across an unbypassed emitter resistance r_e.

ac emitter resistance The ac base-emitter voltage divided by the ac emitter current. This value is normally listed as r'_e and can be calculated by $r'_e = \dfrac{25 \text{ mV}}{I_E}$. This value is important when determining the input impedance and gain of a BJT amplifier.

ac equivalent circuit All that remains when you reduce the dc sources to zero and short all capacitors.

ac ground A node that is bypassed to ground through a capacitor. Such a node will show no ac voltage when it is probed by an oscilloscope, but it will indicate a dc voltage when it is measured with a voltmeter.

ac load line The locus of instantaneous operating points when an ac signal is driving the transistor. This load line is different from the dc load line whenever the ac load resistance is different from the dc load resistance.

ac resistance The resistance of a device to a small ac signal. The ratio of a voltage change to a current change. The key idea here is changes about an operating point.

ac saturation The upper end of the ac load line. At this point, the transistor goes into saturation and clips the ac signal.

ac short A coupling capacitor or bypass capacitor can be treated as an ac short if its capacitive reactance X_C is less than 1/10 of the resistance R. This can be stated mathematically as $X_C < 0.1R$.

active current gain The current gain in the active region of a transistor. That is what you usually find on a data sheet and what most people mean when they talk about current gain. (See *saturated current gain*.)

active filter In the good old days, filters were made out of passive components like inductors and capacitors. Some filters are still made this way. The problem is that at low frequencies, inductors become very large in passive filter designs. Op amps give another way to build filters and eliminate the problem of bulky inductors at low frequencies. Any filter using an op amp is called an active filter.

active half-wave rectifier An op amp circuit with the ability to rectify signals with input voltages less than 0.7 V. This circuit makes use of the very large open-loop gain of an op amp and is also known as a precision rectifier.

active loading This refers to using a bipolar or MOS transistor as a resistor. It's done to save space or to get resistances that are difficult with passive resistors.

active–load resistor A FET with its gate connected to the drain. The resulting two-terminal device is equivalent to a resistor.

active peak detector An op amp circuit used to detect low-level signals.

active positive clamper An op amp circuit used to add a positive dc component to an input signal.

active positive clipper An adjustable op amp circuit used to precisely control the level of positive output voltage.

active region Sometimes called the *linear region*. It refers to that part of the collector curve that is approximately horizontal. A transistor operates in the active region when it is used as an amplifier. In the active region, the emitter diode is forward-biased, the collector diode is reverse-biased, the collector current almost equals the emitter current, and the base current is much smaller than either the emitter or collector current.

all-pass filter A specialized filter having the ideal ability to pass all frequencies between zero and infinity. This filter is also called a *phase filter* because of its ability to shift the phase of the output signal without changing the magnitude.

ambient temperature The temperature of the immediate surrounding air around a device.

amplifier A circuit that can increase the peak-to-peak voltage, current, or power of a signal.

amplitude The size of a signal, usually its peak value.

analog The branch of electronics dealing with infinitely varying quantities. Often referred to as *linear electronics*.

analogy A likeness in some ways between dissimilar things that are otherwise unlike. The analogy between bipolar transistors and JFETs is an example. Because the devices are similar, many of the equations for them are identical except for a change of subscripts.

anode The element of an electronic device that receives the flow of electron current.

approximation A way to retain your sanity with semiconductor devices. Exact answers are tedious and time-consuming and almost never justified in the real world of electronics. On the other hand, approximations give us quick answers, usually adequate for the job at hand.

Armstrong oscillator A circuit distinguished by its use of transformer coupling for the feedback signal.

astable A digital switching circuit with no stable states. This circuit is also referred to as a *free-running* circuit.

attenuation A reduction in signal intensity, normally expressed in decibels. The signal reduction value is compared to the signal level at the midband of the filter. Mathematically, it is expressed as attenuation $= \dfrac{v_{\text{out}}}{v_{\text{out(mid)}}}$ and decibel attenuation $= 20 \log$ attenuation.

audio amplifier Any amplifier designed for the audio range of frequencies, 20 Hz to 20 kHz.

automatic gain control (AGC) A circuit designed to correct the gain of an amplifier according to the amplitude of the incoming signal.

avalanche effect A phenomenon that occurs for large reverse voltages across a *pn* junction. The free electrons are accelerated to such high speeds that they can dislodge valence electrons. When this happens, the valence electrons become free electrons that dislodge other valence electrons.

averager An op amp circuit constructed to provide an output voltage equal to the average of all input voltages.

B

back diode A diode with properties that enable it to conduct better in the reverse direction than in the forward direction. Commonly used in the rectification of weak signals.

bandpass filter A filter capable of passing a range of input frequencies with minimum attenuation while blocking all frequencies below and above the lower and upper cutoff frequencies f_1 and f_2.

bandstop filter A filter capable of rejecting a range of input frequencies while effectively passing all frequencies below and above the cutoff frequencies f_1 and f_2. This filter is also referred to as a notch filter.

bandwidth The difference between the two dominant critical frequencies of an amplifier. If the amplifier has no lower critical frequency, the bandwidth equals the upper critical frequency.

barrier potential The voltage across the depletion layer. This voltage is built into the *pn* junction because it is the difference of potential between the ions on both sides of the junction. It equals approximately 0.7 V for a silicon diode.

base The middle part of a transistor. It is thin and lightly doped. This permits electrons from the emitter to pass through it to the collector.

base bias The worst way to bias a transistor for use in the active region. This type of bias sets up a fixed value of base current.

Bessel filter Filter providing the desired frequency response but with a constant time delay in the passband.

BIFET op amp An IC op amp that combines FETs and bipolar transistors, usually with FET source followers at the front end of the device, followed by bipolar stages of gain.

bipolar junction transistor (BJT) A transistor where both free electrons and holes are necessary for normal operation.

biquadratic filter An active filter, also known as a *TT (Tow-Thomas) filter,* with the ability to independently tune its voltage gain, center frequency, and bandwidth using separate resistors.

Bode plot A graph showing the gain or phase performance of an electronic circuit at various frequencies.

boost regulator The basic topology for a switching regulator circuit in which the output voltage is higher than the input voltage.

bootstrapping A "following" function in which the inverting input voltage immediately increases or decreases the same amount as the noninverting input voltage initiates.

breakdown region For a diode or transistor, it is the region where either avalanche or the zener effect occurs. With the exception of the zener diode, operation in the breakdown region is to be avoided under all circumstances because it usually destroys the device.

breakdown voltage The maximum reverse voltage a diode can withstand before avalanche or the zener effect occurs.

breakover When a transistor breaks down, the voltage across it remains high. But with a thyristor, breakdown turns into saturation. In other words, breakover refers to the way a thyristor breaks down and then immediately goes into saturation.

bridge rectifier The most common type of rectifier circuit. It has four diodes, two of which are conducting at the same time. For a given transformer, it produces the largest dc output voltage with the smallest ripple.

buck–boost regulator The basic topology for a switching regulator circuit in which a positive input voltage produces a negative output voltage.

buck regulator The basic topology for a switching regulator circuit in which the output voltage is lower than the input voltage.

buffer A unity gain amplifier (voltage follower) having a high input impedance and a low output impedance used primarily to provide isolation between two parts of a circuit.

buffer amplifier This is an amplifier that you use to isolate two other circuits when one is overloading the other. A buffer amplifier usually has a very high input impedance, a very low output impedance, and a voltage gain of 1. These qualities mean that the buffer amplifier will transmit the output of the first circuit to the second circuit without changing the signal.

bulk resistance The ohmic resistance of the semiconductor material.

Butterworth filter This is a filter designed to produce as flat a response as possible up to the cutoff frequency. In other words, the output voltage remains constant almost all the way to the cutoff frequency. Then it decreases at $20n$ dB per decade, where n is the number of poles in the filter.

bypass capacitor A capacitor used to ground a node.

C

capacitive coupling The use of a capacitor to pass the ac signal from one stage to another while blocking the dc component of the waveform.

capacitor input filter Nothing more than a capacitor across the load resistor. This type of passive filter is the most common.

capture range The range of input frequencies in which a phase-locked-loop (PLL) circuit can lock on to the input signal.

carrier The high-frequency output signal of a transmitter that is caused to vary in amplitude, frequency, or phase by a modulating signal.

cascaded stages Connecting two or more stages so that the output of one stage is the input to the next.

case temperature This is the temperature of the transistor case or package. When you pick up a transistor, you are in contact with the case. If the case is warm, you are feeling the case temperature.

cathode The element of an electronic device that provides the flow of electron current.

CB amplifier An amplifier configuration in which the input signal is fed into the emitter terminal and the output signal is taken from the collector terminal.

CC amplifier An amplifier configuration in which the input signal is fed into the base terminal and the output signal is taken from the emitter terminal. Also called *emitter follower.*

CE amplifier The most widely applied amplifier configuration, in which the input signal is fed into the base terminal and the output signal is taken from the collector circuit.

channel The *n*-type or *p*-type semiconductor material that provides the main current path between the source and drain of a field effect transistor.

Chebyshev filter A filter with extremely good selectivity. The attenuation rate is much higher than that of Butterworth filters. The main problem with this filter is the ripple in the passband.

chip This has two meanings. First, an IC manufacturer produces hundreds of circuits on a large wafer of semiconductor material. Then the wafer is cut into individual chips, each containing one monolithic circuit. In this case, no leads have been connected to the chip. The chip is still an isolated piece of semiconductor material. Second, after the chip has been put inside a package and external leads have been connected to it, you have a finished IC. This finished IC is also referred to as a *chip.* For instance, we can call a 741C a chip.

chopper A JFET circuit that uses either a shunt or a series switch to convert a dc input voltage to a square-wave output.

clamper A circuit for adding a dc component to an ac signal. Also known as a *dc restorer.*

Clapp oscillator A series-tuned Colpitts configuration noted for its good frequency stability.

class A operation This means the transistor is conducting throughout the ac cycle without going into saturation or cutoff.

class AB operation A power amplifier biased so that each transistor conducts slightly more that $180°$ of the input signal to reduce crossover distortion.

class B operation Biasing of a transistor in such a way that it conducts for only half of the ac cycle.

class C operation Biasing a transistor amplifier so that collector current flows for less that $180°$ of the ac input cycle.

class–D amplifier An amplifier configuration where the output transistors are driven to saturation and cutoff. The two-state output waveform varies its duty cycle based on an input signal level and is essentially pulse width modulated. This results in very low power dissipation by the output transistors and high efficiencies.

clipper A circuit that removes some part of a signal. Clipping may be undesirable in a linear amplifier or desirable in a circuit such as a limiter.

closed-loop quantity The value of any quantity such as the voltage gain, input impedance, and output impedance that is changed by negative feedback.

closed-loop voltage gain Designated as $A_{V(CL)}$ or A_{CL}, this specification designates the voltage gain of an op amp with a feedback path between the output and input.

CMOS inverter A circuit with complementary MOS transistors. The input voltage is either low or high, and the output voltage is either high or low.

cold-solder joint A poor solder connection resulting from insufficient heat applied during the soldering process. The cold-solder joint may act as an intermittent connection or as no connection at all.

collector The largest part of a transistor. It is called the collector because it collects or gathers the carriers sent into the base by the emitter.

collector cutoff current The small collector current that exists when the base current is zero in a CE connection. Ideally, there should be no collector current. But there is because of the minority carriers and the surface leakage current of the collector diode.

collector diode The diode formed by the base and collector of a transistor.

collector-feedback bias An attempt to stabilize the Q point of a transistor circuit by connecting a resistor between the collector and base leads.

Colpitts oscillator One of the most widely used LC oscillators. It consists of a bipolar transistor or FET and an LC resonant circuit. You can recognize it because it has two capacitors in the tank circuit. They act as a capacitive voltage divider that produces the feedback voltage.

common-anode A circuit configuration in a seven-segment indicator where each of the anodes are tied together and connected to a common positive dc supply.

common base (CB) An amplifier configuration in which the input signal is fed into the emitter terminal and the output signal is taken from the collector terminal.

common-cathode A circuit configuration in a seven-segment indicator where all of the cathodes are tied together and connected to a common negative dc supply.

common-collector amplifier This is an amplifier whose collector is at ac ground. The signal goes into the base and comes out of the emitter.

common-emitter circuit A transistor circuit where the emitter is common or grounded.

common-mode rejection ratio (CMRR) The ratio of differential gain to common-mode gain in an amplifier. It is a measure of the ability to reject a common-mode signal and is usually expressed in decibels.

common-mode signal A signal that is applied with equal strength to both inputs of a diff amp or an op amp.

common-source (CS) amplifier A JFET amplifier in which the signal is coupled directly into the gate and all of the ac input voltage appears between the gate and the source, producing an amplified and inverted ac output voltage.

comparator A circuit or device that detects when the input voltage is greater than a predetermined limit. The output is either a low or a high voltage. The predetermined limit is called the *trip point.*

compensating capacitor A capacitor inside an op amp that prevents oscillations. Also, any capacitor that stabilizes an amplifier with a negative-feedback path. Without this capacitor, the amplifier will oscillate. The compensating capacitor produces a low critical frequency and decreases the voltage gain at a rate of 20 dB per decade above the midband. At the unity-gain frequency, the phase shift is in the vicinity of 270°. When the phase shift reaches 360°, the voltage gain is less than 1 and oscillations are impossible.

compensating diodes These are the diodes used in a class B push-pull emitter follower. These diodes have current-voltage curves that match those of the emitter diodes. Because of this, the diodes compensate for changes in temperature.

complementary Darlington A Darlington connection composed of *npn* and *pnp* transistors.

complementary MOS (CMOS) A method of reducing the current drain of a digital circuit by combining *n*-channel and *p*-channel MOSFETs.

conduction angle The angle or number of electrical degrees between the start and end of conduction for a thyristor with an ac waveform applied.

conduction band An energy band in a semiconductor in which electrons are free to move. This energy band is one level higher than the valence band.

correction factor A number used to describe how much one quantity differs from another. This value can be useful when comparing the emitter current to the collector current and determining the percent error that could result.

coupling capacitor A capacitor used to transmit an ac signal from one node to another.

coupling circuit A circuit that couples a signal from a generator to a load. The capacitor is in series with the Thevenin resistance of the generator and the load resistance.

covalent bond The shared electrons between the silicon atoms in a crystal represent covalent bonds because the adjacent silicon atoms pull on the shared electrons, just as two tug-of-war teams pull on a rope.

critical frequency Also known as the *cutoff frequency, break frequency, corner frequency,* etc. This is the frequency where the total

resistance of an RC circuit equals the total capacitive reactance.

crossover distortion The output distortion of a class B emitter follower amplifier resulting from biasing the transistors at cutoff. This distortion occurs during the period when one transistor cuts off and the other transistor comes on. This distortion can be reduced by biasing the transistors slightly above cutoff or class AB.

crowbar The metaphor used to describe the action of an SCR when it is used to protect a load against supply overvoltage.

crystal The geometric structure that occurs when silicon atoms combine. Each silicon atom has four neighbors, and this results in a special shape called a *crystal.*

current amplifier An amplifier configuration in which an input current produces a higher output current. An op amp ICIS circuit has the characteristics of a very low input impedance and high output impedance.

current booster A device, usually a transistor, that increases the maximum allowable load current of an op-amp circuit.

current-controlled current source (ICIS) A type of negative feedback amplifier in which the input current is amplified to get a larger output current, ideal because of stabilized current gain, zero input impedance, and infinite output impedance.

current-controlled voltage source (ICVS) Sometimes called a *transresistance amplifier,* this type of negative feedback amplifier has the input current controlling the output voltage.

current drain The total dc current I_{dc} supplied to an amplifier by the dc voltage source. This current is the combination of the biasing current and the collector current through the transistor.

current feedback This is a type of feedback where the feedback signal is proportional to the output current.

current gain Abbreviated as Ai, this value represents the ratio of output current divided by the input current.

current limiting Electronically reducing the supply voltage so that the current does not exceed a predetermined limit. This is necessary to protect the diodes and transistors, which usually blow out faster than the fuse under shorted-load conditions.

current mirror A circuit that acts as a current source whose value is a reflection of current through a biasing resistor and a diode.

current-regulator diode A special type of diode that holds the current constant through it with changing voltage applied.

current-sensing resistor A small resistor value, placed in series with a pass transistor, used to control the maximum output current of a series voltage regulator. This resistor develops a voltage drop proportional to the load current. If the load current becomes excessive, the voltage drop will turn on an active device that will limit the output current.

current source Ideally, this is an energy source that produces a constant current through a load resistance of any value. To a second approximation, it includes a very high resistance in parallel with the energy source.

current–source bias An FET biasing method using a bipolar junction transistor, configured as a constant current source, to control the drain current.

current–to–voltage converter A circuit that takes an input current value and develops a corresponding output voltage. In op amp circuits, this is also known as a transresistance amplifier or ICVS circuit.

curve tracer An electronic device for drawing characteristic curves on a cathode-ray tube.

cutoff frequency Identical to the critical frequency. The name *cutoff* is preferred when you are discussing filters because that's what most people use.

cutoff point Approximately the same as the lower end of the load line. The exact cutoff point occurs where base current equals zero. At this point, there is a small collector leakage current, which means the cutoff point is slightly above the lower end of the dc load line.

cutoff region The region where the base current is zero in a *CE* connection. In this region, the emitter and collector diodes are nonconducting. The only collector current is the very small current produced by minority carriers and surface leakage current.

D

damping factor The ability of a filter to reduce resonant peaks at its output. The damping factor α is inversely proportional to the circuit's *Q*.

Darlington connection The connection of two transistors producing an overall current gain equal to the product of the individual current gains. This transistor connection can have a very high input impedance and can produce large output currents.

Darlington pair Two transistors connected in a Darlington configuration. The pair can consist of individual transistors or a Darlington pair inside a single case.

Darlington transistor Two transistors connected to get a very high value of β. The emitter of the first transistor drives the base of the second transistor.

dc alpha (α_{dc}) The dc collector current divided by the dc emitter current.

dc amplifier An amplifier that is capable of amplifying signals at very low frequencies, including dc. This amplifier is also known as a direct coupled amplifier.

dc beta (β_{dc}) The ratio of the dc collector current to the dc base current.

dc equivalent circuit What remains after you open all capacitors.

dc return This refers to a path for direct current. Many transistor circuits won't work unless a dc path exists between all three terminals and ground. A diff amp and an op amp are examples of devices that must have dc return paths from their input pins to ground.

dc-to-ac converter A circuit having the ability to convert dc current, generally from a battery, into ac current. This circuitry is also know as an inverter and is the basis for an *uninterruptible power supply*.

dc-to-dc converter A circuit that converts dc voltage of one value to dc voltage at another value. Usually, the dc input voltage is chopped up or changed to a rectangular voltage. This is then stepped up or down as needed, rectified, and filtered to get the output dc voltage.

dc value The same as the average value. For a time-varying signal, the dc value equals the average value of all the points on the waveform. A dc voltmeter reads the average value of a time-varying voltage.

decade A factor of 10. Often used with frequency ratios of 10, as in a decade of frequency referring to a 10:1 change in frequency.

decibel power gain The ratio of output power to input power. Mathematically defined as

$$A_{p(dB)} = 10 \log \frac{P_{out}}{P_{in}}.$$

decibel voltage gain This is a defined voltage gain given by 20 times the logarithm of the ordinary voltage gain.

defining formula A formula or an equation used to define or give the mathematical meaning of a new quantity. Before the defining formula is used for the first time, the quantity does not appear in any other formula.

definition A formula invented for a new concept based on scientific observation.

delay equalizer An all-pass active filter used to compensate for the time delay of another filter.

depletion layer The region at the junction of *p*- and *n*-type semiconductors. Because of diffusion, free electrons and holes recombine at the junction. This creates pairs of oppositely charged ions on each side of the junction. This region is depleted of free electrons and holes.

depletion-mode MOSFET A FET with an insulated gate that relies on the action of a depletion layer to control the drain current.

derating factor A value that tells you how much to reduce the power rating for each degree above the reference temperature given on the data sheet.

derivation A formula produced with mathematics from other formulas.

derived formula A formula or an equation that is a mathematical rearrangement of one or more existing equations.

diac A silicon bilateral device used to gate other devices such as triacs.

diff amp A two-transistor circuit whose ac output is an amplified version of the ac input signal between the two bases.

differential input The difference between the two input signals at the noninverting and inverting input terminals of a diff amp.

differential input voltage The desired input voltage of a differential amplifier as opposed to the common-mode input voltage.

differential output The output voltage value of a differential amplifier equaling the difference between the two collectors.

differential voltage gain The amount of amplification for the desired input signal of a differential amplifier as opposed to the common-mode input voltage.

differentiator An electronic circuit, either active or passive, whose output is proportional to the time rate of change of its input signal. This circuit has the ability to perform a calculus operation called differentiation.

digital A signal level that is found in two distinct states. Digital content is useful in the storage, processing, and transmission of information.

digital-to-analog (D/A) converter A circuit or device used to convert a digital signal into its two input terminals.

diode A *pn* crystal. A device that conducts easily when forward-biased and poorly when reverse-biased.

direct coupling Using a direct wire connection instead of a coupling capacitor between stages. For this to succeed, the designer has to make sure the dc voltages of the two points being connected are approximately equal before the direct connection is made.

discrete circuit A circuit whose components, such as resistors, transistors, etc., are soldered or otherwise connected mechanically.

distortion An undesirable change in the shape or phase of a waveform or signal. When this happens in an amplifier, the output waveform is not a true replica of the input waveform.

dominant capacitors The capacitors that are the main factors in determining a circuit's low- and high-frequency cutoff points.

donor A pentavalent atom, one that has five valence electrons. Each pentavalent atom produces one free electron in a silicon crystal.

doping Adding an impurity element to an intrinsic semiconductor to change its conductivity. Pentavalent or donor impurities increase the number of free electrons, and trivalent or acceptor impurities increase the number of holes.

drain The terminal of a field effect transistor that corresponds to the collector of a bipolar junction transistor.

drain-feedback bias An FET biasing method where a resistor is connected between the drain and gate leads of the transistor. An increase or decrease in drain current results in a corresponding decrease or increase in drain voltage. This voltage is feedback to the gate that stabilizes the transistor's Q point.

driver stage An amplifier designed to provide the proper input signal level to a power amplifier.

dropout voltage The minimum headroom voltage needed for proper regulation of an IC voltage regulator.

duality principle For any theorem in electrical circuit analysis, there is a dual (opposite) theorem in which one replaces the original quantities with dual quantities. This principle can be applied to Thevenin's and Norton's theorems.

duty cycle The width of a pulse divided by the period between pulses. Usually, you multiply by 100 percent to get the answer as a percentage.

E

Ebers–Moll model An early ac model of a transistor also known as the *T model*.

edge frequency The highest frequency in the passband of a low-pass filter. Because it is on the edge of the passband, it is also referred to as the cutoff frequency. The attenuation value at the edge frequency can be specified to be less than 3 dB.

efficiency The ac load power divided by the dc power supplied to the circuit multiplied by 100 percent.

electromagnetic interference (EMI) A form of interference resulting from the radiation of high-frequency energy.

eliptical approximation An active filter with a very steep roll-off in the transition region, but produces ripples in the passband and stopband.

emitter The part of a transistor that is the source of carriers. For *npn* transistors, the emitter sends free electrons into the base. For *pnp* transistors, the emitter sends holes into the base.

emitter bias The best way to bias a transistor for operation in the active region. The key idea is to set up a fixed value of the emitter current.

emitter diode The diode formed by the emitter and base of a transistor.

emitter-feedback bias Stabilizing the *Q* point of a base-bias circuit by adding an emitter resistor. The emitter resistor provides negative feedback.

emitter follower Identical to a *CC amplifier*. The name *emitter follower* caught on because it better describes the action. The ac emitter voltage follows the ac base voltage.

enhancement-mode MOSFET A FET with an insulated gate that relies on an inversion layer to control its conductivity.

epitaxial layer A thin, deposited crystal layer that forms a portion of the electrical structure of certain semiconductors and integrated circuits.

error voltage The voltage between the two input terminals of an op amp. It is identical to the differential input voltage of the op amp.

experimental formula A formula or an equation discovered through experiment or observation. It represents an existing law in nature.

extrinsic Refers to a doped semiconductor.

F

feedback attenuation factor An indication of how much the output voltage is attenuated before the feedback signal reaches the input.

feedback capacitor A capacitor located between the input and output terminals of an amplifier. This capacitor feeds a portion of the output signal back to the input and impacts the voltage gain and frequency response of the amplifier.

feedback fraction B The feedback voltage divided by the output voltage in a VCVS or noninverting amplifier configuration. This value is also known as the feedback attenuation factor *B*.

feedback resistor A resistor placed in a circuit for the purpose of developing a negative feedback signal across it. This resistor is used to control the gain and stability of an amplifier.

FET Colpitts oscillator An FET oscillator in which the feedback signal is applied to the gate.

field effect The control of the depletion layer width existing between the gate and channel of a field effect transistor. The width of this field controls the amount of drain current.

field-effect transistor A transistor that depends on the action of an electric field to control its conductivity.

filter An electronic network designed to either pass or reject a range or band of frequencies.

firing angle The electrical degree point or angle at which a thyristor fires and begins to conduct with an ac input waveform applied.

firm voltage divider A voltage divider whose loaded output voltage is within 10 percent of its unloaded output voltage.

first-order response The frequency response of a passive or active filter that has a 20 dB per decade roll-off.

555 timer A widely used circuit that can run in either of two modes: monostable and astable. In monostable, it can produce accurate time delays, and in astable it can produce rectangular waves with a variable duty cycle.

flag A voltage that indicates an event has taken place. Typically, a low voltage means the event has not occurred, while a high voltage means that it has. The output of a comparator is an example of a flag.

floating load This is a load that has nonzero node voltages on each end. You can spot it on a schematic diagram by the fact that neither end of the load is grounded.

FM demodulator A phase-locked loop (PLL) used as a circuit that recovers the modulating signal from the FM wave.

foldback current limiting Simple current limiting allows the load current to reach a maximum value while the load voltage is reduced to zero. Foldback current limiting takes this one step further. It allows the current to reach a maximum value. Then further decreases in the load resistance reduce both the load current and the load voltage.

The main advantage of foldback limiting is less power dissipation in the pass transistor under shorted-load conditions.

formula A rule that relates quantities. The rule may be an equation, equality, or other mathematical description.

forward bias Applying an external voltage to overcome the barrier potential.

four-layer diode A semiconductor component consisting of an interconnected four-layer *pnpn* structure. This diode allows current to flow through it in only one direction when a specific breakover voltage is reached. Once on, it will remain on until the current through it falls below its holding current I_H value.

free electron One that is loosely held by an atom. Also known as a *conduction-band electron* because it travels in a large orbit, equivalent to a high energy level.

frequency modulation (FM) A basic electronic communication technique in which an input intelligence signal (modulating signal) causes the output (carrier signal) to vary in frequency.

frequency response The graph of voltage gain versus frequency for an amplifier.

frequency scaling factor (FSF) The formula used to scale pole frequencies in direct proportion; cutoff frequency divided by 1 kHz.

frequency-shift keying A modulation technique, used in the transmission of binary data, in which an input signal causes the output signal to vary in one of two distinct output frequencies.

full-wave rectifier A rectifier with a center-tapped secondary winding and two diodes that act as back-to-back half-wave rectifiers. One diode supplies one-half of the output, and the other diode supplies the other half. The output is a full-wave rectified voltage.

fundamental frequency The lowest frequency that a crystal can effectively vibrate and produce an output. This frequency is dependent on the crystal's material constant K and its thickness t where $f = \dfrac{K}{t}$.

G

gain-bandwidth product (GBP) The high frequency at which the gain of an amplifier is 0 dB (unity).

gate The terminal of a field-effect transistor that controls drain current. Also, the terminal of a thyristor used to turn on the device.

gate-bias A simplified method of biasing an FET by connecting a voltage source, through a source resistor, to the gate lead. This biasing method is not suitable for active-region biasing due to the large spread of FET parameters. This biasing method is most applicable for biasing the FET in the ohmic region.

gate-source cutoff voltage The voltage between the gate and the source that reduces the drain current of a depletion-mode device to approximately zero.

gate trigger current I_{GT} The minimum gate current specified to turn on an SCR.

geometric average The center frequency f_0 of a bandpass filter found mathematically by $f_0 = \sqrt{f_1 f_2}$.

germanium One of the first semiconductor materials to be used. Like silicon, it has four valence electrons.

go/no-go test A test or measurement where the readings are distinctly different, really high or really low.

ground loop If you use more than one ground point in a multistage amplifier, the resistance between the ground points will produce small unwanted feedback voltages. This is a ground loop. It can cause unwanted oscillations in some amplifiers.

guard driving Minimizing the effects of cable leakage current and cable capacitance by actively bootstrapping the shield to the common-mode potential.

H

***h* parameters** An early mathematical method for representing transistor action. Still used on data sheets.

half-power frequencies The frequencies at which the load power is reduced to half of its maximum value. This is also referred to as cutoff frequencies because the voltage gain equals 0.707 of the maximum value at this point.

half-wave rectifier A rectifier with only one diode in series with the load resistor. The output is a half-wave rectified voltage.

hard saturation Operating a transistor at the upper end of the load line with a base current that is one-tenth of the collector current. The reason for the overkill is to make sure the transistor remains saturated under all operating conditions, temperature conditions, transistor replacement, etc.

harmonic distortion A form of distortion caused by a signal passing through or being amplified by a nonlinear system resulting in an output signal containing multiples of the fundamental frequency.

harmonics A sine wave whose frequency is some integer multiple of a fundamental sine wave.

Hartley oscillator A circuit distinguished by its inductively tapped tank circuit.

headroom voltage The difference between the input and output voltage of a transistor series voltage regulator or a three-terminal IC voltage regulator.

heat sink A mass of metal attached to the case of a transistor to allow the heat to escape more easily.

high-frequency border The frequency above which a capacitor acts as an ac short. Also, the frequency where the reactance is one-tenth of the total series resistance.

high-pass filter A filter capable of blocking a range of frequencies from zero to a specified

cutoff frequency f_c and passing all frequencies above the cutoff frequency.

holding current The minimum current through a thyristor that can keep it latched in the conducting stage.

hole A vacancy in the valence orbit. For instance, each atom of a silicon crystal normally has eight electrons in the valence orbit. Heat energy may dislodge one of the valence electrons, producing a hole.

hybrid IC A high-power integrated circuit consisting of two or more monolithic ICs in one package or the combination of thin- and thick-film circuits. Hybrid ICs are often used in high-power audio-amplifier applications.

hysteresis The difference between the two trip points of a Schmitt trigger. When used elsewhere, hysteresis refers to the difference between the two trip points on the transfer characteristic.

I

IC voltage regulator An integrated circuit designed to maintain an almost constant output voltage under varying input voltage and load currents.

ideal approximation The simplest equivalent circuit of a device. It includes only a few basic features of the device and ignores many others of less importance.

ideal diode The first approximation of a diode. The viewpoint is to visualize the diode as an intelligent switch that closes when forward-biased and opens when reverse-biased.

ideal transistor The first approximation of a transistor. It assumes a transistor has only two parts: an emitter diode and a collector diode. The emitter diode is treated as an ideal diode, while the collector diode is a controlled current source. The current through the emitter diode controls the collector current source.

initial slope of sine wave The earliest part of a sine wave is a straight line. The slope of this line is the initial slope of the sine wave. This slope depends on the frequency and peak value of the sine wave.

input bias current The average of the two input currents to a diff amp or an op amp.

input offset current The difference of the two input currents to a diff amp or an op amp.

input offset voltage If you ground both inputs of an op amp, you will still have an output offset voltage. The input offset voltage is defined as the input voltage needed to eliminate the output offset voltage. The cause of input offset voltage is the difference in the V_{BE} curves of the two input transistors.

input transducer A device that converts a nonelectrical quantity, such as light, temperature, or pressure into an electrical quantity.

instrumentation amplifier This is a differential amplifier with high input impedance and high CMRR. You find this type of amplifier as the input stage of measuring instruments like oscilloscopes.

insulated-gate bipolar transistor (IGBT) A hybrid semiconductor device constructed with FET characteristics on the input side and FET characteristics on the output side. This device is primarily used in high-power switching control applications.

insulated-gate FET (IGFET) Another name for *MOSFET,* which has a gate that is insulated from the channel, producing a smaller gate current than in a JFET.

integrated circuit A device that contains its own transistors, resistors, and diodes. A complete IC using these microscopic components can be produced in the space occupied by a discrete transistor.

integrator A circuit that performs the mathematical operation of integration. One popular application is generating ramps from rectangular pulses. This is how the time base is generated in oscilloscopes.

interface An electronic component or circuit that enables one type of device or circuit to communicate with or control another device or circuit.

internal capacitance The internal capacitance values between the *pn* junctions of a transistor. These values can normally be neglected under low-frequency conditions, but will provide bypass paths and loss of voltage gain for an ac signal at high frequencies.

intrinsic Refers to a pure semiconductor. A crystal that has nothing but silicon atoms is pure or intrinsic.

inverse Chebyshev approximation An active filter capable of producing a flat basspand response and a fast roll-off. It has the disadvantage of producing ripples in the stopband.

inverting input The input to a diff amp or an op amp that produces an inverted output.

inverting voltage amplifier As the name implies, the amplified output voltage is inverted with respect to the input voltage.

J

junction The border where *p*- and *n*-type semiconductors meet. Unusual things happen at a *pn* junction such as the depletion layer, the barrier potential, etc.

junction temperature The temperature found inside a semiconductor at the *pn* junction. This temperature is normally higher than the ambient temperature due to electron-hole pair recombination.

junction transistor A transistor having three alternate sections of *p*-type and *n*-type materials. These sections are in either a *p-n-p* or *n-p-n* arrangement.

K

knee voltage The point or area on a graph of diode current versus voltage where the forward current suddenly increases. It is approximately equal to the barrier potential of the diode.

L

lag circuit Another name for a bypass circuit. The word *lag* refers to the angle of the output phasor voltage, which is negative with respect to angle of the input phase voltage. The phase angle may vary from 0 to –90° (lagging).

large signal operation An amplifier in which the ac input peak-to-peak signal causes the transistor to use all or most of its ac load line.

laser diode A semiconductor laser device with the acronym for *l*ight *a*mplification by *s*timulated *e*mission of *r*adiation. This active electron device converts input power into a very narrow, intense beam of coherent visible or infrared light.

laser trimming Obtaining very precise resistor values by burning off resistance areas on a semiconductor chip using a laser.

latch Two transistors connected with positive feedback to simulate the action of a thyristor.

law A summary of a relationship that exists in nature and can be verified with an experiment.

lead circuit Another name for a coupling circuit. The word *lead* refers to the angle of the output phasor voltage, which is positive with respect to angle of the input phase voltage. The phase angle may vary from 0 to +90° (leading).

lead-lag circuit A circuit that combines a coupling and a bypass circuit. The angle of the output phasor voltage may be positive or negative with respect to the input phasor voltage. The phase angle may vary from –90 (lagging) to +90° (leading).

leakage current Often used for the total reverse current of a diode. It includes thermally produced current as well as the surface leakage current.

leakage region The graphed region on a reverse-biased zener diode between zero current and breakdown.

LED driver A circuit that can produce enough current through a LED to get light.

lifetime The average amount of time between the creation and recombination of a free electron and a hole.

light-emitting diode A diode that radiates colored light such as red, green, yellow, etc. or invisible light such as infrared.

linear Usually refers to the graph of current versus voltage for a resistor.

linear op-amp circuit This is a circuit where the op amp never saturates under normal operating conditions. This implies that the amplified output has the same shape as the input.

linear phase shift The response of a filter circuit where the phase shift increases linearly with frequency. One such filter is a Bessel filter.

linear regulator The series regulator is an example of a linear regulator. The thing that makes a linear regulator is the fact that the pass transistor operates in the active or linear region. Another example of a linear regulator is the shunt regulator. In this type of regulator, a transistor is shunted across the load. Again, the transistor operates in the active region, so the regulator is classified as a linear regulator.

line regulation A power supply specification indicating how much the output voltage will change for a given input line voltage variation.

line voltage The voltage from the power line. It has a nominal value of 115 V rms. In some places, it may be as low as 105 or as high as 125 V rms.

Lissajous pattern The pattern appearing on an oscilloscope when harmonically related signals are applied to the horizontal and vertical inputs.

load line A tool used to find the exact value of diode current and voltage.

load power The ac power in the load resistor.

load regulation The change in the regulated load voltage when the load current changes from its minimum to its maximum specified value.

lock range The range of input frequencies over which a voltage-controlled-oscillator (VCO) can remain locked on to the input frequency. The lock range is normally specified as a percentage of the VCO frequency.

logarithmic scale A scale where various points are plotted according to the logarithm of the number with which the point is labeled. This scale compresses very large values and allows the plotting of data over many decades.

loop gain The product of the differential voltage gain A and the feedback fraction B. The value of this product is usually very large. If you pick any point in an amplifier with a feedback path, the voltage gain starting from this point and going around the loop is the loop gain. The loop gain is usually made up of two parts: the gain of the amplifier (greater than 1) and the gain of the feedback circuit (less than 1). The product of these two gains is the loop gain.

low-current drop-out The switching of a semiconductor latching circuit from on to off as a result of the latching current dropping low enough to bring the transistors out of saturation.

lower trip point (LTP) One of the two input voltages at which the output voltage changes states. LTP $= -BV_{sat}$.

low-pass filter A filter capable of passing a range of frequencies from zero to a specified cutoff frequency f_c.

LSI Large-scale integration. Integrated circuits with more than 100 integrated components.

M

majority carrier Carriers are either free electrons or holes. If the free electrons outnumber the holes, the electrons are the majority carriers. If the holes outnumber the free electrons, the holes are the majority carriers.

maximum forward current The maximum amount of current that a forward-biased diode can withstand before burning out or being seriously degraded.

measured voltage gain The voltage gain that you calculate from the measured values of input and output voltage.

metal-oxide semiconductor FET (MOSFET) Often used in switching amplifier applications, this transistor provides extremely low power dissipation even with high currents.

midband We have defined this as $10f_1$ to $0.1f_2$. In this range of frequencies, the voltage gain is within 0.5 percent of the maximum voltage gain.

Miller's theorem It says a feedback capacitor is equivalent to two new capacitances, one across the input and the other across the output. The most significant thing is that the input capacitance is equal to the feedback capacitance times the voltage gain of an amplifier. This assumes an inverting amplifier.

minority carrier The carriers that are in the minority. (See the definition of *majority carrier.*)

mixer An op-amp circuit that can have a different voltage gain for each of several input signals. The total output signal is a superposition of the input signals.

modulating signal The low-frequency or intelligence input signal (usually voice or data) used to control the amplitude, frequency, phase, or other condition of an output signal.

monolithic IC An integrated circuit that is entirely on a single chip.

monostable A digital switching circuit with one stable state. This circuit is also referred to as a *one-shot* and is used in timing circuits.

monotonic A description of a filter which has no ripples in the stopband.

motorboating A low-pitched putt-putt sound that comes out of a loudspeaker. It indicates that an amplifier is oscillating at a low frequency. The cause is usually the power supply having too large a Thevenin impedance.

mounting capacitance The equivalent capacitance C_m of a crystal when it is not vibrating. Due to its physical construction, the crystal is essentially two metal plates separated by a dielectric.

MPP value Also called the *output voltage swing.* This is the maximum unclipped peak-to-peak output of an amplifier. With an op amp, the MPP value is ideally equal to the difference of the two supply voltages.

MSI Medium-scale integration. Circuits with 10 to 100 integrated components.

multiple feedback (MFB) An active filter design using more than one feedback path. The feedback paths generally are applied to the op amp's inverting input through a separate resistor and capacitor.

multiplexing A technique that allows more than one signal to be transmitted concurrently over a single medium.

multistage amplifier An amplifier configuration that consists of two or more individual stages of amplification cascaded together. The output of the first stage drives the input of the second stage. The output of the second stage can be used as the input to the third stage.

multivibrator A circuit with positive feedback and two active devices, designed so that one device conducts while the other cuts off. There are three types: a free-running multivibrator, a flip-flop, and a one-shot. The free-running or astable multivibrator produces a rectangular output, similar to a relaxation oscillator.

N

***n*-type semiconductor** A semiconductor where there are more free electrons than holes.

narrowband amplifier An amplifier constructed to operate over a small frequency range. This type of amplifier is often used in RF communications circuits.

narrowband filter A bandpass filter with a Q greater than 1 and effectively passes a small range of frequencies.

natural logarithm The logarithm of a number to the base e. Natural logarithms can be used when analyzing the charging and discharging of capacitors.

negative feedback Feeding a signal back to the input of an amplifier that is proportional to the output signal. The returning signal has a phase that opposes the input signal.

negative resistance The property of an electronic component where an increase in forward voltage produces a decrease in forward current over part of its V/I characteristic curve.

noninverting input The input to a diff amp or an op amp that produces an in-phase output.

nonlinear circuit An amplifier circuit where part of the input signal drives the amplifier into saturation or cutoff. The resulting output waveform has a different shape than the input waveform.

nonlinear device A device that has a graph of current versus voltage that is not a straight line. A device that cannot be treated as an ordinary resistor.

normalized variable A variable that has been divided by another variable with the same units or dimensions.

Norton's theorem Derived from the duality principle, the Norton theorem states that the load voltage equals the Norton current times the Norton resistance in parallel with the load resistance.

notch filter A filter that blocks a signal with at most one frequency.

nulling circuit An external op amp circuit used to reduce the effect of input offset current and input offset voltage. This circuit is used when the output error cannot be ignored.

O

octave A factor of 2. Often used with frequency ratios of 2, as in an octave of frequency referring to a 2:1 change in frequency.

ohmic region The part of the drain curves that starts at the origin and ends at the proportional pinchoff voltage.

op amp A high-gain dc amplifier that provides usable voltage gain for frequencies from 0 to over 1 MHz.

open Refers to a component or connecting wire that has an open circuit, equivalent to a high resistance approaching infinity.

open-collector comparator An op amp comparator circuit that requires the use of an external *pullup* resistor. An open-collector configuration enables higher output switching speeds and allows for the interfacing of circuits with different voltage levels.

open device A device that has infinite resistance resulting in zero current flow through it.

open-loop bandwidth The frequency response of an op amp without a feedback path between the output and input. The cutoff frequency $f_{2(OL)}$ is normally very low due to the internal compensating capacitor.

open-loop voltage gain Designated as A_{VOL} or A_{OL}, this specification designates the maximum voltage gain of an op amp without feedback.

optimum Q point The point where the ac load line has equal maximum signal swings on both half-cycles.

optocoupler A combination of a LED and a photodiode. An input signal to the LED is converted to varying light which is detected by the photodiode. The advantage is very high isolation resistance between the input and output.

optoelectronics A technology that combines optics and electronics, including many devices based on the action of a *pn* junction. Examples of optoelectronic devices are LEDs, photodiodes, and optocouplers.

order of a filter A basic description of the effectiveness of a filter. Generally, the higher the order of a filter, the closer it will be to achieving an ideal response. The order of a passive filter depends on the number of inductors and capacitors. The order of an active filter is determined by the number of RC circuits or poles it has.

oscillations The death of an amplifier. When an amplifier has positive feedback, it may break into oscillations, which is unwanted high-frequency signal. This signal is unrelated to the amplified input signal. Because of this, oscillations interfere with the desired signal. Oscillations make an amplifier useless. This is why a compensating capacitor is used with an op amp; it prevents the oscillations from occurring.

outboard transistor A transistor placed in parallel with a regulating circuit to increase the amount of load current that the overall circuit can regulate. The outboard transistor kicks in at a predetermined current level and

supplies the extra current needed by the load.

output error voltage The output voltage from an op amp circuit when the input voltage is zero. This value should ideally be zero.

output impedance Another term used for the Thevenin impedance of an amplifier. It means the amplifier has been Thevenized, so that the load sees only a single resistance in series with a Thevenin generator. This single resistance is the Thevenin or output impedance.

output offset voltage Any deviation or difference of the output voltage from the ideal output voltage.

output transducer A device that converts an electrical quantity into a nonelectrical quantity such as temperature, sound, pressure, or light.

overloading Using a load resistance so small that it decreases the voltage gain of an amplifier by a noticeable amount. In terms of the Thevenin theorem, overloading occurs when the load resistance is small compared to the Thevenin resistance.

P

parasitic oscillations Oscillations of a very high frequency that cause all sorts of strange things to happen. Circuits act erratically, oscillators may produce more than one output frequency, op amps will have unaccountable offsets, supply voltage will have unexplainable ripples, video displays will contain snow, etc.

passband The range of frequencies that can be effectively passed with minimum attenuation.

passive filter A filter built using resistors, capacitors, and inductors without using amplification devices.

pass transistor The main current-carrying transistor found in a discrete series voltage regulator. This transistor is in effect in series with the load. Therefore, it must pass the entire load current.

peak detector The same as a rectifier with a capacitor input filter. Ideally, the capacitor charges to the peak of the input voltage. This peak voltage is then used for the output voltage of the peak detector, which is why the circuit is called a peak detector.

peak inverse voltage The maximum reverse voltage across a diode in a rectifier circuit.

peak value The largest instantaneous value of a time-varying voltage.

periodic An adjective that describes a waveform that repeats the same basic shape for cycle after cycle.

phase detector The circuit in a phase-locked-loop (PLL) that produces an output voltage proportional to the phase difference between two input signals.

phase-locked loop An electronic circuit that uses feedback and a phase comparator to control frequency or speed.

phase shift The difference in phase angle between phasor voltages at points A and B. For an oscillator, the phase shift around the amplifier and feedback loop at the resonant frequency must equal 360°, equivalent to 0°, for the oscillator to work.

phase splitter A circuit that produces two voltages of the same amplitude but opposite phase. It is useful for driving class B push-pull amplifiers. If you will visualize a swamped CE amplifier with a voltage gain of 1, then you will have a phase splitter because the ac voltages across the collector and emitter resistances are equal in magnitude and opposite in phase.

photodiode A reverse-biased diode that is sensitive to incoming light. The stronger the light, the larger the reverse minority-carrier current.

phototransistor A transistor with a collector junction that is exposed to light, producing more sensitivity to light than a photodiode does.

Pierce crystal oscillator A popular oscillator configuration that uses field-effect transistors, favored because of its simplicity.

piezoelectric effect Vibration that occurs when a crystal is excited by an ac signal across its plates.

Π model An ac model of a transistor shaped like the Greek symbol Π.

pinchoff voltage The border between the ohmic region and the current-source region of a depletion-mode device when the gate voltage is zero.

PIN diode A diode consisting of an intrinsic semiconductor material placed between n-type and p-type materials. When reverse biased, the PIN diode acts like a fixed capacitor and a current-controlled resistance when reverse biased.

pn junction The border between p-type and n-type semiconductors.

pnp transistor A semiconductor sandwich. It contains an n region between two p regions.

pole frequency A special frequency used in the calculations of higher-order active filters.

poles The number of RC circuits in an active filter. The number of poles in an active filter determines the order and response of the filter.

positive clamper A circuit that produces a positive dc shift of a signal by moving all the input signal upward until the negative peaks are at zero and the positive peaks are at $2V_p$.

positive feedback Feedback where the returning signal aids or increases the effect of the input voltage.

positive limiter A circuit that clips off the positive parts of the input signal.

power amplifier A large-signal amplifier designed to produce from a few hundred milliwatts up to several hundred watts of output power.

power bandwidth The highest frequency that an op amp can handle without distorting the output signal. The power bandwidth is inversely proportional to the peak value.

power dissipation The product of voltage and current in a resistor or other nonreactive device. Rate at which heat is produced within a device.

power FET An E-MOSFET designed to handle the necessary current levels for controlling motors, lamps, and switching power supplies as compared to a low-power E-MOSFET used in digital circuits.

power gain The ratio of output power to input power.

power rating The maximum power that can be dissipated in a component or device that is operated according to a manufacturer's specifications.

power supply The section of an electronic system that converts the incoming ac line voltage to dc voltage. This section also provides for the necessary filtering and voltage regulation requirements of the system.

power-supply rejection ratio (PSRR) PSRR equals the change in the input offset voltage divided by the change in the supply voltages.

power transistor A transistor that can dissipate more than 0.5 W. Power transistors are physically larger than small-signal transistors.

preamp An amplifier designed to operate with low-level signals applied. Its main functions are to provide the necessary input impedance values and to produce the output signal value required by the next amplifier stage.

predicted voltage gain The voltage gain you calculate from the circuit values on a schematic diagram. For a CE stage, it equals the ac collector resistance divided by the ac resistance of the emitter diode.

predistortion Decreasing the design value of Q to compensate for op amp bandwidth limitations.

preregulator The first of two zener diodes used to drive a zener regulator circuit configuration. The preregulator provides the proper dc input to the regulator.

programmable unijunction transistor (PUT) A semiconductor device with switching characteristics similar to a UJT, except its intrinsic standoff ratio can be determined (programmed) by external circuitry.

proportional pinchoff voltage The border between the ohmic region and the current-source region for any gate voltage.

prototype A basic circuit that a designer can modify to get more advanced circuits.

p-type semiconductor A semiconductor where there are more holes than free electrons.

pullup resistor A resistor that the user has to add to an IC device to make it work properly. One end of the pullup resistor is connected to the device, and the other end is connected to the positive supply voltage.

pulse-position modulation A procedure in which the pulses change position according to the amplitude of the analog signal.

pulse-width modulation Controlling the width of rectangular waves for the purpose of adding intelligence or to control the average dc value.

push-pull connection Use of two transistors in a connection that makes one of them conduct for half a cycle while the other is turned off. In this way, one of the transistors amplifies the first half-cycle, and the other amplifies the second half-cycle.

Q

quartz-crystal oscillator A very stable and accurate oscillator circuit that uses the piezoelectric effect of a quartz crystal to establish its oscillator frequency.

quiescent point (Q point) The operating point found by plotting the collector current and voltage.

R

r′ parameters One way to characterize a transistor. This model uses quantities like β and r'_e.

radio-frequency (RF) amplifier Also known as a *preselector*, this amplifier provides some initial gain and selectivity.

radio-frequency interference (RFI) Interference from high-frequency electromagnetic waves emanating from electronic devices.

rail-to-rail op amp An op amp whose output voltage can swing all the way to the positive and negative supply voltages. In most op amps, the output swing is limited to 1–2 V less than either supply voltage.

RC differentiator An *RC* circuit used to differentiate an input signal of a rectangular pulse into a series of positive and negative spikes.

recombination The merging of a free electron and a hole.

rectifiers Circuits within a power supply that allow current to flow in only one direction. These circuits convert an ac input waveform to a pulsating dc output waveform.

rectifier diode A diode optimized for its ability to convert ac to dc.

reductio ad absurdum A trick used when a device may be operating as a current source or as a resistor. You assume a current source and proceed with the calculations. If any contradictory answers turn up, you know your original assumption was wrong. Then you can change to the resistor model and finish off the calculations. Reductio ad absurdum usually works whenever you have a two-state system and don't know which state it is in.

reference voltage Usually, a very precise and stable voltage derived from a zener diode with a breakdown voltage between 5 to 6 V. In this range, the temperature coefficient of the zener diode is approximately zero, which means its zener voltage is stable over a large temperature range.

relaxation oscillator A circuit that creates or generates an ac output signal without an ac input signal. This type of oscillator depends on the charging and discharging of a capacitor through a resistor.

resonant frequency The frequency of a lead-lag circuit or the frequency of an LC tank circuit where the voltage gain and phase shift are suitable for oscillations.

reverse-bias Applying an external voltage across a diode to aid the barrier potential. The result is almost zero current. The only exception is when you can exceed the breakdown voltage. If the reverse voltage is large enough, it can produce breakdown through either avalanche or the zener effect.

reverse saturation current The same as the minority-carrier current in a diode. This current exists in the reverse direction.

ripple With a capacitor input filter, this is the fluctuation in load voltage caused by the charging and discharging of the capacitor.

ripple rejection Used with voltage regulators. It tells you how well the voltage regulator rejects or attenuates the input ripple. Data sheets usually list it in decibels, where each 20 dB represents a factor-of-10 decrease in ripple.

risetime The time it takes for a waveform to increase from 10 percent to 90 percent of its maximum value. Abbreviated T_R, risetime can be applied to frequency response using the equation $f_2 = \dfrac{0.35}{T_R}$.

rms value Used with time-varying signals. Also known as the *effective value* and the *heating value*. This is the equivalent value of a dc source that would produce the same amount of heat or power over one complete cycle of the time-varying signal.

RS flip-flop An electronic circuit with two states. Also known as a *multivibrator*. May be free-running (as in an oscillator) or may exhibit one or two stable states.

R/2R ladder A digital-to-analog converter circuit using two basic resistor values arranged in a ladder configuration to reduce resistor value count, improve the accuracy of conversion, and minimize loading effects.

S

safety factor The leeway between the actual operating current, voltage, etc. and the maximum rating specified on a data sheet.

Sallen-Key equal-component filter A VCVS active filter designed using two equal resistor values and two equal capacitor values. The Q of the circuit is effected by the circuit's voltage gain and is determined by $Q = \dfrac{1}{3 - A_v}$.

Sallen-Key low-pass filter An active filter circuit configuration using an op amp connected as a voltage-controlled voltage source (VCVS). This filter has the ability to implement basic Butterworth, Chebyshev, and Bessel low-pass approximations.

Sallen-Key second-order notch filter A VCVS active bandstop filter with the capability of

achieving very steep roll-offs. The circuit's Q is dependent on the voltage gain and is found by $Q = \dfrac{0.5}{2 - A_v}$.

saturation current The current in a reverse-biased diode caused by thermally produced minority carriers.

saturated current gain The current gain of a transistor in the saturation region. This value is less than the active current gain. For soft saturation, the current gain is slightly less than the active current gain. For hard saturation, the current gain is approximately 10.

saturation point Approximately the same as the upper end of the load line. The exact location of the saturation point is slightly lower because the collector-emitter voltage is not quite zero.

saturation region The part of the collector curves that starts at the origin and slopes upward to the right until it reaches the beginning of the active or horizontal region. When a transistor operates in the saturation region, the collector-emitter voltage is typically only a few tenths of a volt.

sawtooth generator A circuit capable of producing a waveform characterized by a slow, linear rise time and a virtually instantaneous fall time.

Schmitt trigger A comparator with hysteresis. It has two trip points. This makes it immune to noise voltages, provided their peak-to-peak values are less than the hysteresis.

Schockley diode Another name for a *four-layer diode, pnpn diode,* and *silicon unilateral switch (SUS)* as named by its inventor.

Schottky diode A special-purpose diode with no depletion layer, extremely short reverse recovery time, and the ability to rectify high-frequency signals.

second approximation An approximation that adds a few more features to the ideal approximation. For a diode or transistor, this approximation includes the barrier potential in the model of the device. For silicon diodes or transistors, this means 0.7 V is included in the analysis.

self-bias The bias you get with a JFET because of the voltage produced across the source resistor.

semiconductor A broad category of materials having four valence electrons and electrical properties between those of conductors and insulators.

series regulator This is the most common type of linear regulator. It uses a transistor in series with the load. The regulation works because a control voltage to the base of the transistor changes its current and voltage as needed to keep the load voltage almost constant.

series switch A type of JFET analog switch where the JFET is in series with the load resistor.

seven-segment display A display containing seven rectangular LEDs.

short One of the common troubles that may occur. A short occurs when an extremely small resistance is approaching zero. Because of this, the voltage across a short approaches zero, but the current may be very large. A component may be internally shorted, or it may be externally shorted by a solder splash or miswire.

short-circuit output current The maximum output current that an op amp can produce for a load resistor of zero.

short-circuit protection A feature of most modern power supplies. It usually means the power supply has some form of electronic current limiting that prevents excessive load currents under shorted-load conditions.

shorted device A device that has zero ohms of resistance resulting in zero voltage dropped across it.

shunt regulator A voltage regulating circuit in which the regulating device is placed in parallel with the load. This could be a simple zener diode, zener/transistor, or zener/transistor/op amp combination configuration.

shunt switch A type of JFET analog switch where the JFET is in shunt with the load resistor.

sign changer An op amp circuit that can be adjusted for a voltage gain from −1 to 1. Mathematically expressed as $-1 < A_v < 1$.

silicon The most widely used semiconductor material. It has 14 protons and 14 electrons in orbit. An isolated silicon atom has four electrons in the valence orbit. A silicon atom that is part of a crystal has eight electrons in the valence orbit because the four neighbors share one of the electrons.

silicon controlled rectifier A thyristor with three external leads called the *anode, cathode,* and *gate.* The gate can turn the SCR on, but not off. Once the SCR is on, you have to reduce the current to less than the holding current to shut off the SCR.

silicon unilateral switch (SUS) Another name for a *Schockley diode.* This device lets current flow in only one direction.

single-ended The output voltage of a differential amplifier taken from one of the collectors in respect to ground.

sink If you visualize water disappearing down a kitchen sink, you will have the general idea of what engineers or technicians mean when they talk about a current sink. It is the point that allows current to flow into ground or out of ground.

slew rate The maximum rate that the output voltage of an op amp can change. It causes distortion for high-frequency large-signal operation.

small-signal amplifier This type of amplifier is used at the front end of receivers because the signal coming in is very weak. (The peak-to-peak emitter current is less than 10 percent of the dc emitter current.)

small-signal operation This refers to an input voltage that produces only small fluctuations in the current and voltage. Our rule for small-signal transistor operation is a peak-to-peak emitter current less than 10 percent of the dc emitter value.

small-signal transistor A transistor that can dissipate 0.5 W or less.

soft saturation Operation of the transistor at the upper end of the load line with just enough base current to produce saturation.

solder bridge An undesirable splash of solder connecting two conducting lines or circuit paths.

source The terminal of a field effect transistor comparable to the emitter of a bipolar junction transistor.

source follower The leading JFET amplifier. You see it used more than any other JFET amplifier.

source regulation The change in the regulated output voltage when the input or source voltage changes from its minimum to its maximum specified voltage.

speed-up capacitor A capacitor used to increase the switching speed of a circuit.

squelch circuit A special circuit used in communications systems where the output signal is automatically quieted with the absence of an input signal.

SSI Small-scale integration. Refers to integrated circuits with 10 or fewer integrated components.

stage A functional part in which a circuit containing one or more active devices can be divided.

state-variable filter A tunable active filter that maintains a constant Q when the center frequency is varied.

stepdown transformer A transformer with more primary turns than secondary turns. This results in less secondary voltage than primary voltage.

step-recovery diode A diode having the properties of reverse snap-off due to lighter doping density near its junction. This diode is often used in frequency multiplier applications.

stiff current source A current source whose internal resistance is at least 100 times larger than the load resistance.

stiff voltage divider A voltage divider whose loaded output voltage is within 1 percent of its unloaded output voltage.

stiff voltage source A voltage source whose internal resistance is at least 100 times smaller than the load resistance.

stopband The range of frequencies that is effectively blocked or not allowed to pass from into to output.

stray wiring capacitance The unwanted capacitance between connecting wires and ground.

substrate A region in a depletion mode MOSFET located opposite from the gate, forming a channel through which electrons flowing from source to drain must pass.

summer An op-amp circuit whose output voltage is the sum of the two or more input voltages.

superposition When you have several sources, you can determine the effect produced by each source acting alone and then add the individual effects, to get the total effect of all sources acting simultaneously.

surface-leakage current A reverse current that flows along the surface of a diode. It increases when you increase the reverse voltage.

surface-mount transistors A transistor package style that enables it to be soldered to the circuit board on the component side instead of using through-hole technology. Surface-mount technology (SMT) allows for densely populated circuit boards.

surge current The large initial current that flows through the diodes of a rectifier. It is the direct result of charging the filter capacitor, which initially is uncharged.

swamped amplifier A CE stage with a feedback resistor in the emitter circuit. This feedback resistor is much larger than the ac resistance of the emitter diode.

swamp out The use of a resistor or other component to nullify the effect of another circuit component. An unbypassed emitter resistor is commonly used to negate the effects of a transistor's r_e' value.

switching circuit A circuit that operates a transistor in either the saturation or cutoff regions. Two distinct operating regions enable the device to be used in digital and computer circuits along with output power control applications.

switching regulator A linear regulator uses a transistor that operates in the linear region. A switching regulator uses a transistor that switches between saturation and cutoff. Because of this, the transistor operates in the active region only during the short time that it is switching states. This implies that power dissipation of the pass transistor is much smaller than in a linear regulator.

T

tail current The current through the common emitter resistor R_E of a differential amplifier. When the transistors are perfectly matched, the individual emitter currents will be equal and can be found by $I_E = \frac{I_T}{2}$.

temperature coefficient The rate of change of a quantity with respect to the temperature.

theorem A derivation, in a statement form, that can be proved mathematically.

thermal energy Random kinetic energy possessed by semiconductor materials at a finite temperature.

thermal noise Noise generated by the random motion of free electrons inside a resistor or other component. This is also called Johnson noise.

thermal resistance A heat transfer characteristic quantity used by designers to determine semiconductor case temperatures and heat sink requirements.

thermal runaway As a transistor heats, its junction temperature increases. This increases the collector current, which forces the junction temperature to increase further, producing more collector current, etc., until the transistor is destroyed.

thermal shutdown A feature found in modern three-terminal IC regulators. When the regulator exceeds a safe operating temperature, the pass transistor is cut off and the output voltage goes to zero. When the device cools, the pass transistor is again turned on. If the original cause of the excessive temperature is still present, the device again shuts off. If the cause has been removed, the device works normally. This feature makes the regulator almost indestructible.

thermistor A device whose resistance experiences large changes with temperature.

Thevenin's theorem A fundamental theorem that says any circuit driving a load can be converted to a single generator and series resistance.

third approximation An accurate approximation of a diode or transistor. Used for designs that need to take into account as many details as possible.

threshold The trip point or input voltage value of a comparator that causes the output voltage to change states.

threshold voltage The voltage that turns on an enhancement-mode MOSFET. At this voltage, an inversion layer connects the source to the drain.

thyristor A four-layer semiconductor device that acts as a latch.

T model An ac model of a transistor looking like a T on its side. The emitter diode acts like an ac resistance and the collector diode acts like a current source.

topology A term used to describe the technique or fundamental layout of a switching-regulator circuit. Common switching-regulator topologies are the buck regulator, boost regulator, and buck-boost regulator.

total voltage gain The overall voltage gain of an amplifier determined by the product of the individual stage gains. Mathematically found by $A_v = (A_{v1})(A_{v2})(A_{vx})$.

transconductance The ratio of ac output current to ac input voltage. A measure of how effectively the input voltage controls the output current.

transconductance amplifier An amplifier with the transfer characteristic where an input voltage controls an output current. This is also known as a voltage-to-current converter or VCIS circuit.

transconductance curve A graph displaying the relationship of I_D versus V_{GS} for a field effect transistor. This graph demonstrates the nonlinear characteristic of a FET and how it follows a *square-law* equation.

transfer characteristic The input/output response of a circuit. The transfer characteristic demonstrates the effectiveness of how the input controls the output.

transfer function The inputs and outputs of an op-amp circuit may be voltages, currents, or a combination of the two. When you use complex numbers for the input and output quantities, the ratio of output to input becomes a function of the frequency. The name for the ratio is the transfer function.

transformer coupling The use of a transformer to pass the ac signal from one stage to another while blocking the dc component of the waveform. The transformer also has the ability to match impedances between stages.

transition The roll-off region of a filter's frequency response between its cutoff frequency f_c and the beginning of the stopband f_s.

transresistance amplifier An amplifier with the transfer characteristic where an input current controls an output voltage. This is also known as a current-to-voltage converter or ICVS circuit.

triac A thyristor that can conduct in both directions. Because of this, it is useful for controlling alternating current. It is equivalent to two SCRs in parallel with opposite polarities.

trial and error Suppose you have a problem involving two simultaneous equations. Instead of solving this in the usual left-brain mathematical way, you can guess an answer and then calculate all the unknowns. One of the calculated unknowns is the very answer that you guessed. You compare the calculated and guessed answers to see how different they are. Then you guess another answer that will close the gap between the guessed and calculated answers. After several trials, the gap becomes small enough that you have an approximate answer.

trigger A sharp pulse of voltage and current that is used to turn on a thyristor or other switching device.

trigger current The minimum current needed to turn on a thyristor.

trigger voltage The minimum voltage needed to turn on a thyristor.

trip point The value of the input voltage that switches the output of a comparator or Schmitt trigger.

troubleshooting A method of determining a circuit fault using an acquired knowledge of electronics theory and electronics measurement equipment.

tuned RF amplifier A type of narrowband amplifier normally using high-Q resonant tank circuits.

tunnel diode A diode having the properties of negative resistance. This diode has a voltage breakdown that occurs at 0 V. Used in high-frequency oscillator circuits.

twin-T oscillator An oscillator that receives the positive feedback to the noninverting input through a voltage divider and the negative feedback through the twin-T filter.

two-stage feedback A circuit configuration where a portion of the second stage's output signal is feedback into the first stage for controlling the overall gain and stability.

two-state output This is the output voltage from a digital or switching circuit. It is referred to as two-state because the output has only two stable states: low and high. The region between the low and high voltages is unstable because the circuit cannot have any value in this range except temporarily when switching between states.

two-supply emitter bias (TSEB) A power supply that produces both positive and negative supply voltages.

U

ultra-large-scale integration (ULSI) The placing of more than 1 million components on a single chip.

unidirectional load current Current that flows through a load in only one direction as the result of a half- or full-wave rectifier.

unijunction transistor Abbreviated UJT, this low-power thyristor is useful in electronic timing, waveshaping, and control applications.

uninterruptible power supply (UPS) A device containing a battery and a dc-to-ac converter to be used during a power failure.

unity-gain frequency The frequency where the voltage gain of an op amp is 1. It indicates the highest usable frequency. It is important because it equals the gain-bandwidth product.

universal curve A solution in the form of a graph that solves a problem for a whole class of circuits. The universal curve for self-biased JFETs is an example. In this universal curve I_D/I_{DSS} is graphed for R_D/R_{DS}.

unwanted bypass circuit A circuit that appears in the base or collector sides of a transistor because of internal transistor capacitances and stray wiring capacitances.

up-down analysis A method of analyzing an electronic circuit using independent and dependent variables. When an independent variable (such as a voltage source) is increased or decreased, the resulting dependent variable (resistor voltage drop or current) change is predicted.

upper trip point (UTP) One of the two input voltages in which the output voltage changes states. $UTP = BV_{sat}$.

upside-down *pnp* bias When you have a positive power supply and a *pnp* transistor, it is customary to draw the transistor upside-down. This is especially helpful when the circuit uses both *npn* and *pnp* transistors.

V

varactor A diode optimized for a reverse capacitance. The larger the reverse voltage, the smaller the capacitance.

varistor A device that acts like two back-to-back zener diodes. Used across the primary winding of a power transformer to prevent line spikes from entering the equipment.

very-large-scale integration (VLSI) The placing of thousands or hundreds of thousands of components on a single chip.

vertical MOS (VMOS) A power MOSFET with a V-shaped groove channel geometry enabling the transistor to handle high currents and block high voltages.

virtual ground A type of ground that appears at the inverting input of an op amp that uses negative feedback. It's called virtual ground because it has some of, but not all, the effects of a mechanical ground. Specifically, it is ground for voltage but not for current. A node that is a virtual ground has 0 V with respect to ground, but the node has no path for current to ground.

virtual short Ideally, because of the large internal voltage gain and extremely high-input impedance of the op amp, the voltage across the inputs, v_1-v_2, is zero and I_{in} is zero for both inputs. A virtual short is a short for voltage, but an open for current. Therefore, an op amp circuit can be analyzed on the input side as having a virtual short between its noninverting and inverting inputs.

voltage amplifier An amplifier that has its circuit values selected to produce a maximum voltage gain.

voltage-controlled current source (VCIS) Sometimes called a *transconductance amplifier,* this type of negative feedback amplifier has input current controlling output voltage.

voltage-controlled device A device like a JFET or MOSFET whose output is controlled by an input voltage.

voltage-controlled oscillator (VCO) An oscillator circuit in which the output frequency is a function of a dc control voltage; also called a *voltage-to-frequency converter.*

voltage-controlled voltage source (VCVS) The ideal op amp, having infinite voltage gain, infinite unity-gain frequency, infinite input impedance, and infinite CMRR, as well as zero output resistance, zero bias, and zero offsets.

voltage-divider bias (VDB) A biasing circuit in which the base circuit contains a voltage divider that appears stiff to the input resistance of the base.

voltage feedback This is a type of feedback where the feedback signal is proportional to the output voltage.

voltage follower An op-amp circuit that uses noninverting voltage feedback. The circuit has a very high input impedance, a very low output impedance, and a voltage gain of 1. It is ideal for use as a buffer amplifier.

voltage gain This is defined as the output voltage divided by the input voltage. Its value indicates how much the signal is amplified.

voltage multiplier Direct current power-supply circuit used to provide transformerless step-up of ac line voltage.

voltage reference A circuit that produces an extremely accurate and stable output voltage. This circuit is often packaged as a special-function IC.

voltage regulator A device or circuit that holds the load voltage almost constant, even though the load current and source voltage are changing. Ideally, a voltage regulator is a stiff voltage source with an output or Thevenin resistance that approaches zero.

voltage source Ideally, an energy source that produces a constant load voltage for any value of the load resistance. To a second approximation, it includes a small internal resistance in series with the source.

voltage step A sudden input voltage change or transition applied to an amplifier. The output response will depend on the amplifier's rate of output voltage change per unit time, also known as its slew rate.

voltage-to-current converter A circuit that is equivalent to a controlled current source. The input voltage controls the current. The current is then constant and independent of the load resistance.

voltage-to-frequency converter A circuit with the ability for an input voltage to control an output frequency. This circuit is also known as a *voltage-controlled oscillator*.

W

wafer A thin slice of a crystal used as a chassis for integrated components.

wideband amplifier An amplifier constructed to operate over a large frequency range. This type of amplifier is generally untuned using resistive loads.

wideband filter A bandpass filter with a Q less than 1 and effectively passes a large range of frequencies.

Wien-bridge oscillator An *RC* oscillator consisting of an amplifier and a Wien bridge. This is the most widely used low-frequency oscillator. It is ideal for generating frequencies from 5 Hz to 1 MHz.

window comparator A circuit used to detect when the input voltage is between two preset limits.

Z

zener diode A diode designed to operate in reverse breakover with a very stable voltage drop.

zener effect Sometimes called *high-field emission,* this occurs when the intensity of the electric field becomes high enough to dislodge valence electrons in a reverse-biased diode.

zener follower A circuit consisting of a zener regulator and an emitter follower. The transistor allows the zener to handle much less current as compared to an ordinary zener regulator. This circuit also has a low output impedance characteristic.

zener regulator A circuit consisting of a power supply or dc input voltage connected to a series resistor and a zener diode. The output voltage of this circuit is less than the output of the power supply. This circuit is also known as a zener voltage regulator.

zener resistance The bulk resistance of a zener diode. It is very small compared to the current-limiting resistance in series with the zener diode.

zener voltage The breakdown voltage of a zener diode. This is the approximate voltage out of a zener voltage regulator.

zero-crossing detector A comparator circuit where an input voltage is compared to a zero-volt reference voltage.

Answers

Odd-Numbered Problems

CHAPTER 1

1-1. $R_L \geq 10\ \Omega$

1-3. $R_L \geq 5\ \text{k}\Omega$

1-5. 0.1 V

1-7. $R_L \leq 100\ \text{k}\Omega$

1-9. 1 kΩ

1-11. 4.80 mA and not stiff

1-13. 6 mA, 4 mA, 3 mA, 2.4 mA, 2 mA, 1.7 mA, 1.5 mA

1-15. V_{TH} is unchanged, and R_{TH} doubles

1-17. $R_{TH} = 10\ \text{k}\Omega$; $V_{TH} = 100\ \text{V}$

1-19. Shorted

1-21. The battery or interconnecting wiring

1-23. 0.08 Ω

1-25. Disconnect the resistor and measure the voltage.

1-27. Thevenin's theorem makes it much easier to solve problems for which there could be many values of a resistor.

1-29. $R_S > 100\ \text{k}\Omega$. Use a 100 V battery in series with 100 kΩ.

1-31. $R_1 = 30\ \text{k}\Omega$, $R_2 = 15\ \text{k}\Omega$

1-33. First, measure the voltage across the terminals—this is the Thevenin voltage. Next, connect a resistor across the terminals. Next, measure the voltage across the resistor. Then, calculate the current through the load resistor. Then, subtract the load voltage from the Thevenin voltage. Then, divide the difference voltage by the current. The result is the Thevenin resistance.

1-35. 1. R_1 shorted

2. R_1 open or R_2 shorted

3. R_3 open

4. R_3 shorted

5. R_2 open or open at point C

6. R_4 open or open at point D

7. Open at point E

8. R_4 shorted

CHAPTER 2

2-1. −2

2-3. *a.* Semiconductor;

b. conductor;

c. semiconductor;

d. conductor

2-5. *a.* 5 mA; *b.* 5 mA; *c.* 5 mA

2-7. Minimum = 0.60 V, maximum = 0.75 V

2-9. 100 nA

CHAPTER 3

3-1. 27.3 mA

3-3. 400 mA

3-5. 10 mA

3-7. 12.8 mA

3-9. 19.3 mA, 19.3 V, 372 mW, 13.5 mW, 386 mW

3-11. 24 mA, 11.3 V, 272 mW, 16.8 mW, 289 mW

3-13. 0 mA, 12 V

3-15. 9.65 mA

3-17. 12 mA

3-19. Open

3-21. The diode is shorted or the resistor is open.

3-23. The <2.0 V reverse diode reading indicates a leaky diode.

3-25. Cathode, toward

3-27. 1N914: forward $R = 100\ \Omega$, reverse $R = 800\ \text{M}\Omega$; 1N4001: forward $R = 1.1\ \Omega$, reverse $R = 5\ \text{M}\Omega$; 1N1185: forward $R = 0.095\ \Omega$, reverse $R = 21.7\ \text{k}\Omega$

3-29. 23 kΩ

3-31. 4.47 µA

3-33. During normal operation, the 15-V power supply is supplying power to the load. The left diode is forward-biased, allowing the 15-V power supply to supply current to the load. The right diode is reverse-biased because 15 V is applied to the cathode and only 12 V is applied to the anode; this blocks the 12-V battery. Once the 15-V power supply is lost, the right diode is no longer reverse-biased and the 12-V battery can supply current to the load. The left diode will become reverse-biased, preventing any current from going into the 15-V power supply.

3-35. The source voltage does not change, but all other variables decrease.

3-37. V_A, V_B, V_C, I_1, I_2, P_1, P_2; since R is so large it has no effect on the voltage divider; therefore the variables associated with the voltage divider do not change.

CHAPTER 4

4-1. 70.7 V, 22.5 V, 22.5 V

4-3. 70.0 V, 22.3 V, 22.3 V

4-5. 20 Vac, 28.3 Vpk

4-7. 21.21 V, 6.74 V

4-9. 15 Vac, 21.2 Vpk, 15 Vac

4-11. 11.42 V, 7.26 V

4-13. 19.81 V, 12.60 V

4-15. 0.5 V

4-17. 21.2 V, 752 mV

4-19. The ripple value will double.

4-21. 18.85 V, 334 mV

4-23. 18.85 V

4-25. 17.8 V; 17.8 V; no; higher

4-27. *a.* 2.12 mA; *b.* 2.76 mA

4-29. 11.99 V

4-31. The capacitor will be destroyed.

4-33. 0.7 V, −50 V

4-35. 1.4 V, −1.4 V

4-37. 2.62 V

4-39. 0.7 V, −89.7 V

4-41. 3393.6 V

4-43. 4746.4 V

4-45. 10.6 V, −10.6 V

4-47. Find the sum of each voltage value in 1° steps, then divide the total voltage by 180.

4-49. Approximately 0 V. Each capacitor will charge up to an equal voltage but opposite polarity.

CHAPTER 5

5-1. 19.2 mA

5-3. 53.2 mA

5-5. $I_S = 19.2$ mA, $I_L = 10$ mA, $I_Z = 9.2$ mA

5-7. 43.2 mA

5-9. $V_L = 12$ V, $I_Z = 12.2$ mA

5-11. 15.05 V to 15.16 V

5-13. Yes, 167 Ω

5-15. 784 Ω

5-17. 0.1 W

5-19. 14.25 V, 15.75 V

5-21. *a.* 0 V; *b.* 18.3 V; *c.* 0 V; *d.* 0 V

5-23. A short across R_S

5-25. 5.91 mA

5-27. 13 mA

5-29. 15.13 V

5-31. Zener voltage is 6.8 V and R_S is less than 440 Ω.

5-33. 24.8 mA

5-35. 7.98 V

5-37. Trouble 5: Open at *A*; Trouble 6: Open at R_L; Trouble 7: Open at *E*; Trouble 8: Zener is shorted.

CHAPTER 6

6-1. 0.05 mA

6-3. 4.5 mA

6-5. 19.8 μA

6-7. 20.8 μA

6-9. 350 mW

6-11. Ideal: 12.3 V, 27.9 mW
Second: 12.7 V, 24.7 mW

6-13. −55 to +150°C

6-15. Possibly destroyed

6-17. *a.* Increase; *b.* increase; *c.* increase; *d.* decrease; *e.* increase; *f.* decrease

6-19. 165.67

6-21. 463 kΩ

6-23. 3.96 mA

6-25. An increase in V_{BB} causes the base current to increase, and since the transistor is controlled by base current, all other dependent variables increase except V_{CE}, which decreases because the transistor is further into conduction.

6-27. I_C, I_B, and all power dissipations decreased. The power dissipations decreased because of the drop in current ($P = IV$). The base current decreased because the voltage drop across it did not change and the resistance increased ($I = V/R$). The collector current decreased because the base current decreased ($I_C = I_B \beta_{dc}$).

6-29. The only variable that decreases is V_C. With an increase in β_{dc}, the same base current will cause a greater collector current, which will create a greater voltage drop across the collector resistor. This leaves less voltage to drop across the transistor.

CHAPTER 7

7-1. 30

7-3. 6.06 mA, 20 V

7-5. The left side of the load line would move down and the right side would remain at the same point.

7-7. 10.64 mA, 5 V

7-9. The left side of the load line will decrease by half, and the right will not move.

7-11. Minimum: 10.79 V; maximum: 19.23 V

7-13. 4.55 V

7-15. Minimum: 3.95 V; maximum: 5.38 V

7-17. *a.* Not in saturation; *b.* not in saturation; *c.* in saturation; *d.* not in saturation

7-19. 4.99995 V, 0.2 V

7-21. 13.2 V

7-23. 3.43 V

7-25. 8.34 V

7-27. 11 mA, 3 V

7-29. I_B will increase; I_C will decrease; V_C will increase.

7-31. I_B will not change; I_C will not change; more voltage at V_C

7-33. I_E will decrease; I_C will decrease; V_C will increase

7-35. I_E will not change; I_C will not change; more voltage at V_C

7-37. V_{BB}, V_{CC}

7-39. R_C could be shorted; the transistor could be open collector-emitter; R_B could be open, keeping the transistor in cutoff; R_E could be open; open in the base circuit; open in the emitter circuit

7-41. Shorted transistor collector-emitter, since the emitter voltage should be 1.1 V; open collector resistor; loss of V_{CC}

7-43. Hand-selecting the components for mass production is not very efficient; instead try using feedback to make the gain independent of the β_{dc} of the transistor.

7-45. 4.94 V

7-47. 7.2 μA

7-49. 22.6 mA

7-51. 1.13 V

7-53. Approximately 0.7 V

7-55. 2 kΩ

7-57. V_B, V_E, I_E, I_C, I_B, and P_E show no change. Since the base voltage did not change, V_B and V_E will not change. Since these voltages do not change, all the currents do not change.

CHAPTER 8

8-1. 3.81 V, 11.28 V

8-3. 1.63 V, 5.21 V

8-5. 4.12 V, 6.14 V

8-7. 3.81 mA, 7.47 V

8-9. 31.96 μA, 3.58 V

8-11. 27.08 μA, 37.36 μA

8-13. 1.13 mA, 6.69 V

8-15. 6.13 V, 7.19 V

8-17. *a.* Decreases; *b.* increases; *c.* decreases; *d.* increases; *e.* increases; *f.* remains the same

8-19. *a.* 0 V; *b.* 7.83 V; *c.* 0 V; *d.* 10 V; *e.* 0 V

8-21. −4.94 V

8-23. −6.04 V, −1.1 V

8-25. The transistor will be destroyed.

8-27. Short R_1, increase the power supply value.

8-29. 9.0 V, 8.97 V, 8.43 V

8-31. 8.8 V

8-33. 27.5 mA

8-35. R_1 shorted

8-37. Trouble 3: R_C is shorted; trouble 4: transistor terminals are shorted together

8-39. Trouble 7: open R_E; trouble 8: R_2 is shorted

8-41. Trouble 11: power supply is not working; trouble 12: emitter-base diode of the transistor is open

CHAPTER 9

9-1. 3.39 Hz

9-3. 1.59 Hz

9-5. 4.0 Hz

9-7. 18.8 Hz

9-9. 0.426 mA

9-11. 150

9-13. 40 μA

9-15. 11.7 Ω

9-17. 2.34 kΩ

9-19. Base: 207 Ω, collector: 1.02 kΩ

9-21. Min h_{fe} = 50; max h_{fe} = 200; current is 1 mA; temperature is 25° C.

9-23. The capacitor has a certain amount of leakage current that will flow through the resistor and create a voltage drop across the resistor.

9-25. 9.09 Hz

9-27. 5.68 kΩ, 2.27 kΩ

9-29. 2700 μF

CHAPTER 10

10-1. 234 mV

10-3. 212 mV

10-5. 39.6 mV

10-7. 0.625 mV, 21.6 mV, 2.53 V

10-9. 3.71 V

10-11. 713 mV

10-13. 14.7

10-15. 12.5 kΩ

10-17. Since there is a voltage at the second stage input, the cause is most likely in the second stage. Some of the possible causes are open transistor, open emitter resistor, open collector resistor, open output coupling capacitor.

10-19. 72.6 mV

10-21. 3.6 kΩ

10-23. Trouble 5: C_2 open; trouble 6: open R_2; trouble 7: open bypass capacitor C_3; trouble 8: collector resistor is open

CHAPTER 11

11-1. 154 kΩ, 1.09 kΩ

11-3. 0.995, 0.951 V

11-5. 2.18 kΩ, 0.956 V

11-7. 0.558 V

11-9. 3.9 Ω

11-11. 351

11-13. It remains at approximately 351.

11-15. 1.6 MΩ

11-17. 100 kΩ

11-19. 6.8 V, 7.5 mA

11-21. 16.4 V

11-23. 650 μA

11-25. 37.8 Ω, 3.3 kΩ

11-27. 63.8 mV

11-29. V_B = 4.48 V, V_E = 3.78 V, V_C = 11.22 V, I_E = 3.78 mA, I_C = 3.78 mA, I_B = 25.2 μA

11-31. With the control voltage at 0 V, the output is 1.79 V. With the control voltage at 5 V, the output is 0 V.

11-33. 21.5 W, the transistor will be destroyed.

11-35. 2 Ω

11-37. 0 V

11-39. T4: open C_3; T5: open between B and C; T6: open C_2; T7: open Q_2

CHAPTER 12

12-1. 680 Ω, 1.76 mA

12-3. 10.62 V

12-5. 10.62 V

12-7. 50 Ω, 274 mA

12-9. 100 Ω

12-11. 500

12-13. 15.84 mA

12-15. 2.2 percent

12-17. 237 mA

12-19. 3.3 percent

12-21. 1.1 A

12-23. 34 Vpp

12-25. 7.03 W

12-27. 31.5 percent

12-29. 1.13 W

12-31. 9.36

12-33. 1679

12-35. 10.73 MHz

12-37. 15.92 MHz

12-39. 31.25 mW

12-41. 15 mW

12-43. 85.84 kHz

12-45. 250 mW

12-47. 72.3 W

12-49. Electrically, it would be safe to touch, but it may be hot and cause a burn.

12-51. No, the collector could have an inductive load.

12-53. Increase, decrease, increase, decrease, no change

12-55. Increase, decrease, decrease, decrease, increase

12-57. Increase, decrease, no change, decrease, increase

12-59. Decrease, no change, no change, no change, decrease

12-61. Increase, no change, no change, no change, increase

CHAPTER 13

13-1. 15 GΩ

13-3. 20 mA, −4 V, 200 Ω

13-5. 500 Ω, 1.1 kΩ

13-7. −2 V, 2.5 mA

13-9. 1.5 mA, 0.849 V

13-11. 0.198 V

13-13. 20.45 V

13-15. 14.58 V

13-17. 7.43 V, 1.01 mA

13-19. −1.5 V, 11.2 V

13-21. −2.5 V, 0.55 mA

13-23. −1.5 V, 1.5 mA

13-25. −5 V, 3200 μS

13-27. 3 mA, 3000 μS

13-29. 7.09 mV

13-31. 3.06 mV

13-33. 0 mVpp, 24.55 mVpp, ∞

13-35. 8 mA, 18 mA

13-37. 8.4 V, 16.2 mV

13-39. 2.94 mA, 0.59 V, 16 mA, 30 V

13-41. Open R_1

13-43. Open R_D

13-45. Open G-S

13-47. Open C_2

CHAPTER 14

14-1. 2.25 mA, 1 mA, 250 μA
14-3. 3 mA, 333 μA
14-5. 381 Ω, 1.52, 152 mV
14-7. 1 MΩ
14-9. *a*. 0.05 V; *b*. 0.1 V; *c*. 0.2 V; *d*. 0.4 V
14-11. 0.23 V
14-13. 0.57 V
14-15. 19.5 mA, 10 A
14-17. 12 V, 0.43 V
14-19. A square-wave +12 V to 0.43 V
14-21. 12 V, 0.012 V
14-23. 1.2 mA
14-25. 1.51 A
14-27. 30.5 W
14-29. 0 A, 0.6 A
14-31. 20 S, 2.83 A
14-33. 24 mS, 3.14, 157 mV
14-35. 187.5 mS, 8.9, 446 mV

14-37. 1.81 W
14-39. 10.5 μA
14-41. 3 V

CHAPTER 15

15-1. 4.7 V
15-3. 0.1 msec, 10 kHz
15-5. 12 V, 0.6 ms
15-7. 7.3 V
15-9. 34.5 V, 1.17 V
15-11. 11.9 ms, 611 Ω
15-13. +10°, +83.7°
15-15. 10.8 V
15-17. 12.8 V
15-19. 22.5 V
15-21. 30.5 V
15-23. 10 V
15-25. 10 V
15-27. 980 Hz, 50 kHz
15-29. T1: *DE* open; T2: no supply voltage; T3: transformer; T4: fuse is open.

CHAPTER 16

16-1. 196, 316
16-3. 19.9, 9.98, 4, 2
16-5. −3.98, −6.99, −10, −13
16-7. −3.98, −13.98, −23.98
16-9. 46 dB, 40 dB
16-11. 31.6, 398
16-13. 50.1
16-15. 41 dB, 23 dB, 18 dB
16-17. 100 mW
16-19. 14 dBm, 19.7 dBm, 36.9 dBm
16-21. 2
16-23. See Figure 1.
16-25. See Figure 2.
16-27. See Figure 3.
16-29. See Figure 4.
16-31. 1.4 MHz
16-33. 222 Hz
16-35. 284 Hz
16-37. 5 pF, 25 pF, 15 pF
16-39. gate: 30.3 MHz; drain: 8.61 MHz

Figure 1

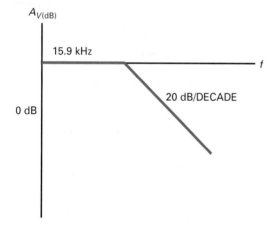

Figure 2

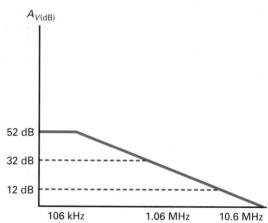

Figure 3

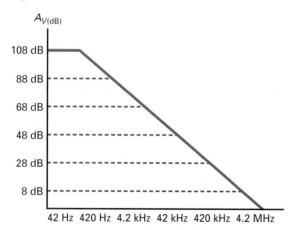

Figure 4

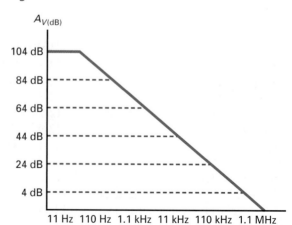

16-41. 40 dB

16-43. 0.44 µS

CHAPTER 17

17-1. 55.6 µA, 27.8 µA, 10 V

17-3. 60 µA, 30 µA, 6 V (right), 12 V (left)

17-5. 518 mV, 125 kΩ

17-7. −207 mV, 125 kΩ

17-9. 4 V, 1.75 V

17-11. 286 mV, 2.5 mV

17-13. 45.4 dB

17-15. 237 mV

17-17. Output will be high; needs a current path to ground for both bases.

17-19. C

17-21. 0 V

17-23. 2 MΩ

17-25. 10.7 Ω, 187

17-27. I_{B1}: increase, no change, increase, increase, no change; I_{B2}: no change, increase, increase, increase, no change

17-29. Increase, increase, no change, no change, increase

CHAPTER 18

18-1. 170 µV

18-3. 19,900, 2000, 200

18-5. 1.59 MHz

18-7. 10, 2 MHz, 250 mVpp, 49 mVpp; See Figure 5.

18-9. 40 mV

18-11. 22 mV

18-13. 50 mVpp, 1 MHz

18-15. 1 to 51, 392 kHz to 20 MHz

18-17. 188 mV/µs, 376 mV/µs

18-19. 38 dB, 21 V, 1000

18-21. 214, 82, 177

18-23. 41, 1

18-25. 1, 1 MHz, 1, 500 kHz

18-27. Go to positive or negative saturation.

18-29. 2.55 Vpp

18-31. I_{B1}: increase, no change, increase, increase, no change, no change; for I_{B2}: no change, increase, increase, no change, no change

18-33. No change, no change, no change, no change, no change, increase

CHAPTER 19

19-1. 0.038, 26.32, 0.10 percent, 26.29

19-3. 0.065, 15.47

19-5. 470 MΩ

19-7. 0.0038 percent

19-9. −0.660 Vpk

19-11. 185 mA$_{rms}$, 34.2 mW

19-13. 106 mA$_{rms}$, 11.2 mW

19-15. 834 mA$_{pp}$, 174 mW

19-17. 2 kHz

19-19. 15 MHz

19-21. 100 kHz, 796 mVpk

19-23. 1 V

19-25. 510 mV, 30 mV, 15 mV

19-27. 110 mV, 14 mV, 11 mV

19-29. 200 mV

19-31. 2 kΩ

19-33. 0.1 V to 1 V

19-35. T1: open between C and D; T2: shorted R_2; T3: shorted R_4

19-37. T7: open between A and B; T8: shorted R_3; T9: R_4 open

CHAPTER 20

20-1. 2, 10

20-3. −18, 712 Hz, 38.2 kHz

20-5. 42, 71.4 kHz, 79.6 Hz

20-7. 510 mV

20-9. 4.4 mV, 72.4 mV

20-11. 0, −10

20-13. 15, −15

20-15. −20, ±0.004

20-17. No

20-19. −200 mV, 10,000

20-21. 1 V

20-23. 19.3 mV

20-25. −3.125 V

20-27. −3.98 V

20-29. 24.5, 2.5 A

20-31. 0.5 mA, 28 kΩ

20-33. 0.3 mVa, 40 kΩ

20-35. 0.02, 10

20-37. −0.018, −0.99

20-39. 11, f_1: 4.68 Hz; f_2: 4.82 Hz; f_3: 32.2 Hz

20-41. 102, 98

20-43. 1 mA

20-45. T4: K-B open; T5: C-D open; T6: J-A open

CHAPTER 21

21-1. 7.36 kHz, 1.86 kHz, 0.25, wideband

21-3. *a.* Narrowband; *b.* narrowband; *c.* narrowband; *d.* narrowband

21-5. 200 dB/decade, 60 dB/octave

21-7. 503 Hz, 9.5

21-9. 39.3 Hz

21-11. −21.4, 10.3 kHz

21-13. 3, 36.2 kHz

21-15. 15 kHz, 0.707, 15 kHz

21-17. 21.9 kHz, 0.707, 21.9 kHz

21-19. 19.5 kHz, 12.89 kHz, 21.74 kHz, 0.8

21-21. 19.6 kHz, 1.23, 18.5 kHz, 18.5 kHz, 14.8 kHz

21-23. −1.04, 8.39, 16.2 kHz

21-25. 1.5, 1, 15.8 Hz, 15.8 Hz

21-27. 127°

21-29. 24.1 kHz, 50, 482 Hz (max and min)

21-31. 48.75 kHz, 51.25 kHz

21-33. 60 dB, 120 dB, 200 dB

21-35. 148 pF, 9.47 nF

CHAPTER 22

22-1. 100 µV

22-3. ±7.5 V

Figure 5

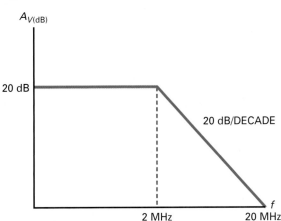

22-5. Zero, between 0.7 V and –9 V

22-7. –4 V, 31.8 Hz

22-9. 40.6 percent

22-11. 1.5 V

22-13. 0.292 V, –0.292V, 0.584 V

22-15. Output voltage is low when the input voltage is between 3.5 and 4.75 V.

22-17. 5 mA

22-19. 1 V, 0.1 V, 10 mV, 1.0 mV

22-21. 0.782 Vpp triangular waveform

22-23. 0.5, 0

22-25. 923 Hz

22-27. 196 Hz

22-29. 135 mVpp

22-31. 106 mV

22-33. –106 mV

22-35. 0 V to 200 mV peak

22-37. 20,000

22-39. Make the 3.3 kΩ resistor variable.

22-41. 1.1 Hz, 0.001 V

22-43. 0.529 V

22-45. Use different capacitors of 0.05 μF, 0.5 μF, and 5 μF, plus an inverter.

22-47. Increase R_1 to 3.3 kΩ.

22-49. Use a comparator with hysteresis and a light dependent resistor in a voltage divider as the input.

22-51. 228,780 mi

22-53. T3: relaxation oscillator circuit; T4: peak detector circuit; T5: positive clamper circuit

22-55. T8: peak detector circuit; T9: integrator circuit; T10: comparator circuit

CHAPTER 23

23-1. 9 Vrms

23-3. *a.* 33.2 Hz, 398 Hz; *b.* 332 Hz, 3.98 kHz; *c.* 3.32 kHz, 39.8 kHz; *d.* 33.2 kHz, 398 kHz

23-5. 3.98 MHz

23-7. 398 Hz

23-9. 1.67 MHz, 0.10, 10

23-11. 1.18 MHz

23-13. 7.34 MHz

23-15. 0.030, 33

23-17. Frequency will increase by 1 percent.

23-19. 517 μs

23-21. 46.8 kHz

23-23. 100 μs, 5.61 μs, 3.71 μs, 8.66 μs, 0.0371, 0.0866

23-25. 10.6 V/ms, 6.67 V, 0.629 ms

23-27. Triangular waveform, 10 kHz, 5 Vpk

23-29. *a.* decrease
b. increase
c. same
d. same
e. same

23-31. One of many possible designs is C = 0.22 μF, 0.022 μF, and 0.0022 μF. Change the 2 kΩ in Fig. 23-53 to 3.3 kΩ and use a 50 kΩ potentiometer.

Use a 1 kΩ potentiometer in place of the 1 kΩ in series with the lamp. Adjust 1 kΩ to get an output of 5 Vrms.

23-33. –360°

CHAPTER 24

24-1. 3.45 percent

24-3. 2.5 percent

24-5. 18.75 V, 284 mA, 187.5 mA, 96.5 mA

24-7. 18.46 V, 798 mA, 369 mA, 429 mA

24-9. 84.5 percent

24-11. 30.9 mA

24-13. 50 Ω, 233 mA

24-15. 421 μV

24-17. 83.3 percent, 60 percent

24-19. 3.84 A

24-21. 6 V

24-23. 14.1 V

24-25. 3.22 kΩ

24-27. 11.9 V

24-29. 0.1 Ω

24-31. 2.4 Ω

24-33. 22.6 kHz

24-35. T1: Triangle-to-pulse converter

24-37. T3: Q_1

24-39. T5: Relaxation oscillator

24-41. T7: Triangle-to-pulse converter

24-43. T9: Triangle-to-pulse converter

Index

common-mode signal, 641
common-source (CS) amplifiers, 448
comparators
     with hysteresis, 856–861
     linear region of, 847
     with nonzero references, 851–856
     with zero reference, 844–850
compensating capacitors, 666
compensating diodes, 399, 648
complementary MOS (CMOS), 497–498
complimentary Darlington, 359
compliment of Q, 916
component level troubleshooting, 554
component substitution, 510
conduction angle, 408–409, 539
conduction band, 45
conductors, 30–31
constant bandwidth, 825
constant time delay, 794
continuity testers, 168
conventional full-wave rectifier, 101
converters, 706
core, 30
corner frequency, 582
correction factor, 239
Coulomb's law, 5
coupling capacitors, 288, 378–379
covalent bonds, 33
critical rate of rise, 541
crossover distortion, 395–396
crowbar
     integrated, 536
     SCR, 534–537
     triac, 546
crystal oscillators, 907–908, 911
crystals, 32, 909
crystal slab, 908–909
crystal stability, 910
current
     derivations of, 194
     and temperature, 224
current amplifier, 719
current boosters, 762–764, 973
current-controlled current source (ICIS), 706
current controlled voltage source (ICVS), 706
current dip, 407
current drain, 386
current gain
     ac, 298
     changes in, 238
     on data sheets, 309
     h parameters, 211
     minor effect of, 238–239
     in saturation region, 233
     of transistors, 194
     variations in, 224
current hogging, 500
current interruption, 530
current limiting, 463, 959–961
current-limiting resistor, 144
current mirror, 648–650
current-regulator diodes, 177
current-sensing resistor, 960
current-source bias, 442–443
current sources, 10
current sourcing, 462
current-to-voltage converter, 706, 715–717

curve tracer, 200
cutoff frequencies, 565, 582, 595, 676
cutoff point, 79, 227
cutoff region, 200
cutoff test, 247

damped response, 802
damping factor, 801–802
Darlington connections, 356–359, 959
Darlington pair, 356
Darlington transistors, 356, 357
data sheets
     ac quantities on, 309
     collection of, 994–1045
     for Darlington transistors, 357
     described, 74
     of E-MOSFETs, 488–489
     for IGBT, 549–550
     for junction field-effect transistors (JFETs),
          464–466
     SCRs, 528–529
     for transistors, 207
     for triacs, 543–544
     of zener diodes, 158–160
dc alpha, 193
dc amplifiers, 379, 454, 457, 565–567
dc analysis of diff amps, 623–627
dc beta, 194
dc clamping of input signal, 404
dc current source, 10
dc equivalent current, 304
dc forward current, 62
dc load lines, 380, 394
dc load voltage, 108
dc resistance
     vs. bulk resistance, 79
     of diodes, 78
dc-to-ac converters, 501–502
dc-to-dc converters, 502, 974–976
dc value of a signal, 91
dc value of half-wave signal, 91
dc voltage source ac effect of, 305
deadband, 858
decades, 581
decibel attenuation, 788
decibel gain, 576
decibel power gain, 569–572
decibels (dBs)
     3-dB frequency, 808
     above reference, 578
     defined, 570
     mathematics of, 569–570
decibel voltage gain, 573–575, 581, 586
definition, 4
delay equalizers, 832
dependent variables, 244
depletion-layer isolation, 647
depletion layers, 39, 41
depletion-mode devices, 485
depletion mode MOSFET (D-MOSFET)
          amplifiers, 482–484
depletion mode MOSFET (D-MOSFET)
          curves, 480
depletion mode MOSFETs (D-MOSFETs),
          478, 480
derating curve, 411
derating factors, 161, 210, 411, 413, 414

derivation, 5
diacs, 541
differential amplifiers (diff amps)
     ac analysis of, 628–634
     construction of, 747–753
     dc analysis of, 623–627
     differential-output gain of, 629
     ideal analysis of, 623
     input impedance of, 631
     loaded, 650
     operation and function of, 620–621
     second approximation of, 624
     single-ended output gain of, 629
     voltage gain of, 628
     voltage gains for, 631
differential input, 620, 621
differential-input configurations, 630–631
differential input voltage, 748
differential output, 620, 630
differential-output gain, 629
differential voltage gain, 748
differentiation, 877
differentiator, 877–879
diffusion, 39
digital circuits, 493–494
digital multimeter (DMM), 15
digital signals, 494
digital switching, 493
digital-to-analog (D/A) converter, 759–762
diode bias, 399
diode case styles, 60
diode circuits, 60
diode clamps, 122–123, 846
diode current, 116
diodes
     back diodes, 178
     current-regulator diodes, 177
     dc resistance of, 78
     PIN diodes, 179
     snap, 178
     step-recovery, 177–178
     surface-mount, 80–81
     troubleshooting of, 71
     tunnel, 178–179
     unbiased, 38
dipole, 39
dips, 176
direct coupled signal, 354
direct coupling, 378
discrete devices, 499
discrete negative feedback amplifier, 713
discrete vs. integrated circuits, 295
distortion, 296, 712–713
distortion analyzer, 712
D-MOSFET amplifiers, 482–484
D-MOSFET curves, 480
D-MOSFETs, 478, 480
dominant capacitor, 565
dominant cutoff frequency, 592
donor atoms, 36
donor impurities, 36
doping, 34, 36–37
doping levels, 190
double-ended limit detection, 861
double scripts notation, 196
double-stage amplifier, 336–338
drain, 426